ON
GROWTH AND FORM

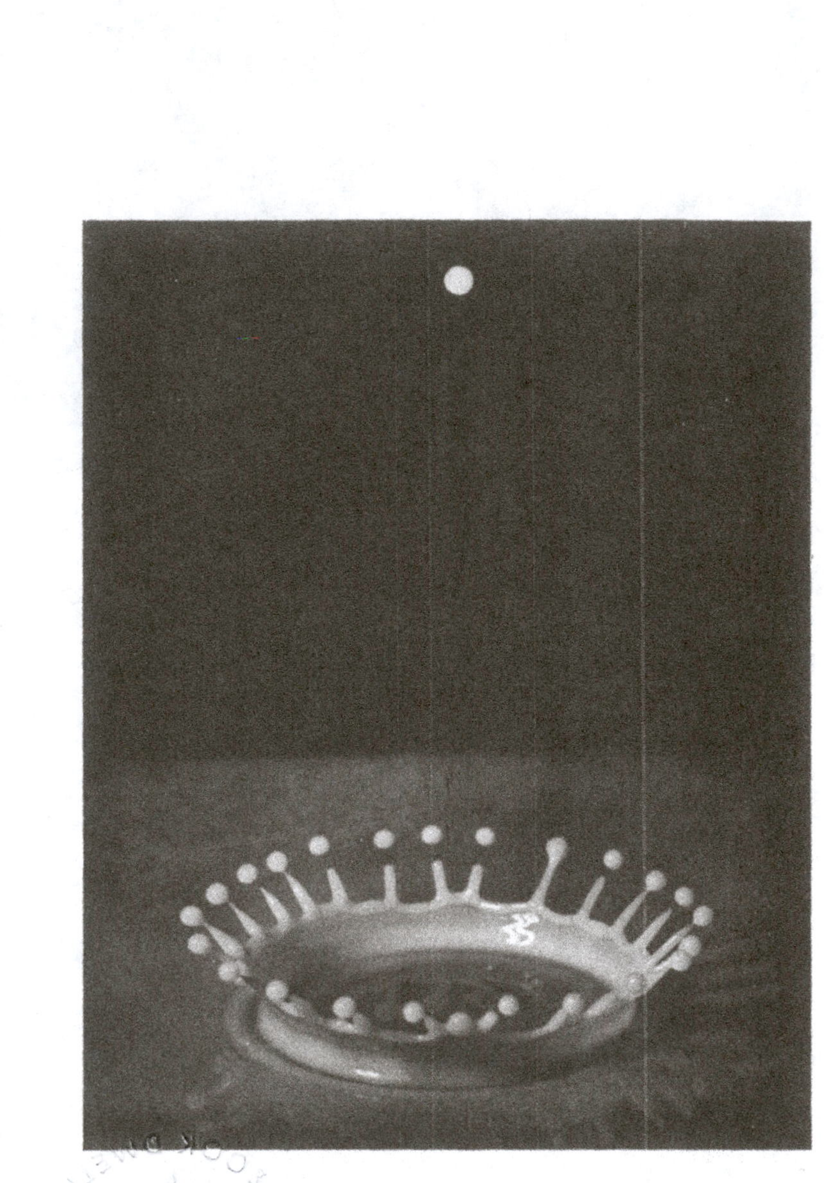

An instantaneous photograph of a 'splash' of milk. From Harold E. Edgerton, Massachusetts Technical Institution

See p. 390

ON GROWTH AND FORM

THE COMPLETE REVISED EDITION

D'Arcy Wentworth Thompson

DOVER PUBLICATIONS, INC.
New York

"The reasonings about the wonderful and intricate operations of Nature are so full of uncertainty, that, as the Wise-man truly observes, *hardly do we guess aright at the things that are upon earth, and with labour do we find the things that are before us.*" Stephen Hales, *Vegetable Staticks* (1727), p. 318, 1738.

"Ever since I have been enquiring into the works of Nature I have always loved and admired the Simplicity of her Ways." Dr George Martine (a pupil of Boerhaave's), in *Medical Essays and Observations*, Edinburgh, 1747.

This Dover edition, first published in 1992, is an unabridged, unaltered republication of the work first published by the Cambridge University Press ("Cambridge at the University Press"), Cambridge, England, 1942, under the title *On Growth and Form: A New Edition.*

Library of Congress Cataloging-in-Publication Data

Thompson, D'Arcy Wentworth, 1860–1948.
On growth and form / D'Arcy Wentworth Thompson.
p. cm.
"The complete revised edition."
Originally published: New ed. Cambridge : University Press, 1942.
Includes bibliographical references and index.
ISBN-13: 978-0-486-67135-2 (pbk.)
ISBN-10: 0-486-67135-6 (pbk.)
1. Growth. 2. Morphology (Animals) I. Title.
QP84.T4 1992
591.1—dc20 92-741
CIP

Manufactured in the United States by Courier Corporation
67135611 2015
www.doverpublications.com

PREFATORY NOTE

THIS book of mine has little need of preface, for indeed it is "all preface" from beginning to end. I have written it as an easy introduction to the study of organic Form, by methods which are the common-places of physical science, which are by no means novel in their application to natural history, but which nevertheless naturalists are little accustomed to employ.

It is not the biologist with an inkling of mathematics, but the skilled and learned mathematician who must ultimately deal with such problems as are sketched and adumbrated here. I pretend to no mathematical skill, but I have made what use I could of what tools I had; I have dealt with simple cases, and the mathematical methods which I have introduced are of the easiest and simplest kind. Elementary as they are, my book has not been written without the help—the indispensable help—of many friends. Like Mr Pope translating Homer, when I felt myself deficient I sought assistance! And the experience which Johnson attributed to Pope has been mine also, that men of learning did not refuse to help me.

I wrote this book in wartime, and its revision has employed me during another war. It gave me solace and occupation, when service was debarred me by my years.

Few are left of the friends who helped me write it, but I do not forget the debt I owe them all. Let me add another to these kindly names, that of Dr G. T. Bennett, of Emmanuel College, Cambridge; he has never wearied of collaboration with me, and his criticisms have been an education to receive.

D. W. T.

1916–1941.

CONTENTS

Plates

"The mathematicians are well acquainted with the difference between pure science, which has to do only with ideas, and the application of its laws to the use of lifė, in which they are constrained to submit to the imperfections of matter and the influence of accident." Dr Johnson, in the fourteenth *Rambler*, May 5, 1750.

"Natural History...is either the beginning or the end of physical science." Sir John Herschel, in *The Study of Natural Philosophy*, p. 221, 1831.

"I believe the day must come when the biologist will—without being a mathematician—not hesitate to use mathematical analysis when he requires it." Karl Pearson, in *Nature*, January 17, 1901.

CHAPTER I

INTRODUCTORY

Of the chemistry of his day and generation, Kant declared that it was a science, but not Science—*eine Wissenschaft, aber nicht Wissenschaft*—for that the criterion of true science lay in its relation to mathematics*. This was an old story: for Roger Bacon had called mathematics *porta et clavis scientiarum*, and Leonardo da Vinci had said much the same†. Once again, a hundred years after Kant, Du Bois Reymond, profound student of the many sciences on which physiology is based, recalled the old saying, and declared that chemistry would only reach the rank of science, in the high and strict sense, when it should be found possible to explain chemical reactions in the light of their causal relations to the velocities, tensions and conditions of equilibrium of the constituent molecules; that, in short, the chemistry of the future must deal with molecular mechanics by the methods and in the strict language of mathematics, as the astronomy of Newton and Laplace dealt with the stars in their courses. We know how great a step was made towards this distant goal as Kant defined it, when van't Hoff laid the firm foundations of a mathematical chemistry, and earned his proud epitaph—*Physicam chemiae adiunxit*‡.

We need not wait for the full realisation of Kant's desire, to apply to the natural sciences the principle which he laid down. Though chemistry fall short of its ultimate goal in mathematical mechanics§, nevertheless physiology is vastly strengthened and enlarged by

* "Ich behaupte nur dass in jeder besonderen Naturlehre nur so viel eigentliche Wissenschaft angetroffen könne als darin Mathematik anzutreffen ist": *Gesammelte Schriften*, IV, p. 470.

† "Nessuna humana investigazione si può dimandare vera scienzia s'essa non passa per le matematiche dimostrazione."

‡ Cf. also Crum Brown, On an application of Mathematics to Chemistry, *Trans. R.S.E.* XXIV, pp. 691–700, 1867.

§ Ultimate, for, as Francis Bacon tells us: Mathesis philosophiam naturalem *terminare* debet, non generare aut procreare.

making use of the chemistry, and of the physics, of the age. Little by little it draws nearer to our conception of a true science with each branch of physical science which it brings into relation with itself: with every physical law and mathematical theorem which it learns to take into its employ*. Between the physiology of Haller, fine as it was, and that of Liebig, Helmholtz, Ludwig, Claude Bernard, there was all the difference in the world†.

As soon as we adventure on the paths of the physicist, we learn to *weigh* and to *measure*, to deal with time and space and mass and their related concepts, and to find more and more our knowledge expressed and our needs satisfied through the concept of *number*, as in the dreams and visions of Plato and Pythagoras; for modern chemistry would have gladdened the hearts of those great philosophic dreamers. Dreams apart, numerical precision is the very soul of science, and its attainment affords the best, perhaps the only criterion of the truth of theories and the correctness of experiments‡. So said Sir John Herschel, a hundred years ago; and Kant had said that it was Nature herself, and not the mathematician, who brings mathematics into natural philosophy.

But the zoologist or morphologist has been slow, where the physiologist has long been eager, to invoke the aid of the physical or mathematical sciences; and the reasons for this difference lie deep, and are partly rooted in old tradition and partly in the diverse minds and temperaments of men. To treat the living body as a mechanism was repugnant, and seemed even ludicrous, to Pascal§; and Goethe, lover of nature as he was, ruled mathematics out of place in natural history. Even now the zoologist has scarce begun to dream of defining in mathematical language even the simplest organic forms. When he meets with a simple geometrical

* "Sine profunda Mechanices Scientia nil veri vos intellecturos, nil boni prolaturos aliis": Boerhaave, *De usu ratiocinii Mechanici in Medicina*, 1713.

† It is well within my own memory how Thomson and Tait, and Klein and Sylvester had to lay stress on the mathematical aspect, and urge the mathematical study, of physical science itself!

‡ Dr Johnson says that "to count is a modern practice, the ancient method was to guess"; but Seneca was alive to the difference—"magnum esse solem philosophus probabit, quantus sit mathematicus."

§ Cf. *Pensées*, XXIX, "Il faut dire, en gros, celà se fait par figure et mouvement, car celà est vrai. Mais de dire quels, et composer la machine, celà est ridicule, car celà est inutile, et incertain, et pénible."

construction, for instance in the honeycomb, he would fain refer it to psychical instinct, or to skill and ingenuity, rather than to the operation of physical forces or mathematical laws; when he sees in snail, or nautilus, or tiny foraminiferal or radiolarian shell a close approach to sphere or spiral, he is prone of old habit to believe that after all it is something more than a spiral or a sphere, and that in this "something more" there lies what neither mathematics nor physics can explain. In short, he is deeply reluctant to compare the living with the dead, or to explain by geometry or by mechanics the things which have their part in the mystery of life. Moreover he is little inclined to feel the need of such explanations, or of such extension of his field of thought. He is not without some justification if he feels that in admiration of nature's handiwork he has an horizon open before his eyes as wide as any man requires. He has the help of many fascinating theories within the bounds of his own science, which, though a little lacking in precision, serve the purpose of ordering his thoughts and of suggesting new objects of enquiry. His art of classification becomes an endless search after the blood-relationships of things living and the pedigrees of things dead and gone. The facts of embryology record for him (as Wolff, von Baer and Fritz Müller proclaimed) not only the life-history of the individual but the ancient annals of its race. The facts of geographical distribution or even of the migration of birds lead on and on to speculations regarding lost continents, sunken islands, or bridges across ancient seas. Every nesting bird, every ant-hill or spider's web, displays its psychological problems of instinct or intelligence. Above all, in things both great and small, the naturalist is rightfully impressed and finally engrossed by the peculiar beauty which is manifested in apparent fitness or "adaptation"—the flower for the bee, the berry for the bird.

Some lofty concepts, like space and number, involve truths remote from the category of causation; and here we must be content, as Aristotle says, if the mere facts be known*. But natural history deals with ephemeral and accidental, not eternal nor universal

* *οὐκ ἀπαιτητέον δ' οὐδὲ τὴν αἰτίαν ὁμοίως, ἀλλ' ἱκανὸν ἔν τισι τὸ ὅτι δειχθῆναι καλῶς* *Eth. Nic.* 1098*a*, 33. Teleologist as he was at heart, Aristotle realised that mathematics was on another plane to teleology: *τὰς δὲ μαθηματικὰς οὐθένα ποιεῖσθαι λόγον περὶ ἀγαθῶν καὶ κακῶν.* *Met.* 996*a*, 35.

things; their causes and effects thrust themselves on our curiosity, and become the ultimate relations to which our contemplation extends*.

Time out of mind it has been by way of the "final cause," by the teleological concept of end, of purpose or of "design," in one of its many forms (for its moods are many), that men have been chiefly wont to explain the phenomena of the living world; and it will be so while men have eyes to see and ears to hear withal. With Galen, as with Aristotle†, it was the physician's way; with John Ray‡, as with Aristotle, it was the naturalist's way; with Kant, as with Aristotle, it was the philosopher's way. It was the old Hebrew way, and has its splendid setting in the story that God made "every plant of the field before it was in the earth, and every herb of the field before it grew." It is a common way, and a great way; for it brings with it a glimpse of a great vision, and it lies deep as the love of nature in the hearts of men.

The argument of the final cause is conspicuous in eighteenth-century physics, half overshadowing the "efficient" or physical cause in the hands of such men as Euler§, or Fermat or Maupertuis, to whom Leibniz|| had passed it on. Half overshadowed by the mechanical concept, it runs through Claude Bernard's *Leçons sur les phénomènes de la Vie*¶, and abides in much of modern physiology**.

* "All reasonings concerning matters of fact seem to be founded on the relation of Cause and Effect. By means of that relation alone we go beyond the evidence of our memory and senses": David Hume, *On the Operations of the Understanding*.

† E.g. "In the works of Nature *purpose*, not accident, is the main thing": τὸ γὰρ μὴ τυχόντως, ἀλλ' ἕνεκά τινος, ἐν τοῖς τῆς φύσεως ἔργοις ἐστὶ καὶ μάλιστα. *PA*, 645*a*, 24.

‡ E.g. "Quaeri fortasse a nonnullis potest, Quis Papilionum usus? Respondeo, ad ornatum Universi, et ut hominibus spectaculo sint." Joh. Raii, *Hist. Insectorum*, p. 109.

§ "Quum enim Mundi universi fabrica sit perfectissima, atque a Creatore sapientissimo absoluta, nihil omnino in Mundo contingit in quo non maximi minimive ratio quaepiam eluceat; quamobrem dubium prorsus est nullum quin omnes Mundi effectus ex causis finalibus, ope Methodi maximorum et minimorum, aeque feliciter determinari queant atque ex ipsis causis efficientibus." *Methodus inveniendi*, etc., 1744, p. 245 (*cit.* Mach, *Science of Mechanics*, 1902, p. 455).

|| Cf. *Opera* (ed. Erdmann), p. 106, "Bien loin d'exclure les causes finales... c'est de là qu'il faut tout déduire en Physique": in sharp contrast to Descartes's teaching, "Nullas unquam res naturales *à fine*, quem Deus aut Natura in iis faciendis sib iproposuit, desumemus, etc." *Princip.* I, 28.

¶ Cf. p. 162. "La force vitale dirige des phénomènes qu'elle ne produit pas: les agents physiques produisent des phénomènes qu'ils ne dirigent pas."

** It is now and then conceded with reluctance. Thus Paolo Enriques, a learned and philosophic naturalist, writing "dell' economia di sostanza nelle osse cave"

Inherited from Hegel, it dominated Oken's *Naturphilosophie* and lingered among his later disciples, who were wont to liken the course of organic evolution not to the straggling branches of a tree, but to the building of a temple, divinely planned, and the crowning of it with its polished minarets*.

It is retained, somewhat crudely, in modern embryology, by those who see in the early processes of growth a significance "rather prospective than retrospective," such that the embryonic phenomena must "be referred directly to their usefulness in building up the body of the future animal†":—which is no more, and no less, than to say, with Aristotle, that the organism is the τέλος, or final cause, of its own processes of generation and development. It is writ large in that Entelechy‡ which Driesch rediscovered, and which he made known to many who had neither learned of it from Aristotle, nor studied it with Leibniz, nor laughed at it with Rabelais and Voltaire. And, though it is in a very curious way, we are told that teleology was "refounded, reformed and rehabilitated" by Darwin's concept of the origin of species§; for, just as the older naturalists held (as Addison|| puts it) that "the make of every kind of animal is different from that of every other kind; and yet there is not the least turn in the muscles, or twist in the fibres of any one, which does not render them more proper for that particular animal's way of life than any other cut or texture of them would have been": so, by the theory of natural selection, "every variety of form and colour was urgently and absolutely called upon to produce its title

(*Arch. f. Entw. Mech.* xx, 1906), says "una certa impronta di teleologismo qua e là è rimasta, mio malgrado, in questo scritto."

* Cf. John Cleland, On terminal forms of life. *Journ. Anat. and Physiol.* XVIII, 1884.

† Conklin, Embryology of Crepidula, *Journ. of Morphol.* XIII, p. 203, 1897; cf. F. R. Lillie, Adaptation in cleavage, *Wood's Hole Biol. Lectures*, 1899, pp. 43–67.

‡ I am inclined to trace back Driesch's teaching of Entelechy to no less a person than Melanchthon. When Bacon (*de Augm.* IV, 3) states with disapproval that the soul "has been regarded rather as a function than as a substance," Leslie Ellis points out that he is referring to Melanchthon's exposition of the Aristotelian doctrine. For Melanchthon, whose view of the peripatetic philosophy had great and lasting influence in the Protestant Universities, affirmed that, according to the true view of Aristotle's opinion, the soul is not a substance but an ἐντελέχεια, or *function*. He defined it as *δύναμις quaedam ciens actiones*—a description all but identical with that of Claude Bernard's "*force vitale.*"

§ Ray Lankester, art. Zoology, *Encycl. Brit.* (9th edit.), 1888, p. 806.

|| *Spectator*, No. 120.

to existence either as an active useful agent, or as a survival" of such active usefulness in the past. But in this last, and very important case, we have reached a teleology without a τέλος, as men like Butler and Janet have been prompt to shew, an "adaptation" without "design," a teleology in which the final cause becomes little more, if anything, than the mere expression or resultant of a sifting out of the good from the bad, or of the better from the worse, in short of a process of mechanism. The apparent manifestations of purpose or adaptation become part of a mechanical philosophy, "une forme méthodologique de connaissance*," according to which "la Nature agit toujours par les moyens les plus simples†," and "chaque chose finit toujours par s'accommoder à son milieu," as in the Epicurean creed or aphorism that Nature *finds a use* for everything‡. In short, by a road which resembles but is not the same as Maupertuis's road, we find our way to the very world in which we are living, and find that, if it be not, it is ever tending to become, "the best of all possible worlds§."

But the use of the teleological principle is but one way, not the whole or the only way, by which we may seek to learn how things came to be, and to take their places in the harmonious complexity of the world. To seek not for ends but for antecedents is the way of the physicist, who finds "causes" in what he has learned to recognise as fundamental properties, or inseparable concomitants, or unchanging laws, of matter and of energy. In Aristotle's parable, the house is there that men may live in it; but it is also there because the builders have laid one stone upon another. It is as a *mechanism*, or a mechanical construction, that the physicist looks upon the world; and Democritus, first of physicists and one of the greatest of the Greeks, chose to refer all natural phenomena to mechanism and set the final cause aside.

* So Newton, in the Preface to the *Principia*: "Natura enim simplex est, et rerum causis superfluis non luxuriat"; "Nature is pleased with simplicity, and affects not the pomp of superfluous causes." Modern physics finds the perfection of mathematical beauty in what Newton called the perfection of simplicity.

† Janet, *Les Causes Finales*, 1876, p. 350.

‡ "Nil ideo quoniam natumst in corpore ut uti Possemus sed quod natumst id procreat usum." Lucret. IV, 834.

§ The phrase is Leibniz's, in his *Théodicée*: and harks back to Aristotle—If one way be better than another, that you may be sure is Nature's way; *Nic. Eth.* 1099*b*, 23 *et al.*

Still, all the while, like warp and woof, mechanism and teleology are interwoven together, and we must not cleave to the one nor despise the other; for their union is rooted in the very nature of totality. We may grow shy or weary of looking to a final cause for an explanation of our phenomena; but after we have accounted for these on the plainest principles of mechanical causation it may be useful and appropriate to see how the final cause would tally with the other, and lead towards the same conclusion*. Maupertuis had little liking for the final cause, and shewed some sympathy with Descartes in his repugnance to its application to physical science. But he found at last, taking the final and the efficient causes one with another, that "l'harmonie de ces deux attributs est si parfaite que sans doute tous les effets de la Nature se pourroient déduire de chacun pris séparément. Une Mécanique aveugle et nécessaire suit les dessins de l'Intelligence la plus éclairée et la plus libre†." Boyle also, the Father of Chemistry, wrote, in his latter years, a *Disquisition about the Final Causes of Natural Things: Wherein it is Inquir'd Whether, And (if at all) With what Cautions, a Naturalist should admit Them?* He found "that all consideration of final cause is not to be banished from Natural Philosophy..."; but on the other hand "that the naturalist who would deserve that name must not let the search and knowledge of final causes make him neglect the industrious indagation of efficients‡." In our own day the philosopher neither minimises nor unduly magnifies the mechanical aspect of the Cosmos; nor need the naturalist either exaggerate or belittle the mechanical phenomena which are profoundly associated with Life, and inseparable from our understanding of Growth and Form.

* "S'il est dangereux de se servir des causes finales à priori pour trouver les lois des phénomènes, il est peut-être utile et il est au moins curieux de faire voir comment le principe des causes finales s'accorde avec les lois des phénomènes, pourvu qu'on commence par déterminer ces lois d'après les principes de mécanique clairs et incontestables." (D'Alembert, Art. Causes finales, *Encyclopédie*, II, p. 789, 1751.)

† See his essay on the "*Accord des différentes lois de la Nature.*"

‡ Cf. also Leibniz (*Discours de la Métaphysique: Lettres inédites*, ed. de Careil, 1857, p. 354), "L'un et l'autre est bon, l'un et l'autre peut être utile...et les auteurs qui suivent ces deux routes différentes ne devraient pas se maltraiter." Or again in the *Monadologie*, "Les âmes agissent selon les causes finales....Les corps agissent selon les lois des causes efficientes ou des mouvements. Et les deux règnes, celui des causes efficientes et des causes finales sont harmonieux entre eux."

Nevertheless, when philosophy bids us hearken and obey the lessons both of mechanical and of teleological interpretation, the precept is hard to follow: so that oftentimes it has come to pass, just as in Bacon's day, that a leaning to the side of the final cause "hath intercepted the severe and diligent enquiry of all real and physical causes," and has brought it about that "the search of the physical cause hath been neglected and passed in silence." So long and so far as "fortuitous variation*" and the "survival of the fittest" remain engrained as fundamental and satisfactory hypotheses in the philosophy of biology, so long will these "satisfactory and specious causes" tend to stay "severe and diligent enquiry... to the great arrest and prejudice of future discovery." Long before the great Lord Keeper wrote these words, Roger Bacon had shewn how easy it is, and how vain, to survey the operations of Nature and idly refer her wondrous works to chance or accident, or to the immediate interposition of God†.

The difficulties which surround the concept of ultimate or "real" causation, in Bacon's or Newton's sense of the word, the insuperable difficulty of giving any just and tenable account of the relation of cause and effect from the empirical point of view, need scarcely hinder us in our physical enquiry. As students of mathematical and experimental physics we are content to deal with those antecedents, or concomitants, of our phenomena without which the phenomenon does not occur—with causes, in short, which, *aliae ex aliis aptae et necessitate nexae*, are no more, and no less, than conditions *sine qua non*. Our purpose is still adequately fulfilled: inasmuch as we are still enabled to correlate, and to equate, our particular phenomena with more and more of the physical phenomena around, and so to weave a web of connection and interdependence which shall serve our turn, though the metaphysician withhold from that interdependence the title of causality‡. We come in touch

* The reader will understand that I speak, not of the "severe and diligent enquiry" of variation or of fortuity, but merely of the easy assumption that these phenomena are a sufficient basis on which to rest, with the all-powerful help of natural selection, a theory of definite and progressive evolution.

† *Op. tert.* (ed. Brewer, p. 99). "Ideo mirabiles actiones naturae, quae tota die fiunt in nobis et in rebus coram oculis nostris, non percipimus; sed aestimamus eas fieri vel per specialem operationem divinam...vel à casu et fortuna."

‡ Cf. Fourier's phrase, in his *Théorie de la Chaleur*, with which Thomson and Tait prefaced their *Treatise on Natural Philosophy*: "Les causes primordiales ne

with what the schoolmen called a *ratio cognoscendi*, though the true *ratio efficiendi* is still enwrapped in many mysteries. And so handled, the quest of physical causes merges with another great Aristotelian theme—the search for relations between things apparently disconnected, and for "similitude in things to common view unlike*." Newton did not shew the cause of the apple falling, but he shewed a similitude ("the more to increase our wonder, with an apple") between the apple and the stars†. By doing so he turned old facts into new knowledge; and was well content if he could bring diverse phenomena under "two or three Principles of Motion" even "though the Causes of these Principles were not yet discovered".

Moreover, the naturalist and the physicist will continue to speak of "causes", just as of old, though it may be with some mental reservations: for, as a French philosopher said in a kindred difficulty: "ce sont là des manières de s'exprimer, et si elles sont interdites il faut renoncer à parler de ces choses."

The search for differences or fundamental contrasts between the phenomena of organic and inorganic, of animate and inanimate, things, has occupied many men's minds, while the search for community of principles or essential similitudes has been pursued by few; and the contrasts are apt to loom too large, great though they may be. M. Dunan, discussing the *Problème de la Vie*‡, in an essay which M. Bergson greatly commends, declares that "les lois physico-chimiques sont aveugles et brutales; là où elles règnent seules, au lieu d'un ordre et d'un concert, il ne peut y avoir qu'incohérence et chaos." But the physicist proclaims aloud that the physical phenomena which meet us by the way have their forms not less beautiful and scarce less varied than those which move us to admira-

nous sont point connues; mais elles sont assujetties à des lois simples et constantes, que l'on peut découvrir par l'observation, et dont l'étude est l'objet de la philosophie naturelle."

* "Plurimum amo analogias, fidelissimos meos magistros, omnium Naturae arcanorum conscios," said Kepler; and Perrin speaks with admiration, in *Les Atomes*, of men like Galileo and Carnot, who "possessed the power of perceiving analogies to an extraordinary degree." Hume declared, and Mill said much the same thing, that all reasoning whatsoever depends on resemblance or analogy, and the power to recognise it. Comparative anatomy (as Vicq d'Azyr first called it), or comparative physics (to use a phrase of Mach's), are particular instances of a sustained search for analogy or similitude.

† As for Newton's apple, see De Morgan, in *Notes and Queries* (2), VI, p. 169, 1858.

‡ *Revue Philosophique*, XXXIII, 1892.

tion among living things. The waves of the sea, the little ripples on the shore, the sweeping curve of the sandy bay between the headlands, the outline of the hills, the shape of the clouds, all these are so many riddles of form, so many problems of morphology, and all of them the physicist can more or less easily read and adequately solve: solving them by reference to their antecedent phenomena, in the material system of mechanical forces to which they belong, and to which we interpret them as being due. They have also, doubtless, their *immanent* teleological significance; but it is on another plane of thought from the physicist's that we contemplate their intrinsic harmony* and perfection, and "see that they are good."

Nor is it otherwise with the material forms of living things. Cell and tissue, shell and bone, leaf and flower, are so many portions of matter, and it is in obedience to the laws of physics that their particles have been moved, moulded and conformed†. They are no exception to the rule that Θεὸς ἀεὶ γεωμετρεῖ. Their problems of form are in the first instance mathematical problems, their problems of growth are essentially physical problems, and the morphologist is, *ipso facto*, a student of physical science. He may learn from that comprehensive science, as the physiologists have not failed to do, the point of view from which her problems are approached, the quantitative methods by which they are attacked, and the wholesome restraints under which all her work is done. He may come to realise that there is no branch of mathematics, however abstract, which may not some day be applied to phenomena of the real

* What I understand by "holism" is what the Greeks called ἁρμονία. This is something exhibited not only by a lyre in tune, but by all the handiwork of craftsmen, and by all that is "put together" by art or nature. It is the "compositeness of any composite whole"; and, like the cognate terms κρᾶσις or σύνθεσις, implies a balance or attunement. Cf. John Tate, in *Class. Review*, Feb. 1939.

† This general principle was clearly grasped by Mr George Rainey many years ago, and expressed in such words as the following: "It is illogical to suppose that in the case of vital organisms a distinct force exists to produce results perfectly within the reach of physical agencies, especially as in many instances no end could be attained were that the case, but that of opposing one force by another capable of effecting exactly the same purpose." (On artificial calculi, *Q.J.M.S.* (*Trans. Microsc. Soc.*), VI, p. 49, 1858.) Cf. also Helmholtz, *infra cit.* p. 9. (Mr George Rainey, a man of learning and originality, was demonstrator of anatomy at St Thomas's; he followed that modest calling to a great age, and is remembered by a few old pupils with peculiar affection.)

world*. He may even find a certain analogy between the slow, reluctant extension of physical laws to vital phenomena and the slow triumphant demonstration by Tycho Brahé, Copernicus, Galileo and Newton (all in opposition to the Aristotelian cosmogony), that the heavens are formed of like substance with the earth, and that the movements of both are subject to the selfsame laws.

Organic evolution has its physical analogue in the universal law that the world tends, in all its parts and particles, to pass from certain less probable to certain more probable configurations or states. This is the second law of thermodynamics. It has been called *the law of evolution of the world*†; and we call it, after Clausius, the Principle of *Entropy*, which is a literal translation of *Evolution* into Greek.

The introduction of mathematical concepts into natural science has seemed to many men no mere stumbling-block, but a very parting of the ways. Bichat was a man of genius, who did immense service to philosophical anatomy, but, like Pascal, he utterly refused to bring physics or mathematics into biology: "On calcule le retour d'un comète, les résistances d'un fluide parcourant un canal inerte, la vitesse d'un projectile, etc.; mais calculer avec Borelli la force d'un muscle, avec Keil la vitesse du sang, avec Jurine, Lavoisier et d'autres la quantité d'air entrant dans le poumon, c'est bâtir sur un sable mouvant un édifice solide par lui-même, mais qui tombe bientôt faute de base assurée‡." Comte went further still, and said that every attempt to introduce mathematics into chemistry must be deemed profoundly irrational, and contrary to the whole spirit of the science§. But the great makers of modern science have all gone the other way. Von Baer, using a bold metaphor, thought that it might become possible "die bildenden Kräfte des thierischen Körpers auf die allgemeinen Kräfte oder *Lebenserscheinungen des Weltganzes* zurückzuführen||." Thomas Young shewed, as Borelli had done, how physics may subserve anatomy; he learned from the heart and arteries that "the mechanical motions which take place in an animal's body are regulated by the same general laws as the motions of

* So said Lobatchevsky.

† Cf. Chwolson, *Lehrbuch*, III, p. 499, 1905; J. Perrin, *Traité de chimie physique*, I, p. 142, 1903; and Lotka's *Elements of Physical Biology*, 1925, p. 26.

‡ *La Vie et la Mort*, p. 81.

§ *Philosophie Positive*, Bk. IV.

|| *Ueber Entwicklung der Thiere: Beobachtungen und Reflexionen*, I, p. 22, 1828.

inanimate bodies*." And Theodore Schwann said plainly, a hundred years ago, "Ich wiederhole übrigens dass, wenn hier von einer physikalischen Erklärung der organischen Erscheinungen die Rede ist, darunter nicht nothwendig eine Erklärung durch die bekannten physikalischen Kräfte...zu verstehen ist, sondern überhaupt eine Erklärung durch Kräfte, die nach strengen Gesetzen der blinden Nothwendigkeit wie die physikalischen Kräfte wirken, mögen diese Kräfte auch in der anorganischen Natur auftreten oder nicht†."

Helmholtz, in a famous and influential lecture, and surely with these very words of Schwann's in mind, laid it down as the fundamental principle of physiology that "there may be other agents acting in the living body than those agents which act in the inorganic world; but these forces, so far as they cause chemical and mechanical influence in the body, must be *quite of the same character* as inorganic forces: in this, at least, that their effects must be ruled by necessity, and must always be the same when acting under the same conditions; and so there cannot exist any arbitrary choice in the direction of their actions." It follows further that, like the other "physical" forces, they must be subject to mathematical analysis and deduction‡.

So much for the physico-chemical problems of physiology. Apart from these, the road of physico-mathematical or dynamical investigation in morphology has found few to follow it; but the pathway is old. The way of the old Ionian physicians, of Anaxagoras§, of Empedocles and his disciples in the days before Aristotle, lay just by that highway side. It was Galileo's and Borelli's way; and Harvey's way, when he discovered the circulation of the blood‖. It was little trodden for long afterwards, but once in a while Swammerdam and Réaumur passed thereby. And of later years Moseley and Meyer, Berthold, Errera and Roux have been among

* Croonian Lecture on the heart and arteries, *Phil. Trans.* 1809, p. 1; *Collected Works*, I, p. 511.

† *Mikroskopische Untersuchungen*, 1839, p. 226.

‡ The conservation of forces applied to organic nature, *Proc. Royal Inst.* April 12, 1861.

§ Whereby he incurred the reproach of Socrates, in the *Phaedo*. See Clerk Maxwell on Anaxagoras as a Physicist, in *Phil. Mag.* (4), XLVI, pp. 453–460, 1873.

‖ Cf. Harvey's preface to his *Exercitationes de Generatione Animalium*, 1651: "Quoniam igitur in Generatione animalium (ut etiam in caeteris rebus omnibus de quibus aliquid scire cupimus), *inquisitio omnis à caussis petenda est, praesertim à materiali et efficiente:* visum est mihi" etc.

the little band of travellers. We need not wonder if the way be hard to follow, and if these wayfarers have yet gathered little. A harvest has been reaped by others, and the gleaning of the grapes is slow.

It behoves us always to remember that in physics it has taken great men to discover simple things. They are very great names indeed which we couple with the explanation of the path of a stone, the droop of a chain, the tints of a bubble, the shadows in a cup. It is but the slightest adumbration of a dynamical morphology that we can hope to have until the physicist and the mathematician shall have made these problems of ours their own, or till a new Boscovich shall have written for the naturalist the new *Theoria Philosophiae Naturalis*.

How far even then mathematics will suffice to describe, and physics to explain, the fabric of the body, no man can foresee. It may be that all the laws of energy, and all the properties of matter, and all the chemistry of all the colloids are as powerless to explain the body as they are impotent to comprehend the soul. For my part, I think it is not so. Of how it is that the soul informs the body, physical science teaches me nothing; and that living matter influences and is influenced by mind is a mystery without a clue. Consciousness is not explained to my comprehension by all the nerve-paths and neurones of the physiologist; nor do I ask of physics how goodness shines in one man's face, and evil betrays itself in another. But of the construction and growth and working of the body, as of all else that is of the earth earthy, physical science is, in my humble opinion, our only teacher and guide.

Often and often it happens that our physical knowledge is inadequate to explain the mechanical working of the organism; the phenomena are superlatively complex, the procedure is involved and entangled, and the investigation has occupied but a few short lives of men. When physical science falls short of explaining the order which reigns throughout these manifold phenomena—an order more characteristic in its totality than any of its phenomena in themselves—men hasten to invoke a guiding principle, an entelechy, or call it what you will. But all the while no physical law, any more than gravity itself, not even among the puzzles of stereochemistry or of physiological surface-action and osmosis, is known to be transgressed by the bodily mechanism.

Some physicists declare, as Maxwell did, that atoms or molecules more complicated by far than the chemist's hypotheses demand, are requisite to explain the phenomena of life. If what is implied be an explanation of psychical phenomena, let the point be granted at once; we may go yet further and decline, with Maxwell, to believe that anything of the nature of physical complexity, however exalted, could ever suffice. Other physicists, like Auerbach*, or Larmor†, or Joly‡, assure us that our laws of thermodynamics do not suffice, or are inappropriate, to explain the maintenance, or (in Joly's phrase) the accelerative absorption, of the bodily energies, the retardation of entropy, and the long battle against the cold and darkness which is death. With these weighty problems I am not for the moment concerned. My sole purpose is to correlate with mathematical statement and physical law certain of the simpler outward phenomena of organic growth and structure or form, while all the while regarding the fabric of the organism, *ex hypothesi*, as a material and mechanical configuration. This is my purpose here. But I would not for the world be thought to believe that this is the only story which Life and her Children have to tell. One does not come by studying living things for a lifetime to suppose that physics and chemistry can account for them all§.

Physical science and philosophy stand side by side, and one upholds the other. Without something of the strength of physics philosophy would be weak; and without something of philosophy's wealth physical science would be poor. "Rien ne retirera du tissu de la science les fils d'or que la main du philosophe y a introduits||." But there are fields where each, for a while at least, must work alone; and where physical science reaches its limitations physical science itself must help us to discover. Meanwhile the appropriate and

* *Ektropismus, oder die physikalische Theorie des Lebens*, Leipzig, 1810.

† Wilde Lecture, *Nature*, March 12, 1908; *ibid.* Sept. 6, 1900; *Aether and Matter*, p. 288. Cf. also Kelvin, *Fortnightly Review*, 1892, p. 313.

‡ The abundance of life, *Proc. Roy. Dublin Soc.* VII, 1890; *Scientific Essays*, 1915, p. 60 *seq.*

§ That mechanism has its share in the scheme of nature no philosopher has denied. Aristotle (or whosoever wrote the *De Mundo*) goes so far as to assert that in the most mechanical operations of nature we behold some of the divinest attributes of God.

|| J. H. Fr. Papillon, *Histoire de la philosophie moderne dans ses rapports avec le développement des sciences de la nature*, I, p. 300, 1876.

legitimate postulate of the physicist, in approaching the physical problems of the living body, is that with these physical phenomena no alien influence interferes. But the postulate, though it is certainly legitimate, and though it is the proper and necessary prelude to scientific enquiry, may some day be proven to be untrue; and its disproof will not be to the physicist's confusion, but will come as his reward. In dealing with forms which are so concomitant with life that they are seemingly controlled by life, it is in no spirit of arrogant assertiveness if the physicist begins his argument, after the fashion of a most illustrious exemplar, with the old formula of scholastic challenge: *An Vita sit? Dico quod non.*

The terms Growth and Form, which make up the title of this book, are to be understood, as I need hardly say, in their relation to the study of organisms. We want to see how, in some cases at least, the forms of living things, and of the parts of living things, can be explained by physical considerations, and to realise that in general no organic forms exist save such as are in conformity with physical and mathematical laws. And while growth is a somewhat vague word for a very complex matter, which may depend on various things, from simple imbibition of water to the complicated results of the chemistry of nutrition, it deserves to be studied in relation to form: whether it proceed by simple increase of size without obvious alteration of form, or whether it so proceed as to bring about a gradual change of form and the slow development of a more or less complicated structure.

In the Newtonian language* of elementary physics, force is recognised by its action in producing or in changing motion, or in preventing change of motion or in maintaining rest. When we deal with matter in the concrete, force does not, strictly speaking, enter into the question, for force, unlike matter, has no independent objective existence. It is energy in its various forms, known or unknown, that acts upon matter. But when we abstract our thoughts from the material to its form, or from the thing moved to its motions, when we deal with the subjective conceptions of form,

* It is neither unnecessary nor superfluous to explain that physics is passing through an empirical phase into a phase of pure mathematical reasoning. But when we use physics to interpret and elucidate our biology, it is the old-fashioned empirical physics which we endeavour, and are alone able, to apply.

or movement, or the movements that change of form implies, then Force is the appropriate term for our conception of the causes by which these forms and changes of form are brought about. When we use the term force, we use it, as the physicist always does, for the sake of brevity, using a symbol for the magnitude and direction of an action in reference to the symbol or diagram of a material thing. It is a term as subjective and symbolic as form itself, and so is used appropriately in connection therewith.

The form, then, of any portion of matter, whether it be living or dead, and the changes of form which are apparent in its movements and in its growth, may in all cases alike be described as due to the action of force. In short, the form of an object is a "diagram of forces," in this sense, at least, that from it we can judge of or deduce the forces that are acting or have acted upon it: in this strict and particular sense, it is a diagram—in the case of a solid, of the forces which *have* been impressed upon it when its conformation was produced, together with those which enable it to retain its conformation; in the case of a liquid (or of a gas) of the forces which are for the moment acting on it to restrain or balance its own inherent mobility. In an organism, great or small, it is not merely the nature of the *motions* of the living substance which we must interpret in terms of force (according to kinetics), but also the *conformation* of the organism itself, whose permanence or equilibrium is explained by the interaction or balance of forces, as described in statics.

If we look at the living cell of an Amoeba or a Spirogyra, we see a something which exhibits certain active movements, and a certain fluctuating, or more or less lasting, form; and its form at a given moment, just like its motions, is to be investigated by the help of physical methods, and explained by the invocation of the mathematical conception of force.

Now the state, including the shape or form, of a portion of matter is the resultant of a number of forces, which represent or symbolise the manifestations of various kinds of energy; and it is obvious, accordingly, that a great part of physical science must be understood or taken for granted as the necessary preliminary to the discussion on which we are engaged. But we may at least try to indicate, very briefly, the nature of the principal forces and the principal properties of matter with which our subject obliges us to deal. Let us imagine, for instance, the case of a so-called "simple" organism, such as Amoeba; and if our short list of its physical properties and conditions be helpful to our further discussion, we need not consider how far it be complete or adequate from the wider physical point of view*.

This portion of matter, then, is kept together by the intermolecular force of cohesion; in the movements of its particles relatively to one another, and in its own movements relative to adjacent matter, it meets with the opposing force of friction—without the help of which its creeping movements could not be performed. It is acted on by gravity, and this force tends (though slightly, owing to the Amoeba's small mass, and to the small difference between its density and that of the surrounding fluid) to flatten it down upon the solid substance on which it may be creeping. Our Amoeba tends, in the next place, to be deformed by any pressure from outside, even though slight, which may be applied to it, and this circumstance shews it to consist of matter in a fluid, or at least semi-fluid, state: which state is further indicated when we observe streaming or current motions in its interior. Like other fluid bodies, its surface†, whatsoever other substance—gas, liquid or solid—it be in contact with, and in varying degree according to the nature of that adjacent substance, is the seat of molecular force exhibiting itself as a surface-tension, from the action of which many important consequences follow, greatly affecting the form of the fluid surface.

While the protoplasm‡ of the Amoeba reacts to the slightest pressure, and tends to "flow," and while we therefore speak of it

* With the special and important properties of *colloidal* matter we are, for the time being, not concerned.

† Whether an animal cell has a membrane, or only a pellicle or *zona limitans*, was once deemed of great importance, and played a big part in the early controversies between the cell-theory of Schwann and the protoplasma-theory of Max Schultze and others. Dujardin came near the truth when he said, somewhat naively, "en niant la présence d'un tégument propre, je ne prétends pas du tout nier l'existence d'une surface."

‡ The word protoplasm is used here in its most general sense, as vaguely as when Huxley spoke of it as the "physical basis of life." Its many changes and shades of meaning in early years are discussed by Van Bambeke in the *Bull. Soc. Belge de Microscopie*, XXII, pp. 1–16, 1896.

as a fluid*, it is evidently far less mobile than such a fluid (for instance) as water, but is rather like treacle in its slow creeping movements as it changes its shape in response to force. Such fluids are said to have a high viscosity, and this viscosity obviously acts in the way of resisting change of form, or in other words of retarding the effects of any disturbing action of force. When the viscous fluid is capable of being drawn out into fine threads, a property in which we know that some Amoebae differ greatly from others, we say that the fluid is also *viscid*, or exhibits viscidity. Again, not by virtue of our Amoeba being liquid, but at the same time in vastly greater measure than if it were a solid (though far less rapidly than if it were a gas), a process of molecular diffusion is constantly going on within its substance, by which its particles interchange their places within the mass, while surrounding fluids, gases and solids in solution diffuse into and out of it. In so far as the outer wall of the cell is different in character from the interior, whether it be a mere pellicle as in Amoeba or a firm cell-wall as in Protococcus, the diffusion which takes place *through* this wall is sometimes distinguished under the term *osmosis*.

Within the cell, chemical forces are at work, and so also in all probability (to judge by analogy) are electrical forces; and the organism reacts also to forces from without, that have their origin in chemical, electrical and thermal influences. The processes of diffusion and of chemical activity within the cell result, by the drawing in of water, salts, and food-material with or without chemical transformation into protoplasm, in *growth*, and this complex phenomenon we shall usually, without discussing its nature and origin, describe and picture as a *force*. Indeed we shall manifestly be inclined to use the term growth in two senses, just indeed as we do in the case of attraction or gravitation, on the one hand as a *process*, and on the other as a *force*.

In the phenomena of cell-division, in the attractions or repulsions of the parts of the dividing nucleus, and in the "caryokinetic" figures which appear in connection with it, we seem to see in operation forces and the effects of forces which have, to say the

* One of the first statements which Dujardin made about protoplasm (or, as he called it, *sarcode*) was that it was *not* a fluid; and he relied greatly on this fact to shew that it was a living, or an organised, structure.

least of it, a close analogy with known physical phenomena: and to this matter we shall presently return. But though they resemble known physical phenomena, their nature is still the subject of much dubiety and discussion, and neither the forms produced nor the forces at work can yet be satisfactorily and simply explained. We may readily admit then, that, besides phenomena which are obviously physical in their nature, there are actions visible as well as invisible taking place within living cells which our knowledge does not permit us to ascribe with certainty to any known physical force; and it may or may not be that these phenomena will yield in time to the methods of physical investigation. Whether they do or no, it is plain that we have no clear rule or guidance as to what is "vital" and what is not; the whole assemblage of so-called vital phenomena, or properties of the organism, cannot be clearly classified into those that are physical in origin and those that are *sui generis* and peculiar to living things. All we can do meanwhile is to analyse, bit by bit, those parts of the whole to which the ordinary laws of the physical forces more or less obviously and clearly and indubitably apply.

But even the ordinary laws of the physical forces are by no means simple and plain. In the winding up of a clock (so Kelvin once said), and in the properties of matter which it involves, there is enough and more than enough of mystery for our limited understanding: "a watchspring is much farther beyond our understanding than a gaseous nebula." We learn and learn, but never know all, about the smallest, humblest thing. So said St Bonaventure: "Si per multos annos viveres, adhuc naturam unius festucae seu muscae seu minimae creaturae de mundo ad plenum cognoscere non valeres*." There is a certain fascination in such ignorance; and we learn (like the Abbé Galiani) without discouragement that Science is "plutôt destiné à étudier qu'à connaître, à chercher qu'à trouver la vérité."

Morphology is not only a study of material things and of the forms of material things, but has its dynamical aspect, under which we deal with the interpretation, in terms of force, of the operations of Energy†. And here it is well worth while to remark that, in dealing

* *Op.* v, p. 541; *cit.* E. Gilson.

† This is a great theme. Boltzmann, writing in 1886 on the second law of thermodynamics, declared that available energy was the main object at stake in the struggle for existence and the evolution of the world. Cf. Lotka, The energetics of evolution, *Proc. Nat. Acad. Sci.* 1922, p. 147.

with the facts of embryology or the phenomena of inheritance, the common language of the books seems to deal too much with the *material* elements concerned, as the causes of development, of variation or of hereditary transmission. Matter as such produces nothing, changes nothing, does nothing; and however convenient it may afterwards be to abbreviate our nomenclature and our descriptions, we must most carefully realise in the outset that the spermatozoon, the nucleus, the chromosomes or the germ-plasma can never *act* as matter alone, but only as seats of energy and as centres of force. And this is but an adaptation (in the light, or rather in the conventional symbolism, of modern science) of the old saying of the philosopher: *ἀρχὴ γὰρ ἡ φύσις μᾶλλον τῆς ὕλης*.

Since this book was written, some five and twenty years ago, certain great physico-mathematical concepts have greatly changed. Newtonian mechanics and Newtonian concepts of space and time are found unsuitable, even untenable or invalid, for the all but infinitely great and the all but infinitely small. The very idea of physical causation is said to be illusory, and the physics of the atom and the electron, and of the quantum theory, are to be elucidated by the laws of probability rather than by the concept of causation and its effects. But the orders of magnitude, whether of space or time, within which these new concepts become useful, or hold true, lie far away. We distinguish, and can never help distinguishing, between the things which are of our own scale and order, to which our minds are accustomed and our senses attuned, and those remote phenomena which ordinary standards fail to measure, in regions where (as Robert Louis Stevenson said) there is no habitable city for the mind of man.

It is no wonder if new methods, new laws, new words, new modes of thought are needed when we make bold to contemplate a Universe within which all Newton's is but a speck. But the world of the living, wide as it may be, is bounded by a familiar horizon within which our thoughts and senses are at home, our scales of time and magnitude suffice, and the Natural Philosophy of Newton and Galileo rests secure.

We start, like Aristotle, with *our own* stock-in-trade of knowledge: *ἀρκτέον ἀπὸ τῶν ἡμῖν γνωρίμων*. And only when we are steeped to the marrow (as Henri Poincaré once said) in the old laws, and in no danger of forgetting them, may we be allowed to learn how they have their remote but subtle limitations, and cease afar off to be more than approximately true*. Kant's axiom of causality, that it is *denknotwendig*—indispensable for thought—remains true however physical science may change. His later aphorism, that all changes take place subject to the law which links cause and effect together—"alle Veränderungen geschehen nach dem Gesetz der Verknüpfung von Ursache und Wirkung"—is still an axiom *à priori*, independent of experience: for experience itself depends upon its truth†.

* So Max Planck himself says somewhere: "In my opinion the teaching of mechanics will still have to begin with Newtonian force, just as optics begins in the sensation of colour and thermodynamics with the sensation of warmth, despite the fact that a more precise basis is substituted later on."

† "Weil er [der Grundsatz das Kausalverhältnisses] selbst der grund der Möglichkeit einer solchen Erfahrung ist": *Kritik d. reinen Vernunft*, ed. Odicke, 1889, p. 221. Cf. also G. W. Kellner, Die Kausalität in der Physik, *Ztschr. f. Physik*, LXIV, pp. 568–580, 1930.

CHAPTER II

ON MAGNITUDE

To terms of magnitude, and of direction, must we refer all our conceptions of Form. For the form of an object is defined when we know its magnitude, actual or relative, in various directions; and Growth involves the same concepts of magnitude and direction, related to the further concept, or "dimension," of Time. Before we proceed to the consideration of specific form, it will be well to consider certain general phenomena of spatial magnitude, or of the extension of a body in the several dimensions of space.

We are taught by elementary mathematics—and by Archimedes himself—that in similar figures the surface increases as the square, and the volume as the cube, of the linear dimensions. If we take the simple case of a sphere, with radius r, the area of its surface is equal to $4\pi r^2$, and its volume to $\frac{4}{3}\pi r^3$; from which it follows that the ratio of its volume to surface, or V/S, is $\frac{1}{3}r$. That is to say, V/S *varies* as r; or, in other words, the larger the sphere by so much the greater will be its volume (or its mass, if it be uniformly dense throughout) in comparison with its superficial area. And, taking L to represent any linear dimension, we may write the general equations in the form

$$S \propto L^2, \qquad V \propto L^3,$$

or
$$S = kL^2, \text{ and } V = k'L^3,$$

where k, k', are "factors of proportion,"

and
$$\frac{V}{S} \propto L, \quad \text{or} \quad \frac{V}{S} = \frac{k}{k'}L = KL.$$

So, in Lilliput, "His Majesty's Ministers, finding that Gulliver's stature exceeded theirs in the proportion of twelve to one, concluded from the similarity of their bodies that his must contain at least 1728 [or 12^3] of theirs, and must needs be rationed accordingly*."

* Likewise Gulliver had a whole Lilliputian hogshead for his half-pint of wine: in the due proportion of 1728 half-pints, or 108 gallons, equal to one pipe or

From these elementary principles a great many consequences follow, all more or less interesting, and some of them of great importance. In the first place, though growth in length (let us say) and growth in volume (which is usually tantamount to mass or weight) are parts of one and the same process or phenomenon, the one attracts our *attention* by its increase very much more than the other. For instance a fish, in doubling its length, multiplies its weight no less than eight times; and it all but doubles its weight in growing from four inches long to five.

In the second place, we see that an understanding of the correlation between length and weight in any particular species of animal, in other words a determination of k in the formula $W = k.L^3$, enables us at any time to translate the one magnitude into the other, and (so to speak) to weigh the animal with a measuring-rod; this, however, being always subject to the condition that the animal shall in no way have altered its form, nor its specific gravity. That its specific gravity or density should materially or rapidly alter is not very likely; but as long as growth lasts changes of form, even though inappreciable to the eye, are apt and likely to occur. Now weighing is a far easier and far more accurate operation than measuring; and the measurements which would reveal slight and otherwise imperceptible changes in the form of a fish—slight relative differences between length, breadth and depth, for instance—would need to be very delicate indeed. But if we can make fairly accurate determinations of the length, which is much the easiest linear dimension to measure, and correlate it with the weight, then the value of k, whether it varies or remains constant, will tell us at once whether there has or has not been a tendency to alteration in the general form, or, in other words, a difference in the rates of growth in different directions. To this subject we shall return, when we come to consider more particularly the phenomenon of *rate of growth*.

double-hogshead. But Gilbert White of Selborne could not see what was plain to the Lilliputians; for finding that a certain little long-legged bird, the stilt, weighed $4\frac{1}{4}$ oz. and had legs 8 in. long, he thought that a flamingo, weighing 4 lbs., should have legs 10 ft. long, to be in the same proportion as the stilt's. But it is obvious to us that, as the weights of the two birds are as 1 : 15, so the legs (or other linear dimensions) should be as the cube-roots of these numbers, or nearly as 1 : $2\frac{1}{2}$. And on this scale the flamingo's legs should be, as they actually are, about 20 in. long.

We are accustomed to think of magnitude as a purely relative matter. We call a thing *big* or *little* with reference to what it is wont to be, as when we speak of a small elephant or a large rat; and we are apt accordingly to suppose that size makes no other or more essential difference, and that Lilliput and Brobdingnag* are all alike, according as we look at them through one end of the glass or the other. Gulliver himself declared, in Brobdingnag, that "undoubtedly philosophers are in the right when they tell us that nothing is great and little otherwise than by comparison": and Oliver Heaviside used to say, in like manner, that there is no absolute scale of size in the Universe, for it is boundless towards the great and also boundless towards the small. It is of the very essence of the Newtonian philosophy that we should be able to extend our concepts and deductions from the one extreme of magnitude to the other; and Sir John Herschel said that "the student must lay his account to finding the distinction of great and little altogether annihilated in nature."

All this is true of *number*, and of *relative magnitude*. The Universe has its endless gamut of great and small, of near and far, of many and few. Nevertheless, in physical science the scale of absolute magnitude becomes a very real and important thing; and a new and deeper interest arises out of the changing ratio of dimensions when we come to consider the inevitable changes of physical relations with which it is bound up. The effect of *scale* depends not on a thing in itself, but in relation to its whole environment or milieu; it is in conformity with the thing's "place in Nature," its field of action and reaction in the Universe. Everywhere Nature works true to scale, and everything has its proper size accordingly. Men and trees, birds and fishes, stars and star-systems, have their appropriate dimensions, and their more or less narrow range of absolute magnitudes. The scale of human observation and experience lies within the narrow bounds of inches, feet or miles, all measured in terms drawn from our own selves or our own doings. Scales which include light-years, parsecs, Ångström units, or atomic

* Swift paid close attention to the arithmetic of magnitude, but none to its physical aspect. See De Morgan, on Lilliput, in *N. and Q.* (2), VI, pp. 123–125, 1858. On relative magnitude see also Berkeley, in his *Essay towards a New Theory of Vision*, 1709.

and sub-atomic magnitudes, belong to other orders of things and other principles of cognition.

A common effect of scale is due to the fact that, of the physical forces, some act either directly at the surface of a body, or otherwise in proportion to its surface or area; while others, and above all gravity, act on all particles, internal and external alike, and exert a force which is proportional to the mass, and so usually to the volume of the body.

A simple case is that of two similar weights hung by two similar wires. The forces exerted by the weights are proportional to their masses, and these to their volumes, and so to the cubes of the several linear dimensions, including the diameters of the wires. But the areas of cross-section of the wires are as the squares of the said linear dimensions; therefore the stresses in the wires *per unit area* are not identical, but increase in the ratio of the linear dimensions, and the larger the structure the more severe the strain becomes:

$$\frac{\text{Force}}{\text{Area}} \propto \frac{l^3}{l^2} \propto l,$$

and the less the wires are capable of supporting it.

In short, it often happens that of the forces in action in a system some vary as one power and some as another, of the masses, distances or other magnitudes involved; the "dimensions" remain the same in our equations of equilibrium, but the relative values alter with the scale. This is known as the "Principle of Similitude," or of dynamical similarity, and it and its consequences are of great importance. In a handful of matter cohesion, capillarity, chemical affinity, electric charge are all potent; across the solar system gravitation* rules supreme; in the mysterious region of the nebulae, it may haply be that gravitation grows negligible again.

To come back to homelier things, the strength of an iron girder obviously varies with the cross-section of its members, and each cross-section varies as the square of a linear dimension; but the weight of the whole structure varies as the cube of its linear dimen-

* In the early days of the theory of gravitation, it was deemed especially remarkable that the action of gravity "is proportional to the quantity of solid matter in bodies, and not to their surfaces as is usual in mechanical causes; this power, therefore, seems to surpass mere mechanism" (Colin Maclaurin, on *Sir Isaac Newton's Philosophical Discoveries*, IV, 9).

sions. It follows at once that, if we build two bridges geometrically similar, the larger is the weaker of the two*, and is so in the ratio of their linear dimensions. It was elementary engineering experience such as this that led Herbert Spencer to apply the principle of similitude to biology†.

But here, before we go further, let us take careful note that increased weakness is no necessary concomitant of increasing size. There are exceptions to the rule, in those exceptional cases where we have to deal only with forces which vary merely with the *area* on which they impinge. If in a big and a little ship two similar masts carry two similar sails, the two sails will be similarly strained, and equally stressed at homologous places, and alike suitable for resisting the force of the same wind. Two similar umbrellas, however differing in size, will serve alike in the same weather; and the expanse (though not the leverage) of a bird's wing may be enlarged with little alteration.

The principle of similitude had been admirably applied in a few clear instances by Lesage‡, a celebrated eighteenth-century physician, in an unfinished and unpublished work. Lesage argued, for example, that the larger ratio of surface to mass in a small animal would lead to excessive transpiration, were the skin as "porous" as our own; and that we may thus account for the hardened or thickened skins of insects and many other small terrestrial animals. Again, since the weight of a fruit increases as the cube of its linear dimensions, while the strength of the stalk increases as the square, it follows that the stalk must needs grow out of apparent due proportion to the fruit: or, alternatively, that tall trees should not bear large

* The subject is treated from the engineer's point of view by Prof. James Thomson, Comparison of similar structures as to elasticity, strength and stability, *Coll. Papers*, 1912, pp. 361–372, and *Trans. Inst. Engineers, Scotland*, 1876; also by Prof. A. Barr, *ibid.* 1899. See also Rayleigh, *Nature*, April 22, 1915; Sir G. Greenhill, On mechanical similitude, *Math. Gaz.* March 1916, *Coll. Works*, VI, p. 300. For a mathematical account, see (e.g.) P. W. Bridgeman, *Dimensional Analysis* (2nd ed.), 1931, or F. W. Lanchester, *The Theory of Dimensions*, 1936.

† Herbert Spencer, The form of the earth, etc., *Phil. Mag.* XXX, pp. 194–6, 1847; also *Principles of Biology*, pt. II, p. 123 *seq.*, 1864.

‡ See Pierre Prévost, *Notices de la vie et des écrits de Lesage*, 1805. George Louis Lesage, born at Geneva in 1724, devoted sixty-three years of a life of eighty to a mechanical theory of gravitation; see W. Thomson (Lord Kelvin), On the ultramundane corpuscles of Lesage, *Proc. R.S.E.* VII, pp. 577–589, 1872; *Phil. Mag.* XLV, pp. 321–345, 1873; and Clerk Maxwell, art. "Atom," *Encycl. Brit.* (9), p. 46.

fruit on slender branches, and that melons and pumpkins must lie upon the ground. And yet again, that in quadrupeds a large head must be supported on a neck which is either excessively thick and strong like a bull's, or very short like an elephant's*.

But it was Galileo who, wellnigh three hundred years ago, had first laid down this general principle of similitude; and he did so with the utmost possible clearness, and with a great wealth of illustration drawn from structures living and dead†. He said that if we tried building ships, palaces or temples of enormous size, yards, beams and bolts would cease to hold together; nor can Nature grow a tree nor construct an animal beyond a certain size, while retaining the proportions and employing the materials which suffice in the case of a smaller structure‡. The thing will fall to pieces of its own weight unless we either change its relative proportions, which will at length cause it to become clumsy, monstrous and inefficient, or else we must find new material, harder and stronger than was used before. Both processes are familiar to us in Nature and in art, and practical applications, undreamed of by Galileo, meet us at every turn in this modern age of cement and steel§.

Again, as Galileo was also careful to explain, besides the questions of pure stress and strain, of the strength of muscles to lift an increasing weight or of bones to resist its crushing stress, we have the important question of *bending moments*. This enters, more or less, into our whole range of problems; it affects the whole form of the skeleton, and sets a limit to the height of a tall tree||.

* Cf. W. Walton, On the debility of large animals and trees, *Quart. Journ. of Math.* IX, pp. 179–184, 1868; also L. J. Henderson, On volume in Biology, *Proc. Amer. Acad. Sci.* II, pp. 654–658, 1916; etc.

† *Discorsi e Dimostrazioni matematiche, intorno à due nuove scienze attenenti alla Mecanica ed ai Muovimenti Locali:* appresso gli Elzevirii, 1638; *Opere*, ed. Favaro, VIII, p. 169 *seq.* Transl. by Henry Crew and A. de Salvio, 1914, p. 130.

‡ So Werner remarked that Michael Angelo and Bramanti could not have built of gypsum at Paris on the scale they built of travertin at Rome.

§ The Chrysler and Empire State Buildings, the latter 1048 ft. high to the foot of its 200 ft. "mooring mast," are the last word, at present, in this brobdingnagian architecture.

|| It was Euler and Lagrange who first shewed (about 1776–1778) that a column of a certain height would merely be compressed, but one of a greater height would be bent by its own weight. See Euler, De altitudine columnarum etc., *Acta Acad. Sci. Imp. Petropol.* 1778, pp. 163–193; G. Greenhill, Determination of the greatest height to which a tree of given proportions can grow, *Cambr. Phil. Soc. Proc.* IV, p. 65, 1881, and Chree, *ibid.* VII, 1892.

We learn in elementary mechanics the simple case of two similar beams, supported at both ends and carrying no other weight than their own. Within the limits of their elasticity they tend to be deflected, or to sag downwards, in proportion to the squares of their linear dimensions; if a match-stick be two inches long and a similar beam six feet (or 36 times as long), the latter will sag under its own weight thirteen hundred times as much as the other. To counteract this tendency, as the size of an animal increases, the limbs tend to become thicker and shorter and the whole skeleton bulkier and heavier; bones make up some 8 per cent. of the body of mouse or wren, 13 or 14 per cent. of goose or dog, and 17 or 18 per cent. of the body of a man. Elephant and hippopotamus have grown clumsy as well as big, and the elk is of necessity less graceful than the gazelle. It is of high interest, on the other hand, to observe how little the skeletal proportions differ in a little porpoise and a great whale, even in the limbs and limb-bones; for the whole influence of gravity has become negligible, or nearly so, in both of these.

In the problem of the tall tree we have to determine the point at which the tree will begin to bend under its own weight if it be ever so little displaced from the perpendicular*. In such an investigation we have to make certain assumptions—for instance that the trunk tapers uniformly, and that the sectional area of the branches varies according to some definite law, or (as Ruskin assumed) tends to be constant in any horizontal plane; and the mathematical treatment is apt to be somewhat difficult. But Greenhill shewed, on such assumptions as the above, that a certain British Columbian pine-tree, of which the Kew flag-staff, which is 221 ft. high and 21 inches in diameter at the base, was made, could not possibly, by theory, have grown to more than about 300 ft. It is very curious that Galileo had suggested precisely the same height (*ducento braccie alta*) as the utmost limit of the altitude of a tree. In general, as Greenhill shewed, the diameter of a tall homogeneous body must increase as the power 3/2 of its height, which accounts for the slender proportions of young trees compared with the squat

* In like manner the wheat-straw bends over under the weight of the loaded ear, and the cat's tail bends over when held erect—not because they "possess flexibility," but because they outstrip the dimensions within which stable equilibrium is possible in a vertical position. The kitten's tail, on the other hand, stands up spiky and straight.

or stunted appearance of old and large ones*. In short, as Goethe says in *Dichtung und Wahrheit*, "Es ist dafür gesorgt dass die Bäume nicht in den Himmel wachsen."

But the tapering pine-tree is but a special case of a wider problem. The oak does not grow so tall as the pine-tree, but it carries a heavier load, and its boll, broad-based upon its spreading roots, shews a different contour. Smeaton took it for the pattern of his lighthouse, and Eiffel built his great tree of steel, a thousand feet high, to a similar but a stricter plan. Here the profile of tower or tree follows, or tends to follow, a logarithmic curve, giving equal strength throughout, according to a principle which we shall have occasion to discuss later on, when we come to treat of form and mechanical efficiency in the skeletons of animals. In the tree, moreover, anchoring roots form powerful wind-struts, and are most developed opposite to the direction of the prevailing winds; for the lifetime of a tree is affected by the frequency of storms, and its strength is related to the wind-pressure which it must needs withstand†.

Among animals we see, without the help of mathematics or of physics, how small birds and beasts are quick and agile, how slower and sedater movements come with larger size, and how exaggerated bulk brings with it a certain clumsiness, a certain inefficiency, an element of risk and hazard, a preponderance of disadvantage. The case was well put by Owen, in a passage which has an interest of its own as a premonition, somewhat like De Candolle's, of the "struggle for existence." Owen wrote as follows‡: "In proportion to the bulk of a species is the difficulty of the contest which, as a living organised whole, the individual of each species has to maintain against the surrounding agencies that are ever tending to dissolve the vital bond, and subjugate the living matter to the ordinary chemical and physical forces. Any changes, therefore, in such external conditions as a species may have been originally adapted

* The stem of the giant bamboo may attain a height of 60 metres while not more than about 40 cm. in diameter near its base, which dimensions fall not far short of the theoretical limits; A. J. Ewart, *Phil. Trans.* CXCVIII, p. 71, 1906.

† Cf. (*int. al.*) T. Petch, On buttress tree-roots, *Ann. R. Bot. Garden, Peradenyia*, XI, pp. 277–285, 1930. Also an interesting paper by James Macdonald, on The form of coniferous trees, *Forestry*, VI, 1 and 2, 1931/2.

‡ *Trans. Zool. Soc.* IV, p. 27, 1850.

to exist in, will militate against that existence in a degree proportionate, perhaps in a geometrical ratio, to the bulk of the species. If a dry season be greatly prolonged, the large mammal will suffer from the drought sooner than the small one; if any alteration of climate affect the quantity of vegetable food, the bulky Herbivore will be the first to feel the effects of stinted nourishment."

But the principle of Galileo carries us further and along more certain lines. The strength of a muscle, like that of a rope or girder, varies with its cross-section; and the resistance of a bone to a crushing stress varies, again like our girder, with its cross-section. But in a terrestrial animal the weight which tends to crush its limbs, or which its muscles have to move, varies as the cube of its linear dimensions; and so, to the possible magnitude of an animal, living under the direct action of gravity, there is a definite limit set. The elephant, in the dimensions of its limb-bones, is already shewing signs of a tendency to disproportionate thickness as compared with the smaller mammals; its movements are in many ways hampered and its agility diminished: it is already tending towards the maximal limit of size which the physical forces permit*. The spindleshanks of gnat or daddy-long-legs have their own factor of safety, conditional on the creature's exiguous bulk and weight; for after their own fashion even these small creatures tend towards an inevitable limitation of their natural size. But, as Galileo also saw, if the animal be wholly immersed in water like the whale, or if it be partly so, as was probably the case with the giant reptiles of the mesozoic age, then the weight is counterpoised to the extent of an equivalent volume of water, and is completely counterpoised if the density of the animal's body, with the included air, be identical (as a whale's very nearly is) with that of the water around†. Under these circumstances there is no longer the same physical barrier to the indefinite growth of the animal. Indeed, in the case of the aquatic animal, there is, as Herbert Spencer pointed out,

* Cf. A. Rauber, Galileo über Knochenformen, *Morphol. Jahrb.* VII, p. 327, 1882.

† Cf. W. S. Wall, *A New Sperm Whale* etc., Sydney, 1851, p. 64: "As for the immense size of Cetacea, it evidently proceeds from their buoyancy in the medium in which they live, and their being enabled thus to counteract the force of gravity."

a distinct advantage, in that the larger it grows the greater is its speed. For its available energy depends on the mass of its muscles, while its motion through the water is opposed, not by gravity, but by "skin-friction," which increases only as the square of the linear dimensions*: whence, other things being equal, the bigger the ship or the bigger the fish the faster it tends to go, but only in the ratio of the square root of the increasing length. For the velocity (V) which the fish attains depends on the work (W) it can do and the resistance (R) it must overcome. Now we have seen that the dimensions of W are l^3, and of R are l^2; and by elementary mechanics

$$W \propto RV^2, \text{ or } V^2 \propto \frac{W}{R}.$$

Therefore $$V^2 \propto \frac{l^3}{l^2} = l, \text{ and } V \propto \sqrt{l}.$$

This is what is known as *Froude's Law*, of the correspondence of speeds—a simple and most elegant instance of "dimensional theory†."

But there is often another side to these questions, which makes them too complicated to answer in a word. For instance, the work (per stroke) of which two similar engines are capable should vary as the cubes of their linear dimensions, for it varies on the one hand with the *area* of the piston, and on the other with the *length* of the stroke; so is it likewise in the animal, where the corresponding ratio depends on the cross-section of the muscle, and on the distance through which it contracts. But in two similar engines, the available horse-power varies as the square of the linear dimensions, and not as the cube; and this for the reason that the actual *energy* developed depends on the heating-surface of the boiler‡. So likewise must

* We are neglecting "drag" or "head-resistance," which, increasing as the cube of the speed, is a formidable obstacle to an unstreamlined body. But the perfect streamlining of whale or fish or bird lets the surrounding air or water behave like a perfect fluid, gives rise to no "surface of discontinuity," and the creature passes through it without recoil or turbulence. Froude reckoned skin-friction, or surface-resistance, as equal to that of a *plane* as long as the vessel's water-line, and of area equal to that of the wetted surface of the vessel.

† Though, as Lanchester says, the great designer "was not hampered by a knowledge of the theory of dimensions."

‡ The analogy is not a very strict or complete one. We are not taking account, for instance, of the thickness of the boiler-plates.

there be a similar tendency among animals for the rate of supply of kinetic energy to vary with the surface of the lung, that is to say (other things being equal) with the *square* of the linear dimensions of the animal; which means that, *caeteris paribus*, the small animal is stronger (having more power per unit weight) than a large one. We may of course (departing from the condition of similarity) increase the heating-surface of the boiler, by means of an internal system of tubes, without increasing its outward dimensions, and in this very way Nature increases the respiratory surface of a lung by a complex system of branching tubes and minute air-cells; but nevertheless in two similar and closely related animals, as also in two steam-engines of the same make, the law is bound to hold that the rate of working tends to vary with the square of the linear dimensions, according to Froude's *law of steamship comparison*. In the case of a very large ship, built for speed, the difficulty is got over by increasing the size and number of the boilers, till the ratio between boiler-room and engine-room is far beyond what is required in an ordinary small vessel*; but though we find lung-space increased among animals where greater rate of working is required, as in general among birds, I do not know that it can be shewn to increase, as in the "over-boilered" ship, with the size of the animal, and in a ratio which outstrips that of the other bodily dimensions. If it be the case then, that the working mechanism of the muscles should be able to exert a force proportionate to the cube of the linear bodily dimensions,

* Let L be the length, S the (wetted) surface, T the tonnage, D the displacement (or volume) of a ship; and let it cross the Atlantic at a speed V. Then, in comparing two ships, similarly constructed but of different magnitudes, we know that $L = V^2$, $S = L^2 = V^4$, $D = T = L^3 = V^6$; also R (resistance) $= S . V^2 = V^6$; H (horse-power) $= R . V = V^7$; and the coal (C) necessary for the voyage $= H/V = V^6$. That is to say, in ordinary engineering language, to increase the speed across the Atlantic by 1 per cent. the ship's length must be increased 2 per cent., her tonnage or displacement 6 per cent., her coal-consumption also 6 per cent., her horse-power, and therefore her boiler-capacity, 7 per cent. Her bunkers, accordingly, keep pace with the enlargement of the ship, but her boilers tend to increase out of proportion to the space available. Suppose a steamer 400 ft. long, of 2000 tons, 2000 H.P., and a speed of 14 knots. The corresponding vessel of 800 ft. long should develop a speed of 20 knots ($1 : 2 :: 14^2 : 20^2$), her tonnage would be 16,000, her H.P. 25,000 or thereby. Such a vessel would probably be driven by four propellers instead of one, each carrying 8000 H.P. See (*int. al.*) W. J. Millar, On the most economical speed to drive a steamer, *Proc. Edin. Math. Soc.* VII, pp. 27–29, 1889; Sir James R. Napier, On the most profitable speed for a fully laden cargo steamer for a given voyage, *Proc. Phil. Soc., Glasgow*, VI, pp. 33–38, 1865.

while the respiratory mechanism can only supply a store of energy at a rate proportional to the square of the said dimensions, the singular result ought to follow that, in swimming for instance, the larger fish ought to be able to put on a spurt of speed far in excess of the smaller one; but the distance travelled by the year's end should be very much alike for both of them. And it should also follow that the curve of fatigue is a steeper one, and the staying power less, in the smaller than in the larger individual. This is the case in long-distance racing, where neither draws far ahead until the big winner puts on his big spurt at the end; on which is based an aphorism of the turf, that "a good big 'un is better than a good little 'un." For an analogous reason wise men know that in the 'Varsity boat-race it is prudent and judicious to bet on the heavier crew.

Consider again the dynamical problem of the movements of the body and the limbs. The work done (W) in moving a limb, whose weight is p, over a distance s, is measured by ps; p varies as the cube of the linear dimensions, and s, in ordinary locomotion, varies as the linear dimensions, that is to say as the length of limb:

$$W \propto ps \propto l^3 \times l = l^4.$$

But the work done is limited by the power available, and this varies as the mass of the muscles, or as l^3; and under this limitation neither p nor s increase as they would otherwise tend to do. The limbs grow shorter, relatively, as the animal grows bigger; and spiders, daddy-long-legs and such-like long-limbed creatures attain no great size.

Let us consider more closely the actual energies of the body. A hundred years ago, in Strasburg, a physiologist and a mathematician were studying the temperature of warm-blooded animals*. The heat lost must, they said, be proportional to the surface of the animal: and the gain must be equal to the loss, since the temperature of the body keeps constant. It would seem, therefore, that the heat lost by radiation and that gained by oxidation vary both alike, as the surface-area, or the square of the linear dimensions, of the animal. But this result is paradoxical; for whereas the heat lost

* MM. Rameaux et Sarrus, *Bull. Acad. R. de Médecine*, III, pp. 1094–1100, 1838–39.

may well vary as the surface-area, that produced by oxidation ought rather to vary as the bulk of the animal: one should vary as the square and the other as the cube of the linear dimensions. Therefore the ratio of loss to gain, like that of surface to volume, ought to increase as the size of the creature diminishes. Another physiologist, Carl Bergmann*, took the case a step further. It was he, by the way, who first said that the real distinction was not between warm-blooded and cold-blooded animals, but between those of constant and those of variable temperature: and who coined the terms *homœothermic* and *poecilothermic* which we use today. He was driven to the conclusion that the smaller animal does produce more heat (per unit of mass) than the large one, in order to keep pace with surface-loss; and that this extra heat-production means more energy spent, more food consumed, more work done†. Simplified as it thus was, the problem still perplexed the physiologists for years after. The tissues of one mammal are much like those of another. We can hardly imagine the muscles of a small mammal to produce more heat (*caeteris paribus*) than those of a large; and we begin to wonder whether it be not nervous excitation, rather than quality of muscular tissue, which determines the rate of oxidation and the output of heat. It is evident in certain cases, and may be a general rule, that the smaller animals have the bigger brains; "plus l'animal est petit," says M. Charles Richet, "plus il a des échanges chimiques actifs, et plus son cerveau est volumineux‡." That the smaller animal needs more food is certain and obvious. The amount of food and oxygen consumed by a small flying insect is enormous; and bees and flies and hawkmoths and humming-

* Carl Bergmann, Verhältnisse der Wärmeökonomie der Tiere zu ihrer Grösse, *Göttinger Studien*, I, pp. 594–708, 1847—a very original paper.

† The metabolic activity of sundry mammals, per 24 hours, has been estimated as follows:

	Weight (kilo.)	Calories per kilo.
Guinea-pig	0·7	223
Rabbit	2	58
Man	70	33
Horse	600	22
Elephant	4000	13
Whale	150000	*circa* 1·7

‡ Ch. Richet, Recherches de calorimétrie, *Arch. de Physiologie* (3), VI, pp. 237–291, 450–497, 1885. Cf. also an interesting historical account by M. Elie le Breton, Sur la notion de "masse protoplasmique active": i. Problèmes posés par la signification de la loi des surfaces, *ibid.* 1906, p. 606.

birds live on nectar, the richest and most concentrated of foods*. Man consumes a fiftieth part of his own weight of food daily, but a mouse will eat half its own weight in a day; its rate of living is faster, it breeds faster, and old age comes to it much sooner than to man. A warm-blooded animal much smaller than a mouse becomes an impossibility; it could neither obtain nor yet digest the food required to maintain its constant temperature, and hence no mammals and no birds are as small as the smallest frogs or fishes. The disadvantage of small size is all the greater when loss of heat is accelerated by conduction as in the Arctic, or by convection as in the sea. The far north is a home of large birds but not of small; bears but not mice live through an Arctic winter; the least of the dolphins live in warm waters, and there are no small mammals in the sea. This principle is sometimes spoken of as *Bergmann's Law*.

The whole subject of the conservation of heat and the maintenance of an all but constant temperature in warm-blooded animals interests the physicist and the physiologist alike. It drew Kelvin's attention many years ago†, and led him to shew, in a curious paper, how larger bodies are kept warm by clothing while smaller are only cooled the more. If a current be passed through a thin wire, of which part is covered and part is bare, the thin bare part may glow with heat, while convection-currents streaming round the covered part cool it off and leave it in darkness. The hairy coat of very small animals is apt to look thin and meagre, but it may serve them better than a shaggier covering.

Leaving aside the question of the supply of energy, and keeping to that of the mechanical efficiency of the machine, we may find endless biological illustrations of the principle of similitude. All through the physiology of locomotion we meet with it in various ways: as, for instance, when we see a cockchafer carry a plate many times its own weight upon its back, or a flea jump many inches high. "A dog," says Galileo, "could probably carry two or three such dogs upon his back; but I believe that a horse could not carry even one of his own size."

* Cf. R. A. Davies and G. Fraenkel, The oxygen-consumption of flies during flight, *Jl. Exp. Biol.* XVII, pp. 402–407, 1940.

† W. Thomson, On the efficiency of clothing for maintaining temperature, *Nature*, XXIX, p. 567, 1884.

Such problems were admirably treated by Galileo and Borelli, but many writers remained ignorant of their work. Linnaeus remarked that if an elephant were as strong in proportion as a stag-beetle, it would be able to pull up rocks and level mountains; and Kirby and Spence have a well-known passage directed to shew that such powers as have been conferred upon the insect have been withheld from the higher animals, for the reason that had these latter been endued therewith they would have "caused the early desolation of the world*."

Such problems as that presented by the flea's jumping powers†, though essentially physiological in their nature, have their interest for us here: because a steady, progressive diminution of activity with increasing size would tend to set limits to the possible growth in magnitude of an animal just as surely as those factors which tend to break and crush the living fabric under its own weight. In the case of a leap, we have to do rather with a sudden impulse than with a continued strain, and this impulse should be measured in terms of the velocity imparted. The velocity is proportional to the impulse (x), and inversely proportional to the mass (M) moved: $V = x/M$. But, according to what we still speak of as "Borelli's law," the impulse (i.e. the work of the impulse) is proportional to the volume of the muscle by which it is produced‡, that is to say (in similarly constructed animals) to the mass of the whole body; for the impulse is proportional on the one hand to the cross-section of the muscle, and on the other to the distance through which it

* *Introduction to Entomology*, II, p. 190, 1826. Kirby and Spence, like many less learned authors, are fond of popular illustrations of the "wonders of Nature," to the neglect of dynamical principles. They suggest that if a white ant were as big as a man, its tunnels would be "magnificent cylinders of more than three hundred feet in diameter"; and that if a certain noisy Brazilian insect were as big as a man, its voice would be heard all the world over, "so that Stentor becomes a mute when compared with these insects!" It is an easy consequence of anthropomorphism, and hence a common characteristic of fairy-tales, to neglect the dynamical and dwell on the geometrical aspect of similarity.

† The flea is a very clever jumper; he jumps backwards, is stream-lined accordingly, and alights on his two long hind-legs. Cf. G. I. Watson, in *Nature*, 21 May 1938.

‡ That is to say, the available energy of muscle, in ft.-lbs. per lb. of muscle, is the same for all animals: a postulate which requires considerable qualification when we come to compare very different kinds of muscle, such as the insect's and the mammal's.

contracts. It follows from this that the velocity is constant, whatever be the size of the animal.

Putting it still more simply, the work done in leaping is proportional to the mass and to the height to which it is raised, $W \propto mH$. But the muscular power available for this work is proportional to the mass of muscle, or (in similarly constructed animals) to the mass of the animal, $W \propto m$. It follows that H is, or tends to be, a constant. In other words, all animals, provided always that they are similarly fashioned, with their various levers in like proportion, ought to jump not to the same relative but to the same *actual* height*. The grasshopper seems to be as well planned for jumping as the flea, and the actual heights to which they jump are much of a muchness; but the flea's jump is about 200 times its own height, the grasshopper's at most 20–30 times; and neither flea nor grasshopper is a better but rather a worse jumper than a horse or a man†.

As a matter of fact, Borelli is careful to point out that in the act of leaping the impulse is not actually instantaneous, like the blow of a hammer, but takes some little time, during which the levers are being extended by which the animal is being propelled forwards; and this interval of time will be longer in the case of the longer levers of the larger animal. To some extent, then, this principle acts as a corrective to the more general one, and tends to leave a certain balance of advantage in regard to leaping power on the side of the larger animal‡. But on the other hand, the question of strength of materials comes in once more, and the factors of stress and strain and bending moment make it more and more difficult for nature to endow the larger animal with the length of lever with which she has provided the grasshopper or the flea. To Kirby and Spence it seemed that "This wonderful strength of insects is doubtless the result of something peculiar in the structure and arrangement of their muscles, and principally their extraordinary

* Borelli, Prop. CLXXVII. Animalia minora et minus ponderosa majores saltus efficiunt respectu sui corporis, si caetera fuerint paria.

† The high jump is nowadays a highly skilled performance. For the jumper contrives that his centre of gravity goes *under* the bar, while his body, bit by bit, goes *over* it.

‡ See also (*int. al.*), John Bernoulli, *De Motu Musculorum*, Basil., 1694; Chabry, Mécanisme du saut, *J. de l'Anat. et de la Physiol.* XIX, 1883; Sur la longueur des membres des animaux sauteurs, *ibid.* XXI, p. 356, 1885; Le Hello, De l'action des organes locomoteurs, etc., *ibid.* XXIX, pp. 65–93, 1893; etc.

power of contraction." This hypothesis, which is so easily seen on physical grounds to be unnecessary, has been amply disproved in a series of excellent papers by Felix Plateau*.

From the *impulse* of the preceding case we may pass to the *momentum* created (or destroyed) under similar circumstances by a given force acting for a given time: $mv = Ft$.

We know that $m \propto l^3$, and $t = l/v$,

so that $l^3 v = Fl/v$, or $v^2 = F/l^2$.

But whatsoever force be available, the animal may only exert so much of it as is in proportion to the strength of his own limbs, that is to say to the cross-section of bone, sinew and muscle; and all of these cross-sections are proportional to l^2, the square of the linear dimensions. The maximal force, F_{max}, which the animal *dare* exert is proportional, then, to l^2; therefore

$$F_{max}/l^2 = \text{constant}.$$

And the maximal speed which the animal can safely reach, namely $V_{max} = F_{max}/l$, is also constant, or independent (*ceteris paribus*) of the dimensions of the animal.

A spurt or effort may be well within the capacity of the animal but far beyond the margin of safety, as trainer and athlete well know. This margin is a narrow one, whether for athlete or racehorse; both run a constant risk of overstrain, under which they may "pull" a muscle, lacerate a tendon, or even "break down" a bone†.

It is fortunate for their safety that animals do not jump to heights proportional to their own. For conceive an animal (of mass m) to jump to a certain altitude, such that it reaches the ground with a velocity v; then if c be the crushing strain at any point of the sectional area (A) of the limbs, the limiting condition is that

$$mv = cA.$$

If the animal vary in magnitude without change in the height to which it jumps (or in the velocity with which it descends), then

$$c \propto \frac{m}{A} \propto \frac{l^3}{l^2}, \quad \text{or } l.$$

The crushing strain varies directly with the linear dimensions of the animal; and this, a dynamical case, is identical with the usual statical limitation of magnitude.

* Recherches sur la force absolue des muscles des Invertébrés, *Bull. Acad. R. de Belgique* (3), VI, VII, 1883–84: see also *ibid.* (2), XX, 1865; XXII, 1866; *Ann. Mag. N.H.* XVII, p. 139, 1866; XIX, p. 95, 1867. Cf. M. Radau, Sur la force musculaire des insectes, *Revue des deux Mondes*, LXIV, p. 770, 1866. The subject had been well treated by Straus-Dürckheim, in his *Considérations générales sur l'anatomie comparée des animaux articulés*, 1828.

† Cf. The dynamics of sprint-running, by A. V. Hill and others, *Proc. R.S.* (B), CII, pp. 29–42, 1927; or *Muscular Movement in Man*, by A. V. Hill, New York, 1927, ch. VI, p. 41.

But if the animal, with increasing size or stature, jump to a correspondingly increasing height, the case becomes much more serious. For the final velocity of descent varies as the square root of the altitude reached, and therefore as the square root of the linear dimensions of the animal. And since, as before,

$$c \propto mv \propto \frac{l^3}{l^2}\,V,$$

$$\therefore \quad c \propto \frac{l^2}{l^2}\cdot\sqrt{l}, \quad \text{or } c \propto l^{\frac{3}{2}}.$$

If a creature's jump were in proportion to its height, the crushing strains would so increase that its dimensions would be limited thereby in a much higher degree than was indicated by statical considerations. An animal may grow to a size where it is unstable dynamically, though still on the safe side statically—a size where it moves with difficulty though it rests secure. It is by reason of dynamical rather than of statical relations that an elephant is of graver deportment than a mouse.

An apparently simple problem, much less simple than it looks, lies in the act of walking, where there will evidently be great economy of work if the leg swing with the help of gravity, that is to say, at a *pendulum-rate*. The conical shape and jointing of the limb, the time spent with the foot upon the ground, these and other mechanical differences complicate the case, and make the rate hard to define or calculate. Nevertheless, we may convince ourselves by counting our steps, that the leg does actually tend to swing, as a pendulum does, at a certain definite rate*. So on the same principle, but to the slower beat of a longer pendulum, the scythe swings smoothly in the mower's hands.

To walk quicker, we "step out"; we cause the leg-pendulum to describe a greater arc, but it does not swing or vibrate faster until we shorten the pendulum and begin to run. Now let two similar individuals, A and B, walk in a similar fashion, that is to say with a similar *angle* of swing (Fig. 1). The *arc* through which the leg swings, or the *amplitude* of each step, will then vary as the length of leg (say as a/b), and so as the height or other linear dimension (l) of the man†. But the time of swing varies inversely as the square

* The assertion that the limb tends to swing in pendulum-time was first made by the brothers Weber (*Mechanik der menschl. Gehwerkzeuge*, Göttingen, 1836). Some later writers have criticised the statement (e.g. Fischer, Die Kinematik des Beinschwingens etc., *Abh. math. phys. Kl. k. Sächs. Ges.* XXV–XXVIII, 1899–1903), but for all that, with proper and large qualifications, it remains substantially true.

† So the stride of a Brobdingnagian was 10 yards long, or just twelve times the 2 ft. 6 in., which make the average stride or half-pace of a man.

root of the pendulum-length, or $\sqrt{a}/\sqrt{b}$. Therefore the velocity, which is measured by amplitude/time, or $a/b \times \sqrt{b}/\sqrt{a}$, will also vary as the square root of the linear dimensions; which is Froude's law over again.

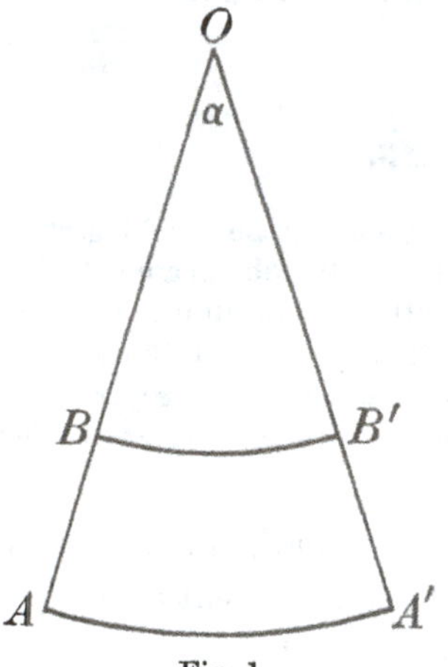

Fig. 1.

The smaller man, or smaller animal, goes slower than the larger, but only in the ratio of the square roots of their linear dimensions; whereas, if the limbs moved alike, irrespective of the size of the animal—if the limbs of the mouse swung no faster than those of the horse—then the mouse would be as slow in its gait or slower than the tortoise. M. Delisle* saw a fly walk three inches in half-a-second; this was good steady walking. When we walk five miles an hour we go about 88 inches in a second, or $88/6 = 14{\cdot}7$ times the pace of M. Delisle's fly. We should walk at just about the fly's pace if our stature were $1/(14{\cdot}7)^2$, or 1/216 of our present height—say 72/216 inches, or one-third of an inch high. Let us note in passing that the number of legs does not matter, any more than the number of wheels to a coach; the centipede runs none the faster for all his hundred legs.

But the leg comprises a complicated system of levers, by whose various exercise we obtain very different results. For instance, by being careful to rise upon our instep we increase the length or amplitude of our stride, and improve our speed very materially; and it is curious to see how Nature lengthens this metatarsal joint, or instep-lever, in horse† and hare and greyhound, in ostrich and in kangaroo, and in every speedy animal. Furthermore, in running we bend and so shorten the leg, in order to accommodate it to a quicker rate of pendulum-swing‡. In short the jointed structure

* Quoted in Mr John Bishop's interesting article in Todd's *Cyclopaedia*, III, p. 443.

† The "cannon-bones" are not only relatively longer but may even be actually longer in a little racehorse than a great carthorse.

‡ There is probably another factor involved here: for in bending and thus shortening the leg, we bring its centre of gravity nearer to the pivot, that is to say to the joint, and so the muscle tends to move it the more quickly. After all,

of the leg permits us to use it as the shortest possible lever while it is swinging, and as the longest possible lever when it is exerting its propulsive force.

The bird's case is of peculiar interest. In running, walking or swimming, we consider the speed which an animal *can attain*, and the increase of speed which increasing size permits of. But in flight there is a certain necessary speed—a speed (relative to the air) which the bird *must attain* in order to maintain itself aloft, and which *must* increase as its size increases. It is highly probable, as Lanchester remarks, that Lilienthal met his untimely death (in August 1896) not so much from any intrinsic fault in the design or construction of his machine, but simply because his engine fell somewhat short of the power required to give the speed necessary for its stability.

Twenty-five years ago, when this book was written, the bird, or the aeroplane, was thought of as a machine whose sloping wings, held at a given angle and driven horizontally forward, deflect the air downwards and derive support from the upward reaction. In other words, the bird was supposed to communicate to a mass of air a downward momentum equivalent (in unit time) to its own weight, and to do so by direct and continuous impact. The downward momentum is then proportional to the mass of air thrust downwards, and to the rate at which it is so thrust or driven: the mass being proportional to the wing-area and to the speed of the bird, and the rate being again proportional to the flying speed; so that the momentum varies as the square of the bird's linear dimensions and also as the square of its speed. But in order to balance its weight, this momentum must also be proportional to the cube of the bird's linear dimensions; therefore the bird's necessary speed, such as enables it to maintain level flight, must be proportional to the square root of its linear dimensions, and the whole work done must be proportional to the power $3\frac{1}{2}$ of the said linear dimensions.

The case stands, so far, as follows: m, the mass of air deflected downwards; M, the momentum so communicated; W, the work done—all in unit time; w, the weight, and V, the velocity of the

we know that the pendulum theory is not the whole story, but only an important first approximation to a complex phenomenon.

bird; l, a linear dimension, the form of the bird being supposed constant.

$$M = w = l^3, \text{ but } M = mV, \text{ and } m = l^2V.$$

Therefore $$M = l^2V^2 = l^3,$$

and therefore $$V = \sqrt{l}$$

and $$W = MV = l^{3\frac{1}{2}}.$$

The gist of the matter is, or seems to be, that the work which *can be done* varies with the available weight of muscle, that is to say, with the mass of the bird; but the work which *has to be done* varies with mass and distance; so the larger the bird grows, the greater the disadvantage under which all its work is done*. The disproportion does not seem very great at first sight, but it is quite enough to tell. It is as much as to say that, every time we double the linear dimensions of the bird, the difficulty of flight, or the work which must needs be done in order to fly, is increased in the ratio of 2^3 to $2^{3\frac{1}{2}}$, or $1 : \sqrt{2}$, or say $1 : 1{\cdot}4$. If we take the ostrich to exceed the sparrow in linear dimensions as $25 : 1$, which seems well within the mark, the ratio would be that between $25^{3\frac{1}{2}}$ and 25^3, or between 5^7 and 5^6; in other words, flight would be five times more difficult for the larger than for the smaller bird.

But this whole explanation is doubly inadequate. For one thing, it takes no account of *gliding flight*, in which energy is drawn from the wind, and neither muscular power nor engine power are employed; and we see that the larger birds, vulture, albatross or solan-goose, depend on gliding more and more. Secondly, the old simple account of the impact of the wing upon the air, and the manner in which a downward momentum is communicated and support obtained, is now known to be both inadequate and erroneous. For the science of flight, or aerodynamics, has grown out of the older science of hydrodynamics; both deal with the special properties of a fluid, whether water or air; and in our case, to be content to think of the air as a body of mass m, to which a velocity v is imparted, is to neglect all its fluid properties. How the

* This is the result arrived at by Helmholtz, Ueber ein Theorem geometrisch-ähnliche Bewegungen flüssiger Körper betreffend, nebst Anwendung auf das Problem Luftballons zu lenken, *Monatsber. Akad. Berlin*, 1873, pp. 501–514. It was criticised and challenged (somewhat rashly) by K. Müllenhof, Die Grösse der Flugflächen etc., *Pflüger's Archiv*, xxxv, p. 407; xxxvi, p. 548, 1885.

fish or the dolphin swims, and how the bird flies, are up to a certain point analogous problems; and *stream-lining* plays an essential part in both. But the bird is much heavier than the air, and the fish has much the same density as the water, so that the problem of keeping afloat or aloft is negligible in the one, and all-important in the other. Furthermore, the one fluid is highly compressible, and the other (to all intents and purposes) incompressible; and it is this very difference which the bird, or the aeroplane, takes special advantage of, and which helps, or even enables, it to fly.

It remains as true as ever that a bird, in order to counteract gravity, must cause air to move downward and obtains an upward reaction thereby. But the air displaced downward beneath the wing accounts for a small and varying part, perhaps a third perhaps a good deal less, of the whole force derived; and the rest is generated above the wing, in a less simple way. For, as the air streams past the slightly sloping wing, as smoothly as the stream-lined form and polished surface permit, it swirls round the front or "leading" edge*, and then streams swiftly over the upper surface of the wing; while it passes comparatively slowly, checked by the opposing slope of the wing, across the lower side. And this is as much as to say that it tends to be compressed below and rarefied above; in other words, that a partial vacuum is formed above the wing and follows it wherever it goes, so long as the stream-lining of the wing and its angle of incidence are suitable, and so long as the bird travels fast enough through the air.

The bird's weight is exerting a downward force upon the air, in one way just as in the other; and we can imagine a barometer delicate enough to shew and measure it as the bird flies overhead. But to calculate that force we should have to consider a multitude of component elements; we should have to deal with the stream-lined tubes of flow above and below, and the eddies round the fore-edge of the wing and elsewhere; and the calculation which was too simple before now becomes insuperably difficult. But the principle of necessary speed remains as true as ever. The bigger the bird

* The arched form, or "dipping front edge" of the wing, and its use in causing a vacuum above, were first recognised by Mr H. F. Phillips, who put the idea into a patent in 1884. The facts were discovered independently, and soon afterwards, both by Lilienthal and Lanchester.

becomes, the more swiftly must the air stream over the wing to give rise to the rarefaction or negative pressure which is more and more required; and the harder must it be to fly, so long as work has to be done by the muscles of the bird. The general principle is the same as before, though the quantitative relation does not work out as easily as it did. As a matter of fact, there is probably little difference in the end; and in aeronautics, the "total resultant force" which the bird employs for its support is said, *empirically*, to vary as the square of the air-speed: which is then a result analogous to Froude's law, and is just what we arrived at before in the simpler and less accurate setting of the case.

But a comparison between the larger and the smaller bird, like all other comparisons, applies only so long as the other factors in the case remain the same; and these vary so much in the complicated action of flight that it is hard indeed to compare one bird with another. For not only is the bird continually changing the incidence of its wing, but it alters the lie of every single important feather; and all the ways and means of flight vary so enormously, in big wings and small, and Nature exhibits so many refinements and "improvements" in the mechanism required, that a comparison based on size alone becomes imaginary, and is little worth the making.

The above considerations are of great practical importance in aeronautics, for they shew how a provision of increasing speed *must* accompany every enlargement of our aeroplanes. Speaking generally, the necessary or minimal speed of an aeroplane varies as the square root of its linear dimensions; if (*ceteris paribus*) we make it four times as long, it must, in order to remain aloft, fly twice as fast as before*. If a given machine weighing, say, 500 lb. be stable at 40 miles an hour, then a geometrically similar one which weighs, say, a couple of tons has its speed determined as follows:

$$W : w :: L^3 : l^3 :: 8 : 1.$$

Therefore $$L : l :: 2 : 1.$$

But $$V^2 : v^2 :: L : l.$$

Therefore $$V : v :: \sqrt{2} : 1 = 1{\cdot}414 : 1.$$

* G. H. Bryan, *Stability in Aviation*, 1911; F. W. Lanchester, *Aerodynamics*, 1909; cf. (*int. al.*) George Greenhill, *The Dynamics of Mechanical Flight*, 1912; F. W. Headley, *The Flight of Birds*, and recent works.

That is to say, the larger machine must be capable of a speed of $40 \times 1{\cdot}414$, or about $56\frac{1}{2}$, miles per hour.

An arrow is a somewhat rudimentary flying-machine; but it is capable, to a certain extent and at a high velocity, of acquiring "stability," and hence of actual flight after the fashion of an aeroplane; the duration and consequent range of its trajectory are vastly superior to those of a bullet of the same initial velocity. Coming back to our birds, and again comparing the ostrich with the sparrow, we find we know little or nothing about the actual speed of the latter; but the minimal speed of the swift is estimated at 100 ft. per second, or even more—say 70 miles an hour. We shall be on the safe side, and perhaps not far wrong, to take 20 miles an hour as the sparrow's minimal speed; and it would then follow that the ostrich, of 25 times the sparrow's linear dimensions, would have to fly (if it flew at all) with a minimum velocity of 5×20, or 100 miles an hour*.

The same principle of *necessary speed*, or the inevitable relation between the dimensions of a flying object and the minimum velocity at which its flight is stable, accounts for a considerable number of

* Birds have an ordinary and a forced speed. Meinertzhagen puts the ordinary flight of the swift at 68 m.p.h., which tallies with the old estimate of Athanasius Kircher (*Physiologia*, ed. 1680, p. 65) of 100 ft. per second for the swallow. Abel Chapman (*Retrospect*, 1928, ch. XIV) puts the gliding or swooping flight of the swift at over 150 m.p.h., and that of the griffon vulture at 180 m.p.h.; but these skilled fliers doubtless far exceed the necessary minimal speeds which we are speaking of. An airman flying at 70 m.p.h. has seen a golden eagle fly past him easily; but even this speed is exceptional. Several observers agree in giving 50 m.p.h. for grouse and woodcock, and 30 m.p.h. for starling, chaffinch, quail and crow. A migrating flock of lapwing travelled at 41 m.p.h., ten or twelve miles more than the usual speed of the single bird. Lanchester, on theoretical considerations, estimates the speed of the herring gull at 26 m.p.h., and of the albatross at about 34 miles. A tern, a very skilful flier, was seen to fly as slowly as 15 m.p.h. A hornet or a large dragonfly may reach 14 or 18 m.p.h.; but for most insects 2–4 metres per sec., say 4–9 m.p.h., is a common speed (cf. A. Magnan, *Vol. des Insectes*, 1834, p. 72). The larger diptera are very swift, but their speed is much exaggerated. A deerfly (*Cephenomyia*) has been said to fly at 400 yards per second, or say 800 m.p.h., an impossible velocity (Irving Langmuir, *Science*, March 11, 1938). It would mean a pressure on the fly's head of half an atmosphere, probably enough to crush the fly; to maintain it would take half a horsepower; and this would need a food-consumption of $1\frac{1}{2}$ times the fly's weight *per second*! 25 m.p.h. is a more reasonable estimate. The naturalist should not forget, though it does not touch our present argument, that the aeroplane is built to the pattern of a beetle rather than of a bird; for the elytra are not wings but planes. Cf. *int. al.*, P. Amans, Géométrie...des ailes rigides, *C.R. Assoc. Franç. pour l'avancem. des Sc.* 1901.

observed phenomena. It tells us why the larger birds have a marked difficulty in rising from the ground, that is to say, in acquiring to begin with the horizontal velocity necessary for their support; and why accordingly, as Mouillard* and others have observed, the heavier birds, even those weighing no more than a pound or two, can be effectually caged in small enclosures open to the sky. It explains why, as Mr Abel Chapman says, "all ponderous birds, wild swans and geese, great bustard and capercailzie, even blackcock, fly faster than they appear to do," while "light-built types with a big wing-area†, such as herons and harriers, possess no turn of speed at all." For the fact is that the heavy birds must fly quickly, or not at all. It tells us why very small birds, especially those as small as humming-birds, and *à fortiori* the still smaller insects, are capable of "stationary flight," a very slight and scarcely perceptible velocity relatively to the air being sufficient for their support and stability. And again, since it is in all these cases velocity relatively to the air which we are speaking of, we comprehend the reason why one may always tell which way the wind blows by watching the direction in which a bird starts to fly.

The wing of a bird or insect, like the tail of a fish or the blade of an oar, gives rise at each impulsion to a swirl or vortex, which tends (so to speak) to cling to it and travel along with it; and the resistance which wing or oar encounter comes much more from these vortices than from the viscosity of the fluid.‡ We learn as a corollary to this, that vortices form only at the edge of oar or wing—it is only the length and not the breadth of these which matters. A long narrow oar outpaces a broad one, and the efficiency of the long, narrow wing of albatross, swift or hawkmoth is so far accounted for. From the length of the wing we can calculate approximately its rate of swing, and more conjecturally the dimensions of each vortex, and finally the resistance or lifting power of the stroke; and the result shews once again the advantages of the small-scale

* Mouillard, *L'empire de l'air; essai d'ornithologie appliquée à l'aviation*, 1881; transl. in *Annual Report of the Smithsonian Institution*, 1892.

† On wing-area in relation to weight of bird see Lendenfeld in *Naturw. Wochenschr.* Nov. 1904, transl. in *Smithsonian Inst. Rep.* 1904; also E. H. Hankin, *Animal Flight*, 1913; etc.

‡ Cf. V. Bjerknes, *Hydrodynamique physique*, II, p. 293, 1934.

mechanism, and the disadvantage under which the larger machine or larger creature lies.

	Weight gm.	Length of wing m.	Beats per sec.	Speed of wing-tip m./s.	Radius of vortex*	Force of wing-beat gm.	Specific force, F/W
(From V. Bjerknes)							
Stork	3500	0·91	2	5·7	1·5	1480	2 : 5
Gull	1000	0·60	3	5·7	1·0	640	2 : 3
Pigeon	350	0·30	6	5·7	0·5	160	1 : 2
Sparrow	30	0·11	13	4·5	0·18	13	2 : 5
Bee	0·07	0·01	200	6·3	0·02	0·2	3½ : 1
Fly	0·01	0·007	190	4·2	0·01	0·04	4 : 1

* Conjectural.

A bird may exert a force at each stroke of its wing equal to one-half, let us say for safety one-quarter, of its own weight, more or less; but a bee or a fly does twice or thrice the equivalent of its own weight, at a low estimate. If stork, gull or pigeon can thus carry only one-fifth, one-third, one-quarter of their weight by the beating of their wings, it follows that all the rest must be borne by *sailing-flight* between the wing-beats. But an insect's wings lift it easily and with something to spare; hence sailing-flight, and with it the whole principle of necessary speed, does not concern the lesser insects, nor the smallest birds, at all; for a humming-bird can "stand still" in the air, like a hover-fly, and dart backwards as well as forwards, if it please.

There is a little group of Fairy-flies (Mymaridae), far below the size of any small familiar insects; their eggs are laid and larvae reared within the tiny eggs of larger insects; their bodies may be no more than ½ mm. long, and their outspread wings 2 mm. from tip to tip (Fig. 2). It is a peculiarity of some of these that their little wings are made of a few hairs or bristles, instead of the continuous membrane of a wing. How these act on the minute quantity of air involved we can only conjecture. It would seem that that small quantity reacts as a viscous fluid to the beat of the wing; but there are doubtless other unobserved anomalies in the mechanism and the mode of flight of these pigmy creatures†.

The ostrich has apparently reached a magnitude, and the moa certainly did so, at which flight by muscular action, according to

† It is obvious that in a still smaller order of magnitude the Brownian movement would suffice to make flight impossible.

the normal anatomy of a bird, becomes physiologically impossible. The same reasoning applies to the case of man. It would be very difficult, and probably absolutely impossible, for a bird to flap its way through the air were it of the bigness of a man; but Borelli, in discussing the matter, laid even greater stress on the fact that a man's pectoral muscles are so much less in proportion than those of a bird, that however we might fit ourselves out with wings, we could never expect to flap them by any power of our own weak muscles. Borelli had learned this lesson thoroughly, and in one of his chapters he deals with the proposition: *Est impossibile ut homines propriis viribus artificiose volare possint**. But gliding flight, where

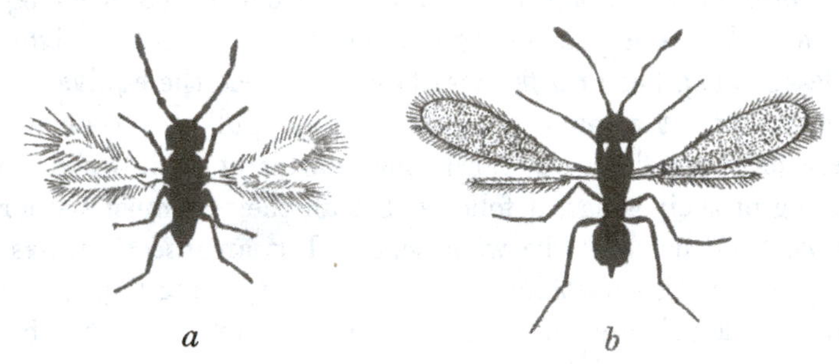

Fig. 2. Fairy-flies (*Mymaridae*): after F. Enock. ×20.

wind-force and gravitational energy take the place of muscular power, is another story, and its limitations are of another kind. Nature has many modes and mechanisms of flight, in birds of one kind and another, in bats and beetles, butterflies, dragonflies and what not; and gliding seems to be the common way of birds, and the flapping flight (*remigio alarum*) of sparrow and of crow to be the exception rather than the rule. But it were truer to say that gliding and soaring, by which energy is captured from the wind, are modes of flight little needed by the small birds, but more and more essential to the large. Borelli had proved so convincingly that we could never hope to fly *propriis viribus*, that all through the eighteenth century men tried no more to fly at all. It was in trying *to glide* that the pioneers of aviation, Cayley, Wenham and Mouillard,

* Giovanni Alfonso Borelli, *De Motu Animalium*, I, Prop. CCIV, p. 243, edit. 1685. The part on *The Flight of Birds* is issued by the Royal Aeronautical Society as No. 6 of its *Aeronautical Classics*.

Langley, Lilienthal and the Wrights—all careful students of birds—renewed the attempt*; and only after the Wrights had learned to glide did they seek to add power to their glider. Flight, as the Wrights declared, is a matter of practice and of skill, and skill in gliding has now reached a point which more than justifies all Leonardo da Vinci's attempts to fly. Birds shew infinite skill and instinctive knowledge in the use they make of the horizontal acceleration of the wind, and the advantage they take of ascending currents in the air. Over the hot sands of the Sahara, where every here and there hot air is going up and cooler coming down, birds keep as best they can to the one, or glide quickly through the other; so we may watch a big dragonfly planing slowly down a few feet above the heated soil, and only every five minutes or so regaining height with a vigorous stroke of his wings. The albatross uses the upward current on the lee-side of a great ocean-wave; so, on a lesser scale, does the flying-fish; and the seagull flies in curves, taking every advantage of the varying wind-velocities at different levels over the sea. An Indian vulture flaps his way up for a few laborious yards, then catching an upward current soars in easy spirals to 2000 feet; here he may stay, effortless, all day long, and come down at sunset. Nor is the modern sail-plane much less efficient than a soaring bird; for a skilful pilot in the tropics should be able to roam all day long at will†.

A bird's sensitiveness to air-pressure is indicated in other ways besides. Heavy birds, like duck and partridge, fly low and apparently take advantage of air-pressure reflected from the ground. Water-hen and dipper follow the windings of the stream as they fly up or down; a bee-line would give them a shorter course, but not so smooth a journey. Some small birds—wagtails, woodpeckers and a few others—fly, so to speak, by leaps and bounds; they fly briskly

* Sir George Cayley (1774–1857), father of British aeronautics, was the first to perceive the capabilities of rigid planes, and to experiment on gliding flight. He anticipated all the essential principles of the modern aeroplane, and his first paper "On Aerial Navigation" appeared in *Nicholson's Journal* for November 1809. F. H. Wenham (1824–1908) studied the flight of birds and estimated the necessary proportion of surface to weight and speed; he held that "the whole secret of success in flight depends upon a proper concave form of the supporting surface." See his paper "On Aerial Locomotion" in the *Report of the Aeronautical Society* 1866.

† Sir Gilbert Walker, in *Nature*, Oct. 2, 1937.

for a few moments, then close their wings and shoot along*. The flying-fishes do much the same, save that they keep their wings outspread. The best of them "taxi" along with only their tails in the water, the tail vibrating with great rapidity, and the speed attained lasts the fish on its long glide through the air†.

Flying may have begun, as in Man's case it did, with short spells of gliding flight, helped by gravity, and far short of sustained or continuous locomotion. The short wings and long tail of Archaeopteryx would be efficient as a slow-speed glider; and we may still see a Touraco glide down from his perch looking not much unlike Archaeopteryx in the proportions of his wings and tail. The small bodies, scanty muscles and narrow but vastly elongated wings of a Pterodactyl go far beyond the limits of mechanical efficiency for ordinary flapping flight; but for gliding they approach perfection‡. Sooner or later Nature does everything which is physically possible; and to *glide* with skill and safety through the air is a possibility which she did not overlook.

Apart from all differences in the action of the limbs—apart from differences in mechanical construction or in the manner in which the mechanism is used—we have now arrived at a curiously simple and uniform result. For in all the three forms of locomotion which we have attempted to study, alike in swimming and in walking, and even in the more complex problem of flight, the general result, obtained under very different conditions and arrived at by different modes of reasoning, shews in every case that speed tends to vary as the square root of the linear dimensions of the animal.

While the rate of progress tends to increase slowly with increasing size (according to Froude's law), and the rhythm or pendulum-rate of the limbs to increase rapidly with decreasing size (according to Galileo's law), some such increase of velocity with decreasing

* Why large birds cannot do the same is discussed by Lanchester, *op. cit.* Appendix IV.

† Cf. Carl L. Hubbs, On the flight of...the Cypselurinae, and remarks on the evolution of the flight of fishes, *Papers of the Michigan Acad. of Sci.* XVII, pp. 575–611, 1933. See also E. H. Hankin, *P.Z.S.* 1920, pp. 467–474; and C. M. Breeder, On the structural specialisation of flying fishes from the standpoint of aerodynamics, *Copeia*, 1930, pp. 114–121.

‡ The old conjecture that their flight was helped or rendered possible by a denser atmosphere than ours is thus no longer called for.

magnitude is true of all the rhythmic actions of the body, though for reasons not always easy to explain. The elephant's heart beats slower than ours*, the dog's quicker; the rabbit's goes pit-a-pat; the mouse's and the sparrow's are too quick to count. But the very "rate of living" (measured by the O consumed and CO_2 produced) slows down as size increases; and a rat lives so much faster than a man that the years of its life are three, instead of threescore and ten.

From all the foregoing discussion we learn that, as Crookes once upon a time remarked†, the forms as well as the actions of our bodies are entirely conditioned (save for certain exceptions in the case of aquatic animals) by the strength of gravity upon this globe; or, as Sir Charles Bell had put it some sixty years before, the very animals which move upon the surface of the earth are proportioned to its magnitude. Were the force of gravity to be doubled our bipedal form would be a failure, and the majority of terrestrial animals would resemble short-legged saurians, or else serpents. Birds and insects would suffer likewise, though with some compensation in the increased density of the air. On the other hand, if gravity were halved, we should get a lighter, slenderer, more active type, needing less energy, less heat, less heart, less lungs, less blood. Gravity not only controls the actions but also influences the forms of all save the least of organisms. The tree under its burden of leaves or fruit has changed its every curve and outline since its boughs were bare, and a mantle of snow will alter its configuration again. Sagging wrinkles, hanging breasts and many another sign of age are part of gravitation's slow relentless handiwork.

There are other physical factors besides gravity which help to limit the size to which an animal may grow and to define the conditions under which it may live. The small insects skating on a pool have their movements controlled and their freedom limited by the surface-tension between water and air, and the measure of that tension determines the magnitude which they may attain. A man coming wet from his bath carries a few ounces of water, and is perhaps 1 per cent. heavier than before; but a wet fly weighs twice as much as a dry one, and becomes a helpless thing. A small

* Say 28 to 30 beats to the minute.

† *Proc. Psychical Soc.* XII, p. 338–355, 1897.

insect finds itself imprisoned in a drop of water, and a fly with two feet in one drop finds it hard to extricate them.

The mechanical construction of insect or crustacean is highly efficient up to a certain size, but even crab and lobster never exceed certain moderate dimensions, perfect within these narrow bounds as their construction seems to be. Their body lies within a hollow shell, the stresses within which increase much faster than the mere scale of size; every hollow structure, every dome or cylinder, grows weaker as it grows larger, and a tin canister is easy to make but a great boiler is a complicated affair. The boiler has to be strengthened by "stiffening rings" or ridges, and so has the lobster's shell; but there is a limit even to this method of counteracting the weakening effect of size. An ordinary girder-bridge may be made efficient up to a span of 200 feet or so; but it is physically incapable of spanning the Firth of Forth. The great Japanese spider-crab, *Macrocheira*, has a span of some 12 feet across; but Nature meets the difficulty and solves the problem by keeping the body small, and building up the long and slender legs out of short lengths of narrow tubes. A hollow shell is admirable for small animals, but Nature does not and cannot make use of it for the large.

In the case of insects, other causes help to keep them of small dimensions. In their peculiar respiratory system blood does not carry oxygen to the tissues, but innumerable fine tubules or tracheae lead air into the interstices of the body. If we imagine them growing even to the size of crab or lobster, a vast complication of tracheal tubules would be necessary, within which friction would increase and diffusion be retarded, and which would soon be an inefficient and inappropriate mechanism.

The vibration of vocal chords and auditory drums has this in common with the pendulum-like motion of a limb that its rate also tends to vary inversely as the square root of the linear dimensions. We know by common experience of fiddle, drum or organ, that pitch rises, or the frequency of vibration increases, as the dimensions of pipe or membrane or string diminish; and in like manner we expect to hear a bass note from the great beasts and a piping treble from the small. The rate of vibration (N) of a stretched string depends on its tension and its density; these being equal, it varies inversely as its own length and as its diameter. For similar strings, $N \propto 1/l^2$, and for a circular membrane, of radius r and thickness e, $N \propto 1/(r^2 \sqrt{e})$.

But the delicate drums or tympana of various animals seem to vary much less in thickness than in diameter, and we may be content to write, once more, $N \propto 1/r^2$.

Suppose one animal to be fifty times less than another, vocal chords and all: the one's voice will be pitched 2500 times as many beats, or some ten or eleven octaves, above the other's; and the same comparison, or the same contrast, will apply to the tympanic membranes by which the vibrations are received. But our own perception of musical notes only reaches to 4000 vibrations per second, or thereby; a squeaking mouse or bat is heard by few, and to vibrations of 10,000 per second we are all of us stone-deaf. Structure apart, mere size is enough to give the lesser birds and beasts a music quite different to our own: the humming-bird, for aught we know, may be singing all day long. A minute insect may utter and receive vibrations of prodigious rapidity; even its little wings may beat hundreds of times a second*. Far more things happen to it in a second than to us; a thousandth part of a second is no longer negligible, and time itself seems to run a different course to ours.

The eye and its retinal elements have ranges of magnitude and limitations of magnitude of their own. A big dog's eye is hardly bigger than a little dog's; a squirrel's is much larger, proportionately, than an elephant's; and a robin's is but little less than a pigeon's or a crow's. For the rods and cones do not vary with the size of the animal, but have their dimensions optically limited by the interference-patterns of the waves of light, which set bounds to the production of clear retinal images. True, the larger animal may want a larger field of view; but this makes little difference, for but a small area of the retina is ever needed or used. The eye, in short, can never be very small and need never be very big; it has its own conditions and limitations apart from the size of the animal. But the insect's eye tells another story. If a fly had an eye like ours, the pupil would be so small that diffraction would render a clear image impossible. The only alternative is to unite a number

* The wing-beats are said to be as follows: dragonfly 28 per sec., bee 190, housefly 330; cf. Erhard, *Verh. d. d. zool. Gesellsch.* 1913, p. 206.

of small and optically isolated simple eyes into a compound eye, and in the insect Nature adopts this alternative possibility*.

Our range of vision is limited to a bare octave of "luminous" waves, which is a considerable part of the whole range of light-heat rays emitted by the sun; the sun's rays extend into the ultra-violet for another half-octave or more, but the rays to which our eyes are sensitive are just those which pass with the least absorption through a watery medium. Some ancient vertebrate may have learned to see in an ocean which let a certain part of the sun's whole radiation through, which part is *our part* still; or perhaps the watery media of the eye itself account sufficiently for the selective filtration. In either case, the dimensions of the retinal elements are so closely related to the wave-lengths of light (or to their interference patterns) that we have good reason to look upon the retina as perfect of its kind, within the limits which the properties of light itself impose; and this perfection is further illustrated by the fact that a few light-quanta, perhaps a single one, suffice to produce a sensation†. The hard eyes of insects are sensitive over a wider range. The bee has two visual optima, one coincident with our own, the other and principal one high up in the ultra-violet‡. And with the latter the bee is able to see that ultra-violet which is so well reflected by many flowers that flower-photographs have been taken through a filter which passes these but transmits no other rays§.

When we talk of light, and of magnitudes whose order is that of a wave-length of light, the subtle phenomenon of colour is near at hand. The hues of living things are due to sundry causes; where they come from chemical pigmentation they are outside our theme, but oftentimes there is no pigment at all, save perhaps as a screen or background, and the tints are those proper to a scale of wave-lengths or range of magnitude. In birds these "optical colours" are of two chief kinds. One kind include certain vivid blues, the

* Cf. C. J. van der Horst, The optics of the insect eye, *Acta Zoolog.* 1933, p. 108.

† Cf. Niels Bohr, in *Nature*, April 1, 1933, p. 457. Also J. Joly, *Proc. R.S.* (B), XCII, p. 222, 1921.

‡ L. M. Bertholf, Reactions of the honey-bee to light, *Journ. of Agric. Res.* XLIII, p. 379; XLIV, p. 763, 1931.

§ A. Kuhn, Ueber den Farbensinn der Bienen, *Ztschr. d. vergl. Physiol.* V, pp. 762–800, 1927; cf. F. K. Richtmeyer, Reflection of ultra-violet by flowers, *Journ. Optical Soc. Amer.* VII, pp. 151–168, 1923; etc.

blue of a blue jay, an Indian roller or a macaw; to the other belong the iridescent hues of mother-of-pearl, of the humming-bird, the peacock and the dove: for the dove's grey breast shews many colours yet contains but one—*colores inesse plures nec esse plus uno*, as Cicero said. The jay's blue feather shews a layer of enamel-like cells beneath a thin horny cuticle, and the cell-walls are spongy with innumerable tiny air-filled pores. These are about $0{\cdot}3\,\mu$ in diameter, in some birds even a little less, and so are not far from the limits of microscopic vision. A deeper layer carries dark-brown pigment, but there is no blue pigment at all; if the feather be dipped in a fluid of refractive index equal to its own, the blue utterly disappears, to reappear when the feather dries. This blue is like the colour of the sky; it is "Tyndall's blue," such as is displayed by turbid media, cloudy with dust-motes or tiny bubbles of a size comparable to the wave-lengths of the blue end of the spectrum. The longer waves of red or yellow pass through, the shorter violet rays are reflected or scattered; the intensity of the blue depends on the size and concentration of the particles, while the dark pigment-screen enhances the effect.

Rainbow hues are more subtle and more complicated; but in the peacock and the humming-bird we know for certain* that the colours are those of Newton's rings, and are produced by thin plates or films covering the barbules of the feather. The colours are such as are shewn by films about $\frac{1}{2}\,\mu$ thick, more or less; they change towards the blue end of the spectrum as the light falls more and more obliquely; or towards the red end if you soak the feather and cause the thin plates to swell. The barbules of the peacock's feather are broad and flat, smooth and shiny, and their cuticular layer splits into three very thin transparent films, hardly more than $1\,\mu$ thick, all three together. The gorgeous tints of the humming-birds have had their places in Newton's scale defined, and the changes which they exhibit at varying incidence have been predicted

* Rayleigh, *Phil. Mag.* (6), XXXVII, p. 98, 1919. For a review of the whole subject, and a discussion of its many difficulties, see H. Onslow, On a periodic structure in many insect scales, etc., *Phil. Trans.* (B), CCXI, pp. 1–74, 1921; also C. W. Mason, *Journ. Physic. Chemistry*, XXVII, XXX, XXXI, 1923–25–27; F. Suffert, *Zeitschr. f. Morph. u. Oekol. d. Tiere*, I, pp. 171–306, 1924 (scales of butterflies); also B. Reusch and Th. Elsasser in *Journ. f. Ornithologie*, LXXIII, 1925; etc.

and explained. The thickness of each film lies on the very limit of microscopic vision, and the least change or irregularity in this minute dimension would throw the whole display of colour out of gear. No phenomenon of organic magnitude is more striking than this constancy of size; none more remarkable than that these fine lamellae should have their tenuity so sharply defined, so uniform in feather after feather, so identical in all the individuals of a species, so constant from one generation to another.

A simpler phenomenon, and one which is visible throughout the whole field of morphology, is the tendency (referable doubtless in each case to some definite physical cause) for mere bodily *surface* to keep pace with *volume*, through some alteration of its form. The development of villi on the lining of the intestine (which increase its surface much as we enlarge the effective surface of a bath-towel), the various valvular folds of the intestinal lining, including the remarkable "spiral valve" of the shark's gut, the lobulation of the kidney in large animals*, the vast increase of respiratory surface in the air-sacs and alveoli of the lung, the development of gills in the larger crustacea and worms though the general surface of the body suffices for respiration in the smaller species—all these and many more are cases in which a more or less constant ratio tends to be maintained between mass and surface, which ratio would have been more and more departed from with increasing size, had it not been for such alteration of surface-form†. A leafy wood, a grassy sward, a piece of sponge, a reef of coral, are all instances of a like phenomenon. In fact, a deal of evolution is involved in keeping due balance between surface and mass as growth goes on.

In the case of very small animals, and of individual cells, the principle becomes especially important, in consequence of the molecular forces whose resultant action is limited to the superficial layer. In the cases just mentioned, action is *facilitated* by increase of surface: diffusion, for instance, of nutrient liquids or respiratory gases is rendered more rapid by the greater area of surface; but

* Cf. R. Anthony, *C.R.* CLXIX, p. 1174, 1919, etc. Cf. also A. Pütter, Studien über physiologische Ähnlichkeit, *Pflüger's Archiv*, CLXVIII, pp. 209–246, 1917.

† For various calculations of the increase of surface due to histological and anatomical subdivision, see E. Babak, Ueber die Oberflächenentwickelung bei Organismen, *Biol. Centralbl.* XXX, pp. 225–239, 257–267, 1910.

there are other cases in which the ratio of surface to mass may change the whole condition of the system. Iron rusts when exposed to moist air, but it rusts ever so much faster, and is soon eaten away, if the iron be first reduced to a heap of small filings; this is a mere difference of degree. But the spherical surface of the rain-drop and the spherical surface of the ocean (though both happen to be alike in mathematical form) are two totally different phenomena, the one due to surface-energy, and the other to that form of mass-energy which we ascribe to gravity. The contrast is still more clearly seen in the case of waves: for the little ripple, whose form and manner of propagation are governed by surface-tension, is found to travel with a velocity which is inversely as the square root of its length; while the ordinary big waves, controlled by gravitation, have a velocity directly proportional to the square root of their wave-length. In like manner we shall find that the form of all very small organisms is independent of gravity, and largely if not mainly due to the force of surface-tension: either as the direct result of the continued action of surface-tension on the semi-fluid body, or else as the result of its action at a prior stage of development, in bringing about a form which subsequent chemical changes have rendered rigid and lasting. In either case, we shall find a great tendency in small organisms to assume either the spherical form or other simple forms related to ordinary inanimate surface-tension phenomena, which forms do not recur in the external morphology of large animals.

Now this is a very important matter, and is a notable illustration of that principle of similitude which we have already discussed in regard to several of its manifestations. We are coming to a conclusion which will affect the whole course of our argument throughout this book, namely that there is an essential difference in kind between the phenomena of form in the larger and the smaller organisms. I have called this book a study of *Growth and Form*, because in the most familiar illustrations of organic form, as in our own bodies for example, these two factors are inseparably associated, and because we are here justified in thinking of form as the direct resultant and consequence of growth: of growth, whose varying rate in one direction or another has produced, by its gradual and unequal increments, the successive stages of development and the final configuration of the whole material structure. But it is by no means true that form and growth are in this direct and simple fashion correlative or complementary in the case of minute portions of living matter. For in the smaller organisms, and in the individual cells of the larger, we have reached an order of magnitude in which the intermolecular forces strive under favourable conditions with, and at length altogether outweigh, the force of gravity, and also those other forces leading to movements of convection which are the prevailing factors in the larger material aggregate.

However, we shall require to deal more fully with this matter in our discussion of the rate of growth, and we may leave it meanwhile, in order to deal with other matters more or less directly concerned with the magnitude of the cell.

The living cell is a very complex field of energy, and of energy of many kinds, of which surface-energy is not the least. Now the whole surface-energy of the cell is by no means restricted to its *outer* surface; for the cell is a very heterogeneous structure, and all its protoplasmic alveoli and other visible (as well as invisible) heterogeneities make up a great system of internal surfaces, at every part of which one "phase" comes in contact with another "phase," and surface-energy is manifested accordingly. But still, the external surface is a definite portion of the system, with a definite "phase" of its own, and however little we may know of the distribution of the total energy of the system, it is at least plain that the conditions which favour equilibrium will be greatly altered by the changed ratio of external surface to mass which a mere change of magnitude produces in the cell. In short, the phenomenon of division of the growing cell, however it be brought about, will be precisely what is wanted to keep fairly constant the ratio between surface and mass, and to retain or restore the balance between surface-energy and the other forces of the system*. But when a germ-cell divides or "segments" into two, it does not increase in mass; at least if there be some slight alleged tendency for the egg to increase in

* Certain cells of the cucumber were found to divide when they had grown to a volume half as large again as that of the "resting cells." Thus the volumes of resting, dividing and daughter cells were as 1 : 1·5 : 0·75; and their surfaces, being as the power 2/3 of these figures, were, roughly, as 1 : 1·3 : 0·8. The ratio of S/V was then as 1 : 0·9 : 1·1, or much nearer equality. Cf. F. T. Lewis, *Anat. Record*, XLVII, pp. 59–99, 1930.

mass or volume during segmentation it is very slight indeed, generally imperceptible, and wholly denied by some*. The growth or development of the egg from a one-celled stage to stages of two or many cells is thus a somewhat peculiar kind of growth; it is growth limited to change of form and increase of surface, unaccompanied by growth in volume or in mass. In the case of a soap-bubble, by the way, if it divide into two bubbles the volume is actually diminished, while the surface-area is greatly increased†; the diminution being due to a cause which we shall have to study later, namely to the increased pressure due to the greater curvature of the smaller bubbles.

An immediate and remarkable result of the principles just described is a tendency on the part of all cells, according to their kind, to vary but little about a certain mean size, and to have in fact certain absolute limitations of magnitude. The diameter of a large parenchymatous cell is perhaps tenfold that of a little one; but the tallest phanerogams are ten thousand times the height of the least. In short, Nature has her materials of predeterminate dimensions, and keeps to the same bricks whether she build a great house or a small. Even ordinary drops tend towards a certain fixed size, which size is a function of the surface-tension, and may be used (as Quincke used it) as a measure thereof. In a shower of rain the principle is curiously illustrated, as Wilding Köller and V. Bjerknes tell us. The drops are of graded sizes, *each twice as big as another*, beginning with the minute and uniform droplets of an impalpable mist. They rotate as they fall, and if two rotate in contrary directions they draw together and presently coalesce; but this only happens when two drops are falling side by side, and since the rate of fall depends on the size it always is a pair of coequal drops which so meet, approach and join together. A supreme instance of constancy or quasi-constancy of size, remote from but yet analogous to the size-limitation of a rain-drop or a cell, is the fact that the stars of heaven (however else one differeth from another), and even the nebulae themselves, are all wellnigh co-equal in *mass*. Gravity draws matter together, condensing it into a world

* Though the entire egg is not increasing in mass, that is not to say that its living protoplasm is not increasing all the while at the expense of the reserve material.

† Cf. P. G. Tait, *Proc. R.S.E.* v, 1866 and VI, 1868.

or into a star; but ethereal pressure is an opponent force leading to disruption, negligible on the small scale but potent on the large. High up in the scale of magnitude, from about 10^{33} to 10^{35} grams of matter, these two great cosmic forces balance one another; and all the magnitudes of all the stars lie within or hard by these narrow limits.

In the living cell, Sachs pointed out (in 1895) that there is a tendency for each nucleus to gather around itself a certain definite amount of protoplasm*. Driesch†, a little later, found it possible, by artificial subdivision of the egg, to rear dwarf sea-urchin larvae, one-half, one-quarter or even one-eighth of their usual size; which dwarf larvae were composed of only a half, a quarter or an eighth of the normal number of cells. These observations have been often repeated and amply confirmed: and Loeb found the sea-urchin eggs capable of reduction to a certain size, but no further.

In the development of *Crepidula* (an American "slipper-limpet," now much at home on our oyster-beds), Conklin‡ has succeeded in rearing dwarf and giant individuals, of which the latter may be five-and-twenty times as big as the former. But the individual cells, of skin, gut, liver, muscle and other tissues, are just the same size in one as in the other, in dwarf and in giant§. In like manner

* *Physiologische Notizen* (9), p. 425, 1895. Cf. Amelung, *Flora*, 1893; Strasbürger, Ueber die Wirkungssphäre der Kerne und die Zellgrösse, *Histol. Beitr.* (5), pp. 95–129, 1893; R. Hertwig, Ueber Korrelation von Zell- und Kerngrösse (Kernplasmarelation), *Biol. Centralbl.* XVIII, pp. 49–62, 108–119, 1903; G. Levi and T. Terni, Le variazioni dell' indice plasmatico-nucleare durante l' intercinesi, *Arch. Ital. di Anat.* X, p. 545, 1911; also E. le Breton and G. Schaeffer, *Variations biochimiques du rapport nucléo-plasmatique*, Strasburg, 1923.

† *Arch. f. Entw. Mech.* IV, 1898, pp. 75, 247.

‡ E. G. Conklin, Cell-size and nuclear size, *Journ. Exp. Zool.* XII, pp. 1–98, 1912; Body-size and cell-size, *Journ. of Morphol.* XXIII, pp. 159–188, 1912. Cf. M. Popoff, Ueber die Zellgrösse, *Arch. f. Zellforschung*, III, 1909.

§ Thus the fibres of the crystalline lens are of the same size in large and small dogs, Rabl, *Z. f. w. Z.* LXVII, 1899. Cf. (*int. al.*) Pearson, On the size of the blood-corpuscles in Rana, *Biometrika*, VI, p. 403, 1909. Dr Thomas Young caught sight of the phenomenon early in last century: "The solid particles of the blood do not by any means vary in magnitude in the same ratio with the bulk of the animal," *Natural Philosophy*, ed. 1845, p. 466; and Leeuwenhoek and Stephen Hales were aware of it nearly two hundred years before. Leeuwenhoek indeed had a very good idea of the size of a human blood-corpuscle, and was in the habit of using its diameter—about 1/3000 of an inch—as a standard of comparison. But though the blood-corpuscles shew no relation of magnitude to the size of the animal, they are related without doubt to its activity; for the corpuscles in the

the leaf-cells are found to be of the same size in an ordinary water-lily, in the great *Victoria regia*, and in the still huger leaf, nearly 3 metres long, of *Euryale ferox* in Japan*. Driesch has laid particular stress upon this principle of a "fixed cell-size," which has, however, its own limitations and exceptions. Among these exceptions, or apparent exceptions, are the giant frond-like cell of a Caulerpa or the great undivided plasmodium of a Myxomycete. The flattening of the one and the branching of the other serve (or help) to increase the ratio of surface to content, the nuclei tend to multiply, and streaming currents keep the interior and exterior of the mass in touch with one another.

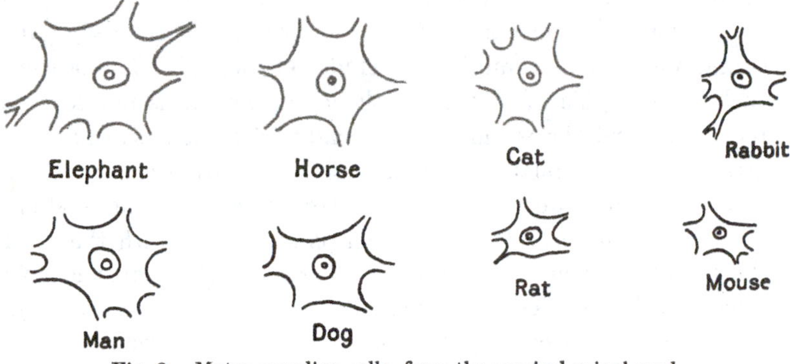

Fig. 3. Motor ganglion-cells, from the cervical spinal cord. From Minot, after Irving Hardesty.

We get a good and even a familiar illustration of the principle of size-limitation in comparing the brain-cells or ganglion-cells, whether of the lower or of the higher animals†. In Fig. 3 we shew certain identical nerve-cells from various mammals, from mouse to elephant, all drawn to the same scale of magnification; and we see that they are all of much the same *order* of magnitude. The nerve-cell of the elephant is about twice that of the mouse in linear

sluggish Amphibia are much the largest known to us, while the smallest are found among the deer and other agile and speedy animals (cf. Gulliver, *P.Z.S.* 1875, p. 474, etc.). This correlation is explained by the surface condensation or adsorption of oxygen in the blood-corpuscles, a process greatly facilitated and intensified by the increase of surface due to their minuteness.

* Okada and Yomosuke, in *Sci. Rep. Tohoku Univ.* III, pp. 271–278, 1928.

† Cf. P. Enriques, La forma come funzione della grandezza: Ricerche sui gangli nervosi degli invertebrati, *Arch. f. Entw. Mech.* xxv, p. 655, 1907–8.

dimensions, and therefore about eight times greater in volume or in mass. But making due allowance for difference of shape, the linear dimensions of the elephant are to those of the mouse as not less than one to fifty; and the bulk of the larger animal is something like 125,000 times that of the less. It follows, if the size of the nerve-cells are as eight to one, that, in corresponding parts of the nervous system, there are more than 15,000 times as many individual cells in one animal as in the other. In short we may (with Enriques) lay it down as a general law that among animals, large or small, the ganglion-cells vary in size within narrow limits; and that, amidst all the great variety of structure observed in the nervous system of different classes of animals, it is always found that the smaller species have simpler ganglia than the larger, that is to say ganglia containing a smaller number of cellular elements*. The bearing of such facts as this upon the cell-theory in general is not to be disregarded; and the warning is especially clear against exaggerated attempts to correlate physiological processes with the visible mechanism of associated cells, rather than with the system of energies, or the field of force, which is associated with them. For the life of the body is more than the *sum* of the properties of the cells of which it is composed: as Goethe said, "Das Lebendige ist zwar in Elemente zerlegt, aber man kann es aus diesen nicht wieder zusammenstellen und beleben."

Among certain microscopic organisms such as the Rotifera (which have the least average size and the narrowest range of size of all the Metazoa), we are still more palpably struck by the small number of cells which go to constitute a usually complex organ, such as kidney, stomach or ovary; we can sometimes number them in a few

* While the difference in cell-volume is vastly less than that between the volumes, and very much less also than that between the surfaces, of the respective animals, yet there *is* a certain difference; and this it has been attempted to correlate with the need for each cell in the many-celled ganglion of the larger animal to possess a more complex "exchange-system" of branches, for intercommunication with its more numerous neighbours. Another explanation is based on the fact that, while such cells as continue to divide throughout life tend to uniformity of size in all mammals, those which do not do so, and in particular the ganglion cells, continue to grow, and their size becomes, therefore, a function of the duration of life. Cf. G. Levi, Studii sulla grandezza delle cellule, *Arch. Ital. di Anat. e di Embriolog.* v, p. 291, 1906; cf. also A. Berezowski, Studien über die Zellgrösse, *Arch. f. Zellforsch.* v, pp. 375–384, 1910.

units, in place of the many thousands which make up such an organ in larger, if not always higher, animals. We have already spoken of the Fairy-flies, a few score of which would hardly weigh down one of the larger rotifers, and a hundred thousand would weigh less than one honey-bee. Their form is complex and their little bodies exquisitely beautiful; but I feel sure that their cells are few, and their organs of great histological simplicity. These considerations help, I think, to shew that, however important and advantageous the subdivision of the tissues into cells may be from the constructional, or from the dynamic, point of view, the phenomenon has less fundamental importance than was once, and is often still, assigned to it.

Just as Sachs shewed there was a limit to the amount of cytoplasm which could gather round a nucleus, so Boveri has demonstrated that the nucleus itself has its own limitations of size, and that, in cell-division after fertilisation, each new nucleus has the same size as its parent nucleus*; we may nowadays transfer the statement to the chromosomes. It may be that a bacterium lacks a nucleus for the simple reason that it is too small to hold one, and that the same is true of such small plants as the Cyanophyceae, or blue-green algae. Even a chromatophore with its "pyrenoids" seems to be impossible below a certain size†.

Always then, there are reasons, partly physiological but in large part purely physical, which define or regulate the magnitude of the organism or the cell. And as we have already found definite limitations to the increase in magnitude of an organism, let us now enquire whether there be not also a lower limit below which the very existence of an organism becomes impossible.

* Boveri, *Zellenstudien*, V: Ueber die Abhängigkeit der Kerngrösse und Zellenzahl von der Chromosomenzahl der Ausgangszellen. Jena, 1905. Cf. also (*int. al.*) H. Voss, Kerngrössenverhältnisse in der Leber etc., *Ztschr. f. Zellforschung*, VII, pp. 187–200, 1928.

† The size of the nucleus may be affected, even determined, by the number of chromosomes it contains. There are giant races of *Oenothera*, *Primula* and *Solanum* whose cell-nuclei contain twice the normal number of chromosomes, and a dwarf race of a little freshwater crustacean, *Cyclops*, has half the usual number. The cytoplasm in turn varies with the amount of nuclear matter, the whole cell is unusually large or unusually small; and in these exceptional cases we see a direct relation between the size of the organism and the size of the cell. Cf. (*int. al.*) R. P. Gregory, *Proc. Camb. Phil. Soc.* XV, pp. 239–246, 1909; F. Keeble, *Journ. of Genetics*, II, pp. 163–188, 1912.

A bacillus of ordinary size is, say, $1\,\mu$ in length. The length (or height) of a man is about a million and three-quarter times as great, i.e. 1·75 metres, or $1{\cdot}75 \times 10^6\,\mu$; and the mass of the man is in the neighbourhood of 5×10^{18} (five million, million, million) times greater than that of the bacillus. If we ask whether there may not exist organisms as much less than the bacillus as the bacillus is less than the man, it is easy to reply that this is quite impossible, for we are rapidly approaching a point where the question of molecular dimensions, and of the ultimate divisibility of matter, obtrudes itself as a crucial factor in the case. Clerk Maxwell dealt with this matter seventy years ago, in his celebrated article *Atom**. Kolli (or Colley), a Russian chemist, declared in 1893 that the head of a spermatozoon could hold no more than a few protein molecules; and Errera, ten years later, discussed the same topic with great ingenuity†. But it needs no elaborate calculation to convince us that the smaller bacteria or micrococci nearly approach the smallest magnitudes which we can conceive to have an organised structure. A few small bacteria are the smallest of visible organisms, and a minute species associated with influenza, *B. pneumosinter*, is said to be the least of them all. Its size is of the order of $0{\cdot}1\,\mu$, or rather less; and here we are in close touch with the utmost limits of microscopic vision, for the wave-lengths of visible light run only from about 400 to 700 $\text{m}\mu$. The largest of the bacteria, *B. megatherium*, larger than the well-known *B. anthracis* of splenic fever, has much the same proportion to the least as an elephant to a guinea-pig‡.

Size of body is no mere accident. Man, respiring as he does, cannot be as small as an insect, nor *vice versa*; only now and then, as in the Goliath beetle, do the sizes of mouse and beetle meet and overlap. The descending scale of mammals stops short at a weight of about 5 grams, that of beetles at a length of about half a millimetre, and every group of animals has its upper and its lower limitations of size. So, not far from the lower limit of our vision, does the long series of bacteria come to an end. There remain still smaller particles which the ultra-microscope in part reveals; and

* *Encyclopaedia Britannica*, 9th edition, 1875.

† Leo Errera, Sur la limite de la petitesse des organismes, *Bull. Soc. Roy. des Sc. méd. et nat. de Bruxelles*, 1903; *Recueil d'œuvres* (*Physiologie générale*), p. 325.

‡ Cf. A. E. Boycott, The transition from live to dead, *Proc. R. Soc. of Medicine*, XXII (*Pathology*), pp. 55–69, 1928.

here or hereabouts are said to come the so-called viruses or "filter-passers," brought within our ken by the maladies, such as hydrophobia, or foot-and-mouth disease, or the mosaic diseases of tobacco and potato, to which they give rise. These minute particles, of the order of one-tenth the diameter of our smallest bacteria, have no diffusible contents, no included water—whereby they differ from every living thing. They appear to be inert colloidal (or even crystalloid) aggregates of a nucleo-protein, of perhaps ten times the diameter of an ordinary protein-molecule, and not much larger than the giant molecules of haemoglobin or haemocyanin*.

Bejerinck called such a virus a *contagium vivum*; "infective nucleo-protein" is a newer name. We have stepped down, by a single step, from living to non-living things, from bacterial dimensions to the molecular magnitudes of protein chemistry. And we begin to suspect that the virus-diseases are not due to an "organism, capable of physiological reproduction and multiplication, but to a mere specific chemical substance, capable of catalysing pre-existing materials and thereby producing more and more molecules like itself. The spread of the virus in a plant would then be a mere autocatalysis, not involving the transport of matter, but only a progressive change of state in substances already there†."

But, after all, a simple tabulation is all we need to shew how nearly the least of organisms approach to molecular magnitudes. The same table will suffice to shew how each main group of animals has its mean and characteristic size, and a range on either side, sometimes greater and sometimes less.

Our table of magnitudes is no mere catalogue of isolated facts, but goes deep into the relation between the creature and its world. A certain range, and a narrow one, contains mouse and elephant, and all whose business it is to walk and run; this is our own world,

* Cf. Svedberg, *Journ. Am. Chem. Soc.* XLVIII, p. 30, 1926. According to the Foot-and-Mouth Disease Research Committee (*5th Report*, 1937), the foot-and-mouth virus has a diameter, determined by graded filters, of 8–12 $m\mu$; while Kenneth Smith and W. D. MacClement (*Proc. R.S.* (B), CXXV, p. 296, 1938) calculate for certain others a diameter of no more than 4 $m\mu$, or less than a molecule of haemocyanin.

† H. H. Dixon, Croonian lecture on the transport of substances in plants, *Proc. R.S.* (B), vol. CXXV, pp. 22, 23, 1938.

with whose dimensions our lives, our limbs, our senses are in tune. The great whales grow out of this range by throwing the burden of their bulk upon the waters; the dinosaurs wallowed in the swamp, and the hippopotamus, the sea-elephant and Steller's great sea-cow pass or passed their lives in the rivers or the sea. The things which

Linear dimensions of organisms, and other objects

	cm.		
(10,000 km.)	10^7	A quadrant of the earth's circumference	
(1000 km.)	10^6	Orkney to Land's End	
	10^5		
	10^4		
		Mount Everest	
(km.)	10^3		
	10^2	Giant trees: *Sequoia*	
		Large whale	
	10^1	Basking shark	
		Elephant; ostrich; man	
(metre)	10^0		
		Dog; rat; eagle	
	10^{-1}		
		Small birds and mammals; large insects	
(cm.)	10^{-2}		
		Small insects; minute fish	
(mm.)	10^{-3}		
		Minute insects	
	10^{-4}		
		Protozoa; pollen-grains	Cells
	10^{-5}		
		Large bacteria; human blood-corpuscles	
(micron, μ)	10^{-6}		
		Minute bacteria	
	10^{-7}		
		Limit of microscopic vision	
		Viruses, or filter-passers	Colloid particles
	10^{-8}	Giant albuminoids, casein, etc.	
		Starch-molecule	
($m\mu$)	10^{-9}		
		Water-molecule	
(Ångström unit)	10^{-10}		

fly are smaller than the things which walk and run; the flying birds are never as large as the larger mammals, the lesser birds and mammals are much of a muchness, but insects come down a step in the scale and more. The lessening influence of gravity facilitates flight, but makes it less easy to walk and run; first claws, then hooks and suckers and glandular hairs help to secure a foothold, until to creep upon wall or ceiling becomes as easy as to walk upon the ground. Fishes, by evading gravity, increase their range of magnitude both above and below that of terrestrial animals. Smaller than all these, passing out of our range of vision and going down to the least dimensions of living things, are protozoa, rotifers, spores, pollen-grains* and bacteria. All save the largest of these float rather than swim; they are buoyed up by air or water, and fall (as Stokes's law explains) with exceeding slowness.

There is a certain narrow range of magnitudes where (as we have partly said) gravity and surface tension become comparable forces, nicely balanced with one another. Here a population of small plants and animals not only dwell in the surface waters but are bound to the surface film itself—the whirligig beetles and pond-skaters, the larvae of gnat and mosquito, the duckweeds (*Lemna*), the tiny *Wolffia*, and *Azolla*; even in mid-ocean, one small insect (*Halobates*) retains this singular habitat. It would be a long story to tell the various ways in which surface-tension is thus taken full advantage of. Gravitation not only limits the magnitude but controls the form of things. With the help of gravity the quadruped has its back and its belly, and its limbs upon the ground; its freedom of motion in a plane perpendicular to gravitational force; its sense of fore-and-aft, its head and tail, its bilateral symmetry. Gravitation influences both our bodies and our minds. We owe to it our sense of the vertical, our knowledge of up-and-down; our conception of the horizontal plane on which we stand, and our discovery of two axes therein, related to the vertical as to one another; it was gravity which taught us to think of three-dimensional space. Our architecture is controlled by gravity, but gravity has less influence over the architecture of the bee; a bee might be excused, might even be commended, if it referred space to four dimensions instead of three!† The plant has its root and its stem; but about this vertical or

* Pollen-grains, like protozoa, have a considerable range of magnitude. The largest, such as those of the pumpkin, are about 200μ in diameter; these have to be carried by insects, for they are above the level of Stokes's law, and no longer float upon the air. The smallest pollen-grains, such as those of the forget-me-not, are about 4½ μ in diameter (Wodehouse).

† Corresponding, that is to say, to the four axes which, meeting in a point, make co-equal angles (the so-called tetrahedral angles) one with another, as do the basal angles of the honeycomb. (See below, chap. VII.)

gravitational axis its radiate symmetry remains, undisturbed by directional polarity, save for the sun. Among animals, radiate symmetry is confined to creatures of no great size; and some form or degree of spherical symmetry becomes the rule in the small world of the protozoon—unless gravity resume its sway through the added burden of a shell. The creatures which swim, walk or run, fly, creep or float are, so to speak, inhabitants and natural proprietors of as many distinct and all but separate worlds. Humming-bird and hawkmoth may, once in a way, be co-tenants of the same world; but for the most part the mammal, the bird, the fish, the insect and the small life of the sea, not only have their zoological distinctions, but each has a physical universe of its own. The world of bacteria is yet another world again, and so is the world of colloids; but through these small Lilliputs we pass outside the range of living things.

What we call mechanical principles apply to the magnitudes among which we are at home; but lesser worlds are governed by other and appropriate physical laws, of capillarity, adsorption and electric charge. There are other worlds at the far other end of the scale, in the uttermost depths of space, whose vast magnitudes lie within a narrow range. When the globular star-clusters are plotted on a curve, apparent diameter against estimated distance, the curve is a fair approximation to a rectangular hyperbola; which means that, to the same rough approximation, the actual diameter is identical in them all*.

It is a remarkable thing, worth pausing to reflect on, that we can pass so easily and in a dozen lines from molecular magnitudes† to the dimensions of a Sequoia or a whale. Addition and subtraction, the old arithmetic of the Egyptians, are not powerful enough for such an operation; but the story of the grains of wheat upon the chessboard shewed the way, and Archimedes and Napier elaborated

* See Harlow Shapley and A. B. Sayer, The angular diameters of globular clusters, *Proc. Nat. Acad. of Sci.* XXI, pp. 593–597, 1935. The same is approximately true of the spiral nebulae also.

† We may call (after Siedentopf and Zsigmondi) the smallest visible particles *microns*, such for instance as small bacteria, or the fine particles of gum-mastich in suspension, measuring 0·5 to 1·0 μ; *sub-microns* are those revealed by the ultra-microscope, such as particles of colloid gold (2–15 mμ), or starch-molecules (5 mμ); amicrons, under 1 mμ, are not perceptible by either method. A water-molecule measures, probably, about 0·1 mμ.

the arithmetic of multiplication. So passing up and down by easy steps, as Archimedes did when he numbered the sands of the sea, we compare the magnitudes of the great beasts and the small, of the atoms of which they are made, and of the world in which they dwell*.

While considerations based on the chemical composition of the organism have taught us that there must be a definite lower limit to its magnitude, other considerations of a purely physical kind lead us to the same conclusion. For our discussion of the principle of similitude has already taught us that long before we reach these all but infinitesimal magnitudes the dwindling organism will have experienced great changes in all its physical relations, and must at length arrive at conditions surely incompatible with life, or what we understand as life, in its ordinary development and manifestation.

We are told, for instance, that the powerful force of surface-tension, or capillarity, begins to act within a range of about 1/500,000 of an inch, or say $0 \cdot 05 \mu$. A soap film, or a film of oil on water, may be attenuated to far less magnitudes than this; the black spots on a soap bubble are known, by various concordant methods of measurement, to be only about 6×10^{-7} cm., or about $6 \, m\mu$ thick, and Lord Rayleigh and M. Devaux have obtained films of oil of $2 \, m\mu$, or even $1 \, m\mu$ in thickness. But while it is possible for a fluid film to exist of these molecular dimensions, it is certain that long before we reach these magnitudes there arise conditions of which we have little knowledge, and which it is not easy to imagine. A bacillus lives in a world, or on the borders of a world, far other than our own, and preconceptions drawn from our experience are not valid there. Even among inorganic, non-living bodies, there comes a certain grade of minuteness at which the ordinary properties become modified. For instance, while under ordinary circumstances crystallisation starts in a solution about a minute solid fragment or crystal

* Observe that, following a common custom, we have only used a logarithmic scale for the round numbers representing powers of ten, leaving the interspaces between these to be filled up, if at all, by ordinary numbers. There is nothing to prevent us from using fractional indices, if we please, throughout, and calling a blood-corpuscle, for instance, $10^{-3 \cdot 2}$ cm. in diameter, a man $10^{2 \cdot 25}$ cm. high, or Sibbald's Rorqual $10^{1 \cdot 48}$ metres long. This method, implicit in that of Napier of Merchiston, was first set forth by Wallis, in his *Arithmetica infinitorum*.

of the salt, Ostwald has shewn that we may have particles so minute that they fail to serve as a nucleus for crystallisation—which is as much as to say that they are too small to have the form and properties of a "crystal." And again, in his thin oil-films, Lord Rayleigh noted the striking change of physical properties which ensues when the film becomes attenuated to one, or something less than one, close-packed layer of molecules, and when, in short, it no longer has the properties of matter *in mass*.

These attenuated films are now known to be "monomolecular," the long-chain molecules of the fatty acids standing close-packed, like the cells of a honeycomb, and the film being just as thick as the molecules are long. A recent determination makes the several molecules of oleic, palmitic and stearic acids measure 10·4, 14·1 and 15·1 cm. in length, and in breadth 7·4, 6·0 and 5·5 cm., all by 10^{-8}: in good agreement with Lord Rayleigh and Devaux's lowest estimates (F. J. Hill, *Phil. Mag.* 1929, pp. 940–946). But it has since been shewn that in aliphatic substances the long-chain molecules are not erect, but inclined to the plane of the film; that the zig-zag constitution of the molecules permits them to interlock, so giving the film increased stability; and that the interlock may be by means of a first or second zig-zag, the measured area of the film corresponding precisely to these two dimorphic arrangements. (Cf. C. G. Lyons and E. K. Rideal, *Proc. R.S.* (A), CXXVIII, pp. 468–473, 1930.) The film may be lifted on to a polished surface of metal, or even on a sheet of paper, and one monomolecular layer so added to another; even the complex protein molecule can be unfolded to form a film one amino-acid molecule thick. The whole subject of monomolecular layers, the nature of the film, whether condensed, expanded or gaseous, its astonishing sensitiveness to the least impurities, and the manner of spreading of the one liquid over the other, has become of great interest and importance through the work of Irving Langmuir, Devaux, N. K. Adam and others, and throws new light on the whole subject of molecular magnitudes*.

The surface-tension of a drop (as Laplace conceived it) is the cumulative effect, the statistical average, of countless molecular attractions, but we are now entering on dimensions where the molecules are few†. The free surface-energy of a body begins to vary with the *radius*, when that radius is of an order comparable to inter-molecular distances; and the whole expression for such energy tends to vanish away when the radius of the drop or particle is less than $0{\cdot}01\,\mu$, or $10\,\mathrm{m}\mu$. The qualities and properties of our

* Cf. (*int. al.*) Adam, *Physics and Chemistry of Surfaces*, 1930; Irving Langmuir, *Proc. R.S.* (A), CLXX, 1939.

† See a very interesting paper by Fred Vles, Introduction à la physique bactérienne, *Revue Scient.* 11 juin 1921. Cf. also N. Rashevsky, Zur Theorie d. spontanen Teilung von mikroskopischen Tropfen, *Ztschr. f. Physik*, XLVI, p. 578, 1928.

particle suffer an abrupt change here; what then can we attribute, in the way of properties, to a corpuscle or organism as small or smaller than, say, 0·05 or 0·03 μ? It must, in all probability, be a homogeneous structureless body, composed of a very small number of albumenoid or other molecules. Its vital properties and functions must be extremely limited; its specific outward characters, even if we could see it, must be *nil*; its osmotic pressure and exchanges must be anomalous, and under molecular bombardment they may be rudely disturbed; its properties can be little more than those of an ion-laden corpuscle, enabling it to perform this or that specific chemical reaction, to effect this or that disturbing influence, or produce this or that pathogenic effect. Had it sensation, its experiences would be strange indeed; for if it could feel, it would regard a fall in temperature as a movement of the molecules around, and if it could see it would be surrounded with light of many shifting colours, like a room filled with rainbows.

The dimensions of a cilium are of such an order that its substance is mostly, if not all, under the peculiar conditions of a surface-layer, and surface-energy is bound to play a leading part in ciliary action. A cilium or flagellum is (as it seems to me) a portion of matter in a state *sui generis*, with properties of its own, just as the film and the jet have theirs. And just as Savart and Plateau have told us about jets and films, so will the physicist some day explain the properties of the cilium and flagellum. It is certain that we shall never understand these remarkable structures so long as we magnify them to another scale, and forget that new and peculiar physical properties are associated with the scale to which they belong*.

As Clerk Maxwell put it, "molecular science sets us face to face with physiological theories. It forbids the physiologist to imagine that structural details of infinitely small dimensions (such as Leibniz assumed, one within another, *ad infinitum*) can furnish an explanation of the infinite variety which exists in the properties and functions of the most minute organisms." And for this reason Maxwell reprobates, with not undue severity, those advocates of pangenesis

* The cilia on the gills of bivalve molluscs are of exceptional size, measuring from say 20 to 120 μ long. They are thin triangular plates, rather than filaments; they are from 4 to 10 μ broad at the base, but less than 1 μ thick. Cf. D. Atkins, *Q.J.M.S.*, 1938, and other papers.

and similar theories of heredity, who "would place a whole world of wonders within a body so small and so devoid of visible structure as a germ." But indeed it scarcely needed Maxwell's criticism to shew forth the immense physical difficulties of Darwin's theory of *pangenesis*: which, after all, is as old as Democritus, and is no other than that Promethean *particula undique desecta* of which we have read, and at which we have smiled, in our Horace.

There are many other ways in which, when we make a long excursion into space, we find our ordinary rules of physical behaviour upset. A very familiar case, analysed by Stokes, is that the viscosity of the surrounding medium has a relatively powerful effect upon bodies below a certain size. A droplet of water, a thousandth of an inch (25μ) in diameter, cannot fall in still air quicker than about an inch and a half per second; as its size decreases, its resistance varies as the radius, not (as with larger bodies) as the surface; and its "critical" or terminal velocity varies as the square of the radius, or as the surface of the drop. A minute drop in a misty cloud may be one-tenth that size, and will fall a hundred times slower, say an inch a minute; and one again a tenth of this diameter (say $0{\cdot}25\mu$, or about twice as big as a small micrococcus) will scarcely fall an inch in two hours*. Not only do dust-particles, spores† and bacteria fall, by reason of this principle, very slowly through the air, but all minute bodies meet with great proportionate resistance to their movements through a fluid. In salt water they have the added influence of a larger coefficient of friction than in fresh‡; and even such comparatively large organisms as the diatoms and the foraminifera, laden though they are with a heavy shell of flint or lime, seem to be poised in the waters of the ocean, and fall with exceeding slowness.

* The resistance depends on the radius of the particle, the viscosity, and the rate of fall (V); the effective weight by which this resistance is to be overcome depends on gravity, on the density of the particle compared with that of the medium, and on the mass, which varies as r^3. Resistance $= krV$, and effective weight $= k'r^3$; when these two equal one another we have the critical or terminal velocity, and $V \propto r^2$.

† A. H. R. Buller found the spores of a fungus (*Collybia*), measuring $5 \times 3\mu$, to fall at the rate of half a millimetre per second, or rather more than an inch a minute; *Studies on Fungi*, 1909.

‡ Cf. W. Krause, *Biol. Centralbl.* I, p. 578, 1881; Flügel, *Meteorol. Ztschr.* 1881, p. 321.

When we talk of one thing touching another, there may yet be a distance between, not only measurable but even large compared with the magnitudes we have been considering. Two polished plates of glass or steel resting on one another are still about 4μ apart—the average size of the smallest dust; and when all dust-particles are sedulously excluded, the one plate sinks slowly down to within $0{\cdot}3\mu$ of the other, an apparent separation to be accounted for by minute irregularities of the polished surfaces*.

The Brownian movement has also to be reckoned with—that remarkable phenomenon studied more than a century ago by Robert Brown†, Humboldt's *facile princeps botanicorum*, and discoverer of the nucleus of the cell‡. It is the chief of those fundamental phenomena which the biologists have contributed, or helped to contribute, to the science of physics.

The quivering motion, accompanied by rotation and even by translation, manifested by the fine granular particle issuing from a crushed pollen-grain, and which Brown proved to have no vital significance but to be manifested by all minute particles whatsoever, was for many years unexplained. Thirty years and more after Brown wrote, it was said to be "due, either directly to some calorical changes continually taking place in the fluid, or to some obscure chemical action between the solid particles and the fluid which is indirectly promoted by heat§." Soon after these words were

* Cf. Hardy and Nottage, *Proc. R.S.* (A), CXXVIII, p. 209, 1928; Baston and Bowden, *ibid.* CXXXIV, p. 404, 1931.

† *A Brief Description of Microscopical Observations...on the Particles contained in the Pollen of Plants; and on the General Existence of Active Molecules in Organic and Inorganic Bodies*, London, 1828. See also *Edinb. New Philosoph. Journ.* v, p. 358, 1828; *Edinb. Journ. of Science*, I, p. 314, 1829; *Ann. Sc. Nat.* XIV, pp. 341–362, 1828; etc. The Brownian movement was hailed by some as supporting Leibniz's theory of Monads, a theory once so deeply rooted and so widely believed that even under Schwann's cell-theory Johannes Müller and Henle spoke of the cells as "organische Monaden"; cf. Emil du Bois Reymond, Leibnizische Gedanken in der neueren Naturwissenschaft, *Monatsber. d. k. Akad. Wiss., Berlin*, 1870.

‡ The "nucleus" was first seen in the epidermis of Orchids; but "this areola, or nucleus of the cell as perhaps it might be termed, is not confined to the epidermis," etc. See his paper on Fecundation in Orchideae and Asclepiadae, *Trans. Linn. Soc.* XVI, 1829–33, also *Proc. Linn. Soc.* March 30, 1832.

§ Carpenter, *The Microscope*, edit. 1862, p. 185.

written it was ascribed by Christian Wiener * to molecular movements within the fluid, and was hailed as visible proof of the atomistic (or molecular) constitution of the same. We now know that it is indeed due to the impact or bombardment of molecules upon a body so small that these impacts do not average out, for the moment, to approximate equality on all sides †. The movement becomes manifest with particles of somewhere about 20μ, and is better displayed by those of about 10μ, and especially well by certain colloid suspensions or emulsions whose particles are just below 1μ in diameter ‡. The bombardment causes our particles to behave just like molecules of unusual size, and this behaviour is manifested in several ways §. Firstly, we have the quivering movement of the particles; secondly, their movement backwards and forwards, in short, straight disjointed paths; thirdly, the particles rotate, and do so the more rapidly the smaller they are: and by theory, confirmed by observation, it is found that particles of 1μ in diameter rotate on an average through 100° a second, while particles of 13μ turn through only 14° a minute. Lastly, the very curious result appears, that in a layer of fluid the particles are not evenly distributed, nor do they ever fall under the influence of gravity to the bottom. For here gravity and the Brownian movement are rival powers, striving for equilibrium; just as gravity is opposed in the atmosphere by the proper motion of the gaseous molecules. And just as equilibrium is attained in the atmosphere when the molecules are so distributed that the density (and therefore the number of molecules per unit volume) falls off in geometrical

* In *Poggendorff's Annalen*, CXVIII, pp. 79–94, 1863. For an account of this remarkable man, see *Naturwissenschaften*, XV, 1927; cf. also Sigmund Exner, Ueber Brown's Molecularbewegung, *Sitzungsber. kk. Akad. Wien*, LVI, p. 116, 1867.

† Perrin, Les preuves de la réalité moléculaire, *Ann. de Physique*, XVII, p. 549, 1905; XIX, p. 571, 1906. The actual molecular collisions are unimaginably frequent; we see only the residual fluctuations.

‡ Wiener was struck by the fact that the phenomenon becomes conspicuous just when the size of the particles becomes comparable to that of a wave-length of light.

§ For a full, but still elementary, account, see J. Perrin, *Les Atomes*; cf. also Th. Svedberg, *Die Existenz der Moleküle*, 1912; R. A. Millikan, *The Electron*, 1917, etc. The modern literature of the Brownian movement (by Einstein, Perrin, de Broglie, Smoluchowski and Millikan) is very large, chiefly owing to the value which the phenomenon is shewn to have in determining the size of the atom or the charge on an electron, and of giving, as Ostwald said, experimental proof of the atomic theory.

progression as we ascend to higher and higher layers, so is it with our particles within the narrow limits of the little portion of fluid under our microscope.

It is only in regard to particles of the simplest form that these phenomena have been theoretically investigated*, and we may take it as certain that more complex particles, such as the twisted body of a Spirillum, would shew other and still more complicated manifestations. It is at least clear that, just as the early microscopists in the days before Robert Brown never doubted but that these phenomena were purely vital, so we also may still be apt to confuse, in certain cases, the one phenomenon with the other. We cannot, indeed, without the most careful scrutiny, decide whether the movements of our minutest organisms are intrinsically "vital" (in the sense of being beyond a physical mechanism, or working model) or not. For example, Schaudinn has suggested that the undulating movements of *Spirochaete pallida* must be due to the presence of a minute, unseen, "undulating membrane"; and Doflein says of the same species that "sie verharrt oft mit eigenthümlich zitternden Bewegungen zu einem Orte." Both movements, the trembling or quivering movement described by Doflein, and the undulating or rotating movement described by Schaudinn, are just such as may be easily and naturally interpreted as part and parcel of the Brownian phenomenon.

While the Brownian movement may thus simulate in a deceptive way the active movements of an organism, the reverse statement also to a certain extent holds good. One sometimes lies awake of a summer's morning watching the flies as they dance under the ceiling. It is a very remarkable dance. The dancers do not whirl or gyrate, either in company or alone; but they advance and retire; they seem to jostle and rebound; between the rebounds they dart hither or thither in short straight snatches of hurried flight, and turn again sharply in a new rebound at the end of each little rush†.

* Cf. R. Gans, Wie fallen Stäbe und Scheiben in einer reibenden Flüssigkeit? *Münchener Bericht*, 1911, p. 191; K. Przibram, Ueber die Brown'sche Bewegung nicht kugelförmiger Teilchen, *Wiener Bericht*, 1912, p. 2339; 1913, pp. 1895–1912.

† As Clerk Maxwell put it to the British Association at Bradford in 1873, "We cannot do better than observe a swarm of bees, where every individual bee is flying furiously, first in one direction and then in another, while the swarm as a whole is either at rest or sails slowly through the air."

Their motions are erratic, independent of one another, and devoid of common purpose*. This is nothing else than a vastly magnified picture, or simulacrum, of the Brownian movement; the parallel between the two cases lies in their complete irregularity, but this in itself implies a close resemblance. One might see the same thing in a crowded market-place, always provided that the bustling crowd had no *business* whatsoever. In like manner Lucretius, and Epicurus before him, watched the dust-motes quivering in the beam, and saw in them a mimic representation, *rei simulacrum et imago*, of the eternal motions of the atoms. Again the same phenomenon may be witnessed under the microscope, in a drop of water swarming with Paramoecia or such-like Infusoria; and here the analogy has been put to a numerical test. Following with a pencil the track of each little swimmer, and dotting its place every few seconds (to the beat of a metronome), Karl Przibram found that the mean successive distances from a common base-line obeyed with great exactitude the "Einstein formula," that is to say the particular form of the "law of chance" which is applicable to the case of the Brownian movement†. The phenomenon is (of course) merely analogous, and by no means identical with the Brownian movement; for the range of motion of the little active organisms, whether they be gnats or infusoria, is vastly greater than that of the minute particles which are passive under bombardment; nevertheless Przibram is inclined to think that even his comparatively large infusoria are small enough for the molecular bombardment to be a stimulus, even though not the actual cause, of their irregular and interrupted movements‡.

* Nevertheless there may be a certain amount of bias or direction in these seemingly random divagations: cf. J. Brownlee, *Proc. R.S.E.* XXXI, p. 262, 1910–11; F. H. Edgeworth, *Metron*, I, p. 75, 1920; Lotka, *Elem. of Physical Biology*, 1925, p. 344.

† That is to say, the mean square of the displacements of a particle, in any direction, is proportional to the interval of time. Cf. K. Przibram, Ueber die ungeordnete Bewegung niederer Tiere, *Pflüger's Archiv*, CLIII, pp. 401–405, 1913; *Arch. f. Entw. Mech.* XLIII, pp. 20–27, 1917.

‡ All that is actually proven is that "pure chance" has governed the movements of the little organism. Przibram has made the analogous observation that infusoria, when not too crowded together, spread or diffuse through an aperture from one vessel to another at a rate very closely comparable to the ordinary laws of molecular diffusion.

George Johnstone Stoney, the remarkable man to whom we owe the name and concept of the *electron*, went further than this; for he supposed that molecular bombardment might be the source of the life-energy of the bacteria. He conceived the swifter moving molecules to dive deep into the minute body of the organism, and this in turn to be able to make use of these importations of energy*.

We draw near the end of this discussion. We found, to begin with, that "scale" had a marked effect on physical phenomena, and that increase or diminution of magnitude might mean a complete change of statical or dynamical equilibrium. In the end we begin to see that there are discontinuities in the scale, defining phases in which different forces predominate and different conditions prevail. Life has a range of magnitude narrow indeed compared to that with which physical science deals; but it is wide enough to include three such discrepant conditions as those in which a man, an insect and a bacillus have their being and play their several roles. Man is ruled by gravitation, and rests on mother earth. A water-beetle finds the surface of a pool a matter of life and death, a perilous entanglement or an indispensable support. In a third world, where the bacillus lives, gravitation is forgotten, and the viscosity of the liquid, the resistance defined by Stokes's law, the molecular shocks of the Brownian movement, doubtless also the electric charges of the ionised medium, make up the physical environment and have their potent and immediate influence on the organism. The predominant factors are no longer those of our scale; we have come to the edge of a world of which we have no experience, and where all our preconceptions must be recast.

* *Phil. Mag.* April 1890.

CHAPTER III

THE RATE OF GROWTH

When we study magnitude by itself, apart from the gradual changes to which it may be subject, we are dealing with a something which may be adequately represented by a number, or by means of a line of definite length; it is what mathematicians call a scalar phenomenon. When we introduce the conception of change of magnitude, of magnitude which varies as we pass from one point to another in space, or from one instant to another in time, our phenomenon becomes capable of representation by means of a line of which we define both the length and the direction; it is (in this particular aspect) what is called a vector phenomenon.

When we deal with magnitude in relation to the dimensions of space, our diagram plots magnitude in one direction against magnitude in another—length against height, for instance, or against breadth; and the result is what we call a picture or outline, or (more correctly) a "plane projection" of the object. In other words, what we call Form is a ratio of magnitudes* referred to direction in space.

When, in dealing with magnitude, we refer its variations to successive intervals of time (or when, as it is said, we equate it with time), we are then dealing with the phenomenon of growth; and it is evident that this term growth has wide meanings. For growth may be positive or negative, a thing may grow larger or smaller, greater or less; and by extension of the concrete signification of the word we easily and legitimately apply it to non-material things, such as temperature, and say, for instance, that a body "grows" hot or cold. When in a two-dimensional diagram we represent a magnitude (for instance length) in relation to time (or "plot" length against time, as the phrase is), we get that kind of vector diagram which is known as a "curve of growth." We see that the phenomenon which we are studying is a *velocity* (whose "dimensions" are space/time, or L/T), and this phenomenon we shall speak of, simply, as a *rate of growth*.

In various conventional ways we convert a two-dimensional into

* In Aristotelian logic, Form is a *quality*. None the less, it is related to *quantity*; and we find the Schoolmen speaking of it as *qualitas circa quantitatem*.

a three-dimensional diagram. We do so, for example, when, by means of the geometrical method of "perspective," we represent upon a sheet of paper the length, breadth and depth of an object in three-dimensional space, but we do it better by means of contour-lines or "isopleths." By contour-lines superposed upon a map of a country, we shew its hills and valleys; and by contour-lines we may shew temperature, rainfall, population, language, or any other "third dimension" related to the two dimensions of the map. *Time* is always implicit, in so far as each map refers to its own date or epoch; but Time as a dimension can only be substituted for one of the three dimensions already there. Thus we may superpose upon our map the successive outlines of the coast from remote antiquity, or of any single isotherm or isobar from day to day. And if in like manner we superpose on one another, or even set side by side, the outlines of a growing organism—for instance of a young leaf and an old, we have a three-dimensional diagram which is a partial representation (limited to two dimensions of *space*) of the organism's gradual change of form, or course of development; in such a case our contours may, for the purposes of the embryologist, be separated by time-intervals of a few hours or days, or, for the palaeontologist, by interspaces of unnumbered and innumerable years*.

Such a diagram represents in two of its three dimensions form, and in two (or three) of its dimensions growth, and we see how intimately the two concepts are correlated or interrelated to one another. In short it is obvious that the *form* of an organism is determined by its rate of *growth* in various directions; hence rate of growth deserves to be studied as a necessary preliminary to the theoretical study of form, and organic form itself is found, mathematically speaking, to be a *function of time*†.

* Sometimes we find one and the same diagram suffice, whether the time-intervals be great or small; and we then invoke "Wolff's law" (or Kielmeyer's), and assert that the life-history of the individual repeats, or recapitulates, the history of the race. This "recapitulation theory" was all-important in nineteenth-century embryology, but was criticised by Adam Sedgwick (*Q.J.M.S.* XXXVI, p. 38, 1894) and many later authors; cf. J. Needham, *Chemical Embryology*, 1931, pp. 1629–1647.

† Our subject is one of Bacon's "Instances of the Course" or studies wherein we "measure Nature by periods of Time." In Bacon's *Catalogue of Particular Histories*, one of the odd hundred histories or investigations which he foreshadows is precisely that which we are engaged on, viz. a "History of the Growth and Increase of the Body, in the whole and in its parts."

At the same time, we need only consider this large part of our subject somewhat briefly. Though it has an essential bearing on the problems of morphology, it is in greater degree involved with physiological problems; also, the statistical or numerical aspect of the question is peculiarly adapted to the mathematical study of variation and correlation. These important subjects we must not neglect; but our main purpose will be served if we consider the characteristics of a rate of growth in a few illustrative cases, and recognise that this rate of growth is a very important specific property, with its own characteristic value in this organism or that, in this or that part of each organism, and in this or that phase of its existence.

The statement which we have just made that "the form of an organism is determined by its rate of growth in various directions," is one which calls for further explanation and for some measure of qualification.

Among organic forms we shall have many an occasion to see that form may be due in simple cases to the direct action of certain molecular forces, among which surface-tension plays a leading part. Now when surface-tension causes (for instance) a minute semifluid organism to assume a spherical form, or gives to a film of protoplasm the form of a catenary or of an elastic curve, or when it acts in various other ways productive of definite contours—just as it does in the making of a drop, a splash or a jet—this is a process of conformation very different from that by which an ordinary plant or animal grows into its specific form. In both cases change of form is brought about by the movement of portions of matter, and in both cases it is ultimately due to the action of molecular forces; but in the one case the movements of the particles of matter lie for the most part within molecular range, while in the other we have to deal with the transference of portions of matter into the system from without, and from one widely distant part of the organism to another. It is to this latter class of phenomena that we usually restrict the term growth; it is in regard to them that we are in a position to study the *rate of action* in different directions and at different times, and to realise that it is on such differences of rate that form and its modifications essentially and ultimately depend.

The difference between the two classes of phenomena is akin to the difference between the forces which determine the form of a raindrop and those which, by the flowing of the waters and the sculpturing of the solid earth, have brought about the configuration of a river or a hill; *molecular* forces are paramount in the one, and *molar* forces are dominant in the other.

At the same time, it is true that *all* changes of form, inasmuch as they necessarily involve changes of actual and relative magnitude, may in a sense be looked upon as phenomena of growth; and it is also true, since the movement of matter must always involve an element of time*, that in all cases the *rate of growth* is a phenomenon to be considered. Even though the molecular forces which play their part in modifying the form of an organism exert an action which is, theoretically, all but instantaneous, that action is apt to be dragged out to an appreciable interval of time by reason of viscosity or some other form of resistance in the material. From the physical or physiological point of view the rate of action may be well worth studying even in such cases as these; for example, a study of the rate of cell-division in a segmenting egg may teach us something about the work done, and the various energies concerned. But in such cases the action is, as a rule, so homogeneous, and the form finally attained is so definite and so little dependent on the time taken to effect it, that the specific rate of change, or rate of growth, does not enter into the morphological problem.

We are dealing with Form in a very concrete way. To Aristotle it was a metaphysical concept; to us it is a quasi-mechanical effect on Matter of the operation of chemico-physical forces†. To

* Cf. Aristotle, *Phys.* VI, 5, 235*a*, 11, ἐπεὶ γὰρ ἅπασα κίνησις ἐν χρόνῳ, κτλ.; he had already told us that natural science deals with magnitude, with motion and with time: ἔστιν ἡ περὶ φύσεως ἐπιστήμη περὶ μέγεθος καὶ κίνησιν καὶ χρόνον. Hence *omnis velocitas tempore durat* became a scholastic aphorism. Bacon emphasised, in like manner, the fact that "all motion or natural action is performed in time: some more quickly, some more slowly, but all in periods determined and fixed in the nature of things. Even those actions which seem to be performed suddenly, and (as we say) in the twinkling of an eye, are found to admit of degree in respect of duration" (*Nov. Organon*, XLVI). That infinitely small motions take place in infinitely small intervals of time is the concept which lies at the root of the calculus. But there is another side to the story.

† Cf. N. K. Koltzoff, Physikalisch-chemische Grundlage der Morphologie, *Biol. Centralbl.* 1928, pp. 345–369.

Aristotle its Form was the essence, the archetype, the very "nature" of a thing, and Matter and Form were an inseparable duality. Even now, when we divide our science into Physiology and Morphology, we are harking back to the old Aristotelian antithesis.

To sum up, we may lay down the following general statements. The form of organisms is a phenomenon to be referred in part to the direct action of molecular forces, in larger part to a more complex and slower process, indirectly resulting from chemical, osmotic and other forces, by which material is introduced into the organism and transferred from one part of it to another. It is this latter complex phenomenon which we usually speak of as "growth."

Every growing organism, and every part of such a growing organism, has its own specific rate of growth, referred to this or that particular direction; and it is by the ratio between these rates in different directions that we must account for the external forms of all save certain very minute organisms. This ratio may sometimes be of a *simple* kind, as when it results in the mathematically definable outline of a shell, or the smooth curve of the margin of a leaf. It may sometimes be a very *constant* ratio, in which case the organism while growing in bulk suffers little or no perceptible change in form; but such constancy seldom endures beyond a season, and when the ratios tend to alter, then we have the phenomenon of morphological "*development*," or steady and persistent alteration of form.

This elementary concept of Form, as determined by varying rates of Growth, was clearly apprehended by the mathematical mind of Haller—who had learned his mathematics of the great John Bernoulli, as the latter in turn had learned his physiology from the writings of Borelli*. It was this very point, the apparently unlimited extent to which, in the development of the chick, inequalities of growth could and did produce changes of form and changes of anatomical structure, that led Haller to surmise that the process was actually without limits, and that all development was but an unfolding or "evolutio," in which no part came into being which

* "Qua in re Incomparabilis Viri Joh. Alph. Borelli vestigiis insistemus." Joh. Bernoulli, *De motu musculorum*, 1694.

had not essentially existed before*. In short the celebrated doctrine of "preformation" implied on the one hand a clear recognition of what growth can do throughout the several stages of development, by hastening the increase in size of one part, hindering that of another, changing their relative magnitudes and positions, and so altering their forms; while on the other hand it betrayed a failure (inevitable in those days) to recognise the essential difference between these movements of masses and the molecular processes which precede and accompany them, and which are characteristic of another order of magnitude.

The general connection between growth and form has been recognised by other writers besides Haller. Such a connection is implicit in the "proportional diagrams" by which Dürer and his brother-artists illustrated the changes in form, or of relative dimensions, which mark the child's growth to boyhood and to manhood. The same connection was recognised by the early embryologists, and appears, as a survival of the doctrine of preformation, in Pander's† study of the development of the chick. And long afterwards, the embryological aspect of the case was emphasised by His‡, who pointed out that the foldings of the blastoderm, by which the neural and amniotic folds are brought into being, were the resultant of unequal rates of growth in what to begin with was a uniform layer of embryonic tissue. If a sheet of paper be made to expand here and contract there, as by moisture or evaporation, the plane surface becomes dimpled, or folded, or buckled, by the said expansions and contractions; and the distortions to which the surface of the "germinal disc" is subject are, as His shewed once and for all, precisely analogous. There are

* Cf. (e.g.) *Elem. Physiologiae*, ed. 1766, VIII, p. 114, "Ducimur autem ad evolutionem potissimum, quando a perfecto animale retrorsum progredimur, et incrementorum atque mutationum seriem relegimus. Ita inveniemus perfectum illud animal fuisse imperfectius, alterius figurae et fabricae, et denique rude et informe: et tamen idem semper animal sub iis diversis phasibus fuisse, quae absque ullo saltu perpetuos parvosque per gradus cohaereant."

† *Beiträge zur Entwickelungsgeschichte des Hühnchens im Ei*, 1817, p. 40. Roux ascribes the same views also to Von Baer and to R. H. Lotze (*Allgem. Physiologie*, 1851, p. 353).

‡ W. His, *Unsere Körperform, und das physiologische Problem ihrer Entstehung*, 1874. See also *Archiv f. Anatomie*, 1894; and cf. C. B. Davenport, Processes concerned in Ontogeny, *Bull. Mus. Comp. Anat.* XXVII, 1895; also G. Dehnel and Jan Tur, *De Embryonum evolutionis progressu inequali: Kosmos* (Lwow), LIII, 1928.

certain Nostoc-algae in which unequal growth, ceasing towards the periphery of a disc and increasing here and there within, gives rise to folds and bucklings curiously like those of our own ears: which indeed owe their shape and characteristic folding to an identical or analogous cause.

An experimental demonstration comparable to the actual case is obtained by making an "artificial blastoderm" of little pills or pellets of dough, which are caused to grow at varying rates by the addition of varying quantities of yeast. Here, as Roux is careful to point out,* it is not only the *growth* of the individual cells, but the *traction* exercised on one another through their mutual interconnections, which brings about foldings, wrinklings and other distortions of the structure. But this again, or such as this, had been in Haller's mind, and formed an essential part of his embryological doctrine. For he has no sooner treated of *incrementum*, or *celeritas incrementi*, than he proceeds to deal with the contributory and complementary phenomena of expansion, traction (*adtractio*)† and pressure, and the more subtle influences which he denominates *vis derivationis et revulsionis*‡: these latter being the secondary and correlated effects on growth in one part, brought about by such changes as are produced, for instance in the circulation, by the growth of another.

We have to do with growth, with exquisitely graded or balanced growth, and with forces subtly exerted by one growing part upon another, in so wonderful a piece of work as the development of the eye: as its primary vesicle expands and then dimples in, as the lens appears and fits into place, as the secondary vesicle closes over to form iris and pupil, and in all the rest of the story.

Let us admit that, on the physiological side, Haller's or His's methods of explanation carry us but a little way; yet even this little way is something gained. Nevertheless, I can well remember

* Roux, *Die Entwickelungsmechanik*, 1905, p. 99.

† *Op. cit.* p. 302, "Magnum hoc naturae instrumentum, etiam in corpore animato evolvendo potenter operatur, etc." The recurrent laryngeal nerve, drawn down as its arch of the aorta descends, is a simple instance of anatomical *traction*. The vitelline and omphalomesenteric arteries lead, by more complicated constraints and tractions, to the characteristic loops of the intestinal blood-vessels, and of the intestine itself. Cf. G. Enbom, *Lunds Univ. Arsskrift*, 1939.

‡ *Ibid.* p. 306, "Subtiliora ista, et aliquantum hypothesi mista, tamen magnam mihi videntur speciem veri habere."

the harsh criticism and even contempt which His's doctrine met with, not merely on the ground that it was inadequate, but because such an explanation was deemed wholly inappropriate, and was utterly disavowed*. Oscar Hertwig, for instance, asserted that, in embryology, when we find one embryonic stage preceding another, the existence of the former is, for the embryologist, an all-sufficient "causal explanation" of the latter. "We consider (he says) that we are studying and explaining a causal relation when we have demonstrated that the gastrula arises by invagination of a blastosphere, or the neural canal by the infolding of a cell-plate so as to constitute a tube†." For Hertwig, then, as Roux remarks, the task of investigating a physical mechanism in embryology—"der Ziel das Wirken zu erforschen"—has no existence at all. For Balfour also, as for Hertwig, the mechanical or physical aspect of organic development had little or no attraction. In one notable instance, Balfour himself adduced a physical, or quasi-physical, explanation of an organic process, when he referred the various modes of segmentation of an ovum, complete or partial, equal or unequal and so forth, to the varying amount or varying distribution of food-yolk associated with the germinal protoplasm of the egg. But in the main, like all the other embryologists of his day, Balfour was engrossed in the

* Cf. His, On the Principles of Animal Morphology, *Proc. R.S.E.* xv, p. 294, 1888: "My own attempts to introduce some elementary mechanical or physiological conceptions into embryology have not generally been agreed to by morphologists. To one it seemed ridiculous to speak of the elasticity of the germinal layers; another thought that, by such considerations, we 'put the cart before the horse'; and one more recent author states, that we have better things to do in embryology than to discuss tensions of germinal layers and similar questions, since all explanations must of necessity be of a phylogenetic nature. This opposition to the application of the fundamental principles of science to embryological questions would scarcely be intelligible had it not a dogmatic background. No other explanation of living forms is allowed than heredity, and any which is founded on another basis must be rejected....To think that heredity will build organic beings without mechanical means is a piece of unscientific mysticism." Even the school of *Entwickelungsmechanik* showed a certain reluctance, or extreme caution, in speaking of the *physical* forces in relation to embryology or physiology. This reluctant caution is well exemplified by Martin Heidenhain, writing on "Formen und Kräfte in der lebendigen Natur" in Roux's *Vorträge*, XXXII, 1923. Speaking of "die Kräfte welche die Entwickelung und den fertigen Zustand der Formen bedingen", he says: "letztere kann man aber *nicht auf dem Felde der Physik* suchen, sondern nur im Umkreis der Lebendigen, obwohl anzunehmen ist, dass diese Kräfte später einmal 'analogienhaft' nach dem Vorbilde der Physik beschreibbar sein werden"

† O. Hertwig, *Zeit- und Streitfragen der Biologie*, II, 1897.

problems of phylogeny, and he expressly defined the aims of comparative embryology (as exemplified in his own textbook) as being "twofold: (1) to form a basis for Phylogeny, and (2) to form a basis for Organogeny, or the origin and evolution of organs*."

It has been the great service of Roux and his fellow-workers of the school of "Entwickelungsmechanik," and of many other students to whose work we shall refer, to try, as His tried, to import into embryology, wherever possible, the simpler concepts of physics, to introduce along with them the method of experiment, and to refuse to be bound by the narrow limitations which such teaching as that of Hertwig would of necessity impose on the work and the thought and the whole philosophy of the biologist.

Before we pass from this general discussion to study some of the particular phenomena of growth, let me give an illustration, from Darwin, of a point of view which is in marked contrast to Haller's simple but essentially mathematical conception of Form.

There is a curious passage in the *Origin of Species*†, where Darwin is discussing the leading facts of embryology, and in particular Von Baer's "law of embryonic resemblance." Here Darwin says: "We are so much accustomed to see a difference in structure between the embryo and the adult that we are tempted to look at this difference as in some necessary manner contingent on growth. But there is no reason why, for instance, the wing of a bat, or the fin of a porpoise, should not have been sketched out with all their parts in proper proportion, as soon as any part became visible." After pointing out various exceptions, with his habitual care, Darwin proceeds to lay down two general principles, viz. "that slight variations generally appear at a not very early period of life," and secondly, that "at whatever age a variation first appears in the parent, it tends to reappear at a corresponding age in the offspring." He then argues that it is with nature as with the fancier, who does not care what his pigeons look like in the embryo so long as the full-grown bird possesses the desired qualities: and that the process of selection takes place when the birds or other animals are nearly

* *Treatise on Comparative Embryology*, I, p. 4, 1881.

† 1st ed. p. 444; 6th ed. p. 390. The student should not fail to consult the passage in question; for there is always a risk of misunderstanding or misinterpretation when one attempts to epitomise Darwin's carefully condensed arguments.

grown up—at least on the part of the breeder, and presumably in nature as a general rule. The illustration of these principles is set forth as follows: "Let us take a group of birds, descended from some ancient form and modified through natural selection for different habits. Then, from the many successive variations having supervened in the several species at a not very early age, and having been inherited at a corresponding age, the young will still resemble each other much more closely than do the adults—just as we have seen with the breeds of the pigeon.... Whatever influence long-continued use or disuse may have had in modifying the limbs or other parts of any species, this will chiefly or solely have affected it when nearly mature, when it was compelled to use its full powers to gain its own living; and the effects thus produced will have been transmitted to the offspring at a corresponding nearly mature age. Thus the young will not be modified, or will be modified only in a slight degree, through the effects of the increased use or disuse of parts." This whole argument is remarkable, in more ways than we need try to deal with here; but it is especially remarkable that Darwin should begin by casting doubt upon the broad fact that a "difference in structure between the embryo and the adult" is "in some necessary matter contingent on growth"; and that he should see no reason why complicated structures of the adult "should not have been sketched out with all their parts in proper proportion, as soon as any part became visible." It would seem to me that even the most elementary attention to form in its relation to growth would have removed most of Darwin's difficulties in regard to the particular phenomena which he is considering here. For these phenomena are phenomena of form, and therefore of relative magnitude; and the magnitudes in question are attained by growth, proceeding with certain specific velocities, and lasting for certain long periods of time. And it seems obvious accordingly that in any two related individuals (whether specifically identical or not) the differences between them must manifest themselves gradually, and be but little apparent in the young. It is for the same simple reason that animals which are of very different sizes when adult differ less and less in size (as well as form) as we trace them backwards to their early stages.

Though we study the visible effects of varying rates of growth throughout wellnigh all the problems of morphology, it is not very often that we can directly measure the velocities concerned. But owing to the obvious importance of the phenomenon to the morphologist we must make shift to study it where we can, even though our illustrative cases may seem sometimes to have little bearing on the morphological problem*.

In a simple spherical organism, such as the single spherical cell of *Protococcus* or of *Orbulina*, growth is reduced to its simplest terms, and indeed becomes so simple in its outward manifestations that it loses interest to the morphologist. The rate of growth is measured by the rate of change in length of a radius, i.e. $V = (R' - R)/T$, and from this we may calculate, as already indicated, the rate in terms of surface and of volume. The growing body remains of constant form, by the symmetry of the system; because, that is to say, on the one hand the pressure exerted by the growing protoplasm is exerted equally in all directions, after the manner of a hydrostatic pressure, which indeed it actually is; while on the other hand the "skin" or surface layer of the cell is sufficiently homogeneous to exert an approximately uniform resistance. Under these simple conditions, then, the rate of growth is uniform in all directions, and does not affect the form of the organism.

But in a larger or a more complex organism the study of growth, and of the rate of growth, presents us with a variety of problems, and the whole phenomenon (apart from its physiological interest) becomes a factor of great morphological importance. We no longer find that growth tends to be uniform in all directions, nor have we any reason to expect it should. The resistances which it meets with are no longer uniform. In one direction but not in others it will be opposed by the important resistance of gravity; within the growing system itself all manner of structural differences come into play, and set up unequal resistances to growth in one direction or another. At the same time the actual sources of growth, the chemical and osmotic forces which lead to the intussusception of new matter, are not uniformly distributed; one tissue or one organ may well increase while another does not; a set of bones, their intervening cartilages and their surrounding muscles, may all be

* "In omni rerum naturalium historia utile est *mensuras definiri et numeros*," Haller, *Elem. Physiol.*, II, p. 258, 1760. Cf. Hales, *Vegetable Staticks*, Introduction.

capable of very different rates of increment. The changes of form which result from these differences in rate are especially manifested during that phase of life when growth itself is rapid: when the organism, as we say, is undergoing its *development*.

When growth in general has slowed down, the differences in rate between different parts of the organism may still exist, and may be made manifest by careful observation and measurement, but the resultant change of form is less apt to strike the eye. Great as are the differences between the rates of growth in different parts of a complex organism, the marvel is that the ratios between them are so nicely balanced as they are, and so capable of keeping the form of the growing organism all but unchanged for long periods of time, or of slowly changing it in its own harmonious way. There is the nicest possible balance of forces and resistances in every part of the complex body; and when this normal equilibrium is disturbed, then we get abnormal growth, in the shape of tumours and exostoses, and other malformations and deformities of every kind.

The rate of growth in man

Man will serve us as well as another organism for our first illustrations of rate of growth, nor can we easily find another which we can better study from birth to the utmost limits of old age. Nor can we do better than go for our first data concerning him to Quetelet's *Essai de Physique Sociale*, an epoch-making book for the biologist. For it is packed with information, some of it unsurpassed, in regard to human growth and form; and it stands out as the first great essay in which social statistics and organic variation are dealt with from the point of view of the mathematical theory of probabilities. How on the one hand Quetelet followed Da Vinci, Luca Pacioli and Dürer in studying the growth and proportions of man: and how on the other he simplified and extended the ideas of James Bernoulli, of d'Alembert, Laplace, Poisson and the rest, is another and a vastly interesting story*.

* Quetelet, *Sur l'Homme, ..., ou Essai de Physique Sociale*, Bruxelles, 1835: trans. Edinburgh, 1842; also *Instructions populaires sur le calcul des probabilités*, 1828; *Lettres...sur la théorie des probabilités appliquée aux sciences morales et politiques*, 1846; and *Anthropométrie*, 1871. For an account of his life and writings, see Lottin's *Quetelet, statisticien et sociologue*, Louvain, 1912; also J. M. Keynes. *Treatise on Probability*, 1921.

The meaning of the word "statistics" is curiously changed. For Shakespeare or for Milton a statist meant (so Dr Johnson says) "a politician, a statesman; one skilled in government." The eighteenth-century *Statistical Account of Scotland* was a description of the State and of its people, its wealth, its agriculture and its trade.

Stature and weight of man (from Quetelet's Belgian data, Essai,* II, pp. 23–43; *Anthropométrie,* p. 346)†

	Stature in metres			Weight in kgm.			$W/L^3 \times 100$	
Age	Male	Female	% F/M	Male	Female	% F/M	Male	Female
0	0·50	0·48	96·0	3·20	2·91	90·9	2·56	2·64
1	0·70	0·69	98·6	10·00	9·30	93·0	2·92	2·83
2	0·80	0·78	97·5	12·00	11·40	95·0	2·35	2·40
3	0·86	0·85	98·8	13·21	12·45	94·2	2·09	2·03
4	0·93	0·91	97·6	15·07	14·18	94·1	1·84	1·88
5	0·99	0·97	98·4	16·70	15·50	92·8	1·89	1·69
6	1·05	1·03	98·6	18·04	16·74	92·8	1·56	1·53
7	1·11	1·10	98·6	20·16	18·45	91·5	1·48	1·39
8	1·17	1·14	97·3	22·26	19·82	89·0	1·39	1·34
9	1·23	1·20	97·8	24·09	22·44	93·2	1·29	1·30
10	1·28	1·25	97·3	26·12	24·24	92·8	1·25	1·24
11	1·33	1·28	96·1	27·85	26·25	94·3	1·18	1·25
12	1·36	1·33	97·6	31·00	30·54	98·5	1·23	1·38
13	1·40	1·39	98·8	35·32	34·65	98·1	1·29	1·29
14	1·49	1·45	97·3	40·50	38·10	94·1	1·21	1·25
15	1·56	1·47	94·6	46·41	41·30	89·0	1·22	1·30
16	1·61	1·52	93·2	53·39	44·44	83·2	1·20	1·32
17	1·67	1·54	92·5	57·40	49·08	85·5	1·23	1·34
18	1·70	1·56	91·9	61·26	53·10	86·7	1·24	1·40
19	1·71	—	—	63·32	—	—	1·20	—
20	1·71	1·57	91·8	65·00	54·46	83·8	1·30	1·41
25	1·72	1·58	91·6	68·29	55·08	80·7	1·39	1·39
30	1·72	1·58	91·7	68·90	55·14	80·0	1·35	1·39
40	1·71	1·56	90·8	68·81	56·65	82·3	1·38	1·49
50	1·67	1·54	91·8	67·45	58·45	86·7	1·45	1·59
60	1·64	1·52	92·5	65·50	56·73	86·6	1·48	1·61
70	1·62	1·51	93·3	63·03	53·72	85·2	1·48	1·58
80	1·61	1·51	93·4	61·22	51·52	84·1	1·46	1·50

This is what Sir William Petty had meant in the seventeenth century by his *Political Arithmetic*, and what Quetelet meant in the nineteenth by his *Physique Sociale*. But "statistics" nowadays are counts and measures of all sorts of things; and statistical science arranges,

* The figures for height and weight given in my first edition were Quetelet's smoothed or adjusted values. I have gone back to his original data.

† This "almost steady growth," from about seven years old to eleven, means that the curve of growth is a nearly straight line during this period: a result already found by Elderton for Glasgow children (*Biometrika*, x, p. 293, 1914–15), by Fessard and Laufer in Paris (*Nouvelles Tables de Croissance*, 1935, p. 13), etc.

explains, and draws deductions from, the resulting series and arrays of numbers. It deals with simple and measurable effects, due to complex and often unknown causes; and when experiment is not at hand to disentangle these causes, statistical methods may still do something to elucidate them.

Now as to the growth of man, if the child be some 20 inches, or say 50 cm., tall at birth, and the man some six feet, or 180 cm., high at twenty, we may say that his average rate of growth had

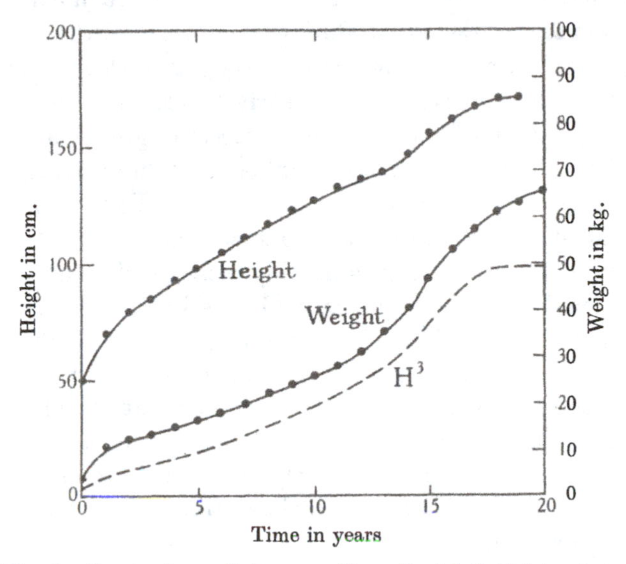

Fig. 4. Curve of growth in man. From Quetelet's Belgian data. The curve H^3 is proportional to the height cubed.

been (180 – 50)/20 cm., or 6·5 cm. per annum. But we well know that this is but a rough preliminary statement, and that growth was surely quick during some and slow during other of those twenty years; we must learn not only the result of growth but the course of growth; we must study it in its *continuity*. This we do, in the first instance, by the method of coordinates, plotting magnitude against time. We measure time along a certain axis (x), and the magnitude in question along a coordinate axis (y); a succession of points defines the magnitudes reached at corresponding epochs, and these points constitute a "*curve of growth*" when we join them together.

Our curve of growth, whether for weight or stature, has a definite form or characteristic curvature: this being a sign that the rate of growth is not always the same but changes as time goes on. Such as it is, the curvature alters in an orderly way; so that, apart from minor and "fortuitous" irregularities, our curves of growth tend to be smooth curves. And the fact that they are so is an instance of that "principle of continuity" which is the foundation of all physical and natural science.

The curve of growth (Fig. 4) for length or stature in man indicates a rapid increase at the outset, during the quick growth of babyhood; a long period of slower but almost steady growth in boyhood; as a rule a marked quickening in his early teens, when the boy comes to the "growing age"; and a gradual arrest of growth as he "comes to his full height" and reaches manhood. If we carried the curve farther, we should see a very curious thing. We should see that a man's full stature endures but for a spell; long before fifty* it has begun to abate, by sixty it is notably lessened, in extreme old age the old man's frame is shrunken and it is but a memory that "he once was tall"; the decline sets in sooner in women than in men, and "a little old woman" is a household word. We have seen, and we see again, that growth may have a negative value, pointing towards an inevitable end. The phenomenon of negative growth extends to weight also; it is largely chemical in origin; the metabolism of the body is impaired, and the tissues keep pace no longer with senile wastage and decay.

We must be very careful, however, how we interpret such a Table as this; for it records the character of a *population*, and we are apt to read in it the life-history of the *individual*. The two things are not necessarily the same. That a man grows less as he grows older all old men know; but it may also be the case, and our Table may indicate it, that the short men live longer than the tall.

Our curve of growth is, by implication, a "time-energy" diagram† or diagram of activity. As man grows he is absorbing energy beyond his daily needs, and accumulating it at a rate depicted in

* Dr Johnson was not far wrong in saying that "life declines from thirty-five"; though the Autocrat of the Breakfast-table declares, like Cicero, that "the furnace is in full blast for ten years longer".

† J. Joly, *The Abundance of Life*, 1915 (1890), p. 86.

our curve; till the time comes when he accumulates no longer, and is constrained to draw upon his dwindling store. But in part, the slow decline in stature is a sign of the unequal contest between our bodily powers and the unchanging force of gravity, which draws us down when we would fain rise up*; we strive against it all our days, in every movement of our limbs, in every beat of our hearts. Gravity makes a difference to a man's height, and no slight one, between the morning and the evening; it leaves its mark in sagging wrinkles, drooping mouth and hanging breasts; it is the indomitable force which defeats us in the end, which lays us on our death-bed and lowers us to the grave†. But the grip in which it holds us is the title by which we live; were it not for gravity one man might hurl another by a puff of his breath into the depths of space, beyond recall for all eternity‡.

Side by side with the curve which represents growth in length, or height or stature, our diagram shews the corresponding curve of weight. That this curve is of a different shape from the former one is accounted for in the main (though not wholly) by the fact—which we have already dealt with—that in similar bodies volume, and therefore weight, varies as the cubes of the linear dimensions; and drawing a third curve to represent the cubes of the corresponding heights, it now resembles the curve of weight pretty closely, but still they are not quite the same. There is a change of direction, or "point of inflection," in the curve of weight at one or two years old, and there are certain other features in our curves which the scale of the diagram does not make clear; and all these differences are due to the fact that the child is changing shape as he grows, that other linear dimensions grow somewhat differently from

* "Lou pes, mèstre de tout (Le poids, maitre de tout), mèstre sènso vergougno, Que te tirasso en bas de sa brutalo pougno." J. H. Fabre, *Oubreto prouvençalo*, p. 61.

† The continuity of the phenomenon of growth, and the natural passage from the phase of increase to that of decrease or decay, are admirably discussed by Enriques, in La Morte, *Rivista di Scienza*, 1907, and in Wachstum und seine analytische Darstellung, *Biol. Centralbl.* June, 1909. Haller (*Elementa*, VII, p. 68) recognised *decrementum* as a phase of growth, not less important (theoretically) than *incrementum*; "*tristis, sed copiosa, haec est materies.*"

‡ Boscovich, *Theoria*, para. 552, "Homo hominem arreptum a Tellure, et utcumque exigua impulsum vi vel uno etiam oris flatu impetitum, ab hominum omnium commercio in infinitum expelleret, nunquam per totam aeternitatem rediturum."

Annual increment of stature (in cm.) from Belgian, French and American statistics

Age	Belgian (Quetelet, *Essai*, II, p. 23)		Paris (Variot et Chaumet, p. 55)				Toronto (Boas, p. 1547)			Worcester, Mass. (Boas, p. 1548)			
			Height		Increment					Increment		Variability	
	Height (Boys)	Ann. increment	Boys	Girls	Boys	Girls	Height (Boys)	Variability of do. (6)	Ann. increment	Boys	Girls	Boys	Girls
0	49·6	—	—	—	—	—	—	—	—	—	—	—	—
1	69·6	20·0	—	—	—	—	—	—	—	—	—	—	—
2	79·7	10·1	74·2	73·6	—	—	—	—	—	—	—	—	—
3	86·0	6·3	82·7	81·8	8·5	8·2	—	—	—	—	—	—	—
4	93·2	7·2	89·1	88·4	6·4	6·6	—	—	—	—	—	—	—
5	99·0	5·8	96·8	95·8	7·7	7·4	105·9	4·4	—	—	—	—	—
6	104·6	5·6	103·3	101·9	6·5	6·1	111·6	4·6	5·7	6·6	5·7	1·6	0·9
7	111·2	6·6	109·9	108·9	6·6	7·0	116·8	4·9	5·2	5·7	5·9	0·7	1·0
8	117·0	5·8	114·4	113·8	4·5	4·9	122·0	5·3	5·2	5·4	5·7	0·9	1·1
9	122·7	5·7	119·7	119·5	5·3	5·7	126·9	5·5	4·9	4·9	5·5	1·0	1·0
10	128·2	5·5	125·0	124·7	5·3	4·8	131·8	5·7	4·9	5·1	6·0	1·0	1·2
11	132·7	4·5	130·3	129·5	5·3	5·2	136·2	6·2	4·4	5·0	6·2	0·9	1·9
12	135·9	3·2	133·6	134·4	3·3	4·9	140·7	6·7	4·5	5·0	7·0	1·3	1·9
13	140·3	4·4	137·6	141·5	4·0	7·1	146·0	7·5	5·3	—	—	—	—
14	148·7	8·4	145·1	148·6	7·5	7·1	152·4	8·5	6·4	—	—	—	—
15	155·9	7·2	153·8	152·9	8·7	4·3	159·7	8·8	7·3	—	—	—	—
16	161·0	5·1	159·6	154·2	5·8	1·3	164·9	7·7	5·2	—	—	—	—
17	167·0	6·0	—	—	—	—	168·9	7·2	4·0	—	—	—	—
18	170·0	3·0	—	—	—	—	171·1	6·7	2·2	—	—	—	—
19	170·6	0·6	—	—	—	—	—	—	—	—	—	—	—
20	171·1	0·5	—	—	—	—	—	—	—	—	—	—	—

length or stature, and in short that infant, boy and man are not *similar figures**. The change of form seems slight and gradual, but behind it lie other and more complex things. The changing ratio between height and weight implies changes in the child's *metabolism*, in the income and expenditure of the body. The infant stores up fat, and the active child "runs it off" again; at four years old or five, bodily metabolism and increase of weight are at a minimum; but a fresh start is made, a new "nutritional period" sets in, and the small schoolboy grows stout and strong†.

Our curve of growth shews at successive epochs of time the height or weight which has been reached by then; it plots changing magnitude (y) against advancing time (x). It is essentially a *cumulative* or *summation* curve; it sums up or "integrates" all the successive magnitudes which have been added in all the foregoing intervals of time. Where the curve is steep it means that growth was rapid, and when growth ceases the curve becomes a horizontal line. It follows that, by measuring the *slope* or steepness of our curve of growth at successive epochs, we shall obtain a picture of the successive *velocities* or *growth-rates*.

The steepness of a curve is measured by its "gradient‡," or we may roughly estimate it by taking for equal intervals of time (strictly speaking, for each infinitesimal interval of time) the increment added during that interval; and this amounts in practice to taking the *differences* between the values given for the successive epochs, or ages, which we have begun by studying. Plotting these successive differences against time, we obtain a curve each point on which represents a certain rate at a certain time; and while the former curve shewed a continuous succession of varying *magnitudes*, this shews a succession of varying *velocities*. The mathematician calls it a *curve of first differences*; we may call it a curve of annual (or other) increments; but we shall not go wrong if we call it a curve of the *rate* (*or rates*) *of growth*, or still more simply, a *velocity-curve*.

* According to Quetelet's data, man's stature is multiplied by 3·4 and his weight by 20·3, between birth and the age of twenty-one. But the cube of 3·4 is nearly 40; so the weight at birth should be multiplied forty times by the age of twenty-one, if infant, boy and man were *similar figures*.

† Cf. T. W. Adams and E. P. Poulton, Heat production in man, *Guy's Hospital Reports* (4), XVII, 1937, and works quoted therein.

‡ That is, by its trigonometrical tangent, referred to the base-line.

We have now obtained two different but closely related curves, based on the selfsame facts or observations, and illustrating them in different ways. One is the inverse of the other; one is the *integral* and one the *differential* of the other; and each makes clear to the eye phenomena which are implicit, but are less conspicuous, in the other. We are using mathematical terms to describe or designate them; but these "curves of growth" are more complicated than the curves with which mathematicians are wont to deal. In our study of growth we may well hope to find curves simpler than these;

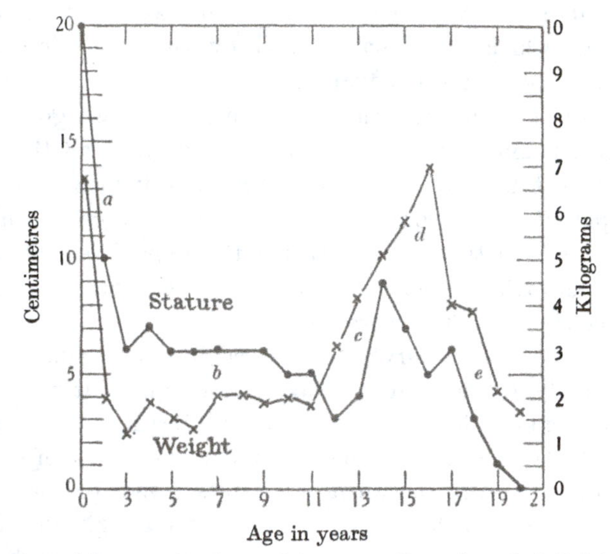

Fig. 5. Annual increments of growth in man. From Quetelet's Belgian data.

but in the successive annual increments of a boy's growth (as Fig. 5 exhibits them) we are dealing with no one continuous operation (such as a mathematical formula might define), but with a *succession of events*, changing as times and circumstances change.

Our curve of increments, or of first differences, for man's stature (Fig. 5) is based, perforce, on annual measurements, and growth alters quickly enough at certain ages to make annual intervals unduly long; nevertheless our curve shews several important things. It suffices to shew, for length or stature, that the growth-rate in early infancy is such as is never afterwards re-attained. From this high early velocity the rate on the whole falls away, until growth itself comes to an end*; but it does so subject to certain important changes and interruptions, which are much the same whether we draw them from Quetelet's Belgian data, or from the British, American and other statistics of later writers. The curve falls fast and steadily during the first couple of years of the child's life (*a*). It runs nearly level during early boyhood, from four or five years old to nine or ten (*b*). Then, after a brief but unmistakable period of depression† during which growth slows down still more (*c*), the boy enters on

Annual increments of stature and of weight in man
(*After Quetelet; see Table*, p. 90)

	Stature (cm.)		Weight (kgm.)	
Age	Male	Female	Male	Female
0– 1	20	21	6·8	6·4
1– 2	10	9	2·0	1·9
2– 3	6	7	1·2	1·1
3– 4	7	6	1·9	1·7
4– 5	6	6	1·6	1·3
5– 6	6	6	1·3	1·2
6– 7	6	7	2·1	1·7
7– 8	6	4	2·1	1·4
8– 9	6	6	1·8	2·6
9–10	5	5	2·0	1·8
10–11	5	3	1·7	2·0
11–12	3	5	3·2	4·3
12–13	4	6	4·3	4·1
13–14	9	6	5·2	3·5
14–15	7	2	5·9	3·2
15–16	5	3	7·0	2·1
16–17	6	4	4·0	4·6
17–18	3	2	3·9	4·0
18–19	1	1	2·1	1·4
19–20	0	0	1·7	—

his teens and begins to "grow out of his clothes"; it is his "growing age", and comes to its height when he is about thirteen or fourteen years old (*d*). The lad goes on growing in stature for some years more, but the rate begins to fall off (*e*), and soon does so with great rapidity.

The corresponding curve of increments in weight is not very different from that for stature, but such differences as there are

* As Haller observed it to do in the chick: "Hoc iterum incrementum miro ordine distribuitur, ut in principio incubationis maximum est; inde perpetuo minuatur" (*Elementa Physiologiae*, VIII, p. 294). Or as Bichat says, "Il y a surabondance de vie dans l'enfant" (*Sur la Vie et la Mort*, p. 1).

† This depression, or slowing down before puberty, seems to be a universal phenomenon, common to all races of men. It is a curious thing that Quetelet's "adjusted figures" (which I used in my first edition) all but smooth out of recognition this characteristic feature of his own observations.

between them are significant enough. There is some tendency for growth in weight to fall off or fluctuate at four or five years old, before the small boy goes to school; but there is, or should be, little retardation of weight when growth in height slows down before he enters on his teens*. The healthy lad puts on weight again more and more rapidly, for some little while after growth in stature has slowed down; and normal increase of weight goes on, more slowly, while the man is "filling out," long after growth in stature has come to an end. But somewhere about thirty he begins losing weight a little; and such subsequent slow changes as men commonly undergo we need not stop to deal with.

The differences in stature and build between one race and another are in like manner a question of growth-rate in the main. Let us take a single instance, and compare the annual increments of growth in Chinese and English boys. The curves are much the same in form, but differ in amplitude and phase. The English boy is growing faster all the while; but the minimal rate and the maximal rate come later by a year or more than in the Chinese curve† (Fig. 6).

Quetelet was not the first to study man's growth and stature, nor was he the first student of social statistics and "demography." The foundations of modern vital statistics had been laid by Graunt and Petty in the seventeenth century‡; the economists developed the subject during the eighteenth§, and parts of it were studied

* That the annual increments of weight in boys are nearly constant, and the curve of growth nearly a straight line at this age, especially from about 8 to 11, has been repeatedly noticed. Cf. Elderton, Glasgow School-children, *Biometrika*, x, p. 283, 1914–15; Fessard and others, *Croissance des Ecoliers Parisiens*, 1934, p. 13. But careful measurements of American children, by Katherine Simmons and T. Wingate Todd, shew steadily increasing increments from four years old till puberty (*Growth*, II, pp. 93–133, 1938).

† For copious bibliography, see J. Needham, *op. cit.*, also Gaston Backman, Das Wachstum der Körperlänge des Menschen, *K. Sv. Vetensk. Akad. Hdlgr.* (3), XIV, 1934.

‡ Cf. John Graunt's *Natural and Political Observations...upon the Bills of Mortality*, London, 1662; *The Economic Writings of Sir William Petty*, ed. by C. H. Hull, 2 vols., Cambridge, Mass., 1927. Concerning Graunt and Petty—two of the original Fellows of the Royal Society—see (*int. al.*) H. Westergaard, *History of Statistics*, 1932, and L. Hogben (and others), *Political Arithmetic*, 1938.

§ Besides the many works of the economists, cf. J. G. Roederer, Sermo de pondere et longitudine recens-natorum, *Comment. Soc. Reg. Sci. Gottingae*, III, 1753; J. F. G. Dietz, De temporum in graviditate et partu aestimatione, Diss., Göttingen, 1757.

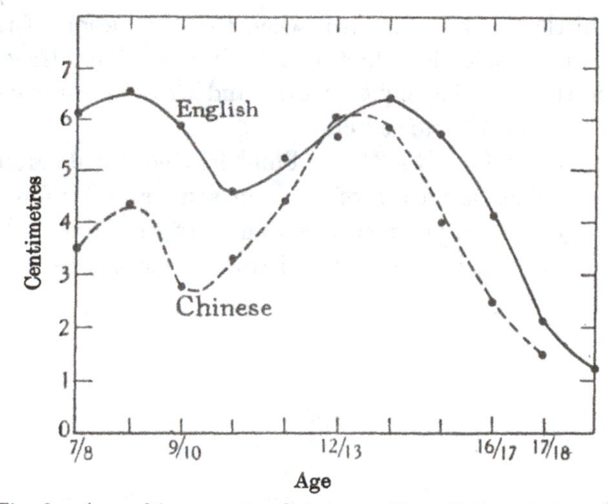

Fig. 6. Annual increments of stature. From Roberts' (English) and Appleton's (Chinese) data.

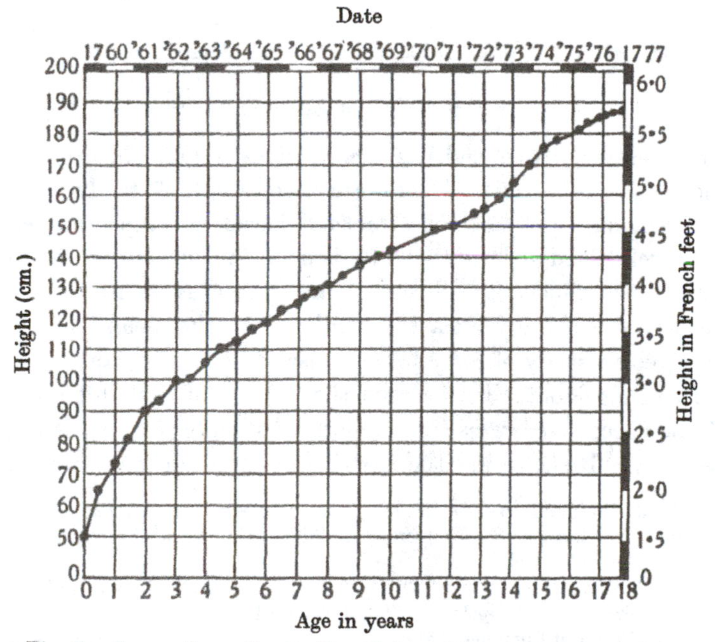

Fig. 7. Curve of growth of a French boy of the eighteenth century. From Scammon, after Buffon.

eagerly in the early nineteenth, when the exhaustion of the armies of France and the evils of factory labour in England drew attention to the stature and physique of man and to the difference between the healthy and the stunted child*.

A friend of Buffon's, the Count Philibert Guéneau de Montbeillard, kept careful measurements of his own son; and Buffon published these in 1777, in a supplementary volume of the *Histoire Naturelle*†. The child was born in April 1759; it was measured every six months

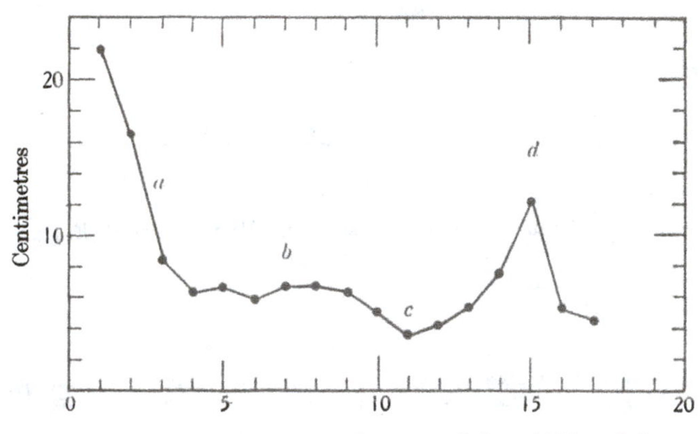

Fig. 8. Annual increments of stature of the said French boy.

for seventeen years, and the record gives a curve of great interest and beauty (Fig. 7). There are two ways of studying such a phenomenon—the statistical method based on large numbers, and the careful study of the individual case; the curve of growth of this one French child is to all intents and purposes identical, save that the boy was throughout a trifle taller, with the mean curve yielded by a recent study of forty-four thousand little Parisians‡.

In young Montbeillard's case the "curve of first differences," or of the successive annual increments of stature (Fig. 8), is clear and beautiful. It shews (*a*) the rapid, but rapidly diminishing, rate of

* Cf. M. Hargenvilliers, *Recherches...sur...le recrutement de l'armée en France*, 1817; J. W. Cowell, Measurements of children in Manchester and Stockport, *Factory Reports*, I; and works referred to by Quetelet.

† See Richard E. Scammon, The first seriatim study of human growth, *Amer. Journ. of Physical Anthropology*, x, pp. 329–336, 1927.

‡ MM. Variot et Chaumet, Tables de croissance, dressées...d'après les mensurations de 44,000 enfants parisiens, *Bull. et Mém. Soc. d'Anthropologie*, III, p. 55, 1906.

growth in infancy; (*b*) the steady growth in early boyhood; (*c*) the period of retardation which precedes, and (*d*) the rapid growth which accompanies, puberty.

Buffon, with his usual wisdom, adds some remarks of his own, which include two notable discoveries. He had observed that a man's stature is measurably diminished by fatigue, and the loss soon made up for in repose; long afterwards Quetelet said, to the same effect, "le lit est favorable à la croissance, et le matin un homme est un peu plus grand que le soir." Buffon asked whether growth varied with the seasons, and Montbeillard's data gave him his reply. Growth was quicker from April to October than during the rest of the year: shewing that "la chaleur, qui agit généralement sur le développement de tous les êtres organisées, influe considérablement sur l'accroissement du corps humain." Between five years old and ten, the child grew seven inches during the five summers, but during the five winters only four; there was a like difference again, though not so great, while the boy was growing quickly in his teens; but there were no seasonal differences at all from birth to five years old, when the child was doubtless sheltered from both heat and cold*.

On rate of growth in man and woman

That growth follows a different course in boyhood and in girlhood is a matter of common knowledge; but differences in the curves of growth are not very apparent on the scale of our diagrams. They are better seen in the annual increments, or first differences; and we may further simplify the comparison by representing the girl's weight or stature as a *percentage* of the boy's.

Taking weight to begin with (Fig. 9), the girl's growth-rate is steady in childhood, from two or three to six or seven years old,

* Growth-rates based on the continuous study of a single individual are rare; we depend mostly on average measurements of many individuals grouped according to their average age. That this is a sound method we take for granted, but we may lose by it as well as gain. (See above, p. 92.) The chief epochs of growth, the chief singularities of the curve, will come out much the same in the individual and in the average curve. But if the individual curves be skew, averaging them will tend to smooth the skewness away; and, more curiously, if they be all more or less diverse, though all symmetrical, a certain skewness will tend to develop in the composite or average curve. Cf. Margaret Merrill, The relationship of individual to average growth, *Human Biology*, III. pp. 37–70, 1931.

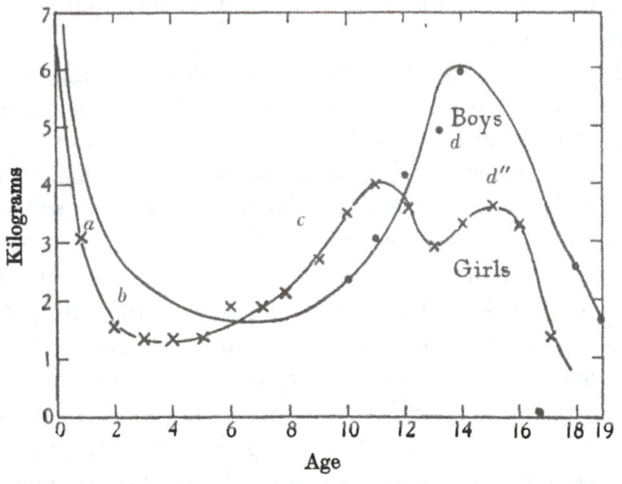

Fig. 9. Annual increase in weight of Belgian boys and girls. From Quetelet's data. (Smoothed curves.)

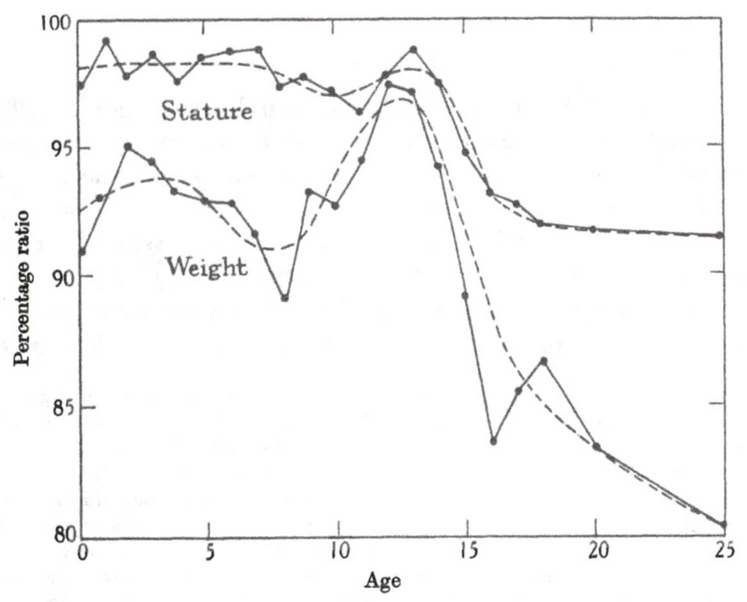

Fig. 10. Percentage ratio of female weight and stature to male. From Quetelet's Belgian data.

just as is the boy's; but her curve stands on a lower level, for the little maid is putting on less weight than the boy (*b*). Later on, her rate accelerates (*c*) sooner than does his, but it never rises quite so high (*d*). After a first maximum at eleven or twelve her rate of growth slows down a little, then rises to a second maximum when

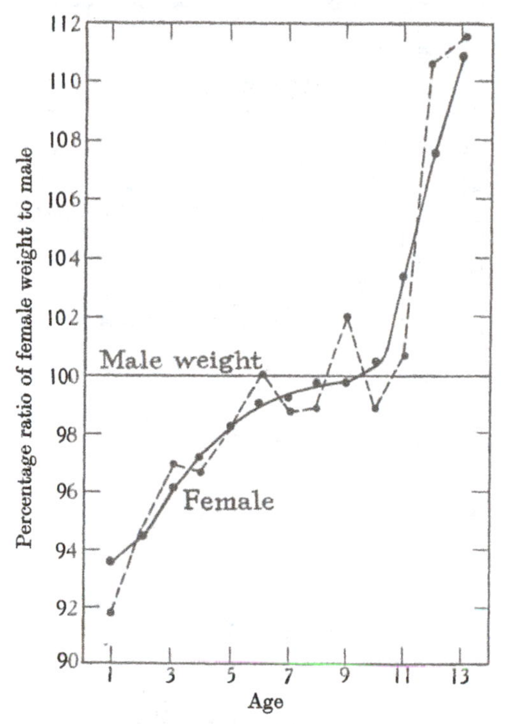

Fig. 11. Relative weight of American boys and girls. From Simmons and Todd's data.

she is sixteen or seventeen, after the boy's phase of quickest growth is over and done. This second spurt of growth, this increase of vigour and of weight in the girl of seventeen or eighteen, Quetelet's figures indicate and common observation confirms. Last of all, while men stop adding to their weight about the age of thirty or before, this does not happen to women. They increase in weight, though slowly, till much later on: until there comes a final phase, in both sexes alike, when weight and height and strength decline together.

Stature of school children, Welsh and English. (From R. M. Fleming's data)

Age	Stature (cm.) Boys	Stature (cm.) Girls	Stature (cm.) Difference	Stature (cm.) Percentage difference	Annual increment (cm.) Boys	Annual increment (cm.) Girls	Annual increment (cm.) Difference	Annual increment (cm.) Percentage difference	Coefficient of variability of stature Boys	Coefficient of variability of stature Girls	Coefficient of variability of stature Difference
3	96·3	96·0	0·3	99·7	6·5	5·6	0·9	86·1	5·44	6·24	−0·80
4	102·8	101·6	1·2	98·7	5·4	6·4	−1·0	118·5	5·27	4·99	0·28
5	108·2	108·0	1·2	99·8	4·8	5·8	−1·0	120·8	5·09	5·16	−0·07
6	113·0	113·8	−0·8	100·7	5·1	4·1	1·0	80·4	5·41	5·17	0·24
7	118·1	117·9	0·2	99·8	5·6	5·5	0·1	98·2	4·97	5·03	−0·06
8	123·7	123·4	0·3	99·7	4·3	4·4	−0·1	102·3	5·32	5·49	−0·17
9	128·0	127·8	0·2	99·9	4·3	4·8	−0·5	111·6	5·45	5·42	0·03
10	132·3	132·6	−0·3	100·2	4·4	5·3	−0·9	120·4	5·58	5·58	0·00
11	136·7	137·9	−1·2	100·8	5·3	6·1	−0·8	115·1	5·70	5·81	−0·11
12	142·0	144·0	−2·0	101·2	6·1	5·6	0·5	91·8	5·55	5·69	−0·14
13	148·1	149·6	−1·5	101·0	7·6	5·1	2·5	67·1	5·81	5·55	0·26
14	155·7	154·7	1·0	99·4	7·6	3·5	4·1	46·0	5·78	4·70	1·08
15	163·3	158·2	5·1	96·9	5·1	1·3	3·8	25·5	5·82	4·18	1·64
16	168·4	159·5	8·9	94·7	3·1	0·8	2·3	25·8	4·96	3·92	1·04
17	171·5	160·3	11·2	93·5	1·0	0·5	0·5	50·0	4·30	3·72	0·58
18	172·5	160·8	11·7	93·3					2·95	3·85	−0·90

Stature and weight of American children (Ohio)
(From Katherine Simmons and T. Wingate Todd's data)

	Stature (cm.)			Weight (lbs.)		
Age	Boys	Girls	% ratio	Boys	Girls	% ratio
3 months	61·3	59·3	96·7	14·4	13·1	91·1
1 year	76·1	74·2	97·5	23·9	21·9	91·8
2 "	87·4	86·2	98·6	29·1	27·5	94·7
3 "	96·2	95·5	99·3	33·5	32·5	96·9
4 "	103·9	103·2	99·3	38·4	37·1	96·7
5 "	110·9	110·3	99·4	43·2	42·3	98·1
6 "	117·2	117·4	100·1	48·5	48·6	100·0
7 "	123·9	123·2	99·4	54·7	54·0	98·8
8 "	130·1	129·3	99·4	62·2	61·5	98·7
9 "	136·0	135·7	99·7	69·5	70·9	102·0
10 "	141·4	140·8	99·6	78·5	77·6	98·8
11 "	146·5	147·8	100·7	86·5	87·0	100·6
12 "	151·1	155·3	102·8	92·7	102·7	110·7
13 "	156·7	159·9	102·0	102·8	114·6	111·4

Mean of observed increments of stature and weight of American children

	Increment of stature (mm.)			Increment of weight (lbs.)		
Age	Boys	Girls	% ratio	Boys	Girls	% ratio
3 m. – 1 yr.	150·4	150·1	99·8	9·32	8·07	93·8
1 yr. – 2 "	123·9	132·0	106·5	4·97	5·56	112·0
2 " – 3 "	88·1	90·0	102·2	4·01	4·18	104·3
3 " – 4 "	73·9	79·1	106·9	4·13	4·46	108·0
4 " – 5 "	69·4	72·2	104·0	4·60	4·55	99·8
5 " – 6 "	67·0	68·0	101·5	4·51	5·08	112·8
6 " – 7 "	64·1	62·6	97·6	5·57	5·40	96·9
7 " – 8 "	61·2	57·8	94·4	6·70	6·65	99·4
8 " —9 "	55·7	60·1	108·0	6·64	7·38	111·1
9 " –10 "	54·9	57·7	105·1	7·92	8·12	104·8
10 " –11 "	51·9	61·3	118·2	8·81	9·58	108·7
11 " –12 "	53·2	66·9	125·6	9·54	11·98	133·1
12 " –13 "	61·0	55·1	89·0	10·90	10·29	94·7

These differences between the two sexes, which are essentially *phase-differences*, cause the *ratio* between their weights to fluctuate in a somewhat complicated way (Figs. 10, 11). At birth the baby girl's weight is about nine-tenths of the boy's. She gains on him for a year or two, then falls behind again; from seven or eight onwards she gains rapidly, and the girl of twelve or thirteen is very little lighter than the boy; indeed in certain American statistics she is by a good deal the heavier of the two. In their teens the boy gains steadily, and the lad of sixteen is some 15 per cent. heavier than the lass. The disparity tends to diminish for a while, when the maid of seventeen has her second spurt of growth; but it increases again, though slowly, until at five-and-twenty the young woman is no more than four-fifths the weight of the man. During middle life she gains on him, and at sixty the difference stands at some 12

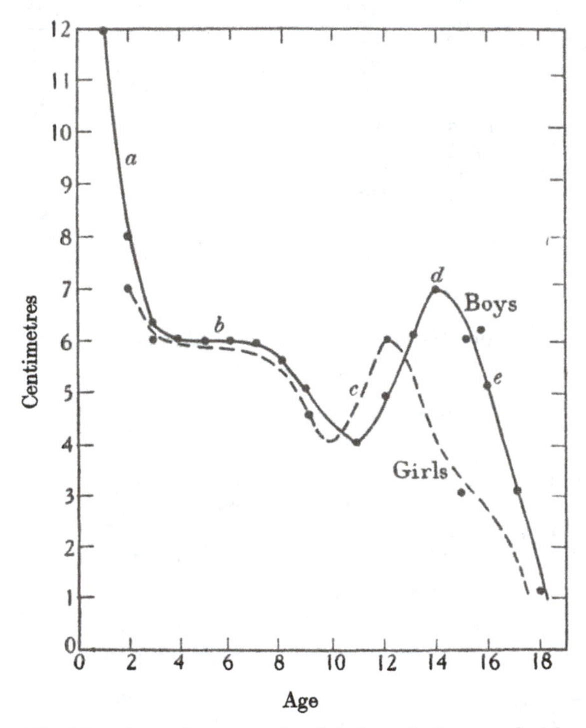

Fig. 12. Annual increments of stature, in boys and girls. From Quetelet's data. (Smoothed curves.)

per cent., not far from the mean for all ages; but the old woman shrinks and dwindles, and the difference tends to increase again.

The rate of increase of stature, like stature itself, differs notably in the two sexes, and the differences, as in the case of weight, are mostly a question of *phase* (Fig. 12). The little girl is adding rather more to her stature than the boy at four years old*, but she grows

* This early spurt of growth in the girl is shewn in English, French and American observations, but not in Quetelet's.

slower than he does for a few years thereafter (*b*). At ten years old the girl's growth-rate begins to rise (*c*), a full year before the boy's; at twelve or thirteen the rate is much alike for both, but it has reached its maximum for the girl. The boys' rate goes on rising, and at fourteen or fifteen they are growing twice as fast as the girls. So much for the annual increments, as a rough measure of the *rates* of growth. In actual stature the baby girl is some 2 or 3 per cent. below the boy at birth; she makes up the difference, and there is

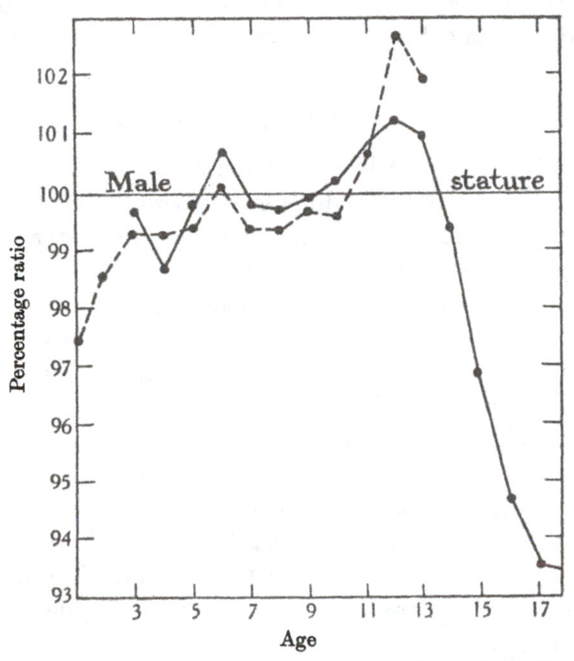

Fig. 13. Ratio of female stature to male. ——— From R. M. Fleming's data. ----- From Simmons and Todd's data.

good evidence to shew that she is by a very little the taller for a while, at about five years old or six. At twelve or thirteen she is very generally the taller of the two, and we call it her "gawky age" (Fig. 13).

Man and woman differ in length of life, just as they do in weight and stature. More baby boys are born than girls by nearly 5 per cent. The numbers draw towards equality in their teens; after twenty the women begin to outnumber the men, and at eighty-five there are twice as many women as men left in the world*.

Men have pondered over the likeness and the unlikeness between the short lifetimes and the long; and some take it to be fallacious to measure all alike by the common timepiece of the sun. Life, they say, has a varying time-scale of its own; and by this modulus the sparrow lives as long as the eagle and the day-fly as the man†. The time-scale of the living has in each case so strange a property of logarithmic decrement that our days and years are long in childhood, but an old man's minutes hasten to their end.

On pre-natal and post-natal growth

The rates of growth which we have so far studied are based on annual increments, or "first differences" between yearly determinations of magnitude. The first increment indicates the *mean rate* of growth during the first year of the infant's life, or (on a further assumption) the mean rate at the mean epoch of six months old; there is a gap between that epoch and the epoch of birth, of which we have learned nothing; we do not yet know whether the very high rate shewn within the first year goes on rising, or tends to fall, as the date of birth is approached. We are accustomed to interpolate freely, and on the whole safely, *between* known points on a curve: "si timide que l'on soit, il faut bien que l'on interpole," says Henri Poincaré; but it is much less safe and seldom justifiable (at least until we understand the physical principle involved and its mathematical expression) to "extrapolate" beyond the limits of our observations.

We must look for more detailed observations, and we may learn much to begin with from certain old tables of Russow's‡, who gives

* Cf. F. E. A. Crew's Presidential Address to Section D of the British Association, 1937.

† Cf. Gaston Backman, Die organische Zeit, *Lunds Universitets Arsskrift*, xxxv, Nr. 7, 1939.

‡ Quoted in Vierordt's *Anatomische...Daten und Tabellen*, 1906, p. 13. See also, among many others, Camerer's data, in Pfaundler and Schlossman's *Hdb. d. Kinderheilkunde*, I, pp. 49, 424, 1908; Variot, *op. cit.*; for pre-natal growth, R. E. Scammon and L. A. Calkins, *Growth in the Foetal Period*, Minneapolis, 1929. Also, on this and many other matters, E. Fauré-Fremiet, *La cinétique du développement*, Paris, 1925; and, not least, J. Needham, *Chemical Embryology*, 1931.

Differences between boy and girl in annual increments of stature

Years of Age ...	1/2	3	4	5	6	7	8	9	10	11	12	13	14	15	16	17	18	19	20
Belgian (Quetelet)	1·0	1·0	1·0	0·0	0·0	1·0	2·0	0·0	0·0	2·0	2·0	2·0	3·0	5·0	2·0	2·0	1·0	—	—
British (Fleming)	—	—	0·9	1·0	1·0	1·0	0·1	0·1	0·5	0·9	0·8	0·5	2·5	4·1	3·8	2·3	—	—	—
Parisian (Variot)	0·3	0·2	0·3	0·1	0·4	0·4	0·4	0·5	0·1	1·6	3·1	0·4	4·4	4·5	—	—	—	—	—
New England (Boas and Wissler)	—	—	—	—	—	—	—	—	—	—	3·4	0·7	4·1	5·8	4·3	2·8	1·8	0·9	0·3

Or we may take (among many others) certain careful measurements of Parisian children*, as follows:—

Months after birth ∴.		0	1	2	3	4	5	6	7	8	9	10	11	12
Stature (cm.)	Boys	49·8	53·1	56·6	58·7	61·1	62·8	64·7	66·0	67·0	68·3	69·5	70·4	71·8
	Girls	49·3	52·9	55·8	57·7	60·5	61·8	63·8	65·2	66·0	67·7	69·1	70·3	71·5
Weight (kgm.)	Boys	3·13	3·62	4·32	4·93	5·71	6·21	6·68	7·15	7·65	8·01	8·53	8·78	9·03
	Girls	3·02	3·55	4·23	4·80	5·40	5·99	6·51	6·92	7·46	7·81	8·30	8·70	8·96
Increments of														
Stature (cm.)	Boys	—	3·3	3·5	2·1	2·4	1·7	1·9	1·3	1·0	1·3	1·2	0·9	1·4
	Girls	—	3·6	2·9	1·9	2·8	1·3	2·0	1·4	0·8	1·7	1·4	1·2	1·2
Weight (kgm.)	Boys	—	4·9	7·0	6·1	7·8	5·0	4·7	4·7	5·0	3·6	5·2	2·5	2·5
	Girls	—	4·3	6·8	5·7	6·0	5·9	5·2	4·1	5·4	3·5	4·9	4·0	2·6
Percentage ratio of boys to girls														
Stature		101·3	100·3	101·4	101·6	101·0	101·5	101·4	101·2	101·4	101·0	100·5	100·1	100·4
Weight		103·6	101·5	102·1	102·7	105·8	103·7	102·6	103·3	102·6	101·5	102·8	101·0	100·7

* From G. Variot, *La Croissance chez le nourisson*, 1925, p. 119.

the stature of the infant, month by month, during the first year of its life, as follows:

Mean growth of an infant, in its first twelve-month
(*After Russow*)

Age (months)	0	1	2	3	4	5	6	7	8	9	10	11	12
Length (cm.)	50	54	58	60	62	64	65	66	67·5	68	69	70·5	72
Monthly increment (cm.)	—	4	4	2	2	2	1	1	1·5	0·5	1	1·5	1·5

From these data of Russow's for German children, rough as indeed they are, from Variot's for little Parisians (Fig. 14), and from

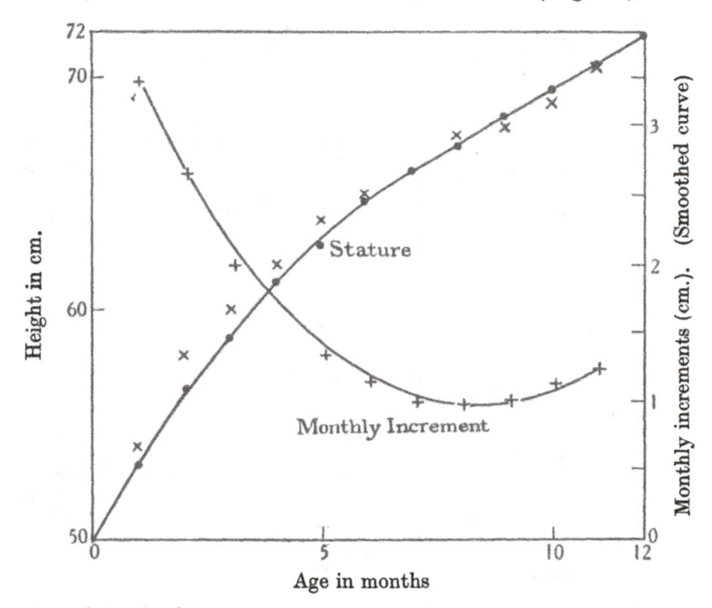

Fig. 14. Growth of Parisian children (boys) from birth to twelve months old. From G. Variot's data; Russow's German data are also shewn, by × × ×.

many more, we see that the rate of growth rises steadily and even rapidly as we pass backwards towards the date of birth. It is never anything like so great again. It is an impressive demonstration of the dynamic potentiality, of the store of energy, in the newborn child.

But birth itself is but an incident, an inconstant epoch, in the life and growth of a viviparous animal. The foal and the lamb are born later than a man-child; the puppy and the kitten are born earlier, and in more helpless case than ours; the mouse comes into the world still earlier and more inchoate, so much so that even the little marsupial is scarcely more embryonic and unformed*. We must take account, so far as each case permits, of pre-natal or intra-uterine growth, if we are to study the curve of growth in its entirety.

According to His†, the following are the mean lengths from month to month of the unborn child:

Months	0	1	2	3	4	5	6	7	8	9	10 (Birth)
Length (mm.)	0	7·5	40	84	162	275	352	402	443	472	490 / 500
Increment per month (mm.)	—	7·5	32·5	44	78	113	77	50	41	29	18 / 28

These data link on very well to those of Russow, which we have just considered; and (though His's measurements for the pre-natal months are more detailed than are those of Russow for the first year of post-natal life) we may draw a continuous curve of growth (Fig. 15) and of increments of growth (Fig. 16) for the combined periods. It will be seen at once that there is a "point of inflection" somewhere about the fifth month of intra-uterine life; up to that date growth proceeds with a continually increasing velocity. After that date, though growth is still rapid, its velocity tends to fall away; the curve, while still ascending, is becoming an **S**-shaped curve (Fig. 15). There is a slight break between our two sets of statistics at the date of birth, an epoch regarding which we should like to have precise and continuous information. But we can see that there is undoubtedly a certain slight arrest of growth, or diminution of the rate of growth, about this epoch; the sudden change of nurture has its inevitable effect, but this slight tem-

* It is part of the story, though by no means all, that (as Minot says) the larger the litter the sooner does birth take place. That the day-old foal or fawn can keep pace with their galloping dams is very remarkable; it is usually explained teleologically, as a provision of Nature, on which their safety and their survival depend. But the fact that they come one at a birth has at least something to do with their comparative maturity.

† *Unsere Körperform und das physiologische Problem ihrer Entstehung*, Leipzig, 1874. On growth in weight of the human embryo, see C. M. Jackson, *Amer. Journ. Anat.* XVII, p. 118, 1909; also J. Needham, *op. cit.* pp. 379–383.

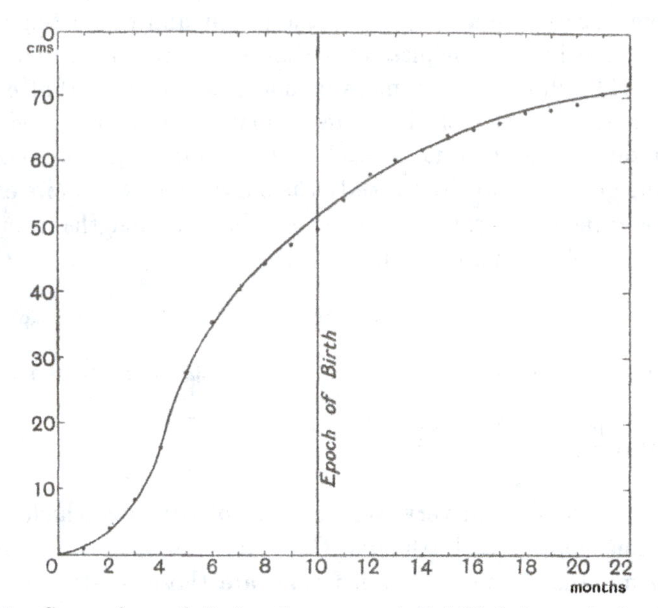

Fig. 15. Curve of growth (in length or stature) of child, before and after birth. From His and Russow's data.

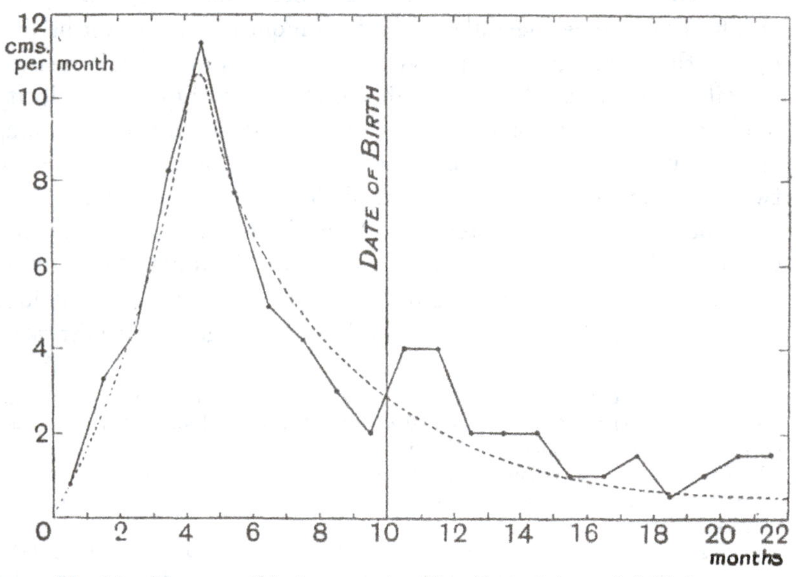

Fig. 16. Mean monthly increments of length or stature of child, in cm. From His and Russow's data.

porary set-back is immediately followed by a secondary, and equally transitory, acceleration*.

Mean weight in grams of American infants during ten days after birth. (From Meredith and Brown)

Age (days)	Weight		Daily increment	
	Male	Female	Male	Female
At birth	3491	3408	—	—
1	3376	3283	-115	-125
2	3294	3207	- 82	- 76
3	3274	3195	- 20	- 12
4	3293	3213	19	17
5	3326	3246	33	34
6	3366	3281	40	35
7	3396	3315	30	34
8	3421	3341	25	26
9	3440	3362	19	21
10	3466	3387	26	25

The set-back after birth of which we have just spoken is better shewn by the child's weight than by any linear measurement. During its first three days the infant loses weight visibly, and it is more than ten days old before it has made up the weight it lost in those first three (Fig. 17).

It is worth our while to illustrate on a larger scale His's careful data for the ten months of pre-natal life (Fig. 18). They give an **S**-shaped curve, beautifully regular, and nearly symmetrical on either side of its point of inflection; and its differential, or curve of monthly increments, is a bell-shaped curve which indicates with the utmost simplicity a rise from a minimal to a maximal rate, and a fall to a minimum again. It has a close family likeness to the well-known "curve of probability," of which we shall presently have much more to say; it is a curve for which we might well hope to find a simple mathematical expression†.

These two curves, then, look more "mathematical," and less merely descriptive, than any others we have yet drawn, and much

* See especially, H. V. Meredith and A. W. Brown, Growth in body-weight during first ten days of postnatal life, *Human Biology*, XI, pp. 24–77, 1939. Also (*int. al.*) T. Brailsford Robertson, Pre- and post-natal growth, etc., *Amer. Journ. Physiol.* XXXVII, pp. 1–42, 74–85, 1915.

† The same is not less true of Friedenthal's more elaborate measurements, in his *Physiologie des Menschenwachstums*, 1914; cf. Needham, *op. cit.* p. 1677.

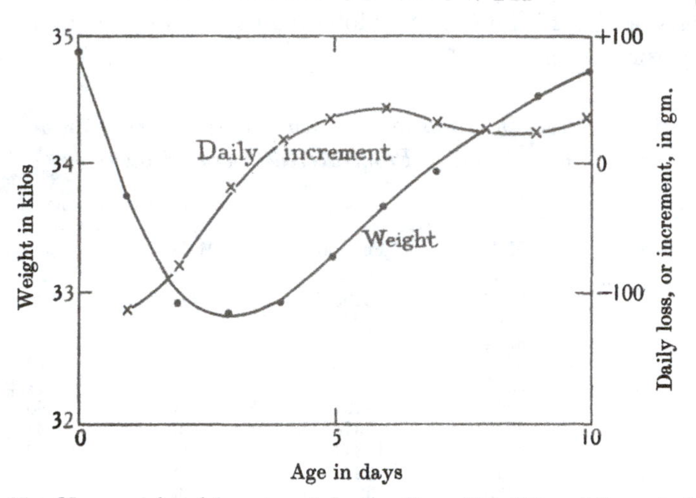

Fig. 17. Mean weight of American infants. From Meredith and Brown's data.

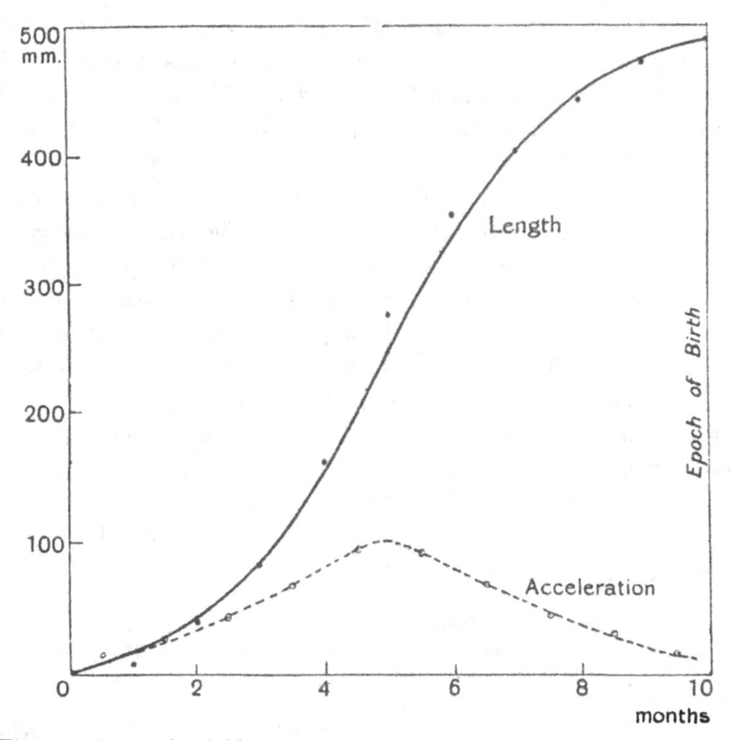

Fig. 18. Curve of a child's pre-natal growth, in length or stature; and corresponding curve of mean monthly increments (mm.). (Smoothed curves.)

the same curves meet us again and again in the growth of other organisms. The pre-natal growth of the guinea-pig is just the same*. We have the same essential features, the same S-shaped curve, in the growth by weight of an ear of maize (Fig. 19), or the growth in length of the root of a bean (Fig. 20); in both we see the same slow beginning, the rate rapidly increasing to a maximum, and the subsequent slowing down or "negative acceleration†." One phase passes into another; so far as these curves go, they exhibit growth as a continuous process, with its beginning, its middle and its end—a continuity which Sachs recognised some seventy years ago, and spoke of as the "grand period of growth‡."

But these simple curves relate to simple instances, to the infant sheltered in the womb, or to plant-growth in the sunny season of the year. They mark a favourable episode, rather than relate the course of a lifetime. A curve of growth to run all life long is only simple in the simplest of organisms, and is usually a very complex affair.

Growth in length of Vallisneria§, and root of bean|| and weight of maize¶

Vallisneria		*Vicia*		*Zea*	
Hours	Inches	Days	Mm.	Days	Gm.
6	0·3	0	1·0	6	1
16	1·7	1	2·8	18	4
42	12·6	2	6·5	30	9
54	15·4	3	24·0	39	17
65	16·1	4	40·5	46	26
77	16·7	5	57·5	53	42
88	17·1	6	72·0	60	62
		7	79·0	74	71
		8	79·0	93	74

It would seem to be a natural rule, that those offspring which are most highly organised at birth are those which are born largest

* See R. L. Draper, *Anat. Record*, XVIII, p. 369, 1920; cf. Needham, *op. cit.*, p. 1672.

† Cf. R. Chodat et A. Monnier; Sur la courbe de croissance chez les végétaux, *Bull. Herbier Boissier* (2), V, p. 615, 1905.

‡ *Arbeiten a. d. bot. Instit. Würzburg*, I, p. 569, 1872.

§ A. Bennett, *Trans. Linn. Soc.* (2), I (Bot.), p. 133, 1880.

|| Sachs, *l.c.*

¶ Stefanowska, *op. cit.*; G. Backman, *Ergebn. d. Physiologie*, XXIII, p. 925, 1931

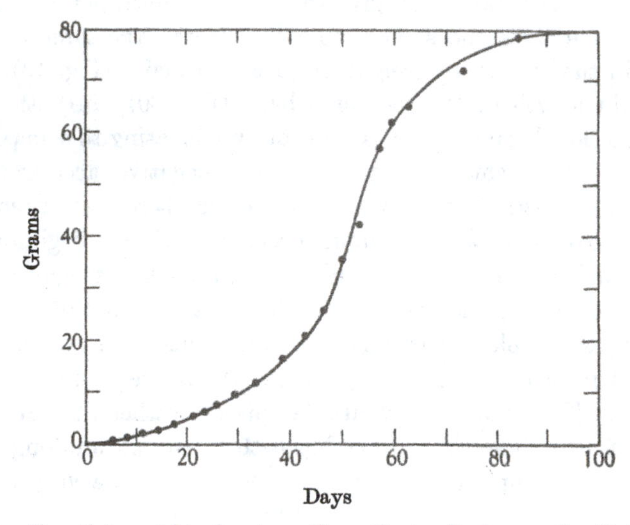

Fig. 19. Growth in weight of maize. From Gustav Backman, after Stefanowska.

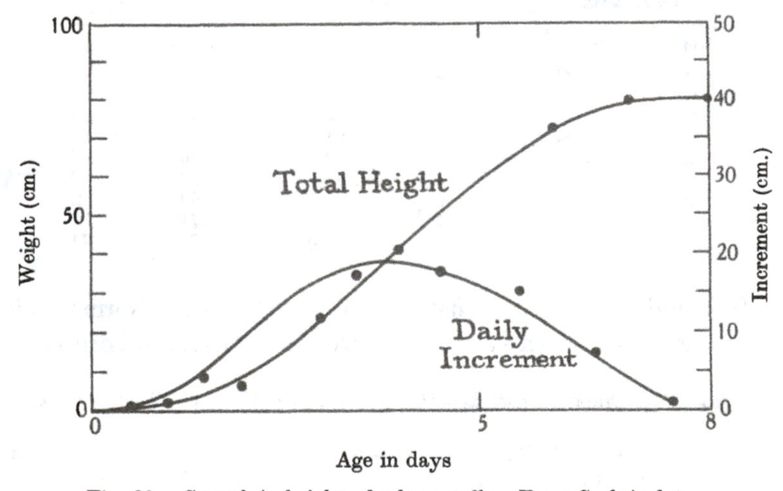

Fig. 20. Growth in height of a beanstalk. From Sachs's data.

relatively to their parents' size. But another rule comes in, which is perhaps less to be expected, that the offspring are born smaller the larger the species to which they belong. Here we shew, roughly, the relative weights of the new-born animal and its mother*:

Bear	1 : 600	Sheep	1 : 14
Lion	160	Ox	13
Hippopotamus	45	Horse	12
Dog	45–50	Rabbit	40
Cat	25	Mouse	10–25
Man	22	Guinea-pig	7

These differences at birth are for the most part made up quickly; in other words, there are great differences in the rate of growth during early post-natal life. Two lion-cubs, studied by M. Anthony, grew as follows:

	Male	Female
Feb. 23 (born)	—	—
28	2·0 kilos	1·7 kilos
Mar. 8	3·0	2·6
15	3·8	3·3
22	4·6	4·0
30	5·3	4·6
Apr. 5	6·1	5·2
12	7·0	6·0
19	8·0	7·0

Thus the lion-cub doubles its weight in the first month, and wellnigh doubles it again in the second; but the newborn child takes fully five months to double its weight, and nearly two years to do so again.

The size finally attained is a resultant of the rate and of the duration of growth; and one or other of these may be the more important, in this case or in that. It is on the whole true, as Minot said, that the rabbit is bigger than the guinea-pig because he grows faster, but man is bigger than the rabbit because he goes on growing for a longer time.

A bantam and a barn-door fowl differ in their rate of growth, which in either case is definite and specific. Bantams have been bred to match almost every variety of fowl; and large size or small, quick growth or slow, is inherited or transmitted as a Mendelian

* Data from Variot, after Anthony.

character in every cross between a bantam and a larger breed. The bantam is not produced by selecting smaller and smaller specimens of a larger breed, as an older school might have supposed; but always by first crossing with bantam blood, so introducing the "character" of smallness or retarded growth, and then segregating the desired types among the dwarfish offspring. In fact, Darwinian selection plays a small and unimportant part in the process*.

From the whole of the foregoing discussion we see that rate of growth is a specific phenomenon, deep-seated in the nature of the organism; wolf and dog, horse and ass, nay man and woman, grow at different rates under the same circumstances, and pass at different epochs through like phases of development. Much the same might be said of mental or intellectual growth; the girl's mind is more precocious than the boy's, and its development is sooner arrested than the man's†.

On variability, and on the curve of frequency or of error

The magnitudes which we are dealing with in this chapter—heights and weights and rates of change—are (with few exceptions) mean values derived from a large number of individual cases. We deal with what (to borrow a word from atomic physics) we may call an *ensemble*; we employ the equalising power of averages, invoke the "law of large numbers‡," and claim to obtain results thereby which are more trustworthy than observation itself§. But in ascertaining a mean value we must also take account of the *amount of variability*, or departure from the mean, among the cases from which the mean value is derived. This leads on far beyond our scope, but we must spare it a passing word; it was this identical phenomenon, in the case of Man, which suggested to Quetelet the

* Cf. Raymond Pearl, The selection problem, *Amer. Naturalist*, 1917, p. 82; R. C. Punnett and P. G. Bailey, *Journ. of Genetics*, IV, pp. 23–39, 1914.

† Cf. E. Devaux, L'allure du développement dans les deux sexes, *Revue génér. des Sci*. 1926, p. 598.

‡ S. D. Poisson, following James Bernoulli's *Ars Conjectandi* (op. posth. 1713), was the discoverer, or inventor, of the law of large numbers. "Les choses de toute nature sont soumises à une loi universelle qu'on peut appeler la loi des grands nombres" (*Recherches*, 1837, pp. 7–12).

§ See p. 137, footnote.

statistical study of Variation, led Francis Galton to enquire into the laws of Natural Inheritance, and served Karl Pearson as the foundation of his science of Biometrics.

When Quetelet tells us that the *mean stature* of a ten-year-old boy is 1·275 metres, this is found to imply, not only that the measurements of all his ten-year-old boys group themselves about this mean value of 1·275 metres, but that they do so *in an orderly way*, many departing little from that mean value, and fewer and fewer departing more and more. In fact, when all the measurements are grouped and plotted, so as to shew the number of instances (y) at each gradation of size (x), we obtain a characteristic

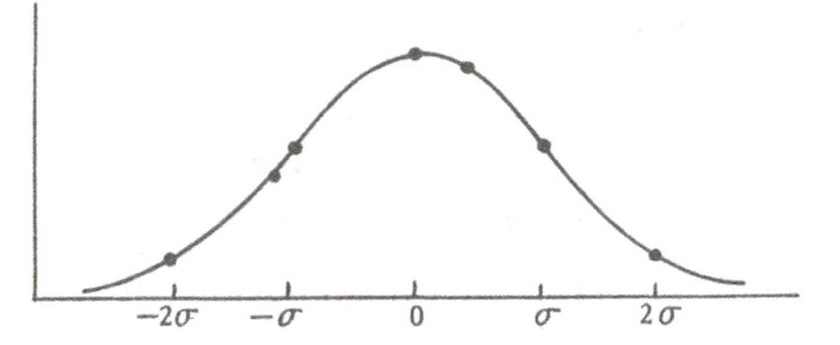

Fig. 21. The normal curve of frequency, or of error. σ, $-\sigma$, the "standard deviation".

configuration, mathematically definable, called the *curve of frequency*, or of *error* (Fig. 21). This is a very remarkable fact. That a "curve of stature" should agree closely with the "normal curve of error" amazed Galton, and (as he said) formed the mainstay of his long and fruitful enquiry into natural inheritance*. The curve is a thing apart, *sui generis*. It depicts no course of events, it is no time or vector diagram. It merely deals with the variability, and variation, of magnitudes; and by magnitudes we mean anything which can be counted or measured, a regiment of men, a basket of

* Stature itself, in a homogeneous population, is a good instance of a normal frequency distribution, save only that the spread or range of variation is unusually low; for one-half of the population of England differs by no more than an inch and a half from the average of them all. Variation is said to be greater among the negroid than among the white races, and it is certainly very great from one race to another: e.g. from the Dinkas of the White Nile with a mean height of 1·8 m. to the Congo pygmies averaging 1·35, or say 5 ft. 11 in. and 4 ft. 6 in. respectively.

nuts, the florets of a daisy, the stripes of a zebra, the nearness of shots to the bull's eye*. It thereby illustrates one of the most far-reaching, some say one of the most fundamental, of nature's laws.

We find the curve of error manifesting itself in the departures from a mean value, which seems itself to be merely accidental—as, for instance, the mean height or weight of ten-year-old English boys; but we find it no less well displayed when a certain definite or normal number is indicated by the nature of the case. For instance the Medusae, or jelly-fishes, have a "radiate symmetry" of eight nodes and internodes. But even so, the number eight is subject to variation, and the instances of more or less group themselves in a Gaussian curve.

Number of "tentaculocysts" in Medusae (Ephyra and Aurelia)
(Data from E. T. Browne, Q.J.M.S. XXXVII, p. 245, 1895)

	5	6	7	8	9	10	11	12	13	14	15
Ephyra (1893)	—	4	8	278	22	18	12	14	3	—	—
„ (1894)	1	6	34	883	75	61	35	17	3	1	—
Aurelia (1894)	—	2	18	296	33	16	18	7	—	—	1
Percentage numbers:											
Ephyra	—	1·1	2·2	77·4	6·1	5·0	3·3	3·9	0·8	—	—
„	—	0·5	3·0	79·0	6·7	5·4	3·1	1·4	0·2	—	—
Aurelia	—	0·5	4·7	77·2	8·6	4·1	2·6	1·8	—	—	—
Mean	—	0·7	3·3	77·9	7·1	4·8	3·0	2·4	0·3	—	—

The curve of error is a "bell-shaped curve," a *courbe en cloche*. It rises to a maximum, falls away on either side, has neither beginning nor end. It is (normally) symmetrical, for lack of cause to make it otherwise; it falls off faster and then slower the farther it departs from the mean or middle line; it has a "point of inflexion," of necessity on either side, where it changes its curvature and from being concave to the middle line spreads out to become convex

* "I know of scarcely anything (says Galton) so apt to impress the imagination as the wonderful form of cosmic order expressed by the Law of Frequency of Error....It reigns with serenity and in complete self-effacement amidst the wildest confusion" (*Natural Inheritance*, p. 62). Observe that Galton calls it the "law of frequency *of* error," which is indeed its older and proper name. Cf. (*int. al.*) P. G. Tait, *Trans R.S.E.* XXIV, pp. 139–145, 1867.

thereto. If we pour a bushel of corn out of a sack, the outline or profile of the heap resembles such a curve; and wellnigh every hill and mountain in the world is analogous (even though remotely) to that heap of corn*. Causes beyond our ken have cooperated to place and allocate each grain or pebble; and we call the result a "random distribution," and attribute it to fortuity, or chance. Galton devised a very beautiful experiment, in which a sloping tray is beset with pins, and sand or millet-seed poured in at the top. Every falling grain has its course deflected again and again; the final distribution is emphatically a random one, and the curve of error builds itself up before our eyes.

The curve as defined by Gauss, *princeps mathematicorum*—who in turn was building on Laplace†—is at once empirical and theoretical‡; and Lippmann is said to have remarked to Poincaré: "Les expérimentateurs s'imaginent que c'est un théorème de mathématique, et les mathématiciens d'être un fait expérimental!" It is theoretical in so far as its equation is based on certain hypothetical considerations: viz. (1) that the arithmetic mean of a number of variants is their best or likeliest average, an axiom which is obviously true in simple cases—but not necessarily in all; (2) that "fortuity" implies the absence of any predominant, decisive or overwhelming cause, and connotes rather the coexistence and joint effect of small, undefined but independent causes, many or few:

* If we pour the corn out carefully through a small hole above, the heap becomes a cone, with sides sloping at an "angle of repose"; and the cone of Fujiyama is an exquisite illustration of the same thing. But in these two instances one predominant cause outweighs all the rest, and the distribution is no longer a random one.

† The Gaussian curve of error is really the "second curve of error" of Laplace. Laplace's first curve of error (which has uses of its own) consists of two exponential curves, joining in a sharp peak at the median value. Cf. W. J. Luyten, *Proc. Nat. Acad. Sci.* XVIII, pp. 360–365, 1932.

‡ The Gaussian equation to the normal frequency distribution or "curve of error" need not concern us further, but let us state it once for all:

$$y=\frac{1}{\sqrt{2\pi}}e^{-\frac{(x_a-x)^2}{2}},$$

where x_a is the abscissa which gives the maximum ordinate, and where the maximum ordinate, $y_0=1/\sqrt{(2\pi)}$. Thus the log of the ordinate is a quadratic function of the abscissa; and a simple property, fundamental to the curve, is that for equally spaced ordinates (starting anywhere) the square of any ordinate divided by the product of its neighbours gives a scalar quantity which is constant all along (G.T.B.).

producing their several variations, deviations or errors; and potent in their combinations, permutations and interferences*.

We begin to see why bodily dimensions lend, or submit, themselves to this masterful law. Stature is no single, simple thing; it is compounded of bones, cartilages and other elements, variable each in its own way, some lengthening as others shorten, each playing its little part, like a single pin in Galton's toy, towards a "fortuitous" resultant. "The beautiful regularity in the statures of a population (says Galton) whenever they are statistically marshalled in the order of their heights, is due to the number of variable and quasi-independent elements of which stature is the sum." In a bagful of pennies fresh from the Mint each coin is made by the single stroke of an identical die, and no ordinary weights and measures suffice to differentiate them; but in a bagful of old-fashioned hand-made nails a slow succession of repeated operations has drawn the rod and cut the lengths and hammered out head, shaft and point of every single nail—and a curve of error depicts the differences between them.

The law of error was formulated by Gauss for the sake of the astronomers, who aimed at the highest possible accuracy, and strove so to interpret their observations as to eliminate or minimise their inevitable personal and instrumental errors. It had its roots also in the luck of the gaming-table, and in the discovery by eighteenth-century mathematicians that "chance might be defined in terms of mathematical precision, or mathematical 'law'." It was Quetelet who, beginning as astronomer and meteorologist, applied the "law of frequency of error" for the first time to biological statistics, with which in name and origin it had nothing whatsoever to do.

The intrinsic significance of the theory of probabilities and the law of error is hard to understand. It is sometimes said that to forecast the future is the main purpose of statistical study, and *expectation*, or expectancy, is a common theme. But all the theory

* "The curve of error would seem to carry the great lesson that the ultimate differences between individuals are simple and few; that they depend on collisions and arrangements, on permutations and combinations, on groupings and interferences, of elementary qualities which are *limited in variety and finite in extent*" (J. M. Keynes). A connection between this law and Mendelian inheritance is discussed by John Brownlee, *P.R.S.E.* XXXI, p. 251, 1910.

in the world enables us to foretell no single unknown thing, not even the turn of a card or the fall of a die. The theory of probabilities is a development of the theory of combinations, and only deals with what occurs, or has occurred, in *the long run*, among large numbers and many permutations thereof. Large numbers simplify many things; a million men are easier to understand than one man out of a million. As David Hume* said: "What depends on a few persons is in a great measure to be ascribed to chance, or to secret and unknown causes; what arises from a great many may often be accounted for by determinate and known causes." Physics is, or has become, a comparatively simple science, just because its laws are based on the statistical averages of innumerable molecular or primordial elements. In that invisible world we are sometimes told that "chance" reigns, and "uncertainty" is the rule; but such phrases as *mere chance*, or *at random*, have no meaning at all except with reference to the knowledge of the observer, and a thing is a "pure matter of chance" when it depends on laws which we do not know, or are not considering†. Ever since its inception the merits and significance of the theory of probabilities have been variously estimated. Some say it touches the very foundations of knowledge‡; and others remind us that "avec les chiffres on peut tout démontrer." It is beyond doubt, it is a matter of common experience, that probability plays its part as a guide to reasoning. It extends, so to speak, the theory of the syllogism, and has been called the "logic of uncertain inference"§.

In measuring a group of natural objects, our measurements are uncertain on the one hand and the objects variable on the other; and our first care is to measure in such a way, and to such a scale, that our own errors are small compared with the natural variations. Then, having made our careful measurements of a group, we want to know more of the *distribution* of the several magnitudes, and

* Essay XIV.

† So Leslie Ellis and G. B. Airy, in correspondence with Sir J. D. Forbes; see his *Life*, p. 480.

‡ Cf. Hans Reichenbach, Les fondements logiques du calcul des probabilités, *Annales de l'inst. Poincaré*, VII, pp. 267, 1937.

§ Cf. J. M. Keynes, *A Treatise on Probability*, 1921; and A. C. Aitken's *Statistical Mathematics*, 1939.

especially to know two important things. We want a *mean value*, as a substitute for the *true value** if there be such a thing; let us use the arithmetic mean to begin with. About this mean the observed values are grouped like a target hit by skilful or unskilful shots; we want some measure of their inaccuracy, some measure of their *spread*, or *scatter*, or dispersion, and there are more ways than one of measuring and of representing this. We do it visibly and graphically every time we draw the curve (or polygon) of frequency; but we want a means of description or tabulation, in words or in numbers. We find it, according to statistical mathematics, in the so-called *index of variability*, or *standard deviation* (σ), which merely means the average deviation from the mean†. But we must take some precautions in determining this average; for in the nature of things these deviations err both by excess and defect, they are partly positive and partly negative, and their mean value is the mean of the variants themselves. Their squares, however, are all positive, and the mean of these takes account of the magnitude of each deviation with no risk of cancelling out the positive and negative terms: but the "dimension" of this average of the squares is wrong. The square root of this average of squares restores the correct dimension, and the result is the useful index of variability, or of deviation, which is called σ‡.

This standard deviation divides the area under the normal curve *nearly* into equal halves, and *nearly* coincides with the point of inflexion on either side; it is the simplest algebraic measure of dispersion, as the mean is the simplest arithmetical measure of position. When we divide this value by the mean, we get a figure

* It is not always obvious what the "errors" are, nor what it is that they depart or deviate from. We are apt to think of the arithmetic mean, and to leave it at that. But were we to try to ascertain the ratio of circumference to diameter by measuring pennies or cartwheels, our "errors" would be found grouped round a mean value which no simple arithmetic could define.

† σ, the standard deviation, was chosen for its convenience in mathematical calculation and formulation. It has no special biological significance; and a simpler index, the "inter-quartile distance," has its advantages for the non-mathematician, as we shall see presently.

‡ That is to say: Square the deviation-from-the-mean of each class or ordinate (ξ); multiply each by the number of instances (or "variates") in that class (f); divide by the total number (N); and take the square-root of the whole: $\sigma^2 = \frac{\Sigma(\xi^2 f)}{N}$.

which is independent of any particular units, and which is called the *coefficient of variability**.

Karl Pearson, measuring the amount of variability in the weight and height of man, found this coefficient to run as follows: In male new-born infants, for weight 15·6, and for stature 6·5; in male adults, for weight 10·8, and for stature 3·6. Here the amount of variability is thrice as great for weight as for stature among grown men, and about 2½ times as great in infancy†. The same curious fact is well brought out in some careful measurements of shell-fish, as follows:

Variability of young Clams (Mactra sp.)‡

	Average size		Coefficient of variability	
Age (years)	1	2	1	2
Number in sample	41	20	41	20
Length (cm.)	3·2	6·3	15·3	6·3
Height	2·3	4·7	14·0	6·7
Thickness	1·3	2·8	9·6	8·3
Weight (gm.)	6·4	59·8	35·4	18·5

The phenomenon is purely mathematical. Weight varies as the product of length, height and depth, or (as we have so often seen) as the cube of any one of these dimensions in the case of similar figures. It is then a mathematical, rather than a biological fact that, for small deviations, the variability of the whole tends to be equal to the *sum* of that of the three constituent dimensions. For if weight, w, varies as height x, breadth y, and depth z, we may write

$$w = c.xyz.$$

Whence, differentiating, $$\frac{dw}{w} = \frac{dx}{x} + \frac{dy}{y} + \frac{dz}{z}.$$

We see that among the shell-fish there is much more variability in the younger than in the older brood. This may be due to

* It is usually multiplied by 100, to make it of a handier amount; and we may then define this coefficient, C, as $= \sigma/M \times 100$.

† Cf. Fr. Boas, Growth of Toronto children, *Rep. of U.S. Comm. of Education, 1896–7*, 1898, pp. 1541–1599; Boas and Clark Wissler, Statistics of growth, *Education Rep. 1904*, 1906, pp. 25–132; H. P. Bowditch, *Rep. Mass. State Board of Health*, 1877; K. Pearson, On the magnitude of certain coefficients of correlation in man, *Proc. R.S.* LXVI, 1900; S. Nagai, Körperkonstitution der Japaner, from Brugsch-Levy, *Biologie d. Person.* II, p. 445, 1928; R. M. Fleming, A study of growth and development, *Medical Research Council, Special Report*, No. 190, 1933.

‡ From F. W. Weymouth, *California Fish Bulletin*, No. 7, 1923.

inequality of age; for in a population only a few weeks old, a few days sooner or later in the date of birth would make more difference than later on. But a more important matter, to be seen in mankind (Fig. 22), is that variability of stature runs *pari passu*, or nearly so, with the rate of growth, or curve of annual increments (cf. Fig. 12). The curve of variability descends when the growth-rate slackens, and rises high when in late boyhood growth is speeded up. In short, the amount of variability in stature or in weight is correlated with, or is a function of, the rate of growth in these magnitudes.

Judging from the evidence at hand, we may say that variability reaches its height in man about the age of thirteen or fourteen, rather earlier in the girls than in the boys, and rather earlier in the case of stature than of weight. The difference in this respect between the boys and the girls is now on one side, now on the other. In infancy variability is greater in the girls; the boys shew it the more at five or six years old; about ten years old the girls have it again. From twelve to sixteen the boys are much the more variable, but by seventeen the balance has swung the other way (Fig. 23).

Coefficient of variability ($\sigma/M \times 100$) *in man, at various ages*

Age ...	5	6	7	8	9	10	11	12	13	14	15	16	17	18
						Stature								
British (Fleming):														
Boys	5·1	5·4	5·0	5·3	5·4	5·6	5·7	5·6	5·8	5·8	5·8	5·0	4·3	3·0
Girls	5·2	5·2	5·0	5·5	5·4	5·6	5·8	5·7	5·6	4·7	4·2	3·9	3·7	3·8
American (Bowditch)	4·8	4·6	4·4	4·5	4·4	4·6	4·7	4·9	5·5	5·8	5·6	5·5	4·6	3·7
Japanese (Nagai):														
Boys	—	4·0	—	4·3	—	4·1	—	4·0	5·0	5·0	4·2	3·2	—	—
Girls	—	4·3	—	4·1	—	4·5	—	4·5	4·6	3·6	3·1	3·0	—	—
Mean	—	4·7	—	4·7	—	4·9	—	5·0	5·3	5·0	4·6	4·1	—	—
						Weight								
American	11·6	10·3	11·1	9·9	11·0	1·6	1·8	13·7	3·6	6·8	15·3	13·3	13·0	10·4
Japanese:														
Boys	—	10·3	—	12·1	—	0·8	—	7·0	5·1	7·0	13·8	10·9	—	—
Girls	—	10·2	—	11·2	—	2·1	—	15·0	5·6	3·4	11·4	11·5	—	—
Mean	—	10·3	—	11·1	—	11·5	—	11·9	14·8	15·7	13·5	11·9	—	—

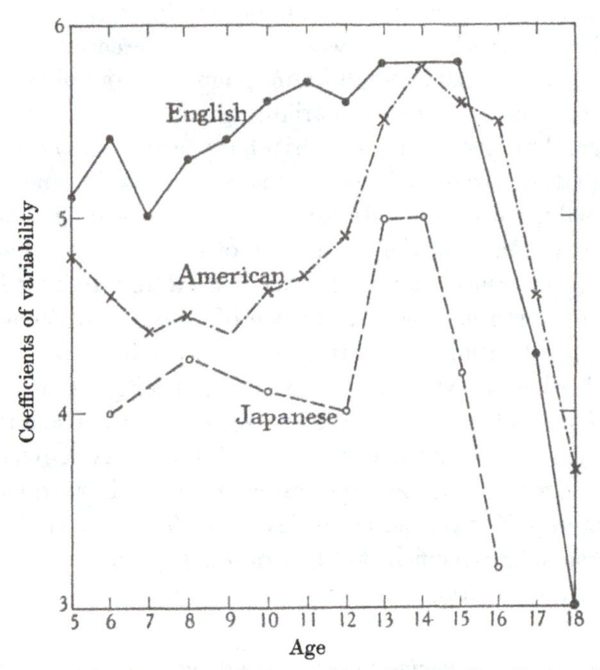

Fig. 22. Variability in stature (boys). After Fleming, Bowditch and Nagai.

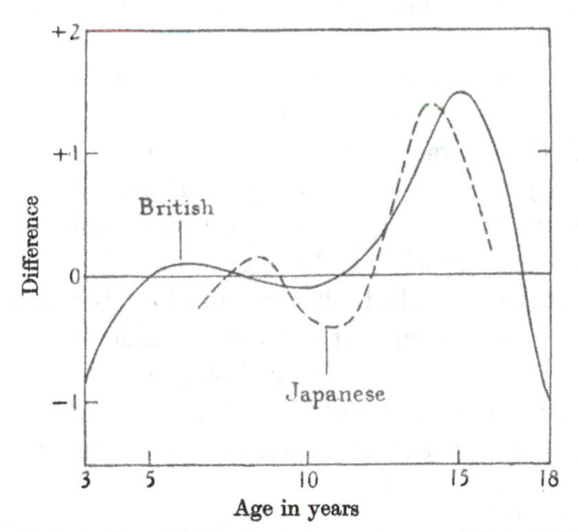

Fig. 23. Coefficient of variability in stature: excess or defect of this coefficient in the boy over the girl. Data from R. M. Fleming, and from Nagai.

The amount of variability is bound to differ from one race or nationality to another, and we find big differences between the Americans and the Japanese, both in magnitude and phase (Fig. 22).

If we take not merely the variability of stature or weight at a given age, but the variability of the yearly *increments*, we find that this latter variability tends to increase steadily, and more and more rapidly, within the ages for which we have information; and this phenomenon is, in the main, easy of explanation. For a great part of the difference between one individual and another in regard to rate of growth is a mere difference of *phase*—a difference in the epochs of acceleration and retardation, and finally a difference as to the epoch when growth comes to an end; it follows that variability will be more and more marked as we approach and reach the period when some individuals still continue, and others have already ceased, to grow. In the following epitomised table, I have taken Boas's determinations* of the standard deviation (σ), converted them into the corresponding coefficients of variability ($\sigma/M \times 100$), and then smoothed the resulting numbers:

Coefficients of variability in annual increments of stature

Age ...	7	8	9	10	11	12	13	14	15
Boys	17·3	15·8	18·6	19·1	21·0	24·7	29·0	36·2	46·1
Girls	17·1	17·8	19·2	22·7	25·9	29·3	37·0	44·8	—

The greater variability in the girls is very marked †, and is explained (in part at least) by the more rapid rate at which the girls run through the several phases of their growth (Fig. 24). To say that children of a given age vary in the rate at which they are growing would seem to be a more fundamental statement than that they vary in the size to which they have grown.

Just as there is a marked difference in phase between the growth-curves of the two sexes, that is to say a difference in the epochs when growth is rapid or the reverse, so also, within each sex, will there be room for similar, but individual, phase-differences. Thus we may have children of accelerated development, who at a given

* *Op. cit.* p. 1548.

† That women are on the whole more variable than men was argued by Karl Pearson in one of his earlier essays: *The Chances of Death and other Studies*, 1897.

epoch after birth are growing rapidly and are already "big for their age"; and others, of retarded development, who are comparatively small and have not reached the period of acceleration which, in greater or less degree, will come to them in turn. In other words, there must under such circumstances be a strong positive "coefficient of correlation" between stature and rate of growth, and also between

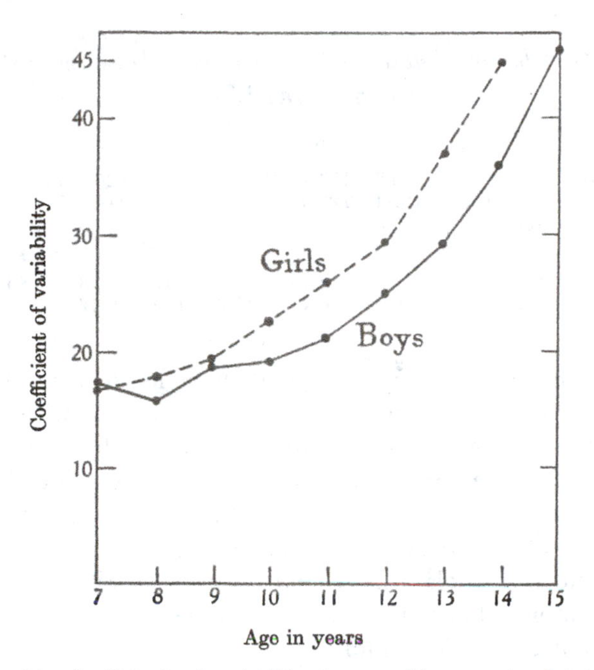

Fig. 24. Coefficients of variability, in annual increments of stature. After Boas.

the rate of growth in one year and the next. But it does not by any means follow that a child who is precociously big will continue to grow rapidly, and become a man or woman of exceptional stature *. On the contrary, when in the case of the precocious or "accelerated" children growth has begun to slow down, the back-

* Some first attempts at analysis seem to shew that the size of the embryo at birth, or of the seed at germination, has more influence than we were wont to suppose on the ultimate size of plant or animal. See (e.g.) Eric Ashby, Heterosis and the inheritance of acquired characters, *Proc. R.S.* (B), No. 833, pp. 431–441, 1937; and papers quoted therein.

ward ones may still be growing rapidly, and so making up (more or less completely) on the others. In other words, the period of high positive correlation between stature and increment will tend to be followed by one of negative correlation. This interesting and important point, due to Boas and Wissler*, is confirmed by the following table:

Correlation of stature and increment in boys and girls
(*From Boas and Wissler*)

Age ...	...	6	7	8	9	10	11	12	13	14	15
Stature	(B)	112·7	115·5	123·2	127·4	133·2	136·8	142·7	147·3	155·9	162·2
	(G)	111·4	117·7	121·4	127·9	131·8	136·7	144·6	149·7	153·8	157·2
Increment	(B)	5·7	5·3	4·9	5·1	5·0	4·7	5·9	7·5	6·2	5·2
	(G)	5·9	5·5	5·5	5·9	6·2	7·2	6·5	5·4	3·3	1·7
Correlation	(B)	0·25	0·11	0·08	0·25	0·18	0·18	0·48	0·29	−0·42	−0·44
	(G)	0·44	0·14	0·24	0·47	0·18	−0·18	−0·42	−0·39	−0·63	0·11

A minor but very curious point brought out by the same investigators is that, if instead of stature we deal with height in the sitting posture (or, practically speaking, with length of trunk or back), then the correlations between this height and its annual increment are throughout negative. In other words, there would seem to be a general tendency for the long trunks to grow slowly throughout the whole period under investigation. It is a well-known anatomical fact that tallness is in the main due not to length of body but to length of limb.

Since growth in height and growth in weight have each their own velocities, and these fluctuate, and even the amount of their variability alters with age, it follows that the *correlation* between height and weight must not only also vary but must tend to fluctuate in a somewhat complicated way. The fact is, this correlation passes through alternate maxima and minima, chief among which are a maximum at about fourteen years of age and a minimum about twenty-one. Other intercorrelations, such as those between height or weight and chest-measurement, shew their periodic variations in like manner; and it is about the time of puberty

* *l.c.* p. 42, and other papers there quoted. Cf. also T. B. Robertson, *Criteria of Normality in the Growth of Children*, Sydney, 1922.

that correlation tends to be closest, or a *norm* to be most nearly approached*.

The whole subject of variability, both of magnitude and rate of increment, is highly suggestive and instructive: inasmuch as it helps further to impress upon us that growth and specific *rate of growth* are the main physiological factors, of which specific magnitude, dimensions and form are the concrete and visible resultant. Nor may we forget for a moment that growth-rate, and growth itself, are both of them very complex things. The increase of the active tissues, the building of the skeleton and the laying up of fat and other stores, all these and more enter into the complex phenomenon of growth. In the first instance we may treat these many factors as though they were all one. But the breeder and the geneticist will soon want to deal with them apart; and the mathematician will scarce look for a simple expression where so many factors are involved. But the problems of variability, though they are intimately related to the general problem of growth, carry us very soon beyond our limitations.

The curve of error

To return to the curve of error.

The normal curve is a symmetrical one. Its middle point, or *median ordinate*, marks the arithmetic *mean* of all the measurements; it is also the *mode*, or class to which the largest number of individual instances belong. *Mean*, *median* and *mode* are three different sorts of average; but they are one and the same in the normal curve.

It is easy to produce a related curve which is not symmetrical, and in which mean, median and mode are no longer the same. The heap of corn will be lop-sided or "skew" if the wind be blowing while the grain is falling: in other words, if some prevailing cause disturb the quasi-equilibrium of fortuity; and there are other ways, some simple, some more subtle, by which asymmetry may be impressed upon our curve.

The Gaussian curve is only one of many similar bell-shaped curves; and the binomial coefficients, the numerical coefficients of $(a + b)^n$, yield a curve so like it that we may treat them as the same. The

* Cf. Joseph Bergson, Growth-changes in physical correlation, *Human Biology*, I, p. 4, 1930.

Gaussian curve extends, in theory, to infinity at either end; and this infinite extension, or asymptotism, has its biological significance. We know that this or that athletic record is lowered, slowly but continually, as the years go by. This is due in part, doubtless, to increasing skill and improved technique; but quite apart from these the record would slowly fall as more and more races are run, owing to the indefinite extension of the Gaussian curve*.

On the other hand, while the Gaussian curve extends in theory to infinity, the fact that variation is always limited and that *extreme* variations are infinitely rare is one of the chief lessons of the law of frequency. If, in a population of 100,000 men, 170 cm. be the mean height and 6 cm. the standard deviation, only 11 per cent., or say 130 men, will exceed 188 cm., only 10 men will be over 191 cm., and only one over 193 cm., or 13½ per cent. above the average. The chance is negligible of a single one being found over 210 cm., or 7 ft. high, or 24 per cent. above the average.

Yet, widely as the law holds good, it is hardly safe to count it as a universal law. Old Parr at 150 years old, or the giant Chang at more than eight feet high, are not so much extreme instances of a law of probability, as exceptional cases due to some peculiar cause or influence coming in†. In a somewhat analogous way, one or two species in a group grow far beyond the average size; the Atlas moth, the Goliath beetle, the ostrich and the elephant, are far-off outliers from the groups to which they belong. A reason is not easy to find. It looks as though variations came at last to be in proportion to the size attained, and so to go on by compound interest or geometrical progression. There may be nothing surprising in this; nevertheless, it is in contradistinction to that summation of small fortuitous differences which lies at the root of the law of error. If size vary in proportion to the magnitude of the variant individuals, not only

* This is true up to a certain extent, but would become a mathematical fiction later on. There will be physical limitations (as there are in quantum mechanics) both to record-breaking, and to the measurement of minute extensions of the record.

† We may indeed treat old Parr's case on the ordinary lines of actuarial probability, but it is "without much actuarial importance." The chance of his record being broken by a modern centenarian is reckoned at $(\frac{1}{2})^{50}$, by Major Greenwood and J. C. Irwin, writing on Senility, in *Human Biology*, XI, pp. 1–23, 1939.

will the frequency curve be obviously skew, but the *geometric mean*, not the arithmetic, becomes the most probable value*. Now the logarithm of the geometric mean of a series of numbers is the arithmetic mean of their logarithms; and it follows that in such cases the logarithms of the variants, and not the variants themselves, will tend to obey the Gaussian law and follow the normal curve of frequency†.

The Gaussian curve, and the standard deviation associated with it, were (as we have seen) invented by a mathematician for the use

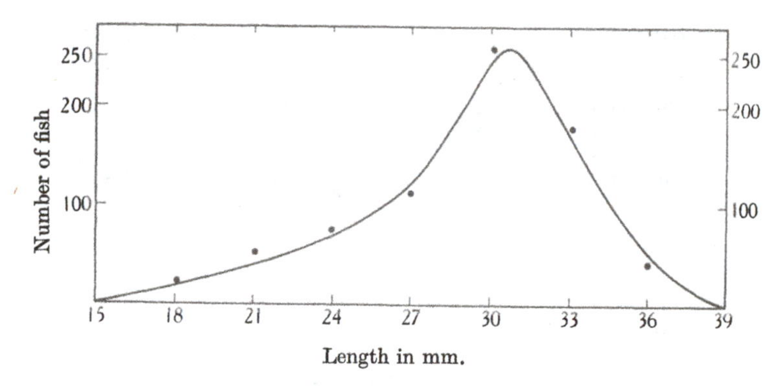

Fig. 25 A. Curve of frequency of a population of minnows.

of an astronomer, and their use in biology has its difficulties and disadvantages. We may do much in a simpler way. Choosing a random example, I take a catch of minnows, measured in 3 mm. groups, as follows (Fig. 25 A):

Size (mm.)	13–15	16–18	19–21	22–24	25–27	28–30	31–33	34–36	37–39
Number	1	22	52	67	114	257	177	41	2

* See especially J. C. Kapteyn, Skew frequency curves in biology and statistics, *Rec. des Trav. Botan. Néerland.*, Groningen, XIII, pp. 105–158, 1916. Also Axel M. Hemmingsen, Statistical analysis of the differences in body-size of related species, *Danske Vidensk. Selsk. Medd.* XCVIII, pp. 125–160, 1934.

† This often holds good. Wealth breeds wealth, hence the distribution of wealth follows a skew curve; but logarithmically this curve becomes a normal one. Weber's law, in physiology, is a well-known instance; on the thresholds of sensations, effects are produced proportional to the magnitudes of those thresholds, and the logs of the thresholds, and not the thresholds themselves, are normally distributed.

Let us sum the same figures up, so as to show the whole number above or below the respective sizes.

Size (mm.)	15	18	21	24	27	30	33	36	39
Number below	1	23	75	142	256	513	690	731	733
Percentage	—	3·1	10·2	19·4	34·9	70·0	94·1	99·6	100

Our first set of figures, the actual measurements, would give us the "courbe en cloche," in the form of an unsymmetrical (or "skew") Gaussian curve: one, that is to say, with a long sloping *talus* on

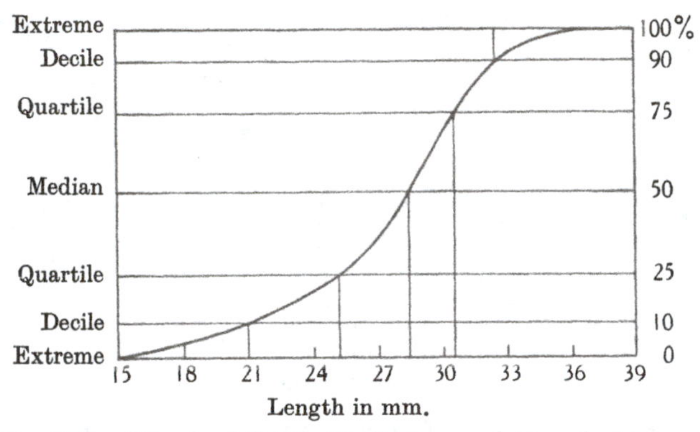

Fig. 25 B. "Curve of distribution" of a population of minnows.

one side of the hill. The other gives us an "**S**-*shaped curve*," apparently limited, but really asymptotic at both ends (Fig. 25 B); and this **S**-shaped curve is so easy to work with that we may at once divide it into two halves (so finding the "median" value), or into quarters and tenths (giving the "quartiles" and "deciles"), or as we please. In short, after drawing the curve to a larger scale, we shall find that we can safely read it to thirds of a millimetre, and so draw from it the following somewhat rough but very useful tabular epitome of our population of minnows, from which the curve can be reconstructed at any time:

	mm.
Extreme	13
First decile	21·0
Lower quartile	25·3
Median	28·6
Upper quartile	30·6
Last decile	32·3
Extreme	39

This **S**-shaped "summation-curve" is what Francis Galton called a *curve of distribution*, and he "liked it the better the more he used it." The spread or "scatter" is conveniently and immediately estimated by the distance between the two quartiles; and it happens that this *very nearly* coincides with the standard deviation of the normal curve.

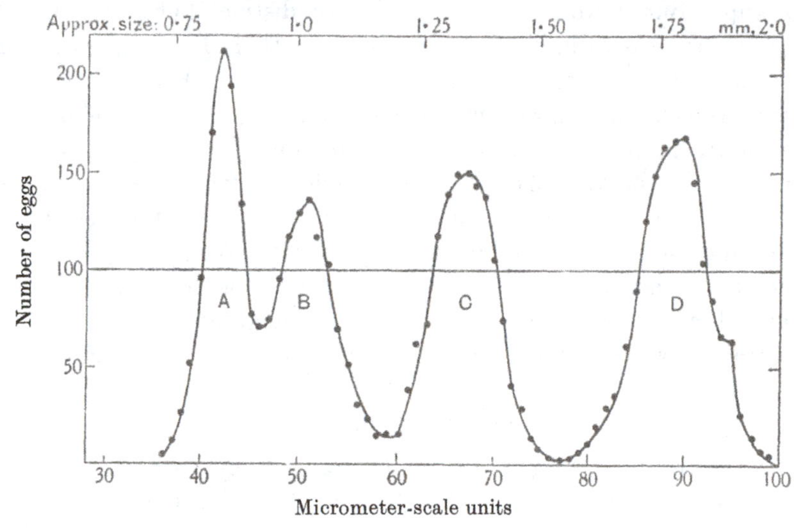

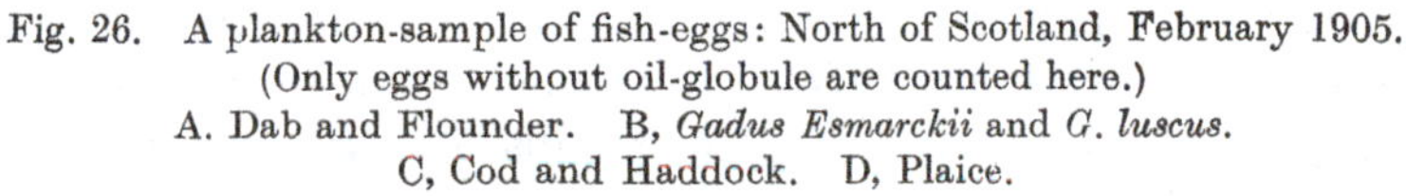

Fig. 26. A plankton-sample of fish-eggs: North of Scotland, February 1905. (Only eggs without oil-globule are counted here.) A. Dab and Flounder. B, *Gadus Esmarckii* and *G. luscus*. C, Cod and Haddock. D, Plaice.

There are biological questions for which we want all the accuracy which biometric science can give; but there are many others on which such refinements are thrown away.

Mathematically speaking, we cannot integrate the Gaussian curve, save by using an infinite series; but to all intents and purposes we are doing so, graphically and very easily, in the illustration we have just shewn. In any case, whatever may be the precise character of each, we begin to see how our two simplest curves of growth, the bell-shaped and the **S**-shaped curve, form a reciprocal pair, *the integral and the differential of one another**—like the distance travelled

* It is of considerable historical interest to know that this *practical* method of summation was first used by Edward Wright, in a Table of Latitudes published in his *Certain Errors in Navigation corrected*, 1599, as a means of virtually integrating sec x. (On this, and on Wright's claim to be the inventor of logarithms, see Florian Cajori, in *Napier Memorial Volume*, 1915, pp. 94–99.)

and the velocity of a moving body. If $y = e^{-x^2}$ be the ordinate of the one, $z = \int e^{-x^2} dx$ is that of the other.

There is one more kind of frequency-curve which we must take passing note of. We begin by thinking of our curve, whether symmetrical or skew, as the outcome of a single homogeneous group. But if we happen to have two distinct but intermingled groups to deal with, differing by ever so little in kind, age, place or circumstance—leaves of both oak and beech, heights of both men and women—this heterogeneity will tend to manifest itself in two separate cusps, or modes, on the common curve: which is then indeed two curves rolled into one, each keeping something of its own individuality. For example, the floating eggs of the food-fishes are much alike, but differ appreciably in size. A random gathering, netted at the surface of the sea, will yield on measurement a multi-modal curve, each cusp of which is recognisable, more or less certainly, as belonging to a particular kind of fish (Fig. 26).

A further note upon curves

A statistical "curve", such as Quetelet seems to have been the first to use*, is a device whose peculiar and varied beauty we are apt, through familiarity, to disregard. The curve of frequency which we have been studying depicts (as a rule) the distribution of magnitudes in a material system (a population, for instance) at a certain epoch of time; it represents a given *state*, and we may call it a *diagram of configuration*†. But we oftener use our curves to compare successive states, or changes of magnitude, as one configuration gives place to another; and such a curve may be called a *diagram of displacement*. An imaginary point moves in imaginary space, the dimensions of which represent those of the phenomenon in question, dimensions which we may further define and measure by a system of "coordinates"; the movements of our point through its figurative space are thus analogous to, and illustrative of, the events which constitute the phenomenon. Time is often represented, and measured, on one of the coordinate axes, and our diagram of "displacement" then becomes a *diagram of velocity*.

* In his *Théorie des probabilités*, 1846.

† See Clerk Maxwell's article "Diagrams," in the *Encyclopaedia Britannica*, 9th edition.

This simple method (said Kelvin) of shewing to the eye the law of variation, however complicated, of an independent variable, is one of the most beautiful results of mathematics*.

We make and use our curves in various ways. We set down on the coordinate network of our chart the points given by a series of observations, and connect them up into a continuous series as we chart the voyage of a ship from her positions day by day; we may "smooth" the line, if we so desire. Sometimes we find our points so crowded, or otherwise so dispersed and distributed, that a line can be drawn not from one to another but *among them all*—a method first used by Sir John Herschel†, when he studied the orbits of the double stars. His delicate observations were affected by errors, at first sight without rhyme or reason, but a curve drawn where the points lay thickest embodied the common lesson of them all; any one pair of observations would have sufficed, whether better or worse, for the calculation of an orbit, but Herschel's dot-diagram obtained "from the whole assemblage of observations taken together, and regarded as a single set of data, a single result in whose favour they all conspire." It put us in possession, said Herschel, of something truer than the observations themselves‡; and Whewell remarked that it enabled us to obtain laws of Nature not only from good but from very imperfect observations§. These are some advantages of the use of "curves," which have made them essential to research and discovery.

It is often helpful and sometimes necessary to *smooth* our curves,

* Kelvin, *Nature*, XXIX, p. 440, 1884.

† *Mem. Astron. Soc.* V, p. 171, 1830; *Nautical Almanack*, 1835, p. 495; etc.

‡ Here a certain distinction may be observed. We take the average height of a regiment, because the men actually vary about a mean. But in estimating the place of a star, or the height of Mont Blanc, we average *results* which only differ by personal or instrumental error. It is this latter process of averaging which leads, in Herschel's phrase, to results more trustworthy than observation itself. Laplace had made a similar remark long before (*Oeuvres*, VII, *Théorie des probabilités*): that we may ascertain the very small effect of a constant cause, by means of a long series of observations the errors of which exceed the effect itself. He instances the small deviation to the eastward which the rotation of the earth imposes on a falling body. In like manner the mean level of the sea may be determined to the second decimal of an inch by observations of high and low water taken roughly to the nearest inch, provided these are faithfully carried out at every tide, for say a hundred years. Cf. my paper on Mean Sea Level, in *Scottish Fishery Board's Sci. Report for* 1915.

§ *Novum Organum Renovatum* (3rd ed.), 1858, p. 20.

whether at free hand or by help of mathematical rules; it is one way of getting rid of non-essentials—and to do so has been called the very key-note of mathematics*. A simple rule, first used by Gauss, is to replace each point by a mean between it and its two or more neighbours, and so to take a "floating" or "running average." In so doing we trade once more on the "principle of continuity"; and recognise that in a series of observations each one is related to another, and is part of the contributory evidence on which our knowledge of all the rest depends. But all the while we feel that Gaussian smoothing gives us a practical or descriptive result, rather than a mathematical one.

Some curves are more elegant than others. We may have to rest content with points in which no order is apparent, as when we plot the daily rainfall for a month or two; for this phenomenon is one whose regularity only becomes apparent over long periods, when average values lead at last to "statistical uniformity." But the most irregular of curves may be instructive if it coincide with another not less irregular: as when the curve of a nation's birth-rate, in its ups and downs, follows or seems to follow the price of wheat or the spots upon the sun.

It seldom happens, outside of the exact sciences, that we comprehend the mathematical aspect of a phenomenon enough to *define* (by formulae and constants) the curve which illustrates it. But, failing such thorough comprehension, we can at least speak of the *trend* of our curves and put into words the character and the course of the phenomena they indicate. We see how this curve or that indicates a uniform velocity, a tendency towards acceleration or retardation, a periodic or non-periodic fluctuation, a start from or an approach to a limit. When the curve becomes, or approximates to, a mathematical one, the types are few to which it is likely to belong†. A straight line, a parabola, or hyperbola, an exponential or a logarithmic curve (like $x = ay^b$), a sine-curve or sinusoid, damped or no, suffice for a wide range of phenomena; we merely modify our scale, and change the names of our coordinates.

* Cf. W. H. Young, The mathematic method and its limitations, *Atti del Congresso dei Matematici, Bologna*, 1928, I, p. 203.

† Hence the engineer usually begins, for his first tentative construction, by drawing one of the familiar curves, catenary, parabola, arc of a circle, or curve of sines.

The curves we mostly use, other than the Gaussian curve, are *time-diagrams*. Each has a beginning and an end; and one and the same curve may illustrate the life of a man, the economic history of a kingdom, the schedule of a train between one station and another. What it then shews is a velocity, an acceleration, and a subsequent negative acceleration or retardation. It depicts a "mechanism" at work, and helps us to see analogous mechanisms in different fields; for Nature rings her many changes on a few simple themes. The same expressions serve for different orders of phenomena. The swing of a pendulum, the flow of a current, the attraction of a magnet, the shock of a blow, have their analogues in a fluctuation of trade, a wave of prosperity, a blow to credit, a tide in the affairs of men.

The same exponential curve may illustrate a rate of cooling, a loss of electric charge, the chemical action of a ferment or a catalyst. The S-shaped population-curve or "logistic curve" of Verhulst (to which we are soon coming) is the hysteresis-curve by which Ewing represented self-induction in a magnetic field; it is akin to the path of a falling body under the influence of friction; and Lotka has drawn a curve of the growing mileage of American railways, and found it to be a typical logistic curve. A few bars of music plotted in wave-lengths of the notes might be mistaken for a tidal record. The periodicity of a wave, the acceleration of gravity, retardation by friction, the role of inertia, the explosive action of a spark or an electric contact—these are some of the modes of action or "forms of mechanism" which recur in limited number, but in endless shapes and circumstances*. The way in which one curve fits many phenomena is characteristic of mathematics itself, which does not deal with the specific or individual case, but *generalises* all the while, and is fond (as Henri Poincaré said) of giving the same name to different things.

Our curves, as we have said, are mostly time-diagrams, and represent a change in time from one magnitude to another; they are diagrams of displacement, in Maxwell's phrase. We may consider four different cases, not equally simple mathematically, but all

* See an admirable little book by Michael Petrovich, *Les mécanismes communs aux phénomènes disparates*, Paris, 1921.

capable of explanation, up to a certain point, without mathematics.

(1) If in our coordinate diagram we have merely to pass from one isolated *point* to another, a *straight line* joining the two points is the shortest—and the likeliest way.

(2) To rise and fall alternately, going to and fro from maximum to minimum, a zig-zag rectilinear path would still be, geometrically, the shortest way; but it would be sharply discontinuous at every turn, it would run counter to the "principle of continuity," it is not likely to be nature's way. A wavy course, with no more change of curvature than is absolutely necessary, is the path which nature follows. We call it a *simple harmonic motion*, and the simplest of

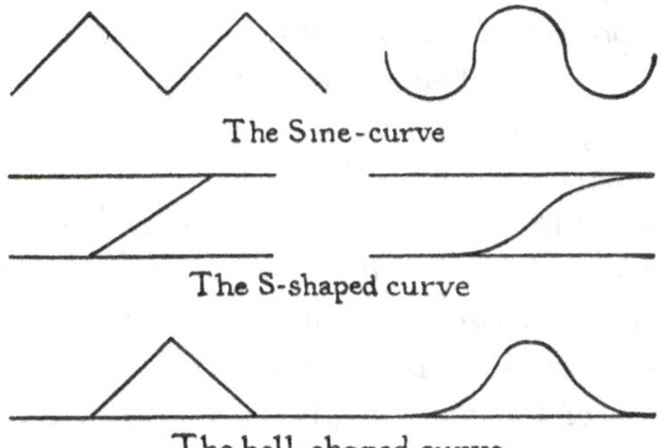

Fig. 27. Simple curves, representing a change from one magnitude to another.

all such wavy curves we call a *sine-curve*. If there be but one maximum and one minimum, which our variant alternates between, the vector pathway may be translated into *polar coordinates*; the vector does what the hands of the clock do, and a *circle* takes the place of the sine-curve.

(3) To pass from a zero-line to a maximum once for all is a very different thing; for now minimum and maximum are both of them continuous states, and the principle of continuity will cause our vector-variant to leave the one gradually, and arrive gradually at the other. The problem is how to go uphill from one level road to another, with the least possible interruption or discontinuity. The path follows an **S**-shaped course; it has an *inflection* midway; and the first phase and the last are represented by horizontal asymptotes. This is an important curve, and a common one. It so far resembles

an "elastic curve" (though it is not mathematically identical with it) that it may be roughly simulated by a watchspring, lying between two parallel straight lines and touching both of them. It has its kinetic analogue in the motion of a pendulum, which starts from rest and comes to rest again, after passing midway through its maximal velocity. It indicates a balance between production and waste, between growth and decay: an approach on either side to a state of rest and equilibrium. It shows the speed of a train between two stations; it illustrates the growth of a simple organism, or even of a population of men. A certain simple and symmetrical case is called the *Verhulst-Pearl curve*, or the *logistic curve*.

(4) Lastly, in order to leave a certain minimum, or zero-line, and return to it again, the simplest way will be by a curve asymptotic to the base-line at both ends—or rather in both directions; it will be a bell-shaped curve, having a maximum midway, and of necessity a point of inflection on either side; it is akin to, and under certain precise conditions it becomes, the *curve of error* or *Gaussian curve*.

Besides the ordinary curve of growth, which is a summation-curve, and the curve of growth-rates, which is its derivative, there are yet others which we may employ. One of these was introduced by Minot*, from a feeling that the rate of growth, or the amount of increment, ought in some way to be equated with the growing structure. Minot's method is to deal, not with the actual increments added in successive periods, but with these successive increments represented as percentages of the amount already reached. For instance, taking Quetelet's values for the height (in centimetres) of a male infant, we have as follows:

Years	0	1	2	3	4
cm.	50·0	69·8	79·1	86·4	92·7

But Minot would state the percentage-growth in each of these four annual periods at 39·6, 13·3, 9·2 and 7·3 per cent. respectively:

Years	0	1	2	3	4
Height (cm.)	50·0	69·8	79·1	86·4	92·7
Increments (cm.)	—	19·8	9·3	7·3	6·3
,, (per cent.)	—	39·6	13·3	9·2	7·3

* C. S. Minot, On certain phenomena of growing old, *Proc. Amer. Assoc.* XXXIX, 1890, 21 pp.; Senescence and rejuvenation, *Journ. Physiol.* XII, pp. 97–153, 1891; etc. Criticised by S. Brody and J. Needham, *op. cit.* pp. 401 *seq.*

Now, in our first curve of growth we plotted length against time, a very simple thing to do. When we differentiate L with respect to T, we have dL/dT, which is rate or velocity, again a very simple thing; and from this, by a second differentiation, we obtain, if necessary, d^2L/dT^2, that is to say, the acceleration.

But when you take percentages of y, you are determining dy/y, and when you plot this against dx, you have

$$\frac{dy/y}{dx}, \quad \text{or} \quad \frac{dy}{y.dx}, \quad \text{or} \quad \frac{1}{y}.\frac{dy}{dx}.$$

That is to say, you are multiplying the thing whose variations you are studying by another quantity which is itself continually varying; and are dealing with something more complex than the original factors*. Minot's method deals with a perfectly legitimate function of x and y, and is tantamount to plotting $\log y$ against x, that is to say, the logarithm of the increment against the time. This would be all to the good if it led to some simple result, a straight line for instance; but it is seldom if ever, as it seems to me, that it does anything of the kind. It has also been pointed out as a grave fault in his method that, whereas growth is a continuous process, Minot chooses an arbitrary time-interval as his basis of comparison, and uses the same interval in all stages of development. There is little use in comparing the percentage increase *per week* of a week-old chick, with that of the same bird at six months old or at six years.

The growth of a population

After dealing with Man's growth and stature, Quetelet turned to the analogous problem of the growth of a population—all the more analogous in our eyes since we know man himself to be a "statistical unit," an assemblage of organs, a population of cells. He had read

* Schmalhausen, among others, uses the same measure of rate of growth, in the form

$$C_v = \frac{\log V - \log V}{k\,(t-t)} \rightarrow \frac{dv}{dt} - \frac{1}{v}:$$

Arch. f. Entw. Mech. CXIII, pp. 462–519, 1928.

Malthus's *Essay on Population** in a French translation, and was impressed like all the world by the importance of the theme. He saw that poverty and misery ensue when a population outgrows its means of support, and believed that multiplication is checked both by lack of food and fear of poverty. He knew that there were, and *must be*, obstacles of one kind or another to the unrestricted increase of a population; and he knew the more subtle fact that a population, after growing to a certain height, oscillates about an unstable level of equilibrium†.

Malthus had said that a population grows by geometrical progression (as 1, 2, 4, 8) while its means of subsistence tend rather to grow by arithmetical (as 1, 2, 3, 4)—that one adds up while the other multiplies‡. A geometrical progression is a natural and a

* T. R. Malthus, *An Essay on the Principle of Population, as it affects the Future Improvement of Society*, etc., 1798 (6th ed. 1826; transl. by P. and G. Prévost, Geneva, 1830, 1845). Among the books to which Malthus was most indebted was *A Dissertation on the Numbers of Mankind in ancient and modern Times*, published anonymously in Edinburgh in 1753, but known to be by Robert Wallace and read by him some years before to the Philosophical Society at Edinburgh. In this remarkable work the writer says (after the manner of Malthus) that mankind naturally increase by successive doubling, and tend to do so thrice in a hundred years. He explains, on the other hand, that "mankind do not actually propagate according to the rule in our tables, or any other constant rule; yet tables of this nature are not entirely useless, but may serve to shew, how much the increase of mankind is prevented by the various causes which confine their number within such narrow limits." Malthus was also indebted to David Hume's *Political Discourse, Of the Populousness of ancient Nations*, 1752, a work criticised by Wallace. See also McCulloch's notes to Adam Smith's *Wealth of Nations*, 1828.

† That the nearest approach to equilibrium in a population is long-continued ebb and flow, a mean level and a tide, was known to Herbert Spencer, and was stated mathematically long afterwards by Vito Volterra. See also Spencer's *First Principles*, ch. 22, sect. 173: "Every species of plant or animal is perpetually undergoing a rhythmical variation in number—now from abundance of food and absence of enemies rising above its average, and then by a consequent scarcity of food and abundance of enemies being depressed below its average....Amid these oscillations produced by their conflict, lies that average number of the species at which its expansive tendency is in equilibrium with surrounding repressive tendencies." Cf. A. J. Lotka, Analytical note on certain rhythmic relations in organic systems, *Proc. Nat. Acad. Sci.* VI, pp. 410–415, 1920; but cf. also his *Elements of Physical Biology*, 1915, p. 90. An analogy, and perhaps a close one, may be found on the Bourse or money market.

‡ That a population will soon outrun its means of subsistence was a natural assumption in Malthus's day, and in his own thickly populated land. The danger may be postponed and the assumption apparently falsified, as by an Argentine cattle-ranch or prairie wheat-farm—but only so long as we enjoy world-wide freedom of import and exchange.

common thing, and, apart from the free growth of a population or an organism, we find it in many biological phenomena. An epidemic declines, or tends to decline, at a rate corresponding to a geometrical progression; the mortality from zymotic diseases declines in geometrical progression among children from one to ten years old; and the chances of death increase in geometrical progression after a certain time of life for us all*.

But in the ascending scale, the story of the horseshoe nails tells us how formidable a thing successive multiplication becomes†. English law forbids the protracted accumulation of compound interest; and likewise Nature deals after her own fashion with the case, and provides her automatic remedies. A fungus is growing on an oaktree—it sheds more spores in a night than the tree drops acorns in a hundred years. A certain bacillus grows up and multiplies by two in two hours' time; its descendants, did they all survive, would number four thousand in a day, as a man's might in three hundred years. A codfish lays a million eggs and more—all in order that *one pair* may survive to take their parents' places in the world. On the other hand, the humming-birds lay only two eggs, the auks and guillemots only one; yet the former are multitudinous in their haunts, and some say that the Arctic auks and auklets outnumber all other birds in the world. Linnaeus‡ shewed that an annual plant would have a million offspring in twenty years, if only two seeds grew up to maturity in a year.

But multiply as they will, these vast populations have their limits. They reach the end of their tether, the pace slows down, and at last they increase no more. Their world is fully peopled, whether it be an island with its swarms of humming-birds, a test-tube with its myriads of yeast-cells, or a continent with its millions of mankind. Growth, whether of a population or an individual, draws to its natural end; and Quetelet compares it, by a bold metaphor, to the motion of a body in a resistant medium. A typical population grows slowly from an asymptotic minimum; it multiplies quickly;

* According to the Law of Gompertz; cf. John Brownlee, in *Proc. R.S.E.* XXXI, pp. 627–634, 1911.

† Herbert Spencer, A theory of population deduced from the general law of animal fertility, *Westminster Review*, April 1852.

‡ In his essay *De Tellure*, 1740.

it draws slowly to an ill-defined and asymptotic maximum. The two ends of the population-curve define, in a general way, the whole curve between; for so beginning and so ending the curve must pass through a point of inflection, it *must be* an **S**-shaped curve. It is just such a curve as we have seen under simple conditions of growth in an individual organism.

This general and all but obvious trend of a population-curve has been recognised, with more or less precision, by many writers. It is implicit in Quetelet's own words, as follows: "Quand une population peut se développer librement et sans obstacles, elle croît selon une progression géométrique; si le développement a lieu au milieu d'obstacles de toute espèce qui tendent à l'arrêter, et qui agissent d'une manière uniforme, c'est à dire si l'état sociale ne change point, la population n'augmente pas d'une manière indéfinie, mais *elle tend de plus en plus à devenir stationnaire**." P. F. Verhulst, a mathematical colleague of Quetelet's, was interested in the same things, and tried to give a mathematical shape to the same general conclusions; that is to say, he looked for a "*fonction retardatrice*" which should turn the Malthusian curve of geometrical progression into the **S**-shaped, or as he called it, the *logistic curve*, which should thus constitute the true "law of population," and thereby indicate (among other things) the limit above which the population was not likely to grow†.

Verhulst soon saw that he could only solve his problem in a preliminary and tentative way; "*la loi de la population nous est inconnue*, parcequ'on ignore la nature de la fonction qui sert de mesure aux obstacles qui s'opposent à la multiplication indéfini de l'espèce humaine." The materials at hand were almost unbelievably scanty and poor. The French statistics were taken from documents "qui ont été reconnus entièrement fictifs"; in England the growth

* *Physique Sociale*, I, p. 27, 1835. But Quetelet's brief account is somewhat ambiguous, and he had in mind a body falling through a resistant medium—which suggests a limiting velocity, or limiting annual increment, rather than a *terminal value*. See Sir G. Udny Yule, The growth of population, *Journ. R. Statist. Soc.* LXXXVIII, p. 42, 1925.

† P. F. Verhulst, Notice sur la loi que la population suit dans son accroissement, *Correspondence math.* etc. publié par M. A. Quetelet, X, pp. 113–121, 1838; Rech. math. sur la loi etc., *Nouv. Mém. de l'Acad. R. de Bruxelles*, XVIII, 38 pp., 1845; deuxième Mém., *ibid.* XX, 32 pp., 1847. The term *logistic curve* had already been used by Edward Wright; see *antea*, p. 135, *footnote*.

of the population was estimated by the number of births, and the births by the baptisms in the Church of England, "de manière que les enfants des dissidents ne sont point portés sur les registres officiels." A law of population, or "loi d'affaiblissement" became a mere matter of conjecture, and the simplest hypothesis seemed to Verhulst to be, to regard "cet affaiblissement comme proportionnel à l'accroissement de la population, depuis le moment où la difficulté de trouver de bonnes terres a commencé à se faire sentir*."

Verhulst was making two assumptions. The first, which is beyond question, is that the rate of increase cannot be, and indeed is not, a constant; and the second is that the rate must somehow depend on (or be *some function* of) the population for the time being. A third assumption, again beyond question, is that the simplest possible function is a *linear function*. He suggested as the simplest possible case that, once the rate begins to fall (or once the struggle for existence sets in), it will fall the more as the population continues to grow; we shall have a *growth-factor* and a *retardation-factor* in *proportion to one another*. He was making early use of a simple differential equation such as Vito Volterra and others now employ freely in the general study of natural selection†.

The point where a struggle for existence first sets in, and where *ipso facto* the rate of increase begins to diminish, is called by Verhulst the *normal level* of the population; he chooses it for the origin of his curve, which is so defined as to be symmetrical on either side of this origin. Thus Verhulst's law, and his logistic curve, owe their form and their precision and all their power to forecast the future to certain hypothetical assumptions; and the tentative solution arrived at is one "sous le point de vue mathématique‡."

* *Op. cit.* p. 8.

† Besides many well-known papers by Volterra, see V. A. Kostitzin, *Biologie mathématique*, Paris, 1937. Cf. also, for the so-called "Malaria equations," Ronald Ross, *Prevention of Malaria*, 2nd ed. 1911, p. 679; Martini, *Zur Epidemiologie d. Malaria*, Hamburg, 1921; W. R. Thompson, *C.R.* CLXXIV, p. 1443, 1922; C. N. Watson, *Nature*, CXI, p. 88, 1923.

‡ Verhulst goes on to say that "une longue série d'observations, non interrompues par de grandes catastrophes sociales ou des révolutions du globe, fera probablement découvrir *la fonction retardatrice* dont il vient d'être fait mention." Verhulst simplified his problem to the utmost, but it is more complicated today than ever; he thought it impossible that a country should draw its bread and meat from overseas: "lors même qu'une partie considérable de la population pourrait être

The mathematics of the Verhulst-Pearl curve need hardly concern us; they are fully dealt with in Raymond Pearl's, Lotka's and other books. Verhulst starts, as Malthus does, with a population growing in geometrical progression, and so giving a logarithmic curve:

$$\frac{dp}{dt} = mp.$$

He then assumes, as his "loi d'affaiblissement," a coefficient of retardation (n) which increases as the population increases:

$$\frac{dp}{dt} = mp - np^2.$$

Integrating, $$p = \frac{m}{n}\frac{1}{e^{-m(t+k)} + 1}.$$

If the point of inflection be taken as the origin, $k = 0$; and again for $t = \infty$, $p = \frac{m}{n} = L$. We may write accordingly:

$$p = L\frac{1}{1 + e^{-mt}}.$$

Malthus had reckoned on a population doubling itself, if unchecked by want or "accident," every twenty-five years*; but fifty years after, Verhulst shewed that this "grande vitesse d'accroissement" was no longer to be found in France or Belgium or other of the older countries†, but was still being realised in the United States (Fig. 28). All over Europe, "le rapport de l'excès annuel des naissances sur les décès, à la population qui l'a fourni, va sans cesse en s'affaiblissant; de manière que l'accroissement annuel, dont la valeur absolue augmente continuellement lorsqu'il y a progression géométrique, paraît suivre une progression tout au plus arithmétique."

nourrie de blès étrangers, jamais un gouvernement sage ne consentira à faire dépendre l'existence de milliers de citoyens du bon vouloir des souverains étrangers." On this and other problems in the growth of a human population, see L. Hogben's *Genetic Problems*, etc., 1937, chap. VII. See also (*int. al.*) Warren S. Thompson and P. K. Whelpton, *Population Trends in the United States*, 1933; F. Lorimer and F. Osborn, *Dynamics of Population*, 1934, etc.

* An estimate based, like the rest of Malthus's arithmetic, on very slender evidence.

† In Quetelet's time the European countries, far from doubling in twenty-five years, were estimated to do so in from sixty years (Norway) to four hundred years (France); see M. Haushofer, *Lehrbuch der Statistik*, 1882.

The "celebrated aphorism" of Malthus was thus, and to this extent, confirmed*. Ir the United States, the Malthusian estimate of unrestricted increase continued to be realised for a hundred years after Malthus wrote; for the 3·93 millions of the U.S. census of 1790 were doubled three times over in the census of 1860, and four times over in that of 1890. A capital which doubles in twenty-five years has grown at 2·85 per cent. per annum, compound interest; the U.S. population did rather more, for it grew at fully 3 per cent. for fifty of those hundred years†.

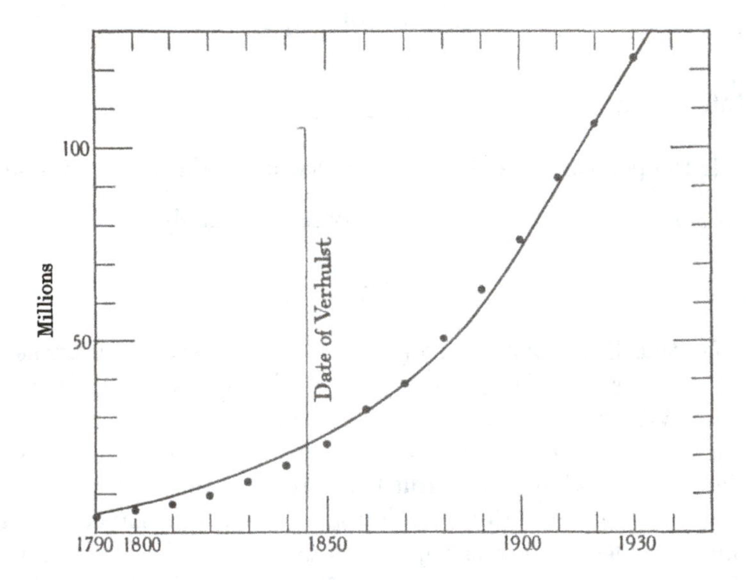

Fig. 28. Population of the United States, 1790–1930.

The population of the whole world and of every continent has increased during modern times, and the increase is large though the rate is low. The rate of increase has been put at about half-a-per-cent per annum for the last three hundred years—a shade more in Europe and a shade less in the rest of the world‡:

* *Op. cit.* 1845, p. 7.

† Verhulst foretold forty millions as the "extreme limit" of the population of France, and $6\frac{1}{2}$ millions as that of Belgium. The latter estimate he increased to 8 millions later on. The actual populations of France and Belgium at the present time are a little more than the ultimate limit which Verhulst foretold.

‡ From A. M. Carr-Saunders' *World Population*, 1936, p. 30.

An estimate of the population of the world
(After W. F. Willcox)

	1650	1750	1800	1850	1900	Mean rate of increase
Europe	100	140	187	266	401	0·52 % per annum
World total	545	728	906	1171	1608	0·49 % „ „

Verhulst was before his time, and his work was neglected and presently forgotten. Only some twenty years ago, Raymond Pearl and L. J. Reed of Baltimore, studying the U.S. population as Verhulst had done, approached the subject in the same way, and came to an identical result; then, soon afterwards (about 1924), Raymond Pearl came across Verhulst's papers, and drew attention to what we now speak of as the Verhulst-Pearl law. Pearl and Reed saw, as Verhulst had done, that a "law of population" which should cover all the ups and downs of human affairs was not to be found; and yet the general form which such a law must take was plain to see. There must be a limit to the population of a region, great or small; and the curve of growth must sooner or later "turn over," approach the limit, and resolve itself into an S-shaped curve. The rate of growth (or annual increment) will depend (1) on the population at the time, and (2) on "the still unutilised reserves of population-support existing" in the available land. Here we have, to all intents and purposes, the growth-factor and retardation-factor of Verhulst, and they lead to the same formula, or the same differential equation, as his*.

A hundred years have passed since Verhulst dealt with the first U.S. census returns, and found them verifying the Malthusian expectation of a doubling every twenty-five years. That "grande vitesse d'accroissement" continued through five decennia; but it ceased some seventy years ago, and a retarding influence has been manifest through all these seventy years (Fig. 29). It is more recently, only after the census of 1910, that the curve seemed to be

* Raymond Pearl and L. J. Reed, on the Rate of growth of the population of the U.S. since 1790, and its mathematical representation, *Proc. Nat. Acad. Sci.* VI, pp. 275–288, 1920; *ibid.* VIII, pp. 365–368, 1922; *Metron*, III, 1923. In the first edition of Pearl's *Medical Biometry and Statistics*, 1923 (2nd ed. 1930), Verhulst is not mentioned. See also his *Studies in Human Biology*, Baltimore, 1924, *Natural History of Population*, 1939, and other works.

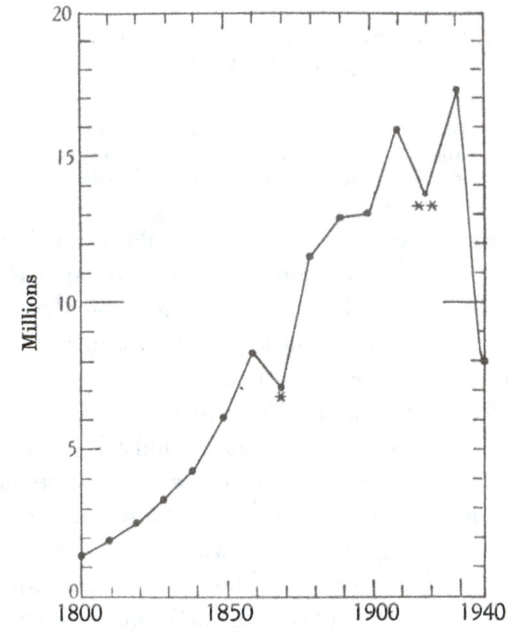

Fig. 29. Decennial increments of the population of the United States.
* The Civil War. ** The "slump".

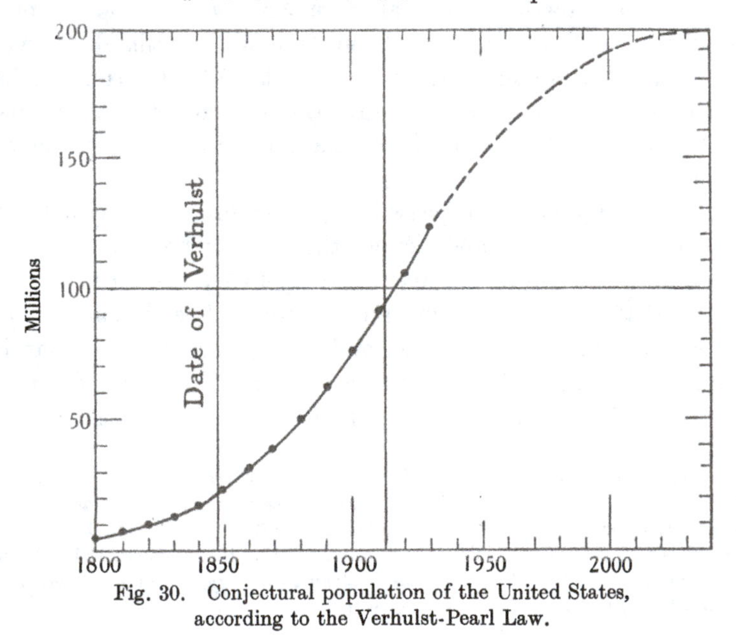

Fig. 30. Conjectural population of the United States, according to the Verhulst-Pearl Law.

finding its turning-point, or point of inflection; and only now, since 1940, can we say with full confidence that it has done so.

A hundred years ago the conditions were still relatively simple, but they are far from simple now. Immigration was only beginning to be an important factor; but immigrants made a quarter of the whole increase of the population of the United States during eighty of these hundred years*. Wars and financial crises have made their mark upon the curve; manners and customs, means and standards of living, have changed prodigiously. But the S-shaped curve makes its appearance through all of these, and the Verhulst-Pearl formula meets the case with surprising accuracy.

Population of the United States

Year	Population × 1000	Calculated by logistic curve (Udny Yule)	In ten years: No. of immigrants landed × 1000	In ten years: Total increase of population × 1000	Percentage increase	Increase by multiplication in 25 years
1790	3,929	3,929	—	—	—	—
1800	5,308	5,336	—	1,379	35·1	—
1810	7,240	7,223	—	1,932	36·4	—
1820	9,638	9,757	250	2,398	33·1	2·08
1830	12,866	13,109	228	3,228	33·5	2·05
1840	17,069	17,506	538	4,203	32·7	2·02
1850	23,192	23,192	1,427	6,123	35·9	2·06
1860	31,443	30,418	2,748	8,251	35·6	2·10
1870	38,558	39,372	2,123	7,115	32·6	1·91
1880	50,156	50,177	2,741	11,598	30·1	1·83
1890	62,948	62,769	5,249	12,792	25·5	1·80
1900	75,995	76,870	3,694	13,047	20·7	1·71
1910	91,972	91,972	8,201	15,977	21·0	1·63
1920	105,711	—	6,347	13,739	14·9	1·52
1930	122,975	—	—	17,264	16·1	1·46
1940	131,669	—	—	8,694	7·1	1·33

A colony of yeast or of bacteria is a population in its simplest terms, and Verhulst's law was rediscovered in the growth of a bacterial colony some years before Raymond Pearl found it in a population of men, by Colonel M'Kendrick and Dr Kesava Pai, who put their case very simply indeed†. The bacillus grows by geometrical

* Without counting the children born to those immigrants after landing, and before the next census return.

† A. G. M'Kendrick and M. K. Pai, The rate of multiplication of micro-organisms: a mathematical study, *Proc. R.S.E.* XXXI, pp. 649–655, 1911. (The period of generation in *B. coli*, answering to Malthus's twenty-five years for men, was found to be 22½ minutes.) Cf. also Myer Coplans, *Journ. of Pathol. and Bacteriol.* XIV, p. 1, 1910 and H. G. Thornton, *Ann. of Applied Biology*, 1922, p. 265.

progression so long as nutriment is enough and to spare; that is to say, the rate of growth is proportional to the number present:

$$\frac{dy}{dt} = by.$$

But in a test-tube colony the supply of nourishment is limited, and the rate of multiplication is bound to fall off. If a be the original concentration of food-stuff, it will have dwindled by time t to $(a - y)$. The rate of growth will now be

$$\frac{dy}{dt} = by\,(a - y),$$

which means that the rate of increase is proportional to the number of organisms present, and to the concentration of the food-supply. It is Verhulst's case in a nutshell; the differential equation so indicated leads to an **S**-shaped curve which further experiment confirms; and Sach's "grand period of growth" is seen to accomplish itself*.

The growth of yeast is studied in the everyday routine of a brewery. But the brewer is concerned only with the phase of unrestricted growth, and the rules of compound interest are all he needs, to find its rate or test its constancy. A population of 1360 yeast-cells grew to 3,550,000 in 35 hours: it had multiplied 2610 times. Accordingly,

$$\frac{\log 2610 = 3{\cdot}417}{35} = 0{\cdot}098 = \log 1{\cdot}254.$$

That is to say, the population had increased at the rate of 25·4 per cent. *per hour*, during the 35 hours.

The time (t_2) required to *double* the population is easily found:

$$t_2 = \frac{\log 2}{\log 1{\cdot}254} = \frac{0{\cdot}301}{0{\cdot}098} = 3{\cdot}07 \text{ hours.}$$

* The sigmoid curve illustrates a theorem which, obvious as it may seem, is of no small philosophical importance, to wit, that a body starting from rest must, in order to attain a certain velocity, pass through all intermediate velocities on its way. Galileo discusses this theorem, and attributes it to Plato: "Platone avendo per avventura avuto concetto non potere alcun mobil passare dalla quiete ad alcun determinato grado di velocità....se non col passare per tutti gli altri gradi di velocità minori, etc."; *Discorsi e dimostrazioni*, ed. 1638, p. 254.

The duplication-period thus determined is known to brewers as the *generation-time*.

Much care is taken to ensure the maximal growth. If the yeast sink to the bottom of the vat only its upper layers enjoy unstinted nutriment; a potent retardation-factor sets in, and the exponential phase of the growth-curve degenerates into a premature horizontal asymptote. Moreover, both the yeast and the bacteria differ in this respect from the typical (or perhaps only simplified) case of man, that they not only begin to suffer want as soon as there comes to be a deficiency of any one essential constituent of their food*, but they also produce things which are injurious to their own growth and in time fatal to their existence. Growth stops long before the food-supply is exhausted; for it does so as soon as a certain balance is reached, depending on the kind or quality of the yeast, between the alcohol and the sugar in the cell†.

If we use the compound-interest law at all, we had better think of Nature's interest as being paid, not once a year nor once an hour as our elementary treatment of the yeast-population assumed, but continuously; and then we learn (in elementary algebra) that in time t, at rate r, a sum P increases to Pe^{rt}, or $P_t = P_0 e^{rt}$.

Applying this to the growth of our sample of 1360 yeast cells, we have

$$\log_e (P_t/P_0) = nr.$$

$P_t/P_0 = 2610$, $\log 2610 = 3{\cdot}417$, which, multiplied by the modulus $2{\cdot}303 = 7{\cdot}868$. Dividing by $n = 35$, the number of hours,

$$7{\cdot}868/35 = 0{\cdot}225 = r.$$

The rate, that is to say, is 22·5 per cent. per hour, continuous compound interest. It becomes a well-defined physiological constant, and we may call it, with V. H. Blackman, an *index of efficiency*.

Our former result, for interest at hourly intervals, was 25·4 per

* According to Liebig's "law of the minimum."

† T. Carlson, Geschwindigkeit und Grösse der Hefevermehrung, *Biochem. Ztschr.* LVII, pp. 313–334, 1913; A. Slator, *Journ. Chem. Soc.* CXIX, pp. 128–142, 1906; *Biochem. Journ.* VII, p. 198, 1913; O. W. Richards, *Ann. of Botany*, XLII, pp. 271–283, 1928; Alf Klem, *Hvalradets Skrifter*, nr. 7, pp. 55–91, Oslo, 1933; Per Ottestad, *ibid.* pp. 30–54. For optimum conditions of temperature, nutriment, *p*H, etc. see Oscar W. Richards, Analysis of growth as illustrated by yeast, *Cold Spring Harbour Symposia*, II, pp. 157–166, 1934.

cent.; there is no great difference between such short intervals and actual continuity, but there is a deal of difference between continuous payment and payment (say) once a year*. Certain sunflowers (*Helianthus*) were found to grow as follows, in thirty-seven days:

	Weight (gm.)		Compound interest rate (%)			
			Continuous		Discontinuous	
	Seedling	Plant	Per day	Per wk.	Per day	Per wk.
Giant sunflower	0·033	17·33	17·0	119	18·5	228 %
Dwarf sunflower	0·035	14·81	16·4	114	17·7	214 %

When the yeast population is allowed to run its course, it yields a simple **S**-shaped curve; and the curve of first differences derived

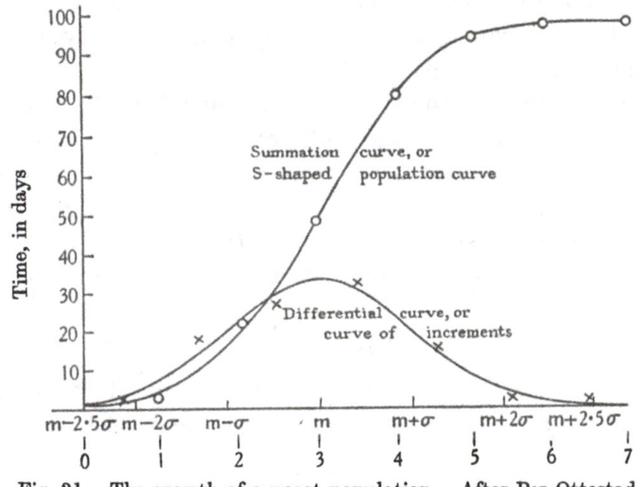

Fig. 31. The growth of a yeast-population. After Per Ottestad.

from this is, necessarily, a bell-shaped curve, so closely resembling the Gaussian curve that any difference between them becomes a delicate matter. Taking the numbers of the population at equal intervals of time from asymptotic start to asymptotic finish, we may treat this series of numbers like any other frequency distribution. Finding in the usual way the mode and standard deviation,

* Cf. V. H. Blackman, The compound interest law and plant growth, *Ann. of Botany*, XXXIII, pp. 353–360, 1919. The first papers on growth by compound interest in plants were by pupils of Noll in Bonn: *e.g.* von Kreusler, Wachstum der Maispflanze, *Landw. JB.* 1877–79; P. Gressler, *Substanz-quotienten von Helianthus*, Diss. Bonn, 1907 etc.

we draw the corresponding Gaussian curve; and the close "fit" between the observed population-curve and the calculated Gaussian curve is sufficiently shewn by Mr Per Ottestad's figure (Fig. 31). This is a very remarkable thing. We began to think of the curve of error as a function with which time had nothing to do, but here we have the same curve (or to all intents and purposes the same) with *time* for one of its coordinates. We might (I think) add one more to the names of the curve of error, and call it the curve of optimum; it represents on either hand the natural passage from best to worst, from likeliest to least likely.

A few flies (*Drosophila*) in a bottle illustrate the rise and fall of a population more complex than yeast, as Raymond Pearl has shewn*. The colony dwindles to extinction if food be withheld; if it be sufficient, the numbers rise in a smooth **S**-shaped curve; if it be plentiful and of the best, they end by fluctuating about an unstable maximum. "The population waves up and down about an average size," as Raymond Pearl says, as Herbert Spencer had foreseen†, and as Vito Volterra's differential equations explain. The growth-rate slackens long before the hunger line is reached; crowding affects the birth-rate as well as the death-rate, and a bottleful of flies produces fewer and fewer offspring *per pair* the more flies we put into the bottle‡. It is true also of mankind, as Dr William Farr was the first to shew, that overcrowding diminishes the birth-rate and shortens the "expectation of life§." It happened so in the United States, *pari passu* with the growth of immigration, incipient congestion acting (or so it seemed) as an obstacle, or a deterrent, to the large families of former days. Nevertheless, children still pullulate in the slums. The struggle for existence is no simple affair, and things happen which no mathematics can foretell.

* Raymond Pearl and S. L. Parker, in *Proc. Nat. Acad. Sci.* VIII, pp. 212–219, 1922; Pearl, *Journ. Exper. Zool.* LXIII, pp. 57–84, 1932.

† "Wherever antagonistic forces are in action, there tends to be alternate predominance.".

‡ In certain insects an optimum density has been observed; a certain amount of crowding accelerates, and a greater amount retards, the rate of reproduction. Cf. D. Stewart Maclagan, Effect of population-density on rate of reproduction, *Proc. R. S.* (B), CXI, p. 437, 1932; W. Goetsch, Ueber wachstumhemmende Factoren, *Zool. Jahrb.* (*Allg. Zool.*), XLV, pp. 799–840, 1928.

§ Dr W. Farr, *Fifth Report of the Registrar-General*, 1843, p. 406 (2nd ed.).

An analogous S-shaped curve, given by the formula $L_x = kg^{c^x}$, was introduced by Benjamin Gompertz in 1825*; it is well known to actuaries, and has been used as a curve of growth by several writers in preference to the logistic curve. It was devised, and well devised, to express a "law of human mortality", and to signify the number surviving at any given age (x), "if the average exhaustions of a man's power to avoid death were such that at the end of infinitely small intervals of time he lost equal portions (i.e. *equal proportions*) of his remaining power to oppose destruction." The principle involved is very important. Death comes by two roads. One is by chance or accident, the other by a steady deterioration, or exhaustion, or growing inability to withstand destruction; and exhaustion comes (roughly speaking) as by the repeated strokes of an air-pump, for the life-tables shew mortality increasing in geometrical progression, at least to a first approximation and over considerable periods of years. Gompertz relied wholly on the experience of "life-contingencies," but the same deterioration of bodily energies is plainly visible as growth itself slows down; for we have seen how growth-rate in infancy is such as is never afterwards attained, and we may speak of growth-energy and its gradual loss or decrement, by an easy but significant alteration of phrase. To deal with the declining growth-rate, as Gompertz did with the falling expectation of life, and so to measure the remaining energy available from time to time, would be a greater thing than to record mere weights and sizes; it raises the problem from mere change of physical magnitudes to an estimation of the falling or fluctuating physiological energies of the body†. We have seen how in only

* Benjamin Gompertz, On the nature of the function expressive of the law of human mortality, *Phil. Trans.* XXXVI, pp. 513–585, 1825. First suggested for use in growth-problems by Sewall Wright, *Journ. Amer. Statist. Soc.* XXI, p. 493, 1926. See also C. P. Winsor, The Gompertz curve as a growth curve, *Proc. Nat. Acad. Sci.* XVIII, pp. 1–8, 1932; cf. (*int. al.*) G. R. Davies, The growth curve, *Journ. Amer. Statist. Soc.* XXII, pp. 370–374, 1927; F. W. Weymouth and S. H. Thompson, Age and growth of the Pacific cockle, *Bull. Bureau Fisheries*, XLVI, pp. 633–641, 1930–31; also Weymouth, McMillen and Rich, in *Journ. Exp. Biol.* VIII, p. 228, 1931.

† A bold attempt to treat the question from the physiological side, and on Gompertz's lines, was made only the other day by P. B. Medawar, The growth, growth-energy and ageing of the chicken's heart, *Proc. R.S.* (B), CXXIX, pp. 332–355, 1940. Cf. James Gray, The kinetics of growth, *Journ. Exp. Biol.* VI, pp. 248–274, 1929.

few and simple cases can a simple curve or single formula be found to represent the growth-rate of an organism; and how our curves mostly suggest cycles of growth, each spurt or cycle enduring for a time, and one following another. Nothing can be more natural from the physiological point of view than that energy should be now added and now withheld, whether with the return of the seasons or at other stages on the eventful journey from childhood to manhood and old age.

The symmetry, or lack of skewness, in the Verhulst-Pearl logistic curve is a weak point rather than a strong; the Gompertz curve is a skew curve, with its point of inflexion not half-way, but about one-third of the way between the asymptotes. But whether in this or in the logistic or any other equation of growth, the precise point of inflexion has no biological significance whatsoever. What we want, in the first instance, is an **S**-shaped curve with a variable, or modifiable, degree of skewness. After all, the same difficulty arises in all the use we make of the Gaussian curve: which has to be eked out by a whole family of skew curves, more or less easily derived from it. We are far from being confined to the Gaussian curve (*sensu stricto*) in our studies of biological probability, or to the logistic curve in the study of population.

Yet another equation has been proposed to the **S**-shaped curve of growth, by Gaston Backman, a very diligent student of the whole subject. The rate of growth is made up, he says, of three components: a constant velocity, an acceleration varying with the time, and a retardation which we may suppose to vary with the square of the time. Acceleration would then tend to prevail in the earlier part of the curve, and retardation in the latter, as in fact they do; and the equation to the curve might be written:

$$\log H = k_0 + k_1 \log T - k_2 \log^2 T.$$

The formula is an elastic one, and can be made to fit many an **S**-shaped curve; but again it is empirical.

The logistic curve, as defined by Verhulst and by Pearl, has doubtless an interest of its own for the mathematician, the statistician and the actuary. But putting aside all its mathematical details and all arbitrary assumptions, the generalised **S**-shaped curve is a very symbol of childhood, maturity and age, of activity which rises to fall again, of growth which has its sequel in decay. The growth of a child or of a nation; the history of a railway*, or the speed between stations of a train; the spread of an epidemic†, or the evolutionary survival of a favoured type‡—all these things run their course, in its beginning, its middle and its end, after the fashion of the **S**-shaped curve. That curve represents a certain common pattern among Nature's "mechanisms," and is (as we have said before), a "mécanisme commun aux phénomènes disparates§."

At the same time—and this is a very interesting part of the story—the **S**-shaped curve is no other than what Galton called a *curve of distribution*, that is to say a curve of integration or summation-curve, whose differential is closely akin to the Gaussian curve of error.

Such, to a first approximation, is our **S**-shaped population-curve, and such are the many phenomena which, to a first approximation, it helps us to compare. But it is *only* to a first approximation that we compare the growth of a population with that of an organism, or for that matter of one organism or one population with another. There are immense differences between a simple and a complex organism, between a primitive and a civilised population. The yeast-plant gives a growth-curve which we can analyse; but we must fain be content with a qualitative description of the growth of a complex organism in its complex world‖.

There is a simplicity in a colony of protozoa and a complexity in a warm-blooded animal, a uniformity in a primitive tribe and a heterogeneity in a modern state or town, which affect all their economies and interchanges, all the relations between *milieu interne* and *externe*, and all the coefficients in any but the simplest equations of growth which we can ever attempt to frame. Every growth-problem becomes at last a specific one, running its own course for its own reasons. Our curves of growth are all alike—but no two are ever

* Raymond Pearl, *Amer. Nat.* LXI, pp. 289–318, 1927.

† Ronald Ross, *Prevention of Malaria* (2nd ed.), 1911, p. 679.

‡ J. B. S. Haldane, *Trans. Camb. Phil. Soc.* XXIII, pp. 19–41, 1924.

§ Cf. (*int. al.*) J. R. Miner's Note on birth-rate and density in a logistic population, *Human Biology*, IV, p. 119, 1932; and cf. Lotka, *ibid.* III, p. 458, 1931.

‖ Cf. (*int. al.*) C. E. Briggs, Attempts to analyse growth-curves, *Proc. R.S.* (B), CII, pp. 280–285, 1928.

the same. Growth keeps calling our attention to its own complexity. We see it in the rates of growth which change with age or season, which vary from one limb to another; in the influence of peace and plenty, of war and famine; not least in those composite populations whose own parts aid or hamper one another, in any form or aspect of the struggle for existence. So we come to the differential equations, easy to frame, more difficult to solve, easy in their first steps, hard and very powerful later on, by which Lotka and Volterra have shewn how to apply mathematics to evolutionary biology, but which lie just outside the scope of this book*.

An important element in a population, and one seldom easy to define, is its *age-composition*. It may vary one way or the other; for the diminution of a population may be due to a decrease in the birth-rate, or to an increasing mortality among the old. A remarkable instance is that of the food-fishes of the North Sea. Their birth-rate is so high that the very young fishes remain, to all appearance, as numerous as ever; those somewhat older are fewer than before, and the old dwindle to a fraction of what they were wont to be.

The rate of growth in other organisms

The rise and fall of growth-rate, the acceleration followed by retardation which finds expression in the **S**-shaped curve, are seen alike in the growth of a population and of an individual, and in most things which have a beginning and an end. But the law of large numbers smooths the population-curve; the individual life draws attention to its own ups and downs; and the characteristic sigmoid curve is only seen in the simpler organisms, or in parts or "phases" of the more complex lives. We see it at its simplest in the simple growth-cycle, or single season, of an annual plant, which cycle draws to its end at flowering; and here not only is the curve simple, but its amplitude may sometimes be very large. The giant *Heracleum* and certain tall varieties of Indian corn grow to twelve feet

* See (*int. al.*) A. J. Lotka, *Elements of Physical Biology*, Baltimore, 1925; *Théorie analytique des associations biologiques*, Paris, 1934; Vito Volterra, *Leçons sur la théorie mathématique de la lutte pour la vie*, 1931; Volterra et U. d'Ancona, *Les associations biologiques au point de vue mathématique*, 1935; V. A. Kostitzin, *op. cit.*; etc.

high in a summer; the kudzu vine (*Pucraria*) may grow twelve inches in twenty-four hours, and some bamboos are said to have grown twenty feet in three days (Figs. 32, 33).

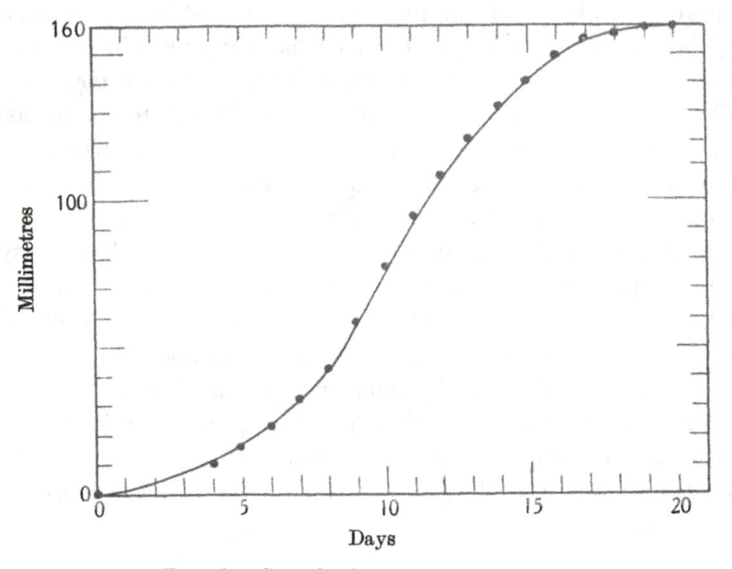

Fig. 32. Growth of Lupine. After Pfeffer.

Growth of Lupinus albus. (*From G. Backman, after Pfeffer*)

Day	Length (mm.)	Difference	Day	Length (mm.)	Difference
4	10·5	—	14	132·3	12·2
5	16·3	5·8	15	140·6	8·3
6	23·3	7·0	16	149·7	9·1
7	32·5	9·2	17	155·6	5·9
8	42·2	9·7	18	158·1	2·5
9	58·7	14·5	19	160·6	2·5
10	77·9	19·2	20	161·4	0·8
11	93·7	15·8	21	161·6	0·2
12	107·4	13·7			
13	120·1	12·7			

In the pre-natal growth of an infant the **S**-shaped curve is clearly seen (Fig. 18); but immediately after birth another phase begins, and a third is implicit in the spurt of growth which precedes puberty. In short, it is a common thing for one wave of growth (or *cycle*, as some call it) to succeed another, whether at special epochs in a lifetime, or as often as winter gives place to spring*.

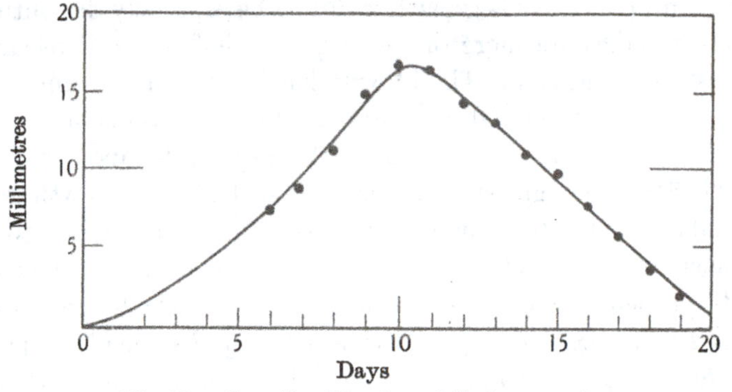

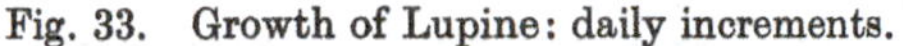
Fig. 33. Growth of Lupine: daily increments.

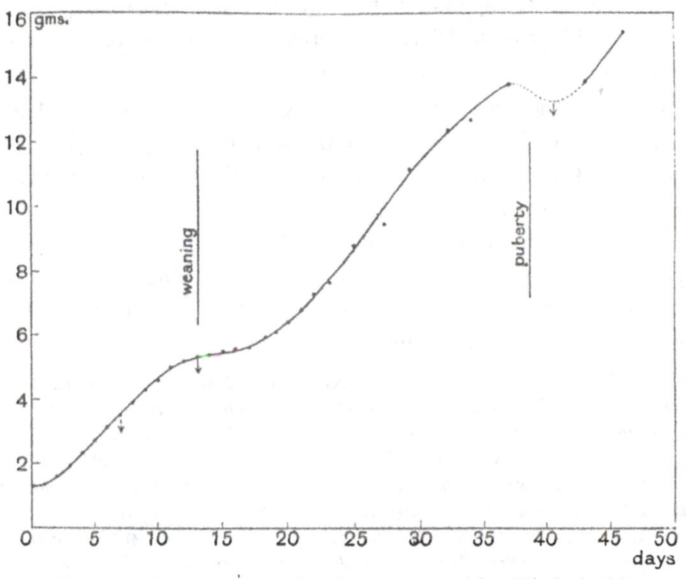

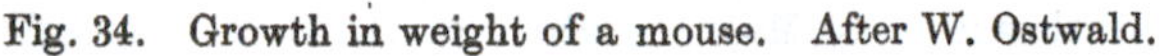
Fig. 34. Growth in weight of a mouse. After W. Ostwald.

In the accompanying curve of weight of the mouse (Fig. 34) we see a slackening of the rate of growth when the mouse is about a fortnight old, at which epoch it opens its eyes, and is weaned soon

* W. Pfeffer, *Pflanzenphysiologie*, 1881, Bd. II, p. 78; A. Bennett, On the rate of growth of the flower-stalk of *Vallisneria spiralis* and of *Hyacinthus*, *Trans. Linn. Soc.* (2), I, Botany, pp. 133, 139, 1880; cited by G. Backman, Das Wachstums-

after. At six weeks old there is another well-marked retardation; it follows on a rapid spurt, and coincides with the epoch of puberty*.

In arthropod animals growth is apt to be especially discontinuous, for their bodies are more or less closely confined until released by the casting of the skin. The blowfly has its striking metamorphoses, yet its growth is wellnigh continuous; for its larval skin is too thin and delicate to impede growth in the usual arthropod way. But in a thick-skinned grasshopper or hard-shelled crab growth goes by fits and starts, by steps and stairs, as Réaumur was the first to shew; for, speaking of insects†, he says: "Peut-être est-il vrai générale-ment que leur accroissement, ou au moins leur plus considérable accroissement, ne se fait que dans le temps qu'ils muent, ou pendant un temps assez court après la mue. Ils ne sont obligés de quitter leur enveloppe que parce qu'elle ne prend pas un accroissement proportionné à celui que prennent les parties qu'elle couvre." All the visible growth of the lobster takes place once a year at moulting-time, but he is growing in weight, more or less, all along. He stores up material for months together; then comes a sudden rush of water to the tissues, the carapace splits asunder, the lobster issues forth, devours his own exuviae, and lies low for a month while his new shell hardens.

The silkworm moults four times, about once a week, beginning on the sixth or seventh day after hatching. There is an arrest or retardation of growth before each moult, but our diagram (Fig. 35) is too small to shew the slight ones which precede the first and

problem, in *Ergebnisse d. Physiologie*, XXXIII, pp. 883–973, 1931. These two cases of *Lupinus* and *Vallisneria*, are among the many which lend themselves easily to Backman's growth-formula, viz. *Lupinus*, $\log p = -2{\cdot}40 + 1{\cdot}48 \log T - 6{\cdot}61 \log^2 T$ and *Vallisneria*, $\log p = +1{\cdot}28 + 4{\cdot}51 \log T - 2{\cdot}62 \log^2 T$. See for an admirable résumé of facts, Wolfgang Ostwald, *Ueber die zeitliche Eigenschaften der Entwicklungsvorgange* (71 pp.), 1908 (in Roux's *Vortrage*, Heft V); and many later works.

* Cf. R. Robertson, Analysis of the growth of the white mouse into its constituent processes, *Journ. Gen. Physiology*, VIII, p. 463, 1926. Also Gustav Backman, Wachstum d. w. Maus, *Lunds Univ. Arsskrift*, XXXV, Nr. 12, 1939, with copious bibliography. Backman analyses the complicated growth-curve of the mouse into one main and three subordinate cycles, two of which are embryonic. Cf. St Loup, Vitesse de croissance chez les souris, *Bull. Soc. Zool. Fr.* XVIII, p. 242, 1893; E. Le Breton and G. Schäfer, *Trav. Inst. Physiol. Strasburg*, 1923; E. C. MacDowell, Growth-curve of the suckling mouse, *Science*, LXVIII, p. 650, 1928; cf. *Journ. Gen. Physiol.* XI, p. 57, 1927; Ph. l'Héritier, Croissance...dans les souris, *Ann. Physiol. et Phys. Chemie*, V, p. i, 1929.

† *Mémoires*, IV, p. 191.

second. Before entering on the pupal or chrysalis stage, when the worm is about seven weeks old, a remarkable process of purgation

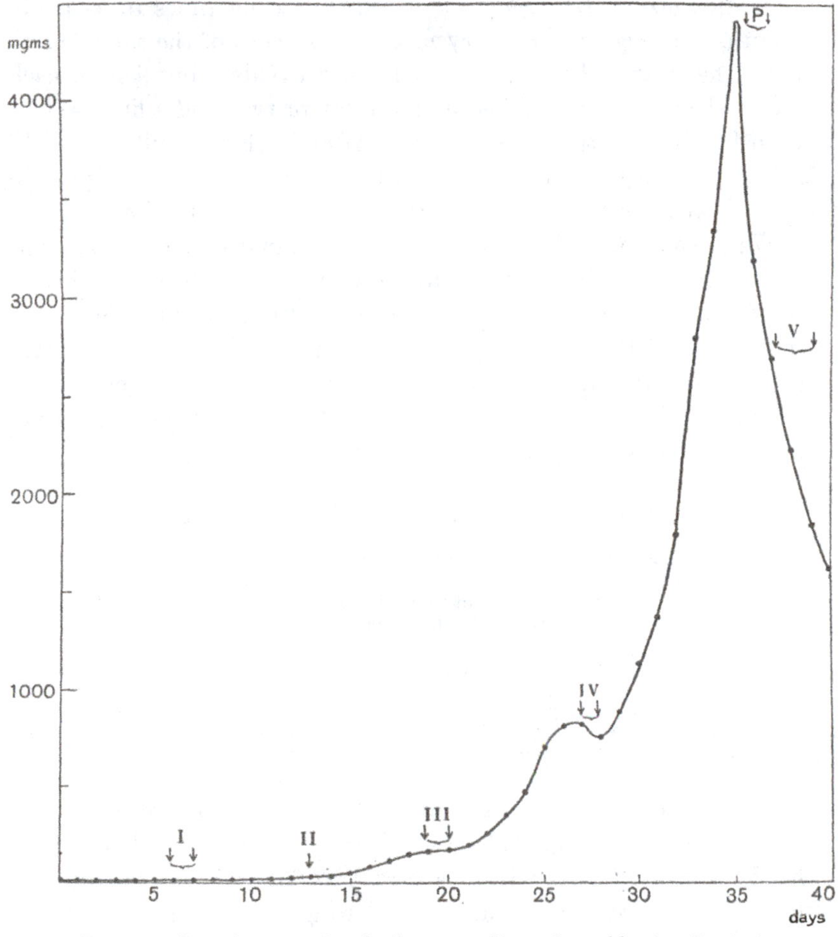

Fig. 35. Growth in weight of silkworm. From Ostwald, after Luciani and Lo Monaco.

takes place, with a sudden loss of water, and of weight, which becomes the most marked feature of the curve*. That the meta-

* Luciani e Lo Monaco, *Arch. Ital. de Biologie*, XXVII, p. 340, 1897; see also Z. Kuwana, Statistics of the body-weight of the silkworm, *Japan. Journ. Zool.* VII, pp. 311–346, 1937. Westwood, in 1838, quoted similar data from Count Dandolo: according to whom 100 silkworms weigh on hatching 1 grain; after the first four moults, 15, 94, 270 and 1085 grains; and 9500 grains when full-grown.

morphoses of an insect are but phases in a process of growth was clearly recognised by Swammerdam, in the *Biblia Naturae**.

A stick-insect (*Dexippus*) moults six or seven times in as many months; it lengthens at every moult, and keeps of the same length until the next. Weight is gained more evenly; but before each moult the creature stops feeding for a day or two, and a little weight is lost in the casting of the skin. After its last moult the stick-insect puts on more weight for a while; but growth soon draws to an end, and the bodily energies turn towards reproduction.

We have careful measurements of the locust from moult to moult, and know from these the *relative* growth-rates of its parts, though we cannot plot these dimensions against *time*. Unlike the metamorphosis of the silkworm, the locust passes through five larval stages (or "instars") all much alike, until in a final moult the "hoppers" become winged. Here are three sets of measurements, of limbs and head, from stage to stage†.

Growth of locust, from one moult to another

	Length (mm.)			Percentage-growth			Ratios		
Stage	Anterior femur	Median femur	Head	Anterior femur	Median femur	Head	Anterior femur	Median femur	Head
I	1·44	3·98	1·44	—	—	—	1	2·76	1·00
II	2·06	5·69	1·94	1·44	1·43	1·35	1	2·76	0·94
III	3·08	8·22	2·70	1·40	1·44	1·39	1	2·67	0·88
IV	4·53	11·94	3·71	1·47	1·45	1·37	1	2·76	0·82
V	6·40	17·22	4·89	1·41	1·44	1·32	1	2·69	0·76
Adult	8·03	22·85	5·59	1·25	1·33	1·14	1	2·84	0·70

As a matter of fact the several parts tend to grow, *for a time*, at a steady rate of compound interest, which rate is not identical for head and limbs, and tends in each case to fall off in the final moult, when material has to be found for the wings. Some fifty years ago, W. K. Brooks found the larva of a certain crab (*Squilla*) increasing at each moult by a quarter of its own length; and soon after H. G. Dyar declared that caterpillars grow likewise, from moult to moult, by geometrical progression‡. This tendency to a compound-

* 1737, pp. 6, 579, etc.

† A. J. Duarte, Growth of the migratory locust, *Bull. Ent. Res.* XXIX, pp. 425–456, 1938.

‡ W. K. Brooks, *Challenger Report on the Stomatopoda*, 1886; H. G. Dyar, Number of moults in lepidopterous larvae, *Psyche*, V, p. 424, 1896.

interest rate in the growth and metamorphosis of insects is known as Dyar's, sometimes as Brooks's, law. According to Przibram, an insect moults as soon (roughly speaking) as cell-division has doubled the number of cells throughout the larval body. That being so, each stage or instar should weigh twice as much as the one before, and each linear dimension should increase by $\sqrt[3]{2}$, or 1·26 times—a measure identical, to all intents and purposes, with Brooks's first estimate. As a *first rough approximation* the rule has a certain value. According to Duarte's measurements the locust's total weight increases from moult to moult by 2·31, 2·16, 2·42, 2·35, 2·21, or a mean increase of 2·29, the cube-root of which is 1·32. Each phase is doubled and more than doubled, in passing to the next*, but Przibram's estimate is not far departed from.

Whatever truth Przibram's law may have in insects, or (as Fowler asserted) in the Ostracods, it would seem to have none in the Cladocera: and this for the sufficient reason that the shell (on which the form of the creature depends) goes on growing all through post-embryonic life without further division or multiplication of its cells, but only by their individual, and therefore collective, enlargement†.

Shells are easily weighed and measured and their various dimensions have been often studied; only in oysters, pearl-oysters and the like, have they been so kept under observation that their actual age is known. The oyster-shell grows for a few weeks in spring just before spawning time, and again in autumn when spawning is over; its growth is imperceptible at other times‡.

* Cf. H. Przibram and F. Megusar, Wachstummessungen an *Sphodromantis*, *Arch. f. Entw. Mech.* XXXIV, pp. 680–741, 1912; etc. How the discrepancy is accounted for, by Bodenheimer and others, need not concern us here. But cf. P. P. Calvert, On rates of growth among...the Odonata, *Proc. Amer. Phil. Soc.* LXVIII, pp. 227–274, 1929, who finds growth faster in nine cases out of ten than Przibram's rule lays down.

Millet asserts, in support of Przibram's law, that in spiders mitotic cell-division is confined to the epoch of the moult, and is then manifested throughout most of the tissues (*Bull. de Biologie* (*Suppl.*), VIII, p. 1, 1926). On the other hand, the rule is rejected by R. Gurney, Rate of growth in Copepoda, *Int. Rev. Hydrobiol.* XXI, pp. 189–27, 1929; Nobumasa Kagi, Growth-curves of insect-larvae, *Mem. Coll. Agric. Kyoto*, No. 1, 1926; and others.

† Cf. W. Rammer, Ueber die Gültigkeit des Brooksschen Wachstumsgesetzes bei den Cladoceren, *Arch. f. Entw. Mech.* CXXI, pp. 111–127, 1930.

‡ Cf. J. H. Orton, Rhythmic periods...in Ostrea, *Journ. Mar. Biol. Assoc.* XV, pp. 365–427, 1928; *Nature*, March 2, 1935, p. 340.

The window-pane oyster in Ceylon (*Placuna placenta*) has been kept under observation for eight years, during which it grows from two inches long to six (Fig. 36). The young grow quickly, and slow down asymptotically towards the end; an **S**-shaped beginning to the growth-curve has not been seen, but would probably be found in the growth of the first year. Changes of shape as growth goes on are hard to see in this and other shells; rather is it characteristic of

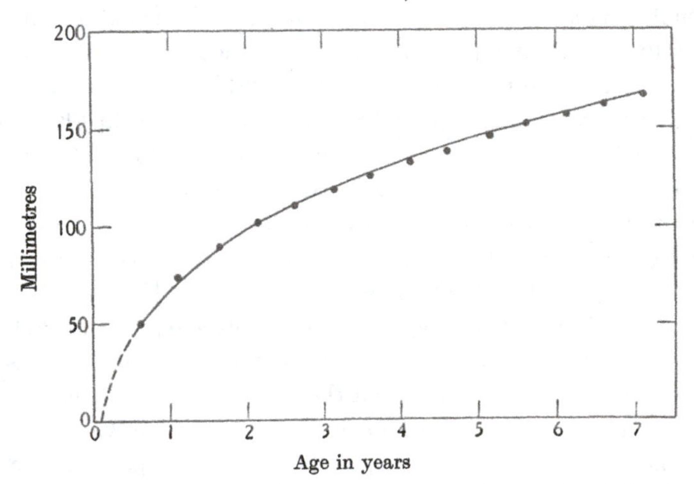

Fig. 36. Growth of the window-pane oyster; short diameter of the shell. From Pearson's data.

them to keep their shape from first to last unchanged. Nevertheless, slight changes are there; in the window-pane oyster the shell grows somewhat rounder; in seven or eight years the one diameter multiplies (roughly speaking) by eleven, and the other by ten*.

Window-pane oysters (Placuna)

Short diameter (mm.)	Long diameter (mm.)	Ratio
15·0	17·6	1·17
65·0	70·5	1·09
102·5	109·7	1·07
132·5	139·9	1·06
167·5	175·2	1·05

The American slipper-limpet has lately and quickly become a pest on English oyster-beds. Its mode of growth is interesting, though

* Joseph Pearson, The growth-rate...of *Placuna placenta*, *Ceylon Bulletin*, 1928.

the actual rate remains unknown. It grows a little longer and narrower with age. Its weight-length coefficient (of which we shall have more to say presently) increases as time goes on, and appears to follow a wavy course which might be accounted for if the shell grew thinner and then thicker again, as if ever so little more lime were secreted at one season than another. The growth of a shell, or the deposition of its calcium carbonate, is much influenced by temperature; clams and oysters enlarge their shells only so long as the temperature stands above a certain specific minimum, and the mean size of the same limpet is very different in Essex and in the United States*. Curious peculiarities of growth have been discovered in slipper-limpets. Young limpets clustered round an old female grow slower than others which live solitary and apart. The solitary forms become in turn male, hermaphrodite and at last female, but the gregarious or clustered forms develop into males, and so remain; development of male characters and duration of the male phase depend on the presence or absence of a female in the near neighbourhood.

Measurements of slipper-limpets
(From J. H. Fraser's data, epitomised)

No. measured	Mean length (mm.)	Breadth (mm.)	Ratio L/B	Weight (gm.)	W/L^3
3	15·3	8·8	1·74	0·33	92
8	17·6	9·8	1·80	0·46	84
9	19·4	10·5	1·85	0·63	88
16	21·5	11·5	1·87	0·77	77
18	23·5	12·5	1·88	1·04	80
41	25·5	13·7	1·86	1·37	85
91	27·4	14·5	1·89	1·81	88
125	39·4	15·4	1·91	2·33	92
98	31·4	16·5	1·90	3·22	104
70	33·6	17·8	1·89	3·61	95
38	35·5	18·6	1·90	4·28	95
10	37·3	19·5	1·91	4·95	95
1	32·1	19·4	2·01	5·35	90
		Mean	1·87		89·3

* Cf. J. H. Fraser, On the size of *Urosalpinx* etc., *Proc. Malacol. Soc.* XIX, pp. 243–254, 1931. Much else is known about the growth of various limpets, their seasonal periodicities, the change of shape in certain species, and other matters; cf. E. S. Russell, Growth of Patella, *P.Z.S.* CXCIX, pp. 235–253; J. H. Orton, *Journ. Mar. Biol. Assoc.* XV, pp. 277–288, 1929; Noboru Abe, *Sci. Rep. Tohoku Imp. Univ. Biol.* VI, pp. 347–363, 1932, and Okuso Hamai, *ibid.* XII, pp. 71–95, 1937.

The growth of the tadpole* is likewise marked by epochs of retardation, and finally by a sudden and drastic change (Fig. 37). There is a slight diminution in weight immediately after the little larva frees itself from what remains of the egg; there is a retardation

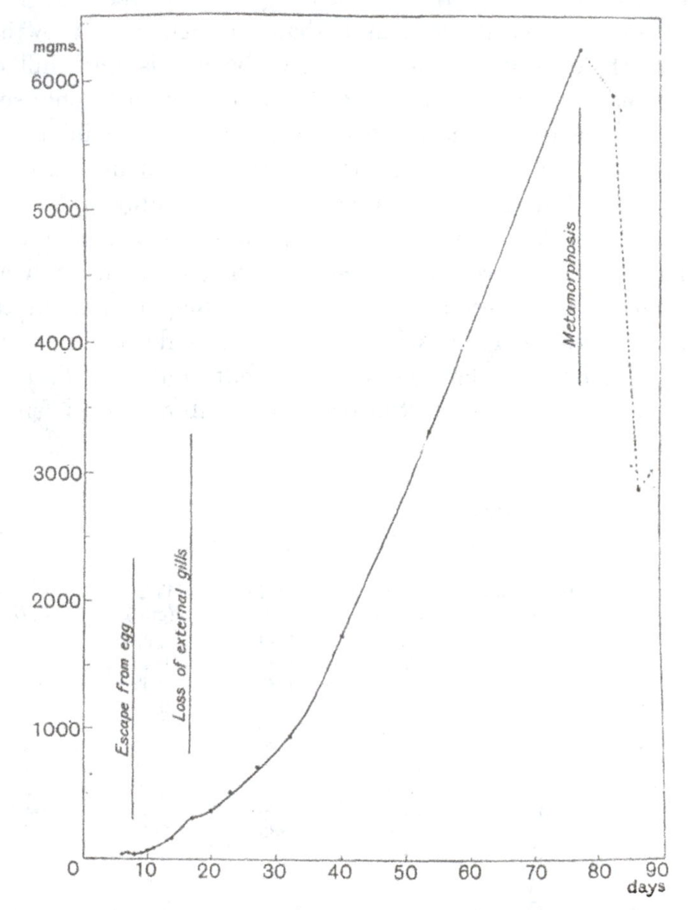

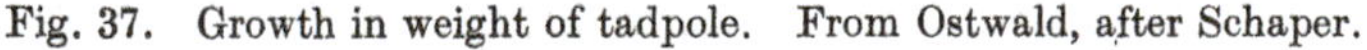
Fig. 37. Growth in weight of tadpole. From Ostwald, after Schaper.

of growth about ten days later, when the external gills disappear; and finally the complete metamorphosis, with the loss of the tail, the growth of the legs and the end of branchial respiration, brings about a loss of weight amounting to wellnigh half the weight of the full-grown

* Cf. (*int. al.*) Barfurth, Versüche über die Verwandlung der Froschlarven, *Arch. f. mikrosk. Anat.* XXIX, 1887.

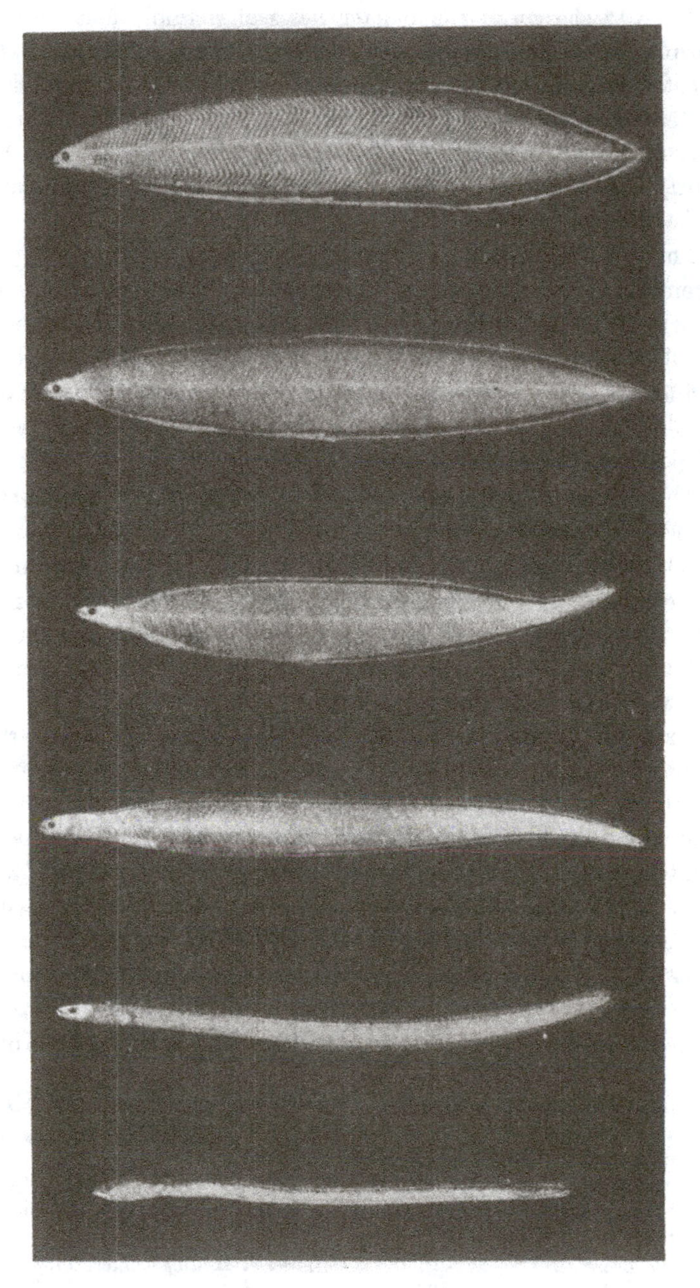

Fig. 38. Development of eel: from *Leptocephalus* larvae to young elver. After Johannes Schmidt.

larva. At the root of the matter lies the simple fact that metamorphosis involves wastage of tissue, increase of oxidation, expenditure of energy and the *doing of work*. While as a general rule the better the animals be fed the quicker they grow and the sooner they metamorphose, Barfurth has pointed out the curious fact that a short spell of starvation, just before metamorphosis is due, appears to hasten the change.

The negative growth, or actual loss of bulk and weight which often, and perhaps always, accompanies metamorphosis, is well shewn in the case of the eel*. The contrast of size is great between the flattened, lancet-shaped *Leptocephalus* larva and the little black, cylindrical, almost thread-like elver, whose magnitude is less than that of the *Leptocephalus* in every dimension, even at first in length (Fig. 38), as Grassi was the first to shew.

The lamprey's case is hardly less remarkable. The larval or Ammocoete stage lasts for three years or more, and metamorphosis, though preceded by a spurt of growth, is followed by an actual decrease in size. The little brook lamprey neither feeds nor grows after metamorphosis, but spawns a few months later and then dies; but the big sea-lampreys become semi-parasitic on other fishes, and live and grow to an unknown age†.

Such fluctuations as these are part and parcel of the general flux of physiological activity, and suggest a finite stock of energy to be spent, now more now less, on growth and other modes of expenditure. The larger fluctuations are special interruptions in a process which is never continuous, but is perpetually varied by rhythms of various kinds and orders. Hofmeister shewed long ago, for instance, that *Spirogyra* grows by fits and starts, in periods of activity and rest alternating with one another at intervals of so many minutes‡ (Fig. 39). And Bose tells us that plant-growth proceeds by tiny and perfectly rhythmical pulsations, at intervals of a few seconds of time.

* Johannes Schmidt, Contributions to the life-history of the eel, *Rapports du Conseil Intern. pour l'exploration de la mer*, v, pp. 137–274, Copenhagen, 1906; and other papers.

† Cf. (*int. al.*) A. Meek, The lampreys of the Tyne, *Rep. Dove Marine Laboratory* (N.S.), VI, p. 49, 1917; cf. L. Hubbs, in *Papers of the Michigan Academy*, IV, p. 587, 1924.

‡ *Die Lehre der Pflanzenzelle*, 1867. Cf. W. J. Koningsberger, *Tropismus und Wachstum* (Thesis), Utrecht, 1922.

A crocus grows, he says, by little jerks, each with an amplitude of about 0·002 mm., every twenty seconds or so, each increment being followed by a partial recoil* (Fig. 40). If this be so we have come

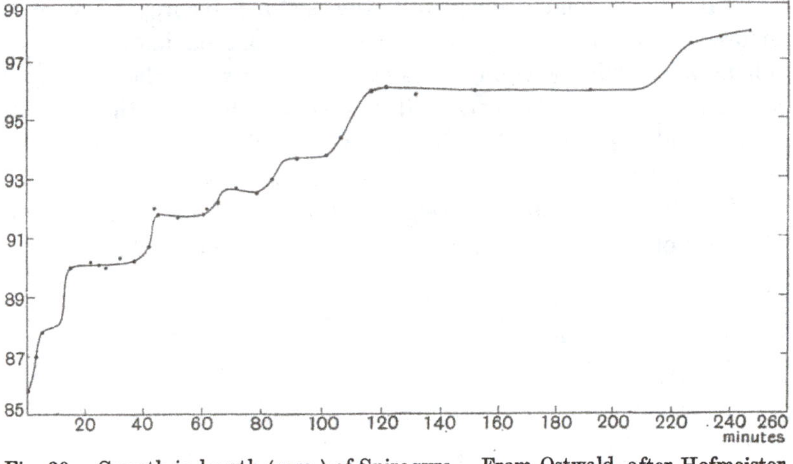

Fig. 39. Growth in length (mm.) of Spirogyra. From Ostwald, after Hofmeister.

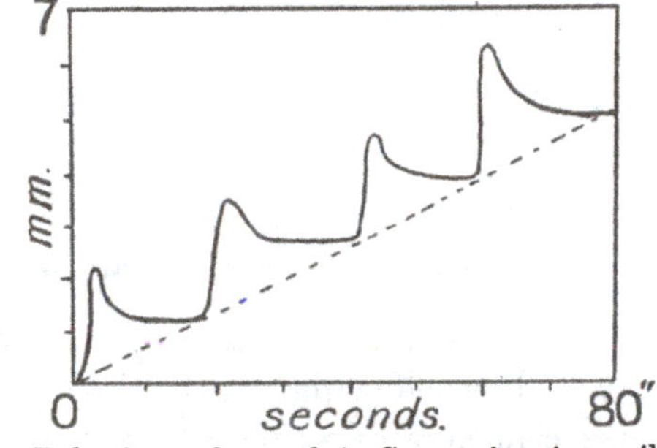

Fig. 40. Pulsations of growth in Crocus, in micro-millimeters. After Bose.

down, so to speak, from a principle of continuity to a principle of discontinuity, and are face to face with what we might call, by rough analogy, "quanta of growth." We seem to be in touch with things of another order than the subject of this book†.

* J. C. Bose, *Plant Response*, 1906, p. 417; *Growth and Tropic Movements of Plants*, 1929.

† There is an apparent and perhaps a real analogy between these periodic phenomena of growth and the well-known phenomenon of periodic, or oscillatory, chemical change, as described by W. Ostwald and others; cf. (e.g.) *Zeitschr. f. phys. Chem.* xxxv, pp. 33, 204, 1900.

We may want now and then to make use of scanty data, and find a rough estimate better than none. The giant tortoises of the Galapagos and the Seychelles grow to a great age, and some have weighed 500 lb. and more; but the scanty records of captive tortoises shew much variation, depending on food and climate as well as age. Ninety young tortoises brought from the Galapagos in 1928 to the southern United States weighed on the average 18½ lb., and grew to 44·3 lb. in two years. Six taken to Honolulu weighed 26½ lb. each in 1929, and 63 lb. each the following year. Another, kept in California, weighed 29 lb. and 360 lb. seven years later, but only gained 65 lb. more in the next seven years. Growth,

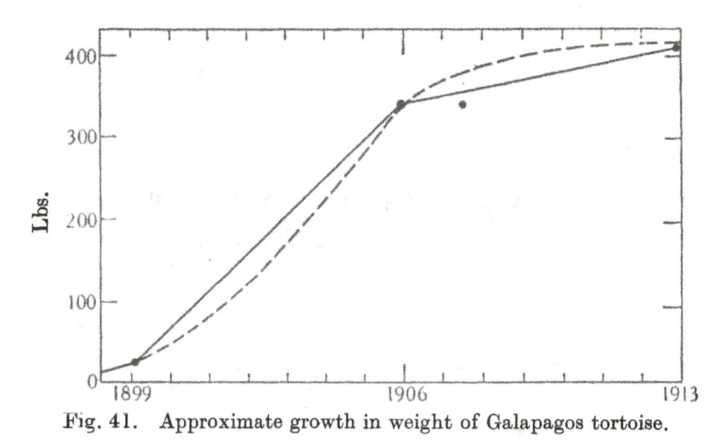

Fig. 41. Approximate growth in weight of Galapagos tortoise.

as usual, is quick to begin with, slower later on, and in the old giants must be slow indeed. If we plot (Fig. 41) the three successive weights of the Californian specimen, at first they help us little; but we can fit an **S**-shaped curve to the three points as a first approximation, and it suggests, with some plausibility, that, at 29 lb. weight the tortoise was from two to three years old. A loggerhead turtle, which reaches a great size, was found to grow from a few grammes to 42 lb. in three years, and to double that weight in another year and a half; these scanty data are in fair accord, so far as they go, with those for the giant tortoises*.

* For these and other data, see C. H. Townsend, Growth and age in the giant tortoises of the Galapagos, *Zoologica*, IX, pp. 459–466, 1931; G. H. Parker, Growth of the loggerhead turtle, *Amer. Naturalist*, LXVII, pp. 367–373, 1929; Stanley F. Flower, Duration of Life in Animals, III, Reptiles, *P.Z.S.* (A), 1937, pp. 1–39.

The horny plates of the tortoise grow, to begin with, a trifle faster than the bony carapace below, and are consequently wrinkled into folds. There is some evidence, at least in the young tortoises, that these folds come once a year, which is as much as to say that there is one season of the year when the growth-rates of bony and horny carapace are especially discrepant. This would give an easy estimate of age; but it is plainer in some species than in others, and it never lasts for long.

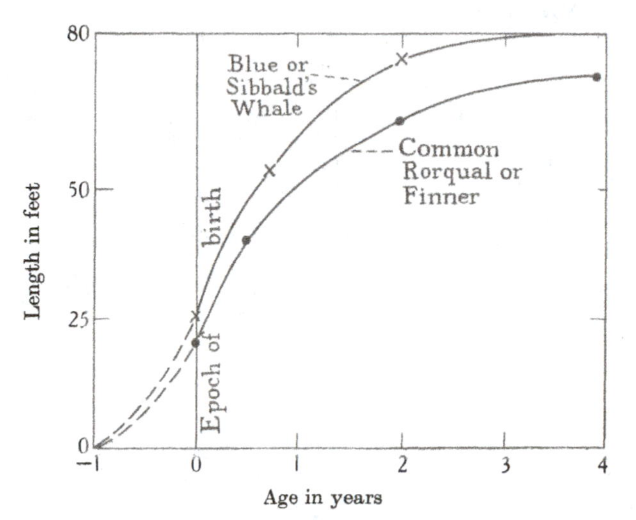

Fig. 42. Growth-rate (approximate) of blue and finner whales.

The blue whale, or Sibbald's rorqual, largest of all animals, grows to 100 ft. long or thereby, the females being a little bigger than the males. The mother goes with young eleven months. The calf measures 22 to 25 ft. at birth, and weighs between three and four tons; it is born big, were it smaller it might lose heat too quickly. It is weaned about nine months later, and is said to be some 16 metres, or say 53 ft., long by then. It is believed to be mature at two years old, by which time it is variously stated to be 60 or even 75 ft. long; the modal size of pregnant females is about 80 ft. or rather more. How long the whale takes to grow the further 15 or 20 feet which bring it to its full size is not known; but, even so far, the rapid growth and early maturity seem very remarkable (Fig. 42). The Norwegian whalers give us statistics, month by month during the Antarctic season, of the sizes of pregnant females and the foetuses they contain; and from these I draw the following averages:

Antarctic blue whales; length of mother and of foetus
(*Season* 1938–39)

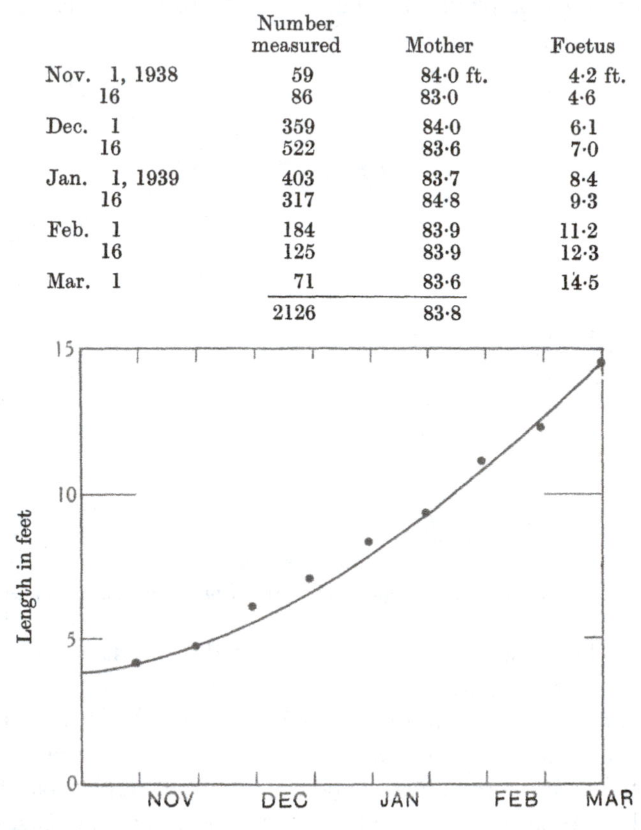

	Number measured	Mother	Foetus
Nov. 1, 1938	59	84·0 ft.	4·2 ft.
16	86	83·0	4·6
Dec. 1	359	84·0	6·1
16	522	83·6	7·0
Jan. 1, 1939	403	83·7	8·4
16	317	84·8	9·3
Feb. 1	184	83·9	11·2
16	125	83·9	12·3
Mar. 1	71	83·6	14·5
	2126	83·8	

Fig. 43. Pre-natal growth of blue whale. Average monthly sizes, from data in *International Whaling Statistics*, XIV, 1940.

The observations are rough but numerous. At the lower end of the scale measurements are few, and the value indicated is probably too high; but on the whole the curve of growth tallies with other estimates, and points to birth about June or July, and to conception about the same time last year (Fig. 43). The mean size of the mother-whales does not alter during the five months in question;

they do not seem to be increasing, though at 84 ft. they still have another 10 feet or more to grow. They may grow slower, and live longer, than is often supposed*.

On the other hand, if we draw from the same official statistics the mean size of mother-whale and foetus at some given epoch of the year (e.g. March 1934), there appears to be a marked correlation between them, such as would indicate very considerable growth of the mother during the months of pregnancy. The matter deserves further study, and the data need confirmation.

Blue whales; length of mother and foetus (*March* 1934)

Number observed	Size (ft.) Mother	Size (ft.) Foetus	Size of foetus (ft.) smoothed in threes
1	74	1·0	—
1	75	7·0	4·4
5	76	5·2	6·4
7	77	7·3	6·4
9	78	6·7	6·9
10	79	6·7	6·8
21	80	7·1	7·2
27	81	7·7	7·5
28	82	7·6	7·9
33	83	8·4	8·3
38	84	8·9	8·6
46	85	8·5	8·6
37	86	8·5	8·8
19	87	9·5	9·3
18	88	9·9	10·3
12	89	11·4	10·9
18	90	11·5	11·1
9	91	10·4	11·1
2	92	11·5	—
341			

On the growth of fishes, and the determination of their age

We may keep a child under observation, and weigh and measure him every day; but more roundabout ways are needed to determine the age and growth of the fish in the sea. A few fish may be caught and marked, on the chance of their being caught again; or a few

* The growth of the finner whale, or common rorqual, is estimated as follows (Hamburg Museum): at birth, 6 m.; at 6 months, 12 m.; at one and two years old, 15 and 19 m.; when full-grown, at 6–8 (?) years old, 21 m. For data, see *Hvalradets Skrifter* and *International Whaling Statistics*, passim; also N. Mackintosh and others in *Discovery Reports*; also Sigmund Rusting, Statistics of whales and whale-foetuses, *Rapports du Conseil Int.* 1928; etc.

more may be kept in a tank or pond and watched as they grow. Both ways are slow and difficult. The advantage of large numbers is not obtained; and it is needed all the more because the rate of growth turns out to be very variable in fishes, as it doubtless is in all cold-blooded or "poecilothermic" animals: changing and fluctuating not only with age and season, but with food-supply, temperature and other known and unknown conditions. Trout in a chalk-stream so differ from those in the peaty water of a highland burn that the former may grow to three pounds weight while the latter only reach four ounces, at three years old or four*.

It is found (and easily verified) that shells on the seashore, kind for kind, do not follow normal curves of frequency in respect of magnitude, but fall into *size-groups* with intervals between, so constituting a *multimodal curve*. The reason is that they are not born all the year round, as we are, but each at a certain annual breeding-season; so that the whole population consists of so many "groups," each one year older, and bigger in proportion, than another. In short we find *size-groups*, and recognise them as *age-groups*. Each group has its own spread or scatter, which increases with size and age; even from the first one group *tends* to overlap another, but the older groups do so more and more, for they have had more time and chance to vary. Hence this way of determining age gets harder and less certain as the years go by; but it is a safe and useful method for short-lived animals, or in the early lifetime of the rest. Aristotle's fishermen used it when they recognised three sorts or sizes of tunnies, the auxids, pelamyds and full-grown fish; and when they found a scarcity of pelamyds in one year to be followed by a failure of the tunny-fishery in the next†.

Shells lend themselves to this method, as Louis Agassiz found when he gathered periwinkles on the New England shore. Winckworth found the *Paphiae* in Madras harbour "of two sizes, one group just under 15 mm. in length, the other nearly all over 30 mm. A small sample, dredged *five months earlier* from the same ground, was intermediate between the other two." When the mean sizes of the two groups were plotted against time, the lesser group being shifted

* Cf. C. A. Wingfield, Effect of environmental factors on the growth of brown trout, *Journ. Exp. Biol.* XVII, pp. 435–448, 1939.

† Aristotle, *Hist. Anim.* VI, 571*a*.

back a year, a growth-curve extending over two seasons was obtained; when extrapolated, it seemed to start from zero about May or June, and this date, at the beginning of the hot season, was in all probability the actual spawning time. Growth stopped in winter, a common thing in our northern climate but surprising at Madras, where the sea-temperature seldom falls below 24° C. Shells over 40 mm. long were rare, and over 50 mm. hardly to be found—an indication that *Paphia* seldom lives over a third season. Here then, though the numbers studied were all too few, the method tells us with little doubt or ambiguity the age of a sample and the growth-rate of the species to which it belongs*.

Dr C. J. G. Petersen of Copenhagen brought this method into use for the study of fishes, and up to a certain point it is safe and trustworthy though seldom easy. For one thing, it is hard to get a "random sample" of fish, for one net catches the big and another the small. The trawl-net takes all the big, but lets more and more of the small ones through. The drift-net catches herring by their heads; if too big, the head fails to catch and the fish goes free, if too small the fish slips through; so the net *selects* a certain modal size according to its mesh, and with no great spread or scatter. When we use Petersen's method and plot the sizes of our catch of fish, the younger age-groups are easily recognised, even though they tend to overlap; but the older fish are few, each size-group has a wider spread, and soon the groups merge together and the modal cusps cease to be recognisable. There is no way, save a rough conjectural one, of analysing the composite curve into the several groups of which it is composed; in short, this method works well for the younger, but fails for the older fish.

Fig. 44 is drawn from a catch of some 500 small cod, or codling, caught one November in the Firth of Forth, in a small-meshed experimental trawl-net. They are too few for the law of large numbers to take full effect; but after smoothing the curve, three peaks are clearly seen, with some sign of a fourth, indicating *about*

* R. Winckworth, Growth of *Paphia undulata*, *Proc. Malacolog. Soc.* XIX, pp. 171–174, 1931. Cf. (*int. al.*) Weymouth, on *Mactra stultorum*, *Bull. Calif. Fish Comm.* VII, 1923; Orton, on *Cardium*, *Journ. Mar. Biol. Assoc.* XIV, 1927, on *Ostrea*, and on *Patella*, *ibid.* XV, 1928; Ikuso Hamai, on Limpets, *Sci. Rep. Tohoku Imp. Univ.* (4), XII, 1937.

11 cm., 26, 44 and 60 cm., as the mean or modal sizes of four successive broods. The dwindling heights of the successive cusps are a first approximation to a "curve of mortality," shewing how the young are many and the old are few. Again, plotting the several sizes against time, we should get our curve of growth for four years, or a first rough approximation to it. Thus we learn from a random sample, caught in a single haul, the mean (or modal) sizes of a fish at several epochs of its life, say at two, three or even more successive intervals of a year; and we learn (to a first approximation) its rate of growth and its actual age, for the slope of the growth-curve, drawing to the base-line, points to the time when growth began.

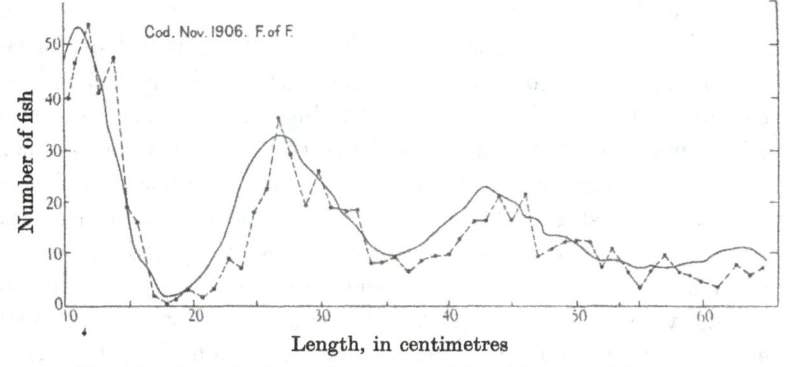

Fig. 44. A catch of cod, shewing a multimodal curve of frequency.

Another haul, soon after, will add new points to the curve, and confirm our first rough approximation.

An experiment in the Moray Firth, a month or two later, shewed the first three annual groups in much the same way; but it also shewed another group, of about 90 cm. long, and others larger still. At first sight these did not seem to fit on to our four successive year-groups, of 11, 26, 44 and 60 cm.; but they did so after all, only *with a gap between*. They were older fish, six and seven years old, which had come back to the Moray Firth to breed after spending a couple of years elsewhere.

It was thought at first that every such experiment should tally with another, and bring us to a more and more accurate knowledge of *the* growth-rate of this fish or that; but there were continual discrepancies, and it was soon found that the rate varied from place to place, from month to month, and from one year to another. The growth-rate of a fish varies far more than does that of a warm-blooded animal. The general character of the curve remains*, save that the fish continues to grow even in extreme old age, but it draws towards its upper asymptote with exceeding slowness.

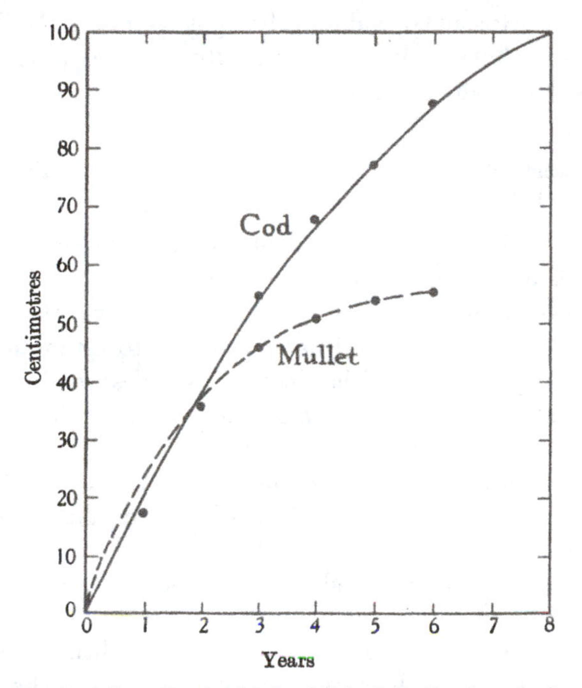

Fig. 45. Growth of cod (after Michael Graham); and of mullet (after C. D. Serbetis).

The following estimate of the mean growth of North Sea cod is based, by Michael Graham, on a great mass of various evidence; and beside it, for comparison, is an estimate for the grey mullet, by C. D. Serbetis. The shape of the curve (Fig. 45) is enough to indicate that at six years old the cod is still growing vigorously†, while the grey mullet has all but ceased to grow. As a matter of

* It is essentially an **S**-shaped curve, as usual; but the conditions of larval life obscure the first beginnings of the **S**.

† Norwegian results, based largely on otoliths, are different. Gunner Rollefsen holds that the spawning cod, or skrei, do not reach maturity, for the most part, till 10 or 11 years old, and grow by no more than 1 to 3 cms. a year (*Fiskeriskrifter*, Bergen, 1933).

fact, 90 cm. is, or was till lately, the median size of cod* in our Scottish trawl-fishery; one-tenth are over a metre long and the largest are in the neighbourhood of 120 cm., with an occasional giant of 150 cm. or even more. But it has come to pass that fish of outstanding size are seen no more save on the virgin fishing grounds; a Greenland halibut, brought home to Hull in 1938, weighed four hundredweight, was nearly two feet thick, and must have been of prodigious age.

Age (years)	1	2	3	4	5	6
Length of cod (cm.)	18	36	55	68	79	89
Length of grey mullet	21	36	46	51	53	55

There are other ways of determining, or estimating, a fish's age. The Greek fishermen shewed Aristotle† how to tell the age of the purple Murex, up to six years old, by counting the whorls and sculptured ridges of the shell, and also how to estimate the age of a scaly fish by the size and hardness of its scales; and Leeuwenhoek saw that a carp's scales‡ bear concentric rings, which increase in number as the fish grows old. In these and other cases, as in the woody rings of a tree, some part of plant or animal carries a record of its own age; and this record may be plain and certain, or may too often be dubious and equivocal.

The scales of most fishes shew concentric rings, sometimes (as in the herring) of a simple kind, sometimes (as in the cod) in a more complex pattern; and the ear-bones, or otoliths, shew opaque concentric zones in their translucent structure. The scales are "read" with apparent ease in herring, haddock, salmon, the otoliths in plaice and hake; but the whole matter is beset with difficulties, and every result deserves to be checked and scrutinised§.

* As distinguished from "codling."

† *Hist. Animalium*, 547*b*, 10; 607*b*, 30.

‡ The carp-breeder is especially interested in the age of his fish; for, like the brewer with his yeast, his profit depends on the rate at which they grow. Leeuwenhoek's and other early observations were brought to light by C. Hoffbauer, Die Alterbestimmung der Karpfen an seiner Schuppen, *Jahresber. d. schles. Fischerei-Vereins*, Breslau, 1899.

§ Thus, for instance, Mr A. Dannevig says (On the age and growth of the cod, *Fiskeridirektorets Skrifter*, 1933, p. 82): "as to the problem of the determination of the age of the cod by means of scales and otoliths, all workers agree that the method is useful. But on a number of fundamental points there are just as many divergences of opinion as there are investigators."

In the following table, we see (*a*) the sizes, and (*b*) the number of scale-rings, in a sample of some 550 herring from the autumn fishery off the east of Scotland.

Rings cm.	3	4	5	6	7	8	9	10	11	12	Total	Mean rings
31	—	—	—	1	1	—	1	—	—	1	4	8·5
30	—	—	—	7	5	6	4	—	—	—	22	7·3
29	—	—	5	18	13	6	6	1	1	1	51	7·0
28	—	3	29	38	11	3	3	1	—	—	88	5·9
27	2	13	41	34	5	5	2	—	—	—	102	5·1
26	7	43	64	29	—	—	1	—	—	—	144	5·0
25	4	36	41	11	—	—	—	—	—	—	92	4·6
24	2	17	15	4	—	—	—	—	—	—	38	4·8
23	—	5	—	—	—	—	—	—	—	—	5	4·0
Total	15	117	195	142	35	20	17	2	1	2		
Mean size	25·6	25·4	26·5	27·4	28·6	28·7	28·8	28·5	—	—		

In this sample, the sizes of the 550 fish are grouped in a somewhat skew curve, about a mode at 26 cm.; and the numbers of scale-rings group themselves in like manner, but with rather more skewness, about a modal number of five rings. Either way we look at it, there is only one "group" of fish; and it is highly characteristic of the herring that a single sample, taken from a single shoal, exhibits a unimodal curve. Accepting in principle the view that scale-rings tend to synchronise with age in years, we may draw this first deduction that our sample consists in part (if not in whole) of five-year old fish, whose average length is about 26 cm.; and this length, of 26 cm. for 5-ringed, or 5-year-old herring, agrees well with many other determinations from the same region. We shall be on the safe side if we deal, after this fashion, with *the one predominant group*, or mode, in each sample of fish; and Fig. 46 shews an approximate curve of growth for our East Coast herring drawn in this way.

But the further assumption is commonly and all but universally made that *each individual herring* carries the record of its age on its scale-rings. If this be so, then our sample of 550 fish is a composite population of some ten separate broods or successive ages, all mixed up in a shoal. And again, if so, the 5-year-olds in the said population average 26·5 cm. in length, the 3-year-olds 25·6 cm., the 10-year-olds 28·5 cm.; but these values do not fit into a normal curve of growth by any means. Still more obvious is it that the several year-classes (if such they be) do not tally with the age-composition of any ordinary population, nor agree with any ordinary curve of mortality. But even if we had ten separate year-groups represented here, which I most gravely doubt, all that we know of the selective action of the drift-net forbids us to assume that we are dealing with a fair random sample of the herring population; so that, even though the number of rings did enable us to distinguish the successive broods, we should still have no right to assume that

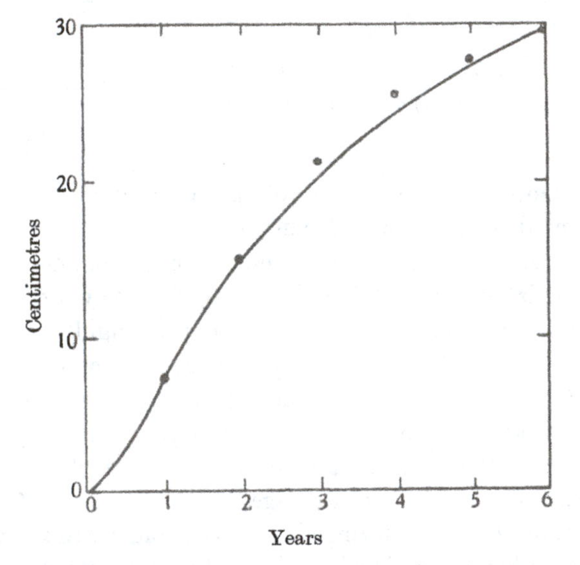

Fig. 46. Mean curve of growth of Scottish (East Coast) herring.

these annual broods actually combine in the proportions shewn, to form the composite population.

It is held by many (in the first instance by Einar Lea) that we may deduce the dimensions of a herring at each stage of its past life from the corresponding dimensions of the rings upon its scales. Some such relation must obviously exist, but it is an approximation of the roughest kind. For it involves the assumption not only that the scales add ring to ring regularly year to year, and that fish and scale grow all the while at corresponding rates or in direct proportion to one another, but also that the scale grows by mere accretion, each annual increment persisting without further change after it is once laid down. This is what happens in a molluscan shell, which is secreted or deposited as mere dead substance or "formed material"; but it is by no means the case in bone, and we have little reason to expect it of the bony mesoblastic tissue of a fish's scale. It is much more likely (though we do not know for sure) that "osteoblasts" and "osteoclasts" continue (as in bone) to play their part in the scale's growth and maintenance, and that some sort of give and take goes on. In any case, it is a matter of

Mean apparent length of one-year-old herring, as deduced by scale-reading from herring of various ages or "year-classes"*

Year-class (or number of rings)	2	3	4	5	6	7	8	9
Estimated length at 1 year old	14·5	13·2	12·7	12·5	12·1	11·8	11·9	11·8

fact and observation that the rings alter in breadth as the fish goes on growing†; that the oldest or innermost rings grow steadily narrower, while the outermost hardly change or even widen a little; that the relative breadths of successive rings alter accordingly; and it follows that when we try to trace the growth of a herring through its lifetime from its scales when it is old, the result is more or less misleading, and the values for the earlier years are apt to be much too small. The whole subject is very difficult, as we might well expect it to be; and I am only concerned to shew some small part of its difficulty‡.

While careful observations on the rate of growth of the higher animals are scanty, they shew so far as they go that the general features of the phenomenon are much the same. Whether the animal be long-lived, as man or elephant, or short-lived like horse §

* From T. Emrys Watkin, The Drift Herring of the S.E. of Ireland, *Rapports du Conseil pour l'Exploration de la Mer*, LXXXIV, p. 85, 1933.

† Cf. (*int. al.*) Rosa M. Lee, Methods of age and growth determination in fishes by means of scales, *Fishery Investigations, Dept. of Agr. and Fisheries*, 1920.

‡ The copious literature of the subject is epitomised, so far, by Michael Graham, in *Fishery Investigations* (2), XI, No. 3, 1928.

§ There is a famous passage in Lucretius (V, 883) where he compares the course of life, or rate of growth, in the horse and his boyish master: *Principio circum tribus actis impiger annis Floret equus, puer hautquaquam*, etc.

or dog, it passes through the same phases of growth; and, to quote Dr Johnson again, "whatsoever is formed for long duration arrives slowly to its maturity*." In all cases growth begins slowly; it attains a maximum velocity somewhat early in its course, and afterwards slows down (subject to temporary accelerations) towards a point where growth ceases altogether. But in cold-blooded animals, as fish or tortoises, the slowing down is greatly protracted, and the size of the creature would seem never to reach, but only to approach asymptotically, to a maximal limit. This, after all, is an important difference. Among certain still lower animals growth ceases early but life goes on, and draws (apparently) to no predetermined end. So sea-anemones have been kept in captivity for sixty or even eighty years, have fed, flourished and borne offspring all the while, but have shewn no growth at all.

The rate of growth of various parts or organs†

That the several parts and organs of the body, within and without, have their own rates of growth can be amply demonstrated in the case of man, and illustrated also, but chiefly in regard to external form, in other animals. There lies herein an endless field for the study of correlation and of variability‡.

In the accompanying table I show, from some of Vierordt's data, the relative weights at various ages, compared with the weight at birth, of the entire body, and of brain, heart and liver; also the changing relation which each of these organs consequently bears, as time goes on, to the weight of the whole body (Fig. 47)§.

* All of which is tantamount to a mere change of scale of the time-curve.

† This phenomenon, of *incrementum inequale*, as opposed to *incrementum in universum*, was most carefully studied by Haller: "Incrementum inequale multis modis fit, ut aliae partes corporis aliis celerius increscant. Diximus hepar minus fieri, majorem pulmonem, minimum thymum, etc." (*Elem.* VIII (2), p. 34.)

‡ See (*int. al.*) A. Fischel, Variabilität und Wachsthum des embryonalen Körpers, *Morphol. Jahrb.* XXIV, pp. 369–404, 1896; Oppel, *Vergleichung des Entwickelungsgrades der Organe zu verschiedenen Entwickelungszeiten bei Wirbelthieren*, Jena, 1891; C. M. Jackson, Pre-natal growth of the human body and the relative growth of the various organs and parts, *Amer. Journ. of Anat.* IX, 1909; and of the albino rat, *ibid.* XV, 1913; L. A. Calkins, Growth of the human body in the foetal period, *Rep. Amer. Assoc. Anat.* 1921. For still more detailed measurements, see A. Arnold, Körperuntersuchungen an 1656 Leipziger Studenten, *Ztschr. f. Konstitutionslehre*, XV, pp. 43–113, 1929.

§ From Vierordt's *Anatomische Tabellen*, pp. 38, 39, much abbreviated.

Weight of various organs, compared with the total weight of the human body (male). (From Vierordt's Anatomische Tabellen)

Age	Wt. (kgm.)	Percentage increase: Body	Brain	Heart	Liver	Percentage of body-wt.: Brain	Heart	Liver
0	3·1	1·0	1·0	1·0	1·0	12·3	0·76	4·6
1	9·0	2·9	2·5	1·8	2·4	10·5	0·46	3·7
2	11·0	3·6	2·7	2·2	3·0	9·3	0·47	3·9
3	12·5	4·0	2·9	2·8	3·4	8·9	0·52	3·9
4	14·0	4·5	3·5	3·1	4·2	9·5	0·53	4·2
5	15·9	5·1	3·3	3·9	3·8	7·9	0·51	3·4
6	17·8	5·7	3·6	3·6	4·3	7·6	0·48	3·5
7	19·7	6·4	3·5	3·9	4·9	6·8	0·47	3·5
8	21·6	7·0	3·6	4·0	4·6	6·4	0·44	3·0
9	23·5	7·6	3·7	4·6	5·0	6·1	0·46	3·0
10	25·2	8·1	3·7	5·4	5·9	5·6	0·51	3·3
11	27·0	8·7	3·6	6·0	6·1	5·0	0·52	3·2
12	29·0	9·4	3·8	(4·1)	6·2	4·9	(0·34)	3·0
13	33·1	10·7	3·9	7·0	7·3	4·5	0·50	3·1
14	37·1	12·0	3·4	9·2	8·4	3·5	0·58	3·2
15	41·2	13·3	3·9	8·5	9·2	3·6	0·48	3·2
16	45·9	14·8	3·8	9·8	9·5	3·2	0·51	3·0
17	49·7	16·0	3·7	10·6	10·5	2·8	0·51	3·0
18	53·9	17·4	3·7	10·3	10·7	2·6	0·46	2·8
19	57·6	18·6	3·7	11·4	11·6	2·4	0·51	2·9
20	59·5	19·2	3·8	12·9	11·0	2·4	0·51	2·6
21	61·2	19·7	3·7	12·5	11·5	2·3	0·49	2·7
22	62·9	20·3	3·5	13·2	11·8	2·2	0·50	2·7
23	64·5	20·8	3·6	12·4	10·8	2·2	0·46	2·4
24	—	—	3·7	13·1	13·0	—	—	—
25	66·2	21·4	3·8	12·7	12·8	2·2	0·46	2·8

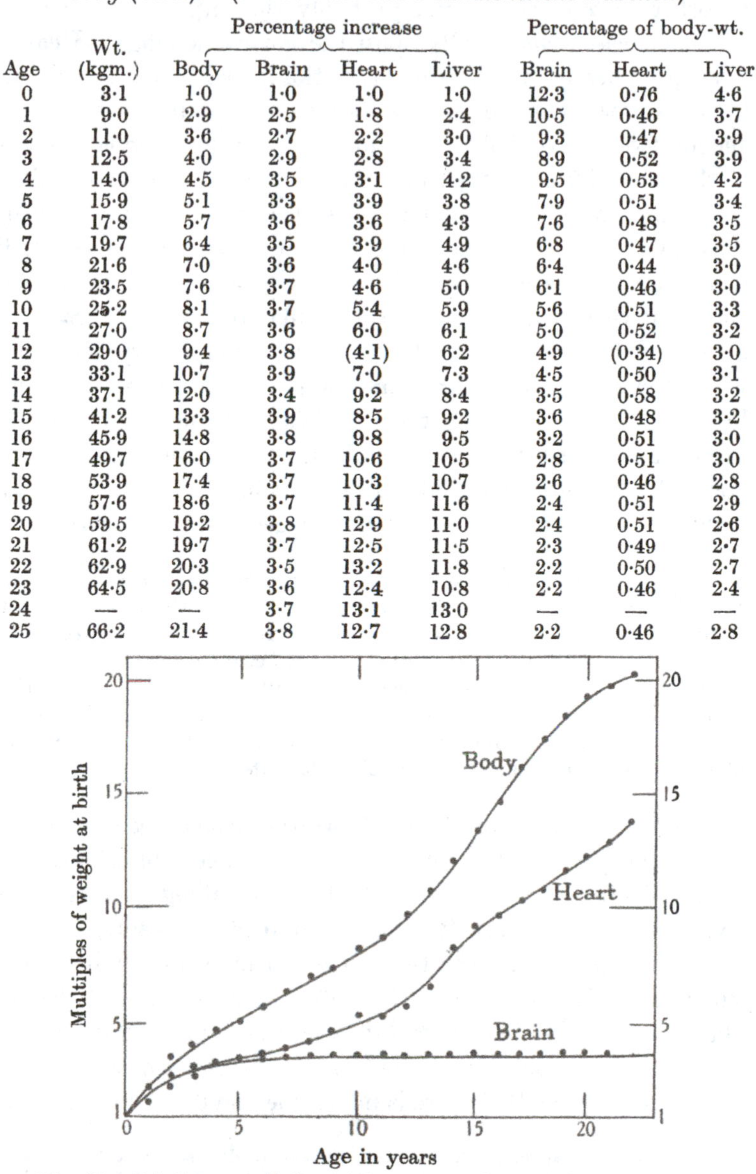

Fig. 47. Relative growth in weight of brain, heart and body of man. From Quetelet's data (smoothed curves).

We see that neither brain, heart nor liver keeps pace by any means with the growing weight of the whole; there must then be other parts of the fabric, probably the muscles and the bones, which increase *more* rapidly than the general average. Heart and liver grow nearly at the same rate, the liver keeping a little ahead to begin with, and the heart making up on it in the end; by the age of twenty-five both have multiplied their original weight at birth about thirteen times, but the body as a whole has multiplied by twenty-one. In contrast to these the brain has only multiplied its weight about three and three-quarter times, and shews but little increase since the child was four or five, and hardly any since it was eight years old. Man and the gorilla are born with brains much of a size; but the gorilla's brain stops growing very soon indeed, while the child's has four years of steady increase. The child's brain grows quicker than the gorilla's, but the great ape's body grows much quicker than the child's; at four years old the young gorilla has reached about 80 per cent. of his bodily stature, and the child's *brain* has reached about 80 per cent. of its full size.

Even during foetal life, as well as afterwards, the *relative* weight of the brain keeps on declining. It is about 18 per cent. of the body-weight in the third month, 16 per cent. in the fourth, 14 per cent. in the fifth; and the ratio falls slowly till it comes to about 12 per cent. at birth, say 10 per cent. a year afterwards, and little more than 2 per cent. at twenty*. Many statistics indicate a further decrease of brain-weight, actual as well as relative. The fact has been doubted and denied; but Raymond Pearl has shewn evidence of a slow decline continuing throughout adult life†.

The latter part of the table shews the decreasing weights of the organs compared with the body as a whole: brain, which was 12 per cent. of the body-weight at birth, falling to 2 per cent. at five-and-twenty; heart from 0·76 to 0·46 per cent.; liver from 4·6 to 2·78 per cent. The thyroid gland (as we know it in the rat) grows for a few weeks, and then diminishes during all the rest of the creature's lifetime; even during the brief period of its own growth it is growing slower than the body as a whole.

It is plain, then, that there is no simple and direct relation, holding

* Cf. J. Ariens Kappers, *Proc. K. Akad. Wetensch., Amsterdam*, XXXIX, No. 7, 1936.

† R. Pearl, Variation and correlation in brain-weight, *Biometrika*, IV, pp. 13–104, 1905.

good throughout life, between the size of the body and its organs; and the ratio of magnitude tends to change not only as the individual grows, but also with change of bodily size from one individual, one race, one species to another. In giant and pigmy breeds of rabbits, the organs have by no means the same ratio to the body-weight; but if we choose individuals of the same weight, then the ratios tend to be identical, irrespective of breed*. The larger breeds of dogs are for the most part lighter and slenderer than the small, and the organs change their proportions with their size. The spleen keeps pace with the weight of the body; but the liver, like the brain, becomes relatively less. It falls from about 6 per cent. of the body-weight in little dogs to rather over 2 per cent. in a great hound†.

The changing ratio with increasing magnitude is especially marked in the case of the brain, which constitutes (as we have just seen) an eighth of the body-weight at birth, and but one-fiftieth at twenty-five. This falling ratio finds its parallel in comparative anatomy, in the general law that the larger the animal the smaller (relatively) is the brain‡. A falling ratio of brain-weight during life is seen in other animals. Max Weber§ tells us that in the lion, at five weeks, four months, eleven months and lastly when full-grown, the brain represents the following fractions of the weight of the body: viz. 1/18, 1/80, 1/184 and 1/546. And Kellicott has shewn that in the dogfish, while certain organs, e.g. pancreas and rectal gland, grow *pari passu* with the body, the brain grows in a diminishing ratio, to be represented (roughly) by a logarithmic curve||.

In the grown man, Raymond Pearl has shewn brain-weight to increase with the stature of the individual and to decrease with his age, both in a straight-line ratio, or *linear regression*, as the

* R. C. Robb, Hereditary size-limitation in the rabbit, *Journ. Exp. Biol.* VI, 1929.

† Cf. H. Vorsteher, *Einfluss d. Gesamtgrösse auf die Zusammensetzung des Körpers*; Diss., Leipzig, 1923.

‡ Oliver Goldsmith argues in his *Animated Nature* as follows, regarding the unlikelihood of dwarfs or giants: "Had man been born a dwarf, he could not have been a reasonable creature; for to that end, he must have a jolt head, and then he would not have body and blood enough to supply his brain with spirits; or if he had a small head, proportionable to his body, there would not be brain enough for conducting life. But it is still worse with giants, etc."

§ *Die Säugethiere*, p. 117.

|| *Amer. Journ. of Anatomy*, VIII, pp. 319–353, 1908.

statisticians call it. Thus the following wholly empirical equations give the required ratios in the case of Swedish males:

$$\text{Brain-weight (gms.)} = 1487{\cdot}8 - 1{\cdot}94 \times \text{age, or}$$
$$= 915{\cdot}06 + 2{\cdot}86 \times \text{stature.}$$

In the two sexes, and in different races, these empirical constants will be greatly changed*; and Donaldson has further shewn that correlation between brain-weight and body-weight is much closer in the rat than in man†.

	Weight of entire animal (gm.) W	Weight of brain (gm.) w	Ratios $w : W$	Ratios $\sqrt[2]{w} : \sqrt[3]{W}$	In $w^n = W$
Marmoset	335	12·5	1 : 26	1 : 2·0	n = 2·30
Spider monkey	1,845	126	15	1·1	1·56
Felis minuta	1,234	23·6	52	1·2	2·25
F. domestica	3,300	31	107	2·4	2·36
Leopard	27,700	164	168	1·2	2·00
Lion	119,500	219	546	1·3	2·17
Dik-dik	4,575	37	124	2·7	2·30
Steinbok	8,600	49·5	173	2·9	2·32
Impala	37,900	148·5	255	2·75	2·11
Wildebeest	212,200	443	479	2·8	2·01
Zebra	255,000	541	472	2·7	1·98
,,	297,000	555	536	2·8	2·00
Rhinoceros	765,000	655	1170	3·6	2·09
Elephant	3,048,000	5,430	560	2·0	1·74
Whale (*Globiocephalus*)	1,000,000	2,511	400	2·0	1·77
			Mean	2·23	2·06

Brandt, a very philosophical anatomist, argued some seventy years ago that the brain, being essentially a hollow structure, a surface rather than a mass, ought to be equated with the surface rather than the mass of the animal. This we may do by taking the square-root of the brain-weight and the cube-root of the body-weight; and while the ratios so obtained do not point to *equality*, they do tend to *constancy*, especially if we limit our comparison to similar or related animals. Or we may vary the method, and ask (as Dubois has done) to what power the brain-weight must be raised

* *Biometrika*, IV, pp. 13–105, 1904.

† H. H. Donaldson, A comparison of the white rat with man, etc., *Boas Memorial Volume*, New York, 1906, pp. 5–26.

to equal the body-weight; and here again we find the same tendency towards uniformity*.

The converse to the unequal growth of organs is found in their unequal loss of weight under starvation. Chossat found, in a well-known experiment, that a starved pigeon had lost 93 per cent. of its fat, about 70 per cent. of liver and spleen, 40 per cent. of its muscles, and only 2 per cent. of brain and nervous tissues†. The salmon spends many weeks in the river before spawning, without taking food. The muscles waste enormously, but the reproductive bodies continue to grow.

As the internal organs of the body grow at different rates, so that their ratios one to another alter as time goes on, so is it with those linear dimensions whose inconstant ratios constitute the changing form and proportions of the body. In one of Quetelet's tables he shews the span of the outstretched arms from year to year, compared with the vertical stature. It happens that height and span are so nearly co-equal in man that direct comparison means little; but the *ratio* of span to height (Fig. 48) undergoes a significant and remarkable change. The man grows faster in stretch of arms than he does in height, and span which was less at birth than stature by about 1 per cent. exceeds it by about 4 per cent. at the age of twenty. Quetelet's data are few for later years, but it is clear enough that span goes on increasing in proportion to stature. How far this is due to actual growth of the arms and how far to increasing breadth of the chest is another story, and is not yet ascertained.

* Cf. A. Brandt, Sur le rapport du poids du cerveau à celui du corps chez différents animaux, *Bull. de la Soc. Imp. des naturalistes de Moscou*, XL, p. 525, 1867; J. Baillanger, De l'étendu de la surface du cerveau, *Ann. Med. Psychol.* XVII, p. 1, 1853; Th. van Bischoff, *Das Hirngewicht des Menschen*, Bonn, 1880 (170 pp.), cf. *Biol. Centralbl.* I, pp. 531–541, 1881; E. Dubois, On the relation between the quantity of brain and the size of the body, *Proc. K. Akad. Wetensch., Amsterdam*, XVI, 1913. Also, Th. Ziehen, Maszverhältnisse des Gehirns, in Bardeleben's *Handb. d. Anatomie des Menschen*; P. Warneke, Gehirn u. Körpergewichtsbestimmungen bei Säugern, *Journ. f. Psychol. u. Neurol.* XIII, pp. 355–403, 1909; B. Klatt, Studien zum Domestikationsproblem, *Bibliotheca genetica*, II, 1921; etc. The case of the heart is somewhat analogous; see Parrot, *Zool. Jahrb.* (System.), VII, 1894; Platt, in *Biol. Centralbl.* XXXIX, p. 406, 1919.

† C. Chossat, Recherches sur l'inanition, *Mém. Acad. des Sci.*, Paris, 1843, p. 438.

The growth-rates of head and body differ still more; for the height of the head is no more than doubled, but stature is trebled,

*Height of the head in man at various ages**
(*After Quetelet, p.* 207, *abbreviated*)

	Men			Women		
Age	Stature m.	Head m.	Ratio	Stature m.	Head m.	Ratio
Birth	0·50	0·11	4·5	0·49	0·11	4·4
1 year	0·70	0·15	4·5	0·69	0·15	4·5
2 years	0·79	0·17	4·6	0·78	0·17	4·5
3 ,,	0·86	0·18	4·7	0·85	0·18	4·7
5 ,,	0·99	0·19	5·1	0·97	0·19	5·1
10 ,,	1·27	0·21	6·2	1·25	0·20	6·2
20 ,,	1·51	0·22	7·0	1·49	0·21	7·0
25 ,,	1·67	0·23	7·3	1·57	0·22	7·1
30 ,,	1·69	0·23	7·4	1·58	0·22	7·1
40 ,,	1·69	0·23	7·4	1·58	0·22	7·1

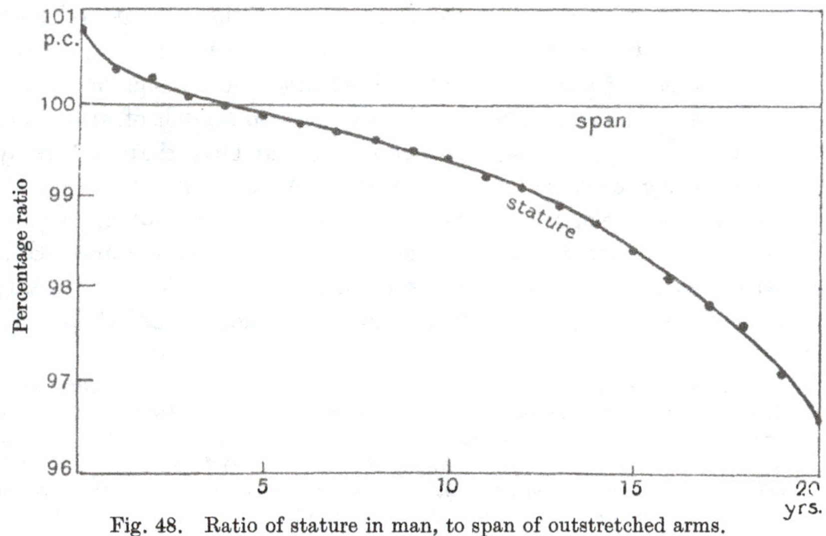

Fig. 48. Ratio of stature in man, to span of outstretched arms. From Quetelet's data.

between infancy and manhood. Dürer studied and illustrated this remarkable phenomenon, and the difference which accompanies and

* A smooth curve, very similar to this, is given by Karl Pearson for the growth in "auricular height" of the girl's head, in *Biometrika*, III, p. 141, 1904.

results from it in the bodily form of the child and the man is easy to see.

The following table shews the relative sizes of certain parts and organs of a young trout during its most rapid development; and so illustrates in a simple way the varying growth-rates in different parts of the body*. It would not be difficult, from a picture of the little trout at any one of these stages, to draw its approximate form at any other by the help of the numerical data here set forth. In like manner a herring's head and tail grow longer, the parts between grow relatively less, and the fins change their places a little; the same changes take place with their specific differences in related fishes, and herring, sprat and pilchard owe their specific characters to their rates of growth or modes of increment†.

Trout (Salmo fario)*: proportionate growth of various organs*
(*From Jenkinson's data*)

Days old	Total length	Eye	Head	1st dorsal	Ventral fin	2nd dorsal	Tail fin	Breadth of tail
40	100	100	100	100	100	100	100	100
63	130	129	148	149	149	108	174	156
77	155	147	189	(204)	(194)	139	258	220
92	173	179	220	(193)	(182)	155	308	272
106	195	193	243	173	165	173	337	288

Sachs studied the same phenomenon in plants, after a method in use by Stephen Hales a hundred and fifty years before. On the growing root of a bean ten narrow zones were marked off, starting from the apex, each zone a millimetre long. After twenty-four hours' growth (at a given temperature) the whole ten zones had grown from 10 to 33 mm., but the several zones had grown very unequally, as shewn in the annexed table‡ (p. 192):

* Cf. J. W. Jenkinson, Growth, variability and correlation in young trout, *Biometrika*, VIII, pp. 444–466, 1912.

† Cf. E. Ford, On the transition from larval to adolescent herring, *Journ. Mar. Biol. Assoc.* XVI, p. 723; XVIII, p. 977, 1930–31. So also in larval eels, tail and body grow at different rates, which rates differ in different species; cf. Johannes Schmidt, *Meddel. Kommiss. Havsundersok.* 1916; L. Bertin, *Bull. Zool. France*, 1926, p. 327.

‡ From Sachs's *Textbook of Botany*, 1882, p. 820.

Graded growth of bean-root

Zone	Increment mm.	Zone	Increment mm.
Apex	1·5	6th	1·3
2nd	5·8	7th	0·5
3rd	8·2	8th	0·3
4th	3·5	9th	0·2
5th	1·6	10th	0·1

"...I marked in the same manner as the Vine, young Honeysuckle shoots, etc....; and I found in them all a gradual scale of unequal extensions, those parts extending most which were tenderest," *Vegetable Staticks*, Exp. cxxiii.

The lengths attained by the successive zones lie very nearly on a smooth curve or gradient; for a certain law, or principle of continuity, connects and governs the growth-rates along the growing axis. This curve has its family likeness to those differential

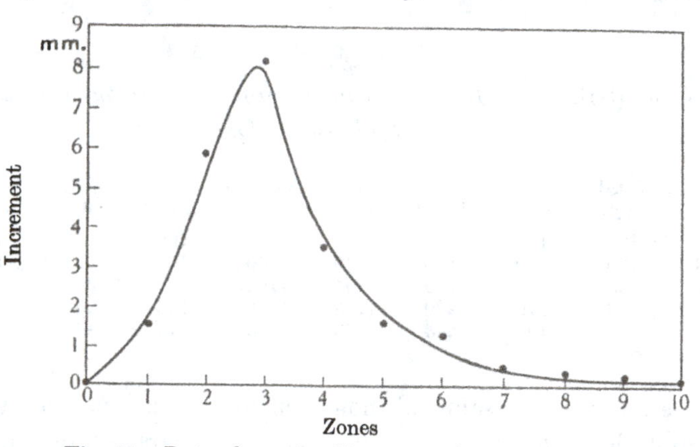

Fig. 49. Rate of growth of bean-root, in successive zones of 1 mm. each, beginning at the tip.

curves which we have already studied, in which rate of growth was plotted against time, as here it is plotted against successive spatial intervals of a growing structure; and its *general* features are those of a curve, a skew curve, of error. Had the several growth-rates been transverse to the axis, instead of being longitudinal and parallel to it, they would have given us a leaf-shaped structure, of which our curve would represent the outline on either side; or again, if growth had been symmetrical about the axis, it might have given us a turnip-shaped solid of revolution. There is always an easy passage from growth to form.

A like problem occurs when we deal with rates of growth in successive natural internodes; and we may then pass from the actual growth of the internodes to the varying number of leaves which they successively produce. Where we have whorls of leaves at each node, as in *Equisetum* or in many water-weeds, then the problem is simplified; and one such case has been studied by Raymond Pearl*. In *Ceratophyllum* the mean number of leaves increases with each successive whorl, but the rate of increase diminishes from whorl to whorl as we ascend. On the main stem the rate of change is very slow; but in the small twigs, or tertiary branches, it becomes rapid, as we see from the following abbreviated table:

Number of leaves per whorl on the tertiary branches of Ceratophyllum

Order of whorl ...	1	2	3	4	5	6
Mean no. of leaves	6·55	8·07	9·00	9·20	9·75	10·00
Smoothed no.	6·5	8·0	9·0	9·5	9·8	10·0

Raymond Pearl gives a logarithmic formula to fit the case; but the main point is that the numbers form a *graded series*, and can be plotted as a simple curve.

In short, a large part of the morphology of the organism depends on the fact that there is not only an average, or aggregate, rate of growth common to the whole, but also a *gradation* of rate from one part to another, tending towards a specific rate characteristic of each part or organ. The least change in the ratio, one to another, of these partial or localised rates of growth will soon be manifested in more and more striking differences of form; and this is as much as to say that the time-element, which is implicit in the idea of *growth*, can never (or very seldom) be wholly neglected in our consideration of form†.

A flowering spray of Montbretia or lily-of-the-valley exemplifies a growth-gradient, after a simple fashion of its own. Along the

* On variation and differentiation in *Ceratophyllum*, *Carnegie Inst. Publications*, No. 58, 1907; see p. 87.

† Herein lies the easy answer to a contention raised by Bergson, and to which he ascribes much importance, that "a mere variation of size is one thing, and a change of form is another." Thus he considers "a change in the form of leaves" to constitute "a profound morphological difference" (*Creative Evolution*, p. 71).

stalk the growth-rate falls away; the florets are of descending age, from flower to bud: their graded differences of age lead to an exquisite gradation of size and form; the time-interval between one and another, or the "space-time relation" between them all, gives a peculiar quality—we may call it phase-beauty—to the whole. A clump of reeds or rushes shews this same phase-beauty, and so do the waves on a cornfield or on the sea. A jet of water is not much, but a fountain becomes a beautiful thing, and the play of many fountains is an enchantment at Versailles.

On the weight-length coefficient, or ponderal index

So much for the visible changes of form which accompany advancing age, and are brought about by a diversity of rates of growth at successive points or in different directions. But it often happens that an animal's change of form may be so gradual as to pass unnoticed, and even careful measurement of such small changes becomes difficult and uncertain. Sometimes one dimension is easily determined, but others are hard to measure with the same accuracy. The length of a fish is easily measured; but the breadth and depth of plaice or haddock are vaguer and more uncertain. We may then make use of that ratio of weight to length which we spoke of in the last chapter: viz. that $W \propto L^3$, or $W = kL^3$, or $W/L^3 = k$, where k, the "ponderal index," is a constant to be determined for each particular case*.

We speak of this k as a "constant," with a mean value specific to each species of animal and dependent on the bodily proportions or form of that animal; yet inasmuch as the animal is continually apt to change its bodily proportions during life, k also is continually subject to change, and is indeed a very delicate index of such

* This relation, and how important it is, were clearly recognised by Herbert Spencer in his *Recent Discussions in Science, etc.*, 1871. The formula has been often, and often independently, employed: first perhaps in the form $\frac{\sqrt[3]{W}}{L} \times 100$, by R. Livi, L'indice ponderale, o rapporto tra la statura e il peso, *Atti Soc. Romana Antropologica*, v, 1897. Values of k for man and many animals are given by H. Przibram, in *Form und Formel*, 1922. On its use as an index to the condition or habit of body of an individual, see von Rhode, in Abderhalden's *Arbeitsmethoden*, IX, 4. The constant k might be called, more strictly, k_l, leaving k_b and k_d for the similar constants to be derived from the *breadth* and *depth* of the fish.

progressive changes: delicate—because our measurements of length are very accurate on the whole, and weighing is a still more delicate method of comparison.

Thus, in the case of plaice, when we deal with mean values for large numbers and with samples so far "homogeneous" that they are taken at one place and time, we find that k is by no means constant, but varies, and varies in an orderly way, with increasing size of the fish. The phenomenon is unexpectedly complex, much more so than I was aware of when I first wrote this book. Fig. 50

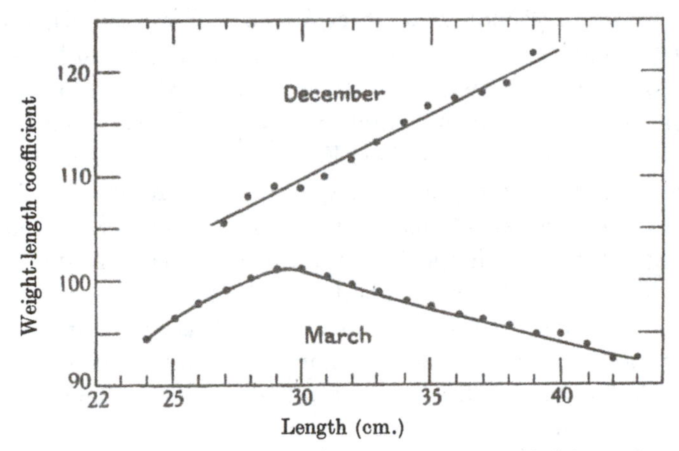

Fig. 50. Changes in the weight-length coefficient of plaice with increasing size; from March and December samples.

shews the weight-length coefficient, or ponderal index, in two large samples, one taken in the month of March, the other in December. In the latter sample k increases steadily as the plaice grow from about 25 to 40 cm. long; weight, that is to say, increases more rapidly than the cube of the length, and it follows that length itself is increasing *less* rapidly than some other linear dimension. In other words, the plaice grow thicker, or bulkier, with length and age. The other sample, taken in the month of March, is curiously different; for now k rises to a maximum when the fish are somewhere about 30 cm. long, and then declines slowly with further increase in size of the fish; and k itself is less in March than in December, the discrepancy being slight in the small fish and great in the large. The "point of inflection" at 30 cm. or thereby marks an epoch in the fish's life; it is about the size when sexual maturity begins, or at least near enough to suggest a connection between the two phenomena*.

A step towards further investigation would be to determine k for the two sexes separately, and to see whether or no the point of inflection occurs, as maturity is known to be reached, at a smaller size in the male. This d'Ancona has done, not for the plaice but for the shad (*Alosa finta*). He finds that the males are the first to reach maturity, first to shew a retardation of the rate of growth, first to reach a maximal value of the ponderal index, and in all probability the first to die†.

Again we may enquire whether, or how, k varies with the time of year; and this correlation leads to a striking result‡. For the ponderal index fluctuates periodically with the seasons, falling steeply to a minimum in March or April, and rising slowly to an annual maximum in December (Fig. 51)§. The main and obvious explanation lies in the process of spawning, the rapid loss of weight thereby, and the slow subsequent rebuilding of the reproductive tissues; whence it follows that, without ever seeing the fish spawn, and without ever dissecting one to see the state of its reproductive system, we may by this statistical method ascertain its spawning season, and determine the beginning and end thereof with considerable accuracy. But all the while a similar fluctuation, of much less amplitude, is to be found in young plaice before the spawning age; whence we learn that the fluctuation is not only due to shedding and replacement of spawn, but in part also to seasonal changes in appetite and general condition.

Returning to our former instance, we now see that the March and December samples of plaice, which shewed such discrepant variations of the ponderal index with increasing size, happen to

* The carp shews still more striking changes than does the plaice in the weight-length coefficient: in other words, still greater changes in bodily shape with advancing age and increasing size; cf. P. H. Struthers, *The Champlain Watershed*, Albany, New York, 1930.

† U. d'Ancona, Il problema dell' accrescimento dei pesci, etc., *Mem. R. Acad. dei Lincei* (6), II, pp. 497–540, 1928.

‡ Cf. Lämmel, Ueber periodische Variationen in Organismen, *Biol. Centralbl.* XXII, pp. 368–376, 1903.

§ When we restrict ourselves, for simplicity's sake, to fish of one particular size, we need not determine the values of k, for changes in weight are obvious enough; but when we have small numbers and various sizes to deal with, the determination of k helps very much.

coincide with the beginning and end of the spawning season; the fish were full of spawn in December, but spent and lean in March. The weight-length ratio was, of necessity, higher at the former season; and the falling-off in condition, and in bulk, which the March sample indicates, is more and more pronounced in the larger and therefore more heavily spawn-laden fish.

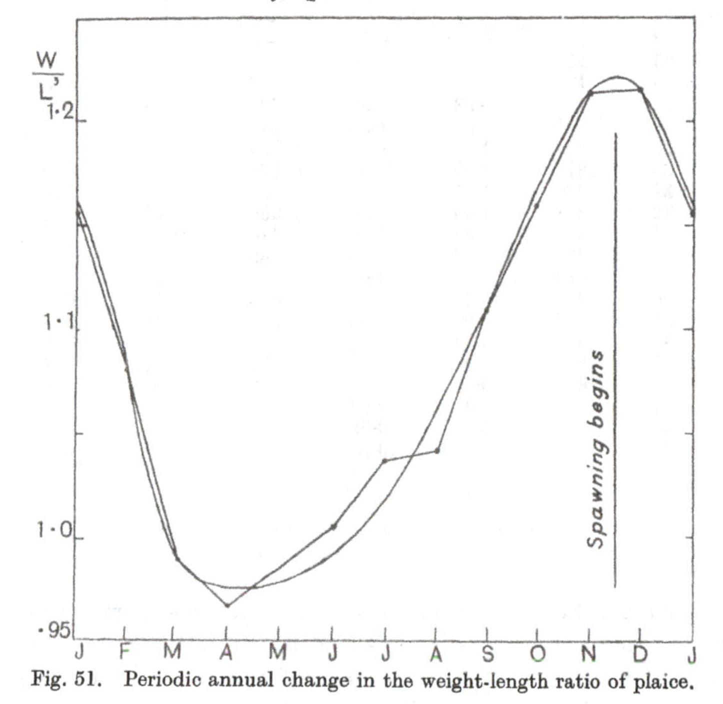

Fig. 51. Periodic annual change in the weight-length ratio of plaice.

Periodic relation of weight to length in plaice of 55 *cm. long*

	Average weight decigrams	W/L^3	W/L^3 (smoothed)
Jan.	204	1·23	1·16
Feb.	174	1·04	1·08
March	162	0·97	0·99
April	159	0·95	0·97
May	162	0·98	0·98
June	171	1·03	1·01
July	169	1·01	1·04
August	178	1·07	1·04
Sept.	173	1·04	1·11
Oct.	203	1·22	1·16
Nov.	203	1·22	1·21
Dec.	200	1·20	1·22
Mean	180	1·08	

Plaice caught in a certain area, March 1907 *and December* 1905.
Variation of k, the weight-length coefficient, with size

	March sample			December sample		
cm.	gm.	W/L^3	Do. smoothed	gm.	W/L^3	Do. smoothed
23	113	0·93	—	—	—	—
24	128	0·93	0·94	—	—	—
25	152	0·97	0·96	—	—	—
26	178	0·96	0·98	177	1·01	—
27	193	0·98	0·99	209	1·06	1·06
28	221	1·01	1·00	241	1·10	1·08
29	250	1·02	1·01	264	1·08	1·09
30	271	1·00	1·01	294	1·09	1·09
31	300	1·01	1·00	325	1·09	1·10
32	328	1·00	1·00	366	1·12	1·12
33	354	0·99	0·99	410	1·14	1·13
34	384	0·98	0·98	449	1·14	1·15
35	419	0·98	0·98	501	1·17	1·17
36	454	0·97	0·97	556	1·19	1·17
37	492	0·95	0·96	589	1·16	1·18
38	529	0·96	0·96	652	1·19	1·19
39	564	0·95	0·95	719	1·21	1·22
40	614	0·96	0·95	809	1·26	—
41	647	0·94	0·94	—	—	—
42	679	0·92	0·93	—	—	—
43	732	0·92	0·93	—	—	—
44	800	0·94	0·94	—	—	—
45	875	0·96	—	—	—	—

These weights and measurements of plaice are taken from the Department of Agriculture and Fisheries' *Plaice-Report*, I, pp. 65, 107, 1908; II, p. 92, 1909.

Japanese goldfish* are exposed to a much wider range of temperature than our plaice are called on to endure; they hibernate in winter and feed greedily in the heat of summer. Their weight is low in winter but rises in early spring, it falls as low as ever at the height of the spawning season in the month of May; so for one weight-length fluctuation which the plaice has, the goldfish has a twofold cycle in the year. The index reaches its second and higher maximum in August, and falls thereafter till the end of the year. That it should begin to fall so soon, and fall so quickly, merely means that late autumn is a time of *growth*; the fish are not losing weight, but growing longer†.

* Cf. Kichiro Sasaki, *Tohoku Sci. Reports* (4), I, pp. 239–260, 1926.

† Much has been written on the weight-length index in fishes. See (*int. al.*) A. Meek, The growth of flatfish, *Northumberland Sea Fisheries Ctee*, 1905, p. 58; W. J. Crozier, Correlations of weight, length, etc., in the weakfish, *Cynoscion*

It is the rule in fishes and other cold-blooded vertebrates that growth is asymptotic and size indeterminate, while in the warm-blooded growth comes, sooner or later, to an end. But the characteristic form is established earlier in the former case, and changes less, save for the minor fluctuations we have spoken of. In the higher animals, such as ourselves, the whole course of life is attended by constant alteration and modification of form; and

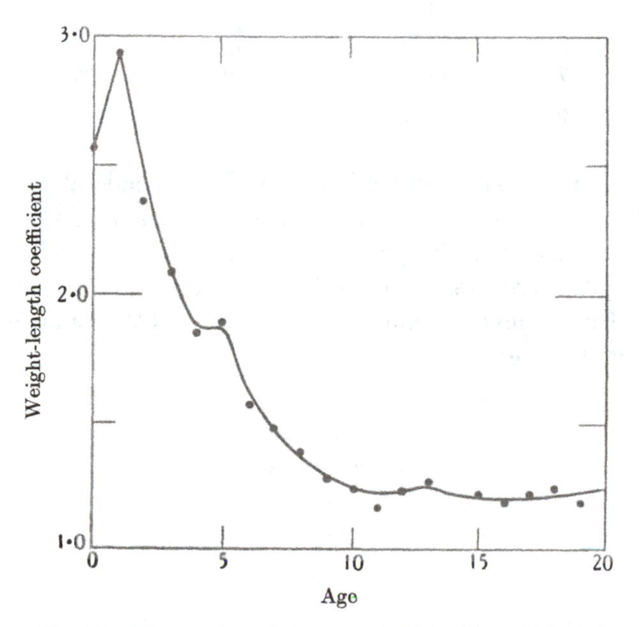

Fig. 52. The ponderal index, or weight-length coefficient, in man. From Quetelet's data.

we may use our weight-length formula, or ponderal index, to illustrate (for instance) the changing relation between height and weight in boyhood, of which we spoke before (Fig. 52).

regalis, *Bull. U.S. Bureau of Fisheries*, XXXIII, pp. 141–147, 1913; Selig Hecht, Form and growth in fishes, *Journ. of Morphology*, XXVII, pp. 379–400, 1916; J. Johnstone (Plaice), *Trans. Liverpool Biolog. Soc.* XXV, pp. 186–224, 1911; J. J. Tesch (Eel), *Journ. du Conseil*, III, 1927; Frances N. Clark (Sardine), *Calif. Fish. Bulletin*, No. 19, 1928 (with full bibliography). For a discussion on statistical lines, apart from any assumptions such as the "law of the cubes," see G. Duncker, Korrelation zwischen Länge u. Gewicht, etc., *Wissensch. Meeresuntersuch. Helgoland*, XV, pp. 1–26, 1923.

The weight-length coefficient, or ponderal index, k, in young Belgians (From Quetelet's figures)

Age (years)	W/L^3	Age (years)	W/L^3
0	2·55	10	1·25
1	2·92	11	1·18
2	2·34	12	1·23
3	2·08	13	1·29
4	1·87	14	1·23
5	1·72	15	1·23
6	1·56	16	1·28
7	1·48	20	1·30
8	1·39	25	1·36
9	1·29		

The infant is plump and chubby, and the ponderal index is at its highest at a year old. As the boy grows, it is in stature that he does so most of all; his ponderal index falls continually, till the growing years are over, and the lad "fills out" and grows stouter again. During prenatal life the index varied little, and less than we might suppose:

Relation between length and weight of the human foetus (From Scammon's data)

Length cm.	Weight gm.	W/L^3
7·7	13	2·9
12·3	41	2·2
17·3	115	2·2
22·3	239	2·2
27·2	405	2·0
32·3	750	2·2
37·2	1163	2·3
42·2	1758	2·3
46·9	2389	2·3
51·7	3205	2·3

As a further illustration of the rate of growth, and of unequal growth in various directions, we have figures for the ox, extending over the first three years of the animal's life, and giving (1) the weight of the animal, month by month, (2) the length of the back, from occiput to tail, and (3) the height to the withers. To these I have added (4) the ratio of length to height, (5) the weight-length coefficient, k, and (6) a similar coefficient, or index-number, k', for the height of the animal. All these ratios change as time goes on. The ratio of length to height increases, at first considerably, for the legs seem disproportionately long at birth in the ox, as in other

Relations between the weight and certain linear dimensions of the ox (Data from Cornevin, abbreviated)*

Age months	Weight kgm.	Length of back m.	Height m.	L/H	$k = W/L^3$	$k' = W/H^3$
0	37	0·78	0·70	1·11	0·78	1·08
1	55	0·94	0·77	1·22	0·66	1·21
2	86	1·09	0·85	1·28	0·67	1·41
3	121	1·21	0·94	1·28	0·69	1·46
4	150	1·31	0·95	1·38	0·66	1·75
5	179	1·40	1·04	1·35	0·65	1·60
6	210	1·48	1·09	1·36	0·64	1·64
7	247	1·52	1·12	1·36	0·70	1·75
8	267	1·58	1·15	1·38	0·68	1·79
9	283	1·62	1·16	1·39	0·66	1·80
10	304	1·65	1·19	1·39	0·68	1·79
11	328	1·69	1·22	1·39	0·67	1·79
12	351	1·74	1·24	1·40	0·67	1·85
13	375	1·77	1·25	1·41	0·68	1·90
14	391	1·79	1·26	1·41	0·69	1·94
15	406	1·80	1·27	1·42	0·69	1·98
16	418	1·81	1·28	1·42	0·70	2·09
17	424	1·83	1·29	1·42	0·69	1·97
18	424	1·86	1·30	1·43	0·66	1·94
19	428	1·88	1·31	1·44	0·65	1·92
20	438	1·88	1·31	1·44	0·66	1·94
21	448	1·89	1·32	1·43	0·66	1·94
22	464	1·90	1·33	1·43	0·68	1·96
23	481	1·91	1·35	1·42	0·69	1·98
24	501	1·91	1·35	1·42	0·71	2·03
25	521	1·92	1·36	1·41	0·74	2·08
26	534	1·92	1·36	1·41	0·75	2·12
27	547	1·93	1·36	1·41	0·76	2·16
28	555	1·93	1·36	1·41	0·77	2·19
29	562	1·93	1·36	1·41	0·78	2·22
30	586	1·95	1·38	1·41	0·79	2·22
31	611	1·97	1·40	1·40	0·80	2·21
32	626	1·98	1·42	1·40	0·80	2·19
33	641	2·00	1·44	1·39	0·81	2·16
34	656	2·01	1·45	1·38	0·81	2·13

ungulate animals; but this ratio reaches its maximum and falls off a little during the third year: so indicating that the beast is growing more in height than length, at a time when growth in both

* Ch. Cornevin, Études sur la croissance, *Arch. de Physiol. norm. et pathol.* (5), IV, p. 477, 1892. Cf. also R. Gärtner, Ueber das Wachstum d. Tiere, *Landwirtsch. Jahresber.* LVII, p. 707, 1922.

dimensions is nearly over*. The ratio W/H^3 increases steadily, and at three years old is double what it was at birth. It is the most variable of the three ratios; and it so illustrates the somewhat obvious but not unimportant fact that k varies most for the dimension which varies least, or grows most uniformly; in other words, that the values of k, as determined at successive epochs for any one dimension, are a measure of the variability of the other two.

The same ponderal index serves as an index of "build," or bodily proportion; and its mean values have been determined for various ages and for many races of mankind. Within one and the

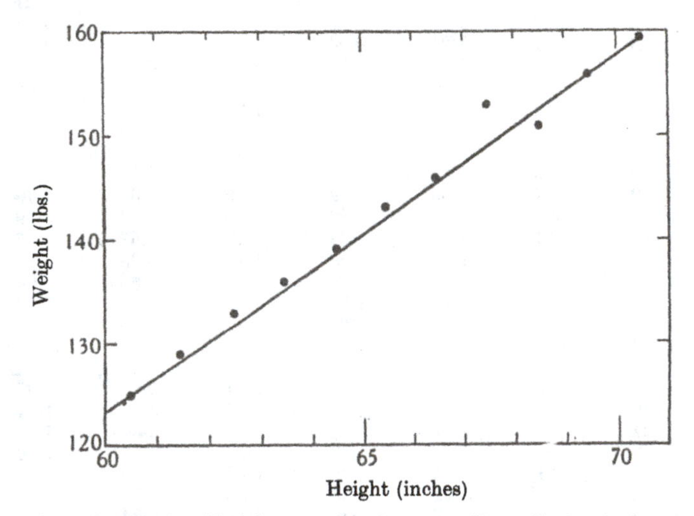

Fig. 53. Ratio of height to weight in man. From Goringe's data.

same race it varies with stature; for tall men, and boys too, are apt to be slender and lean, and short ones to be thickset and strong. And so much does the weight-length ratio change with build or stature that, in the following table of mean heights and weights of men between five and six feet high, it will be seen that weight, instead of varying as the cube of the height, is (within the limits shewn) in nearly simple linear relation to it (Fig. 53)†.

* As a matter of fact, the data shew that the animal grows under 7 per cent. in length, but over 11 per cent. in height, between the twentieth and the thirtieth month of its age.

† Had the weights varied as the cube of the height, the tallest men should have weighed close on 200 lb., instead of 160 lb.

*Ratio of height to weight in man**

No. of instances	Height in.	Weight lb.	W/H	W/H^3
59	60·5	125	2·07	5·62
118	61·5	129	2·13	5·55
220	62·5	133	2·13	5·45
285	63·5	136	2·14	5·30
327	64·5	139	2·15	5·19
386	65·5	143	2·18	5·09
346	66·5	146	2·20	4·97
289	67·5	153	2·27	4·96
220	68·5	151	2·20	4·71
116	69·5	156	2·24	4·64
58	70·5	160	2·27	4·57

The same index may be used as a measure of the condition, even of the quality, of an animal; three Burmese elephants had the following heights, weights, and reputations†:

	Height	Weight	W/H^3	
A	7 ft. 10½ in.	7,511 lb.	1·54	A famous elephant
B	8 1	7,216	1·36	A good elephant
C	7 5	4,756	1·15	A weak, poor elephant

But a great African elephant, 10 ft. 10 in. high, weighed 14,640 lb.‡: whence the weight-height coefficient was no more than 1·15. That is to say, the African elephant is considerably taller than the Indian, and the weight-height ratio is correspondingly less.

Lastly, by means of the same index we may judge, to a first rough approximation, the weight of a large animal such as a whale, where weighing is out of the question. Sigurd Rusting has given us many measurements, and many foetal weights, from the Antarctic whale-fishery: among which, choosing at random, we find that a certain foetus of the blue whale, or Sibbald's rorqual, measured 4 ft. 6 in. long, and weighed 23 kilos, or say 46 lb. A whale of the same kind, 45 ft. long, should then weigh 46×10^3 lb., or about 23 tons; and one of 90 ft., 23×2^3 tons, or over 180 tons. Again in seven young unborn whales, measuring from 39 to 54 inches and weighing from 10 to 23 kilos, the mean value of the index was found

* Data from Sir C. Goringe, *The English Convict*, H.M. Stationery Office, 1913. See also J. A. Harris and others, *The Measurement of Man*, Minnesota, 1930, p. 41.

† Data from A. J. Milroy, *On the management of elephants*, Shillong, 1921.

‡ D. P. Quireng, in *Growth*, III, p. 9, 1939.

to be 15·2, in gramme-inches. From this we calculate the weight of the great rorqual, as follows:

$$\text{At 25 ft., or 300 inches, } W = \frac{15\cdot 2 \times 300^3}{100} = 4{,}100{,}000 \text{ g.}$$
$$= 4{,}100 \text{ kg.}$$
$$= 4 \text{ tons, nearly.}$$

$$\text{At } 50 \text{ ft., } W = 4 \times 2^3 \text{ tons} = 32 \text{ tons.}$$
$$100 \text{ ft.} = 32 \times 2^3 \text{ tons} = 256 \text{ tons.}$$
$$106 \text{ tons (the largest known)} \quad W = 305 \text{ tons, nearly.}$$

The two independent estimates are in close agreement.

Of surface and volume

While the weight-length relation is of especial importance, and is wellnigh fundamental to the understanding of growth and form and magnitude, the corresponding relation of surface-area to weight or volume has in certain cases an interest of its own. At the surface of an animal heat is lost, evaporation takes place, and oxygen may be taken in, all in due proportion as near as may be to the bulk of the animal; and again the bird's wing is a surface, the area of which *must* be in due proportion to the size of the bird. In hollow organs, such as heart or stomach, area is the important thing rather than weight or mass; and we have seen how the brain, an organ not obviously but essentially and developmentally hollow, tends to shew its due proportions when reckoned as a *surface* in comparison with the creature's *mass*.

Surface cannot keep pace with increasing volume in bodies of similar form; wing-area does not and cannot long keep pace with the bird's increasing bulk and weight, and this is enough of itself to set limits to the size of the flying bird. It is the ratio between square-root-of-surface and cube-root-of-volume which should, in theory, remain constant; but as a matter of fact this ratio varies (up to a certain extent) with the circumstances, and in the case of the bird's wing with varying modes and capabilities of flight. The owl, with his silent, effortless flight, capable of short swift spurts of attack, has the largest spread of wings of all; the kite outstrips the other hawks in spread of wing, in soaring, and perhaps in speed.

Stork and seagull have a great expanse of wing; but other skilled and speedy fliers have long narrow wings rather than large ones. The peregrine has less wing-area than the goshawk or the kestrel; the swift and the swallow have less than the lark.

Mean ratio, $\sqrt[2]{S}/\sqrt[3]{W}$, *between wing-area and weight of birds*
(From Mouillard's data)

		Ratio
Owls	1 species	2·2
Hawks	7 ,,	1·7
Gulls	1 ,,	1·7
Waders	3 ,,	1·7
Petrels	2 ,,	1·4
Plovers	3 ,,	1·4
Passeres	4 ,,	1·3
Ducks	2 ,,	1·2

To measure the length of an animal is easy, to weigh it is easier still, but to estimate its surface-area is another thing. Hence we know but little of the surface-weight ratios of animals, and what we know is apt to be uncertain and discrepant. Nevertheless, such data as we possess average down to mean values which are more uniform than we might expect*.

Mean ratio, $\sqrt[2]{S}/\sqrt[3]{W}$, *in various animals (cm. gm. units)*

Ape	11·8	Sheep (shorn)	8
Man	11	Snake	12·5
Dog	10–11	Frog	10·6
Cat, horse	10	Birds	10
Rabbit	9·75	Tortoise	10
Cow, pig, rat	9		

A further note on unequal growth, or heterogony

An organism is so complex a thing, and growth so complex a phenomenon, that for growth to be so uniform and constant in all the parts as to keep the whole shape unchanged would indeed be an unlikely and an unusual circumstance. Rates vary, proportions change, and the whole configuration alters accordingly. In so humble a creature as a medusoid, manubrium and disc grow at different rates, and certain sectors of the disc faster than others, as when the little *Ephyra*-larva "develops" into the great *Aurelia*-jellyfish. Many fishes grow from youth to age with no visible,

* From Fr. G. Benedict, Oberflächenbestimmung verschiedener Tiergattungen, *Ergebnisse d. Physiologie*, XXXVI, pp. 300–346, 1934 (with copious bibliography).

hardly a measurable, change of form*; but the shapes and looks of man and woman go on changing long after the growing age is over, even all their lives long. A centipede has its many pairs of legs alike, to all intents and purposes; they begin alike and grow uniformly. But a lobster has his great claws and his small, his lesser legs, his swimmerets and the broad flaps of his tail; all these begin alike, and diverse rates of growth make up the difference between them. Moreover, we may sometimes watch a single limb growing to an unusual size, perhaps in one sex and not in the other, perhaps on one side and not on the other side of the body: such are the "horns," or mandibles, of the stag-beetle, only conspicuous in the male, and the great unsymmetrical claws of the lobster, or of that extreme case the little fiddler-crab (*Uca pugnax*). For such well-marked cases of differential growth-ratio between one part and another, Julian Huxley has introduced the term *heterogony*†.

Of the fiddler-crabs some four hundred males were weighed, in twenty-five graded samples all nearly of a size, and the weights of the great claw and of the rest of the body recorded separately. To begin with the great claw was about 8 per cent., and at the end about 38 per cent., of the total weight of the unmutilated body. In the female the claw weighs about 8 per cent. of the whole from beginning to end; and this contrast marks the disproportionate, or heterogonic, rate of growth in the male. We know nothing about the actual rate of growth of either body or claw, we cannot plot either against time; but we know the relative proportions, or relative rates of growth of the two parts of the animal, and this is all that matters meanwhile. In Fig. 54, we have set off the successive weights of the body as abscissae, up to 700 mgm., or about one-third of its weight in the adult animal; and the ordinates represent the corresponding weights of the claw. We see that the ratio between the two magnitudes follows a curve, apparently an exponential curve; it does in fact (as Huxley has shewn) follow a compound

* Cf. S. Hecht, Form and growth in fishes, *Journ. Morphology*, XXVII, pp. 379–400, 1916; F. S. and D. W. Hammett, Proportional length-growth of garfish (*Lepidosteus*), *Growth*, III, pp. 197–209, 1939.

† See *Problems of Relative Growth*, 1932, and many papers quoted therein. The term, as Huxley tells us, had been used by Pézard; but it had been used, in another sense, by Rolleston long before to mean an alternation of generations, or production of offspring dissimilar to the parent.

interest law, which (calling y and x the weights of the claw and of the rest of the body) may be expressed by the usual formula for compound interest,

$$y = bx^k, \quad \text{or} \quad \log y = \log b + k \log x;$$

and the coefficients (b and k) work out in the case of the fiddler-crab, to begin with, at

$$y = 0{\cdot}0073\, x^{1{\cdot}62}.$$

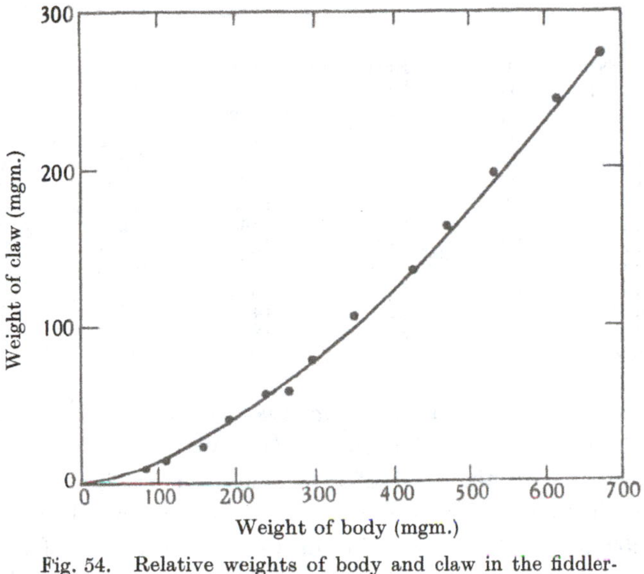

Fig. 54. Relative weights of body and claw in the fiddler-crab (*Uca pugnax*).

But after a certain age, or certain size, these coefficients no longer hold, and new coefficients have to be found. Whether or no, the formula is mathematical rather than biological; there is a lack of either biological or physical significance in a growth-rate which happens to stand, during part of an animal's life, at 62 per cent. compound interest.

Julian Huxley holds, and many hold with him, that the exponential or logarithmic formula, or the compound-interest law, is of general application to cases of differential growth-rates. I do not find it to be so: any more than we have found organ, organism or population to increase by compound interest or geometrical progression, save under exceptional circumstances and in transient phase. Undoubtedly many of Huxley's instances shew increase by compound interest, during a phase of rapid and unstinted growth; but I find many others following a simple-interest rather than a compound-interest law.

Relative weights of claw and body in fiddler-crabs (Uca pugnax). (*Data abbreviated from Huxley*, Problems of Relative Growth, *p.* 12)

Wt. of body less claw (mgm.)	Wt. of claw	Ratio %	Wt. of body	Wt. of claw	Ratio %
58	5	8·6	618	243	39·3
80	9	11·2	743	319	42·9
109	14	12·8	872	418	47·9
156	25	16·0	983	461	46·9
200	38	19·0	1080	537	49·7
238	53	22·3	1166	594	50·9
270	59	21·9	1212	617	50·9
300	78	26·0	1299	670	51·6
355	105	29·7	1363	699	51·3
420	135	32·1	1449	773	53·7
470	165	35·1	1808	1009	55·8
536	196	36·6	2233	1380	61·7

In the common stag-beetle (*Lucanus cervus*) we have the following measurements of mandible and elytron or wing-case: which two organs make up the bulk of, and may for our purpose be held as constituting, the "total length" of the beetle. Here a simple equation meets the case; in other words, the length of elytron or of mandible plotted against total length gives what is to all intents and purposes a straight line, indicating a simple-interest rather than a compound-interest rate of increase.

Measurements of 48 *stag-beetles* (Lucanus cervus)* (*mm.*)

Number of specimens	1	4	5	10	5	7	11	5
Length, total (x)	31·0	38·7	40·5	42·6	45·0	46·9	49·2	53·6
Length of elytron (y)	25·0	30·9	31·5	32·6	33·8	35·1	36·4	39·2
(,, calculated) (y')	26·9	30·8	31·7	32·8	34·0	35·0	36·2	38·5
Length of mandible (z)	6·0	7·8	9·0	10·0	11·2	11·9	12·8	14·4
(,, calculated) (z')	5·9	7·7	9·3	10·1	11·0	11·7	12·6	14·2

* Data, from Julian Huxley, after W. Bateson and H. H. Brindley, in *P.Z.S.* 1892, pp. 585–594.

From the observed data we may solve, by the method of least squares, the simple equations

$$y = a + bx, \quad z = c + dx,$$

or in other words, find the equations of the straight lines in closest agreement with the observed data. The solutions are as follows*:

$$y = 11{\cdot}02 + 0{\cdot}512x, \text{ and } z = -\,5{\cdot}64 + 0{\cdot}368x,$$

the two coefficients 0·368 and 0·512 signifying the difference between the rates of increase of the two organs. The number of samples is not very large, and some deviation is to be expected; nevertheless, the calculated straight lines come close to the observed values.

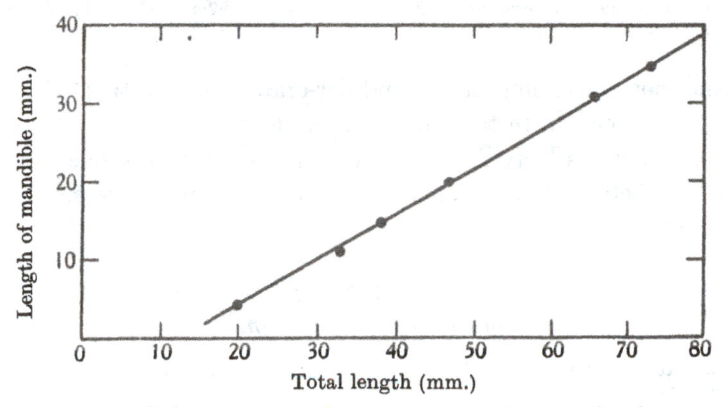

Fig. 55. Relative growth of body and mandible in reindeer-beetle (*Cyclommatus tarandus*).

The reindeer-beetle (*Cyclommatus tarandus*), belonging to the same family, shews much the same thing. The mandible grows in approximately linear ratio to the body, save that it tends to be at first a little above, and later on a little below, this linear ratio (Fig. 55).

Measurements of Cyclommatus tarandus† *(mm.)*

Length of mandible (y)	3·9	10·7	14·1	19·9	24·0	30·7	34·5
Total length (x)	20·4	33·1	38·4	47·3	54·2	66·1	74·0
Total length calculated: $x = 1{\cdot}7y + 13{\cdot}7$	20·3	31·9	37·7	47·5	54·5	65·9	72·4

* As determined for me by Dr A. C. Aitken, F.R.S.

† Data, much abbreviated, from Huxley, after E. Dudich, *Archiv f. Naturgesch.* (A), 1923.

The facial and cranial parts of a dog's skull tend to grow at different rates (Fig. 56); and changes in the ratio between the two go a long way to explain the differences in shape between one dog's skull and another's, between the greyhound's and the pug's. But using Huxley's own data (after Becher) for the sheepdog, I find the ratio between the facial and cranial portions of the skull to be, once again, a simple linear one.

Measurements of skull of sheep-dog (30 *specimens*)* (*mm.*)

Mean length of facial region (y)	22·0	48·3	58·0	73·5	89·1	102·0	112·0
Mean length of cranial region (x)	42·0	65·3	74·5	85·5	99·3	112·6	120·0
Calculated values for cranial region: $x = 22{\cdot}7 + 0{\cdot}88y$	42·1	65·2	73·7	87·4	97·1	112·5	121·2

And now, returning to the fiddler-crab, we find that after the crab has reached a certain size and the first phase of rapid growth is over, claw and body grow in simple linear relation to one another, and the heterogonic or compound-interest formula is no longer required:

Fiddler-crab (Uca pugnax): *ratio of growth-rates, in later stages, of claw and body* (*mgm.*)

Weight of body less claw (x)	872	983	1080	1165	1212	1291	1363	1449
Weight of large claw (y)	418	461	537	594	617	670	699	778
Do., calculated: $y = 0{\cdot}6x - 110$	413	480	538	590	617	665	708	759

* Data from A. Becher, in *Archiv f. Naturgesch.* (A), 1923; see Huxley, *Problems of Relative Growth*, p. 18, and *Biol. Centralbl. loc. cit.* Here, and in the previous case of *Cyclommatus*, the equation has been arrived at in a very simple way. Take any two values, x_1, x_2, and the corresponding values, y_1, y_2. Then let

$$\frac{x - x_1}{x_2 - x_1} = \frac{y - y_1}{y_2 - y_1},$$

$$\text{e.g.}\quad \frac{x - 65{\cdot}3}{112{\cdot}6 - 65{\cdot}3} = \frac{y - 48{\cdot}3}{102{\cdot}0 - 48{\cdot}3},$$

$$\text{or}\quad \frac{x - 65{\cdot}3}{47{\cdot}3} = \frac{y - 48{\cdot}3}{53{\cdot}7},$$

from which $x = 22{\cdot}7 + 0{\cdot}88y$.

We may with advantage repeat this process with other values of x and y; and take the mean of the results so obtained.

Once again we find close agreement between the observed and calculated values, although the observations are somewhat few and the equation is arrived at in a simple way. We may take it as proven that the relation between the two growth-rates is essentially *linear*.

A compound-interest law of growth occurs, as Malthus knew, in cases, and at times, of rapid and unrestricted growth. But unrestricted growth occurs under special conditions and for brief

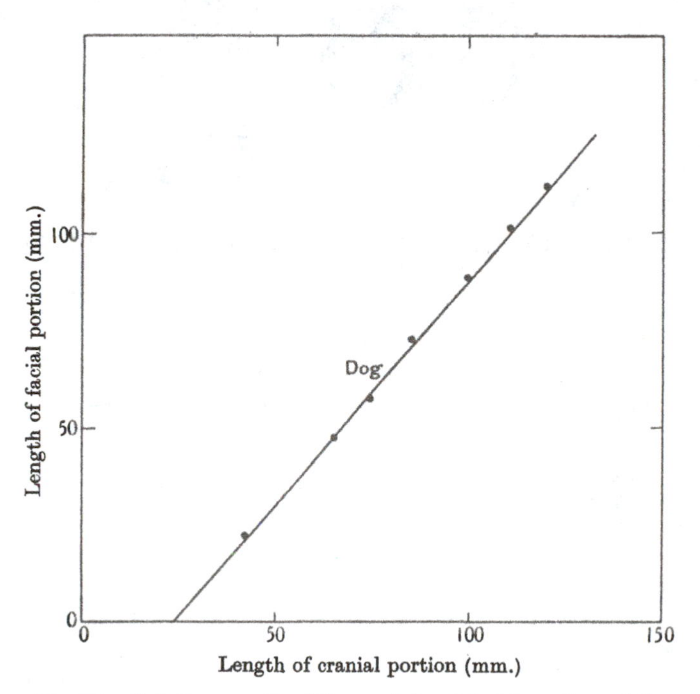

Fig. 56. Relative growth of the cranial and facial portions of the skull in the sheepdog. Cf. Huxley, p. 18, after Becher.

periods; it is the exception rather than the rule, whether in a population or in the single organism. In cases of differential growth the compound-interest law manifests itself, for the same reason, when one of the two growth-rates is rapid and "unrestricted," and when the discrepancy between the two growth-rates is consequently large, for instance in the fiddler-crabs. The compound-interest law is a very natural mode of growth, but its range is limited. A linear relation, or simple-interest law, seems less likely to occur; but the fact is, it does occur, and occurs commonly.

On so-called dimorphism

In a well-known paper, Bateson and Brindley shewed that among a large number of earwigs collected in a particular locality, the males fell into two groups, characterised by large or by small

Fig. 57. Tail-forceps of earwig. From Martin Burr, after Willi Kuhl.

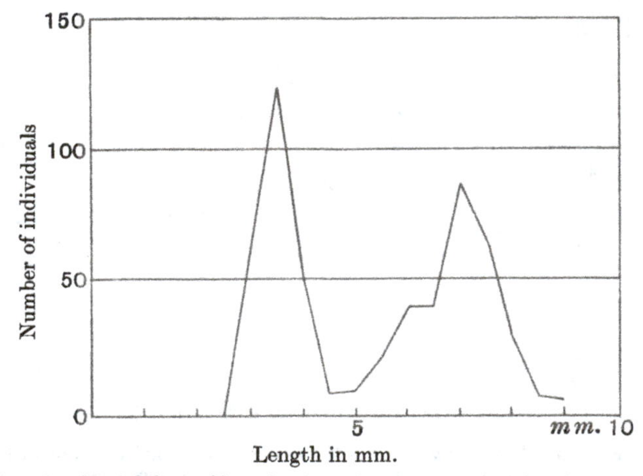

Fig. 58. Variability of length of tail-forceps in a sample of earwigs. After Bateson and Brindley, *P.Z.S.* 1892, p. 588.

tail-forceps (Fig. 57), with few instances of intermediate magnitude*. This distribution into two groups, according to magnitude, is illustrated in the accompanying diagram (Fig. 58); and the

* W. Bateson and H. H. Brindley, On some cases of variation in secondary sexual characters [*Forficula, Xylotrupa*], statistically examined, *P.Z.S.* 1892, pp. 585–594. Cf. D. M. Diakonow, On dimorphic variability of *Forficula, Journ. Genet.* xv, pp. 201–232, 1925; and Julian Huxley, The bimodal cephalic horn of *Xylotrupa, ibid.* xviii, pp. 45–53, 1927.

phenomenon was described, and has been often quoted, as one of dimorphism or discontinuous variation. In this diagram the time-element does not appear; but it looks as though it lay close behind. For the two *size-groups* into which the tails of the earwigs fall look curiously like two *age-groups* such as we have already studied in a fish, where the *ages* and therefore also the *magnitudes* of a random sample form a discontinuous series (Fig. 59). And if, instead of measuring the whole length of our fish, we had confined ourselves to particular parts, such as head, or tail or fin, we should have obtained discontinuous curves of distribution for the magnitudes

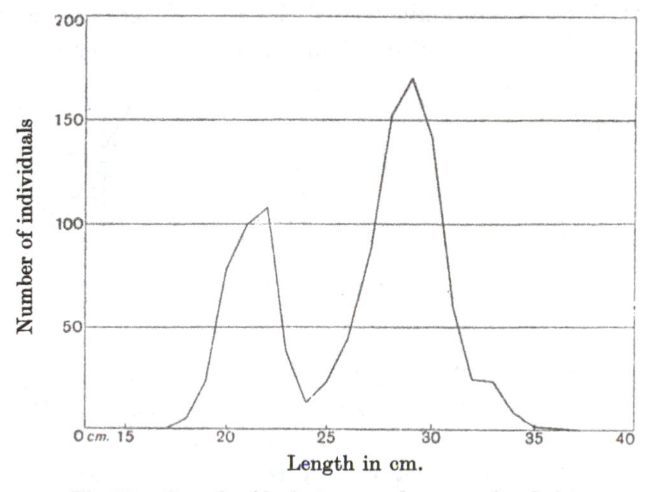

Fig. 59. Length of body in a random sample of plaice.

of these organs, just as for the whole body of the fish, and just as for the tails of Bateson's earwigs. The differences, in short, with which Bateson was dealing were a question of magnitude, and it was only natural to refer these diverse magnitudes to diversities of growth; that is to say, it seemed natural to suppose that in this case of "dimorphism," the tails of the one group of earwigs (which Bateson called the "high males") had either grown faster, or had been growing for a longer period of time, than those of the "low males." If the whole random sample of earwigs were of one and the same age, the dimorphism would appear to be due to two alternative values for the mean growth-rate, individual earwigs varying around one mean or the other. If, on the other hand, the two groups of earwigs were of different ages, or had passed through one moult more or less, the phenomenon would be simple indeed, and there would be no more to be said about it*. Diakonow made the not unimportant observation that in earwigs living in unfavourable conditions only the short-tailed type tended to appear.

In apparent close analogy with the case of the earwigs, and in apparent corroboration of their dimorphism being due to age, Fritz Werner measured large numbers of water-fleas, all apparently adult, found his measurements falling into groups and so giving multimodal curves. The several cusps, or modes, he interpreted without difficulty as indicating differences of age, or the number of moults which the creatures had passed through† (Fig. 60).

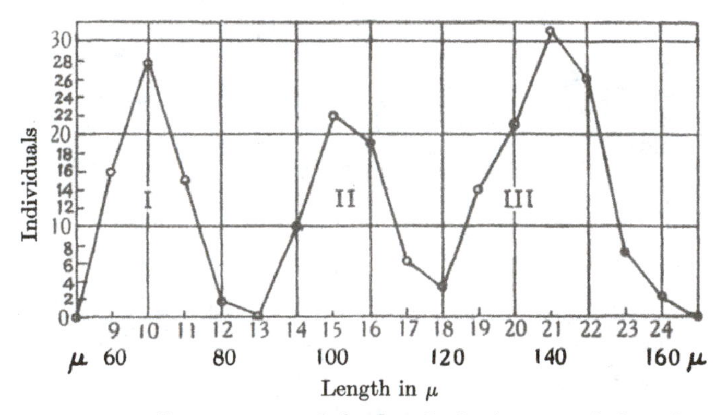

Fig. 60. Measurements of the dorsal edge in a population of *Chydorus sphaericus*, a water-flea. From Fritz Werner.

An apparently analogous but more difficult case is that of a certain little beetle, *Onthophagus taurus*, which bears two "horns" on its head, of variable size or prominence. Linnaeus saw in it a single species, Fabricius saw two; and the question long remained an open one among the entomologists. We now know that there are two "modes," two predominant sizes in a continuous range of

* The number of moults is known to be variable in many species of Orthoptera, and even occasionally in higher insects; and how the number of moults may be influenced by hunger, damp or cold is discussed by P. P. Calvert, *Proc. Amer. Philos. Soc.* LXVIII, p. 246, 1929. On the number of moults in earwigs, see E. B. Worthington, *Entomologist*, 1926, and W. K. Weyrauch, *Biol. Centralbl.* 1929, pp. 543–558.

† Fritz Werner, Variationsanalytische Untersuchungen an Chydoren, *Ztschr. f. Morphologie u. Oekologie d. Tiere*, II, pp. 58–188, 1924.

variation*. In the "complete metamorphosis" of a beetle there is no room for a moult more or less, and the reason for the two modal sizes remains hidden (Fig. 61).

But new light has been thrown on the case of the earwigs, which may help to explain other obscure diversities of shape and size within the class of insects. At metamorphosis, and even in a simple moult, the external organs of an insect may often be seen *to unfold*, as do, for instance, the wings of a butterfly; they then quickly harden, in a form and of a size with which ordinary gradual growth

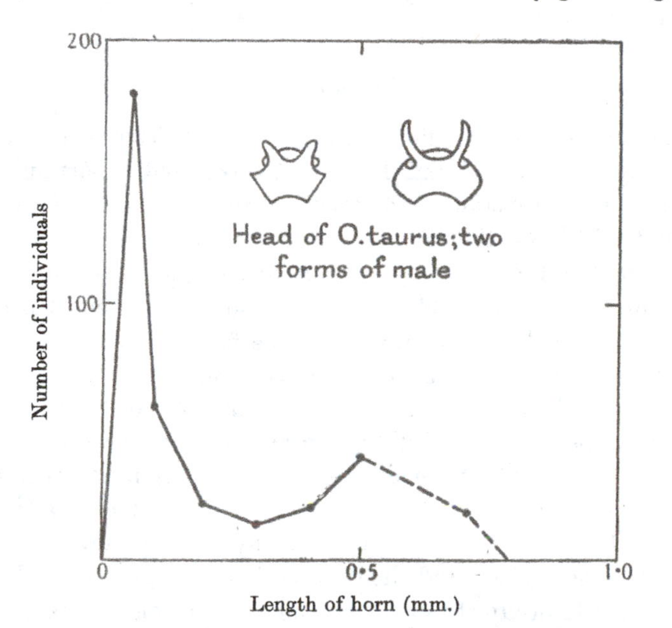

Fig. 61. Two forms of the male, in the beetle *Onthophagus taurus*.

has had nothing directly to do. This is a very peculiar phenomenon, and marks a singular departure from the usual interdependence of growth and form. When the nymph, or larval earwig, is about to shed its skin for the last time, the tail-forceps, still soft and tender, are folded together and wrapped in a sheath; they need to be *distended*, or inflated, by a combined pressure of the body-fluid (or haemolymph) and an intake of respiratory air. If all goes well,

* René Paulian, *Bull. Soc. Zool. Fr.* 1933; also *Le polymorphisme des mâles de Coléoptères*, Paris, 1935, p. 8.

the forceps expand to their full size; if the creature be weak or underfed, inflation is incomplete and the tail-forceps remain small. In either case it is an affair of a few critical moments during the final ecdysis; in ten minutes or less, the chitin has hardened, and shape and size change no more. Willi Kuhl, who has given us this interesting explanation, suggests that the dimorphism observed by Bateson and by Diakonow is not an essential part of the phenomenon; he has found it in one instance, but in other and much larger samples he has found all gradations, but only a single, well-marked unimodal peak*.

The effect of temperature†

The rates of growth which we have hitherto dealt with are mostly based on special investigations, conducted under particular local conditions; for instance, Quetelet's data, so far as we have used them to illustrate the rate of growth in man, are drawn from his study of the Belgian people. But apart from that "fortuitous" individual variation which we have already considered, it is obvious that the normal rate of growth will be found to vary, in man and in other animals, just as the average stature varies, in different localities and in different "races." This phenomenon is a very complex one, and is doubtless a resultant of many undefined contributory causes; but we at least gain something in regard to it when we discover that rate of growth is directly affected by temperature, and doubtless by other physical conditions. Réaumur was the first to shew, and the observation was repeated by Bonnet‡, that the rate of growth or development of the chick was dependent on temperature, being retarded at temperatures below and somewhat

* Willi Kuhl, Die Variabilität der abdominalen Körperanhänge bei *Forficula*, *Ztsch. Morph. u. Oek. d. Tiere*, XII, p. 299, 1924. Cf. Malcolm Burr, *Discovery*, 1939, pp. 340–345.

† The temperature limitations of life, and to some extent of growth, are summarised for a large number of species by Davenport, *Exper. Morphology*, cc. viii, xviii, and by Hans Przibram, *Exp. Zoologie*, IV, c. v.

‡ Réaumur, *L'art de faire éclorre et élever en toute saison des oiseaux domestiques, soit par le moyen de la chaleur du fumée, soit par le moyen de celle du feu ordinaire*, Paris, 1749. He had also studied, a few years before, the effects of heat and cold on growth-rate and duration of life in caterpillars and chrysalids: *Mémoires*, II, p. 1, *de la durée de la vie des crisalides* (1736). See also his Observations du Thermomètre, etc., *Mém. Acad., Paris*, 1735, pp. 345–376.

accelerated at temperatures above the normal temperature of incubation, that is to say the temperature of the sitting hen. In the case of plants the fact that growth is greatly affected by temperature is a matter of familiar knowledge; the subject was first carefully studied by Alphonse De Candolle, and his results and those of his followers are discussed in the textbooks of botany*.

That temperature is only one of the climatic factors determining growth and yield is well known to agriculturists; and a method of "multiple correlation" has been used to analyse the several influences of temperature and of rainfall at different seasons on the future yield of our own crops†. The same joint influence can be recognised in the bamboo; for it is said (by Lock) that the growth-rate of the bamboo in Ceylon is proportional to the humidity of the atmosphere, and again (by Shibata) that it is proportional to the temperature in Japan. But Blackman‡ suggests that in Ceylon temperature conditions are all that can be desired, but moisture is apt to be deficient, while in Japan there is rain in abundance but the average temperature is somewhat low: so that in the one country it is the one factor, and in the other country the other, whose variation is both conspicuous and significant. After all, it is probably rate of evaporation, the joint result of temperature and humidity, which is the crux of the matter§. "Climate" is a subtle thing, and includes a sort of micro-meteorology. A sheltered corner has a climate of its own; one side of the garden-wall has a different climate to the other; and deep in the undergrowth of a wood celandine and anemone enjoy a climate many degrees warmer than what is registered on the screen||.

Among the mould-fungi each several species has its own optimum temperature for germination and growth. At this optimum temperature growth is further accelerated by increase of humidity; and the further we depart from the optimum temperature, the narrower becomes the range of humidity within which growth can proceed¶. Entomologists know, in like manner, how over-abundance of an insect-pest comes, or is apt to come, with a double optimum of temperature and humidity.

* Cf. (*int. al.*) H. de Vries, Matériaux pour la connaissance de l'influence de la température sur les plantes, *Arch. Néerlandaises*, v, pp. 385–401, 1870; C. Linsser, Periodische Erscheinungen des Pflanzenlebens, *Mém. Acad. des Sc., St Pétersbourg* (7), XI, XII, 1867–69; Köppen, Wärme und Pflanzenwachstum, *Bull. Soc. Imp. Nat., Moscou*, XLIII, pp. 41–110, 1871; H. Hoffmann, Thermische Vegetationsconstanten, *Ztschr. Oesterr. Ges. f. Meteorologie*, XVII, pp. 121–131, 1881; Phenologische Studien, *Meteorolog. Ztschr.* III, pp. 113–120, 1886.

† See (*int. al.*) R. H. Hooker, *Journ. Roy. Statist. Soc.* 1907, p. 70; *Journ. Roy. Meteor. Soc.* 1922, p. 46.

‡ F. F. Blackman, *Ann. Bot.* XIX, p. 281, 1905.

§ Száva-Kovátz, in *Petermann's Mitteilungen*, 1927, p. 7.

|| Cf. E. J. Salisbury, On the oecological aspects of Meteorology, *Q.J.R. Meteorol. Soc.* July 1939.

¶ R. G. Tomkins, *Proc. R.S.* (B), CV, pp. 375–401, 1929.

The annexed diagram (Fig. 62), showing growth in length of the roots of some common plants at various temperatures, is a sufficient illustration of the phenomenon. We see that there is always a certain temperature at which the rate is a maximum; while on either side of the optimum the rate falls off, after the fashion of the normal curve of error. We see further, from the data given by Sachs and others, that the optimum is very much the same for all the common plants of our own climate. For these it is somewhere about 26° C.

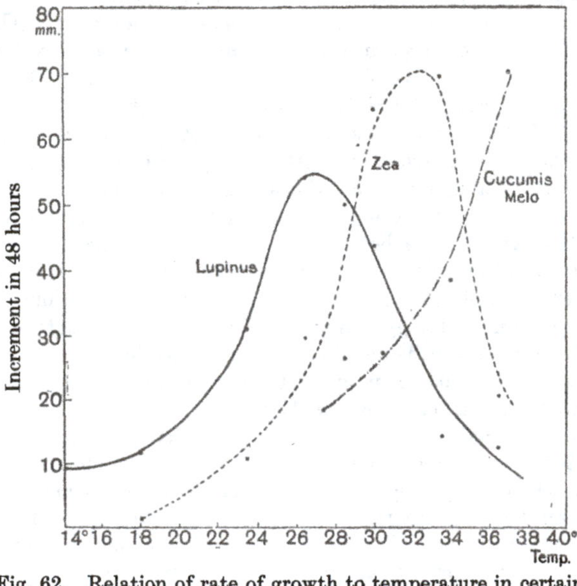

Fig. 62. Relation of rate of growth to temperature in certain plants. From Sachs's data.

(say 77° F.), or about the temperature of a warm summer's day; while it is considerably higher, naturally, in such plants as the melon or the maize, which are at home in warmer countries than our own. The bacteria have, in like manner, their various optima, and sometimes a high one. The tuberculosis-bacillus, as Koch shewed, only begins to grow at about 28° C., and multiplies most rapidly at 37–38°, the body-temperature of its host.

The setting and ripening of fruit is a phase of growth still more dependent on temperature; hence it is a "delicate test of climate," and a proof of its constancy, that the date-palm grows but bears no fruit in Judaea, and the vine bears freely at Eshcol, but not in the hotter country to the south*. Shellfish have their own appropriate spawning-temperatures; it needs a warm summer for the oyster to shed her spat, and *Hippopus* and *Tridacna*, the great clams of the coral-reefs, only do so when the water has reached the high temperature of 30° C. For brown trout, 6° C. is found to be a critical temperature, a minimum short of which they do not grow at all; it follows that in a Highland burn their growth is at a standstill for fully half the year†.

That a rise of temperature accelerates growth is but part of the story, and is not always true. Several insects, experimentally reared, have been found to diminish in size as the temperature increased‡; and certain flies have been found to be larger in their winter than their summer broods. The common copepod, *Calanus finmarchicus*, has spring, summer and autumn broods, which (at Plymouth§) are large, middle-sized and small; but the large spring brood are hatched and reared in the cold "winter" water, and the small autumn-winter brood in the warmest water of the year. In the cold waters of Barents Sea *Calanus* grows larger still; of an allied genus, a large species lives in the Antarctic, a small one in the tropics, a middle-sized is common in the temperate oceans. The large size of many Arctic animals, coelenterates and crustaceans, is well known; and so is that of many tropical forms, like *Fungia* among the corals, or the great Tritons and Tridacnas among molluscs. Another common phenomenon is the increasing number of males in late summer and autumn, as in the Rotifers and in the above-mentioned Calani. All these things seem somehow related to temperature; but other physical conditions enter into the case, for instance the amount of dissolved oxygen in the cold waters, and the physical chemistry of carbonate of lime in the warm||.

The vast profusion of life, both great and small, in Arctic seas, the multitude of individuals and the unusual size to which many species grow, has been often ascribed to a superabundance of dissolved oxygen, but oxygen alone would not go far. The nutrient salts, nitrates and phosphates, are the

* Cf. J. W. Gregory, in *Geogr. Journ.* 1914, and *Journ. R. Geogr. Soc.* Oct. 1930.

† Cf. C. A. Wingfield, *op. cit.* supra, p. 176.

‡ B. P. Uvarow, *Trans. Ent. Soc. Lond.* LXXIX, p. 38, 1931.

§ W. H. Golightly and Ll. Lloyd, in *Nature*, July 22, 1939.

|| Cf. B. G. Bogorow and others, in the *Journ. M.B.A.* XIX, 1933–34.

limiting factor in the growth of that micro-vegetation with which the whole cycle of life begins. The tropical oceans are often very bare of these salts; in our own latitudes there is none too much, and the spring-growth tends to use up the supply. But we have learned from the Discovery Expedition that these salts are so abundant in the Antarctic that plant-growth is never checked for stint of them. Along the Chilean coast and in S.W. Africa, cold Antarctic water wells up from below the warm equatorial current. It is ill-suited for the growth of corals, which build their reefs in the warmer waters of the eastern side; but it teems with nourishment, breeds a plankton-fauna of the richest kind, which feeds fishes preyed on by innumerable birds, the guano of which is sent all over the world. Now and then persistent winds thrust the cold current aside; a new warm current, *el Nino* of the Chileans, upsets the old equilibrium; the fishes die, the water stinks, the birds starve. The same thing happens also at Walfisch Bay, where on such rare occasions dead fish lie piled up high along the shore.

It is curiously characteristic of certain physiological reactions, growth among them, to be affected not merely by the temperature of the moment, but also by that to which the organism has been previously and temporarily exposed. In other words, acclimatisation to a certain temperature may continue for some time afterwards to affect all the temperature relations of the body*. That temporary cold may, under certain circumstances, cause a subsequent acceleration of growth is made use of in the remarkable process known as *vernalisation*. An ingenious man, observing that a winter wheat failed to flower when sown in spring, argued that exposure to the cold of winter was necessary for its subsequent rapid growth; and this he verified by "chilling" his seedlings for a month to near freezing-point, after which they grew quickly, and flowered at the same time as the spring wheat. The economic advantages are great of so shortening the growing period of a crop as to protect it from autumn frosts in a cold climate or summer drought in a hot one; much has been done, especially by Lysenko in Russia, with this end in view†.

The most diverse physiological processes may be affected by temperature. A great astronomer at Mount Wilson, in California, used some idle hours to watch the "trail-running" ants, which run all night and all day. Their speed increases so regularly with the temperature that the time taken to run 30 cm. suffices to tell the

* Cf. Kenneth Mellanby, On temperature coefficients and acclimatisation, *Nature*, 3 August 1940.

† Cf. (*int. al.*) V. H. Blackman, in *Nature*, June 13, 1936.

temperature to 1° C.! Of two allied species, one ran nearly half as fast again as the other, at the same temperature*.

While at low temperatures growth is arrested and at temperatures unduly high life itself becomes impossible, we have now seen that within the range of more or less congenial temperatures growth proceeds the faster the higher the temperature. The same is true of the ordinary reactions of chemistry, and here Van't Hoff and Arrhenius† have shewn that a definite increase in the velocity of the reaction follows a definite increase of temperature, according to an exponential law: such that, for an interval of n degrees the velocity varies as x^n, x being called the "temperature coefficient" for the reaction in question‡. The law holds good throughout a considerable range, but is departed from when we pass beyond certain normal limits; moreover, the value of the coefficient is found to keep to a certain order of magnitude—somewhere about 2 for a temperature-interval of 10° C.—which means to say that the velocity of the reaction is just about doubled, more or less, for a rise of 10° C.

This law, which has become a fundamental principle of chemical mechanics, is applicable (with certain qualifications) to the phenomena of vital chemistry, as Van't Hoff himself was the first to declare; and it follows that, on much the same lines, one may speak of a "temperature coefficient" of growth. At the same time we must remember that there is a very important difference (though we need not call it a *fundamental* one) between the purely physical and the

* Harlow Shapley, On the thermokinetics of Dolichoderine ants, *Proc. Nat. Acad. Sci.* x, pp. 436–439, 1924.

† Van't Hoff and Cohen, *Studien zur chemischen Dynamik*, 1896; Sv. Arrhenius, *Ztschr. f. phys. Chemie*, IV, p. 226.

‡ For various instances of a temperature coefficient in physiological processes, see (e.g.) Cohen, *Physical Chemistry for...Biologists* (English edition), 1903; Kanitz and Herzog in *Zeitschr. f. Elektrochemie*, XI, 1905; F. F. Blackman, *Ann. Bot.* XIX, p. 281, 1905; K. Peter, *Arch. f. Entw. Mech.* XX, p. 130, 1905; Arrhenius, *Ergebn. d. Physiol.* VII, p. 480, 1908, and *Quantitative Laws in Biological Chemistry*, 1915; Krogh in *Zeitschr. f. allgem. Physiologie*, XVI, pp. 163, 178, 1914; James Gray, *Proc. R.S.* (B), XCV, pp. 6–15, 1923; W. J. Crozier, many papers in *Journ. Gen. Physiol.* 1924; J. Belehradek, in *Biol. Reviews*, V, pp. 1–29, 1930. On the general subject, see E. Janisch, Temperaturabhängigkeit biologischer Vorgänge und ihrer kurvenmässige Analyse, *Pflüger's Archiv*, CCIX, p. 414, 1925; G. and P. Hertwig, Regulation von Wachstum...durch Umweltsfaktoren, in *Hdb. d. normal. u. pathol. Physiologie*, XVI, 1930.

physiological phenomenon, in that in the former we study (or seek and profess to study) one thing at a time, while in the living body we have constantly to do with factors which interact and interfere; increase in the one case (or change of any kind) tends to be continuous, in the other case it tends to be brought, or to bring itself, to arrest. This is the simple meaning of that *Law of Optimum*, laid down by Errera and by Sachs as a general principle of physiology; namely that *every* physiological process which varies (like growth itself) with the amount or intensity of some external influence, does so under such conditions that progressive increase is followed by progressive decrease; in other words, the function has its *optimum* condition, and its curve shews a definite *maximum*. In the case of temperature, as Jost puts it, it has on the one hand its accelerating effect, which tends to follow Van't Hoff's law. But it has also another and a cumulative effect upon the organism: "Sie schädigt oder sie ermüdet ihn, und je höher sie steigt desto rascher macht sie die Schädigung geltend und desto schneller schreitet sie voran*." It is this double effect of temperature on the organism which gives, or helps to give us our "optimum" curves, which (like all other curves of frequency or error) are the expression, not of a single solitary phenomenon, but of a more or less complex resultant. Moreover, as Blackman and others have pointed out, our "optimum" temperature is ill-defined until we take account also of the *duration* of our experiment; for a high temperature may lead to a short but exhausting spell of rapid growth, while the slower rate manifested at a lower temperature may be the best in the end. The mile and the hundred yards are won by different runners; and maximum rate of working, and maximum amount of work done, are two very different things†.

In the case of maize, a certain series of experiments shewed that the growth in length of the roots varied with the temperature as follows‡:

* On such limiting factors, or counter-reactions, see Putter, *Ztschr. f. allgem. Physiologie*, XVI, pp. 574–627, 1914.

† Cf. L. Errera, *L'Optimum*, 1896 (*Recueil d'œuvres, Physiologie générale*, pp. 338–368, 1910); Sachs, *Physiologie d. Pflanzen*, 1882, p. 233; Pfeffer, *Pflanzenphysiologie*, II, p. 78, 194; and cf. Jost, Ueber die Reactionsgeschwindigkeit im Organismus, *Biol. Centralbl.* XXVI, pp. 225–244, 1906.

‡ After Köppen, *Bull. Soc. Nat. Moscou*, XLIII, pp. 41–101, 1871.

Temperature °C.	Growth in 48 hours mm.
18·0	1·1
23·5	10·8
26·6	29·6
28·5	26·5
30·2	64·6
33·5	69·5
36·5	20·7

Let us write our formula in the form

$$\frac{V_{(t+n)}}{V_t} = x^n, \quad \text{or } \log V_{(t+n)} - \log V_t = n . \log x.$$

Then choosing two values out of the above experimental series (say the second and the second-last), we have $t = 23{\cdot}5$, $n = 10$, and V, $V' = 10{\cdot}8$ and $69{\cdot}5$ respectively.

Accordingly, $$\frac{\log 69{\cdot}5 - \log 10{\cdot}8}{10} = \log x,$$

or $$\frac{0{\cdot}8414 - 0{\cdot}034}{10} = 0{\cdot}0808,$$

and therefore the temperature-coefficient

$$= \text{antilog } 0{\cdot}0808 = 1{\cdot}204 \text{ (for an interval of } 1^\circ \text{ C.).}$$

This first approximation might be much improved by taking account of all the experimental values, two only of which we have yet made use of; but even as it is, we see by Fig. 63 that it is in very fair accordance with the actual results of observation, within those particular *limits of temperature* to which the experiment is confined.

For an experiment on *Lupinus albus*, quoted by Asa Gray* I have worked out the corresponding coefficient, but a little more carefully. Its value I find to be 1·16, or very nearly identical with that we have just found for the maize; and the correspondence between the calculated curve and the actual observations is now a close one.

Miss I. Leitch has made careful observations of the rate of growth of rootlets of the Pea; and I have attempted a further analysis of her principal results†.

* Asa Gray, *Botany*, p. 387.

† I. Leitch, Some experiments on the influence of temperature on the rate of growth in *Pisum sativum*, *Ann. Bot.* xxx, pp. 25–46, 1916, especially Table III, p. 45. Cf. Priestley and Pearsall, Growth studies, *Ann. Bot.* xxxvi, pp. 224–249, 1922.

In Fig. 64 are shewn the mean rates of growth (based on about a hundred experiments) at some thirty-four different temperatures between 0·8° and 29·3°, each experiment lasting rather less than twenty-four hours. Working out the mean temperature coefficient for a great many combinations of these values, I obtain a value of 1·092 per C.°, or 2·41 for an interval of 10°, and a mean value for the whole series shewing a rate of growth of just about 1 mm. per hour at a temperature of 20°. My curve in Fig. 64 is drawn from these determinations; and it will be seen that, while it is by no means exact at the lower temperatures, and will fail us altogether at very high temperatures, yet it serves as a satisfactory guide to the relations between rate and temperature within the ordinary limits of healthy growth. Miss Leitch

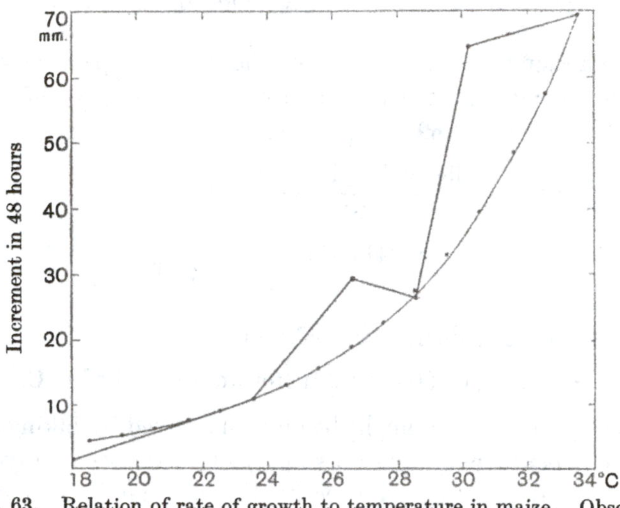

Fig. 63. Relation of rate of growth to temperature in maize. Observed values (after Köppen), and calculated curve.

holds that the curve is *not* a Van't Hoff curve; and this, in strict accuracy, we need not dispute. But the phenomenon seems to me to be one into which the Van't Hoff ratio enters largely, though doubtless combined with other factors which we cannot determine or eliminate.

While the above results conform fairly well to the law of the temperature-coefficient, it is evident that the imbibition of water plays so large a part in the process of elongation of the root or stem that the phenomenon is as much or more a physical than a chemical one: and on this account, as Blackman has remarked, the data commonly given for the rate of growth in plants are apt to be irregular, and sometimes misleading*. We have abundant

* F. F. Blackman, Presidential Address in Botany, *Brit. Assoc.* Dublin, 1908.

illustrations, however, among animals, in which we may study the temperature-coefficient under circumstances where, though the phenomenon is always complicated, true metabolic growth or chemical combination plays a larger rôle. Thus Mlle. Maltaux and Professor Massart* have studied the rate of division in a certain flagellate, *Chilomonas paramoecium*, and found the process to take

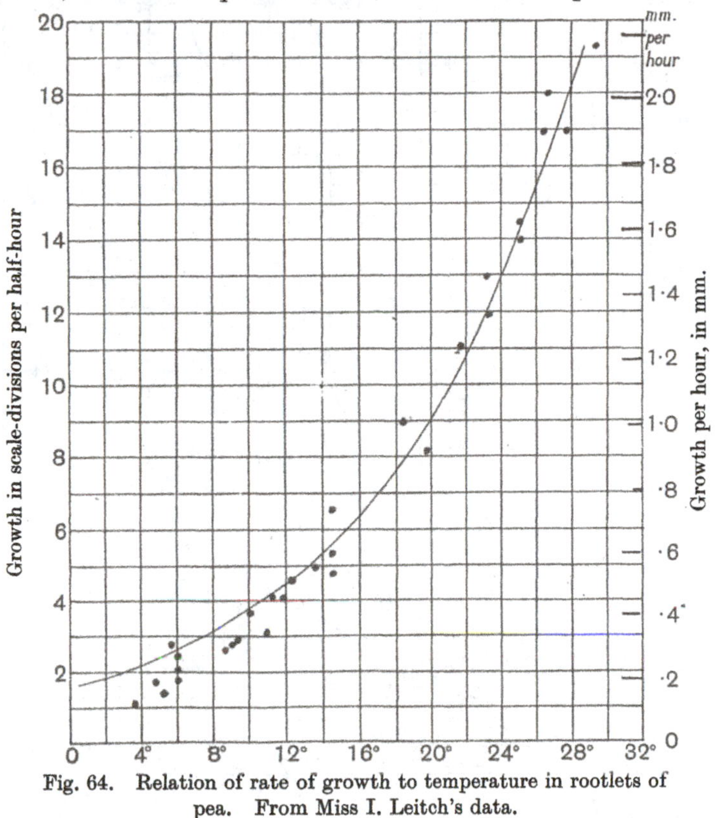

Fig. 64. Relation of rate of growth to temperature in rootlets of pea. From Miss I. Leitch's data.

29 minutes at 15° C., 12 at 25°, and only 5 minutes at 35° C. These velocities are in the ratio of 1 : 2·4 : 5·76, which ratio corresponds precisely to a temperature-coefficient of 2·4 for each rise of 10°, or about 1·092 for each degree centigrade, precisely the same as we have found for the growth of the pea.

By means of this principle we may sometimes throw light on apparently complicated experiments. For instance, Fig. 65 is an

* *Rec. de l'Inst. Bot. de Bruxelles*, VI, 1906.

illustration, which has been often copied, of O. Hertwig's work on the effect of temperature on the rate of development of the tadpole*.

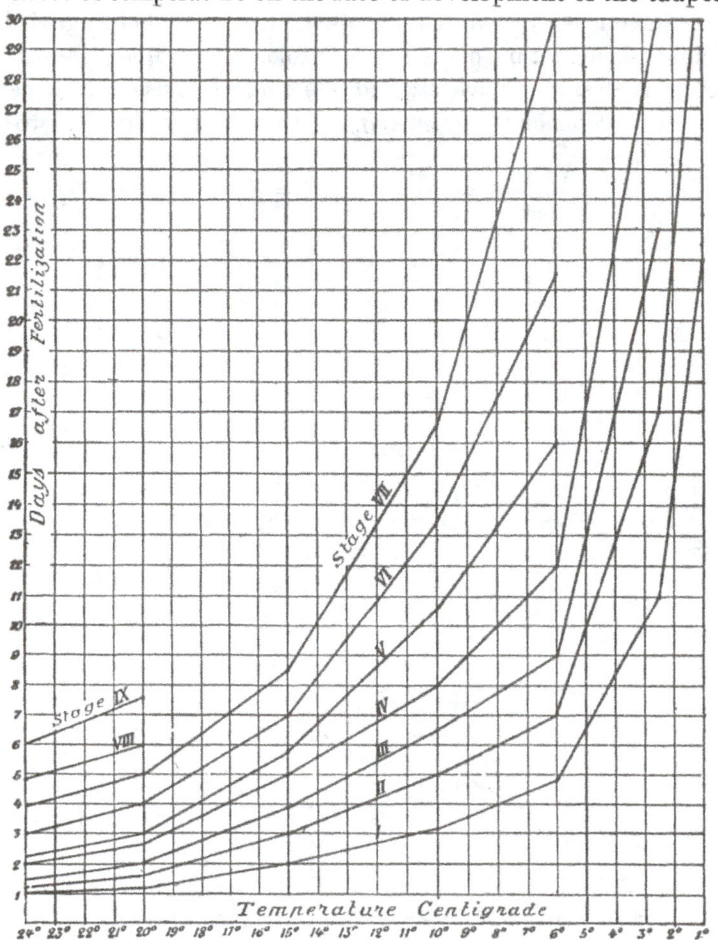

Fig. 65. Diagram shewing time taken (in days), at various temperatures (° C.), to reach certain stages of development in the frog: viz. I, gastrula; II, medullary plate; III, closure of medullary folds; IV, tail-bud; V, tail and gills; VI, tail-fin; VII, operculum beginning; VIII, do. closing; IX, first appearance of hind-legs. From Jenkinson, after O. Hertwig, 1898.

* O. Hertwig, Einfluss der Temperatur auf die Entwicklung von *Rana fusca* und *R. esculenta*, *Arch. f. mikrosk. Anat.* LI, p. 319, 1898. Cf. also K. Bialaszewicz, Beiträge z. Kenntniss d. Wachsthumsvorgänge bei Amphibienembryonen, *Bull. Acad. Sci. de Cracovie*, p. 783, 1908; Abstr. in *Arch. f. Entwicklungsmech.* XXVIII, p. 160, 1909: from which Ernst Cohen determined the value of Q_{10} (*Vorträge üb. physikal. Chemie f. Ärzte*, 1901; English edit. 1903).

From inspection of this diagram, we see that the time taken to attain certain stages of development (denoted by the numbers III–VII) was as follows, at 20° and at 10° C., respectively.

	At 20° C.	At 10° C.
Stage III	2·0	6·5 days
" IV	2·7	8·1 "
" V	3·0	10·7 "
" VI	4·0	13·5 "
" VII	5·0	16·8 "
Total	16·7	55·6 "

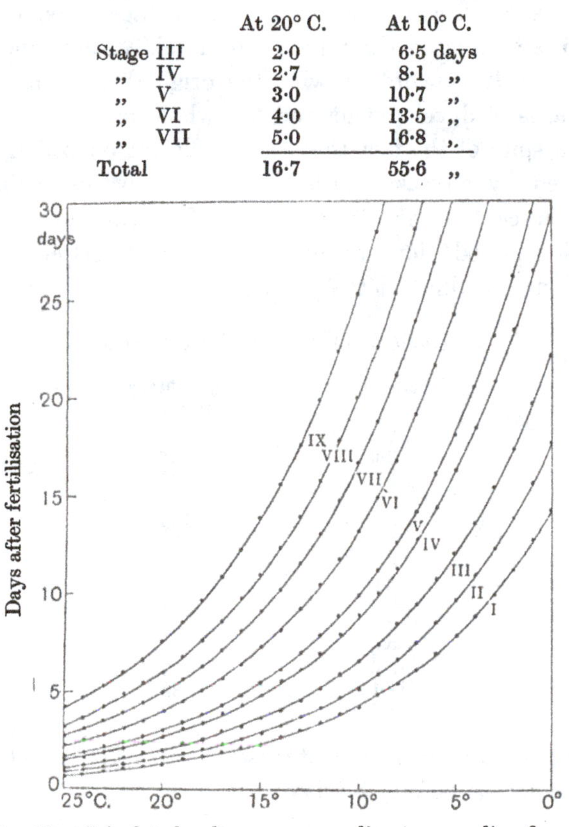

Fig. 66. Calculated values, corresponding to preceding figure.

That is to say, the time taken to produce a given result at 10° was (on the average) somewhere about 55·6/16·7, or 3·33, times as long as was required at 20° C.

We may then put our equation in the simple form,

$$x^{10} = 3{\cdot}33.$$

Or, $$10 \log x = \log 3{\cdot}33 = 0{\cdot}52244.$$

Therefore $$\log x = 0{\cdot}05224,$$

and $$x = 1{\cdot}128.$$

That is to say, between the intervals of 10° and 20° C., if it take m days, at a certain given temperature, for a certain stage of development to be attained, it will take $m \times 1{\cdot}128^n$ days, when the temperature is n degrees less, for the same stage to be arrived at.

Fig. 66 is calculated throughout from this value; and it will be found extremely concordant with the original diagram, as regards all the stages of development and the whole range of temperatures shewn; in spite of the fact that the coefficient on which it is based was derived by an easy method from a very few points on the original curves. In like manner, the following table shews the "incubation period" for trout-eggs, or interval between fertilisation and hatching, at different temperatures*:

Incubation-period of trout-eggs

Temperature ° C.	Days' interval before hatching
2·8	165
3·6	135
3·9	121
4·5	109
5·0	103
5·7	96
6·3	89
6·6	81
7·3	73
8·0	65
9·0	56
10·0	47
11·1	38
12·2	32

Choosing at random a pair of observations, viz. at 3·6° and 10°, and proceeding as before, we have

$$10° - 3{\cdot}6° = 6{\cdot}4°.$$

Then
$$(6{\cdot}4) = \frac{135}{47},$$

or
$$6{\cdot}4 \times \log x = \log 135 - \log 47$$
$$= 2{\cdot}1303 - 1{\cdot}6721 = 0{\cdot}4582$$

and
$$\log x = 0{\cdot}4582 \div 6{\cdot}4 = 0{\cdot}0716,$$
$$x = 1{\cdot}179.$$

* Data from James Gray, The growth of fish, *Journ. Exper. Biology*, VI, p. 126, 1928.

Using three other pairs of observations, we have the following concordant results:

At 12·2°	and 2·8°,	$x = 1·191$
10·0°	3·6°	1·179
9·0°	5·7°	1·178
8·0°	5·0°	1·165
	Mean	1·18

A very curious point is that (as Gray tells us) the young fish which have hatched slowly at a low temperature are bigger than those whose growth has been hastened by warmth.

Again, plaice-eggs were found to hatch and grow to a certain length (4·6 mm.), as follows*:

Temperature (° C.)	Days
4·1	23·0
6·1	18·1
8·0	13·3
10·1	10·3
12·0	8·3

From these we obtain, as before, the following constants:

At 12°	and 8°,	$x = 1·13$
12°	4·1°	1·14
10·1°	6·1°	1·15
8·0°	4·1°	1·15
	Mean	1·14

The value of x is much the same for the one fish as for the other.

Karl Peter†, experimenting on echinoderm eggs, and making use also of Richard Hertwig's experiments on young tadpoles, gives the temperature-coefficients for intervals of 10° C. (commonly written Q_{10}) as follows, to which I have added the corresponding values for Q_1:

Sphaerechinus	$Q_{10} = 2·15$	$Q_1 = 1·08$
Echinus	2·13	1·08
Rana	2·86	1·11

* Data from A. C. Johansen and A. Krogh, Influence of temperature, etc., *Publ. de Circonstance*, No. 68, 1914. The function is here said to be a linear one—which would have been an anomalous and unlikely thing.

† Der Grad der Beschleunigung tierischer Entwicklung durch erhöhte Temperatur, *Arch. f. Entw. Mech.* xx, p. 130, 1905. More recently Bialaszewicz has determined the coefficient for the rate of segmentation in *Rana* as being 2·4 per 10° C.

These values are not only concordant, but are of the same order of magnitude as the temperature-coefficient in ordinary chemical reactions. Peter has also discovered the interesting fact that the temperature-coefficient alters with age, usually but not always decreasing as time goes on*:

Sphaerechinus	Segmentation	$Q_{10} = 2{\cdot}29$	$Q_1 = 1{\cdot}09$
	Later stages	2·03	1·07
Echinus	Segmentation	2·30	1·09
	Later stages	2·08	1·08
Rana	Segmentation	2·23	1·08
	Later stages	3·34	1·13

Furthermore, the temperature-coefficient varies with the temperature itself, falling as the temperature rises—a rule which Van't Hoff shewed to hold in ordinary chemical operations. Thus in *Rana* the temperature-coefficient (Q_{10}) at low temperatures may be as high as 5–6; which is just another way of saying that at low temperatures development is exceptionally retarded.

As the several stages of development are accelerated by warmth, so is the duration of each and all, and of life itself, proportionately curtailed. The span of life itself may have its temperature-coefficient—in so far as Life is a chemical process, and Death a chemical result. In hot climates puberty comes early, and old age (at least in women) follows soon; fishes grow faster and spawn earlier in the Mediterranean than in the North Sea. Jacques Loeb† found (in complete agreement with the general case) that the larval stages of a fly are abbreviated by rise of temperature; that the mean duration of life at various temperatures can be expressed by a temperature-coefficient of the usual order of magnitude; that this coefficient tends, as usual, to fall as the temperature rises; and lastly—what is not a little curious—that the coefficient is very much the same, in fact all but identical, for the larva, pupa and imago of the fly.

* The differences are, after all, of small order of magnitude, as is all the better seen when we reduce the ten-degree to one-degree coefficients.

† J. Loeb and Northrop, On the influence of food and temperature upon the duration of life, *Journ. Biol. Chemistry*, XXXII, pp. 103–121, 1917.

Temperature-coefficients (Q_{10}) *of* Drosophila

	Larva	Pupa	Imago
15–20° C.	1·15	1·17	1·18
20–25° C.	1·06	1·08	1·07

And Japanese students, studying a little fresh-water crustacean, have carried the experiment much beyond the range of Van't Hoff's law, and have found length of life to rise rapidly to a maximum at about 13–14° C., and to fall slowly, in a skew curve, thereafter* (Fig. 67).

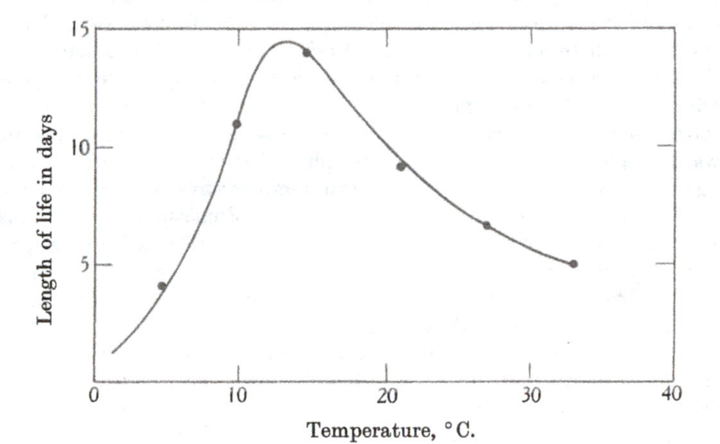

Fig. 67. Length of life, at various temperatures, in a water-flea.

If we now summarise the various temperature-coefficients (Q_1) which we have happened to consider, we are struck by their remarkably close agreement:

Yeast	$Q_1 = 1{\cdot}13$
Lupin	1·16
Maize	1·20
Pea	1·09
Echinoids	1·08
Drosophila (mean)	1·12
Frog, segmentation	1·08
,, tadpole	1·13
Mean	1·12

* A. Terao and T. Tabaka, Duration of life in a water-flea, *Moina* sp.; *Journ. Imp. Fisheries Inst.*, Tokyo, XXV, No. 3, March 1930.

The constancy of these results might tempt us to look on the phenomenon as a simple one, though we well know it to be highly complex. But we had better rest content to see, as Arrhenius saw in the beginning, a general resemblance rather than an identity between the temperature-coefficients in physico-chemical and biological processes*.

It was seen from the first that to extend Van't Hoff's law from physical chemistry to physiology was a bold assumption, to all appearance largely justified, but always subject to severe and cautious limitations. If it seemed to simplify certain organic phenomena, further study soon shewed how far from simple these phenomena were. Living matter is always heterogeneous, and from one phase to another its reactions change; the temperature-coefficient varies likewise, and indicates at the best a summation, or integration, of phenomena. Nevertheless, attempts have been made to go a little further towards a physical explanation of the physiological coefficient. Van't Hoff suggested a viscosity-correction for the temperature-coefficient even of an ordinary chemical reaction; the viscosity of protoplasm varies in a marked degree, inversely with the temperature, and the viscosity-factor goes, perhaps, a long way to account for the aberrations of the temperature-coefficient. It has even been suggested (by Belehradek†) that the temperature-coefficients of the biologist are merely those of protoplasmic viscosity. For instance, the temperature-coefficients of mitotic cell-division have been shewn to alter from one phase to another of the mitotic process, being much greater at the start than at the end‡; and so, precisely, has it been shewn that protoplasmic viscosity is high at the beginning and low at the end of the mitotic process§.

On seasonal growth

There is abundant evidence in certain fishes, such as plaice and haddock, that the ascending curve of growth is subject to seasonal fluctuations or interruptions, the rate during the winter months being always slower than in the months of summer. Thus the Newfoundland cod have their maximum growth-rate in June, and in January–February they cease to grow; it is as though we superimposed a periodic annual sine-curve upon the continuous curve of growth. Furthermore, as growth itself grows less and less from year to year, so will the difference between the summer and the

* Cf. L. V. Heilbronn, *Science*, LXII, p. 268, 1925.

† J. Belehradek, in *Biol. Reviews*, V, pp. 30–58, 1930.

‡ Cf. E. Fauré-Fremiet, *La cinétique du développement*, 1925; also B. Ephrussi, *C.R.* CLXXXII, p. 810, 1926.

§ See (*int. al.*) L. V. Heilbronn, *The Colloid Chemistry of Protoplasm*, 1928.

winter rates grow less and less. The fluctuation in rate represents a vibration which is gradually dying out; the amplitude of the sine-curve diminishes till it disappears; in short our phenomenon is simply expressed by what is known as a "damped sine-curve*."

Growth in height of German military cadets, in half-yearly periods

Number observed	Age	Height (cm.) October	Height (cm.) April	Height (cm.) October	Increment (cm.) Winter ½-year	Increment (cm.) Summer ½-year	Increment (cm.) Year
12	11–12	139·4	141·0	143·3	1·6	2·3	3·9
80	12–13	143·0	144·5	147·4	1·5	2·9	4·4
146	13–14	147·5	149·5	152·5	2·0	3·0	5·0
162	14–15	152·2	155·0	158·5	2·8	3·5	6·3
162	15–16	158·5	160·8	163·8	2·3	3·0	5·3
150	16–17	163·5	165·4	167·7	1·9	2·3	4·2
82	17–18	167·7	168·9	170·4	1·2	1·5	2·7
22	18–19	169·8	170·6	171·5	0·8	0·9	1·7
6	19–20	170·7	171·1	171·5	0·4	0·4	0·8
				Mean	1·6	2·2	

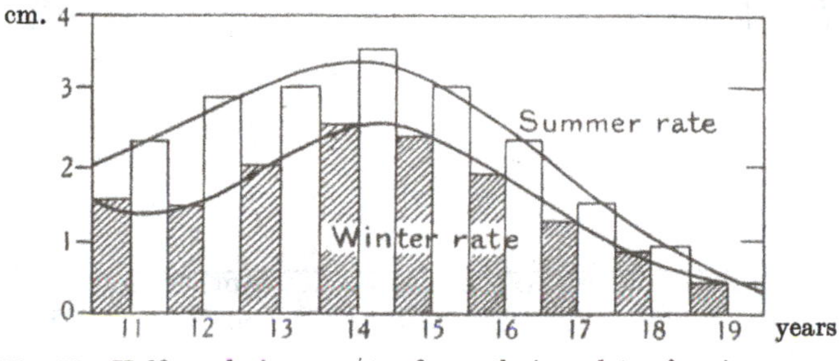

Fig. 68. Half-yearly increments of growth, in cadets of various ages. From Daffner's data.

The same thing occurs in man, though neither in his case nor in that of the fish have we sufficient data for its complete illustration. We can demonstrate the fact, however, by help of certain measurements of the height of German cadets, measured at half-yearly intervals†. In the accompanying diagram (Fig. 68) the half-yearly increments are set forth from the above table, and it will be seen

* The scales, on the other hand, make most of their growth during the intermediate seasons: and with this peculiarity, that a few broad zones are added to the scale in spring, and a larger number of narrow circuli in autumn: see *Contrib. to Canadian Biology*, IV, pp. 289–305, 1929; Ben Dawes, Growth...in plaice, *Journ. M.B.A.* XVII, pp. 103–174, 1930.

† From Daffner, *Das Wachstum des Menschen*, p. 329, 1902.

that they form two even and entirely separate series. Danish schoolboys show just the same periodicity of growth in stature.

The seasonal effect on visible growth-rate is much alike in fishes and in man, in spite of the fact that the bodily temperature of the one varies with the *milieu externe* and that of the other keeps constant to within a fraction of a degree.

While temperature is the dominant cause, it is not the only cause of seasonal fluctuations of growth; for alternate scarcity and abundance of food is often, as in herbivorous animals, the ostensible

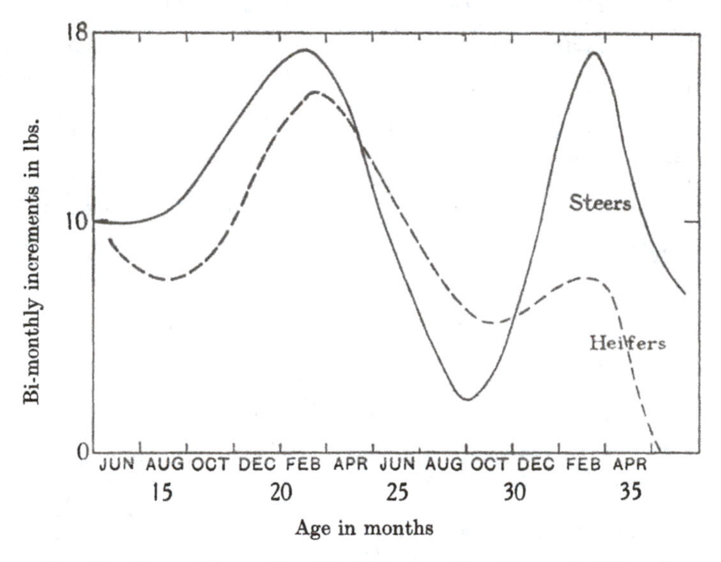

Fig. 69. Seasonal growth of S. African cattle: Sussex half-breeds. After Schütte.

reason. Before turnips came into cultivation in the eighteenth century our own cattle starved for half the year and grew fat the other, and in many countries the same thing happens still. In South Africa the rainy season lasts from November to February; by January the grass is plentiful, by June or July the veldt is parched until rain comes again. Cattle fatten from January to March or April; from July to October they put on little weight, or lose weight rather than put it on*.

* Cf. D. J. Schütte, in *Onderstepoort Journal*, Oct. 1935.

The growth of trees

Some sixty years ago Sir Robert Christison, a learned and versatile Edinburgh professor, was the first to study the "exact measurement" of the girth of trees*; and his way of putting a girdle round the tree, and fitting a recording device to the girdle, is copied in the "dendrographs"† used in forestry today. The Edinburgh beeches begin to enlarge their trunks in late May or June, when in full leaf, and cease growing some three months later; the buds sprout and the leaves begin their work before the cambium wakens to activity. The beech-trees in Maryland do likewise, save that the dates are a little earlier in the year; and walnut-trees on high ground in Arizona shew a like short season of growth, differing somewhat in date or "phase," just as it did in Edinburgh, from one year to another.

Deciduous trees stop growing after the fall of the leaf, but evergreens grow all the year round, more or less. This broad fact is illustrated in the following table, which happens to relate to the

Mean monthly increase in girth of trees at San Jorge, Uruguay: from C. E. Hall's data. Values given in percentages of total annual increment‡

	Jan	Feb.	Mar.	Apr.	May	June	July	Aug.	Sept.	Oct.	Nov.	Dec.
Evergreens	9·1	8·8	8·6	8·9	7·7	5·4	4·3	6·0	9·1	11·1	10·8	10·2
Deciduous trees	20·3	14·6	9·0	2·3	0·8	0·3	0·7	1·3	3·5	9·9	16·7	21·0

southern hemisphere, and to the climate of Uruguay. The measurements taken were those of the girth of the tree, in mm., at three feet from the ground. The evergreens included Pinus, Eucalyptus

* Sir R. Christison, On the exact measurement of trees, *Trans. Edinb. Botan. Soc.* XIV, pp. 164–172, 1882. Cf. also Duhamel du Monceau, *Des semis, et plantation des arbres*, Paris, 1750. On the general subject see (*int. al.*) Pfeffer's *Physiology of Plants*, II, Oxford, 1906; A. Mallock, Growth of trees, *Proc. R.S.* (B), XC, pp. 86–191, 1919. Mallock used an exceedingly delicate optical method, in which interference-bands, produced by two contiguous glass plates, shew a visible displacement on the slightest angular movement of the plates, even of the order of a millionth of an inch.

† W. S. Glock, A. E. Douglass and G. A. Pearson, Principles...of tree-ring analysis, *Carnegie Inst. Washington*, No. 486, 1937; D. T. MacDougal, *Tree Growth*, Leiden, 1938, 240 pp.

‡ *Trans. Edinb. Botan. Soc.* XVIII, p. 456, 1891.

and Acacia; the deciduous trees included Quercus, Populus, Robinia and Melia. The result (Fig. 70) is much as we might expect. The deciduous trees cease to grow in winter-time, and during all the months when the trees are bare; during the warm season the monthly values are regularly graded, approximately in a sine-curve, with a clear maximum (in the southern hemisphere) about the month of December. In the evergreens the amplitude of the

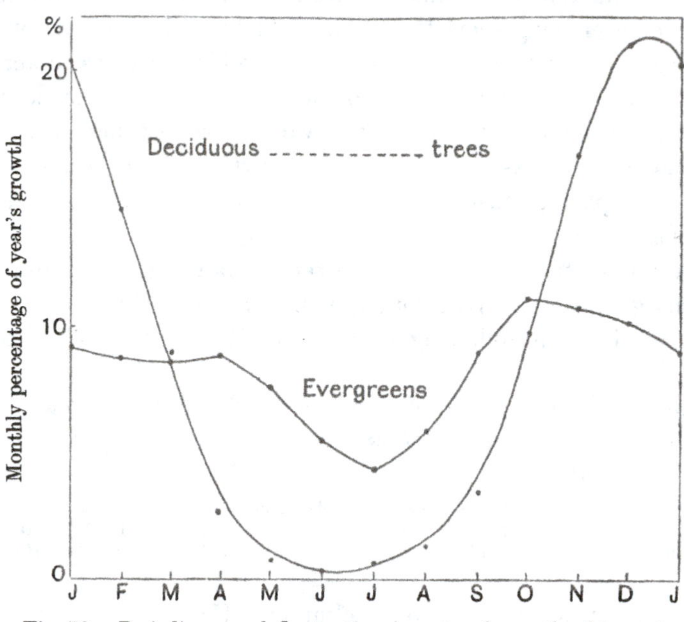

Fig. 70. Periodic annual fluctuation in rate of growth of trees in the southern hemisphere. From C. E. Hall's data.

annual wave is much less; there is a notable amount of growth all the year round, and while there is a marked diminution in rate during the coldest months, there is a tendency towards equality over a considerable part of the warmer season. In short, the evergreens, at least in this case, do not grow the faster as the temperature continues to rise; and it seems probable that some of them, especially the pines, are definitely retarded in their growth, either by a temperature above their optimum or by a deficiency of moisture, during the hottest season of the year.

Fig. 71 shews how a cypress never ceased to grow, but had alternate spells of quicker and slower growth, according to conditions of which we are not informed. Another figure (Fig. 72) illustrates the growth in three successive seasons of the Californian redwood, a near ally of the most gigantic of trees. Evergreen though the redwood is, its growth has periods of abeyance; there is a second minimum about midsummer, and the chief maximum of the year may be that before or after this.

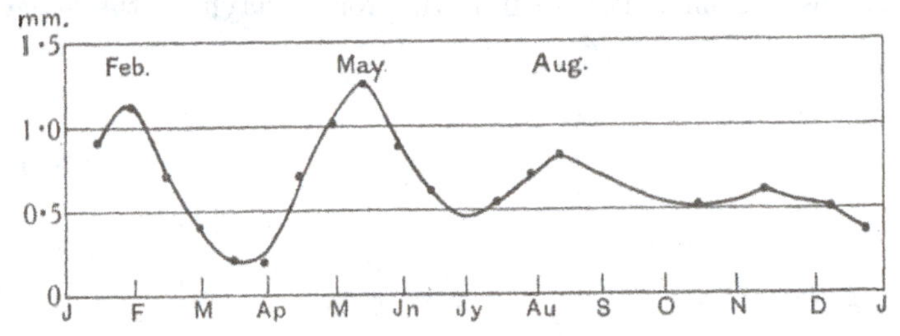

Fig. 71. Growth of cypress (*C. macrocarpa*), shewing seasonal periodicity. From MacDougal's data: smoothed curve.

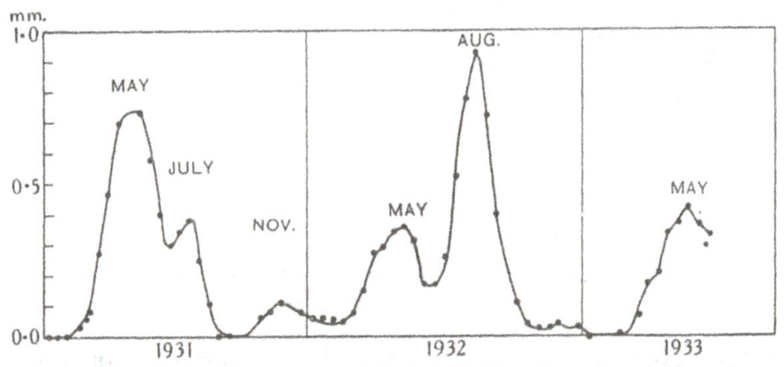

Fig. 72. Fortnightly increase of girth in Californian redwood (*Sequoia sempervirens*), shewing seasonal periodicity. After MacDougal.

In warm countries tree-growth is apt to shew a double maximum, for the cold of winter and the drought of summer are equally antagonistic to it. Trees grow slower—and grow fewer—the farther north we go, till only a few birches and willows remain, stunted and old; it is nearly a hundred years ago since Auguste Bravais* shewed a steadily decreasing growth-rate in the forests between 50° and 70° N.

* Recherches sur la croissance du pin silvestre dans le nord de l'Europe, *Mém. couronnées de l'Acad. R. de Belgique*, xv, 64 pp., 1840–41.

The delicate measuring apparatus now used shews sundry minor but beautiful phenomena. A daily periodicity of growth is a common thing* (Fig. 73). In the tree-cactuses the trunk expands by day and shrinks again after nightfall; for the stomata close in sunlight, and transpiration is checked until the sun goes down. But it is more usual for the trunk to shrink from sunrise until evening and to swell from sunset until dawn; for by daylight the leaves lose

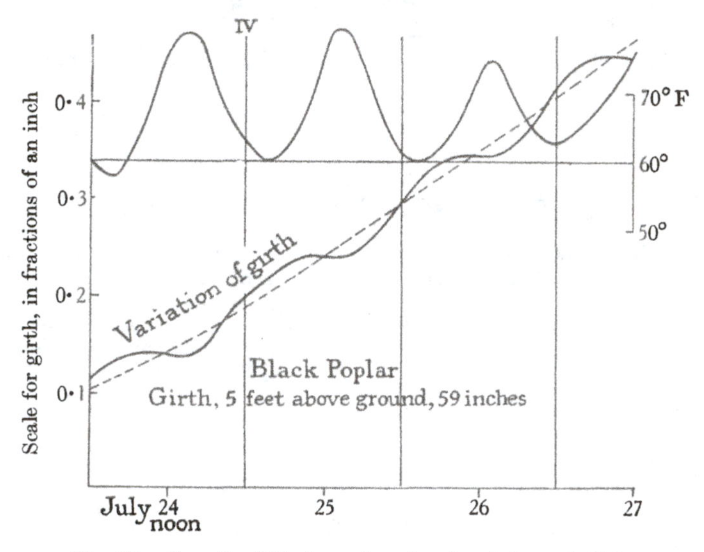

Fig. 73. Growth of black poplar, shewing daily periodicity. After A. Mallock.

water faster, and in the dark they lose it slower, than the roots replace it. The rapid midday loss of water even at the top of a tall *Sequoia* is quickly followed by a measurable constriction of the trunk fifty or even a hundred yards below†.

* The diurnal periodicity is beautifully shewn in the case of the hop by Johannes Schmidt, *C.R. du Laboratoire Carlsberg*, x, pp. 235–248, Copenhagen, 1913.

† This rapid movement is accounted for by Dixon and Joly's "cohesion-theory" of the ascent of sap. The leaves shew innumerable minute menisci, or cup-shaped water-surfaces, in their intercellular air-spaces. As water evaporates from these the little cups deepen, capillarity increases its pull, and suffices to put in motion the strands or columns of water which run continuously through the vessels of wood, and withstand rupture even under a pull of 100–200 atmospheres. See (*int. al.*) H. H. Dixon and J. Joly, On the ascent of sap, *Phil. Trans.* (B), CLXXXVI, p. 563, 1895; also Dixon's *Transpiration and the Ascent of Sap*, 8vo, London, 1914.

In the case of trees, the seasonal periodicity of growth and the direct influence of weather are both so well marked that we are entitled to make use of the phenomenon in a converse way, and to draw deductions (as Leonardo da Vinci did*) as to climate during past years from the varying rates of growth which the tree has recorded for us by the thickness of its annual rings. Mr A. E. Douglass, of the University of California, has made a careful study of this question, and I received from him (through Professor H. H. Turner) some measurements of the average width of the annual rings in Californian redwood, five hundred years old, in which trees the rings are very clearly shewn. For the first hundred years the mean of two trees was used, for the next four hundred years the mean of five; and the means of these (and sometimes of larger numbers) were found to be very concordant. A correction was applied by drawing a nearly straight line through the curve for the whole period, which line was assumed to represent the slowly diminishing mean width of annual ring accompanying the increasing size, or age, of the tree; and the actual growth as measured was equated with this diminishing mean. The figures used give, then, the ratio of the actual growth in each year to the mean growth of the tree at that epoch.

It was at once manifest that the growth-rate so determined shewed a tendency to fluctuate in a long period of between 100 and 200 years. I then smoothed the yearly values in groups of 100 (by Gauss's method of "moving averages"), so that each number thus found represented the mean annual increase during a century: that is to say, the value ascribed to the year 1500 represented the *average annual growth* during the whole period between 1450 and 1550, and so on. These values, so simply obtained, give us a curve of beautiful and surprising smoothness, from which we draw the direct conclusion that the climate of Arizona, during the last five hundred years, has fluctuated with a regular periodicity of almost precisely 150 years. I have drawn, more recently, and also from Mr Douglass's data, a similar curve for a group of pine trees in Calaveras County†. These trees are about 300 years old, and the

* Cf. J. Playfair McMurrich, *Leonardo da Vinci*, 1930, p. 247.

† When this was first written I had not seen Mr Douglass's paper On a method of estimating rainfall by the growth of trees, *Bull. Amer. Geograph. Soc.* XLVI,

data are reduced, as before, to moving averages of 100 years, but without further correction. The agreement between the growth-rate of these pines and that of the great *Sequoias* during the same period is very remarkable (Fig. 74).

We should be left in doubt, so far as these observations go, whether the essential factor be a fluctuation of temperature or an alternation of drought and humidity; but the character of the Arizona climate, and the known facts of recent years, encourage the belief that the latter is the more direct and more important factor. In a New England forest many trees of many kinds were studied after a hurricane; they shewed on the whole no correlation

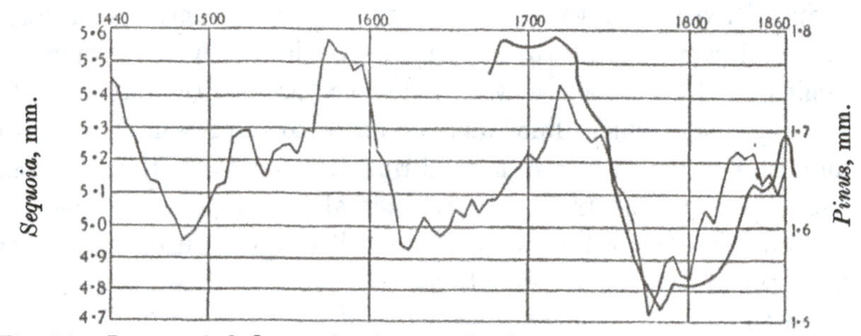

Fig. 74. Long-period fluctuation in growth of Arizona redwood (*Sequoia*), from A.D. 1390 to 1910; and of yellow pine from Calaveras County, from A.D. 1620 to 1920. (Smoothed in 100-year periods.)

between growth-rate and temperature, with the remarkable exception (in the conifers) of a clear correlation with the temperature of March and April, a month or two before the season's growth began. In a cold spring the melting snows and early rains ran off into the rivers, in a warm and early one they sank into the soil*; in other words, humidity was still the controlling factor. An ancient oak tree in Tunis is said to have recorded fifty years of abundant rain,

pp. 321–335, 1914; nor, of course, his great work on Climatic cycles and tree-growth, *Carnegie Inst. Publications*, 1919, 1928, 1936. Mr Douglass does not fail to notice the long period here described, but he is more interested in the sunspot-cycle and other shorter cycles known to meteorologists. See also (*int. al.*) E. Huntingdon, The fluctuating climate of North America, *Geograph. Journ.* Oct. 1912; and Otto Pettersson, Climatic variation in historic and prehistoric time, *Svenska Hydrografisk-Biolog. Skrifter*, v, 1914.

* C. J. Lyon, *Amer. Assoc. Rep.* 1939; *Nature*, Apr. 13, 1940, p. 595.

with short intervals of drought, during the eighteenth century; then, after 1790, longer droughts and shorter spells of rainy seasons*.

It has been often remarked that our common European trees, such as the elm or the cherry, have larger leaves the farther north we go; but the phenomenon is due to the longer hours of daylight throughout the summer, rather than to intensity of illumination or difference of temperature. On the other hand, long daylight, by prolonging vegetative growth, retards flowering and fruiting; and late varieties of soya bean may be forced into early ripeness by artificially shortening their daylight at midsummer†.

The effect of ultra-violet light, or any other portion of the spectrum, is part, and perhaps the chief part, of the same problem. That ultra-violet light accelerates growth has been shewn both in plants and animals‡. In tomatoes, growth is favoured by just such ultra-violet light as comes very near the end of the solar spectrum§, and as happens, also, to be especially absorbed by ordinary green-house glass‖. At the other end of the spectrum, in red or orange light, the leaves become smaller, their petioles longer, the nodes more numerous, the very cells longer and more attenuated. It is a physiological problem, and as such it shews how plant-life is adapted, on the whole, to just such rays as the sun sends; but it also shews the morphologist how the secondary effects of climate may so influence growth as to modify both size and form¶. An analogous case is the influence of light, rather than temperature, in modifying the coloration of organisms, such as certain butterflies.

* Le chêne Zeem d'Ain Draham, *Bull. du Directeur Général*, Tunisie, 1927.

† That the plant grows by turns in darkness and in light, and has its characteristic growth-phases in each, longer or shorter according to species and variety and normal habitat, is a subject now studied under the name of "photoperiodism," and become of great practical importance for the northerly extension of cereal crops in Canada and Russia. Cf. R. G. Whyte and M. A. Oljhovikov, *Nature*, Feb. 18, 1939.

‡ Cf. Kuro Suzuki and T. Hatano, in *Proc. Imp. Acad. of Japan*, III, pp. 94–96, 1927.

§ Withrow and Benedict, in *Bull. of Basic Scient. Research*, III, pp. 161–174, 1931.

‖ Cf. E. C. Teodoresco, Croissance des plantes aux lumières de diverses longueurs d'onde, *Ann. Sc. Nat., Bot.* (8), pp. 141–336, 1929; N. Pfeiffer, *Botan. Gaz.* LXXXV, p. 127, 1929; etc.

¶ See D. T. MacDougal, Influence of light and darkness, etc., *Mem. N.Y. Botan. Garden*, 1903, 392 pp.; Growth in trees, *Carnegie Inst.* 1921, 1924, etc.; J. Wiesner, *Lichtgenuss der Pflanzen*, VII, 322 pp., 1907; Earl S. Johnston, *Smithson. Misc. Contrib.* 18 pp., 1938; etc. On the curious effect of short spells of light and darkness, see H. Dickson, *Proc. R.S.* (B), CXV, pp. 115–123, 1938.

Now if temperature or light affect the rate of growth in strict uniformity, alike in all parts and in all directions, it will only lead to local races or varieties differing in size, as the Siberian goldfinch or bullfinch differs from our own. But if there be ever so little of a discriminating tendency such as to enhance the growth of one tissue or one organ more than another*, then it must soon lead to racial, or even "specific," difference of form.

It is hardly to be doubted that climate has some such discriminating influence. The large leaves of our northern trees are an instance of it; and we have a better instance of it still in Alpine plants, whose general habit is dwarfed though their floral organs suffer little or no reduction†. Sunlight of itself would seem to be a hindrance rather than a stimulant to growth; and the familiar fact of a plant turning towards the sun means increased growth on the shady side, or partial inhibition on the other.

More curious and still more obscure is the moon's influence on growth, as on the growth and ripening of the eggs of oysters, sea-urchins and crabs. Belief in such lunar influence is as old as Egypt; it is confirmed and justified, in certain cases, nowadays, but the way in which the influence is exerted is quite unknown‡.

Osmotic factors in growth

The curves of growth which we have been studying have a twofold interest, morphological and physiological. To the morphologist, who has learned to recognise form as a "function of growth," the most important facts are these: (1) that rate of growth is an orderly phenomenon, with general features common to various organisms, each having its own characteristic rates, or specific constants; (2) that rate of growth varies with temperature, and so with season and with climate, and also with various other physical factors, external and internal to the organism; (3) that it varies in different parts of the body, and along various directions or axes:

* Or as we might say nowadays, have a different "threshold value" in one organ to another.

† Cf. for instance, Nägeli's classical account of the effect of change of habitat on alpine and other plants, *Sitzungsber. Baier. Akad. Wiss.* 1865, pp. 228–284.

‡ Cf. Munro Fox, Lunar periodicity in reproduction, *Proc. R.S.* (B), xcv, pp. 523–550, 1935; also Silvio Ranzi, *Pubblic. Staz. Zool. Napoli*, xi, 1931.

such variations being harmoniously "graded," or related to one another by a "principle of continuity," so giving rise to the characteristic form and dimensions of the organism and to the changes of form which it exhibits in the course of its development. To the physiologist the phenomenon of growth suggests many other considerations, and especially the relation of growth itself to chemical and physical forces and energies.

To be content to shew that a certain rate of growth occurs in a certain organism under certain conditions, or to speak of the phenomenon as a "reaction" of the living organism to its environment or to certain stimuli, would be but an example of that "lack of particularity" with which we are apt to be all too easily satisfied. But in the case of growth we pass some little way beyond these limitations: to this extent, that an affinity with certain types of chemical and physical reaction has been recognised by a great number of physiologists*.

A large part of the phenomenon of growth, in animals and still more conspicuously in plants, is associated with "turgor," that is to say, is dependent on osmotic conditions. In other words, the rate of growth depends (as we have already seen) as much or more on the amount of water taken up into the living cells†, as on the actual amount of chemical metabolism performed by them; and sometimes, as in certain insect-larvae, we can even distinguish between tissues which grow by increase of cell-size, the result of imbibition, and others which grow by multiplication of their constituent cells‡. Of the chemical phenomena which result in the

* Cf. F. F. Blackman, Presidential Address in Botany, *Brit. Assoc.*, Dublin, 1908. The idea was first enunciated by Baudrimont and St Ange, Recherches sur le développement du foetus, *Mém. Acad. Sci.* XI, p. 469, 1851.

† Cf. J. Loeb, *Untersuchungen zur physiologischen Morphologie der Tiere*, 1892; also Experiments on cleavage, *Journ. Morphology*, VII, p. 253, 1892; Ueber die Dynamik des tierischen Wachstums, *Arch. f. Entw. Mech.* XV, p. 669, 1902–3; Davenport, On the rôle of water in growth, *Boston Soc. N.H.* 1897; Ida H. Hyde in *Amer. Journ. Physiology*, XII, p. 241, 1905; Bottazzi, Osmotischer Druck und elektrische Leitungsfähigkeit der Flüssigkeiten der Organismen, in Asher-Spiro's *Ergebnisse der Physiologie*, VII, pp. 160–402, 1908; H. A. Murray in *Journ. Gener. Physiology*, IX, p. 1, 1925; J. Gray, The role of water in the evolution of the terrestrial vertebrates, *Journ. Exper. Biology*, VI, pp. 26–31, 1928; and A. N. J. Heyn, Physiology of cell-elongation, *Botan. Review*, VI, pp. 515–574, 1940.

‡ Cf. C. A. Berger, Carnegie Inst. of Washington, *Contributions to Embryology*, XXVII, 1938.

actual increase of protoplasm we shall speak presently, but the rôle of water in growth deserves a passing word, even in our morphological enquiry.

The lower plants only live and grow in abundant moisture; few fungi continue growing when the humidity falls below 85 per cent. of saturation, and the mould-fungi, such as *Penicillium*, need more moisture still (Fig. 75). Their limit is reached a little below 90 %.

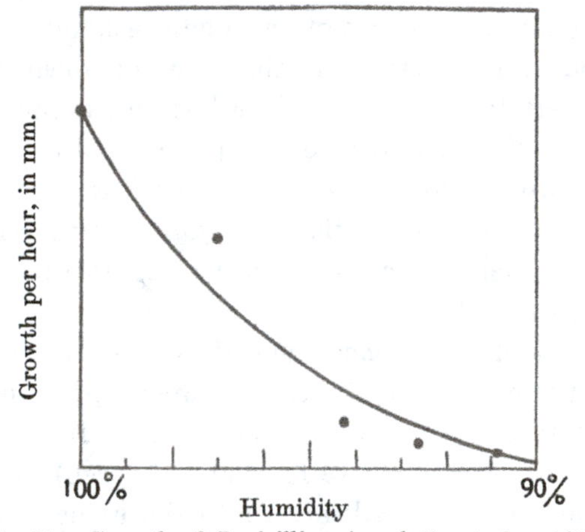

Fig. 75. Growth of *Penicillium* in relation to humidity.

Growth of Penicillium *(at* 25° C.)*

Humidity (% of saturation)	Growth per hour (mm.)
100·0	7·7
97·0	5·0
94·2	1·0
92·6	0·5
90·8	0·3

Among the coelenterate animals growth and ultimate size depend on little more than absorption of water and consequent turgescence, the process shewing itself in simple ways. A sea-anemone may live to an immense age†, but its age and size have little to do with one

* From R. G. Tomkins, Studies of the growth of moulds, *Proc. R.S.* (B), CV, pp. 375–401, 1929.

† Like Sir John Graham Dalyell's famous "Granny," and Miss Nelson's family of *Cereus* (not *Sagartia*) of which one still lives at over 80 years old. Cf. J. H. Ashworth and Nelson Annandale, in *Trans. R. Physical Soc. Edin.* xxv, pp. 1–14 1904.

another. It has an upper limit of size vaguely characteristic of the species, and if fed well and often it may reach it in a year; on stinted diet it grows slowly or may dwindle down; it may be kept at wellnigh what size one pleases. Certain full-grown anemones were left untended in war-time, unfed and in water which evaporated down to half its bulk; they shrank down to little beads, and grew up again when fed and cared for.

Loeb shewed, in certain zoophytes, that not only must the cells be turgescent in order to grow, but that this turgescence is possible only so long as the salt-water in which the cells lie does not overstep a certain limit of concentration: a limit reached, in the case of *Tubularia*, when the salinity amounts to about 3·4 per cent. Sea-water contains some 3·0 to 3·5 per cent. of salts in the open sea, but the salinity falls much below this normal, to about 2·2 per cent., before *Tubularia* exhibits its full turgescence and maximal growth; a further dilution is deleterious to the animal. It is likely enough that osmotic conditions control, after this fashion, the distribution and local abundance of many zoophytes. Loeb has also shewn* that in certain fish-eggs (e.g. of *Fundulus*) an increasing concentration, leading to a lessening water-content of the egg, retards the rate of segmentation and at last arrests it, though nuclear division goes on for some time longer.

The eggs of many insects absorb water in large quantities, even doubling their weight thereby, and fail to develop if drought prevents their doing so; and sometimes the egg has a thin-walled stalk, or else a "hydropyle," or other structure by which the water is taken in†.

In the frog, according to Bialaszewicz‡, the growth of the embryo while within the vitelline membrane depends wholly on absorption of water. The rate varies with the temperature, but the amount of water absorbed is constant, whether growth be fast or slow. Moreover, the successive changes of form correspond to definite quantities of water absorbed, much of which water is intracellular. The solid residue, as Davenport has also shewn, may even diminish

* *Pflüger's Archiv*, LV, 1893.

† Cf. V. B. Wigglesworth, *Insect Physiology*, 1939, p. 2.

‡ Beiträge zur Kenntniss d. Wachstumsvorgänge bei Amphibienembryonen, *Bull. Acad. Sci. de Cracovie*, 1908, p. 783; also A. Drzwina and C. Bohn, De l'action...des solutions salines sur les larves des batraciens, *ibid.* 1906.

notably, while all the while the embryo continues to grow in bulk and weight. But later on, and especially in the higher animals, the water-content diminishes as growth proceeds and age advances; and loss of water is followed, or accompanied, by retardation and cessation of growth. A crab loses water as each phase of growth draws to an end and the corresponding moult approaches; but it absorbs water in large quantities as soon as the new period of growth begins*. Moreover, that water is lost as growth goes on has been shewn by Davenport for the frog, by Potts for the chick, and particularly by Fehling in the case of man. Fehling's results may be condensed as follows:

Age in weeks (man)	6	17	22	24	26	30	35	39
Percentage of water	97·5	91·8	92·0	89·9	86·4	83·7	82·9	74·2

The following illustrate Davenport's results for the frog:

Age in weeks (frog)	1	2	5	7	9	14	41	84
Percentage of water	56·3	58·5	76·7	89·3	93·1	95·0	90·2	87·5

The following table epitomises the drying-off of ripening maize†; it shews how ripening and withering are closely akin, and are but two phases of senescence (Fig. 76):

Days (from August 6)	0	22	35	49	56	63
Percentage of water	87	81	77	68	65	58

The bird's egg provides all the food and all the water which the growing embryo needs, and to carry a provision of water is the special purpose of the white of the egg; the water contained in the albumen at the beginning of incubation is just about what the chick contains at the end. The yolk is not surrounded by water, which would diffuse too quickly into it, nor by a crystalloid solution, whose osmotic value would soon increase; but by a watery albuminous colloid, whose osmotic pressure changes slowly as its charge of water is gradually withdrawn‡.

* Cf. A. Krogh, *Osmotic regulation in aquatic animals*, Cambridge, 1939.

† Henry and Morrison, 1917; quoted by Otto Glaser, on Growth, time and form, *Biolog. Reviews*, XIII, pp. 2–58, 1938.

‡ Cf. James Gray, in *Journ. Exper. Biology*, IV, pp. 214–225, 1926.

Distribution of water in a hen's egg

Day of incubation	Gm. of water contained in Albumen	Embryo	Yolk	Loss by evaporation	Gain by combustion
0	29·9	0·0	8·5	0·0	0·0
6	27·2	0·4	8·45	2·4	0·01
12	20·4	4·6	7·8	5·6	0·27
18	9·2	18·1	2·3	8·8	1·20
20	2·2	27·4	1·0	9·8	2·00

The actual amount of water, compared with the dry solids in the egg, has been determined as follows:

Day of incubation (chick)	5	8	11	14	17	19
Percentage of water	94·7	93·8	92·3	87·7	82·8	82·3

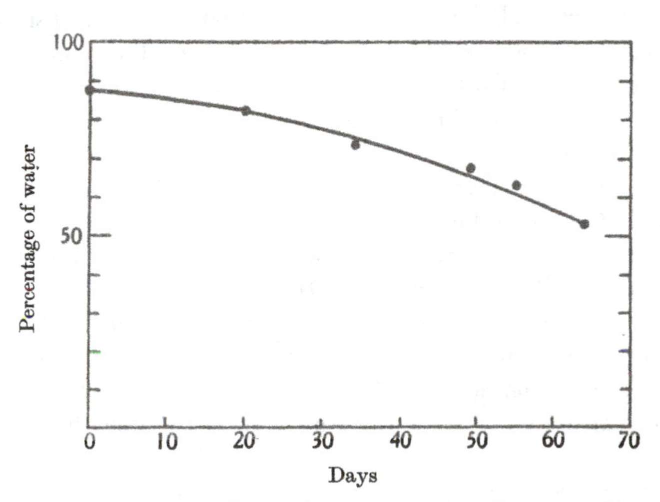

Fig. 76. Percentage of water in ripening maize. From Otto Glaser.

We know very little of the part which all this water plays: how much is mere "reaction-medium," how much is fixed in hydrated colloids, how much, in short, is bound or unbound. But we see that somehow or other water is lost, and lost in considerable amount, as the embryo draws towards completion and ceases for the time being to grow.

All vertebrate animals contain much the same amount of water in their living bodies, say 85 per cent. or thereby, however unequally distributed in the tissues that water may be*. Land animals have evolved from water animals with little change in this respect, though the constant proportion of water is variously achieved. A newt loses moisture by evaporation with the utmost freedom, and regains it by no less rapid absorption through the skin; while a lizard in his scaly coat is less liable to the one and less capable of the other, and must drink to replace what water it may lose.

We are on the verge of a difficult subject when we speak of the rôle of water in the living tissues, in the growth of the organism, and in the manifold activities of the cell; and we soon learn, among other more or less unexpected things, that osmotic equilibrium is neither universal nor yet common in the living organism. The yolk maintains a higher osmotic pressure than the white of the egg—so long as the egg is living; and the watery body of a jellyfish, though not far off osmotic equilibrium, has a somewhat less salinity than the sea-water. In other words, its surface acts to some extent as a semipermeable membrane, and the fluid which causes turgescence of the tissues is less dense than the sea-water outside†.

In most marine invertebrates, however, the body-fluids constituting the *milieu interne* are isotonic with the *milieu externe*, and vary in these animals *pari passu* with the large variations to which sea-water itself is subject. On the other hand, the dwellers in fresh-water, whether invertebrates or fishes, have, naturally, a more concentrated medium within than without. As to fishes, different kinds shew remarkable differences. Sharks and dogfish have an osmotic pressure in their blood and their body fluids little different

* The vitreous humour is nearly all water, the enamel has next to none, the grey matter has some 86 per cent., the bones, say 22 per cent.; lung and kidney take up more than they can hold, and so become excretory or regulatory organs. Eggs, whether of dogfish, salmon, frogs, snakes or birds, are composed, roughly speaking, of half water and half solid matter.

† Cf. (*int. al.*) G. Teissier, Sur la teneur en eau...de Chrysaora, *Bull. Soc. Biol. de France*, 1926, p. 266. And especially A. V. Hill, R. A. Gortner and others, On the state of water in colloidal and living systems, *Trans. Faraday Soc.* XXVI, pp. 678–704, 1930. For recent literature see (e.g.) Homer Smith, in *Q. Rev. Biol.* VII, p. 1, 1932; E. K. Marshall, *Physiol. Rev.* XIV, p. 133, 1934; Lovatt Evans, *Recent Advances in Physiology*, 4th ed., 1930; M. Duval, Recherches...sur le milieu intérieur des animaux aquatiques, *Thèse*, Paris, 1925; Paul Portier, *Physiologie des animaux marins*, Chap. III, Paris, 1938; G. P. Wells and I. C. Ledingham, Effects of a hypotonic environment, *Journ. Exp. Biol.* XVII, pp. 337–352, 1940.

from that of the sea-water outside: but with certain chemical differences, for instance that the chlorides within are much diminished, and the molecular concentration is eked out by large accumulations of urea in the blood. The marine teleosts, on the other hand, have a much lower osmotic pressure within than that of the sea-water outside, and only a little higher than that of their fresh-water allies. Some, like the conger-eel, maintain an all but constant internal concentration, very different from that outside; and this fish, like others, is constantly absorbing water from the sea; it must be exuding or excreting salt continually*. Other teleosts differ greatly in their powers of regulation and of tolerance, the common stickleback (which we may come across in a pool or in the middle of the North Sea) being exceptionally tolerant or "euryhaline†." Physiology becomes "comparative" when it deals with differences such as these, and Claude Bernard foresaw the existence of just such differences: "Chez tous les êtres vivants le milieu intérieur, qui est un produit de l'organisme, conserve les rapports nécessaires d'échange avec le milieu extérieur; mais à mesure que l'organisme devient plus parfait *le milieu organique se spécifie*, et s'isole en quelque sorte de plus en plus au milieu ambiant‡." Claude Bernard was building, if I mistake not, on Bichat's earlier concept, famous in its day, of life as "une alternation habituelle d'action de la part des corps extérieurs, et de réaction de la part du corps vivant": out of which grew his still more famous aphorism, "La vie est l'ensemble des fonctions qui résistent à la mort§."

One crab, like one fish, differs widely from another in its power

* Probably by help of Henle's tubules in the kidney, which structures the dogfish does not possess. But the gills have their part to play as water-regulators, as also, for instance, in the crab.

† The grey mullets go down to the sea to spawn, but may live and grow in brackish or nearly fresh-water. The several species differ much in their adaptability, and Brunelli sets forth, as follows, the range of salinity which each can tolerate:

M. auratus	24–35 per mille
saliens	16–40
chelo	10–40
capito	5–40
cephalus	4–40

‡ *Introduction à l'étude de la médecine expérimentale*, 1855, p. 110. For a discussion of this famous concept see J. Barcroft, "La fixité du milieu intérieur est la condition de la vie libre," *Biol. Reviews*, VIII, pp. 24–87, 1932.

§ *Sur la vie et la mort*, p. 1.

of self-regulation; and these physiological differences help to explain, in both cases, the limitation of this species or that to more or less brackish, or more or less saline, waters. In deep-sea crabs (*Hyas*, for instance) the osmotic pressure of the blood keeps nearly to that of the *milieu externe*, and falls quickly and dangerously with any dilution of the latter; but the little shore-crab (*Carcinus moenas*) can live for many days in sea-water diluted down to one-quarter of its usual salinity. Meanwhile its own fluids dilute slowly, but not near so far; in other words, this crab combines great powers of osmotic regulation with a large capacity for tolerating osmotic gradients which are beyond its power to regulate. How the unequal balance is maintained is yet but little understood. But we do know that certain organs or tissues, especially the gills and the antennary gland, absorb, retain or eliminate certain elements, or certain ions, faster than others, and faster than mere diffusion accounts for; in other words, "ionic" regulation goes hand in hand with "osmotic" regulation, as a distinct and even more fundamental phenomenon*. This at least seems generally true—and only natural—that quickened respiration and increased oxygen-consumption accompany all such one-sided conditions: in other words, the "steady state" is only maintained by the doing of work and the expenditure of energy†.

To the dependence of growth on the uptake of water, and to the phenomena of osmotic balance and its regulation, Höber‡ and also Loeb were inclined to refer the modifications of form which certain phyllopod crustacea undergo when the highly saline waters which they inhabit are further concentrated, or are abnormally diluted. Their growth is retarded by increased concentration, so that individuals from the more saline waters appear stunted and dwarfish; and they become altered or transformed in other ways, suggestive of "degeneration," or a failure to attain full and perfect develop-

* See especially D. A. Webb, Ionic regulation in *Carcinus moenas*, *Proc. R.S.* (B), CXXIX, pp. 107–136, 1940.

† In general the fresh-water Crustacea have a larger oxygen-consumption than the marine. *Stenohaline* and *euryhaline* are terms applied nowadays to species which are confined to a narrow range of salinity, or are tolerant of a wide one. An extreme case of toleration, or adaptability, is that of the Chinese woolly-handed crab, *Eriocheir*, which has not only acclimatised itself in the North Sea but has ascended the Elbe as far as Dresden.

‡ R. Höber, Bedeutung der Theorie der Lösungen für Physiologie und Medizin, *Biol. Centralbl.* XIX, p. 272, 1899.

ment*. Important physiological changes ensue. The consumption of oxygen increases greatly in the stronger brines, as more and more active "osmo-regulation" is required. The rate of multiplication is increased, and parthenogenetic reproduction is encouraged. In the less saline waters male individuals, usually rare, become plentiful, and here the females bring forth their young alive; males disappear altogether in the more concentrated brines, and then the females lay eggs, which, however, only begin to develop when the salinity is somewhat reduced.

The best-known case is the little brine-shrimp, *Artemia salina*, found in one form or another all the world over, and first discovered nearly two hundred years ago in the salt-pans at Lymington. Among many allied forms, one, *A. milhausenii*, inhabits the natron-lakes of Egypt and Arabia, where, under the name of "loul," or "Fezzan-worm," it is eaten by the Arabs†. This fact is interesting, because it indicates (and investigation has apparently confirmed) that the tissues of the creature are not impregnated with salt, as is the medium in which it lives. In short *Artemia*, like teleostean fishes in the sea, lives constantly in a "hypertonic medium"; the fluids of the body, the *milieu interne*, are no more salt than are those of any ordinary crustacean or other animal, but contain only some 0·8 per cent. of NaCl‡, while the *milieu externe* may contain from 3 to 30 per cent. of this and other salts; the skin, or body-wall, of the creature acts as a "semi-permeable membrane," through which the dissolved salts are not permitted to diffuse, though water passes freely. When brought into a lower concentration the animal may grow large and turgescent, until a statical equilibrium, or steady state, is at length attained.

Among the structural changes which result from increased con-

* Schmankewitsch, *Zeitschr. f. wiss. Zool.* XXIX, p. 429, 1877. Schmankewitsch has made equally interesting observations on change of size and form in other organisms, after some generations in a milieu of altered density; e.g. in the flagellate infusorian *Ascinonema acinus* Bütschli.

† These "Fezzan-worms," when first described, were supposed to be "insects' eggs"; cf. Humboldt, *Personal Narrative*, VI, i, 8, note; Kirby and Spence, Letter X.

‡ See D. J. Kuenen, Notes, systematic and physiological, on *Artemia*, *Arch. Néerland. Zool.* III, pp. 365–449, 1939; cf. also Abonyi, *Z. f. w. Z.* CXIV, p. 134, 1915. Cf. Mme. Medwedewa, Ueber den osmotischen Druck der Haemolymph v. *Artemia*; in *Ztsch. f. vergl. Physiolog.* V, pp. 547–554, 1922.

centration of the brine (partly during the life-time of the individual, but more markedly during the short season which suffices for the development of three or four, or perhaps more, successive generations), it is found that the tail comes to bear fewer and fewer bristles, and the tail-fins themselves tend at last to disappear: these changes corresponding to what have been described as the specific characters of *A. milhausenii*, and of a still more extreme form, *A. köppeniana*; while on the other hand, progressive dilution of the water tends to precisely opposite conditions, resulting in forms which have also been described as separate species, and even

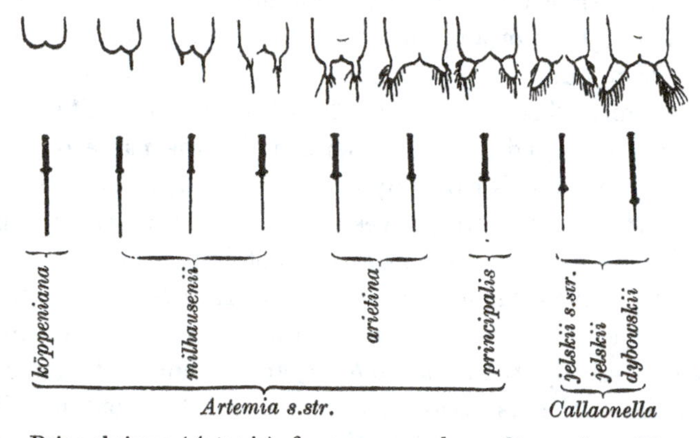

Fig. 77. Brine-shrimps (*Artemia*), from more or less saline water. Upper figures shew tail-segment and tail-fins; lower figures, relative length of cephalothorax and abdomen. After Abonyi.

referred to a separate genus, *Callaonella*, closely akin to *Branchipus* (Fig. 77). *Pari passu* with these changes, there is a marked change in the relative lengths of the fore and hind portions of the body, that is to say, of the cephalothorax and abdomen: the latter growing relatively longer, the salter the water. In other words, not only is the rate of growth of the whole animal lessened by the saline concentration, but the specific rates of growth in the parts of its body are relatively changed. This latter phenomenon lends itself to numerical statement, and Abonyi has shewn that we may construct a very regular curve, by plotting the proportionate length of the creature's abdomen against the salinity, or density, of the water; and the several species of *Artemia*, with all their other correlated specific characters, are then found to occupy successive, more or less well-defined, and more or less extended, regions of the curve (Fig. 78). In short, the density of the water is so clearly "specific," that we might briefly define *Artemia jelskii*, for instance, as the *Artemia* of density 1000–1010 (NaCl), or all but fresh water, and the typical *A. salina* (or *principalis*) as the *Artemia* of density 1018–1025, and so on*.

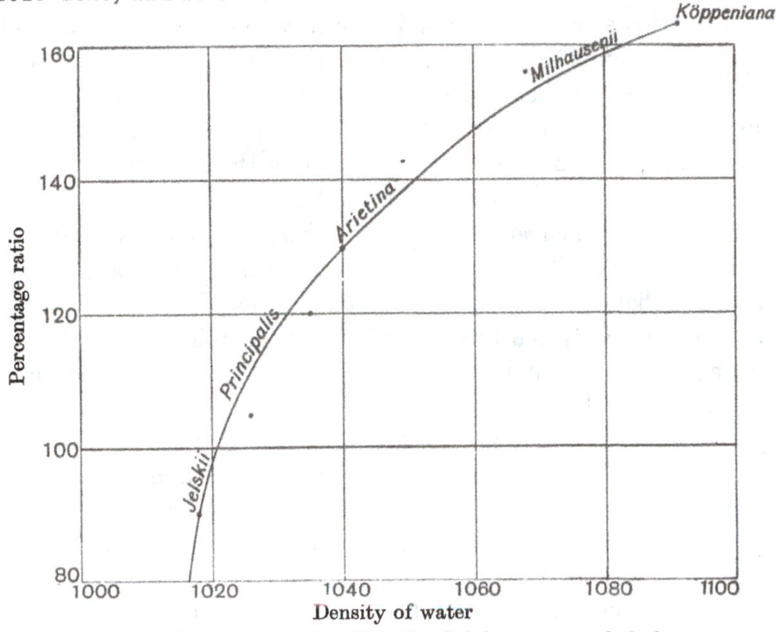

Fig. 78. Percentage ratio of length of abdomen to cephalothorax in brine-shrimps, at various salinities. After Abonyi.

These Artemiae are capable of living in waters not only of great density, but of very varied chemical composition, and it is hard to say how far they are safeguarded by semi-permeability or by specific properties and reactions of the living colloids†. The natron-lakes,

* Different authorities have recognised from one to twenty species of *Artemia*. Daday de Deés (*Ann. sci. nat.* 1910) reduces the salt-water forms to one species with four varieties, but keeps *A. jelskii* in a separate sub-genus. Kuenen suggests two species, *A. salina* and *gracilis*, one for the European and one for the American forms. According to Schmankewitsch every systematic character can be shewn to vary with the external medium. Cf. Professor Labbé on change of characters, specific and even generic, of Copepods according to the *p*H of saline waters at Le Croisic, *Nature*, March 10, 1928.

† We may compare Wo. Ostwald's old experiments on *Daphnia*, which died in a pure solution of NaCl isotonic with normal sea-water. Their death was not to be explained on osmotic grounds; but was seemingly due to the fact that the organic gels do not retain their normal water-content save in the presence of such concentrations of $MgCl_2$ (and other salts) as are present in sea-water.

for instance, contain large quantities of magnesium sulphate; and the Artemiae continue to live equally well in artificial solutions where this salt, or where calcium chloride, has largely replaced the common salt of the more usual habitat. Moreover, such waters as those of the natron-lakes are subject to great changes of chemical composition as evaporation and concentration proceed, owing to the different solubilities of the constituent salts; but it appears that the forms which the Artemiae assume, and the changes which they undergo, are identical, or indistinguishable, whichever of the above salts happen to exist or to predominate in their saline habitat. At the same time we still lack, so far as I know, the simple but crucial experiments which shall tell us whether, in solutions of different chemical composition, it is *at equal densities*, or *at isotonic concentrations* (that is to say, under conditions where the osmotic pressure, and consequently the rate of diffusion, is identical), that the same changes of form and structure are produced and corresponding phases of equilibrium attained.

Sea-water has been described as an instance of the "fitness of the environment*" for the maintenance of protoplasm in an appropriate milieu; but our Artemias suffice to shew how nature, when hard put to it, makes shift with an environment which is wholly abnormal and anything but "fit."

While Höber and others† have referred all these phenomena to osmosis, Abonyi is inclined to believe that the viscosity, or mechanical resistance, of the fluid also reacts upon the organism; and other possible modes of operation have been suggested. But we may take it for certain that the phenomenon as a whole is not a simple one. We should have to look far in organic nature for what the physicist would call simple osmosis‡; and assuredly there is always at work, besides the passive phenomena of intermolecular

* L. H. Henderson, *The Fitness of the Environment*, 1913.

† Cf. Schmankewitsch, *Z. f. w. Zool.* XXV, 1875; XXIX, 1877, etc.; transl. in appendix to Packard's *Monogr. of N. American Phyllopoda*, 1883, pp. 466–514; Daday de Deés, *Ann. Sci. Nat.* (*Zool*), (9), XI, 1910; Samter und Heymons, *Abh. d. K. pr. Akad. Wiss.* 1902; Bateson, *Mat. for the Study of Variation*, 1894, pp. 96–101; Anikin, *Mitth. Kais. Univ. Tomsk*, XIV: *Zool. Centralbl.* VI, pp. 756–760, 1908; Abonyi, *Z. f. w. Zool.* CXIV, pp. 96–168, 1915 (with copious bibliography), etc.

‡ Cf. C. F. A. Pantin, Body fluids in animals, *Biol. Reviews*, VI, p. 4, 1931; J. Duclaux, *Chimie appliquée à la biologie*, 1937, II, chap. 4.

diffusion, some other activity to play the part of a regulatory mechanism*.

On growth and catalytic action

In ordinary chemical reactions we have to deal (1) with a specific velocity proper to the particular reaction, (2) with variations due to temperature and other physical conditions, (3) with variations due to the quantities present of the reacting substances, according to Van't Hoff's "Law of Mass Action," and (4) in certain cases with variations due to the presence of "catalysing agents," as Berzelius called them a hundred years ago†. In the simpler reactions, the law of mass involves a steady slowing-down of the process as the reaction proceeds and as the initial amount of substance diminishes: a phenomenon, however, which is more or less evaded in the organism, part of whose energies are devoted to the continual bringing-up of supplies.

Catalytic action occurs when some substance, often in very minute quantity, is present, and by its presence produces or accelerates a reaction by opening "a way round," without the catalysing agent itself being diminished or used up‡. It diminishes the resistance somehow—little as we know what resistance means

* According to the empirical canon of physiology, that, as Léon Frédéricq expresses it (*Arch. de Zool.* 1885), "L'être vivant est agencé de telle manière que chaque influence perturbatrice provoque d'elle-même la mise en activité de l'appareil compensateur qui doit neutraliser et réparer le dommage." Herbert Spencer had conceived a similar principle, and thought he recognised in it the *vis medicatrix Naturae.* It is the physiological analogue of the "principle of Le Chatelier" (1888), with this important difference that the latter is a rigorous and quantitative law, based on a definite and stable equilibrium. The close relation between the two is maintained by Le Dantec (*La Stabilité de la Vie*, 1910, p. 24), and criticised by Lotka (*Physical Biology*, p. 283 *seq.*).

† In a paper in the *Berliner Jahrbuch* for 1836. This paper was translated in the *Edinburgh New Philosophical Journal* in the following year; and a curious little paper On the coagulation of albumen, and catalysis, by Dr Samuel Brown, followed in the *Edinburgh Academic Annual* for 1840.

‡ Such phenomena come precisely under the head of what Bacon called *Instances of Magic*: "By which I mean those wherein the material or efficient cause is scanty and small as compared with the work or effect produced; so that even when they are common, they seem like miracles, some at first sight, others even after attentive consideration. These magical effects are brought about in three ways...[of which one is] by excitation or invitation in another body, as in the magnet which excites numberless needles without losing any of its virtue, *or in yeast and such-like.*" *Nov. Org.*, cap. li.

in a chemical reaction. But the velocity-curve is not altered in form; for the amount of energy in the system is not affected by the presence of the catalyst, the law of mass exerts its effect, and the rate of action gradually slows down. In certain cases we have the remarkable phenomenon that a body capable of acting as a catalyser is necessarily formed as a product, or by-product, of the main reaction, and in such a case as this the reaction-velocity will tend to be steadily accelerated. Instead of dwindling away, such a reaction continues with an ever-increasing velocity: always subject to the reservation that limiting conditions will in time make themselves felt, such as a failure of some necessary ingredient (the "law of the minimum"), or the production of some substance which shall antagonise and finally destroy the original reaction. Such an action as this we have learned, from Ostwald, to describe as "autocatalysis." Now we know that certain products of protoplasmic metabolism—we call them enzymes—are very powerful catalysers, a fact clearly understood by Claude Bernard long ago*; and we are therefore entitled, to that extent, to speak of an autocatalytic action on the part of protoplasm itself.

Going a little farther in the footsteps of Claude Bernard, Chodat of Geneva suggested (as we are told by his pupil Monnier) that growth itself might be looked on as a catalytic, or autocatalytic reaction: "On peut bien, ainsi que M. Chodat l'a proposé, considérer l'accroissement comme une réaction chimique complexe, dans laquelle le catalysateur est la cellule vivante, et les corps en présence sont l'eau, les sels et l'acide carbonique†."

A similar suggestion was made by Loeb, in connection with the

* "Les diastases contiennent, en définitive, le secret de la vie. Or, les actions diastatiques nous apparaissent comme des phénomènes catalytiques, en d'autres termes, des accélérations de vitesse de réaction." Cf. M. F. Porchet, *Revue Scientifique*, 18th Feb. 1911. For a last word on this subject, see W. Frankenberger, *Katalytische Umsetzungen in homogenen u. enzymatischen Systemen*, Leipzig, 1937.

† Cf. R. Chodat, *Principes de Botanique* (2nd ed.), 1907, p. 133; A. Monnier, La loi d'accroissement des végétaux, *Publ. de l'Inst. de Bot. de l'Univ. de Genève* (7), III, 1905. Cf. W. Ostwald, *Vorlesungen über Naturphilosophie*, 1902, p. 342; Wo. Ostwald, Zeitliche Eigenschaften der Entwicklungsvorgänge, in Roux's *Vorträge*, Heft 5, 1908; Robertson, Normal growth of an individual, and its biochemical significance, *Arch. f. Entw. Mech.* XXV, pp. 581–614; XXVI, pp. 108–118, 1908; S. Hatai, Growth-curves from a dynamical standpoint, *Anat. Record*, V, p. 373, 1911; A. J. Lotka, *Ztschr. f. physikal. Chemie*, LXXII, p. 511, 1910; LXXX, p. 159, 1912; etc.

synthesis of nuclear protoplasm, or *nuclein*; for he remarked that, as in an autocatalysed chemical reaction, the rate of synthesis increases during the initial stage of cell-division in proportion to the amount of nuclear matter already there. In other words, one of the products of the reaction, i.e. one of the constituents of the nucleus, accelerates the production of nuclear from cytoplasmic material. To take one more instance, Blackman said, in the address already quoted, that "the botanists (or the zoologists) speak of *growth*, attribute it to a specific power of protoplasm for assimilation, and leave it alone as a fundamental phenomenon; but they are much concerned as to the distribution of new growth in innumerable specifically distinct forms. While the chemist, on the other hand, recognises it as a familiar phenomenon, and refers it to the same category as his other known examples of autocatalysis."

Later on, Brailsford Robertson upheld the autocatalytic theory with skill and learning*; and knowing well that growth was no simple solitary chemical reaction, he thought that behind it lay some one master-reaction, essentially autocatalytic, by which protoplasmic synthesis was effected or controlled. He adduced at least one curious case, in the growth and multiplication of the Infusoria, which can hardly be described otherwise than as catalytic. Two minute individuals (of *Enchelys* or *Colpodium*) kept in the same drop of water, so enhance each other's rate of asexual reproduction that it may be many times as great when two are together as when one is alone; the phenomenon has been called *allelocatalysis*. When a single infusorian is isolated, it multiplies the quicker the smaller the drop it is in—a further proof or indication that something is being given off, in this instance by the living cells, which hastens growth and reproduction. But even the ordinary multiplication of a bacterium, which doubles its numbers every few minutes till (were it not for limiting factors) those numbers would be all but incalculable in a day, looks like and has been cited as a simple but most striking instance of the potentialities of protoplasmic catalysis.

It is not necessary for us to pursue this subject much further.

* T. B. Robertson, *The Chemical Basis of Growth and Senescence*, 1923; and earlier papers. Cf. his Multiplication of isolated infusoria, *Biochem. Journ.* xv, pp. 598–611, 1921; cf. *Journ. Physiol.* LVI, pp. 404–412, 1921; R. A. Peters, Substances needed for the growth of...*Colpodium*, *Journ. Physiol.* LV, p. 1, 1921.

It is sufficiently obvious that the normal **S**-shaped curve of growth of an organism resembles in its general features the velocity-curve of chemical autocatalysis, and many writers have enlarged on the resemblance; but the **S**-shaped curve of growth of a population resembles it just as well. When the same curve depicts the growth of an individual, and of a population, and the velocity of a chemical reaction, it is enough to shew that the analogy between these is a mathematical and not a physico-chemical one. The sigmoid curve of growth, common to them all, is sufficiently explained as an interference effect, due to opposing factors such as we may use a differential equation to express: a phase of acceleration is followed by a phase of retardation, and the causes of both are in each case complex, uncertain or unknown. Nor are points of difference lacking between the chemical and the biological phenomena. As the chemical reaction draws to a close, it is by the gradual attainment of chemical equilibrium; but when organic growth comes to an end, it is (in all but the lowest organisms) by reason of a very different kind of equilibrium, due in the main to the gradual differentiation of the organism into parts, among whose peculiar properties or functions that of growth or multiplication falls into abeyance.

The analogy between organic growth and chemical autocatalysis is close enough to let us use, or try to use, just such mathematics as the chemist applies to his reactions, and so to reduce certain curves of growth to logarithmic formulae. This has been done by many, and with no little success *in simple cases*. So have we done, partially, in the case of yeast; so the statisticians and actuaries do with human populations; so we may do again, borrowing (for illustration) a certain well-known study of the growing sunflower (Figs. 79, 80). Taking our mathematics from elementary physical chemistry, we learn that:

The velocity of a reaction depends on the concentratior a of the substance acted on: V varies as a,

$$V = Ka.$$

The concentration continually decreases, so that at time t (in a monomolecular reaction),

$$V = \frac{dx}{dt} = k\,(a - x).$$

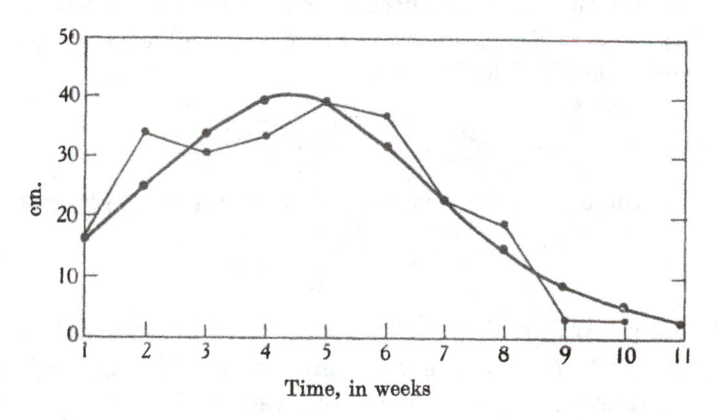

Fig. 79. Growth of sunflower-stem: observed and calculated curves. From Reed and Holland.

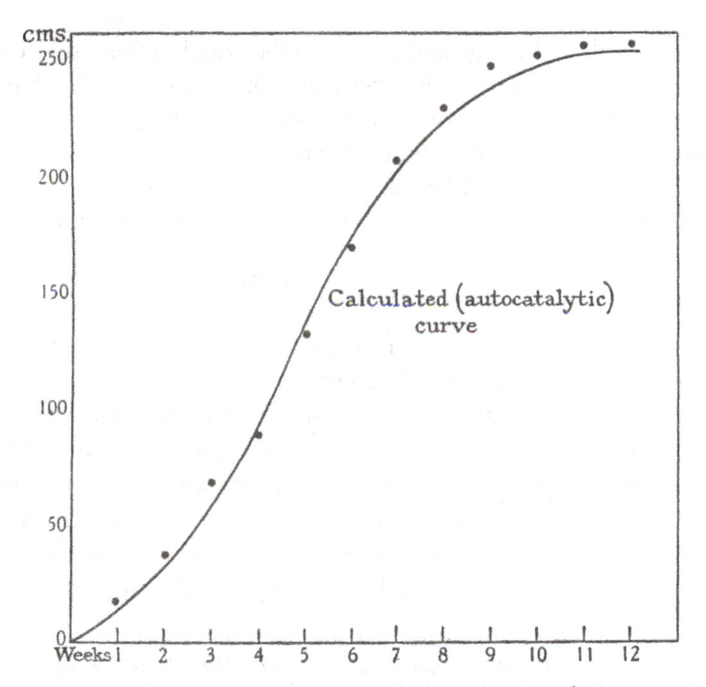

Fig. 80. Growth of sunflower-stem: calculated (autocatalytic) curve. After Reed and Holland.

But if the substance produced exercise a catalytic effect, then the velocity will vary not only as above but will also increase as x increases: the equation becomes

$$V = \frac{dx}{dt} = k'x\,(a - x),$$

which is the elementary equation of autocatalysis. Integrating,

$$\frac{1}{at}\log\frac{ax}{a - x} = k'.$$

In our growth-problem it is sometimes found convenient to choose for our epoch, t', the time when growth is half-completed, as the chemist takes the time at which his reaction is half-way through; and we may then write (with a changed constant)

$$\log\frac{x}{a - x} = K\,(t - t').$$

This is the physico-chemical formula which Reed and Holland apply to the growing sunflower-stem—a simple case*. For a we take the maximum height attained, viz. 254·5 cm.; for t', the epoch when one-half of that height was reached, viz. (by interpolation) about 34·2 days. Taking an observation at random, say that for the 56th day, when the stem was 228·3 cm. high, we have

$$\log\frac{228{\cdot}3}{254{\cdot}5 - 228{\cdot}3} = K\,(56 - 34{\cdot}2).$$

K in this case is found to be 0·043, and the mean of all such determinations† is not far different.

Applying this formula to successive epochs, we get a calculated curve in close agreement with the observed one; and by well-known statistical methods we confirm, and measure, its "closeness of fit." But just as the chemist must vary and develop his fundamental formula to suit the course of more and more complicated reactions, so the biologist finds that only the simplest of his curves

* H. S. Reed and R. H. Holland, The growth-rate of an annual plant, *Helianthus*, *Proc. Nat. Acad. of Sci.* (Washington), v, p. 135, 1919; cf. Lotka, *op. cit.*, p. 74, A similar case is that of a gourd, recorded by A. P. Anderson, *Bull. Survey, Minnesota*, 1895, and analysed by T. B. Robertson, *ibid.* pp. 72–75.

† Better determined, especially in more complex cases, by the method of least squares.

of growth, or only portions of the rest, can be fitted to this simplest of formulae. In a life-time are many ages; and no all-embracing formula covers the infant in the womb, the suckling child, the growing schoolboy, the old man when his work is done. Besides, we need such a formula as a biologist can understand! One which gives a mere coincidence of numbers may be of little use or none, unless it go some way to depict and explain the *modus operandi* of growth. As d'Ancona puts it: "Il importe d'appliquer des formules qui correspondent non seulement au point de vue géométrique, mais soient représentées par des valeurs de signification biologique." A mere curve-diagram is better than an *empirical* formula; for it gives us at least a picture of the phenomenon, and a qualitative answer to the problem.

Growth of sunflower-stem. (*After Reed and Holland*)

Age (days)	Height (cm.) Observed	Height (cm.) Calculated	1st diff.
7	17·9	21·9	
14	34·4	37·7	15·8
21	67·8	62·1	24·4
28	98·1	95·4	33·3
35	131·0	134·6	39·2
42	169·0	173·0	38·4
49	205·5	204·6	31·6
56	228·3	227·2	22·6
63	247·1	241·6	14·4
70	250·5	250·1	8·5
77	253·8	255·0	4·9
84	254·5	257·8	2·8

The chemical aspect of growth

As soon as we touch on such matters as the chemical phenomenon of catalysis we are on the threshold of a subject which, if we were able to pursue it, would lead us far into the special domain of physiology; and there it would be necessary to follow it if we were dealing with growth as a phenomenon in itself, instead of mainly as a help to our study and comprehension of form. The whole question of diet, of overfeeding and underfeeding*, would present

* For example, A. S. Parker has shewn that mice suckled by rats, and consequently much overfed, grow so quickly that in three weeks they reach double their normal weight; but their development is not accelerated; *Ann. Appl. Biol.* XVI, 1929.

itself for discussion*. But without opening up this large subject, we may say one more passing word on the remarkable fact that certain chemical substances, or certain physiological secretions, have the power of accelerating or of retarding or in some way regulating growth, and of so influencing the morphological features of the organism.

To begin with there are numerous elements, such as boron, manganese, cobalt, arsenic, which serve to stimulate growth, or whose complete absence impairs or hampers it; just as there are a few others, such as selenium, whose presence in the minutest quantity is injurious or pernicious. The chemistry of the living body is more complex than we were wont to suppose.

Lecithin was shewn long ago to have a remarkable power of stimulating growth in animals†, and accelerators of plant-growth, foretold by Sachs, were demonstrated by Bottomley and others‡; the several vitamins are either accelerators of growth, or are indispensable in order that it may proceed.

In the little duckweed of our ponds and ditches (*Lemna minor*) the botanists have found a plant in which growth and multiplication are reduced to very simple terms. For it multiplies by budding, grows a rootlet and two or three leaves, and buds again; it is all young tissue, it carries no dead load; while the sun shines it has no lack of nourishment, and may spread to the limits of the pond. In one of Bottomley's early experiments, duckweed was grown (1) in a "culture solution" without stint of space or food, and (2) in the same, with the addition of a little bacterised peat or "auximone." In both cases the little plant spread freely, as in the first, or Malthusian, phase of a population curve; but the peat greatly accelerated the rate, which was not slow before. Without the auximone the population doubled in nine or ten days, and with it in five or six; but in two months the one was seventy-fold the other!

The subject has grown big from small beginnings. We know certain substances, haematin being one, which stimulate the growth of bacteria, and seem to act on them as true catalysts. An obscure but complex body known as "bios" powerfully stimulates the growth of yeast; and the so-called *auxins*, a name which covers numerous bodies both nitrogenous and non-nitrogenous, serve in

* For a brief résumé of this subject see Morgan's *Experimental Zoology*, chap. XVI.

† Hatai, *Amer. Journ. Physiology*, X, p. 57, 1904; Danilewsky, *C.R.* CXXI, CXXII, 1895–96.

‡ W. B. Bottomley, *Proc. R.S.* (B), LXXXVIII, pp. 237–247, 1914, and other papers. O. Haberlandt, *Beitr. z. allgem. Botanik*, 1921.

minute doses to accelerate the growth of the higher plants*. Some of these "growth-substances" have been extracted from moulds or from bacteria, and one remarkable one, to which the name auxin is especially applied, from seedling oats. This last is no enzyme but a stable non-nitrogenous substance, which seems to act by softening the cell-wall and so facilitating the expansion of the cell. Lastly the remarkable discovery has been made that certain indol-compounds, comparatively simple bodies, act to all intents and purposes in the same way as the growth-hormones or natural auxins, and one of these "hetero-auxins," an indol-acetic acid†, is already in common and successful use to promote the growth and rooting of cuttings.

Growth of duckweed, with and without peat-auximone

	Without		With	
Weeks	Obs.	Calc.	Obs.	Calc.
0	20	20	20	20
1	30	33	38	55
2	52	54	102	153
3	77	88	326	424
4	135	155	1,100	1,173
5	211	237	3,064	3,250
6	326·	390	6,723	8,980
7	550	640	19,763	2,490
8	1052	1048	69,350	68,800
Percentage increase, per week		164%		277%

There are kindred matters not less interesting to the morphologist. It has long been known that the pituitary body produces, in its anterior lobe, a substance by which growth is increased and regulated. This is what we now call a "hormone"—a substance produced in one organ or tissue and regulating the functions of another. In this case atrophy of the gland leaves the subject a dwarf, and its hyper-

* The older literature is summarised by Stark, *Ergebn. d. Biologie*, II, 1906; the later by N. Nielsen, *Jb. wiss. Botan.* LXXIII, 1930; by Boyson Jensen, *Die Wuchsstofftheorie*, 1935; by F. W. Went and K. V. Thimann, *Phytohormones*, New York, 1937, and by H. L. Pearse, Plant hormones and their practical importance, *Imp. Bureau of Horticulture*, 1939. Cf. Went, *Rec. d. Trav. Botan. Néerl.* XXV, p. 1, 1928; A. N. J. Heyn, *ibid.* XXVIII, p. 113, 1931.

† Discovered by Kögl and Kostermans, *Ztschr. f. physiol. Chem.* CCXXXV, p. 201, 1934. Cf. (*int. al.*) P. W. Zimmermann and F. W. Wilcox in *Contrib. Boyce-Thompson Instit.* 1935.

trophy or over-activity goes to the making of a giant; the limb-bones of the giant grow longer, their epiphyses get thick and clumsy, and the deformity known as "acromegaly" ensues*. This has become a familiar illustration of functional regulation, by some glandular or "endocrinal" secretion, some enzyme or *harmozone* as Gley called it, or *hormone*† as Bayliss and Starling called it—in the particular case where the function to be regulated is growth, with its consequent influence on form. But we may be sure that this so-called regulation of growth is no simple and no specific thing, but implies a far-reaching and complicated influence on the bodily metabolism‡.

Some say that in large animals the pituitary is apt to be disproportionately large§; and the giant dinosaur Branchiosaurus, hugest of land animals, is reputed to have the largest hypophyseal recess (or cavity for the pituitary body) ever observed.

The thyroid also has its part to play in growth, as Gudernatsch was the first to shew||; perhaps it acts, as Uhlenhorth suggests, by releasing the pituitary hormone. In a curious race of dwarf frogs both thyroid and pituitary were found to be atrophied¶. When tadpoles are fed on thyroid their legs grow out long before the usual time; on the other hand removal of the thyroid delays metamorphosis, and the tadpoles remain tadpoles to an unusual size**.

The great American bull-frog (*R. Catesbeiana*) lives for two or three years in tadpole form; but a diet of thyroid turns the little tadpoles into bull-frogs before they are a month old††. The converse

* Cf. E. A. Schafer, The function of the pituitary body, *Proc. R.S.* (B), LXXXI, p. 442, 1904.

† It is not easy to draw a line between *enzyme* and *vitamin*, or between *hormone* and *enzyme*.

‡ The physiological relations between insulin and the pituitary body might seem to indicate that it is the carbohydrate metabolism which is more especially concerned. Cf. (e.g.) Eric Holmes, *Metabolism of the Living Tissue*, 1937.

§ Van der Horst finds this to be the case in *Zalophus* and in the ostrich, compared with smaller seals or birds; cf. Ariens Kappers, *Journ. Anat.* LXIV, p. 256, 1930.

|| Gudernatsch, in *Arch. f. Entw. Mech.* XXXV, 1912.

¶ Eidmann, *ibid.* XLIX, pp. 510–537, 1921.

** Allen, *Journ. Exp. Zool.* XXIV, p. 499, 1918. Cf. (*int. al.*) E. Uhlenhuth, Experimental production of gigantism, *Journ. Gen. Physiol.* III, p. 347; IV, p. 321, 1921–22.

†† W. W. Swingle, *Journ. Exp. Zool.* XXIV, 1918; XXXVII, 1923; *Journ. Gen. Physiol.* I, II, 1918–19; etc.

experiment has been performed on ordinary tadpoles*; with their thyroids removed they remain normal to all appearance, but the weeks go by and metamorphosis does not take place. Gill-clefts and tail persist, no limbs appear, brain and gut retain their larval features; but months after, or apparently at any time, the belated tadpoles respond to a diet of thyroid, and may be turned into frogs by means of it. The Mexican axolotl is a grown-up tadpole which, when the ponds dry up (as they seldom do), completes its growth and turns into a gill-less, lung-breathing newt or salamander†; but feed it on thyroid, even for a single meal, and its metamorphosis is hastened and ensured‡.

Much has been done since these pioneering experiments, all going to shew that the thyroid plays its active part in the tissue-changes which accompany and constitute metamorphosis. It looks as though more thyroid meant more respiratory activity, more oxygen-consumption, more oxidative metabolism, more tissue-change, hence earlier bodily development§. Pituitary and thyroid are very different things; the one enhances growth, the other retards it. Thyroid stimulates metabolism and hastens development, but the tissues waste.

It is a curious fact, but it has often been observed, that starvation or inanition has, in the long run, a similar effect of hastening metamorphosis||. The meaning of this phenomenon is unknown.

An extremely remarkable case is that of the "galls", brought into existence on various plants in response to the prick of a small insect's ovipositor. One tree, an oak for instance, may bear galls

* Bennett Allen, *Biol. Bull.* XXXII, 1917; *Journ. Exp. Zool.* XXIV, 1918; XXX, 1920; etc.

† Colorado axolotls are much more apt to metamorphose than the Mexican variety.

‡ Babak, Ueber die Beziehung der Metamorphose...zur inneren Secretion, *Centralbl f. Physiol.* X, 1913. Cf. Abderhalden, Studien über die von einzelnen Organen hervorgebrachten Substanzen mit spezifischer Wirkung, *Pflüger's Archiv*, CLXII, 1915.

§ Certain experiments by M. Morse (*Journ. Biol. Chem.* XIX, 1915) seemed to shew that the effect of thyroid on metamorphosis depended on iodine; but the case is by no means clear (cf. O. Shinryo, *Sci. Rep. Tohoku Univ.* III, 1928, and others). The axolotl is said to shew little response to experimental iodine, and its ally *Necturus* none at all (cf. B. M. Allen, in *Biol. Reviews*, XIII, 1939).

|| Cf. Krizensky, Die beschleunigende Einwirkung des Hungerns auf die Metamorphose, *Biol. Centralbl.* XLIV, 1914. Cf. *antea*, p. 170.

of many kinds, well-defined and widely different, each caused to grow out of the tissues of the plant by a chemical stimulus contributed by the insect, in very minute amount; and the insects are so much alike that the galls are easier to distinguish than the flies. The same insect may produce the same gall on different plants, for instance on several species of willow; or sometimes on different parts, or tissues, of the same plant. Small pieces of a dead larva have been used to infect a plant, and a gall of the usual kind has resulted. Beyerinck killed the eggs with a hot wire as soon as they were deposited in the tree, yet the galls grew as usual. Here, as Needham has lately pointed out, is a great field for reflection and future experiment. The minute drop of fluid exuded by the insect has marvellous properties. It is not only a stimulant of growth, like any ordinary auxin or hormone; it causes the growth of a peculiar tissue, and shapes it into a new and specific form*.

Among other illustrations (which are plentiful) of the subtle influence of some substance upon growth, we have, for instance, the growth of the placental decidua, which Loeb shewed to be due to a substance given off by the corpus luteum, lending to the uterine tissues an enhanced capacity for growth, to be called into action by contact with the ovum or even of a foreign body. Various sexual characters, such as the plumage, comb and spurs of the cock, arise in like manner in response to an internal secretion or "male hormone"; and when castration removes the source of the secretion, well-known morphological changes take place. When a converse change takes place the female acquires, in greater or less degree, characters which are proper to the male: as in those extreme cases, known from time immemorial, when an old and barren hen assumes the plumage of the cock†.

The mane of the lion, the antlers of the stag, the tail of the peacock, are all examples of intensified differential growth, or localised and

* Joseph Needham, Aspects nouveaux de la chimie et de la biologie de la croissance organisée, *Folia Morphologica*, Warszawa, VIII, p. 32, 1938. On galls, see (*int. al.*) Cobbold, Ross und Hedicke, *Die Pflanzengallen*, Jena, 1927; etc. And on their "morphogenic stimulus", cf. Herbst, *Biolog. Cblt.*, 1894–5, passim.

† The hen which assumed the voice and plumage of the male was a portent or omen—*gallina cecinit*. The first scientific account was John Hunter's celebrated Account of an extraordinary pheasant, and Of the appearance of the change of sex in Lady Tynte's peahen, *Phil. Trans.* LXX, pp. 527, 534, 1780.

sex-linked hypertrophy; and in the singular and striking plumage of innumerable birds we may easily see how enhanced growth of a tuft of feathers, perhaps exaggeration of a single plume, is at the root of the whole matter. Among extreme instances we may think of the immensely long first primary of the pennant-winged nightjar; of the long feather over the eye in *Pteridophora alberti*,

Fig. 81. A single pair of hypertrophied feathers in a bird-of-paradise, *Pteridophora alberti*.

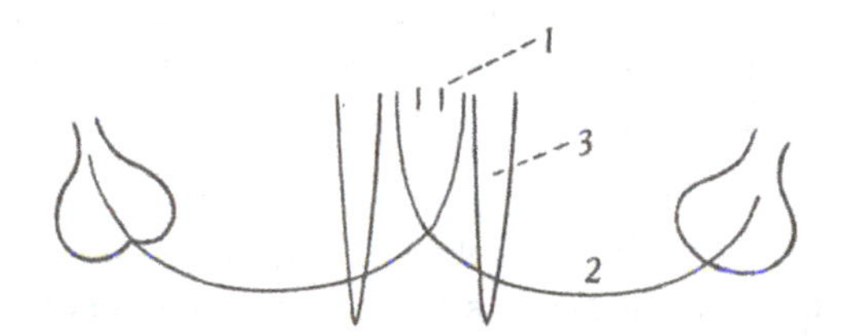

Fig. 82. Unequal growth in the three pairs of tail-feathers of a humming-bird (*Loddigesia*). 1, rudimentary: 2, short and stiff; 3, long and spathulate.

or the six long plumes over or behind the eye in the six-shafted bird-of-paradise; or among the humming-birds, of the long outer rectrix in *Lesbia*, the second outer one in *Aethusa*, or of the extra-ordinary inequalities of the tail-feathers of *Loddigesia mirabilis*, some rudimentary, some short and straight and stiff, and other two immensely elongated, curved and spathulate. The sexual hormones have a potent influence on the plumage of a bird; they serve, somehow, to orientate and regulate the rate of growth from one feather-tract to another, and from one end to another, even from one side to the other, of a single feather. An extreme case is the occasional phenomenon of a "gynandrous" feather, male and female on two sides of the same vane*.

While unequal or differential growth is of peculiar interest to the morphologist, rate of growth pure and simple, with all the agencies which control or accelerate it, remains of deeper importance to the practical man. The live-stock breeder keeps many desirable qualities in view: constitution, fertility, yield and quality of milk or wool are some of these; but rate of growth, with its corollaries of early maturity and large ultimate size, is generally more important than them all. The inheritance of size is somewhat complicated, and limited from the breeder's point of view by the mother's inability to nourish and bring forth a crossbred offspring of a breed larger than her own. A cart mare, covered by a Shetland sire, produces a good-sized foal; but the Shetland mare, crossed with a carthorse, has a foal a little bigger, but not much bigger, than herself (Fig. 83). In size and rate of growth, as in other qualities, our farm animals differ vastly from their wild progenitors, or from the "un-improved" stock in days before Bakewell and the other great breeders began. The improvement has been brought about by "selection"; but what lies behind? Endocrine secretions, especially pituitary, are doubtless at work; and already the stock-raiser and the biochemist may be found hand in hand.

If we once admit, as we are now bound to do, the existence of factors which by their physiological activity, and apart from any direct action of the nervous system, tend towards the acceleration of growth and consequent modification of form, we are led into wide fields of speculation by an easy and a legitimate pathway. Professor Gley carries such speculations a long, long way: for he says† that by these chemical influences "Toute une partie de la construction des êtres paraît s'expliquer d'une façon toute mécanique. La forteresse, si longtemps inaccessible, du vitalisme est entamée. Car la notion morphogénique était, suivant le mot de Dastre‡, comme 'le dernier réduit de la force vitale'."

* See an interesting paper by Frank R. Lillie and Mary Juhn, on The physiology of development of feathers: I, Growth-rate and pattern in the individual feather. *Physiological Zoology*, v, pp. 124–184, 1932, and many papers quoted therein.

† Le Néo-vitalisme, *Revue Scientifique*, March 1911.

‡ *La Vie et la Mort*, 1902, p. 43.

The physiological speculations we need not discuss: but, to take a single example from morphology, we begin to understand the possibility, and to comprehend the probable meaning, of the all but sudden appearance on the earth of such exaggerated and almost monstrous forms as those of the great secondary reptiles and the

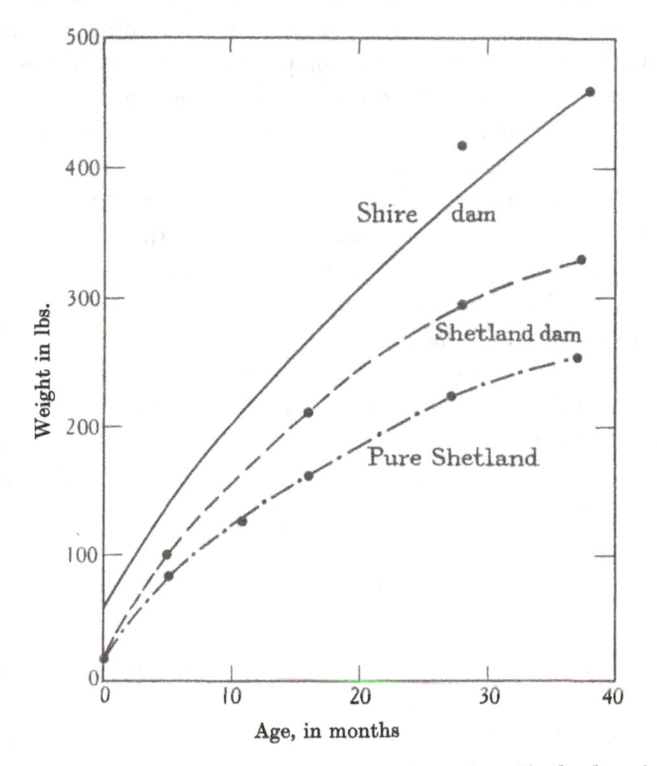

Fig. 83. Effect of cross-breeding on rate of growth in Shetland ponies. From Walton and Hammond's data.*

great tertiary mammals†. We begin to see that it is in order to account not for the appearance but for the disappearance of such forms as these that natural selection must be invoked. And we then, I think, draw near to the conclusion that what is true of these is universally true, and that the great function of natural selection

* Walton and Hammond, *Proc. R.S.* (B), No. 840, p. 317, 1938.

† Cf. also Dendy, *Evolutionary Biology*, 1912, p. 408.

is not to originate* but to remove: *donec ad interitum genus id natura redegit*†.

The world of things living, like the world of things inanimate, grows of itself, and pursues its ceaseless course of creative evolution. It has room, wide but not unbounded, for variety of living form and structure, as these tend towards their seemingly endless but yet strictly limited possibilities of permutation and degree: it has room for the great and for the small, room for the weak and for the strong. Environment and circumstance do not always make a prison, wherein perforce the organism must either live or die; for the ways of life may be changed, and many a refuge found, before the sentence of unfitness is pronounced and the penalty of extermination paid. But there comes a time when "variation," in form, dimensions, or other qualities of the organism, goes further than is compatible with all the means at hand of health and welfare for the individual and the stock; when, under the active and creative stimulus of forces from within and from without, the active and creative energies of growth pass the bounds of physical and physiological equilibrium: and so reach the limits which, as again Lucretius tells us, natural law has set between what may and what may not be,

> et quid quaeque queant per foedera naturai
> quid porro nequeant.

Then, at last, we are entitled to use the customary metaphor, and to see in natural selection an inexorable force whose function is not to create but to destroy—to weed, to prune, to cut down and to cast into the fire‡.

* So said Yves Delage (*L'hérédité*, 1903, p. 397): "La sélection naturelle est un principe admirable et parfaitement juste. Tout le monde est d'accord sur ce point. Mais où l'on n'est pas d'accord, c'est sur la limite de sa puissance et sur la question de savoir si elle peut engendrer des formes spécifiques nouvelles. *Il semble bien démontré aujourd'hui qu'elle ne le peut pas.*"

† Lucret. v, 875. "Lucretius nowhere seems to recognise the possibility of improvement or change of species by 'natural selection'; the animals remain as they were at the first, except that the weaker and more useless kinds have been crushed out. Hence he stands in marked contrast with modern evolutionists." Kelsey's note, *ad loc.*

‡ Even after we have so narrowed its scope and sphere, natural selection is still a hard saying; for the causes of *extinction* are wellnigh as hard to understand as are those of the *origin* of species. If we assert (as has been lightly and too

Of regeneration, or growth and repair

The phenomenon of regeneration, or the restoration of lost or amputated parts, is a particular case of growth which deserves separate consideration. It is a property manifested in a high degree among invertebrates and many cold-blooded vertebrates, diminishing as we ascend the scale, until it lessens down in the warm-blooded animals to that *vis medicatrix* which heals a wound. Ever since the days of Aristotle, and still more since the experiments of Trembley, Réaumur and Spallanzani in the eighteenth century, physiologist and psychologist alike have recognised that the phenomenon is both perplexing and important. "Its discovery," said Spallanzani, "was an immense addition to the riches of organic philosophy, and an inexhaustible source of meditation for the philosopher." The general phenomenon is amply treated of elsewhere*, and we need only deal with it in its immediate relation to growth.

Regeneration, like growth in other cases, proceeds with a velocity which varies according to a definite law; the rate varies with the time, and we may study it as velocity and as acceleration. Let us take, as an instance, Miss M. L. Durbin's measurements of the rate of regeneration of tadpoles' tails: the rate being measured in terms of length, or longitudinal increment†. From a number of tadpoles, whose average length was in one experiment 34 mm., and in another 49 mm., about half the tail was cut off, and the average amounts regenerated in successive periods are shewn as follows:

Days	3	5	7	10	12	14	17	18	24	28	30
Amount regenerated (mm.):											
First experiment	1·4	—	3·4	4·3	—	5·2	—	5·5	6·2	—	6·5
Second ,,	0·9	2·2	3·7	5·2	6·0	6·4	7·1	—	7·6	8·2	8·4

confidently done) that Smilodon perished on account of its gigantic tusks, that Teleosaurus was handicapped by its exaggerated snout, or Stegosaurus weighed down by its intolerable load of armour, we may call to mind kindred forms where similar conditions did not lead to rapid extermination, or where extinction ensued apart from any such apparent and visible disadvantages. Cf. F. A. Lucas, On momentum in variation, *Amer. Nat.* XLI, p. 46, 1907.

* See Professor T. H. Morgan's *Regeneration* (316 pp.), 1901, for a full account and copious bibliography. The early experiments on regeneration, by Vallisneri, Dicquemare, Spallanzani, Réaumur, Trembley, Baster, Bonnet and others, are epitomised by Haller, *Elementa Physiologiae*, VIII, pp. 156 *seq.*

† *Journ. Exper. Zool.* VII, p. 397, 1909.

Both experiments give us fairly smooth curves of growth within the period of the observations; and, with a slight and easy extrapolation, both curves draw to the base-line at zero (Fig. 84). More-

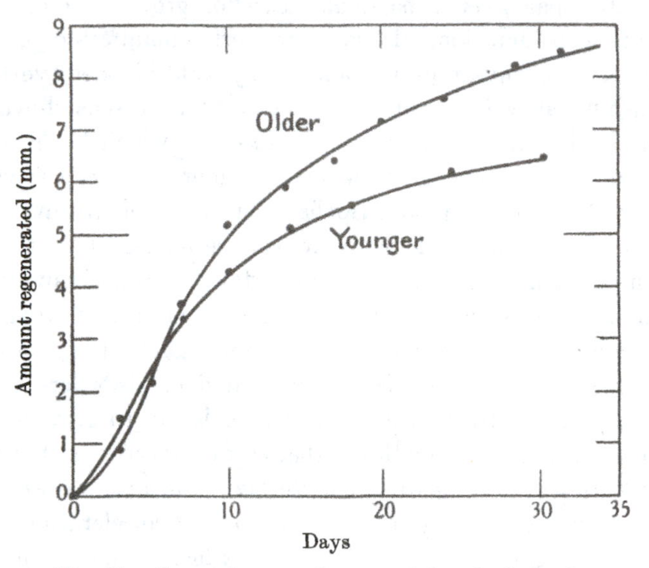

Fig. 84. Curve of regenerative growth in tadpoles' tails. From M. L. Durbin's data.

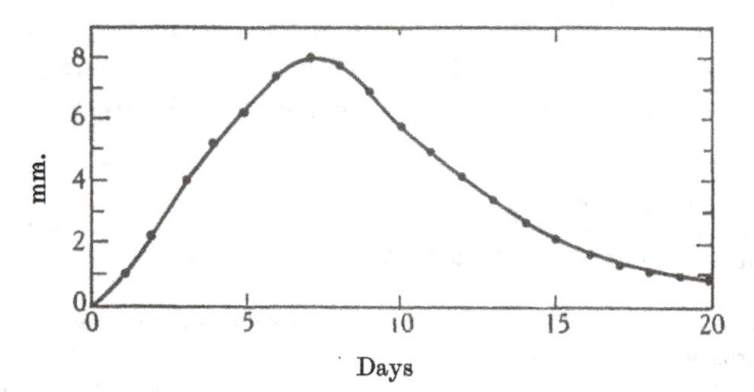

Fig. 85. Tadpoles' tails: amount regenerated daily, in mm. (Smoothed curve).

over, if from the smoothed curves we deduce the daily increments, we get (Fig. 85) a bell-shaped curve similar to (or to all appearance identical with) a skew curve of error. In point of fact, this instance of *regeneration* is a very ordinary example of *growth*, with its

S-shaped curve of integration and its bell-shaped differential curve, just as we have seen it in simple cases, or simple phases, of the growth of a population or an individual.

If we amputate one limb of a pair in some animal with rapid powers of regeneration, we may compare from time to time the dimensions of the regenerating limb with those of its uninjured fellow, and so deal with a relative rather than an absolute velocity. The legs of insect-larvae are easily restored, but after pupation no further growth or regeneration takes place. An easy experiment, then, is to remove a limb in larvae of various ages, and to compare

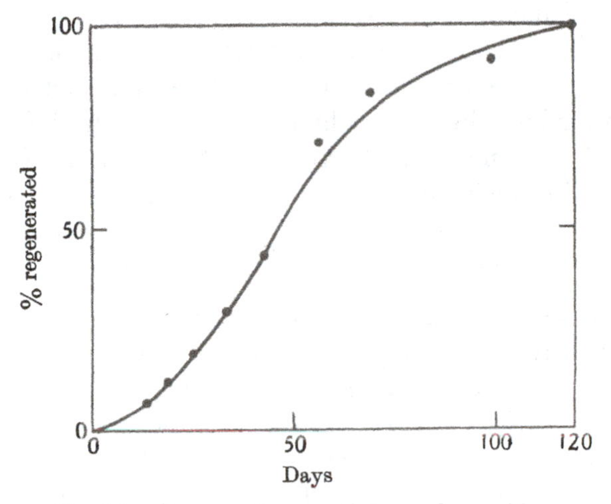

Fig. 86. Regenerative growth in mealworms' legs.

at leisure in the pupa the dimensions of the new limb with the old. The following much-abbreviated table shews the gradual increase of a regenerating limb in a mealworm, up to final equality with the normal limb, the rate varying according to the usual S-shaped curve* (Fig. 86).

Rate of regeneration in the mealworm (Tenebrio molitor, *larva*)

Days after amputation ...	0	16	21	25	34	44	58	70	100	121
% ratio of new limb to old	0	7	11	20	29	42	71	83	91	100

* From J. Krizenecky, Versuch zur statisch-graphischen Untersuchung...der Regenerationsvorgänge, *Arch. f. Entw. Mech.* XXXIX, 1914; XLII, 1917.

Some writers have found the curve of regenerative growth to be different from the curve of ordinary growth, and have commented on the apparent difference; but they have been misled (as it seems to me) by the fact that regeneration is seen from the start or very nearly so, while the ordinary curves of growth, as they are usually presented to us, date not from the beginning of growth, but from the comparatively late, and unimportant, and even fallacious epoch of birth. A complete curve of growth, starting from zero, has the same essential characteristics as the regeneration curve.

Indeed the more we consider the phenomenon of regeneration, the more plainly does it shew itself to us as but a particular case of the general phenomenon of growth*, following the same lines, obeying the same laws, and merely started into activity by the special stimulus, direct or indirect, caused by the infliction of a wound. Neither more nor less than in other problems of physiology are we called upon, in the case of regeneration, to indulge in metaphysical speculation, or to dwell upon the beneficent purpose which seemingly underlies this process of healing and repair.

It is a very general rule, though not a universal one, that regeneration tends to fall somewhat short of a *complete* restoration of the lost part; a certain percentage only of the lost tissues is restored. This fact was well known to some of those old investigators, who, like the Abbé Trembley and like Voltaire, found a fascination in the study of artificial injury and the regeneration which followed it. Sir John Graham Dalyell, for instance, says, in the course of an admirable paragraph on regeneration†: "The reproductive faculty...is not confined to one portion, but may extend over many; and it may ensue even in relation to the regenerated portion more than once. Nevertheless, the faculty gradually weakens, so that in general every successive regeneration is smaller and more imperfect than the organisation preceding it; and at length it is exhausted."

* The experiments of Loeb on the growth of Tubularia in various saline solutions, referred to on p. 245, might as well or better have been referred to under the heading of regeneration, as they were performed on cut pieces of the zoophyte. (Cf. Morgan, *op. cit.* p. 35.)

† *Powers of the Creator*, I, p. 7, 1851. See also *Rare and Remarkable Animals*, II, pp. 17–19, 90, 1847.

In certain minute animals, such as the Infusoria, in which the capacity for regeneration is so great that the entire animal may be restored from a mere fragment, it becomes of great interest to discover whether there be some definite size at which the fragment ceases to display this power. This question has been studied by Lillie*, who found that in *Stentor*, while still smaller fragments were capable of surviving for days, the smallest portions capable of regeneration were of a size equal to a sphere of about 80μ in diameter, that is to say of a volume equal to about one twenty-seventh of the average entire animal. He arrives at the remarkable conclusion that for this, and for all other species of animals, there is a "minimal organisation mass," that is to say a "minimal mass of definite size consisting of nucleus and cytoplasm within which the organisation of the species can just find its latent expression." And in like manner, Boveri† has shewn that the fragment of a sea-urchin's egg capable of growing up into a new embryo, and so discharging the complete functions of an entire and uninjured ovum, reaches its limit at about one-twentieth of the original egg—other writers having found a limit at about one-fourth. These magnitudes, small as they are, represent objects easily visible under a low power of the microscope, and so stand in a very different category to the minimal magnitudes in which life itself can be manifested, and which we have discussed in another chapter.

The Bermuda "life-plant" (*Bryophyllum calycinum*) has so remarkable a power of regeneration that a single leaf, kept damp, sprouts into fresh leaves and rootlets which only need nourishment to grow into a new plant. If a stem bearing two opposite leaves be split asunder, the two co-equal sister-leaves will produce (as we might indeed expect) *equal masses* of shoots in equal times, whether these shoots be many or few; and, if one leaf of the pair have part cut off it and the other be left intact, the amount of new growth

* F. R. Lillie, The smallest parts of *Stentor* capable of regeneration, *Journn. Morphology*, XII, p. 239, 1897.

† Boveri, Entwicklungsfähigkeit kernloser Seeigeleier, etc., *Arch. f. Entw. Mech.* II, 1895. See also Morgan, Studies of the partial larvae of *Sphaerechinus*, *ibid.* 1895; J. Loeb, On the limits of divisibility of living matter, *Biol. Lectures*, 1894; *Pflüger's Archiv*, LIX, 1894, etc. Bonnet studied the same problem a hundred and seventy years ago, and found that the smallest part of the worm *Lumbriculus* capable of regenerating was 1½ lines (3·4 mm.) long. For other references and discussion see H. Przibram, *Form und Formel*, 1922, ch. V.

will be in direct and precise proportion to the mass of the leaf from which it grew. The leaf is all the while a living tissue, manufacturing material to build its own offshoots; and we have a simple case of the law of mass action in the relation between the mass of the leaf with its included chlorophyll and that of its regenerated offshoot*.

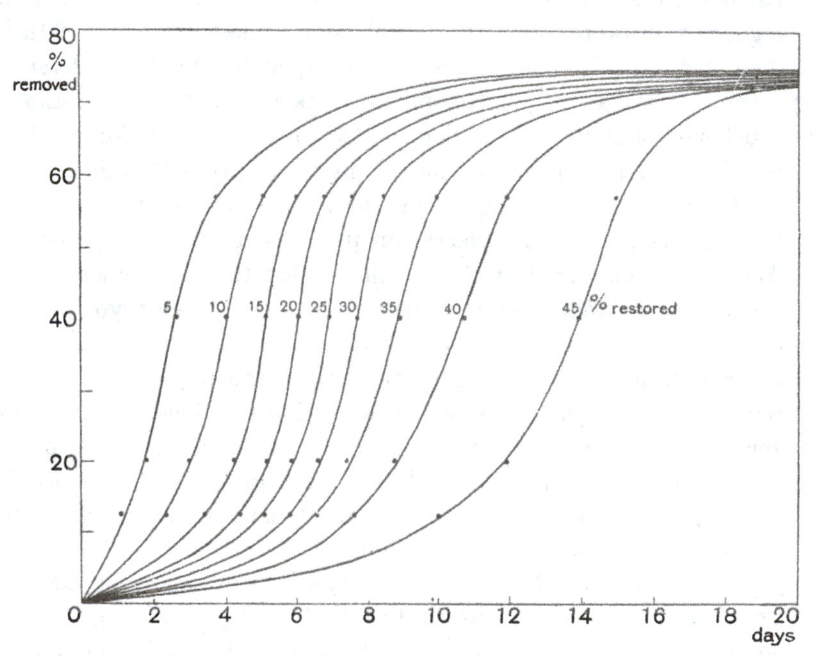

Fig. 87. Relation between the percentage amount of tail removed, the percentage restored, and the time required for its restoration. Constructed from M. M. Ellis's data.

A number of phenomena connected with the linear rate of regeneration are illustrated and epitomised in the accompanying diagram (Fig. 87), which I have constructed from certain data given by Ellis in a paper on the relation of the amount of tail *regenerated* to the amount *removed*, in tadpoles. These data are summarised in the next table. The tadpoles were all very much

* Jacques Loeb, The law controlling the quantity and rate of regeneration, *Proc. Nat. Acad. Sci.* IV, pp. 117–121, 1918; *Journ. Gen. Physiol.* I, pp. 81–96, 1918; *Botan. Gaz.* LXV, pp. 150–174, 1918.

of a size, about 40 mm.; the average length of tail was very near to 26 mm., or 65 per cent. of the whole body-length; and in four series of experiments about 10, 20, 40 and 60 per cent. of the tail were severally removed. The amount regenerated in successive intervals of three days is shewn in our table. By plotting the actual amounts regenerated against these three-day intervals of time, we may interpolate values for the time taken to regenerate definite percentage amounts, 5 per cent., 10 per cent., etc., of the amount removed; and my diagram is constructed from the four sets of values thus obtained, that is to say from the four sets of experiments which differed from one another in the amount of tail amputated. To these we have to add the general result of a fifth series of experiments, which shewed that when as much as 75 per cent. of the tail was cut off, no regeneration took place at all, but the animal presently died. In our diagram, then, each curve indicates the time taken to regenerate n per cent. of the amount removed. All the curves converge towards infinity of time, when the amount removed approaches 75 per cent. of the whole; and all start from zero, for nothing is regenerated where nothing had been destroyed.

The rate of regenerative growth in tadpoles' tails
(*After M. M. Ellis, Journ. Exp. Zool.* VII, *p.* 421, 1909)

Series*	Body length mm.	Tail length mm.	Amount removed mm.	Per cent. of tail removed	% amount regenerated in days						
					3	6	9	12	15	18	32
O	39·575	25·895	3·2	12·36	13	31	44	44	44	44	44
P	40·21	26·13	5·28	20·20	10	29	40	44	44	44	44
R	39·86	25·70	10·4	40·50	6	20	31	40	48	48	48
S	40·34	26·11	14·8	56·7	0	16	33	39	45	48	48

* Each series gives the mean of 20 experiments.

The amount regenerated varies also with the age of the tadpole, and with other factors such as temperature; in short, for any given age or size of tadpole, and for various temperatures, and doubtless for other varying physical conditions, a similar diagram might be constructed†.

The power of reproducing, or regenerating, a lost limb is par-

† Cf. also C. Zeleny, Factors controlling the rate of regeneration, *Illinois Biol. Monographs,* III, p. 1, 1916.

ticularly well developed in arthropod animals, and is sometimes accompanied by remarkable modification of the form of the regenerated limb. A case in point, which has attracted much attention, occurs in connection with the claws of certain Crustacea*.

In many of these we have an asymmetry of the great claws, one being larger than the other and also more or less different in form. For instance in the common lobster, one claw, the larger of the two, is provided with a few great "crushing" teeth, while the smaller claw has more numerous teeth, small and serrated. Though Aristotle thought otherwise, it appears that the crushing-claw may be on the right or left side, indifferently; whether it be on one or the other is a matter of "chance." It is otherwise in many other Crustacea, where the larger and more powerful claw is always left or right, as the case may be, according to the species: where, in other words, the "probability" of the large or the small claw being left or being right is tantamount to certainty†.

As we have already seen, the one claw is the larger because it has grown the faster; it has a higher "coefficient of growth," and accordingly, as age advances, the disproportion between the two claws becomes more and more evident. Moreover, we must assume that the characteristic form of the claw is a "function" of its magnitude; the knobbiness is a phenomenon coincident with growth, and we never, under any circumstances, find the smaller claw with big crushing teeth and the big claw with little serrate ones. There are many other somewhat similar cases where size and form are manifestly correlated, and we have already seen, to some extent, how the phenomenon of growth is often accompanied by such ratios of velocity as lead inevitably to changes of form. Meanwhile, then, we must simply assume that the essential difference between the two claws is one of magnitude, with which a certain differentiation of form is inseparably associated.

* Cf. H. Przibram, Scheerenumkehr bei dekapoden Crustaceen, *Arch. f. Entw. Mech.* XIX, pp. 181–247, 1905; XXV, pp. 266–344, 1907; Emmel, *ibid.* XXII, p. 542, 1906; Regeneration of lost parts in lobster, *Rep. Comm. Inland Fisheries, Rhode Island,* XXXV, XXXVI, 1905–6; *Science* (N.S.), XXVI, pp. 83–87, 1907; Zeleny, Compensatory regulation, *Journ. Exp. Zool.* II, pp. 1–102, 347–369, 1905; etc.

† Lobsters are occasionally found with two symmetrical claws: which are then usually serrated, sometimes (but very rarely) both blunt-toothed. Cf. W. T. Calman, *P.Z.S.* 1906, pp. 633, 634, and *reff.*

If we amputate a claw, or if, as often happens, the crab "casts it off," it undergoes a process of regeneration—it grows anew, and does so with an accelerated velocity which ceases when equilibrium of the parts is once more attained: the accelerated velocity being a case in point to illustrate that *vis revulsionis* of Haller to which we have already referred.

With the help of this principle, Przibram accounts for certain curious phenomena which accompany the process of regeneration. As his experiments and those of Morgan shew, if the large or knobby claw (*A*) be removed, there are certain cases, e.g. the common lobster, where it is directly regenerated. In other cases, e.g. *Alpheus**, the other claw (*B*) assumes the size and form of that which was amputated, while the latter regenerates itself in the form of the lesser and weaker one; *A* and *B* have apparently changed places. In a third case, as in the hermit-crabs, the *A*-claw regenerates itself as a small or *B*-claw, but the *B*-claw remains for a time unaltered, though slowly and in the course of repeated moults it later on assumes the large and heavily toothed *A*-form.

Much has been written on this phenomenon, but in essence it is very simple. It depends upon the respective rates of growth, upon a ratio between the rate of regeneration and the rate of growth of the uninjured limb: that is to say, on the familiar phenomenon of unequal growth, or, as it has been called, *heterogony*†. It is complicated a little, however, by the possibility of the uninjured limb growing all the faster for a time after the animal has been relieved of the other. From the time of amputation, say of *A*, *A* begins to grow from zero, with a high "regenerative" velocity; while *B*, starting from a definite magnitude, continues to increase with its normal or perhaps somewhat accelerated velocity. The ratio between the two velocities of growth will determine whether, by a given time, *A* has equalled, outstripped, or still fallen short of the magnitude of *B*.

That this is the gist of the whole problem is confirmed (if confirmation be necessary) by certain experiments of Wilson's. It is

* E. B. Wilson, Reversal of symmetry in *Alpheus heterocheles*, *Biol. Bull.* IV, p. 197, 1903.

† See p. 205.

known that by section of the nerve to a crab's claw, its growth is retarded, and as the general growth of the animal proceeds the claw comes to appear stunted or dwarfed. Now in such a case as that of *Alpheus*, we have seen that the rate of regenerative growth in an amputated large claw fails to let it reach or overtake the magnitude of the growing little claw: which latter, in short, now appears as the big one. But if at the same time as we amputate the big claw we also sever the nerve to the lesser one, we so far slow down the latter's growth that the other is able to make up to it, and in this case the two claws continue to grow at approximately equal rates, or in other words continue of coequal size.

The phenomenon of regeneration goes some little way towards helping us to comprehend the phenomenon of "multiplication by fission," as it is exemplified in its simpler cases in many worms and worm-like animals. For physical reasons which we shall have to study in another chapter, there is a natural tendency for any tube, if it have the properties of a fluid or semi-fluid substance, to break up into segments after it comes to a certain length*; and nothing can prevent its doing so except the presence of some controlling force, such for instance as may be due to the pressure of some external support, or some superficial thickening or other intrinsic rigidity of its own substance. If we add to this natural tendency towards fission of a cylindrical or tubular worm, the ordinary phenomenon of regeneration, we have all that is essentially implied in "reproduction by fission." And in so far as the process rests upon a physical principle, or natural tendency, we may account for its occurrence in a great variety of animals, zoologically dissimilar; and for its presence here and absence there, in forms which are materially different in a physical sense, though zoologically speaking they are very closely allied.

But the phenomena of regeneration, like all the other phenomena of growth, soon carry us far afield, and we must draw this long discussion to a close.

* A morphological *polarity*, or essential difference between one end and the other of a segment, is important even in so simple a case as the internode of a hydroid zoophyte; and an electrical polarity seems always to accompany it. Cf. A. P. Matthews, *Amer. Journ. Physiology*, VIII, p. 294, 1903; E. J. Lund, *Journ. Exper. Zool.* XXXIV, pp. 477–493; XXXVI, pp. 477–494, 1921–22.

Summary and Conclusion

For the main features which appear to be common to all curves of growth we may hope to have, some day, a simple explanation. In particular we should like to know the plain meaning of that point of inflection, or abrupt change from an increasing to a decreasing velocity of growth, which all our curves, and especially our acceleration curves, demonstrate the existence of, provided only that they include the initial stages of the whole phenomenon: just as we should also like to have a full physical or physiological explanation of the gradually diminishing velocity of growth which follows, and which (though subject to temporary interruption or abeyance) is on the whole characteristic of growth in all cases whatsoever. In short, the characteristic form of the curve of growth in length (or any other linear dimension) is a phenomenon which we are at present little able to explain, but which presents us with a definite and attractive problem for future solution. It would look as though the abrupt change in velocity must be due, either to a change in that pressure outwards from within by which the "forces of growth" make themselves manifest, or to a change in the resistances against which they act, that is to say the *tension* of the surface; and this latter force we do not by any means limit to "surface-tension" proper, but may extend to the development of a more or less resistant membrane or "skin," or even to the resistance of fibres or other histological elements binding the boundary layers to the parts within*. I take it that the sudden arrest of velocity is much more likely to be due to a sudden increase of resistance than to a sudden diminution of internal energies: in other words, I suspect that it is coincident with some notable event of histological differentiation, such as the rapid formation of a comparatively firm skin; and that the dwindling of velocities, or the negative acceleration, which follows, is the resultant or composite effect of waning forces of growth on the one hand, and increasing superficial resistance

* It is natural to suppose the cell-wall less rigid, or more plastic, in the growing tissue than in the full-grown or resting cell. It has been suggested that this plasticity is due to, or is increased by, auxins, whether in the course of nature, or in our stimulation of growth by the use of these bodies. Cf. H. Söding, *Jahrb. d. wiss. Bot.* LXXIV, p. 127, 1931.

on the other. This is as much as to say that growth, while its own energy tends to increase, leads also, after a while, to the establishment of resistances which check its own further increase.

Our knowledge of the whole complex phenomenon of growth is so scanty that it may seem rash to advance even this tentative suggestion. But yet there are one or two known facts which seem to bear upon the question, and to indicate at least the manner in which a varying resistance to expansion may affect the velocity of growth. For instance, it has been shewn by Frazee* that electrical stimulation of tadpoles, with small current density and low voltage, increases the rate of regenerative growth. As just such an electrification would tend to lower the surface-tension, and accordingly decrease the external resistance, the experiment would seem to support, in some slight degree, the suggestion which I have made.

To another important aspect of regeneration we can do no more than allude. The Planarian worms rival *Hydra* itself in their powers of regeneration; and in both cases even small bits of the animal are likely to include endoderm cells capable of intracellular digestion, whereby the fragment is enabled to live and to grow. Now if a Planarian worm be cut in separate pieces and these be suffered to grow and regenerate, they do so in a definite and orderly way; that part of a slice or fragment which had been nearer to the original head will develop a head, and a tail will be regenerated at the opposite end of the same fragment, the end which had been tailward in the beginning; the amputated fragments possess sides and ends, a front end and a hind end, like the entire worm; in short, they *retain their polarity*. This remarkable discovery is due to Child, who has amplified and extended it in various instructive ways. The existence of two poles, positive and negative, implies a "gradient" between them. It means that one part leads and another follows; that one part is dominant, or prepotent over the rest, whether in regenerative growth or embryonic development.

We may summarise, as follows, the main results of the foregoing discussion:

(1) Except in certain minute organisms, whose form (like that

* *Journ. Exper. Zool.* VII, p. 457, 1909.

of a drop of water) is due to the direct action of the molecular forces, we may look upon the form of an organism as a "function of growth," or a direct consequence of growth whose rate varies in its different directions. In a newer language we might call the form of an organism an "event in space-time," and not merely a "configuration in space."

(2) Growth varies in rate in an orderly way, or is subject, like other physiological activities, to definite "laws." The rates differ in degree, or form "gradients," from one point of an organism to another; the rates in different parts and in different directions tend to maintain more or less constant ratios to one another in each organism; and to the regularity and constancy of these relative rates of growth is due the fact that the form of the organism is in general regular and constant.

(3) Nevertheless, the ratio of velocities in different directions is not absolutely constant, but tends to alter in course of time, or to fluctuate in an orderly way; and to these progressive changes are due the changes of form which accompany development, and the slower changes which continue perceptibly in after life.

(4) Rate of growth depends on the age of the organism. It has a maximum somewhat early in life, after which epoch of maximum it slowly declines.

(5) Rate of growth is directly affected by temperature, and by other physical conditions: the influence of temperature being notably large in the case of cold-blooded or "poecilothermic" animals. Growth tends in these latter to be asymptotic, becoming slower but never ending with old age.

(6) It is markedly affected, in the way of acceleration or retardation, at certain physiological epochs of life, such as birth, puberty or metamorphosis.

(7) Under certain circumstances, growth may be *negative*, the organism growing smaller; and such negative growth is a common accompaniment of metamorphosis, and a frequent concomitant of old age.

(8) The phenomenon of regeneration is associated with a large transitory increase in the rate of growth (or *acceleration* of growth) in the region of injury; in other respects regenerative growth is similar to ordinary growth in all its essential phenomena.

In this discussion of growth, we have left out of account a vast number of processes or phenomena in the physiological mechanism of the body, by which growth is effected and controlled. We have dealt with growth in its relation to magnitude, and to that relativity of magnitudes which constitutes form; and so we have studied it as a phenomenon which stands at the beginning of a morphological, rather than at the end of a physiological enquiry. Under these restrictions, we have treated it as far as possible, or in such fashion as our present knowledge permits, on strictly physical lines. That is to say, we rule "heredity" or any such concept out of our present account, however true, however important, however indispensable in another setting of the story, such a concept may be. In physics "on admet que l'état actuel du monde ne dépend que du passé le plus proche, sans être influencé, pour ainsi dire, par le souvenir d'un passé lointain*." This is the concept to which the differential equation gives expression; it is the step which Newton took when he left Kepler behind.

In all its aspects, and not least in its relation to form, the growth of organisms has many analogies, some close, some more remote, among inanimate things. As the waves grow when the winds strive with the other forces which govern the movements of the surface of the sea, as the heap grows when we pour corn out of a sack, as the crystal grows when from the surrounding solution the proper molecules fall into their appropriate places: so in all these cases, very much as in the organism itself, is growth accompanied by change of form, and by a development of definite shapes and contours. And in these cases (as in all other mechanical phenomena), we are led to equate our various magnitudes with time, and so to recognise that growth is essentially a question of rate, or of velocity.

The differences of form, and changes of form, which are brought about by varying rates (or "laws") of growth, are essentially the same phenomenon whether they be episodes in the life-history of the individual, or manifest themselves as the distinctive characteristics of what we call separate species of the race. From one form, or one ratio of magnitude, to another there is but one straight and direct road of transformation, be the journey taken fast or

* Cf. H. Poincaré, La physique générale et la physique mathématique, *Rev. gén. des Sciences*, XI, p. 1167, 1900.

slow; and if the transformation take place at all, it will in all likelihood proceed in the self-same way, whether it occur within the lifetime of an individual or during the long ancestral history of a race. No small part of what is known as Wolff's or von Baer's law, that the individual organism tends to pass through the phases characteristic of its ancestors, or that the life-history of the individual tends to recapitulate the ancestral history of its race, lies wrapped up in this simple account of the relation between growth and form.

But enough of this discussion. Let us leave for a while the subject of the growth of the organism, and attempt to study the conformation, within and without, of the individual cell.

CHAPTER IV

ON THE INTERNAL FORM AND STRUCTURE OF THE CELL

IN the early days of the cell-theory, a hundred years ago, Goodsir was wont to speak of cells as "centres of growth" or "centres of nutrition," and to consider them as essentially "centres of force*". He looked forward to a time when the forces connected with the cell should be particularly investigated: when, that is to say, minute anatomy should be studied in its dynamical aspect. "When this branch of enquiry," he says, "shall have been opened up, we shall expect to have a science of organic forces, having direct relation to anatomy, the science of organic forms." And likewise, long afterwards, Giard contemplated a science of *morphodynamique*—but still looked upon it as forming so guarded and hidden a "territoire scientifique, que la plupart des naturalistes de nos jours ne le verront que comme Moïse vit la terre promise, seulement de loin et sans pouvoir y entrer†."

To the external forms of cells, and to the forces which produce and modify these forms, we shall pay attention in a later chapter. But there are forms and configurations of matter within the cell which also deserve to be studied with due regard to the forces, known or unknown, of whose resultant they are the visible expression.

* *Anatomical and Pathological Observations*, p. 3, 1845; *Anatomical Memoirs*, II, p. 392, 1868. This was a notable improvement on the "kleine wirkungsfähige Zentren oder Elementen" of the *Cellularpathologie*. Goodsir seems to have been seeking an analogy between the living cell and the physical atom, which Faraday, following Boscovich, had been speaking of as a *centre of force* in the very year before Goodsir published his *Observations*: see Faraday's *Speculations concerning Electrical Conductivity and the Nature of Matter*, 1844. For Newton's "molecules" had been turned by his successors into material points; and it was Boscovich (in 1758) who first regarded these material points as mere persistent centres of force. It was the same fertile conception of a *centre of force* which led Rutherford, later on, to the discovery of the nucleus of the atom.

† A. Giard, L'œuf et les débuts de l'évolution, *Bull. Sci. du Nord de la Fr.* VIII, pp. 252–258, 1876.

In the long interval since Goodsir's day, the visible structure, the conformation and configuration, of the cell, has been studied far more abundantly than the purely dynamic problems which are associated therewith. The overwhelming progress of microscopic observation has multiplied our knowledge of cellular and intracellular structure; and to the multitude of visible structures it has been often easier to attribute virtues than to ascribe intelligible functions or modes of action. But here and there nevertheless, throughout the whole literature of the subject, we find recognition of the inevitable fact that dynamical problems lie behind the morphological problems of the cell.

Bütschli pointed out sixty years ago, with emphatic clearness, the failure of morphological methods and the need for physical methods if we were to penetrate deeper into the essential nature of the cell*. And such men as Loeb and Whitman, Driesch and Roux, and not a few besides, have pursued the same train of thought and similar methods of enquiry.

Whitman†, for instance, puts the case in a nutshell when, in speaking of the so-called "caryokinetic" phenomena of nuclear division, he reminds us that the leading idea in the term "*caryokinesis*" is *motion*—"motion viewed as an exponent of forces residing in, or acting upon, the nucleus. It regards the nucleus as a *seat of energy, which displays itself in phenomena of motion*‡."

In short it would seem evident that, except in relation to a dynamical investigation, the mere study of cell structure has but

* *Entwickelungsvorgänge der Eizelle*, 1876; *Investigations on Microscopic Foams and Protoplasm*, p. 1, 1894.

† *Journ. Morphology*, I, p. 229, 1887.

‡ While it has been very common to look upon the phenomena of mitosis as sufficiently explained by the results *towards which* they seem to lead, we may find here and there a strong protest against this mode of interpretation. The following is a case in point: "On a tenté d'établir dans la mitose dite primitive plusieurs catégories, plusieurs types de mitose. On a choisi le plus souvent comme base de ces systèmes des concepts abstraits et téléologiques: répartition plus ou moins exacte de la chromatine entre les deux noyaux-fils suivant qu'il y a ou non des chromosomes (*Dangeard*), distribution particulière et signification dualiste des substances nucléaires (substance kinétique et substance générative ou héréditaire, *Hartmann et ses élèves*), etc. Pour moi tous ces essais sont à rejeter catégoriquement à cause de leur caractère finaliste; de plus, ils sont construits sur des concepts non démontrés, et qui parfois représentent des généralisations absolument erronées." A. Alexeieff, *Archiv für Protistenkunde*, XIX, p. 344, 1913.

little value of its own. That a given cell, an ovum for instance, contains this or that visible substance or structure, germinal vesicle or germinal spot, chromatin or achromatin, chromosomes or centrosomes, obviously gives no explanation of the *activities* of the cell. And in all such hypotheses as that of "pangenesis," in all the theories which attribute specific properties to micellae, chromosomes, idioplasts, ids, or other constituent particles of protoplasm or of the cell, we are apt to fall into the error of attributing to *matter* what is due to *energy* and is manifested in force: or, more strictly speaking, of attributing to material particles individually what is due to the energy of their collocation.

The tendency is a very natural one, as knowledge of structure increases, to ascribe particular virtues to the material structures themselves, and the error is one into which the disciple is likely to fall but of which we need not suspect the master-mind. The dynamical aspect of the case was in all probability kept well in view by those who, like Goodsir himself, first attacked the problem of the cell and originated our conceptions of its nature and functions*.

If we speak, as Weismann and others speak, of an "hereditary *substance*," a substance which is split off from the parent-body, and which hands on to the new generation the characteristics of the old, we can only justify our mode of speech by the assumption that that particular portion of matter is the essential vehicle of a particular charge or distribution of energy, in which is involved the capability of producing motion, or of doing "work." For, as Newton said, to tell us that a thing "is endowed with an occult specific quality†, by which it acts and produces manifest effects, is to tell us nothing; but to derive two or three general principles of motion‡ from

* See also (*int. al.*) R. S. Lillie's papers on the physiology of cell-division in the *Journ. Exper. Physiology*; especially No. VI, Rhythmical changes in the resistance of the dividing sea-urchin egg, *ibid.* XVI, pp. 369–402, 1916.

† Such as the *vertu dormitive* which accounts for the soporific action of opium. We are now more apt, as Le Dantec says, to substitute for this occult quality the hypothetical substance *dormitin*.

‡ This is the old philosophic axiom writ large: *Ignorato motu, ignoratur natura*; which again is but an adaptation of Aristotle's phrase, ἡ ἀρχὴ τῆς κινήσεως, as equivalent to the "Efficient Cause." FitzGerald holds that "all explanation consists in a description of underlying motions" (*Scientific Writings*, 1902, p. 385); and Oliver Lodge remarked, "You can move Matter; it is the only thing you can do to it."

phenomena would be a very great step in philosophy, though the causes of those principles were not yet discovered." The *things* which we see in the cell are less important than the *actions* which we recognise in the cell; and these latter we must especially scrutinise, in the hope of discovering how far they may be attributed to the simple and well-known physical forces, and how far they be relevant or irrelevant to the phenomena which we associate with, and deem essential to, the manifestation of *life*. It may be that in this way we shall in time draw nigh to the recognition of a specific and ultimate residuum.

And lacking, as we still do lack, direct knowledge of the actual forces inherent in the cell, we may yet learn something of their distribution, if not also of their nature, from the outward and inward configuration of the cell and from the changes taking place in this configuration; that is to say from the movements of matter, the kinetic phenomena, which the forces in action set up.

The fact that the germ-cell develops into a very complex structure is no absolute proof that the cell itself is structurally a very complicated mechanism: nor yet does it prove, though this is somewhat less obvious, that the forces at work or latent within it are especially numerous and complex. If we blow into a bowl of soapsuds and raise a great mass of many-hued and variously shaped bubbles, if we explode a rocket and watch the regular and beautiful configuration of its falling streamers, if we consider the wonders of a limestone cavern which a filtering stream has filled with stalactites, we soon perceive that in all these cases we have begun with an initial system of very slight complexity, whose structure in no way foreshadowed the result, and whose comparatively simple intrinsic forces only play their part by complex interaction with the equally simple forces of the surrounding medium. In an earlier age, men sought for the visible embryo, even for the *homunculus*, within the reproductive cells; and to this day we scrutinise these cells for visible structure, unable to free ourselves from that old doctrine of "pre-formation*."

Moreover, the microscope seemed to substantiate the idea (which

* As when Nägeli concluded that the organism is, in a certain sense, "vorgebildet"; *Beitr. zur wiss. Botanik,* II, 1860.

we may trace back to Leibniz* and to Hobbes†), that there is no limit to the mechanical complexity which we may postulate in an organism, and no limit, therefore, to the hypotheses which we may rest thereon. But no microscopical examination of a stick of sealing-wax, no study of the material of which it is composed, can enlighten us as to its electrical manifestations or properties. Matter of itself has no power to do, to make, or to become: it is in energy that all these potentialities reside, energy invisibly associated with the material system, and in interaction with the energies of the surrounding universe.

That "function presupposes structure" has been declared an accepted axiom of biology. Who it was that so formulated the aphorism I do not know; but as regards the structure of the cell it harks back to Brücke, with whose demand for a mechanism, or an organisation, within the cell histologists have ever since been trying to comply‡. But unless we mean to include thereby invisible, and merely chemical or molecular, structure, we come at once on dangerous ground. For we have seen in a former chapter that organisms are known of magnitudes so nearly approaching the molecular, that everything which the morphologist is accustomed to conceive as "structure" has become physically impossible; and recent research tends to reduce, rather than to extend, our conceptions of the visible structure necessarily inherent in living protoplasm§. The microscopic structure which in the last resort

* "La matière arrangée par une sagesse divine doit être essentiellement organisée partout...il y a machine dans les parties de la machine naturelle à l'infini." *Sur le principe de la Vie*, p. 431 (Erdmann). This is the very converse of the doctrine of the Atomists, who could not conceive a condition "*ubi dimidiae partis pars semper habebit Dimidiam partem, nec res praefiniet ulla.*"

† Cf. an interesting passage from the *Elements* (I, p. 445, Molesworth's edit.), quoted by Owen, *Hunterian Lectures on the Invertebrates*, 2nd ed. pp. 40, 41, 1855.

‡ "Wir müssen deshalb den lebenden Zellen, abgesehen von der Molekularstructur der organischen Verbindungen welche sie enthàlt, noch eine andere und in anderer Weise complicirte Structur zuschreiben, und diese es ist welche wir mit dem Namen *Organisation* bezeichnen," Brücke, Die Elementarorganismen, *Wiener Sitzungsber.* XLIV, 1861, p. 386; quoted by Wilson, *The Cell*, etc., p. 289. Cf. also Hardy, *Journ. Physiol.* XXIV, 1899, p. 159.

§ The term *protoplasm* was first used by Purkinje, about 1839 or 1840 (cf. Reichert, *Arch. f. Anat. u. Physiol.* 1841). But it was better defined and more strictly used by Hugo von Mohl in his paper Ueber die Saftbewegung im Inneren der Zellen, *Botan. Zeitung*, IV, col. 73–78, 89–94, 1846.

or in the simplest cases it seems to shew, is that of a more or less viscous colloid, or rather mixture of colloids, and nothing more. Now, as Clerk Maxwell puts it in discussing this very problem, "one material system can differ from another only in the configuration and motion which it has at a given instant*." If we cannot assume differences in structure or configuration, we must assume differences in *motion*, that is to say in *energy*. And if we cannot do this, then indeed we are thrown back upon modes of reasoning unauthorised in physical science, and shall find ourselves constrained to assume, or to "admit, that the properties of a germ are not those of a purely material system."

But we are by no means necessarily in this dilemma. For though we come perilously near to it when we contemplate the lowest orders of magnitude to which life has been attributed, yet in the case of the ordinary cell, or ordinary egg or germ which is going to develop into a complex organism, if we have no reason to assume or to believe that it comprises an intricate "mechanism," we may be quite sure, both on direct and indirect evidence, that, like the powder in our rocket, it is very heterogeneous in its structure. It is a mixture of substances of various kinds, more or less fluid, more or less mobile, influenced in various ways by chemical, electrical, osmotic and other forces, and in their admixture separated by a multitude of surfaces or boundaries, at which these or certain of these forces are made manifest.

Indeed, such an arrangement as this is already enough to constitute a "mechanism"; for we must be very careful not to let our physical or physiological concept of mechanism be narrowed to an interpretation of the term derived from the complicated contrivances of human skill. From the physical point of view, we understand by a "mechanism" whatsoever checks or controls, and guides into determinate paths, the workings of energy: in other words, whatsoever leads in the degradation of energy to its manifestation in some form of *work*, at a stage short of that ultimate degradation which lapses in uniformly diffused heat. This, as Warburg has well explained, is the general effect or function of the physiological machine, and in particular of that part of it which we call "cell-

* Precisely as in the Lucretian *concursus, motus, ordo, positura, figurae*, whereby bodies *mutato ordine mutant naturam*.

structure*." The normal muscle-cell is something which turns energy, derived from oxidation, into work; it is a mechanism which arrests and utilises the chemical energy of oxidation in its downward course; but the same cell when injured or disintegrated loses its "usefulness," and sets free a greatly increased proportion of its energy in the form of heat. It was a saying of Faraday's, that "even a life is but a chemical act prolonged. If death occur, the more rapidly oxygen and the affinities run on to their final state†."

Very great and wonderful things are done by means of a mechanism (whether natural or artificial) of extreme simplicity. A pool of water, by virtue of its surface, is an admirable mechanism for the making of waves; with a lump of ice in it, it becomes an efficient and self-contained mechanism for the making of currents. Music itself is made of simple things—a reed, a pipe, a string. The great cosmic mechanisms are stupendous in their simplicity; and, in point of fact, every great or little aggregate of heterogeneous matter (not identical in "phase") involves, *ipso facto*, the essentials of a mechanism. Even a non-living colloid, from its intrinsic heterogeneity, is in this sense a mechanism, and one in which energy is manifested in the movement and ceaseless rearrangement of the constituent particles. For this reason Graham speaks somewhere or other of the colloid state as "the dynamic state of matter"; in the same philosopher's phrase, it possesses "*energia*‡."

Let us turn then to consider, briefly and diagrammatically, the structure of the cell, a fertilised germ-cell or ovum for instance, not in any vain attempt to correlate this structure with the structure or properties of the resulting and yet distant organism; but merely to see how far, by the study of its form and its changing internal configuration, we may throw light on certain forces which are for the time being at work within it.

We may say at once that we can scarcely hope to learn more of these forces, in the first instance, than a few facts regarding their

* Otto Warburg, Beiträge zur Physiologie der Zelle, insbesondere über die Oxidationsgeschwindigkeit in Zellen; in Asher-Spiro's *Ergebnisse der Physiologie*, XIV, pp. 253–337, 1914 (see p. 315).

† See his *Life* by Bence Jones, II, p. 299.

‡ Both phrases occur, side by side, in Graham's classical paper on Liquid diffusion applied to analysis, *Phil. Trans.* CLI, p. 184, 1861; *Chem. and Phys. Researches* (ed. Angus Smith), 1876, p. 554.

direction and magnitude; the nature and specific identity of the force or forces is a very different matter. This latter problem is likely to be difficult of elucidation, for the reason, among others, that very different forces are often much alike in their outward and visible manifestations. So it has come to pass that we have a multitude of discordant hypotheses as to the nature of the forces acting within the cell, and producing in cell division the "caryokinetic" figures of which we are about to speak. One student may, like Rhumbler, choose to account for them by an hypothesis of mechanical traction, acting on a reticular web of protoplasm*; another, like Leduc, may shew us how in many of their most striking features they may be admirably simulated by salts diffusing in a colloid medium; others, like Lamb and Graham Cannon, have compared them to the stream-lines produced and the field of force set up by bodies vibrating in a fluid; others, like Gallardo† and Rhumbler in his earlier papers‡, insisted on their resemblance to certain phenomena of electricity and magnetism§; while Hartog believed that the force in question is only analogous to these, and has a specific identity of its own‖. All these conflicting views are of secondary importance, so long as we seek only to account for certain *configurations* which reveal the direction, rather than the nature, of a force. One and the same system of lines of force may appear in a field of magnetic or of electrical energy, of the osmotic energy of diffusion, of the gravitational energy of a flowing stream. In short, we may expect to learn something of the pure or abstract dynamics long before we can deal with the special physics of the

* L. Rhumbler, Mechanische Erklärung der Aehnlichkeit zwischen magnetischen Kraftliniensystemen und Zelltheilungsfiguren, *Arch. f. Entw. Mech.* xv, p. 482, 1903.

† A. Gallardo, Essai d'interprétation des figures caryocinétiques, *Anales del Museo de Buenos-Aires* (2), II, 1896; *Arch. f. Entw. Mech.* XXVIII, 1909, etc.

‡ *Arch. f. Entw. Mech.* III, IV, 1896–97.

§ On various theories of the mechanism of mitosis, see (e.g.) Wilson, *The Cell in Development*, etc.; Meves, *Zelltheilung*, in Merkel u. Bonnet's *Ergebnisse der Anatomie*, etc., VII, VIII, 1897–98; Ida H. Hyde, *Amer. Journ. Physiol.* XII, pp. 241–275, 1905; and especially A. Prenant, Théories et interprétations physiques de la mitose, *Journ. de l'Anat. et Physiol.* XLVI, pp. 511–578, 1910. See also A. Conard, *Sur le mécanisme de la division cellulaire, et sur les bases morphologiques de la Cytologie*, Bruxelles, 1939: a work which I find hard to follow.

‖ M. Hartog, Une force nouvelle: le mitokinétisme, *C.R.* 11 Juli 1910; *Arch. f. Entw. Mech.* XXVII, pp. 141–145, 1909; cf. *ibid.* XL, pp. 33–64, 1914.

cell. For indeed, just as uniform expansion about a single centre, to whatsoever physical cause it may be due, will lead to the configuration of a sphere, so will any two centres or foci of potential (of whatsoever kind) lead to the configurations with which Faraday first made us familiar under the name of "lines of force*"; and this is as much as to say that the phenomenon, though physical in the concrete, is in the abstract purely mathematical, and in its very essence is neither more nor less than *a property of three-dimensional space*.

But as a matter of fact, in this instance, that is to say in trying to explain the leading phenomena of the caryokinetic division of the cell, we shall soon perceive that any explanation which is based, like Rhumbler's, on mere mechanical traction, is obviously inadequate, and we shall find ourselves limited to the hypothesis of some polarised and polarising force, such as we deal with, for instance, in magnetism or electricity, or in certain less familiar phenomena of hydrodynamics. Let us speak first of the cell itself, as it appears in a state of rest, and let us proceed afterwards to study the more active phenomena which accompany its division.

Our typical cell is a spherical body; that is to say, the uniform surface-tension at its boundary is balanced by the outward resistance of uniform forces within. But at times the surface-tension may be a fluctuating quantity, as when it produces the rhythmical contractions or "Ransom's waves"† on the surface of a trout's egg; or again, the surface-tension may be locally unequal and variable, giving rise to an amoeboid figure, as in the egg of *Hydra*‡.

Within the cell is a nucleus or germinal vesicle, also spherical,

* The configurations, as obtained by the usual experimental methods, were of course known long before Faraday's day, and constituted the "convergent and divergent magnetic curves" of eighteenth century mathematicians. As Leslie said, in 1821, they were "regarded with wonder by a certain class of dreaming philosophers, who did not hesitate to consider them as the actual traces of an invisible fluid, perpetually circulating between the poles of the magnet." Faraday's great advance was to interpret them as indications of *stress in a medium*—of tension or attraction along the lines, and of repulsion transverse to the lines, of the diagram.

† W. H. Ransom, On the ovum of osseous fishes, *Phil. Trans.* CLVII, pp. 431–502, 1867 (*vide* p. 463 *et. seq.*) (Ransom, afterwards a Nottingham physician, was Huxley's friend and class-fellow at University College, and beat him for the medal in Grant's class of zoology.)

‡ Cf. also the curious phenomenon in a dividing egg described as "spinning" by Mrs G. F. Andrews, *Journ. Morph.* XII, pp. 367–389, 1897.

and consisting of portions of "chromatin," aggregated together within a more fluid drop. The fact has often been commented upon that, in cells generally, there is no correlation of *form* (though there apparently is of *size*) between the nucleus and the "cytoplasm," or main body of the cell. So Whitman* remarks that "except during the process of division the nucleus seldom departs from its typical spherical form. It divides and sub-divides, ever returning to the same round or oval form....How different with the cell. It preserves the spherical form as rarely as the nucleus departs from it. Variation in form marks the beginning and the end of every important chapter in its history." On simple dynamical grounds, the contrast is easily explained. So long as the fluid substance of the nucleus is qualitatively different from, and incapable of mixing with, the fluid or semi-fluid protoplasm surrounding it, we shall expect it to be, as it almost always is, of spherical form. For on the one hand, it has a surface of its own whose surface-tension is presumably uniform, and on the other, it is immersed in a medium which transmits on all sides a uniform fluid or "hydrostatic" pressure†; thus the case of the spherical nucleus is closely akin to that of the spherical yolk within the bird's egg. Again, for a similar reason, the contractile vacuole of a protozoon is spherical‡. It is just a drop of fluid, bounded by a

* Whitman, *Journ. Morph.* II, p. 40, 1889.

† "Souvent il n'y a qu'une séparation *physique* entre le cytoplasme et le suc nucléaire, comme entre deux liquides immiscibles, etc."; Alexeieff, Sur la mitose dite *primitive*, *Arch. f. Protistenk.* XXIX, p. 357, 1913.

‡ The appearance of "vacuolation" is a result of endosmosis, or the diffusion of a less dense fluid into the denser plasma of the cell. But while water is probably taken up at the surface of the cell by purely passive osmotic intake, a definite "vacuole" appears at a place where osmotic work is being actively done. A higher osmotic pressure than that of the external medium is maintained within the cell, but as a "steady state" rather than a condition of equilibrium, in other words by the continual expenditure of energy; and the difference of pressure is at best small. The "contractile vacuole" bursts when it touches the surface of the cell, and bursting may be delayed by manipulating the vacuole towards the interior. It may sometimes burst towards the interior of the cell through inequalities in its own surface-tension, and the collapsing vacuole is then apt to shew a star-shaped figure. The cause of the higher osmotic pressure within the cell is a matter for the colloid chemist, and cannot be discussed here. On the physiology of the contractile vacuole, see (*int. al.*) H. Z. Gow, *Arch. f. Protistenk.* LXXXVII, pp. 185–212, 1936; J. Spek, Einfluss der Salze auf die Plasmkolloide von *Actinosphaerium*, *Acta Zool.* 1921; J. A. Kitching, *Journ. Exp. Biology*, XI, XIII, XV, 1934–38.

uniform surface-tension, and through whose boundary-film diffusion is taking place; but here, owing to the small difference between the fluid constituting and that surrounding the drop, the surface-tension equilibrium is somewhat unstable; it is apt to vanish, and the rounded outline of the drop disappears, like a burst bubble, in a moment.

If, on the other hand, the substance of the cell acquire a greater solidity, as for instance in a muscle-cell, or by reason of mucous accumulations in an epithelium cell, then the laws of fluid pressure no longer apply, the pressure on the nucleus tends to become unsymmetrical, and its shape is modified accordingly. Amoeboid movements may be set up in the nucleus by anything which disturbs the symmetry of its own surface-tension; and where "nuclear material" is scattered in small portions throughout the cell as in many Rhizopods, instead of being aggregated in a single nucleus, the simple explanation probably is that the "phase difference" (as the chemists say) between the nuclear and the protoplasmic substance is comparatively slight, and the surface-tension which tends to keep them separate is correspondingly small*.

Apart from that invisible or ultra-microscopic heterogeneity which is inseparable from our notion of a "colloid," there is a visible heterogeneity of structure within both the nucleus and the outer protoplasm. The former contains, for instance, a rounded nucleolus or "germinal spot," certain conspicuous granules or strands of the peculiar substance called chromatin†, and a coarse meshwork of a protoplasmic material known as "linin" or achromatin; the outer protoplasm, or cytoplasm, is generally believed to consist throughout of a sponge-work, or rather alveolar meshwork, of more and less fluid substances; it may contain "mitochondria," appearing in tissue-cultures as small amoeboid bodies; and lastly, there are generally to be detected (in the animal, rarely in the vegetable kingdom) one or more very minute bodies, usually in the cytoplasm sometimes within the nucleus, known as the centrosome or centrosomes.

* The elongated or curved "macronucleus" of an Infusorian is to be looked upon as a single mass of chromatin, rather than as an aggregation of particles in a fluid drop, as in the case described. It has a shape of its own, in which ordinary surface-tension plays a very subordinate part.

† First so-called by W. Flemming, in his *Zellsubstanz, Kern und Zelltheilung*, 1882.

The morphologist is accustomed to speak of a "polarity" of the cell, meaning thereby a symmetry of visible structure about a particular axis. For instance, whenever we can recognise in a cell both a nucleus and a centrosome, we may consider a line drawn through the two as the morphological axis of polarity; an epithelium cell is morphologically symmetrical about a median axis passing from its free surface to its attached base. Again, by an extension of the term polarity, as is customary in dynamics, we may have a "radial" polarity, between centre and periphery; and lastly, we may have several apparently independent centres of polarity within the single cell. Only in cells of quite irregular or amoeboid form do we fail to recognise a definite and symmetrical polarity. The *morphological* polarity is accompanied by, and is but the outward expression (or part of it) of a true *dynamical* polarity, or distribution of forces; and the lines of force are, or may be, rendered visible by concatenation of particles of matter, such as come under the influence of the forces in action.

When lines of force stream inwards from the periphery towards a point in the interior of the cell, particles susceptible of attraction either crowd towards the surface of the cell or, when retarded by friction, are seen forming lines or "fibrillae" which radiate outwards from the centre. In the cells of columnar or ciliated epithelium, where the sides of the cell are symmetrically disposed to their neighbours but the free and attached surfaces are very diverse from one another in their external relations, it is these latter surfaces which constitute the opposite poles; and in accordance with the parallel lines of force so set up, we very frequently see parallel lines of granules which have ranged themselves perpendicularly to the free surface of the cell (cf. Fig. 149).

A simple manifestation of polarity may be well illustrated by the phenomenon of diffusion, where we may conceive, and may automatically reproduce, a field of force, with its poles and its visible lines of equipotential, very much as in Faraday's conception of the field of force of a magnetic system. Thus, in one of Leduc's experiments*, if we spread a layer of salt solution over a level plate of glass, and let fall into the middle of it a drop of indian ink, or of blood, we shall find the coloured particles travelling

* *Théorie physico-chimique de la Vie*, 1910, p. 73.

outwards from the central "pole of concentration" along the lines of diffusive force, and so mapping out for us a "monopolar field" of diffusion: and if we set two such drops side by side, their lines of diffusion will oppose and repel one another. Or, instead of the uniform layer of salt solution, we may place at a little distance from one another a grain of salt and a drop of blood, representing two opposite poles: and so obtain a picture of a "bipolar field" of diffusion. In either case, we obtain results closely analogous to the morphological, but really *dynamical*, polarity of the organic cell. But in all probability, the dynamical polarity or asymmetry of the cell is a very complicated phenomenon: for the obvious reason that, in any system, one asymmetry will tend to beget another. A chemical asymmetry will induce an inequality of surface-tension, which will lead directly to a modification of form; the chemical asymmetry may in turn be due to a process of electrolysis in a polarised electrical field; and again the chemical heterogeneity may be intensified into a chemical polarity, by the tendency of certain substances to seek a locus of greater or less surface-energy. We need not attempt to grapple with a subject so complicated, and leading to so many problems which lie beyond the sphere of interest of the morphologist. But yet the morphologist, in his study of the cell, cannot quite evade these important issues; and we shall return to them again when we have dealt somewhat with the form of the cell, and have taken account of some of its simpler phenomena.

We are now ready, and in some measure prepared, to study the numerous and complex phenomena which accompany the division of the cell, for instance of the fertilised egg. But it is no easy task to epitomise the facts of the case, and none the easier that of late new methods have shewn us new things, and have cast doubt on not a little that we have been accustomed to believe.

Division of the cell is of necessity accompanied, or preceded, by a change from a radial or monopolar to a definitely bipolar symmetry. In the hitherto quiescent or apparently quiescent cell, we perceive certain movements, which correspond precisely to what must accompany and result from a polarisation of forces within: of forces which, whatever be their specific nature, are at least capable of polarisation, and of producing consequent attraction or repulsion between charged particles. The opposing forces which are distributed in equilibrium throughout the cell become focused in two "centrosomes*," which may or may not be already visible. It generally happens that, in the egg, one of these centrosomes is near to and the other far from the "animal pole," which is both visibly and chemically different from the other, and is where the more conspicuous developmental changes will presently begin.

Between the two centrosomes, in stained preparations, a spindle-shaped figure appears (Fig. 88), whose striking resemblance to the

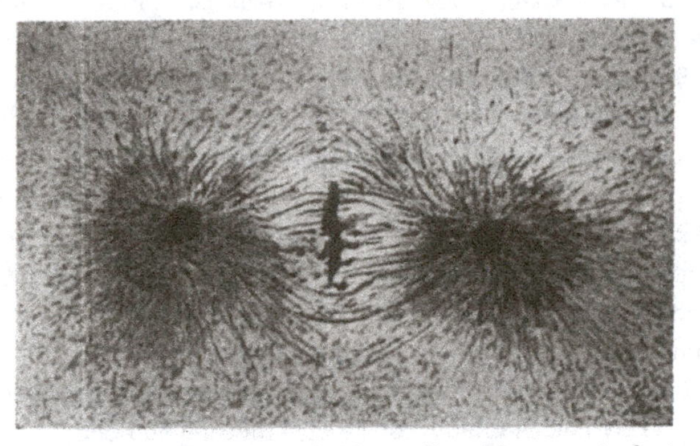

Fig. 88. Caryokinetic figure in a dividing cell (or blastomere) of a trout's egg. After Prenant, from a preparation by Prof. Bouin.

lines of force made visible by iron-filings between the poles of a magnet was at once recognised by Hermann Fol, in 1873, when he witnessed the phenomenon for the first time†. On the farther side of the centrosomes are seen star-like figures, or "asters," in which we seem to recognise the broken lines of force which run externally to those stronger lines which lie nearer to the axis and constitute the "spindle." The lines of force are rendered visible, or materialised, just as in the experiment of the iron-filings, by the fact that, in the heterogeneous substance of the cell, certain portions

* These centrosomes are the two halves of a single granule, and are said (by Boveri) to come from the middle piece of the original spermatozoon.

† He did so in the egg of a medusa (*Geryon*), *Jen. Zeitschr.* VII, p. 476, 1873. Similar ideas have been expressed by Strasbürger, Henneguy, Van Beneden, Errera, Ziegler, Gallardo and others.

of matter are more "permeable" to the acting force than others, become themselves polarised after the fashion of a magnetic or "paramagnetic" body, arrange themselves in an orderly way between the two poles of the field of force, seem to cling to one another as it were in threads*, and are only prevented by the friction of the surrounding medium from approaching and congregating around the adjacent poles.

As the field of force strengthens, the more will the lines of force be drawn in towards the interpolar axis, and the less evident will be those remoter lines which constitute the terminal, or extrapolar, asters: a clear space, free from materialised lines of force, may thus tend to be set up on either side of the spindle, the so-called "Bütschli space" of the histologists†. On the other hand, the lines of force constituting the spindle will be less concentrated if they find a path of less resistance at the periphery of the cell: as happens in our experiment of the iron-filings, when we encircle the field of force with an iron ring. On this principle, the differences observed between cells in which the spindle is well developed and the asters small, and others in which the spindle is weak and the asters greatly developed, might easily be explained by variations in the potential of the field, the large, conspicuous asters being correlated in turn with a marked permeability of the surface of the cell.

The visible field of force, though often called the "nuclear spindle," is formed outside of, but usually near to, the nucleus.

* Whence the name "mitosis" (Greek μίτος, a thread), applied first by Flemming to the whole phenomenon. Kollmann (*Biol. Centralbl.* II, p. 107, 1882) called it *divisio per fila*, or *divisio laqueis implicata.* Many of the earlier students, such as Van Beneden (Rech. sur la maturation de l'œuf, *Arch. de Biol.* IV, 1883), and Hermann Fol (Zur Lehre v. d. Entstehung d. karyokinetischen Spindel, *Arch. f. mikrosk. Anat.* XXXVII, 1891) thought they recognised actual muscular threads, drawing the nuclear material asunder towards the respective foci or poles; and some such view of *Zugkräfte* was long maintained by other writers, by Heidenhain especially, by Boveri, Flemming, R. Hertwig, Rhumbler, and many more. In fact, the existence of contractile threads, or the ascription to the spindle rather than to the poles or centrosomes of the active forces concerned in nuclear division, formed the main tenet of all those who declined to go beyond the "contractile properties of protoplasm" for an explanation of the phenomenon (cf. J. W. Jenkinson, *Q.J.M.S.* XLVIII, p. 471, 1904. See also J. Spek's historical account of the theories of cell-division, *Arch. f. Entw. Mech.* XLIV, pp. 5–29, 1918).

† Cf. O. Bütschli, Ueber die künstliche Nachahmung der karyokinetischen Figur, *Verh. Med. Nat. Ver. Heidelberg*, V, pp. 28–41 (1892), 1897.

Let us look a little more closely into the structure of this body, and into the changes which it presently undergoes.

Within its spherical outline (Fig. 89 A), it contains an "alveolar" meshwork (often described, from its appearance in optical section, as a "reticulum"), consisting of more solid substances with more fluid matter filling up the interalveolar spaces. This phenomenon, familiar to the colloid chemist, is what he calls a "two-phase system," one substance or "phase" forming a continuum through which the other is dispersed; it is closely allied to what we call in

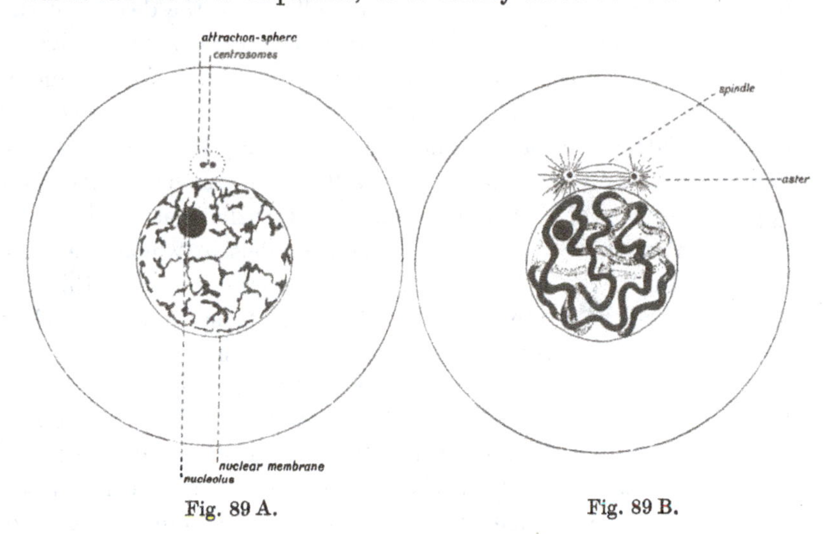

Fig. 89 A. Fig. 89 B.

ordinary language a *froth* or a *foam**, save that in these latter the disperse phase is represented by *air*. It is a surface-tension phenomenon, due to the interaction of two intermixed fluids not very different in density, as they strive to separate. Of precisely the same kind (as Bütschli was the first to shew) are the minute alveolar networks which are to be discerned in the cytoplasm of the cell†,

* Froth and foam have been much studied of late years for technical reasons, and other factors than surface-tension are found to be concerned in their existence and their stability. See (*int. al.*) Freundlich's *Capillarchemie*, and various papers by Sasaki, in *Bull. Chem. Soc. of Japan*, 1936–39.

† Bütschli, *Untersuchungen über mikroskopische Schäume und das Protoplasma*, 1892; *Untersuchungen über Strukturen*, etc., 1898; L. Rhumbler, Protoplasma als physikalisches System, *Ergebn. d. Physiologie*, 1914; H. Giersberg, Plasmabau der Amöben, im Hinblick auf die Wabentheorie, *Arch. f. Entw. Mech.* LI, pp. 150–250, 1922; etc.

and which we now know to be not inherent in the nature of protoplasm nor of living matter in general, but to be due to various causes, natural as well as artificial*. The microscopic honeycomb structure of cast metal under various conditions of cooling is an example of similar surface-tension phenomena.

Such then, in briefest outline, is the typical structure commonly ascribed to a cell when its latent energies are about to manifest themselves in the phenomenon of cell-division. The account is based on observation not of the living cell but of the dead: on the assumption, that is to say, that fixed and stained material gives a true picture of reality. But in Robert Chambers's method of micro-dissection†, the living cell is manipulated with fine glass needles under a high magnification, and shews us many interesting things. Chambers assures us that the spindle fibres never make their appearance as visible structures until coagulation has set in; and that astral rays are, or appear to be, channels in which the more fluid content of the cell flows towards a centrosome‡. Within the bounds to which we are at present keeping, these things are of no great moment; for whether the spindle appear early or late, it still bears witness to the fact that matter has arranged itself along bipolar lines of force; and even if the astral rays be only streams or currents, on lines of force they still approximately lie. Yet the change from the old story to the new is important, and may make a world of difference when we attempt to define the forces concerned. All our descriptions, all our interpretations, are bound to be influenced by our conception of the mechanism before us; and he

* Arrhenius, in describing a typical colloid precipitate, does so in terms that are very closely applicable to the ordinary microscopic appearance of the protoplasm of the cell. The precipitate consists, he says, "en un réseau d'une substance solide contenant peu d'eau, dans les mailles duquel est inclus un fluide contenant un peu de colloide dans beaucoup d'eau....Évidemment cette structure se forme à cause de la petite différence de poids spécifique des deux phases, et de la consistance gluante des particules séparées, qui s'attachent en forme de réseau" (*Rev. Scientifique*, Feb. 1911). This, however, is far from being the whole story: cf. (e.g.) S. C. Bradford, On the theory of gels, *Biochem. Journ.* XVII, p. 230, 1925; W. Seifritz, The alveolar structure of protoplasm, *Protoplasma*, IX, p. 198, 1930; and A. Frey-Wissling, *Submikroskopische Morphologie des Protoplasmas*, Berlin, 1938.

† See R. Chambers, An apparatus...for the dissection and injection of living cells, *Anatom. Record*, XXIV, 19 pp., 1922.

‡ This centripetal flow of fluid was announced by Butschli in his early papers, and confirmed by Rhumbler, though attributed to another cause.

who sees threads where another sees channels is likely to tell a different story about neighbouring and associated things.

It has also been suggested that the spindle is somehow due to a re-arrangement of protein macromolecules or micelles; that such changes of orientation of large colloid particles may be a widespread phenomenon; and that coagulation itself is but a polymerisation of larger and larger macromolecules*.

But here we have touched the brink of a subject so important that we must not pass it by without a word, and yet so contentious that we must not enter into its details. The question involved is simply whether the great mass of recorded observations and accepted beliefs with regard to the visible structure of protoplasm and of the cell constitute a fair picture of the actual *living cell*, or be based on appearances which are incident to death itself and to the artificial treatment which the microscopist is accustomed to apply. The great bulk of histological work is done by methods which involve the sudden killing of the cell or organism by strong reagents, the assumption being that death is so rapid that the visible phenomena exhibited during life are retained or "fixed" in our preparations.

Hermann Fol struck a warning note full sixty years ago: "Il importe à l'avenir de l'histologie de combattre la tendance à tirer des conclusions des images obtenues par des moyens artificiels et à leur donner une valeur intrinsèque, sans que ces images aient été controlées sur le vivant†." Fol was thinking especially of cell-membranes and the delimitation of cells; but still more difficult and precarious is the interpretation of the minute internal networks, granules, etc., which represent the alleged structure of protoplasm. A colloid body, or colloid solution, is *ipso facto* heterogeneous; it has after some fashion a *structure* of its own. And this structure chemical action, under the microscope, may demonstrate, or emphasise, or alter and disguise. As Hardy put it, "It is notorious that the various fixing reagents are coagulants of organic colloids, and that the figure varies according to the reagent used."

A case in point is that of the vitreous humour, to which some histologists have ascribed a fairly complex structure, seeing in it a framework of fibres with the meshes filled with fluid. But it is really *a true gel*, without any structure in the usual sense of the word. The "fibres" seen in ordinary microscopic preparations are due to the coagulation of micellae by the fixative employed. Under the ultra-microscope the vitreous is optically empty to begin with; then innumerable minute fibrillae appear in the beam of light, criss-crossing one another. Soon these break down into strings of beads, and

* Cf. J. D. Bernal, on Molecular architecture of biological systems, *Proc. Roy. Inst.*, 1938; H. Staudinger, *Nature*, Aug. 1, 1939.

† H. Fol, *Recherches sur la fécondation et le commencement de l'hénogénie chez divers animaux*, Genève, 1879, pp. 241–242. Cf. A. Dalcq, in *Biol. Reviews*, III, p. 24, 1928: "Il serait désirable de nous débarrasser de l'idée que tout ce qu'il y a d'important dans la cellule serait providentiellement colorable par l'hématoxyline, la safranine ou le violet de gentiane."

finally only separate dots are seen*. Other sources of error arise from the optical principles concerned in microscopic vision; for the diffraction-pattern which we call the "image" may, under certain circumstances, be very different from the actual object†. Furthermore, the optical properties of living protoplasm are especially complicated and imperfectly known, as in general those of colloids may be said to be; the minute aggregates of the "disperse phase" of gels produce a scattering action on light, leading to appearances of turbidity etc., with no other or more real basis‡.

So it comes to pass that some writers have altogether denied the existence in the living cell-protoplasm of a network or alveolar "foam"; others have cast doubts on the main tenets of recent histology regarding nuclear structure; and Hardy, discussing the structure of certain gland-cells, declared that "there is no evidence that the structure discoverable in the cell-substance of these cells after fixation has any counterpart in the cell when living." "A large part of it" he went on to say "is an artefact. The profound difference in the minute structure of a secretory cell of a mucous gland according to the reagent which is used to fix it would, it seems to me, almost suffice to establish this statement in the absence of other evidence§."

Nevertheless, histological study proceeds, especially on the part of the morphologists, with but little change in theory or in method, in spite of these and many other warnings. That certain visible structures, nucleus, vacuoles, "attraction-spheres" or centrosomes, etc., are actually present in the living cell we know for certain; and to this class belong the majority of structures with which we are at present concerned. That many other alleged structures are artificial has also been placed beyond a doubt; but where to draw the dividing line we often do not know.

The following is a brief epitome of the visible changes undergone by a typical cell, subsequent to the resting stage, leading up to the act of segmentation, and constituting the phenomenon of mitosis or caryokinetic division. In the fertilised egg of a sea-urchin we see with almost diagrammatic completeness, in fixed and stained specimens, what is set forth here||.

* W. S. Duke-Elder, *Journ. Physiol.* LXVIII, pp. 154–165, 1930; cf. Baurmann, *Arch. f. Ophthalm.* 1923, 1926; etc.

† Abbé, *Arch. f. mikrosk. Anat.* IX, p. 413, 1874; *Gesammelte Abhandl.* I, p. 45, 1904.

‡ Cf. Rayleigh, On the light from the sky, *Phil. Mag.* (4) XLI, p. 107, 1871.

§ W. B. Hardy, On the structure of cell protoplasm, *Journ. Physiol.* XXIV, pp. 158–207, 1889; also Höber, *Physikalische Chemie der Zelle und der Gewebe*, 1902; W. Berg, Beiträge zur Theorie der Fixation, etc., *Arch. f. mikr. Anat.* LXII, pp. 367–440, 1903. Cf. (*int. al.*) Flemming, *Zellsubstanz, Kern und Zelltheilung*, 1882, p. 51; etc.

|| My description and diagrams (Figs. 89–93) are mostly based on those of the late Professor E. B. Wilson.

1. The chromatin, which to begin with had been dimly seen as granules on a vague achromatic reticulum (Figs. 89, 90)—perhaps no more than an histological artefact—concentrates to form a skein or *spireme*, often looked on as a continuous thread, but perhaps discontinuous or fragmented from the first. It, or its several fragments, will presently split asunder; for it is essentially double, and may even be seen as a double thread, or pair of *chromatids*, from an early stage. The *chromosomes* are portions of this double thread, which shorten down to form little rods, straight or curved, often

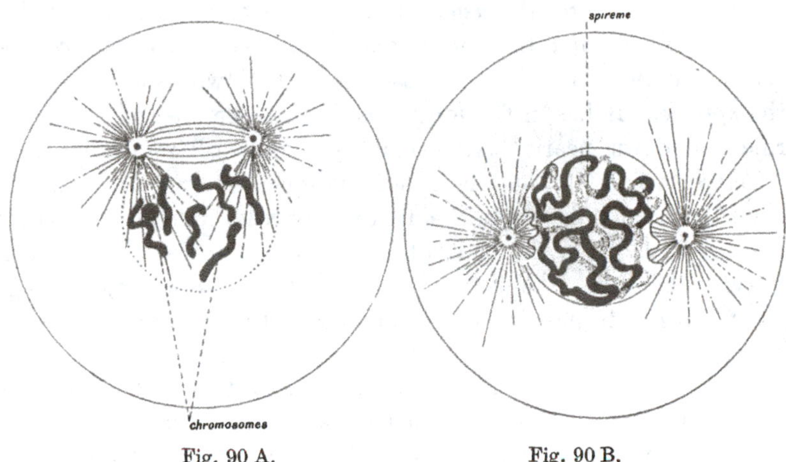

Fig. 90 A. Fig. 90 B.

bent into a **V**, sometimes ovoid, round or even annular, and which in the living cell are frequently seen in active, writhing movement, "like eels in a box"*; they keep apart from one another, as by some repulsion, and tend to move outward towards the nuclear membrane. Certain deeply staining masses, the nucleoli, may be present in the resting nucleus, but take no part (at least as a rule) in the formation of the chromosomes; they are either cast out of the nucleus and dissolved in the cytoplasm, or else fade away *in situ*.

* T. S. Strangeways, *Proc. R.S.* (B), XCIV, p. 139, 1922. The tendency of the chromatin to form spirals, large or small, while the nucleus is issuing from its resting-stage, is very remarkable. The tensions to which it is due may be overcome, and the chromosomes made to uncoil, by treatment with ammonia or acetic acid vapour. See Y. Kuwada, *Botan. Mag.* Tokyo, XLVI, p. 307, 1932; and C. D. Darlington, Mechanical aspects of nuclear division, *Sci. Journ. R. Coll. of Sci.* IV, p. 94, 1934.

But this rule does not always hold; for they persist in many protozoa, and now and then the nucleolus remains and becomes itself a chromosome, as in the spermogonia of certain insects.

2. Meanwhile a certain deeply staining granule (here extra-nuclear), known as the *centrosome**, has divided into two. It is all but universally visible, save in the higher plants; perhaps less stress is laid on it than at one time, but Bovery called it the "dynamic centre" of the cell†. The two resulting granules travel to opposite poles of the nucleus, and there each becomes surrounded by a starlike figure, the *aster*, of which we have spoken already; immediately around the centrosome is a clear space, the *centrosphere.* Between the two centrosomes, or the two asters, stretches the *spindle.* It lies in the long axis, if there be one, of the cell, a rule laid down nearly sixty years ago, and still remembered as "Hertwig's Law"‡; but the rule is as much and no more than to say that the spindle sets in the direction of least resistance. Where the egg is laden with food-yolk, as often happens, the latter is heavier than the cytoplasm; and gravity, by orienting the egg itself, thus influences, though only indirectly, the first planes of segmentation§.

3. The definite nuclear outline is soon lost; for the chemical "phase-difference" between nucleus and cytoplasm has broken down, and where the nucleus was, the chromosomes now lie (Figs. 90, 91). The lines of the spindle become visible, the chromosomes arrange themselves midway between its poles, to form the *equatorial plate,* and are spaced out evenly around the central spindle, again a simple result of mutual repulsion.

4. Each chromosome separates longitudinally into two||: usually at this stage—but it is to be noted that the splitting may have taken place as early as the spireme stage (Fig. 92).

* The centrosome has a curious history of its own, none too well ascertained. The ovum has a centrosome, and in self-fertilised eggs this is retained; but when a sperm-cell enters the egg the original centrosome degenerates, and its place is taken by the "middle-piece" of the spermatozoon.

† The stages 1, 2, 5 and 6 are called by embryologists the *prophase, metaphase, anaphase* and *telophase.*

‡ C. Hertwig, *Jenaische Ztschr.* XVIII, 1884.

§ See James Gray, The effect of gravity on the eggs of Echinus, *Jl. Exp. Zool.* v, pp. 102–11, 1927.

|| A fundamental fact, first seen by Flemming in 1880.

5. The halves of the split chromosomes now separate from and apparently repel one another, travelling in opposite directions towards the two poles* (Fig. 92 B), for all the world as though they were being pulled asunder by actual threads.

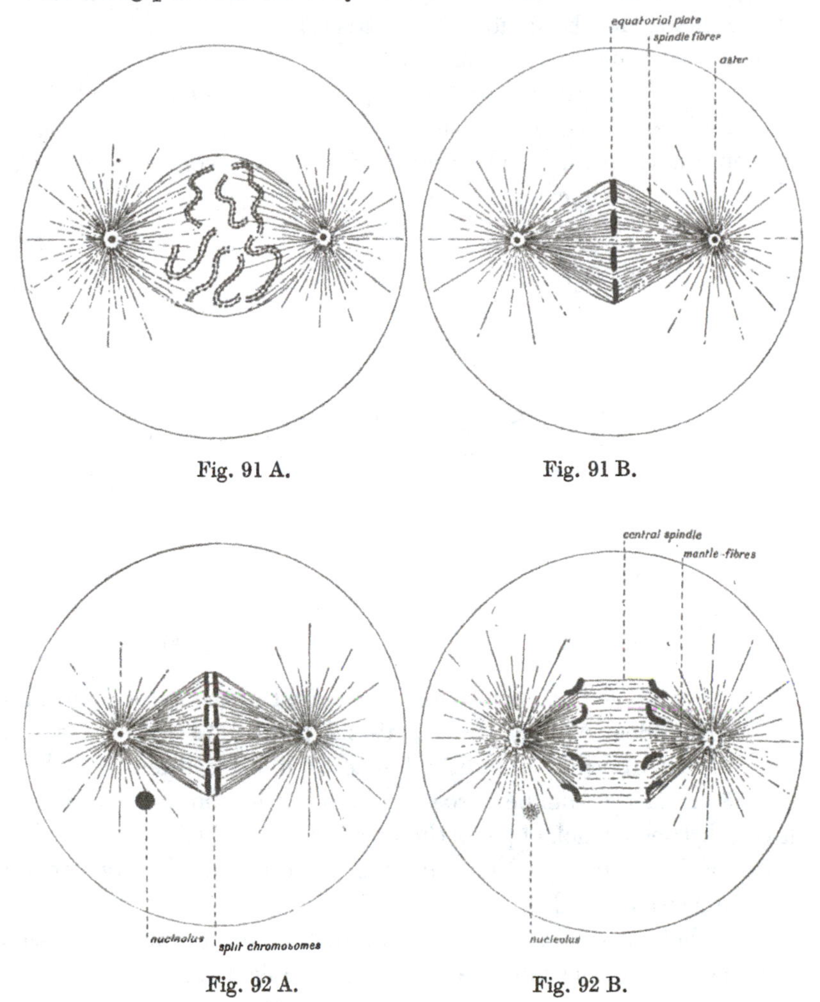

Fig. 91 A. Fig. 91 B.

Fig. 92 A. Fig. 92 B.

6. Presently the spindle itself changes shape, lengthens and contracts, and seems as it were to push the two groups of daughter-

* Cf. K. Belar, Beiträge zur Causalanalyse der Mitose, *Ztschr. f. Zellforschung*, x, pp. 73–124, 1929.

chromosomes into their new places* (Figs. 92, 93); and its chromosomes form once more an alveolar reticulum and may occasionally form another spireme at this stage. A boundary-surface, or at least a recognisable phase-difference, now develops round each reconstructed nuclear mass, and the spindle disappears (Fig. 93 B). The centrosome remains, as a rule, outside the nucleus.

7. On the central spindle, in the position of the equatorial plate, a "cell-plate," consisting of deeply staining thickenings, has made its appearance during the migration of the chromosomes. This cell-plate is more conspicuous in plant-cells.

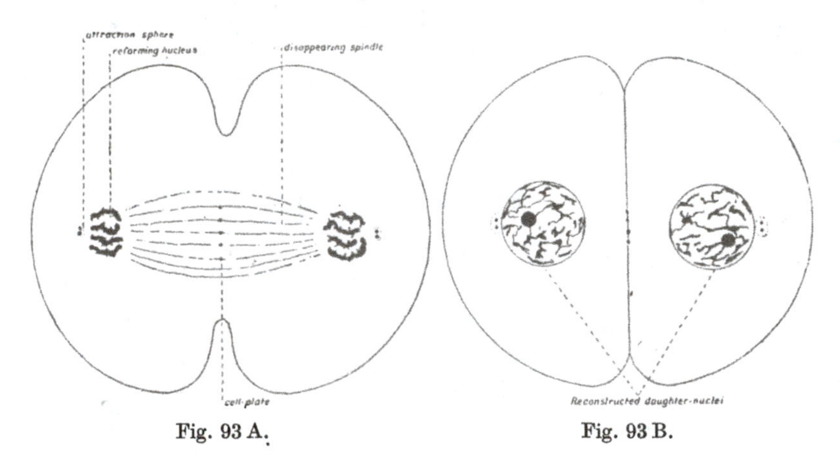

Fig. 93 A. Fig. 93 B.

8. Meanwhile a constriction has appeared in the cytoplasm, and the cell divides through the equatorial plane. In plant-cells the line of this division is foreshadowed by the "cell-plate," which extends from the spindle across the entire cell, and splits into two layers, between which appears the membrane by which the daughter-cells are cleft asunder. In animal cells the cell-plate does not attain such dimensions, and no cell-wall is formed.

The whole process takes from half-an-hour to an hour; and this extreme slowness is not the least remarkable part of the phenomenon, from a physical point of view. The two halves of the

* The spindle has no actual threads or fibres, for Robert Chambers's micro-needles pass freely through it without disturbing the chromosomes: nor is it visible at all in living cells *in vitro*. It seems to be due to partial gelation of the cytoplasm, under conditions which, whether they be mechanical or chemical, are not easy to understand.

dividing centrosome, while moving apart, take some twenty minutes to travel a distance of 20μ, or at the rate, say, of two years to a yard. It is a question of inertia, and the inertia of the system must be very large.

The beautiful technique of cell-culture *in vitro* has of late years let this whole succession of phenomena, once only to be deduced from sections, be easily followed as it proceeds within the living tissue or cell. The vivid accounts which have been given of this spectacle add little to the older account as we have related it: save that, when the equatorial constriction begins and the halves of the split chromosomes drift apart, the protoplasm begins to show a curious and even violent activity. The cytoplasm is thrust in and out in bulging pustules or "balloons"; and the granules and fat-globules stream in and out as the pustules rise and fall away. At length the turmoil dies down; and now each half of the cell (not an ovum but a tissue-cell or "fibroplast") pushes out large pseudopodia, flattens into an amoeboid phase, the connecting thread of protoplasm snaps in the divided cell, and the daughter-cells fall apart and crawl away. The two groups of chromosomes, on reaching the poles of the spindle, turn into bunches of short thick rods; these grow diffuse, and form a network of chromatin within a nucleus; and at last the chromosomes, having lost their identity, disappear entirely, and two or more nucleoli are all that is to be seen within the cell.

The whole, or very nearly the whole, of these nuclear phenomena may be brought into relation with some such polarisation of forces in the cell as a whole as is indicated by the "spindle" and "asters" of which we have already spoken: certain particular phenomena, directly attributable to surface-tension and diffusion, taking place in more or less obvious and inevitable dependence upon the polar system. At the same time, in attempting to explain the phenomena, we cannot say too clearly, or too often, that all that we are meanwhile justified in doing is to try to shew that such and such actions lie *within the range* of known physical actions and phenomena, or that known physical phenomena produce effects similar to them. We feel that the whole phenomenon is not *sui generis*, but is somehow or other capable of being referred to dynamical laws, and to the general principles of physical science. But when we speak of some particular force or mode of action, using it as an illustrative hypothesis, we stop far short of the implication that this or that force is necessarily the very one which is actually at work within the living cell; and certainly we need not attempt the formidable task of trying to reconcile, or to choose between, the various hypotheses which have already been enunciated, or the several assumptions on which they depend.

Many other things happen within the cell, especially in the germ-cell both before and after fertilisation. They also have a physical element, or a mechanical aspect, like the phenomena of cell-division which we are speaking of; but the narrow bounds to which we are keeping hold difficulties enough*.

Any region of space within which action is manifested is a field of force; and a simple example is a bipolar field, in which the action is symmetrical with reference to the line joining two points, or poles, and with reference also to the "equatorial" plane equidistant from both. We have such a field of force in the neighbourhood of the centrosome of the ripe cell or ovum, when it is about to divide; and by the time the centrosome has divided, the field is definitely a bipolar one.

The *quality* of a medium filling the field of force may be uniform, or it may vary from point to point. In particular, it may depend upon the magnitude of the field; and the quality of one medium may differ from that of another. Such variation of quality, within one medium, or from one medium to another, is capable of diagrammatic representation by a variation of the direction or the strength of the field (other conditions being the same) from the state manifested in some uniform medium taken as a standard. The medium is said to be *permeable* to the force, in greater or less degree than the standard medium, according as the variation of the density of the lines of force from the standard case, under otherwise identical conditions, is in excess or defect. *A body placed in the medium will tend to move towards regions of greater or less force according as its*

* Cf. C. D. Darlington, *Recent Advances in Cytology*, 1932, and other well-known works.

*permeability is greater or less than that of the surrounding medium**. In the common experiment of placing iron-filings between the two poles of a magnetic field, the filings have a very high permeability; and not only do they themselves become polarised so as to attract one another, but they tend to be attracted from the weaker to the stronger parts of the field, and as we have seen, they would soon gather together around the nearest pole were it not for friction or some other resistance. But if we repeat the same experiment with such a metal as bismuth, which is very little permeable to the magnetic force, then the conditions are reversed, and the particles, being repelled from the stronger to the weaker parts of the field, tend to take up their position as far from the poles as possible. The particles have become polarised, but in a sense opposite to that of the surrounding, or adjacent, field.

Now, in the field of force whose opposite poles are marked by the centrosomes, we may imagine the nucleus to act as a more or less permeable body, as a body more permeable than the surrounding medium, that is to say the "cytoplasm" of the cell. It is accordingly attracted by, and drawn into, the field of force, and tries, as it were, to set itself between the poles and as far as possible from both of them. In other words, the centrosome-foci will be apparently drawn over its surface, until the nucleus as a whole is involved within the field of force which is visibly marked out by the "spindle" (Fig. 90 B).

If the field of force be electrical, or act in a fashion analogous to an electrical field, the charged nucleus will have its surface-tensions diminished†: with the double result that the inner alveolar meshwork will be broken up (par. 1), and that the spherical boundary of the whole nucleus will disappear (par. 2). The break-up of the alveoli (by thinning and rupture of their partition walls)

* If the word *permeability* be deemed too directly suggestive of the phenomena of magnetism, we may replace it by the more general term of *specific inductive capacity*. This would cover the particular case, which is by no means an improbable one, of our phenomena being due to a "surface charge" borne by the nucleus itself and also by the chromosomes: this surface charge being in turn the result of a difference in inductive capacity between the body or particle and its surrounding medium.

† On the effect of electrical influences in altering the surface-tensions of the colloid particles, see Bredig, *Anorganische Fermente*, pp. 15, 16, 1901.

leads to the formation of a net, and the further break-up of the net may lead to the unravelling of a thread or "spireme".

Here there comes into play a fundamental principle which, in so far as we require to understand it, can be explained in simple words. The effect (and we might even say the *object*) of drawing the more permeable body in between the poles is to obtain an "easier path" by which the lines of force may travel; but it is obvious that a longer route through the more permeable body may at length be found less advantageous than a shorter route through the less permeable medium. That is to say, the more permeable body will only tend to be drawn into the field of force until a point is reached where (so to speak) the way *round* and the way *through* are equally advantageous. We should accordingly expect that (on our hypothesis) there would be found cases in which the nucleus was wholly, and others in which it was only partially, and in greater or less degree, drawn in to the field between the centrosomes. This is precisely what is found to occur in actual fact. Figs. 90 A and B represent two so-called "types," of a phase which follows that represented in Fig. 89. According to the usual descriptions we are told that, in such a case as Fig. 90 B, the "primary spindle" disappears* and the centrosomes diverge to opposite poles of the nucleus; such a condition being found in many plant-cells, and in the cleavage-stages of many eggs. In Fig. 90 A, on the other hand, the primary spindle persists, and subsequently comes to form the main or "central" spindle; while at the same time we see the fading away of the nuclear membrane, the breaking up of the spireme into separate chromosomes, and an ingrowth into the nuclear area of the "astral rays"—all as in Fig. 91 A, which represents the next succeeding phase of Fig. 90 B. This condition, of Fig. 91 A, occurs in a variety of cases; it is well seen in the epidermal cells of the salamander, and is also on the whole characteristic of the mode of formation of the "polar bodies†." It is clear and obvious that the two "types" correspond to mere differences of degree,

* The spindle is potentially there, even though (as Chambers assures us) it only becomes visible after post-mortem coagulation. It is also said to become visible under crossed nicols: W. J. Schmidt, *Biodynamica*, XXII, 1936.

† These were first observed in the egg of a pond-snail (*Limnaea*) by B. Dumortier, *Mém. sur l'embryogénie des mollusques*, Bruxelles, 1837.

and are such as would naturally be brought about by differences in the relative permeabilities of the nuclear mass and of the surrounding cytoplasm, or even by differences in the magnitude of the former body.

But now an important change takes place, or rather an important difference appears; for, whereas the nucleus as a whole tended to be drawn in to the *stronger* parts of the field, when it comes to break up we find, on the contrary, that its contained spireme-thread or separate chromosomes tend to be repelled to the *weaker* parts. Whatever this difference may be due to—whether, for instance, to actual differences of permeability, or possibly to differences in "surface-charge" or to other causes—the fact is that the chromatin substance now *behaves* after the fashion of a "diamagnetic" body, and is repelled from the stronger to the weaker parts of the field. In other words, its particles, lying in the inter-polar field, tend to travel towards the equatorial plane thereof (Figs. 91, 92), and further tend to move outwards towards the periphery of that plane, towards what the histologist calls the "mantle-fibres," or outermost of the lines of force of which the spindle is made up (par. 5, Fig. 91 B). And if this comparatively non-permeable chromatin substance come to consist of separate portions, more or less elongated in form, these portions, or separate "chromosomes," will adjust themselves longitudinally, in a peripheral equatorial circle (Figs. 92 A, B). This is precisely what actually takes place. Moreover, before the breaking up of the nucleus, long before the chromatin material has broken up into separate chromosomes, and at the very time when it is being fashioned into a "spireme," this body already lies in a polar field, and must already have a tendency to set itself in the equatorial plane thereof. But the long, continuous spireme thread is unable, so long as the nucleus retains its spherical boundary wall, to adjust itself in a simple equatorial annulus; in striving to do so, it must tend to coil and "kink" itself, and in so doing (if all this be so), it must tend to assume the characteristic convolutions of the "spireme."

After the spireme has broken up into separate chromosomes, these bodies come to rest in the equatorial plane, somewhere near its periphery; and here they tend to set themselves in a symmetrical arrangement (Fig. 94), such as makes for still better equilibrium.

The particles may be rounded or linear, straight or bent, sometimes annular; they may be all alike, or one or more may differ from the rest. Lying as they do in a semi-fluid medium, and subject (doubtless) to some symmetrical play of forces, it is not to be wondered at that they arrange themselves in a symmetrical configuration; and the field of force seems simple enough to let us predict, to some extent, the symmetries open to them. We do not know, we cannot safely surmise, the nature of the forces involved. In discussing Brauer's observations on the splitting of the chromatic filament, and on the symmetrical arrangement of the separate granules, in *Ascaris megalocephala*, Lillie* remarks: "This behaviour

Fig. 94. Chromosomes, undergoing splitting and separation. After Hatschek and Flemming, diagrammatised.

is strongly suggestive of the division of a colloidal particle under the influence of its surface electrical charge, and of the effects of mutual repulsion in keeping the products of division apart." It is probable that surface-tensions between the particles and the surrounding protoplasm would bring about an identical result, and would sufficiently account for the obvious, and at first sight very curious symmetry. If we float a couple of matches in water, we know that they tend to approach one another till they lie close together, side by side; and if we lay upon a smooth wet plate four matches, half broken across, a similar attraction brings the four matches together in the form of a symmetrical cross. Whether one of these, or yet another, be the explanation of the phenomenon,

* R. S. Lillie, Conditions determining the disposition of the chromatic filaments, etc., in mitosis; *Biol. Bulletin*, VIII, 1905.

it is at least plain that by some physical cause, some mutual attraction or common repulsion of the particles, we must seek to account for the symmetry of the so-called "tetrads," and other more or less familiar configurations. The remarkable annular chromosomes, shewn in Fig. 95, can be closely imitated by loops of thread upon a soapy film, when the film within the annulus is broken or its tension reduced; the balance of forces is here a simple one, between the uniform capillary tension which tends to widen out the ring and the uniform cohesion of its particles which keeps it together.

We may find other cases, at once simpler and more varied, where the chromosomes are bodies of rounded form and more or less

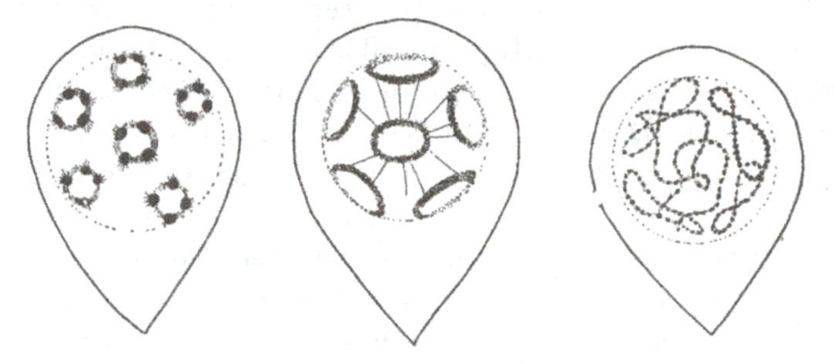

Fig. 95. Annular chromosomes, formed in the spermatogenesis of the mole-cricket. From Wilson, after Vom Rath.

uniform size. These also find their way to an equatorial plate; we gather (and Lamb assures us) that they are repelled from the centrosomes. They may go near the equatorial periphery, but they are not driven there; and we infer that some bond of mutual attraction holds them together. If they be free to move in a fluid medium, subject both to some common repulsion and some mutual attraction, then their circumstances are much like those of Mayer's well-known experiment of the floating magnets. A number of magnetised needles stuck in corks, all with like poles upwards, are set afloat in a basin; they repel one another, and scatter away to the sides. But bring a strong magnet (of unlike pole) overhead, and the little magnets gather in under its common attraction, while still keeping asunder through their own mutual repulsion. The symmetry of forces leads to a symmetrical configuration, which is the mathematical expression of a physical equilibrium—and is the not too remote counterpart of the arrangement of the electrons in an atom. Be that as it may, it is found that a group of three, four or five little magnets arrange themselves at the corners of an equilateral triangle, square or pentagon; but a sixth passes within the ring, and comes to rest in the centre of symmetry of the pentagon. If there be seven magnets, six form the ring, and the seventh occupies the centre; if there be ten, there is a ring of eight and two within it; and so on, as follows*:

Number of magnets	5	6	7	8	9	10	11	12	13	14	15	16
Do. in outer ring	5	5	6	7	8	8	8	9	10	10	10	11
Do. in inner ring	0	1	1	1	1	2	3	3	3	4	5	5

When we choose from the published figures cases where the chromosomes are as nearly as possible alike in size and form—the condition necessary for our parallel to hold—then, as Lillie predicted and as Doncaster and Graham Cannon have shewn, their congruent arrangement agrees, even to a surprising degree, with what we are led to expect by theory and analogy (Fig. 96).

The break-up of the nucleus, already referred to and ascribed to a diminution of its surface-tension, is accompanied by certain diffusion phenomena which are sometimes visible to the eye; and we are reminded of Lord Kelvin's view that diffusion is implicitly associated with surface-tension changes, of which the first step is a minute puckering of the surface-skin, a sort of interdigitation with the surrounding medium. For instance, Schewiakoff has observed in *Euglypha*† that, just before the break-up of the nucleus, a system of rays appears, concentred about it, but having nothing to do with the polar asters: and during the existence of this striation the nucleus enlarges very considerably, evidently by imbibition of fluid from the surrounding protoplasm. In short, diffusion is at work, hand in hand with, and as it were in opposition to, the surface-tensions which define the nucleus. By diffusion, hand in hand with surface-tension, the alveoli of the nuclear meshwork are formed, enlarged and finally ruptured: diffusion sets up the movements

* H. Graham Cannon, On the nature of the centrosomal force, *Journ. Genetics*, XIII, p. 55, 1923.

† Schewiakoff, Ueber die karyokinetische Kerntheilung der *Euglypha alveolata*, *Morph. Jahrb.* XIII, pp. 193–258, 1888 (see p. 216).

which give rise to the appearance of rays, or striae, around the nucleus: and through increasing diffusion and weakening surface-tension the rounded outline of the nucleus finally disappears.

As we study these manifold phenomena in the individual cases of particular plants and animals, we recognise a close identity of type coupled with almost endless variation of specific detail; and

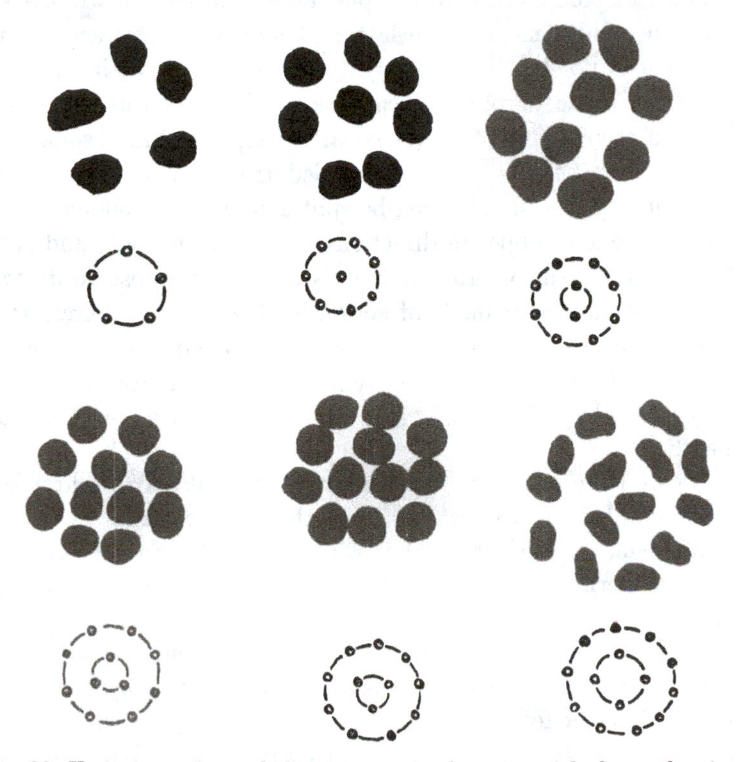

Fig. 96. Various numbers of chromosomes in the equatorial plate: the ring-diagrams give the arrangements predicted by theory. From Graham Cannon.

in particular, the order of succession in which certain of the phenomena occur is variable and irregular. The precise order of the phenomena, the time of longitudinal and of transverse fission of the chromatin thread, of the break-up of the nuclear wall, and so forth, will depend upon various minor contingencies and "interferences." And it is worthy of particular note that these variations in the order of events and in other subordinate details, while doubtless attributable to specific physical conditions, would seem to be without any obvious classificatory meaning or other biological significance.

So far as we have now gone, there is no great difficulty in pointing to simple and familiar examples of a field of force which are similar, or comparable, to the phenomena which we witness within the cell. But among these latter phenomena there are others for which it is not so easy to suggest, in accordance with known laws, a simple mode of physical causation. It is not at once obvious how, in any system of symmetrical forces, the chromosomes, which had at first been apparently repelled from the poles towards the equatorial plane, should then be split asunder, and should presently be attracted in opposite directions, some to one pole and some to the other. Remembering that it is not our purpose to *assert* that some one particular mode of action is at work, but merely to shew that there do exist physical forces, or distributions of force, which are capable of producing the required result, I give the following suggestive hypothesis, which I owe to my colleague Professor W. Peddie.

As we have begun by supposing that the nuclear or chromosomal matter differs in *permeability* from the medium, that is to say the cytoplasm, in which it lies, let us now make the further assumption that its permeability is variable, and depends upon the *strength of the field*.

In Fig. 97, we have a field of force (representing our cell), consisting of a homogeneous medium, and including two opposite poles: lines of force are indicated by full lines, and *loci of constant magnitude of force* are shewn by dotted lines, these latter being what are known as Cayley's equipotential curves*.

Let us now consider a body whose permeability (μ) depends on the strength of the field F. At two field-strengths, such as F_a, F_b, let the permeability of the body be equal to that of the medium, and let the curved line in Fig. 98 represent generally its permeability at other field-strengths; and let the outer and inner dotted curves in Fig. 97 represent respectively the loci of the field-strengths F_b

* *Phil. Trans.* XIV, p. 142, 1857. Cf. also F. G. Teixeira, *Traité des Courbes*, I, p. 372, Coimbra, 1908.

and F_a. The body if it be placed in the medium within either branch of the inner curve, or outside the outer curve, will tend to move into the neighbourhood of the adjacent pole. If it be placed

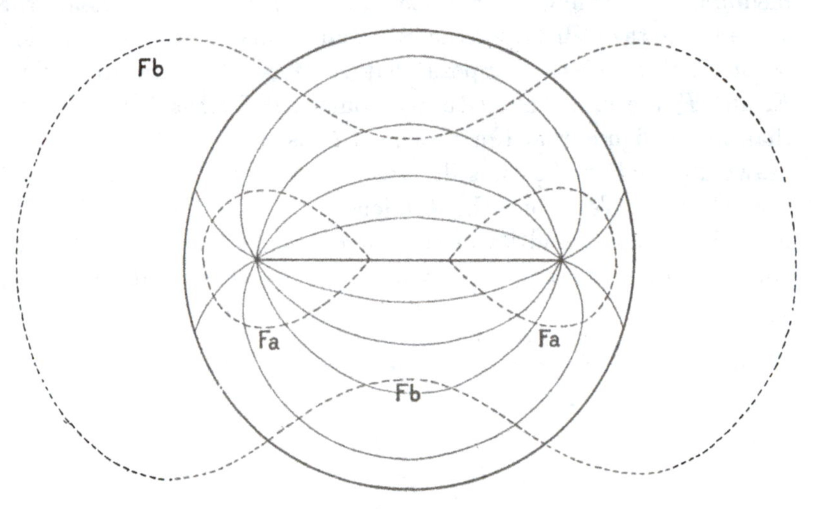

Fig. 97.

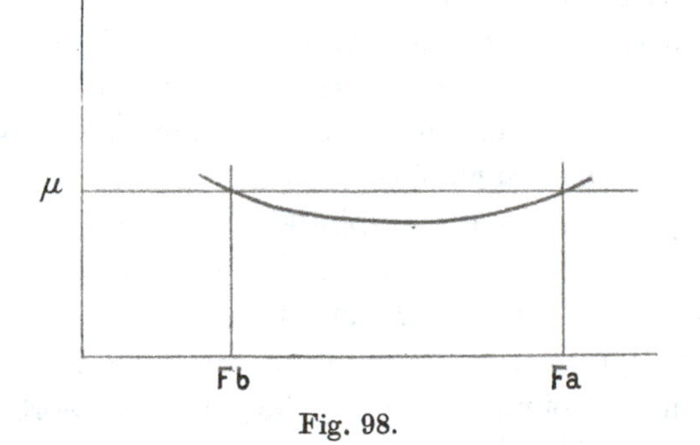

Fig. 98.

in the region intermediate to the two dotted curves, it will tend to move towards regions of weaker field-strength.

The locus F_b is therefore a locus of stable position, towards which the body tends to move; the locus F_a is a locus of unstable position, from which it tends to move. If the body were placed across F_a, it might be torn asunder into two portions, the split coinciding with the locus F_a.

Suppose a number of such bodies to be scattered throughout the medium. Let at first the regions F_a and F_b be entirely outside the space where the bodies are situated: and, in making this supposition we may, if we please, suppose that the loci which we are calling F_a and F_b are meanwhile situated somewhat farther from the axis than in our figure, that (for instance) F_a is situated where we have drawn F_b, and that F_b is still farther out. The bodies then tend towards the poles; but the tendency may be very small if, in Fig. 98, the curve and its intersecting straight line do not diverge very far from one another beyond F_a; in other words, if, when

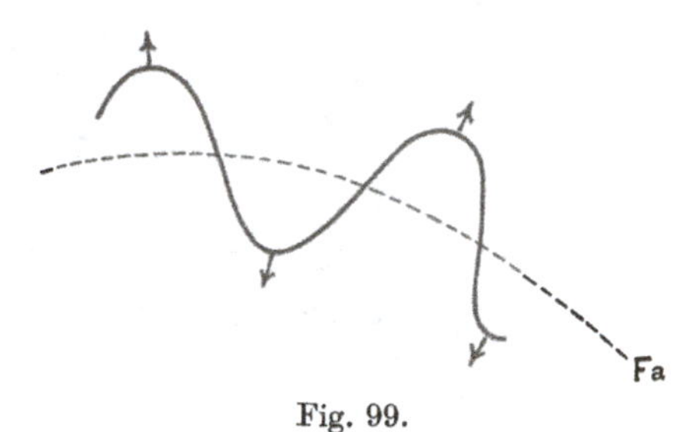

Fig. 99.

situated in this region, the permeability of the bodies is not very much in excess of that of the medium.

Let the poles now tend to separate farther and farther from one another, the strength of each pole remaining unaltered; in other words, let the centrosome-foci recede from one another, as they actually do, drawing out the spindle-threads between them. The loci F_a, F_b will close in to nearer relative distances from the poles. In doing so, when the locus F_a crosses one of the bodies, the body may be torn asunder; if the body be of elongated shape, and be crossed at more points than one, the forces at work will tend to exaggerate its foldings, and the tendency to rupture is greatest when F_a is in some median position (Fig. 99).

When the locus F_a has passed entirely over the body, the body tends to move towards regions of weaker force; but when, in turn, the locus F_b has crossed it, then the body again moves towards regions of stronger force, that is to say, towards the nearest pole.

And, in thus moving towards the pole, it will do so, as appears actually to be the case in the dividing cell, along the course of the outer lines of force, the so-called "mantle-fibres" of the histologist*

Such considerations as these give general results, easily open to modification in detail by a change of any of the arbitrary postulates which have been made for the sake of simplicity. Doubtless there are other assumptions which would meet the case; for instance, that during the active phase of the chromatin molecule (when it decomposes and sets free nucleic acid) it carries a charge opposite to that which it bears during its resting, or alkaline phase; and that it would accordingly move towards different poles under the influence of a current, wandering with its negative charge in an alkaline fluid during its acid phase to the anode, and to the kathode during its alkaline phase. A whole field of speculation is opened up when we begin to consider the cell not merely as a polarised electrical field, but also as an electrolytic field, full of wandering ions. Indeed it is high time we reminded ourselves that we have perhaps been dealing too much with ordinary physical analogies: and that our whole field of force within the cell is of an order of magnitude where these grosser analogies may fail to serve us, and might even play us false, or lead us astray. But our sole object meanwhile, as I have said more than once, is to demonstrate, by such illustrations as these, that, whatever be the actual and as yet unknown *modus operandi*, there are physical conditions and distributions of force which *could* produce just such phenomena of movement as we see taking place within the living cell. This, and no more, is precisely what Descartes is said to have claimed for his description of the human body as a "mechanism†."

While it can scarcely be too often repeated that our enquiry is not directed towards the solution of physiological problems, save only in so far as they are inseparable from the problems presented by the visible configurations of form and structure, and while we try, as far as possible, to evade the difficult question of what

* We have not taken account in the above paragraphs of the obvious fact that the supposed symmetrical field of force is distorted by the presence in it of the more or less permeable bodies; nor is it necessary for us to do so, for to that distorted field the above argument continues to apply, word for word.

† Michael Foster, *Lectures on the History of Physiology*, 1901, p. 62.

particular forces are at work when the mere visible forms produced are such as to leave this an open question, yet in this particular case we have been drawn into the use of electrical analogies, and we are bound to justify, if possible, our resort to this particular mode of physical action. There is an important paper by R. S. Lillie, on the "Electrical convection of certain free cells and nuclei*," which, while I cannot quote it in direct support of the suggestions which I have made, yet gives just the evidence we need in order to shew that electrical forces act upon the constituents of the cell, and that their action discriminates between the two species of colloids represented by the cytoplasm and the nuclear chromatin. And the difference is such that, in the presence of an electrical current, the cell substance and the nuclei (including sperm-cells) tend to migrate, the former on the whole with the positive, the latter with the negative stream: a difference of electrical potential being thus indicated between the particle and the surrounding medium, just as in the case of minute suspended particles of various kinds in various feebly conducting media†. And the electrical difference is doubtless greatest, in the case of the cell constituents, just at the period of mitosis: when the chromatin is invariably in its most deeply staining, most strongly acid, and therefore, presumably, in its most electrically negative phase. In short, Lillie comes easily to the conclusion that "electrical theories of mitosis are entitled to more careful consideration than they have hitherto received."

* *Amer. J. Physiol.* VIII, pp. 273–283, 1903 (*vide supra*, p. 314); cf. *ibid.* XV, pp. 46–84, 1905; XXII, p. 106, 1910; XXVII, p. 289, 1911; *Journ. Exp. Zool.* XV, p. 23, 1913; etc.

† In like manner Hardy shewed that colloid particles migrate with the negative stream if the reaction of the surrounding fluid be alkaline, and *vice versa*. The whole subject is much wider than these brief allusions suggest, and is essentially part of Quincke's theory of Electrical Diffusion or Endosmosis: according to which the particles and the fluid in which they float (or the fluid and the capillary wall through which it flows) each carry a charge: there being a discontinuity of potential at the surface of contact and hence a field of force leading to powerful tangential or shearing stresses, communicating to the particles a velocity which varies with the density per unit area of the surface charge. See W. B. Hardy's paper on Coagulation by electricity, *Journ. Physiol.* XXIV, pp. 288–304, 1899; also Hardy and H. W. Harvey, Surface electric charges of living cells, *Proc. R.S.* (B), LXXXIV, pp. 217–226, 1911, and papers quoted therein. Cf. also E. N. Harvey's observations on the convection of unicellular organisms in an electric field (Studies on the permeability of cells, *Journ. Exp. Zool.* X, pp. 508–556, 1911).

Among other investigations all leading towards the same general conclusion, namely that differences of electric potential play their part in the phenomena of cell division, I would mention a noteworthy paper by Ida H. Hyde*, in which the writer shews (among other important observations) that not only is there a measurable difference of potential between the animal and vegetative poles of a fertilised egg (*Fundulus*, toad, turtle, etc.), but also that this difference fluctuates, or actually reverses its direction, periodically, at epochs coinciding with successive acts of segmentation or other

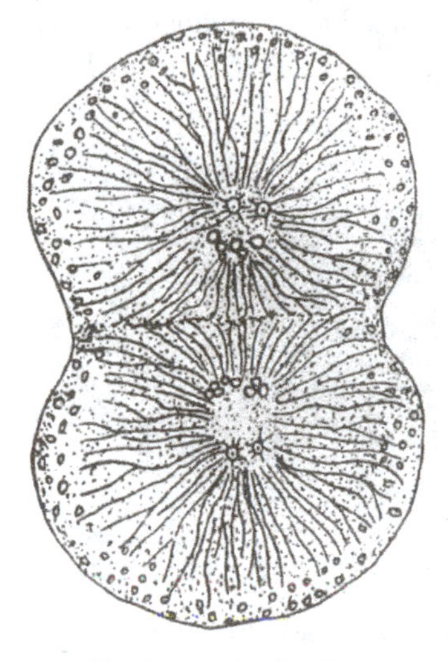

Fig. 100. Final stage in the first segmentation of the egg of *Cerebratulus*. From Prenant, after Coe*.

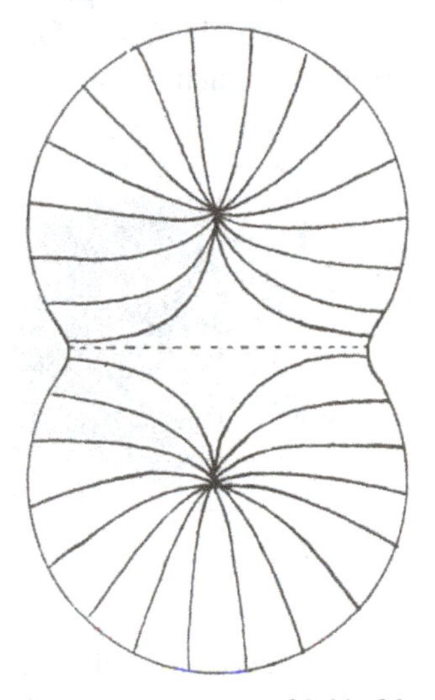

Fig. 101. Diagram of field of force with two similar poles.

important phases in the development of the egg†; just as other physical rhythms, for instance, in the production of CO_2, had already been shewn to do. Hence we need not be surprised to find that the "materialised" lines of force, which in the earlier stages form the

* On differences in electrical potential in developing eggs, *Amer. Journ. Physiol.* XII, pp. 241–275, 1905. This paper contains an excellent summary, for the time being, of physical theories of the segmentation of the cell.

† Gray has demonstrated a temporary increase of electrical conductivity in sea-urchin eggs during the process of fertilisation, and ascribes the changes in resistance to polarisation of the surface: Electrical conductivity of echinoderm eggs, etc., *Phil. Trans.* (B), CCVII, pp. 481–529, 1916.

convergent curves of the spindle, are replaced in the later phases of caryokinesis by divergent curves, indicating that the two foci, which are marked out in the field by the divided and reconstituted nuclei, are now alike in their polarity* (Figs. 100, 101).

The foregoing account is based on the provisional assumption that the phenomena of caryokinesis are analogous to those of a bipolar electrical field—a comparison which seems to offer a helpful and instructive series of analogies. But there are other forces which lead to similar configurations. For instance, some of Leduc's diffusion-experiments offer very remarkable analogies to the diagrammatic phenomena of caryokinesis, as shewn in Fig. 102†.

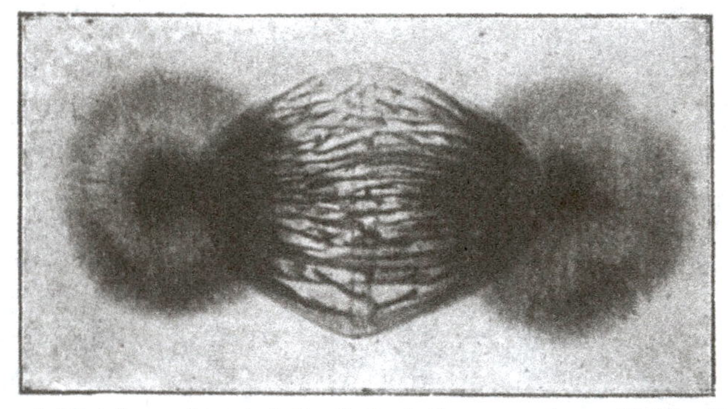

Fig. 102. Artificial caryokinesis (after Leduc), for comparison with Fig. 88, p. 299.

Here we have two identical (not opposite) poles of osmotic concentration, formed by placing a drop of indian ink in salt water, and then on either side of this central drop, a hypertonic drop of salt solution more lightly coloured. On either side the pigment of the central drop has been drawn towards the focus nearest to it; but in the middle line, the pigment is drawn in opposite directions by equal forces, and so tends to remain undisturbed, in the form of an "equatorial plate."

To account for the same mitotic phenomena an elegant hypothesis has been put forward by A. B. Lamb‡, and developed by Graham

* W. R. Coe, Maturation and fertilisation of the egg of *Cerebratulus*, *Zool. Jahrbücher* (*Anat. Abth.*), XII, pp. 425–476, 1899.

† *Op. cit.* pp. 110 and 91.

‡ A. B. Lamb, A new explanation of the mechanism of mitosis, *Journ. Exp. Zool.* V, pp. 27–33, 1908.

Cannon*. It depends on certain investigations of the Bjerknes, father and son†, which prove that bodies pulsating or oscillating‡ in a fluid set up a field of force precisely comparable with the lines of force in a magnetic field. Certain old and even familiar observations had pointed towards this phenomenon. Guyot had noticed that bits of paper were attracted towards a vibrating tuning-fork; and Schellbach found that a sounding-board so acts on bodies in its neighbourhood as to attract those which are heavier and repel those which are lighter than the surrounding medium; in air bits of paper are attracted and a gas-flame is repelled. To explain these simple observations, Bjerknes experimented with little drums attached to an automatic bellows. He found that two bodies in a fluid field, synchronously pulsating or synchronously oscillating, repel one another when their oscillations are in the same phase, or their pulsations are in opposite phase; and *vice versa*: while other particles, floating passively in the same fluid, tend (as Schellbach had observed before) to be attracted or repulsed according as they are heavier or lighter than the fluid medium. The two bodies behave towards one another like two electrified bodies, or like two poles of a magnet; we are entitled to speak of them as "hydrodynamic poles," we might even call them "hydrodynamic magnets"; and pursuing the analogy, we may call the heavy bodies paramagnetic, and the light ones diamagnetic with regard to them. Lamb's hypothesis then, and Cannon's, is that the centrosomes act as "hydrodynamic magnets." The explanation depends on oscillations which have never been seen, in centrosomes which are not always to be discovered. But it brings together certain curious analogies, and these, where we know so little, may be worth reflecting on.

If we assume that each centrosome is endowed with a vibratory motion as it floats in the semi-fluid colloids, or hydrosols (to use Graham's word) of the cell, we may take it that the visible intracellular phenomena will be much the same as those we have

* *Op. cit.* Cf. also Gertrud Woken, Zur Physik der Kernteilung, *Z. f. allg. Physiol.* XVIII, pp. 39–57, 1918.

† V. Bjerknes, *Vorlesungen über hydrodynamische Fernkräfte, nach C. A. Bjerknes' Theorie*, Leipzig, 1900.

‡ A body is said to pulsate when it undergoes a rhythmic change of volume; it oscillates when it undergoes a rhythmic change of place.

described under an electrical hypothesis; the lines of force will have the same distribution, and such movements as the chromosomes undergo, and such symmetrical configurations as they assume, may be accounted for under the one hypothesis pretty much as under the other. There are however other phenomena accompanying mitosis, such as Chambers's astral currents and certain local changes in the viscosity of the egg, which are more easily explained by the hydrodynamic theory.

We may assume that the cytoplasm, however complex it may be, is but a sort of microscopically homogeneous emulsion of high dispersion, that is to say one in which the minute particles of one phase are widely scattered throughout, and freely mobile in, the other; and this indeed is what is meant by calling it a *hydrosol*. Let us assume also that the particles are a little less dense than the continuous phase in which they are dispersed; and assume lastly (it is not the easiest of our assumptions) that these ultra-minute particles will be affected, just as are the grosser ones, by the forces of the hydrodynamic field.

All this being so, the disperse particles will be repelled from the oscillating centrosome, with a force which falls off very rapidly, for Bjerknes tells us that it varies inversely as the seventh power of the distance; a round clear field, like a drop or a bubble, will be formed round the centrosome; and the disperse particles, expelled from this region, will tend to accumulate in a crowded spherical zone immediately beyond it. Outside of this again they will continue to be repulsed, but slowly, and we may expect a second and lesser concentration at the periphery of the cell. A clear central mass, or "centrosphere," will thus come into being; and the surrounding cytoplasm will be rendered denser and more viscous, especially close around the centrosphere and again peripherally, by condensation of the disperse particles. Moreover, all outward movements of these lighter particles entail inward movements of the heavier, which (by hypothesis) are also the more fluid; stream-lines or visible currents will flow towards the centre, giving rise to the star-shaped "aster," and the best accounts of the sea-urchin's egg* tally well with what is thus deduced from the hydrodynamic

* Cf. R. Chambers, in *Journ. Exp. Zool.* XXIII, p. 483, 1917; *Trans. R.S. Canada*, XII, 1918; *Journ. Gen. Physiol.* II, 1919.

hypothesis. The round drop of clear fluid which forms the centre of the aster grows as the aster grows, fluid streaming towards it from all parts of the cell along the channels of the astral rays. The cytoplasm between the rays is in the *gel* state, but gradually passes into a *sol* beyond the confines of the aster. Seifritz asserts that the substance of the centrosphere is "not much more viscous than water," but that the wedges of cytoplasm between the inwardly directed streams are stiff and viscous*.

After the centrosome divides we have two oscillating bodies instead of one; they tend to repel one another, and pass easily through the fluid centrosphere to the denser layer around. But now the new centrosomes, on opposite sides of the centrosphere, repel, each on its own side, the disperse particles of the denser zone; and two new asters are formed, their rays marked by the streams coursing inwards to the centrosome-foci. Thus the *amphiaster* comes into being; it is not that the old aster divides, as a definite entity; but the old aster ceases to exist when its focus is disturbed, and about the new foci new asters are necessarily and automatically developed. Again this hypothetic account tallies well with Chambers's description.

The same attractions and repulsions should be manifested, perhaps better still, in whatsoever bodies lie or float within the cell, whether liquid or solid, oil-globules, yolk-particles, mitochondria, chromosomes or what not. A zoned, concentric arrangement of yolk-globules is often seen in the egg, with the centrosome as focus; and in certain sea-urchin eggs the mitochondria gather around the centrosome while the amphiaster is forming, collecting together in that very zone to which Chambers ascribes a semi-rigid or viscous consistency†. The Golgi bodies found in various germ-cells are at first black rod-like bodies embedded in the centrosphere; they undergo changes and complex movements, now scattering through the cytoplasm and anon crowding again around the centrosome. Some periodic change in the density of these bodies compared with

* Cf. W. Seifritz, Some physical properties of protoplasm, *Ann. Bot.* XXXV, 1921. Wo. Ostwald and M. H. Fischer had thought that the astral rays were due to local changes of the plasma-sol into a gel, Zur physikal. chem. Theorie der Befruchtung, *Pflüger's Archiv*, CVI, pp. 229–266, 1905.

† Cf. F. Vejdovsky and A. Mrazek, Umbildung des Cytoplasma während der Befruchtung und Zelltheilung, *Arch. f. mikr. Anat.* LXII, 431–579, 1903.

that of the medium in which they lie seems all that is required to account for their excursions; and such changes of density are not only of likely occurrence during the active chemical operations associated with fertilisation and division, but are in all probability inseparable from the changes in viscosity which are known to occur*. The movements and arrangements of the chromosomes, already described, may be easily accounted for if we postulate, in addition to their repulsion from the oscillating centrosomes, induced oscillations in themselves such as to cause them to attract one another.

The well-defined length of the spindle and the position of equilibrium in which it comes to rest may be conceived as resultants of the several mutual repulsions of the centrosomes by one another, by the chromosomes or other lighter material of the equatorial plate, and again by such lighter material as may have accumulated at the periphery of the egg; the first two of these will tend to lengthen the spindle, the last to shorten it; and the last will especially affect its position and direction. When Chambers amputated part of an amphiastral egg, the remains of the amphiaster disappeared, and then came into being again in a new and more symmetrical position; it or its centrosomal focus had been symmetrically repelled, we may suppose, by the fresh surface. Hertwig's law that the spindle-axis tends to lie in the direction of the largest mass of protoplasm, in other words to point where the cell-surface lies farthest off and its repulsion is least felt, may likewise find its easy explanation.

Between these hypotheses we may choose one or other (if we choose at all), according to our judgment. As Henri Poincaré tells us, we never know that any one physical hypothesis is *true*, we take the simplest we can find; and this we call the guiding principle of simplicity! In this case, the hydrodynamic hypothesis is a simple one; but it all rests on a hypothetic oscillation of the centrosomes, which has never been witnessed. Bayliss has shewn that precisely such reversible states of gelation as we have been speaking of as

* Cf. G. Odquist, Viscositätsänderungen des Zellplasmas während der ersten Entwicklungsstufen des Froscheies, *Arch. f. Entw. Mech.* LI, pp. 610–624, 1922; A. Gurwitsch, Prämissen und anstossgebende Faktoren der Furchung und Zelltheilung, *Arch. f. Zellforsch.* II, pp. 495–548, 1909; L. V. Heilbrunn, Protoplasmic viscosity-changes during mitosis, *Journ. Exp. Zool.* XXXIV, pp. 417–447, 1921; *ibid.* XLIV, pp. 255–278, 1926; E. Leblond, Passage de l'état de gel à l'état de sol dans le protoplasme vivant, *C.R. Soc. Biol.* LXXXII, p. 1150; cf. *ibid.* p. 1220; etc.

"periodic changes in viscosity" may be induced in living protoplasm by electrical stimulation*. On the other hand, the fact that the hydrodynamic forces fall off as fast as they do with increasing distance limits their efficacy; and the minute disperse particles must, under Stokes's law, be slow to move. Lastly, it may well be (as Lillie has urged) that such work as his own, or Ida Hyde's, or Gray's, on change of potential in developing eggs, taken together with that of many others on the behaviour of colloid particles in an electrical field, has not yet been followed out in all its consequences, either on the physical or the physiological side of the problem.

But to return to our general discussion.

As regards the actual mechanical division of the cell into two halves, we shall see presently that, in certain cases, such as that of a long cylindrical filament, surface-tension, and what is known as the principle of "minimal areas," go a long way to explain the mechanical process of division; and in all cells whatsoever, the process of division must somehow be explained as the result of a conflict between surface-tension and its opposing forces. But in such a case as our spherical cell, it is none too easy to see what physical cause is at work to disturb its equilibrium and its integrity.

The fact that when actual division of the cell takes place, it does so at right angles to the polar axis and precisely in the direction of the equatorial plane, would lead us to suspect that the new surface formed in the equatorial plane sets up an annular tension, directed inwards, where it meets the outer surface layer of the cell itself. But at this point the problem becomes more complicated. Before we can hope to comprehend it, we shall have not only to enquire into the potential distribution at the surface of the cell in relation to that which we have seen to exist in its interior, but also to take account of the differences of potential which the material arrangements along the lines of force must themselves tend to produce. Only thus can we approach a comprehension of the balance of forces which cohesion, friction, capillarity and electrical distribution combine to set up.

The manner in which we regard the phenomenon would seem to

* W. M. Bayliss, Reversible gelation in living protoplasm, *Proc. R.S.* (B), XCI, pp. 196–201, 1920.

turn, in great measure, upon whether or no we are justified in assuming that, in the liquid surface-film of a minute spherical cell, local and symmetrically localised differences of surface-tension are likely to occur. If not, then changes in the conformation of the cell such as lead immediately to its division must be ascribed not to local changes in its surface-tension, but rather to direct changes in internal pressure, or to mechanical forces due to an induced surface-distribution of electrical potential. We have little reason to be sceptical; in fact we now know that the cell is so far from being chemically and physically homogeneous that local variations in its surface-tension are more than likely, they are certain to occur.

Bütschli suggested more than sixty years ago that cell-division was brought about by an increase of surface-tension in the equatorial region of the cell; and the suggestion was the more remarkable that it was (I believe) the very first attempt to invoke surface-tension as a factor in the physical causation of a biological phenomenon*. An increase of equatorial tension would cause the surface-area there to diminish, and the equator to be pinched in; but the total surface-area of the cell would be increased thereby, and the two effects would strike a balance†. But, as Bütschli knew very well, the surface-tension change would not stand alone; it would bring other phenomena in its train, currents would tend to be set up, and tangential strains would be imposed on the cell-membrane or cell-surface as a whole. The secondary if not the direct effects of increased equatorial tension might, after all, suffice for the division of the cell. It was Loeb, in 1895, who first shewed that streaming went on from the equator towards the divided nuclei. To the violence of these streaming movements he attributed the phenomenon of division, and many other physiologists have adopted this hypothesis‡. The currents of which Loeb spoke call for counter-currents

* O. Bütschli, Über die ersten Entwicklungsvorgänge der Eizelle, *Abh. Senckenberg. naturf. Gesellsch.* x, 1876; Über Plasmaströmungen bei der Zelltheilung, *Arch. f. Entw. Mech.* x, p. 52, 1900. Ryder ascribed the caryokinetic figures to surface-tension in his *Dynamics in Evolution*, 1894.

† A relative, not positive, increase of surface-tension, was part of Giardina's hypothesis: Note sul mecanismo della divisione cellulare, *Anat. Anz.* XXI. 1902.

‡ J. Loeb, *Amer. Journ. Physiol.* VI, p. 432, 1902; E. G. Conklin, Protoplasmic movements as a factor in differentiation, *Wood's Hole Biol. Lectures*, p. 69, etc., 1898–99; J. Spek, Oberflächenspannungsdifferenzen als eine Ursache der Zellteilung, *Arch. f. Entw. Mech.* XLIV, pp. 54–73, 1918.

towards the equator, in or near the surface of the cell; and theory and observation both indicate that precisely such currents are bound to be set up by the surface-energy involved in the increase of equatorial tension.

An opposite view has been held by some, and especially by T. B. Robertson*. Quincke had shewn that the formation of soap at the surface of an oil-droplet lowers the surface-tension of the latter, and that if the saponification be local, that part of the surface tends to enlarge and spread out accordingly. Robertson, in a very curious experiment, found that by laying a thread, moistened with dilute caustic alkali or merely smeared with soap, across a drop of olive oil afloat in water, the drop at once divided into two. A vast amount of controversy has arisen over this experiment, but Spek seems to have shewn conclusively that it is an exceptional case.

In a drop of olive-oil, balanced in water† and touched *anywhere* with an alkali, there is so copious a formation of lighter soaps that differences of density tend to drag the drop in two. But in the case of other oils (and especially the thinner oils, such as oil of bergamot) the saponified portion bulges, as theory directs; and when the alkali is applied to two opposite poles the equatorial region is pinched in, as McClendon‡, in opposition to Robertson, had found it to do. Conversely, if an alkaline thread be looped around the drop, the zone of contact bulges, and instead of dividing at the equator the drop assumes a lens-like form.

We may take it then as proven that a relative increase of equatorial surface-tension, whether in oil-drops, mercury-globules or living cells, does lead, or tend to lead, to an equatorial constriction. In all cases a system of surface-currents is set up among the fluid drops towards the zone of increased tension; and an axial counter-current flows towards the pole or poles of lowered tension. Precisely such currents have been observed to run in various eggs (especially of

* T. B. Robertson, Note on the chemical mechanics of cell-division, *Arch. f. Entw. Mech.* XXVII, p. 29, 1909; XXXII, p. 308, 1911; XXXV, p. 402, 1913. Cf. R. S. Lillie, *Journ. Exp. Zool.* XXI, pp. 369–402, 1916; McClendon, *loc. cit.*; etc.

† In these experiments, and in many of Quincke's, a little chloroform is added to the oil, in order to bring its density as near as may be to that of water.

‡ J. F. McClendon, Note on the mechanics of cell-division, *Arch. f. Entw. Mech.* XXXIV, pp. 263–266, 1912.

certain Nematodes) during division of the cell; but if the process be slow, more than 7 or 8 minutes long, the slow currents become hard to see. Various contents of the cell are transported by these currents, and clear, yolk-free polar caps and equatorial accumulations of yolk and pigment are among the various manifestations of the phenomenon. The extrusion of a polar body, at a small and sharply defined region of lowered tension, is a particular case of the same principle*.

But purely chemical changes are not of necessity the fundamental cause of alteration in the surface-tension of the egg, for the action of electrolytes on surface-tension is now well known and easily demonstrated. So, according to other views than those with which we have been dealing, electrical charges are sufficient in themselves to account for alterations of surface-tension, and in turn for that protoplasmic streaming which, as so many investigators agree, initiates the segmentation of the egg†. A great part of our difficulty arises from the fact that in such a case as this the various phenomena are so entangled and apparently concurrent that it is hard to say which initiates another, and to which this or that secondary phenomenon may be considered due. Of recent years the phenomenon of *adsorption* has been adduced (as we have already briefly said) in order to account for many of the events and appearances which are associated with the asymmetry, and lead towards the division, of the cell. But our short discussion of this phenomenon may be reserved for another chapter.

However, we are not directly concerned here with the phenomena of segmentation or cell-division in themselves, except only in so far as visible changes of form are capable of easy and obvious correlation with the play of force. The very fact of "development" indicates that, while it lasts, the equilibrium of the egg is never complete‡. And the gist of the matter is that, if you have caryokinetic figures developing inside the cell, that of itself indicates that the dynamic system and the localised forces arising from it are in

* J. Spek, *loc. cit.* pp. 108–109.

† Cf. D'Arsonval, Relation entre la tension superficielle et certains phénomènes électriques d'origine animale, *Arch. de Physiol.* I, pp. 460–472, 1889; Ida H. Hyde, *op. cit.* p. 242.

‡ Cf. Plateau's remarks (*Statique des liquides*, II, p. 154) on the *tendency* towards equilibrium, rather than actual equilibrium, in many of his systems of soap-films.

gradual alteration; and changes in the outward configuration of the system are bound, consequently, to take place.

Perhaps we may simplify the case still more. We have learned many things about cell-division, but we do not know much in the end. We have dealt, perhaps, with too many related phenomena, and failed because we tried to combine and account for them all. A physical problem, still more a mathematical one, wants reducing to its simplest terms, and Dr Rashevsky has simplified and generalised the problem of cell-division (or division of a drop) in a series of papers, which still outrun by far the elementary mathematics of this book. If we cannot follow him in all he does, we may find useful lessons in his way of doing it. Cells are of many kinds; they differ in size and shape, in visible structure and chemical composition. Most have a nucleus, some few have none; most need oxygen, some few do not; some metabolise in one way, some in another. What small residuum of properties remains common to them all? A living cell is a little fluid (or semi-fluid) system, in which work is being done, physical forces are in operation and chemical changes are going on. It is in such intimate relation with the world outside—its own *milieu interne* with the great *milieu externe*—that substances are continually entering the cell, some to remain there and contribute to its growth, some to pass out again with loss of energy and metabolic change. The picture seems simplicity itself, but it is less simple than it looks. For on either side of the boundary-wall, both in the adjacent medium and in the living protoplasm within, there will be no uniformity, but only degrees of activity, and *gradients of concentration*. Substances which are being absorbed and consumed will diminish from periphery to centre; those which are diffusing outwards have their greatest concentration near the centre, decrease towards the periphery, and diminish further with increasing distance in the near neighbourhood of the system. Size, shape, diffusibility, permeability, chemical properties of this and that, may affect the gradients, but in the living cell the interchanges are always going on, and the *gradients* are always there*.

* Outward diffusion makes one of the many contrasts between cell-growth and crystal-growth. But the diffusion-gradients round a growing crystal are far more complicated than was once supposed. Cf. W. F. Berg, Crystal growth from solutions, *Proc. R.S.* (A), CLXIV, pp. 79–95, 1938.

If the cell be homogeneous, taking in and giving out at a constant rate in a uniform way, its shape will be spherical, the concentration-field of force, or concentration-field, will likewise have a spherical symmetry, and the resultant force will be zero. But if the symmetry be ever so little disturbed, and the shape be ever so little deformed, then there will be forces at work tending to increase the deformation, and others tending to equalise the surface-tension and restore the spherical symmetry, and it can be shewn that such agencies are within the range of the chemistry of the cell. Since surface-tension becomes more and more potent as the size of the drop diminishes, it follows that (under fluid conditions) the smallest solitary cells are least likely to depart from a spherical shape, and that cell-division is only likely to occur in cells above a certain critical order of magnitude; and using such physical constants as are available, Rashevsky finds that this critical magnitude tallies fairly well with the average size of a living cell. The more important lesson to learn, however, is this, that, merely *by virtue of its metabolism*, every cell contains within itself factors which may lead to its division after it reaches a certain critical size.

There are simple corollaries to this simple setting of the case. Since unequal concentration-gradients are the chief cause which renders non-spherical shapes of cell possible, and these last only so long as the cell lives and metabolises, it follows that, as soon as the gradients disappear, whether in death or in a "resting-stage", the cell reverts to a spherical shape and symmetry. Again, not only is there a critical size above which cell-division becomes possible, and more and more probable, but there must also be a size beyond which the cell is not likely to grow. For the "specific surface" decreases, the metabolic exchanges diminish, the gradients become less steep, and the rate of growth decreases too; there must come a stage where anabolism just balances katabolism, and growth ceases though life goes on. When streaming currents are visible within the cell, they seem to complicate the problem; but after all, they are part of the result, and proof of the existence, of the gradients we have described. In any further account of Rashevsky's theories the mathematical difficulties very soon begin. But it is well to realise that pure theory often carries the mathematical physicist a long way; and that higher and higher powers of the microscope, and greater and greater histological skill are not the one and only way to study the physical forces acting within the cell*.

As regards the phenomena of fertilisation, of the union of the spermatozoon with the "pronucleus" of the egg, we might study these also in illustration, up to a certain point, of the forces which are more or less manifestly at work. But we shall merely take, as a single illustration, the paths of the male and female pronuclei, as they travel to their ultimate meeting-place.

The spermatozoon, when within a very short distance of the egg-cell, is attracted by it, the same attraction being further manifested in a small conical uprising of the surface of the egg†. The nature of the attractive force has been much disputed. Loeb found the spermatozoon to be equally attracted by other substances, even by a bead of glass. It has been held also that the attraction is chemotropic, some substance being secreted by the egg which drew the sperm towards it: just as Pfeffer, having shewn that malic acid has an attraction for fern-antheridia, supposed this substance to play its attractive part within the mucus of the archegonia. Again, the chemical secretion may be neither attractive nor directive, but yet play a useful part in activating the spermatozoa. However that may be, Gray has shewn reason to believe that an electromotive force is developed in the contact between active spermatozoon and inactive ovum; and that it is the electrical change so set up, and almost instantaneously propagated, which precludes the entry of another spermatozoon‡. Whatever the force may be, it is one which acts normally to the surface of the ovum, and after entry the

* Cf. N. Rashevsky, *Mathematical Biophysics*, Chicago, 1938; and many earlier papers. E.g. Physico-mathematical aspects of cellular multiplication and development, *Cold Spring Harbor Symposia*, II, 1934; The mechanism of division of small liquid systems which are the seat of physico-chemical reactions, *Physics*, III, pp. 374–379, 1934; papers in *Protoplasma*, XIV–XX, 1931–33, etc.

† With the classical account by H. Fol, *C.R.* LXXXIII, p. 667, 1876; *Mém. Soc. Phys. Genève*, XXVI, p. 89, 1879, cf. Robert Chambers, The mechanism of the entrance of sperm into the star-fish egg, *Journ. Gen. Physiol.* V, pp. 821–829, 1923. Here a delicate filament is said to run out from the fertilisation-cone and drag the spermatozoon in; but this is disputed and denied by E. Just, *Biol. Bull.* LVII, pp. 311–325, 1929.

‡ But, under artificial conditions, "polyspermy" may take place, e.g. under the action of dilute poisons, or of an abnormally high temperature, these being doubtless also conditions under which the surface-tension is diminished.

spermatozoon points straight towards the centre of the egg. From the fact that other spermatozoa, subsequent to the first, fail to effect an entry, we may safely conclude that an immediate consequence of the entry of the spermatozoon is an increase in the surface-tension of the egg: this being but one of the complex reactions exhibited by the surface, or cortex of the cell*. Somewhere or other, within the egg, near or far away, lies its own nuclear body, the so-called female pronucleus, and we find that after a while this has fused with the "male pronucleus" or head of the spermatozoon, and that the body resulting from their fusion has come to occupy the centre of the egg. This *must* be due (as Whitman pointed out many years ago) to a force of attraction acting between the two bodies, and another force acting upon one or other or both in the direction of the centre of the cell. Did we know the magnitude of these several forces, it would be an easy task to calculate the precise path which the two pronuclei would follow, leading to conjugation and to the central position. As we do not know the magnitude, but only the direction, of these forces, we can only make a general statement: (1) the paths of both moving bodies will lie wholly within a plane triangle drawn between the two bodies and the centre of the cell; (2) unless the two bodies happen to lie, to begin with, precisely on a diameter of the cell, their paths until they meet one another will be curved paths, the convexity of the curve being towards the straight line joining the two bodies; (3) the two bodies will meet a little before they reach the centre; and, having met and fused, will travel on to reach the centre in a straight line. The actual study and observation of the path followed is not very easy, owing to the fact that what we usually see is not the path itself, but only a *projection* of the path upon the plane of the microscope; but the curved path is particularly well seen in the frog's egg, where the path of the spermatozoon is marked by a little streak of brown pigment, and the fact of the meeting of the pronuclei before reaching the centre has been repeatedly seen by many observers†.

* See Mrs Andrews' beautiful observations on "Some spinning activities of protoplasm in starfish and echinoid eggs," *Journ. Morphol.* XII, pp. 307–389, 1897.

† W. Pfeffer, Locomotorische Richtungsbewegungen durch chemische Reize, *Unters. a. d. Botan. Inst. Tübingen*, I, 1884; *Physiology of Plants*, III, p. 345, Oxford, 1906; W. J. Dakin and M. G. C. Fordham, *Journ. Exp. Biol.* I, pp. 183–200, 1924. Cf. J. Loeb, *Dynamics of Living Matter*, 1906, p. 153.

The problem recalls the famous problem of three bodies, which has so occupied the astronomers; and it is obvious that the foregoing brief description is very far from including all possible cases. Many of these are particularly described in the works of Fol, Roux, Whitman and others*.

The intracellular phenomena of which we have now spoken have assumed great importance in biological literature and discussion during the last fifty years; but it is open to us to doubt whether they will be found in the end to possess more than a secondary, even a remote, biological significance. Most, if not all of them, would seem to follow immediately and inevitably from certain simple assumptions as to the physical constitution of the cell, and from an extremely simple distribution of polarised forces within it. We have already seen that how a thing grows, and what it grows into, is a dynamic and not a merely material problem; so far as the material substance is concerned, it is so only by reason of the chemical, electrical or other forces which are associated with it. But there is another consideration which would lead us to suspect that many features in the structure and configuration of the cell are of secondary biological importance; and that is, the great variation to which these phenomena are subject in similar or closely related organisms, and the apparent impossibility of correlating them with the peculiarities of the organism as a whole. In a broad and general way the phenomena are always the same. Certain structures swell and contract, twine and untwine, split and unite, advance and retire; certain chemical changes also repeat themselves. But Nature rings the changes on all the details. "Comparative study has shewn that almost every detail of the processes (of mitosis) described above is subject to variation in different forms of cells†." A multitude of cells divide to the accompaniment of caryokinetic phenomena; but others do so without any visible caryokinesis at all. Sometimes the polarised field of force is within,

* H. Fol, *Recherches sur la fécondation*, 1879; W. Roux, Beiträge zur Entwickelungsmechanik des Embryos, *Arch. f. Mikr. Anat.* XIX, 1887; C. O. Whitman, Oökinesis, *Journ. Morph.* I, 1887; E. Giglio-Tos, Entwicklungsmechanische Studien, I, *Arch. f. Entw. Mech.* LI, p. 94, 1922. See also Frank R. Lillie, *Problems of Fertilisation*, Chicago, 1919.

† Wilson, *The Cell*, p. 77; cf. 3rd ed. (1925), p. 120.

sometimes it is adjacent to, and at other times it lies remote from, the nucleus. The distribution of potential is very often symmetrical and bipolar, as in the case described; but a less symmetrical distribution often occurs, with the result that we have, for a time at least, numerous centres of force, instead of the two main correlated poles: this is the simple explanation of the numerous stellate figures,

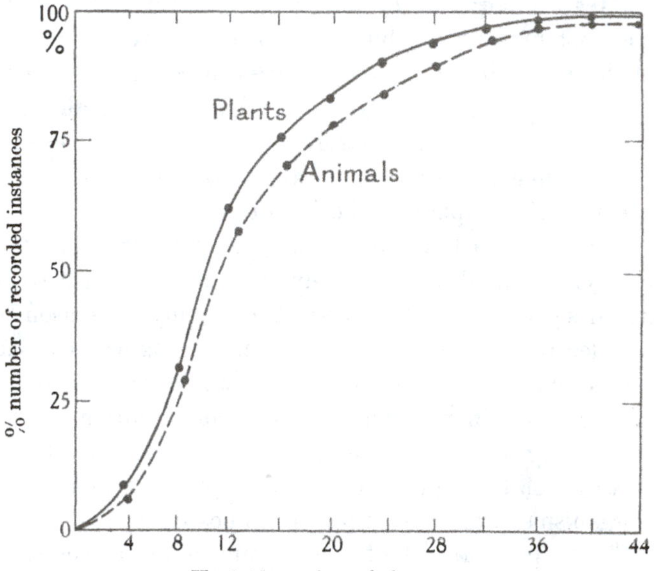

Fig. 103. Summation diagram shewing the % number of instances (among 2,415 phanerogams and 1,070 metazoa), in which the chromosomes do not exceed a given number. Data from M. J. D. White.

or "Strahlungen," which have been described in certain eggs, such as those of *Chaetopterus*. The number of chromosomes may be constant within a group, as in the tailed Amphibia, with 12; or very variable, as in sedges, and in grasshoppers*; in one and the same species of worm (*Ascaris megalocephala*), one group or two groups of chromosomes may be present. And remarkably constant, in general, as the number in any one species undoubtedly is, yet we must not forget that, in plants and animals alike, the whole range of observed numbers is but a small one (Fig. 103); for (as regards

* There are varieties of *Artemia salina* which hardly differ in outward characters, but differ widely in the number of their chromosomes.

the germ-nuclei) few have less than six chromosomes, and few have more than twenty*. In closely related animals, such as various species of Copepods, and even in the same species of worm or insect, the form of the chromosomes and their arrangement in relation to the nuclear spindle have been found to differ in ways alluded to above; while only here and there, as among the chrysanthemums, do related species or varieties shew their own characteristic chromosome numbers. In contrast to the narrow range of the chromosome numbers, we may reflect on the all but infinite possibilities of chemical variability. Miescher shewed that a molecule containing 40 C-atoms would admit (arithmetically though not necessarily chemically) of a million possible isomers; and changes in position of the N-atoms of a protein, for instance, might vastly increase that prodigious number. In short, we cannot help perceiving that many nuclear phenomena are not specifically related to the particular organism in which they have been observed, and that some are not even specially and indisputably connected with the organism as such. They include such manifestations of the physical forces, in their various permutations and combinations, as may also be witnessed, under appropriate conditions, in non-living things.

When we attempt to separate our purely morphological or "purely

* The commonest numbers of (haploid) chromosomes, both in plants and animals, are 8, 12 and 16. The median number is 12 in both, and the lower quartile is 8, likewise in both; but the upper quartile is 24 or thereby in animals, and in the neighbourhood of 16 in plants. If we may judge by the long lists given by E. B. Wilson (*The Cell*, 3rd ed. pp. 855–865), by M. Ishikawa in *Botan. Mag. Tokyo*, XXX, 1916, by M. J. D. White in his book on *Chromosomes*, or by Tischler in *Tabulae Biologicae* (1927), fully 60 per cent. of the observed cases lie between 6 and 16. As Wilson says (p. 866) "the number of chromosomes is *per se* a matter of secondary importance"; and (p. 868) "We must admit the present inadequacy of attempts to reduce the chromosome numbers to any single or consistent arithmetical rules." Clifford Dobell had said the same thing: "Nobody nowadays will be prepared to argue that chromosome numbers, as such, have any quantitative or qualitative relation to the characters exhibited by their owners. Complexity of bodily structure is certainly not correlated in any way with multiplicity of chromosomes"; *La Cellule*, XXXV, p. 188, 1924. On the other hand, Tischler stoutly maintains that chromosome-numbers give useful evidence of phylogenetic affinity (*Biol. Centralbl.* XLVIII, pp. 321–345, 1928); and there are a few well-known cases, such as the chrysanthemums, where, undoubtedly, the numbers are constant and specific. Again in certain cases, the number of the chromosomes may differ in different *races* (diploid and tetraploid) of the same plant; and the difference is accompanied by differences in cell-size, in rate of growth, and even in the shape of the fruit (cf. Sinnott and Blakeslee, *Nat. Acad. of Sci.* 1938, p. 476).

embryological" studies from physiological and physical investigations, we tend *ipso facto* to regard each particular structure and configuration as an attribute, or a particular "character," of this or that particular organism. From this assumption we are easily led to the framing of theories as to the ancestral history, the classificatory position, the natural affinities of the several organisms: in fact, to apply our embryological knowledge to the study of *phylogeny*. When we find, as we are not long of finding, that our phylogenetic hypotheses become complex and unwieldy, we are nevertheless reluctant to admit that the whole method, with its fundamental postulates, is at fault; and yet nothing short of this would seem to be the case, in regard to the earlier phases at least of embryonic development. All the evidence at hand goes, as it seems to me, to shew that embryological data, prior to and even long after the epoch of segmentation, are essentially a subject for physiological and physical investigation and have but the slightest link, if any, with the problems of zoological classification. Comparative embryology has its own facts to classify, and its own methods and principles of classification. We may classify eggs according to the presence or absence, the paucity or abundance, of their associated food-yolk, the chromosomes according to their form and their number, the segmentation according to its various "types"—radial, bilateral, spiral, and so forth. But we have little right to expect, and in point of fact we shall very seldom and (as it were) only accidentally find, that these embryological categories coincide with the lines of "natural" or "phylogenetic" classification which have been arrived at by the systematic zoologist.

The efforts to explain "heredity" by help of "genes" and chromosomes, which have grown up in the hands of Morgan and others since this book was first written, stand by themselves in a category which is all their own and constitutes a science which is justified of itself. To weigh or criticise these explanations would lie outside my purpose, even were I fitted to attempt the task. When these great discoveries began to be made, Bateson crossed the ocean to see and hear for himself what Morgan and his pupils had to shew and to tell. He came home convinced, and humbly marvelling. And I leave this great subject on one side not because I doubt for a moment the facts nor dispute the hypotheses nor decry the importance of one or other; but because we are so much in the dark as to the mysterious field of force in which the chromosomes lie, far from the visible horizon of physical science, that the matter lies (for the present) beyond the range of problems which this book professes to discuss, and the trend of reasoning which it endeavours to maintain.

The cell*, which Goodsir spoke of as a *centre of force*, is in reality a *sphere of action* of certain more or less localised forces; and of these, surface-tension is the particular force which is especially responsible for giving to the cell its outline and its morphological individuality. The partially segmented differs from the totally segmented egg, the unicellular Infusorian from the minute multi-

* The "cell-theory" began early and grew slowly. In a curious passage which Mr Clifford Dobell has shewn me (*Nov. Org.* II, 7, *ad fin.*), Bacon speaks of "cells" in the human body: of a "collocatio spiritus per corpoream molem, eiusque pori, meatus, venae *et cellulae*, et rudimenta sive tentamenta corporis organici." It is "surely one of the most strangely prophetic utterances which even Bacon ever made." Apart from this the story begins in the seventeenth century, with Robert Hooke's well-known figure of the "cells" in a piece of cork (1665), with Grew's "bladders" or "bubbles" in the parenchyma of young beans, and Malpighi's "utriculi" or "sacculi" in the parenchyma or "utriculorum substantia" of various plants. Christian Fr. v. Wolff conceived, about the same time, a hypothetical "cell-theory," on the analogy of Leibniz's Monads; but the first clear idea of a cellular parenchyma, or *contextus cellularis*, came from C. Gottlieb Ludwig (1742), and from K. Fr. Wolff, who spoke freely of cells or *cellulae*. Fontana, author of a curious *Traité sur le venin de la vipère* (1781), described various histological elements, caught a glimpse of the nucleus, and experimented with reagents, using syrup of violets for a stain. Early in the eighteenth century the vessels of the plant played an important rôle, under Kurt Sprengel and Treviranus; but it was not till 1831 that Hugo v. Mohl recognised that they also arose from "cells." About this time Robert Brown discovered, or re-discovered, the nucleus (1833), which Schleiden called the *cytoblast*, or "cell-producer." It was Schleiden's idea, and a far-seeing one, that the cell lived a double life, a life of its own and the life of the plant to which it belonged: "jede Zelle führt nun ein zweifaches Leben: ein selbstständiges, nur ihrer eigenen Entwicklung angehörigen, und ein anderes mittelbares, insofern sie integrierender Theil einer Pflanze geworden ist" (*Phytogenesis*, 1838, p. 1). The cell-theory, so long a-building, may be said to have been launched, and christened, with Schwann's *Mikroskopische Untersuchungen* of 1839. Within the next five years Martin Barry shewed how cell-division starts with the nucleus, Henle described the budding of certain cells, and Goodsir declared that all cells originate in pre-existing cells, a doctrine at once accepted by Remak, and made famous in pathology by Virchow. (Cf. (*int. al.*) J. G. McKendrick, On the modern cell-theory, etc., *Proc. Phil. Soc. Glasgow*, XIX, pp. 1–55, 1887; J. Stephenson, Robert Brown...and the cell-theory, *Proc. Linn. Soc.* 1931–2, pp. 45–54; M. Möbius, Hundert Jahre Zellenlehre, *Jen. Ztschr.* LXXI, pp. 313–326, 1938.)

cellular Turbellarian, in the intensity and the range of those surface-tensions which in the one case succeed and in the other fail to form a visible separation between the cells. Adam Sedgwick used to call attention to the fact that very often, even in eggs that appear to be totally segmented, it is yet impossible to discover an actual separation or cleavage, through and through, between the cells which on the surface of the egg are so clearly delimited; so far and no farther have the physical forces effectuated a visible "cleavage." The vacuolation of the protoplasm in *Actinophrys* or *Actinosphaerium* is due to localised surface-tensions, quite irrespective of the multi-nuclear nature of the latter organism. In short, the boundary walls due to surface-tension may be present or may be absent, with or without the delimination of the other specific fields of force which are usually correlated with these boundaries and with the independent individuality of the cells. What we may safely admit, however, is that one effect of these circumscribed fields of force is usually such a separation or segregation of the protoplasmic constituents, the more fluid from the less fluid and so forth, as to give a field where surface-tension may do its work and bring a visible boundary into being. When the formation of a "surface" is once effected, its physical condition, or phase, will be bound to differ notably from that of the interior of the cell, and under appropriate chemical conditions the formation of an actual cell-wall, cellulose or other, is easily intelligible. To this subject we shall return again, in another chapter.

From the moment that we enter on a dynamical conception of the cell, we perceive that the old debates were vain as to what visible portions of the cell were active or passive, living or non-living: For the manifestations of force can only be due to the *interaction* of the various parts, to the transference of energy from one to another. Certain properties may be manifested, certain functions may be carried on, by the protoplasm apart from the nucleus; but the interaction of the two is necessary, that other and more important properties or functions may be manifested. We know, for instance, that portions of an Infusorian are incapable of regenerating lost parts in the absence of a nucleus, while nucleated pieces soon regain the specific form of the organism: and we are told that reproduction by fission cannot be *initiated*, though apparently all its later steps can be carried on, independently of nuclear action. Nor, as Verworn pointed out, can the nucleus possibly be regarded as the "sole vehicle of inheritance," since only in the conjunction of cell and nucleus do we find the essentials of cell-life. "Kern und Protoplasma sind nur *vereint* lebensfähig," as Nussbaum said. Indeed we may, with E. B. Wilson, go further, and say that "the terms 'nucleus' and 'cell-body' should probably be regarded as only topographical expressions denoting two differentiated areas in a common structural basis."

Endless discussion has taken place regarding the centrosome, some holding that it is a specific and essential structure, a permanent corpuscle derived from a similar pre-existing corpuscle, a "fertilising element" in the spermatozoon, a special "organ of cell-division," a material "dynamic centre" of the cell (as Van Beneden and Boveri call it); while on the other hand, it is pointed out that many cells live and multiply without any visible centrosomes, that a centrosome may disappear and be created anew, and even that under artificial conditions abnormal chemical stimuli may lead to the formation of new centrosomes. We may safely take it that the centrosome, or the "attraction sphere," is essentially a "centre of force," and that this dynamic centre may or may not be constituted by (but will be very apt to produce) a concrete and visible concentration of matter.

It is far from correct to say, as is often done, that the cell-wall, or cell-membrane, belongs "to the passive products of protoplasm rather than to the living cell itself"; or to say that in the animal cell, the cell-wall, because it is "slightly developed," is relatively unimportant compared with the important rôle which it assumes in plants. On the contrary, it is quite certain that, whether visibly differentiated into a semi-permeable membrane or merely constituted by a liquid film, the surface of the cell is the seat of important forces, capillary and electrical, which play an essential part in the dynamics of the cell. Even in the thickened, largely solidified cellulose wall of the plant-cell, apart from the mechanical resistances which it affords, the osmotic forces developed in connection with it are of essential importance.

But if the cell acts, after this fashion, as a whole, each part interacting of necessity with the rest, the same is certainly true of the entire multicellular organism: as Schwann said of old, in very precise and adequate words, "the whole organism subsists only by means of the *reciprocal action* of the single elementary parts*." As Wilson says again, "the physiological autonomy of the individual cell falls into the background...and the apparently composite character which the multicellular organism may exhibit is owing to a secondary distribution of its energies among local centres of action†." It is here that the homology breaks down which is so often drawn, and overdrawn, between the unicellular organism and the individual cell of the metazoon‡.

Whitman, Adam Sedgwick§, and others have lost no opportunity of warning us against a too literal acceptation of the cell-theory, against the view that the multicellular organism is a colony (or, as Haeckel called it, in the case of the plant, a "republic") of independent units of life||. As Goethe said long ago, "Das lebendige ist zwar in Elemente zerlegt, aber man kann es aus diesen nicht wieder zusammenstellen und beleben"; the dictum of the *Cellularpathologie* being just the opposite, "Jedes Thier erscheint als eine Summe vitaler Einheiten, von denen *jede den vollen Charakter des Lebens an sich trägt.*"

Hofmeister and Sachs have taught us that in the plant the growth

* *Theory of Cells*, p. 191.

† *The Cell in Development*, etc., p. 59; cf. 3rd ed. (1925), p. 102.

‡ E.g. Brücke, *Elementarorganismen*, p. 387: "Wir müssen in der Zelle einen kleinen Thierleib sehen, und dürfen die Analogien, welche zwischen ihr und den kleinsten Thierformen existiren, niemals aus den Augen lassen."

§ C. O. Whitman, The inadequacy of the cell-theory, *Journ. Morphol.* VIII, pp. 639–658, 1893; A. Sedgwick, On the inadequacy of the cellular theory of development, *Q.J.M.S.* XXXVII, pp. 87–101, 1895; XXXVIII, pp. 331–337, 1896. Cf. G. C. Bourne, *ibid.* XXXVIII, pp. 137–174, 1896; Clifford Dobell, The principles of Protistology, *Arch. f. Protistenk.* XXIII, p. 270, 1911.

|| Cf. O. Hertwig, *Die Zelle und die Gewebe*, 1893, p. 1: "Die Zellen, in welche der Anatom die pflanzlichen und thierischen Organismen zerlegt, sind die Träger der Lebensfunktionen; sie sind, wie Virchow sich ausgedrückt hat, die 'Lebenseinheiten.' Von diesem Gesichtspunkt aus betrachtet, erscheint der Gesammtlebensprozess eines zusammengesetzten Organismus nichts Anderes zu sein als das höchst verwickelte Resultat der einzelnen Lebensprozesse seiner zahlreichen, verschieden functionirenden Zellen." But in 1920 Doncaster (*Cytology*, p. 1) declared that "the old idea of discrete and independent cells is almost abandoned," and that the word *cell* was coming to be used "rather as a convenient descriptive term than as denoting a fundamental concept of biology"; and James Gray (*Experimental Cytology*, p. 2) said, in 1931, that "we must be careful to avoid any tacit assumption that the cell is a natural, or even legitimate, unit of life and function."

of the mass, the growth of the organ, is the primary fact, that "cell formation is a phenomenon very general in organic life, but still only of secondary significance." "Comparative embryology," says Whitman, "reminds us at every turn that the organism dominates cell-formation, using for the same purpose one, several, or many cells, massing its material and directing its movements and shaping its organs, as if cells did not exist*." So Rauber declared that, in the whole world of organisms, "das Ganze liefert die Theile, nicht die Theile das Ganze: letzteres setzt die Theile zusammen, nicht diese jenes†." And on the botanical side De Bary has summed up the matter in an aphorism, "Die Pflanze bildet Zellen, nicht die Zelle bildet Pflanzen."

Discussed almost wholly from the concrete, or morphological point of view, the question has for the most part been made to turn on whether actual protoplasmic continuity can be demonstrated between one cell and another, whether the organism be an actual reticulum, or syncytium‡. But from the dynamical point of view the question is much simpler. We then deal not with material continuity, not with little bridges of connecting protoplasm, but with a continuity of forces, a comprehensive field of force, which runs through and through the entire organism and is by no means restricted in its passage to a protoplasmic continuum. And such a continuous field of force, somehow shaping the whole organism, independently of the number, magnitude and form of the individual cells, which enter like a froth into its fabric, seems to me certainly and obviously to exist. As Whitman says, "the fact that physiological unity is not broken by cell-boundaries is confirmed in so many ways that it must be accepted as one of the fundamental truths of biology§."

* *Journ. Morph.* VIII, p. 653, 1893.

† Neue Grundlegungen zur Kenntniss der Zelle, *Morph. Jahrb.* VIII, pp. 272, 313, 333, 1883.

‡ Cf. e.g. Ch. van Bambeke, A propos de la délimitation cellulaire, *Bull. Soc. belge de Microsc.* XXIII, pp. 72–87, 1897.

§ *Journ. Morph.* II, p. 49, 1889.

CHAPTER V

THE FORMS OF CELLS

PROTOPLASM, as we have already said, is a fluid* or a semi-fluid substance, and we need not try to describe the particular properties of the colloid or jelly-like substances to which it is allied, or rather the characteristics of the "colloidal state" in which it and they exist; we should find it no easy matter†. Nor need we appeal to precise theoretical definitions of fluidity, lest we come into a debatable land. It is in the most general sense that protoplasm is "fluid." As Graham said (of colloid matter in general), "its softness *partakes of fluidity*, and enables the colloid to become a vehicle for liquid diffusion, like water itself‡." When we can deal with protoplasm in sufficient quantity we see it *flow*§; particles move freely through it, air-bubbles and liquid droplets shew round or spherical within it; and we shall have much to say about other phenomena manifested by its own surface, which are those especially characteristic of liquids. It may encompass and contain solid bodies, and it may "secrete" solid substances within or around itself; and it often happens in the complex living organism that these solid substances, such as shell or nail or horn or feather, remain when the protoplasm which formed them is dead and gone. But the protoplasm itself is fluid or semi-fluid, and permits of free (though not necessarily rapid) *diffusion* and easy *convection* of particles within itself, which simple fact is of elementary importance

* Cf. W. Kühne, *Ueber das Protoplasma*, 1864.

† Sand, or a heap of millet-seed, may in a sense be deemed a "fluid," and such the learned Father Boscovich held them to be (*Theoria*, p. 427), but at best they are fluids without a surface. Galileo had drawn the same comparison; but went on to contrast the continuity, or infinite subdivision, of a fluid with the finite, discontinuous subdivision of a fine powder. Cf. Boyer, *Concepts of the Calculus*, 1939, p. 291.

‡ *Phil. Trans.* CLI, p. 183, 1861; *Researches*, ed. Angus Smith, 1877, p. 553. We no longer speak, however, of "colloids" in a specific sense, as Graham did; for any substance can be brought into the "colloidal state" by appropriate means or in an appropriate medium.

§ The copious protoplasm of a Myxomycete has been passed unharmed through filter-paper with a pore-size of about 1 μ, or 0·001 mm.

in connection with form, throwing light on what seem to be common characteristics and peculiarities of the forms of living things.

Much has been done, and more said, about the nature of protoplasm since this book was written. Calling *cytoplasm* the cell-protoplasm after deduction of chloroplasts and other gross inclusions, we find it to contain fats, proteins, lecithin and some other substances combined with much water (up to 97 per cent.) to form a sort of watery *gel*. The microscopic structures attributed to it, alveolar, granular or fibrillar, are inconstant or invalid; but it does appear to possess an invisible or submicroscopic structure, distinguishing it from an ordinary colloid gel, and forming a quasi-solid framework or reticulum. This framework is based on proteid macromolecules, in the form of polypeptide chains, of great length and carrying in side-chains other organic constituents of the cytoplasm*. The polymerised units represent the micellae† which the genius of Nägeli predicted or postulated more than sixty years ago; and we may speak of a "micellar framework" as representing in our cytoplasm the dispersed phase of an ordinary colloid. In short, as the cytoplasm is neither true fluid not true solid, neither is it true colloid in the ordinary sense. Its micellar structure gives it a certain rigidity or tendency to retain its shape, a certain plasticity and tensile strength, a certain ductility or capacity to be drawn out in threads; but yet leaves it with a permeability (or semi-permeability), a capacity to swell by imbibition, above all an ability to stream and flow, which justify our calling it "fluid or semi-fluid," and account for its exhibition of surface-tension and other capillary phenomena.

The older naturalists, in discussing the differences between organic and inorganic bodies, laid stress upon the circumstance that the latter grow by "agglutination," and the former by what they termed "intussusception." The contrast is true; but it applies rather to solid or crystalline bodies as compared with colloids of all kinds, whether living or dead. But it so happens that the great majority of colloids are of organic origin; and out of them our bodies, and our food, and the very clothes we wear, are almost wholly made.

A crystal "grows" by deposition of new molecules, one by one and layer by layer, each one superimposed on the solid substratum

* See (*int. al.*) A. Frey-Wyssling, *Submikroskopische Morphologie des Protoplasmas*, Berlin, 1938; cf. *Nature*, June 10, 1939, p. 965; also A. R. Moore, in *Scientia*, LXII, July 1, 1937. On the nature of viscid fluid threads, cf. Larmor, *Nature*, July 11, 1936, p. 74.

† *Micella*, or *micula*, diminutive of *mīca*, a crumb, grain or morsel—*mica panis, salis, turis*, etc. Nägeli used the word to mean an aggregation of molecules, as the molecule is an aggregation of atoms; the one, however, is a physical and the other a chemical concept. Roughly speaking, we may think of micellae as varying from about 1 to 200 $\mu\mu$; they play a corresponding part in the "disperse phase" of a colloid to that played by the molecules in an ordinary solution. The macromolecules of modern chemistry are sometimes distinguished from these as still larger aggregates. See Carl Nägeli, *Das Mikroskop* (2nd ed.), 1877; *Theorie der Gahrung*, 1879.

already formed. Each particle would seem to be influenced only by the particles in its immediate neighbourhood, and to be in a state of freedom and independence from the influence, either direct or indirect, of its remoter neighbours. So Lavoisier was the first to say. And as Kelvin and others later on explained the formation and the resulting forms of crystals, so we believe that each added particle takes up its position in relation to its immediate neighbours already arranged, in the holes and corners that their arrangement leaves, and in closest contact with the greatest number*; hence we may repeat or imitate this process of arrangement, with great or apparently even with precise accuracy (in the case of the simpler crystalline systems), by piling up spherical pills or grains of shot. In so doing, we must have regard to the fact that each particle must drop into the place where it can go most easily, or where no easier place offers. In more technical language, each particle is free to take up, and does take up, its position of least potential energy relative to those already there: in other words, for each particle motion is induced until the energy of the system is so distributed that no tendency or resultant force remains to move it more. This has been shewn to lead to the production of *plane* surfaces† (in all cases where, by the limitation of material, surfaces *must* occur); where we have planes, there straight edges and solid angles must obviously occur also, and, if equilibrium is to follow, must occur symmetrically. Our piling up of shot to make mimic crystals gives us visible demonstration that the result is actually to obtain, as in the natural crystal, plane surfaces and sharp angles symmetrically disposed.

* Cf. Kelvin, On the molecular tactics of a crystal, *The Boyle Lecture*, Oxford, 1893; *Baltimore Lectures*, 1904, pp. 612–642. Here Kelvin was mainly following Bravais's (and Frankenheim's) theory of "space-lattices," but he had been largely anticipated by the crystallographers. For an account of the development of the subject in modern crystallography, by Sohncke, von Fedorow, Schönfliess, Barlow and others, see (e.g.) Tutton's *Crystallography*, and the many papers by W. E. Bragg and others.

† In a homogeneous crystalline arrangement, *symmetry* compels a locus of one property to be a plane or set of planes; the locus in this case being that of least surface potential energy. Crystals "seem to be, as it were, the Elemental Figures, or the A B C of Nature's working, the reason of whose curious Geometrical Forms (if I may so call them) is very easily explicable" (Robert Hooke, *Posthumous Works*, 1745, p. 280).

But the living cell grows in a totally different way, very much as a piece of glue swells up in water, by "imbibition," or by interpenetration into and throughout its entire substance. The semi-fluid colloid mass takes up water, partly to combine chemically with its individual molecules*; partly by physical diffusion into the interstices between molecules or micellae, and partly, as it would seem, in other ways; so that the entire phenomenon is a complex and even an obscure one†. But, so far as we are concerned, the net result is very simple. For the equilibrium, or tendency to equilibrium, of fluid pressure in all parts of its interior while the process of imbibition is going on, the constant rearrangement of its fluid mass, the contrast in short with the crystalline method of growth where each particle comes to rest to move (relatively to the whole) no more, lead the mass of jelly to swell up very much as a bladder into which we blow air, and so, by a *graded* and harmonious distribution of forces, to assume everywhere a rounded and more or less bubble-like external form‡. So, when the same school of older naturalists called attention to a new distinction or contrast of form between organic and inorganic objects, in that the contours of the former tended to roundness and curvature, and those of the latter to be bounded by straight lines, planes and sharp angles, we see that this contrast was not a new and different one, but only another aspect of their former statement, and an immediate consequence of the difference between the processes of agglutination and intussusception§.

So far then as growth goes on undisturbed by pressure or other external force, the fluidity of the protoplasm, its mobility internal

* This is what Graham called the *water of gelatination*, on the analogy of *water of crystallisation*; *Chem. and Phys. Researches*, p. 597.

† On this important phenomenon, see J. R. Katz, *Gesetze der Quellung*, Dresden, 1916. Swelling is due to "concentrated solution," and is accompanied by increase of volume and liberation of energy, as when the Egyptians split granite by the swelling of wood.

‡ Here, in a non-crystalline or random arrangement of particles, symmetry ensures that the potential energy shall be the same per unit area of all surfaces; and it follows from geometrical considerations that the total surface energy will be least if the surface be spherical.

§ Intussusception has its shades of meaning; it is excluded from the idea of a crystalline body, but not limited to the ordinary conception of a colloid one. When new micellar strands become interwoven in the micro-structure of a cellulose cell-wall, that is a special kind of "intussusception."

and external*, and the way in which particles move freely hither and thither within, all manifestly tend to the production of swelling, rounded surfaces, and to their great predominance over plane surfaces in the contours of the organism. These rounded contours will tend to be preserved for a while, in the case of naked protoplasm by its viscosity, and in presence of a cell-wall by its very lack of fluidity. In a general way, the presence of curved boundary surfaces will be especially obvious in the unicellular organisms, and generally in the external form of all organisms, and wherever mutual pressure between adjacent cells, or other adjacent parts, has not come into play to flatten the rounded surfaces into planes.

The swelling of any object, organic or inorganic, living or dead, is bound to be influenced by any lack of structural symmetry or homogeneity†. We may take it that all elongated structures, such as hairs, fibres of silk or cotton, fibrillae of tendon and connective tissue, have by virtue of their elongation an invisible as well as a visible polarity. Moreover, the ultimate fibrils are apt to be invested by a protein different from the "collagen" within, and liable to swell more or to swell less. In ordinary tendons there is a "reticular sheath," which swells less, and is apt to burst under pressure from within; it breaks into short lengths, and when the strain is relieved these roll back, and form the familiar *annuli*. Another instance is the tendency to swell of the "macro-molecules" of many polymerised organic bodies, proteins among them.

But the rounded contours which are assumed and exhibited by a piece of hard glue when we throw it into water and see it expand as it sucks the water up, are not near so regular nor so beautiful as are those which appear when we blow a bubble, or form a drop, or even pour water into an elastic bag. For these curving contours depend upon the properties of the bag itself, of the film or membrane which contains the mobile gas, or which contains or bounds the mobile liquid mass. And hereby, in the case of the fluid or semifluid mass, we are introduced to the subject of *surface-tension*: of which indeed we have spoken in the preceding chapter, but which we must now examine with greater care.

* The protoplasm of a sea-urchin's egg has a viscosity only about four times, and that of various plants not more than ten to twenty times, that of water itself. See, for a general discussion, L. V. Heilbrunn, *Colloid Symposium Monograph*, 1928.

† D. Jordan Lloyd and R. H. Marriott, The swelling of structural proteins, *Proc. R.S.* (B), No. 810, pp. 439–445, 1935.

Among the forces which determine the forms of cells, whether they be solitary or arranged in contact with one another, this force of surface-tension is certainly of great, and is probably of paramount, importance. But while we shall try to separate out the phenomena which are directly due to it, we must not forget that, in each particular case, the actual conformation which we study may be, and usually is, the more or less complex resultant of surface-tension acting together with gravity, mechanical pressure, osmosis, or other physical forces. The peculiar beauty of a soap-bubble, solitary or in collocation, depends on the absence (to all intents and purposes) of these alien forces from the field; hence Plateau spoke of the films which were the subject of his experiments as "lames fluides *sans pesanteur*." The resulting form is in such a case so pure and simple that we come to look on it as wellnigh a mathematical abstraction.

Surface-tension, then, is that force by which we explain the form of a drop or of a bubble, of the surfaces external and internal of a "froth" or collocation of bubbles, and of many other things of like nature and in like circumstances*. It is a property of liquids (in the sense at least with which our subject is concerned), and it is manifested at or very near the surface, where the liquid comes into contact with another liquid, a solid or a gas. We note here that the term *surface* is to be interpreted in a wide sense; for wherever we have solid particles embedded in a fluid, wherever we have a non-homogeneous fluid or semi-fluid, or a "two-phase colloid" such as a particle of protoplasm, wherever we have the presence of "impurities" as in a mass of molten metal, there we have always to bear in mind the existence of *surfaces* and of surface-phenomena, not only on the exterior of the mass but also throughout its interstices, wherever like and unlike meet.

* The idea of a "surface-tension" in liquids was first enunciated by Segner, and ascribed by him to forces of attraction whose range of action was so small "ut nullo adhuc sensu percipi potuerat" (*De figuris superficierum fluidarum*, in *Comment. Soc. Roy. Göttingen*, 1751, p. 301). Hooke, in the *Micrographia* (1665, Obs. VIII, etc.), had called attention to the globular or spherical form of the little morsels of steel struck off by a flint, and had shewn how to make a powder of such spherical grains, by heating fine filings to melting point. "This Phaenomenon" he said "proceeds from a propriety which belongs to all kinds of fluid Bodies more or less, and is caused by the Incongruity of the Ambient and included Fluid, which so acts and modulates each other, that they acquire, as neer as is possible, a *spherical* or *globular* form...."

A liquid in the mass is devoid of structure; it is homogeneous, and without direction or polarity. But the very concept of surface-tension forbids this to be true of the surface-layer of a body of liquid, or of the "interphase" between two liquids, or of any film, bubble, drop, or capillary jet or stream. In all these cases, and more emphatically in the case of a "monolayer," even the liquid has a structure of its own; and we are reminded once again of how largely the living organism, whether high or low, is composed of colloid matter in precisely such forms and structural conditions.

Surface-tension is due to molecular force*: to force, that is to say, arising from the action of one molecule upon another; and since we can only ascribe a small "sphere of action" to each several molecule, this force is manifested only within a narrow range. Within the interior of the liquid mass we imagine that such molecular interactions negative one another; but at and near the free surface, within a layer or film approximately equal to the range of the molecular force—or to the radius of the aforesaid "sphere of action" —there is a lack of equilibrium and a consequent manifestation of force.

The action of the molecular forces has been variously explained. But one simple explanation (or mode of statement) is that the molecules of the surface-layer are being constantly attracted into the interior by such as are just a little more deeply situated; the surface shrinks as molecules keep quitting it for the interior, and this *surface-shrinkage* exhibits itself as a *surface-tension*. The process continues till it can go no farther, that is to say until the surface itself becomes a "minimal area†." This is a sufficient description of the phenomenon in cases where a portion of liquid is subject to no other than its own molecular forces, and (since the sphere has,

* While we *explain* certain phenomena of the organism by reference to atomic or molecular forces, the following words of Du Bois Reymond's seem worth recalling: 'Naturerkennen ist Zurückführen der Veränderungen in der Körperwelt auf Bewegung von Atomen, die durch deren von der Zeit unabhängige Centralkräfte bewirkt werden, oder Auflösung der Naturkräfte in Mechanik der Atome. Es ist eine psychologische Erfahrungstatsache dass, wo solche Auflösung gelängt, unser Causalbedürfniss vorlaüfig sich befriedigt fühlt" (*Ueber die Grenzen des Naturerkennens*, Leipzig, 1873).

† There must obviously be a certain kinetic energy in the molecules within the drop, to balance the forces which are trying to contract and diminish the surface.

of all solids, the least surface for a given volume) it accounts for the spherical form of the raindrop*, of the grain of shot, or of the living cell in innumerable simple organisms†. It accounts also, as we shall presently see, for many much more complicated forms, manifested under less simple conditions.

Let us note in passing that surface-tension is a comparatively small force and is easily measurable: for instance that between water and air is equivalent to but a few grains per linear inch, or a few grammes per metre. But this small tension, when it exists in a *curved* surface of great curvature, such as that of a minute drop, gives rise to a very great pressure, directed inwards towards the centre of curvature. We may easily calculate this pressure, and so satisfy ourselves that, when the radius of curvature approaches molecular dimensions, the pressure is of the order of thousands of atmospheres—a conclusion which is supported by other physical considerations.

The contraction of a liquid surface, and the other phenomena of surface-tension, involve the doing of work, and the power to do work is what we call Energy. The whole energy of the system is diffused throughout its molecules, as is obvious in such a simple case as we have just considered; but of the whole stock of energy only the part residing at or very near the surface normally manifests itself in work, and hence we speak (though the term be open to

* Raindrops must be spherical, or they would not produce a rainbow; and the fact that the upper part of the bow is the brightest and sharpest shews that the higher raindrops are more truly spherical, as well as smaller than the lower ones. So also the smallest dewdrops are found to be more iridescent than the large, shewing that they also are the more truly spherical; cf. T. W. Backhouse, in *Monthly Meteorol. Mag.* March, 1879. Mercury has a high surface-tension, and its globules are very nearly round.

† That the offspring of a spherical cell (whether it be raindrop, plant or animal) should be also a spherical cell, would seem to need no other explanation than that both are of identical substance, and each subject to a similar equilibrium of surface-forces; but the biologists have been apt to look for a subtler reason. Giglio-Tos, speaking of a sea-urchin's dividing egg, asks why the daughter-cells are spherical like the mother-cell, and finds the reason in "heredity": "Wenn also die letztere (d. i. die Mutterzelle) eine sphärische Form besass, so nehmen auch die Töchterzellen dieselbe ein; wäre ursprünglich eine kubische Form vorhanden, so wurden also auch die Töchterzellen dieselbe auch aneignen. Die Ursache warum die Töchterzellen die sphärische Form anzunehmen trachten liegt darin, *dass diese die Ur- und Grundform aller Zellen ist, sowohl bei Tieren wie bei den Pflanzen*" (*Arch. f. Entw. Mech.* LI, p. 115, 1922).

some objection) of a specific *surface-energy*. Surface-energy, and the way it is increased and multiplied by the multiplication of surfaces due to the subdivision of the tissues into cells, is of the highest interest to the physiologist; and even the morphologist cannot pass it by. For the one finds surface-energy present, often perhaps paramount, in every cell of the body; and the other may find, if he will only look for it, the form of every solitary cell, like that of any other drop or bubble, related to if not controlled by capillarity. The theory of "capillarity," or "surface-energy," has been set forth with the utmost possible lucidity by Tait and by Clerk Maxwell, on whom the following paragraphs are based: they having based their teaching on that of Gauss*, who rested on Laplace.

Let E be the whole potential energy of a mass M of liquid; let e_0 be the energy per unit mass of the interior liquid (we may call it the *internal energy*); and let e be the energy per unit mass for a layer of the skin, of surface S, of thickness t, and density ρ (e being what we call the *surface-energy*). It is obvious that the total energy consists of the internal *plus* the surface-energy, and that the former is distributed through the whole mass, minus its surface layers. That is to say, in mathematical language,

$$E = (M - S \,.\, \Sigma t\rho)\, e_0 + S \,.\, \Sigma t\rho e.$$

But this is equivalent to writing:

$$= Me_0 + S \,.\, \Sigma t\rho\, (e - e_0);$$

and this is as much as to say that the total energy of the system may be taken to consist of two portions, one uniform throughout the whole mass, and another, which is proportional on the one hand to the amount of surface, and on the other hand to the difference between e and e_0, that is to say to the difference between the unit values of the internal and the surface energy.

It was Gauss who first shewed how, from the mutual attractions between all the particles, we are led to an expression for what we

* See Gauss's *Principia generalia Theoriae Figurae Fluidorum in statu equilibrii*, Göttingen, 1830. The historical student will not overlook the claims to priority of Thomas Young, in his Essay on the cohesion of fluids, *Phil. Trans.* 1805; see the account given in his *Life* by Dean Peacock, 1855, pp. 199–210.

now call the *potential energy** of the system; and we know, as a fundamental theorem of dynamics, as well as of molecular physics, that the potential energy of the system tends to a minimum, and finds in that minimum its stable equilibrium.

We see in our last equation that the term Me_0 is irreducible, save by a reduction of the mass itself. But the other term may be diminished (1) by a reduction in the area of surface, S, or (2) by a tendency towards equality of e and e_0, that is to say by a diminution of the specific surface energy, e.

These then are the two methods by which the energy of the system will manifest itself in work. The one, which is much the more important for our purposes, leads always to a diminution of surface, to the so-called "principle of minimal areas"; the other, which leads to the lowering (under certain circumstances) of surface tension, is the basis of the theory of Adsorption, to which we shall have some occasion to refer as the *modus operandi* in the development of a cell-wall, and in a variety of other histological phenomena. In the technical phraseology of the day, the "capacity factor" is involved in the one case, and the "intensity factor" in the other†.

Inasmuch as we are concerned with the *form* of the cell, it is the former which becomes our main postulate: telling us that the energy-equations of the surface of a cell, or of the free surfaces of cells in partial contact, or of the partition-surfaces of cells in contact with one another, all indicate a minimum of potential energy in the system, by which minimal condition the system is brought, *ipso facto*, into equilibrium. And we shall not fail to observe, with something more than mere historical interest and curiosity, how

* The word *Energy* was substituted for the old *vis viva* by Thomas Young early in the nineteenth century, and was used by James Thomson, Lord Kelvin's brother, about 1852, to mean, more generally, "capacity for doing work." The term *potential*, or *latent*, in contrast to *actual* energy, in other words the distinction between "energy of activity and energy of configuration," was proposed by Macquorn Rankine, and suggested to him by Aristotle's use of δύναμις and ἐνέργεια; see Rankine's paper On the general law of the transformation of energy, *Phil. Soc. Glasgow*, Jan. 5, 1853, cf. *ibid.* Jan. 23, 1867, and *Phil. Mag.* (4), XXVII, p. 404, 1864. The phrase *potential energy* was at once adopted, but *kinetic* was substituted for *actual* by Thomson and Tait.

† The capacity factor, inasmuch as it leads to diminution of surface, is responsible for the concrescence of droplets into drops, of microcrystals into larger units, for the flocculation of colloids, and for many other similar "changes of state."

deeply and intrinsically there enter into this whole class of problems the method of maxima and minima discovered by Fermat, the "loi universelle de repos" of Maupertuis, "dont tous les cas d'équilibre dans la statique ordinaire ne sont que des cas particuliers", and the *lineae curvae maximi minimive proprietatibus gaudentes* of Euler, by which principles these old natural philosophers explained correctly a multitude of phenomena, and drew the lines whereon the foundations of great part of modern physics are well and truly laid. For that physical laws deal with *minima* is very generally true, and is highly characteristic of them. The hanging chain so hangs that the height of its centre of gravity is a minimum; a ray of light takes the path, however devious, by which the time of its journey is a minimum; two chemical substances in reaction so behave that their thermodynamic potential tends to a minimum, and so on. The natural philosophers of the eighteenth century were engrossed in minimal problems; and the differential equations which solve them nowadays are among the most useful and most characteristic equations in mathematical physics.

"Voici," said Maupertuis, "dans un assez petit volume à quoi je reduis mes ouvrages mathématiques!" And when Lagrange, following Euler's lead*, conceived the principle of least action, he regarded it not as a metaphysical principle but as "un resultat simple et général des lois de la mécanique†." The principle of least action‡ explains nothing, it tells us nothing of causation, yet it illuminates a host of things. Like Maxwell's equations and other such flashes of genius it clarifies our knowledge, adds weight to our observations, brings order into our stock-in-trade of facts. It embodies and extends that "law of simplicity" which Borelli was the first to lay down: "Lex perpetua Naturae est ut agat minimo labore, mediis et modis simplicissimis, facillimis, certis et

* Euler, *Traité des Isopérimètres*, Lausanne, 1744.

† Lagrange, *Mécanique Analytique* (2), II, p. 188; ed. in 4to, 1788.

‡ This profound conception, not less metaphysical in the outset than physical, began in the seventeenth century with Fermat, who shewed (in 1629) that a ray of light followed the quickest path available, or, as Leibniz put it, *via omnium facillima*; it was over this principle that Voltaire quarrelled with Euler and Maupertuis. The mathematician will think also of Hamilton's restatement of the principle, and of its extension to the theory of probabilities by Boltzmann and Willard Gibbs. Cf. (*int. al.*) A. Mayer, *Geschichte des Prinzips der kleinsten Action*, 1877.

tutis: evitando, quam maxime fieri potest, incommoditates et prolixitates." The principle of least action grew up, and grew quickly, out of cruder, narrower notions of "least time" or "least space or distance." Nowadays it is developing into a principle of "least action in space-time," which shall still govern and predict the motions of the universe. The infinite perfection of Nature is expressed and reflected in these concepts, and Aristotle's great aphorism that "Nature does nothing in vain" lies at the bottom of them all.

In all cases where the principle of maxima and minima comes into play, as it conspicuously does in films at rest under surface-tension, the configurations so produced are characterised by obvious and remarkable *symmetry**. Such symmetry is highly characteristic of organic forms, and is rarely absent in living things—save in such few cases as *Amoeba*, where the rest and equilibrium on which symmetry depends are likewise lacking. And if we ask what physical equilibrium has to do with formal symmetry and structural regularity, the reason is not far to seek, nor can it be better put than in these words of Mach's†: "In every symmetrical system every deformation that tends to destroy the symmetry is complemented by an equal and opposite deformation that tends to restore it. In each deformation, positive and negative work is done. One condition, therefore, though not an absolutely sufficient one, that a maximum or minimum of work corresponds to the form of equilibrium, is thus supplied by symmetry. Regularity is successive symmetry; there is no reason, therefore, to be astonished that the forms of equilibrium are often symmetrical and regular."

A crystal is the perfection of symmetry and of regularity; symmetry is displayed in its external form, and regularity revealed in its internal lattices. Complex and obscure as the attractions, rotations, vibrations and what not within the crystal may be, we rest assured that the configuration, repeated again and again, of

* On the mathematical side, cf. Jacob Steiner, Einfache Beweise der isoperimetrischen Hauptsätze, *Abh. k. Akad. Wiss. Berlin*, XXIII, pp. 116–135, 1836 (1838). On the biological side, see (*int. al.*) F. M. Jaeger, *Lectures on the Principle of Symmetry, and its application to the natural sciences*, Amsterdam, 1917; also F. T. Lewis, Symmetry...in evolution, *Amer. Nat.* LVII, pp. 5–41, 1923.

† *Science of Mechanics*, 1902, p. 395; see also Mach's article Ueber die physikalische Bedeutung der Gesetze der Symmetrie, *Lotos*, XXI, pp. 139–147, 1871.

the component atoms is precisely that for which the energy is a minimum; and we recognise that this minimal distribution is of itself tantamount to symmetry and to stability.

Moreover, the principle of least action is but a setting of a still more universal law—that the world and all the parts thereof tend ever to pass from less to more probable configurations; in which the physicist recognises the principle of Clausius, or second law of thermodynamics, and with which the biologist must somehow reconcile the whole "theory of evolution."

As we proceed in our enquiry, and especially when we approach the subject of *tissues*, or agglomerations of cells, we shall have from time to time to call in the help of elementary mathematics. But already, with very little mathematical help, we find ourselves in a position to deal with some simple examples of organic forms.

When we melt a stick of sealing-wax in the flame, surface-tension (which was ineffectively present in the solid but finds play in the now fluid mass) rounds off its sharp edges into curves, so striving towards a surface of minimal area; and in like manner, by merely melting the tip of a thin rod of glass, Hooke made the little spherical beads which served him for a microscope*. When any drop of protoplasm, either over all its surface or at some free end, as at the extremity of the pseudopodium of an amoeba, is seen likewise to "round itself off," that is not an effect of "vital contractility," but, as Hofmeister shewed so long ago as 1867, a simple consequence of surface-tension; and almost immediately afterwards Engelmann† argued on the same lines, that the forces which cause the contraction of protoplasm in general may "be just the same as those which tend to make every non-spherical drop of fluid become spherical." We are not concerned here with the many theories and speculations which would connect the phenomena of surface-tension with contractility, muscular movement, or other special *physiological* func-

* Similarly, Sir David Brewster and others made powerful lenses by simply dropping small drops of Canada balsam, castor oil, or other strongly refractive liquids, on to a glass plate: *On New Philosophical Instruments* (Description of a new fluid microscope), Edinburgh, 1813, p. 413. See also Hooke's *Micrographia*, 1665; and Adam's *Essay on the Microscope*, 1798, p. 8: "No person has carried the use of these globules so far as Father Torre of Naples, etc." Leeuwenhoek, on the other hand, *ground* his lenses with exquisite skill.

† Beiträge zur Physiologie des Protoplasma, *Pflüger's Archiv*, II, p. 307, 1869.

tions, but we find ample room to trace the operation of the same cause in producing, under conditions of rest and equilibrium, certain definite and inevitable forms.

It is of great importance to observe that the living cell is one of those cases where the phenomena of surface-tension are by no means limited to the *outer* surface; for within the heterogeneous emulsion of the cell, between the protoplasm and its nucleus and other contents, and in the "alveolar network" of the cytoplasm itself (so far as that alveolar structure is actually present in life), we have a multitude of interior surfaces; and, especially among plants, we may have large internal "interfacial contacts" between the protoplasm and its included granules, or its vacuoles filled with the "cell-sap." Here we have a great field for surface-action; and so long ago as 1865, Nägeli and Schwendener shewed that the streaming currents of plant cells might be plausibly explained by this phenomenon. Even ten years earlier, Weber had remarked upon the resemblance between the protoplasmic streamings and the currents to be observed in certain inanimate drops for which no cause but capillarity could be assigned*. What sort of chemical changes lead up to, or go hand in hand with, the variations of surface-tension in a living cell, is a vastly important question. It is hardly one for us to deal with; but this at least is clear, that the phenomenon is more complicated than its first investigators, such as Bütschli and Quincke, ever took it to be. For the lowered surface-tension which leads, say, to the throwing out of a pseudopodium, is accompanied first by local acidity, then by local adsorption of proteins, lastly and consequently by gelation; and this last is tantamount to the formation of "ectoplasm"—a step in the direction of encystment†.

The elementary case of *Amoeba* is none the less a complicated one. The "amoeboid" form is the very negation of rest or of equilibrium;

* *Poggendorff's Annalen*, XCIV, pp. 447–459, 1855. Cf. Strethill Wright, *Phil. Mag.* Feb. 1860; *Journ. Anat. and Physiol.* I, p. 337, 1867.

† Cf. C. J. Pantin, *Journ. Mar. Biol. Assoc.* XIII, p. 24, 1923; *Journ. Exp. Biol.* 1923 and 1926; S. O. Mast, *Journ. Morph.* XLI, p. 347, 1926; and O. W. Tiegs, Surface tension and the theory of protoplasmic movement, *Protoplasma*, IV, pp. 88–139, 1928. See also (*int. al.*) N. K. Adam, *Physics and Chemistry of Surfaces*, 1930; also Discussion on colloid science applied to biology (*passim*), *Trans. Faraday Soc.* XXVI, pp. 663 *seq.*, 1930.

the creature is always moving, from one protean configuration to another; its surface-tension is never constant, but continually varies from here to there. Where the surface tension is greater, that portion of the surface will contract into spherical or spheroidal forms; where it is less, the surface will correspondingly extend. While generally speaking the surface-energy has a minimal value, it is not necessarily constant. It may be diminished by a rise of temperature; it may be altered by contact with adjacent substances*, by the transport of constituent materials from the interior to the surface, or again by actual chemical and fermentative change; for within the cell, the surface-energies developed about its heterogeneous contents will continually vary as these contents are affected by chemical metabolism. As the colloid materials are broken down and as the particles in suspension are diminished in size the "free surface-energy" will be increased, but the osmotic energy will be diminished†. Thus arise the various fluctuations of surface-tension, and the various phenomena of amoeboid form and motion, which Bütschli and others have reproduced or imitated by means of the fine emulsions which constitute their "artificial amoebae."

A multitude of experiments shew how extraordinarily delicate is the adjustment of the surface-tension forces, and how sensitive they are to the least change of temperature or chemical state. Thus,

* Haycraft and Carlier pointed out long ago (*Proc. R.S.E.* xv, pp. 220–224, 1888) that the amoeboid movements of a white blood-corpuscle are only manifested when the corpuscle is in contact with some solid substance: while floating freely in the plasma or serum of the blood, these corpuscles are spherical, that is to say they are at rest and in equilibrium. The same fact was recorded anew by Ledingham (On phagocytosis from an adsorptive point of view, *Journ. Hygiene*, XII, p. 324, 1912). On the emission of pseudopodia as brought about by changes in surface tension, see also (*int. al.*) J. A. Ryder, *Dynamics in Evolution*, 1894; Jensen, Ueber den Geotropismus niederer Organismen, *Pflüger's Archiv*, LIII, 1893. Jensen remarks that in Orbitolites, the pseudopodia issuing through the pores of the shell first float freely, then as they grow longer bend over till they touch the ground, whereupon they begin to display amoeboid and streaming motions. Verworn indicates (*Allg. Physiol.* 1895, p. 429), and Davenport says (*Exper. Morphology*, II, p. 376), that "this persistent clinging to the substratum is a 'thigmotropic' reaction, and one which belongs clearly to the category of 'response'." Cf. Pütter, Thigmotaxis bei Protisten, *Arch. f. Physiol.* 1900, Suppl. p. 247; but it is not clear to my mind that to account for this simple phenomenon we need invoke other factors than gravity and surface-action.

† Cf. Pauli, *Allgemeine physikalische Chemie d. Zellen u. Gewebe*, in Asher-Spiro's *Ergebnisse der Physiologie*, 1912; Przibram, *Vitalität*, 1913, p. 6.

on a plate which we have warmed at one side a drop of alcohol runs towards the warm area, a drop of oil away from it; and a drop of water on the glass plate exhibits lively movements when we bring into its neighbourhood a heated wire, or a glass rod dipped in ether*. The water-colour painter makes good use of the surface-tension effect of the minutest trace of ox-gall. When a plasmodium of *Aethalium* creeps towards a damp spot or a warm spot, or towards substances which happen to be nutritious, and creeps away from solutions of sugar or of salt, we are dealing with phenomena too often ascribed to 'purposeful' action or adaptation, but every one of which can be paralleled by ordinary phenomena of surface-tension†. The soap-bubble itself is never in equilibrium: for the simple reason that its film, like the protoplasm of *Amoeba* or *Aethalium*, is exceedingly heterogeneous. Its surface-energies vary from point to point, and chemical changes and changes of temperature increase and magnify the variation. The surface of the bubble is in continual movement, as more concentrated portions of the soapy fluid make their way outwards from the deeper layers; it thins and it thickens, its colours change, currents are set up in it and little bubbles glide over it; it continues in this state of restless movement as its parts strive one with another in their interactions towards unattainable equilibrium‡. On reaching a certain tenuity the bubble bursts: as is bound to happen when the attenuated film has no longer the properties of *matter in mass*.

* So Bernstein shewed that a drop of mercury in nitric acid moves towards, or is "attracted by," a crystal of potassium bichromate; *Pflüger's Archiv*, LXXX, p. 628, 1900.

† The surface-tension theory of protoplasmic movement has been denied by many. Cf. (e.g.) H. S. Jennings, Contributions to the behaviour of the lower organisms, *Carnegie Instit.* 1904, pp. 130–230; O. P. Dellinger, Locomotion of Amoebae, etc., *Journ. Exp. Zool.* III, pp. 337–357, 1906; also various papers by Max Heidenhain, in Merkel u. Bonnet's *Anatomische Hefte*; etc.

‡ These motions of a liquid surface, and other still more striking movements, such as those of a piece of camphor floating on water, were at one time ascribed by certain physicists to a peculiar force, *sui generis*, the *force épipolique* of Dutrochet: until van der Mensbrugghe shewed that differences of surface-tension were enough to account for this whole series of phenomena (Sur la tension superficielle des liquides, considérée au point de vue de certains mouvements observés à leur surface, *Mém. Cour. Acad. de Belgique*, XXXIV, 1869, *Phil. Mag.* Sept. 1867; cf. Plateau, *Statique des Liquides*, p. 283). An interesting early paper is by Dr G. Carradini of Pisa, Dell' adesione o attrazione di superficie, *Mem. di Matem. e di Fisica d. Soc. Ital. d. Sci.* (*Modena*), XI, p. 75, XII, p. 89, 1804–5.

The film becomes a mere bimolecular, or even a monomolecular, layer; and at last we may treat it as a simple "surface of discontinuity." So long as the changes due to imperfect equilibrium are taking place *very slowly*, we speak of the bubble as "at rest"; it is then, as Willard Gibbs remarks, that the characters of a film are most striking and most sharply defined*.

So also, and surely not less than the soap-bubble, is every cell-surface a complex affair. Face and interface have a molecular orientation of their own, depending both on the partition-membrane and on the phases on either side. It is a variable orientation, changing at short intervals of space and time; it coincides with inconstant fields of force, electrical and other; it initiates, and controls or catalyses, chemical reactions of great variety and importance. In short we acknowledge and confess that, in simplifying the surface phenomena of the cell, for the time being and for our purely morphological ends, we may be losing sight, or making abstraction, of some of its most specific physical and physiological characteristics.

In the case of the naked protoplasmic cell, as the amoeboid phase is emphatically a phase of freedom and activity, of unstable equilibrium, of chemical and physiological change, so on the other hand does the spherical form indicate a phase of stability, of inactivity, of rest. In the one phase we see unequal surface-tensions manifested in the creeping movements of the amoeboid body, in the rounding-off of the ends of its pseudopodia, in the flowing out of its substance over a particle of "food," and in the current-motions in the interior of its mass; till, in the alternate phase, when internal homogeneity and equilibrium have been as far as possible attained and the potential energy of the system is at a minimum, the cell assumes a rounded or spherical form, passes into a state of "rest," and (for a reason which we shall presently consider) becomes at the same time encysted†.

* On the equilibrium of heterogeneous substances, *Collected Works*, I, pp. 55–353; *Trans. Conn. Acad.* 1876–78.

† We still speak of the *naked protoplasm* of *Amoeba*; but short, and far short, of "encystment," there is always a certain tendency towards adsorptive action, leading to a surface-layer, or "plasma-membrane," still semi-fluid but less fluid than before, and different from the protoplasm within; it was one of the first and chief things revealed by the new technique of "micro-dissection." Little is known of

In their amoeboid phase the various Amoebae are just so many varying distributions of surface-energy, and varying amounts of surface-potential*. An ordinary floating drop is a figure of equilibrium under conditions of which we shall soon have something to say; and if both it and the fluid in which it floats be homogeneous it will be a round drop, a "figure of revolution." But the least chemical heterogeneity will cause the surface-tension to vary here and there, and the drop to change its form accordingly. The little swarm-spores of many algae lose their flagella as they settle down, and become mere drops of protoplasm for the time being; they "put out pseudopodia"—in other words their outline changes; and presently this amoeboid outline grows out into characteristic lobes or lappets, a sign of more or less symmetrical heterogeneity in the cell-substance.

In a budding yeast-cell (Fig. 103 A), we see a definite and restricted change of surface-tension. When a "bud" appears, whether with or without actual growth by osmosis or otherwise of the mass, it does so because at a certain part of the cell-surface the tension has diminished, and the area of that portion expands accordingly; but in turn the surface-tension of the expanded or extruded portion makes itself felt, and the bud rounds itself off into a more or less spherical form.

Fig. 103 A.

The yeast-cell with its bud is a simple example of an important principle. Our whole treatment of cell-form in relation to surface-tension depends on the fact (which Errera was the first to give clear expression to) that the *incipient* cell-wall retains with but little impairment the properties of a liquid film†, and that the growing cell, in spite of the wall by which it has begun to be surrounded,

the physical nature of this so-called membrane. It behaves more or less like a fluid lipoid envelope, immiscible with its surroundings. It is easily injured and easily repaired, and the well-being of the internal protoplasm is said to depend on the maintenance of its integrity. Robert Chambers, *Physical Properties of Protoplasm*, 1926; The living cell as revealed by microdissection, *Harvey Lectures*, Ser. XXII, 1926–27; *Journ. Gen. Physiol.* VIII, p. 369, 1926; etc.

* See (*int. al.*) Mary J. Hogue, The effect of media of different densities on the shape of Amoebae, *Journ. Exp. Zool.* XXII, pp. 565–572, 1917. Scheel had said in 1889 that *A. radiosa* is only an early stage of *A. proteus* (*Festschr. z. 70. Geburtstag C. V. Kupffer*).

† Cf. *infra*, p. 561.

behaves very much like a fluid drop. So, to a first approximation, even the yeast-cell shews, by its ovoid and non-spherical form, that it has acquired its shape under some influence other than the uniform and symmetrical surface-tension which makes a soap-bubble into a sphere. This oval or any other asymmetrical form, once acquired, may be retained by virtue of the solidification and consequent rigidity of the membrane-like wall of the cell; and, unless rigidity ensue, it is plain that such a conformation as that of the yeast-cell with its attached bud could not be long retained as a figure of even partial equilibrium. But as a matter of fact, the cell in this case is not in equilibrium at all; it is *in process* of budding, and is slowly altering its shape by rounding off its bud. In like manner the developing egg, through all its successive phases of form, is never in complete equilibrium: but is constantly responding to slowly changing conditions, by phases of partial, transitory, unstable and conditional equilibrium.

There are innumerable solitary plant-cells, and unicellular organisms in general, which, like the yeast-cell, do not correspond to any of the simple forms which may be generated under the influence of simple and homogeneous surface-tension; and in many cases these forms, which we should expect to be unstable and transitory, have become fixed and stable by reason of some comparatively sudden solidification of the envelope. This is the case, for instance, in the more complicated forms of diatoms or of desmids, where we are dealing, in a less striking but even more curious way than in the budding yeast-cell, not with one simple act of formation, but with a complicated result of successive stages of localised growth, interrupted by phases of partial consolidation. The original cell has acquired a certain form, and then, under altering conditions and new distributions of energy, has thickened here or weakened there, and has grown out, or tended (as it were) to branch, at particular points. We can often trace in each particular stage of growth, or at each particular temporary growing point, the laws of surface tension manifesting themselves in what is for the time being a fluid surface; nay more, even in the adult and completed structure we have little difficulty in tracing and recognising (for instance in the outline of such a desmid as *Euastrum*) the rounded lobes which have successively grown or flowed out from the original rounded and flattened cell. What we see in a many chambered foraminifer, such as *Globigerina* or *Rotalia*, is the same thing, save that the stages are more separate and distinct, and the whole is carried out to greater completeness and perfection. The little organism as a whole is not a figure of equilibrium nor of minimal area; but each new bud or separate chamber is such a figure, conditioned by the forces of surface-tension, and superposed upon the complex aggregate of similar bubbles after these latter have become consolidated one by one into a rigid system.

Let us now make some enquiry into the forms which a fluid surface can assume under the mere influence of surface-tension. In doing so we are limited to conditions under which other forces are relatively unimportant, that is to say where the surface energy is a considerable fraction of the whole energy of the system; and in general this will be the case when we are dealing with portions of liquid so small that their dimensions come within or near to what we have called the molecular range, or, more generally, in which the "specific surface" is large. In other words it is the small or minute organisms, or small cellular elements of larger organisms, whose forms will be governed by surface-tension; while the forms of the larger organisms are due to other and non-molecular forces. A large surface of water sets itself level because here gravity is predominant; but the surface of water in a narrow tube is curved, for the reason that we are here dealing with particles which lie within the range of each other's molecular forces. The like is the case with the cell-surfaces and cell-partitions which we are about to study, and the effect of gravity will be especially counteracted and concealed when the object is immersed in a liquid of nearly its own density.

We have already learned, as a fundamental law of "capillarity," that a liquid film *in equilibrium* assumes a form which gives it a minimal area under the conditions to which it is subject. These conditions include (1) the form of the boundary, if such exist, and (2) the pressure, if any, to which the film is subject: which pressure is closely related to the volume of air, or of liquid, that the film (if it be a closed one) may have to contain. In the simplest of cases, as when we take up a soap-film on a plane wire ring, the film is exposed to equal atmospheric pressure on both sides, and it obviously has its minimal area in the form of a plane. So long as our wire ring lies in one plane (however irregular in outline), the film stretched across it will still be in a plane; but if we bend the ring so that it lies no longer in a plane, then our film will become curved into a surface which may be extremely complicated, but is still the smallest possible surface which can be drawn continuously across the uneven boundary.

The question of pressure involves not only external pressures acting on the film, but also that which the film itself is capable of exerting. For we have seen that the film is always contracting to the utmost; and when the film is curved, this leads to a pressure directed inwards—perpendicular, that is to say, to the surface of the film. In the case of the soap-bubble, the uniform contraction of whose surface has led to its spherical form, this pressure is balanced by the pressure of the air within; and if an outlet be given for this air, then the bubble contracts with perceptible force until it stretches across the mouth of the tube, for instance across the mouth of the pipe through which we have blown the bubble. A precisely similar pressure, directed inwards, is exercised by the surface layer of a drop of water or a globule of mercury, or by the surface pellicle on a portion or "drop" of protoplasm. Only we must always remember that in the soap-bubble, or the bubble which a glass-blower blows, there is a twofold pressure as compared with that which the surface-film exercises on the drop of liquid of which it is a part; for the bubble consists (unless it be so thin as to consist of a mere layer of molecules*) of a liquid layer, with a free surface within and another without, and each of these two surfaces exercises its own independent and coequal tension and its corresponding pressure†.

If we stretch a tape upon a flat table, whatever be the tension of the tape it obviously exercises no pressure upon the table below. But if we stretch it over a *curved* surface, a cylinder for instance, it does exercise a downward pressure; and the more curved the surface the greater is this pressure, that is to say the greater is this share of the entire force of tension which is resolved in the down-

* Or, more strictly speaking, unless its thickness be less than twice the range of the molecular forces.

† It follows that the tension of a bubble, depending only on the surface-conditions, is independent of the thickness of the film.

ward direction. In mathematical language, the pressure (p) varies directly as the tension (T), and inversely as the radius of curvature (R): that is to say, $p = T/R$, per unit of surface.

If instead of a cylinder, whose curvature lies only in one direction, we take a case of curvature in two dimensions (as for instance a sphere), then the effects of these two curvatures must be added together to give the resulting pressure p: which becomes equal to $T/R + T/R'$, or

$$p = \frac{1}{R} + \frac{1}{R'}*.$$

And if in addition to the pressure p, which is due to surface-tension, we have to take into account other pressures, p', p'', etc., due to gravity or other forces, then we may say that the total pressure

$$P = p' + p'' + T\left(\frac{1}{R} + \frac{1}{R'}\right).$$

We may have to take account of the extraneous pressures in some cases, as when we come to speak of the shape of a bird's egg; but in this first part of our subject we are able for the most part to neglect them.

Our equation is an equation of equilibrium. The resistance to compression—the pressure outwards—of our fluid mass is a constant quantity (P); the pressure inwards, $T(1/R + 1/R')$, is also constant; and if the surface (unlike that of the mobile amoeba) be homogeneous, so that T is everywhere equal, it follows that $1/R + 1/R' = C$ (a constant), throughout the whole surface in question.

Now equilibrium is reached after the surface-contraction has done its utmost, that is to say when it has reduced the surface to the least possible area. So we arrive at the conclusion, from the physical side, that a surface such that $1/R + 1/R' = C$, in other words a surface which has the same *mean curvature* at all points, is equivalent to a surface of minimal area for the volume enclosed†;

* This simple but immensely important formula is due to Laplace (*Mécanique Céleste*, Bk x, suppl. *Théorie de l'action capillaire*, 1806).

† A surface may be "minimal" in respect of the area occupied, or of the volume enclosed: the former being such as the surface which a soap-film forms when it fills up a ring, whether plane or no. The geometers are apt to restrict the term "minimal surface" to such as these, or, more generally, to all cases where the mean curvature is *nil*; the others, being only minimal with respect to the volume contained, they call "surfaces of constant mean curvature."

and to the same conclusion we may also come by ways purely mathematical. The plane and the sphere are two obvious examples of such surfaces, for in both the radius of curvature is everywhere constant.

From the fact that we may extend a soap-film across any ring of wire, however fantastically the wire be bent, we see that there is no end to the number of surfaces of minimal area which may be constructed or imagined*. While some of these are very complicated indeed, others, such as a spiral or helicoid screw, are relatively simple. But if we limit ourselves to *surfaces of revolution* (that is to say, to surfaces symmetrical about an axis), we find, as Plateau was the first to shew, that those which meet the case are few in number. They are six in all, namely the plane, the sphere, the cylinder, the catenoid, the unduloid, and a curious surface which Plateau called the nodoid.

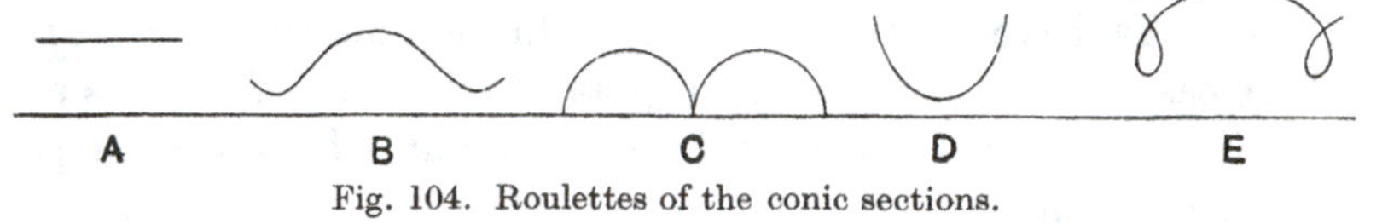

Fig. 104. Roulettes of the conic sections.

These several surfaces are all closely related, and the passage from one to another is generally easy. Their mathematical interrelation is expressed by the fact (first shewn by Delaunay†, in 1841) that the plane curves by whose revolution they are generated are themselves generated as "roulettes" of the conic sections.

Let us imagine a straight line, or axis, on which a circle, ellipse or other conic section rolls; the focus of the conic section will describe a line in some relation to the fixed axis, and this line (or roulette), when we rotate it around the axis, will describe in space one or another of the six surfaces of revolution of which we are speaking.

If we imagine an ellipse so to roll on a base-line, either of its foci will describe a sinuous or wavy line (Fig. 104, *B*) at a distance

* To fit a minimal surface to the boundary of any given closed curve in space is a problem formulated by Lagrange, and commonly known as the "problem of Plateau," who solved it with his soap-films.

† Sur la surface de révolution dont la courbure moyenne est constante, *Journ. de M. Liouville*, VI, p. 309, 1841. Cf. (*int. al.*) J. Clerk Maxwell, On the theory of rolling curves, *Trans. R.S.E.* XVI, pp. 519–540, 1849; J. K. Wittemore, Minimal surfaces of rotation, *Ann. Math.* (2), XIX, 1917, *Amer. Journ. Math.* XL, p. 69, 1918; Gino Loria, *Courbes planes spéciales, théorie et histoire*, Milan, 574 pp., 1930.

alternately maximal and minimal from the axis; this wavy line, by rotation about the axis, becomes the meridional line of the surface which we call the *unduloid*, and the more unequal the two axes are of our ellipse, the more pronounced will be the undulating sinuosity of the roulette. If the two axes be equal, then our ellipse becomes a circle; the path described by its rolling centre is a straight line parallel to the axis (A), and the solid of revolution generated therefrom will be a *cylinder*: in other words, the cylinder is a "limiting case" of the unduloid. If one axis of our ellipse vanish, while the other remains of finite length, then the ellipse is reduced to a straight line with its foci at the two ends, and its roulette will appear as a succession of semicircles touching one another upon the axis (C); the solid of revolution will be a series of equal *spheres*. If as before one axis of the ellipse vanish, but the other be infinitely long, then the roulette described by the focus of this ellipse will be a circular arc at an infinite distance; i.e. it will be a straight line normal to the axis, and the surface of revolution traced by this straight line turning about the axis will be a *plane*. If we imagine one focus of our ellipse to remain at a given distance from the axis, but the other to become infinitely remote, that is tantamount to saying that the ellipse becomes transformed into a parabola; and by the rolling of this curve along the axis there is described a catenary (D), whose solid of revolution is the *catenoid*.

Lastly, but this is more difficult to imagine, we have the case of the hyperbola. We cannot well imagine the hyperbola rolling upon a fixed straight line so that its focus shall describe a continuous curve. But let us suppose that the fixed line is, to begin with, asymptotic to one branch of the hyperbola, and that the rolling proceeds until the line is now asymptotic to the other branch, that is to say touching it at an infinite distance; there will then be mathematical continuity if we recommence rolling with this second branch, and so in turn with the other, when each has run its course. We shall see, on reflection, that the line traced by one and the same focus will be an "elastic curve," describing a succession of kinks or knots (E), and the solid of revolution described by this meridional line about the axis is the so-called *nodoid*.

The physical transition of one of these surfaces into another can be experimentally illustrated by means of soap-bubbles, or better still, after another method of Plateau's, by means of a large globule of oil, supported when necessary by wire rings, and lying in a fluid of specific gravity equal to its own.

To prepare a mixture of alcohol and water of a density precisely equal to that of the oil-globule is a troublesome matter, and a method devised by Mr C. R. Darling is a great improvement on Plateau's*. Mr Darling used the oily liquid orthotoluidene, which does not mix with water, has a beautiful and conspicuous red colour, and has precisely the same density as water when both are kept at a temperature of 24° C. We have therefore only to run the liquid into water at this temperature in order to produce beautifully spherical drops of any required size: and by adding a little salt to the lower layers of water, the drop may be made to rest or float upon the denser liquid.

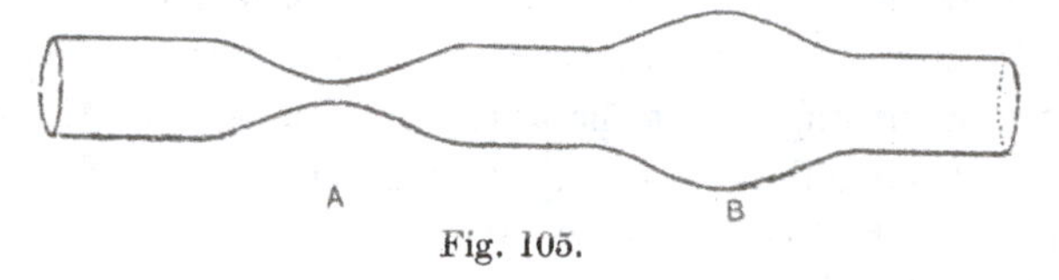

Fig. 105.

We have seen that the soap-bubble, spherical to begin with, is transformed into a plane when we release its internal pressure and let the film shrink back upon the orifice of the pipe. If we blow a bubble and then catch it up on a second pipe, so that it stretches between, we may draw the two pipes apart, with the result that the spheroidal surface will be gradually flattened in a longitudinal direction, and the bubble will be transformed into a cylinder. But if we draw the pipes yet farther apart, the cylinder narrows in the middle into a sort of hour-glass form, the increasing curvature of its transverse section being balanced by a gradually increasing *negative* curvature in the longitudinal section; the cylinder has, in turn, been converted into an unduloid. When we hold a soft glass tube in the flame and "draw it out," we are in the same identical fashion converting a cylinder into an unduloid (Fig. 105, *A*); when on the other hand we stop the end and blow, we again convert the cylinder into an unduloid (*B*), but into one which is now positively, while the former was negatively, curved. The two figures are

* See *Liquid Drops and Globules*, 1914, p. 11. Robert Boyle used turpentine in much the same way; for other methods see Plateau, *op. cit.* p. 154.

essentially the same, save that the two halves of the one change places in the other.

That spheres, cylinders and unduloids are of the commonest occurrence among the forms of small unicellular organisms or of individual cells in the simpler aggregates, and that in the processes of growth, reproduction and development transitions are frequent from one of these forms to another, is obvious to the naturalist*, and we shall deal presently with a few of these phenomena. But before we go further in this enquiry we must consider, to some small extent at least, the *curvatures* of the six different surfaces, so far as to determine what modification is required, in each case, of the general equation which applies to them all. We shall find that with this question is closely connected the question of the *pressures* exercised by or impinging on the film, and also the very important question of the limiting conditions which, from the nature of the case, set bounds to the extension of certain of the figures. The whole subject is mathematical, and we shall only deal with it in the most elementary way.

We have seen that, in our general formula, the expression $1/R + 1/R' = C$, a constant; and that this is, in all cases, the condition of our surface being one of minimal area. That is to say, it is always true for one and all of the six surfaces which we have to consider; but the constant C may have any value, positive, negative or nil.

In the case of the plane, where R and R' are both infinite, $1/R + 1/R' = 0$. The expression therefore vanishes, and our dynamical equation of equilibrium becomes $P = p$. In short, we can only have a plane film, or we shall only find a plane surface in our cell, when on either side thereof we have equal pressures or no pressure at all; a simple case is the plane partition between two equal and similar cells, as in a filament of *Spirogyra*.

In the sphere the radii are all equal, $R = R'$; they are also positive, and $T(1/R + 1/R')$, or $2T/R$, is a positive quantity, involving a constant positive pressure P, on the other side of the equation.

In the cylinder one radius of curvature has the finite and positive value R; but the other is infinite. Our formula becomes T/R, to

* They tend to reappear, no less obviously, in those precipitated structures which simulate organic form in the experiments of Leduc, Herrera and Lillie.

which corresponds a positive pressure P, supplied by the surface-tension as in the case of the sphere, but evidently of just half the magnitude.

In plane, sphere and cylinder the two principal curvatures are constant, separately and together; but in the unduloid the curvatures change from one point to another. At the middle of one of the swollen "beads" or bubbles, the curvatures are both positive; the expression $(1/R + 1/R')$ is therefore positive, and it is also finite. The film exercises (like the cylinder) a positive pressure inwards, to be compensated by an equivalent outward pressure from within. Between two adjacent beads, at the middle of one of the narrow necks, there is obviously a much stronger curvature in the transverse direction; but the total pressure is unchanged, and we now see that a negative curvature *along* the unduloid balances the increased curvature in the transverse direction. The sum of the two must remain positive as well as constant; therefore the convex or positive curvature must always be greater than the concave or negative curvature at the same point, and this is plainly the case in our figure of the unduloid.

The catenoid, in this respect a limiting case of the unduloid, has its curvature in one direction equal and opposite to its curvature in the other, this property holding good for all points of the surface; $R = -R'$; and the expression becomes

$$(1/R + 1/R') = (L/R - 1/R) = 0.$$

That is to say, the mean curvature is zero, and the catenoid, like the plane itself, has *no curvature*, and exerts no pressure. None of the other surfaces save these two share this remarkable property; and it follows that we may have at times the plane and the catenoid co-existing as parts of one and the same boundary system, just as the cylinder or the unduloid may be capped by portions of spheres. It follows also that if we stretch a soap-film between two rings, and so form an annular surface open at both ends, that surface is a catenoid: the simplest case being when the rings are parallel and normal to the axis of the figure*.

* A topsail bellied out by the wind is not a catenoid surface, but in vertical section it is everywhere a catenary curve; and Dürer shews beautiful catenary curves in the wrinkles under an Old Man's eyes. A simple experiment is to invert

The nodoid is, like the unduloid, a continuous curve which keeps altering its curvature as it alters its distance from the axis; but in this case the resultant pressure inwards is negative instead of positive. But this curve is a complicated one, and its full mathematical treatment is too hard for us.

In one of Plateau's experiments, a bubble of oil (protected from gravity by a fluid of equal density to its own) is balanced between annuli; and by adjusting the distance apart of these, it may be brought to assume the form of Fig. 106, that is to say, of a cylinder with spherical ends; there is then everywhere a pressure inwards on the fluid contents of the bubble a pressure due to the convexity

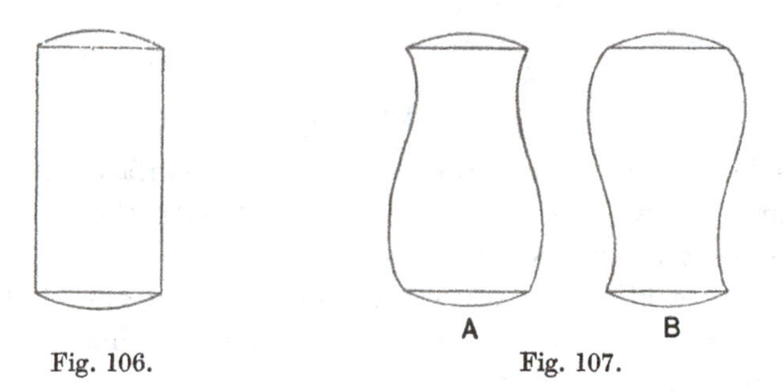

Fig. 106. Fig. 107.

of the surface film. This cylinder may be converted into an unduloid, either by drawing the rings farther apart or by abstracting some of the oil, until at length rupture ensues, and the cylinder breaks up into two spherical drops. Or again, if the surrounding liquid be made ever so little heavier or lighter than that which constitutes the drop, then gravity comes into play, the conditions of equilibrium are modified accordingly, and the cylinder becomes part of an unduloid, with its dilated portion above or below as the case may be (Fig. 107).

In all cases the unduloid, like the original cylinder, is capped by spherical ends, the sign and the consequence of a positive pressure produced by the curved walls of the unduloid. But if our initial cylinder, instead of being tall, be a flat or dumpy one

a small funnel in a large one, wet them with soap-solution, and draw them apart; the film which develops between them is a catenoid surface, set perpendicularly to the two funnels. On this and other geometrical illustrations of the fact that a soap-film sets itself at right angles to a solid boundary, see an elegant paper by Mary E. Sinclair, in *Annals of Mathematics*, VIII, 1907.

(with certain definite relations of height to breadth), then new phenomena may occur. For now, if oil be cautiously withdrawn from the mass by help of a small syringe, the cylinder may be made to flatten down so that its upper and lower surfaces become plane: which is of itself a sufficient indication that the pressure inwards is now *nil*. But at the very moment when the upper and lower surfaces become plane, it will be found that the sides curve inwards, in the fashion shewn in Fig. 108 B. This figure is a catenoid, which,

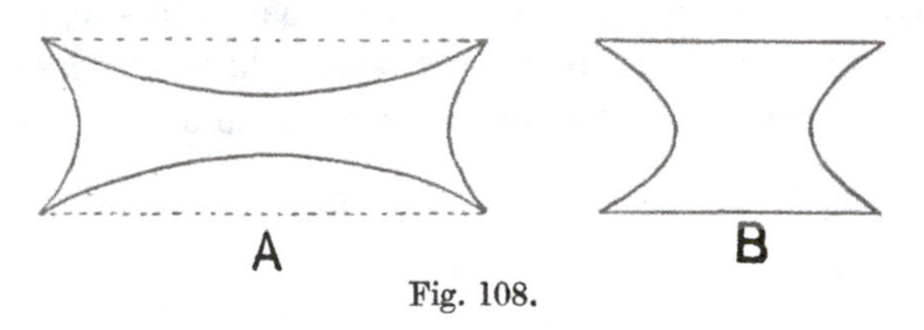

Fig. 108.

as we have seen, is, like the plane itself, a surface exercising no pressure, and which therefore may coexist with the plane as part of one and the same system.

We may continue to withdraw more oil from our bubble, drop by drop, and now the upper and lower surfaces dimple down into concave portions of spheres, as the result of the *negative* internal pressure; and thereupon the peripheral catenoid surface alters its form (perhaps, on this small scale, imperceptibly), and becomes a portion of a nodoid. It represents, in fact, that portion of the nodoid which in Fig. 109 lies between such points as O, P. While it is easy to draw the outline, or meridional section, of the nodoid, it is obvious that the solid of revolution to be derived from it can never be realised in its entirety: for one part of the solid figure would cut, or entangle with, another. All that we can ever do, accordingly, is to realise isolated portions of the nodoid*.

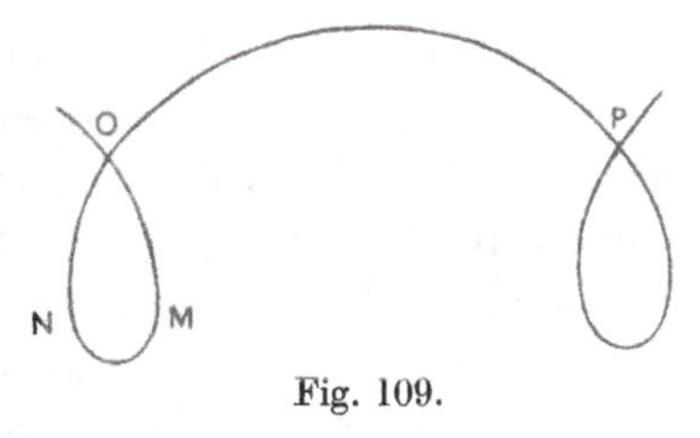

Fig. 109.

* This curve resembles the looped Elastic Curve (see Thomson and Tait, II, p. 148, fig. 7), but has its axis on the other side of the curve. The nodoid was represented upside-down in the first edition of this book, a mistake into which others have fallen, including no less a person than Clerk Maxwell, in his article "Capillarity" in the *Encycl. Brit.* 9th ed.

In all these cases the ring or annulus is not merely a means of mechanical restraint, controlling the form of the drop or bubble; it also marks the boundary, or "locus of discontinuity," between one surface and another.

If, in a sequel to the preceding experiment of Plateau's, we use solid discs instead of annuli, we may exert pressure on our oil-globule as we exerted traction before. We begin again by adjusting the pressure of these discs so that the oil assumes the form of a cylinder: our discs, that is to say, are adjusted to exercise a mechanical pressure just equal to what in the former case was supplied by the surface-tension of the spherical caps or ends of the bubble. If we now increase the pressure slightly, the peripheral walls become convexly curved, exercising a precisely corresponding pressure; the form assumed by the sides of our figure is now that of a portion of an unduloid. If we increase the pressure, the peripheral surface of oil will bulge out more and more, and will presently constitute a portion of a sphere. But we may continue the process yet further, and find within certain limits the system remaining perfectly stable. What is this new curved surface which has arisen out of the sphere, as the latter was produced from the unduloid? It is no other than a portion of a nodoid, that part which in Fig. 109 lies between M and N. But this surface, which is concave in both directions towards the surface of the oil within, is exerting a pressure upon the latter, just as did the sphere out of which a moment ago it was transformed; and we had just stated, in considering the previous experiment, that the pressure inwards exerted by the nodoid was a negative one. The explanation of this seeming discrepancy lies in the simple fact that, if we follow the outline of our nodoid curve in Fig. 109, from OP, the surface concerned in the former case, to MN, that concerned in the present, we shall see that in the two experiments the surface of the liquid is not the same, but lies on the positive side of the curve in the one case, and on the negative side in the other.

These capillary surfaces of Plateau's form a beautiful example of the "materialisation" of mathematical law. Theory leads to certain equations which determine the position of points in a system, and these points we may then plot as curves on a coordinate diagram; but a drop or a bubble may realise in an instant the whole result of our calculations, and materialise our whole apparatus of curves. Such a case is what Bacon calls a "collective instance," bearing witness to the fact that one common law is obeyed by every point or particle of the system. Where the underlying equations are unknown to us, as happens in so many natural configurations, we may still rest assured that kindred mathematical laws are being automatically followed, and rigorously obeyed, and sometimes half-revealed.

Of all the surfaces which we have been describing, the sphere is the only one which can enclose space of itself; the others can only help to do so, in combination with one another or with the sphere. Moreover, the sphere is also, of all possible figures, that which encloses the greatest volume with the least area of surface*; it is strictly and absolutely the surface of minimal area, and it is, *ipso facto*, the form which will be assumed by a unicellular organism (just as by a raindrop), if it be practically homogeneous and if, like *Orbulina* floating in the ocean, its surroundings be likewise homogeneous and its field of force symmetrical†. It is only relatively speaking that the rest of these configurations are surfaces *minimae areae*; for they are so under conditions which involve various pressures or restraints. Such restraints are imposed by the pipe or annulus which supports and confines our oil-globule or soap-bubble; and in the case of the organic cell, similar restraints are supplied by solidifications partial or complete, or other modifications local or general, of the cell-surface or cell-wall.

One thing we must not fail to bear in mind. In the case of the soap-bubble we look for stability or instability, equilibrium or non-equilibrium, in its several configurations. But the living cell is seldom in equilibrium. It is continually using or expending energy; and this ceaseless flow of energy gives rise to a "steady state," taking the place of and simulating equilibrium. In like manner the

* On the circle and sphere as giving the smallest boundary for a given content, see (e.g.) Jacob Steiner, Einfache Beweisen der isoperimetrischen Hauptsätze, *Berlin. Abhandlungen*, 1836, pp. 123–132.

† The essential conditions of homogeneity and symmetry are none too common, and a spherical organism is only to be looked for among simple things. The floating (or pelagic) eggs of fishes, the spores of red seaweeds, the oospheres of *Fucus* or *Oedogonium*, the plasma-masses escaping from the cells of *Vaucheria*, are among the instances which come to mind.

hardly changing outline of a jet or waterfall is but in pseudo-equilibrium; it is in a steady state, dynamically speaking. Many puzzling and apparent paradoxes of physiology, such (to take a single instance) as the maintenance of a constant osmotic pressure on either side of a cell-membrane, are accounted for by the fact that energy is being spent and work done, and a *steady state* or pseudo-equilibrium maintained thereby.

Before we pass to biological illustrations of our surface-tension figures we have still another matter to deal with. We have seen from our description of two of Plateau's classical experiments, that at some particular point one type of surface gives place to another; and again we know that, when we draw out our soap-bubble into a cylinder, and then beyond, there comes a certain point at which the bubble breaks in two, and leaves us with two bubbles of which each is a sphere or a portion of a sphere. In short there are certain limits to the *dimensions* of our figures, within which limits equilibrium is stable, but at which it becomes unstable, and beyond which it breaks down. Moreover, in our composite surfaces, when the cylinder for instance is capped by two spherical cups or lenticular discs, there are well-defined ratios which regulate their respective curvatures and their respective dimensions. These two matters we may deal with together.

Let us imagine a liquid drop which in appropriate conditions has been made to assume the form of a cylinder; we have already seen that its ends will be capped by portions of spheres. Since one and the same liquid film covers the sides and ends of the drop (or since one and the same delicate membrane encloses the sides and ends of the cell), we assume the surface-tension (T) to be everywhere identical; and it follows, since the internal fluid-pressure is also everywhere identical, that the expression $(1/R + 1/R')$ for the cylinder is equal to the corresponding expression, which we may call $(1/r + 1/r')$, in the case of the terminal spheres. But in the

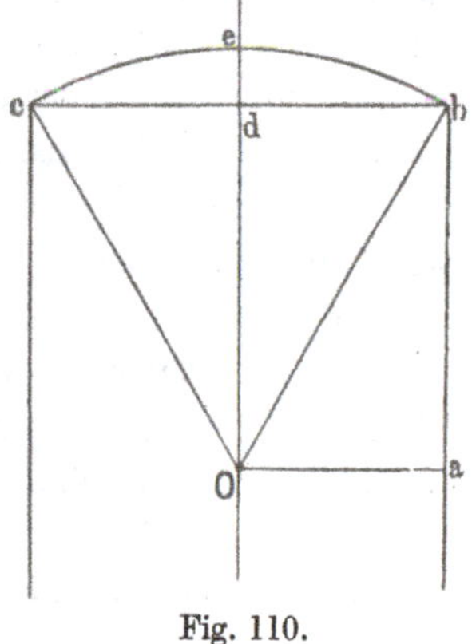

Fig. 110.

cylinder $1/R' = 0$, and in the sphere $1/r = 1/r'$. Therefore our relation of equality becomes $1/R = 2/r$, or $r = 2R$; which means that the sphere in question has just twice the radius of the cylinder of which it forms a cap.

And if Ob, the radius of the sphere, be equal to twice the radius (Oa) of the cylinder, it follows that the angle aOb is an angle of 60°, and bOc is also an angle of 60°; that is to say, the arc bc is equal to $\frac{1}{3}\pi$. In other words, the spherical disc which (under the given conditions) caps our cylinder is not a portion taken at haphazard, but is neither more nor less than that portion of a sphere which is subtended by a cone of 60°. Moreover, it is plain that the height of the spherical cap, de, $= Ob - ab = R(2 - \sqrt{3}) = 0{\cdot}27R$, where R is the radius of our cylinder, or one-half the radius of our spherical cap: in other words the normal height of the spherical cap over the end of the cylindrical cell is just a very little more than one-eighth of the diameter of the cylinder, or of the radius of the sphere. And these are the proportions which we recognise, more or less, under normal circumstances, in such a case as the cylindrical cell of *Spirogyra*, when one end is free and capped by a portion of a sphere*.

Among the many theoretical discoveries which we owe to Plateau, one to which we have just referred is of peculiar importance: namely that, with the exception of the sphere and the plane, the surfaces with which we have been dealing are only in complete equilibrium within certain dimensional limits, or in other words, have a certain definite limit of stability; only the plane and the sphere, or any portion of a sphere, are perfectly stable, because they are perfectly symmetrical, figures.

Perhaps it were better to say that their symmetry is such that any small disturbance will probably readjust itself, and leave the plane or spherical surface as it was before, while in the other configurations the chances are that a disturbance once set up will travel in one direction or another, increasing as it goes. For equilibrium and probability (as Boltzman told us) are nearly allied:

* The conditions of stability of the cylinder, and also of the catenoid, are explained with the utmost simplicity by Clerk Maxwell, in his article, already quoted, on "Capillarity." On the catenoids, see A. Terquem. *C.R.* XCII, pp. 407–9, 1881.

so nearly that that state of a system which is most likely to occur, or most likely to endure, is precisely that which we call the state of equilibrium.

For experimental demonstration, the case of the cylinder is the simplest. If we construct a liquid cylinder, either by drawing out a bubble or by supporting a globule of oil between two rings, the experiment proceeds easily until the length of the cylinder becomes just about three times as great as its diameter. But soon afterwards instability begins, and the cylinder alters its form; it narrows at

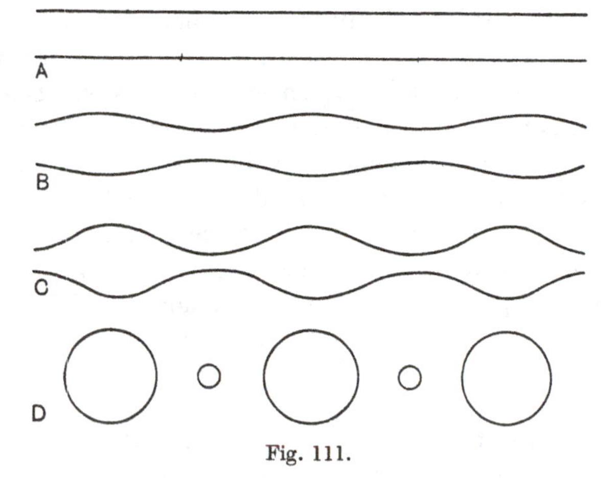

Fig. 111.

the waist, so passing into an unduloid, and the deformation progresses quickly until our cylinder breaks in two, and its two halves become portions of spheres. This physical change of one surface into another corresponds to what the mathematicians call a "discontinuous solution" of a problem of minima. The theoretical limit of stability, according to Plateau, is when the length of the cylinder is equal to its circumference, that is to say, when $L=2\pi r$, or when the ratio of length to diameter is represented by π.

The fact is that any small disturbance takes the form of a wave, and travels along the cylinder. Short waves do not affect the stability of the system; but waves whose length exceeds that of the circumference tend to grow in amplitude: until, contracting here, expanding there, the cylinder turns into a pronounced unduloid, and soon breaks into two parts or more. Thus the cylinder is a stable figure until it becomes longer than its own circumference, and then the risk of rupture may be said to begin. But Rayleigh shewed that still longer waves, leading to still greater instability, are needed to break down material resistance*. For, as Plateau knew well, his was a theoretical result, to be departed from under material conditions; it is affected largely by viscosity, and, as in the case of a flowing cylinder or jet, by inertia. When inertia plays a leading part, viscosity being small, the node of maximum instability corresponds to nearly half as much again as in the simple or theoretical case: and this result is very near to what Plateau himself had deduced from Savart's experiments on jets of water†. When the fluid is external (as when the cylinder is of air) the wave-length of maximal instability is longer still. Lastly, when viscosity is very large, and becomes paramount, then the wave-length between regions of maximal instability may become very long indeed: so that (as Rayleigh put it) "long (viscid) threads do not tend to divide themselves into drops at mutual distances comparable with the diameter of the cylinder, but rather to give way by attenuation at few and distant places." It is this that renders possible the making of long glass tubes, or the spinning of threads of "viscose" and like materials; but while these latter preserve their continuity, the principle of Plateau tends to give them something of a wavy, unduloid surface, to the great enhancement of their beauty. We are prepared, then, to find that such cylinders and unduloids as occur in organic nature seldom approach in regularity to those which theory prescribes or a soap-film may be made to shew; but rather exhibit all manner of gradations, from something exquisitely neat and regular to a coarse and distant approximation to the ideal thing‡.

The unduloid has certain peculiar properties as regards its limitations of stability, but we need mention two facts only: (1) that when the unduloid, which we produce with our soap-bubble or our

* Rayleigh, On the instability of fluid surfaces, *Sci. Papers*, III, p. 594.

† Cf. E. Tylor, *Phil. Mag.* XVI, pp. 504–518, 1933.

‡ Cf. F. Savart, Sur la constitution des veines liquides lancées par des orifices, etc., *Ann. de Chimie*, LIII, pp. 337–386, 1833. Rayleigh, On the instability of a cylinder of viscous liquid, etc., *Phil. Mag.* (5), XXXIV, 1892, or *Sci. Papers*, I, p. 361. See also Larmor, On the nature of viscid fluid threads, *Nature*, July 11, 1936, p. 74.

oil-globule, consists of the figure containing a complete constriction, then it has somewhat wide limits of stability; but (2) if it contain the swollen portion, then equilibrium is limited to the case of the figure consisting of one complete unduloid, no less nor more; that is to say when the ends of the figure are constituted by the narrowest portions, and its middle by the widest portion of the entire curve. The theoretical proof of this is difficult; but if we take the proof for granted, the fact itself will serve to throw light on what we have learned regarding the stability of the cylinder. For, when we remember that the meridional section of our unduloid is generated by the rolling of an ellipse upon a straight line in its own plane, we easily see that the length of the entire unduloid is equal to the circumference of the generating ellipse. As the unduloid becomes less and less sinuous in outline it approaches, and in time reaches, the form of the cylinder, as a "limiting case"; and *pari passu*, the ellipse which generated it passes into a circle, as its foci come closer and closer together. The cylinder of a length equal to the circumference of its generating circle is homologous to an unduloid whose length is equal to the circumference of its generating ellipse; and this is just what we recognise as constituting one complete segment of the unduloid.

The cylinder turns so easily into an unduloid, and the unduloid is capable of assuming so many graded differences of form, that we may expect to find it abundantly and variously represented among the simpler living things. For the same reason it is the very stand-by of the glass-blower, whose flasks and bottles are, of necessity, unduloids*. The blown-glass bottle is a true unduloid, and the potter's vase a close approach to an unduloid; but the alabaster bottle, turned on the lathe, is another story. It may be an imitation, or a reminiscence, of the potter's or the glass-blower's work; but it is no unduloid nor any surface of minimal area at all.

The catenoid, as we have seen, is a surface of zero pressure, and as such is unlikely to form part (unless momentarily) of the closed boundary of a cell. It forms a limiting case between unduloid and nodoid, and, were it realised, it would seldom be visibly different from the other two. In *Trichodina pediculus*, a minute infusorian para-

* Unless, that is to say, their shape be cramped and their mathematical beauty annihilated, by compression in a mould.

site of the freshwater polype, we have a circular disc bounded (apparently) by two parallel rings of cilia, with a pulley-like groove between. The groove looks very like that catenoid surface which we have produced from two parallel and opposite annuli; and the fact that the lower surface of the little creature is practically plane, where it creeps over the smooth body of the Hydra, looks like confirming the catenoid analogy. But the upper surface of the infusorian, with its ciliated "gullet," gives no assurance of a zero pressure; and we must take it that the equatorial groove of Trichodina resembles, or approaches, but is not mathematically identical with, a catenoid surface.

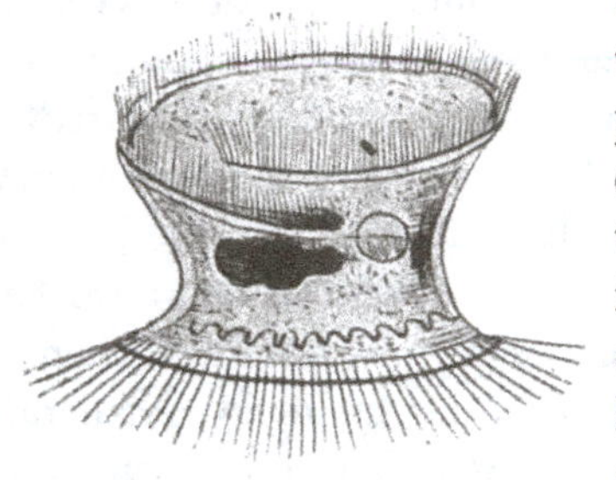

Fig. 112. *Trichodina pediculus*

While those figures of equilibrium which are also surfaces of revolution are only six in number, there is an infinite number of other figures of equilibrium, that is to say of surfaces of constant mean curvature, which are not surfaces of revolution; and it can be shewn mathematically that any given contour can be occupied by a finite portion of some one such surface, in stable equilibrium. The experimental verification of this theorem lies in the simple fact (already noted) that however we bend a wire into a closed curve, plane or not plane, we may always fill the entire area with a continuous film. No more interesting problem has ever been propounded to mathematicians as the outcome of experiment than the general problem so to describe a minimal surface passing through a closed contour; and no complete solution, no general method of approach, has yet been discovered*.

Of the regular figures of equilibrium, or surfaces of constant mean curvature, apart from the surfaces of revolution which we have discussed, the helicoid spiral is the most interesting to the biologist.

* Partial solutions, closely connected with recent developments of mathematical analysis, are due to Riemann, Weierstrass and Schartz. Cf. (*int. al.*) G. Darboux, *Théorie des surfaces*, 1914, pp. 490–601; T. Bonneson, *Problèmes des isopérimètres et des isépiphanes*, Paris, 1929; Hilbert's *Anschauliche Geometrie*, 1932, p. 237 *seq.*; a good account also in G. A. Bliss's *Calculus of Variations*, Chicago, 1925. See also (*int. al.*) Tibor Rado, *Mathem. Ztschr.* XXXII, 1930; Jesse Douglas, *Amer. Math. Journ.* XXXIII, 1931, *Journ. Math. Phys.* XV, 1936.

This is a helicoid generated by a straight line perpendicular to an axis, about which it turns at a uniform rate, while at the same time it slides, also uniformly, along this same axis. At any point in this surface, the curvatures are equal and of opposite sign, and the sum of the curvatures is accordingly nil. Among what are called "ruled surfaces," or surfaces capable of being defined by a system of stretched strings*, the plane and the helicoid are the only two whose mean curvature is null, while the cylinder is the only one whose curvature is finite and constant. As this simplest of helicoids corresponds, in three dimensions, to what in two dimensions is merely a plane (the latter being generated by the rotation of a straight line about an axis without the superadded gliding motion which generates the helicoid), so there are other and much more complicated helicoids which correspond to the sphere, the unduloid and the rest of our figures of revolution, the generating planes of these latter being supposed to wind spirally about an axis. In the case of the cylinder it is obvious that the resulting figure is indistinguishable from the cylinder itself. In the case of the unduloid we obtain a grooved spiral, and we meet with something very like it in nature (for instance in Spirochaetes, *Bodo gracilis*, etc.); but in point of fact, the screw motion given to an unduloid or catenary curve fails to give a minimal screw surface, as we might have expected it to do.

The foregoing considerations deal with a small part only of the theory of surface-tension, or capillarity: with that part, namely, which relates to the surfaces capable of subsisting in equilibrium under the action of that force, either of itself or subject to certain simple constraints. And as yet we have limited ourselves to the case of a single surface, or of a single drop or bubble, leaving to another occasion a discussion of the forms assumed when such drops or vesicles meet and combine together. In short, what we have said may help us to understand the form of a *cell*—considered, as with certain limitations we may legitimately consider it, as a liquid drop or liquid vesicle; the conformation of a *tissue* or cell-aggregate must be dealt with in the light of another series of theoretical considerations. In both cases, we can do no more than touch on the fringe of a large and difficult subject. There are many forms

* Or rather, surfaces such that through every point there runs a straight line which lies wholly in the surface.

capable of realisation under surface-tension, and many of them doubtless to be recognised among organisms, which we cannot deal with in this elementary account. The subject is a very general one; it is, in its essence, more mathematical than physical; it is part of the mathematics of surfaces, and only comes into relation with surface-tension because this physical phenomenon illustrates and exemplifies, in a concrete way, the simple and symmetrical conditions with which the mathematical theory is capable of dealing. And before we pass to illustrate the physical phenomena by biological examples, we must repeat that the simple physical conditions which we presuppose will never be wholly realised in the organic cell. Its substance will never be a perfect fluid, and hence equilibrium will be slowly reached; its surface will seldom be perfectly homogeneous, and therefore equilibrium will seldom be perfectly attained; it will very often, or generally, be the seat of other forces, symmetrical or unsymmetrical; and all these causes will more or less perturb the surface-tension effects*. But we shall find that, on the whole, these effects of surface-tension though modified are not obliterated nor even masked; and accordingly the phenomena to which I have devoted the foregoing pages will be found manifestly recurring and repeating themselves among the phenomena of the organic cell.

In a spider's web we find exemplified several of the principles of surface-tension which we have now explained. The thread is spun out of a glandular secretion which issues from the spider's body as a semi-fluid cylinder, the force of expulsion giving it its length and that of surface-tension giving it its circular section. It is too viscid, and too soon hardened on exposure to the air, to break up into drops or spherules; but it is otherwise with another sticky secretion which, coming from another gland, is simultaneously poured over the

* That "every particular that worketh any effect is a thing compounded more or less of diverse single natures, more manifest and more obscure" is a point made and dwelt on by Bacon. Of the same principle a great astronomer speaks as follows: "It is one of the fundamental characteristics of natural science that we *never get beyond an approximation*...Nature *never* offers us simple and undivided phenomena to observe, but always infinitely complex compounds of many different phenomena. Each single phenomenon can be described mathematically in terms of the accepted fundamental laws of Nature:...but we can never be sure that we have carried the analysis to its full exhaustion, and have isolated one single simple phenomenon." W. de Sitter, in *Nature*, Jan. 21, 1928, p. 99.

slacker cross-threads as they issue to form the spiral portion of the web. This latter secretion is more fluid than the first, and only dries up after several hours*. By capillarity it "wets" the thread, spreading over it in an even film or liquid cylinder. As such it has its limits of stability, and tends to disrupt at points more distant than the theoretical wave-length, owing to the imperfect fluidity of the viscous film and still more to the frictional drag of the inner thread with which it is in contact. Save for this qualification the cylinder disrupts in the usual manner, passing first into the wavy outline of an unduloid, whose swollen internodes swell more and more till the necks between them break asunder, and leave a row of spherical drops or beads strung like dewdrops at regular intervals along the thread. If we try to varnish a thin taut wire we produce automatically the same identical result†; unless our varnish be such as to dry almost instantaneously it gathers into beads, and do what we will we fail to spread it smooth. It follows that, according to the drying qualities of our varnish, the process may stop at any point short of the formation of perfect spherules; and as our final stage we may only obtain half-formed beads or the wavy outlines of an unduloid. The beads may be helped to form by jerking the stretched thread, and so disturbing the unstable equilibrium of the viscid cylinder. This the spider has been said to do, but Dr G. T. Bennett assures me that she does nothing of the kind. She only draws her thread out a little, and leaves it a trifle slack; if the gum should break into droplets, well and good, but it matters little. The web with its sticky threads is not improved thereby. Another curious phenomenon here presents itself.

In Plateau's experimental separation of a cylinder of oil into two spherical halves, it was noticed that, when contact was nearly broken, that is to say when the narrow neck of the unduloid had become very thin, the two spherical bullae, instead of absorbing the fluid out of the narrow neck into themselves as they had done with the preceding portion, drew out this small remaining part of

* When we see a web bespangled with dew of a morning, the dewdrops are not drops of pure water, but of water mixed with the sticky, gummy fluid of the cross-threads; the radii seldom if ever shew dewdrops. See F. Strehlke, Beobachtungen an Spinnengewebe, *Poggendorff's Annalen*, XL, p. 146, 1937.

† Felix Plateau recommends the use of a weighted thread or plumb-line, to be drawn up slowly out of a jar of water or oil; *Phil. Mag.* XXXIV, p. 246, 1867.

the liquid into a thin thread as they completed their spherical form and receded from one another: the reason being that, after the thread or "neck" has reached a certain tenuity, internal friction prevents or retards a rapid exit of the fluid from the thread to the adjacent spherule. It is for the same reason that we are able to draw a glass rod or tube, which we have heated in the middle, into a long and uniform cylinder or thread by quickly separating the two ends. But in the case of the glass rod the long thin thread quickly cools and solidifies, while in the ordinary separation of a liquid cylinder the corresponding intermediate cylinder remains liquid; and therefore, like any other liquid cylinder, it is liable to

Fig. 113. Dew-drops on a spider's web.

break up, provided that its dimensions exceed the limit of stability. And its length is generally such that it breaks at two points, thus leaving two terminal portions continuous and confluent with the spheres, and one median portion which resolves itself into a tiny spherical drop, midway between the original and larger two. Occasionally, the same process of formation of a connecting thread repeats itself a second time, between the small intermediate spherule and the large spheres; and in this case we obtain two additional spherules, still smaller in size, and lying one on either side of our first little one. This whole phenomenon, of equal and regularly interspaced beads, often with little beads regularly interspaced between the larger ones, and now and then with a third order of still smaller beads regularly intercalated, may be easily observed in a spider's web, such as that of *Epeira*, very often with beautiful regularity—sometimes interrupted and disturbed by a slight want of homogeneity in the secreted fluid; and the same phenomenon is repeated on a grosser scale when the web is bespangled with dew, and its threads bestrung with pearls innumerable. To the older naturalists, these regularly arranged and beautifully formed globules on the spider's web were a frequent source of wonderment. Blackwall, counting some twenty globules in a tenth of an inch, calculated that a large garden-spider's web should comprise about 120,000 globules; the net was spun and finished in about forty minutes, and Blackwall was filled with admiration of the skill and quickness with which the spider manufactured these little beads. And no wonder, for according to the above estimate they had to be made at the rate of about 50 per second*.

Here we see exemplified what Plateau told us of the law of minimal areas transforming the cylinder into the unduloid and disrupting it

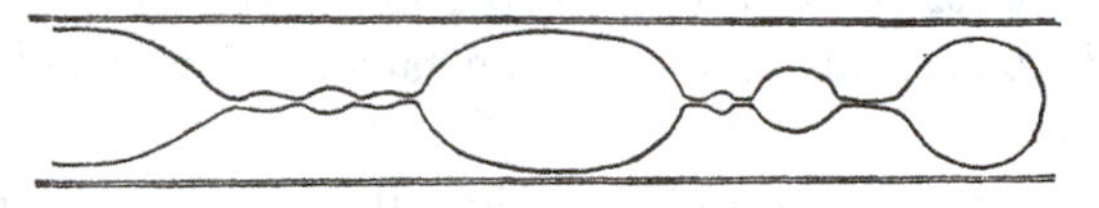

Fig. 114. Root-hair of *Trianea*, in glycerine. After Berthold.

into spheres. The little delicate beads which stud the long thin pseudopodia of a foraminifer, such as *Gromia*, or which appear in like manner on the film of protoplasm coating the long radiating spicules of *Globigerina*, represent an identical phenomenon. Indeed we may study in a protoplasmic filament the whole process of formation of such beads: if we squeeze out on a slide the viscid contents of a mistletoe-berry, the long sticky threads into which the substance runs shew the whole phenomenon particularly well. True, many long cylindrical cells, such as are common in plants, shew no sign of beading or disruption; but here the cell-walls are never fluid but harden as they grow, and the protoplasm within is kept in place and shape by its contact with the cell-wall. It was noticed many years ago by Hofmeister†, and afterwards explained by Berthold, that if we dip the long root-hairs of certain water-plants, such as *Hydrocharis* or *Trianea*, in a denser fluid (a little sugar-solution or

* J. Blackwall, *Spiders of Great Britain* (Ray Society), 1859, p. 10; *Trans. Linn. Soc.* XVI, p. 477, 1833. On the strength and elasticity of the spider's web, see J. R. Benton, *Amer. Journ. Science*, XXIV, pp. 75–78, 1907.

† *Lehrbuch von der Pflanzenzelle*, p. 71; cf. Nägeli, *Pflanzenphysiologische Untersuchungen* (*Spirogyra*), III, p. 10.

dilute glycerine), the cell-sap tends to diffuse outwards, the protoplasm parts company with its surrounding and supporting wall, and then lies free as a protoplasmic cylinder in the interior of the cell. Thereupon it soon shews signs of instability, and commences to disrupt; it tends to gather into spheres, which however, as in our illustration, may be prevented by their narrow quarters from assuming the complete spherical form; and in between these spheres, we have more or less regularly alternate ones, of smaller size*. We could not wish for a better or a simpler proof of the *essential fluidity* of the protoplasm†. Similar, but less regular, beads or droplets may be caused to appear, under stimulation by an alternating current, in the protoplasmic threads within the living cells of the hairs of *Tradescantia*; the explanation usually given is, that the viscosity of the protoplasm is reduced, or its fluidity increased; but an increase of the surface-tension would seem a more likely reason‡.

In one of Robert Chambers's delicate experiments, a filament of protoplasm is drawn off, by a micro-needle, from the fluid surface of a starfish-egg. If drawn too far it breaks, and part returns within the protoplasm while the other rounds itself off on the needle's point. If drawn out less far, it looks like a row of beads or chain of droplets; if yet more relaxed, the droplets begin to fuse until the whole filament is withdrawn; if drawn out anew the process repeats itself. The whole story is a perfect description of the behaviour of a fluid jet or cylinder, of varying length and thickness§.

We may take note here of a remarkable series of phenomena, which, though they seem at first sight to be of a very different order,

* The intermediate spherules appear with great regularity and beauty whenever a liquid jet breaks up into drops. So a bursting soap-bubble scatters a shower of droplets all around, sometimes all alike, but often with a beautiful alternation of great and small. How the breaking up of thread or jet into drops may be helped, regularised, and sometimes complicated, by external vibrations is another and by no means unimportant story.

† Though doubtless to speak of the viscid thread as a fluid is but a first approximation; cf. Larmor, in *Nature*, July 11, 1936.

‡ Kühne, *Untersuchungen über das Protoplasma*, 1864, p. 75, etc.

§ Cf. R. Chambers in *Colloid Chemistry, theoretical and applied*, II, cap. 24, 1928; also *Ann. de Physiol.* VI, p. 234, 1930; etc.

are closely related to those which attend and which bring about the breaking-up of a liquid cylinder or thread.

In Mr Worthington's beautiful experiments on splashes*, it was found that the fall of a round pebble into water from a height first formed a dip or hollow in the surface, and then caused a filmy "cup" of water to rise up all round, opening out trumpet-fashion

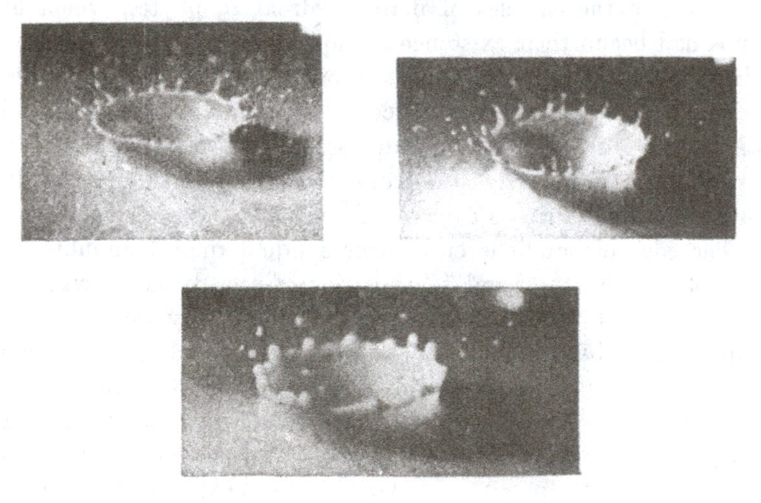

Fig. 115. Phases of a splash. From Worthington.

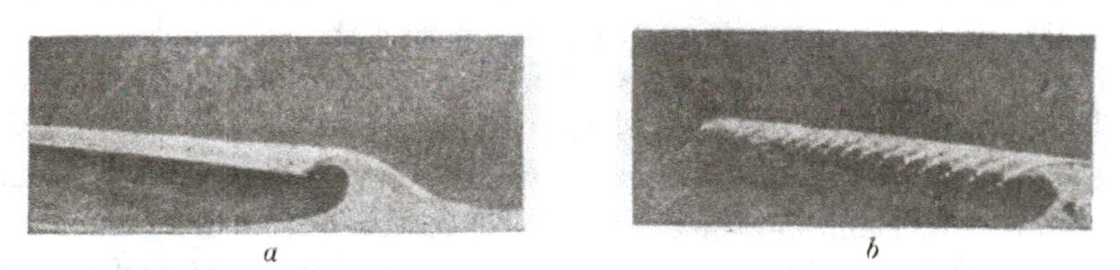

a b

Fig. 116. A wave breaking into spray.

or closing in like a bubble, according to the height from which the pebble fell. The cup or "crater" tends to be fluted in alternate ridges and grooves, its edges get scolloped into corresponding lobes and notches, and the projecting lobes or prominences tend to break off or break up into drops or beads (Fig. 115). A similar appearance is seen on a great scale in the edge of a breaking wave: for the smooth

* *A Study of Splashes*, 1908, p. 38, etc.; also various papers in *Proc. R.S.* 1876–1882, and *Phil. Trans.* (A), 1897 and 1900.

edge becomes notched or sinuous, and the surface near by becomes ribbed or fluted, owing to the internal flow being helped here and hindered there by a viscous shear; and then all of a sudden the uneven edge shoots out an array of tiny jets, which break up into the countless droplets which constitute "spray" (Fig. 116). The naturalist may be reminded also of the beautifully symmetrical notching of the calycles of many hydroid zoophytes, which little cups had begun their existence as liquid or semi-liquid films before they became stiff and rigid. The next phase of the splash (with which we are less directly concerned) is that the crater subsides, and where it stood a tall column rises up, which also tends, if it be tall enough, to break up into drops. Lastly the column sinks down in its turn, and a ripple runs out from where it stood.

The edge of our little cup forms a liquid ring or annulus, comparable on the one hand to the edge of an advancing wave, and on the other to a liquid thread or cylinder if only we conceive the thread to be bent round into a ring; and accordingly, just as the thread segments first into an unduloid and then into separate spherical drops, so likewise will the edge of cup or annulus tend to do. This phase of notching, or beading, of the edge of the splash is beautifully seen in many of Worthington's experiments*, and still more beautifully in recent work (Frontispiece†). In the second place the fact that the crater rises up means that liquid is flowing in from below; the segmentation of the rim means that channels of easier flow are being created, along which the liquid is led or driven into the protuberances; and these last are thereby exaggerated into the jets or streams which become conspicuous at the edge of the crater. In short any film or film-like fluid or semi-fluid cup will be unstable; its instability will tend to show itself in a fluting of the surface and a notching of the edge; and just such a fluting and notching are conspicuous features of many minute organic cup-like structures. In the hydroids (Fig. 117), we see that these common features of the

* Cf. *A Study of Splashes*, pp. 17, 77. The same phenomenon is often well seen in the splash of an oar. It is beautifully and continuously evident when a strong jet of water from a tap impinges on a curved surface and then shoots off again.

† We owe this picture to the kindness of Mr Harold E. Edgerton, of the Massachusetts Institute of Technology. It shews the splash caused by a drop falling into a thin layer of milk; a second drop of milk is seen above, following the first. The exposure-time was 1/50,000 of a second.

The latter phase of a splash: the crater has subsided, a columnar jet has risen up, and the jet is dividing into droplets. From Harold E. Edgerton, Massachusetts Technical Institution

cup and the annulation of the stem are phenomena of the same order. A cord-like thickening of the edge of the cup is a variant of the same order of phenomena; it is due to the checking at the rim of the flow of liquid from below, and a similar thickening is to be seen, not only in some hydroid calycles but also in many Vorticellae (cf. Fig. 124) and other cup-shaped organisms. And these are by no means the only manifestations of surface-tension in a splash which shew resemblances and analogies to organic form*.

The phenomena of an ordinary liquid splash are so swiftly transitory that their study is only rendered possible by photography:

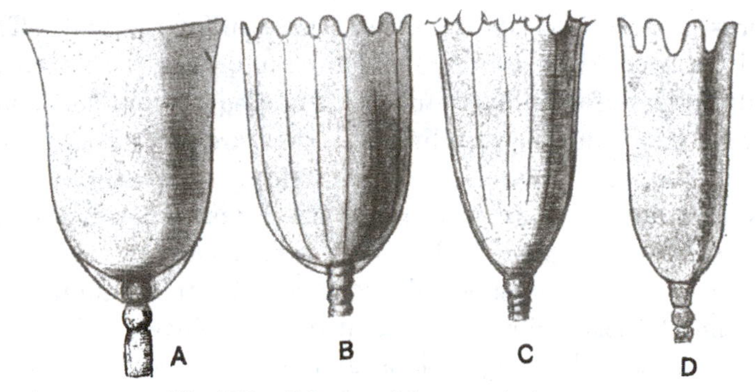

Fig. 117. Calycles of Campanularia spp.

but this excessive rapidity is not an essential part of the phenomenon. For instance, we can repeat and demonstrate many of the simpler phenomena, in a permanent or quasi-permanent form, by splashing water on to a surface of dry sand†, or by firing a bullet into a soft metal target. There is nothing, then, to prevent a slow and lasting manifestation, in a viscous medium such as a protoplasmic organism, of phenomena which appear and disappear with

* The same phenomena are modified in various ways, and the drops are given off much more freely, when the splash takes place in an electric field—all owing to the general instability of an electrified liquid surface; and a study of this aspect of the subject might suggest yet more analogies with organic form. Cf. J. Zeleny, *Phys. Rev.* x, 1917; J. P. Gott, *Proc. Cambridge Philos. Soc.* xxxi, 1935; etc.

† We find now and then in certain brick-clays of glacial origin, hard, quoit-shaped rings, each with an equally indurated, round or flattened ball resting on it. These may be precisely imitated by splashing large drops of water on a smooth surface of fine dry sand. The ring corresponds, apparently, to the crater of the splash, and the ball (or its water content) to the pillar rising in the middle.

evanescent rapidity in a more mobile liquid. Nor is there anything peculiar in the splash itself; it is simply a convenient method of setting up certain motions or currents, and producing certain surface-forms, in a liquid medium—or even in such an imperfect fluid as a bed of sand. Accordingly, we have a large range of possible conditions under which the organism might conceivably display configurations analogous to, or identical with, those which Mr Worthington has shewn us how to exhibit by one particular experimental method.

To one who has watched the potter at his wheel, it is plain that the potter's thumb, like the glass-blower's blast of air, depends for its efficacy upon the physical properties of the clay or "slip" it works on, which for the time being is essentially a fluid. The cup and the saucer, like the tube and the bulb, display (in their simple and primitive forms) beautiful surfaces of equilibrium as manifested under certain limiting conditions. They are neither more nor less than glorified "splashes," formed slowly, under conditions of restraint which enhance or reveal their mathematical symmetry. We have seen, and we shall see again before we are done, that the art of the glass-blower is full of lessons for the naturalist as also for the physicist: illustrating as it does the development of a host of mathematical configurations and organic conformations which depend essentially on the establishment of a constant and uniform pressure within a *closed* elastic shell or fluid envelope or bubble. In like manner the potter's art illustrates the somewhat obscurer and more complex problems (scarcely less frequent in biology) of a figure of equilibrium which is an *open* surface of revolution. The two series of problems are closely akin; for the glass-blower can make most things which the potter makes, by cutting off *portions* of his hollow ware; besides, when this fails and the glass-blower, ceasing to blow, begins to use his rod to trim the sides or turn the edges of wineglass or of beaker, he is merely borrowing a trick from the still older craft of the potter.

It would seem venturesome to extend our comparison with these liquid surface-tension phenomena from the cup or calycle of the hydrozoon to the little hydroid polyp within: and yet there is something to be learned by such a comparison. The cylindrical body of the tiny polyp, the jet-like row of tentacles, the beaded annulations which these tentacles exhibit, the web-like film which sometimes (when they stand a little way apart) conjoins their bases, the thin annular film of tissue which surrounds the little organism's mouth, and the manner in which this annular "peristome" contracts*, like a shrinking soap-bubble, to close the aperture, are every one of them features to which we may find a singular and striking parallel in the surface-tension phenomena of the splash†.

Some seventy years ago much interest was aroused by Helmholtz's work (and also Kirchhoff's) on "discontinuous motions of a fluid‡"; that is to say, on the movements of one body of fluid within another, and the resulting phenomena due to friction at the surfaces between. What Kelvin§ called Helmholtz's "admirable discovery of the law of vortex-motion in a perfect fluid" was the chief result of this investigation; and was followed by much experimental work, in order to illustrate and to extend the mathematical conclusions.

The drop, the bubble and the splash are parts of a long story; and a "falling drop," or a drop moving through surrounding fluid, is a case deserving to be considered. A drop of water, tinged with fuchsin, is gently released (under a pressure of a couple of millimetres) at the bottom of a glass of water‖. Its momentum enables it to rise through a few centimetres of the surrounding water, and in doing so it communicates motion to the water around. In front the rising drop *thrusts* its way through, almost like a solid body; behind it tends to *drag* the surrounding water after it, by fluid friction¶; and these two motions together give rise to beautiful verticoid configurations, the *Strömungspilze* or *Tintenpilze* of their first discoverers (Fig. 119). Under a higher and more continuous pressure

* See a *Study of Splashes*, p. 54.

† There is little or no difference between a *splash* and a *burst bubble*. The craters of the moon have been compared with, and explained by, both of these.

‡ Helmholtz, in *Berlin. Monatsber.* 1868, pp. 215–228; Kirchhoff, in *Crelle's Journal*, LXX, pp. 289–298, LXXI, 237–273, 1869–70.

§ W. Thomson, in *Proc. R.S.E.* VI, p. 94, 1867.

‖ See A. Overbeck, Ueber discontinuirliche Flüssigkeitsbewegungen, *Wiedemann's Annalen*, II, 1877; W. Bezold, Ueber Strömungsfiguren in Flüssigkeiten, *ibid.* XXIV, pp. 569–593, 1885; P. Czermak, *ibid.* L, p. 329, 1893; etc.

¶ The frictional drag on the hinder part of the drop is felt alike in the ship, the bird and the aeroplane, and tends to produce retarding vortices in them all. It is always minimised in one way or another, and it is automatically minimised in the present instance, as the drop thins off and tapers down.

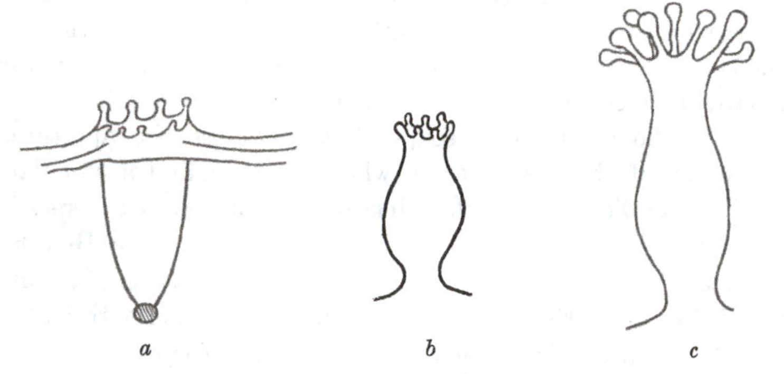

Fig. 118. *a*, *b*. More phases of a splash, after Worthington. *c*. A hydroid polype, after Allman.

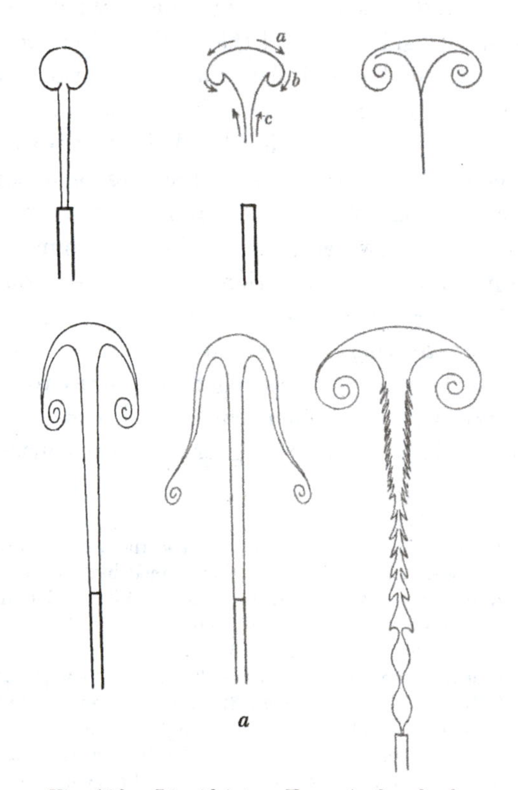

Fig. 119. Liquid jets. From A. Overbeck.

the drop becomes a jet; the form of the vortex is modified thereby, and may be further modified by slight differences of temperature (i.e. of density), or by interrupting the rate of flow. To let a drop of ink fall into water is a simple and most beautiful experiment*. The effect is more violent than in the former case. The descending

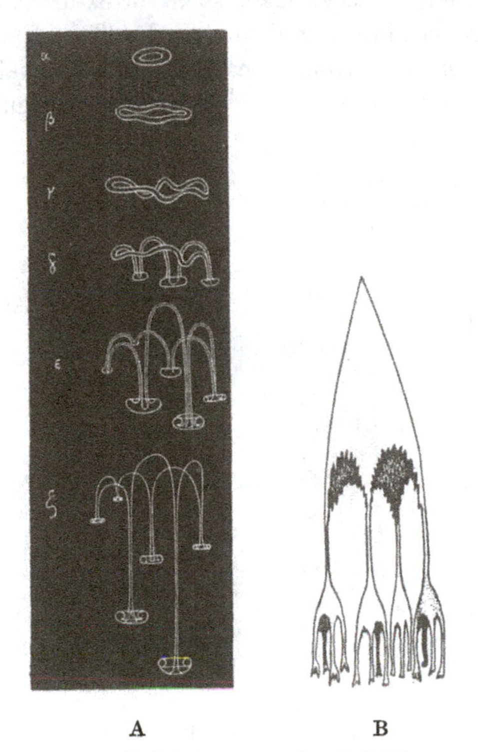

Fig. 120. Falling drops. A, ink in water, after J. J. Thomson and Newall. B, fusel oil in paraffin, after Tomlinson.

drop turns into a complete vortex-ring; it expands and attenuates; it waves about, and the descending loops again turn into incipient vortices (Fig. 120).

Lastly, instead of letting our drop rise or fall freely, we may use a hanging drop, which, while it sinks, remains suspended to the surface. Thus it cannot form a complete annulus, but only a

* J. J. Thomson and H. F. Newall, On the formation of vortex-rings by drops, *Proc. R.S.* XXXIX, pp. 417–436, 1885. Emil Hatschek, On forms assumed by a gelatinising liquid in various coagulating solutions, *ibid.* (A) XCIV, pp. 303–316, 1918.

partial vortex suspended by a thread or column—just as in Overbeck's jet-experiments; and the figure so produced, in either case, is closely analogous to that of a medusa or jellyfish, with its bell or "umbrella," and its clapper or "manubrium" as well. Some years ago Emil Hatschek made such vortex-drops as these of liquid gelatine dropped into a hardening fluid. These "artificial medusae" sometimes show a symmetrical pattern of radial "ribs", due to shrinkage, and this to dehydration by the coagulating fluid. An

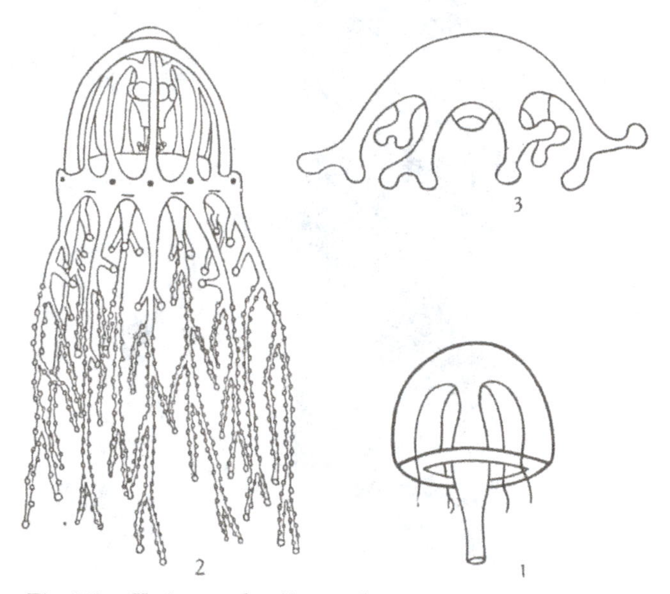

Fig. 121. Various medusoids: 1, *Syncoryne*; 2, *Cordylophora*; 3, *Cladonema* (after Allman).

extremely curious result of Hatschek's experiments is to shew how sensitive these vorticoid drops are to physical conditions. For using the same gelatine all the while, and merely varying the density of the fluid in the third decimal place, we obtain a whole range of configurations, from the ordinary hanging drop to the same with a ribbed pattern, and then to medusoid vortices of various graded forms.

The living medusa has a geometrical symmetry so marked and regular as to suggest a physical or mechanical element in the little creature's growth and construction. It has, to begin with, its vortex-like bell or umbrella, with its cylindrical handle or manubrium. The bell is traversed by radial canals, four or in multiples of four; its edge is beset with tentacles, smooth or often beaded, at regular intervals and of graded sizes; and certain sensory structures, including solid concretions or "otoliths," are also symmetrically interspaced. No sooner made, than it begins to pulsate; the little bell begins to "ring."

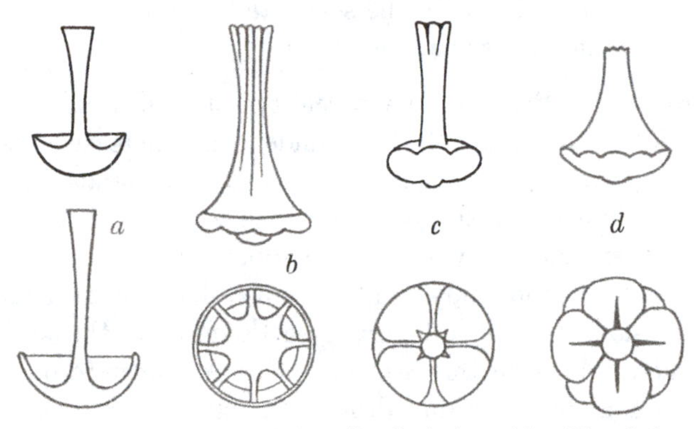

Fig. 121*b*. "Medusoid drops", of gelatin. After Hatschek.

Buds, miniature replicas of the parent-organism, are very apt to appear on the tentacles, or on the manubrium or sometimes on the edge of the bell; we seem to see one vortex producing others before our eyes. The development of a medusoid deserves to be studied without prejudice, from this point of view. Certain it is that the tiny medusoids of *Obelia*, for instance, are budded off with a rapidity and a complete perfection which suggest an automatic and all but instantaneous act of conformation, rather than a gradual process of growth.

Moreover, not only do we recognise in a vorticoid drop a "schema" or analogue of medusoid form, but we seem able to discover various actual phases of the splash or drop in the all but innumerable living types of jellyfish; in *Cladonema* we seem to see an early stage of a breaking drop, and in *Cordylophora* a beautiful picture of incipient vortices. It is hard indeed to say how much or little all these analogies imply. But they indicate, at the very least, how certain simple organic forms might be naturally assumed by one fluid mass within another, when gravity, surface tension and fluid friction play

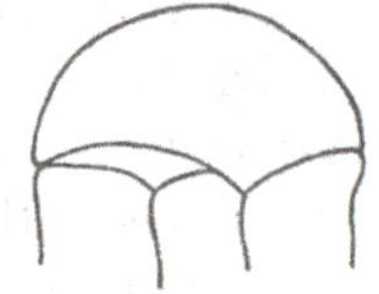

Fig. 122. *Medusachloris*, a ciliate infusoria.

their part, under balanced conditions of temperature, density and chemical composition.

A little green infusorian from the Baltic Sea is, as near as may be, a medusa in miniature*. It is curious indeed to find the same medusoid, or as we may now call it vorticoid, configuration occurring in a form so much lower in the scale, and so much less in order of magnitude, than the ordinary medusae.

According to Plateau, the viscidity of the liquid, while it retards the breaking up of the cylinder and increases the length of the segments beyond that which theory demands, has nevertheless less influence in this direction than we might have expected. On the other hand any external support or adhesion, or mere contact with a solid body, will be equivalent to a reduction of surface-tension and so will very greatly increase the stability of our cylinder. It is for this reason that the mercury in our thermometers seldom separates into drops: though it sometimes does so, much to our inconvenience. And again it is for this reason that the protoplasm in a long tubular or cylindrical cell need not divide into separate cells and internodes until the length of these far exceeds the theoretical limits.

An interesting case is that of a viscous drop immersed in another viscous fluid, and drawn out into a thread by a shearing motion of the latter. The thread seems stable at first, but when left to rest it breaks up into drops of a very definite and uniform size, the size of the drops, or wave-length of the unduloid of which they are made, depending on the relative viscosities of the two threads†.

Plateau's results, though discovered by way of experiment and though (as we have said) they illustrate the "materialisation" of mathematical law, are nevertheless essentially theoretical results approached rather than realised in material systems. That a liquid cylinder begins to be unstable when its length exceeds $2\pi r$ is all but mathematically true of an all but immaterial soap-bubble; but very far from true, as Plateau himself was well aware, in a flowing jet, retarded by viscosity and by inertia. The principle is true and universal; but our living cylinders do not follow the abstract laws

* *Medusachloris phiale*, of A. Pascher, *Biol. Centralbl.* XXXVII, pp. 421–429, 1917.

† See especially Rayleigh, *Phil. Mag.* XXXIV, p. 145, 1892, by whom the subject is carried much further than where Plateau left it. See also (*int. al.*) G. I. Taylor, *Proc. R.S.* (A), CXLVI, p. 501, 1934; S. Tomotika, *ibid.* CL, p. 322, 1935; etc.

of mathematics, any more than do the drops and jets of ordinary fluids or the quickly drawn and quickly cooling tubes in the glass-worker's hands.

Plateau says that in most liquids the influence of viscosity is such as to cause the cylinder to segment when its length is about four times, or even six times, its diameter, instead of a fraction over three times, as theory would demand of a perfect fluid. If we take it at four times, the resulting spherules would have a diameter of about 1·8 times, and their distance apart would be about 2·2 times, the original diameter of the cylinder; and the calculation is not difficult which would shew how these dimensions are altered in the case of a cylinder formed around a solid core, as in the case of a spider's web. Plateau also observed that the *time* taken in the division of the cylinder is directly proportional to its diameter, while varying with the nature of the liquid. This question, of the time taken in the division of a cell or filament in relation to its dimensions, has not so far as I know been enquired into by biologists.

From the simple fact that the sphere is of all configurations that whose surface-area for a given volume is an absolute minimum, we have seen it to be the one figure of equilibrium assumed by a drop or vesicle when no disturbing factor is at hand; but such freedom from counter-influences is likely to be rare, and neither does the rain-drop nor the round world itself retain its primal sphericity. For one thing, gravity will always be at hand to drag and distort our drop or bubble, unless its dimensions be so minute that gravity becomes insignificant compared with capillarity. Even the soap-bubble will be flattened or elongated by gravity, according as we support it from below or from above; and the bubble which is thinned out almost to blackness will, from its small mass, be the one which remains most nearly spherical*.

Innumerable new conditions will be introduced, in the shape of complicated tensions and pressures, when one drop or bubble becomes associated with another, and when a system of intermediate films or partition-walls is developed between them. This subject we shall discuss later, in connection with cell-aggregates or tissues, and we shall find that further theoretical considerations are

* Cf. Dewar, On soap-bubbles of long duration, *Proc. Roy. Inst.* Jan. 19, 1929.

needed as a preliminary to any such enquiry. Meanwhile let us consider a few cases of the forms of cells, either solitary, or in such simple aggregates that their individual form is little disturbed thereby. Let us clearly understand that the cases we are about to consider are those where the perfect symmetry of the sphere is replaced by another symmetry, less complete, such as that of an ellipsoidal or cylindrical cell. The cases of asymmetrical deformation or displacement, such as are illustrated in the production of a bud or the development of a lateral branch, are much simpler; for here we need only assume a slight and localised variation of surface-tension, such as may be brought about in various ways through the heterogeneous chemistry of the cell. But such diffused and graded asymmetry as brings about for instance the ellipsoidal shape of a yeast-cell is another matter.

If the sphere be the one surface of complete symmetry and therefore of independent equilibrium, it follows that in every cell which is otherwise conformed there must be some definite cause of its departure from sphericity; and if this cause be the obvious one of resistance offered by a solidified envelope, such as an egg-shell or firm cell-wall, we must still seek for the deforming force which was in action to bring about the given shape prior to the assumption of rigidity. Such a cause may be either external to, or may lie within, the cell itself. On the one hand it may be due to external pressure or some form of mechanical restraint, as when we submit our bubble to the partial restraint of discs or rings or more complicated cages of wire; on the other hand it may be due to intrinsic causes, which must come under the head either of differences of internal pressure, or of lack of homogeneity or isotropy in the surface or its envelope*.

* A case which we have not specially considered, but which may be found to deserve consideration in biology, is that of a cell or drop suspended in a liquid of *varying* density, for instance in the upper layers of a fluid (e.g. sea-water) at whose surface condensation is going on, so as to produce a steady density-gradient. In this case the normally spherical drop will be flattened into an oval form, with its maximum surface-curvature lying at the level where the densities of the drop and the surrounding liquid are just equal. The sectional outline of the drop has been shewn to be not a true oval or ellipse, but a somewhat complicated quartic curve. (Rice, *Phil. Mag.* Jan. 1915.) A more general case, which also may well deserve consideration by the biologist, is that of a charged bubble in (for instance) a uniform field of force: which will expand or elongate in the direction of the lines of force, and become a spheroidal surface in continuous transformation with the original sphere.

Our formula of equilibrium, or equation to an elastic surface, is $P = p_e + (T/R + T'/R')$, where P is the internal pressure, p_e any extraneous pressure normal to the surface, R, R' the radii of curvature at a point, and T, T' the corresponding tensions, normal to one another, of the envelope.

Now in any given form which we seek to account for, R, R' are known quantities; but all the other factors of the equation are subject to enquiry. And somehow or other, by this formula, we must account for the form of any solitary cell whatsoever (provided always that it be not formed by successive stages of solidification), the cylindrical cell of *Spirogyra*, the ellipsoidal yeast-cell, or (as we shall see in another chapter) even the egg of any bird. In using this formula hitherto we have taken it in a simplified form, that is to say we have made several limiting assumptions. We have assumed that P was the uniform hydrostatic pressure, equal in all directions, of a body of liquid; we have assumed likewise that the tension T was due to surface-tension in a homogeneous liquid film, and was therefore equal in all directions, so that $T = T'$; and we have only dealt with surfaces, or parts of a surface, where extraneous pressure, p_n, was non-existent. Now in the case of a bird's egg the external pressure p_n, that is to say the pressure exercised by the walls of the oviduct, will be found to be a very important factor; but in the case of the yeast-cell or the *Spirogyra*, wholly immersed in water, no such external pressure comes into play. We are accordingly left in such cases as these last with two hypotheses, namely that the departure from a spherical form is due to inequalities in the internal pressure P, or else to inequalities in the tension T, that is to say to a difference between T and T'. In other words, it is theoretically possible that the oval form of a yeast-cell is due to a greater internal pressure, a greater "tendency to grow" in the direction of the longer axis of the ellipse, or alternatively, that with equal and symmetrical tendencies to growth there is associated a difference of external resistance in respect of the tension, and implicitly the molecular structure, of the cell-wall. Now the former hypothesis is not impossible. Protoplasm is far from being a perfect fluid; it is the seat of various internal forces, sometimes manifestly polar, and it is quite possible that the forces, osmotic and other, which lead to an increase of the content of the cell and are manifested in pressure outwardly directed upon its wall may be unsymmetrical, and such as to deform what would otherwise be a simple sphere. But while this hypothesis is not impossible, it is not very easy of acceptance. The protoplasm, though not a perfect fluid, has yet on the whole the properties of a fluid; within the small compass of the cell there is little room for the development of unsymmetrical pressures; and in such a case as *Spirogyra*, where most part of the cavity is filled by watery sap, the conditions are still more obviously, or more nearly, those under which a uniform hydrostatic pressure should be displayed. But in variations of T, that is to say of the specific surface-tension per unit area, we have an ample field for all the various deformations with which we shall have to deal. Our condition now is, that $(T/R + T'/R') =$ a constant; but it no longer follows, though it may still often be the case, that this will represent a surface of absolute minimal area. As soon as T and T' become unequal, we are no longer dealing with a perfectly liquid surface film; but its departure from perfect fluidity may be of all degrees, from that of a slight non-isotropic viscosity to the state of a firm elastic membrane*; and it matters little whether this viscosity or semi-rigidity be manifested in the self-same layer which is still a part of the protoplasm of the cell, or in a layer which is completely differentiated into a distinct and separate membrane. As soon as, by secretion or adsorption, the molecular constitution of the surface-layer is altered, it is clearly conceivable that the alteration, or the secondary chemical changes which follow it, may be such as to produce an anisotropy, and to render the molecular forces less capable in one direction than another of exerting that contractile force by which they are striving to reduce to a minimum the surface area of the cell. A slight inequality in two opposite directions will produce the ellipsoid cell, and a great inequality will give rise to the cylindrical cell.

I take it therefore, that the cylindrical cell of *Spirogyra*, or any other cylindrical cell which grows in freedom from any manifest external restraint, has assumed that particular form simply by reason of the molecular constitution of its developing wall or

* Indeed any non-isotropic *stiffness*, even though T remained uniform, would simulate, and be indistinguishable from, a condition of non-stiffness and non-isotropic T.

membrane; and that this molecular constitution was anisotropous, in such a way as to render extension easier in one direction than another. Such a lack of homogeneity or of isotropy in the cell-wall is often rendered visible, especially in plant-cells, in the form of concentric lamellae, annular and spiral striations, and the like. But there exists yet another heterogeneity, to help us account for the long threads, hairs, fibres, cylinders, which are so often formed. Carl Nägeli said many years ago that organised bodies, starch-grains, cellulose and protoplasm itself, consisted of invisible particles, each an aggregate of many molecules—he called them *micellae*; and these were isolated, or "dispersed" as we should say, in a watery medium. This theory was, to begin with, an attempt to account for the colloid state; but at the same time, the particles were supposed to be so ordered and arranged as to render the substance anisotropic, to confer on it vectorial properties as we say nowadays, and so to account for the polarisation of light by a starch-grain or a hair. It was so criticised by Bütschli and von Ebner that it fell into disrepute, if not oblivion; but a great part of it was true. And the micellar structure of wool, cotton, silk and similar substances is now rendered clearly visible by the same X-ray methods as revealed the molecular orientation, or lattice-structure, of a crystal to von Laue.

It is now well known that the cell-wall has in many cases a definite structure which depends on molecular assemblages in the material of which it is composed, and is made visible by X-rays in the form of "diffraction patterns". The green alga *Valonia* has very large bubbly cells, 2–3 centimetres long, with cell-walls formed, as usual, of cellulose; this substance is a polysaccharide, with long-chain molecules some 500 Ångström-units, or say $0{\cdot}05\,\mu$ long, bound together sideways to form a multiple sheet or three-dimensional lattice. In the cell-wall of *Valonia* one set of chains runs round in a left-handed spiral, another forms meridians from pole to pole, and these two layers are superposed alternately to build the wall. Hemp has two layers, both running in right-handed spirals; flax two layers, crossing and recrossing in spirals of opposite sign. Even the cytoplasm and its contents seem to be influenced by molecular "*lignes directrices*," corresponding to the striae of the cell-wall. Analogous but still more complicated results of molecular structure are to be found in wool, cotton and other fibres*.

* Cf. R. D. Preston, *Phil. Trans.* (B), CCXXIV, p. 131, 1934; Preston and Astbury, *Proc. R.S.* (B), CXXII, pp. 76–97, 1937; and many other important papers by Astbury, van Iterson, Heyn, and others. We are brought by them to a borderland

But this phenomenon, while it brings about a certain departure from complete symmetry, is still compatible with, and coexistent with, many of the phenomena which we have seen to be associated with surface-tension. The symmetry of tensions still leaves the cell a solid of revolution, and its surface is still a surface of equilibrium. The fluid pressure within the cylinder still causes the film or membrane which caps its ends to be of a spherical form. And in the young cell, where the surface pellicle is absent or but little differentiated, as for instance in the oögonium of *Achlya* or in the young zygospore of *Spirogyra*, we see the tendency of the entire structure towards a spherical form reasserting itself: unless, as in the latter case, it be overcome by direct compression within the cylindrical mother-cell. Moreover, in those cases where the adult filament consists of cylindrical cells we see that the young germinating spore, at first spherical, very soon assumes with growth an elliptical or ovoid form—the direct result of an incipient anisotropy of its envelope, which when more developed will convert the ovoid into a cylinder. We may also notice that a truly cylindrical cell is comparatively rare, for in many cases what we call a cylindrical cell shews a distinct bulging of its sides; it is not truly a cylinder, but a portion of a spheroid or ellipsoid.

Unicellular organisms in general—protozoa, unicellular cryptogams, various bacteria and the free isolated cells, spores, ova, etc. of higher organisms—are referable for the most part to a small number of typical forms; but there are many others in which either no symmetry is to be recognised, or in which the form is clearly not one of equilibrium. Among these latter we have *Amoeba* itself and all manner of amoeboid organisms, and also many curiously shaped cells such as the Trypanosomes and various aberrant Infusoria. We shall return to the consideration of these; but in the meanwhile it will suffice to say (and to repeat) that, inasmuch as their surfaces are not equilibrium-surfaces, so neither are the living cells themselves in any stable equilibrium. On the contrary, they are in continual flux and movement, each portion of the

between chemical and histological structure, where micellae and long-chain molecules enlarge and alter our conceptions not only of cellulose and keratin, but of pseudopodia and cilia, of bone and muscle, and of the naked surface of the cell. See L. E. R. Picken, The fine structure of biological systems, *Biol. Reviews*, xv, pp. 133–67, 1940.

surface constantly changing its form, passing from one phase to another of an equilibrium which is never stable for more than a moment, and which death restores to the stable equilibrium of a sphere. The former class, which rest in stable equilibrium, must fall (as we have seen) into two classes—those whose equilibrium arises from liquid surface-tension alone, and those in whose conformation some other pressure or restraint has been superimposed upon ordinary surface-tension.

To the fact that all these organisms belong to an order of magnitude in which form is mainly, if not wholly, conditioned and controlled by molecular forces is due the limited range of forms which they actually exhibit. They vary according to varying physical conditions. Sometimes they do so in so regular and orderly a way that we intuitively explain them as "phases of a life-history," and leave physical properties and physical causation alone: but many of their variations of form we treat as exceptional, abnormal, decadent or morbid, and are apt to pass these over in neglect, while we give our attention to what we call a typical or "characteristic" form or attitude. In the case of the smallest organisms, bacteria, micrococci, and so forth, the range of form is especially limited, owing to their minuteness, the powerful pressure which their highly curved surfaces exert, and the comparatively homogeneous nature of their substance. But within their narrow range of possible diversity these minute organisms are protean in their changes of form. A certain species will not only change its shape from stage to stage of its little "cycle" of life; but it will be remarkably different in outward form according to the circumstances under which we find it, or the histological treatment to which we subject it. Hence the pathological student, commencing the study of bacteriology, is early warned to pay little heed to differences of *form*, for purposes of recognition or specific identification. Whatever grounds we may have for attributing to these organisms a permanent or stable specific identity (after the fashion of the higher plants and animals), we can seldom safely do so on the ground of definite and always recognisable *form*: we may often be inclined, in short, to ascribe to them a physiological (sometimes a "pathogenic") rather than a morphological specificity.

Many unicellular forms, and a few other simple organisms, are spherical, and serve to illustrate in the simplest way the point at issue. Unicellular algae, such as *Protococcus* or *Halisphaera*, the innumerable floating eggs of fishes, the floating unilocular foraminifer *Orbulina*, the lovely green multicellular *Volvox* of our ponds, all these in their several grades of simplicity or complication are so many round drops, spherical because no alien forces have deformed or mis-shapen them. But observe that, with the exception of *Volvox*, whose spherical body is covered wholly and uniformly with minute cilia, all the above are passive or inactive forms; and in a "resting" or encysted phase the spherical form is common and general in a great range of unicellular organisms.

Conversely, we see that those unicellular forms which depart markedly from sphericity—excluding for the moment the amoeboid forms and those provided with skeletons—are all ciliate or flagellate. Cilia and flagella are *sui generis*; we know nothing of them from the physical side, we cannot reproduce or imitate them in any non-living drop or fluid surface. But we can easily see that they have an influence on *form*, besides serving for locomotion. When our little *Monad* or *Euglena* develops a flagellum, that is in itself an indication of asymmetry or "polarity," of non-homogeneity of the little cell; and in the various flagellate types the flagellum or its analogues always stand on prominent points, or ends, or edges of the cell—on parts, that is to say, where curvature is high and surface-tension may be expected to be low—for the product of surface-tension by mean curvature tends to be constant.

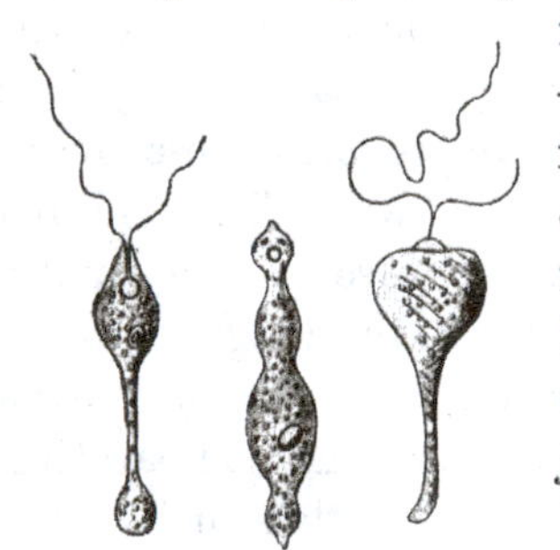

Fig. 123. A flagellate "monad," *Distigma proteus* Ehr. After Saville Kent.

The minute dimensions of a cilium or a flagellum are such that the molecular forces leading to surface-tension must here be under peculiar conditions and restraints; we cannot hope to understand them by comparison with a whip-lash, or through any other analogy drawn from a different order of magnitude. I suspect that a ciliary surface is always electrically charged, and that a point-charge is formed or induced in each cilium or flagellum. Just as we learn the properties of a drop or a jet as phenomena proper to their scale of magnitude, so some day we shall learn the very different physical, but microcosmic, properties of these minute, mobile, pointed, fluid or semi-fluid threads.*

Cilia, like flagella, tend to occupy positions, or cover surfaces, which would otherwise be unstable; and often indeed (as in a trochosphere larva or even in a Rotifer) a ring of cilia seems to play the very part of one of Plateau's wire rings, supporting and steadying the semi-fluid mass in its otherwise unstable configuration. Let us note here (in passing) what seems to be an analogous phenomenon. Chitinous hairs, spines or bristles are common and characteristic structures among the smaller Crustacea, and more or less generally among the Arthropods. We find them at every exposed point or corner; they fringe the sharp edge or border of a limb; as we draw the creature, we seem to know where to put them in! In short, they tend to occur, as the flagella do, just where the surface-tension would be lowest, if or when the surface was in a fluid condition.

Of the other surfaces of Plateau, we find cylinders enough and to spare in *Spirogyra* and a host of other filamentous algae and fungi. But it is to the vegetable kingdom that we go to find them, where a cellulose envelope enables the cylinder to develop beyond its ordinary limitations.

The unduloid makes its appearance whenever sphere or cylinder begin to give way. We see the transitory figure of an unduloid in the normal fission of a simple cell, or of the nucleus itself; and we have already seen it to perfection in the incipient beadings of a spider's web, or of a pseudopodial thread of protoplasm. A large number of infusoria have unduloid contours, in part at least; and this figure appears and reappears in a great variety of forms. The cups of various Vorticellae (Fig. 124), below the ciliated ring, look like a beautiful series of unduloids, in every gradation of form, from what is all but cylindrical to all but a perfect sphere; moreover successive phases in their life-history appear as mere graded changes

* It is highly characteristic of a cilium or a flagellum that neither is ever seen motionless, unless the cell to which it belongs is moribund. "I believe the motion to be ceaseless, unconscious and uncontrolled, a direct function of the chemical and physical environment"; George Bidder, in *Presidential Address to Section D, British Association*, 1927. Cf. also James Gray, *Proc. R.S.* (B), xcix, p. 398, 1926.

of unduloid form. It has been shewn lately, in one or two instances at least, that species of *Vorticella* may "metamorphose" into one another: in other words, that contours supposed to characterise species are not "specific". These Vorticellid unduloids are

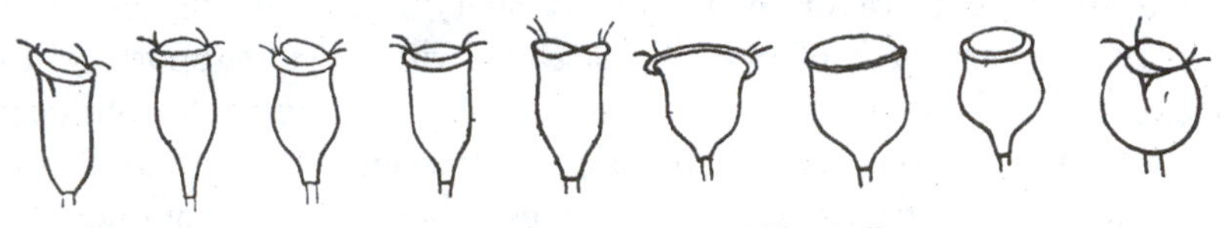

Fig. 124. Various species of *Vorticella*.

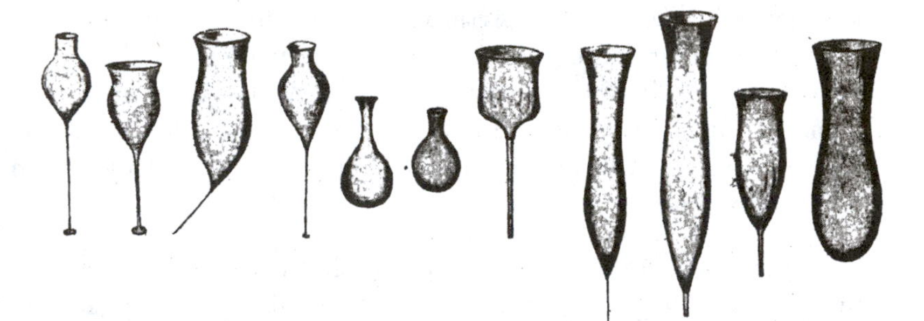

Fig. 125. Various species of *Salpingoeca*.

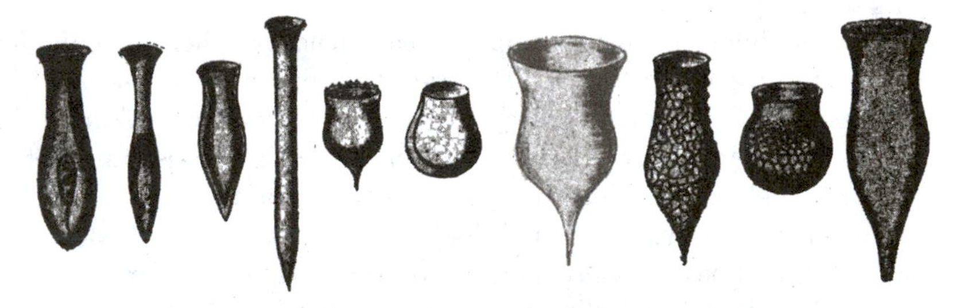

Fig. 126. Various species of *Tintinnus*, *Dinobryon* and *Codonella*. After Saville Kent and others.

not fully symmetrical; rather are they such unduloids as develop when we suspend an oil-globule between two unequal rings, or blow a bubble between two unequal pipes. For our Vorticellid bell hangs by two terminal supports, the narrow stalk to which it is attached below, and the thickened ring from which spring its circumoral cilia; and it is most interesting to see how, when the bell leaves its stalk (as sometimes happens) and swims away, a new ring of cilia comes into being, to encircle and support its narrow end.

Similar unduloids may be traced in even greater variety among other families or genera of the Infusoria. Sometimes, as in *Vorticella* itself, the unduloid is seen in the contour of the soft semifluid body of the living animal. At other times, as in *Salpingoeca*, *Tintinnus*, and many other genera, we have a membranous cup containing the animal, but originally secreted by, and moulded upon, its semifluid living surface. Here we have an excellent illustration of the contrast between the different ways in which such a structure may be regarded and interpreted. The teleological explanation is that it is developed for the sake of protection,

Fig. 127. *Vaginicola.*

Fig. 128. *Folliculina.*

as a domicile and shelter for the little organism within. The mechanical explanation of the physicist (seeking after the "efficient," not the "final" cause) is that it owes its presence, and its actual conformation, to certain chemico-physical conditions: that it was inevitable, under the given conditions, that certain constituent substances present in the protoplasm should be drawn by molecular forces to its surface layer; that under this adsorptive process, the conditions continuing favourable, the particles accumulated and concentrated till they formed (with the help of the surrounding medium) a pellicle or membrane, thicker or thinner as the case might be; that this surface pellicle or membrane was inevitably bound, by molecular forces, to contract into a surface of the least possible area which the circumstances permitted; that in the present case the symmetry and "freedom" of the system permitted, and *ipso facto* caused, this surface to be a surface of revolution; and that of the few surfaces of revolution which, as being also surfaces *minimae areae*, were available, the unduloid was manifestly the one permitted, and *ipso facto* caused, by the dimensions of the organism and other circumstances of the case. And just as the thickness or thinness of the pellicle was obviously a subordinate matter, a mere matter of degree, so we see that the actual outline of this or that particular unduloid is also a very subordinate matter, such as physico-chemical variants of a minor order would suffice to bring about; for between the various unduloids which the various species of *Vorticella* represent, there is no more real difference than that difference of ratio or degree which exists between two circles of different diameter, or two lines of unequal length.

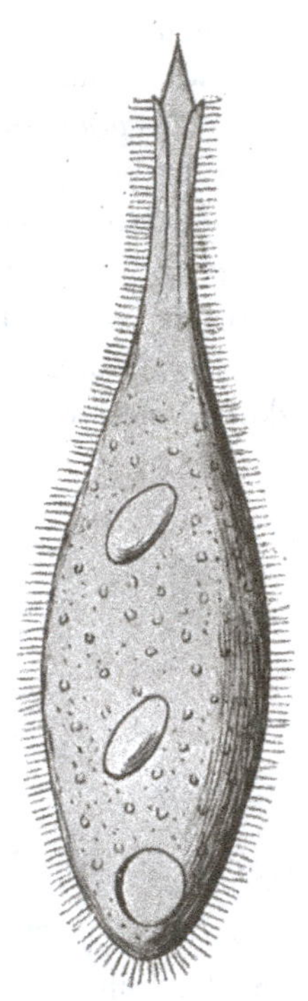

Fig. 129. *Trachelophyllum*. After Wreszniowski.

In many cases (of which Fig. 129 is an example) we have a more or less unduloid form exhibited not by a surrounding pellicle or shell, but by the soft protoplasmic body of a ciliated organism; in such cases the form is mobile, and changes continually from one to another unduloid contour according to the movements of the animal.* We are dealing here with no stable equilibrium, but possibly with a subtle problem of "stream-lines," as in the difficult but beautiful problems suggested by the form of a fish. But this whole class of cases, and of problems, we merely take note of here; we shall speak of them again, but their treatment is hard.

In considering such series of forms as these various unduloids we are brought sharply up (as in the case of our bacteria or micrococci) against the biological concept of organic *species*. In the intense classificatory activity of the last hundred years it has come about that every form which is apparently characteristic, that is to say which is capable of being described or portrayed, and of being

* Doflein lays stress, in like manner, on the fact that *Spirochaete*, unlike *Spirillum*, "ist nicht von einer starren Membran umhüllt," and that waves of contraction may be seen passing down its body.

recognised when met with again, has been recorded as a *species*—for we need not concern ourselves with the occasional discussions, or individual opinions, as to whether such and such a form deserves "specific rank," or be "only a variety." And this secular labour is pursued in direct obedience to the precept of the *Systema Naturae*—"*ut sic in summa confusione rerum apparenti, summus conspiciatur Naturae ordo.*" In like manner the physicist records, and is entitled to record, his many hundred "species" of snow-crystals*, or of crystals of calcium carbonate. Indeed the snow-crystal illustrates to perfection how Nature rings the changes on every possible variation and permutation and combination of form: subject only to the condition (in this instance) that a snow-crystal shall be a plane, symmetrical, rectilinear figure, with all its external angles those of a regular hexagon. We may draw what we please on a sheet of "hexagonal paper," keeping to its lines; and when we repeat our drawing, kaleidoscope-fashion, about a centre, the stellate figure so obtained is sure to resemble one or another of the many recorded species of snow-crystals. And this endless beauty of crystalline form is further enhanced when the flakes begin to thaw, and all their feathery outlines soften. But regarding these "species" of his, the physicist makes no assumptions: he records them *simpliciter*; he notes, as best he can, the circumstances (such as temperature or humidity) under which each occurs, in the hope of elucidating the conditions which determine their formation†; but above all, he

* The case of the snow-crystals is a particularly interesting one; for their "distribution" is analogous to what we find, for instance, among our microscopic skeletons of Radiolarians. That is to say, we may one day meet with myriads of some one particular form or species, and another day with myriads of another; while at another time and place we may find species intermingled in all but inexhaustible variety. Cf. e.g. J. Glaisher, *Illustrated London News*, Feb. 17, 1855; *Q.J.M.S.* III, pp. 179–185, 1855; Sir Edward Belcher, *Last of the Arctic Voyages*, II, pp. 288–306 (4 plates), 1855; William Scoresby, *An Account of the Arctic Regions*, Edinburgh, 1820; G. Hellmann, *Schneekrystalle*, Berlin, 1893; Bentley and Humphreys, *Snow Crystals*, New York, 1931; and the especially beautiful figures of Nakaya and Hasikura in *Journ. Fac. Sci. Hokkaido*, Dec. 1934.

† Every snow-crystal tells, more or less plainly, the story of its own development. The cold upper air is saturated with water-vapour, but this is scanty and rarefied compared with the space in which snow-crystallisation is going on. Hence crystallisation tends to proceed only along the main axes, or cardinal framework, of the crystalline structure of ice; in so doing it gives a visible picture or actual embodiment of the trigonal-hexagonal space-lattice, in the endless permutations and combinations of its constituent elements.

does not introduce the element of time, and of succession, or discuss their origin and affiliation as an *historical* sequence of events. But in biology, the term species carries with it many large though often vague assumptions; though the doctrine or concept of the "permanence of species" is dead and gone, yet a certain quasi-permanency is still connoted by the term. If a tiny foraminiferal shell, a *Lagena* for instance, be found living to-day, and a shell indistinguishable from it to the eye be found fossil in the Chalk or some still more remote geological formation, the assumption is deemed legitimate that that species has "survived," and has handed down its minute specific character or characters from generation to generation, unchanged for untold millions of years*. If the ancient forms be like to rather than identical with the recent, we still assume an unbroken descent, accompanied by the hereditary transmission of common characters and progressive variations. And if two identical forms be discovered at the ends of the earth, still (with occasional slight reservations on the score of possible "homoplasy") we build hypotheses on this fact of identity, taking it for granted that the two appertain to a common stock, whose dispersal in space must somehow be accounted for, its route traced, its epoch determined, and its causes discussed or discovered. In short, the naturalist admits no exception to the rule that a *natural* classification can only be a *genealogical* one, nor ever doubts that "*The fact that we are able to classify organisms at all in accordance with the structural characteristics which they present is due to the fact of their being related by descent*†." But this great and valuable and even fundamental generalisation sometimes carries us too far. It may be safe and sure and helpful and illuminating when we apply it to such complex entities—such thousand-fold resultants of the combination and permutation of many variable characters—as a horse, a lion or an eagle; but (to my mind) it has a very different look, and a far less firm foundation, when we attempt to extend it to minute organisms whose specific characters are few and simple, whose simplicity

* Cf. Bergson, *Creative Evolution*, p. 107: "Certain Foraminifera have not varied since the Silurian epoch. Unmoved witnesses of the innumerable revolutions that have upheaved our planet, the Lingulae are today what they were at the remotest times of the palaeozoic era."

† Ray Lankester, *A.M.N.H.* (4), XI, p. 321, 1873.

becomes more manifest from the point of view of physical and mathematical analysis, and whose form is referable, or largely referable, to the direct action of a physical force. When we come to the minute skeletons of the Radiolaria we shall again find ourselves dealing with endless modifications of form, in which it becomes more and more difficult to discern, and at last vain and hopeless to apply, the guiding principle of affiliation or "*phylogeny*."

Among the Foraminifera we have an immense variety of forms, which, in the light of surface-tension and of the principle of minimal area, are capable of explanation and of reduction to a small number of characteristic types. Many of them are composite structures, formed by the successive imposition of cell upon cell, and these we shall deal with later on; let us glance here at the simpler conformations exhibited by the single chambered or "monothalamic" genera, and perhaps one or two of the simplest composites.

We begin with forms like *Astrorhiza* (Fig. 320, p. 703), which are large, coarse and highly irregular, and end with others which are minute and delicate, and which manifest a perfect and mathematical regularity. The broad difference between these two types is that the former are characterised, like *Amoeba*, by a variable surface-tension, and consequently by unstable equilibrium; but the strong contrast between these and the regular forms is bridged over by various transition-stages, or differences of degree. Indeed, as in all other Rhizopods, the very fact of the emission of pseudopodia, which are especially characteristic of this group of animals, is a sign of unstable surface-equilibrium; and we must therefore consider, or may at least suspect, that those forms whose shells indicate the most perfect symmetry and equilibrium have secreted these during periods when rest and uniformity of surface-conditions contrasted with the phases of pseudopodial activity. The irregular forms are in almost all cases arenaceous, that is to say they have no solid shells formed by steady adsorptive secretion, but only a looser covering of sand grains with which the protoplasmic body has come in contact and cohered. Sometimes, as in *Ramulina*, we have a calcareous shell combined with irregularity of form; but here we can easily see a partial and as it were a broken regularity, the regular forms of sphere and cylinder being repeated in various parts of the ramified mass. When we look more closely at the arenaceous forms, we find the same thing true of them; they represent, in whole or part, approximations to the surfaces of equilibrium, spheres, cylinders and so forth. In *Aschemonella* we have a precise replica of the calcareous *Ramulina*; and in *Astrorhiza* itself, in the forms distinguished by naturalists as *A. crassatina*, what is described as the "subsegmented interior*" seems to shew

Fig. 130. Various species of *Lagena*. After Brady.

the natural, physical tendency of the long semifluid cylinder of protoplasm to contract at its limit of stability into unduloid constrictions, as a step towards the breaking up into separate spheres: the completion of which process is restrained or prevented by contact with the unyielding arenaceous covering.

Passing to the typical calcareous Foraminifera, we have the most symmetrical of all possible types in the perfect sphere of *Orbulina*; this is a pelagic organism, whose floating habitat gives it a field of

* Brady, *Challenger Monograph*, pl. xx, p. 233.

force of perfect symmetry. Save for one or two other forms which are also spherical, or approximately so, like *Thurammina*, the rest of the monothalamic calcareous Foraminifera are all comprised by naturalists within the genus *Lagena*. This large and varied genus consists of "flask-shaped" shells, whose surface is that of an unduloid, or, like that of a flask itself, an unduloid combined with a portion of a sphere. We do not know the circumstances under which the shell of *Lagena* is formed, nor the nature of the force by which, during its formation, the surface is stretched out into the unduloid form; but we may be pretty sure that it is suspended vertically in the sea, that is to say in a position of symmetry as regards its vertical axis, about which the unduloid surface of revolution is symmetrically formed. At the same time we have other types of the same shell in which the form is more or less flattened; and these are doubtless the cases in which such symmetry of position was not present, or was replaced by a broader, lateral contact with the surface pellicle*.

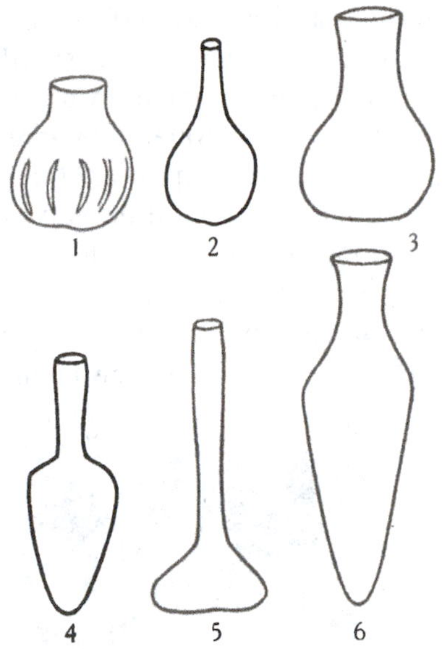

Fig. 131. Roman pottery, for comparison with species of *Lagena*. E.g., 1, 2, with *L. sulcata*; 3, *L. orbignyana*; 4, *L. striata*; 5, *L. crenata*; 6, *L. stelligera*.

While *Orbulina* is a simple spherical drop, *Lagena* suggests to our minds a hanging drop, drawn out to a longer or shorter neck by

* That the Foraminifera not only can but do hang from the surface of the water is confirmed by the following apt quotation which I owe to Mr E. Heron-Allen: "Quand on place, comme il a été dit, le dépôt provenant du lavage des fucus dans un flacon que l'on remplit de nouvelle eau, on voit au bout d'une heure environ les animaux [*Gromia dujardinii*] se mettre en mouvement et commencer à grimper. Six heures après ils tapissent l'extérieur du flacon, de sorte que les plus élevés sont à trente-six ou quarante-deux millimètres du fond; le lendemain beaucoup d'entre eux, *après avoir atteint le niveau du liquide, ont continué à ramper à sa surface, en se laissant pendre au-dessous* comme certains mollusques gastéropodes." (F. Dujardin, Observations nouvelles sur les prétendus céphalopodes microscopiques, *Ann. des Sci. Nat.* (2), III, p. 312, 1835.)

its own weight, aided by the viscosity of the material. Indeed the various hanging drops, such as Mr C. R. Darling shews us, are the most beautiful and perfect unduloids, with spherical ends, that it is possible to conceive. A suitable liquid, a little denser than water and incapable of mixing with it (such as ethyl benzoate), is poured on a surface of water. It spreads over the surface and gradually forms a hanging drop, approximately hemispherical; but as more liquid is added the drop sinks or rather stretches downwards, still adhering to the surface film; and the balance of forces between gravity and surface-tension results in the unduloid contour, as the increasing weight of the drop tends to stretch it out and finally break it in two. At the moment of rupture, by the way, a tiny

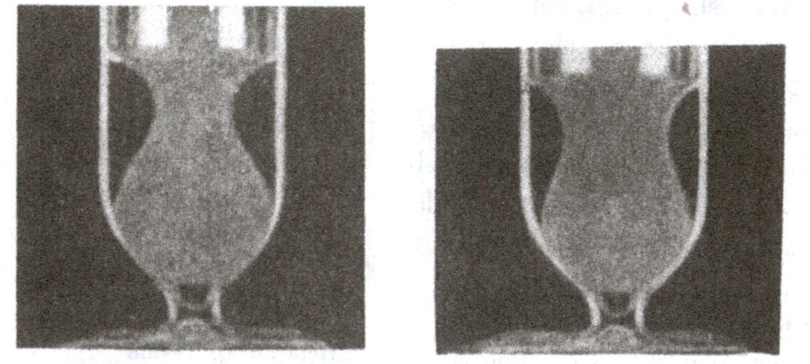

Fig. 132. Large "hanging drops" of oil. After Darling.

droplet is formed in the attenuated neck, such as we described in the normal division of a cylindrical thread.

The thin, fusiform, pointed, non-globular *Lagenas* are less easily explained. Surface-tension, which tends to keep the drop spherical, is overmastered here, and the elongate shape suggests the viscous drag of a shearing fluid*.

To pass to a more highly organised class of animals, we find the unduloid beautifully exemplified in the little flask-shaped shells of certain Pteropod mollusca, e.g. *Cuvierina*†. Here again the symmetry of the figure would at once lead us to suspect that the creature lived in a position of symmetry to the surrounding forces, as for instance if it floated in the ocean in an erect position, that is to say with its long axis coincident with the direction of gravity; and this we know to be actually the mode of life of the little Pteropod.

* Cf. G. I. Taylor, The formation of emulsions in definable fields of flow, *Proc. R.S.* (A), No. 858, p. 501, 1934.

† Cf. Boas, *Spolia Atlantica*, 1886, pl. 6.

Many species of *Lagena* are complicated and beautified by a pattern, and some by the superaddition to the shell of plane extensions or "wings." These latter give a secondary, bilateral symmetry to the little shell, and are strongly suggestive of a phase or period of growth in which it lay horizontally on the surface, instead of hanging vertically from the surface-film: in which, that is to say, it was a floating and not a hanging drop. The pattern is of two kinds. Sometimes it consists of a sort of fine reticulation, with rounded or more or less hexagonal interspaces: in other cases it is produced by a symmetrical series of ridges or folds, usually longitudinal, on the body of the flask-shaped cell, but occasionally transversely arranged upon the narrow neck. The reticulated and folded patterns we may consider separately. The netted pattern is very similar to the wrinkled surface of a dried pea, or to the more regular wrinkled patterns on poppy and other seeds and even pollen-grains. If a spherical body after developing a "skin" begin to shrink a little, and if the skin have so far lost its elasticity as to be unable to keep pace with the shrinkage of the inner mass, it will tend to fold or wrinkle; and if the shrinkage be uniform, and the elasticity and flexibility of the skin be also uniform, then the amount of foldings will be uniformly distributed over the surface. Little elevations and depressions will appear, regularly interspaced, and separated by concave or convex folds. These being of equal size (unless the system be otherwise perturbed), each one will tend to be surrounded by six others; and when the process has reached its limit, the intermediate boundary-walls, or folds, will be found converted into a more or less regular pattern of hexagons. To these symmetrical wrinkles or shrinkage-patterns we shall return again.

But the analogy of the mechanical wrinkling of the coat of a seed is but a rough and distant one; for we are dealing with molecular rather than with mechanical forces. In one of Darling's experiments, a little heavy tar-oil is dropped on to a saucer of water, over which it spreads in a thin film shewing beautiful interference colours after the fashion of those of a soap-bubble. Presently tiny holes appear in the film, which gradually increase in size till they form a cellular pattern or honeycomb, the oil gathering together in the meshes or walls of the cellular net. Some action of this sort is in all probability at work in a surface-film of protoplasm covering the shell. As a physical phenomenon the actions involved are by no means fully understood, but surface-tension, diffusion and cohesion play their respective parts therein*. The very perfect cellular patterns obtained by Leduc (to which we shall have occasion to refer in a subsequent chapter) are diffusion patterns on a larger scale, but not essentially different.

The folded or pleated pattern is doubtless to be explained, in a general way, by the shrinkage of a surface-film under certain conditions of viscous or frictional restraint. A case which (as it seems to me) is closely allied to that of our foraminiferal shells is described by Quincke†, who let a film of chromatised gelatin or of resin set and harden upon a surface of quicksilver, and found that the little solid pellicle had been thrown into a pattern of symmetrical folds, as fine as a diffraction grating. If the surface thus folded or wrinkled be a cylinder, or any other figure with one principal axis

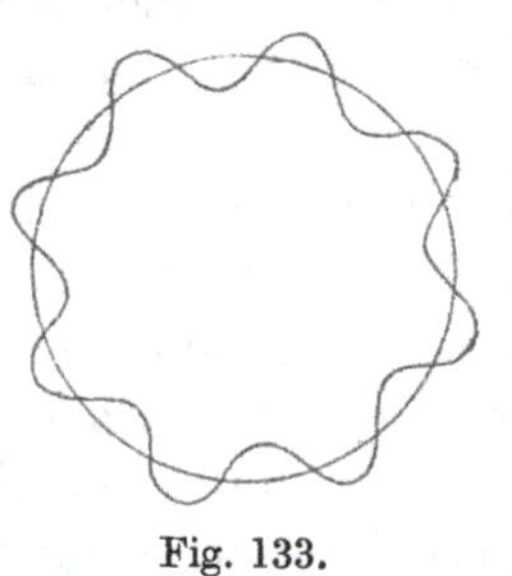

Fig. 133.

* This cellular pattern would seem to be related to the "cohesion figures" described by Tomlinson in various surface-films (*Phil. Mag.* 1861–70); to the "tesselated structure" on liquid surfaces described by James Thomson in 1882 (*Collected Papers*, p. 136); and (more remotely) to the *tourbillons cellulaires* of Bénard, *Ann. de Chimie* (7), XXIII, pp. 62–144, 1901; (8), XXIV, pp. 563–566, 1911, *Rev. génér. des Sci.* XI, p. 1268, 1900; cf. also E. H. Weber, Mikroskopische Beobachtungen sehr gesetzmässiger Bewegungen welche die Bildung von Niederschlagen harziger Körper aus Weingeist begleiten, *Poggend. Ann.* XCIV, pp. 447–459, 1855; etc. Some at least of Tomlinson's cohesion-figures arise, according to van Mensbrugghe, from the disengagement of minute bubbles of gas, when a fluid holding gases in solution comes in contact with a fluid of lower surface-tension. The whole phenomenon is of great interest and various appearances have been referred to it, in biology, geology, metallurgy and even astronomy: for the flocculent clouds in the solar photosphere shew an analogous configuration. (See letters by Kerr Grant, Larmor, Wager and others, in *Nature*, April 16 to June 11, 1914; also Rayleigh, *Phil. Mag.* XXXII, p. 529, 1916; G. T. Walker, Clouds, natural and artificial, *Royal Inst.* 8 Feb. 1935; etc.) In many instances, marked by strict symmetry or regularity, it is very possible that the interference of waves or ripples may play its part in the phenomenon. But in the majority of cases, it is fairly certain that localised centres of action, or of diminished tension, are present, such as might be provided by dust-particles in the case of Bénard's experiment (cf. *infra*, p. 503).

† Quincke, Ueber physikalische Eigenschaften dünner fester Lamellen, *Sitzungsb. Berlin. Akad.* 1888, p. 789; Ueber ansichtbare Eigenschaften, etc., *Ann. d. Physik*, 1920, p. 653. Quincke found that "sehr kleine Menge fremder Substanz haben eine grosse Einfluss auf die Bildung der Schaumwände."

of symmetry, such as an ellipsoid or unduloid, the folds will tend to be related to the axis of symmetry, and we may expect accordingly to find regular longitudinal, or regular transverse wrinkling. Now as a matter of fact we almost invariably find in *Lagena* the former condition: that is to say, in our ellipsoid or unduloid shell, the puckering takes the form of the vertical fluting on a column, rather than that of the transverse pleating of an accordion; and further, there is often a tendency for such longitudinal flutings to be more or less localised at the end of the ellipsoid, or in the region where the unduloid merges into its spherical base*. In the latter region we often meet with a regular series of short longitudinal folds, as in the forms denominated *L. semistriata.* All these various forms of surface can be imitated, or precisely reproduced, by the art of the glass-blower; and they can be seen in a contracting bubble of saponin, though not in the more fluid soap-bubble. They remind one of the ribs or flutings in the film or sheath which splashes up to envelop a smooth pebble dropped into a liquid, as Mr Worthington has so beautifully shewn.

In Mr Worthington's experiment there appears to be something of the nature of a viscous drag in the surface-pellicle; but whatever be the actual cause of variation of tension, it is not difficult to see that there must be in general a tendency towards *longitudinal* puckering or "fluting" in the case of a thin-walled cylindrical or other elongated body, rather than a tendency towards transverse puckering, or "pleating." For let us suppose that some change takes place involving an increase of surface-tension in some small area of the curved wall, and leading therefore to an increase of pressure: that is to say let T become $T + t$, and P become $P + p$. Our new equation of equilibrium, then, in place of $P = T/r + T/r'$, becomes

$$P + p = \frac{T + t}{r} + \frac{T + t}{r'},$$

* Certain palaeontologists (e.g. Haeusler and Spandel) have asserted that in each family or genus the plain smooth-shelled forms are primitive and ancient, and that the ribbed and otherwise ornamented shells make their appearance at later dates in the course of advancing evolution (cf. Rhumbler, *Foraminiferen der Plankton-Expedition*, 1911, p. 21). If this were true it would be of fundamental importance: but this book of mine would not deserve to be written.

and by subtraction,

$$p = t/r + t/r'.$$

Now if $r < r'$, $t/r > t/r'$.

Therefore, in order to produce the small increment of pressure p, it is easier to do so by increasing t/r than t/r'; that is to say, the easier way is to alter or diminish r. And the same will hold good if the tension and pressure be diminished instead of increased.

This is as much as to say that, when corrugation or "rippling" of the walls takes place owing to small changes of surface-tension, and consequently of pressure, such corrugation is more likely to take place in the plane of r—that is to say, *in the plane of greatest curvature*. And it follows that in such a figure as an ellipsoid, wrinkling will be most likely to take place not only in a longitudinal direction but near the extremities of the figure, that is to say again in the region of greatest curvature.

The longitudinal wrinkling of the flask-shaped bodies of our Lagenae, and of the more or less cylindrical cells of many other Foraminifera (Fig. 134), is in complete accord with the above considerations; but nevertheless, we soon find that our result is not a general one but is defined by certain limiting conditions, and is accordingly subject to what are, at first sight, important exceptions For instance, when we turn to the narrow neck of the *Lagena* we see at once that our theory no longer holds; for the wrinkling which was invariably longitudinal in the body of the cell is as invariably transverse in the narrow neck. The reason for the difference is not far to seek. The conditions in the neck are very different from those in the expanded portion of the cell: the main difference being that the thickness of the wall is no longer insignificant, but is of considerable magnitude as compared with the diameter, or circumference, of the neck. We must accordingly take it into account in considering the *bending moments* at any point in this region of the shell-wall. And it is at once obvious that, in any portion of the narrow neck, *flexure* of a wall in a transverse direction will be very difficult, while flexure in a longitudinal direction will be comparatively easy; just as, in the case of a long narrow strip of iron, we may easily bend it into folds running transversely to its long axis, but not the other way. The manner in which our little *Lagena*-shell tends to fold or wrinkle, longitudinally in its wider part and transversely or annularly in its narrow neck, is thus completely explained.

An identical phenomenon is apt to occur in the little flask-shaped gonangia, or reproductive capsules, of some of the hydroid zoophytes. In the annexed drawings of these gonangia in two species of *Campanularia*, we see that in one case the little vesicle has the flask-shaped or unduloid configuration of a *Lagena*; and here the walls of the flask are longitudinally fluted, just after the manner we have witnessed in the latter genus. In the other Campanularian the vesicles are long, narrow and tubular, and here a transverse folding

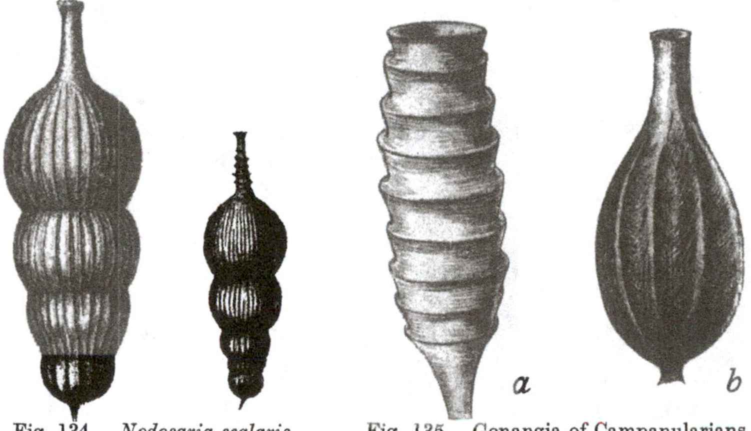

Fig. 134. *Nodosaria scalaris* Batsch.

Fig. 135. Gonangia of Campanularians. (*a*) *C. gracilis*; (*b*) *C. grandis*. After Allman.

or pleating takes the place of the longitudinally fluted pattern; and the very form of the folds or pleats is enough to suggest that we are not dealing here with a simple phenomenon of surface-tension, but with a condition in which surface-tension and *stiffness* are both present, and play their parts in the resultant form.

An everted rim, or short neck, may arise in various ways apart from the phenomenon of the hanging drop. To make a "thistle-head" the glassblower blows a bubble, and from that another one; after blowing the latter up large and thin he crushes it to pieces, and melting down what is left of it he forms the rim. I take it that the neck or rim of the shell in *Difflugia* is formed in an analogous way, in connection with the growth of a new individual at the mouth of the first. There is a very neat expanded orifice in the cyst of *Chromulina* (Fig. 136); it is doubtless fashioned in just as simple a way, but how I know not.

Passing from the solitary flask-shaped cell of *Lagena*, but without leaving the Foraminifera, we find in *Nodosaria*, *Rheophax* or *Sagrina* constricted cylinders, or successive unduloids, such as are represented in Fig. 137. In some of these, as in the arenaceous genus

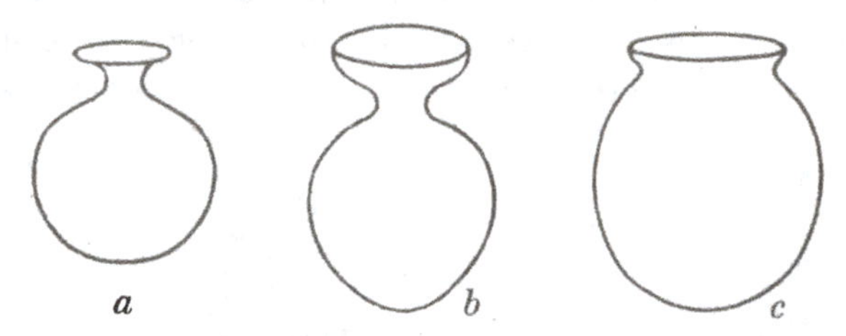

Fig. 136. Flask-shaped shells or cysts. *a*, *b*, *Chromulina* and *Deropyxis* (Flagellata); *c*, *Difflugia*.

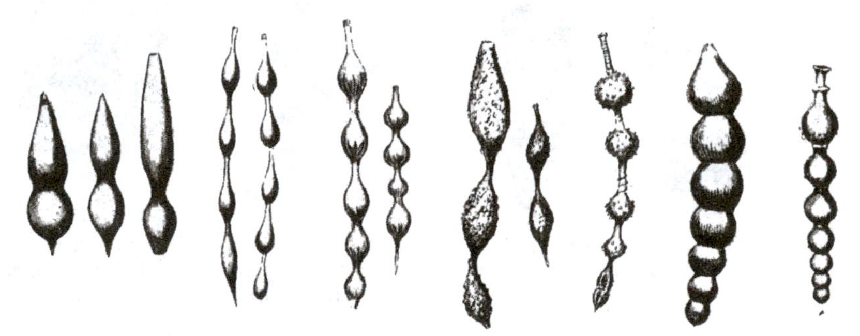

Fig. 137. Various species of *Nodosaria*, *Rheophax*, *Sagrina*. After Brady.

Rheophax, we have to do with the ordinary phenomenon of a partially segmenting cylinder. But in others, the structure is not developed out of a continuous protoplasmic cylinder, but, as we can see by examining the interior of the shell, it has been formed in successive stages, beginning with a simple unduloid "*Lagena*," about which, after it solidified, another drop of protoplasm accumulated, and in turn assumed the unduloid or lagenoid form. The chains of interconnected bubbles which Morey and Draper made many years ago of melted resin are a similar if not identical phenomenon*.

* See *Silliman's Journal*, II, p. 179, 1820; and cf. Plateau, *op. cit.* II, pp. 134, 461.

Fishes shew a vast though limited variety of form; and some of the strangest shapes are found in the great depths of the ocean. Here, in unchanging temperature, in darkness save for a few phosphorescent rays, above all in unruffled stillness and eternal

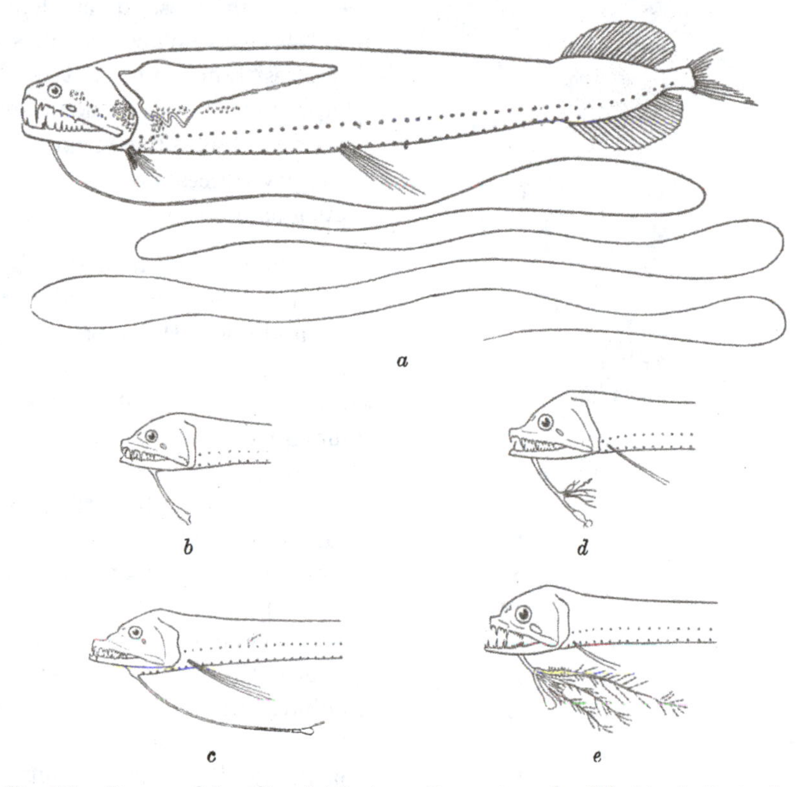

Fig. 138. Deep-sea fishes (Stomiatidae). *a*, *Lamprotoxus flagellibarbis*; *b*, *Eustomias dactylobus*; *c*, *E. parri*; *d*, *E. schmidti*; *e*, *E. silvescens*. After Tate Regan and Trewavas.

calm*, the conditions of life are strange indeed. In deep-sea fishes length and attenuation are common characters of the body and of its parts. A barbel below the lip may grow to ten times the whole length of the fish; it ends, commonly, in a little bulb or blob; it may give off threadlike branches, and these last slender filaments

* In Overbeck's jet-experiments (*supra*, p. 394) the water into which the jet is led must first stand for many hours, till all internal movements and temperature-differences are eliminated.

are sometimes finely beaded*; and slight differences in the beading and branching are said to characterise allied species of fish. Such a barbel looks like a jet or branching stream of one fluid falling through another. It may indeed be that in these quiet depths growth easily follows its lines of least resistance, and that in the shaping of these peculiar outgrowths hydrodynamical and capillary forces are taking the upper hand.

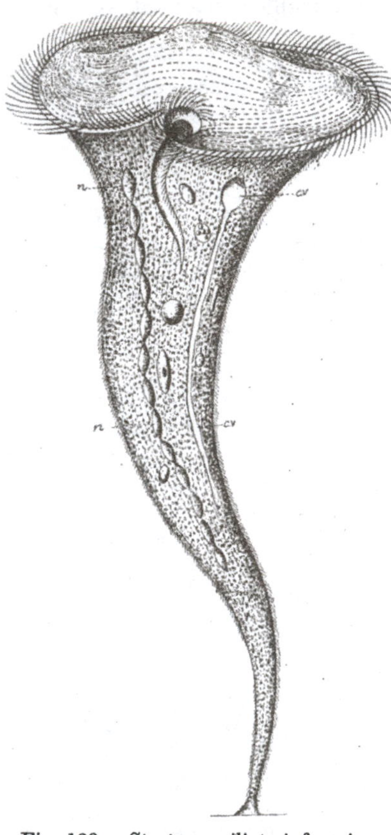

Fig. 139. *Stentor*, a ciliate infusorian: from Savile Kent.

We have found it easy to illustrate the sphere, the cylinder and the unduloid, three of the six "surfaces of Plateau," all with an endless wealth of illustration among the simplest of organisms. The plane we need hardly look for among the finite outlines of a fluid body; and the catenoid, also a surface of zero mean curvature, can likewise only be a rare and transitory configuration. One last surface still remains, namely the nodoid; and there also remains one very common but most remarkable Protozoan configuration, that of the ciliate Infusoria, to the most characteristic feature of which we have not so far found a physical analogue. Here the curved contour seems to enter, re-enter, and disappear within the substance of the body, so bounding a deep and twisted space or passage, which merges with the fluid contents and vanishes within the cell, and is called by naturalists the "gullet." This very peculiar and complicated structure is only kept in equilibrium, and in existence, by the constant activity of cilia over the general surface

* See, for instance, C. Tate Regan and E. Trewavas on *The Stomiatidae of the Dana Expedition*, 1930; W. Beebe, Deep-sea Stomiatoids, *Copeia*, Dec. 1833.

of the body and very especially in the said gullet or re-entrant portion of the surface. Now we have seen the nodoid to be a curved surface, re-entering on itself and endless; no method of support, by wire-rings or otherwise, enables us to construct or realise more than a small portion of it. But the typical ciliate, such as *Paramoecium*, looks just like what we might expect a nodoid surface to be, if we could only realise it (or a single segment of it) in a drop of fluid, and imagine it to be kept in quasi-equilibrium by continual ciliary activity. I suspect, indeed, that here is nothing more, and nothing less, than a partial realisation of the nodoid itself; that the so-called gullet is but the characteristic inversion or "kink" in that curve; and that the cilia, which normally clothe the surface and always line the gullet, are needed to realise and to maintain the unstable equilibrium of the figure. If this be so—it is a suggestion and no more—we shall have found among our simple organisms the complete realisation, in varying abundance, of each and all of the six surfaces of Plateau. On each and all of them we have a host of beautiful "patterns" of various sorts; all of them so beautiful and so symmetrical that they *ought* to be capable of geometric representation—and all waiting for their interpreter!

From all these configurations, which the law of minimal area controls and dominates, *Amoeba* stands aloof and alone. The rest are all figures of equilibrium, unstable though it may sometimes be. But *Amoeba* is the characteristic case of a fluid surface without an equilibrium; it is the very negation of stability. In composition it is neither constant nor homogeneous; its chemistry is in constant flux, its surface energies vary from here to there, its fluid substance is drawn hither and thither; within and without it is never still, be its motions swift or be they slow. The heterogeneity of its system points towards a maximal surface-area, rather than a minimal one; only here and there, in small portions of its heterogeneous substance, do we see the rounded contours of a fluid drop, in token of temporary equilibrium. Only when its heterogeneous reactions quieten down and the little living speck enters on its "resting-stage," does the protoplasmic body withdraw itself into a sphere and the law of area minima come into its own. Physically analogous is the case of such complicated pseudopodia, or "axopodia", as we find among the Foraminifera and Heliozoa: where the whole fabric is in a flux, and currents flow and granules are carried hither and thither. Here again there is no statical equilibrium; but surface tension varies, as does the chemistry of the protoplasm, from one spot to another.

The great oceanic group of the Radiolaria, and the highly complicated skeletons which they construct, give us many beautiful illustrations of physical phenomena, among which the effects of surface-tension are as usual prominent. But we shall deal later on with these little skeletons under the head of spicular concretions.

In a simple and typical Heliozoan, such as the sun-animalcule, *Actinophrys sol*, we have a "drop" of protoplasm, contracted by its surface tension into a spherical form. Within this heterogeneous protoplasm are more fluid portions, and a similar surface-tension causes these also to assume the form of spherical "vacuoles," or of little clear drops within the big one; unless indeed they become numerous and closely packed, in which case they run together and constitute a "froth," such as we shall study in the next chapter. One or more of such clear spaces may be what is called a "contractile vacuole": that is to say, a droplet whose surface-tension is in unstable equilibrium and is apt to vanish altogether, so that the definite outline of the vacuole suddenly disappears*. Again, within the protoplasm are one or more nuclei, whose own surface-tension draws them in turn into the shape of spheres. Outwards through the protoplasm, and stretching far beyond the spherical surface of the cell, run stiff linear threads of modified protoplasm, reinforced in some cases by delicate siliceous needles. In either case we know little or nothing about the forces which lead to their production, and we do not hide our ignorance when we ascribe their development to a "radial polarisation" of the cell. In the case of the protoplasmic filament, we may (if we seek for a hypothesis) suppose that it is somehow comparable to a viscid stream, or "liquid vein," thrust or spirted out from the body of the cell. But when it is once formed, this long and comparatively rigid filament is separated by a distinct surface from the neighbouring

* The presence or absence of the contractile vacuole or vacuoles is one of the chief distinctions, in systematic zoology, between the Heliozoa and the Radiolaria. As we have seen on p. 295 (footnote), it is probably no more than a physical consequence of the different conditions of existence in fresh and in salt water.

protoplasm, that is to say, from the more fluid surface-protoplasm of the cell; and the latter begins to creep up the filament, just as water would creep up the interior of a glass tube, or the sides of a glass rod immersed in the liquid. It is the simple case of a balance between three separate tensions: (1) that between the filament and the adjacent protoplasm, (2) that between the filament and the adjacent water, and (3) that between the water and the protoplasm. Calling these tensions respectively T_{fp}, T_{fw}, and T_{wp}, equilibrium will be attained when the angle of contact between the fluid protoplasm and the filament is such that $\cos \alpha = \frac{T_{fw} - T_{wp}}{T_{fp}}$. It is evident in this case that the angle is a very small one. The precise form of the curve is somewhat different from that which, under ordinary circumstances, is assumed by a liquid which creeps up a solid surface, as water in contact with air creeps up a surface of glass; the difference being due to the fact that here, owing to the density of the protoplasm being all but identical with that of the surrounding medium, the whole system is practically immune from gravity. Under normal circumstances the curve is part of the "elastic curve" by which that surface of revolution is generated which we have called, after Plateau, the nodoid; but in the present case it is apparently a catenary. Whatever curve it be, it obviously forms a surface of revolution around the filament.

Since this surface-tension is symmetrical around the filament, the latter will be pulled equally in all directions; in other words the filament will tend to be set normally to the surface of the sphere, that is to say radiating directly outwards from the centre. If the distance between two adjacent filaments be considerable, the curve will simply meet the filament at the angle α already referred to; but if they be sufficiently near together, we shall have a continuous catenary curve forming a hanging loop between one filament and the other. And when this is so, and the radial filaments are more or less symmetrically interspaced, we may have a beautiful system of honeycomb-like depressions over the surface of the organism, each cell of the honeycomb having a strictly defined geometric configuration (cf. p. 710).

In the simpler Radiolaria, the spherical form of the entire organism is equally well marked; and here, as also in the more complicated

Heliozoa (such as *Actinosphaerium*), the organism is apt to be differentiated into layers, so constituting sphere within sphere, whose inter-surfaces become the seat of adsorption, and the locus of skeletal secretion. One layer at least is close-packed with vacuoles, forming an "alveolar meshwork," with the configurations of which we shall attempt in another chapter to correlate certain characteristic types of skeleton. In *Actinosphaerium* the radial filaments pass through the outer layer, and seem to rest on but do not penetrate the layer below; this must happen if the surface-energy between the one plasma-layer and the other be less than that between the filament and the water around*.

A very curious conformation is that of the vibratile "collar," found in *Codosiga* and the other "Choanoflagellates," and which we also meet with in the "collar-cells" which line the interior cavities of a sponge. Such collar-cells are always very minute, and the collar is constituted of a very delicate film which shews an undulatory or rippling motion. It is a surface of revolution, and as it maintains itself in equilibrium (though a somewhat unstable and fluctuating one) it must be, under the restraining circumstances of its case, a surface of minimal area. But it is not so easy to see what these special circumstances are, and it is obvious that the collar, if left to itself, must shrink or shrivel towards its base and become confluent with the general surface of the cell; for it has no longitudinal supports and no strengthening ring at its periphery. But in all these collar-cells, there stands within the annulus of the collar a large and powerful cilium or flagellum, in constant movement; and by the action of this flagellum, and doubtless in part also by the intrinsic vibrations of the collar itself, there is set up a constant steady current in the surrounding water, whose direction would seem to be such that it passes up the outside of the collar, down its inner side, and out in the middle in the direction of the flagellum; and there is a distinct eddy, in which foreign particles tend to be caught, around the peripheral margin of the collar†.

* Cf. N. K. Koltzoff, *Anat. Anzeiger*, XLI, p. 190, 1912.

† The very minute size of *Codosiga*, whose collar and flagellum measure about 30–40 μ, and of all such collar-cells, make the apparently complex current-system all the harder to comprehend. Cf. G. Lepage, Notes on *C. botrytis*, *Q.J.M.S.* LXIX, pp. 471–508, 1925.

When the cell dies, that is to say when motion ceases, the collar immediately shrivels away and disappears. It is notable, by the way, that the edge of this little mobile cup is always smooth, never notched or lobed as in the cases we have discussed on p. 390: this latter condition being the outcome of a definite instability, marking the close of a period of equilibrium. But the vibratile collar of *Codosiga* is in "a steady state," its equilibrium, such as it is, being constantly renewed and perpetuated, like that of a juggler's pole, by the motions of the system. Somehow its existence is due to the current motions and to the traction exerted upon it through the frictien of the stream which is constantly passing by. In short, I think that it is formed very much in the same way as the cup-like ring of streaming ribbons, which we see fluttering and vibrating in the air-current of a ventilating fan. If we turn once more to Mr Worthington's *Study of Splashes*, we may find a curious suggestion of analogy in the beautiful craters encircling a central jet (as the collar of *Codosiga* encircles the flagellum), which we see produced in the later stages of the splash of a pebble.

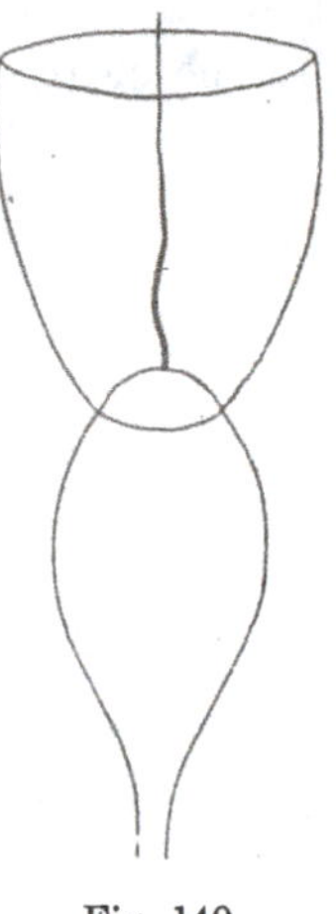

Fig. 140.

Another exceptional form of cell, and beautiful manifestation of capillarity, occurs in Trypanosomes, those tiny parasites of the blood which are associated with sleeping-sickness and certain other dire maladies of beast and man. These minute organisms consist of elongated solitary cells down one side of which runs a very delicate frill, or "undulating membrane," the free edge of which is seen to be slightly thickened, and the whole of which undergoes rhythmical and beautiful wavy movements. When certain Trypanosomes are artificially cultivated (for instance *T. rotatorium*, from the blood of the frog), phases of growth are witnessed in which the organism has no undulating membrane, but possesses a long cilium or "flagellum," springing from near the front end, and exceeding the whole body in length*. Again, in *T. lewisii*, when it reproduces by "multiple fission," the products of this division are likewise

* Cf. Doflein, *Lehrbuch der Protozoenkunde*, 1911, p. 422.

devoid of an undulating membrane, but are provided with a long free flagellum*. It is a plausible assumption to suppose that, as the flagellum waves about, it comes to lie near and parallel to the body of the cell, and that the frill or undulating membrane is formed by the clear, fluid protoplasm of the surface layer springing up in

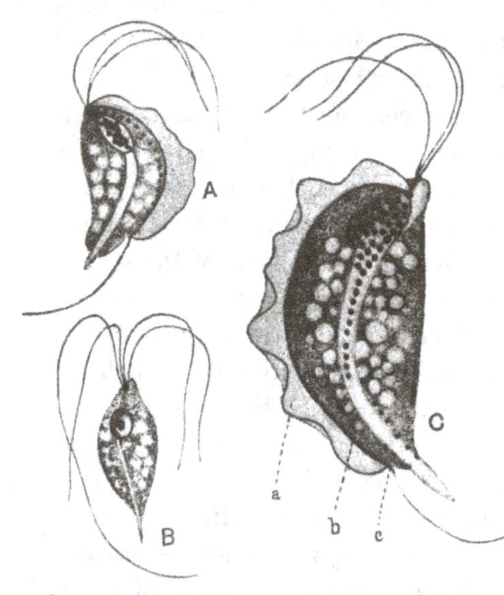

Fig. 141. *A*, *Trichomonas muris* Hartmann; *B*, *Trichomastix serpentis* Dobell; *C*, *Trichomonas angusta* Alexeieff. After Kofoid.

Fig. 142. A Trypanosome.

a film to run up and along the flagellum, just as a soap-film would form under similar circumstances.

This mode of formation of the undulating membrane or frill appears to be confirmed by the appearances shewn in Fig. 141. Here we have three little organisms closely allied to the ordinary Trypanosomes, of which one, *Trichomastix* (*B*), possesses four flagella, and the other two, *Trichomonas*, apparently three only:

* Cf. Minchin, *Introduction to the Study of the Protozoa*, 1914, p. 293, Fig. 127.

the two latter possess the frill, which is lacking in the first*. But it is impossible to doubt that when the frill is present (as in *A* and *C*), its outer edge is constituted by the apparently missing flagellum *a*, which has become *attached* to the body of the creature at the point *c*, near its posterior end; and all along its course the superficial protoplasm has been drawn out into a film, between the flagellum *a* and the adjacent surface or edge of the body *b*.

Moreover, this mode of formation has been actually witnessed and described, though in a somewhat exceptional case. The little flagellate monad *Herpetomonas* is normally destitute of an undulating membrane, but possesses a single long terminal flagellum. According to Prof. D. L. Mackinnon, the cytoplasm in a certain stage of growth becomes somewhat "sticky," a phrase which we may in all probability interpret to mean that its surface-tension is being reduced. For this stickiness is shewn in two ways. In the first place, the long body, in the course of its various bending movements, is apt to adhere head to tail (so to speak), giving a rounded or sometimes annular form to the organism, such as has also been described in certain species or stages of Trypanosomes. But again, the long flagellum, if it get bent backwards upon the body, tends to adhere to its surface. "Where the flagellum was pretty long and active, its efforts to continue movement under these abnormal conditions resulted in the gradual lifting up from the cytoplasm of the body of a sort of *pseudo*-undulating membrane (Fig. 143). The movements of this structure were so exactly those of a true undulating membrane that it was difficult to believe one was not dealing with a small, blunt Trypanosome"*. This in short is a precise

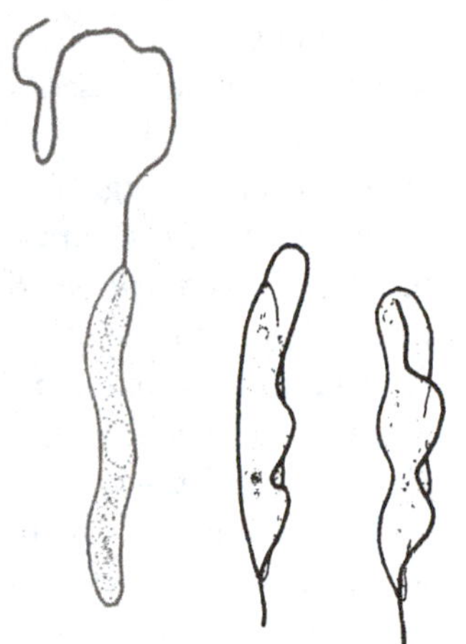

Fig. 143. *Herpetomonas* assuming the undulatory membrane of a Trypanosome. After D. L. Mackinnon.

* Cf. C. A. Kofoid and Olive Swezy, On Trichomonad flagellates, etc., *Pr. Amer. Acad. of Arts and Sci.* LI, pp. 289–378, 1915. Also C. H. Martin and Muriel Robertson, *Q.J.M.S.* LVII, pp. 53–81, 1912.

† D. L. Mackinnon, Herpetomonads from the alimentary tract of certain dungflies, *Parasitology*, III, p. 268, 1910.

description of the mode of development which, from theoretical considerations alone, we should conceive to be the natural if not the only possible way in which the undulating membrane could come into existence.

There is a genus closely allied to *Trypanosoma*, viz. *Trypanoplasma*, which possesses one free flagellum, together with an undulating membrane; and it resembles the neighbouring genus *Bodo*, save that the latter has two flagella and no undulating membrane. In like manner, *Trypanosoma* so closely resembles *Herpetomonas* that, when individuals ascribed to the former genus exhibit a free flagellum only, they are said to be in the "Herpetomonas stage." In short, all through the order, we have pairs of genera which are presumed to be separate and distinct, viz..*Trypanosoma-Herpetomonas*, *Trypanoplasma-Bodo*, *Trichomastix-Trichomonas*, in which one differs from the other mainly if not solely in the fact that a free flagellum in the one is replaced by an undulating membrane in the other. We can scarcely doubt that the two structures are essentially one and the same.

The undulating membrane of a Trypanosome, then, according to our interpretation of it, is a liquid film and must obey the law of constant mean curvature. It is under curious limitations of freedom: for by one border it is attached to the comparatively motionless body, while its free border is constituted by a flagellum which retains its activity and is being constantly thrown, like the lash of a whip, into wavy curves. It follows that the membrane, for every alteration of its longitudinal curvature, must at the same instant become curved in a direction perpendicular thereto; it bends, not as a tape bends, but with the accompaniment of beautiful but tiny waves of double curvature, all tending towards the establishment of an "equipotential surface", which indeed, as it is under no pressure on either side, is really a surface of no curvature at all; and its characteristic undulations are not originated by an active mobility of the membrane but are due to the molecular tensions which produce the very same result in a soap-film under similar circumstances. Some of the larger Spirochaetes possess a structure so like to the undulating membrane of the Trypanosomes that it has led some persons to include these peculiar allies of the bacteria among the flagellate protozoa; but it would seem (according to the weight of evidence) that the Spirochaete membrane does not undulate, and possesses no thickened border or marginal filament (*Randfade*)*. It forms a "screw-surface," or helicoid, and, though we might think that nothing could well be more curved, yet its mathematical properties are such that it constitutes a "ruled surface" whose mean curvature is everywhere *nil*. Precisely such a surface, and of exquisite beauty, may be produced by bending a wire upon itself so that part forms an axial rod and part winds spirally round the axis, and then dipping the whole into a soapy solution.

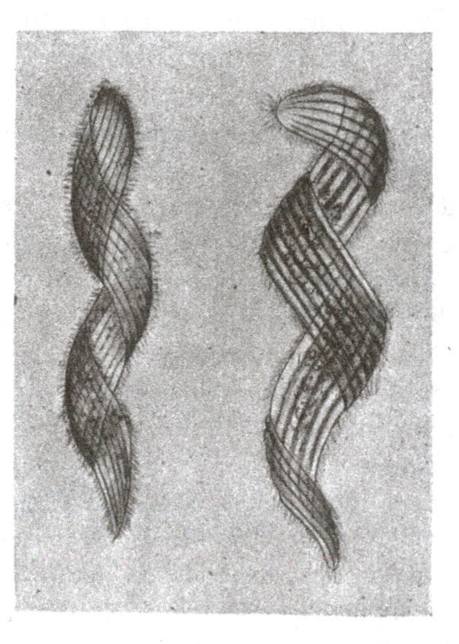

Fig. 144. *Dinenympha gracilis* Leidy.

A peculiar type is the flattened spiral of *Dinenympha*†, which reminds us of the cylindrical spiral of a *Spirillum* among the bacteria. Here we have a symmetrical figure, whose two opposite surfaces each constitute a surface of constant mean curvature; it is evidently

* For a discussion of this obscure lamella, and of the *crista* which seems to correspond with it in other species, see Doflein, *Probleme der Protistenkunde*, II, Die Natur der Spirochaeten, Jena, 1911; see also Clifford Dobell, *Arch. f. Protistenkunde*, 1912.

† Leidy, Parasites of the termites, *Journ. Nat. Sci., Philadelphia*, VIII, pp. 425–447, 1874–81; cf. Savile-Kent's *Infusoria*, II, p. 551.

a figure of equilibrium under certain special conditions of restraint. The cylindrical coil of the *Spirillum*, on the other hand, is a surface of constant mean curvature, and therefore of equilibrium, as truly, and in the same sense, as the cylinder itself.

A very beautiful "saddle-shaped" surface, of constant mean curvature, is to be found in the little diatom *Campylodiscus*, and others, a little more complicated, in the allied genus *Surirella**.

These undulating and helicoid surfaces are exactly reproduced among certain forms of spermatozoa. The tail of a spermatozoon consists normally of an axis surrounded by clearer and more fluid protoplasm, and the axis sometimes splits up into two or more slender filaments. To surface-tension operating between these and the surface of the fluid protoplasm (just as in the case of the flagellum of the Trypanosome), I ascribe the formation of the undulating membrane which we find, for instance, in the spermatozoa of the newt or salamander; and of the helicoid membrane, wrapped in a far closer and more beautiful spiral than that which we saw in Spirochaeta, which is characteristic of the spermatozoa of many birds. The undulatory membrane which certain *ciliate* infusoria exhibit is, seemingly, a different thing. It is not based on a single marginal flagellum, but consists of a row of fine cilia fused together. The membrane can be broken up by certain reagents into fibrillae, and—what is more remarkable—a touch of the micro-dissection needle may split it into a multitude of cilia, all active but beating out of time; a moment more and they unite again, all but disappearing from view as they fuse into the optically homogeneous membrane. They unite as quickly and as intimately as though they were so many liquid jets, and they manifestly "partake of fluidity." Neither they, nor cilia in general, have received, nor seem likely to receive, a simple explanation†. Nevertheless, we may see a little light in the darkness after all.

It would be overbold to seek for every form of living cell a parallel configuration due to simple capillary forces, as manifested in drop or bubble or jet. And yet, if the simple cases of sphere or cylinder be the beginning of the story, they assuredly are not the end. The

* Van Heurck, *Synopsis des Diatomées de Belgique*, pls. lxxiv, 6; lxxvii, 4.

† H. N. Maier, Der feinere Bau der Wimperapparate der Infusorien, *Arch. f. Protistenk.* II, p. 73, 1903; R. Chambers and J. A. Dawson, Structure of the undulating membrane in the ciliate *Blepharisma*, *Biol. Bull.* XLVIII, p. 240, 1925.

pointed and flagellate cell of a Monad, one of the least and commonest of micro-organisms, is far removed from a simple "drop," and all its characters, to the microscopist's eye, are both generally and specifically those of a living thing. But a drop of water falling through an electric field, as in a thunderstorm, is found to lengthen out to three or four times as long as it is broad; and then, if the strength of the field increase a little, the prolate drop becomes unstable, it grows spindle-shaped, and suddenly from one of its two pointed spindle-ends (the positive end especially) a long and slender filament shoots out, to the accompaniment of an electrical discharge. We need not assert that the phenomena are identical, nor that the forces in action are absolutely the same. Yet it is no small thing to have learned that the peculiar conformation of the little flagellate Monad has its *analogue* in an electrified drop, and is not unique after all*.

Before we pass from the subject of the conformation of the solitary cell we must take some account of certain other exceptional forms, less easy of explanation, and still less perfectly understood. Such is the case, for instance, of the red blood-corpuscles of man and other vertebrates; and among the sperm-cells of the decapod crustacea we find forms still more aberrant and not less perplexing. These are among the comparatively few cells or cell-like structures whose form *seems* to be incapable of explanation by theories of surface-tension.

In all the mammalia (save a very few) the red blood-corpuscles are flattened circular discs, dimpled in upon their two opposite sides. This configuration closely resembles that of an india-rubber ball when we pinch it tightly between finger and thumb†.

The form of the corpuscle is symmetrical; it is a solid of revolution, but its surface is not a surface of constant mean curvature. From the surface-tension point of view, the blood-corpuscle is not a surface of equilibrium; in other words, it is not a fluid drop poised

* Cf. W. A. Macky, On the deformation of water-drops in strong electric fields, *Proc. R.S.* (A), CXXXIII, pp. 565–587, 1931.

† On this analogy we might expect the double concavity to pass, with no great difficulty, into the single hollow of a cup or bell, and such a shape the blood-corpuscles are said sometimes to assume. Cf. Weidenreich, *Arch. f. mikr. Anat.* LXVII, 1902; and cf. Clerk-Maxwell on "dimples" in *Tr. R.S.E.* XXVI, p. 11, 1870.

in another liquid. Some other force or forces must be at work to conform it, and the simple effect of mechanical pressure is excluded, because the corpuscle exhibits its characteristic shape while floating freely in the blood. It has been suggested that the corpuscle is perhaps comparable to a solid of revolution described about one of Cayley's equipotential curves*, such as we have spoken of briefly on p. 318. Were the corpuscle a sphere, or a thin plate, a gas diffusing inwards would reach all parts equally soon; but the surface would be small in the one case and the volume in the other. In so far as the corpuscle resembles or approaches the equipotential form, we might look on it as a compromise; but however advantageous such a shape might be, and however interesting physiologically, we should be as far as ever from understanding how it was produced. In all other vertebrates, from fishes to birds, sluggish or active, warm-blooded or cold, the blood-corpuscles have the simpler form of a flattened oval disc, with somewhat sharp edges and ellipsoidal surfaces, and this again is manifestly not a surface of fluid equilibrium. But there is nothing to choose between the one type and the other in the way of physiological efficiency, nor any apparent need for a refinement of adaptive form in either of them.

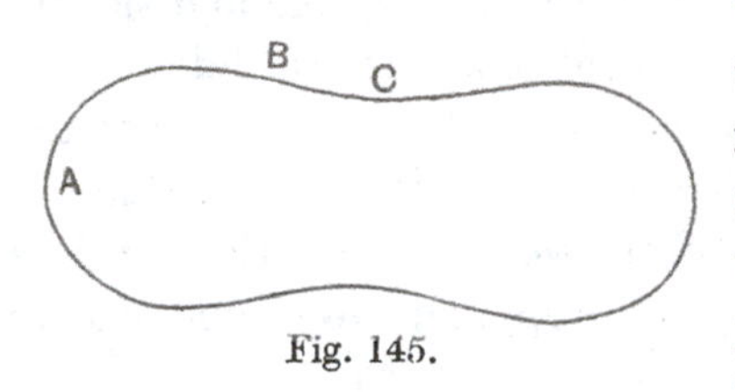

Fig. 145.

Two facts are noteworthy in connection with the form of the mammalian blood-corpuscle. In the first place its form is only maintained, that is to say it is only in equilibrium, in specific relation to the medium in which it floats. If we add water to the blood, the corpuscle becomes a spherical drop, a true surface of minimal area and stable equilibrium; if, conversely, we add a little salt, or a drop of glycerine, the corpuscle shrinks, and its surface becomes puckered and uneven. So far, it merely obeys the laws of diffusion; but the phenomenon is more complex than this†. For the spherical form is assumed just as well in various *isotonic* solutions, leaving

* Cf. H. Hartridge, *Journ. Physiol.* LII, p. lxxxi, 1919–20; Eric Ponder, *Journ. Gen. Physiol.* IX, pp. 197–204, 625–629, 1925–26.

† Cf. A. Gough, On the assumption of a spherical form by human blood-corpuscles, *Biochem. Journ.* XVIII, p. 202, 1924.

the volume unchanged; a little ammonium oxalate impedes or inhibits the change of form, a little serum brings the spherical corpuscles back to biconcave discs again. We are no longer dealing with simple diffusion, but with phenomena of a very subtle kind.

Secondly, the form of the corpuscle can be imitated artificially by means of other colloid substances. Many years ago Norris made the interesting observation that drops of glue in an emulsion assumed a biconcave form closely resembling that of the mammalian corpuscles*; the glue was impure and doubtless contained lecithin. Waymouth Reid made similar emulsions of cholesterin oleate, in which the same conformation of the drops or particles is beautifully shewn; and Emil Hatschek has made somewhat similar biconcave bodies by dropping gelatine containing potassium ferrocyanide into copper sulphate or a tannin solution. Here Hatschek believes that his biconcave drops are half-formed vortex-rings, arrested by the formation of a semi-permeable membrane; but the explanation does not seem to fit the blood-corpuscle. The cholesterin bodies in Waymouth Reid's experiment are such as have a place of their own among Lehmann's "fluid crystals"†; and it becomes at least conceivable that obscure forces akin to those of crystallisation may be playing their part along with surface-energy in these strange but familiar conformations. The case is a hard one in every way. From the physiological point of view it is difficult and complex enough. For the surface of the corpuscle is equivalent to a semi-permeable membrane‡, through which certain substances pass freely but not others—for the most part anions and not cations§; and accordingly we have here in life a steady state of osmotic inequilibrium, of negative osmotic tension within, and to this comparatively simple cause the imperfect distension of the corpuscle may be due.

* *Proc. R.S.* XII, pp. 251–257, 1862–63.

† Cf. (*int. al.*) Lehmann, Ueber scheinbar lebende Kristalle und Myelinformen, *Arch. f. Entw. Mech.* XXVI, p. 483, 1908; *Ann. d. Physik*, XLIV, p. 969, 1914.

‡ That no "true membrane" exists has long been known; cf. (*int. al.*) Röhring, *Koll. Chem. Beihefte*, VIII, pp. 337–398, 1916. On the other hand the surface of the corpuscle is defined by a monolayer, and very probably by the still more stable condition of two "interpenetrating" monolayers, a proteid and a lipoid. Cf. Eric Ponder, *Phys. Rev.* XVI, p. 19, 1936; and on "interpenetration," Schulman and Rideal in *Proc. R.S.* (B), CXXII, pp. 29–57, 1937.

§ Cf. Hamburger, *Z. f. physikal. Chem.* LXIX, p. 663, 1909; *Pflüger's Archiv*, 1902, p. 442; etc.

Whatever the forces are which make and keep the man's corpuscle a dimpled disc and the frog's a flattened ellipsoid, they seem to be of a powerful kind. When we submit either to great hydrostatic pressure, it tends to become spherical at last, the natural result of uniform pressure over its whole surface; but the pressure necessary to bring this result about is very great indeed*. Since the form of the blood-corpuscle cannot, then, be rated as a figure of equilibrium, we must be content to regard it as a "steady state"; and this, moreover, is all we can say of its physico-chemical condition. The red blood-corpuscle, especially the non-nucleated one, is in no ordinary sense *alive*†. It has no power of movement, of reproduction or of repair; it is a mere haemoglobin-freighted drop of protein; its own metabolism, apart from its alternate give and take of oxygen, is slight indeed or absent altogether. But all the same, chemical change is continually going on; anions (like HCO_3) pass freely through its walls, simple cations (like Na, K) find it impermeable; and so, between plasma and corpuscles the conditions are fulfilled for that steady osmotic state known as a "Donnan equilibrium." Somehow, but we know not how, a steady state is maintained alike in the corpuscle's osmotic equilibrium and in its form.

In mammalian blood, the running together of the round biconcave corpuscles into "rouleaux" gives a well-known and characteristic picture. When cold, rouleaux are formed slowly, in warmed plasma they form quickly and well, in salt-solution they do not form at all.

* The whole phenomenon would become simple and mechanical if we might postulate a stiffer peripheral region to the corpuscle, in the form (for instance) of an elastic ring. Such an annular stiffening, like the "collapse-rings" which an engineer inserts in a boiler or the whalebone ring which a Breton fisherman fits into his *beret*, has been repeatedly asserted to exist; by Dehler, *Arch. f. mikr. Anat.* XLVI, 1895; by Meves, *ibid.* LXXVII, 1911; and especially by J. Rünnstrom, Was bedingt die Form und die Formveränderungen der Saügetiererythrocyten, *Arch. f. Entw. Mech.* I, pp. 391–409, 1922. It has been denied at least as often; but the remarkable statement has been lately made that in a corpuscle which has been swollen up and then brought back to its biconcave form, the dimples reappear on the same sides as before: apparently in "strong evidence for some sort of fixed cellular structure"; see R. F. Furchgott and Eric Ponder in *Journ. Exp. Biol.* XVII, pp. 30–44, 117–127, 1940. See also, on the whole subject, Eric Ponder, *The Mammalian Red Cell*, Berlin, 1934.

† Cf. A. V. Hill, *Trans. Faraday Soc.* XXVI, p. 667, 1930; *Proc. R.S.* (B), 1930; K. R. Dixon, in *Current Sci.* VII, p. 169, 1938; etc.

The phenomenon, though a purely physical one, is none too clear. There is a difference of electrical potential between corpuscles and plasma, and the charged corpuscles tend to repel one another; but they also tend to adhere together, all the more when they meet broadside on, whether by actual stickiness or through surface-energy. The attractive forces then overcome the repulsive, and the rouleau is formed. But if the potential be reduced, and mutual repulsion reduced with it, then the corpuscles stick together just as they happen to meet; rouleaux are no longer formed, and ordinary "agglutination" takes place. Whatever be the precise nature of the phenomenon, the number of rouleaux and the mean number of

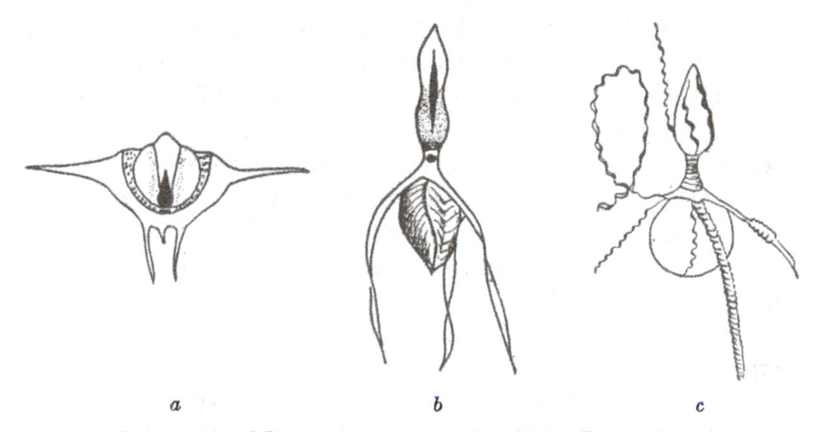

Fig. 146. Sperm-cells of Decapod crustacea (after Koltzoff). *a, Inachus scorpio*; *b, Galathea squamifera*; *c*, do. after maceration, to shew spiral fibrillae.

corpuscles in each is found, after a given time, to obey a certain law (Smoluchowsky's Law), defining the number of contacts of floating bodies under ordinary physical conditions*.

The sperm-cells of the Decapod crustacea exhibit various singular shapes. In the crayfish they are flattened cells with stiff curved processes radiating outwards like St Catherine's wheel; in *Inachus* there are two such circles of stiff processes; in *Galathea* we have a still more complicated form, with long and slightly twisted processes.

* Smoluchowsky, *Ztschr. f. physik. Chemie*, XCIX, p. 129, 1917; Eric Ponder, On Rouleaux-formation, *Q. Journ. Exp. Physiol.* XVI, pp. 173–194, 1926.

In all these cases, just as in the case of the blood-corpuscle, the structure alters, and finally loses, its characteristic form when the constitution of the surrounding medium is changed*.

Here again, as in the blood-corpuscle, we have to do with the important force of osmosis, manifested under conditions similar to those of Pfeffer's classical experiments on the plant-cell†. The surface of the cell acts as a semi-permeable membrane, permitting the passage of certain dissolved substances (or their ions), and including or excluding others: and thus rendering manifest and measurable the existence of a definite "osmotic pressure." Again, in the hen's egg a delicate yolk-membrane separates the yolk from the white. The morphologist looks on it but as the cell-wall of a vast yolk-laden germ-cell; the physiologist sees in it a semi-permeable membrane, the seat of many complex activities. The end and upshot of these last is that a steady difference of osmotic pressure, the equivalent of some two atmospheres, is maintained between yolk and white; and yet there is no current flowing through. Somewhere or other in the system there is a constant metabolic flux, a continuous liberation of energy, a continual doing of work, all leading to the maintenance of a steady dynamical state, which is not "equilibrium‡."

In the case of the sperm-cells of *Inachus*, certain quantitative experiments have been performed. The sperm-cell exhibits its characteristic conformation while lying in the serous fluid of the animal's body, in ordinary sea-water, or in a 5 per cent. solution of potassium nitrate, these three fluids being all "isotonic" with one another. As we alter the concentration of potassium nitrate, the cell assumes certain definite forms corresponding to definite concentrations of the salt; and, as a further and final proof that the phenomenon is entirely physical, it is found that other salts produce an identical effect when their concentration is proportionate to their molecular weight, and whatever identical effect is produced

* Cf. N. K. Koltzoff, Studien über die Gestalt der Zelle, *Arch. f. mikrosk. Anat.* LXVII, pp. 365–572, 1905; *Biol. Centralbl.* XXIII, pp. 680–696, 1903; XXVI, pp. 854–863, 1906; XLVIII, pp. 345–369, 1928; *Arch. f. Zellforschung,* II, pp. 1–65, 1908; VII, pp. 344–423, 1911; *Anat. Anzeiger,* XLI, pp. 183–206, 1912.

† W. Pfeffer, *Osmotische Untersuchungen,* Leipzig, 1877.

‡ Cf. J. Straub, Der Unterschied in osmotischer Konzentration zwischen Eigelb und Eiklar, *Rec. Trav. Chim. du Pays-Bas,* XLVIII, p. 49, 1929.

by various salts in their respective concentrations, a similarly identical effect is produced when these concentrations are doubled or otherwise proportionately changed.

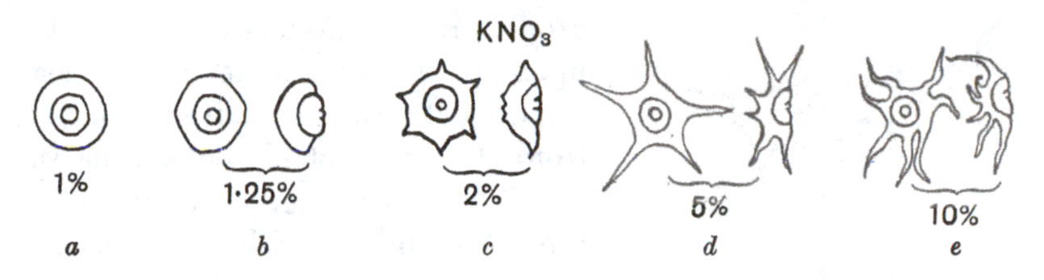

Fig. 147. Sperm-cells of *Inachus*, as they appear in saline solutions of varying density. After Koltzoff.

Thus the following table shews the percentage concentrations of certain salts necessary to bring the cell into the forms *a* and *c* of Fig. 147; in each case the quantities are proportional to the molecular weights, and in each case twice the quantity is necessary to produce the effect of *c*, compared with that which gives rise to the all but spherical form of *a*.

	% concentration of salts in which the sperm-cell of *Inachus* assumes the form of	
	a	*c*
Sodium chloride	0·6	1·2
Sodium nitrate	0·85	1·7
Potassium nitrate	1·0	2·0
Acetic acid	2·2	4·5
Cane sugar	5·0	10·0

If we look then upon the spherical form of this cell as its true condition of symmetry and of equilibrium, we see that what we call its normal appearance is just one of many intermediate phases of shrinkage, brought about by the abstraction of fluid from its interior as the result of an osmotic pressure greater outside than inside the cell, and where the shrinkage of *volume* is not kept pace with by a contraction of the *surface-area*. In the case of the blood-corpuscle, the shrinkage is of no great amount, and the resulting deformation is symmetrical; such structural inequality as may be necessary to account for it need be but small. But in the case of the sperm-cells, we must have, and we actually do find, a somewhat complicated arrangement of more or less rigid or elastic structures in the wall of the cell, which, like the wire framework in Plateau's experiments, restrain and modify the forces acting on the drop. In one form of Plateau's experiments, instead of supporting his drop on rings or frames of wire, he laid upon its surface one or more elastic coils; and then, on withdrawing oil from the centre of his globule, he saw its uniform shrinkage counteracted by the spiral springs, with the result that the centre of each elastic coil seemed to shoot out into a prominence. Just such spiral coils are figured (after Koltzoff) in Fig. 148*; and they may be regarded as closely akin to those local thickenings or striations, spiral and other, which are common in vegetable cells.

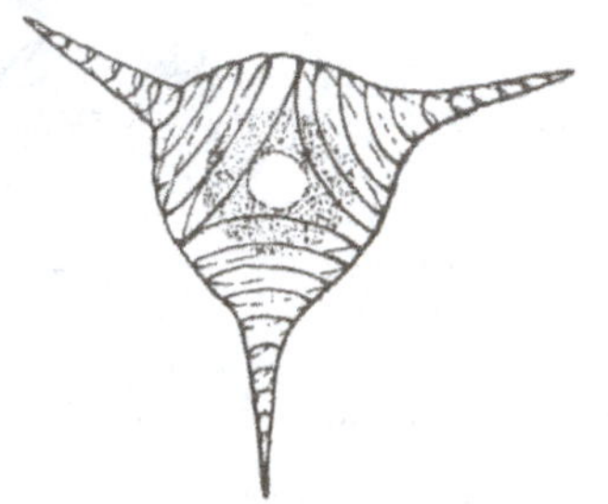

Fig. 148. Sperm-cell of *Dromia*. After Koltzoff.

Physically speaking, the protoplasmic colloids are neither simple nor uniform. We begin by thinking of our cell as a *drop* of a homogeneous fluid and on this bold simplifying assumption we account for its form to a first, and often to a near, approximation. For the cell is largely composed of fluid "hydrosols," which are still fluid however viscous they may be, and still tend towards rounded, drop-like configurations. But it has also its "hydrogels," which shew a certain tenacity, a certain elasticity, a certain reluctance to let their particles move on one another; and of these are formed the scarce distinguishable fibrillae within a host of highly specialised cells, the elastic fibres of a tendon, the incipient cell-walls of a plant, the rudiments of many axial and skeletal structures.

The cases which we have just dealt with lead us to another consideration. In a semi-permeable membrane, through which water passes freely in and out, the conditions of a liquid surface are greatly modified; in the ideal or ultimate case, there is neither surface nor surface-tension at all. And this would lead us some-

* As Bethe points out (Zellgestalt, Plateausche Flüssigkeitsfigur und Neurofibrille, *Anat. Anz.* XL, p. 209, 1911), the spiral fibres of which Koltzoff speaks must lie *in the surface*, and not within the substance, of the cell whose conformation is affected by them.

what to reconsider our position, and to enquire whether the true surface-tension of a liquid film is actually responsible for *all* that we have ascribed to it, or whether certain of the phenomena which we have assigned to that cause may not in part be due to the contractility of definite and elastic membranes. But to investigate this question, in particular cases, is rather for the physiologist: and the morphologist may go his way, paying little heed to what is no great difficulty. For in surface-tension we have the production of a film with the properties of an elastic membrane, and with the special peculiarity that contraction continues with the same energy however far the process may have already gone; while the ordinary elastic membrane contracts to a certain extent, and contracts no more. But within wide limits the essential phenomena are the same in both cases. Our fundamental equations apply to both cases alike. And accordingly, so long as our purpose is *morphological*, so long as what we seek to explain is regularity and definiteness of form, it matters little if we should happen, here or there, to confuse surface-tension with elasticity, the contractile forces manifested at a liquid surface with those which come into play at the complex internal surfaces of an elastic solid.

CHAPTER VI

A NOTE ON ADSORPTION

An important corollary to, or amplification of, the theory of surface-tension is to be found in the chemico-physical doctrine of Adsorption; which means, in a word, the concentration of a substance *at a surface*, by reason of that *surface-energy* of which we have had so much to say*. Charcoal, with its vast internal surface-area of carbonised cell-walls, is the commonest and most familiar of adsorbents, and of it Du Bois Reymond first used the name. In its full statement this subject becomes very complicated, and involves physical conceptions and mathematical treatment which go far beyond our range. But it is necessary for us to take account of the phenomenon, even though it be in the most elementary way.

In the brief account of the theory of surface-tension with which our last chapter began, it was shewn that, in a drop of liquid, the potential energy of the system could be diminished, and work manifested accordingly, in two ways. In the first place we saw that, at our liquid surface, surface-tension tends to set up an equilibrium of form, in which the surface is reduced or contracted either to the absolute minimum of a sphere, or at any rate to the least possible area which is permitted by the various circumstances and conditions; and if the two bodies which comprise our system, namely the drop of liquid and its surrounding medium, be simple substances, and the system be uncomplicated by other distributions of force, then the energy of the system will have done its work when this equilibrium of form, this minimal area of surface, is once attained. This phenomenon of the production of a minimal surface-area we have now seen to be of fundamental importance in the external

* Some define adsorption as surface-condensation, without reference to the forces which produce it; in other words they recognize chemical, electrical and other forces, including cohesion, as producing analogous or indistinguishable results: cf. A. P. Mathews, in *Physiological Reviews*, I, pp. 553–597, 1921.

morphology of the cell, and especially (so far as we have yet gone) of the solitary cell or unicellular organism.

But we also saw, according to Gauss's equation, that the potential energy of the system will be diminished (and its diminution will accordingly be manifested in work) if from any cause the specific surface-energy be diminished, that is to say if it be brought more nearly to an equality with the specific energy of the molecules in the interior of the liquid mass. This latter is a phenomenon of great moment in physiology, and, while we need not attempt to deal with it in detail, it has a bearing on cell-form and cell-structure which we cannot afford to overlook.

A diminution of the surface-energy may be brought about in various ways. For instance, it is known that every isolated drop of fluid has, under normal circumstances, a surface-charge of electricity: in such a way that a positive or negative charge (as the case may be) is inherent in the surface of the drop, while a corresponding charge, of contrary sign, is inherent in the immediately adjacent molecular layer of the surrounding medium. Now the effect of this distribution, by which all the surface molecules of our drop are similarly charged, is that by virtue of the charge they tend to repel one another, and possibly also to draw other molecules, of opposite charge, from the interior of the mass; the result being in either case to antagonise or cancel, more or less, that normal tendency of the surface molecules to attract one another which is manifested in surface-tension. In other words, an increased electrical charge concentrating at the surface of a drop tends, whether it be positive or negative, to *lower* the surface-tension.

Again, a rise of temperature diminishes surface-tension, and consequently facilitates the formation of a bubble or a froth. It follows (from the principle of Le Chatelier) that foam is warmer than the fluid of which it is made, and the difference is all the greater the lower the concentration of the foaming (or capillary-active) substance*.

But a still more important case has next to be considered. Let us suppose that our drop consists no longer of a single chemical substance, but contains other substances either in suspension or in solution. Suppose (as a very simple case) that it be a watery

* Cf. Fr. Schütz, in *Nature*, April 10, 1937. In the case of 0·01 per cent. solution of saponin, the temperature-difference is no less than 3·3° C.

fluid, exposed to air, and containing droplets of oil: we know that the specific surface-tension of oil in contact with air is much less than that of water, and it follows that, if the watery surface of our drop be replaced by an oily surface the specific surface-energy of the system will be notably diminished. Now under these circumstances it is found that (quite apart from gravity, which might cause it to *float* to the surface) the oil has a tendency to be *drawn* to the surface; and again this phenomenon of molecular attraction or adsorption represents work done, equivalent to the diminished potential energy of the system*. In more general terms, if a liquid be a chemical mixture, some one constituent in which, if it entered into or increased in amount in the surface layer, would have the effect of diminishing its surface-tension, then that constituent will have a tendency to accumulate or concentrate at the surface: the surface-tension may be said, as it were, to exercise an attraction on this constituent substance, drawing it into the surface-layer, and this tendency will proceed until at a certain "surface-concentration" equilibrium is reached, its opponent being that osmotic force which tends to keep the substance in uniform solution or diffusion. In other words, in any "two-phase" system, a change of concentration at the boundary-surface and a diminution of surface-tension there accompany one another of necessity; positive adsorption means negative surface-tension, and *vice versa*. Furthermore, the lowering of surface-tension (as by saponin) will permit (*caeteris paribus*) an extension of surface, manifesting itself in "froth." Thus the production of a froth and the concentration of appropriate substances therein are two sides of one and the same phenomenon.

In the complex mixtures which constitute the protoplasm of the living cell, this phenomenon of adsorption has abundant play: for many of its constituents, such as fats, soaps, proteins, lecithin, etc., possess the required property of diminishing surface-tension.

* The first instance of what we now call an adsorptive phenomenon was observed in soap-bubbles. Leidenfrost was aware that the outer layer of the bubble was covered by an "oily" layer (*De aquae communis nonnullis qualitatibus tractatus*, Duisburg, 1756). A hundred years later Dupré shewed that in a soap-solution the soap tends to concentrate at the surface, so that the surface-tension of a very weak solution is very little different from that of a strong one (*Théorie mécanique de la chaleur*, 1869, p. 376; cf. Plateau, II, p. 100).

Moreover, the more a substance has the power of lowering the surface-tension of the liquid in which it happens to be dissolved, the more will it tend to displace another and less effective substance from the surface-layer. Thus we know that protoplasm always contains fats, not only in visible drops, but also in the finest suspension or "colloidal solution"; and if under any impulse, such for instance as might arise from the Brownian movement, a droplet of oil be brought close to the surface, it is at once drawn into that surface and tends to spread itself in a thin layer over the whole surface of the cell. But a soapy surface (for instance) in contact with the surrounding water would have a surface-tension even less than that of the film of oil: and consequently, if soap be present in the water it will in turn be adsorbed, and will tend to displace the oil from the surface pellicle*. And all this is as much as to say that the molecules of the dissolved or suspended substance or substances will so distribute themselves throughout the drop as to lead towards an equilibrium, for each small unit of volume, between the superficial and internal energy; or, in other words, so as to reduce towards a minimum the potential energy of the system. This tendency to concentration at a surface of any substance within the cell by which the surface-tension tends to be diminished, or *vice versa*, constitutes, then, the phenomenon of *adsorption*; and the general statement by which it is defined is known as the Willard-Gibbs, or Gibbs-Thomson law†, and was arrived at not by experimental but by theoretical and hydrodynamical methods.

An assemblage of drops or droplets offers a great extension of surface, but so also does an assemblage of equally minute cells or

* This identical phenomenon was the basis of Quincke's theory of amoeboid movement (Ueber periodische Ausbreitung von Flüssigkeitsoberflächen, etc., *SB. Berlin. Akad.* 1888, pp. 791–806; cf. *Pflüger's Archiv*, 1879, p. 136). We must bear in mind that to describe an amoeboid cell as "naked" does not imply that its outer layer is identical with its internal substance.

† J. Willard Gibbs, Equilibrium of heterogeneous substances, *Tr. Conn. Acad.* III, pp. 380–400, 1876, also in *Collected Papers*, I, pp. 185–218, London, 1906; J. J. Thomson, *Applications of Dynamics to Physics and Chemistry*, 1888 (Surface tension of solutions), p. 190. See also (*int. al.*) various papers by C. M. Lewis, *Phil. Mag.* (6), XV, p. 499, 1908; XVII, p. 466, 1909; *Zeitschr. f. physik. Chemie*, LXX, p. 129, 1910; Milner, *Phil. Mag.* (6), XIII, p. 96, 1907; A. B. Macallum, The role of surface-tension in determining the distribution of salts in living matter, *Trans. 15th Int. Congress on Hygiene, etc.*, Washington, 1912; etc.

pores; both alike are "two-phase" systems, and in both alike the phenomenon of adsorption has free play. The occlusion of gases, including water-vapour, by charcoal is a familiar phenomenon of adsorption, and is due to the minuteness of the pores only in so far as surface-area is increased and multiplied thereby. For surface-energy is surface-strain or surface-tension × surface-area, and is vastly increased by minute subdivision. And surface-energy is such that, whenever a substance is introduced into a two-phase system—which merely means two things in touch (or surface-contact) with one another—it is apt to concentrate itself *on the surface* where the two phases meet. *Absorption* implies uniform distribution, as when a gas is absorbed by a liquid; *adsorption* implies a heterogeneous field, and a concentration localised on the surfaces therein.

Among the many important physical features or concomitants of this phenomenon, let us take note at present that we need not conceive of a strictly superficial distribution of the adsorbed substance, that is to say of its direct association with the surface-layer of molecules such as we imagined in the case of an electrical charge; but rather of a progressive tendency to concentrate more and more, the nearer the surface is approached. Indeed we may conceive the colloid or gelatinous precipitate in which, in the case of our protoplasmic cell, the dissolved substance tends often to be thrown down, to constitute one boundary layer after another, the general effect being intensified and multiplied by the repetition of these new surfaces.

Moreover, it is not less important to observe that the process of adsorption, in the neighbourhood of the surface of a heterogeneous liquid mass, is a process which *takes time*; the tendency to surface concentration is a gradual and progressive one, and will fluctuate with every minute change in the composition of our substance and with every change in the area of its surface. In other words, it involves (in every heterogeneous substance) a continual instability: and a constant manifestation of motion, sometimes in the mere invisible transfer of molecules, but often in the production of visible currents, or manifest alterations in the form or outline of the system.

Cellular activity is of necessity associated with cellular structure, even in our simplest interpretation thereof, as a mere increase of surface due to the existence and the multiplication of cells. In the chemistry of the tissues there may be substances (catalysts and others) which exhibit their proper reactions even though the cells containing them be disintegrated or destroyed; but other processes, oxidation itself among them, are essentially surface-actions, based on adsorption at the vast cell-surface of the tissue*. The breaking-down of the cell-walls, the disintegration of cellular structure in a tissue, brings about "a biochemical chaos, a medley of reactions†." Cells are not merely there because the tissue has grown by their multiplication; there are physico-chemical reasons, even of an elementary kind, which render the morphological phenomenon of the cell indispensable to physiological action.

The physiologist deals with the surface-phenomena of the cell in ways undreamed of when I began to write this book. To begin with, the concept of a *surface* (in the old mathematical or quasi-mathematical sense) no longer suffices to describe the boundary conditions of even a "naked" protoplasmic cell. As Rayleigh foretold, and as Irving Langmuir has proved, the "boundary-state" consists of a layer of complex molecules, each one a long array of atoms, all set side by side in an orderly and uniform way. There is not merely a boundary-surface between two phases (as the older colloid chemistry supposed) but a *boundary-layer*, which itself constitutes a third phase, or interphase, and which part of the surface-energy has gone to the making of.

Surface-energy plays a leading part in modern theories of muscular contraction, and has indeed done so ever since FitzGerald and d'Arsonval indicated a connection between them some sixty years or more ago‡. It plays its part handsomely (we may be sure) in the electric pile of the Torpedo, where two million tiny discs present a

* Many surface-active substances are known to be among the most active pharmacologically; cf. Michaelis and Rona, *Physikal. Chemie*, 1930.

† A. V. Hill, *Proc. R.S.* (B), CIII, p. 138; cf. also M. Penrose and J. H. Quastel, on Cell structure and cell activity, *ibid.* CVII, p. 168.

‡ Cf. G. F. FitzGerald, On the theory of muscular contraction, *Brit. Ass. Rep.* 1878; also in *Scientific Writings*, ed. Larmor, 1902, pp. 34, 75. A. d'Arsonval, Relations entre l'électricité animale et la tension superficielle, *C.R.* CVI, p. 1740, 1888; A. Imbert, Le mécanisme de la contraction musculaire, déduit de la considération des forces de tension superficielle, *Arch. de Phys.* (5), IX, pp. 289–301, 1897; A. J. Ewart, *Protoplasmic Streaming in Plants*, Oxford, 1903, pp. 112–119.

vast aggregate of interfacial contact. It gives us a new conception, as Wolfgang Ostwald was the first to shew, of the relation of oxygen to the red corpuscles of the blood*. But many more and still more complicated "film-reactions" are started or intensified by the oriented molecules of the monolayer. The catalytic action of living ferments (a subject vast indeed) is largely a question of modified adsorption, or of surface-action. The range of bodies so adsorbed is extremely limited; the specific reactions, which depend on the bacterium engaged, are fewer still; and sometimes a whole class of substances may be adsorbed, and only one of them thrown specifically into action†. The physiological, and sometimes lethal, actions of various substances are examples of similar effects. The chemistry of the surface-layer in this cell or that may be elucidated by its reactions to various "penetrants," and depends somehow on the molecular orientation of the surfaces, and on the potentials associated with the characteristic electric fields which we may suppose to correspond to the particular molecular arrangements‡.

It is the dynamic aspect of the case, the ingresses, egresses and metabolic changes associated with the boundary-layer, which interest the physiologist. He finds the monolayer acting in ways not known in a homogeneous liquid—and adsorption is one of these ways. We keep as much as may be to the morphological side of the case rather than to the physiological, to the static side rather than to the dynamic, to the equilibrium attained rather than to the energies to which it is due. We continue to speak of surface, and of surface-

* Ueber die Natur der Bindung der Gase im Blut und in seinen Bestandteilen, *Kolloid Ztschr.* II, pp. 264–272, 1908; cf. Loewy, Dissociationsspannung des Oxyhaemoglobin im Blut, *Arch. f. Anat. u. Physiol.* 1904, p. 231. Arrhenius remarked long ago that the forces which produce adsorption are of the same order, and of the same nature, as those which cause the mutual attractions of the molecules of a gas. Hence the *order* is constant in which various gases are adsorbed by different adsorbents. The question of the inner mechanism of the forces which result in surface-tension, adsorption and allied phenomena, and their relation to electric charge on particles or ions, belongs to the highest parts of physical chemistry. Besides countless recent papers, M. v. Smoluchowski's Versuch einer mathematischen Theorie der Koagulationskritik, *Z. f. physik. Chemie,* XCII, pp. 129–168, 1918, is still interesting.

† Cf. N. K. Adams, *Physics and Chemistry of Surfaces*, 1930. Also (*int. al.*) J. H. Quastel, Mechanism of bacterial action, *Trans. Faraday Soc.* XXVI, pp. 831–861, 1930.

‡ This is Loeb's so-called "membrane-effect," cf. *Journ. Biol. Chemistry*, XXXII, p. 147, 1917; and J. Gray, *Journ. Physiol.* LIV, pp. 68–78, 1920.

energy and of adsorptive phenomena, in a somewhat old-fashioned way; but even with this simplifying limitation we find them helpful, throwing light upon our subject.

In the first place our preliminary account, such as it is, is already tantamount to a description of the process of development of a cell-membrane, or cell-wall. The so-called "secretion" of this cell-wall is nothing more than a sort of exudation, or striving towards the surface, of certain constituent molecules or particles within the cell; and the Gibbs-Thomson law formulates, in part at least, the conditions under which they do so. The adsorbed material may range from an almost unrecognisable pellicle to the distinctly differentiated "ectosarc" of a protozoon, and again to the development of a fully-formed cell-wall, as in the cellulose partitions of a vegetable tissue. In such cases, the dissolved and adsorbtive material has not only the property of lowering the surface-tension, and hence of itself accumulating at the surface, but has also the property of increasing the viscosity and mechanical rigidity of the material in which it is dissolved or suspended, and so of constituting a visible and tangible "membrane*." The "zoogloea" around a group of bacteria is probably a phenomenon of the same order. In the superficial deposition of inorganic materials we see the same process abundantly exemplified. Not only do we have the simple case of the building of a shell or "test" upon the outward surface of a living cell, as for instance in a Foraminifer, but in a subsequent chapter, when we come to deal with spicules and spicular skeletons such as those of the sponges and of the Radiolaria, we shall see how highly characteristic it is of the whole process of

* We may trace the first steps in the study of this phenomenon to Melsens, who found that thin films of white of egg become firm and insoluble (Sur les modifications apportées à l'albumine...par l'action purement mécanique, *C.R.* XXXIII, p. 247; *Ann. de chimie et de physique* (3), XXXIII, p. 170, 1851); and Harting made similar observations about the same time. Ramsden investigated the same subject, and also the more general phenomenon of the formation of albuminoid and fatty membranes by adsorption, and found (*int. al.*) that on shaking white of egg practically all the albumin passes gradually into the froth; cf. his Koagulierung der Eiweisskörper auf mechanischer Wege, *Arch. f. Anat. u. Phys.* (*Phys. Abth.*), 1894, p. 517; Abscheidung fester Körper in Oberflächenschichten, *Z. f. phys. Chem.* XLVII, p. 341, 1902; *Proc. R.S.* LXXII, p. 156, 1904. For a general review of the whole subject see H. Zangger, Ueber Membranen und Membranfunktionen, in Asher-Spiro's *Ergebnisse der Physiologie*, VII, pp. 99–160, 1908.

spicule-formation for the deposits to be laid down just in the "interfacial" boundaries between cells or vacuoles, and how the form of the spicular structures tends in many cases to be regulated and determined by the arrangement of these boundaries. The so-called *collenchyma*, in which an excess of cellulose is laid down around the angles of contact of adjacent cells, in a kind of exaggerated "bourrelet," is another case in point *.

No pure liquid ever forms a froth or foam. White of egg is no exception to the rule; for the albumin is somehow changed, or "denatured," and becomes a quasi-solid when we beat it up. But in the frothing liquid there must always be some admixture present to concentrate on, or be adsorbed by, the surfaces and interfaces of the other; and this dispersion must go on completely and uniformly, so as to leave the whole system homogeneous. The resulting diminution of surface-tension facilitates the subdivision of the bubbles and dispersion of the air; and the adsorbed surface-layer gives firmness and stability to the system. The sudden increase of surface diminishes, for the moment, the concentration, or "thickness" of the surface-layer; the tension rises accordingly, and the cycle of operations begins anew †.

In physical chemistry, a distinction is usually drawn between adsorption and *pseudo-adsorption*, the former being a *reversible*, the latter an irreversible or permanent phenomenon. That is to say, adsorption, strictly speaking, implies the surface-concentration of a dissolved substance, under circumstances which, if they be altered or reversed, will cause the concentration to diminish or disappear. But pseudo-adsorption includes cases, doubtless originating in adsorption proper, where subsequent changes leave the concentrated substance incapable of re-entering the liquid system. It is obvious that many (though not all) of our biological illustrations, for instance the formation of spicules or of permanent cell-membranes, belong to the class of so-called pseudo-adsorption phenomena. But the apparent contrast between the two is in the main a secondary one, and however important to the chemist is of little consequence to us.

While this brief sketch of the theory of membrane-formation is cursory and inadequate, it is enough to shew that the physical theory of adsorption tends in part to overturn, in part to simplify

* Cf. G. Haberlandt, Zelle u. Elementarorgane, *Biol. Centralbl.* 1925, p. 263.

† Cf. F. G. Donnan, Some aspects of the physical chemistry of interfaces, *Brit. Ass. Address* (Section B), 1923; *Nature*, Dec. 15, 22, 1923.

enormously, the older histological descriptions. We can no longer be content with such statements as that of Strasbürger, that membrane-formation in general is associated with the "activity of the kinoplasm," or that of Harper that a certain spore-membrane arises directly from the astral rays*. In short, we have easily reached the general conclusion that the formation of a cell-wall or cell-membrane is a chemico-physical phenomenon, which the purely objective methods of the biological microscopist do not suffice to interpret.

Having reached this conclusion we may wait patiently, and confidently, for more. But when the physico-chemical nature of these phenomena is admitted, and their dependence on adsorption recognised, or at least assumed, we have still to remember that the chemist himself is none too certain of his ground. He still finds it hard, now and then, to tell how far adsorption and direct chemical action go their way together, what parts they severally play, what shares they take in their intimate cooperation†.

If the process of adsorption, on which the formation of a membrane depends, be itself dependent on the power of the adsorbed substance to lower the surface-tension, it is obvious that adsorption can only take place when the surface-tension already present is greater than zero. It is for this reason that films or threads of creeping protoplasm shew little tendency, or none, to cover themselves with an encysting membrane; and that it is only when, in an altered phase, the protoplasm has developed a positive surface-tension, and has accordingly gathered itself up into a more or less spherical body, that the tendency to form a membrane is manifested, and the organism develops its "cyst" or cell-wall. The holes in a Globigerina-shell are there "to let the pseudopodia through." They may also be described as due to unequal distribution of surface-energy, such as to prevent shell-substance from being adsorbed here and there, and at the same time inducing a pseudopodium to emerge.

* Strasbürger, Ueber Cytoplasmastrukturen, etc., *Jahrb. f. wiss. Bot.* XXX, 1897; R. A. Harper, Kerntheilung und freie Zellbildung im Ascus, *ibid.*; cf. Wilson, *The Cell in Development, etc.*, pp. 53–55.

† The "adsorption theory" of dyeing is a case in point, where the precise mode, or modes, of action seem still far from settled.

It is found that a rise of temperature greatly reduces the adsorbability of a substance, and this doubtless comes, either in part or whole, from the fact that a rise of temperature is itself a cause of the lowering of surface-tension. We may in all probability ascribe to this fact and to its converse, or at least associate with it, such phenomena as the encystment of unicellular organisms at the approach of winter, or the frequent formation of strong shells or membranous capsules in "winter-eggs."

Again, since a film or a froth (which is a system of films) can only be maintained by virtue of a certain viscosity or rigidity of the liquid, it may be quickly caused to disappear by the presence in its neighbourhood of some substance capable of materially reducing the surface-tension; for this substance, being adsorbed, may displace from the surface-layer a material to which was due the rigidity of the film. In this way a "bathytonic" substance, such as ether, causes most foams to subside, and the pouring oil on troubled waters not only calms the waves but still more quickly dissipates the foam of the breakers. In a very different order of things, the breaking up of an alveolar network, as at a certain stage in the nuclear division of the cell, may be due in part to just such a cause, as well as to the direct lowering of surface-tension by electrical agency.

Our last illustration has led us back to the subject of a previous chapter, namely to the visible configuration of the interior of the cell, in so far (at least) as it represents a "dispersed system," coarse enough to be visible; and in connection with this wide subject there are many phenomena on which light is apparently thrown by our knowledge of adsorption, of which we took little or no account in our former discussion. One of these phenomena is nothing less than that visible or concrete "polarity," which we have seen to be in some way associated with a dynamical polarity of the cell.

This morphological polarity may be of a very simple kind, as when it is manifested, in an epithelial cell, by the outward shape of the elongated or columnar cell itself, by the essential difference between its free surface and its attached base, or by the presence in the neighbourhood of the former of mucus or other products of the cell's activity. But in a great many cases, this polarised symmetry is supplemented by the presence of various fibrillae, or of linear arrangements of particles, which in the elongated or "monopolar" cell run parallel with its axis, but tend to a radial arrangement in the more or less rounded or spherical cell. Of late years great importance has been attached to these various linear or fibrillar arrangements, as they are seen (*after staining*) in the cell-substance of intestinal epithelium, of spermatocytes, of ganglion cells, and most abundantly and frequently of all in gland cells. Various functions have been assigned, and hard names given to them; for these structures include your mitochondria* and your chondriokonts (both of these being varieties of chondriosomes), your Altmann's granules, your microsomes, pseudo-chromosomes, epi-

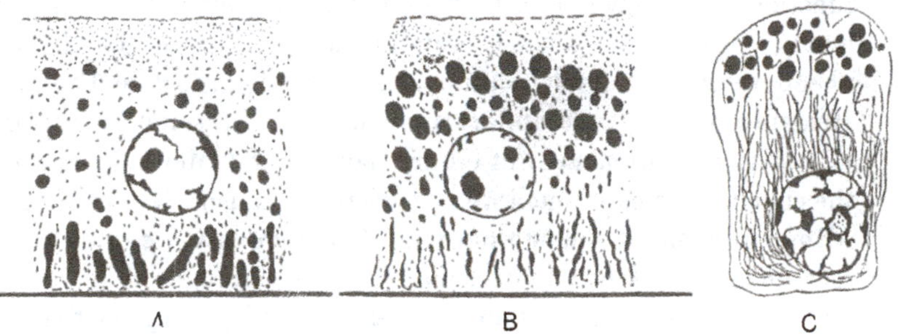

Fig. 149. A, B, Chondriosomes in kidney-cells, prior to and during secretory activity (after Barratt); C, do. in pancreas of frog (after Mathews).

dermal fibrils and basal filaments, your archeoplasm and ergastoplasm, and probably your idiozomes, plasmosomes, and many other histological minutiae†.

The position of these bodies with regard to the other cell-structures is carefully described. Sometimes they lie in the neighbourhood of the nucleus itself, that is to say in proximity to the fluid boundary surface which separates the nucleus from the

* Mitochondria are threads which move slowly through the protoplasm, sometimes break in two, and often tend to radiate from the centrosphere or division-centre of the cell. The nucleoli are two or more opaque bodies within the nucleus, which keep shifting their position; within the cytoplasm many small fatty bodies likewise move about, and display the Brownian oscillation.

† Cf. A. Gurwitsch, *Morphologie und Biologie der Zelle*, 1904, pp. 169–185; Meves, Die Chondriosomen als Träger erblicher Anlagen, *Arch. f. mikrosk. Anat.* 1908, p. 72; J. O. W. Barratt, Changes in chondriosomes, etc., *Q.J.M.S.* LVIII, pp. 553–566, 1913, etc.; A. P. Mathews, Changes in structure of the pancreas cell, etc., *Journ. Morph.* XV (Suppl.), pp. 171–222, 1899.

cytoplasm; and in this position they often form a somewhat cloudy sphere which constitutes the *Nebenkern*. In the majority of cases, as in the epithelial cells, they form filamentous structures, and rows of granules, whose main direction is parallel to the axis of the cell; and which may, in some cases, and in some forms, be conspicuous at the one end, and in some cases at the other end of the cell. But I seldom find the histologists attempting to explain, or to correlate with other phenomena, the tendency of these bodies to lie parallel with the axis, and perpendicular to the extremities of the cell; it is merely noted as a peculiarity, or a specific character, of these particular structures. Extraordinarily complicated and diverse functions have been ascribed to them. Engelmann's "Fibrillenkonus," which was almost certainly another aspect of the same phenomenon, was held by him and by cytologists like Breda and Heidenhain to be an apparatus connected in some unexplained way with the mechanism of ciliary movement. Meves looked upon the chondriosomes as the actual carriers or transmitters of heredity. Altmann invented a new aphorism, *Omne granulum e granulo*, as a refinement of Virchow's (or Remak's) *omnis cellula e cellula**: and many other histologists, more or less in accord, accepted the chondriosomes as important entities, *sui generis*, intermediate in grade between the cell itself and its ultimate molecular components. The extreme cytologists of the Munich school, Popoff, Goldschmidt and others, following Richard Hertwig, declaring these structures to be identical with "chromidia" (under which name Hertwig ranked all extra-nuclear chromatin), would assign them complex functions in maintaining the balance between nuclear and cytoplasmic material; and the "chromidial hypothesis," as every reader of cytological literature knows, has become a very abstruse and complicated thing†. With the help of the "binuclearity hypothesis" of Schaudinn and his school, it has given us the chromidial net, the

* Virchow, *Arch. f. pathol. Anat.* VIII, p. 23, 1855; but used, implicitly, by Remak, in his paper Ueber extracelluläre Entstehung thierischer Zellen und über die Vermehrung derselben durch Theilung, *Müller's Archiv*, 1852, pp. 47–57. That cells come, and only come, from pre-existing cells seems to have been clearly understood by John Goodsir, in 1846; see his *Anatomical Memoirs*, II, pp. 90, 389.

† Cf. Clifford Dobell, Chromidia and the binuclearity hypotheses; a review and a criticism, *Q.J.M.S.* LIII, pp. 279–326, 1909; A. Prenant, Les Mitochondries et l'Ergastoplasme, *Journ. de l'Anat. et de la Physiol.* XLVI, pp. 217–285, 1910 (both with copious bibliography).

chromidial apparatus, the trophochromidia, idiochromidia, gametochromidia, the protogonoplasm, and many other novel and original conceptions. There is apt to be confusion between important and unimportant things; and the very names are apt to vary somewhat in significance from one writer to another.

The outstanding fact, as it seems to me, is that physiological science has been heavily burdened in this matter, with a jargon of names and a thick cloud of hypotheses; but from the physical point of view we see but little mystery in the whole phenomenon. For, on the one hand, it is likely enough that these various bodies, by vastly extending the intra-cellular surface-area, may serve to increase the physico-chemical activities of the cell; and, on the other hand, we ascribe their very existence, in all probability and in general terms, to the "clumping" together under surface-tension of various constituents of the heterogeneous cell-contents, and to the drawing out of the little clumps along the axis of the cell towards one extremity or the other, in relation to osmotic currents as these are set up in turn in direct relation to the phenomena of surface-energy and of adsorption*. And all this implies that the study of these minute structures, even if it taught us nothing else, at least surely and certainly reveals the presence of a definite field of force, and a dynamical polarity within the cell†.

* Traube in particular has maintained that in differences of surface-tension we have the origin of the active force productive of osmotic currents, and that herein we find an explanation, or an approach to an explanation, of many phenomena which were formerly deemed peculiarly "vital" in their character. "Die Differenz der Oberflächenspannungen oder der Oberflächendruck eine Kraft darstellt, welche als treibende Kraft der Osmose, an die Stelle des nicht mit dem Oberflächendruck identischen osmotischen Druckes zu setzen ist, etc." (Oberflächendruck und seine Bedeutung im Organismus, *Pflüger's Archiv*, CV, p. 559, 1904.) There is, moreover, good reason to believe that physiological "osmosis" is not a general phenomenon common to this or that colloid membrane or dialyser, but depends (*int. al.*) on a specific affinity between the particular membrane (or the particular material it is moistened with) and the substance dialysed. This statement, made by Kahlenberg in 1906 (*Journ. Phys. Chem.* X, p. 141; also *Nature*, LXXV, p. 430, 1907), has been confirmed (e.g.) by R. Brinkmann and A. von Szent-Gyorgyi in *Biochem. Ztschr.* CXXXIX, pp. 261–273, 1923.

† C. E. Walker, in an interesting paper on Artefacts as a guide to the chemistry of the cell, *Proc. R.S.* (B), CIII, pp. 397–403, 1928, tells how he took mixtures of albumen, gelatine and lipins, with droplets of methyl myristate (with or without phosphorus) to act as nuclei; and found on treating with osmic acid that the lipins had separated out and arranged themselves very much as do Golgi bodies and other structural elements in ordinary histological preparations.

Our next and last illustration of the effects of adsorption, which we owe to the work of the late Professor A. B. Macallum* of Montreal, is of great importance; for it introduces us to phenomena in regard to which we seem to stand on firmer ground than in some of the foregoing cases, albeit the whole story has not been told. In our last chapter we were restricted mainly, though not entirely, to a consideration of figures of equilibrium, such as the sphere, the cylinder or the unduloid; and we began at once to find ourselves in difficulties when we were confronted by departures from symmetry, even in such a simple case as the ellipsoidal yeast-cell and the production of its bud. We found the cylindrical cell of *Spirogyra*, with its plane partitions or its spherical ends, a simple matter to understand; but when this uniform cylinder puts out a lateral outgrowth in the act of conjugation, we have a new and very different system of forces to account for and explain. The analogy of the soap-bubble, or of the simple liquid drop, was apt to lead us to suppose that surface-tension was, on the whole, uniform over the surface of the cell; and that its departures from symmetry of form were due to variations in external resistance. But if we have been inclined to make such an assumption we must now reconsider it, and be prepared to deal with important localised variations in the surface-tension of the cell. For, as a matter of fact, the simple case of a perfectly symmetrical drop, with uniform surface, at which adsorption takes place with similar uniformity, is probably rare in physics, and rarer still (if it exist at all) in the fluid or fluid-containing system which we call in biology a cell. We have more to do with cells whose general heterogeneity of substance leads to qualitative differences of surface, and hence to varying distributions of surface-tension. We must accordingly investigate the case of a cell which displays some definite and regular heterogeneity of its liquid surface, just as *Amoeba* displays a heterogeneity which is complex, irregular and continually fluctuating in amount and distribution. Such heterogeneity as we are speaking of must be essentially chemical, and the preliminary problem is to devise methods of "microchemical" analysis, which shall reveal *localised* accumulations of particular substances within

* See his Methoden u. Ergebnisse der Mikrochemie in der biologischen Forschung; Asher-Spiro's *Ergebnisse*, VII, 1908.

the narrow limits of a cell, in the hope that, their normal effect on surface-tension being ascertained, we may then correlate with their presence and distribution the actual indications of varying surface-tension which the form or movement of the cell displays. In theory the method is all that we could wish, but in practice we must be content with a very limited application of it; for the substances which have such action as we are looking for, and which are also actual or possible constituents of the cell, are very numerous, while the means are very seldom at hand to demonstrate their precise distribution and localisation. But in one or two cases we have such means, and the most notable is in connection with the element potassium. As Macallum has shewn, this element can be revealed in very minute quantities by means of a certain salt, a nitrite of cobalt and sodium*. This salt penetrates readily into the tissues and into the interior of the cell; it combines with potassium to form a sparingly soluble nitrite of cobalt, sodium and potassium; and this, on subsequent treatment with ammonium sulphide, is converted into a characteristic black precipitate of cobaltic sulphide†.

By this means Macallum demonstrated, years ago, the unexpected presence of potassium (i.e. of chlorides or other potassium salts) accumulated in particular parts of various cells, both solitary cells and tissue cells‡; and he arrived at the conclusion that the localised accumulations in question were simply evidences of *concentration* of the dissolved potassium salts, formed and localised in accordance with the Gibbs-Thomson Law. For potassium (as we now know) has a much higher ionic velocity than sodium; and accordingly the

* On the distribution of potassium in animal and vegetable cells, *Journ. Physiol.* XXXII, p. 95, 1905. (The only substance at all likely to be confused with potassium in this reaction is *creatine*.)

† The reader will recognise a fundamental difference, and contrast, between such experiments as those of Macallum's and the ordinary staining processes of the histologist. The latter are (as a general rule) merely empirical, while the former endeavour to reveal the true microchemistry of the cell. "On peut dire que la microchimie n'est encore qu'à la période d'essai, et que l'avenir de l'histologie et spécialement de la cytologie est tout entier dans la microchimie": A. Prenant, Méthodes et résultats de la microchimie, *Journ. de l'Anat. et de la Physiol.* XLVI, pp. 343–404, 1910. There is an interesting paper by Brunswick, on the Limitations of microchemical methods in biology, in *Die Naturwissenschaften*, Nov. 2, 1923.

‡ It is always conspicuously absent, as are chlorides and phosphates in general, from the nuclear substance.

K-ions reach and occupy the adsorbing surfaces of the cell-membranes out of all proportion to their abundance in the external media*. And we may take it also that our potassium salts, like inorganic substances in general, tend to raise the surface-tension, and will be found concentrated, therefore, at a portion of the surface where the tension is weak†.

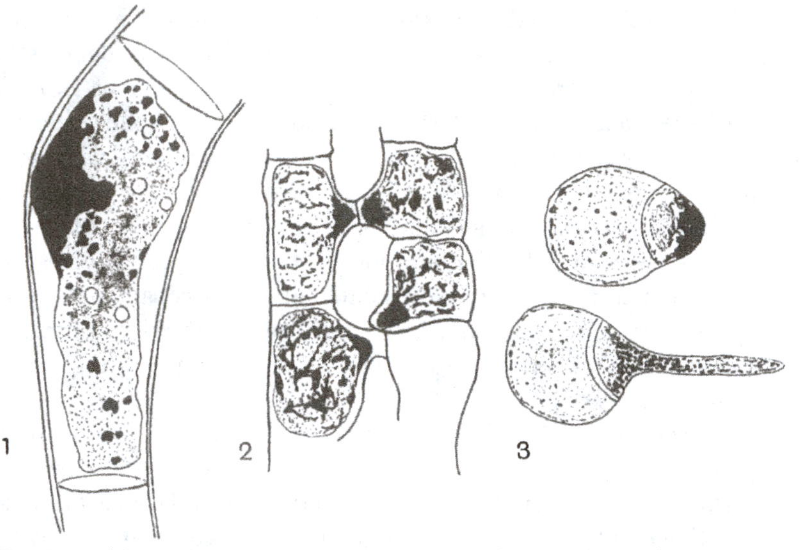

Fig. 150. Adsorptive concentration of potassium salts in (1) a cell of *Pleurocarpus* about to conjugate; (2) conjugating cells of *Mesocarpus*; (3) sprouting spores of *Equisetum*. After Macallum.

In Professor Macallum's figure (Fig. 150, 1) of the little green alga *Pleurocarpus*, we see that one side of the cell is beginning to bulge out in a wide convexity. This bulge is, in the first place, a sign of weakened surface-tension on one side of the cell, which as a whole had hitherto been a symmetrical cylinder; in the second place, we see that the bulging area corresponds to the position of a great concentration of the potassium salt; while in the third place,

* Cf. A. B. Macallum, Address to Section I, *Brit. Ass.* 1910; Oberflächenspannung und Lebenserscheinungen, in Asher-Spiro's *Ergebnisse der Physiologie*, XI, pp. 598–688, 1911; also his important paper on Ionic mobility as a factor in influencing the distribution of potassium in living matter, *Proc. R.S.* (B), CIV, pp. 440–458, 1929; cf. E. F. Burton, *Trans. Faraday Soc.* XXVI, p. 677, 1930.

† In accordance with the "principle of Le Chatelier," which is in fact a corollary to the Gibbs-Thomson Law.

from the physiological point of view, we call the phenomenon the first stage in the process of conjugation. In the figure of *Mesocarpus* (a close ally of *Spirogyra*), we see the same phenomenon admirably exemplified in a later stage. From the adjacent cells distinct outgrowths are being emitted, where the surface-tension has been weakened: just as the glass-blower warms and softens a small part of his tube to blow out the softened area into a bubble or diverticulum; and in our *Mesocarpus* cells (besides a certain amount of potassium rendered visible over the boundary which separates the green protoplasm from the cell-sap), there is a very large accumulation precisely at the point where the tension of the originally cylindrical cell is weakening to produce the bulge. But in a still later stage, when the boundary between the two conjugating cells is lost and the cytoplasm of the two cells becomes fused together, then the signs of potassium concentration quickly disappear, the salt becoming generally diffused through the now symmetrical and spherical "zygospore."

In a spore of *Equisetum*, while it is still a single cell, no localised concentration of potassium is to be discerned; but as soon as the spore has divided by an internal partition into two cells, the potassium salt is found to be concentrated in the smaller one, and especially towards its outer wall which is marked by a pronounced convexity. As this convexity (which corresponds to one pole of the now asymmetrical, or quasi-ellipsoidal spore) grows out into the root-hair, the potassium salt accompanies its growth and is concentrated under its wall. The concentration is, accordingly, a concomitant of the diminished surface-tension which is manifested in the altered configuration of the system.

The Acinete protozoa obtain their food through suctorial tentacles extruded from the surface of the cell: their extrusion being doubtless due to a local diminution of surface-tension. A dense concentration of potassium reveals itself, accordingly, in the surface-film of each tiny tentacle. As the tentacles are withdrawn their potassium diffuses into the cytoplasm; when retraction is complete it is again found in surface-concentration, but the surface-films on which it now concentrates are the surfaces of the protein-spherules (or "food-vacuoles") within the body of the cell.

In the case of ciliate or flagellate cells, there is to be found a characteristic accumulation of potassium at and near the base of the cilia. The relation of ciliary movement to surface-tension* lies beyond our range, but the fact which we have just mentioned throws light upon the frequent or general presence of a little protuberance of the cell-surface just where a flagellum is given off (cf. p. 406), and of a little projecting ridge or fillet at the base of an isolated row of cilia, such as we find in *Vorticella*.

Yet another of Professor Macallum's demonstrations, though its interest is mainly physiological, will help us somewhat further to comprehend what is implied in our phenomenon. In a normal cell of *Spirogyra*, a concentration of potassium is revealed along the whole surface of the spiral coil of chlorophyll-bearing, or "chromatophoral," protoplasm, the rest of the cell being wholly destitute of that substance: the inference being that at this particular boundary, between chromatophore and cell-sap, the surface-tension is small in comparison with any other interfacial surface within the system. And again, in certain minute *Chytridia*-like fungi, parasitic on *Spirogyra* and the like, the potassium-reaction helps to trace the delicate haustoria of the parasite in their course within the host-cell —a clear indication of low surface-tension at the surface between.

Now as Macallum points out, the presence of potassium is known to be a factor, in connection with the chlorophyll-bearing protoplasm, in the synthetic production of starch from CO_2 under the influence of sunlight; but we are left in some doubt as to the consecutive order of the phenomena. For the lowered surface-tension, indicated by the presence of the potassium, may be itself a cause of the carbohydrate synthesis; while on the other hand, this synthesis may be attended by the production of substances (e.g. formaldehyde) which lower the surface-tension, and so conduce to the concentration of potassium. All we know for certain is that the several phenomena are associated with one another, as apparently inseparable parts or inevitable concomitants of a certain complex action†.

* Cf. J. Gray, The mechanism of ciliary movement, *Proc. R.S.* (B), 1922–24.

† The distribution of potassium within plant-cells is more complicated than it seemed at first to be; but it is still the general if not the invariable rule to find it associated (by adsorption) with one boundary-surface or another. Cf. E. S. Dowding, Regional and seasonal distribution of potassium in plant tissues, *Ann. Bot.* XXXIX, pp. 459–476, 1925. The whole question, first adumbrated by Macallum,

And now to return, for a moment, to the question of cell-form. When we assert that the form of a cell (in the absence of mechanical pressure) is essentially dependent on surface-tension, and even when we make the preliminary assumption that protoplasm is essentially a fluid, we are resting our belief on a general consensus of evidence, rather than on compliance with any one crucial definition. The simple fact is that the agreement of cell-forms with the forms which physical experiment and mathematical theory assign to liquid surfaces under the influence of surface-tension is so frequently and often so typically manifested that we are led, or driven, to accept the surface-tension hypothesis as generally applicable and as equivalent to a universal law. The occasional difficulties or apparent exceptions are such as to call for further enquiry, but fall short of throwing doubt on the hypothesis. Macallum's researches introduce a new element of certainty, a "nail in a sure place," when they demonstrate that in certain movements or changes of form which we should naturally attribute to weakened surface-tension, a chemical concentration which would naturally accompany such weakening actually takes place. They further teach us that in the cell a chemical heterogeneity may exist of a very marked kind, certain substances being accumulated here and absent there, within the narrow bounds of the system.

Such localised accumulations can as yet only be demonstrated in the case of a very few substances, and of a single one in particular; and these few are substances whose presence does not produce, but whose concentration tends to follow, a weakening of surface-tension. The physical cause of the localised inequalities of surface-tension remains unknown. We may assume, if we please, that they are due to the prior accumulation, or local production, of bodies which have this direct effect; though we are by no means limited to this hypothesis. But in spite of some remaining difficulties and uncertainties, we have arrived at the conclusion, as regards unicellular organisms, that not only their general configuration but also *their*

is part of the general subject of *ionic regulation*, which has since become a matter of great physiological importance; cf. (*int. al.*) D. A. Webb, Ionic regulation in *Carcinus moenas*, *Proc. R.S.* (B), CXXIX, pp. 107–136, 1940, and many works quoted therein. It is curious and interesting that Macallum's first work on unequal ionic distribution in the tissues and Donnan's fundamental conception of the Donnan equilibrium (*Journ. Chem. Soc.* XCIX, p. 1554, 1911) came just at the same time.

departures from symmetry may be correlated with the molecular forces manifested in their fluid or semi-fluid surfaces.

Looking at the physiological side, rather than at the morphological which is more properly our own, we see how very important a *cellular system* is bound to be, even in respect of its surface-area alone. The order of magnitude of the cells which constitute our tissues is such as to give a relation of surface to volume far beyond anything in all the structures or mechanisms devised and fabricated by man. At this extensive surface, capillary energy, a form of energy scarcely utilised by man, plays a large predominant part in the energetics of the organism. Even the warm-blooded animal is not in reality a heat-engine; working as it does at almost constant temperatures its output of energy is bound, by the principle of Carnot, to be small. Nor is it an electrostatic machine, nor yet an electrodynamic one. It is a mechanism in which chemical energy turns into surface-energy, and, working hand in hand, the two are transformed into mechanical energy, by steps which are for the most part unknown*.

We are led on by these considerations to reflect on the molecular, rather than the histological, structure of the cell. We have already spoken in passing of "monomolecular layers," such as Henri Devaux imagined some thirty years ago, and afterwards obtained†, and such as Irving Langmuir has lately made his own. The free surface of every liquid (provided the form and symmetry of its molecules permit) presents a single layer of oriented molecules. Such a surface is no mere limit or simple boundary; it becomes a region of great importance and peculiar activity in certain cases, when, for instance, protein molecules of vast complexity are concerned. It is then a morphological field with a molecular structure of its own, and a dynamical field with energetics of its own. It becomes a frontier where this alien molecule may be excluded and that other be passed through: where some must submit to mere adsorption, and others suffer chemical change. In a word, we begin to look on a surface-layer or membrane, visible or invisible, as a vastly important thing, a place of delicate operations, and a field of peculiar and potent activity.

* Lippmann imagined a *moteur électrocapillaire*, unique in the history of mechanical invention. Cf. Berthelot, *Rev. Sci.* Dec. 7, 1913.

† Cf. (*int. al.*) H. Devaux, La structure moléculaire de la cellule végétale, *Bull. Soc. Bot. de France*, LXXV, p. 88, 1928.

CHAPTER VII

THE FORMS OF TISSUES OR CELL-AGGREGATES

We pass from the solitary cell to cells in contact with one another—to what we may call in the first instance "cell-aggregates," through which we shall be led ultimately to the study of complex tissues. In this part of our subject, as in the preceding chapters, we shall have to consider the effect of various forces; but, as in the case of the solitary cell, we shall probably find, and we may at least begin by assuming, that the agency of surface-tension is especially manifest and important. The effect of this surface-tension will manifest itself in surfaces *minimae areae*: where, as Plateau was always careful to point out, we must understand by this expression not an absolute but a relative minimum, an area, that is to say, which approximates to an absolute minimum as nearly as the circumstances and material exigencies of the case permit.

There are certain fundamental principles, or fundamental equations, besides those we have already considered, which we shall need in our enquiry; for instance, the case which we briefly touched on (on p. 426) of the angle of contact between the protoplasm and the axial filament in a Heliozoan, we shall now find to be but a particular case of a general and elementary theorem.

Let us re-state as follows, in terms of *Energy*, the general principle which underlies the theory of surface-tension or capillarity*.

When a fluid is in contact with another fluid, or with a solid or with a gas, a portion of the total energy of the system (that, namely, which we call *surface energy*) is proportional to the area of the surface of contact; it is also proportional to a coefficient which is specific for each particular pair of substances and is constant for these, save only in so far as it may be modified by changes of temperature or of electrical charge. Equilibrium, which is the condition of *minimum potential energy* in the system, will accordingly

* See Clerk Maxwell's famous article on "Capillarity" in the ninth edition of the *Encyclopedia Britannica*, revised by Lord Rayleigh in the tenth edition.

be obtained, *caeteris paribus*, by the utmost possible reduction of the surfaces in contact.

When we have three bodies in contact with one another the same is true, but the case becomes a little more complex. Suppose a drop of some fluid, A, to float on another fluid, B, while both are exposed to air, C. Here are three surfaces of contact, that of the drop with the fluid on which it floats, and those of air with the one and other of these two; and the whole surface-energy, E, of the system consists of three parts resident in these three surfaces,

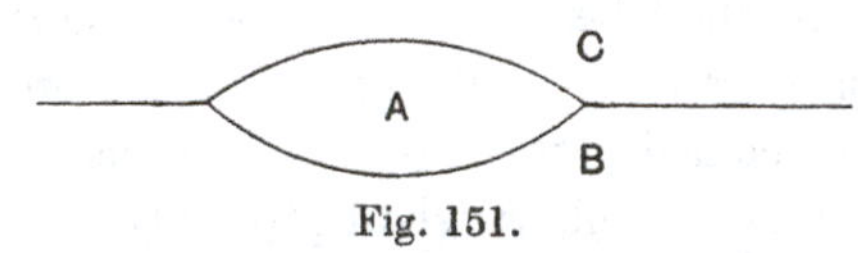

Fig. 151.

or of three specific energies, E_{AB}, E_{AC}, E_{BC}. The condition of equilibrium, or minimal potential energy, will be reached by contracting those surfaces whose specific energy happens to be large and extending those where it is small—contraction leading to the production of a "drop," and extension to a spreading "film." Floating on water, turpentine gathers into a drop, olive-oil spreads out in a film; and these, according to the several specific energies, are the ways by which the total energy of the system is diminished and equilibrium attained.

A drop will continue to exist provided its own two surface-energies exceed, per unit area, the specific energy of the water-air surface around: that is to say, provided (Fig. 151)

$$E_{AB} + E_{AC} > E_{BC}.$$

But if the one fluid happen to be oil and the other water, then the combined energy per unit-area of the oil-water and the oil-air surfaces together is less than that of the water-air surface:

$$E_{wa} > E_{oa} + E_{ow}.$$

Hence the oil-air and oil-water surfaces increase, the air-water surface contracts and disappears, the oil spreads over the water, and the "drop" gives place to a "film." In both cases the total surface-area is a minimum under the circumstances of the case, and always provided that no external force, such as gravity, complicates the situation.

The surface-energy of which we are speaking here is manifested in that contractile force, or tension, of which we have had so much to say*. In any part of the free water-surface, for instance, one surface-particle attracts another surface-particle, and the multitudinous attractions result in equilibrium. But a water-particle in the immediate neighbourhood of the drop may be pulled outwards,

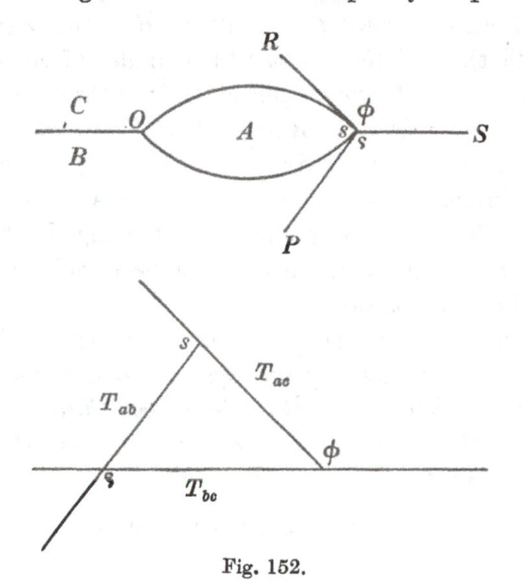

Fig. 152.

so to speak, by another water-particle, but find none on the other side to furnish the counter-pull; the pull required for equilibrium must therefore be provided by tensions existing in the *other two* surfaces of contact. In short, if we imagine a single particle placed at the very point of contact, it will be drawn upon by three different forces, whose directions lie in the three surface-planes and whose

* It can easily be proved (by equating the increase of energy stored in an increased surface with the work done in increasing that surface), that the tension measured per unit breadth, T_{ab}, is equal to the energy per unit area, E_{ab}. Surface-tensions are very diverse in magnitude, but all are positive; Clerk Maxwell conceived the existence of *negative* surface-tensions, but could not point to any certain instance. When blood-serum meets a solution of common salt, the two fluids hasten to mix, long streamers of the one running into the other; this remarkable phenomenon, first observed by Almroth Wright (*Proc. R.S.* (B), XCII, 1921) and called by him "pseudopodial intertraction," was described by Schoneboom (*ibid.* (A), CI, 1922) as a case of negative surface-tension. But it is a diffusion-phenomenon rather than a capillary one.

magnitudes are proportional to the specific tensions characteristic of the three "interfacial" surfaces. Now for three forces acting at a point to be in equilibrium they must be capable of representation, in magnitude and direction, by the three sides of a triangle taken in order, in accordance with the theorem of the Triangle of Forces. So, if we know the form of our drop as it floats on the surface (Fig. 152), then by drawing tangents P, R, from O (the point of mutual contact), we determine the three angles of our triangle, and know therefore the relative magnitudes of the three surface-tensions proportional to its sides. Conversely, if we know the three tensions acting in the directions P, R, S (viz. T_{ab}, T_{ac}, T_{bc}) we know the three sides of the triangle, and know from its three angles the form of the section of the drop. All points round the edge of the drop being under similar conditions, the drop must be circular and its figure that of a solid of revolution*.

The principle of the triangle of forces is expanded, as follows, in an old seventeenth-century theorem, called Lamy's Theorem:

If three forces acting at a point be in equilibrium, each force is proportional to the sine of the angle contained between the directions of the other two. That is to say (in Fig. 152)

$$P : R : S = \sin \phi : \sin \rho : \sin \varsigma,$$

or

$$\frac{P}{\sin \phi} = \frac{R}{\sin \rho} = \frac{S}{\sin \varsigma}.$$

And from this, in turn, we derive the equivalent formulae by which each force is expressed in terms of the other two and of the angle between them: viz.

$$P^2 = R^2 + S^2 + 2RS \cos \phi, \text{ etc.}$$

From this and the foregoing, we learn the following important and useful deductions:

(1) The three forces can only be in equilibrium when each is less

* Bubbles have many beautiful properties besides the more obvious ones. For instance, a floating bubble is always part of a sphere, but never more than a hemisphere; in fact it is always rather less, and a very small bubble is considerably less, than a hemisphere. Again, as we blow up a bubble, its thickness varies inversely as the square of its diameter; the bubble becomes a hundred and fifty times thinner as it grows from an inch in diameter to a foot. In an actual calculation we must always take account of the tensions *on both surfaces* of each film or membrane.

than the sum of the other two; otherwise the triangle is impossible. In the case of a drop of olive-oil on a clean water-surface, the relative magnitudes of the three tensions (at 15° C.) are nearly as follows:

Water-air surface		59
Oil-air	,,	25
Oil-water	,,	16

No triangle having sides of these relative magnitudes is possible, and no such drop can remain in existence*.

(2) The three surfaces may be all alike: as when two soap-bubbles are joined together on either side of a partition-film. The three tensions then are all co-equal, and the three angles are co-equal; that is to say, when three similar liquid surfaces, or films, meet together, they always do so at identical angles of 120°. Whether our two conjoined soap-bubbles be equal or unequal, this is still the invariable rule; because the specific tension of a particular surface is independent of form or magnitude.

(3) If all three surfaces be different, as when a fluid drop lies between water and air, the three surface-tensions will (in all likelihood) be different, and the two surfaces of the drop will differ in their amount of curvature.

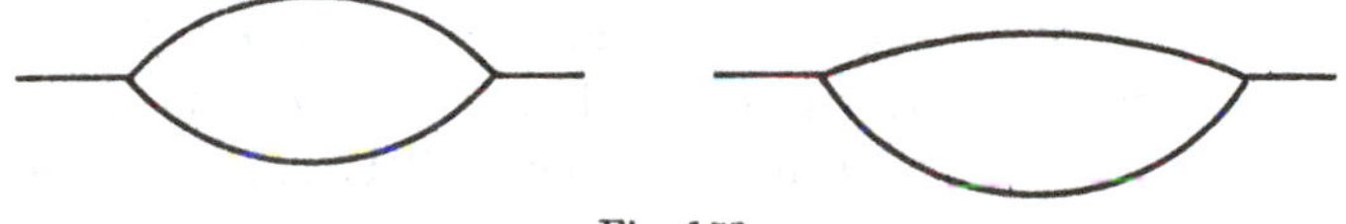

Fig. 153.

(4) If two only of the surfaces be alike, then two of the angles will be alike and the other will be unlike; and this last will be the difference between 360° and the sum of the other two. A particular case is when a film is stretched between solid and parallel walls, like a soap-film within a cylindrical tube. Here, so long as no external pressure is applied to either side, so long as both ends of the tube are open or closed, the angles on either side of the film will be equal, that is to say the film will set itself at right angles to the sides. Many years ago Sachs laid it down as a principle, which

* Nevertheless, if the water-surface be contaminated by ever so thin a film of oil, the oil-drop may be made to float upon it. See Rayleigh on Foam, *Collected Works*, III, p. 351.

has become celebrated in botany under the name of Sachs's Rule, that one cell-wall always tends to set itself at right angles to another cell-wall. But this rule only applies to the case we have just illustrated; and such validity as it possesses is due to the fact that among plant-tissues it commonly happens that one cell-wall has become solid and rigid before another partition-wall impinges upon it.

(5) Another important principle arises, not out of our equations but out of the general considerations which led to them. We saw in the soap-bubble that at and near the point of contact between our several surfaces, there is a continued balance of forces, carried (so to speak) across the interval; in other words, there is *physical continuity* between one surface and another and it follows that the surfaces merge one into another by a continuous curve. Whatever

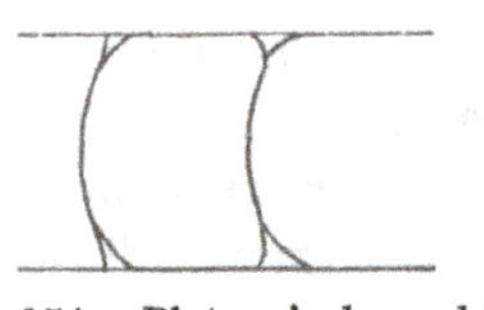

Fig. 154. Plateau's *bourrelet*, in an algal filament. After Berthold.

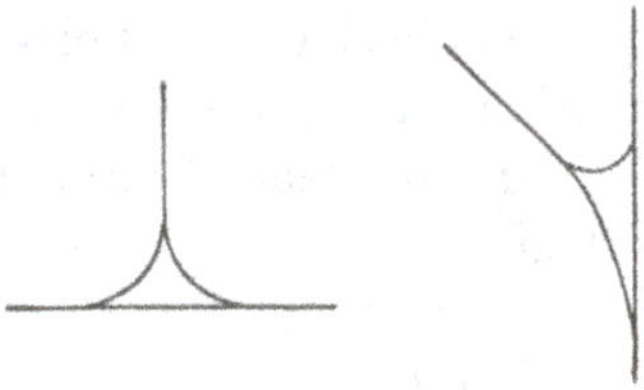

a *b*

Fig. 155.

be the form of our surfaces and whatever the angle between them, a small intervening curved surface is always there to bridge over the line of contact; and this little fillet, or "bourrelet," as Plateau called it, is big enough to be a common and conspicuous feature in the microscopy of tissues (Fig. 154). A similar "bourrelet" is clearly seen at the boundary between a floating bubble and the liquid on which it floats: in which case it constitutes a "masse annulaire," whose mathematical properties and relation to the form of the *nearly* hemispherical bubble have been investigated by van der Mensbrugghe*. The superficial vacuoles in *Actinophrys* or *Actinosphaerium* present an identical phenomenon.

(6) It is a curious effect, or consequence, of the bourrelet that a "horizontal" soap-film is never either horizontal or plane. For the bourrelet at its edge is deformed by gravity, and the film is correspondingly inclined upwards where it meets it (Fig. 155 *b*).

* Cf. Plateau, *op. cit.* p. 366.

(7) The bourrelet, a fluid mass connected with a fluid film, is no mere passive phenomenon but has its active influence or dynamical effect. This was pointed out by Willard Gibbs*, and Plateau's bourrelet is more often called, nowadays, "Gibbs's Ring." The ring is continuous in phase with the interior of the film, and fluid is sucked into it from the latter, which thins rapidly; and this, becoming a more potent factor of unrest than gravity itself, leads presently to the rupture of the film. Plateau's explanation of his bourrelet as a "surface of continuity" is thus but a part, and a small part of the story.

(8) In the succulent, or parenchymatous, tissue of a vegetable, the cells have their internal corners rounded off (Fig. 156) in a way which might suggest the bourrelet, but comes of another cause.

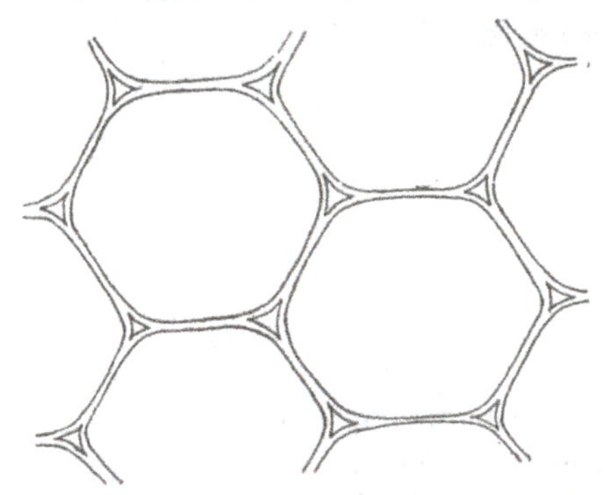

Fig. 156. Parenchyma of maize; shewing intercellular spaces.

Where the angles are rounded off the cell-walls tend to split apart from one another, and each cell seems tending to withdraw, as far as it can, into a sphere; and this happens, not when the tissue is young and the cell-walls tender and quasi-fluid, but later on, when cellulose is forming freely at the surface of the cell. The cell-walls no longer meet as fluid films, but are stiffening into pellicles; the cells, which began as an association of bubbles, are now so many balls, in solid contact or partial detachment; and flexibility and elasticity have taken the place of the capillary forces of an earlier and more liquid phase†.

* *Collected Works*, I, p. 309.

† J. H. Priestley, Cell-growth...in the flowering plant, *New Phytologist*, XXVIII, pp. 54–81, 1929.

(9) Statically though not dynamically, that is to say as a line or surface of continuity in Plateau's sense, our bourrelet is analogous to the accumulation of sand seen where two nodal lines cross in a Chladni figure: "Vers les endroits où des lignes nodales se coupent, elles s'élargissent toujours, de sorte que la forme des parties vibrantes près de ces endroits n'est pas angulaire mais plus ou moins arrondie, souvent en forme d'hyperbole*." And in somewhat remoter analogy, we may look on the three *corpora Arantii* as so many *bourrelets*, helping to fill the angles where three semilunar valves meet at the base of the great arteries.

We may now illustrate some of the foregoing principles, constantly bearing in mind the principles set forth in our chapter on the Forms of Cells, and especially those relating to the pressure exercised by a curved film.

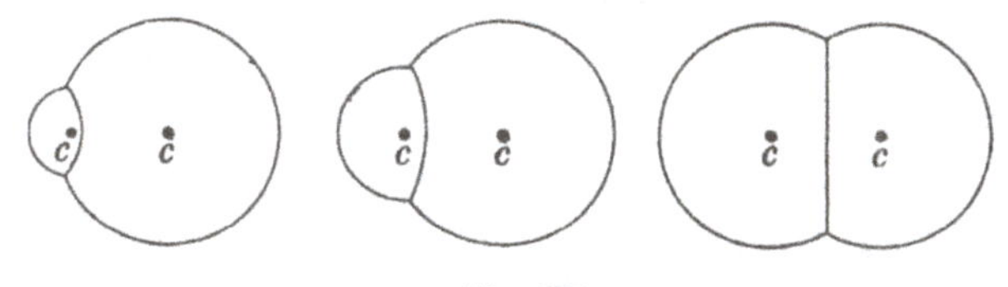

Fig. 157.

Let us look for a moment at the case presented by the partition-wall in a double soap-bubble. As we have just seen, the three films in contact (viz. the outer walls of the two bubbles and the partition-wall between) being all composed of the same substance and being all alike in contact with air, the three tensions must be equal, and the three films must, in all cases, meet at co-equal angles of 120°. But unless the two bubbles be of precisely equal size, and therefore of equal curvature, the tangents to the spheres will not meet the plane of their circle of contact at equal angles, and the partition-wall will of necessity be a curved, and indeed a spherical, surface; it is only plane when it divides two equal and symmetrical cells. It is obvious, from the symmetry of the figure, that the centres of the two bubbles and of the partition between are all on one and the same straight line.

The two bubbles exert a pressure inwards which is inversely

* E. F. F. Chladni, *Traité d'acoustique*, 1809, p. 127.

proportional to their radii: that is to say, $p : p' :: 1/r : 1/r'$; and the partition-wall must, for equilibrium, exert a pressure (P) which is equal to the difference between these two pressures, that is to say, $P = 1/R = 1/r' - 1/r = (r - r')/rr'$. It follows that the curvature of the partition must be just such as is capable of exerting this pressure, that is to say, $R = rr'/(r - r')$. The partition, then, is a portion of a spherical surface, whose radius is equal to the product, divided by the difference, of the radii of the two bubbles; if the two bubbles be equal, the radius of curvature of the partition is infinitely great, that is to say the partition is (as we have already seen) a plane surface.

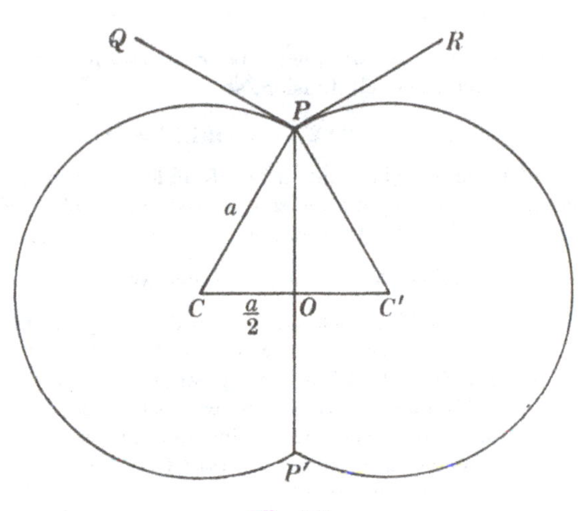

Fig. 158.

In the typical case of an evenly divided cell, such as a double and co-equal soap-bubble (Fig. 158), where partition-wall and outer walls are identical with one another and the same air is in contact with them all, we can easily determine the form of the system. For, at any point of the boundary of the partition, P, the tensions being equal, the angles QPP', RPP', QPR are all equal, and each is, therefore, an angle of 120°. But PQ, PR being tangents, the centres of the two spheres (or circular arcs in the figure) lie on lines perpendicular to them; therefore the radii CP, $C'P$ meet at an angle of 60°, and CPC' is an equilateral triangle. That is to say, the centre of each circle lies on the circumference of the other; the partition lies midway between the two centres; and the diameter of the partition-wall, PP', is $\frac{OP}{CP} = \sin 60° = \frac{\sqrt{3}}{2} = 0{\cdot}866$ times the diameter of each of the two cells. This gives us, then, the form of a combination of two co-equal spherical cells under uniform conditions.

By integrating between the known values of the meridian section and the plane partition, we should find each half of the double cell (or soap-bubble) to be equal to 27/32 of a complete sphere. Therefore the radius of curvature of each half of the divided bubble is greater than that of a sphere of equal volume in the ratio of:

$$\sqrt[3]{32} : \sqrt[3]{27} = 2.\sqrt[3]{4} : 3 = 1{\cdot}058 : 1 = 1 : 0{\cdot}945.$$

And the radius of the original sphere, before division, is to the radius of each half, or each product of cell-division, as

$$\sqrt[3]{54} : \sqrt[3]{32} = 3.\sqrt[3]{2} : 2.\sqrt[3]{4} = 1{\cdot}191 : 1 = 1 : 0{\cdot}84.$$

In the case of three co-equal and united bubbles (to which case we shall presently return), each is approximately five-sevenths of a whole sphere: and their radii, therefore, are to the radius of the whole sphere as

$$\sqrt[3]{7} : \sqrt[3]{5} = 1 : 0{\cdot}893 = 1 : (0{\cdot}945)^2.$$

When two co-equal bubbles coalesce, the internal pressure, due to the tension of the wall and varying inversely as its radius of curvature, is diminished in the ratio of 1 : 0·945, or say 5½ per cent. And we begin to see, in the case of three bubbles, that the process proceeds in a geometrical progression, each new coalescence increasing the radius of curvature and diminishing the internal pressure, by a constant fraction of the whole. This and other simple corollaries may perchance, some day, be found useful to the biologist.

In the case of unequal bubbles, the curvature of their partition-wall is easily determined, and is shewn in Fig. 159. The three films meeting in P being (as before) identical films, the three tangents, PQ, PR, PS, meet at co-equal angles of 120°, and PS produced bisects the angle QPR. PQ, PR are tangents perpendicular to the radii CP, $C'P$; and $C''P$, the radius of the spherical partition PP', is found by drawing a perpendicular to PS in P. The centre C'' is, by the symmetry of the figure, in a straight line with C, C'.

Whether the partition be or be not a plane surface, it is obvious that its *line of junction* with the rest of the system lies in a plane, and is at right angles to the axis of symmetry. The actual curvature of the partition-wall is easily seen in optical section; but in surface view the line of junction is *projected* as a plane (Fig. 160), perpendicular to the axis, and this appearance has helped to lend support and authority to "Sachs's Rule."

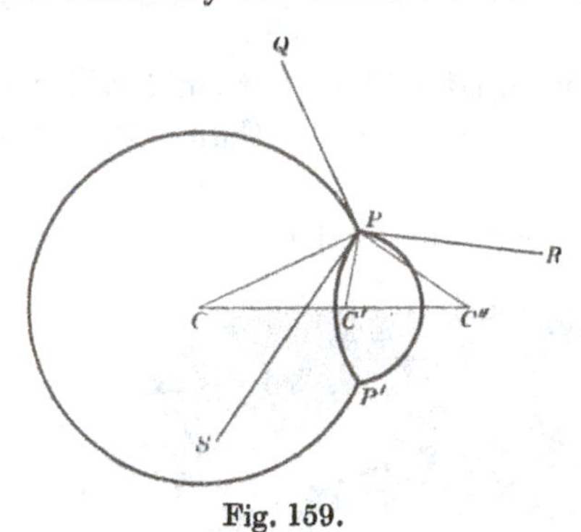

Fig. 159.

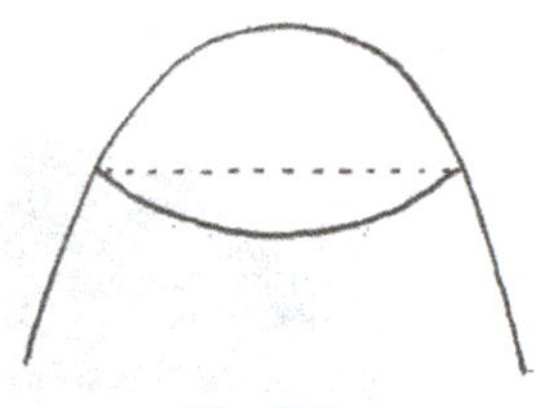

Fig. 160.

As soon as the tensions of the cell-walls become unequal, whether from changes in their own substance or in the substances with which they are in contact, then the form alters. If the tension along the partition P diminishes, the partition itself enlarges and the angle QPR increases: until, when the tension p is very small compared with q or r, the whole figure becomes a sphere, and the partition-wall, dividing it into two hemispheres, stands at right angles to the outer wall. This is the case when the outer wall of the cell is practically solid. On the other hand, if p begins to

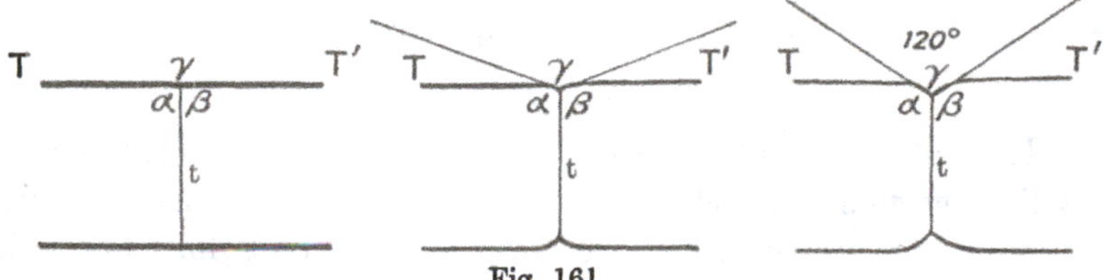

Fig. 161.

increase relatively to q and r, then the partition-wall contracts, and the two adjacent cells become larger and larger segments of a sphere, until at length the system becomes divided into two separate cells.

To put the matter still more simply, let the annexed diagrams (Fig. 161) represent a system of three films, one being a partition-wall running between the other two; and where the partition t meets the outer wall TT', let the several tensions, or the tractions exerted on a point at their meeting-place, be proportional to T, T' and t. Let α, β, γ be, as in the figure, the opposite angles. Then:

(1) If T be equal to T', and t be relatively insignificant, the angles α, β will be of 90°.

(2) If $T = T'$, but be a little greater than t, then t will exert an appreciable traction, and α, β will be more than 90°, say for instance, 100°.

(3) If $T = T' = t$, then α, β, γ will all equal 120°.

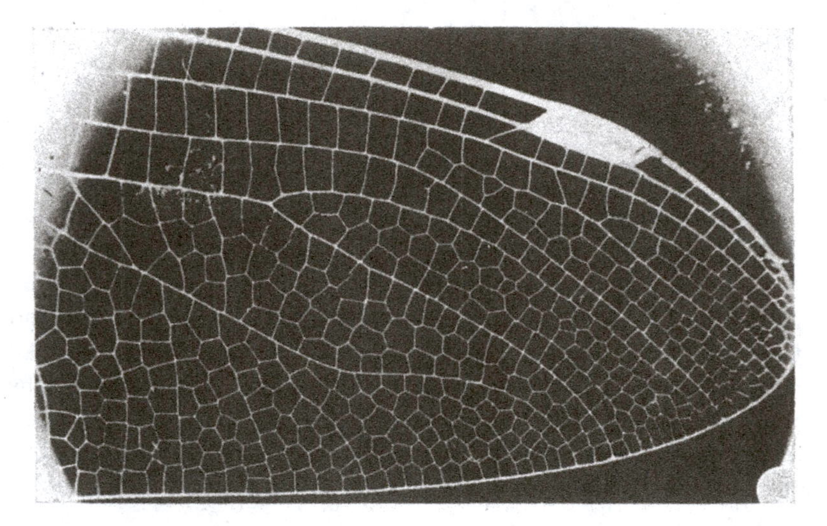

Fig. 162. Part of a dragonfly's wing.

The outer walls of the two cells on either side of the partition will be straight, as well as continuous, in the first case, and more or less curved in the other two. We have a vivid illustration (if a somewhat crude one) of the first case in a section of honey: where the waxen walls, which meet one another at 120°, meet the wooden sides of the box at 90°.

The wing of a dragon-fly shews a seemingly complicated system of veins which the foregoing considerations help much to simplify. The wing is traversed by a few strong "veins," or ribs, more or less parallel to one another, between which finer veins make a meshwork of "cells," these lesser veins being all much of a muchness, and exerting tensions insignificant compared with those of the greater veins. Where (*a*) two ribs run so near together that only one row of cells lies between, these cells are quadrangular in form, their thin partitions meeting the ribs at right angles on either side. Where (*b*) two rows of cells are intercalated between a pair of ribs, one row fits into the other by angles of 120°, the result of co-equal tensions; but both meet the ribs at right angles, as in the former case. Where (*c*) the cell-rows are numerous, all their angles in common tend to be co-equal angles of 120°, and the cells resolve, consequently, into a hexagonal meshwork.

Many spherical cells, such as *Protococcus*, divide into two equal halves, separated by a plane partition. Among other lower Algae akin to *Protococcus*, such as the Nostocs and Oscillatoriae, in which the cells are embedded in a gelatinous matrix, we find a series of forms such as are represented in Fig. 163, which various conditions depend, according to what we have already learned, upon the relative magnitudes of the tensions at the surface of the cells and the boundary between them. In some cases (Fig. 163, B) the cells remain spherical, because they are merely embedded in the matrix, with no other physical continuity between them; even two soap-bubbles do not tend to unite, unless their surfaces be moist or we put a drop of soap-solution between them. In certain other cases, the system consists of a relatively thick-walled tube, subdivided by more delicate partitions, which latter then tend (as in D) to become plane septa, set at right angles to the walls. Or again, side-walls and septa may be all alike, or nearly so; and then the configuration (as in C, on Fig. 163) is that of a linear cluster of soap-bubbles*.

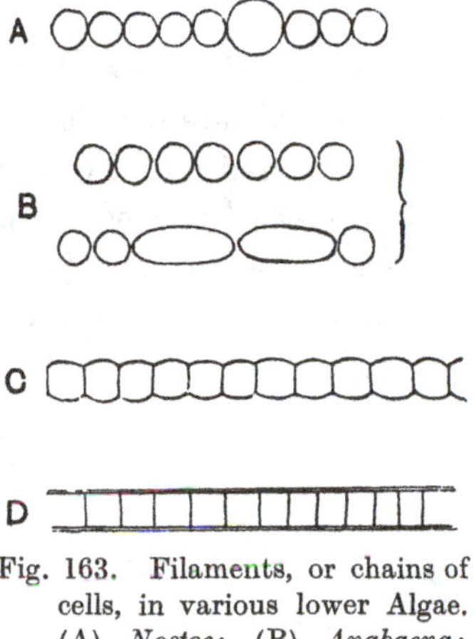

Fig. 163. Filaments, or chains of cells, in various lower Algae. (A) *Nostoc*; (B) *Anabaena*; (C) *Rivularia*; (D) *Oscillatoria*.

In the spores of liverworts, such as *Pellia*, the first partiti

* Cf. Dewar, Studies on liquid films, *Proc. Roy. Inst.* 1918, p. 359.

(the equatorial partition in Fig. 165 *a*) divides the spore into two equal halves, and is therefore a plane surface normal to the surface of the cell. But the next partitions arise near to either end of the original spherical or elliptical cell, and each of these latter will likewise tend to set itself normally to the cell-wall—at least the

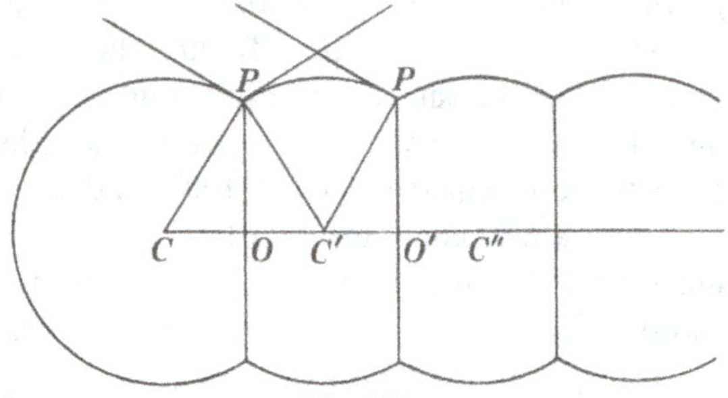

Fig. 164.

angles on either side of the partition will tend to be identical, and their magnitude will depend on the relative tensions of the cell-wall and the partition. The angles will be right angles if the cell-wall is solid or nearly so when the partition is formed; but they will be somewhat greater, if (in all probability) rigidity of the cell-wall has not been quite attained. In either case the partition itself will

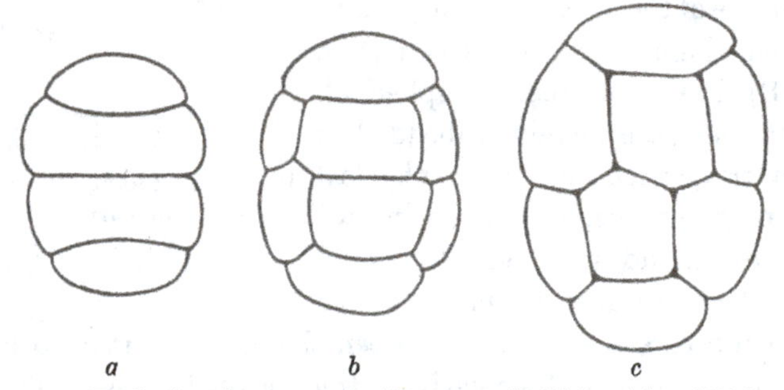

Fig. 165. Early development of a liverwort (*Pellia*). After Wildeman.

be part of a spherical surface, whose curvature will now correspond to the difference of pressures in the two chambers (or cells) which it serves to separate.

We have innumerable cases, near the tip of a growing filament for instance, where in like manner the partition-wall which cuts off the terminal, more or less conical, cell constitutes a spherical lens-shaped surface, set normally to the adjacent walls; and the centre of curvature is the meeting-point of two tangents to the cone. We find such a lenticular partition at the tips of the branches of many Florideae; in *Dictyota dichotoma*, as figured by Reinke, we have a succession of them. And by the way, where, in such cases as these, the tissues happen to be very transparent, we often have a puzzling confusion of lines (Fig. 166); one being the optical section

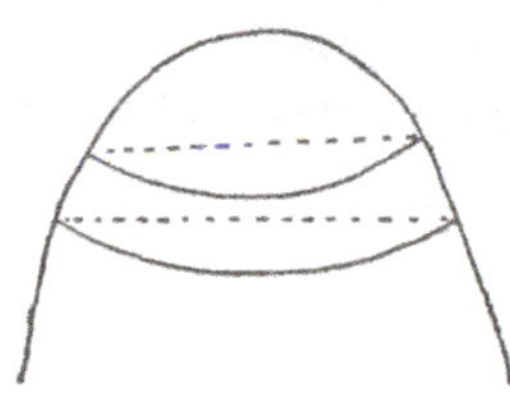

Fig. 166. Cells of *Dictyota*. After Reinke.

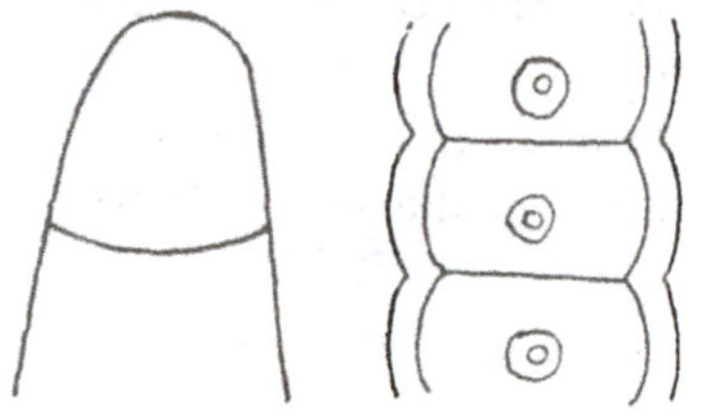

Fig. 167. Terminal and other cells of *Chara*.

of the curved partition-wall, the other being the straight linear projection of its outer edge to which we have already referred. In the conical terminal cell of *Chara*, we have the same lens-shaped curve; but a little lower down, where the sides of the shoot are approximately parallel, we have flat transverse partitions, and the form of the cells is, more or less, what we have been led to expect in the simple case of successive transverse partitions (Fig. 167).

In the young antheridia of *Chara* (Fig. 168), and in the geometrically similar case of the sporangium (or conidiophore) of *Mucor*, we easily recognise the hemispherical form of the septum which shuts off the large spherical cell from the cylindrical filament. Here, in the first phase of development, we should have to take into consideration the different pressures exerted by the single curvature of the cylinder and the double curvature of its spherical cap (p. 371); and we should find that the partition would have a somewhat low curvature, with a radius *less* than the diameter of the cylinder, which it would have exactly equalled but for the additional pressure inwards which it receives from the curvature of the large surrounding sphere. But as the latter continues to

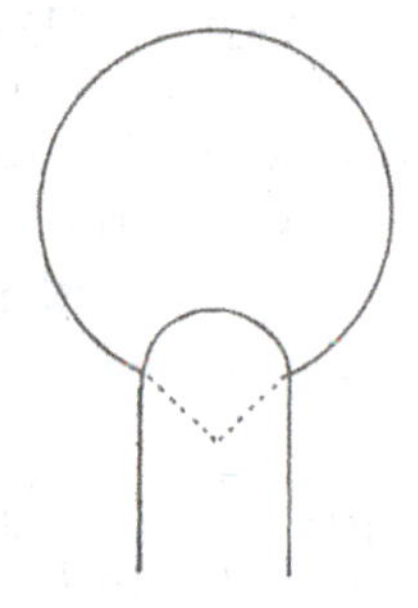

Fig. 168. Young antheridium of *Chara*.

grow its curvature decreases, and so likewise does the inward pressure of its surface; and accordingly the little convex partition bulges out more and more.

In the ordinary meristematic tissue of a plant, the new partition-wall within a dividing cell will generally meet the old walls at right angles to begin with, because its tension is usually small compared to what theirs has become. But as the system grows and the old wall strengthens, the tensions of all three walls become approximately the same; and they tend towards a new position of equilibrium, in which (as seen in optical section) they meet as before, at co-equal angles of 120°*.

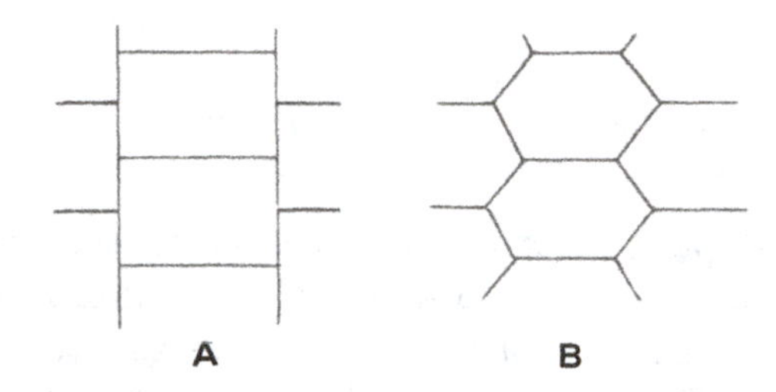

Fig. 169. Cambium cells after division, altering from *A* to *B*.

The biological facts which the foregoing considerations go far to explain and account for have been the subject of much argument and discussion on the part of the botanists. Let me recapitulate, in a very few words, the history of this long discussion.

Some seventy years ago, Hofmeister laid it down as a general, but purely empirical, law that "The partition-wall stands always perpendicular to what was previously the principal direction of growth in the cell"—or, in most cases, perpendicular to the long axis of the cell†. This contains an important truth; for it is as much as to say that the cell tends to be divided by the smallest

* J. H. Priestley, Studies...of cambium activity, *New Phytologist*, XXIX, p. 101, 1930. Cf. also J. J. Beijer, Vermehrung der radialen Reihen in Cambium, *Rec. de trav. bot. Néerl.* XXIV, pp. 631–786, 1927.

† Hofmeister, *Pringsheim's Jahrb.* III, p. 272, 1863; *Hdb. d. physiol. Bot.* I, p. 129, 1867; etc. Hofmeister adds the somewhat curious qualification: "Wohlbemerkt, nicht senkrecht zum grössten Durchmesser der Zelle, der mit der Richtung des stärksten Wachstums nicht zusammenfallen braucht, und in sehr viel Fällen in der That auch nicht mit ihr zusammenfällt."

partition capable of doing so. Ten years later, Sachs formulated his rule of "rectangular section," declaring that in all tissues, however complex, the cell-walls cut one another (at the time of their formation) at right angles*. Years before, Schwendener had found in the final results of cell-division a universal system of "orthogonal trajectories†"; and this idea Sachs further developed, introducing complicated systems of confocal ellipses and hyperbolae, and distinguishing between periclinal walls whose curves approximate to the peripheral contours, radial partitions which cut these at an angle of 90°, and finally anticlines, which stand at right angles to the other two.

Reinke (in 1880) was the first to throw doubt upon this explanation. He pointed out cases where the angle was not a right angle, but very definitely an acute one; and he saw in the commoner rectangular symmetry merely what he called a necessary, but *secondary*, result of growth‡.

Within the next few years a number of botanical writers were content to point out further exceptions to Sachs's rule§, and in some cases to show that the *curvatures* of the partition-walls, especially such cases of lenticular curvature as we have described, were by no means accounted for by either Hofmeister or Sachs; while within the same period, Sachs himself, and also Rauber, attempted to extend the main generalisation to animal tissues¶. The simple fact is that Sachs's rule is limited to those many cases where one cell-wall grows stiff or solid before another

* Sachs, Ueber die Anordnung d. Zellen in jüngsten Pflanzentheilen, *Verh. phys.-med. Gesellsch. Würzburg*, XI, pp. 219–242, 1877; Ueber Zellenanordnung u. Wachstum, *ibid.* XII, 1878; cf. *Arb. bot. Inst. Würzburg*, II, 1882; Ueber die durch Wachstum bedingte Verschiebung kleinster Theilchen in trajectorischen Curven, *Monatsb. k. Akad. Wiss. Berlin*, 1880; *Physiology of Plants*, chap. XXVII, Oxford, 1887.

† Schwendener, Bau u. Wachstum des Flechtenthallus, *Naturf. Gesellsch. Zürich*, 1860, pp. 272–296.

‡ Reinke, *Lehrbuch d. Botanik*, 1880, p. 519; Kienitz-Gerloff, *Botan. Ztg.* 1878, p. 58, had already shewn some exceptions to Sachs's rules, and ascribed them, vaguely, to "heredity." It was a time when *heredity* overruled everything, and when Sachs himself spoke of the difficulty of demonstrating the causes of any morphological phenomenon in any other way than "genetically": *Textbook*, 1882, p. 201.

§ E.g., Leitgeb, *Untersuchungen über die Lebermoose*, II, p. 4, Graz, 1881.

¶ Rauber, Neue Grundlegungen zur Kenntniss der Zelle, *Morphol. Jahrb.* VIII, pp. 279, 334, 1882.

impinges upon it; and, subject to this limitation, the rule is strictly true.

While these writers regarded the form and arrangement of the cell-walls as a biological phenomenon, with little if any direct relation to ordinary physical laws, or with but a vague reference to "mechanical conditions," the physical side of the case was soon urged by others, with more or less force and cogency. Indeed the general resemblance between a cellular tissue and a "froth" had been pointed out long before. Robert Hooke described the cells within the shaft of a feather as forming "a kind of solid or hardened froth, or a congeries of very small bubbles," and Grew described a parenchyma as made by "fermentation", "as we see Bread in Baking", and again as being "much the same thing, as to its construction which the froth of beer or eggs is." Later on, within the days of the cell-theory, Melsens made an "artificial tissue" by blowing into a solution of white of egg*.

In 1886, Berthold published his *Protoplasmamechanik*, in which he definitely adopted the principle of "minimal areas," and, following on the lines of Plateau, compared the forms of many cells and the arrangement of their partitions with those assumed under surface-tension by a system of "weightless films." But, as Klebs† pointed out, in reviewing the book, Berthold was so cautious as to stop short of attributing the biological phenomena to a mechanical cause. They remained for him, as they had done for Sachs, so many "phenomena of growth," or "properties of protoplasm."

In the same year, but while still unacquainted, apparently, with Berthold's work, Leo Errera published a short but very striking article‡ in which he definitely ascribed to the cell-wall (as Hofmeister had already done) the properties of a semi-liquid film, and drew from this as a logical consequence the deduction that it *must* assume the various configurations which the law of minimal areas imposes on the soap-bubble. So what we may call *Errera's Law* is formulated as follows: A cell-wall, at the moment of its

* *C.R.* XXXIII, p. 247, 1851; *Ann. de chimie et de phys.* (3), XXXIII, p. 170, 1851; *Bull. R. Acad. Belg.* XXIV, p. 531, 1857.

† Georg Klebs, *Biol. Centralbl.* VII, pp. 193–201, 1887.

‡ L. Errera, Sur une condition fondamentale d'équilibre des cellules vivantes, *C.R.* CIII, p. 822, 1886; *Bull. Soc. Belge de Microscopie*, XIII, Oct. 1886; *Recueil d'œuvres* (*Physiologie générale*), 1910, pp. 201–205.

formation, tends to assume the form which would be assumed under the same conditions by a liquid film destitute of weight*.

Soon afterwards Chabry†, discussing the segmentation of the Ascidian egg, indicated many ways in which cells and cell-partitions repeat the surface-tension phenomena of the soap-bubble. He came to the conclusion that some, at least, of the embryological phenomena were purely physical, and the same line of investigation and thought was pursued and developed by Robert‡ in connection with the embryology of the Mollusca. Driesch also, in a series of papers, continued to draw attention to capillary phenomena in the segmenting cells of various embryos, and came to the conclusion, startling to the embryologists of the time, that the mode of segmentation was of little importance as regards the final result§. Lastly de Wildeman¶, in a somewhat wider but also vaguer generalisation than Errera's, declared that "The form of the cellular framework of plants and also of animals depends, in its essential features, upon the forces of molecular physics."

Let us return to our problem of the arrangement of partition films. When we have three bubbles in contact, instead of two as in the case already considered, the phenomenon is strictly analogous to the former case. The three bubbles are separated by three partition surfaces, whose curvature will depend upon the relative size of the spheres, and which will be plane if the latter are all of equal size; but whether plane or curved, the three partitions will meet one another at angles of 120°, in an axial line. Various pretty geometrical corollaries accompany this arrangement. For instance, if Fig. 170 represent the three associated bubbles in a

* There was no lack of hearty antagonism to Berthold and Errera's views. Cf. (e.g.) Zimmermann, *Beitr. z. Morphologie und Physiologie der Pflanzenzelle*, Tübingen, 1891; Jost, *Vorlesungen uber Pflanzenphysiologie*, 1904, p. 329, etc.; Giesenhagen, *Studien über Zelltheilungen im Pflanzenreiche*, 1905. Cf. also K. Habermehl, *Die mechanische Ursache für die regelmässige Anordnung der Teilungswände in Pflanzenzellen* (Inaug. Diss.), Kaiserslautern, 1909.

† L. Chabry, Embryologie des Ascidiens, *J. Anat. et Physiol.* XXIII, p. 266, 1887.

‡ H. Robert, Embryologie des Troques, *Arch. de Zool. expér. et gén.* (3), X, 1892.

§ "Dass der Furchungsmodus etwas für das Zukunftige unwesentliches ist," *Z. f. w. Z.* LV, 1893, p. 37. With this statement compare, or contrast, that of Conklin, quoted on p. 5; cf. also p. 287 (footnote).

¶ E. de Wildeman, Études sur l'attache des cloisons cellulaires, *Mém. Couronn. de l'Acad. R. de Belgique*, LIII, 84 pp., 1893–94.

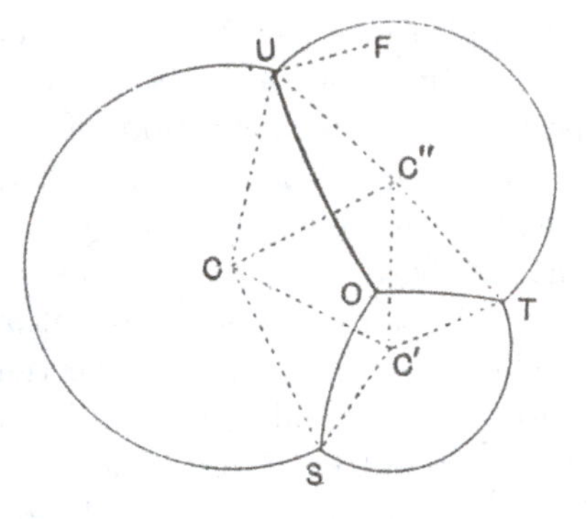

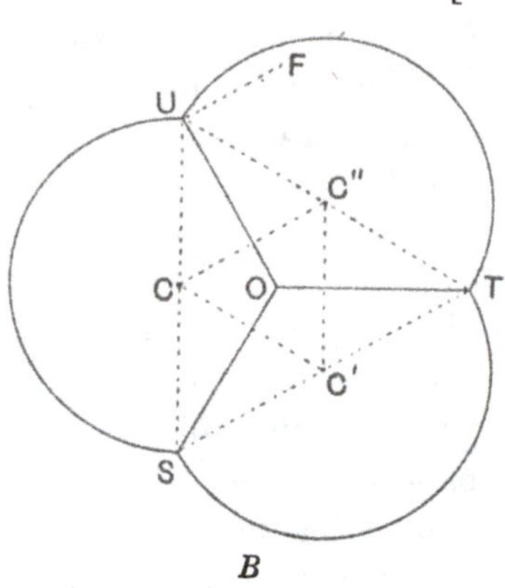

A Fig. 170. *B*

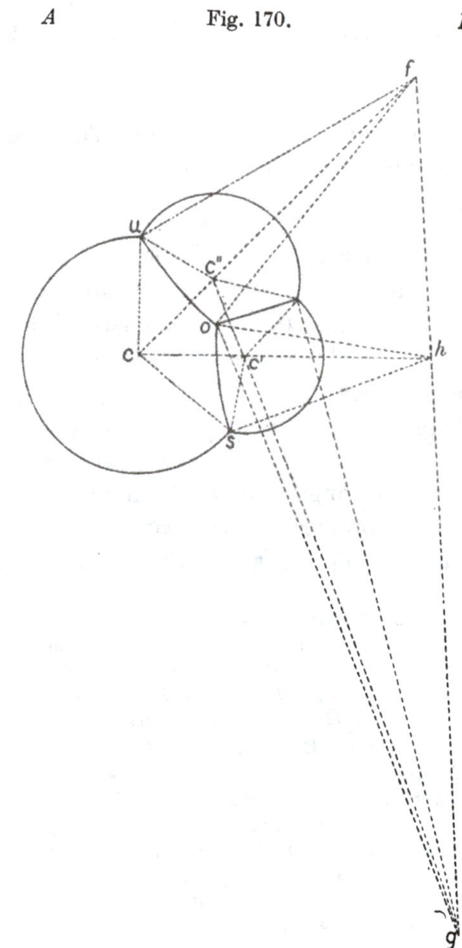

Fig. 171.

plane drawn through their centres, c, c', c'' (or what is the same thing, if it represent the base of three bubbles resting on a plane), then the lines uc, uc'', or sc, sc', etc., drawn to the centres from the points of intersection of the circular arcs, will always enclose an angle of 60°. Again (Fig. 171), if we make the angle $c''uf$ equal to 60°, and produce uf to meet cc'' in f, f will be the centre of the circular arc which constitutes the partition Ou; and further, the three points f, g, h, successively determined in this manner, will lie on one and the same straight line. In the case of three co-equal bubbles (as in Fig. 170, B), it is obvious that the lines joining their centres form an equilateral triangle: and consequently, that the centre of each circle (or sphere) lies on the circumference of the other two; it is also obvious that uf is now parallel to cc'', and accordingly that the centre of curvature of the partition is now infinitely distant, or (as we have already said) that the partition itself is plane.

The mathematician will find a more elegant way of dealing with our spherical bubbles and their associated interfaces by the method of spherical inversion. (i) Take three planes through a line, cutting one another at 60°, and invert from any point, and you have the case of two spherical bubbles fused, with their interface also spherical. (ii) Take the six planes projecting the edges of a regular tetrahedron from its centre, and you get by inversion the case of the three unequal bubbles and their three interfaces. (iii) Take these same planes with a bubble added centrally (thus adding a spherical tetrahedron), and inversion gives the general case of four fused bubbles and their six spherical partitions.

When we have four bubbles meeting in a plane (Fig. 172), they would seem capable of arrangement in two symmetrical ways: either (a) with four partition-walls intersecting at right angles, or (b) with five partitions meeting, three and three, at angles of 120°. The latter arrangement is strictly analogous to the arrangement of *three* bubbles in Fig. 170. Now, though both of these figures might seem, from their apparent symmetry, to be figures of equilibrium, yet in point of fact the latter turns out to be of stable and the former of unstable equilibrium. If we try to bring four bubbles into the form (a), that arrangement endures only for an instant; the partitions glide upon one another, an intermediate wall springs into existence, and the system assumes the form (b), with its two triple, instead of one quadruple, conjunction. In like manner, when four billiard-balls are packed close upon a table, two tend to come together and separate the other two.

Let us epitomise the Law of Minimal Areas and its chief clauses or corollaries in the particular case of an assemblage of fluid films, as was first done by Lamarle*. Firstly and in general: In every liquid system of thin films in stable equilibrium, the sum of the areas of the films is a minimum. From observation and experience, rather than by demonstration, it follows that (2) the area of *each* is a minimum under its own limiting conditions; and further that (3) the mean curvature of any film is constant throughout its whole area, null when the pressures are equal on either side and in other

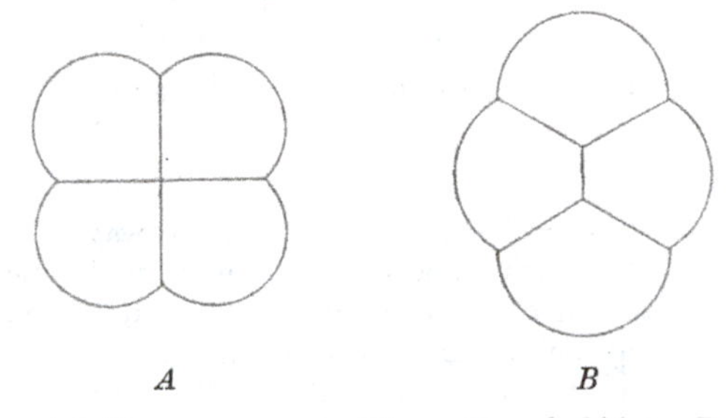

Fig. 172. A, an unstable arrangement of four cells or bubbles. B, the normal and stable configuration, showing the polar furrow.

cases proportional to their difference. Less obvious, very important, and likewise subject (but none too easily) to rigorous mathematical proof, are the next two propositions, both of which had been laid down empirically by Plateau: (4) the films meeting in any one edge are three in number; (5) the crests or edges meeting in any one corner are four in number, neither more nor less. Lastly, and following easily from these: (6) the three films meeting in a crest or edge do so at co-equal angles, and the same is true of the four edges meeting in a corner.

Wherever we have a true cellular complex, an arrangement of cells in actual physical contact by means of their intervening boundary walls, we find these general principles in force; we must only bear in mind that, for their easy and perfect recognition, we

* Ernest Lamarle, Sur la stabilité des systèmes liquides en lames minces, *Mém. de l'Acad. R. de Belgique*, XXXV, XXXVI, 1864–67.

must be able to view the object in a plane at right angles to the boundary walls. For instance, in any ordinary plane section of a vegetable parenchyma, we recognise the appearance of a "froth," precisely resembling that which we can construct by imprisoning a mass of soap-bubbles in a narrow vessel with flat sides of glass; in both cases we see the cell-walls everywhere meeting, by threes, at angles of 120°, irrespective of the size of the individual cells: whose relative size, on the other hand, determines the *curvature* of the partition-walls. On the surface of a honey-comb we have precisely the same conjunction, between cell and cell, of three boundary walls, meeting at 120°. In embryology, when we examine a segmenting egg, of four (or more) segments, we find in like manner, in the majority of cases if not in all, that the same principle is still exemplified. The four segments do not meet in a common centre, but each cell is in contact with two others; and the three, and only three, common boundary walls meet at the normal angle of 120°. A so-called *polar furrow**, the visible edge of a vertical partition-wall, joins (or separates) the two triple contacts, precisely as in Fig. 172, B, and so gives rise to a diamond-shaped figure, which was recognised more than a hundred years ago (in a newt or salamander) by Rusconi, and called by him a *tetracitula*.

That four cells, contiguous in a plane, tend to meet in a lozenge with three-way junctions and a "polar furrow" between the cells, is a geometrical theorem of wide bearing. The first four cells in a wasp's nest shew it neither better nor worse than do those of a segmenting ovum, or the ambulacral plates of a sea-urchin or the oosphere of *Oedogonium* giving birth to its four zoospores†. Going farther afield for an illustration, we find it in the molecules of a viscous liquid under shear: where a group of four

* It was so termed by Conklin in 1897, in his paper on Crepidula (*Journ. Morph.* XIII, 1897). It is the *Querfurche* of Rabl (*Morph. Jahrb.* v, 1879); the *Polarfurche* of O. Hertwig (*Jen. Zeitschr.* XIV, 1880); the *Brechungslinie* of Rauber (Neue Grundlage zur Kenntniss der Zelle, *Morph. Jahrb.* VIII, 1882); and the *cross-line* of T. H. Morgan (1897). It is carefully discussed by Robert, *op. cit.* p. 307 *seq.*

† Speaking of the complicated polygonal patterns in the test of the protozoon genus Peridinium, Barrows says: "In the experience of the writer no case has been found in which four sutures actually meet at one point. Cases which at first sight appeared as such, upon closer analysis in a favourable position have been resolved into two junction-points of three sutures each, etc." On skeletal variation in the genus Peridinium, *Univ. Calif. Publ.* 1918, p. 463.

molecules is supposed to slip from one lozenge-configuration to an opposite one, passing on the way through the simple cross or square —a configuration of "higher energy" and less stable equilibrium*.

The solid geometry of this four-celled figure is not without interest. If the two polar furrows (the one above and the other below) run criss-cross, the whole is a more or less flattened and distorted spherical tetrahedron. If they run parallel, then it is a four-sided lozenge with two curved quadrilateral faces, and two bilateral faces each bounded by two curved edges, like the "liths"

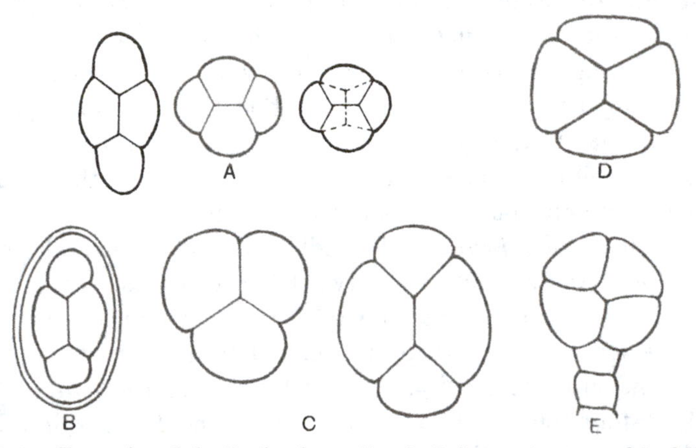

Fig. 173. Examples of the "polar furrow". A, Pollen-grains (tetrads) of *Neottia*. B, Egg of hookworm (*Ankylostoma*). C, First cells of a wasp's nest (*Polistes*). (From Packard, after Saussure.) D, Four-celled stage of Volvox: from Janet. E, Hair of Salvia, after Hanstein.

of an orange†. In either case the lozenge-configuration is under some restraint to keep its four cells in a plane; for a tetrahedral pile, or pyramid, of four spheres would be the simplest arrangement of all.

The polar furrow and the partition of which it forms an edge are, like all the edges and partitions in our associated cells, perfectly definite in dimensions and position; and to draw them to scale, in projection, is a simple matter. Taking the simplest case, when the radii of all four cells are equal to one another, let c, c', c'' and c''' be the centres of the four cells, Fig. 174. The centres of

* Cf. J. D. Bernal, *Proc. R.S.* (A), No. 914, p. 321, 1937.

† The geometer seldom takes account of such two-sided surfaces or facets; but in groups of cells or bubbles they are of common occurrence, and in the theory of polyhedra they fit in without difficulty with the rest (cf. *infra*, p. 737).

any two are related precisely as though two cells only were conjoined; the centres of three *contiguous* cells (as c, c' and c'') are related as though three only were concerned; and the centres of two opposite cells are situated symmetrically to one another. This is as much as to say that if there be two bubbles in contact the addition of a third does not disturb their symmetry; and if there be three in contact, the addition of a fourth leaves the first three likewise *in statu quo*. Thus the triangle $cc'c''$ is equilateral, as we already know. The partition so bisects the side cc'', and the angle $cc'c''$; and the point o is the centre of gravity of the triangle. Therefore $op = \frac{1}{2}oc''$ and $oo' = \frac{1}{3}c''c'''$.

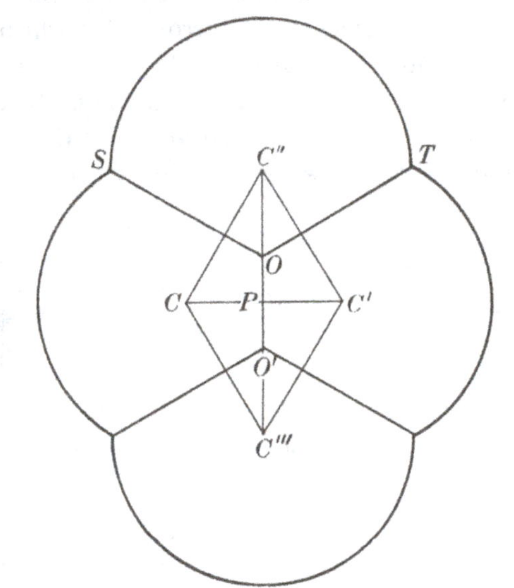

Fig. 174. The geometric symmetry of a system of four cells.

Again, in the triangle cpc'', where $cc'' = r$, $pc'' = \frac{\sqrt{3}}{2} r$, and oo' (the polar furrow) $= \frac{1}{\sqrt{3}} r$. Once again, in the triangle soc'', $sc = r$; and so (one of the partitions) $= \frac{2}{\sqrt{3}} r =$ twice oo'. The length of the polar furrow, then, as seen in vertical projection in a system of four co-equal cells, is (theoretically) just one-half that of the four intercellular partitions, and very nearly three-fifths that of a cell-radius.

It is worth while to remark that the universal phenomenon of a polar furrow gives *an appearance* of bilateral symmetry to every egg or embryo in its four-celled stage, no matter to what kind or class or organism it belongs.

In the four-celled stage of the frog's egg, Rauber (an exceptionally careful observer) shews us three alternative modes in which the four cells may be found to be conjoined (Fig. 175). In A we have the commonest arrangement, which is that which we have just studied and found to be the simplest theoretical one; that namely where a straight polar furrow intervenes, and where the partition-walls are conjoined at its extremities, three by three. In B, we have again a polar furrow, which is now seen to be a portion of the first "segmentation-furrow" by which the egg was originally divided into two; the four-celled stage being reached by the appearance of the two transverse furrows. In this case, the polar furrow is seen to be sinuously curved, and Rauber tells us that its curvature gradually alters; as a matter of fact, it, or rather the

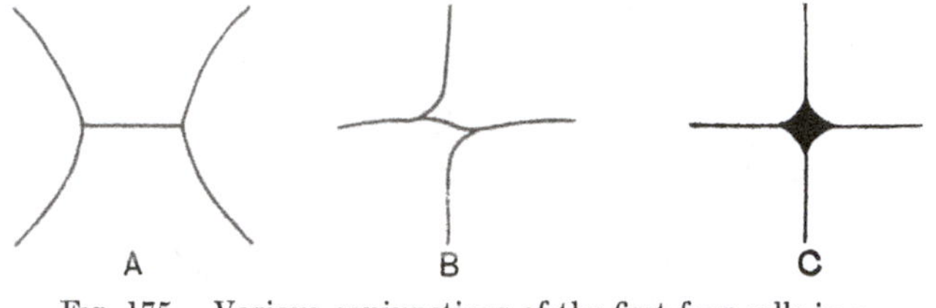

Fig. 175. Various conjunctions of the first four cells in a frog's egg. After Rauber.

partition-wall corresponding to it, is gradually setting itself into a position of equilibrium, that is to say of equiangular contact with its neighbours, which position is already attained or nearly so in A. In C we have a very different condition, with which we shall deal in a moment.

The polar furrow may be longer or shorter, and it may be so minute as to be not easily discernible; but it is quite certain that no simple and homogeneous system of fluid films such as we are dealing with is in equilibrium without its presence. In the accounts given, however, by embryologists of the segmentation of the egg, while the polar furrow is depicted in the great majority of cases, there are others in which it has not been seen and some in which its absence is definitely asserted*. The cases where four cells lying

* Thus Wilson declared (*Journ. Morph.* VIII, 1895) that in *Amphioxus* the polar furrow was occasionally absent, and Driesch took occasion to criticise and to throw doubt upon the statement (*Arch. f. Entw. Mech.* I, p. 418, 1895).

in one plane meet *in a point*, such as were frequently figured by the older embryologists, are hard to verify and sometimes not easy to believe. Considering the physical stability of the other arrangement, the great preponderance of cases in which it is known to occur, the difficulty of recognising the polar furrow in cases where it is very small and unless it be specially looked for, and the natural tendency of the draughtsman to make an all but symmetrical structure appear wholly so, I was wont to attribute to error or imperfect observation all those cases where the junction-lines of four cells are represented (after the manner of Fig. 172, A) as a simple cross*. As a matter of fact, the simple cross is no very rare phenomenon, even in the frog's egg; but it is a transitory one, and unstable. Viscosity and friction may enable it to endure for a while, but the partitions inevitably shift into the stable, three-way, configuration. In such a case, the polar furrow manifests itself slowly and as it were laboriously; but in the more fluid soap-bubble it does so in the twinkling of an eye.

While a true four-rayed intersection, or simple cross, is theoretically impossible save as a transitory and unstable condition, there is another configuration which may closely simulate it, and which is common enough. There are plenty of faithful representations of segmenting eggs in which, instead of the triple junctions and polar furrow, the four cells (and also their more numerous successors) are represented as *rounded off*, and separated from one another by an empty space, or by a little drop of extraneous fluid, evidently not directly miscible with the fluid surface of the cells. Such is the case in the obviously accurate figure which Rauber gives (Fig. 175, C) of his third mode of conjunction in the four-celled stage of the frog's egg. Here Rauber is most careful to point out that the furrows do not simply "cross," or meet in a point, but are separated by a little space, which he calls the *Polgrübchen*, and asserts to be constantly present whensoever the polar furrow, or *Brechungslinie*, is not to be discerned. This little interposed space with its contained drop of fluid materially alters the case, and implies a new

* The same remark was made long ago by Driesch: "Das so oft schematisch gezeichnete Vierzellenstadium mit zwei sich in zwei Punkten scheidende Medianen kann man wohl getrost aus der Reihe des Existierenden streichen" (Entw. mech. Studien, *Z. f. w. Z.* LIII, p. 166, 1892). Cf. also his *Math. mechanische Bedeutung morphologischer Probleme der Biologie*, Jena, 59 pp., 1891.

condition of theoretical and actual equilibrium. For on the one hand, we see that now the four intercellular partitions do not meet *one another at all*; but really impinge upon four new and separate partitions, which constitute interfacial contacts not between cell and cell, but between the respective cells and the intercalated drop. And secondly, the angles at which these four little surfaces meet the four cell-partitions will be determined, in the usual way, by the balance between the respective tensions of these several surfaces. In an extreme case (as in some pollen-grains) it may be found that the cells under the observed circumstances are not truly in surface contact: that they are so many drops which touch but do not "wet" one another, and which are merely held together

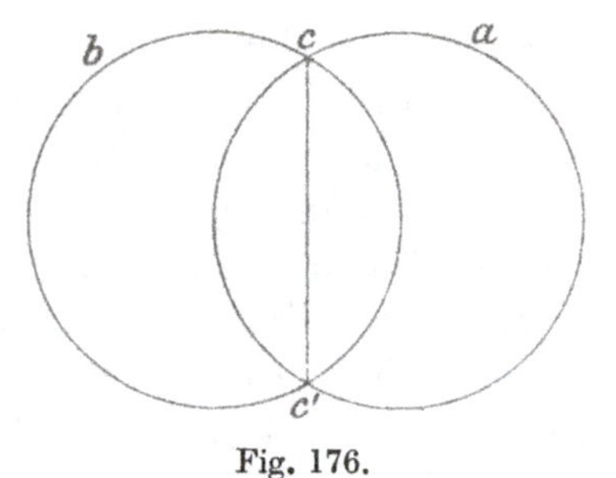

Fig. 176.

by the pressure of the surrounding envelope. But even supposing that they are in actual fluid contact, the case from the point of view of surface-tension presents no difficulty. In the case of the conjoined soap-bubbles, we were dealing with *similar* contacts and with *equal* surface-tensions throughout the system; but in the system of protoplasmic cells which constitute the segmenting egg we must make allowance for *inequality* of tensions, between the surfaces where cell meets cell and where on the other hand cell-surface is in contact with the surrounding medium—generally water or one of the fluids of the body. Remember that our general condition is that, in our entire system, the *sum of the surface energies* is a minimum; and, while this is attained by the *sum of the surfaces* being a minimum in the case where the energy is uniformly distributed, it is not necessarily so under non-uniform conditions. In the diagram (Fig. 176), if the energy per unit area be greater along the contact surface *cc′*, where cell meets cell, than along *ca* or *cb*, where cell-surface is in contact with the surrounding medium, these latter surfaces will tend to increase and the surface of cell-contact to diminish. In short there will be the usual balance of forces between the tension along the surface *cc'*, and the two opposing tensions along *ca* and *cb*. If the former be greater than either of the other two, the outside angle will be less than 120°; and if the tension along the surface *cc'* be as much or more than the sum of the other two, then the drops will merely touch one another, save for the possible effect of external pressure. This is the explanation, in general terms, of the peculiar conditions obtaining in *Nostoc* and its allies (p. 477), and it also leads us to a consideration of the general properties and characters of a superficial or "epidermal" layer*.

While the inner cells of the honeycomb are symmetrically situated, sharing with their neighbours in equally distributed pressures or tensions, and therefore all tending closely to identity of form, the case is obviously different with the cells at the borders of the system. So it is with our froth of soap-bubbles†. The bubbles, or cells, in the interior of the mass are all alike in general character, and if they be equal in size are alike in every respect: as we see them in projection their sides are uniformly flattened, and tend to meet at equal angles of 120°. But the bubbles which constitute the outer layer retain their spherical surfaces (just as in the cells of a honeycomb), and these still tend to meet the partition-walls connected with them at constant angles of 120°. This outer layer of bubbles, which forms the surface of our froth, constitutes after a fashion what we should call in botany an "epidermal" layer. But in our froth of soap-bubbles we have, as

* A surface-layer always tends to have, *ipso facto*, a character of its own: a "skin" has such and such characteristics just because it is a skin. The "Beilby layer" on a metallic surface is, in its own special way, a consequence of its own externality.

† A froth is a collocation of bubbles containing air; or in the language of colloid chemistry, an emulsion with air for its disperse phase. The power of forming a froth is not the same as that of forming isolated bubbles; for some liquids, such as a solution of saponin, of gum arabic, of albumin itself, give a copious and lasting froth, but we find it hard to blow even a single tiny bubble with any of them. Something more than surface-tension seems necessary for the production and maintenance of a film: perhaps a certain amount of viscosity, to resist the tendency of surface-tension to tear the film asunder.

a rule, the same kind of contact (that is to say, contact with *air*) both within and without the bubbles; while in our living cell, the outer wall of the epidermal cell is exposed to air on the one side, but is in contact with the protoplasm of the cell on the other: and this involves a difference of tensions, so that the outer walls and their adjacent partitions need no longer meet at precisely equal angles of 120°. Moreover a chemical change, due perhaps to oxidation or possibly also to adsorption, is very apt to affect the external wall and lead to the formation of a "cuticle"; and this process, as we have seen, is tantamount to a large increase of tension in that outer wall, and will cause the adjacent partitions to impinge upon it at angles more and more nearly approximating to 90°: the bubble-like, or spherical, surfaces of the individual cells being more

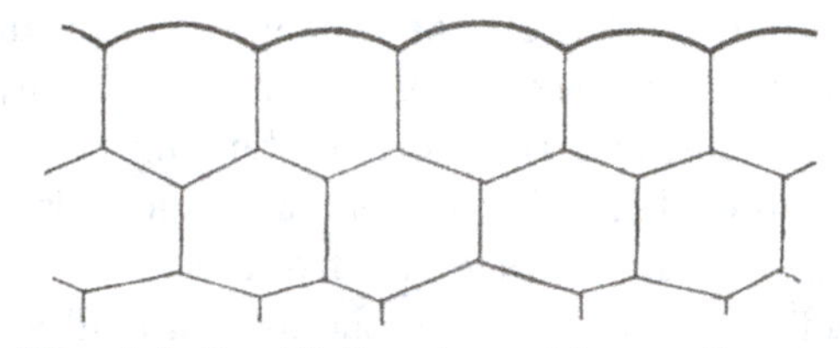

Fig. 177. A froth, with its outer and inner cells or vesicles.

and more flattened in consequence. Lastly, the chemical changes which affect the outer walls of the superficial cells may extend in greater or less degree to their inner walls also: with the result that these cells will tend to become more or less rectangular throughout, and will cease to dovetail into the interstices of the next subjacent layer. These then are the general characters which we recognise in an epidermis; and we now perceive that its fundamental character simply is that it lies outside, and that its physical characteristics follow, as a matter of course, from the position which it occupies and from the various consequences which that situation entails.

In the young shoot or growing point of a flowering plant botanists (following Hanstein) find three cell-layers, and call them *dermatogen*, *periblem* and *plerome*. The first is an epidermis, such as we have just described. Its cells grow long as the shoot grows long; new partitions cross the lengthening cell and tend to lie at right angles to its hardening walls; and this epidermis, once formed, remains a single superficial layer. The next few layers, the so-called periblem, are compressed and flattened between the epidermis with its tense cuticle and the growing mass within; and under this restraint the cell-layers of the periblem also continue to divide in their own plane or planes. But the cells of the inner mass or plerome, lying in a more homogeneous field, tend to form "space-filling" polyhedra, twelve- or perhaps fourteen-sided according to the freedom which they enjoy. In a well-known passage Sachs declares that the behaviour of the cells in the growing point is determined not by any specific characters or properties of their own, but by their position and the forces to which they are subject in the system of which they are a part*. This was a prescient utterance, and is abundantly confirmed†.

We have hitherto considered our cells, or our bubbles, as lying in a plane of symmetry, and have only considered their appearance as projected on that plane; but we must also begin to consider them as solids, whether they lie in a plane (like the four cells in Fig. 172), or are heaped on one another, like a froth of bubbles or a pile of cannon-balls. We have still much to do with the study of more complex partitioning in a plane, and we have the whole subject to enter on of the solid geometry of bodies in "close packing," or three-dimensional juxtaposition.

The same principles which account for the development of hexagonal symmetry hold true, as a matter of course, not only of *cells* (in the biological sense), but of any bodies of uniform size and originally circular outline, close-packed in a plane; and hence the hexagonal pattern is of very common occurrence, under widely varying circumstances. The curious reader may consult Sir Thomas Browne's quaint and beautiful account, in the *Garden of Cyrus*, of hexagonal, and also of quincuncial, symmetry in plants and animals, which "doth neatly declare how nature Geometrizeth, and observeth order in all things."

We come back to very elementary geometry. The first and simplest of all figures in plane geometry (with which for that reason Euclid begins his book) is the equilateral triangle; because three straight lines are the least number which enclose two-dimensional

* *Lectures on the Physiology of the Plant*, Oxford, 1887, p. 460, etc.

† Cf. J. H. Priestley in *Biol. Reviews*, III, pp. 1–20, 1928; U. Tetley in *Ann. Bot.* L, pp. 522–557, 1936; etc.

space, and three equal sides make the simplest of triangles. But it by no means follows that equilateral, or any other, triangles combine to form the simplest of polygonal associations or patterns. On the other hand, three straight lines meeting in a point are the least number by which we can subdivide or partition two-dimensional space; the simplest case of all is when the three partitions meet at co-equal angles, and a pattern of hexagons, so produced, is, geometrically speaking, the simplest of all ways in which a surface can be subdivided—the simplest of all two-dimensional "space-filling" patterns. So it comes to pass that we meet with a pattern of hexagons here and there and again and again, in all sorts of plane symmetrical configurations, from a soapy froth to the retinal pigment, from the cells of the honeycomb to the basaltic columns of Staffa and the Giant's Causeway.

We pass to solid geometry, and arrive by similar steps at an analogous result. Four plane sides are now the least number which enclose space, and (next to the sphere itself) the regular tetrahedron is the first and simplest of solids; but its simplicity is that of a solitary or isolated figure, and tetrahedra do not combine to fill space at all. But as the partitioning of an equilateral triangle was the first step towards the symmetrical partitioning of two-dimensional space, so we draw from the regular tetrahedron a first lesson in the partitioning of space of three dimensions; and as three lines meeting in a point were needed to partition two-dimensional space, so here, for three-dimensional space, we need four. The simplest case is, as before, when these meet at co-equal angles, but we do not see quite so easily what those four co-equal angles are.

For as the centre of symmetry of our equilateral triangle was defined by three lines bisecting its three angles and meeting one another in a point at co-equal angles of 120°, so in our regular tetrahedron four straight lines, running symmetrically inwards from the four corners, meet in a point at co-equal angles, and again define the centre of symmetry. If we make (as Plateau made) a wire tetrahedron, and dip it into soap-solution, we find that a film has attached itself to each of the six wires which constitute the little tetrahedral cage; that these six films meet, three by three, in four edges; and that these four edges meet at co-equal angles in a point, which is the *centroid*, or centre of symmetry, or centre of gravity, of the system.

This is the centre of symmetry not only for our tetrahedron, but for any close-packed tetrahedral aggregate of co-equal spheres; we meet with it over and over again, in a pile of cannon-balls, a froth of soap-suds, a parenchyma of cells, or the interior of the honeycomb. Moreover, in the actual demonstration by soap-films of this tetrahedral symmetry, we see realised all the main criteria laid down by Plateau and by Lamarle for a system *minimae areae*: three films and no more meet in an edge; four fluid edges and no more meet in a point, just as three wire edges and one fluid edge met in a point at each corner of the experimental figure. Lastly, the symmetry of the whole configuration is such that the three fluid films

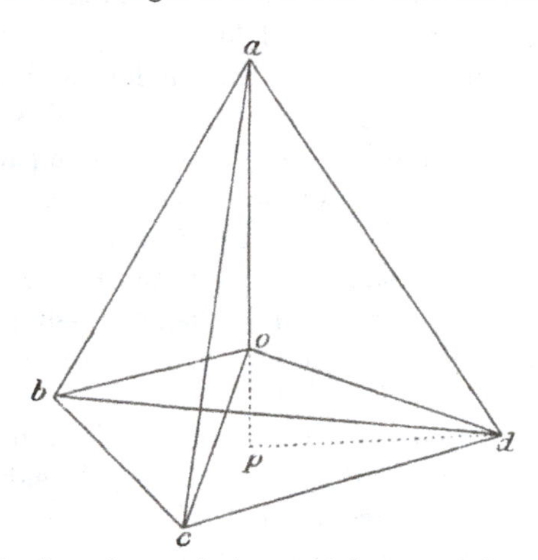

Fig. 178. A regular tetrahedron, with its centre of symmetry.

meeting in an edge, or the four fluid edges meeting in a point, all do so at co-equal angles.

In the plane configuration we saw without more ado that the angles of symmetry were the co-equal angles of 120°; but the four co-equal angles between the four edges which meet at the centre of our tetrahedron require a little more consideration. If in our figure of a regular tetrahedron (Fig. 178) o be the centroid, and we produce ao to p, the centre of the opposite side, bcd, it may be shewn that the line ap is so divided that $ao = 3op$ and $ao = bo = co = do$. For let four equal weights be put at the four corners of the tetrahedron, a, b, c, d. The resultant of the three at b, c, d is equivalent to $3W$ at p, the centre of symmetry of the equilateral triangle.

The resultant of all four is equal to the resultant of W at a, and $3W$ at p; it lies, therefore, on the straight line ap, and at the point o, such that $ao = 3op$. Therefore, in the triangle pod, as in the other three similar triangles in the figure, $\cos pod = 1/3$, and $\cos aod = -1/3$. Our tables tell us that the angle pod, whose trigonometrical value is the very simple one of $\cos pod = 1/3$, has, in degrees and minutes to the nearest second, the seemingly less simple value of $70° 31' 43''$; and its supplement, the angle aod, has the corresponding value of $109° 28' 16''$.

This latter angle, then, of $109° 28' 16''$, or very nearly 109 degrees and a half, is the angle at which, in this and throughout *every other three-dimensional system* of liquid films, the edges of the partition-walls meet one another. It is the fundamental angle in simple homogeneous partitioning of three-dimensional space. It is an angle of statical equilibrium, an angle of close-packing, an angle of repose. In the simplest of carbon-compounds, the molecule of marsh-gas (CH_4), we may be sure that this angle governs the arrangement of the H-atoms; it determines the relation of the carbon-atoms one to another in a diamond—simplest of crystal-lattices; it defines the intersections of the bubbles in a froth, and of the cells in the honeycomb of the bee.

It is sometimes called the "tetrahedral angle"; it might be better called (for a reason we shall see presently) "Maraldi's angle." The whole story is less a physical than a mathematical one; for the phenomena do not depend on surface tension nor on any other physical force, but on such relations between surface and volume as are involved in the properties of space. If we take four little elastic balloons, half fill them with air, smear them with glycerine to lessen friction, place them in a bottle and exhaust the air therein, they will expand, adjust themselves together, and group themselves in a tetrahedral configuration, whose partition walls, edges and centre of symmetry are just those of our experiment of the soap-films.

This characteristic angle, though it leads in ordinary angular measurement to an endless decimal of a second, is nevertheless a very simple and perfectly definite magnitude. It is a strange property of Number that it fails to express certain simple and definite magnitudes, such as π, or $\sqrt{2}$, or $\sqrt{3}$, or this four-fold angle made by four lines meeting symmetrically in three-dimensional space. It is not these magnitudes that are peculiar, it is Number itself that is so! In all of these cases we have to import a new symbol; and in this case, when we draw it not from arithmetic but from trigonometry, and define our angle as cos $-\frac{1}{3}$, nothing can or need be more precise or simpler. We may put the same thing a little differently, and say that Number itself fails us, now and then, to express what we want, although we have all the ten digits and their apparently endless permutations at our command. In such a deadlock, we have only to bring one new symbol, one new quantity, into use; and at once a wide new field is open to us.

Out of these two angles—the Maraldi angle of 109° etc., and the plane angle of 120°—we may construct a great variety of figures,

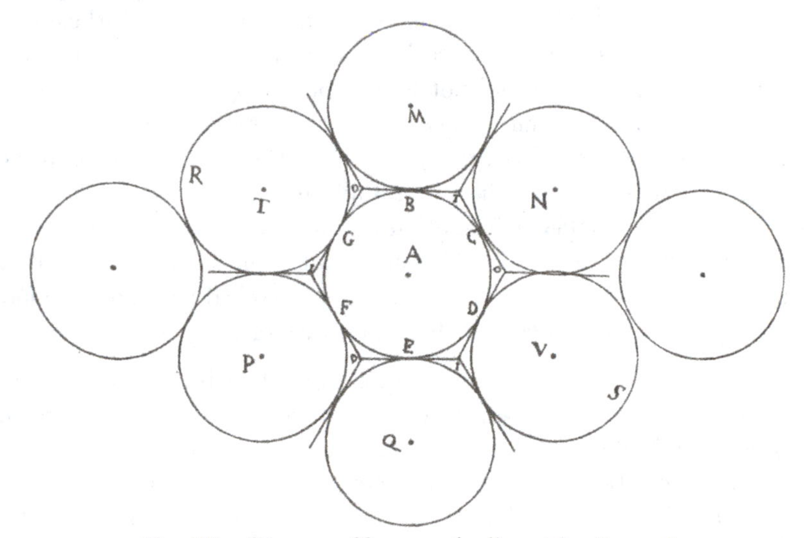

Fig. 179. Diagram of hexagonal cells. After Bonanni.

plane and solid, which become still more complex and varied when we consider associations of unequal as well as of co-equal cells, and thereby admit curved as well as plane intercellular partitions. Let us consider some examples of these, beginning with such as we need only consider in reference to a plane.

Let us imagine a system of equal cylinders, or equal spheres, in contact with one another in a plane, and represented in section by the equal and contiguous circles of Fig. 179. I borrow my figure from an old Italian naturalist, Bonanni (a contemporary of Borelli, of Ray and Willoughby, and of Martin Lister), who dealt with this matter in a book chiefly devoted to molluscan shells*.

* A. P. P. Bonanni, *Ricreatione dell' occhio e della mente, nell' Osservatione delle Chiocciole*, Roma, 1681.

It is obvious, as a simple geometrical fact, that each of these co-equal circles is in contact with six others around. Imagine the whole system under some uniform stress—of pressure caused by growth or expansion within the cells, or due to some uniformly applied constricting pressure from without. In these cases the six *points of contact* between the circles in the diagram will be extended into *lines*, representing *surfaces* of contact in the actual spheres or cylinders; and the equal circles of our diagram will be converted into regular and co-equal hexagons. The result is just the same so far as form is concerned—so long as we are concerned only with a morphological result and not with a physiological process—whatever be the force which brings the bodies together. For instance, the cells of a segmenting egg, lying within their vitelline membrane or within some common film or ectoplasm, are pressed together as they grow, and suffer deformation accordingly; their surface tends towards an *area minima*, but we need not even enquire, in the first instance, whether it be surface-tension, mechanical pressure, or what not other physical force, which is the cause of the phenomenon*.

The production by mutual interaction of polygons, which become regular hexagons when conditions are perfectly symmetrical, is beautifully illustrated by Bénard's *tourbillons cellulaires*, and also in some of Leduc's diffusion experiments. In these latter, a solution of gelatine is allowed to set on a plate of glass, and little drops of weak potassium ferrocyanide are then let fall at regular intervals upon the gelatine. Immediately each little drop becomes the centre of a system of diffusion currents, and the several systems conflict with and repel one another; so that presently each little area becomes the seat of a to-and-fro current system, outwards and back again, until the concentration of the field becomes equalised and the currents cease. When equilibrium is attained, and when the gelatin-layer is allowed to dry, we have an artificial tissue of

* The following is one of many curious corollaries to the principle of close-packing here touched upon. A circle surrounded by six similar circles, the whole bounded by a circle of three times the radius of the original one, forms a unit, so to speak, next in order after the circle itself. A round pea or grain of shot will pass through a hole of its own size; but peas or shot will not *run out* of a vessel through a hole less than *three times* their own diameter. There can be no freedom of motion among the close-packed grains when confronted by a smaller orifice. Cf. K. Takahasi, *Sci. Papers Inst. Chem., etc.*, Tokio, XXVI, p. 19. 1935.

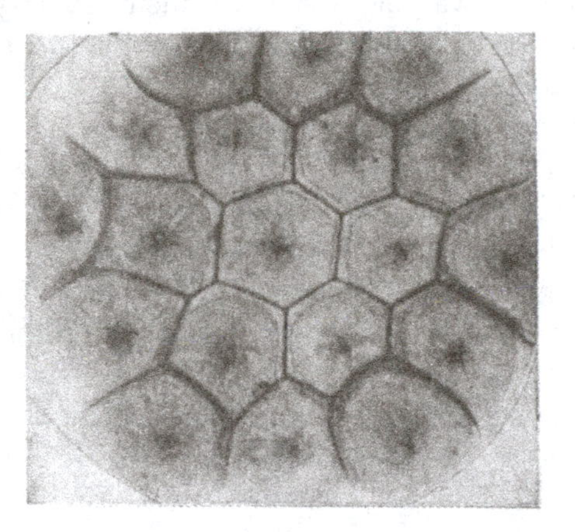

Fig. 180. An "artificial tissue," formed by coloured drops of sodium chloride solution diffusing in a less dense solution of the same salt. After Leduc.

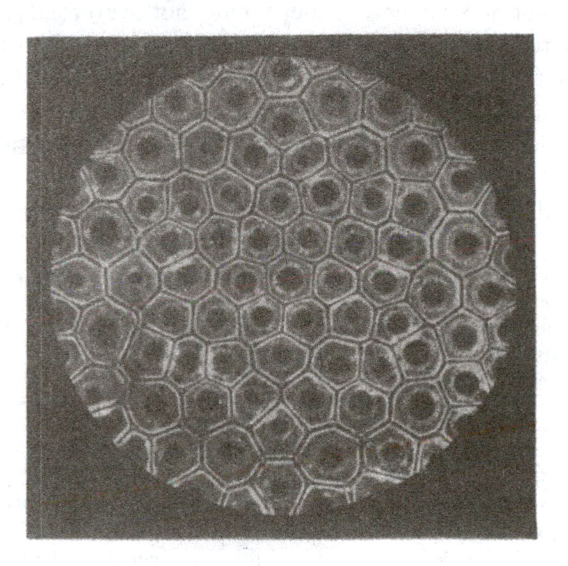

Fig. 181. An artificial cellular tissue, formed by the diffusion in gelatine of drops of a solution of potassium ferrocyanide. After Leduc.

hexagonal "cells," which simulate an organic parenchyma very closely; and by varying the experiment in ways which Leduc describes, we may imitate various forms of tissue, and produce cells with thick walls or with thin, cells in close contact or with wide intercellular spaces, cells with plane or with curved partitions, and so forth.

James Thomson (Kelvin's elder brother) had observed nearly sixty years ago a curious "tesselated structure" on a liquid surface, to wit, the soapy water of a wash-tub. The eddies and streaks of swirling water settled down into a cellular configuration, which continued for hours together to alter its details; small areoles disappeared, large ones grew larger, and subdivided into small ones again. With few and transitory exceptions three partitions and no more met at every node of the meshwork; and (as it seems to me) the subsequent changes were all due to such shifting of the lines as tended to make the three adjacent angles more and more nearly co-equal with one another: the obvious effect of this being to make the pattern more and more regularly hexagonal*.

In a not less homely experiment, hot water is poured into a shallow tin and a layer of milk run in below; on blowing gently to cool the water, holes, more or less close-packed and evenly interspaced, appear in the milk. They shew how cooling has taken place, so to speak, in spots, and the cooled water has descended in isolated columns†.

Bénard's "tourbillons cellulaires"‡, set up in a thin liquid layer,

* James Thomson, On a changing tesselated structure in certain liquids, *Proc. Glasgow Phil. Soc.* 1881–82; *Coll. Papers*, p. 136—a paper with which M. Bénard was not acquainted, but see Bénard's later note in *Ann. de Chim.* Dec. 1911.

† See Graham's paper, quoted below.

‡ H. Bénard, Les tourbillons cellulaires dans une nappe liquide, *Rev. génér. des Sciences*, XII, pp. 1261–1271, 1309–1328, 1900; *Ann. Chimie et Physique* (7), XXIII, pp. 62–144, 1901; *ibid.* 1911. Quincke had seen much the same long before: *Ann. d. Phys.* CXXXIX, p. 28, 1870. The "figures of de Heen" are an analogous electrical phenomenon; cf. P. de Heen, Les tourbillons et les projections de l'éther, *Bull. Acad. de Bruxelles* (3) XXXVII, p. 589, 1899; A. Lafay, *Ann. de Physique* (10), XIII, pp. 349–394, 1930. These various phenomena, all leading to a pattern of hexagons, have often been studied mathematically: cf. Rayleigh, *Phil. Mag.* XXXII, pp. 529–546, 1916, *Coll. Papers*, VI, p. 48; also Ann Pellew and R. V. Southwell, *Proc. R.S.* (A), CLXXVI, pp. 312–343, 1940. The hexagonal pattern is a particular case of stability, but not necessarily the simplest; it is only by experiment that we know it to be the permanent condition in an unlimited field.

are similar to but more elegant than James Thomson's tesselated patterns, and both of them are in their own way still more curious than M. Leduc's; for the latter depend on centres of diffusion artificially inserted into the system and determining the number and position of the "cells," while in the others the cells make themselves. In Bénard's experiment a thin layer of liquid is warmed in a copper dish. The liquid is under peculiar conditions of instability, for the least fortuitous excess of heat here or there would suffice to start a current, and we should expect the whole system to be highly unstable and unsymmetrical. But if all be kept carefully uniform, small disturbances appear at random all over the system; a current ascends in the centre of each; and a "steady state," if not a stable equilibrium, is reached in time, when the descending currents, impinging on one another, mark out a "cellular system." If we set the fluid gently in motion to begin with, the first "cell-divisions" will be in the direction of the flow; long tubes appear, or "vessels," as the botanist would be apt to call them. As the flow slows down new cell-boundaries appear, at right angles to the first and at even distances from one another; parallel rows of cells arise, and this transitory stage of partial equilibrium or imperfect symmetry is such as to remind the botanist of his *cambium* tissues, which are, so to speak, a temporary phase of histological equilibrium. If the impressed motion be not longitudinal but rotary, the first lines of demarcation are spiral curves, followed by orthogonal intersections.

Whether we start with liquid in motion or at rest, symmetry and uniformity are ultimately attained. The cells draw towards uniformity, but four, five or seven-sided cells are still to be found among the prevailing hexagons. The larger cells grow less, the smaller enlarge or disappear; where four partition-walls happen to meet, they shift till only three converge; the sides adjust themselves to equal lengths, the angles also to equality. In the final stage the cells are hexagonal prisms of definite dimensions, which depend on temperature and on the nature and thickness of the liquid layer; molecular forces have not only given us a definite cellular pattern, but also a "fixed cell-size."

Solid particles in the fluid come to rest in symmetrical positions. If they be heavier they accumulate in little isolated heaps, each in the focus or axis of a cell; if they be lighter they drift to the boundaries, then towards the nodes, where they tend to form tri-radiate figures like so many "tri-radiate spicules". But if they be in very fine suspension a curious thing happens: for as they are carried round in the vortex, the lowermost layer of liquid, next to the solid floor, keeps free of particles; and this "dust-free coat*", rising in the axis of the cell and descending at its boundary-walls, surrounds an inner vortex to which the suspended particles are confined. The cell-contents have, so to speak, become differentiated into an "ectoplasm" and an "endoplasm"; and an analogy appears

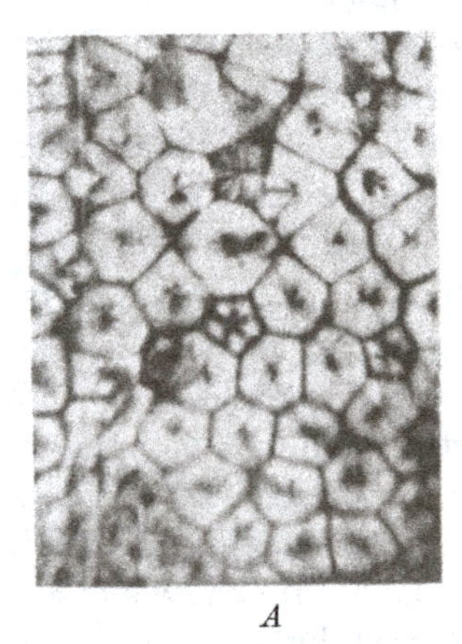

A

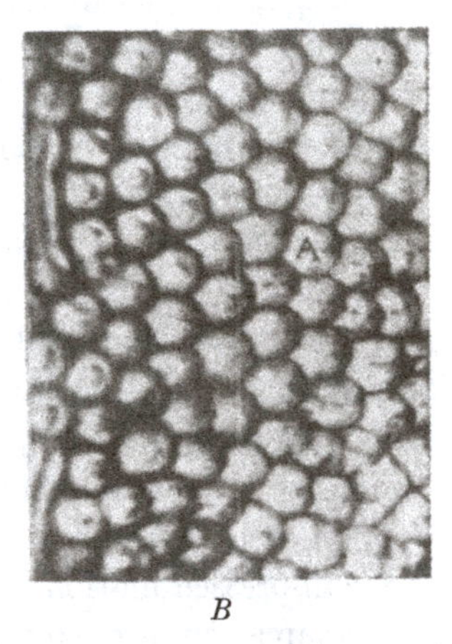

B

Fig. 182. Bénard patterns in smoke: *A*, at rest; *B*, under shear. After K. Chandra.

with the phenomenon of protoplasmic "rotation," where the outer layer of a cell tends to be free from granules. When bright glittering particles are used for the suspension (such as graphite or butterfly-scales) beautiful optical effects are obtained, deep shadows marking the outlines and the centres of the cells. Lastly, and this is by no means the least curious part of the phenomenon, the free surface of the liquid is not plane; but each little cell is found to be dimpled in the centre and raised at the edges, in a surface of very complex curvature†, and there is a curious pulsation in the flow, especially when waxes are used.

* Cf. Tyndall, *Proc. Roy. Inst.* VI, p. 3, 1870.

† The differences of level are of a very small order of magnitude, say 1 μ in a layer of spermaceti 1 mm. thick.

Ringing the changes on Bénard's experiment, we may use unstable layers of various liquids or gases, or even a thin layer of smoke between a hot plate and a cold (Fig. 182). The smoke will form waves and folds and rolling clouds; then, with increasing and more and more symmetrical instability, polygonal or hexagonal prisms; and all these configurations we may deform or "shear" by sliding one plate over the other. Familiar cloud-patterns, as of a dappled or mackerel sky, can be imitated in this way. When the hexagonal prisms have been developed it is found that a steady shear deforms them into a well-known curvilinear tesselated pattern (Fig. 183)

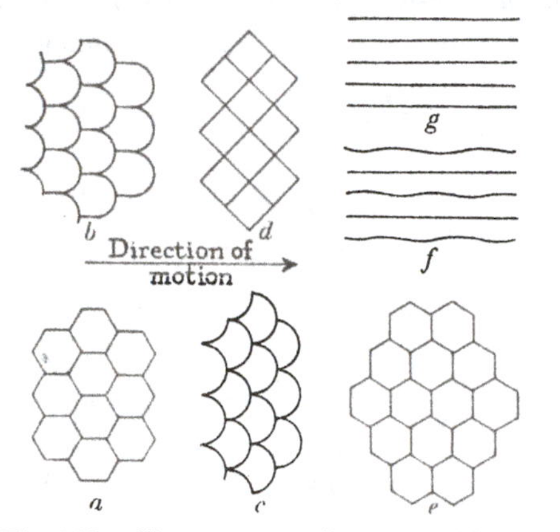

Fig. 183. Shear-patterns in an unstable layer of air or smoke. After Graham.

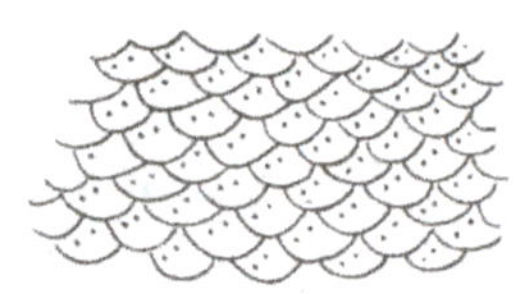

Fig. 184. Ambulacral plates of a sea-urchin (*Lepidesthes*), to illustrate the "shearing" of a pattern normally hexagonal. After Hawkins*.

and this may be sheared again into hexagons, oriented in an opposite direction to the first†. The very same pattern occurs now and then in organisms, as a deformation of what is normally a pattern of hexagons (Fig. 184).

* From H. L. Hawkins, *Phil. Trans.* (B), CCIX, p. 383, 1920.

† Cf. A. Graham, Shear patterns in an unstable layer of air, *Phil. Trans.* (A), No. 714, 1933; Gilbert Walker and Phillips, *Q.J.R. Met. Soc.* LVIII, p. 23, 1932; Mals, *Beitr. Phys. frei. Atmosph.* XVII, p. 45, 1930; H. Jeffreys, *Proc. R.S.* (A), CXVIII, p. 195, 1928; Krishna Chandra, *ibid.* CLXIV, pp. 231–242, 1938. After a steady state is reached it is found that in air or smoke the centre of each polygon is a funnel of descent, but it is an ascending column if the layer be liquid. Now the viscosity of a gas increases, and that of a liquid decreases, with rise of temperature, and the greater the viscosity the more stable is the layer. Accordingly, for a gas the upper, and for a liquid the lower layer is the more unstable, and it is there that in each case the flow begins.

We learn from these experiments of Bénard and others how similar distributions of force, and identical figures of equilibrium, may arise through different physical agencies. We see that patterns closely analogous to those of living cells and tissues may be due to very different causes; and we may be led to scrutinise anew, with an open mind, various histological configurations whose origin is doubtful or obscure. The chitinous shells of certain water-fleas (*Cladocera*) are beset with a roughly hexagonal pattern, and each little chitinous polygon is supposed to correspond to, and to be formed by, an underlying "hypodermis" cell*. But we presently discover that the existence of these hypodermis-cells is merely deduced from the polygons themselves and from a coincident

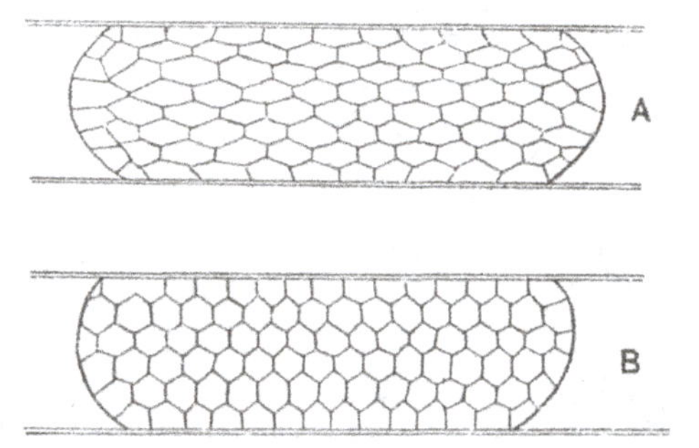

Fig. 185. Soap-froth under pressure. After Rhumbler.

distribution of pigment; it might not be amiss to look again into the development of the pattern, with an open mind as to the possibility of its being a purely physical phenomenon. Nor need we by any means assume that the calcareous prisms of a molluscan shell are necessarily derived from, or associated with, a like number of histological elements.

In a soap-froth imprisoned between two glass plates we have a symmetrical system of cells which appear in optical section (Fig. 185, B) as regular hexagons; but if we press the plates a little closer together the hexagons become deformed and flattened. The

* Cf. F. Claus, Zur Kenntniss...des feineren Baues der Daphniden, *Ztschr. f. wiss. Zool.* XXIII, XXVII, pp. 362–402, 1876; Ernest Warren, Relationship between size of cell and size of body in *Daphnia magna*, *Biometrika*, (3) II, pp. 255–259, 1902; Fritz Werner, Die Veränderung der Schalenform und der Zellenaufbau bei *Scapholeberis*, *Int. Revue der ges. Hydrobiologie*, pp. 1–20, 1923.

change is from a more to a less probable configuration—the entropy is diminished; and if we apply no further pressure the tension of the films adjusts itself again, and the system recovers its former symmetry*.

The epithelial lining of the blood-vessels shews a curious and beautiful pattern. The cells seem diamond-shaped, but looking closer we see that each is in contact (usually) with six others; they are not rhombs, or diamonds, but elongated hexagons, pulled out long by the growth of the vessel and the elastic traction of its walls. The sides of each cell are curiously waved, and a simple experiment explains this phenomenon. If we make a froth of white-of-egg upon a stretched sheet of rubber, the cells of the froth will tend to

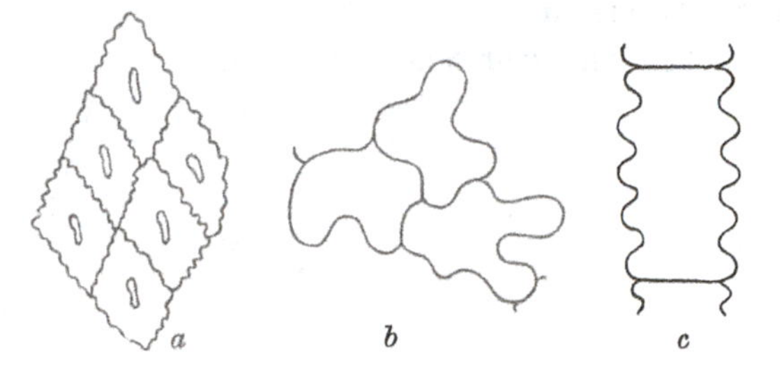

Fig. 186. Sinuous outlines of epithelial cells. *a*, endothelium of a blood-vessel; *b*, epidermis of *Impatiens*; *c*, epidermal cells of a grass (*Festuca*).

assume their normal hexagonal pattern; but relax the elastic membrane, and the cell-walls are thrown into beautiful sinuous or wavy folds. The froth-cells cannot contract as the rubber does which carries them, nor can the epithelial cells contract as does the muscular coat of the blood-vessel; in both cases alike the cell-walls are obliged to fold or wrinkle up, from lack of power to shorten. The epithelial cells on the gills of a mussel† are wrinkled after the same fashion; but the more coarsely sinuous outlines of the epithelium in many plants is another story, and not so easily accounted for.

The hexagonal pattern is illustrated among organisms in countless cases, but those in which the pattern is perfectly regular, by

* That *everything* is passing all the while towards a "*more probable state*" is known as the "principle of Carnot," and is the most general of all physical laws or aphorisms.

† Cf. James Gray's *Experimental Cytology*, p. 252.

reason of perfect uniformity of force and perfect equality of the individual cells, are not so numerous. The hexagonal cells of the pigmented epithelium of the retina are a good example. Here we have a single layer of uniform cells, reposing on the one hand upon a basement membrane, supported behind by the solid wall of the sclerotic, and exposed on the other hand to the uniform fluid pressure of the vitreous humour. The conditions all point, and lead, to a symmetrical result: the cells, uniform in size, are flattened out to a uniform thickness by uniform pressure, and their reaction one upon another converts each flattened disc into a regular hexagon. An equally symmetrical case, one of the first-known examples of an "epithelium," is to be found on the inner wall of the amnion, where, as Theodor Schwann remarked, "die sechseckige Plättchen sind sehr schön und gross*."

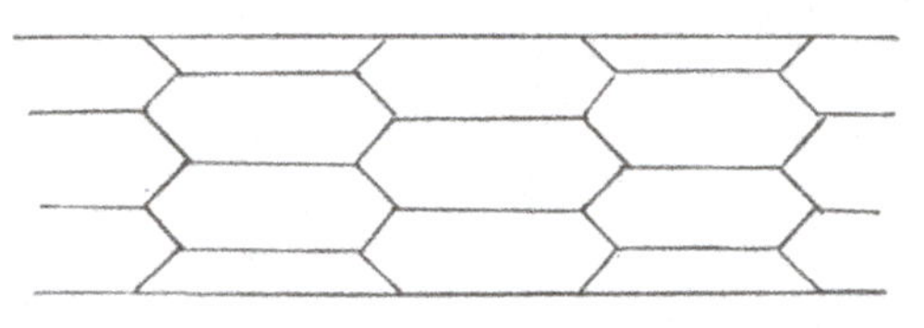

Fig. 187. Epidermis of *Girardia*. After Goebel.

In an ordinary columnar epithelium, such as that of the intestine, again the columnar cells are compressed into hexagonal prisms; but here the cells are less uniform in size, small cells are apt to be intercalated among the larger, and the perfect symmetry is lost accordingly. But obviously, wherever we have, in addition to the forces which tend to produce the regular hexagonal symmetry, some other component arising asymmetrically from growth or traction, then our regular hexagons will be distorted in various simple ways. Thus in the delicate epidermis of a leaf or young shoot we begin with hexagonal cells of exquisite regularity: on which, however, subsequent longitudinal growth may impose an equally simple and symmetrical deformation or polarity (Fig. 187).

In the growth of an ordinary dicotyledonous leaf, we see reflected in the form of its cells the tractions, irregular but on the whole longitudinal, which growth has superposed on the tensions of the

* *Untersuchungen*, p. 84; cf. Sydenham Society's translation, p. 75.

partition walls (Fig. 188). In the narrow elongated leaf of a monocotyledon, such as a hyacinth, the elongated, apparently quadrangular cells of the epidermis appear as a necessary consequence of the simpler laws of growth which gave its simple form to the leaf as a whole. In all these cases alike, however, the rule still holds that only three partitions (in surface view or plane projection) meet in a point; and near their point of meeting the walls are manifestly curved for a little way, so as to permit the triple conjunction to take place at or near the co-equal angles of 120°, after the fashion described above.

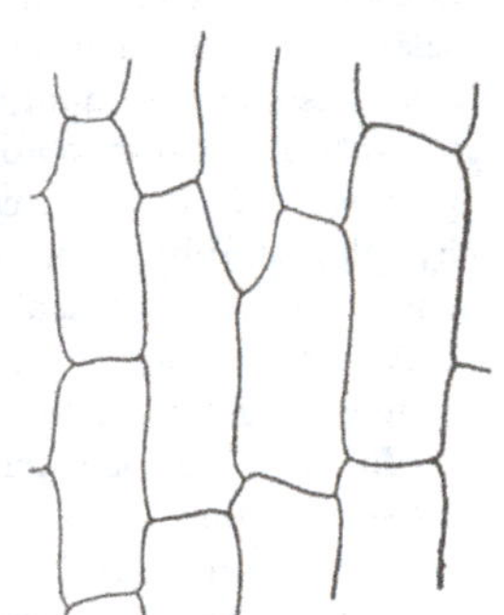

Fig. 188. Epidermal cells from leaf of *Elodeá canadensis*. After Berthold.

Briefly speaking, wherever we have a system of cylinders or spheres, associated together with sufficient mutual interaction to bring them into complete surface contact, there, in section or in surface view, we tend to get a pattern of hexagons.

In thickened cells or fibres of bast or wood, the "sclerenchyma" of vegetable histology, the hexagonal pattern is all but lost, and we see in cross-section the more or less *circular* transverse outlines of elongated and tapering cells. Looking closer we see that the primitive cell-walls preserve their angular contours, and shew much as usual an hexagonal pattern, with only such irregularities as follow from the unequal sizes of the associated cells. But when these primary walls are once laid down, the secondary deposits which follow them are under different conditions; and these obey the law of minimal areas in their own way, by filling up the angles of the primary cell and by continuing to grow inwards in concentric and more and more nearly circular rings.

While the formation of an hexagonal pattern on the basis of ready-formed and symmetrically arranged material units is a very common, and indeed the general way, it does not follow that there are not others by which such a pattern can be obtained. For instance, if we take a little triangular dish of mercury and set it vibrating (either by help of a tuning-fork, or by simply tapping on the sides) we shall have a series of little waves or ripples starting inwards from each of the three faces; and the intercrossing, or interference of these three sets of waves produces crests and hollows, and intermediate points of no disturbance, whose *loci* are seen as a beautiful pattern of minute hexagons. It is possible that the very minute and astonishingly regular pattern of hexagons which we see on the surface of many diatoms (Fig. 189) may be a phenomenon of this order*. The same may be the case also in Arcella, where an apparently hexagonal pattern is found not to consist of simple hexagons, but of "straight lines in three sets of parallels, the lines of each set making an angle of sixty degrees with those of the other two sets†." We must also bear in mind, in the case of the minuter forms, the large possibilities of optical illusion. For instance, in one of Abbe's "diffraction-plates," a pattern of dots, set at equal interspaces, is reproduced on a very minute scale by photography; but under certain conditions of microscopic illumination and focusing, these isolated dots appear as a pattern of hexagons.

A symmetrical arrangement of hexagons, such as we have just been studying, suggests various geometrical corollaries, of which the following may be a useful one. We sometimes desire to estimate the number of hexagonal areas or facets in some structure where these are numerous, such for instance as the cornea of an insect's eye, or in the minute pattern of hexagons on many diatoms. An approximate enumeration is easily made as follows.

For the area of a hexagon (if we call δ the short diameter, that namely which bisects two of the opposite sides) is $\delta^2 \times \sqrt{3}/2$, the area of a circle being $d^2 . \pi/4$. Then, if the diameter (d) of a circular area include n hexagons, the area of that circle equals $(n . \delta)^2 \times \pi/4$. And, dividing this number by the area of a single hexagon, we obtain for the number of areas in the circle, each equal to a hexagonal facet, the expression $n^2 \times \pi/4 \times 2/\sqrt{3} = 0{\cdot}907n^2$, or $9/10 . n^2$, nearly.

This calculation deals, not only with the complete facets, but with the areas of the broken hexagons at the periphery of the circle. If we neglect these latter, and consider our whole field as consisting of successive rings of hexagons about a central one, we obtain a simpler rule. For obviously, around our central hexagon there stands a zone of six, and around these

* Cf. some of J. H. Vincent's photographs of ripples, in *Phil. Mag.* 1897–99; or those of F. R. Watson, in *Phys. Review*, 1897, 1901, 1916. The appearance will depend on the rate of the wave, and in turn on the surface-tension; with a low tension one would probably see only a moving "jabble." Cf. also Faraday, On the crispations of fluids resting upon a vibrating support, *Phil. Mag.* 1831, p. 299; and Rayleigh, *Sound*, II, p. 346, 1896. FitzGerald thought diatom-patterns might be due to electromagnetic vibrations (*Works*, p. 503, 1902); with which cf. W. D. Dye, Vibration-patterns of quartz plates, *Proc. R.S.* (A), CXXXVIII, p. 1, 1932. Dye's Fig. 17, which he calls "one of the most beautiful types of minor vibration met with in discs", is closely akin to the diatom *Orthoneis splendida*. In both cases two nodal systems, conjugate to one another, are based on two foci near the ends of an elliptical plate; but bands in the experimental plate are further broken up into rows of dots in the diatom. See also Max Schultze, Die Struktur der Diatomeenschale verglichen mit gewissen aus Fluorkiesel künstlich darstellbaren Kieselhäuten, *Verh. naturh. Ver. Bonn*, XX, pp. 1–42, 1863; *Trans. Microsc. Soc.* (N.S.), XI, pp. 120–136, 1863; H. J. Slack, *Monthly Microsc. Journ.* 1870, p. 183.

† J. A. Cushman and W. P. Henderson, *Amer. Nat.* XL, pp. 797–802, 1906.

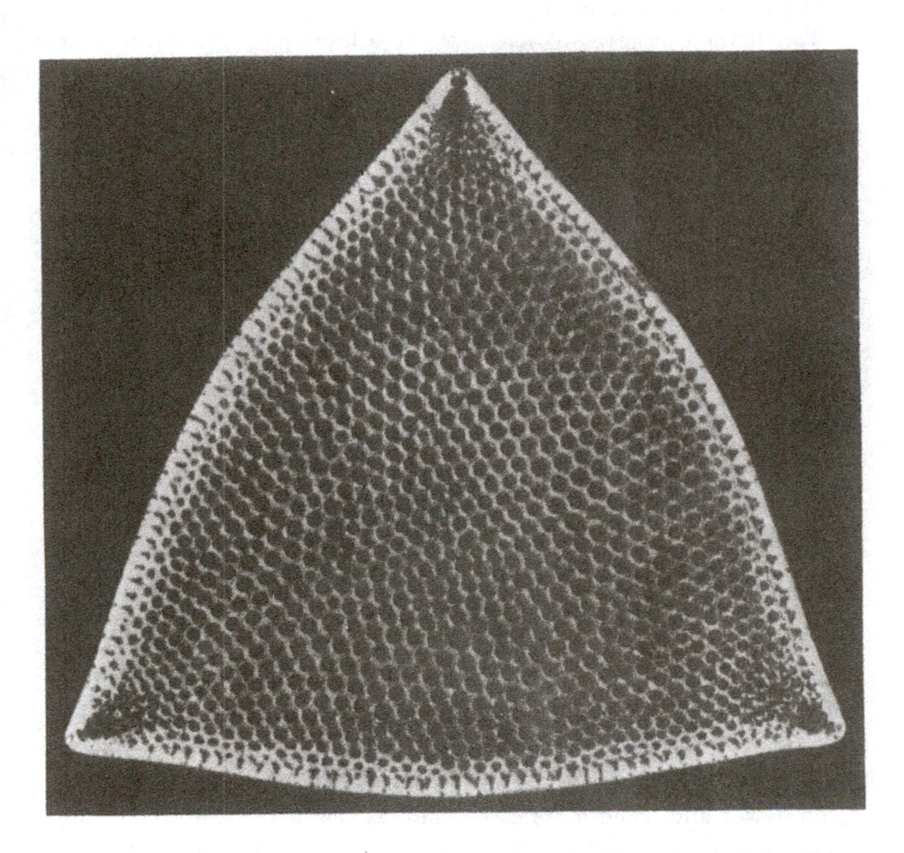

Fig. 189. A diatom, *Triceratium* sp., shewing pattern of hexagons; ×300. From O. Prochnow, *Formenkunst der Natur*.

a zone of twelve, and around these a zone of eighteen, and so on. And the total number, excluding the central hexagon, is accordingly:

For one zone	6	$=3\times 1\times 2$	$=6\times 1$
„ two zones	18	2×3	3
„ three zones	36	3×4	6
„ four zones	60	4×5	10
„ five zones	90	5×6	15

and so forth. If N be the number of zones, and if we add one to the above numbers for the odd central hexagon, then the rule is that the total number $H=3N(N+1)+1$. Thus, if in a preparation of a fly's cornea I can count twenty-five facets in a line from a central one, the total number in the entire field is $(3\times25\times26)+1=1951$ *.

The electrical engineer is dealing with the selfsame problem when he finds he can pack $6+18+36+\ldots$ wires around a central wire, to form a multiple cable of 1, 2 and 3 concentric strands. He counts them by the same formula, in the simpler form of $6t+1$: where t is a "triangular number," 1, 3, 6, 10, etc., corresponding to the number of strands. Thus $1951=6\times325+1$; 325 being the triangular number of 25, $1+2+3+\ldots+25$.

We have many varied examples of this principle among corals, wherever the polypes are in close juxtaposition, with neither empty space nor accumulations of matrix between their adjacent walls. *Favosites gothlandica*, for instance, furnishes us with an excellent example. In the great genus Lithostrotion we have some species which are "massive" and others which are "fasciculate." In other words, in some the long cylindrical corallites are closely packed together, and in others they are separate and loosely bundled (Fig. 190); in the former the corallites are squeezed into hexagonal prisms, while in the latter they retain their cylindrical form. Where the polypes are comparatively few, and so have room to spread, the mutual pressure ceases to work or only tends to push them asunder, letting them remain circular in outline (e.g. *Thecosmilia*). Where they vary gradually in size, as for instance in *Cyathophyllum hexagonum*, they are more or less hexagonal but are not regular hexagons; and where there is greater and more irregular variation in size, the cells will be *on the average* hexagonal, but some will have fewer and some more sides than six, as in the annexed figure of *Arachnophyllum* (Fig. 192). Where larger and smaller cells, corresponding to two different kinds of zooids, are mixed together, we may get various results. If the larger cells are numerous enough

* This estimate neglects not merely the broken hexagons, but all those whose centres lie between the circle and a hexagon inscribed in it. The discrepancy is considerable, but a correction is easily made. It will be found that the numbers arrived at by the two methods are approximately as 6 : 5. For more detailed calculations see a paper by H.M. (? H. Munro) in *Q.J.M.S.* VI, p. 83, 1858. The methods of enumeration used by older writers, especially by Leeuwenhoek, are sometimes curious and interesting; cf. Hooke, *Micrographia*, 1665, p. 176; Leeuwenhoek, *Arcana naturae*, 1695, p. 477; *Phil. Trans.* 1698, p. 169; *Epist. physiolog.* 1719, p. 342; Swammerdam, *Biblia Naturae*, 1737, p. 490. Leeuwenhoek found, or estimated, 3181 facets on the cornea of a scarab, and 8000 on that of a fly; M. Puget, about the same time, found 17,325 in that of a butterfly. See also Karl Leinemann, *Die Zahl der Facetten in den...Coleopteren*, Hildesheim, 1904.

to be more or less in contact with one another (e.g. various Monticuliporae) they will be irregular hexagons, while the smaller cells between them will be crushed into all manner of irregular angular

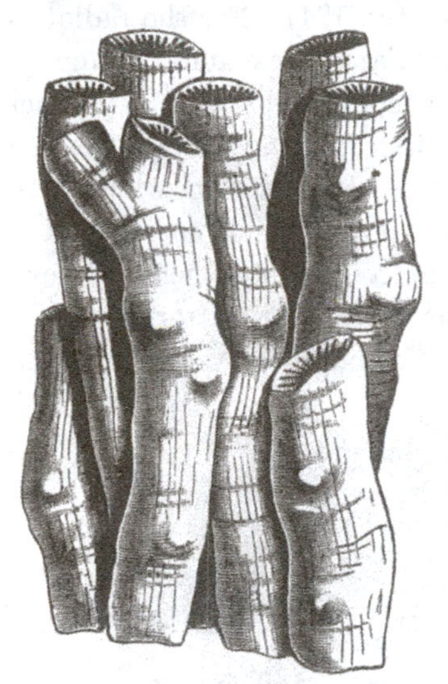

Fig. 190. *Lithostrotion Martini.* After Nicholson.

Fig. 191. *Cyathophyllum hexagonum.* From Nicholson, after Zittel.

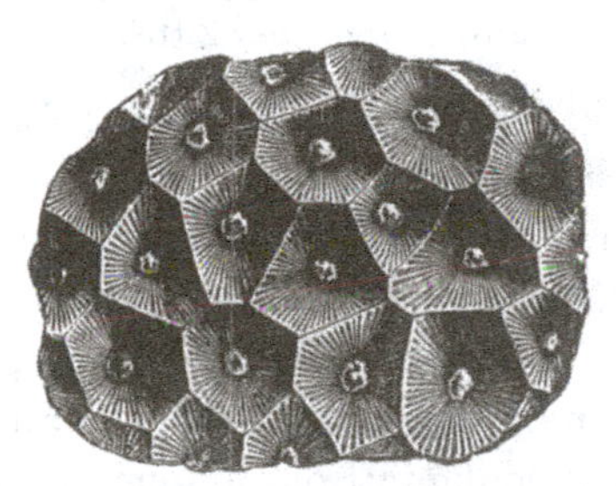

Fig. 192. *Arachnophyllum pentagonum.* After Nicholson.

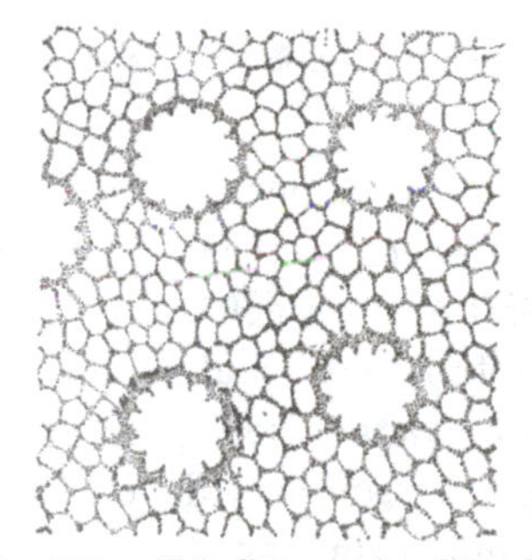

Fig. 193. *Heliolites.* After Woods.

forms. If on the other hand the large cells are comparatively few and are large and strong-walled compared with their smaller neighbours, then the latter alone will be squeezed into hexagons while the larger ones will tend to retain their circular outline undisturbed (e.g. *Heliopora*, *Heliolites*, etc. (Fig. 193)).

When, as happens in certain corals, the peripheral walls or "thecae" of the individual polypes remain undeveloped but the radiating septa are formed and calcified, then we obtain new and beautiful mathematical configurations (Fig. 194). For the radiating septa are no longer confined to the circular or hexagonal bounds of a polypite, but tend to meet and become confluent with their neighbours on every side; and, tending to assume positions of equilibrium, or of minimum, under the restraints to which they are subject, they fall into congruent curves, which correspond in a striking manner to lines running in a common field of force between a number of secondary centres. Similar patterns may be produced in various ways by the play of osmotic or magnetic forces;

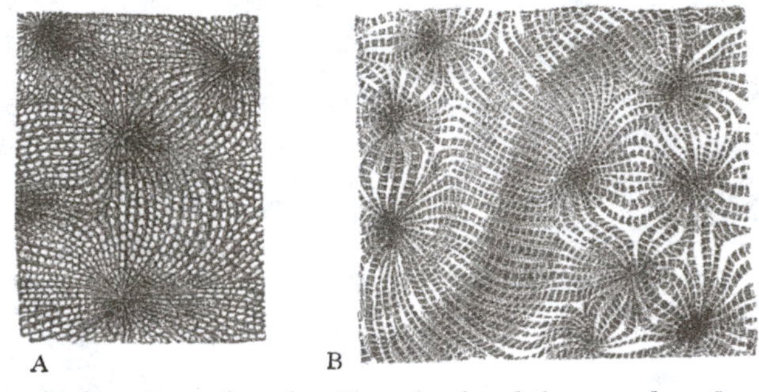

Fig. 194. Surface-views of corals with undeveloped thecae and confluent septa. A, *Thamnastraea*; B, *Comoseris*. From Nicholson, after Zittel.

and a very curious case is to be found in those complicated forms of nuclear division known as triasters, polyasters, etc., whose relation to a field of force Hartog in part explained*. It is obvious that in our corals these curving septa are all orthogonal to the non-existent hexagonal boundaries; and, as the phenomenon is due to the imperfect development, or non-existence, of a thecal wall, it is not surprising that we find identical configurations among various corals, or families of corals, not otherwise related to one another. We find the same or very similar patterns displayed, for instance, in Synhelia (*Oculinidae*), in Phillipsastraea (*Rugosa*), in Thamnastraea (*Fungida*), and in many more.

* Cf. M. Hartog, The dual force of the dividing cell, *Science Progress* (N.S.), I, Oct. 1907, and other papers. Also Baltzer, *Mehrpolige Mitosen bei Seeeigeleiern*, Diss., 1908.

A beautiful hexagonal pattern is seen in the male and female cones of *Zamia*, where the scales which bear the pollen-sacs or the ovules are crowded together, and are so formed and circumstanced that they cannot protrude and overlap. They become compressed accordingly into regular hexagons, smaller and more regular in the male cone than in the female, in which latter the cone as a whole has tended to grow more in breadth than in length, and the hexagons are somewhat broader than they are long. In a cob of maize the hexagonal form of the grains, such as should result from close-packing and mutual compression, is exhibited faintly if at all; for growth and elongation of the spike itself has relieved, or helped to relieve, the mutual pressure of the grains.

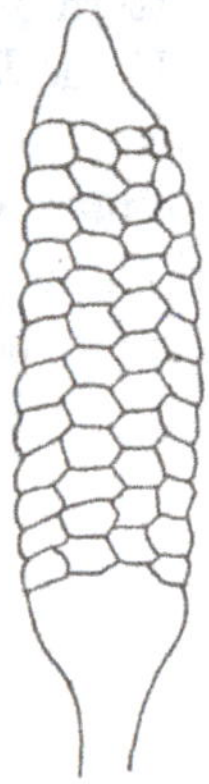

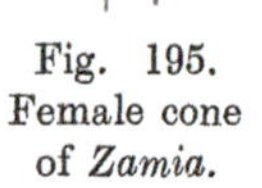

Fig. 195. Female cone of *Zamia*.

The pine-cone shews a simple, but unusual mode of close-packing. The spiral arrangement causes each scale to lie, to right and to left, on two principal spirals; it has close neighbours on four sides, and mutual compression leads to a square or rhomboidal, instead of an hexagonal, configuration*. On the other hand, the scales of the larch-cone overlap: therefore they are not subject to compression, but grow more freely into leaf-like curves.

The story of the hexagon leads us far afield, and in many directions, but it begins with something simpler even than the hexagon. We have seen that in a soapy froth three films, and three only, meet in an edge, a phenomenon capable of explanation by the law of *areae minimae*. But the conjunction, three by three, of almost any assemblage of partitions, of cracks in drying mud, of varnish on an old picture, of the various cellular systems we have described, is a general tendency, to be explained more simply still. It would be a complex pattern indeed, and highly improbable, were all the cracks (for instance) to meet one another six by six; four by four would be less so, but still too much; and three by three is nature's way, simply because it is the simplest and the least. When the partitions meet three by three, the angles by which they do so may vary indefinitely, but their *average* will be 120°; and if all be *on the*

* In some small, few-scaled cones the packing remains incomplete, and the scales are four-, five-, or six-sided, as the case may be.

average angles of 120°, the polygonal areas must, on the average, be *hexagonal*. This, then, is the simple *geometrical* explanation, apart from any physical one, of the widespread appearance of the pattern of hexagons.

If the law of minimal areas holds good in a "cellular" structure, as in a froth of soap-bubbles or in a vegetable parenchyma, then not merely on the average, but actually at every node, three partition-walls (in plane projection) meet together. Under perfect symmetry they do so at co-equal angles of 120°, and the assemblage consists

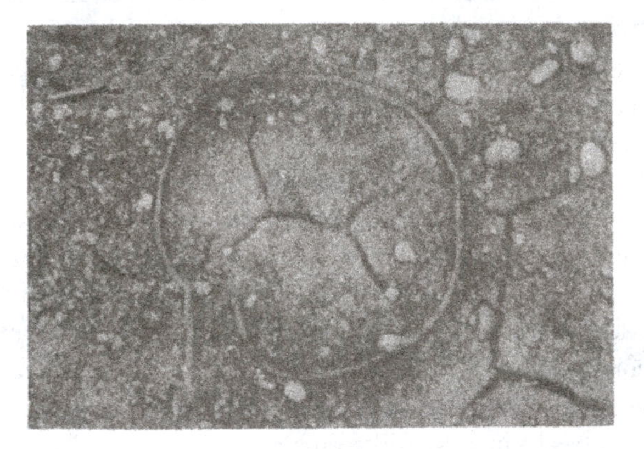

Fig. 196. Cracks in drying mud; a thread encircles and marks out a "polar furrow"; cf. p. 487. From R. H. Wodehouse.

(in plane projection) of co-equal hexagons; but the angles may vary, the cells be unequal, and the hexagons interspersed with other polygonal figures. Nevertheless, so long as three partition-walls and only three meet together, the cells are, *ipso facto*, on the average hexagonal*.

We may count the cells if we please. A section of Cycas-petiole gave the following numbers:

Number of sides	3	4	5	6	7	8	9
,, ,, instances	0	8	97	207	96	9	0

Mean number of sides: 6·00.

The fine emulsion of an Agfa plate shews a beautiful polygonal pattern which obeys the law of the triple node; and a patch of a

* A more elaborate proof is given by W. C. Goldstein, On the average number of sides of polygons of a net, *Ann. Math.* (2), XXXII, pp. 149–153, 1931.

thousand cells in such a plate has been found, like the Cycas-petiole, to average out at six sides each, precisely*.

The cracking of a fine varnish may illustrate (as in Fig. 197) the same phenomenon. A little water in the varnish tends to accumulate between the cells, interferes with their close-packing, and complicates the arrangement of their partitions.

The horny plates which form the carapace of a tortoise (different as the case may seem) still obey the two guiding principles (1) that the polygonal boundaries meet in three-way nodes, and (2) that the three angles tend towards equality, always provided that no alien influences interfere. These principles are of the widest application; the carapace of a Eurypterid, the dermal armour of an Old Red

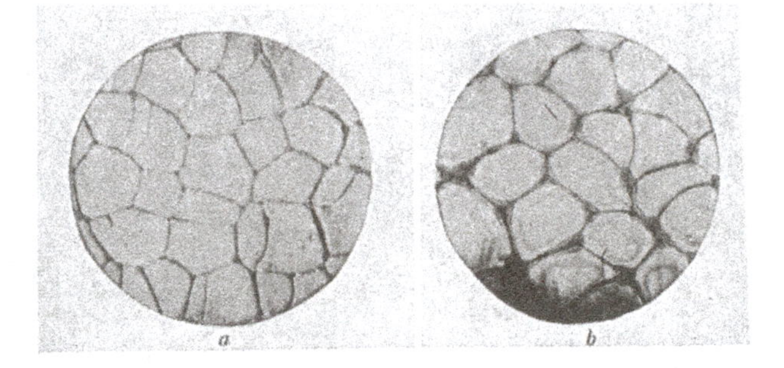

Fig. 197. Cellular patterns in varnish. *a*, dissolved in dry acetone; *b*, containing a little water.

Sandstone fish like Hugh Miller's *Asterolepis*, exhibit them at a glance†. The carapace of our tortoise is formed of a bony framework of ribs and vertebrae, overlaid by superficial plates of horn or tortoiseshell; it is these latter with which we are about to deal. They are arranged in three rows down the back, and a marginal row of smaller plates surrounds the others; there are (normally) twenty-four plates in the marginal series, and five large ones in the median longitudinal row. With these few facts, and our general principles

* F. T. Lewis, Polygons in a film...and the pattern of simple epithelium, *Anat. Record*, L, pp. 235–265, 1931.

† The all but universal law of the triple corner, or triradiate suture, is now and then enough to give a deceptive likeness to very different things. When Cope marred his brilliant classification of the Ostracoderm fishes by seeing in the carapace of Bothriolepis a likeness to the dorsal plates of the tunicate Chelyosoma, it was this and this alone which led him astray.

in hand, how far can we go towards depicting the carapace? The five plates of the median row *must* alternate with their lateral neighbours, and four plates in each lateral row will, accordingly, be the simplest case or most probable number. If at each three-way junction between median and lateral plates the angles tend to equality, it follows that the median plates become converted into more or less regular, or at least symmetrical, hexagons. As to the twenty-four marginal plates, let us put one in front* and one behind, leaving eleven for each side; and let us see to it carefully

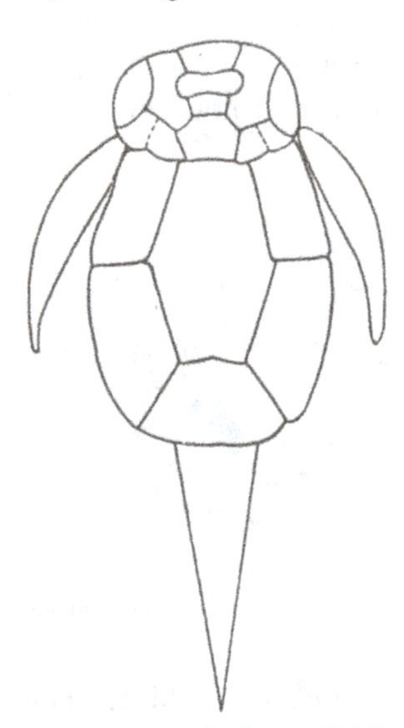

Fig. 198. *Asterolepis*: an Old Red Sandstone fish. After Traquair.

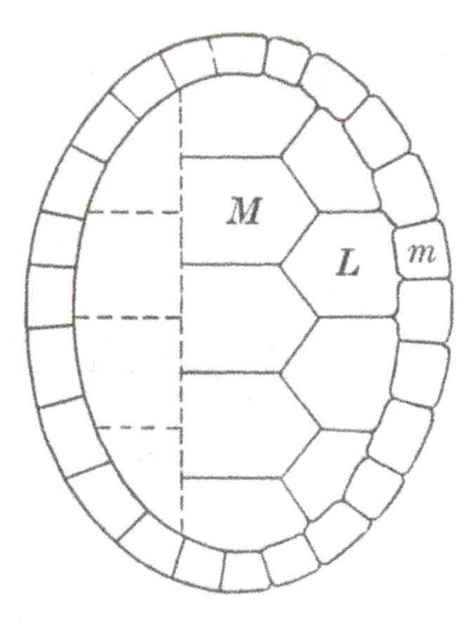

Fig. 199. Horny carapace of a tortoise; diagrammatic.

that the sutures between these do not coincide but *alternate* with those of the lateral row. Here we begin to meet with conditions of restraint analogous to those of the surface layer of a froth, for the long marginal cells must remain marginal, and their sides must continue to run more or less parallel, or more or less perpendicular, to the edge of the shell; only in the immediate

* The Old World tortoises have twenty-five marginal plates, those of the New World lack the anterior median, or "nuchal" plate. This difference is a biological accident, it has neither mathematical interest nor functional significance; it exemplifies the aphorism that whatsoever is possible Nature will sooner or later do.

neighbourhood of each corner will the sides tend to curve in, so forming a notch whose curved sides have tangents approximately 120° apart, again just as in our projection of the surface of a froth (p. 494). Already these considerations lead us to a fair sketch of, or first approximation to, the carapace of a tortoise; but we may go on a little further. The horny plates and the bony carapace below must grow at such a rate as to keep pace, more or less exactly, with one another; but it does not follow that they will keep time precisely. If the horny plates grow ever so little faster than the bones below, they will fail to fit, will overcrowd one another, and will be forced to bulge or wrinkle. Both of these things they often, and even characteristically, do; the wrinkles appear in orderly, parallel folds, pointing to alternate periods, or spurts, of faster and slower growth; and the characteristic patterns which ensue are the visible expression of these differential growth-rates.

In all this we assume that the plates are lying in one and the same plane or even surface, abutting against one another as they grow, and so crowding and squeezing one another into the form of straight-edged polygons. The result will be very different if they overlap, after the manner of slates on a roof: the difference is what we have seen to exist between the cones of *Pinus* and of *Larix*. The overlapping edges will be free to grow into natural, rounded curves; each plate, uncrowded and unconstrained, will stay smooth and unwrinkled; the number and order of the plates will be the same as before—but the shell will be no longer that of a tortoise, but of the turtle from which "tortoise-shell" is obtained.

A snow-crystal is a very beautiful example of hexagonal symmetry. It belongs to another order of things to those we have been speaking of: for in substance it is a solid, and in form it is a crystal, and its own intrinsic molecular forces build it up in its own way. But (as we have mentioned once before) it is an exquisite illustration of Nature's way of producing infinite variety from the permutations and combinations of a single type. The snowflake is a crystal formed by sublimation, that is to say by precipitation from a vapour without passing through a liquid phase. It begins as a tiny hexagon, the making of which tends to use up the vapour near by; the angles of the hexagon jut out, so to speak, into regions of greater, or less depleted, vapour-pressure, and at these corners further crystallisation will next set in. The hexagon will tend to grow out into a six-rayed star; and later and more slowly the material for further crystallisation will make its way between the rays, and begin to build side-growths on them*.

The basaltic column

Hexagonal patterns are by no means confined to the organic world. The basalt of Staffa and the Giant's Causeway shews a wonderful array of prismatic columns of irregular size and form,

Fig. 200. Basalt at Giant's Causeway. By Mr R. Welch, Belfast.

but mostly hexagonal; so also does the frozen soil of Spitzbergen; starch sets on cooling into analogous prisms, but in a ruder fashion as on a smaller scale; and all these are due to simple forces in a simple field, namely to tension, or shrinkage, in a horizontal mass or layer. Imagine a sheet or "sill" of intrusive basalt, thrust in as a molten mass between older rocks. It is gradually chilled by the cold air above or by the rocks on either side, and its inner mass, cooling slower than the outer layer, contracts slowly. Nothing hinders its vertical contraction, rather is this helped by its own weight and by the load above; but no further lateral contraction can take place without splitting the mass, once the basalt sets hard. Con-

* Cf. Gerald Seligmann, *Nature*, 26 June, 1937, p. 1090.

traction, however, does take place, irresistibly, and it may be that long cracks appear; the strain being so far relieved, the next cracks will tend to take place at right angles to the first. But more commonly rupture is delayed until considerable strain-energy has been stored up; once started, it proceeds explosively from a number of centres, and shatters the whole mass into prismatic fragments. However quickly and explosively the cracks succeed one another each relieves an existing tension, and the next crack will give relief in a different direction to the first. When one crack meets another it will seldom cross it, for the strain which led to the former fracture does not extend into the new field. In short the cracks will be found to meet one another three by three, and therefore at angles *on the average* of 120°, and the columns will be *on the average* hexagonal. For the making of a prismatic structure all that is required is more or less uniform tensile strain in the two dimensions of a horizontal plane; uniform tension in three dimensions would have given rise to a cellular structure, of which the hexagonal "causeway" is the two-dimensional analogue.

The columnar structure is accompanied by sundry secondary phenomena. The vertical columns tend to break across on further contraction, and exhibit rounded or basin-shaped ends, fitting together in a shallow ball-and-socket; this beautiful configuration has only lately been explained*. When cooling has caused the mass to split into vertical columns, air or it may be water enters the rifts and further cools or quenches the now solid but still glowing basalt. Each column tends to be chilled all round while still hot within; but the hot unshrunken mass within checks or hinders the contraction of the cooler outer layers. Thus unequal cooling causes vertical as well as horizontal tensions; and just as these last are relieved by the existing cracks, so new rifts appear crosswise to the column, and relieve the vertical tensions.

If the cooling come downward from above, then, at any given level, the column will always be cooler above it than below; the

* F. W. Preston, On ball-and-socket jointing in basalt prisms, *Proc. R.S.* (B), CVI, pp. 87–92, 1930; A study of the rupture of glass, *Trans. Soc. of Glass Technology*, 1926, p. 263. On the general subject of prismatic structure in igneous rocks, see also Robert Mallet, *Phil. Mag.* (4), L, pp. 122–135, 201–226, 1875; James Thomson, *Trans. Geol. Soc. Glasgow*, March, 1877; *Coll. Works*, p. 422; R. B. Sosman, *Journ. Geology*, XXIV, p. 215, 1916.

part above will shrink the more; and this will set up a horizontal shear in the interior of the column, in addition to the existing vertical tension. The principal stress, compounded of shear and tension, will be neither vertical nor horizontal, but inclined obliquely between the two. Now in a brittle substance, such as glass or basalt, an advancing fracture tends to advance at right angles to the principal tension. Where the surface of the column meets the cool air the tension is parallel to the face, and the fissure enters at right angles to the face, that is to say, horizontally; it is for this reason that the ball-and-socket joint is found to have *a square lip*. Once inside the boundary, however, the advancing rift finds itself in a region where the principal stress is inclined,

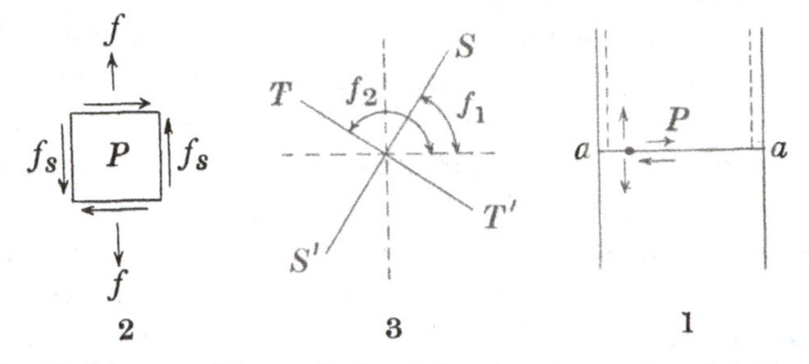

Fig. 201. (1) Diagram of the vertical and shearing stresses in a shrinking column of basalt. (2) The same in the neighbourhood of the point P. (3) SS' resultant stress, and TT' direction of rupture. After F. W. Preston.

slightly, to the vertical; the crack consequently bends down, the downward tilt increases for a short distance and then approaches the horizontal again, and the opposing surfaces of "ball and socket" are thus defined. The crack often fails to complete its journey, and leaves a core of rock unbroken in the middle of the bowl.

The curved sides of the basin, its square lip, its flattened centre, often incomplete, are thus all explained; and whether it be convex or concave, dome or basin, merely depends on whether the cooling or quenching came from above or from below.

The basin, or bowl, will always be a shallow one. For if at a point P, within the column, the vertical tension be f, and the horizontal shear-stress be f_s, then the direction of the principal planes will be $2\phi = \tan^{-1}(-2f_s/f)$; so that, since f and f_s are both positive, ϕ_1 lies between 45° and 90°, while ϕ_2 lies between 135°

and 180°. Hence the rift, dipping down at the angle TT', will never dip at a greater angle than 45°, and generally much less. Now a spherical bowl whose lip is at 45° to the horizontal has a depth of $\frac{\sqrt{2}-1}{2}$ times its lip diameter, or approximately one-fifth. And one-fifth of the diameter of the column is, approximately, the depth of the deepest bowl.

The hexagonal pattern and the three-way corners on which it is based are characteristic (as we have already explained) of a condition of symmetry or uniformity under which a partition is as likely to arise in any one direction as in any other, and no series of partitions has precedence over the rest. We have seen,

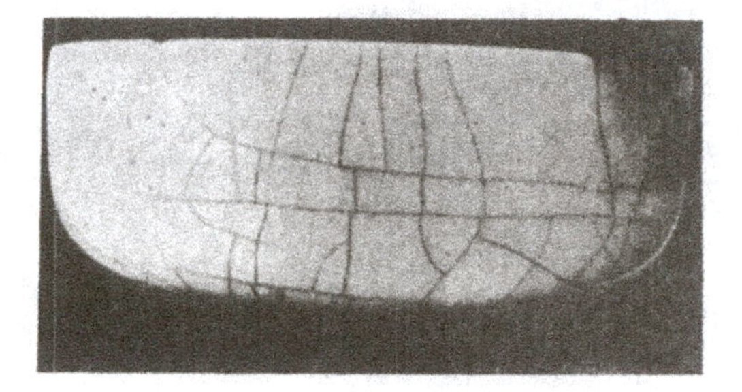

Fig. 202. Crackles on a porcelain bowl. From H. Hukusima.

in the dragon-fly's wing and in cambium-tissue, how different is the result when primary partitions are first established and consolidated, to be followed by a secondary and a weaker set. The "crackles" on a porcelain bowl look somewhat like a cellular epithelium; but the porcelain has been under strain in more ways than one. The plastic clay was first shaped upon a wheel, and potential stress-energy so acquired is stored up even in the finished ware; again, as the ware cools and shrinks after it is drawn from the kiln the glaze is apt to cool quicker and shrink more than the paste below, and tension-energy is stored up till the glaze ruptures and the cracks appear. Various rates of cooling and of contraction, the nature of paste and glaze, the shape of the ware, and even the way in which the potter worked the clay, may all influence the pattern of the crackle. The primary crack will be perpendicular to the main tension; secondaries will tend to be more or less orthogonal to the primary cracks; two secondaries on opposite sides of a primary will tend to be near together, though not opposite; and spiral cracks are often to be seen, the remote effect of plasticity under the potter's wheel. The net result is that certain primary cracks appear, related (more or less clearly) to the circular or spiral shaping of the clay upon the wheel; and each primary crack so relieves the tension in one direction that the secondaries tend to follow in a direction at right angles to the first. While co-equal angles of 120° were likely to occur in certain symmetrical cases, leading to a simultaneous pattern of hexagons, in other cases suc-

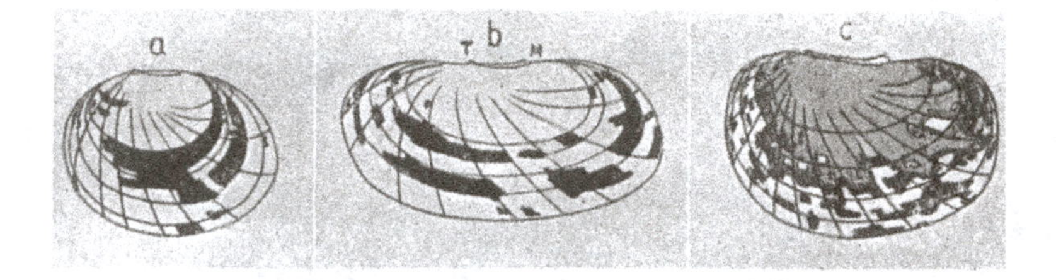

Fig. 203. Colour-patterns of kidney-beans with diagrammatic contour-lines added; *a*, *b*, Japanese "quail-beans"; *c*, scarlet-runner. After M. Hirata.

cessive partitions or cracks tend to be at right angles to one another; and Sachs's Law becomes truer of the porcelain than of the plant*.

In this latter case, and doubtless in many more, we are dealing not with a random pattern, but with one *based* on systematic and predetermined lines. The apparently confused or random pattern of a kidney-bean comes under the same class of configurations, inasmuch as it also is based on an underlying polarity, whose centre of symmetry is in the stalk or "hilus." For simplicity's sake, imagine the bean round like a pea, and its surface mapped out, orthogonally, by two sets of boundary-lines, radial and concentric. Then suppose an asymmetry of growth to be introduced so that the round pea grows into the ellipsoid of a bean; and suppose that the whole system of boundary-lines is subject to the same conformal transformation—which elliptic functions might help

* See H. Hukusima, Cracks upon the glazed surface of ceramic wares, *Sci. Papers, Inst. of Chem. and Phys. Research*, Tokyo, XXVII, pp. 235–243. 1935.

us to define. The colour-pattern of the bean will then be found following the direction of the boundary-lines, and occupying areas or patches corresponding to parts of the orthogonal system. The lines are equipotential lines, or akin thereto. If we varnish an elastic bag, dry it and expand it, the varnish will tend to crack along the same orthogonal boundaries*.

The bee's cell

The most famous of all hexagonal conformations, and one of the most beautiful, is the bee's cell. As in the basalt or the coral, we have to deal with an assemblage of co-equal cylinders, of circular section, compressed into regular hexagonal prisms; but in this case we have two layers of such cylinders or prisms, one facing one way and one the other, and a new problem arises in connection with their inner ends. We may suppose the original cylinders to have spherical ends†, which is their normal and symmetrical way of terminating; then, for closest packing, it is obvious that the end of any one cylinder in the one layer will touch, and fit in between, the ends of three cylinders in the other. It is just as when we pile round-shot in a heap; we begin with three, a fourth fits into its nest between the three others, and the four form a "tetrad," or regular tetragonal arrangement.

Just as it was obvious, then, that by mutual pressure from the *sides* of six adjacent cells any one cell would be squeezed into a hexagonal prism, so is it also obvious that, by mutual pressure against the *ends* of three opposite neighbours, the end of each and every cell will be compressed into a trihedral pyramid. The three sides of this pyramid are set, *in plane projection*, at co-equal angles of 120° to one another; but the three apical angles (as in the analogous case already described of a system of soap-bubbles) are,

* M. Hirata, Coloured patches in kidney-beans, *Sci. Papers, Inst. Chem. Research, Tokyo*, XXVI, pp. 122–135, 1936.

† In the combs of certain tropical bees the hexagona structure is imperfect and the cells are not far removed from cylinders. They are set in tiers, not contiguous but separated by little pillars of wax, and the base of each cell is a portion of a sphere. They differ from the ordinary honeycomb in the same sort of way as the fasciculate from the massive corals, of which we spoke on p. 512. Cf. Leonard Martin, Sur les Mélipones de Brésil, *La Nature*, 1930, pp. 97–100.

by the geometry of the case*, co-equal angles of 109° and so many minutes and seconds.

If we experiment, not with cylinders but with spheres, if for instance we pile bread-pills together and then submit the whole to a uniform pressure, as we shall presently find that Buffon did: each ball (like the seeds in a pomegranate, as Kepler said) will be in contact with *twelve* others—six in its own plane, three below and three above, and under compression it will develop twelve plane surfaces. It will repeat, above and below, the conditions to which the bee's cell is subject at one end only; and, since the sphere is symmetrically situated towards its neighbours on all sides, it follows that the twelve plane sides to which its surface has been reduced will be all similar, equal and similarly situated. Moreover, since we have produced this result by squeezing our original spheres close together, it is evident that the bodies so formed completely fill space. The regular solid which fulfils all these conditions is the *rhombic dodecahedron*. The bee's cell is this figure incompletely formed; it represents, so to speak, one-half of that figure, with its apex and the six adjacent corners proper to the rhombic dodecahedron, but six sides continued, as a hexagonal prism, to an open or unfinished end†.

The bee's comb is vertical and the cells nearly horizontal, but sloping slightly downwards from mouth to floor; in each prismatic cell two sides stand vertically, and two corners lie above and below. Thus for every honeycomb or "section" of honey, there is one and only one "right way up"; and the work of the hive is so far controlled by gravity. Wasps build the other way, with the cells upright and the combs horizontal; in a hornet's nest, or in that of *Polistes*, the cells stand upright like the wasp's, but their mouths look downwards in the hornet's nest and upwards in the wasp's.

What Jeremy Taylor called "the discipline of bees and the rare fabric of honeycombs" must have attracted the attention and excited the admiration of mathematicians from time immemorial. "Ma maison est construite," says the bee in the *Arabian Nights*, "selon

* The dihedral angle of 120° is, physically speaking, the essential thing; the Maraldi angle, of 109°, etc., is a geometrical consequence. Cf. G. Césaro, Sur la forme de l'alvéole de l'abeille, *Bull. Acad. R. Belgique (Sci.)*, 1920, p. 100.

† See especially Haüy, the crystallographer; Sur le rapport des figures qui existe entre l'alvéole des abeilles et le grenat dodécaèdre, *Journ. d'hist. naturelle*, II, p. 47, 1792.

les lois d'une sévère architecture; et Euclidos lui-même s'instruirait en admirant la géométrie de ses alvéoles*." Ausonius speaks of the *geometrica forma favorum*, and Pliny tells of men who gave a lifetime to its study.

Pappus the Alexandrine has left us an account of its hexagonal plan, and drew from it the conclusion that the bees were endowed with "a certain geometrical forethought"†. "There being, then, three figures which of themselves can fill up the space round a point, viz. the triangle, the square and the hexagon, the bees have wisely selected for their structure that which contains most angles, suspecting indeed that it could hold more honey than either of the

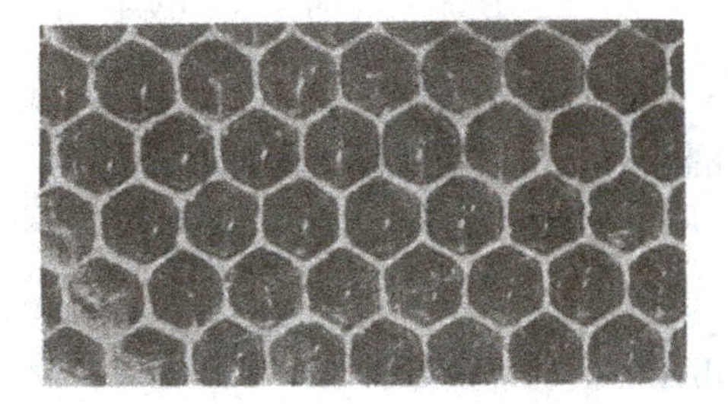

Fig. 204. Portion of a honeycomb. After Willem.

other two‡." Erasmus Bartholin was apparently the first to suggest that the hypothesis of "economy" was not warranted, and that the hexagonal cell was no more than the necessary result of equal pressures, each bee striving to make its own little circle as large as possible.

The investigation of the ends of the cell was a more difficult matter than that of its sides, and came later. In general terms the arrangement was doubtless often studied and described: as for

* Ed. Mardrus, XV, p. 173.

† φυσικὴν γεωμετρικὴν πρόνοιαν. Pappus, Bk. V; cf. Heath, *Hist. of Gk. Math.* II, p. 589. St Basil discusses τὴν γεωμετρίαν τῆς σοφωτάτης μελίσσης: *Hexaem.* VIII, p. 172 (Migne); Virgil speaks of the *pars divinae mentis* of the bee, and Kepler found the bees *animâ praeditas et geomet. riae suo modo capaces.*

‡ This was according to the "theorem of Zenodorus." The use by Pappus of "economy" as a guiding principle is remarkable. For it means that, like Hero with his mirrors, he had a pretty clear adumbration of that principle of *minima*, which culminated in the *principle of least action*, which guided eighteenth-century physics, was generalised (after Fermat) by Lagrange, inspired Hamilton and Maxwell, and reappears in the latest developments of wave-mechanics.

instance, in the *Garden of Cyrus*: "And the Combes themselves so regularly contrived that their mutual intersections make three Lozenges at the bottom of every Cell; which severally regarded make three Rows of neat Rhomboidall Figures, connected at the angles, and so continue three several chains throughout the whole comb." Or as Réaumur put it, a little later on: "trois cellules accolées laissent un vuide pyramidal, précisément semblable à celui de la base d'une autre cellule tournée en sens contraire."

Kepler had deduced from the space-filling symmetry of the honeycomb that its angles must be those of the rhombic dodecahedron; and Swammerdam also recognised the same geometrical figure in the base of the cell*. But Kepler's discovery passed unnoticed, and Maraldi the astronomer, Cassini's nephew, has the credit of ascertaining for the first time the shape of the rhombs and of the solid angle which they bound, while watching the bees in "les ruches vitrées dans le jardin de M. Cassini attenant l'Observatoire de Paris†." The angles of the rhomb, he tells us, are 110° and 70°: "Chaque base d'alvéole est formée de trois rombes presque toujours égaux et semblables, qui, *suivant les mesures que nous avons prises*, ont les deux angles obtus chacun de 110 degrès, et par conséquent les deux aigus chacun de 70 degrès." Further on (p. 312), he observes that on the magnitude of the angles of the three rhombs at the base of the cell depends that of the basal angles of the six trapezia which form its sides; and it occurs to him to ask what must these angles be, if those of the floor and those of the sides be equal one to another. The solution of this problem is that "les angles aigus des rombes étant de 70 degrès 32 minutes, et les obtus de 109 degrès 28 minutes, ceux des trapèzes qui leur sont contigus doivent être aussi de la même grandeur." And lastly: "Il résulte de cette grandeur d'angle non seulement une plus grande facilité et simplicité dans la construction, à cause que par cette manière les abeilles n'employent que deux sortes d'angles, mais il en résulte encore une plus belle simétrie dans la disposition et dans la figure

* Kepleri *Opera omnia*, ed. Fritsch, v, pp. 115, 122, 178, vii, p. 719, 1864; Swammerdam, *Tractatus de apibus* (observations made in 1673).

† Obs. sur les abeilles, *Mém. Acad. R. Sciences* (1712), 1731, pp. 297–331. Sir C. Wren had used "transparent bee-hives" long before; see his Letter concerning that pleasant and profitable invention, etc., in S. Hartlib's *Reformed Common-Wealth of Bees*, 1655.

de l'Alvéole." In short, Maraldi takes the two principles of simplicity and mathematical beauty as his sure and sufficient guides.

The next step was that which had been foreshadowed long before by Pappus. Though Euler had not yet published his famous dissertation on curves *maximi minimive proprietate gaudentes*, the idea of *maxima* and *minima* was in the air as a guiding postulate, an heuristic method, to be used as Maraldi had used his principle of simplicity. So it occurred to Réaumur, as apparently it had not done to Maraldi, that a minimal configuration, and consequent economy of material in the waxen walls of the cell, might be at the root of the matter: and that, just as the close-packed hexagons gave the minimal extent of boundary in a plane, so the figure determined by Maraldi, namely the rhombic dodecahedron, might be that which employs the minimum of surface for a given content: or which, in other words, should hold the most honey for the least wax. "Convaincu que les abeilles employent le fond pyramidal qui mérite d'être préféré, j'ai soupçonné que la raison, ou une des raisons, qui les avoit décidées était l'épargne de la cire; qu'entre les cellules de même capacité et à fond pyramidal, celle qui pouvait être faite avec moins de matière ou de cire étoit celle dont chaque rhombe avoit deux angles chacun d'environ 110 degrès, et deux chacun d'environ 70°." He set the problem to Samuel Koenig, a young Swiss mathematician: Given an hexagonal cell terminated by three similar and equal rhombs, what is the configuration which requires the least quantity of material for its construction? Koenig confirmed Réaumur's conjecture, and gave 109° 26′ and 70° 34′ as the angles which should fulfil the condition; and Réaumur then sent him the Mémoires de l'Académie for 1712, where Koenig was "agreeably surprised" to find: "que les rombes que sa solution avait déterminé, avait à deux minutes près* les angles que M. Maraldi avait trouvés *par des mesures actuelles* à chaque rhombe des cellules d'abeilles....Un tel accord entre la solution et les mesures actuelles a assurément de quoi surprendre." Koenig asserted that the bees had solved a problem beyond the reach of

* The discrepancy was due to a mistake of Koenig's, doubtless misled by his tables, in the determination of $\sqrt{2}$; but Koenig's own paper, sent to Réaumur, remained unpublished and his method of working is unknown. An abridged notice appears in the *Mém. de l'Acad.* 1739, pp. 30–35.

the old geometry and requiring the methods of Newton and Leibniz. Whereupon Fontenelle, as Secrétaire Perpétuel, summed up the case in a famous judgment, in which he denied intelligence to the bees but nevertheless found them blindly using the highest mathematics by divine guidance and command*.

When Colin Maclaurin studied the honeycomb in Edinburgh, a few years after Maraldi in Paris, he proceeded to solve the problem without using "any higher Geometry than was known to the Antients," and he began his account by saying: "These bases are formed from Three equal Rhombus's, the obtuse angles of which are found to be the doubles of an Angle that often offers itself to mathematicians in Questions relating to Maxima and Minima.†" It was an angle of 109° 28′ 16″, with its supplement of 70° 31′ 44″. And this angle of the bee's cell determined by Maraldi, Koenig and Maclaurin in their several ways, this angle which has for its cosine 1/3 and is double of the angle which has for its tangent $\sqrt{2}$, is on the one hand an angle of the rhombic dodecahedron, and on the other is that very angle of simple tetrahedral symmetry which the soap-films within the tetrahedral cage spontaneously assume, and whose frequent appearance and wide importance we have already touched upon‡.

That "the true theoretical angles were 109° 28′ and 70° 32′, *precisely corresponding with the actual measurement of the bee's cell*," and that the bees had been "proved to be right and the mathematicians wrong," was long believed by many. Lord Brougham

* La grande merveille est que la détermination de ces angles passe de beaucoup les forces de la Géométrie commune, et n'appartient qu'aux nouvelles Méthodes fondées sur la Théorie de l'Infini. Mais à la fin les Abeilles en sçauraient trop, et l'excès de leur gloire en est la ruine. Il faut remonter jusqu'à une Intelligence infinie, qui les fait agir aveuglément sous ses ordres, sans leur accorder de ces lumières capables de s'accroître et de se fortifier par elles-mêmes, qui font l'honneur de notre Raison." *Histoire de l'Académie Royale*, 1739, p. 35.

† Colin Maclaurin, On the bases of the cells wherein the bees deposit their honey, *Phil. Trans.* XLII, pp. 561–571, 1743; also in the Abridgement, VIII, pp. 709–713, 1809; it was characteristic of Maclaurin to use geometrical methods for wellnigh everything, even in his book on Fluxions, or in his famous essay on the equilibrium of spinning planets. Cf. also Lhuiller, Mémoire sur le minimum du cire des alvéoles des Abeilles, et en particulier sur un *minimum minimorum* relatif à cette matière, *Nouv. Mém. de l'Acad. de Berlin*, 1781 (1783), pp. 277–300. Cf. Castillon, *ibid.* (commenting on Lhuiller); also Ettore Carruccio, Notizie storiche sulla geometria delle api, *Periodico di Mathematiche*, (4) XVI, pp. 35–54, 1936.

‡ *Supra*, p. 497. The faces of a regular octahedron meet at the same angle.

helped notably to spread these and other errors, and his writings on the bee's cell contain, according to Glaisher, "as striking examples of bad reasoning as are often to be met with in writings relating to mathematical subjects." The fact is that, were the angles and facets of the honeycomb as sharp and smooth, and as constant and uniform, as those of a quartz-crystal, it would still be a delicate matter to measure the angles within a minute or two of arc, and a technique unknown in Maraldi's day would be required to do it. The minute-hand of a clock (if it move continuously) moves through one degree of arc in ten seconds of time, and through an angle of two minutes in one-third of a second;—and this last is the angle which Maraldi is supposed to have measured. It was eighty years after Maraldi had told Réaumur what the angle was that Boscovich pointed out for the first time that to ascertain the angle to the nearest minute by direct admeasurement of the waxen cell was utterly impossible. Yet Réaumur had certainly believed, and apparently had persuaded Koenig, that Maraldi's determinations, first and last, were the result of measurement; and Fontenelle, the historian of the Academy, epitomising Koenig's paper, speaks of "les mesures actuelles de M. Maraldi," of the bees being in error to the trifling extent of 2′, and of the *grande merveille* of their so nearly solving a problem belonging to the higher geometry. Boscovich, in a long-forgotten note, rediscovered by Glaisher, puts the case in a nutshell: "Mirum sane si Maraldus ex observatione angulum aestimasset intra minuta, quod in tam exigua mole fieri utique non poterat. At is (ut satis patet ex ipsa ejus determinatione) affirmat se invenisse angulos circiter 110° et 70°, nec minuta eruit ex observatione sed ex equalitate angulorum pertinentium ad rhombos et ad trapezia; ad quam habendam Geometria ipsa docuit requiri illa minuta*." Indeed he goes on to say the wonder is that the angles could be measured even within a few *degrees*, variable and irregular as they are seen to be, and as even Réaumur† knew

* In his note *De apium cellulis*, appended to the philosophical poem of Benedict Stay, II, pp. 498–504, Romae, 1792.

† *Op. cit.* V, p. 382. Several authors recognised that the cells are far from identical, and do no more than approximate to an average or ideal angle: e.g. Swammerdam in the *Biblia Naturae*, II, p. 379; G. S. Klügel, Grösstes u. Kleinstes, in *Mathem. Wörterb.* 1803; Castillon, *op. cit.*; and especially Jeffries Wyman, Notes on the cells of the bee, *Proc. Amer. Acad. Sci. and Arts*, VII, 1868.

well they were. The old misunderstanding was at last explained and corrected by Leslie Ellis; and better still by Glaisher, in a little-known but very beautiful paper*. For these two mathematicians shewed that, though Maraldi's account of his "measurements" led to misunderstanding, yet he had really done well and scientifically when he eked out a rough observation by finer theory, and deemed himself entitled thereby to discuss the cell and its angles in the same precise terms that he would use as a mathematician in speaking of its geometrical prototype†.

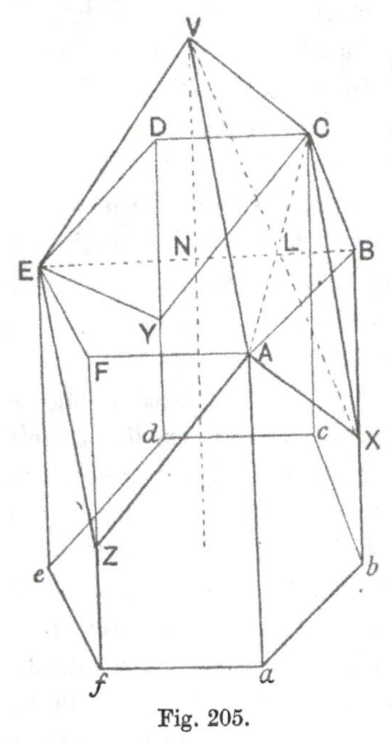

Fig. 205.

Many diverse proofs‡ have been given of the minimal character of the bee's cell, some few, like Maclaurin's, purely geometrical, others arrived at by help of the calculus. The following seems as simple as any:

ABCDEF, *abcdef*, is a right prism upon a regular hexagonal base. The corners *B*, *D*, *F* are cut off by planes through the lines *AC*, *CE*, *EA*, meeting in a point *V* on the axis *VN* of the prism, and intersecting *Bb*, *Dd*, *Ff*, in *X*, *Y*, *Z*. The volume of the figure thus formed is the same as that of the original prism with its hexagonal ends: for, if the axis cut the hexagon *ABCDEF* in *N*, the volumes *ACVN*, *ACBX* are equal.

It is required to find the inclination to the axis of the faces forming the

* Leslie Ellis, On the form of bees' cells, in *Mathematical and other Writings*, 1863, p. 353; J. W. L. Glaisher, do., *Phil. Mag.* (4), XLVI, pp. 103–122, 1873.

† The learned and original Kieser, in his *Mémoire sur l'organisation des plantes*, 1812, p. iv, gives advice to the same effect: "Il est indispensable de se former, avant de dessiner, une idée de l'objet dans sa plus grande perfection, et de dessiner selon cette idée, et non pas l'objet plus ou moins imparfait, plus ou moins altéré par le scalpel. Voilà la méthode qu'ont suivi Haller, Albinus et tous les autres grands anatomistes....Mais il faut employer pour cela la plus grande précaution, la circonspection la plus tranquille pour l'observation, etc."

‡ Cf. Koenig, Lhuiller and Boscovich, *opp. cit.*; H. Hennessy, *Proc. R.S.* XXXIX, p. 253, 1885; XLI, pp. 442, 443, 1886; XLII, pp. 176, 177, 1887.

trihedral angle at V, such that the surface of the whole figure may be a minimum.

Let the angle NVX, which is the inclination of the plane of the rhombus to the axis of the prism, $= \theta$; the side of the hexagon, as AB, $= s$; and the height, as Aa, $= h$.

Then $AC = 2s \cos 30°$, $= s\sqrt{3}$. And, from inspection of the triangle LXB,

$$VX = \frac{s}{\sin\theta}.$$

Therefore the area of the rhombus

$$VAXC = \frac{s^2\sqrt{3}}{2\sin\theta}.$$

And the area of $\quad AabX = \frac{s}{2}(2h - \tfrac{1}{2}VX\cos\theta),$

$$= \frac{s}{2}(2h - \tfrac{1}{2}s\cot\theta).$$

Therefore the total area of the figure

$$= \text{the hexagonal base } abcdef + 3s(2h - \tfrac{1}{2}s\cot\theta) + 3\frac{s^2\sqrt{3}}{2\sin\theta}.$$

Therefore $\quad \dfrac{d\,(\text{area})}{d\theta} = \dfrac{3s^2}{2}\left(\dfrac{1}{\sin^2\theta} - \dfrac{\sqrt{3}\cos\theta}{\sin^2\theta}\right).$

But this expression vanishes, or $\dfrac{d\,(\text{area})}{d\theta} = 0$, when $\cos\theta = \dfrac{1}{\sqrt{3}}$,

that is to say, when $\quad \theta = 54° \, 44' \, 8'' = \tfrac{1}{2}(109° \, 28' \, 16'').$

Such then are the conditions under which the total area of the figure has its minimal value.

The following is, in substance, Maclaurin's elementary but somewhat lengthy proof of the minimal properties of the bee's cell, using "no higher Geometry than was known to the Antients."

Let $ABCD$, $abcd$, represent one-half of a right prism on a regular hexagonal base; and let $AabE$, $EbcC$ be the trapezial portions of two adjacent sides, to which one of the three rhombs, $AECe$, is fitted.

Let O be the centre of the hexagon, of which AB, BC are adjacent sides; join AC and OB, intersecting in P. Then, because $AOC = ABC$, and $BE = Oe$, the solid $AECB = AeCO$; whence it appears that the solid *content* of the whole cell will be the same, wherever the point E be taken in Bb, and will in fact be identical with the content of the hexagonal prism. We have then to enquire where E is to be taken in Bb, in order that the combined *surfaces* of the rhomb and of the two trapezia may be a minimum.

Because Ee is perpendicular to AC in P, the area of the rhombus $= PE.AC$; and the area of the two trapezia $= (Aa + Eb) \times BC$. The total area in question, then, is $PE.AC + 2BC.Bb - BC.BE$. But $BC.Bb$ is constant; so the question remains, When is $PE.AC - BC.BE$ a minimum?

Let a point L be so taken in Bb that $BL : PL :: BC : AC$. From the centre P, in the plane PBE, describe the circular arc ER, meeting PL in R; and on PL let fall the perpendiculars ES, BT.

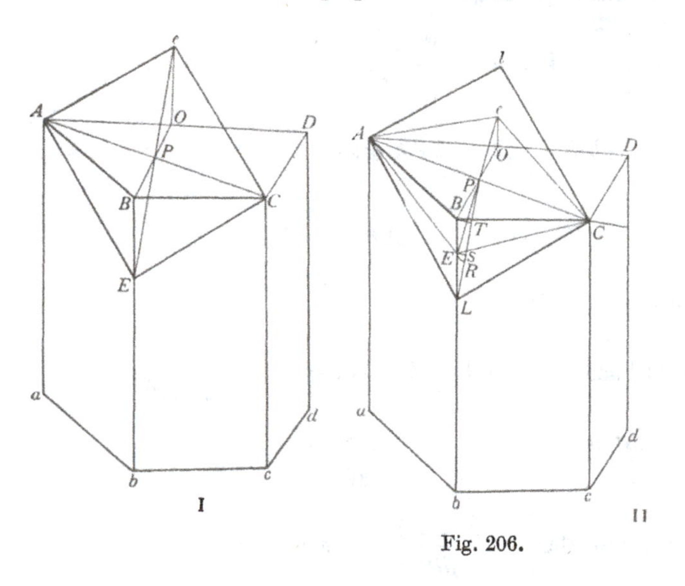

Fig. 206.

The triangles LES, LBT, LPB are all similar. Therefore

$$LS : LE :: LT : LB :: LB : LP,$$

and (by hypothesis) $$:: BC : AC.$$

Hence $$(LT - LS) : (LB - LE) :: BC : AC,$$

i.e. $$ST : BE :: BC : AC.$$

Therefore $$ST.AC = BE.BC$$

and consequently, $$PE.AC - BE.BC = PE.AC - ST.AC$$
$$= AC\,(PE - ST).$$

But $PE = PR$; therefore $AC\,(PR - ST) = AC\,(PT + RS)$.

But AC and PT do not vary, while RS varies with the position of E. Accordingly, $AC\,(PT+RS)$, or $PE.AC-BE.BC$, is a minimum when RS vanishes: that is to say, when E coincides with L.

"Therefore $ALCl$ is the *Rhombus* of the most advantageous Form in respect of Frugality, when BL is to PL as BC to AC."

Again, since $OB=BC$, and $OP=PB$, $BC^2=4PB^2$, and $PC^2=3PB^2$, and $AC=2PC=2\sqrt{3}.PB$.

Therefore $$BC : AC :: 2PB : 2\sqrt{3}.PB :: 1 : \sqrt{3},$$

and, by hypothesis, $$BL : PL :: BC : AC$$
$$=1 : \sqrt{3}$$

or $$PL : PB :: \sqrt{3} : \sqrt{2},$$

and $$PB : PC :: 1 : \sqrt{3}.$$

Therefore $$PL : PC :: 1 : \sqrt{2}.$$

"That is, the angle CLP is that whose Tangent is to the *Radius* as $\sqrt{2}$ is to 1, or as 1·4142135 to 1·0000000; and therefore is of 54° 44′ 08″, and consequently the Angle of the *Rhombus* of the Best Form is that of 109° 28′ 16″."

When we have thus ascertained that the characteristic angles of the rhombs are 109° 28′ 16″ and its supplement 70° 31′ 44″, the cosine of which latter angle is 1/3, the construction of a model is of the easiest.

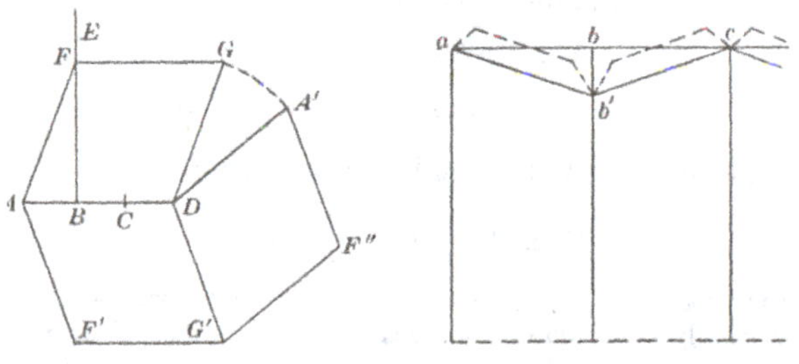

Fig. 207. Construction of a model of the bee's cell

On AD make $AB = BC = CD$. Let $AF = AD$ meet the perpendicular BE in F. Then the angle BAF (whose cosine = 1/3, or whose tangent = $2\sqrt{2}$) = 70° 31′ 44″. Complete the rhomb $ADGF$, and repeat three times as indicated. Make a developed hexagonal prism with sides ab, bc, = BF. Cut away angles $bb'a$, $bb'c$, etc., = BAF. Fold, and attach together.

A soap-bubble, or soap-film, assumes a minimal configuration instantaneously*, however small the saving of surface-area may be. But after learning that the bee's cell has undoubted minimal properties, we should like to know what saving is actually obtained by substituting a rhomboidal pyramid for a plane base in the hexagonal prism. It turns out, after all, to be a small matter! The calculation was first made by Maclaurin and by Lhuiller, in both cases briefly but correctly. Lhuiller stated that the whole amount used in the bee's cell was to that required for a flat-topped prismatic cell of equal volume as $25 + \sqrt{6}$ (or 27·45) to 28; the saving being thus a little more than 2 per cent. of the whole quantity of wax required†. Glaisher recalculated the values, taking the cell part by part. Assuming, with Lhuiller, that the radius of the inscribed circle of the hexagon is to the depth of the prismatic cell, when the latter has the same capacity as the real cell, as $1\frac{1}{5}$ to 5, then, taking the side of the hexagon as unity, we have for the same depth (viz. the longest side of the trapezium in the real cell) the value $\frac{25\sqrt{3}}{12}$; and then (to three places of decimals):

Area of the three rhombs, $\frac{9}{4}\sqrt{2}$	=	3·182	(i).
,, ,, ,, six triangles, $\frac{3}{4}\sqrt{2}$	=	1·061	(ii).
,, ,, ,, six sides of the equivalent prismatic cell, $\frac{25}{2}\sqrt{3}$	=	21·651	(iii).
,, ,, ,, hexagonal base, $\frac{3}{2}\sqrt{3}$	=	2·598	(iv).

The whole surface of the real cell, accordingly,

$$= \text{(i)} + \text{(iii)} - \text{(ii)} = 23{\cdot}772;$$

* For the most part *instantaneously*; but sometimes, when there are two positions of nearly equal potential energy, the film "creeps" from the less to the more advantageous of the two.

† We must take into account the depth of the cell, or assume a value for it, if we are to estimate the percentage saving of wax on the whole construction. But (as Dr G. T. Bennett says) the whole saving is on the roof, and the height of the house does not matter; the question rather is, what is saved on the rhomboidal sloping roof compared with a flat one? If the short axis of the rhombs be 2 units (the edge of the cube), then 3 rhombs have area $6\sqrt{2}$, the wall-saving is $2\sqrt{2}$, while the flat hexagonal top is $4\sqrt{3}$. So the actual saving is the difference between $4\sqrt{2}$ and $4\sqrt{3}$—which looks much less negligible! But it is only on a small portion of the work.

and that of the flat-bottomed hexagonal prism

$$= \text{(iii)} + \text{(iv)} = 24{\cdot}249;$$

and $$\frac{24{\cdot}249}{23{\cdot}772} = \frac{102}{100}, \text{ or a little more;}$$

so that the saving in the former case amounts to about 2 per cent., as Lhuiller had found it to be.

Glaisher sums up the matter as follows: "As the result of a tolerably careful examination of the whole question, I may be permitted to say that I agree with Lhuiller in believing that the economy of wax has played a very subordinate part in the determination of the form of the cell; in fact I should not be surprised if it were acknowledged hereafter that the form of the cell had been determined by other considerations, into which the saving of wax did not enter (that is to say did not enter sensibly; of course I do not mean that the amount of wax required was a matter of absolute indifference to the bees). The fact of all the dihedral angles being 120° is, it is not unlikely, the cause that determined the form of the cell." This last fact, that in such a cell every plane cuts every other plane at an angle of 120°, was known both to Klügel and to Boscovich; it is no mere corollary, but the root of the matter. It is, as Glaisher indicates, the fundamental physical principle of construction from which the apical angles of 109° follow as a geometrical corollary. And it is curious indeed to see how the obtuse angle of the rhomb, and its cosine $-\frac{1}{3}$, drew attention all the while; but the dihedral angle of 120° of the rhombohedron, and the inclination of its three short diagonals at 90° to one another, got rare and scanty notice.

Darwin had listened too closely to Brougham and the rest when he spoke of the bee's architecture as "the most wonderful of known instincts"; and when he declared that "beyond this stage of perfection in architecture natural selection could not lead; for the comb of the hive-bee, as far as we can see, is *absolutely perfect* in economising labour and wax."

The minimal properties of the cell and all the geometrical reasoning in the case postulate cell-walls of uniform tenuity and edges which are mathematically straight. But the walls, and still more their edges, are always thickened; the edges are never accurately straight, nor the cells strictly horizontal. The base is always thicker than the side-walls; its solid angles are by no means sharp, but filled up with curving surfaces of wax, after the fashion, but more coarsely, of Plateau's bourrelet. Hence the Maraldi angle is seldom or never attained; the mean value (according to Vogt) is no more than 106·7° for the workers, and 107·3° for the drones. The hexagonal angles of the prism are fairly constant; about 4° is the limit of departure, and about 1·8° the mean error, on either side.

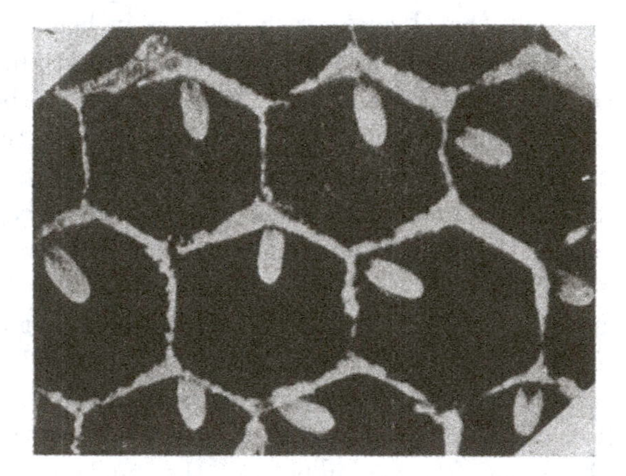

Fig. 208. Brood-comb, with eggs.

The bee makes no economies; and whatever economies lie in the theoretical construction, the bee's handiwork is not fine nor accurate enough to take advantage of them*.

The cells vary little in size, so little that Thévenot, a friend of Swammerdam's, suggested using their dimensions as a modulus or standard of length; but after all, the constancy is not so great as has been supposed. Swammerdam gives measurements which work out at 5·15 mm. for the mean diameter of the worker-cells, and 7 mm. for those of the drones; Jeffries Wyman found mean values for the worker-cells from 5·1 to 5·2 mm.; Vogt, after many careful measurements, found a mean of 5·37 mm. for the worker-

* All this Heinrich Vogt has abundantly shewn, in part by making casts of the interior of the cells, as Castellan had done a hundred years before. See his admirable paper on the Geometrie und Oekonomie der Bienenzelle, in *Festschrift d. Universität Breslau*, 1911, pp. 27–274.

cells, with an insignificant difference in the various diameters, and a mean of 6·9 for the drone-cells, with their horizontal diameter somewhat in excess, and averaging 7·1 mm. A curious attempt has been made of late years by Italian bee-keepers to let the bees work on a larger foundation, and so induce them to build larger cells; and some, but by no means all, assert that the young bees reared in the larger cells are themselves of larger stature*.

That the beautiful regularity of the bee's architecture is due to some automatic play of the physical forces, and that it were fantastic to assume (with Pappus and Réaumur) that the bee intentionally seeks for a method of economising wax, is certain; but the precise manner of this automatic action is not so clear. When the hive-bee builds a solitary cell, or a small cluster of cells, as it does for those eggs which are to develop into queens, it makes but a rude construction. The queen-cells are lumps of coarse wax hollowed out and roughly bitten into shape, bearing the marks of the bee's jaws like the marks of a blunt adze on a rough-hewn log.

Omitting the simplest of all cases, when (among some humble-bees) the old cocoons are used to hold honey, the cells built by the "solitary" wasps and bees are of various kinds. They may be formed by partitioning off little chambers in a hollow stem; they may be rounded or oval capsules, often very neatly constructed out of mud or vegetable fibre or little stones, agglutinated together with a salivary glue; but they shew, except for their rounded or tubular form, no mathematical symmetry. The social wasps and many bees build, usually out of vegetable matter chewed into a paste with saliva, very beautiful nests of "combs"; and the close-set papery cells which constitute these combs are just as regularly hexagonal as are the waxen cells of the hive-bee. But in these cases (or nearly all of them) the cells are in a single row; their sides are regularly hexagonal, but their ends, for want of opponent forces, remain simply spherical.

In *Melipona domestica* (of which Darwin epitomises Pierre Huber's description) "the large waxen honey-cells are nearly spherical, nearly equal in size, and are aggregated into an irregular mass."

* Cf. (*int. al.*) H. Gontarsi, Sammelleistungen von Bienen aus vergrösserten Brutzellen, *Arch. f. Bienenkunde*, XVI, p. 7, 1935; A. Ghetti, Celli ed api piu grandi, *IV Congresso nazion. della S.A.I.* 1935.

But the spherical form is only seen on the outside of the mass; for inwardly each cell is flattened into "two, three or more flat surfaces, according as the cell adjoins two, three or more other cells. When one cell rests on three other cells, which from the spheres

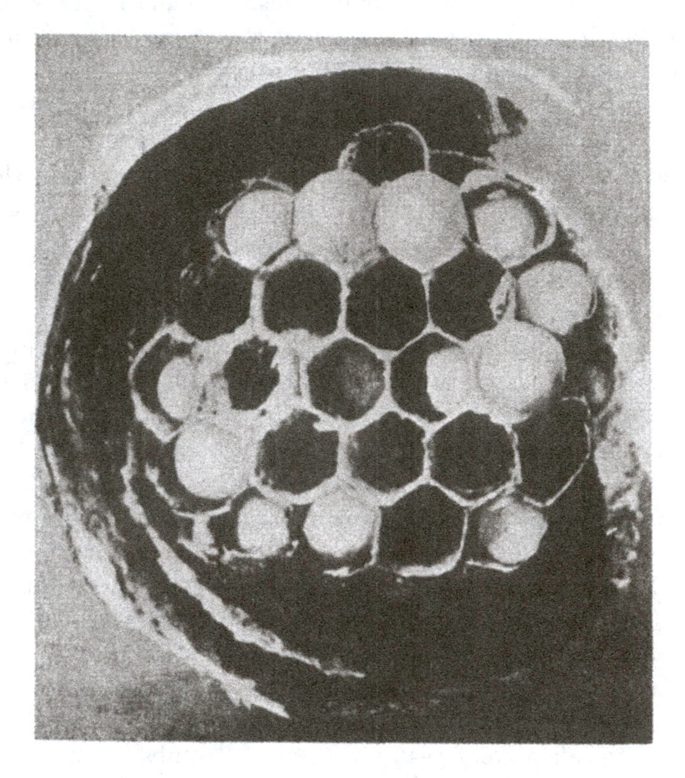

Fig. 209. An early stage of a wasp's nest. Observe the spherical caps, and the irregular shape of the peripheral cells. After R. Bott.

being nearly of the same size is very frequently and necessarily the case, the three flat surfaces are united into a pyramid; and this pyramid, as Huber has remarked, is manifestly a gross imitation of the three-sided pyramidal base of the cell of the hive-bee*."

* *Origin of Species*, ch. VIII (6th ed., p. 221). The cells of various bees, humble-bees and social wasps have been described and mathematically investigated by K. Müllenhoff, *Pflüger's Archiv*, XXXII, p. 589, 1883; but his many interesting results are too complex to epitomise. For figures of various nests and combs see (e.g.) von Büttel-Reepen, *Biol. Centralbl.* XXXIII, pp. 4, 89, 129, 183, 1903.

We had better be content to say that it depends on the same elementary geometry.

The question is, To what particular force are we to ascribe the plane surfaces and definite angles which define the sides of the cell in all these cases, and the ends of the cell in cases where one row meets and opposes another? We have seen that Bartholin suggested, and it is still commonly believed, that this result is due to mere physical pressure, each bee enlarging as much as it can the cell which it is a-building, and nudging its wall outwards till it fills every intervening gap, and presses hard against the similar efforts of its neighbour in the cell next door*.

That the bee, if left to itself, "works in segments of circles," or in other words builds a rounded and roughly spherical cell, is an old contention† which some recent experiments of M. Victor Willem amplify and confirm‡. M. Willem describes vividly how each cell begins as a little hemispherical basin or "cuvette," how the workers proceed at first with little apparent order and method, laying on the wax roughly like the mud when a swallow builds; how presently they concentrate their toil, each burying its head in its own cuvette, and slowly scraping, smoothing and ramming home; how those on the other side gradually adjust themselves

* Darwin had a somewhat similar idea, though he allowed more play to the bee's instinct or conscious intention. Thus, when he noticed certain half-completed cell-walls to be concave on one side and convex on the other, but to become perfectly flat when restored for a short time to the hive, he says: "It was absolutely impossible, from the extreme thinness of the little plate, that they could have effected this by gnawing away the convex side; and I suspect that the bees in such cases stand on opposite sides and push and bend the ductile and warm wax (which as I have tried is easily done) into its proper intermediate plane, and thus flatten it." Huber thought the difference in form between the inner and the outer cells a clear proof of intelligence; it is really a direct proof of the contrary. And while cells differ when their situations and circumstances differ, yet over great stretches of comb extreme uniformity, unbroken by any sign of individual differences, is the strikingly mechanical characteristic of the cells.

† It is so stated in the *Penny Cyclopedia*, 1835, Art. "Bees"; and is expounded by Mr G. H. Waterhouse (*Trans. Entom. Soc., London*, II, p. 115, 1864) in an article of which Darwin made good use. Waterhouse shewed that when the bees were given a plate of wax, the separate excavations they made therein remained hemispherical, or were built up into cylindrical tubes; but cells in juxtaposition with one another had their party-walls flattened, and their forms more or less prismatic.

‡ Victor Willem, L'architecture des abeilles, *Bull. Acad. Roy. de Belgique* (5), XIV, pp. 672–705, 1928.

to their opposite neighbours; and how the rounded ends of the cells fashion themselves into the rhomboidal pyramids, "à la suite de l'amincissement progressif des cloisons communes, et des pressions antagonistes exercées sur les deux faces de ces cloisons."

Among other curious and instructive observations, M. Willem has watched the bees at work on the waxen "foundations" now commonly used, on which a rhomboidal pattern is impressed with a view to starting the work and saving the labour of the bees. The bees (he says) disdain these half-laid foundations of their cells; they hollow out the wax, erase the rhombs, and turn the pyramidal hollows into hemispherical "cuvettes" in their usual way; and the vertical walls which they raise, more or less on the lines laid down for them, are not hexagonal but cylindrical to begin with. "La forme plane, en facettes, tant de prismes que des fonds, n'est obtenue que plus tard, progressivement, comme résultat de retouches, d'enlèvements et de pressions exercées sur les cloisons qui s'amincissent, par des groupes d'ouvrières opérant face à face, de manière antagoniste."

But when all is said and done, it is doubtful whether such *retouches*, *enlèvements* and *pressions antagonistes*, such mechanical forces intermittently exercised, could produce the nearly smooth surfaces, the all but constant angles and the close approach to a minimal configuration which characterise the cell, whether it be constructed by the bee of wax or by the wasp of papery pulp. We have the properties of the material to consider; and it seems much more likely to me that we have to do with a true tension effect: in other words, that the walls assume their configuration when in a semi-fluid state, while the watery pulp is still liquid or the wax warm under the high temperature of the crowded hive. In the first few cells of a wasp's comb, long before crowding and mutual pressure come into play, we recognise the identical configurations which we have seen exhibited by a group of three or four soap-bubbles, the first three or four cells of a segmenting egg. The direct efforts of the wasp or bee may be supposed to be limited, at this stage, to the making of little hemispherical cups, as thin as the nature of the material permits, and packing these little round cups as close as possible together. It is then conceivable, and indeed probable, that the symmetrical tensions of the semi-fluid films should suffice (however retarded by viscosity) to bring the whole system into equilibrium, that is to say into the configuration which the comb actually assumes.

The remarkable passage in which Buffon discusses the bee's cell and the hexagonal configuration in general is of such historical importance, and tallies so closely with the whole trend of our enquiry, that before we leave the subject I will quote it in full*: "Dirai-je encore un mot: ces cellules des abeilles, tant vantées, tant admirées, me fournissent une preuve de plus contre l'enthousiasme et l'admiration; cette figure, toute géométrique et toute régulière qu'elle nous paraît, et qu'elle est en effet dans la spéculation, n'est ici qu'un résultat mécanique et assez imparfait qui se trouve souvent dans la nature, et que l'on remarque même dans les productions les plus brutes; les cristaux et plusieurs autres pierres, quelques sels, etc., prennent constamment cette figure dans leur formation. Qu'on observe les petites écailles de la peau d'une roussette, on verra qu'elles sont hexagones, parce que chaque écaille croissant en même temps se fait obstacle et tend à occuper le plus d'espace qu'il est possible dans un espace donné: on voit ces mêmes hexagones dans le second estomac des animaux ruminans, on les trouve dans les graines, dans leurs capsules, dans certaines fleurs, etc. Qu'on remplisse un vaisseau de pois, ou plûtot de quelque autre graine cylindrique, et qu'on le ferme exactement après y avoir versé autant d'eau que les intervalles qui restent entre ces graines peuvent en recevoir; qu'on fasse bouillir cette eau, tous ces cylindres deviendront de colonnes à six pans. On y voit clairement la raison, qui est purement mécanique; chaque graine, dont la figure est cylindrique, tend par son renflement à occuper le plus d'espace possible dans un espace donné, elles deviennent donc toutes nécessairement hexagones par la compression réciproque. Chaque abeille cherche à occuper de même le plus d'espace possible dans un espace donné, il est donc nécessaire aussi, puisque le corps des abeilles est cylindrique, que leurs cellules sont hexagones—par la même raison

* Buffon, *Histoire naturelle*, IV, p. 99, Paris, 1753. Bonnet criticised Buffon's explanation, on the ground that his description was incomplete; for Buffon took no account of the Maraldi pyramids. Not a few others discovered impiety in his hypotheses, and some dismissed them with the remark that "philosophical absurdities are the most difficult to refute"; cf. W. Smellie, *Philosophy of Natural History*, Edinburgh, 1790, p. 424.

des obstacles réciproques. On donne plus d'esprit aux mouches dont les ouvrages sont les plus réguliers; les abeilles sont, dit-on, plus ingénieuses que les guêpes, que les frélons, etc., qui savent aussi l'architecture, mais dont les constructions sont plus grossières et plus irrégulières que celles des abeilles: on ne veut pas voir, ou l'on ne se doute pas, que cette régularité, plus ou moins grande, dépend uniquement du nombre et de la figure, et nullement de l'intelligence de ces petites bêtes; plus elles sont nombreuses, plus il y a des forces qui agissent également et s'opposent de même, plus il y a par conséquent de contrainte mécanique, de régularité forcée, et de perfection apparente dans leurs productions*."

Of parenchymatous cells

Just as Bonanni and other early writers sought, as we have seen, to explain hexagonal symmetry on mechanical principles, so other early naturalists, relying more or less on the analogy of the bee's cell, endeavoured to explain the cells of vegetable parenchyma; and to refer them to the rhombic dodecahedron or garnet-form, which solid figure, in close-packed association, was believed in their time, and long afterwards, to enclose space with a minimal extent of surface.

* Among countless papers on the bee's cell, see John Barclay and others in *Ann. of Philosophy*, IX, X, 1817; Henry Lord Brougham, in *Dissertations... connected with Natural Theology*, app. to Paley's Works, I, pp. 218–368, 1839; *C.R. Acad. Sci. Paris*, XLVI, pp. 1024–1029, 1858; *Tracts, Mathematical and Physical*, 1860, pp. 103–121, etc.; E. Carruccio, Note storiche sulla geometria delle api, *Periodico di Matem.* (4), XVI, 20 pp., 1936; G. Césaro, Sur la forme de l'alvéole des abeilles, *Bull. Acad. Roy. Belg.* (Sci.), Avril 10, 1929; Sam. Haughton, On the form of the cells made by various wasps and by the honey-bee, *Proc. Nat. Hist. Soc. Dublin*, III, pp. 128–140, 1863; *Ann. Mag. Nat. Hist.* (3), XI, pp. 415–429, 1863; A. R. Wallace, Remarks on the foregoing paper, *ibid.* XII, p. 33; J. O. Hennum, *Arch. f. Math. u. Vidensk.*, Christiania, IX, p. 301, 1884; F. Huber, *Nouv. obs. sur les abeilles*, II, p. 475, 1814; F. W. Hultmann, *Tidsskr. f. Math.*, Uppsala, I, p. 197, 1868; John Hunter, Observations on bees, *Phil. Trans.* 1792, pp. 128–195; Jacob, *Nouv. Ann. de Math.* II, p. 160, 1843; G. S. Klügel, Mathem. Betrachtungen ub. d. kunstreichen Bau d. Bienenzellen, *Hannoversches Mag.* 1772, pp. 353–368; Léon Lalanne, Note sur l'architecture des abeilles, *Ann. Sc. Nat. Zool.* (2), XIII, pp. 358–374, 1840; B. Powell, *Proc. Ashmol. Soc.* I, p. 10, 1844; K. H. Schellbach, *Mathem. Lehrstunde: Lehre v. Grossten u. Kleinsten*, 1860, pp. 35–37; Sam. Sharpe, *Phil. Mag.* IV, pp. 19–21, 1828; J. E. Siegwart, Die Mathematik im Dienste d. Bienenzucht, *Schw. Bienenzeitung*, III, 1880; O. Terquem, *Nouv. Ann. de Math.* XV, p. 176, 1856; C. M. Willick, On the angle of dock-gates and the bee's cell, *Phil. Mag.* (4) XVIII, p. 427, 1859; *C.R.* LI, p. 633, 1860; Chauncy Wright, *Proc. Amer. Acad. Arts and Sci.* IV, p. 432, 1860.

We have mentioned both Hooke and Grew*, and we have just heard Buffon engaged in such speculations; but the matter was more elaborately treated near the beginning of last century by Dieterich George Kieser†, an ingenious friend and colleague of the celebrated Lorenz Oken. Kieser clearly understood that the cell has not a shape of its own, but merely one impressed on it by physical forces and defined by mathematical laws. In his *Mémoire sur l'organisation des plantes*, he gives an admirable historical account of the work of Malpighi, Hooke, Grew, John Hill and other early microscopists; and then he says "La forme des cellules est variée dans les plantes différentes, mais il y a des formes principales, fondées sur les lois des mathématiques, que la nature suit toujours dans ses formations....La forme la plus commune est celle que prennent nécessairement des globules rondes ou allongées, pressées ensemble, celle des corps hexagonaux à parois quadrilatérales, ou d'une colonne très courte hexagone, coupée horizontalement d'en haut et d'en bas." Here we have, briefly described and sufficiently accounted for, the configuration of what we call a "pavement epithelium," or other simple association of cells in a single layer.

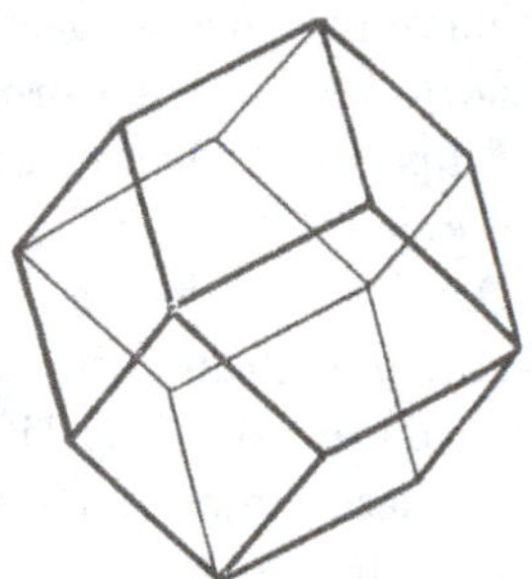

Fig. 210. A rhombic dodecahedron.

But another passage (from the same author's *Phytotomie*) is worth quoting at length, where he deals with cells in the mass, that is to say with the three-dimensional problem. "Die nach mathematische Gesetzen bestimmte als nothwendige Grundform der Zelle der vollkommenen Zellengewebe ist das langgezogene Rhombododekaheder. ...Mathematisch liegt das Beweis dass diese Figur die Grundform der vollkommenen Zellengewebe sei darin, dass unter allen mathematischen Körpern welche durch Zusammensetzung einen soliden Körper ohne Zwischenräume bilden, das Rhombododekaheder die einzige ist welche mit der wenigsten Masse des Umkreises den grössten Raum einschliesst. Sollte also aus dem Globus—dem

* R. Hooke, *Micrographia*, 1665, pp. 115–116; Nehemiah Grew, *Anatomy of Plants*, 1682, pp. 64, 76, 120.

† D. G. Kieser, *Mémoire*, etc., Haarlem, 1814, p. 89; *Phytotomie, oder Grundzüge der Anatomie der Pflanzen*, Jena, 1815, p. 4.

ursprünglichsten Schleimbläschen der Pflanzenzelle—ein eckiger Körper gebildet werden, so musste dieser das Rhombododekaheder sein, weil dieser im Hinsicht des Minimums der Masse zu dem Maximum des eingeschlossenen Raumes dem Globus am nächsten liegt. Als die Urform der Pflanzenzelle ist nicht Globus sondern Ellipsoide, daher muss das Dodekaheder, welche die Grundform der eckigen Pflanzenzelle ist, auch aus dem Ellipsoide entstanden sein. Das Rhombododekaheder wird also vom unten nach oben gestreckt, und die Grundform der eckigen Pflanzenzelle ist das in perpendiculärer Richtung längsgestreckte Rhombododekaheder."

These views and speculations of Kieser's, now all but forgotten, were by no means neglected in their day. Oken accepted them, and taught them*; Schleiden remarks that "the form of cells frequently passes into that of the rhombic dodecahedron, so beautifully determined, *à priori*, by Kieser†"; and De Candolle thought it necessary to warn his readers that cells are not as geometrically regular as published figures might lead one to believe‡.

The same principles apply to various orders of magnitude, and close-packing may be seen even in the inner contents of a cell. In vitally stained "goblet-cells," the mucin gathers into clumps or droplets, of which each appears in optical section to be surrounded by six more. When fixed they draw together, appear in optical section to be hexagonal, and we may take it that they have become, to a first approximation, rhombic dodecahedra§.

These then, and such as these, were the not unimportant speculations on the forms of cells by men who early grasped the fact that form had a physical cause and a mathematical significance. But their conception of the phenomenon was of necessity limited to the play of the mechanical forces; for Plateau's *Statique des Liquides* had not yet shewn what the capillary forces can do, nor opened a way thereby for Berthold and for Errera.

A very beautiful hexagonal symmetry as seen in section, or dodecahedral as viewed in the solid, is presented by the pith of certain rushes (e.g. *Juncus effusus*), and somewhat less diagram-

* Oken, *Physiophilosophy* (Ray Society), 1847, p. 209.

† *Müller's Archiv*, 1838, p. 146.

‡ *Organogénie végétale*, I, p. 13, 1827.

§ E. S. Duthie, in *Proc. R.S.* (B), CXIII, pp. 459–463, 1933.

matically by the pith of the banana. The cells are stellate, and the tissue has the appearance in section of a network of six-rayed stars (Fig. 211), linked together by the tips of the rays, and separated by symmetrical, air-filled intercellular spaces, which give its snow-like whiteness to the pith. In thick sections, the solid twelve-rayed "star-dodecahedra" may be very beautifully seen under the binocular microscope. They are not difficult to understand. Imagine, as before, a system of equal spheres in close contact, each one touching its twelve neighbours, six of them in the equatorial

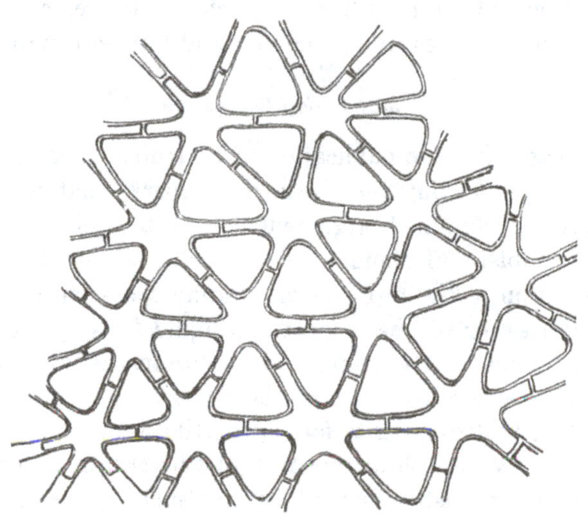

Fig. 211. Stellate cells in pith of *Juncus*.

plane; and let the cells be not only in contact, but become attached at the points of contact. Then, instead of each cell expanding so as to encroach on and fill up the intercellular spaces, let each tend to shrink or shrivel up by the withdrawal of fluid from its interior. The result will be to enlarge the intercellular spaces; the attachments of each cell to its neighbours will remain fixed, but the walls between these points of attachment will be withdrawn in a symmetrical fashion towards the centre. As the final result we have the star-dodecahedron, which appears in plane section as a six-rayed figure. It is necessary not only that the pith-cells should be attached to one another, but also that the outermost should be attached to a boundary wall, to preserve the symmetry of the system. What actually occurs in the rush is tantamount to this, but not absolutely identical. It is not so much the pith-cells which tend to shrink within a boundary of constant size, but rather the boundary wall which continues to expand after the pith-cells which it encloses have ceased to grow or to multiply. The points of attachment on the surface of each little pith-cell are drawn asunder, but the content of the cell does not correspondingly increase; and the remaining portions of the surface shrink inwards, accordingly, and gradually constitute the complicated figure which Kepler called a star-dodecahedron, which is still a symmetrical figure, and is still a surface of minimal area under the new and altered conditions.

The tetrakaidekahedron

A few years after the publication of Plateau's book, Lord Kelvin shewed, in a short but very beautiful paper*, that we must not hastily assume from such arguments as the foregoing that a close-packed assemblage of rhombic dodecahedra will be the true and general solution of the problem of dividing space with a minimum partitional area, or will be present in a liquid "foam," in which the general problem is completely and automatically solved. The general mathematical solution of the problem (as we have already indicated) is, that every interface or partition-wall must have constant mean curvature throughout; that where these partitions meet in an edge, they must intersect at angles such that equal forces, in planes perpendicular to the line of intersection, shall balance; that no more than three such interfaces may meet in a line or edge, whence it follows (for symmetry) that the angle of intersection of all surfaces or facets must be 120°; and that neither more nor less than four edges meet in a point or corner. An assemblage of rhombic dodecahedra goes far to meet the case. It fills space; its surfaces or interfaces are planes, and therefore surfaces of constant curvature throughout; and they meet together at angles of 120°. Nevertheless, the proof that the rhombic dodecahedron (which we find exemplified in the bee's cell) is a figure of minimal area is not a comprehensive proof; it is limited to certain conditions, and

* Sir W. Thomson, On the division of space with minimum partitional area, *Phil. Mag.* (5), XXIV, pp. 503–514, Dec. 1887; cf. *Baltimore Lectures*, 1904, p. 615; *Molecular tactics of a crystal* (Robert Boyle Lecture), 1894, pp. 21–25.

practically amounts to no more than this, that of the ordinary space-filling solids with all sides plane and similar, this one has the least surface for its solid content.

The rhombic dodecahedron has six tetrahedral angles and eight trihedral angles. At each of the latter three, and at each of the former six, dodecahedra meet in a point in close packing; and four edges meet in a point in the one case and eight in the other. This is enough to shew that the conditions for minimal area are not rigorously met. In one of Plateau's most beautiful experiments*, a wire cube is dipped in soap-solution. When lifted out, a film is seen to pass inwards from each of the twelve edges of the cube, and these twelve films meet, three by three, in eight edges, running inwards from the eight corners of the cube; but the twelve films and their eight edges do not meet in a point, but are grouped around a small central quadrilateral film (Fig. 212). Two of the eight edges run to each corner of the little square, and, with the two sides of the square itself, make up the four edges meeting in a point which the theory of area minima requires. We may substitute (by a second dip) a little cube for the little square; now an edge from each corner of the outer cube runs to the corresponding corner of the inner one, and with the three adjacent edges of the little cube itself the number four is still maintained. Twelve films, and eight edges meeting in a point, were essentially unstable; but the introduction of the little square or cube meets most of the conditions of stability which Plateau was the first to lay down. One more condition has to be met, namely the equality of angles at which the four edges meet in each conjunction. These co-equal "Maraldi angles" at each corner of the square can only be constructed by help of a slight curvature of the sides, and the little square is seen to have its sides curved into circular arcs accordingly; moreover its size and shape, as that of all the other films in the system, are perfectly definite. It is all one, according to the

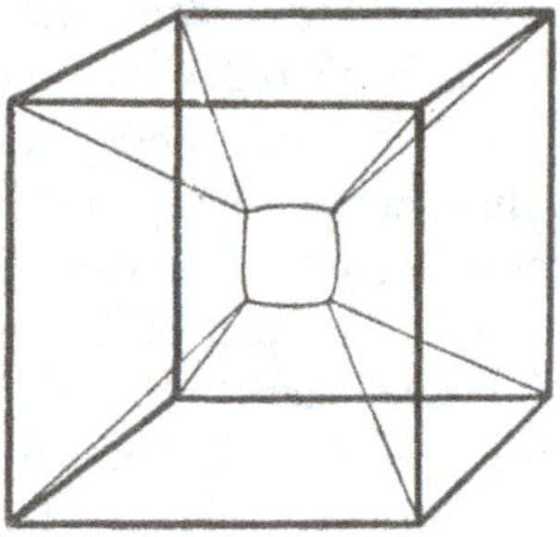

Fig. 212.

* Also discovered independently by Sir David Brewster, *Trans. R.S.E.* XXIV, p. 505, 1867; XXV, p. 115, 1869.

symmetry of the figure, to which side of the skeleton cube the square lies parallel; wherever it may be, if we blow gently on it, then (as M. Van Rees discovered) it alters its place and sets itself parallel now to one and now to another of the paired faces of the cube.

The skeleton cube, like the tetrahedron which we have already studied, is only one of many interesting cases; for we may vary the shape of our wire cages and obtain other and not less beautiful configurations. An hexagonal prism, if its sides be square or nearly so, gives us six vertical triangular films, whose apices meet the corners of a horizontal hexagon*; also six pairs of truncated triangles, which link the top and bottom edges of the cage to the sides of the median hexagon. But if the height of the hexagonal prism be increased, the six vertical films become curvilinear triangles, with sides concave towards the apex; and the twelve remaining films, which spring from the top and bottom of the hexagon, are curved surfaces, looking like a sort of hexagonal hourglass†.

There is a deal of elegant geometry in these various configurations. Lamarle shewed that if, in a figure represented by our wire cage, we suppress (in imagination) one face and all the other faces adjoining it, then the faces which remain are those which appear in the centre of the figure after the cage has been withdrawn from the soap-solution. Thus, in a cube, we suppress one face and the four adjacent to it; only one remains, and it reappears as the central square in the middle of the new configuration; in the tetrahedron, when we have suppressed one face and the three adjacent to it, there is nothing left—save a median point, corresponding to the opposite corner. In a regular dodecahedron, if we suppress one pentagonal face and its five neighbours, the other half of the whole figure remains; and the dodecahedral cage, after immersion in the soap, shews a central and symmetrical group of six pentagons‡.

Moreover, while the cage is carrying its configuration of films, we may blow a bubble within it, and so insert a new polyhedron

* The angles of a hexagon are too big, as those of a square were too small, to form the Maraldi angles of symmetry; hence the sides of the hexagon are found to be concave, as those of the square bulged out convexly.

† Cf. Dewar, *op. cit.* 1918.

‡ That is to say, if nF_m be a polyhedron (of n m-faced sides), the corresponding wire cage will exhibit $(n - m + 1)\ F_m$ as central fenestrae.

within the old, and set it in place of the former fenestra. The inner polyhedral bubble so produced may be of any dimensions, but it resembles the outer polyhedral cage precisely, except in the curvature of its sides; it has all its faces spherical, and all of equal radius of curvature; its edges are either arcs of circles or straight lines. Later on, we shall see that there is no small biological interest attaching to these configurations.

Lord Kelvin made the remarkable discovery that the square fenestra with the four quadrilateral films impinging on its sides, in Plateau's experiment, represented the one-sixth part of a symmetrical figure; that this figure when complete was bounded by six squares and eight hexagons; that by means of an assemblage of these

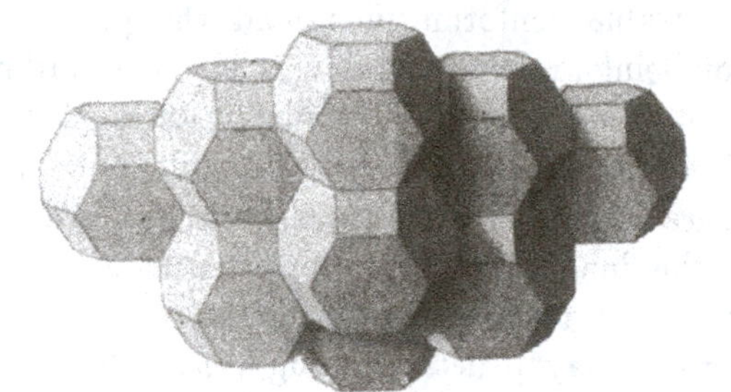

Fig. 213. A set of 14-hedra, to shew close-packing. From F. T. Lewis.

fourteen-sided figures, or "tetrakaidekahedra," space is filled and homogeneously partitioned—into equal, similar and similarly situated cells—with an economy of surface in relation to volume even greater than in an assemblage of rhombic dodecahedra*.

The tetrakaidekahedron, in its most generalised case, is bounded by three pairs of equal and opposite quadrilateral faces, and four pairs of equal and opposite hexagonal faces, neither the quadrilaterals nor the hexagons being necessarily plane. In its simplest case, with all its facets plane and equilateral, it is Kelvin's "orthotetrakaidekahedron"; and also (though Kelvin was unaware of the fact) one of the thirteen semi-regular and isogonal polyhedra, or "Archimedean bodies." In a particular case, the quadrilaterals are plane surfaces with curved edges, but the hexagons are slightly

* Kelvin, *Boyle Lecture* and *Baltimore Lectures*. In the first of these Kelvin described the plane-faced tetrakaidekahedron; in the second he shewed how that figure must have its faces warped and edges curved to fulfil all the conditions of minimal area.

curved "anticlastic" surfaces; and these latter have at every point equal and opposite curvatures, and are surfaces of minimal curvature for a boundary of six curved edges. This figure has the remarkable property that, like the plane rhombic dodecahedron, it so partitions space that three faces meeting in an edge do so everywhere at co-equal angles of 120°; and, unlike the rhombic dodecahedron, four edges meet in each point or corner at co-equal angles of 109°.28′*.

We may take it as certain that, in a homogeneous system of fluid films like the interior of a froth of soap-bubbles, where the films are perfectly free to glide or turn over one another and are of approximately co-equal size, the mass is actually divided into cells of this remarkable conformation: and the possibility of such a configuration being present even in the cells of an ordinary vegetable parenchyma was suggested in the first edition of this book. It is all a question of *restraint*, of degrees of mobility or fluidity. If we squeeze a mass of clay pellets together, like Buffon's peas, they come out, or all the inner ones do, in neat garnet-shape, or rhombic dodecahedra. But a young student once shewed me (in Yale) that if you wet these clay pellets thoroughly, so that they slide easily on one another and so acquire a sort of pseudo-fluidity in the mass, they no longer come out as regular dodecahedra, but with square and hexagonal facets recognisable as those of ill-formed or half-formed tetrakaidekahedra.

Dr F. T. Lewis has made a long and careful study of various vegetable parenchymas, by simple maceration, wax-plate recon-

* Von Fedorow had already described (in Russian), unaware that Archimedes had done so, the same figure under the name of cubo-octahedron, or hepta-parallelohedron, limited however to the case where all the faces are plane and regular. This cubo-octahedron, together with the cube, the hexagonal prism, the rhombic dodecahedron and the "elongated dodecahedron," constitute the five plane-faced, parallel-sided figures by which space is capable of being completely filled and uniformly partitioned; the series so forming the foundation of Von Fedorow's theory of crystalline structure—though the space-fillers are not all, and cannot all be, crystalline forms. All of these figures, save the hexagonal prism, are related to and derivable from the cube; so we end by recognising two principal types, cubic and hexagonal. We have learned to recognise the dodecahedron, and we may find in still closer packing the cubo-octahedron, in a parenchyma; the elongated dodecahedron is, essentially, the figure of the bee's cell; the cube we have, in essence, in cambium-tissue; the hexagonal prism, dwarf or tall, simple or recognisably deformed, we see in every epithelium.

struction and otherwise, and has succeeded in shewing that the tetrakaidekahedral form is closely approached, or even attained, in certain simple and homogeneous tissues. After reconstructing a large model of the cells of elder-pith, he finds that the fourteen-sided figure clearly manifests itself as the characteristic or typical form to which the cells approximate, in spite of repeated cell-divisions and consequent inequalities of size. Counting in a hundred cells the number of contacts which each made with its neighbours, that is to say the total number both of actual and potential facets, Lewis found that 74 per cent. of the cells were either 12, 13, 14, 15 or 16-sided, 56 per cent. either 13, 14 or 15-sided, and that the average

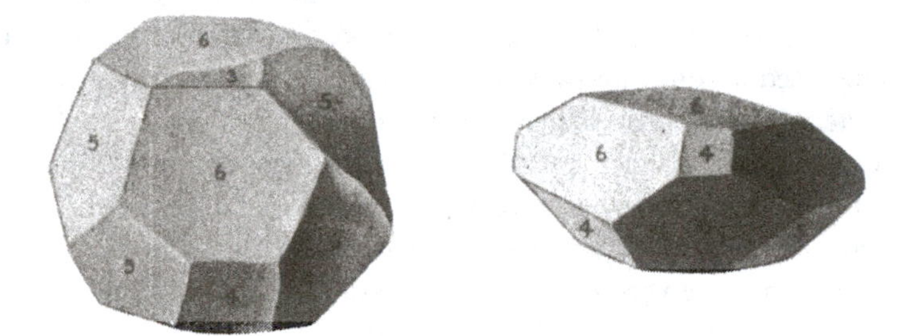

Fig. 214. Reconstructed models of cells of elder-pith, shewing a certain approximation to 14-hedral form. From F. T. Lewis.

number of facets or contacts was, in this instance, just 13·96. These figures indicate the general symmetry of the cells, their departure from the dodecahedral, and their tendency towards the tetrakaidekahedral, form*.

But after all, the geometry of the 14-hedron, displayed to perfection by our soap-films in the twinkling of an eye, is only roughly developed in an organic structure, even one so delicate as elder-pith; the conditions are no longer simple, for friction, viscosity and

* F. T. Lewis, The typical shape of polyhedral cells in vegetable parenchyma, and the restoration of that shape following cell-division, *Proc. Amer. Acad. of Arts and Sci.* LVIII, pp. 537–552, 1923, and other papers. See also (*int. al.*) J. W. Marvin, The aggregation of orthis-tetrakaidekahedra, *Science*, LXXXIII, p. 188, 1936; E. B. Metzger, An analysis of the orthotetrakaidekahedron, *Bull. Torrey Bot. Club*, LIV, pp. 341–348, 1927. Professor van Iterson of Delft tells me that *Asparagus Sprengeri* (a common greenhouse plant) is a good subject for shewing the 14-hedral cells.

solidification have vastly complicated the case. We get a curious and an unexpected variant of the same phenomenon in the microscopic foam-like structure assumed as molten metal cools. If these foam-cells were again 14-hedra, their facets would all be either squares or hexagons; but pentagonal facets are commoner than either, and the cells often approach closely to the form of a regular *pentagonal* dodecahedron! The edges of this figure meet at angles of 108°, not far from the characteristic Maraldi angle of 109° 28′; and the faces meet at an angle not far removed from 120°. A slight curvature of the sides is enough to turn our pentagonal dodecahedron into a possible figure of equilibrium for a foam-cell. We cannot close-pack pentagonal dodecahedra, whether equal or unequal, so as to fill space; but still the figure may be, and seems to be, common, interspersed among the polyhedra of various shapes and sizes which are packed together in a metallic foam*.

A somewhat similar result, and a curious one, was found by Mr J. W. Marvin, who compressed leaden small-shot in a steel cylinder, as Buffon compressed his peas; but this time the pressure on the plunger ran from 1000 to 35,000 lb. or nearly twenty tons to the square inch. When the shot was introduced carefully, so as to lie in ordinary close packing, the result was an assemblage of regular rhombic dodecahedra, as might be expected and as Buffon had found. But the result was very different when the shot was poured at random into the cylinder, for the average number of facets on each grain now varied with the pressure, from about 8·5 at 1000 lb. to 12·9 at 10,000 lb., and to no less than 14·16 facets or contacts after all interstices were eliminated, which took the full pressure of 35,000 lb. to do. An average of just over fourteen facets might seem to indicate a tendency to the production of tetrakaidekahedra, just as in the froth of soap-bubbles; but this is not so. The squeezed grains are irregular in shape, and pentagonal facets are much the commonest, just as we found them to be in the microscopic structure of a once-molten metal. At first sight it might seem that, though the experiment has something to teach us about random packing in a limited space, it has no biological significance; but it is curious to find that the pith-cells of

* Cf. Cecil H. Desch, The solidification of metals from the liquid state, *Journ. Inst. of Metals*, XXII, p. 247, 1919.

Eupatorium have a similar average configuration, with the same predominance of pentagonal facets*.

We learn, in short, from Lewis and from Marvin that the mechanical result of mutual pressure, even in an assemblage of co-equal spheres, is more varied and more complex than we had supposed. The two simple and homogeneous configurations—the rhombo-dodecahedral and tetrakaidekahedral assemblages—are easily and commonly produced, the one by the compression of solid spheres in ordinary close-packing, the other when a liquid system of spheres or bubbles is free to slide and glide into a packing which is closer still. Between these two configurations there is no other symmetrical or homogeneous arrangement possible; but random packing and degrees of compression leave their random effects, among which are traces here and there of regular shape and symmetry.

As a froth has its histological lessons for us, throwing light on the structure of a parenchyma, so may we draw an illustration or two from the analogous characteristics of an emulsion. Both alike are "states of aggregation"; both are "two-phase systems," one phase being dispersed and the other the medium of dispersion. Both phases are liquid in the emulsion, in the froth the dispersed phase is a gas; our living tissue is, so far, more likely to be an emulsion than a froth. The concept widens. A colony of bacteria, the blood corpuscles in their plasma, the filaments of an alga, the heterogeneous texture of any ordinary tissue, may all be brought under the general concept of "phase systems," and share the common character that one phase exposes a large "interface" to the other. If we take milk as a simple emulsion, we see its liquid oil-globules dispersed in a watery medium and rounded by surface tension into spheres. The watery medium, as is usual in such emulsions, contains dissolved substances which tend to lower the interfacial tension; for were that tension high the globules would tend to be larger and their aggregate surface less. Suppose the "phase-ratio" to alter, the globules becoming more numerous and the disperse medium less and less, the globules will be close-packed

* J. W. Marvin, The shape of compressed lead-shot, etc.; *Amer. Journ. of Botany*, XXVI, pp. 280–288, 1939; Cell-shape studies in the pith of Eupatorium, *ibid.* pp. 487–504.

at last. Then each (provided they be of equal size) will be in touch with twelve neighbours; and if the spheres were solid—were the system not an emulsion but a "suspension"—the matter would end here. But our liquid globules are capable of deformation, and the points of contact are flattened in still closer packing into planes. They become polyhedral, and tend to take the form of rhombic dodecahedra, or it may be even of 14-hedra, and the dispersion-medium is reduced to mere films or pellicles between. At the stage of mere twelve-point contact, the spherules constitute about 74 per cent., and the disperse medium 26 per cent., of the whole. But in the final stage the phase-ratio has so altered that the disperse-medium is but a small fraction of the whole, the thin film to which it has been reduced has the appearance of a cell-membrane separating the cells, and the microscopic structure of the whole corresponds to the cellular configuration of a parenchymatous tissue*.

Of certain groupings of cells

It follows from all that we have said that the problems connected with the conformation of cells, and with the manner in which a given space is partitioned by them, soon become complex; and while this is so even when all our cells are equal and symmetrically placed, it becomes vastly more so when cells varying even slightly in size, in hardness, rigidity or other qualities, are packed together. The mathematics of the case very soon become too hard for us, but in its essence the phenomenon remains the same. We have little reason to doubt, and no just cause to disbelieve, that the whole configuration, for instance of an egg in the advanced stages of segmentation, is accurately determined by simple physical laws: just as much as in the early stages of two or four cells, during which early stages we are able to recognise and demonstrate the forces and their effects. But when mathematical investigation has become too difficult, physical experiment can often reproduce the phenomena which Nature exhibits, and which we are striving to comprehend. In an admirable research, M. Robert not only shewed some years ago that the early segmentation of the egg of *Trochus* (a marine univalve mollusc) proceeded in accordance with the laws

* Cf. E. Hatschek, Homogeneous partitionings, etc., *Phil. Mag.* XXXIII, p. 83, 1917.

of surface-tension, but he also succeeded in imitating by means of soap-bubbles one stage after another of the developing egg.

M. Robert carried his experiments as far as the stage of sixteen cells, or bubbles. It is not easy to carry the artificial system quite so far, but in the earlier stages the experiment is easy; we have merely to blow our bubbles in a little dish, adding one to another, and adjusting their sizes to produce a symmetrical system. One of the simplest and prettiest parts of his investigation concerned the "polar furrow" of which we have spoken on p. 489. On blowing four little contiguous bubbles he found (as we may all find with the greatest ease) that they form a symmetrical system, two in contact with one another by a laminar film, and two which are elevated a little above the others and are separated by the length of the aforesaid lamina. The bubbles are thus in contact three by three, their partition-walls making with one another equal angles of 120°. The upper and lower edges of the intermediate lamina (the lower one visible through the transparent system) constitute the two polar furrows of the embryologist (Fig. 215, 1–3). The lamina itself is plane when the system is symmetrical, but it responds by a corresponding curvature to the least inequality of the bubbles on either side. In the experiment, the upper polar furrow is usually a little shorter than the lower, but parallel to it; that is to say, the lamina is of trapezoidal form: this lack of perfect symmetry being due (in the experimental case) to the lower portion of the bubbles being somewhat drawn asunder by the tension of their attachments to the sides of the dish (Fig. 215, 4). A similar phenomenon is usually found in *Trochus*, according to Robert, and many other observers have likewise found the upper furrow to be shorter than the one below. In the various species of the genus *Crepidula*, Conklin asserts that the two furrows are equal in *C. convexa*, that the upper one is the shorter in *C. fornicata*, and that the upper one all but disappears in *C. plana*; but we may well be permitted to doubt, without the evidence of very special investigations, whether these slight physical differences are actually characteristic of, and constant in, particular *species*. Returning to the experimental case, Robert found that by withdrawing a little air from, and so diminishing the bulk of the two terminal bubbles (i.e. those at the ends of the intermediate lamina), the upper polar furrow was caused to elongate, till it became equal in length to the lower; and by continuing the process it became the longer in its turn. These two conditions have again been described by investigators as characteristic of this embryo or that; for instance in *Unio*, Lillie has described the two furrows as gradually altering their respective lengths*; and Wilson (as Lillie

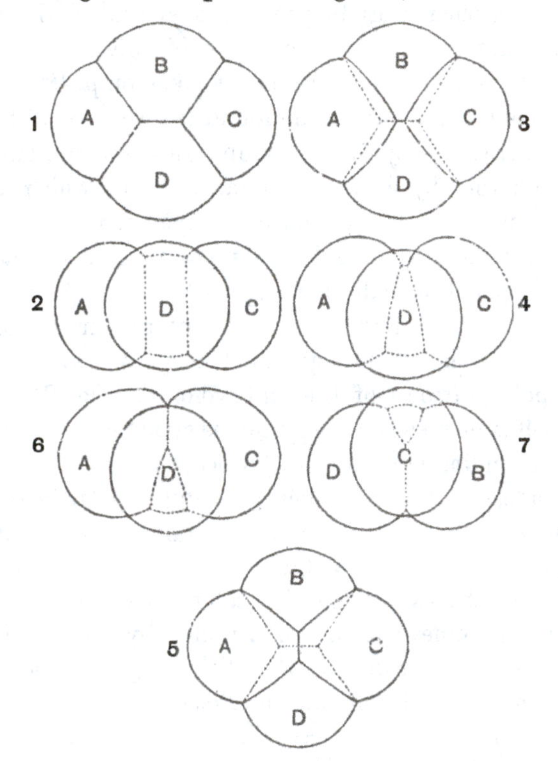

Fig. 215. Aggregations of four soap-bubbles, to shew various arrangements of the intermediate partition and polar furrows. After Robert.

remarks) had already pointed out that "the reduction of the apical cross-furrow, as compared with that at the vegetative pole in molluscs and annelids, 'stands in obvious relation to the different size of the cells produced at the two poles†'."

When the two lateral bubbles are gradually reduced in size, or the two terminal ones enlarged, the upper furrow becomes shorter

* F. R. Lillie, Embryology of the Unionidae, *Journ. Morph.* x, p. 12, 1895.

† E. B. Wilson, The cell-lineage of Nereis, *Journ. Morph.* vi, p. 452, 1892.

and shorter; and at the moment when it is about to vanish, a new furrow makes its instantaneous appearance in a direction perpendicular to the old one; but the inferior furrow, constrained by its attachment to the base, remains unchanged, and it looks as though our two polar furrows, which were formerly parallel, were now at right angles to one another. But in fact, the geometry of the whole system is entirely altered. Before, two furrows left each end of one polar furrow for *the same end* of the other polar furrow, and the two cells at either end were shaped like "liths" of an orange. Under the new arrangement, two furrows leave each end of one for *the two ends* of another. The figure is now divided by six similar furrows into four similar curvilinear triangles*; it has become (approximately) a spherical tetrahedron, and the four cells into which it is divided are four similar and symmetrical figures, also tetrahedral, all meeting in a point at the centroid of the figure. Such a four-celled embryo, described as having two polar furrows arranged in a cross, has often been seen and figured by the embryologists. Robert himself found this condition in *Trochus*, as an occasional or exceptional occurrence: it has been described as normal in *Asterina* by Ludwig, in *Branchipus* by Spangenberg, and in *Podocoryne* and *Hydractinia* by Bunting.

So, by slight and delicate modifications, we pass through many, and perhaps through *all*, of the possible arrangements of external furrows and internal partitions which divide the four cells from one another in a four-celled egg or embryo; and many, or most, or possibly *all* of these arrangements have been more or less frequently observed in the four-celled stages of various embryos. And all these configurations, which the embryologists have witnessed and described, belong to that large class of phenomena whose distribution among embryos, or among organisms in general, bears no relation to the boundaries of zoological classification; through molluscs, worms, coelenterates, vertebrates and what not, we meet with now one and now another, in a medley which defies classification. They are not "vital phenomena," or "functions" of the organism, or special characteristics of this organism or that, but purely physical

* That the sphere can be symmetrically divided into four equilateral triangles, after the manner of these embryos (or of many pollen-grains), is an elementary fact of great importance in geometry and trigonometry.

phenomena. The kindred but more complicated phenomena analogous to the polar furrow, which arise when a larger number of cells than four are associated together, we shall deal with in the next chapter.

Having shewn that the capillary phenomena are patent and unmistakable during the earlier stages of embryonic development, but soon become more obscure and less capable of experimental reproduction in the later stages when the cells have increased in number, various writers including Robert himself have been inclined to argue that the physical phenomena die away, and are overpowered and cancelled by agencies of a different order. Here we pass into a region where observation and experiment are not at hand to guide us, and where a man's trend of thought, and way of judging the whole evidence in the case, must shape his philosophy. We must always remember that even in a froth of soap-bubbles we can apply an exact analysis only to the simplest cases and conditions; we cannot describe, but can only imagine, the forces which in such a froth control the respective sizes, positions and curvatures of the innumerable bubbles and films of which it consists; but our knowledge is enough to leave us assured that what we have learned by investigation of the simplest cases includes the principles which determine the most complex. In the case of the growing embryo we know from the beginning that surface-tension is only one of the physical forces at work; and that other forces, including those displayed within the interior of each living cell, play their part in the determination of the system. But we have no evidence whatsoever that at this point, or that point, or at any, the dominion of the physical forces over the material system gives place to a new condition where agencies at present unknown to the physicist impose themselves on the living matter, and become responsible for the conformation of its material fabric.

Before we leave for the present the subject of the segmenting egg, we may take brief note of two associated problems: viz. (1) the formation and enlargement of the segmentation cavity, or central interspace around which the cells tend to group themselves in a single layer, and (2) the formation of the gastrula, that is to say (in a typical case) the conversion by "invagination," of the one-layered ball into a two-layered cup. Neither problem is free from difficulty, and all we can do meanwhile is to state them in general terms, introducing some more or less plausible assumptions.

The former problem is comparatively easy, as regards the tendency of a segmentation cavity to *enlarge*, when once it has been established. We may then assume that subdivision of the cells is due to the appearance of a new-formed septum within each cell, that this septum has a tendency to shrink under surface-tension, and that these changes will be accompanied on the whole by a diminution of surface-energy in the system. This being so, it may be shewn that the volume of the divided cells must be less than it was prior to division, or in other words that part of their contents must exude during the process of segmentation*. Accordingly, the case where the segmentation cavity enlarges and the embryo developes into a hollow blastosphere may, under the circumstances, be simply described as the case where that outflow or exudation from the cells of the blastoderm is directed on the whole inwards.

The physical forces involved in the invagination of the cell-layer to form the gastrula have been repeatedly discussed†, but the several explanations are conflicting, and are far from clear. There is, however, a certain homely phenomenon which goes some way, perhaps a long way, to explain this remarkable configuration. An ordinary gelatine lozenge, or jujube, has (like the developing gastrula) a more or less spherical form, depressed or dimpled at one side; this is a very noteworthy conformation, and it arises, automatically, by the shrinkage of a sphere. Were the initial sphere of gelatine perfectly homogeneous, and so situated as to shrink with absolute uniformity, it would merely shrink into a smaller sphere; it does nothing of the kind. There is always some part or other which shrinks *a little more* than the rest‡; and the dimple so formed goes on increasing, until at last a very perfect cup-shaped figure is formed. I imagine that the gastrula is formed in much the

* Professor Peddie has given me this interesting result, but the mathematical reasoning is too lengthy to be set forth here.

† Cf. Bütschli, *Arch. f. Entw. Mech.* v, p. 592, 1897; Rhumbler, *ibid.* XIV, p. 401, 1902; Assheton, *ibid.* XXXI, p. 46, 1910.

‡ Just as there may be some small part which shrinks a little *less*. But this we should not distinguish from the common case where one small part *grows a little more*, and so "produces a *bud*," as in the yeast-cell on p. 363.

same way, save only that the initial dimple, instead of being fortuitous, has its constant place, determined by the physico-chemical heterogeneity of the embryo. We may even go one step further, and see (or imagine we see) in the formation of the gastrula a physico-chemical or physiological turning-point, the segmentation cavity being due (as we have seen) to an inward flow, and a reversal of the current leading to that shrinkage which produces the gastrula.

Fig. 216. Effect of shrinkage on a globule of gelatine. After E. Hatschek.

A note on shrinkage

We have dealt much with *growth*, but the fact is that negative growth, or shrinkage, is also an important matter; and just as we find a whole series of phenomena to be based on the extension or expansion of bubbles, vesicles, etc., so there is another series, physically alike and mathematically identical, which depend on the shrinkage of a solid or semi-fluid mass. After all, growth and its converse go hand in hand, and a special case of shrinkage is that surface-tension to which all the Plateau configurations are due. One clear case, the gastrula, we have touched on, and we have discussed another which led to the stellate dodecahedra of the Rush.

As a cube of gelatine, or of paraffin, dries, and shrinks, it alters its shape in a remarkable way*. Its corners become more salient, its sides become concave; its cross-section has the form of a four-

* Emil Hatschek, *Kolloid Ztschr.* xxxv, pp. 67–76, 1924; *Nature*, 1st Nov. 1924.

rayed star with rounded angles. The block has dried unequally; its corners and its edges were naturally the first to dry*, and the twelve dried and hardened edges began to play the part of the wire frame in Plateau's soap-bubble experiment. The shrinking cube is tending towards the identical configuration shewn in Fig. 212; it is a minimal configuration, partially realised in a coarse material, but realisable to perfection in a film.

A shrinking cylinder (as Plateau knew) shews various phenomena, depending on its proportions. A low, squat cylinder begins to show a pulley-like groove—a catenoid—around its periphery, precisely like the soap-film between its two wire rings in Fig. 108; and as the groove deepens, the plane surfaces of the cylinder also begin

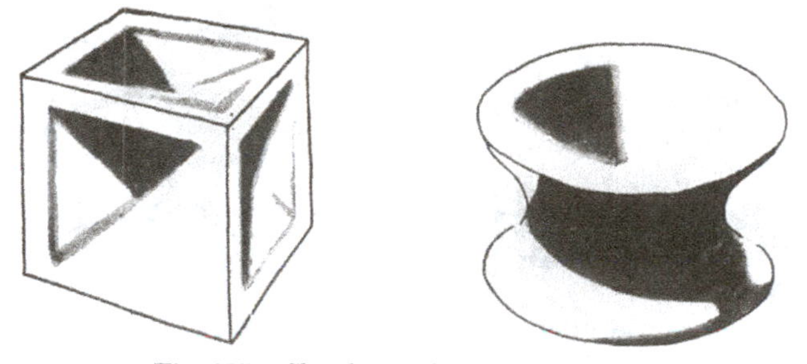

Fig. 217. Shrinkage of cube and cylinder.

to dimple in. They become spherical as they grow more concave, and the deepening groove of the pulley passes from a catenoid to a nodoid curve—so at least theory tells us†, for, beautiful as the experimental configurations are, they hardly lend themselves to precise measurements of curvature. But this shrunken cylinder is now wonderfully like the "amphicoelous" vertebra of a cartilaginous fish, the simplest and most "primitive" vertebra of all. A series of cracks, or splits, around the circular groove in the vertebra seem to be a final result of irregular shrinkage, not shewn in the more homogeneous gelatine.

A long cylinder, or thread, of gelatine tends to become fluted, with three or more ribs or folds, and it is in this way that threads

* Just as, conversely, the prominent parts of a crystal tend to grow more rapidly than the rest in a super-saturated solution, and to dissolve more rapidly in one below saturation; cf. O. Lehmann, Ueber das Wachstum der Krystalle, *Ztschr. f. Krystallogr.* I, p. 453.

† See p. 369.

of viscose, or artificial silk, tend likewise to have a ridged or fluted structure, and gain in lustre thereby. The subject is new, and hardly ripe for full discussion; but it holds out promise (as it seems to me) of many biological lessons and illustrations.

We glanced in passing at such "shrinkage-patterns" as are found, for instance, on the little shells of *Lagena*, or on those other hanging drops which constitute Emil Hatschek's artificial medusae; it is no small subject. A stretched elastic membrane, circular or spherical, remains spherical or circular when we let its tension relax; but if, to begin with, we coat the rubber with a pliant but non-elastic material such as wax, the waxen layer, failing to con-

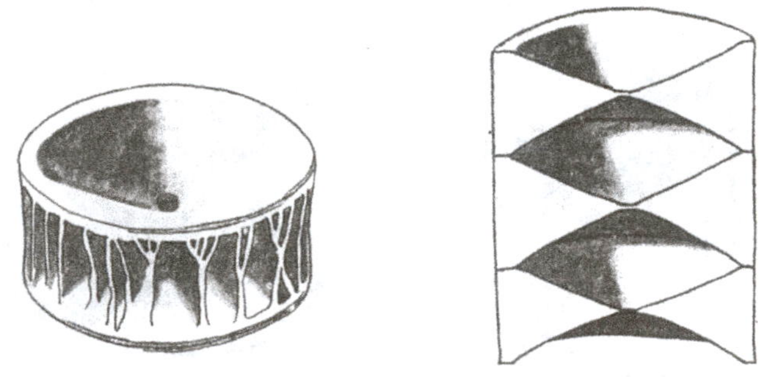

Fig. 218. Amphicoelous vertebrae of a shark.

tract, is thrown into more or less characteristic folds. In a dried pea the seed has shrunken through loss of moisture, and the loose outer coat wrinkles up*. The pretty pattern of a poppy-seed arises in the same way; but so do the wrinkles on an old man's withered skin. When our experimental elastic with its non-contractile coat is suffered to contract, the first sign of the coat's inability to keep pace is the appearance of little domes, or hummocks, or blisters; and soon from each of these there run out folds, which tend to fork, and the angles between the three branches tend to equalise. They tend, in simple and symmetrical cases, to form a pattern of hexagons, with occasional pentagons or quadrilaterals between; but where the surface is larger and the coat more flexible the folds form an irregular network, still with the various anticlines mostly meeting in three-

* The difference between a smooth and a wrinkled pea, familiar to Mendelians, merely depends, somehow, on amount and rate of shrinkage.

way nodes*. Nature will ring the changes on the resultant patterns, according as the surface be plane or curved, spherical or cylindrical, coarse or fine, fragile or tough. But on these general lines very many structures, both regular and irregular, spines, bristles, ridges, tubercles and wrinkled patterns, bid fair to find their physical or mechanical interpretation; and it is in the more or less hardened parts of plant or animal that we find them one and all displayed. On the egg of a butterfly, on the grooved and dotted elytron of a beetle, on the notched forehead of a scarab, in the saw-like teeth on a grasshopper's leg, in the little lines of dotted tubercles on the shell of a *Rissoa*, more crudely in the lozenged bark of elm or pine, we see a very few of this innumerable class of "shrinkage-patterns."

* The fact that such triplets of divergent ridges or crests are not a feature in the topography of mountain-ranges is a strong argument against the view that general shrinkage accounts for the pattern of the earth's crust. Cf. A. J. Bull, The pattern of a contracting earth, *Geolog. Mag.* LXIX, pp. 73–75, 1932; A. E. B. de Chancourtois, *C.R.* LIX, p. 348, 1903.

CHAPTER VIII

THE FORMS OF TISSUES OR CELL-AGGREGATES (*continued*)

THE problems which we have been considering, and especially that of the bee's cell, belong to a class of "isoperimetrical" problems, which deal with figures whose surface is a minimum for a definite content or volume. Such problems soon become difficult*, but we may find many easy examples which lead us towards the explanation of biological phenomena; and the particular subject which we shall find most easy of approach is that of the division, in definite proportions, of some definite portion of space, by a partition-wall of minimal area. The theoretical principles so arrived at we shall then attempt to apply, after the manner of Berthold and Errera, to the biological phenomena of cell-division.

This investigation may be approached in two ways: by considering the partitioning off from some given space or area of one-half (or some other fraction) of its content; or again, by dealing with the partitions necessary for the breaking up of a given space into a definite number of compartments.

If we begin with the simple case of a cubical cell, it is obvious that, to divide it into two halves, the smallest partition-wall is one which runs parallel to, and midway between, two of its opposite sides. If we call a the length of one of the edges of the cube, then a^2 is the area, alike of one of its sides and of the partition which we have interposed parallel thereto. But if we now consider the bisected cube, and wish to divide the one-half of it again, it is obvious that another partition parallel to the first, so far from being the smallest possible, is twice the size of a cross-partition perpendicular to it; for the area of this new partition is $a \times a/2$. And again, for a third bisection, our next partition must be perpendicular to the other two, and is obviously a little square, with an area of $(\frac{1}{2}a)^2 = \frac{1}{4}a^2$.

* Minkowski and others have shewn how hard it is, for instance, to prove the seemingly obvious proposition that the sphere, of all figures, has the greatest volume for a given surface; cf. (e.g.) T. Bonneson, *Les problèmes des isopérimètres et des isépiphanes*, Paris, 1929. For a historical account of this class of problems, see G. Enestrom, in *Bibl. Math.* 1888.

From this we may draw the simple rule that, for a rectangular body or parallelepiped to be bisected by means of a partition of minimal area, (1) the partition must cut across the longest axis of the figure; and (2) in successive bisections, each partition must run at right angles to its immediate predecessor.

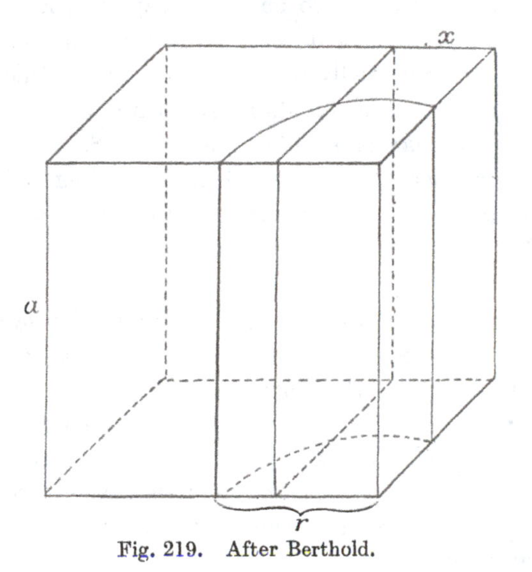

Fig. 219. After Berthold.

We have already spoken of "Sachs's Rules," which are an empirical statement of the method of cell-division in plant-tissues; and we may now set them forth as follows:

(1) The cell tends to divide into two co-equal parts.

(2) Each new plane of division tends to intersect the preceding plane of division at right angles.

The first of these rules is a statement of physiological fact, not without its exceptions, but so generally true that it will justify us in limiting our enquiry for the most part to cases of equal subdivision. That it is by no means universally true for cells generally is shewn, for instance, by such well-known cases as the unequal segmentation of the frog's egg. It is true, when the dividing cell is homogeneous and under the influence of symmetrical forces; but it ceases to be true when the field is no longer dynamically symmetrical, as when the parts differ in surface tension or internal pressure, or, speaking generally, in their chemico-physical properties and conditions. This latter condition, of asymmetry of field, is frequent in segmenting eggs*, and it then covers or includes the principle upon which Balfour laid stress as leading to "unequal" or to "partial" segmentation of the egg—viz. the unequal or asymmetrical distribution of protoplasm and of food-yolk.

The second rule, which also has its exceptions, is true in a large number of cases, and owes its validity, as we may judge from the illustration of the repeatedly bisected cube, to the guiding principle of minimal areas. It is in short subordinate to a much more important and fundamental rule, due not to Sachs but to Errera; that (3) the incipient partition-wall of a dividing cell tends to be such that *its area is the least possible by which the given space-content can be enclosed.*

Let us return to the case of our cube, and suppose that, instead of bisecting it, we desire to shut off some small portion only of its volume. It is found in the course of experiments upon soap-films, that if we try to bring a partition-film too near to one side of a cubical (or rectangular) space it becomes unstable, and is then easily shifted to a new position in which it constitutes a curved cylindrical wall cutting off one corner of the cube. It still meets the sides of the cube at right angles (for reasons which we have already considered); and, as we may see from the symmetry of the case, it constitutes one-quarter of a cylinder. Our plane transverse partition had always the same area, wherever it was placed, viz. a^2; and it is obvious that a cylindrical wall, if it cut off a small corner, may be much less than this. We want, accordingly, to determine what volume might be partitioned off with equal economy of wall-space in one way as the other, that is to say, what area of cylindrical

* M. Robert (*loc. cit.* p. 305) has compiled a long list of cases among the molluscs and the worms, where the initial segmentation of the egg proceeds by equal or unequal division. The two cases are about equally numerous. But like most other writers of his time, he would ascribe this equality or inequality rather to a provision for the future than to a direct effect of immediate physical causation: "Il semble assez probable, comme on l'a dit souvent, que la plus grande taille d'un blastomère est liée à l'importance et au développement précoce des parties du corps qui doivent en naître: il y aurait là une sorte de reflet des stades postérieures du développement sur les premières phénomènes, ce que M. Ray Lankester appelle *precocious segregation.* Il faut avouer pourtant qu'on est parfois assez embarrassé pour assigner une cause à pareilles différences."

wall would be neither more nor less than the area a^2. The calculation is easy:

The *surface-area* of a cylinder of length a is $2\pi r \,.\, a$, and that of our quarter-cylinder is, therefore, $a \,.\, \pi r/2$; and this being, by hypothesis, $= a^2$, we have $a = \pi r/2$, or $r = 2a/\pi$.

The *volume* of a cylinder of length a is $a\pi r^2$, and that of our quarter-cylinder is $a \,.\, \pi r^2/4$, which (by substituting the value of r) is equal to a^3/π.

Now precisely this same volume is, obviously, shut off by a transverse partition of area a^2 if the third side of the rectangular space be equal to a/π; and this fraction, if we take $a = 1$, is equal to 0·318..., or rather less than one-third. And, as we have just seen, the radius, or side, of the corresponding quarter-cylinder will be twice that fraction, or equal to 0·636 times the side of the cubical cell.

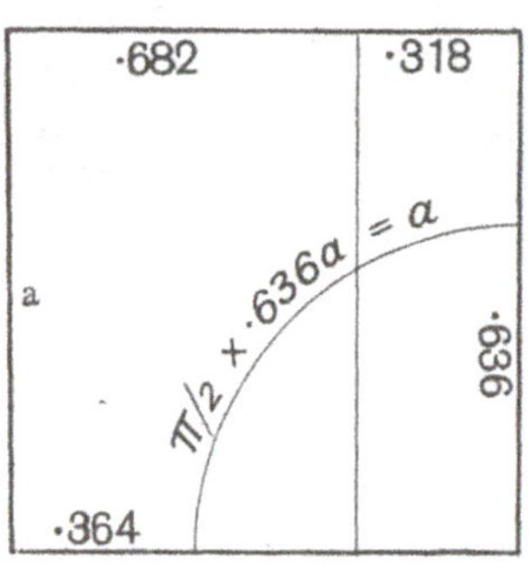

Fig. 220.

If then, in the process of division of a cubical cell, it so divide that the two portions be not equal in volume but that one portion be anything less than about three-tenths of the whole or three-sevenths of the other portion, there will be a tendency for the cell to divide, not by means of a plane transverse partition, but by means of a curved, cylindrical wall cutting off one corner of the original cell; and the part so cut off will be one-quarter of a cylinder.

By a similar calculation we can shew that a *spherical* wall, cutting off one solid angle of the cube and constituting an octant of a sphere, would likewise be of less area than a plane partition as soon as the volume to be enclosed was not greater than about one-quarter of the original cell*. But while both the cylindrical wall and the

* The principle is well illustrated in an experiment of Sir David Brewster's (*Trans. R.S.E.* xxv, p. 111, 1869). A soap-film is drawn over the rim of a wine-glass, and then covered by a watch-glass. The film is inclined or shaken till it becomes attached to the glass covering, and it then immediately changes place, leaving its transverse position to take up that of a spherical segment extending from one side of the wine-glass to its cover, and so enclosing the same volume of air as formerly but with a great economy of surface, precisely as in the case of our spherical partition cutting off one corner of a cube.

spherical wall would be of less area than the plane transverse partition after that limit (of one-quarter volume) was passed, the cylindrical would still be the better of the two up to a further limit. It is only when the volume to be partitioned off is no greater than

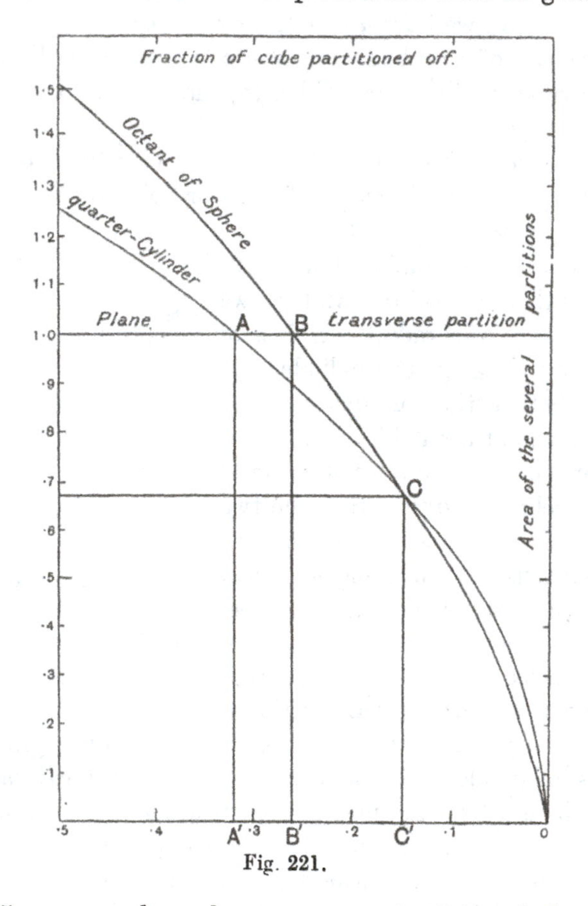

Fig. 221.

about 0·15, or somewhere about one-seventh of the whole, that the spherical cell-wall in a corner of the cubical cell, that is to say the octant of a sphere, is definitely of less area than the quarter-cylinder. In the accompanying diagram (Fig. 221) the relative areas of the three partitions are shewn for all fractions, less than one-half, of the divided cell.

In this figure, we see that the plane transverse partition, whatever fraction of the cube it cut off, is always of the same dimensions, that is to say is always equal to a^2, or $=1$. If one-half of the cube have to be cut off, this plane transverse partition is much the best, for we see by the diagram that a cylindrical partition cutting off an equal volume would have an area about 25 per cent. and a spherical partition would have an area about 50 per cent. greater. The point A in the diagram corresponds to the point where the cylindrical partition would begin to have an advantage over the plane, that is to say (as we have seen) when the fraction to be cut off is about one-third, or 0·318 of the whole. In like manner, at B the spherical octant begins to have an advantage over the plane; and it is not till we reach the point C that the spherical octant becomes of less area than the quarter-cylinder.

The case we have dealt with is of little practical importance to the biologist, because the cases in which a cubical, or rectangular, cell divides unequally and unsymmetrically are apparently few; but we can find, as Berthold pointed out, a few examples, as in the hairs within the reproductive "conceptacles" of certain Fuci (*Sphacelaria*, etc., Fig. 222), or in the "paraphyses" of mosses (Fig. 226). But it is of great theoretical importance: as serving to introduce us to a large class of cases in which, under the guiding principle of minimal areas, the shape and relative dimensions of the original cavity lead to cell-division in very definite and sometimes unexpected ways. It is not easy, nor indeed possible, to give a general account of these cases, for the limiting conditions are somewhat complex and the mathematical treatment soon becomes hard. But it is easy to comprehend a few simple cases, which carry us a good long way; and which will go far to persuade the student that, in other cases which we cannot fully master, the same guiding principle is at the root of the matter.

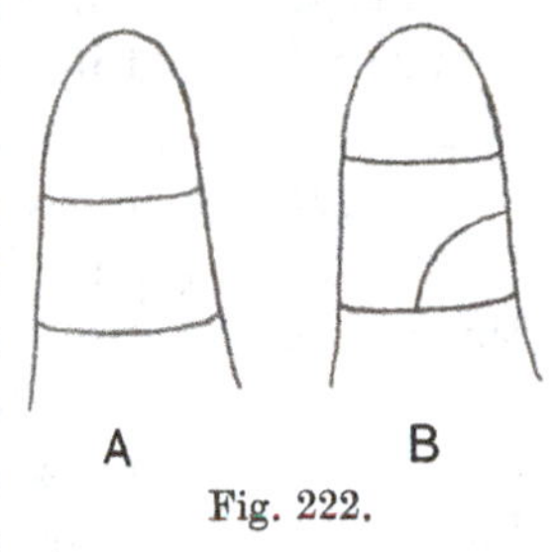

Fig. 222.

The bisection of a solid (or its subdivision in other definite proportions) soon leads us into a geometry which, if not necessarily difficult, is apt to be unfamiliar; but in such problems we can go some way, and often far enough for our purpose, if we merely consider the plane geometry of a side or section of our figure. For instance, in the case of the cube which we have just been considering, and in the case of the plane and cylindrical partitions by which it has been divided, it is obvious, since these two partitions extend symmetrically from top to bottom of our cube, that we need only have considered the manner in which they subdivide the *base* of the cube; in short the problem of the solid, up to a certain point, is contained in our plane diagram of Fig. 221. And when our particular solid is a solid of revolution, then it is equally obvious that a study of its plane of symmetry (that is to say any plane passing through its axis of rotation) gives us the solution of the whole problem. The right cone is a case in point, for here the investigation of its modes of symmetrical subdivision is completely met by an examination of the isosceles triangle which constitutes its plane of symmetry.

The bisection of an isosceles triangle by a line which shall be the shortest possible is an easy problem; for it is obvious that, if the triangle be low, a vertical partition will be shortest; if it be high, a horizontal one; if it be equilateral, the partition may run parallel to any side; and if it be right-angled, the partition may bisect the right angle or run parallel to either side equally well.

Let ABC be an isosceles triangle of which A is the apex; it may be shewn that, for its shortest line of bisection, we are limited to three cases: viz. to a vertical line AD, bisecting the angle at A and the side BC; to a transverse line parallel to the base BC; or to an oblique line parallel to AB or to AC. The lengths of these partition lines follow at once from the magnitudes of the angles of our triangle. We know, to begin with, since the areas of similar figures vary as the squares of their linear dimensions, that, in order to bisect the area, a line parallel to one side of our triangle must always have a length equal to $1/\sqrt{2}$ of that side. If then, we take our base, BC, in all cases of a length $= 2$, the transverse partition, EF, drawn parallel to it will always have a length equal to $2/\sqrt{2}$, or $= \sqrt{2}$. The vertical partition, AD, since $BD = 1$, will always equal $\tan\beta$; and the oblique partition, GH, being equal to $AB/\sqrt{2}$, $= 1/\sqrt{2}\cos\beta$. If then we call our vertical, transverse and oblique partitions V, T, and O, we have $V = \tan\beta$; $T = \sqrt{2}$; and $O = 1/\sqrt{2}\cos\beta$, or

$$V : T : O = \tan\beta/\sqrt{2} : 1 : 1/2\cos\beta.$$

And, working out these equations for various values of β, we soon see that the vertical partition (V) is the least of the three until $\beta = 45°$, at which limit V and O are each equal to $1/\sqrt{2} = 0{\cdot}707$; that O then becomes the least of the three, and remains so until

$\beta = 60°$, when $\cos\beta = 0{\cdot}5$, and $O = T$; after which T (whose value always $= 1$) is the shortest of the three partitions. And, as we have

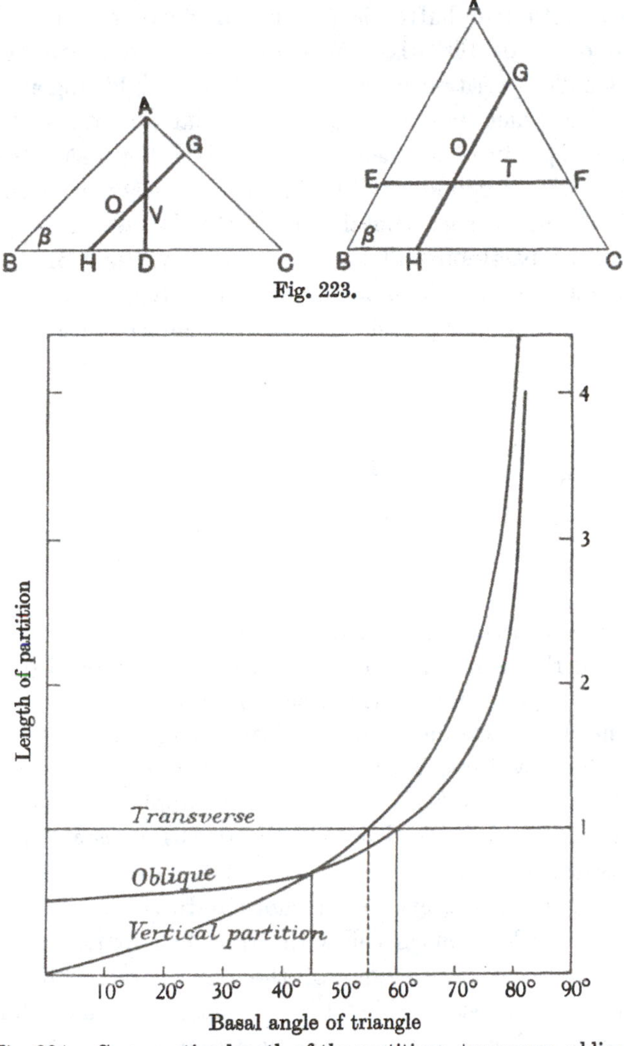

Fig. 223.

Fig. 224. Comparative length of the partitions, transverse, oblique or vertical, bisecting an isosceles triangle.

seen, these results are at once applicable, not only to the case of the plane triangle, but also to that of the conical or pyramidal cell.

In like manner, if we have a spheroidal body less than a hemisphere, such for instance as a low, watchglass-shaped cell (Fig. 225, A), it is obvious that the smallest partition by which we can divide it into two halves is (as in our flattened disc) a median vertical one; and likewise, the hemisphere itself can be bisected by no smaller partition meeting the walls at right angles than that median one which divides it into two similar quadrants of a sphere. But if we produce our hemisphere into a more elevated conical body, or into a cylinder with spherical cap, there comes a point where a transverse horizontal partition will bisect the figure with less area of partition-wall than a median vertical one (C). And furthermore, there will be an intermediate region, a region where height and base have their relative dimensions nearly equal (as

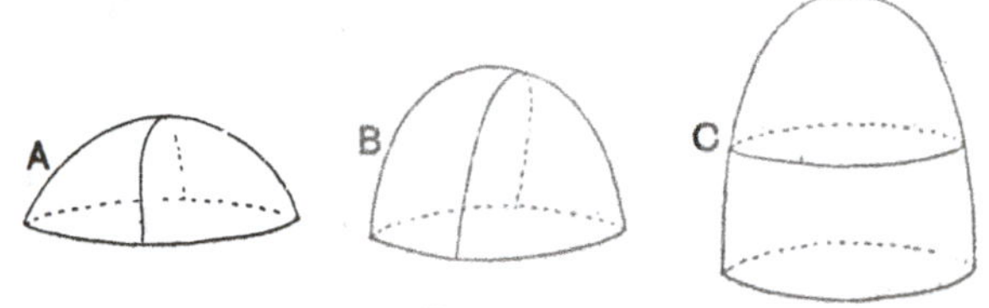

Fig. 225.

in B), where an oblique partition will be better than either the vertical or the transverse; though here the analogy of our triangle does not suffice to give us the precise limiting values.

We need not examine these limitations in detail, but we must look at the curvatures which accompany the several conditions. We have seen that a film tends to set itself at equal angles to the surface which it meets, and therefore, when that surface is a solid, to meet it (or its tangent) at right angles. Our *vertical* partition is, therefore, a plane surface, everywhere normal to the original cell-walls. But in the taller, conical cell with transverse partition, the latter still meets the opposite sides of the cell at right angles, and it follows that it must itself be curved; moreover, since the tension, and therefore the curvature, of the partition is everywhere uniform, it follows that its curved surface must be a portion of a sphere, concave towards the apex of the original cell. In the intermediate case, where we have an oblique partition meeting both the base and the curved sides of the mother-cell, the contact must still be everywhere at right angles: provided we continue to suppose that the walls of the mother-cell (like those of our diagrammatic cube) have become practically rigid before the partition appears, and are therefore not affected and deformed by the tension of the latter. In such a case, and especially when the cell is elliptical in cross-section or still more complicated in form, the partition may have to assume a complex curvature in order to remain a surface of minimal area.

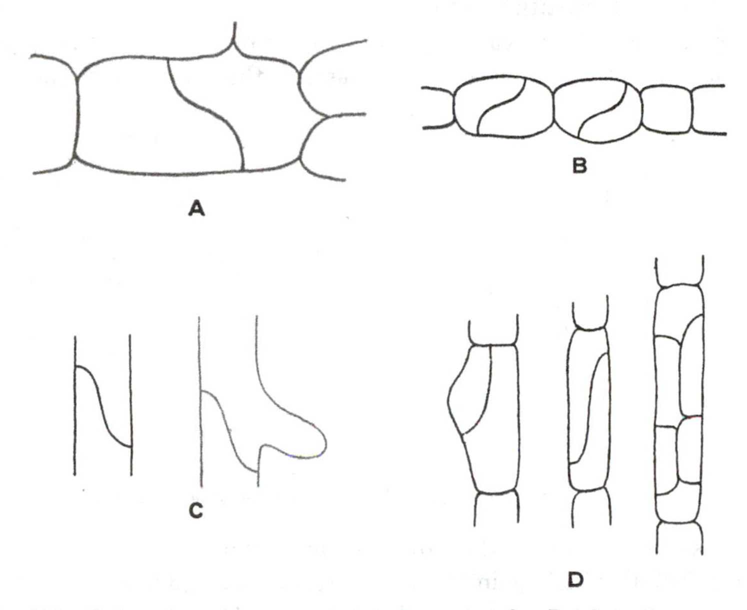

Fig. 226. S-shaped partitions: A, *Taonia atomaria* (after Reinke); B, paraphyses of *Fucus*; C, rhizoids of moss; D, paraphyses of *Polytrichum*.

While in very many cases the partitions (like the walls of the original cell) will be either plane or spherical, a more complex curvature will sometimes be assumed. It will be apt to occur when the mother-cell is irregular in shape, and one particular case of such asymmetry will be that in which (as in Fig. 227) the cell has begun to branch before division takes place. And again, whenever we have a marked internal asymmetry of the cell, leading to irregular and anomalous modes of division, in which the cell is not necessarily divided into two equal halves and in which the partition-wall may assume an oblique position, then equally anomalous curvatures will tend to make their appearance*.

Suppose an oblong cell to divide by means of an oblique partition (as may happen through various causes or conditions of asymmetry), such a partition will still have a tendency to set itself at right angles to the rigid walls of the mother-cell: and it follows that our oblique partition, throughout its whole extent, will assume the form of a complex, saddle-shaped or anticlastic surface.

Many such partitions of complex or double curvature exist, but they are not always easy of recognition, nor do they often appear in a *terminal* cell. We may see them in the roots (or rhizoids) of

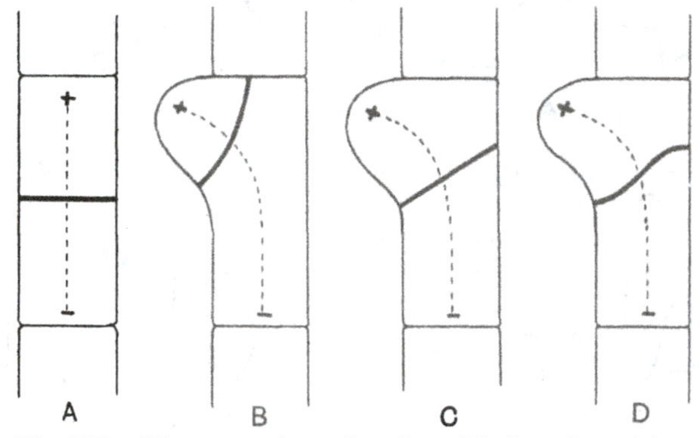

Fig. 227. Diagrammatic explanation of S-shaped partition.

mosses, especially at the point of development of a new rootlet (Fig. 226, C); and again among mosses, in the "paraphyses" of the male plants (e.g. in *Polytrichum*), we find more or less similar partitions (D). They are frequent also among Fuci, as in the hairs or paraphyses of *Fucus* itself (B). In *Taonia atomaria*, as figured in Reinke's memoir on the Dictyotaceae of the Gulf of Naples†, we see, in like manner, oblique partitions, which on more careful examination are seen to be curves of double curvature (Fig. 226, A).

The physical cause and origin of these S-shaped partitions is somewhat obscure, but we may attempt a tentative explanation. When we assert a tendency for the cell to divide transversely to its long axis, we are not only stating empirically that the partition

* Cf. Wildeman, *Attache des cloisons*, etc., pls. 1, 2.

† *Nova Acta K. Leop. Akad.* XI, 1, pl. IV.

tends to appear in a small, rather than a large cross-section of the cell: but we are also ascribing to the cell a longitudinal *polarity* (Fig. 227, A), and implicitly asserting that it tends to divide (just as the segmenting egg does), by a partition transverse to its polar axis. Such a polarity may conceivably be due to a chemical asymmetry, or anisotropy, such as we have learned of (from Macallum's experiments) in our chapter on Adsorption. Now if the chemical concentration, on which this anisotropy or polarity (by hypothesis) depends, be unsymmetrical, one of its poles being as it were deflected to one side where a little branch or bud is being (or about to be) given off—all in precise accordance with the adsorption phenomena described on p. 460—then our "polar axis" would necessarily be a curved axis, and the partition, being constrained (again *ex hypothesi*) to arise transversely to the polar axis, would lie obliquely to the *apparent* axis of the cell (as in B or C). And if the oblique partition be so situated that it has to meet the *opposite* walls (as in C), then, in order to do so symmetrically (i.e. either perpendicularly, as when the cell-wall is already solidified, or at least at equal angles on either side), it is evident that the partition, in its course from one side of the cell to the other, must necessarily assume a more or less **S**-shaped curvature (D).

The complex curvature of the partition-walls in such cases as these may be illustrated by the following experiment. Set two plates of glass (as in Fig. 228) in a wire frame, so that they may lie parallel or at any angle to one another; and dip the whole thing in soap-solution, so that a sheet of film is formed between the two plates and is framed by the two wires which carry them. The film is, of course, a surface of minimal area; its mean curvature is constant everywhere, and (since the film is an open surface with identical pressure on both sides) the mean curvature is everywhere *nil*. A related condition is that the film must meet its solid framework, glass or wire, everywhere at right angles or "orthogonally"; and this last constraint leads to curvatures of extreme complexity, which continually vary as we rotate one plate on the plane of the other.

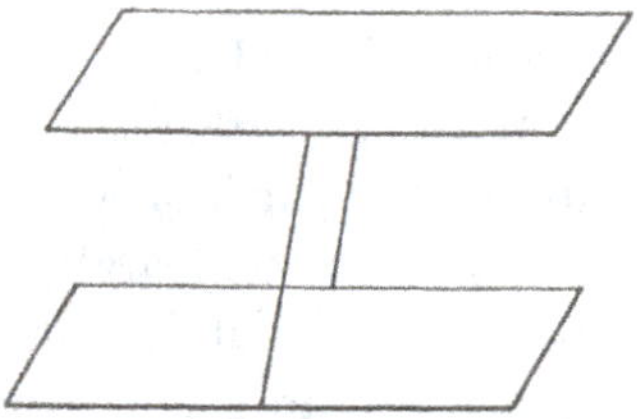

Fig. 228.

As a matter of fact, while we have abundant simple illustrations of the principles which we have now begun to study, apparent exceptions to this simplicity, due to an asymmetry of the cell itself or of the system of which the single cell is a part, are by no means rare. We know that in cambium-cells division often takes place parallel to the long axis of the cell, though a partition of much less area would suffice if it were set cross-ways: and it is only when a considerable disproportion has been set up between the length and breadth of the cell that the balance is in part redressed by the appearance of a transverse partition. It was owing to such exceptions that Berthold was led to qualify and even to depreciate the importance of the law of minimal areas as a factor in cell-division, after he himself had done so much to demonstrate and elucidate it*. He was deeply and rightly impressed by the fact that other forces besides surface tension, both external and internal to the cell, play their part in determining its partitions, and that the answer to our problem is not to be given in a word. How fundamentally important it is, however, in spite of all conflicting tendencies and apparent exceptions, we shall see better and better as we proceed.

But let us leave the exceptions and consider the simpler and more general phenomena. And let us leave the case of the cubical, quadrangular or cylindrical cell, and examine that of a spherical cell and of its successive divisions, or the still simpler case of a circular, discoidal cell.

When we attempt to investigate mathematically the place and form of a partition of minimal area, it is plain that we shall be dealing with comparatively simple cases wherever even one dimension of the cell is much less than the other two. Where two dimensions are small compared with the third, as in a thin cylindrical filament like that of *Spirogyra*, we have the problem at its simplest; for it is obvious, then, that the partition must lie transversely to the long axis of the thread. But even where one dimension only is relatively small, as for instance in a flattened plate, our problem

* Cf. *Protoplasmamechanik*, p. 229: "Insofern liegen also die Verhaltnisse hier wesentlich anders als bei der Zertheilung hohler Körperformen durch flüssige Lamellen. Wenn die Membran bei der Zelltheilung die von dem Prinzip der kleinsten Flächen geforderte Lage und Krümmung annimmt, so werden wir den Grund dafür in andrer Weise abzuleiten haben."

is so far simplified that we see at once that the partition cannot be parallel to the extended plane, but most cut the cell, somehow, at right angles to that plane. In short, the problem of dividing a much flattened solid becomes identical with that of dividing a simple *surface* of the same form.

There are a number of small algae growing in the form of small flattened discs, and consisting (for a time at any rate) of but a single layer of cells, which, as Berthold shewed, exemplify this comparatively simple problem; and we shall find presently that it is admirably illustrated in the cell-divisions which occur in the egg of a frog or a sea-urchin, when it is flattened out under artificial pressure. These same little algae which serve to exemplify the partitioning of a disc also illustrate, now and then, a curious feature of its contour. Such a small green alga as *Castagna* (Fig. 229) shews, and many Desmids shew just as well, a sinuous border running out into rounded crenations or lobes. This is a surface-tension phenomenon. A little milk poured over an apple-pie gives a homely illustration of the same sinuous outlines; a drop on a greasy plate spreads in the same uneven way, and does so indeed unless the utmost care be taken to ensure absolute cleanliness and surface equilibrium*.

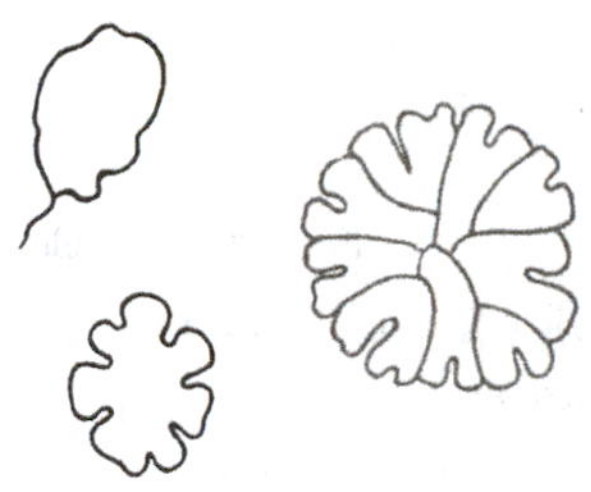

Fig. 229. *Castagna polycarpa.* Swarm-spore and young plants. After Berthold.

Fig. 230† represents younger and older discs of the little alga *Erythrotrichia discigera*; and it will be seen that in all stages save the first we have an arrangement of cell-partitions which looks somewhat complex, but into which we must attempt to throw some light and order. Starting with the original single, and flattened, cell, we have no difficulty with the first two cell-divisions; for we know that no bisecting partitions can possibly be shorter than the two diameters, which divide the cell into halves and into quarters. We have only to remember that, for the sum total of partitions to

* Cf. Quincke's "Ausbreitungserscheinungen," in *Poggendorff's Annalen*, CXXXIX, p. 37, 1870; also Tomlinson's papers in *Phil. Mag.* VIII–XXXIX; and Van der Mensbrugghe, *Mém. Cour. de l'Acad. R. Belgique*, XXXIV, 1870; XXXVII, 1873.

† From Berthold's *Monograph of the Naples Bangiaceae*, 1882.

be a minimum, three only must meet in a point; and therefore, the four quadrantal walls must shift a little, producing the usual little median partition, or cross-furrow, instead of one common central point of junction. This intermediate partition, however,

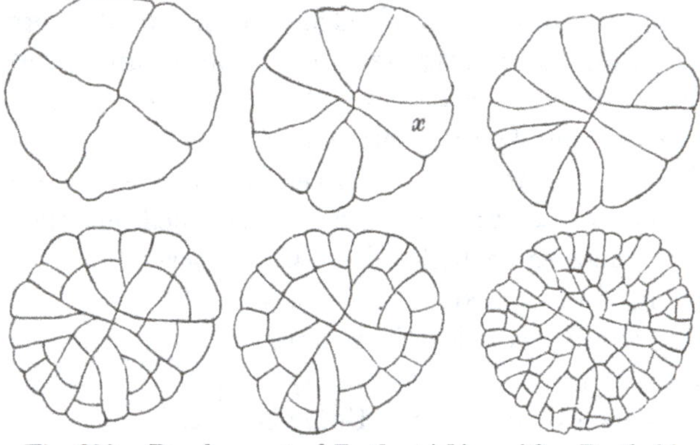

Fig. 230. Development of *Erythrotrichia*. After Berthold.

will be small, and to all intents and purposes we may deal with the case as though we had now to do with four equal cells, each one of them a perfect quadrant; so our problem is, to find the shortest line which shall divide the quadrant of a circle into two halves of

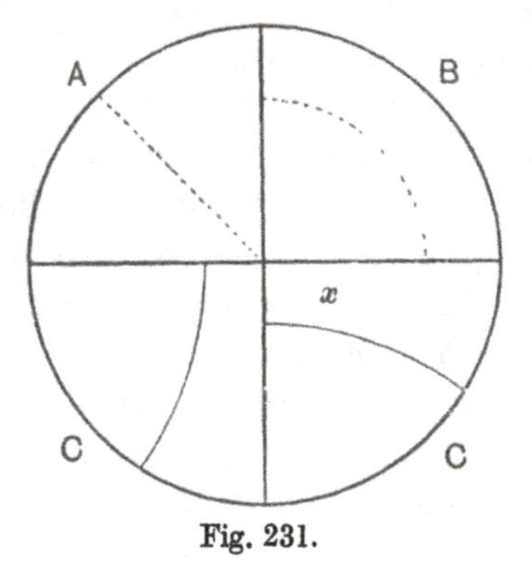

Fig. 231.

equal area. A radial partition (Fig. 231, A), starting from the apex of the quadrant, is at once excluded, for the reason just referred to; our choice must lie between two modes of division such as are illustrated in Fig. 231, where the partition is either (as in B) concentric with the outer border of the cell, or else (as in C) cuts that outer border; in other words, our partition may (B) cut *both* radial walls, or (C) may cut *one* radial wall and the periphery. These are the two methods of division which Sachs called, respectively, (B) *periclinal*, and (C) *anticlinal**. We may either treat the walls of the dividing quadrant as already solidified, or at least as having a tension compared with which that of the incipient partition film is inconsiderable; in either case the new partition must meet the old wall, on either side, at right angles, and (its own tension and curvature being everywhere uniform) must take the form of a circular arc.

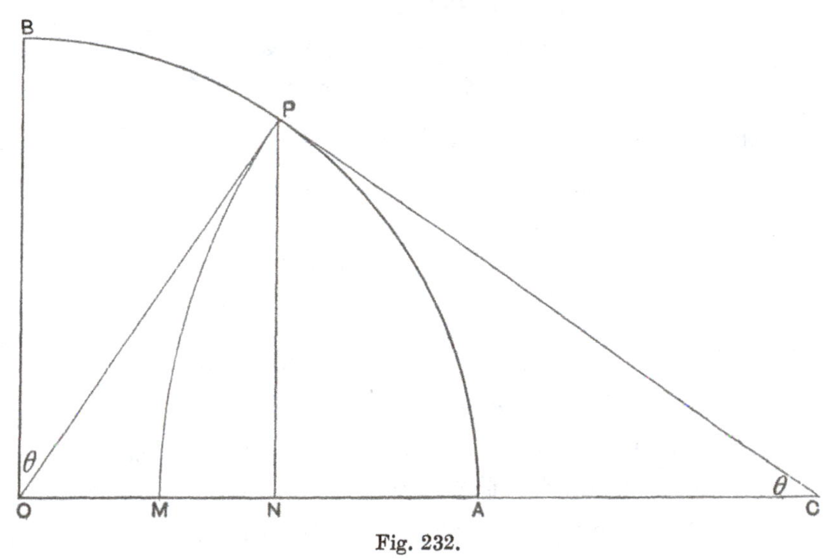

Fig. 232.

We find that a flattened cell which is approximately a quadrant of a circle invariably divides after the manner of Fig. 231, C, that is to say, by an approximately circular, *anticlinal* wall, and this we now recognise in the eight-celled stage of *Erythrotrichia* (Fig. 230); let us then consider that Nature has solved our problem, and let us work out the actual geometric conditions.

Let the quadrant OAB (in Fig. 232) be divided into two parts of equal area, by the circular arc MP. It is required to determine

* There is, I think, some ambiguity or disagreement among botanists as to the use of this latter term: the sense in which I am using it, viz. for any partition which meets the outer or peripheral wall at right angles (the strictly *radial* partition being for the present excluded), is, however, clear.

(1) the position of P upon the arc of the quadrant, that is to say the angle BOP; (2) the position of the point M on the side OA; and (3) the length of the arc MP in terms of a radius of the quadrant.

(1) Draw OP; also PC a tangent, meeting OA in C; and PN, perpendicular to OA. Let us call a a radius; and θ the angle at C, which is equal to OPN, or POB. Then

$$CP = a \cot \theta;\ PN = a \cos \theta;\ NC = CP \cos \theta = a \,.\, \cos^2 \theta / \sin \theta.$$

The area of the portion PMN

$$\begin{aligned} &= \tfrac{1}{2} CP^2 \,\theta - \tfrac{1}{2} PN \,.\, NC \\ &= \tfrac{1}{2} a^2 \,\theta \cot^2 \theta - \tfrac{1}{2} a \cos \theta \,.\, a \cos^2 \theta / \sin \theta \\ &= \tfrac{1}{2} a^2 (\theta \cot^2 \theta - \cos^3 \theta / \sin \theta). \end{aligned}$$

And the area of the portion PNA

$$\begin{aligned} &= \tfrac{1}{2} a^2 (\pi/2 - \theta) - \tfrac{1}{2} ON \,.\, NP \\ &= \tfrac{1}{2} a^2 (\pi/2 - \theta) - \tfrac{1}{2} a \sin \theta \,.\, a \cos \theta \\ &= \tfrac{1}{2} a^2 (\pi/2 - \theta - \sin \theta \,.\, \cos \theta). \end{aligned}$$

Therefore the area of the whole portion PMA

$$\begin{aligned} &= a^2/2\,(\pi/2 - \theta + \theta \cot^2 \theta - \cos^3 \theta/\sin \theta - \sin \theta \,.\, \cos \theta) \\ &= a^2/2\,(\pi/2 - \theta + \theta \cot^2 \theta - \cot \theta), \end{aligned}$$

and also, by hypothesis, $= \tfrac{1}{2}$. area of the quadrant, $= \pi a^2/8$.

Hence θ is defined by the equation

$$a^2/2\,(\pi/2 - \theta + \theta \cot^2 \theta - \cot \theta) = \pi a^2/8,$$

or

$$\pi/4 - \theta + \theta \cot^2 \theta - \cot \theta = 0.$$

We may solve this equation by constructing a table (of which the following is a small portion) for various values of θ.

θ	$\pi/4$	$-\theta$	$-\cot\theta$	$+\theta\cot^2\theta$	$=x$
34° 34′	0·7854	− 0·6033	− 1·4514	+ 1·2709 =	0·0016
35′	0·7854	0·6036	1·4505	1·2700	0·0013
36′	0·7854	0·6039	1·4496	1·2690	0·0009
37′	0·7854	0·6042	1·4487	1·2680	0·0005
38′	0·7854	0·6045	1·4478	1·2671	0·0002
39′	0·7854	0·6048	1·4469	1·2661	−0·0002
40′	0·7854	0·6051	1·4460	1·2652	−0·0005

We see accordingly that the equation is solved (as accurately as need be) when θ is an angle somewhat over 34° 38′, or say

34° 38½′. That is to say, a quadrant of a circle is bisected by a circular arc cutting the side and the periphery of the quadrant at right angles, when the arc is such as to include (90° − 34° 38′), i.e. 55° 22′ of the quadrantal arc. This determination of ours is practically identical with that which Berthold arrived at by a rough and ready method, without the use of mathematics. He simply tried various ways of dividing a quadrant of paper by means of a circular arc, and went on doing so till he got the weights of his two pieces of paper approximately equal. The angle, as he thus determined it, was 34·6°, or say 34° 36′.

(2) The position of M on the side of the quadrant OA is given by the equation $OM = a \operatorname{cosec} \theta - a \cot \theta$; the value of which expression, for the angle which we have just discovered, is 0·3028. That is to say, the radius (or side) of the quadrant will be divided by the new partition into two parts, in the proportions, nearly, of three to seven.

(3) The length of the arc MP is equal to $a\theta \cot \theta$; and the value of this for the given angle is 0·8751. This is as much as to say that the curved partition-wall which we are considering is shorter than a radial partition in the proportion of 8¾ to 10, or seven-eighths, almost exactly.

But we must also compare the length of this curved anticlinal partition-wall (MP) with that of the concentric, or periclinal, one (RS, Fig. 233) by which the quadrant might also be bisected. The length of this partition is obviously equal to the arc of the quadrant (i.e. the peripheral wall of the cell) divided by $\sqrt{2}$; or, in terms of the radius, $= \pi/2\sqrt{2} = 1{\cdot}111$. So that, not only is the anticlinal partition (such as we actually find in nature) notably the best, but the periclinal one, when it comes to dividing an entire quadrant, is very considerably larger even than a radial partition.

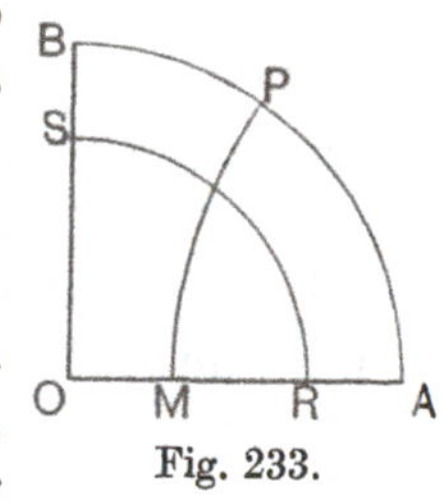

Fig. 233.

The two cells into which our original quadrant is now divided are equal in volume, but of very different shapes; the one is a triangle (MAP) with circular arcs for two of its sides, and the other is a four-sided figure ($MOBP$), which we may call approximately oblong. How will they continue to divide? We cannot say as yet how the triangular portion ought to divide; but it is obvious that the least possible partition-wall which shall bisect the other must run across the long axis of the oblong, that is to say periclinally. This is precisely what tends actually to take place. In the following diagrams (Fig. 234) of a frog's egg dividing under pressure, that is to say when reduced to the form of a flattened plate, we see, firstly (A), the division into four quadrants (by the partitions 1, 2); secondly (B), the division of each quadrant by means of an anticlinal circular arc (3, 3), cutting the peripheral wall of the quadrant approximately in the proportions of three to seven; and thirdly

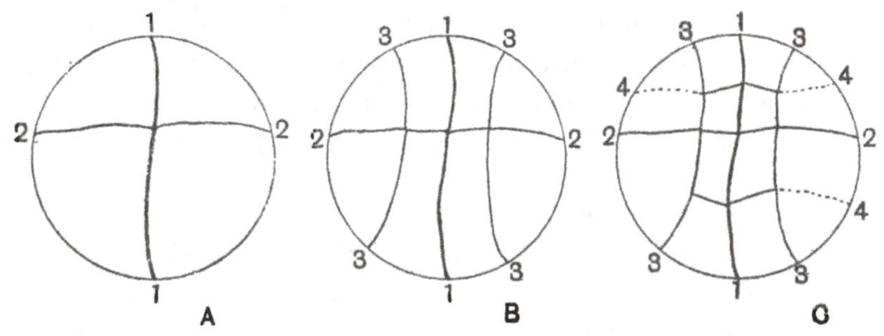

Fig. 234. Segmentation of frog's egg, under artificial compression. After Roux.

(C), we see that of the eight cells (four triangular and four oblong) into which the whole egg is now divided, the four which we have called oblong now proceed to divide by partitions transverse to their long axes, or roughly parallel to the periphery of the egg.

The question how the other, or triangular, portion of the divided quadrant will next divide leads us to a well-defined problem which is only a slight extension, making allowance for the circular arcs, of that elementary problem of the triangle we have already considered. We know now that an entire quadrant (in order that its bisecting wall shall have the least possible area) must divide by means of an anticlinal partition, but how about any smaller sectors of circles? It is obvious in the case of a small prismatic sector, such as that shewn in Fig. 235, that a *periclinal* partition is the least by which we can bisect the cell; we want, accordingly, to know the limits below which the periclinal partition is always the best, and above which the anticlinal arc has the advantage, as in the case of the whole quadrant.

This may be easily determined; for the preceding investigation is a perfectly general one, and the results hold good for sectors of any other arc, as well as for the quadrant, or arc of 90°. That is to say, the length of the partition-wall MP is always determined by the angle θ, according to our equation $MP = a\theta \cot \theta$; and the angle θ has a definite relation to α, the angle of arc.

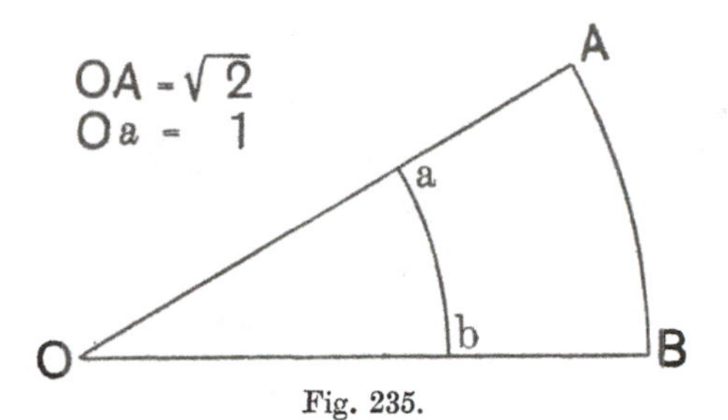

Fig. 235.

Moreover, in the case of the periclinal boundary, RS (Fig. 233) (or ab, Fig. 235), we know that, if it bisects the cell,

$$RS = a \cdot \alpha/\sqrt{2}.$$

Accordingly, the arc RS will be just equal to the arc MP when

$$\theta \cot \theta = \alpha/\sqrt{2}.$$

When $\theta \cot \theta > \alpha/\sqrt{2}$, or $MP < RS$,

then division will take place as in RS, or periclinally.

When $\theta \cot \theta < \alpha/\sqrt{2}$, or $MP > RS$,

then division will take place as in MP, or anticlinally.

In the accompanying diagram (Fig. 236), I have plotted the various magnitudes with which we are concerned, in order to exhibit the several limiting values. Here we see, in the first place, the curve marked α, which shews on the (left-hand) vertical scale the various possible magnitudes of that angle (viz. the angle of arc of the whole sector which we wish to divide), and on the horizontal scale the corresponding values of θ, or the angle which determines the point on the periphery where it is cut by the partition-wall,

MP. Two limiting cases are to be noticed here: (1) at 90° (point *A* in diagram), because we are at present only dealing with arcs no greater than a quadrant; and (2), the point (*B*) where the angle θ

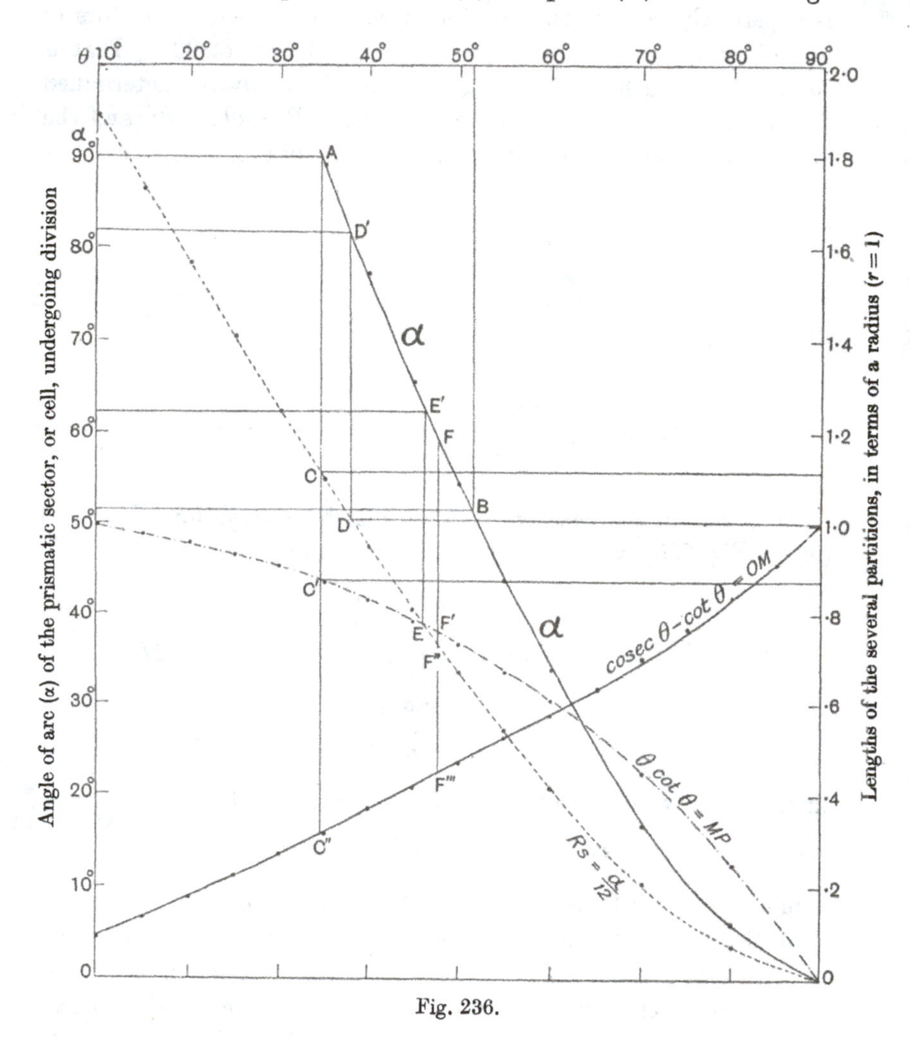

Fig. 236.

comes to equal the angle α, for after that point the construction becomes impossible, since an anticlinal bisecting partition-wall would be partly outside the cell. The only partition which, after that point, can possibly exist is a periclinal one; and this point, as our diagram shews us, occurs when the two angles (α and θ) are both rather under 52°.

Next I have plotted on the same diagram, and in relation to the same scale of angles, the corresponding lengths of the two partitions, viz. RS and MP, their lengths being expressed (on the right-hand side of the diagram) in terms of the radius of the circle (a), that is to say the side wall, OA, of our cell.

The limiting values here are (1), C, C', where the angle of arc is 90°, and where, as we have already seen, the two partition-walls have the relative magnitudes of $MP : RS = 0{\cdot}875 : 1{\cdot}111$: (2) the point D, where RS equals unity, that is to say where the periclinal partition has the same length as a radial one; this occurs when α is rather under 82° (cf. the points D, D'): (3) the point E, where RS and MP intersect, that is to say the point at which the two partitions, periclinal and anticlinal, are of the same magnitude; this is the case, according to our diagram, when the angle of arc is just over $62\frac{1}{2}$°. We see from this that what we have called an anticlinal partition, as MP, is only likely to occur in a triangular or prismatic cell whose angle of arc lies between 90° and $62\frac{1}{2}$°; in all narrower or more tapering cells the periclinal partition will be of less area, and will therefore be more and more likely to occur.

The case (F) where the angle α is just 60° is of some interest. Here, owing to the curvature of the peripheral border, and the consequent fact that the peripheral angles are somewhat greater than the apical angle α, the periclinal partition has a very slight and almost imperceptible advantage over the anticlinal, the relative proportions being about as $MP : RS :: 0{\cdot}73 : 0{\cdot}72$. But if the triangle be a plane equiangular triangle, bounded by circular arcs, then we see that there is no longer any distinction at all between our two partitions; MP and RS are now identical.

On the same diagram, I have inserted the curve for values of $\operatorname{cosec}\theta - \cot\theta = OM$, that is to say the distances from the centre, along the side of the cell, of the starting-point (M) of the anticlinal partition. The point C'' represents its position in the case of a quadrant, and shews it to be (as we have already said) about 3/10 of the length of the radius from the centre. If on the other hand our cell be an equilateral triangle, then we have to read off the point on this curve corresponding to $\alpha = 60°$; and we find it at the point F'''' (vertically under F), which tells us that the partition now starts 45/100, or nearly halfway, along the radial wall.

The foregoing considerations carry us a long way in our investigation of the simpler forms of cell-division. Strictly speaking they are limited to the case of flattened cells, in which we can treat the problem as though we were partitioning a plane surface. But it is obvious that, though they do not teach us the whole conformation of the partition which divides a more complicated solid into two halves, yet, even in such a case they so far enlighten us as to tell us the appearance presented in one plane of the actual solid. And, as this is all that we see in a microscopic section, it follows that the results we have arrived at will help us greatly in the interpretation of microscopic appearances, even in comparatively complex cases of cell-division.

Let us now return to our quadrant cell ($OAPB$), which we have found to be divided into a triangular and a quadrilateral portion, as in Figs. 233 or 237; and let us now suppose the whole system to grow, in a uniform fashion, as a prelude to further subdivision. The whole quadrant, growing uniformly (or with equal radial increments), will still remain a quadrant, and it is obvious, therefore, that for every new increment of size, more will be added to the margin of its triangular portion than to the narrower margin of the quadrilateral; and the increments will be in proportion to the angles of arc, viz. 55° 22′ : 34° 38′, or as 0·96 : 0·60, i.e. as 8 : 5. Accordingly, if we may assume (and the assumption is a very plausible one), that, just as the quadrant itself divided into two halves after it got to a certain size, so each of its two halves will reach the same size before again dividing, it is obvious that the triangular portion will be doubled in size, and therefore ready to divide, a considerable time before the quadrilateral part. To work out the problem in detail would lead us into troublesome mathe-

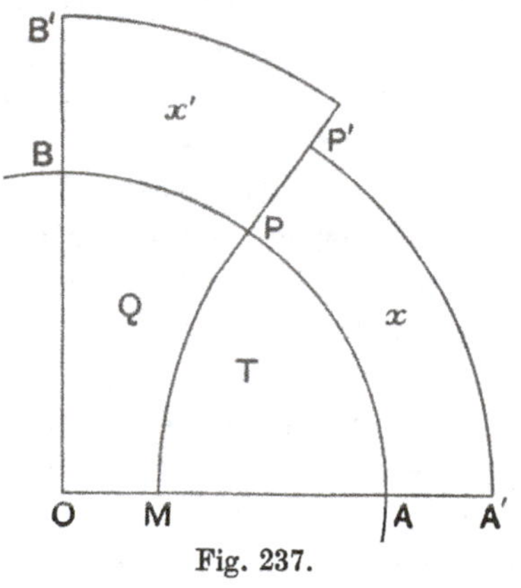

Fig. 237.

matics; but if we simply assume that the increments are proportional to the increasing radii of the circle, we have the following equations:

Call the triangular cell T, and the quadrilateral Q (Fig. 237); let the radius, OA, of the original quadrantal cell $= a = 1$; and let the increment which is required to add on a portion equal to T (such as $PP'A'A$) be called x, and let that required, similarly, for the doubling of Q be called x'.

Then we see that the area of the original quadrant

$$= T + Q = \tfrac{1}{4}\pi a^2 = 0{\cdot}7854a^2,$$

while the area of T $= Q = 0{\cdot}3927a^2$.

The area of the enlarged sector, $P'OA'$,

$$= (a + x)^2 \times (55^\circ\ 22') \div 2 = 0{\cdot}4831\ (a + x)^2,$$

and the area OPA

$$= a^2 \times (55^\circ\ 22') \div 2 = 0{\cdot}4831a^2.$$

Therefore the area of the added portion, T',

$$= 0{\cdot}4831\ \{(a + x)^2 - a^2\}.$$

And this, by hypothesis,

$$= T = 0{\cdot}3927a^2.$$

We get, accordingly, since $a = 1$,

$$x^2 + 2x = 0{\cdot}3927/0{\cdot}4831 = 0{\cdot}810,$$

and, solving,

$$x + 1 = \sqrt{1{\cdot}81} = 1{\cdot}345, \text{ or } x = 0{\cdot}345.$$

Working out x' in the same way, we arrive at the approximate value, $x' + 1 = 1{\cdot}517$.

This is as much as to say that, supposing each cell tends to divide into two halves when (and not before) its original size is doubled, then, in our flattened disc, the triangular cell T will tend to divide when the radius of the disc has increased by about a third (from 1 to 1·345), but the quadrilateral cell, Q, will not tend to divide until the linear dimensions of the disc have increased by fully a half (from 1 to 1·517).

The case here illustrated is of no small importance. For it shews us that a uniform and symmetrical growth of the organism (symmetrical, that is to say, under the limitations of a plane surface, or plane section) by no means involves a uniform or symmetrical growth of the individual cells, but may under certain conditions actually lead to inequality among these; and this phenomenon (or to be quite candid, this hypothesis, which is due to Berthold) is independent of any change or variation in surface tensions, and is essentially different from that unequal segmentation (studied by Balfour) to which we have referred on p. 568.

After this fashion we might go on to consider the manner, and the order of succession, in which subsequent cell-divisions should tend to take place, as governed by the principle of minimal areas.

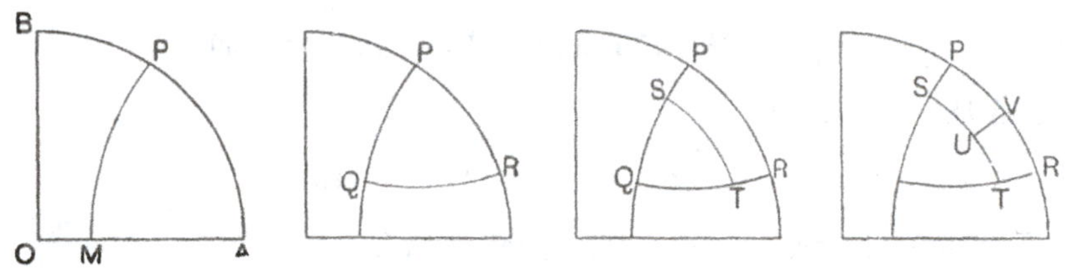

Fig. 238.

The calculations would grow more difficult, and the results got by simple methods would grow less and less exact; at the same time some of the results would be of great interest, and well worth our while to obtain. For instance, the precise manner in which our triangular cell, T, would next divide would be interesting to know, and a general solution of this problem is certainly troublesome to calculate. But in this particular case we see that the width of the triangular cell near P (Fig. 238) is so obviously less than that near either of the other two angles, that a circular arc cutting off that angle is bound to be the shortest possible bisecting line; and that, in short, our triangular cell will tend to subdivide, just like the original quadrant, into a triangular and a quadrilateral portion.

But the case will be different next time, because in this new triangle, PRQ, the least width is near the innermost angle, that at Q; and the bisecting circular arc will therefore be opposite to Q, or (approximately) parallel to PR. The importance of this fact is at once evident; for it means to say that there comes a time when, whether by the division of triangles or of quadrilaterals, we find only quadrilateral cells adjoining the periphery of our circular disc. In the subsequent division of these quadrilaterals, the partitions will arise transversely to their long axes, that is to say, *radially* (as U, V); and we shall consequently have a superficial or peripheral layer of quadrilateral cells, with sides approximately parallel, that is to say what we are accustomed to call *an epidermis*. And this epidermis or superficial layer will be in clear contrast with the more irregularly shaped cells, the products of triangles and quadrilaterals, which make up the deeper, underlying layers of tissue.

In following out these theoretic principles, and others like to them, in the actual division of living cells, we must bear in mind certain conditions and qualifications. In the first place, the law of minimal area and the other rules which we have arrived at are not absolute but relative: they are links, and very important links, in a chain of physical causation; they are always at work, but their effects may be overridden and concealed by the operation of other forces. Secondly, we must remember that, in most cases, the cell-system which we have in view is constantly increasing in magnitude by active growth; and by this means the form and also the proportions of the cells are continually altering, of which phenomenon we have already had an example. Thirdly, we must carefully remember that, until our cell-walls become absolutely solid and rigid, they are always apt to be modified in form owing to the tension of the adjacent walls; and again, that so long as our partition films are fluid or semifluid, their points and lines of contact with one another may shift, like the shifting outlines of a system of soap-bubbles. This is the physical cause of the movements frequently seen among segmenting cells, like those to which Rauber called attention in the segmenting ovum of the frog, and like those more striking movements or accommodations which give rise to a so-called "spiral" type of segmentation.

Bearing in mind these considerations, let us see what our flattened disc is likely to look like, after a few successive cell-divisions. In Fig. 239 *a*, we have a diagrammatic representation of our disc, after it has divided into four quadrants, and each quadrant into a triangular and a quadrilateral portion; but as yet, this figure has scarcely anything like the normal look of an aggregate of living cells. But let us go a little further, still limiting ourselves to the consideration of the eight-celled stage. Wherever one of our radiating partitions meets the peripheral wall, there will (as we know) be a mutual tension between the three convergent films, which will tend to set their edges at equal angles to one another, angles that is to say of 120°. In consequence of this, the outer wall of each individual cell will (in this surface view of our disc) be an arc of a circle of which we can determine the centre by the method used on p. 485; and, furthermore, the narrower cells, that is to say the quadrilaterals, will have this outer border somewhat more

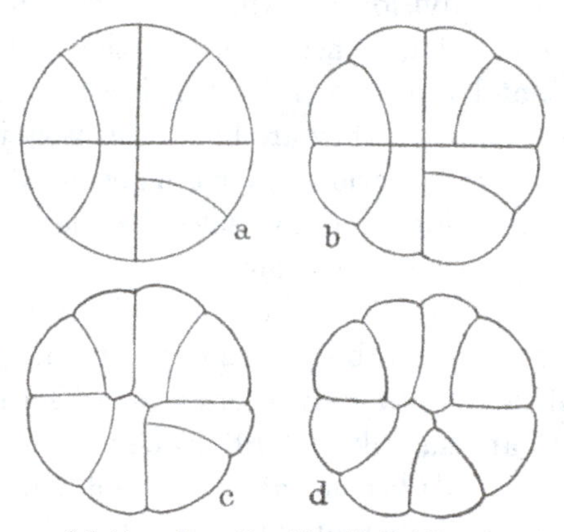

Fig. 239. Diagram of flattened or discoid cell dividing into octants: to shew gradual tendency towards a position of equilibrium.

curved than their broader neighbours. We arrive, then, at the condition shewn in Fig. 239 *b*. Within the cell, also, wherever wall meets wall, the angle of contact must tend, in every case, to be an angle of 120°; in no case may more than three films (as seen in section) meet in a point (*c*); and this condition, of the partitions meeting three by three and at co-equal angles, will involve the curvature of some, if not all, of the partitions (*d*) which to begin with we treated as plane. To solve this problem in a general way is no easy matter; but it is a problem which Nature solves in every case where, as in the case we are considering, eight bubbles or eight cells meet together in a plane or curved surface. An approximate solution has been given in *d*; and it will at once be recognised that this figure has vastly more resemblance to an aggregate of living cells than had the diagram of *a*, with which we began.

Just as we have constructed in this case a series of purely diagrammatic or schematic figures, so will it be possible as a rule to diagrammatise, with but little alteration, the complicated appearances presented by any ordinary aggregate of cells. The accompanying little figure (Fig. 240), of a germinating spore of a Liverwort (*Riccia*), after a drawing of D. H. Campbell's, scarcely needs further explanation: for it is well-nigh a typical diagram of the method of space-partitioning which we are now considering. The same is equally true of any one of Hanstein's figures of the hairs on a leaf-bud*, or Berthold's of the small discoid algae. Let us look again at our figures of *Erythrotrichia* or *Chaetopeltis* from Berthold's *Monograph*, and redraw some of the earlier stages.

Fig. 240.

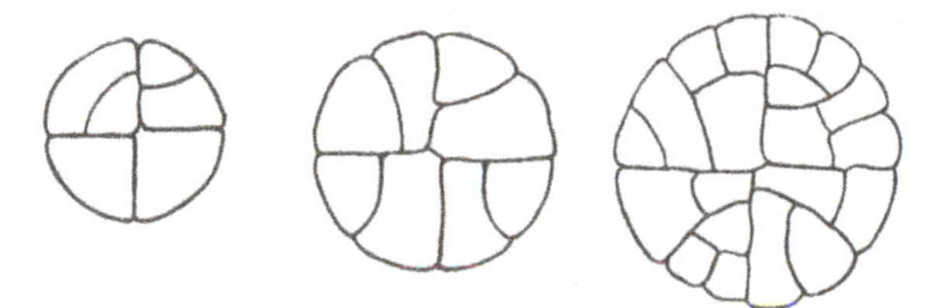

Fig. 241. Embryo-stages of *Chaetopeltis orbicularis*. After Berthold.

In the following diagrams (Fig. 242) the new partitions, or those just about to form, are in each case outlined; and in the next succeeding stage they are shewn after settling down into position, and after exercising their respective tractions on the walls previously laid down. It is clear, I think, that these four diagrammatic figures represent all that is shewn in the first five stages drawn by Berthold from the plant itself; but the correspondence cannot in this case be precisely accurate, for the reason that Berthold's figures are taken from different individuals, and so are not strictly and consecutively continuous. The last of the six drawings in Fig. 230 is already too complicated for diagrammatisation, that is to say it is too complicated for us to decipher with certainty the order of appearance of the numerous partitions which it contains. But in Fig. 243 I shew one more diagrammatic figure, of a disc which has

* *Bot. Zeitung*, XXVI, p. 11, xi, xii, 1868.

divided, according to the theoretical plan, into about sixty-four cells; and making due allowance for the changes which mutual tensions and tractions bring about, increasing in complexity with each succeeding stage, we can see, even at this advanced and

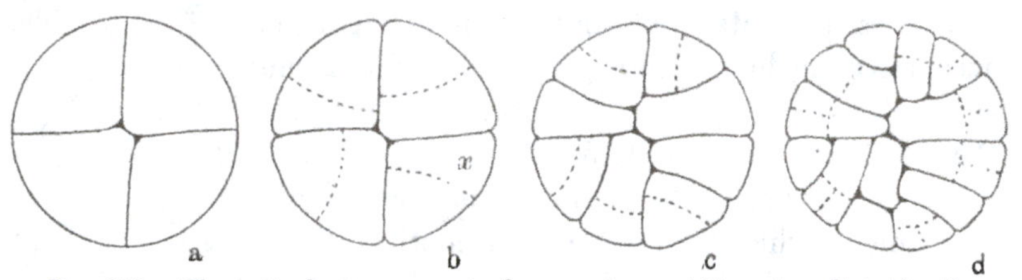

Fig. 242. Theoretical arrangement of successive partitions in a discoid cell; for comparison with Figs. 230 and 241.

complicated stage, a very considerable resemblance between the actual picture and the diagram which we have here constructed in obedience to a few simple rules.

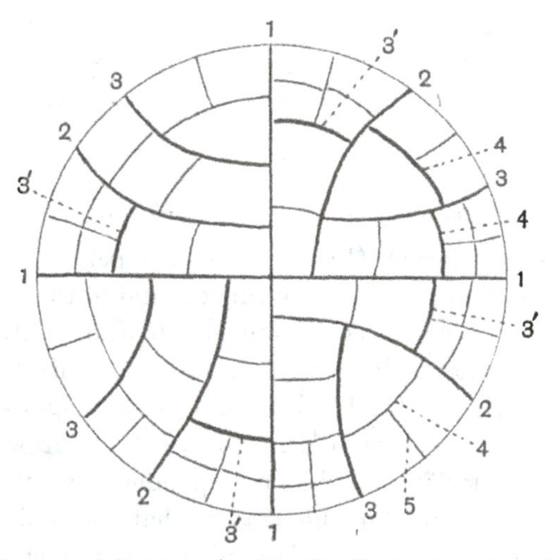

Fig. 243. Theoretical division of a discoid cell into sixty-four chambers: no allowance being made for the mutual tractions of the cell-walls.

In like manner, in the annexed figures representing sections through a young embryo of a moss, we have little difficulty in discerning the successive stages which must have intervened between the two stages shewn: so as to lead from the just divided or dividing quadrants (*a*), to the stage (*b*) in which a well-marked epidermal layer surrounds an at first sight irregular agglomeration of "fundamental tissue".

In the last paragraph but one, I have spoken of the difficulty of so arranging the meeting-places of a number of cells that at each junction only three cell-walls shall meet in a point, and all three shall meet at equal angles of 120°. As a matter of fact, the problem is soluble in a number of ways; that is to say, when we have a number of cells enclosed in a common boundary, say eight as in the case considered, there are various ways in which their walls may meet internally, three by three, at equal angles; and these differences will entail differences also in the curvature of the walls, and consequently in the shape of the cells. The question is some-

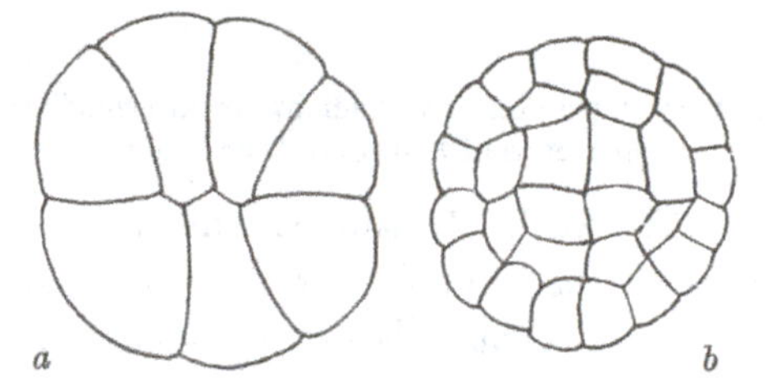

Fig. 244. Sections of embryo of a moss. After Kienitz-Gerloff.

what complex; it has been dealt with by Plateau, and treated mathematically by M. Van Rees*.

If within our boundary we have only three cells all meeting internally, they must meet in a point; furthermore, they tend to do so at equal angles of 120°, and there is an end of the matter. If we have four cells, then, as we have already seen, the conditions are satisfied by interposing a little intermediate wall, the two extremities of which constitute the meeting-points of three cells each, and the upper edge of which marks the "polar furrow." In the case of five cells, we require *two* little intermediate walls, and two polar furrows; and we soon arrive at the rule that, for n cells, we require $n - 3$ little longitudinal partitions (and corresponding polar furrows), connecting the triple junctions of the cells†; and these little walls, like all the rest within the system, tend to

* *Cit.* Plateau, *Statique des Liquides*, I, p. 358.

† There is an obvious analogy between this rule for the number of internal partitions within a polygonal system, and Lamarle's rule for the number of "free films" within a polyhedron. *Vide supra*, p. 550.

incline to one another at angles of 120°. Where we have only one such wall (in the case of four cells), or only two (in the case of five cells), there is no room for ambiguity. But where we have three little connecting-walls, in the case of six cells, we can arrange them in three different ways, as Plateau* found his six

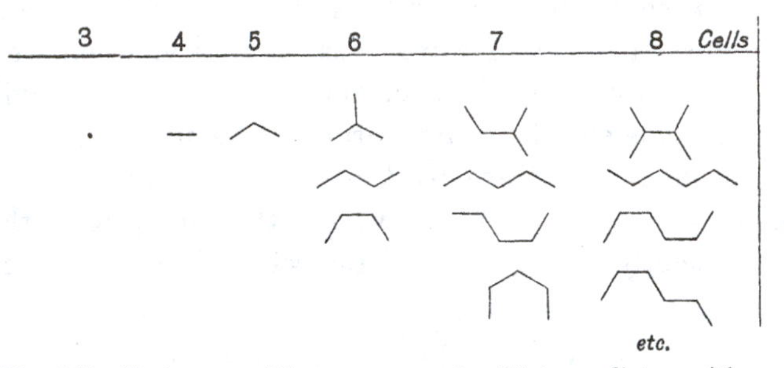

Fig. 245. Various possible arrangements of intermediate positions, in groups of 3, 4, 5, 6, 7 or 8 cells.

soap-films to do (Fig. 246). In the system of seven cells, the four partitions can be arranged in four ways; and the five partitions required in the case of eight cells can be arranged in no less than

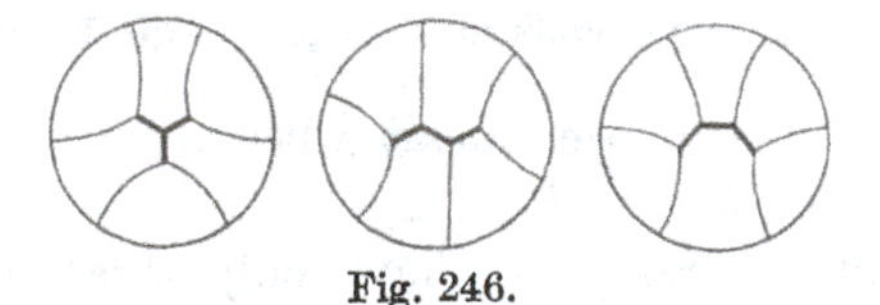

Fig. 246.

twelve different ways†. It does not follow that these various arrangements are all equally good; some are known to be more stable than others, and some are hard to realise in actual experiment.

Examples of these various arrangements meet us at every turn, in all sorts of partitioning, whether there be actual walls or mere

* Plateau experimented with a wire frame or "cage", in the form of a low hexagonal prism. When this was plunged in soap-solution and withdrawn upright, a vertical film occupied its six quadrangular sides and nothing more. But when it was drawn out sideways, six films starting from the six vertical edges met somehow in the middle, and divided the hexagon into six cells. Moreover the partition-films automatically solved the problem of meeting one another three-by-three, at co-equal angles of 120°; and did so in more ways than one, which could be controlled more or less, according to the manner and direction of lifting the cage.

† Plateau, on Van Rees's authority, says thirteen; but this is wrong—unless he meant to include the case where one cell is wholly surrounded by the seven others.

rifts and cracks in a broken surface. The phenomenon is in the first instance mathematical, in the second physical; and the limited number of possible arrangements appear and reappear in the most diverse fields, and are capable of representation by the same diagrams. We have seen in Fig. 196 how the cracks in drying mud exhibit to perfection the polar furrow joining two three-way nodes, which is the characteristic feature of the four-celled stage of a segmenting egg.

The possible arrangements of the intermediate partitions becomes a question of permutations. Let us call the flexure between two consecutive furrows *a* or *b*, according to its direction, right or left; and let a triple conjunction be called *c*. Then the three possible arrangements in a system of six cells are *aa*, *ab*, *c*; the four in a system of seven cells are *aaa*, *aab*, *aba*, *ac*; and the twelve possible arrangements in a system of eight cells are as follows*:

a	*aac*	*f*	*aabb*
b	*abc*	*g*	*aaba*
c	*acb*	*h*	*aaab*
d	*aca*	*i*	*aaaa*
e	*bcb*	*j*	*abab*
		k	*abba*
	l	*cc*	

We may classify, and may denote or symbolise, these several arrangements in various ways. In the following table we see: A, the twelve arrangements of the five intermediate partitions which are necessary to enable all the boundary walls of a plane assemblage of eight cells (none being "insular") to meet in three-way junctions; B, the literal permutations which symbolise the same; C, the number of sides (other than the external boundary) which in each case each cell possesses, i.e. the number of contacts each makes with its neighbours. The total number of contacts (as we shall see presently) is 26, and the mean number 3·25; if we take the departures from the mean, and sum them irrespective of sign, the sum is shewn under D.

* I believe that Kirkman, in a paper of more than 80 years ago, said that the number of 8-sided convex, Eulerian polyhedra, with trihedral corners, was thirteen.

	A	B	C	D
a		*aac*	222 33 44 6	8·5
e		*aca*	222 33 44 6	8·5
c	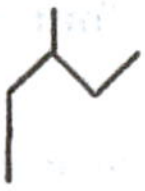	*acb*	222 33 4 55	8·5
d	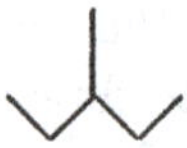	*bcb*	222 33 4 55	8·5
b	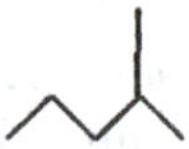	*abc*	222 3 444 5	8·0
j		*abab*	22 33 4444	6·0
f		*aabb*	22 3333 55	7·0
g		*aaba*	22 333 44 5	6·5
k		*abba*	22 333 44 5	6·5
h		*aaab*	22 3333 4 6	7·0
i		*aaaa*	22 33333 7	7·5
l		*cc*	2222 44 55	10·0

Nine cells may be arranged in twenty-seven ways. In higher series the numbers increase very rapidly, but the cells will tend to overlap, and so introduce a new complication*.

We may draw help from the theory of polyhedra (in an elementary way) if we treat our group of eight cells (none of them "insular") as part of a polyhedron, to be completed by one eight-sided cell,

* Max Brückner states the number of possible arrangements of thirteen cells, with trihedral junctions, as nearly 50,000; of sixteen, nearly 30 millions; and of eighteen, "bereits über einige Billionen" (*Proc. Math. Congress*, Bologna, 1930, vol. IV, p. 11). It is plain that the study of "cell-lineage," or the mapping out in detail of the cell-arrangements after repeated cell-divisions, is only possible under severe limitations.

serving (so to speak) as a base under all the rest. Then, calling F_3, F_4, etc., the number of three-sided and four-sided facets, we reclassify our twelve configurations as follows:

Polyhedral arrangements of eight cells (*none of them "insular"*), *considered as part of a nine-faced polyhedron, whose ninth face is octagonal.*

	F_3	F_4	F_5	F_6	F_7	F_8
l	4	—	2	2	—	1
ae	3	2	2	—	1	1
cd	3	2	1	2	—	1
b	3	1	3	1	—	1
i	2	5	—	—	—	2
h	2	4	1	—	1	1
f	2	4	—	2	—	1
gk,	2	3	2	1	—	1
j	2	2	4	—	—	1

It is of interest, and of more than mere mathematical interest, to know, not only that these possible arrangements are few, but that they are strictly defined as to the number and form of the respective faces. For we know that we are limited to three-way corners or nodes; and, that being so, the following simple rule holds for the facets—a rule which we shall use later on in still more curious circumstances, and which may be easily verified in any line of the foregoing table:

$$3F_3 + 2F_4 + F_5 \pm 0 . F_6 - F_7 - 2F_8 - \text{etc.} = 12.$$

We may produce and illustrate all these configurations by blowing bubbles in a dish and here (Fig. 247) is the complete series, up to seven cells. They correspond precisely to the diagrams shewn on p. 596, and their resemblance to embryological diagrams is only cloaked a little by the circular outline, the artificial boundary of the system. Of the twelve eight-celled arrangements, four seem unstable; these include the one case (*i*) where one cell of the eight is in contact with all seven others, and the three cases (*a*, *e*, *h*) where one is in contact with six others. The reason of this instability is, I imagine, that the internal angles cannot be angles of 120°, as equilibrium demands, unless the sides be curved, and convex inwards; but this implies a combined pressure from without on the large cell in the middle. While it adjusts its walls, then, to the required angles, the large cell tends to close up, to lose hold of the boundary of the system, and to become an island-cell entirely surrounded by the rest.

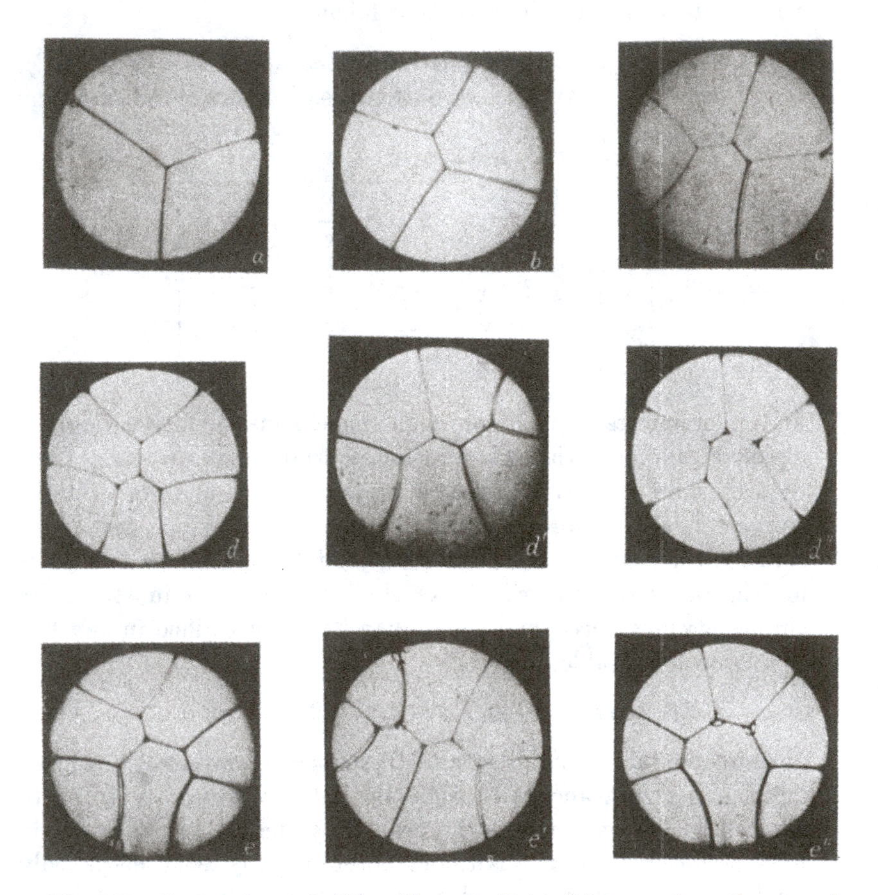

Fig. 247. Group of soap-bubbles, blown in Petri dishes. *a*, *b*, *c*, the normal partitioning of groups of three, four or five cells or bubbles. *d*, the three ways of partitioning a group of six cells or bubbles. *e*, three of the four ways of partitioning a group of seven cells.

Among the published figures of embryonic stages and other cell aggregates, we only discern the little intermediate partitions in cases where the investigator has drawn carefully just what lay before him, without any preconceived notions as to radial or other symmetry; but even in other cases we can often recognise, without much difficulty, what the actual arrangement was whereby the cell-walls met together in equilibrium. I suspect that a leaning towards Sachs's Rule, that one cell-wall tends to set itself at right angles to another cell-wall (a rule whose strict limitations and narrow range of application we have already considered) is responsible for many inaccurate or incomplete representations of the mutual arrangement of associated cells.

In the accompanying series of figures (Figs. 248–255) I have set forth a few aggregates of eight cells, mostly from drawings of

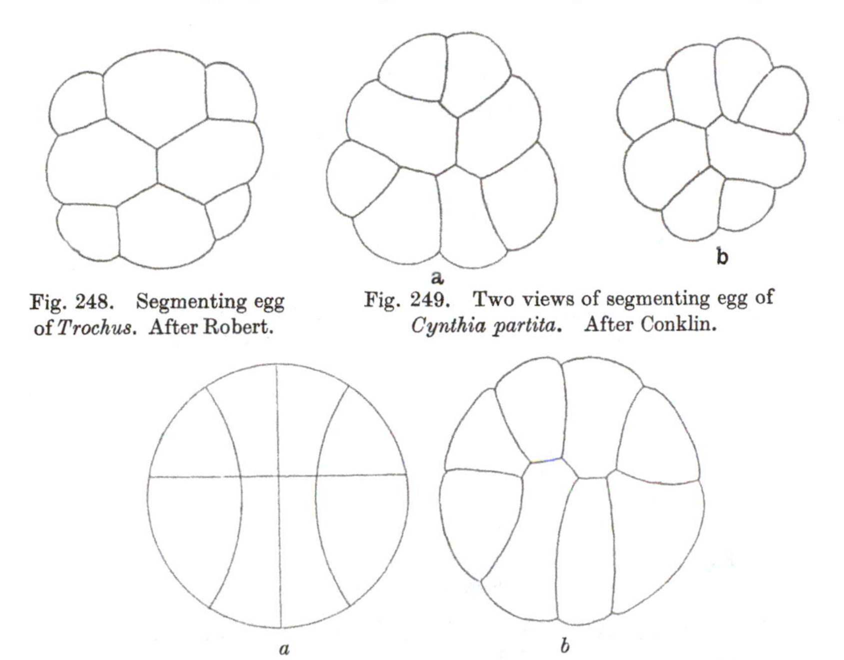

Fig. 248. Segmenting egg of *Trochus*. After Robert.

Fig. 249. Two views of segmenting egg of *Cynthia partita*. After Conklin.

Fig. 250. (*a*) Section of apical cone of *Salvinia*. After Pringsheim*. (*b*) Diagram of probable actual arrangement.

segmenting eggs. In some cases they shew clearly the manner in which the cell-walls meet one another, always by three-way junctions, at angles of about 120°, more or less, and always with the help of five intermediate boundary walls within the eight-celled system; in other cases I have added a slightly altered drawing, so as to shew, with as little change as possible, the arrangement of boundaries

* This, like many similar figures, is manifestly drawn under the influence of Sachs's theoretical views, or assumptions, regarding orthogonal trajectories, coaxial circles, confocal ellipses, etc.

which may have existed, and given rise to the appearance which the observer drew. These drawings may be compared with the diagrams on p. 598, in which the twelve possible arrangements of five intermediate partitions for a system of eight cells have been set forth.

It will be seen that Robert-Tornow's figure of the segmenting egg of *Trochus* (Fig. 248) clearly shews the cells grouped after the fashion of *l*; while Conklin's figure of the ascidian egg (*Cynthia*) shews equally clearly the arrangement *e*. A sea-urchin egg segmenting

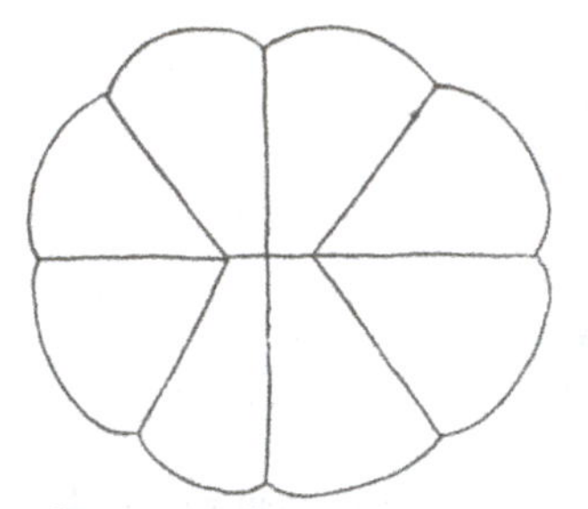

Fig. 251. Egg of *Pyrosoma*. After Korotneff.

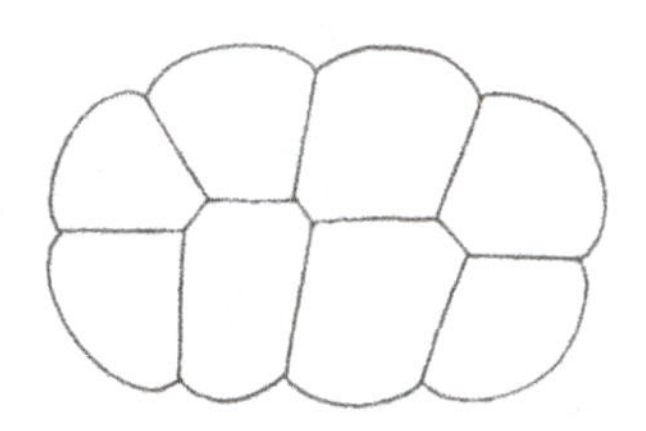

Fig. 252. Egg of *Echinus*, segmenting under pressure. After Driesch.

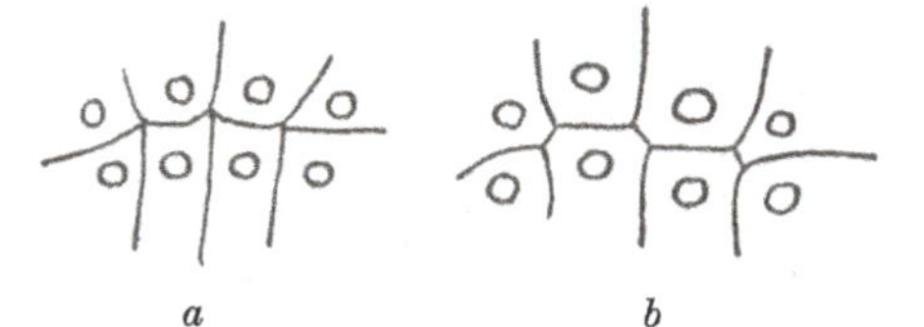

Fig. 253. (*a*) Part of segmenting egg of Cephalopod (after Watase); (*b*) probable actual arrangement.

under pressure, as figured by Driesch, scarcely wants any modification of the drawing to appear in one case as type *f*, in another as *g*. Turning to a botanical illustration, we have a figure of Pringsheim's shewing an eight-celled stage in the apex of the young cone of *Salvinia*: it is ill drawn, but may be referable, as in my diagram, to type *f*; after it is figured a very different object, a segmenting egg of the ascidian *Pyrosoma*, after Korotneff, also, but still more doubtfully, referred to *f*. In the cuttlefish egg there is again some uncertainty, but it is probably referable to *g*. Lastly, I have copied from Roux a curious figure of the frog's egg, viewed from the animal pole; it is obviously inaccurate, but may perhaps belong to type *e*. Of type *i*, in which the five partitions form four re-entrant angles, that is to say a figure representing the five sides of a hexagon, and one cell is in touch with seven others, I have found no examples among published figures of segmenting eggs. It is obvious enough, without more ado, that these phenomena

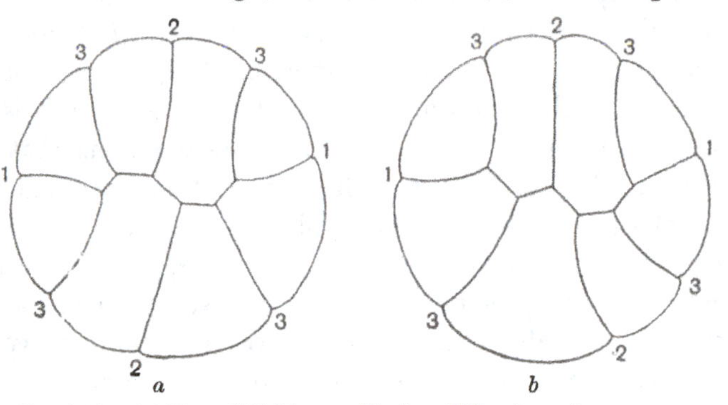

Fig. 254. (*a*) Egg of *Echinus*; (*b*) do. of *Nereis*, under pressure. After Driesch.

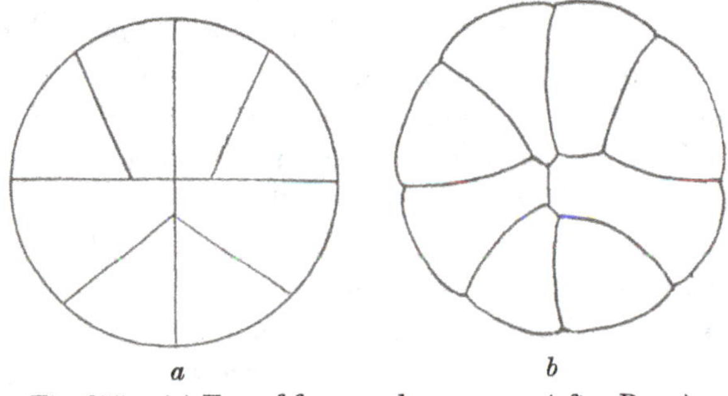

Fig. 255. (*a*) Egg of frog, under pressure (after Roux); (*b*) probable actual arrangement.

are in the strictest and completest way common to both plants and animals, in which respect they tally with, and further extend, the fundamental conclusions laid down by Schwann wellnigh a hundred years ago, in his *Mikroskopische Untersuchungen über die Uebereinstimmung in der Struktur und dem Wachsthum der Thiere und Pflanzen**.

But now that we have seen how a certain limited number of types of eight-celled segmentation (or of arrangements of eight cell-partitions) appear and reappear here and there throughout the whole world of organisms, there still remains the very important

* Berlin, 1839; Sydenham Society, 1847.

question, whether *in each particular organism* the conditions are such as to lead to one particular arrangement being predominant, characteristic, or even invariable. In short, is a particular arrangement of cell-partitions to be looked upon (as the published figures of the embryologist are apt to suggest) as a *specific character*, or at least a constant or normal character, of the particular organism? The answer to this question is a direct negative, but it is only in the work of the most careful and accurate observers that we find it revealed. Rauber (whom we have more than once had occasion to quote) was one of those embryologists who recorded just what he saw, without prejudice or preconception; as Boerhaave said of Swammerdam, *quod vidit id asseruit*. Now Rauber has put on record a considerable number of variations in the arrangement of the first eight cells, which form a discoid surface about the dorsal (or "animal") pole of the frog's egg. In a certain number of cases these figures are identical with one another in type, identical (that is to say) save for slight differences in magnitude, relative proportions, or orientation. But I have selected (Fig. 256) six diagrammatic figures, which are all *essentially different*, and these diagrams seem to me to bear intrinsic evidence of their accuracy: the curvatures of the partition-walls and the angles at which they meet agree closely with the requirements of theory, and when they depart from theoretical symmetry they do so only to the slight extent which we might expect in a material system*. Of these six illustrations, two are exceptional. In Fig. 256, 5, we observe that one of the eight cells is *insular*, and surrounded by the other seven. This is a perfectly natural condition, and represents, like the rest, a phase of partial or conditional equilibrium; but it is not included in the series we are now considering, which is restricted to the case of eight cells extending outwards to a common boundary. The condition shewn in Fig. 256, 6, is

* Such preconceptions as Rauber entertained were all in a direction likely to lead him away from such phenomena as he has faithfully depicted. Rauber had no idea whatsoever of the principles by which we are guided in this discussion, nor does he introduce at all the analogy of surface-tension, or any other purely physical concept; but he was deeply under the influence of Sachs's rule of rectangular intersection, and he was accordingly disposed to look upon the configuration represented above in Fig. 256, 6, as the most typical or primitive. His articles on *Thier und Pflanze*, in *Biol. Cbt.* IV, 1881, tell us much about this and other biological theories of his time.

again peculiar, and is probably rare, but it is included under the cases considered on p. 491, in which the cells are not in complete fluid contact but are separated by little droplets of extraneous matter; it needs no further comment. But the other four cases are beautiful diagrams of space-partitioning, similar to those we have just been considering, but so exquisitely clear that they need no modification, no "touching-up," to exhibit their mathematical regularity. It will easily be recognised that in Fig. 256, 1 and 2,

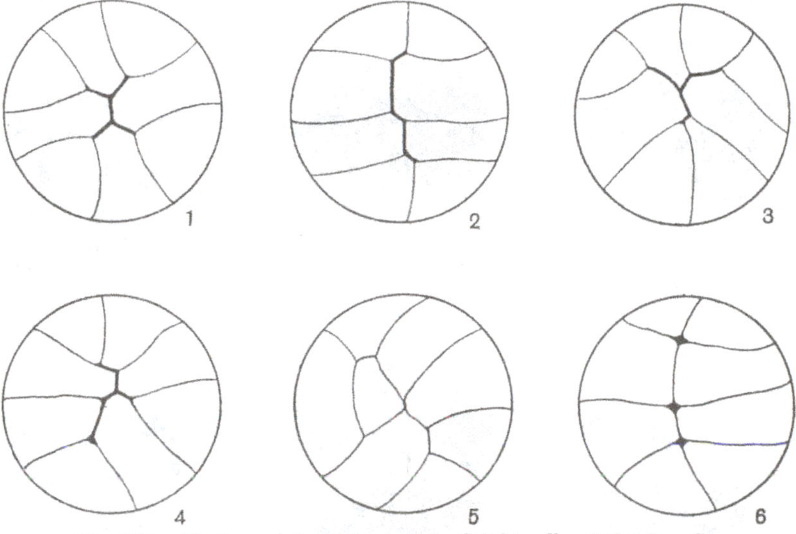

Fig. 256. Various modes of grouping of eight cells, at the dorsal or epiblastic pole of the frog's egg. After Rauber.

we have the arrangements corresponding to *l* and *g*, and in 3 and 4 to *c* in our table on p. 598. One thing stands out as very certain indeed: that the elementary diagram of the frog's segmenting egg given in textbooks of embryology—in which the cells are depicted as uniformly symmetrical and more or less quadrangular bodies—is entirely inaccurate and grossly misleading*.

* Cf. Rauber, Neue Grundlegungen z. K. der Zelle, *Morphol. Jahrb.* VIII, p. 273, 1883: "Ich betone noch, dass unter meinen Figuren diejenige gar nicht enthalten ist, welche zum Typus der Batrachierfurchung gehörig am meisten bekannt ist....Es haben so ausgezeichnete Beobachter sie als vorhanden beschrieben, dass es mir nicht einfallen kann, sie überhaupt nicht anzuerkennen." See also O. Hertwig, Ueber den Werth d. erste Furchungszelle für die Organbildung des Embryo, *Arch. f. Anat.* XLIII, 1893; here O. Hertwig maintains that there is no such thing as "cellular homology."

We begin to realise the remarkable fact, which may even appear a startling one to the biologist, that all possible groupings or arrangements whatsoever of eight cells in a single layer or surface (none being submerged or wholly enveloped by the rest) are referable to one or another of twelve types or patterns; and that all the thousands and thousands of drawings which diligent observers have made of such eight-celled embryos or blastoderms, or other eight-celled structures, animal or vegetable, anatomical, histological or

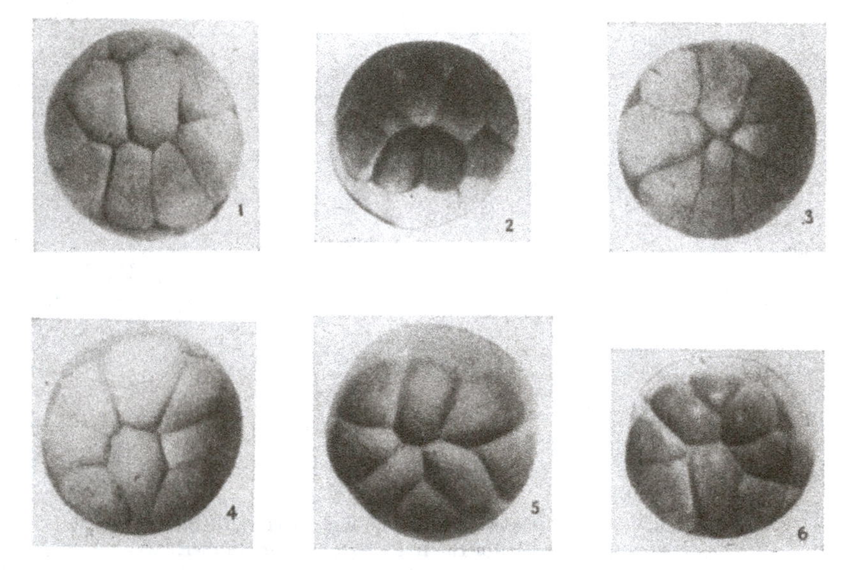

Fig. 257. Photographs of frogs' eggs, shewing various arrangements, or partitionings, of the first eight cells.

embryological, are one and all of them representations of some one or other of these twelve types—or rather for the most part of less than the whole twelve; for a certain small number are essentially unstable, and have at best but a transitory and evanescent existence. But that even the unstable cases should now and then be seen is not to be wondered at: when viscidity and friction, and in general the imperfect fluidity of the system, retard the adjustment of the cells and delay the advent of equilibrium.

As soon as we realise that the number of cell-patterns, for instance in a segmenting egg, is strictly limited, we want to know how many patterns actually occur and in what proportions they do so in a random sample of identical eggs. Some years ago Mr Martin Adamson photographed more than a thousand frogs' eggs in my laboratory, all at the stage shewing an eight-celled group of epiblastic cells: with the remarkable result that every one of the twelve possible arrangements was found to occur, but some were common and some rare, and the following were their comparative frequencies:

Type	Frequency	Type	Frequency
c	19·0 %	*d*	6·7 %
j	17·0	*h*	6·6
b	12·8	*f*	5·1
g	10·3	*k*	3·0
a	7·8	*l*	2·4
e	6·9	*i*	1·8

In six separate batches of eggs (combined in the above list) one or other of the first two types (*c* or *j*) was always the commonest; and the first four taken together made up from 50 to 80 per cent. of each separate sample. On the other hand, when Roux, many years ago, shewed how various cell-configurations might be simulated by oil-drops*—as we have done by means of soap-bubbles—he found that the type *i* was essentially unstable, the large drop with its seven contacts easily slipping into the centre of the system, and there taking up a stable position of equilibrium. That the latter is the more stable, and therefore the more probable, configuration, seems obvious enough; and indeed type *i* seems so obviously unstable that we are not surprised to find it at the bottom of Martin Adamson's list of frequencies. The order in which the rest occur is by no means so easy of explanation.

There is a point worth considering in regard to the number of contacts between cell and cell. In a system of eight cells, all reaching the boundary and all with three-way junctions, there are, besides the eight peripheral boundary-walls, thirteen internal partitions, or $2(n-2)+1$; the number of interfacial *contacts* is double that number, or twenty-six; and the mean number of contacts for each cell is 26/8, or 3·25. But, looking at the diagrams in Fig. 259 (which represent three out of our twelve possible arrangements of

* Roux's experiments were performed with drops of paraffin suspended in dilute alcohol, to which a little calcium acetate was added to form a soapy pellicle over the drops and prevent them from reuniting with one another.

eight cells), we see that, in type *j*, two cells are each in contact with two others, two with three others, and four each with four other cells; in type *l*, four cells are each in contact with two, two with

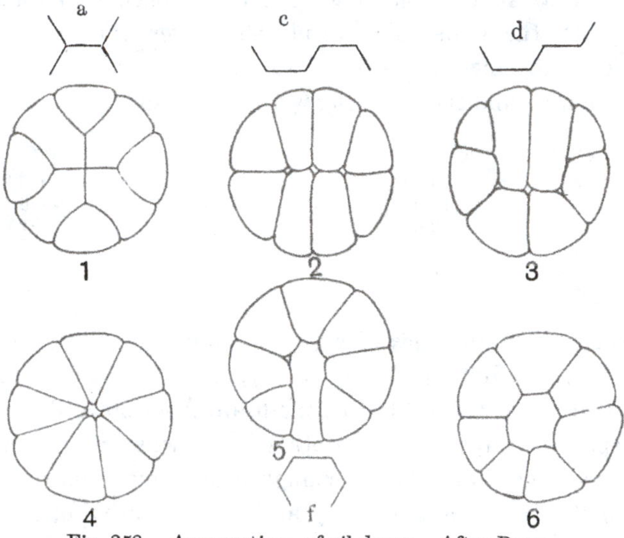

Fig. 258. Aggregations of oil-drops. After Roux.
Nos. 5, 6 represent successive changes in a single system.

four and two with five; and in type *i*, two are in contact with two, four with three and one with no less than seven. And if we sum up, irrespective of sign, the differences from the mean in these three cases, the sum amounts in *j* to 6, in *i* to 7·5, and in *l* to no less than

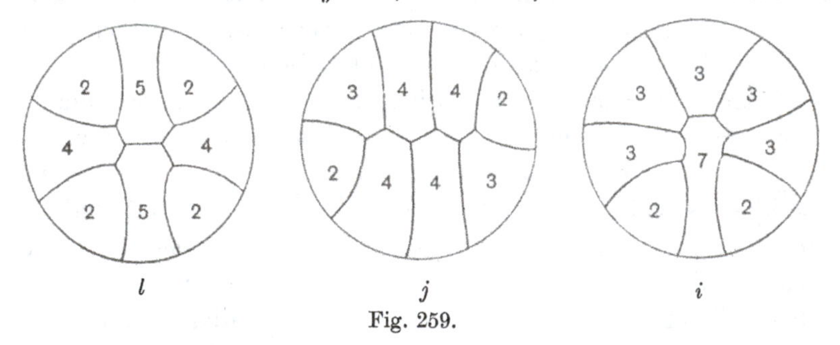

Fig. 259.

10. We might expect to find in such arrangements, that the commonest and most stable types were those in which the cell-contacts were most evenly distributed, and the fact that *j* is (according to Martin Adamson's results) one of the commonest, and *l* one of the

rarest of all looks like supporting the conjecture. Moreover, in all the commonest types we have a more or less equable division; but on the other hand, the number of contacts in type f is just the same as in b, but the latter occurred thrice as often, and k, which is as equable as any, was one of the least frequent of all. Coincidences are weighed down by discrepancies, and we are left pretty much in the dark as to why some types are much commoner than others.

The rules and principles which we have arrived at from the point of view of surface tension have a much wider bearing than is at once suggested by the problems to which we have applied them; for in this study of a segmenting egg we are on the verge of a subject adumbrated by Leibniz, studied more deeply by Euler, and greatly developed of recent years. It is the *Geometria Situs* of Gauss, the *Analysis Situs* of Riemann, the Theory of Partitions of Cayley, of Spatial Complexes or Topology of Johann Benedict Listing*. It begins with regions, boundaries and neighbourhoods, but leads to abstruse developments in modern mathematics. Leibniz had pointed out† that there was room for an analysis of mere position, apart from magnitude: "je croy qu'il nous faut encor une autre analyse, qui nous exprime directement *situm*, comme l'Algèbre exprime *magnitudinem*." There were many things to which the new *Geometria Situs* could be applied. Leibniz used it to explain the game of solitaire, Euler to explain the knight's move on the chess-board, or the routes over the bridges of a town. Vandermonde created a *géometrie de tissage*‡, which Leibniz himself had foreseen, to describe the intricate complexity of interwoven threads in a satin or a brocade§. Listing, in a famous paper||, admired by Maxwell, Cayley and Tait, gave a new name to this new "algorithm," and shewed its application to the curvature of a twining stem or tendril,

* Cf. Clerk Maxwell, On reciprocal figures, *Trans. R.S.E.* XXVI, p. 9, 1870.

† In a letter to Huygens, Sept. 8, 1679; see *Hugenii Exercitationes math. et philos.*, etc., ed. Uylenbroeck, p. 9, 1833.

‡ Remarques sur les problèmes de situation, *Mém. Acad. Sci. Paris* (1771), 1774, p. 566.

§ A problem developed by many eminent mathematicians, and which Edouard Lucas shewed to be intimately related to the construction of Magic Squares: *Récréations mathém.* I, p. xxii, 1891.

|| Vorstudien über Topologie, *Gottinger Studien*, I, pp. 811–75, 1847; Der Census räumlicher Complexe, *ibid.* X, pp. 97 *seq.*, 1861.

the aestivation of a flower, the spiral of a snail-shell, the scales on a fir-cone, and many other common things. The theory of "spatial complexes," as illustrated especially by knots, is a large part of the subject.

Topological analysis seems somewhat superfluous here; but it may come into use some day to describe and classify such complicated, and diagnostic, patterns as are seen in the wings of a butterfly or a fly. Let us look for a moment at how the topologist might begin to study one of our groups of cells; he would probably

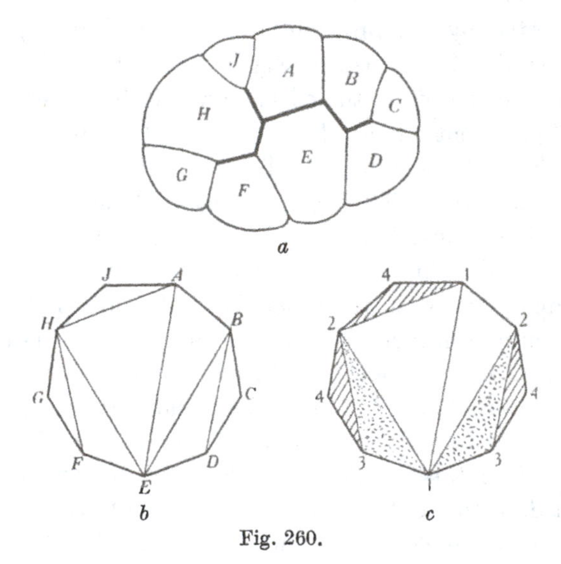

Fig. 260.

call it an island divided into n counties, all maritime (i.e. none encircled by the rest), and having inland none but three-way junctions*. Here (in Fig. 260 *a*) is an island with nine counties; and here (*b*) is a 9-gon, whose corners represent the same counties, and the lines connecting these (whether sides or chords) represent the contacts between. The polygon is now divided by six chords into seven triangles. Three of these are peripheral, *BCD*, *FGH*, *HJA*; mark their vertices, *C*, *G*, *J*, each with the symbol 4, and obliterate these three triangles (as in *c*). The remaining polygon has two peripheral triangles *BDE*, *EFH*; obliterate these, after marking

* This, like many another thing, comes from my good friend Dr G. T. Bennett.

their vertices with the symbol 3. There remains the quadrilateral *ABEH*, containing two peripheral triangles *ABE*, *EHA*; mark *B* and *H* each with the symbol 2. The residual points *A*, *E* are to be marked 1 and 1. The polygon *ABCDEFGHJ* may now be read off: 1 2 4 3 1 3 4 2 4; and this formula, resulting from the triangulation, defines completely the system of chords and the topology of the "island." This is one of the twenty-seven cases of a nine-celled arrangement; and here are our twelve arrangements of eight cells, recatalogued under the new method:

a	1 1 2 3 1 3 2 3	*g*	1 2 3 4 1 2 4 3
b	1 1 3 2 1 3 2 3	*h*	1 2 3 4 1 3 4 2
c	1 2 3 1 2 3 1 3	*i*	1 2 3 4 1 4 3 2
d	1 2 3 1 3 1 3 2	*j*	1 2 4 3 1 2 4 3
e	1 2 3 1 3 2 1 3	*k*	1 2 4 3 1 3 4 2
f	1 2 3 4 1 2 3 4	*l*	1 3 2 3 1 3 2 3

The crucial point for the biologist to comprehend is, that in a closed surface divided into a number of faces, the arrangement of all the faces, lines and points in the system is capable of analysis, and that, when the number of faces or areas is small, the number of possible arrangements is small also. This is the simple reason why we meet in such a case as we have been discussing (viz. the arrangement of a group or system of eight cells) with the same few types recurring again and again in all sorts of organisms, plants as well as animals, and with no relation to the lines of biological classification: and why, further, we find similar configurations occurring to mark the symmetry, not of cells merely, but of the parts and organs of entire animals. The phenomena are not "functions," or specific characters, of this or that tissue or organism, but involve general principles, even "properties of space," which lie within the province of the mathematician.

The theory of space-partitioning, to which the segmentation of the egg gives us an easy practical introduction, is illustrated in innumerable ways, some simple, some extremely complicated, in other fields of natural history; and some serve the better to illustrate the mathematical, and others the physical groundwork of the phenomenon.

Very beautiful instances are to be found in insects' wings. In the dragonfly's wing (which we have already spoken of on p. 476) we see at first sight a vague assemblage of reticulate cells; but their arrangement is both orderly and simple. The long narrow wing is stiffened by longitudinal "veins," which in front lie near and parallel, for reasons well known to the student of aerodynamics*, but become remote and divergent over the rest of the wing; finer veinlets, running between the veins, break up the surface into cells or areolae. Where two large veins run parallel, and so near together that there is only room for one row of cells between, the walls of these meet the large veins at right angles, for the reason that the

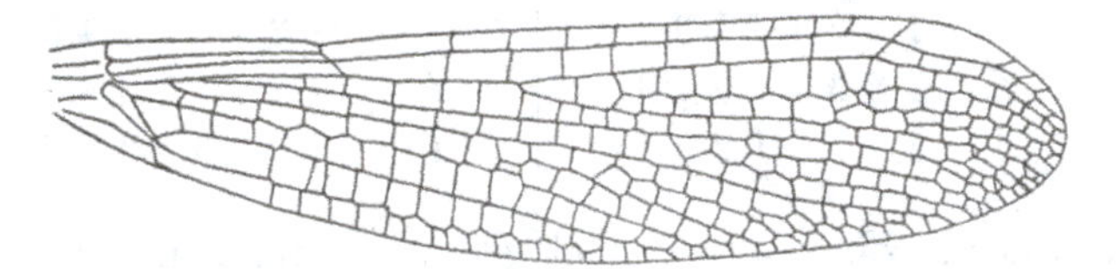

Fig. 261. Wing of "demoiselle" dragonfly (*Agrion*).

tension in these latter is much greater than their own; and this happens nearly all over the delicate wings of the little dragonflies called "demoiselles." But in the big dragonflies (*Aeschna*), and in general wherever there is space enough between two strong veins to hold a double row of cells, the walls of these intercalate with one another at co-equal angles of 120°, while still impinging at right angles on the strong longitudinal partitions. Wherever, as in the hinder parts of the wing, the great veins are few, the cells numerous, and their walls equally delicate, then the reticulum of cells becomes an hexagonal network of all but perfect regularity. In a cicada and in many others there is less contrast between great veins and small; the cells are few, the veins meet neither orthogonally nor at co-equal angles, and the shape of the cells suggests a common deformation under strain. In this last case, and generally in flies, bees and butterflies, the few cells form a complex space-arrangement, simplified

* Sir George Cayley was the first to shew that in a sail—or wing—set at an acute angle to the wind, the centre of pressure lay near the front edge, which had, therefore, to be supported or stiffened (*Nicholson's Journal*, XXV, 1810).

only by the condition that the walls impinge on one another three by three; and this being so, the assemblage includes a number of small intermediate partitions analogous to the "polar furrows" of embryology. We have seen how complex such configurations become as the cells increase in number; and another source of complexity comes in when the veins are of varying thickness and unequal tension, and hence meet one another at varying angles.

The entomologist is much concerned with the number and arrangement of these veins. In Fig. 263 we shew three forewings of a certain stonefly, which serve first to shew how constantly the veins meet in three-way junctions; and then we notice how the

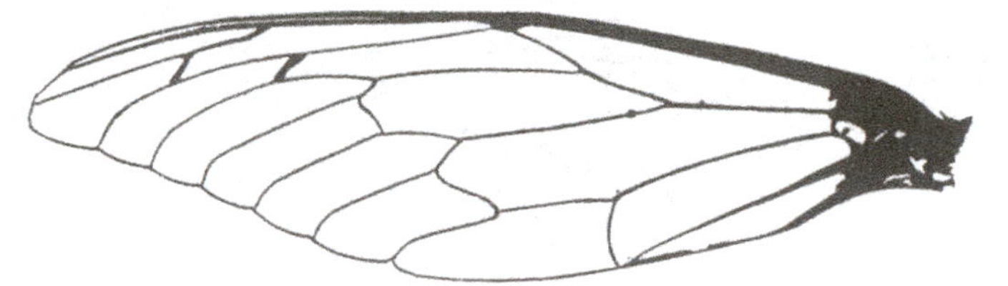

Fig. 262. Forewing of cicada.

three wings are not exactly alike, for all their close resemblance, because in two of them there is one cell less than in the third, *a* being confluent with *b* in one of these cases and with *c* in the other*. In other words, the veinlet *ab* has gone amissing in the one, and *ac* in the other, and in each case the remaining veinlet has sprung into a position of equilibrium. This is one of more than two hundred variations which have been recorded in the wing-veins of this one insect; it might seem superfluous to look for more.

The lower algae shew us many beautiful patterns or collocations of cells, sometimes very complicated, as in *Volvox* or *Hydrodictyon*. A simpler case is that of *Gonium*. Of its sixteen cells four commonly form a square, being so thrust apart out of closer packing by accumulated intercellular substance; the other twelve are grouped around these four, and obey the rule that three cells and no more meet at each node or point of contact (Fig. 264). The twelve cells

* From Arthur Willey, Graded mutations in wings of a stonefly (*Allocapnia pygmaea* Burm.), *Nature*, July 17, 1937.

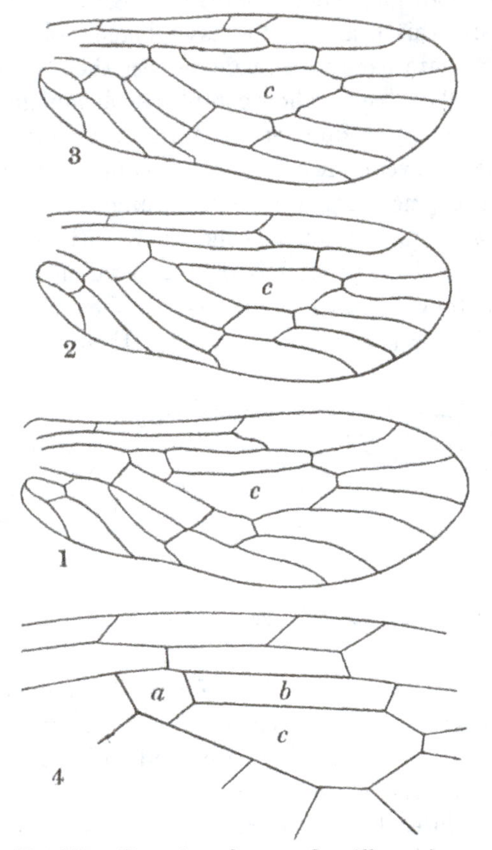

Fig. 263. Forewing of a stonefly, *Allocapnia* sp. 1–3 from A. Willey.

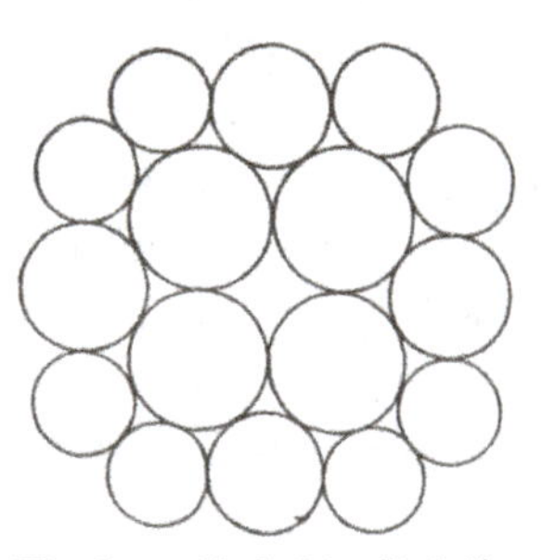

Fig. 264. The four-sided, 16-celled disc of *Gonium*, a minute algae. After Harper, diagrammatic.

are thus in three series, four in touch with six neighbours each, four with four, and four with three; and this arrangement leads to curved contour-lines which may be seen in some of Cohn's figures. Sometimes the interstitial substance is not enough to keep the four cells apart; then they come together in the usual rhomb or lozenge, and the rest group themselves around in the simplest and most symmetrical way. That the whole arrangement is as compact as possible under the conditions, that it is in accord with the principle of minimal areas, and is "such as would result from surface-tension and adhesion between viscous colloidal globules" is now well known to botanists*.

The tiny plates which form the microscopic shell of a Peridinian illustrate over again by their various collocations the principles which we have been studying in the partition-walls of the segmenting egg; and if the one case has shewn us pitfalls in the way of the embryologist, the other shews how the systematist, in his endless task of describing the forms and patterns of things, may sometimes base distinctions on what seem trivial differences from the physical or mathematical point of view.

On the upper half (or *epitheca*) of the globular test of a Peridinium we have fourteen little plates, or fourteen "cells," to use the word in a mathematical rather than a histological sense, whose boundary-walls always meet in three-way nodes. We may reproduce the identical arrangement of the Peridinial plates by blowing bubbles in a saucer; but to deal with so many bubbles at once needs more patience than do the other similar experiments which we have described. That the cells are fourteen in number is, from the physical point of view, the merest accident, but from the zoologist's it is a criterion of the genus; when there are more cells or fewer, the organism is called by another name. The number of possible arrangements of fourteen polygonal cells, linked by three-way nodes, is very large; but the "characters of the genus" exclude many of the variants. Many of them occur—it is quite possible that all occur—in Nature; but they are not called *Peridinium*. The following arrangement defines the genus. There is a central or apical cell, around which are grouped six others; of these six, one extends to the boundary of the figure, that is, to the equator of the globular

* R. A. Harper, The colony in *Gonium*, *Trans. Amer. Microsc. Soc.* XXXI, pp. 65–84, 1912; cf. F. Cohn, in *Nova Acta Acad. C.L.C.*, XXIV, p. 101, 1854.

shell, and seven other "equatorial cells" complete the boundary. In other words, the apical hexagon is surrounded by two concentric rows, originally of six cells each, whereof one cell of the inner row has (as it were) burst through a cell of the outer row, so reaching the boundary itself, and so dividing into two the equatorial cell which it encroached on and bisected.

In any such collocation as this, the number of sides and the number of nodes or corners are strictly determined; there are here fourteen cells, all conjoined by three-way nodes, and it follows that there are just 39 separate walls or edges, and just 26 nodes or corners. Many of these last are already defined for us; for six of them are the corners of the central hexagon, eight lie on the equator, and six more are at the inner ends of the radial partitions which separate the equatorial cells from one another. Six remain to be determined, those, namely, where the partitions running outwards from the apical cell meet the walls of the equatorial cells. The diagram (Fig. 265) shews us two sets of radiating partitions, six running inwards from the equator (a, b, c, d, e, f) and six running outwards from the central hexagon ($A \ldots F$), those of the one set being nearly opposite to those of the other; but near as they may be they never meet, for to do so would be to make a four-way node, which theory forbids and which observation tells us does not occur. In every case, one partition must be slewed a little to one side or other of its opposite neighbour; and the whole range of possible variations depends on whether the shift be to the one side or to the other. We have six pairs of partitions, and in each of the six there is this possible alternative of right or left; there are therefore 2^6 or 64 possible variations in all. Whether all six may vary, I do not know; there is no obvious reason why they should not. But alternative variation does occur in the two anterior and two posterior pairs of partitions; and these four give us 2^4 or 16 possible arrangements.

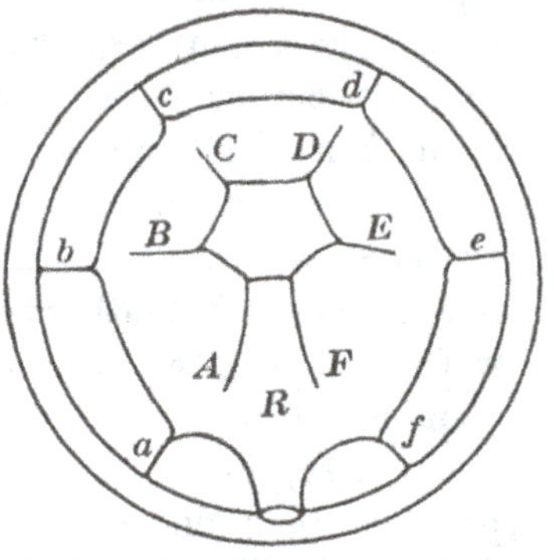

Fig. 265. Dorsal view of a *Peridinium*: diagrammatic.

When the partitions A, F meet the equatorial cells on the *hither* side of the partitions a, f, then, obviously, the large cell R is an irregular hexagon, and is in contact with two equatorial cells only; such an arrangement is said to define the genus *Orthoperidinium*. When A, F happen to fall on the *farther* sides of a, f, the large cell has eight sides, and is in contact with four equatorial cells; we have the new genus *Paraperidinium*. When A falls within a, but F falls beyond f, we have the genus *Metaperidinium*. There remains one alternative case, the converse of the last, to which the systematists have not given a name.

The same physical phenomenon occurs at the opposite pole of the disc, where the partitions C, D may fall within or without, or one within and one without the positions of c, d; where, in other words, the intercalated cell CD is in contact with one, with three, or with two equatorial cells. Jörgensen, seeing these three types occurring both among the Orthoperidinia and the Metaperidinia, draws the conclusion that this character is more primitive, or more ancestral, than that by which Ortho- and Metaperidinia are separated from one another, a phylogenetic deduction concerning which topology has nothing to say*. Within the restricted genus *Peridinium* we have at present two sub-genera and seven sub-groups of these, this being the number of the 64 possible arrangements so far recognised and named. These may have a certain constancy or stability; and trivial as their differences may seem to the physicist, they may still be worth the naturalist's while to study and record.

Another case, geometrically akin but biologically very different, is to be found in the little diatoms of the genus *Asterolampra*, and their immediate congeners†. In *Asterolampra* we have a little disc, in which we see (as it were) radiating spokes of one material alternating with intervals occupied on the flattened wheel-like disc by another (Fig. 266). The spokes vary in number, but the general appearance is in a high degree suggestive of the Chladni figures produced by the vibration of a circular plate. The spokes broaden out towards the centre, and interlock by visible junctions, which

* E. Jörgensen, Ueber Planktonproben, *Svenska Hydrogr. Biol. Komm. Skrifter*, IV, 1913.

† See K. R. Greville, Monograph of the genus *Asterolampra*, *Q.J.M.S.* VIII, (Trans.), pp. 102–124, 1860; cf. *ibid.* (n.s.), II, pp. 41–55, 1862.

generally obey the rule of triple intersection, and accordingly exemplify the partition-figures with which we have been dealing. But whereas we have found the particular arrangement in which one cell is in contact with all the rest to be unstable, according to Roux's oil-drop experiments, and to be conspicuous by its absence from our diagrams of segmenting eggs, here in *Asterolampra*, on the other hand, it occurs frequently, and is indeed the commonest arrangement (Fig. 266, B). In all probability, we are entitled to consider this marked difference natural enough. For we may suppose that in *Asterolampra* (unlike the case of the segmenting egg) the tendency is to perfect radial symmetry, all the spokes emanating from a point in the centre: such a condition would be

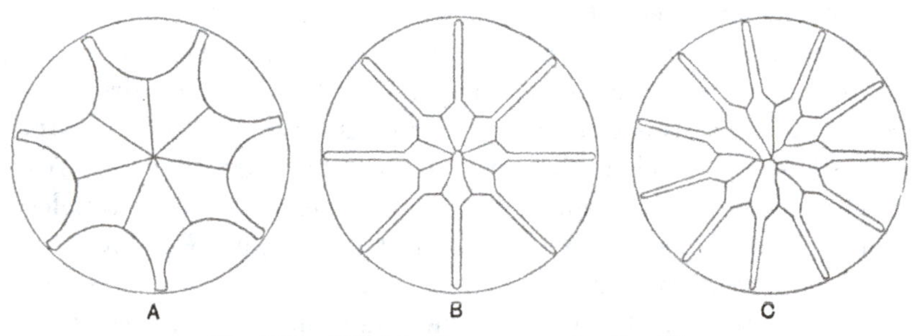

Fig. 266. (A) *Asterolampra marylandica* Ehr.; (B, C) *A. variabilis* Grev. After Greville.

eminently unstable, and would break down under the least asymmetry. A very simple, perhaps the simplest case, would be that one single spoke should differ slightly from the rest, and should so tend to be drawn in amid the others, these latter remaining similar and symmetrical among themselves. Such a configuration would be vastly less unstable than the original one in which all the boundaries meet in a point; and the fact that further progress is not made towards other configurations of still greater stability may be sufficiently accounted for by viscosity, rapid solidification, or other conditions of restraint. A perfectly stable condition would of course be obtained if, as in the case of Roux's oil-drop (Fig. 257, 6), one of the cellular spaces passed into the centre of the system, the other partitions radiating outwards from its circular wall to the periphery of the whole system. Precisely such a condition occurs among our diatoms; but when it does so, it is looked upon as the mark and characterisation of the *allied genus Arachnoidiscus*.

A simple case, introductory to others of a more complex kind, is that of the radial canals of the Medusae. Here, in certain cases (e.g. *Eleutheria*), the usual arrangement of eight radial canals is not seldom modified, as for example, when two or more of them arise

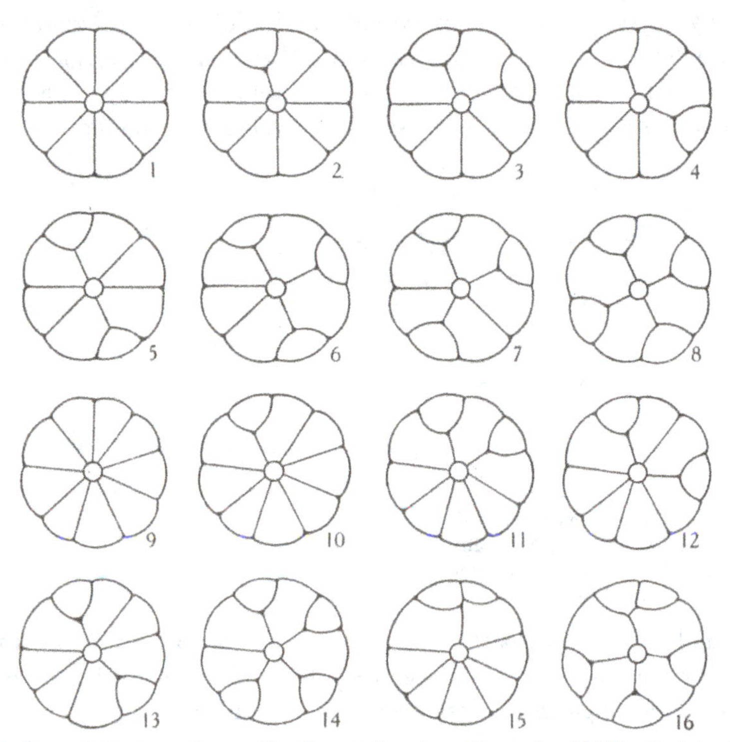

Fig. 267. Variations observed in the canal-system of a medusoid (*Eleutheria*); after Hans Lengerich. 1–8, the eight possible arrangements of eight radial canals; 9–16, some observed instances of nine radial canals.

not separately but by bifurcation*. We then have just eight possible arrangements, as shewn in Fig. 267, 1–8, and of these eight no less than six have been actually observed. The other two are just as likely to occur, and we may take it that they also will in due time be recorded. It is yet another simple illustration of the

* Hans Lengerich. Verzweigungsarten der Radialkanäle bei *Eleutheria*, *Zool. Jahrbuch*, 1922, p. 325.

aphorism that whatsoever is possible, that Nature does, all in her own time and way; what things Nature does not do are the things which are mathematically impossible or are barred by physical conditions. It is possible for our little Medusa to develop nine radial canals instead of eight, and so at times she does; again they may be simple or branched, and trifurcations as well as bifurcations may appear. So we may extend our list of possible permutations and combinations, and find as before that a fair proportion of these possible arrangements have been observed already. There are many other Medusae (e.g. *Willsia*), where the number of radial

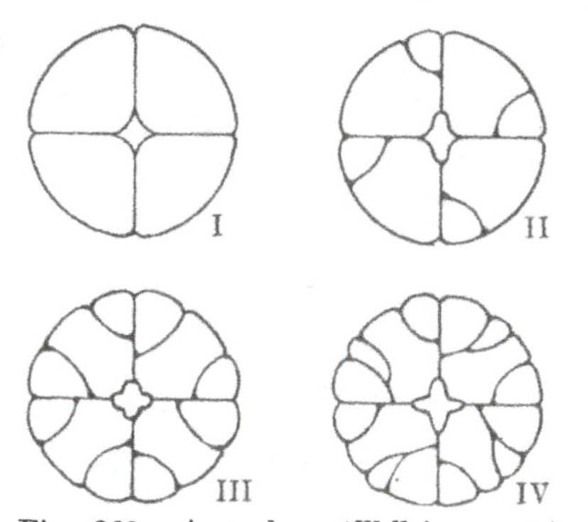

Fig. 268. A medusa (*Willsia ornata*) shewing, diagrammatically, the order of development of the numerous radial canals. After Mayor.

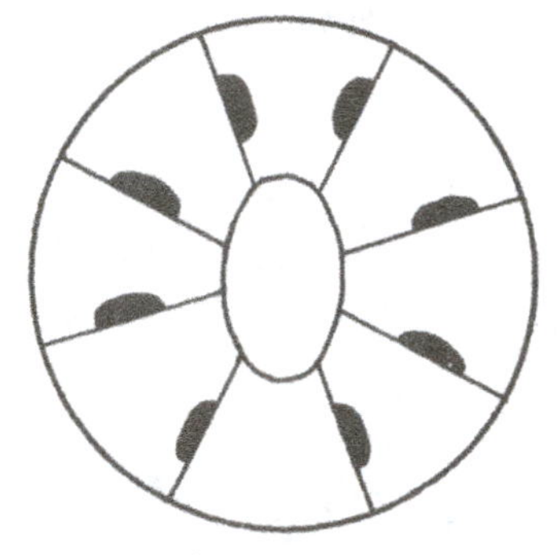

Fig. 269. Section of Alcyonarian polyp.

canals much exceeds the simple symmetry of four or eight; and in these we may sometimes see, very beautifully, how the successive canals arrange themselves according to the same principles which we have now studied in so many diverse cases of partitioning (Fig. 268)*.

In a diagrammatic section of an Alcyonarian polyp (Fig. 269), we have eight chambers set, symmetrically, about a ninth, which constitutes the "stomach." In this arrangement there is no difficulty, for it is obvious that, throughout the system, three boundaries meet (in plane section) in a point. In many corals we have as

* Such branching canals are characteristic of the Dendrostaurinae, a subfamily of the Oceanidae, a family of Anthomedusae; and very much the same occur in a certain subfamily of Leptomedusae. See Mayor's *Medusae of the World*, I, p. 190.

simple or even simpler conditions, for the radiating calcified partitions either converge upon a central chamber, or fail to meet it and end freely. But in a few cases, the partitions or "septa" converge to meet *one another*, there being no central chamber on which they may impinge; and here the manner in which contact is effected becomes complicated, and involves problems identical with those which we are now studying.

In the great majority of corals we have as simple or even simpler conditions than those of *Alcyonium*; for as a rule the calcified partitions or septa of the coral either converge upon a central chamber (or central "columella"), or else fail to meet it and end freely. In the latter case the problem of space-partitioning does not arise; in the former, however numerous the septa be, their separate contacts with the wall of the central chamber comply with our fundamental rule according to which three lines and no more meet in a point, and from this simple and symmetrical arrangement there is little tendency to variation. But in a few cases, the septal partitions converge to meet *one another*, there being no central chamber on which they may impinge; and here the manner in which contact is effected becomes complicated, and involves problems of space-partitioning identical with those which we are now studying. In the genus *Heterophyllia* and in a few allied forms we have such conditions, and students of the Coelenterata have found them very puzzling. McCoy*, their first discoverer, pronounced these corals to be "totally unlike" any other group, recent or fossil; and Professor Martin Duncan, writing a memoir on *Heterophyllia* and its allies†, described them as "paradoxical in their anatomy."

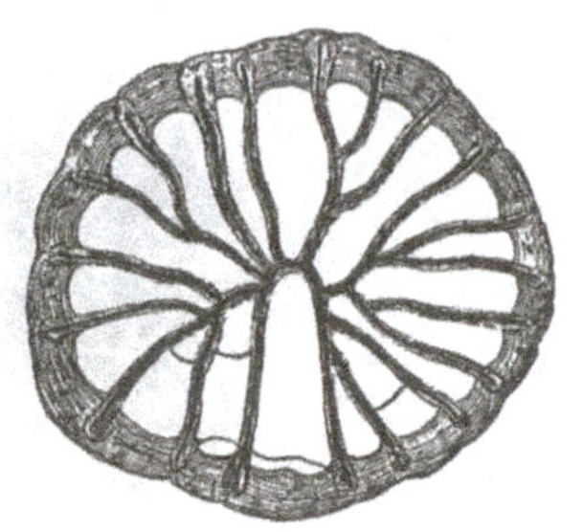

Fig. 270. *Heterophyllia angulata.* After Nicholson.

The simplest or youngest Heterophylliae known have six septa (as in Fig. 271, *A*); in the case figured, four of these septa are conjoined two and two, thus forming the usual triple junctions together with their intermediate partition-walls; and in the case of the other two we may fairly assume that their proper and original

* *Ann. Mag. N.H.* (2), III, p. 126, 1849.
† *Phil. Trans.* CLVII, pp. 643–656, 1867.

arrangement was that of our type 6 *b* (Fig. 245), though the central intermediate partition has been crowded out by partial coalescence. When with increasing age the septa become more numerous, their arrangement becomes exceedingly variable; for the simple reason that, from the mathematical point of view, the number of possible arrangements, of 10, 12 or more cellular partitions in triple contact, tends to increase with great rapidity, and there is little to choose among many of them in regard to symmetry and equilibrium. But while, mathematically speaking, each particular case among the

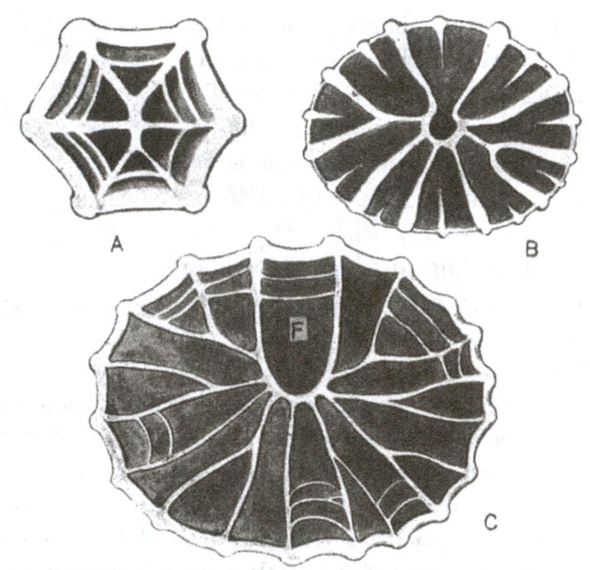

Fig. 271. *Heterophyllia* sp. After Martin Duncan.

multitude of possible cases is an orderly and definite arrangement, from the purely biological point of view on the other hand no law or order is recognisable; and so McCoy described the genus as being characterised by the possession of septa "destitute of any order of arrangement, but irregularly branching and coalescing in their passage from the solid external walls towards some indefinite point near the centre where the few main lamellae irregularly anastomose."

In the two examples figured (Fig. 271 B, C), both comparatively simple ones, it will be seen that, of the main chambers, one is in each case an unsymmetrical one; that is to say, there is one chamber which is in contact with a greater number of its neighbours than any other, and which at an earlier stage must have had contact with them all; this was the case of our type *i*, in the eight-celled system (p. 598). Such an asymmetrical chamber (which may occur in a system of any number of cells greater than six) constitutes what is known to students of the Coelenterata as a "fossula"; and we may recognise it not only here, but also in *Zaphrentis* and its allies, and in a good many other corals besides. Moreover, certain corals are described as having more than one fossula: this appearance being naturally produced under certain of the other asymmetrical variations of normal space-partitioning. Where a single fossula occurs, we are usually told that it is a symptom of "bilaterality"; and this is in turn interpreted as an indication of a higher grade of organisation than is implied in the purely "radial symmetry" of the commoner types of coral. The mathematical aspect of the case gives no warrant for this interpretation.

Let us carefully notice (lest we run the risk of confusing two distinct problems) that the space-partitioning of *Heterophyllia* by no means agrees with the details of that which we have studied in (for instance) the case of the developing disc of *Erythrotrichia*: the difference simply being that *Heterophyllia* illustrates the general case of cell-partitioning as Plateau and Van Rees studied it, while in *Erythrotrichia*, and in our other embryological and histological instances, we have found ourselves justified in making the additional assumption that each new partition divided a cell into *co-equal parts*. No such law holds in *Heterophyllia*, whose case is essentially different from the others: inasmuch as the chambers whose partition we are discussing in the coral are mere empty spaces (empty save for the mere access of sea-water); while in our histological and embryological instances, we were speaking of the division of a cellular unit of living protoplasm. Accordingly, among other differences, the "transverse" or "periclinal" partitions, which were bound to appear at regular intervals and in definite positions, when co-equal bisection was a feature of the case, are comparatively few and irregular in the earlier stages of *Heterophyllia*, though they begin to appear in numbers after the main, more or less radial, partitions have become numerous, and when accordingly these radiating partitions come to bound narrow and almost parallel-sided interspaces; then it is that the transverse or periclinal partitions begin to come in, and form what the student of the Coelenterata calls the "dissepiments" of the coral. We need go no further into the configuration and anatomy of the corals; but it seems to me beyond a doubt that the whole question of the complicated arrangement of septa and dissepiments throughout the group (including the curious vesicular or bubble-like tissue of the Cyathophyllidae and the general structural plan of the *Tetracoralla*, such as *Streptoplasma* and its allies) is well worth investigation from the physical and mathematical point of view, after the fashion which is here slightly adumbrated.

The method of dividing a circular, or spherical, system into eight parts, equal as to their areas but unequal in their peripheral boundaries, is probably of wide biological application; that is to say, without necessarily supposing it to be rigorously followed, the typical configuration which it yields seems to recur again and

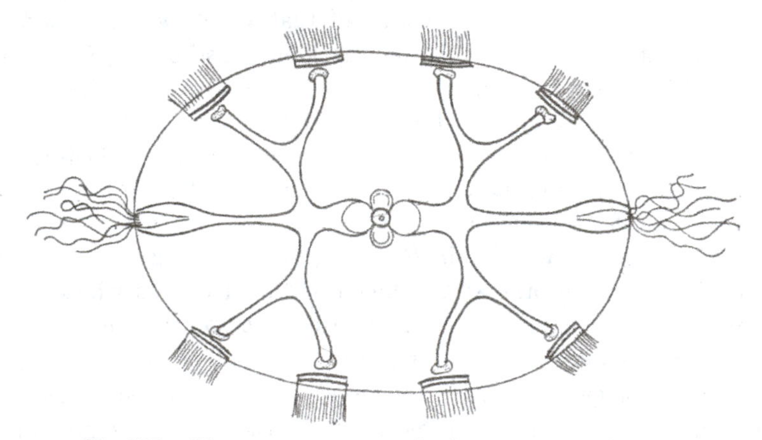

Fig. 272. Diagrammatic section of a Ctenophore (*Eucharis*).

again, with more or less approximation to precision, and under widely different circumstances. I am inclined to think, for instance, that the unequal division of the surface of a Ctenophore by its meridian-like ciliated bands is a case in point (Fig. 272). Here, if we imagine each quadrant to be twice bisected by a curved anticline, we shall get what is apparently a close approximation to the actual position of the ciliated bands. The case however is complicated by the fact that the sectional plan of the organism is never quite circular, but always more or less elliptical. One point, at least, is clearly seen in the symmetry of the Ctenophores; and that is that the radiating canals which pass outwards to correspond in position with the ciliated bands have no common centre, but diverge from one another by repeated bifurcations, in a manner comparable to the conjunctions of our cell-walls.

In the early development of the shell (or "test") of a sea-urchin*, each interambulacral area consists of a lozenge of four plates, in the familiar configuration assumed by four cells or bubbles, the polar furrow lying in the direction of a "radius" of the shell. A fifth plate, or "cell," presently fits itself in between the third and fourth, that is to say between the terminal plate and one of the lateral ones, and in doing so thrusts the former to one side; a sixth intercalates itself between the fourth and fifth, and so on alternately. An ambulacrum consists of two columns of calcareous plates, which fit into one another in the usual way by sutures set at angles of 120°. Each plate consists of three platelets; a "primary" plate (*a*) is succeeded by a smaller and narrower secondary plate (*b*); the squarish primary has one corner cut off by a curved partition, to form a "demiplate", and the whole is called by students of this group an "echinoid triad". Though we do not know precisely how the partitions arise, nor can we prove by measurement their obedience to the laws of maxima and minima, yet their general analogy to the principles we have explained is sufficiently obvious.

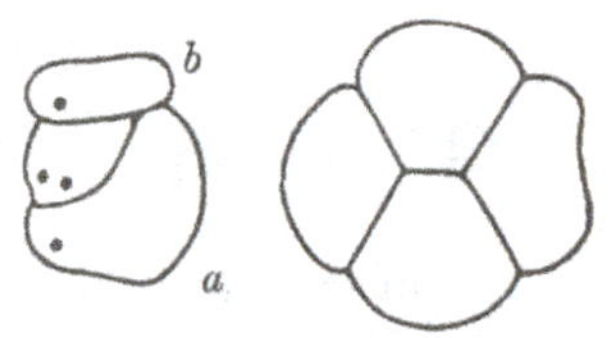

Fig. 273. A "triad" and a "lozenge": stages in the development of the ambulacral and interambulacral plates of a sea-urchin. After I. Gordon.

I am even inclined to think that the same principle helps us to understand the arrangement of the skeletal rods of a larval Echinoderm, and the complex conformation of the larva which is brought about by the presence of these long, slender skeletal

* Isabella Gordon, The development of the calcareous test of *Echinus miliaris*, *Phil. Trans.* (B), No. 214, p. 282, 1926; etc.

radii*. In Fig. 274 I have divided a circle into its four quadrants, and have bisected each quadrant by a circular arc (*BC*), passing from radius to periphery, as in the foregoing cases of cell-division (*e.g.* p. 590); and I have again bisected, in a similar way, the triangular

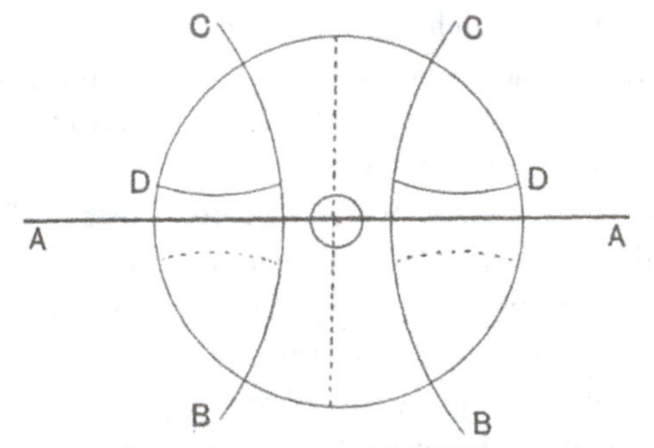

Fig. 274. Diagrammatic arrangement of partitions, represented by skeletal rods, in a larval Echinoderm (*Ophiura*).

halves of each quadrant (*D*, *D*). I have also inserted a small circle in the middle of the figure, concentric with the large one. If now we imagine the partition-lines in the figure to be replaced by solid rods, we shall have at once the frame-work of an Ophiurid (Pluteus)

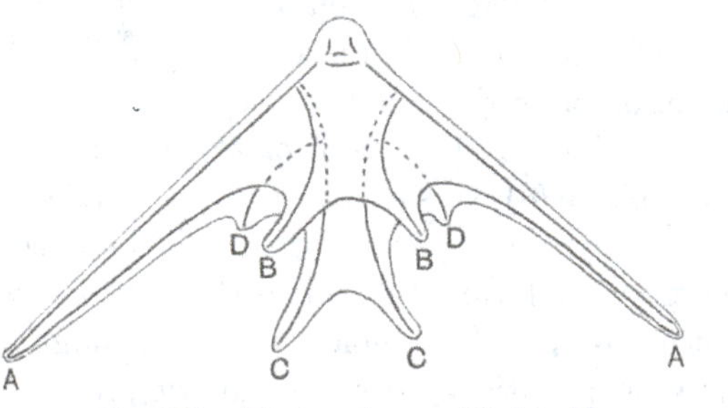

Fig. 275. Pluteus-larva of Ophiurid.

larva. Let us imagine all these arms to be bent symmetrically downwards, so that the plane of the paper is transformed into a conical surface with curved sides; let a membrane be spread, umbrella-like, between the outstretched skeletal rods, and let

* J. Loeb has shewn (*Amer. Journ. Physiol.* VII, p. 441, 1900) that the sea-urchin's egg can be reared for a time in a balanced solution of sodium, potassium and calcium chloride, developing no spicules and so forming no pluteus larva; on adding sodium carbonate the spicules are laid down and the pluteus larva takes shape accordingly.

its margin loop from rod to rod in curves which are possibly catenaries but are more probably portions of an "elastic curve," and the outward resemblance to a Pluteus larva is now complete. By various slight modifications, by altering the relative lengths of the rods, by modifying their curvature or by replacing the curved rod by a tangent to itself, we can ring the changes which lead us from one known type of Pluteus to another. The case of the Bipinnaria larvae of Echinids is certainly analogous, but it becomes very much more complicated; we have to do with a more complex partitioning of space, and I confess that I am not yet able to represent the more complicated forms in so simple a way.

There are a few notable exceptions (besides the various unequally segmenting eggs) to the general rule that in cell-division the mother-cell tends to divide into equal halves; and one of these exceptional cases is to be found in connection with the development of "stomata" in the leaves of plants*. The epidermal cells by which the leaf is covered may be of various shapes; sometimes, as in a hyacinth, they are oblong, but more often they have an irregular shape in which we can recognise, more or less clearly, a distorted or imperfect hexagon. In the case of the oblong cells, a transverse partition will be the least possible, whether the cell be equally or unequally divided, unless (as we have already seen) the space to be cut off be a very small one, not more than about three-tenths the area of a square based on the *short* side of the original rectangular cell. As the portion usually cut off is not nearly so small as this, we get the form of partition shewn in Fig. 276, and the cell so cut off is next bisected by a partition at right angles to the first; this latter partition splits, and the two last-formed cells constitute the so-called "guard-cells" of the stoma. In other cases, as in Fig. 277, there will come a point where the minimal partition necessary to cut off the required fraction of the cell-content is no longer a

* We know more about the physical activities of the stomata than about the mechanics of their development. It is known that the rate of gaseous diffusion through apertures *of their order of magnitude* is inversely proportional to the diameters of the apertures; and this law, by which the sufficient entry of carbonic acid through the stomata is fully accounted for, is (like Pfeffer's work on natural semi-permeable membranes) one of the notable cases where physiology has enlarged the boundaries of physical science. Cf. Horace T. Brown, Some recent work on diffusion, *Proc. Roy. Instit.* March, 1901.

transverse one, but is a portion of a cylindrical wall (2) cutting off one corner of the mother-cell. The cell so cut off is now a certain segment of a circle, with an arc of approximately 120°; and its next division will be by means of a curved wall cutting it into a

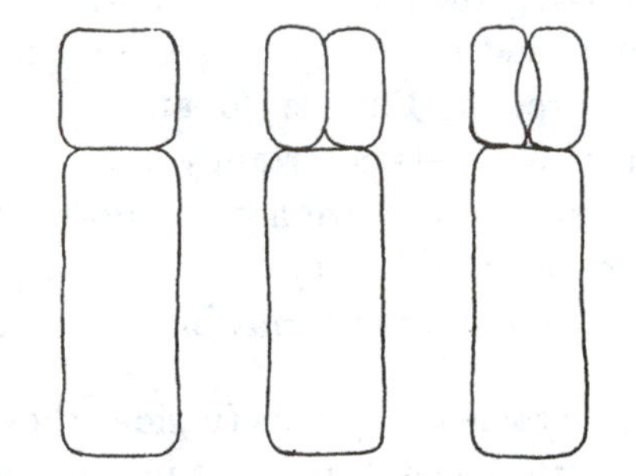

Fig. 276. Diagrammatic development of stomata in hyacinth.

triangular and a quadrangular portion (3). The triangular portion will continue to divide in a similar way (4, 5), and at length (for a reason which is not yet clear) the partition wall between the new-formed cells splits, and again we have the phenomenon of a

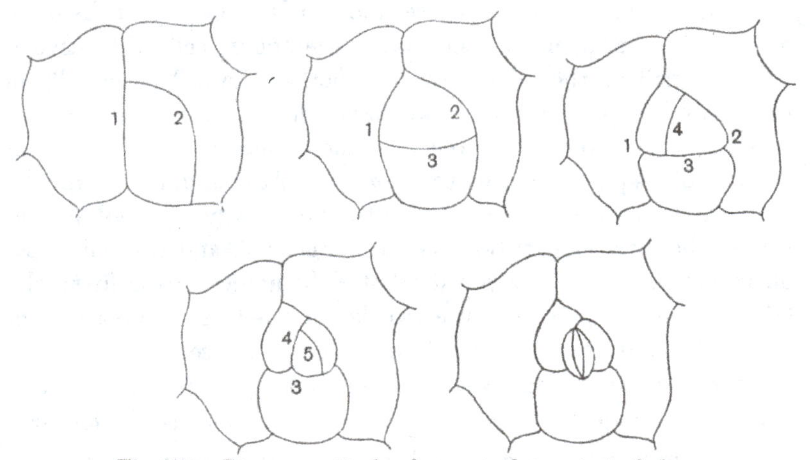

Fig. 277. Diagrammatic development of stomata in *Sedum*. (Cf. fig. in Sachs's *Botany*, 1882, p. 103.)

"stoma" with its attendant guard-cells. In Fig. 277 are shewn the successive stages of division, and the changing curvatures of the various walls which ensue as each subsequent partition appears, and introduces a new tension into the system. Among the oblong cells of the epidermis in the hyacinth the stomata will be found arranged in regular rows, while they will be irregularly distributed over the surface of the leaf in such a case as we have depicted in *Sedum*.

As I have said, the mechanical cause of the split which constitutes the orifice of the stoma is not quite clear. It may be directly due to the subepidermal air-space which the stoma communicates with, for an air-surface on both sides of the delicate epidermis might well cause such an alteration of tensions that the two halves of the dividing cell would tend to part company. In Professor Macallum's experiments, which we have briefly discussed in our short chapter on Adsorption, it was found that large quantities of potassium gathered together along the outer walls of the guard-cells of the stoma, thereby indicating a low surface-tension along these outer walls. The tendency of the guard-cells to bulge outwards is so far explained, and it is possible that, under the existing conditions of restraint, we may have here a force tending, or helping, to split the two cells asunder. It is clear enough, however, that the last stage in the development of a stoma is, from the physical point of view, not yet properly understood*. It is noteworthy, and Nägeli took note of it wellnigh a hundred years ago, that the stomatal mother-cells remain small while the others grow, and also that they only divide once for all, while their neighbours divide and divide again, to produce the lateral or accessory guard-cells.

In all our foregoing examples of the development of a "tissue" we have seen that the process consists in the *successive* division of cells, each act of division being accompanied by the formation of a boundary-surface, which, whether it become at once a solid or semi-solid partition or whether it remain semi-fluid, exercises in all cases an effect on the position and the form of the boundary which comes into being with the next act of division. In contrast to this general process stands the phenomenon known as "free cell-formation," in which, out of a common mass of protoplasm, a number of separate cells are *simultaneously*, or all but simultaneously, differentiated; and the case is all the more interesting

* Botanische Beiträge, *Linnaea*, XVI, p. 238, 1842. Cf. Garreau, Mém. sur les stomates, *Ann. Sc. Nat., Bot.* (4), I, p. 213, 1854.

when the daughter-cells remain, for a time at least, within the envelope of the mother-cell. It sometimes happens, to begin with, that a number of mother-cells are formed simultaneously, and that the content of each divides, by successive divisions, into four "daughter-cells." These daughter-cells tend to group themselves, just as would four soap-bubbles, into a "tetrad," the four cells forming a spherical tetrahedron. For the system of four bodies is in perfect symmetry. The four cells are closely packed within the cell-wall of the mother-cell; their outer walls divide the sphere into four equiangular triangles; their inner walls meet three-by-three in an edge, and the four edges converge in the geometrical centre of the system; and these partition walls and their respective edges meet one another everywhere at co-equal angles. This is the typical mode of development of pollen-grains, common among monocotyledons and all but universal among dicotyledonous plants. By a loosening of the surrounding tissue and an expansion of the cavity, or anther-cell, in which they lie, the pollen-grains afterwards fall apart, and their individual form will depend upon whether or no their walls have solidified before this liberation takes place. For if not, then the separate grains will be free to assume a spherical form as a consequence of their own individual and unrestricted growth; but if they become set or rigid prior to the separation of the tetrad, then they will conserve more or less completely the plane interfaces and sharp angles of the elements of the tetrahedron. The latter is apparently the case in the pollen-grains of *Epilobium* (Fig. 278, 1) and in many others. In the passion-flower (2) we have an intermediate condition: in which we can still see an indication of the facets where the grains abutted on one another in the tetrad, but the plane faces have been swollen by growth into spheroidal or spherical surfaces. In heaths and in azaleas the four cells of the tetrad remain attached together, and form a compound tetrahedral pollen-grain. Six furrows correspond to the six edges of the tetrahedron, and each is continued across a pair of cells; they are formed (I take it) along lines of weakness at the edges of the tetrahedron, and they make three furrows upon each one of the four coherent grains, just as we see them on a large number of ordinary separate and non-coherent pollen-grains. On the other hand, there may easily be cases where the tetrads of daughter-cells fail to assume, even temporarily, the tetrahedral form: cases, in a general way, where the four cells escape from the confinement of their envelope, and fall into a looser, less close-packed arrangement*. The figures given by Goebel of the development of the pollen of *Neottia* (3, *a–e*: all the figures referring to grains taken from a single anther) illustrate this to perfection, and it will be seen that,

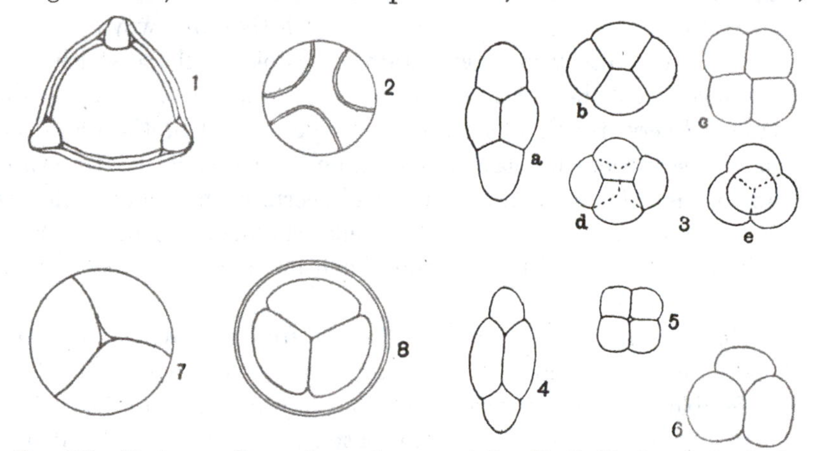

Fig. 278. Various pollen-grains and spores (after Berthold, Campbell, Goebe and others). (1) *Epilobium*; (2) *Passiflora*; (3) *Neottia*; (4) *Periploca graeca*; (5) *Apocynum*; (6) *Erica*; (7) spore of *Osmunda*; (8) tetraspore of *Callithamnion*.

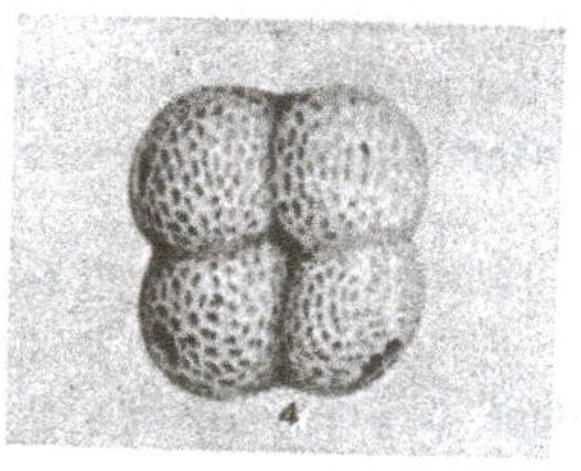

Fig. 279. Pollen of bulrush (*Typha*). After Wodehouse.

when the four cells lie in a plane, they conform exactly to our typical diagram of the first four cells in a segmenting ovum; physically, as well as biologically, the tetrads *a–d* and the tetrad *e* are "allelomorphs" of one another. Again in the bulrush (Fig. 279),

* Cf. C. Nägeli, *Zur Entwicklungsgeschichte des Pollens bei den Phanerogamen*, 36 pp., Zürich, 1842; Hugo Fischer, *Vergleichende Morphologie der Pollenkörner*, Berlin, 1890; see also, for many and varied illustrations, R. P. Wodehouse's beautiful book on *Pollen*, 574 pp., New York, 1935, and earlier papers.

the four cells remain attached to one another, and lie upon a level with a "polar furrow" well displayed. Occasionally, though the four cells lie in a plane, the diagram seems to fail us, for the cells appear to meet in a simple cross (as in 5); but here we soon perceive that the cells are not in complete interfacial contact, but are kept apart by a little intervening drop of fluid or bubble of air. The spores of ferns (7) for the most part develop in much the same way as pollen-grains; they also very often retain traces of the shape which they assumed as members of a tetrahedral figure, and the same is equally true of liverworts. Among the "tetraspores" (8) of the Florideae, or red seaweeds, we have a condition which is in every respect analogous. The same thing happens in certain simple algae allied to *Protococcus*: where four daughter-cells, confined within a mother-cell, form a spherical tetrahedron, much like a spore of *Osmunda* on a smaller scale*.

Here again it is obvious that, apart from differences in actual magnitude, and apart from superficial or "accidental" differences (referable to other physical phenomena) in the way of colour, texture and minute sculpture or pattern, a very small number of diagrammatic figures will sufficiently represent the outward forms of all the tetraspores, four-celled pollen-grains, and other four-celled aggregates which are known or are even capable of existence. And it is equally obvious that the resemblance of these things, to this extent, is a matter of physical and mathematical symmetry, and carries no proof of near relationship or common ancestry.

We have been dealing hitherto (save for some slight exceptions) with the partitioning of cells on the assumption that the system either remains unaltered in size or else that growth has proceeded uniformly in all directions. But we extend the scope of our enquiry greatly when we begin to deal with *unequal growth*, with cells so growing and dividing as to produce a greater extension along some one axis than another. And here we come close in touch with that great and still (as I think) insufficiently appreciated generalisation of Sachs, that the manner in which the cells divide is *the result*, and not the cause, of the form of the dividing structure: that the form of the mass is caused by its growth as a whole, and

* Cf. A. Pascher, *Arch. f. Protozoenk.* LXXVI, p. 409; LXXVII, p. 195, 1932.

is not a resultant of the growth of the cells individually considered*. Such asymmetry of growth may be easily imagined, and may conceivably arise from a variety of causes. In any individual cell, for instance, it may arise from molecular asymmetry of the structure of the cell-wall, giving it greater rigidity in one direction than another, while all the while the hydrostatic pressure within the cell remains constant and uniform. In an aggregate of cells, it may very well arise from a greater chemical, or osmotic, activity in one than another, leading to a localised increase in the fluid pressure, and to a corresponding bulge over a certain area of the external surface. It might conceivably occur as a direct result of preceding cell-divisions, when these are such as to produce many peripheral or concentric walls in one part and few or none in another, with the obvious result of strengthening the boundary wall here and weakening it there; that is to say, in our dividing quadrant, if its quadrangular portion subdivide by periclines, and the triangular portion by oblique anticlines (as we have seen to be the natural tendency), then we might expect that external growth would be more manifest over the latter than over the former areas. As a direct and immediate consequence of this we might expect a tendency for special outgrowths, or "buds," to arise from the triangular rather than from the quadrangular cells; and this turns out to be not merely a tendency towards which theoretical considerations point, but a widespread and important factor in the morphology of the cryptogams. But meanwhile, without enquiring further into this complicated question, let us simply take it that, if we start from such a simple case as a round cell which has divided into two halves or four quarters (as the case may be), we shall at once get bilateral symmetry about a main axis, and other secondary results arising therefrom, as soon as one of the halves, or one of the quarters, begins to shew a rate of growth in advance of the others; for the more rapidly growing cell, or the peripheral wall common to two or more such rapidly growing cells, will bulge out, and may finally extend into a cylinder with rounded end. This latter very simple case is illustrated in the development of a

* Sachs, *Pflanzenphysiologie* (*Vorlesung* XXIV), 1882; cf. Rauber, Neue Grundlegungen zur Kenntniss der Zelle, *Morphol. Jahrb.* VIII, p. 303 *seq.*, 1883; E. B. Wilson, Cell-lineage of *Nereis*, *Journ. Morph.* VI, p. 448, 1892; etc.

pollen-tube, where the rapidly growing cell develops into the elongated cylindrical tube, and the slow-growing or quiescent part remains behind as the so-called "vegetative" cell or cells.

Just as we have found it easier to study the segmentation of a circular disc than that of a spherical cell, so let us begin in the same way, by enquiring into the divisions which will ensue if the disc tend to grow, or elongate, in some one particular direction instead of in radial symmetry. The figures which we shall then obtain will not only apply to the disc, but will also represent, in all essential features, a projection or longitudinal section of a solid body, spherical to begin with, preserving its symmetry as a solid of revolution, and subject to the same general laws as we study in the disc*.

(1) Suppose, in the first place, that the axis of growth lies symmetrically in one of the original quadrantal cells of a segmenting disc; and let this growing cell elongate with comparative rapidity before it subdivides. When it does divide, it will necessarily do so by a transverse partition, concave towards the apex of the cell: and, as further elongation takes place, the cylindrical structure which will be developed thereby will tend to be again and again subdivided by similar transverse partitions (Fig. 280). If at any time, through this process of concurrent elongation and subdivision, the apical cell become equivalent to, or less than, a hemisphere, it will next divide by means of a longitudinal, or vertical partition; and similar longitudinal partitions will arise in the other segments of the cylinder, as soon as it comes about that their length (in the direction of the axis) is less than their breadth.

But when we think of this structure in the solid, we at once perceive that each of these flattened segments, into which our cylinder divided to begin with, is equivalent to a flattened circular disc; and its further division will accordingly tend to proceed like

* In the following account I follow closely on the lines laid down by Berthold; *Protoplasmamechanik*, cap. vii. Many botanical phenomena identical and similar to those here dealt with are elaborately discussed by Sachs in his *Physiology of Plants* (chap. xxvii, pp. 431–459, Oxford, 1887), and in his earlier papers, Ueber die Anordnung der Zellen in jüngsten Pflanzentheilen, and Ueber Zellenanordnung und Wachsthum (*Arb. d. botan. Inst. Würzburg*, 1877/78). But Sachs's treatment differs entirely from that which I adopt and advocate here: his explanations being based on his "law" of rectangular succession, and involving complicated systems of confocal conics, with their orthogonally intersecting ellipses and hyperbolas.

any other flattened disc, namely into four quadrants, and afterwards by anticlines and periclines in the usual way. A section across the cylinder, then, will tend to shew us precisely the same arrangements as we have already so fully studied in connection with the typical division of a circular cell into quadrants, and of these quadrants into triangular and quadrangular portions, and so on.

But there are other possibilities to be considered, in regard to the mode of division of the elongating quasi-cylindrical portion, as it gradually develops out of the growing and bulging quadrantal cell; for the manner in which this latter cell divides will simply depend upon the form it has assumed before each successive act

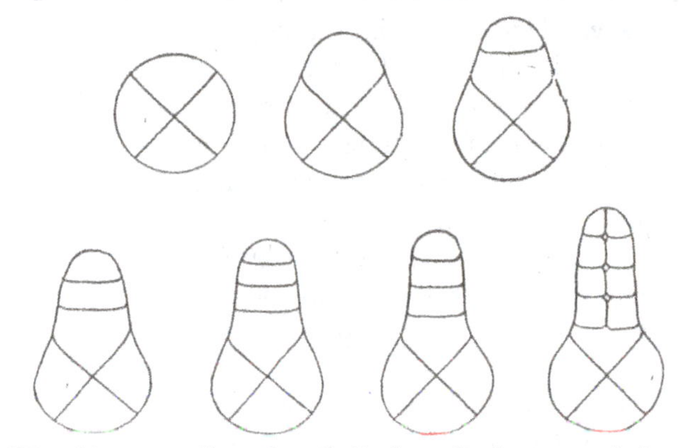

Fig. 280. Diagrammatic, or hypothetical, result of asymmetrical growth.

of division takes place, that is to say upon the ratio between its rate of growth and the frequency of its successive divisions. For, as we have already seen, if the growing cell attain a markedly oblong or cylindrical form before division ensues, then the partition will arise transversely to the long axis; if it be but a little more than a hemisphere, it will divide by an oblique partition; and if it be less than a hemisphere (as it may come to be after successive transverse divisions) it will divide by a vertical partition, that is to say by one coinciding with its axis of growth. An immense number of permutations and combinations may arise in this way, and we must confine our illustrations to a small number of cases. The important thing is not so much to trace out the various conformations which may arise, but to grasp the fundamental principle: which is, that the forces which dominate the *form* of each cell regulate the manner of its subdivision, that is to say the form of the new cells into which it subdivides; or in other words, the form of the growing organism regulates the form and number of the cells which eventually constitute it. The complex cell-network is not the cause but the result of the general configuration, which latter has its essential cause in whatsoever physical and chemical processes have led to a varying velocity of growth in one direction as compared with another.

In the annexed figure of an embryo of *Sphagnum* we see a mode of development almost precisely corresponding to the hypothetical case which we have just described—the case, that is to say, where one of the four original quadrants of the mother-cell is the chief agent in future growth and development. We see at the base of our first figure (*a*), the three stationary, or undivided quadrants, one of which has further slowly divided in the stage *b*. The active quadrant has grown quickly into a cylindrical structure, which inevitably divides, in the next place, into a series of transverse partitions; and accordingly, this mode of development carries with it the presence of a single "apical cell," whose lower wall is a spherical surface with its convexity downwards. Each cell of the subdivided cylinder now appears as a more or less flattened disc, whose mode of further subdivision we may prognosticate according to our former investigation, to which subject we shall presently return.

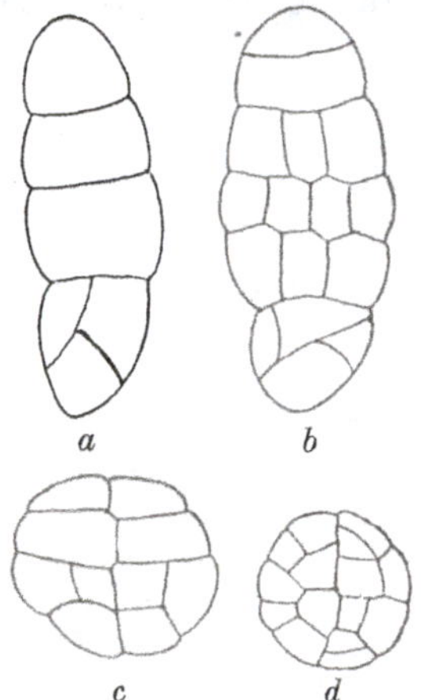

Fig. 281. Development of *Sphagnum*. After Campbell.

(2) In the next place, still keeping to the case where only one of the original quadrant-cells continues to grow and develop, let us suppose that this growing cell falls to be divided when by growth it has become just a little greater than a hemisphere; it will then divide, as in Fig. 282, 2, by an oblique partition, in the usual way, whose precise position and inclination to the base will depend entirely on the configuration of the cell itself, save only, of course, that we may have also to take into account the possibility of the division being into two unequal halves. By our hypothesis, the

growth of the whole system is mainly in a vertical direction, which is as much as to say that the more actively growing protoplasm, or at least the strongest osmotic force, will be found near the apex; where indeed there is obviously more external surface for osmotic action. It will therefore be that one of the two cells which contains, or constitutes, the apex which will grow more rapidly than the other, and which therefore will be the first to divide; and indeed in any case, it will usually be this one of the two which will tend to divide first, inasmuch as the triangular and not the quadrangular half is bound to constitute the apex*. It is obvious that (unless

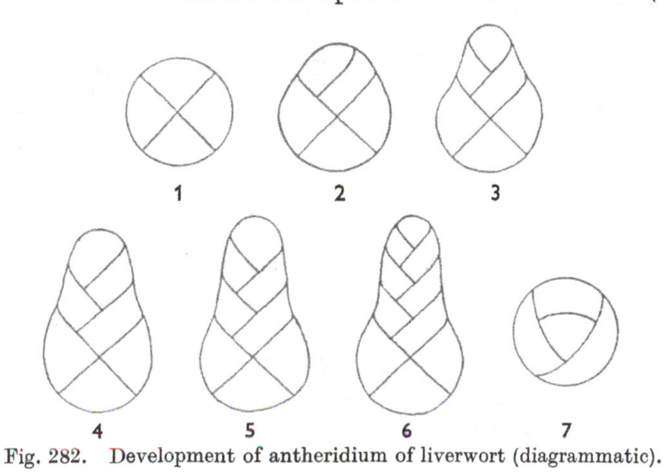

Fig. 282. Development of antheridium of liverwort (diagrammatic).

the act of division be so long postponed that the cell has become quasi-cylindrical) it will divide by another oblique partition, starting from, and running at right angles to, the first. And so division will proceed by oblique alternate partitions, each one tending to be, at first, perpendicular to that on which it is based and also to the peripheral wall; but all these points of contact soon tending, by reason of the equal tensions of the three films or surfaces which meet there, to form angles of 120°. There will always be a single apical cell, of a triangular form. The developing antheridium of a liverwort (*Riccia*) is a typical example of such a case. In Fig. 283 which represents a "gemma" of a moss, we see just the same thing; with this addition, that here the lower of the two original cells has grown even more quickly than the other, constituting a long cylin-

* Cf. p. 590.

drical stalk, and dividing in accordance with its shape by means of transverse septa. In all such cases the cells may continue to subdivide, and the manner in which they do so must depend upon their own proportions; and in all cases there will sooner or later be a tendency to the formation of periclinal walls, cutting off an epidermal layer of cells, as Fig. 284 illustrates very well.

The method of division by means of oblique partitions is a common one in the case of "growing points"; for it evidently includes all cases in which the act of cell-division does not lag far behind that elongation which is determined by the specific rate of

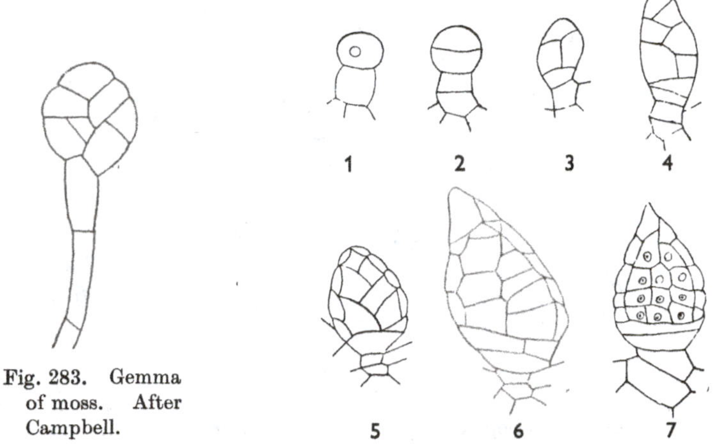

Fig. 283. Gemma of moss. After Campbell.

Fig. 284. Development of antheridium of *Riccia*. After Campbell.

growth. And it is also obvious that, under a common type, there must here be included a variety of cases which will, at first sight, present a very different appearance one from another. For instance, in Fig. 285 which represents a growing shoot of *Selaginella*, and somewhat less diagrammatically in the young embryo of *Jungermannia* (Fig. 286), we have the appearance of an almost straight vertical partition running up in the axis of the system, and the primary cell-walls are set almost at right angles to it—almost transversely, that is to say, to the outer walls and to the long axis of the structure. We soon recognise, however, that the difference is merely a difference of degree. The more remote the partitions are, that is to say the greater the velocity of growth relatively to that of division, the less abrupt will be the alternate kinks or curvatures of the portions which lie along the axis, and the more will these portions appear to constitute a single unbroken wall.

(3) But an appearance nearly, if not quite, indistinguishable from this may be got in another way, namely, when the original growing cell is so nearly hemispherical that it is actually divided

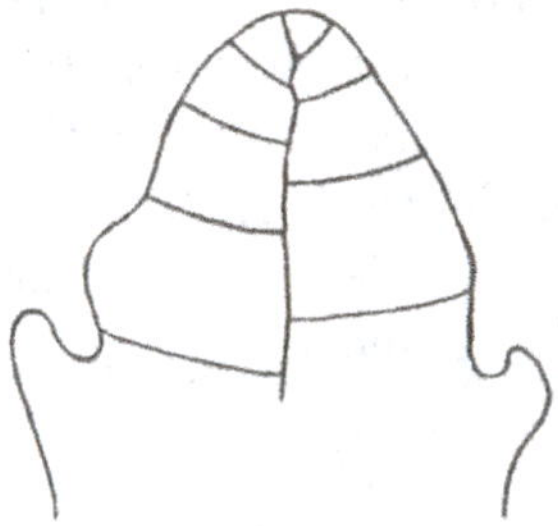

Fig. 285. Section of growing shoot of *Selaginella*, diagrammatic.

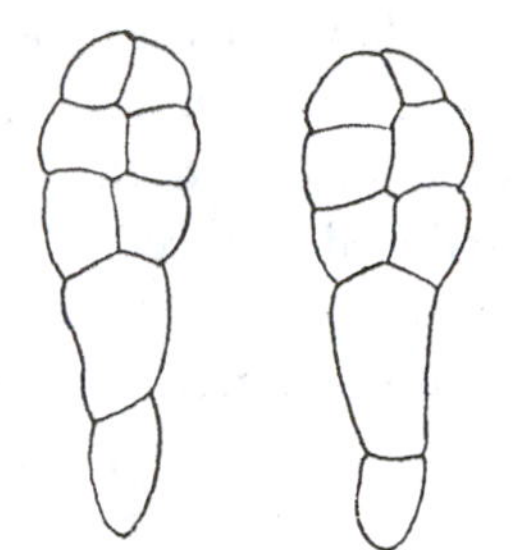

Fig. 286. Embryo of *Jungermannia*. After Kienitz-Gerloff.

by a vertical partition into two quadrants, and when from this vertical partition, as it elongates, lateral partition-walls arise on either side. Then, by the tensions exercised by these, the vertical partition

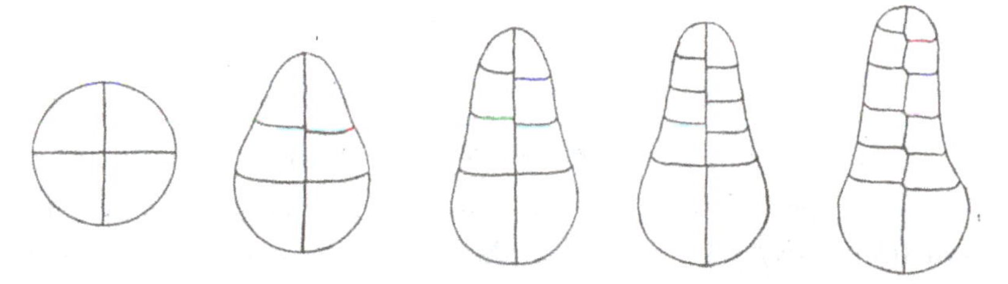

Fig. 287.

will be bent into little portions set at 120° one to another, and the whole will come to look just like that which, in the former case, was made up of parts of many successive oblique partitions (Fig. 287).

Let us now, in one or two cases, follow out a little further the stages of cell-division whose beginnings we have studied in the last paragraphs. In the antheridium of *Riccia*, after successive oblique partitions have produced the longitudinal series of cells shewn in Fig. 284, 4, it is plain that the next partitions will arise periclinally, that is to say parallel to the outer wall, which coincides with the short axis of the oblong cells. The effect is to produce an epidermal layer, whose cells subdivide further by partitions perpendicular to the surface, that is to say crossing the flattened cells by their shortest diameter. The inner mass consists of cells which are still more or less oblong, or which become so in process of growth; and these again divide, parallel to their short axes, into squarish cells, which as usual, by the mutual tension of their walls, become hexagonal as seen in a plane section. There is a clear distinction, then, in form as well as in position, between the outer covering-cells and those which lie within this envelope; the latter are reduced to a condition which fulfils the mechanical function of a protective coat, while the former undergo less modification, and become the actively living, reproductive elements.

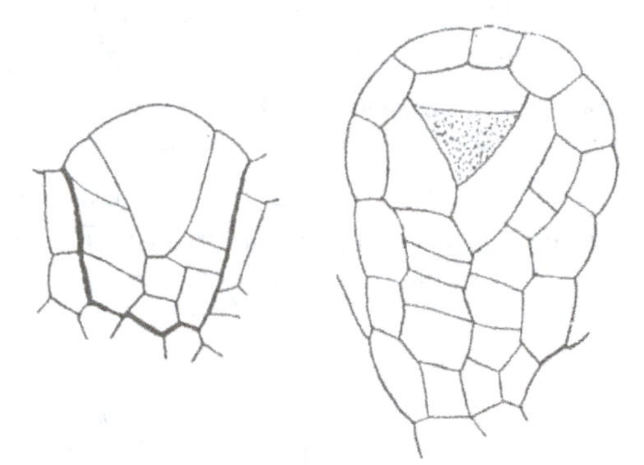

Fig. 288. Development of sporangium of *Osmunda*. After Bower.

In Fig. 288 is shewn the development of the sporangium of a fern (*Osmunda*). We may trace here the common phenomenon of a series of oblique partitions, built alternately on one another, and cutting off a conspicuous triangular apical cell. Over the whole system an epidermal layer is formed, in the manner we have described; and in this case it covers the apical cell also, owing to the fact that it was of such dimensions that, at one stage of growth, a periclinal partition wall, cutting off its outer end, was indicated as of less area than an anticlinal one. This periclinal wall cuts down the apical cell to the proportions, very nearly, of an equilateral triangle, but the solid form of the cell is obviously that of a tetrahedron with curved faces; and accordingly, the least possible partitions by which further subdivision can be effected will run successively parallel to its four sides (or its three sides when we confine ourselves to the appearances as seen in section). The effect is to cut off on each side of the apical cell a characteristically flattened cell, oblong as seen in section, still leaving a triangular (or strictly speaking, a tetrahedral) one in the centre. The oblong cells, which constitute no specific structure and perform no specific physiological function*, but which merely represent certain directions in space towards which the whole system of partitioning has gradually led, are called by botanists the "tapetum." The active growing tetrahedral cell which lies between them, and from which in a sense every other cell in the system has been either directly or indirectly segmented off, still manifests its vigour and activity, and becomes, by internal subdivision, the mother-cell of the spores.

In all these cases, for simplicity's sake, we have merely considered the appearances presented in a single longitudinal plane of optical section. But it is not difficult to interpret from these appearances what would be seen in another plane, for instance in a transverse section. In our first example, for instance, that of the developing embryo of *Sphagnum* (Fig. 281 *c*, *d*), we see that, at appropriate levels, the cells of the original cylindrical row have divided into transverse rows of four, and then of eight cells. We may be sure that the four cells represent, approximately, quadrants of a cylindrical disc, the four cells, as usual, not meeting in a point, but intercepted by a small intermediate partition. Again, where we have a plate of eight cells, we may well imagine that the eight octants are arranged in what we have found to be the way naturally resulting from the division of four quadrants, that is to say into alternately triangular and quadrangular portions; and this is found by means of sections to be the case. The figure is precisely comparable to our previous diagrams of the arrangement of eight cells in a dividing disc, save only that, in two cases, the cells have already undergone a further subdivision.

It follows that we are apt to meet with this characteristic figure, in one or other of its possible and strictly limited variations, in the cross-sections of many growing structures, just as we have already

* This is not to say that Nature makes *no use* of the tapetal cells. In the end they break down and contribute to the growth of the spore-mother-cell: very much as the "superfluous" eggs in a fly's ovary contribute yolk-material to the developing ovum.

seen it appear in cases where the entire system consists of eight cells only. For example, we have it in a section of a young embryo of a moss (*Phascum*), and again, in a section of an embryo of a fern (*Adiantum*). In Fig. 290, shewing a section through a growing frond of a sea-weed (*Girardia*), we have a case where the partitions forming the eight octants have conformed to the usual type; but instead of the usual division by periclines of the four quadrangular spaces, these latter are dividing by means of oblique septa, apparently owing to the fact that the cell is not dividing into two equal, but into two unequal portions. In this last figure we have a peculiar look of stiffness or formality, such that it appears at first to bear little resemblance to the rest. The explanation is of the simplest. The mode of partitioning differs little (except to some slight extent

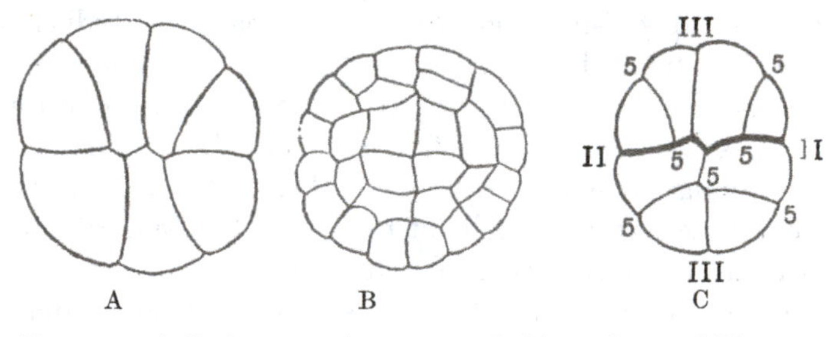

Fig. 289. (A, B) Sections of younger and older embryos of *Phascum*; (C) do. of *Adiantum*. After Kienitz-Gerloff.

in the way already mentioned) from the normal type; but in this case the partition walls are so thick and become so soon comparatively solid and rigid, that the secondary curvatures due to their successive mutual tractions are here imperceptible.

A curious and beautiful case, apparently aberrant but which would doubtless be found conforming strictly to physical laws if only we clearly understood the actual conditions, is indicated in the development of the antheridium of a fern, as described by Strasbürger. Here the antheridium develops from a single cell, which (Fig. 291) has grown to something more than a hemisphere; and the first partition, instead of stretching transversely across the cell, as we should expect it to do if the cell were actually spherical, has as it were sagged down to come in contact with the base, and so to develop into an annular partition, running round the lower margin of the cell. The phenomenon is precisely identical to that bisection of a quadrant by means of a circular arc, of which we spoke on p. 581, and the annular film is very easy to reproduce by means of a soap-bubble in the bottom of a cylindrical dish or beaker. The next partition is a periclinal one, concentric with the outer surface of the young antheridium; and this in turn is followed by a concave partition which cuts off the apex of the original cell: but which becomes connected with the second, or periclinal partition in precisely the same annular fashion as the first partition did with the base of the little antheridium. The result is that, at this stage, we have four cell-cavities in the little antheridium: (1) a central

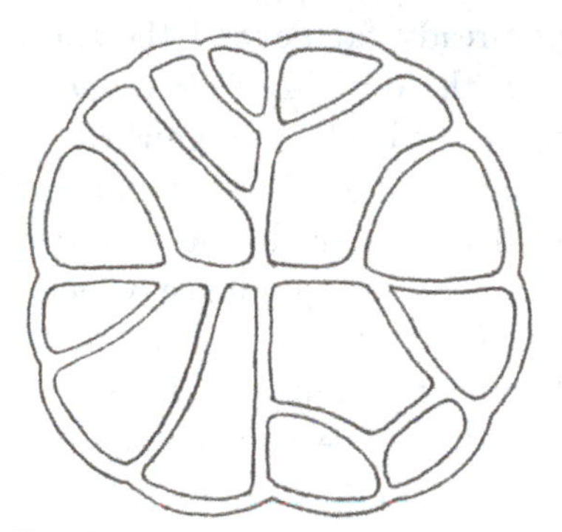

Fig. 290. Section through frond of *Girardia sphacelaria*. After Goebel.

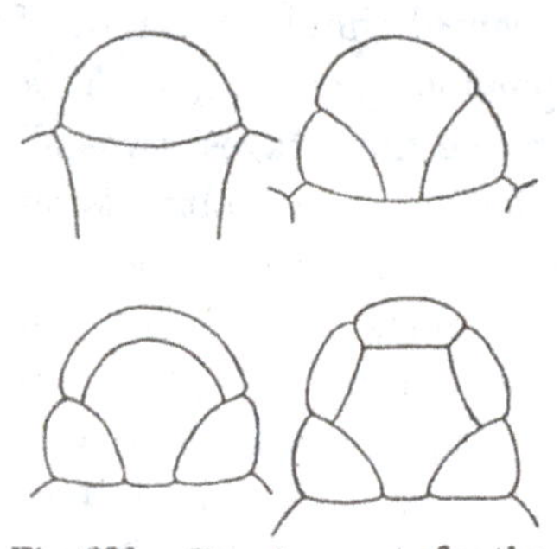

Fig. 291. Development of antheridium of *Pteris*. After Strasbürger.

cavity; (2) an annular space around the lower margin; (3) a narrow annular or cylindrical space around the sides of the antheridium; and (4) a small terminal or apical cell. It is evident that the tendency, in the next place, will be to subdivide the flattened external cells by means of anticlinal partitions, and so to convert the whole structure into a single layer of epidermal cells, surrounding a central cell within which, in course of time, the antherozoids are developed.

The foregoing account deals only with a few elementary phenomena, and may seem to fall far short of an attempt to deal in general with "the forms of tissues." But it is the principle involved, and not its ultimate and very complex results, that we can alone attempt to grapple with. The stock-in-trade of mathematical physics, in all the subjects with which that science deals, is for the most part made up of simple, or simplified, cases of phenomena which in their actual and concrete manifestations are usually too complex for mathematical analysis; hence, even in physics, the full mechanical explanation of a phenomenon is seldom if ever more than the "cadre idéal" towards which our never-finished picture extends. When we attempt to apply the same methods of mathematical physics to our biological and histological phenomena, we need not wonder if we be limited to illustrations of a simple kind, which cover but a small part of the phenomena with which histology has to do. But yet it is only relatively that these phenomena to which we have found the method applicable are to be deemed simple and few. They go already far beyond the simplest phenomena of all, such as we see in the dividing *Protococcus*, and in the first stages, two-celled or four-celled, of the segmenting egg. They carry us into stages where the cells are already numerous, and where the whole conformation has become by no means easy to depict or visualise, without the help and guidance which the phenomena of surface-tension, the laws of equilibrium and the principle of minimal areas are at hand to supply. And so far as we have gone, and so far as we can discern, we see no sign of the guiding principles failing us, or of the simple laws ceasing to hold good.

CHAPTER IX

ON CONCRETIONS, SPICULES, AND SPICULAR SKELETONS

THE deposition of inorganic material in the living body, usually in the form of calcium salts or of silica, is a common phenomenon. It begins by the appearance of small isolated particles, crystalline or non-crystalline, whose form has little relation or none to the structure of the organism; it culminates in the complex skeletons of the vertebrate animals, in the massive skeletons of the corals, or in the polished, sculptured and mathematically regular molluscan shells. Even among very simple organisms, such as diatoms, radiolarians, foraminifera or sponges, the skeleton displays extraordinary variety and beauty, whether by reason of the intrinsic form of its elementary constituents or the geometric symmetry with which these are interconnected and arranged.

With regard to the form of these various structures (and this is all that immediately concerns us here), we have to do with two distinct problems, which merge with one another though they are theoretically distinct. For the form of the spicule or other skeletal element may depend solely on its chemical nature, as for instance, to take a simple but not the only case, when it is purely crystalline; or the inorganic material may be laid down in conformity with the shapes assumed by cells, tissues or organs, and so be, as it were, moulded to the living organism; and there may well be intermediate stages in which both phenomena are simultaneously at work, the molecular forces playing their part in conjunction with the other forces inherent in the system.

So far as the problem is a purely chemical one we must deal with it very briefly indeed: all the more because special investigations regarding it have as yet been few, and even the main facts of the case are very imperfectly known. This at least is clear, that the phenomena with which we are about to deal go deep into the subject of colloid chemistry, and especially that part of the science which deals with colloids in connection with surface phenomena. It is to the special student of the chemistry and physics of the colloids that we must look for the elucidation of our problem*.

In the first and simplest part of our subject, the essential problem is the problem of crystallisation in presence of colloids. In the cells of plants true crystals are found in comparative abundance, and consist, in the majority of cases, of calcium oxalate. In the stem and root of the rhubarb for instance, in the leaf-stalk of *Begonia* and in countless other cases, sometimes within the cell, sometimes in the substance of the cell-wall, we find large and well-formed crystals of this salt; their varieties of form, which are extremely numerous, are simply the crystalline forms proper to the salt itself, and belong to the two systems, cubic and monoclinic, in one or other of which, according to the amount of water of crystallisation, this salt is known to crystallise. When calcium oxalate crystallises according to the latter system (as it does when its molecule is combined with two molecules of water), the microscopic crystals have the form of fine needles, or "raphides"; these are very common in plants, and may be artificially produced when the salt is crystallised out in presence of glucose or of dextrin†.

Calcium carbonate, on the other hand, when it occurs in plant-cells, as it does abundantly (for instance in the "cystoliths" of the Urticaceae and Acanthaceae, and in great quantities in *Melobesia* and the other calcareous or "stony" algae), appears in the form of fine rounded granules, whose inherent crystalline structure is only revealed (like that of a molluscan shell) under polarised light. Among animals, a skeleton of carbonate of lime occurs under a multitude of forms, of which we need only mention a few of the most conspicuous. The spicules of the calcareous sponges are triradiate, occasionally quadriradiate, bodies, with pointed rays, not crystalline in outward form but with a definitely crystalline internal

* There is much information regarding the chemical composition and mineralogical structure of shells and other organic products in H. C. Sorby's Presidential Address to the Geological Society (*Proc. Geol. Soc.* 1879, pp. 56–93); but Sorby failed to recognise that association with "organic" matter, or with colloid matter whether living or dead, introduced a new series of purely physical phenomena.

† Julien Vesque, Sur la production artificielle de cristaux d'oxalate de chaux semblables à ceux qui se forment dans les plantes, *Ann. Sc. Nat.* (*Bot.*) (5) XIX, pp. 300–313, 1874.

structure; we shall return again to these, and find for them what would seem to be a satisfactory explanation of their form. Among the Alcyonarian zoophytes we have a great variety of spicules*, which are sometimes straight and slender rods, sometimes flattened and more or less striated plates, and still more often disorderly aggregations of micro-crystals, in the form of rounded or branched concretions with rough or knobby surfaces† (Figs. 292, 298). A third type, presented by several very different things, such as a

Fig. 292. Alcyonarian spicules: *Siphonogorgia* and *Anthogorgia*. After Studer.

pearl or the ear-bone of a bony fish, consists of a more or less rounded body, sometimes spherical, sometimes flattened, in which the calcareous matter is laid down in concentric zones, denser and clearer layers alternating with one another. In the development of the molluscan shell and in the calcification of a bird's egg or a crab's shell, small spheroidal bodies with similar concentric striation make their appearance; but instead of remaining separate they become

* Cf. Kölliker, *Icones Histologicae*, 1864, p. 119, etc.

† In rare cases, these shew a single optic axis and behave as individual crystals: W. J. Schmidt, *Arch. f. Entw. Mech.* LI, pp. 509–551, 1922.

crowded together, and in doing so are apt to form a pattern of hexagons. In some cases the carbonate of lime, on being dissolved away by acid, leaves behind it a certain small amount of organic residue; in many cases other salts, such as phosphates of lime, ammonia or magnesia, are present in small quantities; and in most cases, if not all, the developing spicule or concretion is somehow so associated with living cells that we are apt to take it for granted that it owes its form to the constructive or plastic agency of these.

The appearance of direct association with living cells, however, is apt to be fallacious; for the actual *precipitation* takes place, as a rule, not in actively living, but in dead or at least inactive tissue*; that is to say in the "formed material" or matrix which accumulates round the living cells, or in the interspaces between these latter, or, as often happens, in the cell-wall or cell-membrane rather than within the substance of the protoplasm itself. We need not go the length of asserting that this is a rule without exception; but, so far as it goes, it is of great importance and to its consideration we shall presently return†.

Cognate with this is the fact that, at least in some cases, the organism can go on, in apparently unimpaired health, when stinted or even wholly deprived of the material of which it is wont to make its spicules or its shell. Thus the eggs of sea-urchins reared in lime-free water develop, in apparent health and comfort, into larvae which lack the usual skeleton of calcareous rods: and in which, accordingly, the long arms of the Pluteus larva, which the rods should support and extend, are entirely absent‡. Again, when foraminifera are kept for generations in water from which they gradually exhaust the lime, their shells grow hyaline and transparent, and dwindle to a mere chitinous pellicle; on the other hand,

* In an interesting paper by Robert Irvine and Sims Woodhead on the Secretion of carbonate of lime by animals (*Proc. R.S.E.* XV, pp. 308–316; XVI, pp. 324–351, 1889–90) it is asserted (p. 351) that "lime salts, of whatever form, are deposited *only* in vitally inactive tissue."

† The tube of *Teredo* shews no trace of organic matter, but consists of irregular prismatic crystals: the whole structure "being identical with that of small veins of calcite, such as are seen in thin sections of rocks" (Sorby, *Proc. Geol. Soc.* 1879, p. 58). This, then, would seem to be a somewhat exceptional case of a shell laid down completely outside of the animal's external layer of organic substance.

‡ Cf. Pouchet and Chabry, *C.R. Soc. Biol. Paris* (9), I, pp. 17–20, 1889; *C.R. Acad. Sci.* CVIII, pp. 196–198, 1889.

in the presence of excess of lime their shells become much altered, are strengthened with various ridges or "ornaments," and come to resemble other varieties and even "species*."

The crucial experiment, then, is to attempt the formation of similar spicules or concretions apart from the living organism. But however feasible the attempt may be in theory, we must be prepared to encounter many difficulties; and to realise that, though the reactions involved may be well within the range of physical chemistry, yet the actual conditions of the case may be so complex, subtle and delicate that only now and then, and only in the simplest of cases, has it been found possible to imitate the natural objects successfully. Such an attempt is part of that wide field of enquiry through which Stéphane Leduc and other workers have sought to produce, by synthetic means, forms similar to those of living things; but it is a circumscribed and well-defined part of that wider investigation†.

When we find ourselves investigating the forms assumed by chemical compounds under the peculiar circumstances of association with a living body, and when we find these forms to be characteristic or recognisable, and somehow different from those which the same substance is wont to assume under other circumstances, an analogy, captivating though perhaps remote, presents itself to our minds between this subject of ours and certain synthetic problems of the organic chemist. There is doubtless an essential difference, as well as a difference of scale, between the visible form of a spicule or con-

* Cf. Heron-Allen, *Phil. Trans.* (B), CCVI, p. 262, 1915.

† Leduc's artificial growths were mostly obtained by introducing salts of the heavy metals or alkaline earths into solutions which form with them a "precipitation-membrane"—as when we introduce copper sulphate into a ferrocyanide solution. See his *Mechanism of Life*, 1911, ch. X, for copious references to other works on the "artificial production of organic forms." Closely related to Leduc's experiments are those of Denis Monnier and Carl Vogt, Sur la fabrication artificielle des formes des éléments organiques, *Journ. de l'Anat.* XVIII, pp. 117–123, 1882; cf. Moritz Traube, Zur Geschichte der mechanischen Theorie des Wachstums der organischen Zelle, *Botan. Ztg.* XXXVI, 1878. Cf. also A. L. Herrera, Sur les phénomènes de vie apparente observés dans les émulsions de carbonate de chaux dans la silice gélatineuse, *Mem. Soc. Alzate, Mexico,* XXVI, 1908; Los Protobios, *Boll. de la Dir. de Estud. Biolog., Mexico,* I, pp. 607–631, and other papers. Also (*int. al.*) R. S. Lillie and E. N. Johnston, Precipitation-structures simulating organic growth, *Biol. Bull.* XXXIII, p. 135, 1917; XXXVI, pp. 225–272, 1919; *Scientific Monthly,* Feb. 1922, p. 125; H. W. Morse, C. H. Warren and J. D. H. Donnay, Artificial spherulites, etc., *Amer. Jl. of Sci.* (5) XXIII, pp. 421–439, 1932.

cretion and the hypothetical form of an individual molecule. But molecular form is a very important concept; and the chemist has not only succeeded, since the days of Wöhler, in synthetising many substances which are characteristically associated with living matter, but his task has included the attempt to account for the molecular *forms* of certain "asymmetric" substances—glucose, malic acid and many more—as they occur in Nature. These are bodies which, when artificially synthetised, have no optical activity, but which, as we actually find them in organisms, turn (when *in solution*) the plane of polarised light in one direction rather than the other; thus dextroglucose and laevomalic acid are common products of plant metabolism, but dextromalic acid and laevoglucose do not occur in Nature at all. The optical activity of these bodies depends, as Pasteur shewed eighty years ago*, upon the form, right-handed or left-handed, of their molecules, which molecular asymmetry further gives rise to a corresponding right- or left-handedness (or enantiomorphism) in the crystalline aggregates. It is a distinct problem in organic or physiological chemistry, and by no means without its interest for the morphologist, to discover how it is that Nature, for each particular substance, habitually builds up, or at least selects, its molecules in a one-sided fashion, right-handed or left-handed as the case may be. It will serve us no better to assert that this phenomenon has its origin in "fortuity" than to repeat the Abbé Galiani's saying, "*les dés de la nature sont pipés.*"

The problem is not so closely related to our immediate subject that we need discuss it at length; but it has its relation, such as it is, to the general question of *form* in relation to vital phenomena, and it has its historic interest as a theme of long-continued discussion. According to Pasteur, there lay in the molecular asymmetry of the natural bodies and their symmetry when artificially produced, one of the most deep-seated differences between vital and non-vital phenomena: he went further, and declared that "this was perhaps the *only* well-marked line of demarcation that can at present [1860] be drawn between the chemistry of dead and of living matter." Nearly forty years afterwards the same theme was pursued and

* Lectures on the molecular asymmetry of natural organic compounds, *Chemical Soc. of Paris*, 1860; also in Ostwald's *Klassiker d. exact. Wiss.* No. 28, and in *Alembic Club Reprints*, No. 14, Edinburgh, 1897; cf. G. M. Richardson, *Foundations of Stereochemistry*, New York, 1901.

elaborated by Japp in a celebrated lecture*, and the distinction still has its weight, I believe, in the minds of many chemists. "We arrive at the conclusion," said Professor Japp, "that the production of single asymmetric compounds, or their isolation from the mixture of their enantiomorphs, is, as Pasteur firmly held, the prerogative of life. Only the living organism, or the living intelligence with its conception of asymmetry, can produce this result. Only asymmetry can beget asymmetry." In these last words (which, so far as the chemist and the biologist are concerned, we may acknowledge to be true†) lies the crux of the difficulty.

Observe that it is only the first beginnings of chemical asymmetry that we need discover; for when asymmetry is once manifested, it is not disputed that it will continue "to beget asymmetry." A plausible suggestion is at hand, which if it were confirmed and extended would supply or at least sufficiently illustrate the kind of explanation that is required. We know that when ordinary non-polarised light acts upon a chemical substance, the amount of chemical action is proportionate to the amount of light absorbed. We know in the second place‡ that light circularly polarised is absorbed in certain cases in different amounts by the right-handed or left-handed varieties of an asymmetric substance. And thirdly, we know that a portion of the light which comes to us from the sun is already plane-polarised light, which becomes in part circularly polarised, by reflection (according to Jamin) at the surface of the sea, and then rotated in a particular direction under the influence of terrestrial magnetism. We only require to be assured that the relation between absorption of light and chemical activity will continue to hold good in the case of circularly polarised light; that is to say that the formation of some new substance or other, under the influence of light so polarised, will proceed asymmetrically in consonance with the asymmetry of the light itself; or conversely,

* F. R. Japp, Stereochemistry and vitalism, *Brit. Ass. Rep.* (Bristol), 1898, p. 813; cf. also a voluminous discussion in *Nature*, 1898–99.

† They represent the general theorem of which particular cases are found, for instance, in the asymmetry of the ferments (or *enzymes*) which act upon asymmetrical bodies, the one fitting the other, according to Emil Fischer's well-known phrase, as lock and key. Cf. his Bedeutung der Stereochemie für die Physiologie, *Z. f. physiol. Chemie*, v, p. 60, 1899, and various papers in the *Ber. d. d. chem. Ges.* from 1894.

‡ Cf. Cotton, *Ann. de Chim. et de Phys.* (7), VIII, pp. 347–432 (cf. p. 373), 1896.

that the asymmetrically polarised light will tend to more rapid decomposition of those molecules by which it is chiefly absorbed. This latter proof is said to be furnished by Byk*, who asserts that certain tartrates become unsymmetrical under the continued influence of the asymmetric rays. Here then we seem to have an example, of a particular kind and in a particular instance, an example limited but yet crucial if confirmed, of an asymmetric force, non-vital in its origin, which might conceivably be the starting-point of that asymmetry which is characteristic of so many organic products.

The mysteries of organic chemistry are great, and the differences between its processes or reactions as they are carried out in the organism and in the laboratory are many†; the actions, catalytic and other, which go on in the living cell are of extraordinary complexity. But the contention that they are different in kind from ordinary chemical operations, or that in the production of single asymmetric compounds there is actually, as Pasteur maintained, a "prerogative of life," would seem to be no longer tenable. Our historic interest in the whole question is increased by the fact, or the great probability, that "the tenacity with which Pasteur fought against the doctrine of spontaneous generation was not unconnected with his belief that chemical compounds of one-sided symmetry could not arise save under the influence of life‡." But the question whether spontaneous generation be a fact or not does not depend upon theoretical considerations; our negative response is based, and is soundly based, on repeated failures to demonstrate its occurrence. Many a great law of physical science, not excepting gravitation itself, has no higher claim on our acceptance.

Let us return from this digression to the general subject of the forms assumed by certain chemical bodies when deposited or precipitated within the organism, and to the question of how far these forms may be artificially imitated or theoretically explained.

* A. Byk, Zur Frage der Spaltbarkeit von Racemverbindungen durch zirkularpolarisiertes Licht, ein Beitrag zur primaren Entstehung optisch-activer Substanzen, *Zeitsch. f. physikal. Chemie*, XLIX, pp. 641–687, 1904. It must be admitted that positive evidence on these lines is still awanting.

† Cf. (*int. al.*) Emil Fischer, *Untersuchungen über Aminosäuren, Proteine*, etc. Berlin, 1906.

‡ Japp, *loc. cit.* p. 828.

Mr George Rainey, of St Thomas's Hospital (of whom we have spoken before), and Professor P. Harting, of Utrecht, were the first to deal with this specific problem. Rainey published, between 1857 and 1861, a series of valuable and thoughtful papers to shew that shell and bone and certain other organic structures were formed "by a process of molecular coalescence, demonstrable in certain artificially formed products*." Harting, after thirty years of experimental work, published in 1872 a paper, which has become classical, entitled *Recherches de morphologie synthétique, sur la production artificielle de quelques formations calcaires organiques*†; his aim was to pave the way for a "morphologie synthétique," as Wöhler had laid the foundations of a "chimie synthétique" by his classical discovery forty years before.

Rainey and Harting used similar methods—and these were such as other workers have continued to employ—partly with the direct object of explaining the genesis of organic forms and partly as an integral part of what is now known as Colloid Chemistry. The gist of the method was to bring some soluble salt of lime, such as the chloride or nitrate, into solution within a colloid medium, such as gum, gelatine or albumin; and then to precipitate it out in the form of some insoluble compound, such as the carbonate or oxalate. Harting found that, when he added a little sodium or potassium carbonate to a concentrated solution of calcium chloride in albumin, he got at first a gelatinous mass, or "colloid precipitate": which slowly transformed by the appearance of tiny microscopic particles,

* George Rainey, On the elementary formation of the skeletons of animals, and other hard structures formed in connection with living tissue, *Brit. and For. Med. Ch. Rev.* XX, pp. 451–476, 1857; published separately with additions, 8vo, London, 1858. For other papers by Rainey on kindred subjects see *Q.J.M.S.* VI (*Tr. Microsc. Soc.*), pp. 41–50, 1858; VII, pp. 212–225, 1859; VIII, pp. 1–10, 1860; I (n.s.), pp. 23–32, 1861. Cf. also W. Miller Ord, *On the influence exercised by colloids upon crystalline form*, pp. x, 179, 1874; cf. also *Q.J.M.S.* XII, pp. 219–239, 1872; also the early but still interesting observations of Mr Charles Hatchett, Chemical experiments on zoophytes; with some observations on the component parts of membrane, *Phil. Trans.* 1800, pp. 327–402. For early references to sclerites formed in cells, see (e.g.) L. Selenka, *Z.f.w.Z.* XXXIII, p. 45, 1879 and R. Semon, *Mitth. Zool. St. Neapel*, VII, p. 288, 1886 (both in holothurians); Blochmann, *Die Epithelfrage bei Cestoden u. Trematoden*, Hamburg, 1896; also Leger's Observations on crystals of calcium oxalate in the cysts of *Lithocystis Schneideri*, *A.M.N.H.* (6), XVIII, p. 479, 1895.

† Cf. *Q.J.M.S.* XII, pp. 118–123, 1872.

shewing, as they grew larger, the typical Brownian movement. So far, much the same phenomena were witnessed whether the solution were albuminous or not, and similar appearances indeed had been witnessed and recorded by Gustav Rose, so far back as 1837*; but in the later stages the presence of albuminoid matter made a great difference. Now, after a few days, the calcium carbonate was seen to be deposited in the form of large rounded concretions, each with a more or less distinct central nucleus and with a surrounding structure at once radiate and concentric; the presence of concentric zones or lamellae, alternately dark and clear, was especially characteristic. These round "calcospherites" shewed a tendency to aggregate in layers, and then to assume polyhedral, often regularly hexagonal, outlines. In this latter condition they closely resemble the early stages of calcification in a molluscan (Fig. 296), or still more in a crustacean shell†; while in their isolated condition they

* Cf. Quincke, Ueber unsichtbare Flüssigkeitsschichten, etc., *Ann. der Physik* (4), VII, pp. 631–682, 701–744, 1902.

† See for instance other excellent illustrations in Carpenter's article "Shell," in Todd's *Cyclopædia*, IV, pp. 556–571, 1847–49. According to Carpenter, the shells of the mollusca (and also of the crustacea) are "essentially composed of *cells*, consolidated by a deposit of carbonate of lime in their interior." That is to say, Carpenter supposed that the spherulites or calcospherites of Harting were, to begin with, just so many living protoplasmic cells. Soon afterwards, however, Huxley pointed out that the mode of formation, while at first sight "irresistibly suggesting a cellular structure...is in reality nothing of the kind," but "is simply the result of the concretionary manner in which the calcareous matter is deposited"; *ibid.* art. "Tegumentary organs," V, p. 487, 1859. Quekett (*Lectures on Histology*, II, p. 393, 1854, and *Q.J.M.S.* XI, pp. 95–104, 1863) supported Carpenter; but Williamson (Histological features in the shells of the Crustacea, *Q.J.M.S.* VIII, pp. 35–47, 1860) amply confirmed Huxley's view, which in the end Carpenter himself adopted (*The Microscope*, 1862, p. 604). A like controversy arose later in regard to corals. Mrs Gordon (M. M. Ogilvie) asserted that the coral was built up "of successive layers of calcified cells, which hang together at first by their cell-walls, and ultimately, as crystalline changes continue, form the individual laminae of the skeletal structures" (*Phil. Trans.* CLXXXVII, p. 102, 1896): whereas von Koch had figured the coral as formed out of a mass of "Kalkconcremente" or "crystalline spheroids," laid down outside the ectoderm, and precisely similar both in their early rounded and later polygonal stages (though von Koch was not aware of the fact) to the calcospherites of Harting (Entw. d. Kalkskelettes von Astroides, *Mitth. Zool. St. Neapel*, III, pp. 284–290, pl. XX, 1882). Lastly, W. H. Bryan finds all ordinary corals (*Hexacoralla*) to be mineral aggregates formed by "spherulitic crystallisation," due in turn to the presence of a colloid matrix secreted by certain areas of ectoderm; see *Proc. R.S. Queensland*, LII, pp. 41–53, 1940; *Univ. of Queensland Papers, Geology*, II, 4 and 5, 1941. Cf. J. E. Duerden, On *Siderastraea*, *Carnegie Inst. Washington*, 1904, p. 34.

Fig. 293. Calcospherites, or concretions of calcium carbonate, deposited in white of egg. After Harting.

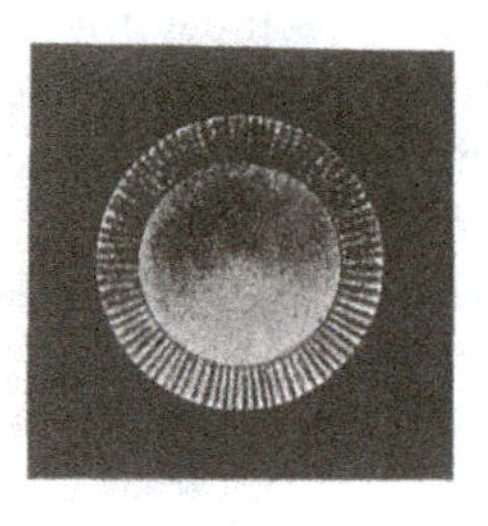

Fig. 294. A single calcospherite, with central "nucleus," and striated, iridescent border. After Harting.

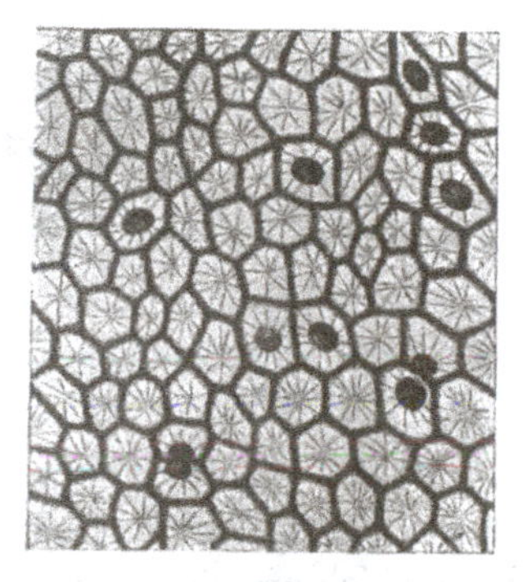

Fig. 295. Later stages in the same experiment.

Fig. 296. A, Section of shell; B, Section of hinge-tooth of *Mya*. After Carpenter.

closely resemble the little calcareous bodies in the tissues of a trematode or a cestode worm, or in the oesophageal glands of an earthworm*.

When the albumin was somewhat scanty, or when it was mixed with gelatine, and especially when a little phosphate of lime was added to the mixture, the spheroidal globules tended to become rough, by an outgrowth of spinous or digitiform projections; and

Fig. 297. Large irregular calcareous concretions, or spicules, deposited in a piece of dead cartilage, in presence of calcium phosphate. After Harting.

in some cases, but not without the presence of the phosphate†, the result was an irregularly shaped knobby spicule, precisely similar to those which are characteristic of the *Alcyonaria*‡.

The rough spicules of the *Alcyonaria* are extraordinarily variable in shape and size, as, looking at them from the chemist's or the physicist's point of view, we should expect them to be. Partly upon the form of these spicules, and partly on the general form or mode of branching of the entire colony of polyps, a vast number of separate "species" have been based by systematic

* Cf. Claparède, *Z.f.w.Z.* XIX, p. 604, 1869. On the structure of the molluscan shell, see O. B. Boggild, *K. Vidensk. Selsk. Skr.*, Kjöbenh., (9) II, 1930. On nacre, or mother-of-pearl, see Brewster, *Treatise on Optics*, 1853, p. 137; Schmidt, *Die Bausteine der Tierkörper in polarisirtem Licht*, Bonn, 1924. Also S. Ruma Swamy, *Proc. Ind. Acad. Sci.* (A), I, p. 871, 1935; P. S. Srinivasam, *ibid.* V, pp. 464–483, 1937; and, on the specific qualities of the nacre in the several divisions of the Mollusca, Sir C. V. Raman, *ibid.* pp. 559, etc., 1935.

† On the deposition of phosphates in organisms, cf. Pauli u. Samec, *Biochem. Ztschr.* XVII, p. 235, 1909; *Wiener mediz. Wochenschr.* 1910, pp. 2287–2292.

‡ Spicules much like those of the *Alcyonaria* occur also in a few sponges; cf. (e.g.), Vaughan Jennings, *Journ. Linn. Soc.* XXIII, p. 531, pl. 13, fig. 8, 1891.

zoologists. But it is now admitted that even in specimens of a single species, from one and the same locality, the spicules may vary immensely in shape and size: and Professor S. J. Hickson declared that after many years of laborious work in striving to determine species of these animal colonies, he felt "quite convinced that we have been engaged in a more or less fruitless task*."

The formation of a tooth is a phenomenon of the same order. That is to say, "calcification in both dentine and enamel is in great part a physical phenomenon; the actual deposit in both tissues occurs in the form of calcospherites, and the process in mammalian tissue is identical in every point with

Fig. 298. Additional illustrations of alcyonarian spicules: *Eunicea*. After Studer.

the same process occurring in lower organisms†." The ossification of bone, we may be sure, is in the same sense and to the same extent a physical phenomenon.

The typical structure of a calcospherite is no other than that of a pearl, nor does it differ essentially from that of the otolith of a mollusc or of a bony fish. (The otoliths of the elasmobranch fishes, like those of reptiles and birds, are not developed after this fashion, but are true crystals of calc-spar.)

The effect of surface-tension is manifest throughout these phenomena. It is by surface-tension that ultra-microscopic particles are brought together in the first floccular precipitate or coagulum; by

* *Mem. Manchester Lit. and Phil. Soc.* LX, p. 11, 1916.

† J. H. Mummery, On calcification in enamel and dentine, *Phil. Trans.* (B), CCV, pp. 95–111, 1914.

the same agency the coarser particles are in turn agglutinated into visible lumps; and the form of the calcospherites, whether it be that of the solitary spheres or that assumed in various stages of aggregation (e.g. Fig. 300)*, is likewise due to the same agency.

From the point of view of colloid chemistry the whole phenomenon is important and significant; and not the least significant part is this tendency of the solidified deposits to assume the form of "spherulites" and other rounded contours. In the phraseology

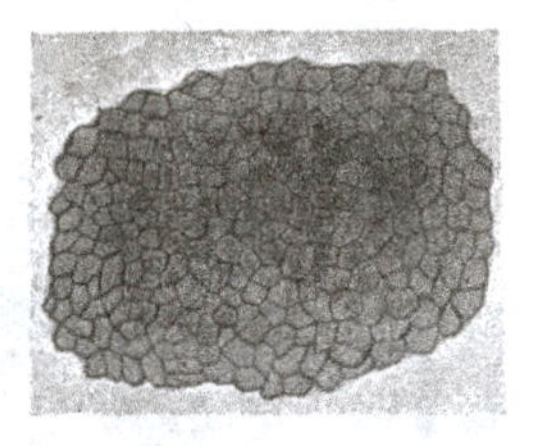

Fig. 299. A "crust" of close-packed calcareous concretions, precipitated at the surface of an albuminous solution. After Harting.

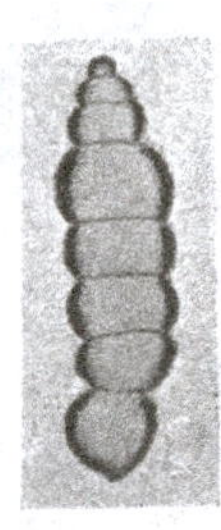

Fig. 300. Aggregated calcospherites. After Harting.

of that science, we are dealing with a *two-phase* system, which finally consists of solid particles in suspension in a liquid—a *disperse phase* in a *dispersion medium.* In accordance with a rule first recognised by Ostwald, when a substance begins to separate out from a solution, so making its appearance as a *new phase*, it always makes its appearance first as a liquid†. Here is a case in point. The minute quantities of material, on their way from a state of solution to a state of "suspension," pass through a liquid to a solid form; their temporary sojourn in the former leaves its impress in the rounded contours which surface-tension brought about while the little aggregate was still labile or fluid: while coincidently with this surface-tension effect, crystallisation tends to take place throughout the little liquid mass, or in such portions of it as have not yet consolidated and crystallised.

* The artificial concretion represented in Fig. 300 is identical in appearance with the concretions found in the kidney of *Nautilus*, as figured by Willey (*Zoological Results*, p. lxxvi, Fig. 2, 1902).

† This rule, undreamed of by Errera, supports and justifies his cardinal assumption (of which we have had so much to say in discussing the forms of cells and tissues) that the *incipient* cell-wall behaves as, and indeed actually is, a liquid film (cf. p. 482).

Where we have simple aggregates of two or three calcospherites the resulting figure is that of so many contiguous soap-bubbles. In other cases composite forms result which are not so easily explained, but which, if we could only account for them, would be of very great interest to the biologist. For instance, when smaller calcospheres seem, as it were, to invade the substance of a larger one, we get curious conformations which somewhat resemble the outlines of certain diatoms (Fig. 301). Another curious formation, which Harting calls a "conostat," is of frequent occurrence, and in it we

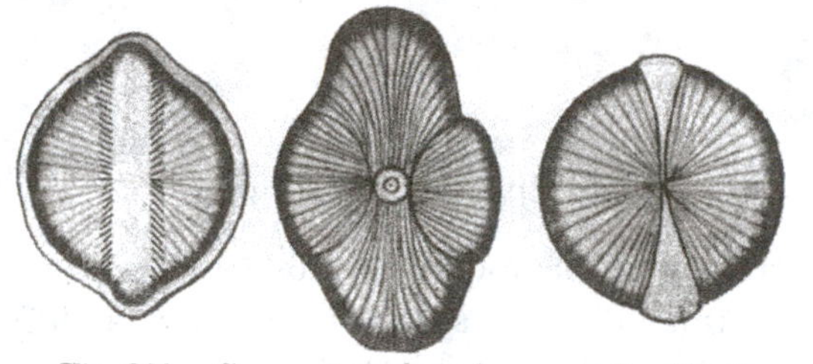

Fig. 301. Composite calcospheres. After Harting.

see at least a suggestion of analogy with the configuration which, in a protoplasmic structure, we have spoken of as a "collar-cell." The conostats, which are formed in the surface layer of the solution, consist of a portion of a spheroidal calcospherite, whose upper part is continued into a thin spheroidal collar of somewhat larger radius than the solid sphere; but the precise manner in which the collar is formed, possibly around a bubble of gas, possibly about a vortex-like diffusion-current, is not obvious.

Among these various phenomena, the concentric striation of the calcospherite has acquired a special interest and importance*. It is part of a phenomenon now widely known under the name of "Liesegang's Rings†."

* Cf. Harting, *op. cit.* pp. 22, 50: "J'avais cru d'abord que ces couches concentriques étaient produites par l'alternance de la chaleur ou de la lumière, pendant le jour et la nuit. Mais l'expérience, expressément instituée pour examiner cette question, y a répondu négativement."

† R. E. Liesegang, *Ueber die Schichtungen bei Diffusionen*, Leipzig, 1907, and earlier papers. A periodic precipitate is said to have been first noticed (on filter-paper) by Runge, in 1885; cf. Quincke, Ueber unsichtbare Flüssigkeitsschichten, *Ann. d. Physik* (4), VII, pp. 643–7, 1902. On a very minute periodicity in the so-called Hookham's crystals, formed by crystallising copper sulphate and salicin in strong syrup, see Rayleigh, *Collected Papers*, VI, p. 661: "There is much here," says Rayleigh, "to excite admiration and perplexity."

If we dissolve, for instance, a little potassium bichromate in gelatine, pour it on to a glass plate, and after it is set pour upon it a drop of silver nitrate solution, there appears in the course of a few hours the phenomenon of Liesegang's rings. At first the silver

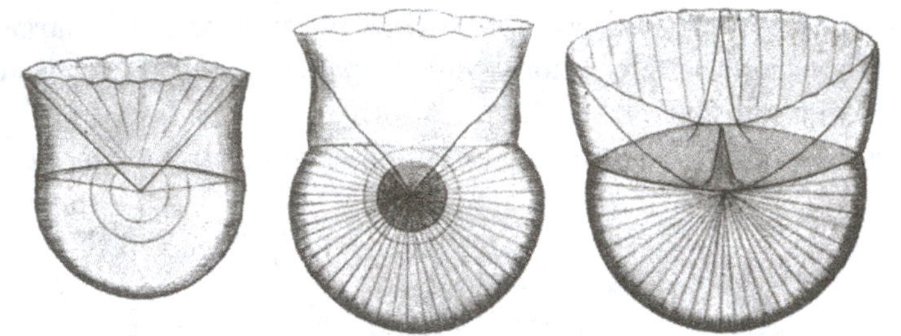

Fig. 302. Conostats. After Harting.

forms a central patch of abundant reddish-brown chromate precipitate; but around this, as the silver nitrate diffuses slowly through the gelatine, the precipitate no longer comes down continuously, but forms a series of concentric rings or zones, beautifully

Fig. 303. Liesegang's rings. After Leduc.

regular, which alternate with clear interspaces of jelly and stand farther and farther apart in a definite ratio as they recede from the centre*. For a discussion of the *raison d'être* of this phenomenon, the student will consult the textbooks of physical and colloid chemistry. But, speaking generally, we may say that the appearance

* It is now known that periodic precipitation may be exhibited even in aqueous solutions, and that what the gel does is to enlarge the intervals, and to enhance the phenomenon, by affecting the rate or relative rates of diffusion. Cf. H. W. Morse, *Journ. Phys. Chem.* 1931.

of Liesegang's rings is but a particular case of a more general phenomenon, namely the influence on crystallisation of the presence of foreign bodies or "impurities," represented in this case by the gel or colloid matrix. F. S. Beudant had shewn in a fine paper, more than a hundred years ago, that impurities were the chief cause of variation of crystal habit*. Faraday proved that to diffusion in presence of slight impurities, not to actual stratification or alternate deposition, could be ascribed the banded structure of ice,

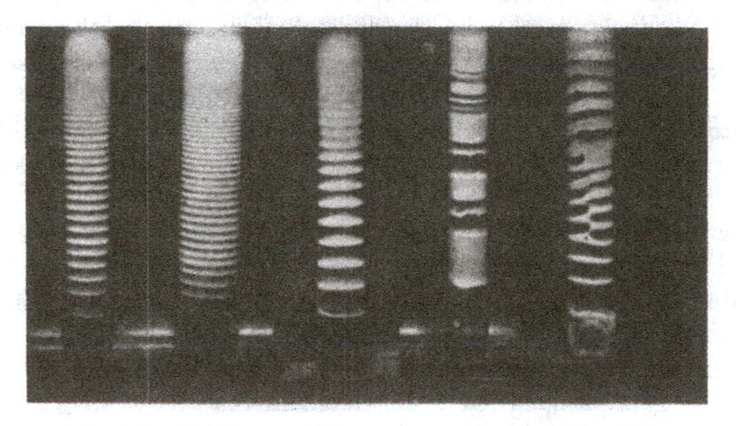

Fig. 304. The Liesegang phenomena. After Emil Hatschek.

of agate or of onyx; and Quincke and Tomlinson added to our scanty knowledge of this remarkable phenomenon†. Ruskin, who knew a great deal about agates, spoke of the perpetual difficulty of distinguishing "between concretionary separation and successive deposition." And Rayleigh shewed how to such a *periodic*, but

* F. S. Beudant, Recherches sur les causes qui peuvent varier les formes crystallines d'une même substance minérale, *Ann. de Chimie*, VIII, pp. 5–52, 1818. See also his Mémoire sur les parties solides des Mollusques, *Mém. du Muséum*, XV, pp. 66–75, 1810.

† Cf. Faraday, On ice of irregular fusibility, *Phil. Trans.* 1858, p. 228; *Researches in Chemistry, etc.*, 1859, p. 374; Canon Moseley, On the veined structure of the ice of glaciers, *Phil. Mag.* (4), XXXIX, p. 241, 1870; R. Weber, in *Poggend. Ann.* CIX, p. 379, 1860; Tyndall, *Forms of Water*, 1872, p. 178; C. Tomlinson, On some effects of small quantities of foreign matter on crystallisation, *Phil. Mag.* (5) XXXI, p. 393, 1891, and other papers. Cf. Liesegang, *Centralbl. f. Mineralogie*, XVI, p. 497, 1911; E. S. Hedges and J. E. Myers, *The problem of physico-chemical periodicity*, London, 1926; W. F. Berg, Crystal growth from solutions, *Proc. R.S.* (A), CLXIV, pp. 79–95, 1938.

unstratified structure all the colours of the opal and the iridescence of ancient glass are alike due.

Besides the tendency to rhythmic action, as manifested in Liesegang's rings, the association of colloid matter with a crystalloid in solution may lead to other well-marked effects. These include*: (1) the total prevention of crystallisation; (2) suppression of certain of the lines of crystal growth; (3) extension of the crystal to abnormal proportions, with a tendency to become compound; (4) a curving or gyrating of the crystal or its parts.

It would seem that, if the supply of material to the growing crystal begin to run short (as may well happen in a colloid medium for lack of convection-currents), then growth will follow only the strongest lines of crystallising force, and will be suppressed or partially suppressed along other axes. The crystal will have a tendency to become filiform, or "fibrous"; and the raphides of our plant-cells, and the needle-like "oxyotes" of sponges, are cases in point. Again, the long slender crystal so formed, pushing its way into new material, may start a new centre of crystallisation: whereby we get the phenomenon known as a "relay," along the principal lines of force and sometimes along subordinate axes as well. This phenomenon is illustrated in the accompanying figure of common salt crystallising in a colloid medium; and it may be that we have here an explanation, or part of an explanation, of the compound siliceous spicules of the Hexactinellid sponges. Lastly, when the crystallising force is nearly equalled by the resistance of the viscous medium, the crystal takes the line of least resistance, with very various results. One of these results would seem to be a gyratory course, giving to the crystal a curious wheel-like shape, as in Fig. 306; and other results are the feathery, fern-like or arborescent shapes so frequently seen in microscopic crystallisation.

To return to Liesegang's rings, the typical appearance of concentric rings upon a plate of gelatine may be modified in various experimental ways. For instance, if our gelatinous medium be placed in a capillary tube immersed in a solution of the precipitating salt, we obtain (Fig. 304) a vertical succession of bands or zones regularly

* Cf. J. H. Bowman, A study in crystallisation, *Journ. Soc. of Chem. Industry*, XXV, p. 143, 1906.

interspaced: the result being very closely comparable to the banded pigmentation which we see in the hair of a rabbit or a rat. In the ordinary plate preparation, the free surface of the gelatine is under different conditions to the layers below and especially to the lowest layer of all in contact with the glass; and so we often obtain a

Fig. 305. Relay-crystals of common salt. After Bowman.

double series of rings, one deep and the other superficial, which by occasional blending or interlacing may produce a netted pattern. Sometimes, when only the inner surface of our capillary tube is covered with a layer of gelatine, there is a tendency for the deposit

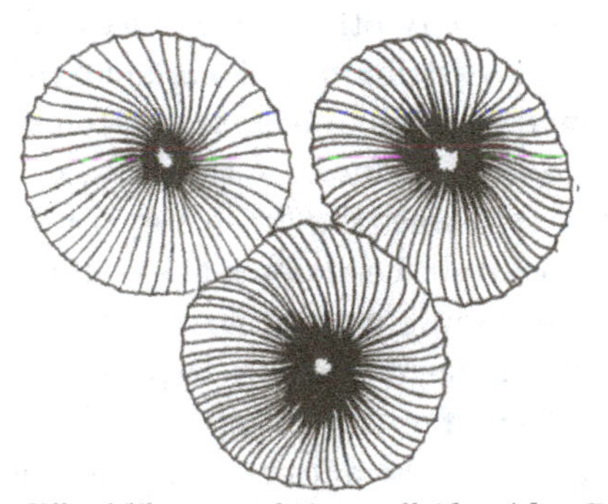

Fig. 306. Wheel-like crystals in a colloid. After Bowman.

to take place in a continuous spiral, rather than in concentric and separate zones. By such means, according to Küster*, various forms of annular, spiral and reticulated thickenings in the vascular tissue of plants may be closely imitated; and he and certain other writers have been inclined to carry the same chemico-physical

* E. Küster, Ueber die Schichtung der Stärkekörner, *Ber. d. botan. Gesellsch.* XXXI, pp. 339–346, 1913; Ueber Zonenbildung in kolloidalen Medien, *Koll. Ztschr.* XIII, pp. 192–194; XIV, pp. 307–319, 1913–14.

phenomenon a very long way, in the explanation of various banded, striped, and other rhythmically successional types of structure or pigmentation. The striped leaves of many plants (such as *Eulalia japonica*), the striped or clouded colouring of many feathers or of a cat's skin, the patterns of many fishes, such for instance as the brightly coloured tropical Chaetodonts and the like, are all regarded by him as so many instances of "diffusion-figures" closely related to the typical Liesegang phenomenon. Gebhardt* declares that the banded wings of *Papilio podalirius* are analogous to or even closely imitated in Liesegang's experiments; that the finer markings on the wings of the goatmoth shew a double rhythm, alternately coarse and fine, such as is manifested in certain experimental cases of the same kind; that the alternate banding of the antennae (for instance in *Sesia spheciformis*), a pigmentation not concurrent with the antennal joints, is explicable in the same way; and that the ocelli on the wings of the Emperor moth are typical illustrations of the common concentric type. Darwin's well-known disquisition on the ocellar pattern of the feathers of the Argus pheasant, as a result of sexual selection, will occur to the reader's mind, in striking contrast to this or to any other direct physical explanation†.

To turn from the distribution of pigment to more deeply seated structural characters, Leduc has argued, for instance, that the laminar structure of the cornea or the lens is, or may be, a similar phenomenon. In the lens of the fish's eye, we have a very curious appearance, the consecutive lamellae being roughened or notched by close-set, interlocking sinuosities; and the same appearance, save that it is not quite so regular, is presented in one of Küster's figures as the effect of precipitating a little sodium phosphate in a gelatinous medium. Biedermann has studied, from the same

* *Verh. d. d. zool. Gesellsch.* p. 179, 1912.

† As a matter of fact, the phenomena associated with the development of an "ocellus" are or may be of great complexity, inasmuch as they involve not only a graded distribution of pigment, but also, in "optical" coloration, a symmetrical distribution of structure or form. The subject therefore deserves very careful discussion, such as Bateson gives to it (*Variation*, chap. XII). This, by the way, is one of the very rare cases in which Bateson appears inclined to suggest a purely physical explanation of an organic phenomenon: "The suggestion is strong that the whole series of rings (in *Morpho*) may have been formed by some one central disturbance, somewhat as a series of concentric waves may be formed by the splash of a stone thrown into a pool." Cf. Darwin, *Descent of Man*, II, p. 132, 1871.

point of view, the structure and development of the molluscan shell, the problem which Rainey had first attacked more than fifty years before*; and Liesegang himself has applied his results to the formation of pearls, and, as Bechhold has also done, to the development of bone†.

The presence of concentric rings or zones in slow-growing structures is evidently after some fashion a function of the time, and an indication of periodic acceleration or variation of growth; it is apt to be referred, rightly or wrongly, to the seasons of the year, and to be interpreted (with or without confirmation and proof) as a sure mark and measure of the creature's age. This is the case, for instance, with the scales, bones and otoliths of fishes; and a kindred phenomenon in starch-grains has given rise, in like manner, to the belief that they indicate a diurnal and nocturnal periodicity of activity and rest‡ on the part of the cell wherein they grew.

That this is actually the case in growing starch-grains is often if not generally believed, on the authority of Meyer§; but while under certain circumstances a marked alternation of growing and resting periods may occur, and may leave its impress on the structure of the grain, there is now more reason to believe that, apart from such external influences, the internal phenomena of diffusion may, just as in the typical Liesegang experiment, produce the well-known concentric rings. The spherocrystals of inulin, in like manner, shew, like the calcospherites of Harting (Fig. 307), a concentric structure which in all likelihood has had no causative impulse save from within.

The striation, or concentric lamellation, of the scales and otoliths of fishes has been much employed, not as a mere indication, but

* Cf. also Sir D. Brewster, On optical properties of mother of pearl, *Phil. Trans.* 1814, p. 397; and J. F. W. Herschel, in *Edin. Phil. Journ.* II, p. 116, 1819.

† W. Biedermann, Ueber die Bedeutung von Kristallisationsprozessen der Skelette wirbelloser Thiere, namentlich der Molluskenschalen, *Z. f. allg. Physiol.* I, p. 154, 1902; Ueber Bau und Entstehung der Molluskenschale, *Jen. Zeitschr.* XXXVI, pp. 1–164, 1902. Cf. also Steinmann, Ueber Schale und Kalksteinbildungen, *Ber. Naturf. Ges. Freiburg i. Br.* IV, 1889; Liesegang, *Naturw. Wochenschr.* 1910. p. 641; *Arch. f. Entw. Mech.* XXXIV, p. 452, 1912; H. Bechhold, *Ztschr. f. phys. Chem.* LII, p. 185, 1905.

‡ Cf. Bütschli, Ueber die Herstellung künstlicher Stärkekörner oder von Sphärokrystallen der Stärke, *Verh. nat. med. Ver. Heidelberg*, V, pp. 457–472, 1896.

§ *Untersuchungen über die Stärkekörner*, Jena, 1905.

as a trustworthy and unmistakeable measure of the fish's age (see *ante*, p. 180). There are some difficulties in the way of accepting this hypothesis, not the least of which is the fact that the otolith-zones, for instance, are extremely well marked even in the case of some fishes which spend their lives in deep water, where temperature and other physical conditions shew little or no appreciable fluctuation with the seasons of the year. There are, on the other hand, phenomena which seem strongly confirmatory of the hypothesis: for instance, the fact (if it be fully established) that in such a fish as the cod, zones of growth, *identical in number*, are found both on

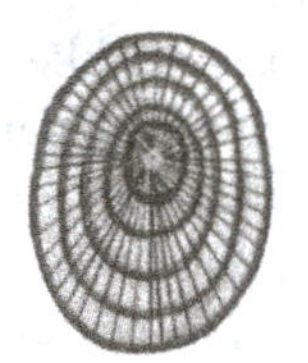

Fig. 307. A sphero-crystal of inulin.

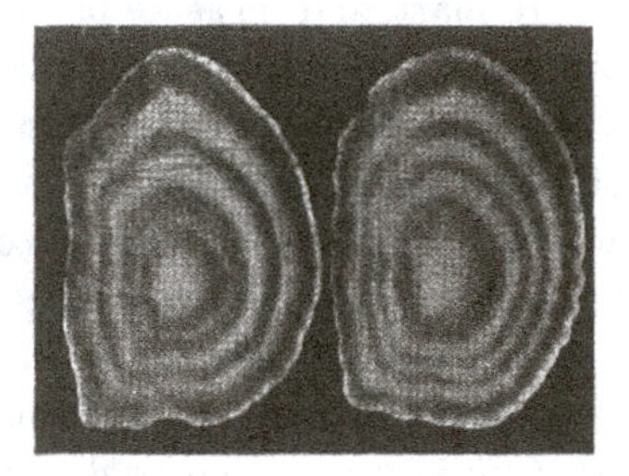

Fig. 308. Otoliths of plaice, shewing four zones or "age-rings." After Wallace.

the scales and in the otoliths*. The subject is as difficult as it is important, but it is at least certain, with the Liesegang phenomenon in view, that we have no right to *assume*, without proof and confirmation, that rhythm and periodicity in structure and growth are necessarily bound up with, and indubitably brought about by, a periodic or seasonal recurrence of particular *external* conditions†.

But while in the ordinary Liesegang phenomenon rhythmic

* Cf. Winge, *Meddel. fra Komm. for Havundersögelse* (*Fiskeri*), IV, p. 20, Copenhagen, 1915.

† A. W. Morosow strongly supports the view—uncertain as it seems to be—that the concentric pattern of a fish's scale is due to the Liesegang phenomenon; he produces an "artificial scale," with its "summer and winter rings," by precipitating sodium carbonate and calcium chloride in gelatin: Zur Frage über die Natur des Schuppenwachstums bei Fischen (and in Russian), *Nation. Comm. Agriculture: Rep. Sci. Inst. Fisheries*, I, Moscow, 1924; abstract in Michael Graham's Studies of age-determination in fish, *Rep. Ministry of Agr. and Fisheries, Fishery Investigations*, (2) XI, no. 3, p. 28, 1928.

precipitation depends only on forces intrinsic to the system, and is independent of any corresponding rhythmic changes in external conditions, we have not far to seek for analogous chemico-physical phenomena where rhythmic alternations of structure are produced in close relation to periodic fluctuations of temperature. The banding, or "varving," of Swedish and Irish glacial clays is a remarkable instance. A well-known and a simple case is that of the Stassfurt deposits, where the rock-salt alternates with thin layers of "anhydrite," or (in another series of beds) with "polyhalite*": and where these zones are commonly regarded as marking years, and their alternate bands as due to the seasons. A discussion, however, of this remarkable and significant phenomenon, and of how the chemist explains it, by help of the "phase-rule," in connection with temperature conditions, would lead us far beyond our scope.

We may turn aside to touch, for a single moment, on certain forms and patterns not easy to classify: some of which depend on the molecular structure of a colloid matrix, while others are of a coarser and more mechanical grade. So many organic forms and patterns await explanation that we cannot seek too widely for examples, nor for explanations, of such things. For instance, a drop of dried egg-albumin shews beautiful radial cracks, with cross-lines here and there; and a drop of blood drying on a glass plate shews a complete system of radial fissures, in series after series, sometimes with and sometimes without a clear central space. The general resemblance to the cross-section of a stem, with its pith and its primary and secondary medullary rays, is striking enough to have led some even to look upon a tree as one great complicated but symmetrical colloid mass†. We may compare also the beautiful radiating structure which Bütschli observed long ago around small

* The anhydrite is sulphate of lime ($CaSO_4$); the polyhalite is a triple sulphate of lime, magnesia and potash ($2CaSO_4 . MgSO_4 . K_2SO_4 + 2H_2O$).

† Cf. H. Wislicenus, *Ztschr. f. Chemie u. Kolloide*, VI, 1910; A. Lingelsheim, Pflanzenanatomische Strukturbilder in trocknenden Kolloiden, *Arch. f. Entw. Mech.* XLII, pp. 117–125, 1917. Cf. also Liesegang, Trocknungserscheinungen bei Gelen, *Ztschr. f. Ch. u. K.* X, p. 229 *sq.*, 1912; Bütschli, *Verh. n. h. Ver. Heidelberg*, VII, p. 653, 1904. Also (*int. al.*) Norman Stuart, on Spiral growths in silica gel, *Nature*, Oct. 2, 1937, p. 589.

bubbles in chrome-gelatine, and which he used in one of his early (and none too fortunate) speculations on the nature of the nuclear spindle.

We see that the methods by which we attempt to study the chemico-physical characteristics of an inorganic concretion or spicule within the body of an organism soon introduce us to a multitude of phenomena of which our knowledge is extremely scanty, and which we must not attempt to discuss at greater length. As regards our main point, namely the formation of spicules and other elementary skeletal forms, we have seen that some of them may be safely ascribed to precipitation or crystallisation of inorganic materials in ways modified by the presence of albuminous or other colloid substances. The effect of these latter is found to be much greater in the case of some crystallisable bodies than in others. For instance Harting, and Rainey also, found that calcium oxalate was much less affected by a colloid medium than was calcium carbonate; it shewed in their hands no tendency to form rounded concretions or "calcospherites" in presence of a colloid, but continued to crystallise, either normally or with a tendency to form needles or raphides. It is doubtless for this reason that, as we have seen, *crystals* of calcium oxalate are so common in the tissues of plants, while those of other calcium salts are rare; but true calcospherites, or spherocrystals, even of the oxalate are occasionally found, for instance in certain Cacti, and Bütschli* has succeeded in making them artificially in Harting's usual way, that is to say by crystallisation in a colloid medium. If the nature of the salt has a marked specific effect, so also has the gel: silver chromate is thrown down in rings in gelatin but not in agar; replace the silver by lead, and the rings come in agar but not in gelatin; while neither lead nor silver produce them in silicic acid gel.

There link on to such observations as Harting's, and to the statement already quoted that calcareous deposits are associated with the dead residua, or "formed materials," rather than with the living cells of the organism, certain very interesting facts in regard to the *solubility* of salts in colloid media, which go far to account for the presence (apart from the form) of calcareous pre-

* Spharocrystalle von Kalkoxalat bei Kakteen, *Ber. d. d. Bot. Gesellsch.* p. 178, 1885.

cipitates within the organism*. It has been shewn, in the first place, that the presence of albumin has a notable effect on the solubility in a watery solution of calcium salts, increasing the solubility of the phosphate in a marked degree and that of the carbonate in still greater proportion; but the sulphate is only very little more soluble in presence of albumin than in pure water, and the rarity of its occurrence within the organism is accounted for thereby. On the other hand, the bodies derived from the breaking down of the albumins—their "catabolic" products, such as the peptones, etc.—dissolve the calcium salts to a much less degree than albumin itself; and phosphate of lime is scarcely more soluble in them than in water. The probability is, therefore, that the actual precipitation of the calcium salts is not due to the direct action of carbonic acid on a more soluble salt (as was at one time believed); but to catabolic changes in the proteids of the organism, which throw down salts that had been already formed, but had remained hitherto in albuminous solution. The very slight solubility of calcium phosphate under such circumstances accounts for its predominance in mammalian bone†; and, in short, wherever a supply of this salt has been available to the organism.

To sum up, we see that, whether from food or from sea-water, calcium sulphate will tend to pass but little into solution in the albuminoid substances of the body: that calcium carbonate will enter more freely, but a considerable part of it will tend to remain in solution: while calcium phosphate will pass into solution in considerable amount, but will be almost wholly precipitated again as the albumin becomes broken down in the normal process of metabolism. We have still to wait for a similar and equally illuminating study of the solution and precipitation of *silica* in presence of organic colloids.

When carbonate of lime is secreted or precipitated by living organisms, to form bone, shell, egg-shell, coral and what not, its mineralogical form may vary, but the causes which determine it

* W. Pauli u. M. Samec, Ueber Löslichkeitsbeeinflüssung von Elektrolyten durch Eiweisskörper, *Biochem. Zeitschr.* XVII, p. 235, 1910. Some of these results were known much earlier; cf. Fokker in *Pfluger's Archiv*, VII, p. 274, 1873; also Robert Irvine and Sims Woodhead, *op. cit.* p. 347.

† Which, in 1000 parts of ash, contains about 840 parts of phosphate and 76 parts of calcium carbonate.

are all but unknown. It is amorphous in our bones. It has the form of calcite in an oyster, a starfish, a *Gorgonia*, a *Globigerina*; but of aragonite in most molluscs and in all ordinary corals. It is of calcite in a bird's egg, of aragonite in a tortoise's; of the one in *Argonauta*, of the other in *Nautilus*; of the one in an Ammonite, and the other in its *Aptychus*-lid; of the one in *Ostrea*, the other in *Unio*; of the one in the outer and the other in the inner layers of a limpet or a mussel-shell. Physical chemistry has little to say of the formation of these two, of the parts played by temperature, by the presence of sulphate of lime, or of magnesia or of various impurities; it leaves us in the dark as to what brings the one form or the other into being in the organism*.

Organic fibres, animal and vegetable, proteid and non-proteid, hair and wool, silk, cotton and the rest, may be mentioned here in passing: because, as formed material, they have a certain analogy to the spicular formations with which we are concerned. A hair or a wool-fibre may shew upon its surface the scaly or scurfy remnants of the living cells among which its substance was laid down; but the wool itself is by no means living, but is so nearly crystalline as to shew, in an X-ray photograph, the Laue interference-figures well known to physicists. Moreover, the same identical figure is obtained from such diverse sources as human hair, merino-wool and porcupine's quill. But if we stretch the thread, whether of hair or wool, the first Laue diagram changes to another; one crystalline arrangement has shifted over into a new form of molecular equilibrium. We are dealing with a crystalline, or crystal-like, form of *keratin*, the substance of which hoof and horn, nail, scale and feather are made; and this remarkable substance turns out to be a comparatively simple substance after all, with no very high or protein-like molecule†.

From the comparatively small group of inorganic formations which, arising within living organisms, owe their form to precipitation or to crystallisation, that is to say to chemical or other molecular

* Cf. Marcel Prenant, Les formes minéralogiques du calcaire chez les êtres vivants, *Biol. Reviews*, II, pp. 365–393, 1927.

† The study of wool and other fibres has much technical importance, and has gone far during the last few years; cf. W. T. Astbury, in *Phil. Trans.* (A), CCXXX, pp. 75–100, 1931, and other papers.

forces, we shall presently pass to that other and larger group which appears to be conformed in direct relation to the forms and the arrangement of cells or other protoplasmic elements*. The two principles of conformation are both illustrated in the spicular skeletons of the sponges.

In a considerable number but withal a minority of cases, the form of the sponge-spicule may be deemed sufficiently explained on the lines of Harting's and Rainey's experiments, that is to say as the direct result of chemical or physical phenomena associated with the deposition of lime or of silica in presence of colloids†. This is the case, for instance, with various small spicules of a globular or spheroidal form, consisting of amorphous silica, concentrically striated within, and often developing irregular knobs or tiny tubercles over their surfaces. In the aberrant sponge *Astrosclera*‡, we have, to begin with, rounded, striated discs or globules, which in like manner are nothing more nor less than the calcospherites of Harting's experiments; and as these grow they become closely aggregated together (Fig. 309), and assume an angular, polyhedral form, once more in complete accordance with the results of experiment§. Again, in many monaxonid sponges, we have irregularly shaped, or branched spicules, roughened or tuberculated by secondary superficial deposits, and reminding one of the spicules of the *Alcyonaria*. These also must be looked upon as the simple result of chemical deposition, the form of the deposit being somewhat modified in conformity with the surrounding tissues: just as in the simple experiment the form of the concretionary precipitate is affected by the heterogeneity, visible or invisible, of the matrix. Lastly, the simple needles of amorphous

* Cf. Fr. Dreyer, Die Principien der Gerüstbildung bei Rhizopoden, Spongien und Echinodermen, *Jen. Zeitschr.* XXVI, pp. 204–468, 1892.

† In a very anomalous Australian sponge, described by Professor Dendy (*Nature*, May 18, 1916, p. 253) under the name of *Collosclerophora*, the spicules are "gelatinous," consisting of a gel of colloid silica with a high percentage of water. It is not stated whether an organic colloid is present together with the silica. These gelatinous spicules arise as exudations on the outer surface of cells, and come to lie in intercellular spaces or vesicles.

‡ J. J. Lister, in Willey's *Zoological Results*, pt IV, p. 459, 1900.

§ The peculiar spicules of *Astrosclera* are said to consist of spherules, or calcospherites, of aragonite, spores of a certain red seaweed forming the nuclei or starting-points of the concretions (R. Kirkpatrick, *Proc. R.S.* (B), LXXXIV, p. 579, 1911).

silica which constitute one of the commonest types of spicule call for little in the way of explanation; they are accretions or deposits about a linear axis, or fine thread of organic material, just as the ordinary rounded calcospherite is deposited about some minute point or centre of crystallisation, and as ordinary crystallisation may be started by a particle of dust; in some cases they also, like the others, are apt to be roughened by more irregular secondary

Fig. 309. Close-packed calcospherites, or so-called "spicules," of *Astrosclera*. After Lister.

deposits, which probably, as in Harting's experiments, assume this irregular form when material runs short.

Our few foregoing examples, diverse as they are in look and kind, from the spicules of *Astrosclera* or *Alcyonium* to the otoliths of a fish, seem all to have their free origin in some larger or smaller fluid-containing space or cavity of the body: pretty much as Harting's calcospheres made their appearance in the albuminous content of a dish. But we come at last to a much larger class of spicular and skeletal structures, for whose regular and often complex forms some other explanation than the intrinsic forces of crystallisation or molecular adhesion is required. As we enter on this subject, which is certainly no small nor easy one, it may conduce to simplicity and to brevity if we make a rough classification, by way of forecast, of the conditions we are likely to meet with.

Just as we look upon animals as constituted, some of a great number of cells, others of a single cell or of but few, and just as the shape of the former has no longer a visible relation to the individual shapes of its constituent cells while in the latter it is cell-form which dominates or is actually equivalent to the form of the organism, so shall we find it to be, with more or less exact analogy, in the case of the skeleton. For example, our own skeleton consists of bones, in the formation of each of which a vast number of minute living cellular elements are necessarily concerned; but the form and even the arrangement of these bone-forming cells or corpuscles are monotonously simple, and give no physical explanation of the outward and visible configuration of the bone. It is as part of a far larger field of force—in which we must consider gravity, the action of various muscles, the compressions, tensions and bending moments due to variously distributed loads, the whole interaction of a very complex mechanical system—that we must explain (if we are to explain at all) the configuration of a bone.

In contrast to these massive skeletons we have other skeletal elements whose whole magnitude is commensurate with that of a living cell, or (as comes to very much the same thing) is comparable to the range of action of the molecular forces. Such is the case with the ordinary spicules of a sponge, with the delicate skeleton of a radiolarian, or with the denser and robuster shells of the foraminifera. The effect of *scale*, then, of which we had so much to say in our introductory chapter on Magnitude, is bound to be apparent in the study of skeletal fabrics, and to lead to essential differences between the big and the little, the massive and the minute, in regard to their controlling forces and resultant forms. And if all this be so, and if the range of action of the molecular forces be now the important and fundamental thing, then we may somewhat extend our statement of the case, and include among our directive or constructive influences not only association with the living cellular elements of the body, but also association with any bubbles, drops, vacuoles or vesicles which may be comprised within the bounds of the organism, and which are (as their names and characters connote) of the order of magnitude of which we are speaking.

Proceeding a little farther in our classification, we may conceive each little skeletal element to be associated with, and developed by, a single cell or vesicle, or alternatively a cluster or "system" of consociated cells. In either case there are various possibilities. For instance, the calcified or other skeletal material may tend to overspread the entire outer surface of the cell or cluster of cells, and so tend to assume a configuration comparable to the surface of a fluid drop or aggregation of drops; this, in brief, is the gist and essence of our story of the foraminiferal shell. Another common but very different condition will arise if, in the case of the cell-aggregates, the skeletal material tends to accumulate in the interstices *between* the cells, in the partition-walls which separate them, or in the still more restricted edges, or junctions between these partition-walls; conditions such as these will go a long way to help us to understand many sponge-spicules and an immense variety of radiolarian skeletons. And lastly (for the present), there is a possible and very interesting case of a skeletal element associated with the surface of a cell, not so as to cover it like a shell, but only so as to pursue a course of its own within it, and subject to the restraints imposed by such confinement to a curved and limited surface. With this curious condition we shall deal immediately.

This preliminary and much simplified classification of the lesser skeletal, or micro-skeletal, forms does not pretend (as is evident enough) to completeness. It leaves out of account some conformations and configurations with which we shall attempt to deal, and others which we must perforce omit. But nevertheless it may help to clear or mark our way towards the subjects which this chapter has to consider, and the conditions by which they are at least partially defined.

Among the possible, or conceivable, types of microscopic skeletons let us begin with the case of a spicule, more or less simply linear as far as its *intrinsic* powers of growth are concerned, but which owes its more complicated form to a restraint imposed by the cell to which it is confined, and within whose bounds it is generated.

The conception of a spicule developed under such conditions came from that very great mathematical physicist, G. F. FitzGerald. Many years ago, Sollas pointed out that if a spicule begin to grow in some particular way, presumably under the control or constraint imposed by the organism, it continues to grow by further chemical deposition in the same form or direction even after it has got beyond the boundaries of the organism or its cells. This phenomenon is what we see in, and this imperfect explanation goes so far to account for, the continued growth in straight lines of the long calcareous spines of *Globigerina* or *Hastigerina*, or the similarly radiating but siliceous spicules of many Radiolaria. In physical language, if our crystalline structure has once begun to be laid down in a definite orientation, further additions tend to accrue in a like regular fashion and in an identical direction: corresponding to the phenomenon of so-called "orientirte Adsorption," as described by Lehmann.

In *Globigerina* or in *Acanthocystis* the long needles grow out freely into the surrounding medium, with nothing to impede their rectilinear growth and approximately radiate symmetry. But let us consider some simple cases to illustrate the forms which a spicule will tend to assume when, striving (as it were) to grow straight, it comes under some simple and constant restraint or compulsion.

If we take any two points on a smooth curved surface, such as that of a sphere or spheroid, and imagine a string stretched between them, we obtain what is known in mathematics as a "geodesic" curve. It is the shortest line which can be traced between the two points upon the surface itself, and it has always the same direction upon the surface to which it is confined; the most familiar of all cases, from which the name is derived, is that curve, or "rhumb-line," upon the earth's surface which the navigator learns to follow in the practice of "great-circle sailing," never altering his direction nor departing from his nearest road. Where the surface is spherical, the geodesic is literally a "great circle," a circle, that is to say, whose centre is the centre of the sphere. If instead of a sphere we be dealing with a spheroid, whether prolate or oblate (that is to say a figure of revolution in which an ellipse rotates about its long or its short axis), then the system of geodesics becomes more complicated. For in it the elliptic meridians are all geodesics, and so is the circle of the equator; though the circles of latitude are not so, any more than in the sphere. But a line which crosses the equator at an oblique angle, if it is to be geodesic, will go on so far and then turn back again, winding its way in a continual figure-of-eight curve between two extreme latitudes, as when we wind a ball of wool. To say, as we have done, that the geodesic is the shortest line between two points upon the surface, is as much as to say that it is a *trace* of some particular straight line upon the surface in question; and it follows that, if any linear body be confined to that surface, while retaining a tendency to grow (save only for its confinement to that surface) in a straight line, the resultant form which it will assume will be that of a geodesic.

Let us now imagine a spicule whose natural tendency is to grow into a straight linear element, either by reason of its own molecular anisotropy or because it is deposited about a thread-like axis, and let us suppose that it is confined either within a cell-wall or in adhesion thereto; its line of growth will be a geodesic to the surface of the cell. And if the cell be an imperfect sphere, or a more or less regular ellipsoid, the spicule will tend to grow into one or other of three forms: either a plane curve of nearly circular arc; or, more commonly, a plane curve which is a portion of an ellipse; or, most commonly of all, a curve which is a portion of a spiral in space. In the latter case, the number of turns of the spiral will depend not only on the length of the spicule, but on the relative dimensions of the ellipsoidal cell, as well as on the angle by which the spicule is inclined to the ellipsoid axes; but a very common case will probably be that in which the spicule looks at first sight to be a plane **C**-shaped figure, but is discovered, on more careful inspection, to lie not in one plane but in a more complicated twist. This investigation includes a series of forms which are abundantly represented among actual sponge-spicules, as illustrated in Figs. 310 and 311.

Growth or motion, when confined to some particular curved surface, may appear in various forms and in unexpected places. An amoeba, creeping along the inside or the outside of a glass tube, was found in either case to follow a winding, spiral path: it was really doing its best to go straight—in other words it was following a geodesic or loxodromic path, determined by whatsoever angle of obliquity to the axis of the tube it had chanced to start out upon.

The spiral bands of chlorophyll in *Spirogyra*, set at varying angles of helicoid obliquity, are (I take it) very beautiful examples of continuous growth under the restraint of a cylindrical surface.

Fig. 310. Sponge and holothurian spicules.

To return to our sponge-spicules. If the spicule be not restricted to linear growth, but have a tendency to expand, or to branch out from a main axis, we shall obtain a series of more complex figures, all related to the geodesic system of curves. A notable case will arise where the spicule occupies, in the first instance, the axis of the containing cell, and then, on reaching its boundary, tends to branch or spread outwards. We shall now get various figures, in some of

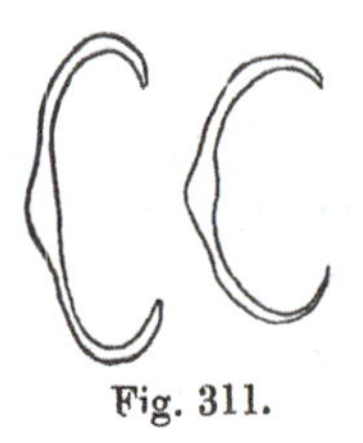

Fig. 311.

Fig. 312. An "amphidisc" of *Hyalonema*.

which the spicule will appear as an axis expanding into a disc or wheel at either end; and in other cases, the terminal disc will be replaced by rays or spokes with a reflex curvature, corresponding to the spherical or ellipsoid curvature of the cell. Such spicules as these are exceedingly common among various sponges (Fig. 312).

Furthermore, if these mechanical methods of conformation, and others like to these, be the true cause of the shapes which the spicules assume, it is plain that the production of these spicular shapes is not a specific function of the sponge, but that we should expect the same or similar spicules to occur in other organisms, wherever the conditions of inorganic secretion within closed cells were very much the same. As a matter of fact, in the sea-cucumbers, where the formation of intracellular spicules is a characteristic feature of the group, all the principal types of conformation which we have just described can be closely paralleled; indeed, in many cases, the forms of the holothurian spicules are identical and indistinguishable from those of the sponges*. But the holothurian spicules are composed of calcium carbonate while those which we have just described in the case of sponges are siliceous: this being just another proof of the fact that in such cases as these the form of the spicule is not due to its chemical nature or molecular structure, but to the external forces to which it is subjected.

The broad fact that the skeleton is calcareous in certain large groups of animals and calcareous in others is as remarkable as its causes are obscure. I for one have no idea why some sponge-skeletons are of the one and some the other, with never the least admixture of the two; or why the diatoms and radiolarians are all the one, and the molluscs and corals and foraminifera are all the other†.

So much for that small class of sponge-spicules whose forms seem due to the fact that they are developed within, or under the restraint imposed by, the surface of a single cell or vesicle. Such spicules are usually of small size as well as of simple form; and they are greatly outstripped in number, in size, and in supposed importance as guides to zoological classification, by another class of spicules. These are the many and various cases which we explain on the

* See for instance the plates in Théel's Monograph of the Challenger *Holothuroidea*; also Sollas's *Tetractinellida*, p. lxi. Cf. also E. Merke, Studien am Skelet der Echinodermen, *Zool. Jahrbücher (Abth. f. allgem. Zoologie)*, 1916–19.

† The particles of lime and silica tend to bear opposite charges; siliceous organisms seem to flourish in the colder waters as the calcareous certainly do in warmer seas. And such facts, or tendencies, as these may help some day to explain the phenomenon.

assumption that they develop in association (of some sort or another) with the *lines of junction*, or boundary-edges, of contiguous cells. They include the triradiate spicules of the calcareous sponges, the quadriradiate or "tetractinellid" spicules which occur sometimes in the same group but more characteristically in certain siliceous sponges known as the Tetractinellidae, and perhaps (though these last are somewhat harder to understand) the six-rayed spicules of the Hexactinellids. We shall come later on to more complicated skeletons of the same type among the Radiolaria.

The spicules of the calcareous sponges are commonly triradiate, and the three radii are usually inclined to one another at *nearly* equal angles; in certain cases, two of the three rays are nearly in a straight line, and at right angles to the third*. They are not always in a plane, but are often inclined to one another in a trihedral angle, not easy of precise measurement under the microscope. The three rays are often supplemented by a fourth, which is set tetrahedrally, making *nearly* co-equal angles with the other three. The calcareous spicule consists mainly of carbonate of lime in the form of calcite, with (according to von Ebner) some admixture of soda and magnesia, of sulphates and of water. According to the same writer there is no organic matter in the spicule, either in the form of an axial filament or otherwise, and the appearance of stratification, often simulating the presence of an axial fibre, is due to "mixed crystallisation" of the various constituents. The spicule is a true crystal, and therefore its existence and its form are *primarily* due to the molecular forces of crystallisation; moreover it is a single crystal and not a group of crystals, as is seen by its behaviour in polarised light. But its axes are not crystalline axes, its angles are variable and indefinite, and its form neither agrees with, nor in any way resembles, any one of the countless, polymorphic forms in which calcite is capable of crystallising. It is as though it were carved out of a solid crystal; it is, in fact, a crystal under restraint, a crystal growing, as it were, in an artificial mould, and this mould is constituted by the surrounding cells or structural vesicles of the sponge.

* For very numerous illustrations of the triradiate and quadriradiate spicules of the calcareous sponges, see (*int. al.*), papers by Dendy (*Q.J.M.S.* xxxv, 1893), Minchin (*P.Z.S.* 1904), Jenkin (*P.Z.S.* 1908), etc.

We have already studied in an elementary way, but enough for our purpose, the manner in which three, four or more cells, or bubbles, meet together under the influence of surface-tension, in configurations geometrically similar to what may be brought about by a uniform distribution of mechanical pressure. And we have seen how surface-energy leads to the *adsorption* of certain chemical substances, first at the corners, then at the edges, lastly in the partition-walls, of such an assemblage of cells. A spicule formed in the interior of such a mass, starting at a corner where four cells meet and extending along the adjacent edges, would then (in theory) have the characteristic form which the geometry of the bee's cell has taught us, of four rays radiating from a point, and set at co-equal angles to one another of 109°, approximately. Precisely such "tetractinellid" spicules are often formed.

But when we confine ourselves to a plane assemblage of cells, or to the outer surface of a mass, we need only deal with the simpler geometry of the hexagon. In such a plane assemblage we find the cells meeting one another in threes; when the cells are uniform in size the partitions are straight lines, and combine to form regular hexagons; but when the cells are unequal, the partitions tend to be curved, and to combine to form other and less regular polygons. Accordingly, a skeletal secretion originating in a layer or surface of cells will begin at the corners and extend to the edges of the cells, and will thus take the form of triradiate spicules, whose rays (in a typical case) will be set at co-equal angles of 120° (Fig. 313, F). This latter condition of inequality will be open to modification in various ways. It will be modified by any inequality in the specific tensions of adjacent cells; as a special case, it will be apt to be greatly modified at the surface of the system, where a spicule happens to be formed in a plane perpendicular to the cell-layer, so that one of its three rays lies between two adjacent cells and the other two are associated with the surface of contact between the cells and the surrounding medium; in such a case (as in the cases considered in connection with the forms of the cells themselves on p. 494), we shall tend to obtain a spicule with two equal angles and one unequal (Fig. 313, A, C); in the last case, the two outer, or superficial rays, will tend to be markedly curved. Again, the equiangular condition will be departed from, and more or less curvature will be imparted to the rays, wherever the cells of the system cease to be uniform in size, and when the hexagonal symmetry of the system is lost accordingly. Lastly, although we speak of the rays as meeting at certain definite angles, this statement applies to their *axes* rather than to the rays themselves. For if the triradiate spicule be developed in the *interspace* between three juxtaposed cells it is obvious that its sides will tend to be concave, because the space between three contiguous equal circles is an equilateral, curvilinear triangle; and even if our

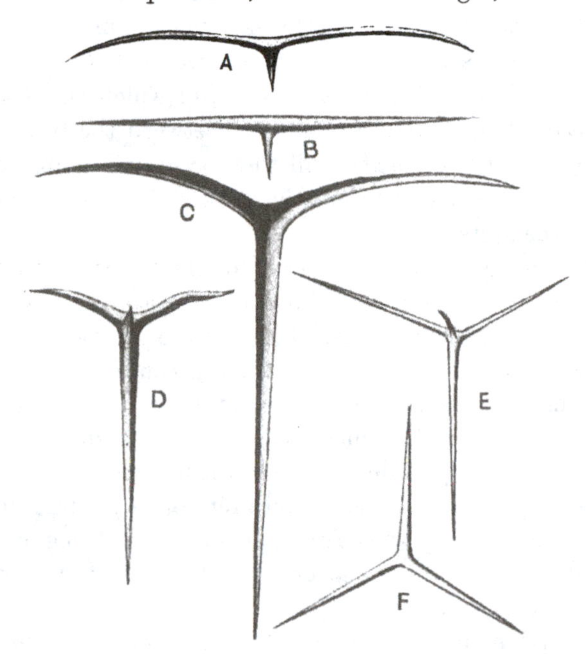

Fig. 313. Spicules of *Grantia* and other calcareous sponges. After Haeckel.

spicule be deposited, not in the space between our three cells, but in the mere thickness of an intervening wall, then we may recollect that the several partitions never actually meet at sharp angles, but the angle of contact is always bridged over by an accumulation of material (varying in amount according to its fluidity) whose boundary takes the form of a circular arc, and which constitutes the "bourrelet" of Plateau. In any sample of the triradiate spicules of *Grantia*, or in any series of careful drawings, such as Haeckel's, we shall find all these various configurations severally and completely illustrated.

The tetrahedral, or rather tetractinellid, spicule needs no further explanation in detail (Fig. 313, D, E). For just as a triradiate spicule corresponds to the case of three cells in mutual contact, so does the four-rayed spicule to that of a solid aggregate of four cells: these latter tending to meet one another in a tetrahedral system, shewing four edges, at each of which three facets or partitions meet, their edges being inclined to one another at equal angles of about 109°—the "Maraldi" angle. And even in the case of a single layer, or superficial layer, of cells, if the skeleton originate in connection with all the edges of mutual contact, we shall (in complete and typical cases) have a four-rayed spicule, of which one straight limb will correspond to the line of junction between the three cells, and the other three limbs (which will then be curved limbs) will correspond to the three edges where the three cells meet in pairs on the surface of the system.

But if such a physical explanation of the forms of our spicules is to be accepted, we must seek for some physical agency to explain the presence of the solid material just at the junctions or interfaces of the cells, and for the forces by which it is confined to, and moulded to the form of, these intercellular or interfacial contacts. We owe to Dreyer the physical or mechanical theory of spicular conformation which I have just described—a theory which ultimately rests on the form assumed, under surface-tension, by an aggregation of cells or vesicles. But this fundamental point being granted, we have still several possible alternatives by which to explain the details of the phenomenon.

Dreyer, if I understand him aright, was content to assume that the solid material, secreted or excreted by the organism, accumulates in the interstices between the cells, and is there subjected to mechanical pressure or constraint as the cells get crowded together by their own growth and that of the system generally. As far as the general form of the spicules goes such explanation is not inadequate, though under it we might have to renounce some of our assumptions as to what takes place at the surface of the system. But where a few years ago the concept of secretion seemed precise enough, we turn now-a-days to the phenomenon of adsorption as a further stage towards the elucidation of our facts, and here we have a case in point. In the tissues of our sponge, wherever two cells meet, there we have a definite *surface* of contact, and there accordingly we have a manifestation of surface-energy; and the concentration of surface-energy will tend to be a maximum at the *lines* or *edges* whereby such surfaces are conjoined. Of the micro-chemistry of the sponge-cells our ignorance is great; but (without venturing on any hypothesis involving the chemical details of the process) we may safely assert that there is an inherent probability that certain substances will tend to be concentrated and ultimately deposited just in these lines of intercellular contact and conjunction. In other words, adsorptive concentration, under osmotic pressure, at and in the surface-film which bounds contiguous cells, and especially in the *edges* where these films meet and intersect, emerges as an alternative (and, as it seems to me, a highly preferable alternative) to Dreyer's conception of an accumulation under mechanical pressure in the vacant spaces left between one cell and another.

But a purely chemical, or purely molecular, adsorption is not the only form of the hypothesis on which we may rely. For from the purely physical point of view, angles and edges of contact between adjacent cells will be *loci* in the field of distribution of surface-energy, and any material particles whatsoever will tend to undergo a diminution of freedom on entering one of those boundary regions. Let us imagine a couple of soap bubbles in contact with one another; over the surface of each bubble tiny bubbles and droplets glide in every direction; but as soon as these find their way into the groove or re-entrant angle between the two bubbles, there their freedom of movement is so far restrained, and out of that groove they have little tendency, or little freedom, to emerge. A cognate phenomenon is to be witnessed in microscopic sections of steel or other metals. Here, together with its crystalline structure, the metal develops a cellular structure by reason of its lack of homogeneity; for in the molten state one constituent tends to separate out into drops, while the other spreads over these and forms a filmy reticulum between —the *disperse* phase and the *continuous* phase of the colloid chemists. In a polished section we easily observe that the little particles of graphite and other foreign bodies common in the matrix have tended to aggregate themselves in the walls and at the angles of the polygonal cells—this being a direct result of the diminished freedom which they undergo on entering one of these boundary regions. And the same phenomenon is turned to account in the various "separation-processes" in which metallic particles are caught up in the interstices of a froth, that is to say in the walls of the *foam-cells* or *Schaumkammern**.

It is by a combination of these two principles, chemical adsorption on the one hand and physical quasi-adsorption or concentration of grosser particles on the other, that I conceive the substance of the sponge-spicule to be concentrated and aggregated at the cell boundaries; and the forms of the triradiate and tetractinellid spicules are in precise conformity with this hypothesis. A few general matters, and a few particular cases, remain to be considered. It matters little or not at all for the phenomenon in question, what is the histological nature or "grade" of the vesicular structures on which it depends. In some cases (apart from sponges), they may be no-more than little alveoli of an intracellular protoplasmic network, and this would seem to be the case at least in the protozoan *Entosolenia aspera*, within the vesicular protoplasm of whose single cell Möbius has described tiny spicules in the shape of little tetrahedra with sunken or concave sides. It is probably the case also in the small beginnings of Echinoderm spicules, which are likewise intracellular and are of similar shape. Among the sponges we have many varying conditions. In some cases there is reason to believe that the spicule is formed at the boundaries of true cells or histological units; but in the case of the larger triradiate or tetractinellid spicules they far surpass in size the actual "cells." We find them lying, regularly and symmetrically arranged, between the "pore-canals" or "ciliated chambers," and it is in conformity with the shape and arrangement of these large rounded or spheroidal structures that their shape is assumed.

Again, it is not at variance with our hypothesis to find that, in the adult sponge, the larger spicules may greatly outgrow the bounds not only of actual cells but also of the ciliated chambers, and may even appear to project freely from the surface of the sponge. For we have already seen that the spicule is capable of

* The crystalline composition of iron was recognised by Hooke in the *Micrographia* (1665); and the cellular or polyhedral structure of the metal was clearly recognised by Réaumur, in his *Art de convertir le fer forgé en acier*, 1722.

growing, without marked change of form, by further deposition, or crystallisation, of layer upon layer of calcareous molecules, even in an artificial solution; and we are entitled to believe that the same process may be carried on in the tissues of the sponge, without greatly altering the symmetry of the spicule, long after it has established its characteristic but non-crystalline form of a system of slender trihedral or tetrahedral rays.

Neither is it of great importance to our hypothesis whether the rayed spicule necessarily arises as a single structure, or does so from separate minute centres of aggregation. Minchin has shewn that, in some cases at least, the latter is the case; the spicule begins, he tells us, as three tiny rods, separate from one another, each developed in the interspace between two sister-cells, which are themselves the results of the division of one of a little trio of cells; and the little rods meet and fuse together while still very minute, when the whole spicule is only about $\frac{1}{200}$ of a millimetre long. At this stage, it is interesting to learn that the spicule is non-crystalline; but the new accretions of calcareous matter are soon deposited in crystalline form.

This observation threw difficulties in the way of former mechanical theories of the conformation of the spicule, and was quite at variance with Dreyer's theory, according to which the spicule was bound to begin from a central nucleus coinciding with the meeting-place of three contiguous cells, or rather the interspace between them. But the difficulty is removed when we import the concept of adsorption; for by this agency it is natural enough, or conceivable enough, that deposition should go on at separate parts of a common system of surfaces; and if the cells tend to meet one another by their interfaces before these interfaces extend to the angles and so complete the polygonal cell, it is again only natural that the spicule should first arise in the form of separate and detached limbs or rays.

Among the "tetractinellid" sponges, whose spicules are composed of amorphous silica or opal, all or most of the above-described main types of spicule occur, and, as the name of the group implies, the four-rayed, tetrahedral spicules are especially represented. A somewhat frequent type of spicule is one in which one of the four rays is greatly developed, and the other three constitute small prongs diverging at equal angles from the main or axial ray. In all probability, as Dreyer suggests, we have here had to do with a group of four vesicles, of which three were large and co-equal, while a fourth and very much smaller one lay above and between the other three. In certain cases where we have likewise one large and three much smaller rays, the latter are recurved, as in Fig. 314, *a–c*. This type, save for the constancy of the number of rays and the limitation of the terminal ones to three, and save also for the more important difference that they occur only at one and not at both ends of the long axis, is similar to the type of spicule illustrated in Fig. 312, which we have explained as being probably developed within an oval cell, by whose walls its branches have been cabined and confined. But it is more probable that we have here to do with a spicule developed in the midst of a group of three co-equal and more or less elongated or cylindrical cells or vesicles, the long axial ray corresponding to their common edge or line of contact, and the three short rays having each lain in the surface furrow between two out of the three adjacent cells.

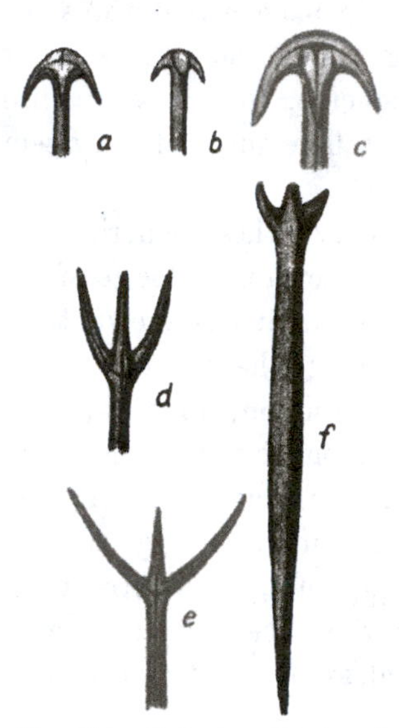

Fig. 314. Spicules of tetractinellid sponges (after Sollas). *a–e*, anatriaenes; *d–f*, protriaenes.

Just as in the case of the little **S**-shaped spicules formed within the bounds of a single cell, so also in the case of the larger tetractinellid types do we find the same configurations reproduced among the holothuroids as we have dealt with in the sponges. The holothurian spicules are a little less neatly formed, a little rougher, than the sponge-spicules, and certain forms occur among the former group which do not present themselves among the latter; but for the most part a community of type is obvious and striking (Fig. 315).

The very peculiar spicules of the holothurian *Synapta*, where a tiny anchor is pivoted or hinged on a perforated plate, are a puzzle indeed; but we may at least solve part of the riddle. How the hinge is formed, I do not know; the anchor gets its shape, perhaps, in some such way as we have supposed the "amphidiscs" of *Hyalo-*

nema to acquire their reflexed spokes, but the perforated plate is more comprehensible. Each plate starts in a little clump of cells in whose boundary-walls calcareous matter is deposited, doubtless by adsorption, the holes in the finished plate thus corresponding to the cells which formed it. Close-packing leads to an arrangement of six cells round a central one, and the normal pattern of the plate displays this hexagonal configuration. The calcareous plate begins as a little rod whose ends fork, and then fork again: in the same inevitable trinodal pattern which includes the "polar furrow" of the embryologists. The anchor had been first formed, and the

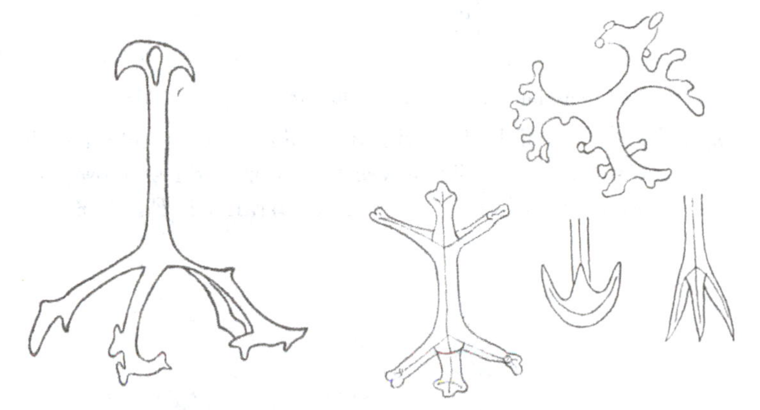

Fig. 315. Various holothurian spicules. After Théel.

little plate is added on beneath it. The first spicular rudiment of the plate may lie parallel to the stock of the anchor or it may lie athwart* it. From the physical point of view it would seem to be a mere matter of chance which way the cluster of cells happens to lie; but this difference of direction will cause a certain difference in the symmetry of the resulting plate. It is this very difference which systematic zoologists at one time seized upon to distinguish *S. Buskii* from our two commoner "species." The two latter

* Cf. S. Becher, Nicht-funktionelle Korrelation in der Bildung selbständiger Skeletelemente, *Zool. Jahrbücher* (*Physiol.*), XXXI, pp. 1–189, 1912; Hedwiga Wilhelmi, Skeletbildung der füsslosen Holothurien, *ibid.* XXXVII, pp. 493–547, 1920; *Arch. f. Entw. Mech.* XLVI, pp. 210–258, 1920. See also W. Woodland, Studies in spicule-formation, *Q.J.M.S.* XLIX, pp. 535–559, 1906; LI, pp. 483–509, 1907 and R. Semon, Naturgeschichte der Synaptiden, *Mitth. Zool. St. Neapel*, VII, pp. 272–299, 1886. On the common species of *Synapta*, see Koehler, *Faune de France, Echinodermes*, 1921, pp. 188–9.

(*S. inhaerens* and *S. digitata*) are mainly distinguished from one another by the number of holes in the plate, that is to say, by the average number of cells in the little cluster of which the plate or spicule was formed. In many or perhaps most other holothurians

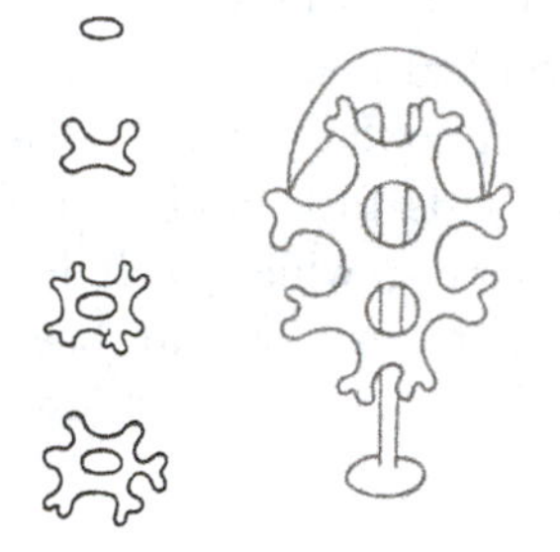

Fig. 316. Development of anchor-plate in *Synapta*. After Semon.

the spicules consist of little perforated plates or baskets, developed in the same way, about cells or vesicles more or less close-packed, and therefore more or less symmetrically arranged (Fig. 316).

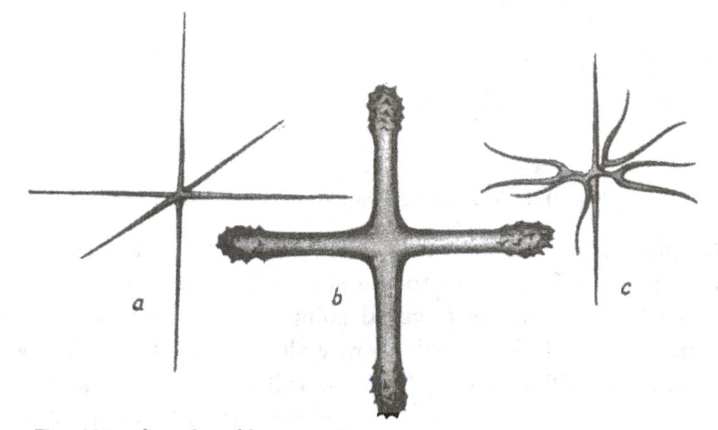

Fig. 317. Spicules of hexactinellid sponges. After F. E. Schultze.

The six-rayed siliceous spicules of the hexactinellid sponges, while they are perhaps the most regular and beautifully formed spicules to be found within the entire group, have been found very difficult to explain, and Dreyer has confessed his complete inability to account for their conformation*. But, though it may only be

* Cf. Albr. Schwan, Ueber die Funktion des Hexactinellidenskelets, u. seine Vergleichbarkeit mit dem Radiolarienskelet, *Zool. Jb., Abth. allg. Zool. u. Physiol.* XXXIII, pp. 603–616, 1913; cf. V. Hacker, Bericht über d. Tripyleenausbeute d. d. Tiefsee-Exped. *Verh. d. zool. Ges.* 1904.

throwing the difficulty a little further back, we may so far account for them by considering that the cells or vesicles by which they are conformed are not arranged in what is known as "closest packing," but in linear series; so that in their arrangement, and by their mutual compression, we tend to get a pattern not of hexagons but of squares: or, looking to the solid, not of dodecahedra but of cubes or parallelepipeda. This indeed appears to be the case, not with the individual cells (in the histological sense), but with the larger units or vesicles which make up the body of the hexactinellid. And this being so, the spicules formed between the linear, or cubical series of vesicles, will have the same tendency towards a "hexactinellid" shape, corresponding to the angles and adjacent edges of a system of cubes, as in our former case they had to a triradiate or a tetractinellid form, when developed in connection with the angles and edges of a system of hexagons, or a system of rhombic dodecahedra.

However the hexactinellid spicules be arranged (and this is none too easy to determine) in relation to the tissues and chambers of the sponge, it is at least clear that, whether they lie separate or be fused together in a composite skeleton, they effect a symmetrical partitioning of space according to the cubical system, in contrast to that closer packing which is represented and effected by the tetrahedral system*.

Histologically, the case is illustrated by a well-known phenomenon in embryology. In the segmenting ovum, there is a tendency for the cells to be budded off in linear series; and so they often remain, in rows side by side, at least for a considerable time and during the course of several consecutive cell divisions. Such an arrangement constitutes what the embryologists call the "radial type" of segmentation†. But in what is described as the "spiral type" of segmentation, it is stated that, as soon as the first horizontal furrow has divided the cells into an upper and a lower layer, those of "the upper layer are shifted in respect to the lower layer,

* *Chall. Rep.*, *Hexactinellida*, pls. xvi, liii, lxxvi, lxxxviii.

† See, for instance, the figures of the segmenting egg of Synapta (after Selenka), in Korschelt and Heider's *Vergleichende Entwicklungsgeschichte*. On the spiral type of segmentation as a secondary derivative, due to mechanical causes, of the "radial" type of segmentation, see E. B. Wilson, Cell-lineage of *Nereis*, *Journ. Morph.* VI, p. 450, 1892.

by means of a rotation about the vertical axis*." It is, of course, evident that the whole process is merely that which is familiar to physicists as "close packing." It is a very simple case of what Lord Kelvin used to call "a problem in tactics." It is a mere question of the rigidity of the system, of the freedom of movement on the part of its constituent cells, whether or at what stage this tendency to slip into the closest propinquity, or position of *minimum potential*, will be found to manifest itself.

Lastly, a curious case is presented by the so-called "chessman" spicules of *Latrunculia* and of a few other sponges, where the spicular shaft is thickened at regular intervals, and the thickenings grow into whorled and flattened lobes. Dendy suggested that the developing spicule is in a state of vibration (due perhaps to the water-currents of the sponge), and that the whorls correspond to nodes, or *loci* of comparative rest, where the formative cells tend to settle down and do their work undisturbed. The position of the nodes and internodes will depend on many circumstances, on whether the spicule be a fixed rod or a free one, straight or curved, uniform in section or tapering towards either end. In the free bar there should tend, in any case, to be a node in the middle, and two more at definite distances from either end. It so happens that in the forms investigated there are only two whorls, the median and one other; but J. W. Nicholson has calculated the positions of these according to the vibration theory, and the theoretical results are found to agree with those of observation very closely indeed. That one of the whorls should be lacking might seem to imperil the proof; but on the other hand among large numbers of spicules no one was found to have its whorls in a position inconsistent with the theory, and there was the required agreement between the shape of the spicule and the position of the whorls. The absence of a third whorl is explained as due to a lack of the necessary formative cells at that part of the spicule†. The theory is in a way supported by recent work (by R. W. Wood of Baltimore and others) on "supersonic vibrations," showing excessively rapid

* Korschelt and Heider, p. 16.

† A. Dendy and J. W. Nicholson, On the influence of vibration upon the form of certain sponge-spicules, *Proc. R.S.* (B), LXXXIX, pp. 573–587, 1917; A. Dendy, The chessman spicules of the genus *Latrunculia*, etc., *Journ. Quekett Microsc. Club*, XIII, pp. 1–16, 1917.

vibrations in quartz rods, more rapid even than Dendy's hypothesis would seem to require. But on the other hand, it is only to few and even exceptional spicules that the theory would seem to apply.

This question of the origin and causation of the forms of sponge-spicules, with which we have now sought to deal, is all the more important and all the more interesting because it has been discussed time and again, from points of view which are characteristic of very different schools of thought in biology. Haeckel found in the form of the sponge-spicule a typical illustration of his theory of "bio-crystallisation"; he considered that these "biocrystals" represented something midway—*ein Mittelding*—between an inorganic crystal and an organic secretion; that there was a "compromise between the crystallising efforts of the calcium carbonate and the formative activity of the fused cells of the syncytium"; and that the semi-crystalline secretions of calcium carbonate "were utilised by natural selection as 'spicules' for building up a skeleton, and afterwards, by the interaction of adaptation and heredity, became modified in form, and differentiated in a vast variety of ways, in the struggle for existence*." What Haeckel precisely meant by these words is not clear to me.

F. E. Schultze, perceiving that identical forms of spicule were developed whether the material were crystalline or non-crystalline, abandoned all theories based upon crystallisation; he simply saw in the form and arrangement of the spicules something which was "best fitted" for its purpose, that is to say for the support and strengthening of the porous walls of the sponge, and finding clear evidence of "utility" in the specific characters of these skeletal elements, had no difficulty in ascribing them to natural selection.

Sollas and Dreyer, as we have seen, introduced in various ways the conception of physical causation—as indeed Haeckel himself had done in regard to one particular, when he supposed the *position* of the spicules to be due to the constant passage of the water-

* "Hierbei nahm der kohlensaure Kalk eine halb-krystallinische Beschaffenheit an, und gestaltete sich unter Aufnahme von Krystallwasser und in Verbindung mit einer geringen Quantität von organischer Substanz zu jenen individuellen, festen Körpern, welche durch die natürliche Züchtung als *Spicula* zur Skeletbildung benützt, und späterhin durch die Wechselwirkung von Anpassung und Vererbung im Kampfe ums Dasein auf das Vielfältigste umgebildet und differenziert wurden." *Die Kalkschwämme*, I, p. 377, 1872; cf. also pp. 482, 483.

currents; though even here, by the way, if I understand Haeckel aright, he was thinking not of a direct or immediate physical causation, but rather of one manifesting itself through the agency of natural selection*. Sollas laid stress upon the "path of least resistance" as determining the direction of growth; while Dreyer dealt in greater detail with the tensions and pressures to which the growing spicule was exposed, amid the alveolar or vesicular structure which was represented alike by the chambers of the sponge, by the constituent cells, or by the minute structure of the intracellular protoplasm. But neither of these writers, so far as I can discover, was inclined to doubt for a moment the received canon of biology, which sees in such structures as these the characteristics of true organic *species*, the indications of blood-relationship and family likeness, and the evidence by which evolutionary descent throughout geologic time may be deduced or deciphered.

Minchin, in a well-known paper†, took sides with F. E. Schultze, and gave his reasons for dissenting from such mechanical theories as those of Sollas and of Dreyer. For example, after pointing out that all protoplasm contains a number of "granules" or microsomes, contained in an alveolar framework and lodged at the nodes of a reticulum, he argued that these also ought to acquire a form such as the spicules possess, if it were the case that these latter owed their form to their similar or identical position. "If vesicular tension cannot in any other instance cause the granules at the nodes to assume a tetraxon form, why should it do so for the sclerites?" The answer is not far to seek. If the force which the "mechanical" hypothesis has in view were simply that of *mechanical pressure*, as between solid bodies, then indeed we should expect that *any* substances lying between the impinging spheres would tend to assume the quadriradiate or "tetraxon" form; but this conclusion does not follow at all, in so far as it is to *surface-energy* that we ascribe the phenomenon. Here the specific nature of the substance makes all the difference. We cannot argue from one

* *Op. cit.* p. 483. "Die geordnete, oft so sehr regelmässige und zierliche Zusammensetzung des Skeletsystems ist zum grössten Theile unmittelbares Product der Wasserströmung; die characteristische Lagerung der Spicula ist von der constanten Richtung des Wasserstroms hervorgebracht; zum kleinsten Theile ist sie die Folge von Anpassungen an untergeordnete äussere Existenzbedingungen."

† Materials for a Monograph of the Ascones, *Q.J.M.S.* XL, pp. 469–587, 1898.

substance to another; adsorptive attraction shews its effect on one and not on another; and we have no reason to be surprised if we find that the little granules of protoplasmic material, which as they lie bathed in the more fluid protoplasm have (presumably, and as their shape indicates) a strong surface-tension of their own, behave towards the adjacent vesicles in a very different fashion to the incipient aggregations of calcareous or siliceous matter in a colloid medium. "The ontogeny of the spicules," says Professor Minchin, "points clearly to their regular form being a *phylogenetic adaptation, which has become fixed and handed on by heredity, appearing in the ontogeny as a prophetic adaptation.*" And again, "The forms of the spicules are the result of adaptation to the requirements of the sponge as a whole, produced by *the action of natural selection upon variation in every direction.*" It would scarcely be possible to illustrate more briefly and more cogently than by these few words (or the similar words of Haeckel quoted on p. 691), the fundamental difference between the Darwinian conception of the causation and determination of Form, and that which is based on, and characteristic of, the physical sciences.

Last of all, Dendy took a middle course. While admitting that the majority of sponge-spicules are "the outcome of conditions which are in large part purely physical," he still saw in them "a very high taxonomic value," as "indications of phylogenetic history" all on the ground that "it seems impossible to account in any other way for the fact that we can actually arrange the different forms in such well-graduated series." At the same time he believed that "the vast majority of spicule-characters appear to be non-adaptive," "that no one form of spicule has, as a rule, any greater survival-value than another," and that "the natural selection of favourable varieties can have had very little to do with the matter*."

The quest after lines and evidences of descent dominated morphology for many years, and preoccupied the minds of two or three generations of naturalists. We find it easier to see than they did that a graduated or consecutive series of forms may be based on physical causes, that forms mathematically akin may belong to

* Cf. A. Dendy, The Tetraxonid sponge-spicule: a study in evolution, *Acta Zoologica*, 1921, pp. 136, 146, etc. Cf. also Bye-products of organic evolution, *Journ. Quekett Microscop. Club*, XII, pp. 65–82, 1913.

organisms biologically remote, and that, in general, mere formal likeness may be a fallacious guide to evolution in time and to relationship by descent and heredity.

If I have dealt comparatively briefly with the inorganic skeletons of sponges, in spite of the interest of the subject from the physical point of view, it has been owing to several reasons. In the first place, though the general trend of the phenomena is clear, it must be admitted that many points are obscure, and could only be discussed at the cost of a long argument. In the second place, the physical theory is too often (as I have shewn) in conflict with the accounts given by embryologists of the development of the spicules, and with the current biological theories which their descriptions embody; it is beyond our scope to deal with such descriptions in detail. Lastly, we find ourselves able to illustrate the same physical principles with greater clearness and greater certitude in another group of animals, namely the Radiolaria.

The group of microscopic organisms known as the Radiolaria is extraordinarily rich in diverse forms or "species." I do not know how many of such species have been described and defined by naturalists, but some fifty years ago the number was said to be over four thousand, arranged in more than seven hundred genera*; of late years there has been a tendency to reduce the number. But apart from the extraordinary multiplicity of forms among the Radiolaria, there are certain features in this multiplicity which arrest our attention. Their distribution in space is curious and vague; many species are found all over the world, or at least every here and there, with no evidence of specific limitations of geographical habitat; some occur in the neighbourhood of the two poles, some are confined to warm and others to cold currents of the ocean. In time their distribution is not less vague: so much so that it has been asserted of them that "from the Cambrian age downwards, the families and even genera appear identical with those now living." Lastly, except perhaps in the case of a few large "colonial forms," we seldom if ever find, as is usual in most animals, a local predominance of one particular species. On the

* Haeckel, in his *Challenger Monograph*, p. clxxxviii (1887), estimated the number of known forms at 4314 species, included in 739 genera. Of these, 3508 species were described for the first time in that work.

contrary, in a little pinch of deep-sea mud or of some fossil "radiolarian earth," we shall probably find scores, and it may be even hundreds, of different forms. Moreover, the radiolarian skeletons are of quite extraordinary delicacy and complexity, in spite of their minuteness and the comparative simplicity of the "unicellular" organisms within which they grow; and these complex conformations have a wonderful and unusual appearance of geometric regularity. All these general considerations seem such as to prepare us for some physical hypothesis of causation. The little skeletons remind us of such things as snow-crystals (themselves almost endless in their diversity), rather than of a collection of animals, constructed in accordance with functional needs and distributed in accordance with their fitness for particular situations. Nevertheless, great efforts have been made to attach "a biological meaning" to these elaborate structures, and "to justify the hope that in time their utilitarian character will be more completely recognised*."

As Ernst Haeckel described and figured many hundred "species" of radiolarian skeletons, so have the physicists depicted snow-crystals in several thousand different forms†. These owe their multitudinous variety to symmetrical repetitions of one simple crystalline form—a beautiful illustration of Plato's *One among the Many*, τὸ ἓν παρὰ τὰ πολλά. On the other hand, the radiolarian skeleton rings its endless changes on combinations of certain facets, corners and edges within a filmy and bubbly mass. The broad difference between the two is very plain and instructive.

Kepler studied the snowflake with care and insight, though he said that to care for such a trifle was like Socrates measuring the hop of a flea. The first drawings I know are by Dominic Cassini; and if that great astronomer was content with them they shew how the physical sciences lagged behind astronomy. They date from the time when Maraldi, Cassini's nephew, was studying the bee's cell;

* Cf. Gamble, *Radiolaria* (Lankester's *Treatise on Zoology*), I, p. 131, 1909. Cf. also papers by V. Häcker, in *Jen. Zeitschr.* XXXIX, p. 581, 1905; *Z. f. wiss. Zool.* LXXXIII, p. 336, 1905; *Arch. f. Protistenkunde*, IX, p. 139, 1907; etc.

† See above, p. 411; and see (besides the works quoted there) Kepler, De nive sexangula (1611), *Opera*, ed. Fritsch, VII, pp. 715–730; Erasmus Bartholin, *De figura nivis, Diss.*, Hafniae, 1661; Dom. Cassini, Obs. de la figure de la neige (Abstr.), *Mém. Acad. R. des Sciences* (1666–1699), x, 1730; J. C. Wilcke, Om de naturliga snö-figurers, *K. V. Akad. Handl.* XXII, 1761.

and they shew once more how very rough his measurements of the honeycomb are bound to have been.

Crystals lie outside the province of this book; yet snow-crystals, and all the rest besides, have much to teach us about the variety, the beauty and the very nature of form. To begin with, the snow-crystal is a regular hexagonal plate or thin prism; that is to

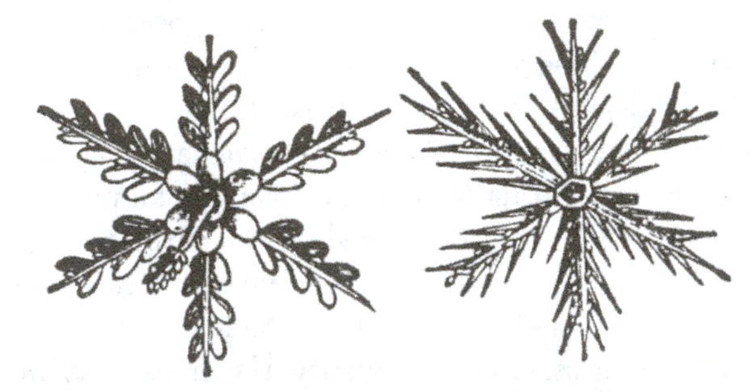

Fig. 318 *a*. Snow-crystals, or "snow-flowers." From Dominic Cassini (*c*. 1600).

say, it shews hexagonal faces above and below, with edges set at co-equal angles of 120°. Ringing her changes on this fundamental form, Nature superadds to the primary hexagon endless combinations of similar plates or prisms, all with identical angles but varying lengths of side; and she repeats, with an exquisite symmetry,

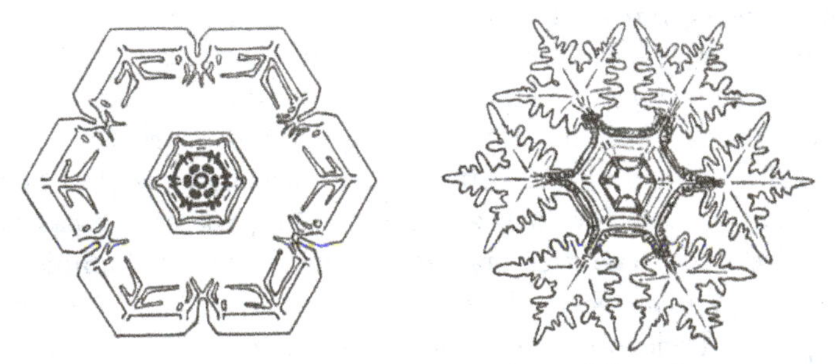

Fig. 318 *b*. Snow-crystals. From Bentley and Humphreys, 1931.

about all three axes of the hexagon, whatsoever she may have done for the adornment and elaboration of one. These snow-crystals seem (as Tutton says) to give visible proof of the space-lattice on which their structure is framed.

The beauty of a snow-crystal depends on its mathematical regularity and symmetry; but somehow the association of many variants of a single type, all related but no two the same, vastly increases our pleasure and admiration. Such is the peculiar beauty which a Japanese artist sees in a bed of rushes or a clump of bamboos, especially when the wind's ablowing; and such (as we saw before) is the phase-beauty of a flowering spray when it shews every gradation from opening bud to fading flower.

The snow-crystal is further complicated, and its beauty is notably enhanced, by minute occluded bubbles of air or drops of water, whose symmetrical form and arrangement are very curious and not always easy to explain*. Lastly, we are apt to see our snow-crystals after a slight thaw has rounded their sharp edges, and has heightened their beauty by softening their contours.

In the majority of cases, the skeleton of the Radiolaria is composed, like that of so many sponges, of silica; in one large family, the Acantharia, and perhaps in some others, it is made of a very unusual constituent, namely strontium sulphate†. There is no important morphological character in which the shells made of these two constituents differ from one another; and in no case can the chemical properties of these inorganic materials be said to influence the form of the complex skeleton or shell, save only in this general way that, by their hardness, toughness and rigidity, they give rise to a fabric more slender and delicate than we find among calcareous organisms.

A slight exception to this rule is found in the presence of true crystals, which occur within the central capsules of certain Radiolaria, for instance the genus *Collosphaera*‡. Johannes Müller (whose knowledge and insight never fail to astonish us§) remarked

* We may find some suggestive analogies to these occlusions in Emil Hatschek's paper, Gestalt und Orientirung von Gasblasen in Gelen, *Kolloid. Ztschr.* xx, pp. 226–234, 1914.

† Bütschli, Ueber die chemische Natur der Skeletsubstanz der Acantharia, *Zool. Anz.* xxx, p. 784, 1906.

‡ For figures of these crystals see Brandt, *F. u. Fl. d. Golfes von Neapel*, XIII, *Radiolaria*, 1885, pl. v. Cf. Johannes Müller, Ueber die Thalassicollen, etc., *Abh. K. Akad. Wiss. Berlin*, 1858.

§ It is interesting to think of the lesser discoveries or inventions, due to men famous for greater things. Johannes Müller first used the tow-net, and Edward Forbes first borrowed the oyster-man's dredge. When we watch a living polyp under the microscope in its tiny aquarium of a glass-cell, we are doing what John Goodsir was the first to do; and the microtome itself was the invention of that best of laboratory-servants, "old Stirling," Goodsir's right-hand man.

that these were identical in form with crystals of celestine, a sulphate of strontium and barium; and Bütschli's discovery of sulphates of strontium and of barium in kindred forms renders it all but certain that they are actually true crystals of celestine*.

In its typical form, the radiolarian body consists of a spherical mass of protoplasm, around which, and separated from it by some sort of porous "capsule," lies a frothy protoplasm, bubbled up into a multitude of alveoli or vacuoles, filled with a fluid which can scarcely differ much from sea-water†. According to their surface-tension conditions, these vacuoles may appear more or less isolated and spherical, or joined together in a "froth" of polyhedral cells; and in the latter, which is the commoner condition, the cells tend to be of equal size, and the resulting polygonal meshwork beautifully regular. In some cases a large number of such simple individual organisms are associated together, forming a floating colony; and it is probable that many others, with whose scattered skeletons we are alone acquainted, had likewise formed part of a colonial organism.

In contradistinction to the sponges, in which the skeleton always begins as a loose mass of isolated spicules, which only in a few exceptional cases (such as *Euplectella* and *Farrea*) fuse into a continuous network, the characteristic feature of the radiolarians lies in the production of a continuous skeleton, of netted mesh or perforated lacework, sometimes replaced by and oftener associated with minute independent spicules. Before we proceed to treat of the more complex skeletons, we may begin by dealing with those comparatively few simple cases where the skeleton is represented by loose, separate spicules or aciculae, which seem, like the spicules of *Alcyonium*, to be isolated formations or deposits, precipitated in the colloid matrix, with no relation to cellular or vesicular boundaries. These simple acicular spicules occupy a definite position in the organism. Sometimes, as for instance among the fresh-water Heliozoa (e.g. *Raphidiophrys*), they lie on the outer surface of the organism, and not infrequently (when few in number) they tend to

* Celestine, or celestite, is $SrSO_4$ with some BaO replacing SrO.

† With the colloid chemists, we may adopt (as Rhumbler has done) the terms *spumoid* or *emulsoid* to denote an agglomeration of fluid-filled vesicles, restricting the name *froth* to such vesicles when filled with air or some other gas.

collect round the bases of the pseudopodia, or around the larger radiating spicules or axial rays in cases where these latter are present. When the spicules are thus localised around some prominent centre, they tend to take up a position of symmetry in regard to it; instead of forming a tangled or felted layer, they come to lie side by side, in a radiating cluster round the focus. In other cases (as for instance in the well-known radiolarian *Aulacantha scolymantha*) the felted layer of aciculae lies at some depth below the surface, forming a sphere concentric with the entire spherical organism. In either case, whether the layer of spicules be deep or be superficial, it tends to mark a "surface of discontinuity," a meeting place either between two distinct layers of protoplasm or between the protoplasm and the water around; and it is evident that, in either case, there are manifestations of surface-energy at the boundary, which cause the spicules to be retained there and to take up their position in its plane. The case is analogous to that of a cirrus cloud, which marks a surface of discontinuity in a stratified atmosphere.

We have, then, to enquire what are the conditions, apart from gravity, which confine an extraneous body to a surface-film; and we may do this very simply, by considering the surface-energy of the entire system. In Fig. 319 we have two fluids in contact with one another (let us call them water and protoplasm), and a body (b) which may be immersed in either, or may be restricted to the boundary between. We have here three possible "interfacial contacts," each with its own specific surface-energy per unit of surface area: namely, that between our particle and the water (let us call it α), that between the particle and the protoplasm (β), and that between water and protoplasm (γ). When the body lies in the boundary of the two fluids, let us say half in one and half in the other, the surface-energies concerned are equivalent to $(S/2)\,\alpha + (S/2)\,\beta$; but we must also remember that, by the presence

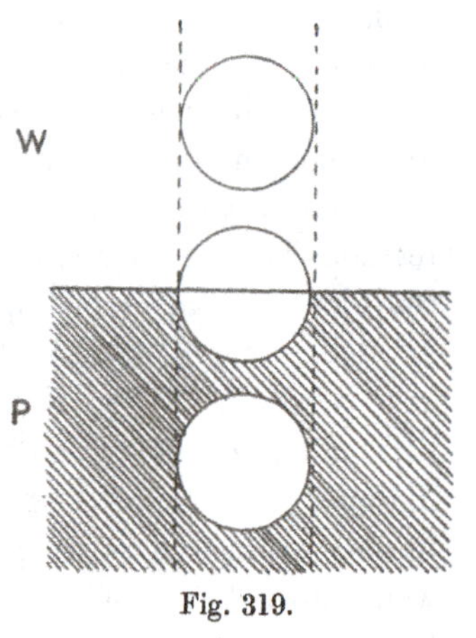

Fig. 319.

of the particle, a small portion (equal to its sectional area s) of the original contact-surface between water and protoplasm has been obliterated, and with it a proportionate quantity of energy, equivalent to $s\gamma$, has been set free. When, on the other hand, the body lies entirely within one or other fluid, the surface-energies of the system (so far as we are concerned) are equivalent to $S\alpha + s\gamma$, or $S\beta + s\gamma$, as the case may be. Accordingly as α be less or greater than β, the particle will have a tendency to remain immersed in the water or in the protoplasm; but if $(S/2)(\alpha + \beta) - s\gamma$ be less than either $S\alpha$ or $S\beta$, then the condition of minimal potential will be found when the particle lies, as we have said, in the boundary zone, half in one fluid and half in the other; and, if we were to attempt a more general solution of the problem, we should have to deal with possible conditions of equilibrium under which the necessary balance of energies would be attained by the particle rising or sinking in the boundary zone, so as to adjust the relative magnitudes of the surface-areas concerned. This principle may, in certain cases, help us to explain the position even of a *radial* spicule, which is just a case where the surface of the solid spicule is distributed between the fluids with a minimal disturbance, or minimal replacement, of the original surface of contact between the one fluid and the other.

In like manner we may provide for the case (a common and an important one) where the protoplasm "creeps up" the spicule, covering it with a delicate film, and forming catenary curves or festoons between one spicule and another; and a less fluid or more tenacious thread of protoplasm may serve, like a solid spicule, to extend the more fluid film, as we see, for instance, in *Chlamydomyxa* or in *Gromia*. When the spicules are numerous and close-set, the surface-film of protoplasm stretching between them will tend to look like a layer of concave or inverted bubbles; and this honeycombed bubbly surface is sometimes beautifully regular, to judge from a well-known figure of the living *Globigerina**. In *Acanthocystis* we have yet another special case, where the radial spicules plunge only a certain distance into the protoplasm of the cell, being arrested at a boundary-surface between an inner and an outer layer of

* See H. B. Brady's *Challenger Monograph*, pl. lxxvii; and see the figure of *Chlamydomyxa* in Doflein's *Protozoenkunde*, p. 374.

cytoplasm; here we have only to assume that there is a tension at this surface, between the two layers of protoplasm, sufficient to balance the tensions which act directly on the spicule*.

In various Acanthometridae, besides such typical characteristics as the radial symmetry, the concentric layers of protoplasm, and the capillary surfaces in which the outer vacuolated protoplasm is festooned upon the projecting radii, we have another curious feature. On the surface of the protoplasm where it creeps up the sides of the long radial spicules, we find a number of elongated bodies, forming in each case one or several little groups, and lying neatly arranged in parallel bundles. A Russian naturalist, Schewiakoff, whose views have been accepted in the text-books, tells us that these are muscular structures, serving to raise or lower the conical masses of protoplasm about the radial spicules, which latter serve as so many "tent-poles" or masts, on which the protoplasmic membranes are hoisted up; and the little elongated bodies are dignified with various names, such as "myonemes" or "myophriscs," in allusion to their supposed muscular nature†. This explanation is by no means convincing. To begin with, we have precisely similar festoons of protoplasm in a multitude of other cases where the "myonemes" are lacking; from their minute size (0·006–0·012 mm.) and the amount of contraction they are said to be capable of, the myonemes can hardly be very efficient instruments of traction; and further, for them to act (as is alleged) for a specific purpose, namely the "hydrostatic regulation" of the organism giving it power to sink or to swim, would seem to imply a mechanism of action and of coordination not easy to conceive in these minute and simple organisms. The fact is that the whole explanation is unnecessary. Just as the supposed "hauling up" of the protoplasmic festoons may be at once explained by capillary phenomena, so also (in all probability) may the position and arrangement of the little elongated bodies. Whatever the actual nature of these bodies may be, whether they be truly portions of differentiated protoplasm, or whether they be foreign bodies or spicular structures (as bodies occupying a similar position in other cases undoubtedly

* Cf. Koltzoff, Zur Frage der Zellgestalt, *Anat. Anzeiger*, XLI, p. 190, 1912.

† *Mém. de l'Acad. des Sci., St Pétersbourg*, XII, Nr. 10, 1902.

are), we can explain their situation on the surface of the protoplasm, and their arrangement around the radial spicules, all by the principles of capillarity.

This last case is not of the simplest; and I do not forget that my explanation of it, which is wholly theoretical, implies a doubt of Schewiakoff's statements, founded on his personal observation. This I am none too willing to do; but whether it be justly done in this case or not, I hold that it is in principle justifiable to look with suspicion upon all such statements where the observer has obviously left out of account the physical aspect of the phenomenon, and all the opportunities of simple explanation which the consideration of that aspect might afford.

Whether it be applicable to this particular and complex case or no, our general theorem of the localisation and arrestment of solid particles in a surface-film is of great biological significance; for on it depends the power displayed by many little naked protoplasmic organisms of covering themselves with an "agglutinated" shell. Sometimes, as in *Difflugia*, *Astrorhiza* (Fig. 320) and others, this covering consists of sand-grains picked up from the surrounding medium, and sometimes, on the other hand, as in *Quadrula*, it consists of solid particles said to arise as inorganic deposits or concretions within the protoplasm itself, and to find their way outwards to a position of equilibrium in the surface-layer; and in both cases, the mutual capillary attractions between the particles, confined to the boundary-layer but enjoying a certain measure of freedom therein, tends to the orderly arrangement of the particles one with another, and even to the appearance of a regular "pattern" as the result of this arrangement.

The "picking up" by the protoplasmic organism of a solid particle with which "to build its house" (for it is hard to avoid this customary use of figures of speech, misleading though it be) is a physical phenomenon akin to that by which an amoeba "swallows" a particle of food. This latter process has been reproduced or imitated in various pretty experimental ways. For instance, Rhumbler has shewn that if a splinter of glass be covered with shellac and brought near a drop of chloroform suspended in water, the drop takes in the spicule, robs it of its shellac covering,

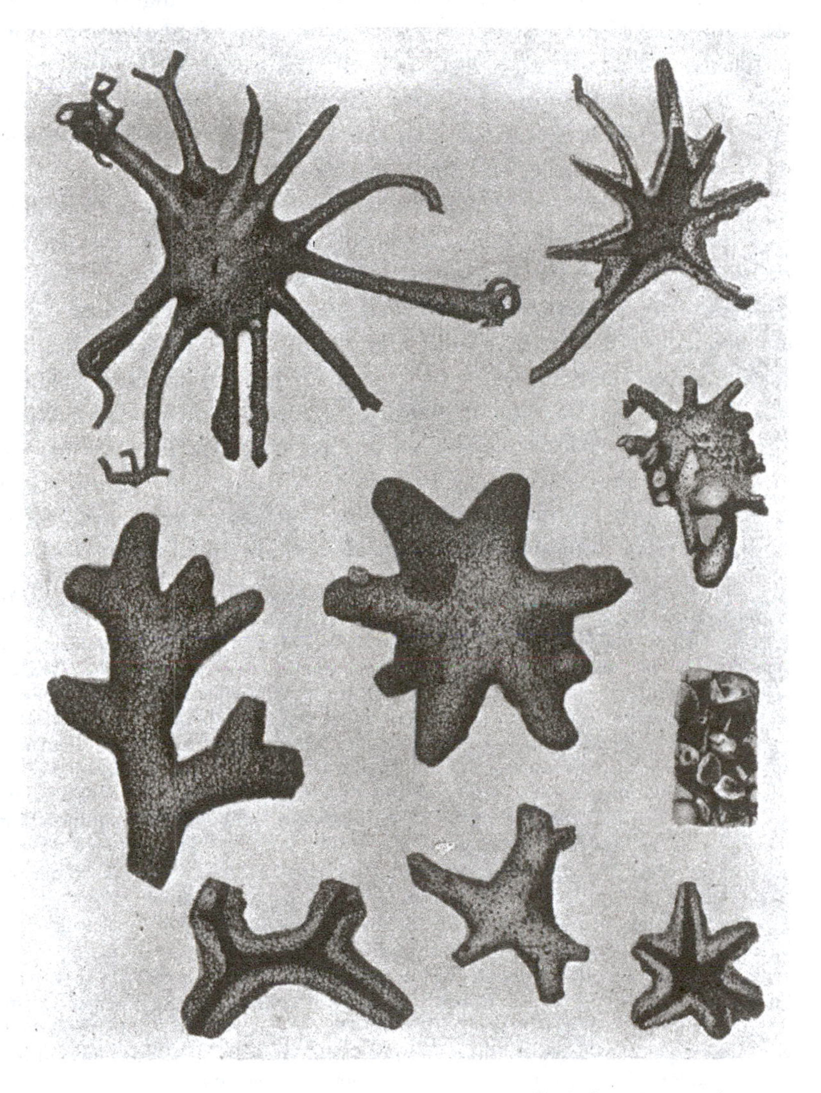

Fig. 320. Arenaceous Foraminifera; *Astrorhiza limicola* and *arenaria*. From Brady's *Challenger Monograph*.

and then passes it out again*. In another case a thread of shellac, laid on a drop of chloroform, is drawn in and coiled within it: precisely as we may see a filament of Oscillatoria ingested by an Amoeba, and twisted and coiled within its cell. It is all a question of relative surface-energies, leading to different degrees of "adhesion" between the chloroform and the splinter of glass or its shellac covering. Thus it is that the Amoeba takes in the diatom, dissolves off its proteid covering, and casts out the shell.

Furthermore, as the whole phenomenon depends on a distribution of surface-energy, the amount of which is specific to certain particular substances in contact with one another, we have no difficulty in understanding the *selective action* which is very often a conspicuous feature in the phenomenon†. Just as some caddis-worms make their houses of twigs, and others of shells and again others of stones, so some Rhizopods construct their agglutinated "test" out of stray sponge-spicules, or frustules of diatoms, or again of tiny mud particles or of larger grains of sand. In all these cases, we have to deal with specific surface-energies, and also doubtless with differences in the total available amount of surface-

* Rhumbler, Physikalische Analyse von Lebenserscheinungen der Zelle, *Arch. f. Entw. Mech.* VII, p. 250, 1898.

† The whole phenomenon has been described as a "surprising exhibition of constructive and selective activity," and ascribed, in varying phraseology, to intelligence, skill, purpose, psychical activity, or "microscopic mentality": that is to say, to Galen's τεχνικὴ φύσις, or "artistic creativeness" (cf. Brock's *Galen*, 1916, p. xxix); cf. Carpenter, *Mental Physiology*, 1874, p. 41; Norman, Architectural achievements of Little Masons, etc., *Ann. Mag. Nat. Hist.* (5), I, p. 284, 1878; Heron-Allen, Contributions...to the study of the Foraminifera, *Phil. Trans.* (B), CCVI, pp. 227–279, 1915; Theory and phenomena of purpose and intelligence exhibited by the Protozoa, as illustrated by selection and behaviour in the Foraminifera, *Journ. R. Microsc. Soc.* 1915, pp. 547–557; *ibid.*, 1916, pp. 137–140. Sir J. A. Thomson (*New Statesman*, Oct. 23, 1915) describes a certain little foraminifer, whose protoplasmic body is overlaid by a crust of sponge-spicules, as "a psycho-physical individuality, whose experiments in self-expression include a masterly treatment of sponge-spicules, and illustrate that organic skill which came before the dawn of Art." Sir Ray Lankester finds it "not difficult to conceive of the existence of a mechanism in the protoplasm of the Protozoa which selects and rejects building-material, and determines the shapes of the structures built, comparable to that mechanism which is assumed to exist in the nervous system of insects and other animals which 'automatically' go through wonderfully elaborate series of complicated actions." And he agrees with "Darwin and others [who] have attributed the building up of these inherited mechanisms to the age-long action of Natural Selection, and the survival of those individuals possessing qualities or 'tricks' of life-saving value," *Journ. R. Microsc. Soc.* April, 1916, p. 136.

energy in relation to gravity or other extraneous forces. In my early student days, Wyville Thomson used to tell us that certain deep-sea "Difflugias," after constructing a shell out of particles of the black volcanic sand common in parts of the North Atlantic, finished it off with "a clean white collar" of little grains of quartz. Even this phenomenon may be accounted for on surface-tension principles, if we may assume that the surface-energy ratios have tended to change, either with the growth of the protoplasm or by reason of external variation of temperature or the like; we are by no means obliged to attribute even this phenomenon to a manifestation of volition, or taste, or aesthetic skill, on the part of the microscopic organism. Nor, when certain Radiolaria tend more than others to attract into their own substance diatoms and suchlike foreign bodies, is it scientifically correct to speak, as some text-books do, of species "in which diatom-selection has become *a regular habit.*" To do so is an exaggerated misuse of anthropomorphic phraseology.

The formation of an "agglutinated" shell is thus seen to be a purely physical phenomenon, and indeed a special case of a more general physical phenomenon which has important consequences in biology. For the shell to assume the solid and permanent character which it acquires in *Difflugia*, we have only to make the further assumption that small quantities of a cementing substance are secreted by the animal, and that this substance flows or creeps by capillary attraction through all the interstices of the little quartz grains, and ends by binding them together. Rhumbler* has shewn us how these agglutinated tests of spicules or of sand-grains can be precisely imitated, and how they are formed with greater or less ease and greater or less rapidity according to the nature of the materials employed, that is to say according to the specific surface-tensions which are involved. If we mix up a little powdered glass with chloroform, and set a drop of the mixture in water, the glass particles gather neatly round the surface of the drop so quickly that the eye cannot follow the operation. If we do the same with oil and fine sand, dropped into 70 per cent. alcohol, a still more

* Rhumbler, Beiträge z. Kenntniss d. Rhizopoden, I–V, *Z. f. w. Z.* 1891–5; *Das Protoplasma als physikalisches System*, Jena, p. 591, 1914; also in *Arch. f. Entwickelungsmech.* VII, pp. 279–335, 1898; *Biol. Centralbl.* XVIII, 1898; etc.

beautiful artificial Rhizopod-shell is formed, but it takes some three hours to do. Where the action is quick the little test forms as the droplet exudes from the pipette: precisely as in the living *Difflugia* when new protoplasm, laden with solid particles, is being extruded from the mouth of the parent-cell. The experiment can be varied, simply and easily. Instead of a spherical drop a pear-shaped one may easily be formed, so exactly like the common *Difflugia pyriformis* that Rhumbler himself was unable, sometimes, to tell under the microscope the real from the artefact. Again he found that, when the alcohol dissolved the oily substance of the drop and shrinkage took place accordingly, the surface-layer with its solid particles got kinked or folded in—and reproduced in doing so, with startling accuracy, a little shell of common occurrence, known by the generic name of *Lesqueureusia*, or *Difflugia spiralis*. The peculiar shape of this little twisted and bulging shell has been taken to shew that "it had enlarged after its first formation, a very rare occurrence in this group*"; the very opposite is the case. Neither here nor in any allied form does the agglutinated test, once set in order by capillary forces, yield scope for intercalation and enlargement.

At the very time when Rhumbler was thus demonstrating the physical nature of the Difflugian shell, Verworn, a very notable person, was studying the same and kindred organisms from the older standpoint of an incipient psychology†. But as Rhumbler himself admits, Verworn (unlike many another) was doing his best not to over-estimate the appearance of volition, or selective choice, in the little organism's use of materials to construct its dwelling.

This long parenthesis has led us away, for the time being, from the subject of the radiolarian skeleton, and to that subject we must now return. Leaving aside, then, the loose and scattered spicules, which we have sufficiently discussed, the more perfect radiolarian skeletons consist of a continuous and regular structure; and the siliceous (or other inorganic) material of which this framework is composed tends to be deposited in one or other of two ways or in both combined: (1) in long radial spicules, emanating symmetrically from, and usually conjoined at, the centre of the protoplasmic body;

* Cf. *Cambridge Natural History*, Protozoa, p. 55.

† Max Verworn, *Psycho-physiologische Protisten-Studien*, Jena, 1889 (219 pp.); Biologische Protisten-Studien, *Z. f. wiss. Z.* L, pp. 445–467, 1890.

(2) in the form of a crust, developed either on the outer surface of the organism or in relation to one or more of the internal surfaces which separate its concentric layers or its component vesicles. Not infrequently, this superficial skeleton comes to constitute a spherical shell, or a system of concentric spheres.

We have already seen that a great part of the body of the Radiolarian, and especially that outer portion to which Haeckel has given the name of the "calymma," is built up of a mass of "vesicles," forming a sort of stiff froth, and equivalent in the physical though not necessarily in the biological sense to "cells," inasmuch as the little vesicles have their own well-defined boundaries, and their own surface phenomena. In short, all that we have said of cell-surfaces and cell-conformations in our discussion of cells and of tissues will apply in like manner, and under appropriate conditions, to these. In certain cases, even in so common and so simple a one as the vacuolated substance of an *Actinosphaerium*, we may see a close resemblance, or formal analogy, to a cellular or parenchymatous tissue in the close-packed arrangement and consequent configuration of these vesicles, and even at times in a slight membranous hardening of their walls. Leidy has figured* some curious little bodies like small masses of consolidated froth, which seem to be nothing else than the dead and empty husks, or filmy skeletons, of *Actinosphaerium*; and Carnoy† has demonstrated in certain cell-nuclei an all but precisely similar framework, of extreme minuteness and tenuity, formed by adsorption or partial solidification of interstitial matter in a close-packed system of alveoli (Fig. 321). In short, we are again dealing or about to deal with a network or basketwork, whose meshes correspond to the boundary lines between associated cells or vesicles. It is just in those boundary walls or films, still more in their edges or at their corners, that surface-energy will be concentrated and adsorption will be hard at work; and the whole arrangement will follow, or tend to follow, the rules of *areae minimae*—the partition-walls meeting at co-equal angles, three by three in an edge, and their edges meeting four by four in a corner.

Let us suppose the outer surface of our Radiolarian to be covered

* J. Leidy, *Fresh-water Rhizopods of North America*, 1879, p. 262, pl. xli, figs. 11, 12.
† Carnoy, *Biologie Cellulaire*, p. 244, fig. 108; cf. Dreyer, *op. cit.* 1892, fig. 185.

by a layer of froth-like vesicles, uniform in size or nearly so. We know that their mutual tensions will *tend* to conform them into the fashion of a honeycomb, or regular meshwork of hexagons, and that the free end of each hexagonal prism will be a little spherical cap. Suppose now that it be at the outer surface of the protoplasm (in contact with the surrounding sea-water) that the siliceous particles have a tendency to be secreted or adsorbed; the distribution of surface-energy will lead them to accumulate in the grooves which separate the vesicles, and the result will be the development

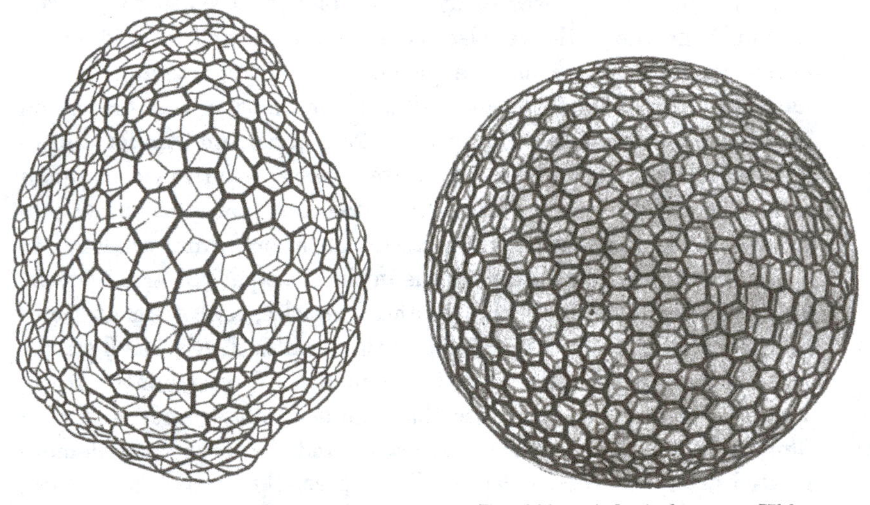

Fig. 321. "*Reticulum plasmatique.*" After Carnoy.

Fig. 322. *Aulonia hexagona* Hkl.

of a delicate sphere composed of tiny rods arranged, or apparently arranged, in a hexagonal network after the fashion of Carnoy's *reticulum plasmatique*, only more solid, and still more neat and regular. Just such a spherical basket, looking like the finest imaginable Chinese ivory ball, is found in the siliceous skeleton of *Aulonia*, another of Haeckel's Radiolaria from the Challenger.

But here a strange thing comes to light. *No system of hexagons can enclose space;* whether the hexagons be equal or unequal, regular or irregular, it is still under all circumstances mathematically impossible. So we learn from Euler: the array of hexagons may be extended as far as you please, and over a surface either plane or curved, but *it never closes in.* Neither our *reticulum plasmatique* nor what seems the very perfection of hexagonal symmetry in *Aulonia* are as we are wont to conceive them; hexagons indeed predominate in both, but a certain number of facets are and must be other than hexagonal. If we look carefully at Carnoy's careful drawing we see that both pentagons and heptagons are shewn in his reticulum, and Haeckel actually states, in his brief description of his *Aulonia hexagona*, that a few square or pentagonal facets are to be found among the hexagons.

Such skeletal conformations are common: and Nature, as in all her handiwork, is quick to ring the changes on the theme. Among

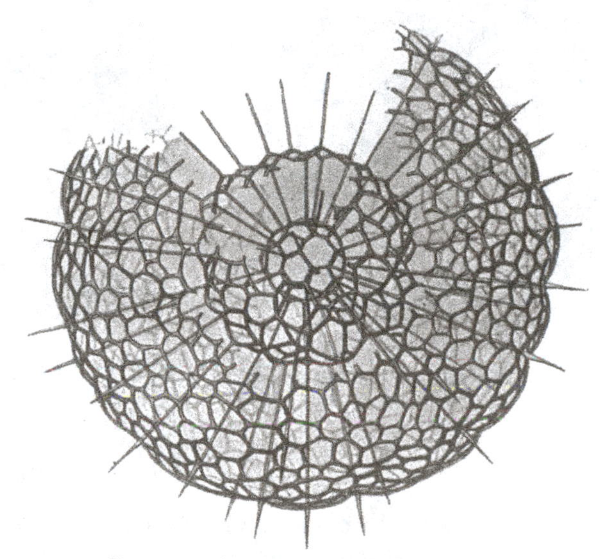

Fig. 323. *Actinomma arcadophorum* Hkl.

its many variants may be found cases (e.g. *Actinomma*) where the vesicles have been less regular in size; and others in which the meshwork has been developed not on an outer surface only but at successive levels, producing a system of concentric spheres. If the siliceous material be not limited to the linear junctions of the cells but spread over a portion of the outer spherical surfaces or caps, then we shall have the condition represented in Fig. 324 (*Ethmosphaera*), where the shell appears perforated by circular instead of hexagonal apertures and the circular pores are set on slight spheroidal eminences; and, interconnected with such types as this, we have others in which the accumulating pellicles of skeletal matter have extended from the edges into the substances of the boundary walls and have so produced a system of films, normal to the surface of the sphere, constituting a very perfect honeycomb, as in *Cenosphaera favosa* and *vesparia**.

In one or two simple forms, such as the fresh-water *Clathrulina*, just such a spherical perforated shell is produced out of some

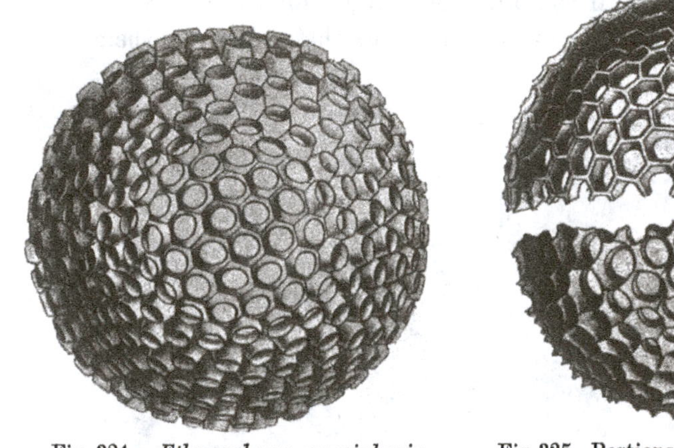

Fig. 324. *Ethmosphaera conosiphonia* Hkl.

Fig. 325. Portions of shells of two "species" of *Cenosphaera*: upper figure, *C. favosa*; lower, *C. vesparia* Hkl.

organic, acanthin-like substance; and in some examples of *Clathrulina* the chitinous lattice-work of the shell is just as regular and delicate, with the meshes for the most part as beautifully hexagonal as in the siliceous shells of the oceanic Radiolaria. This is only another proof (if proof be needed) that the peculiar form and character of these little skeletons are due not to the material of which they are composed, but to the moulding of that material upon an underlying vesicular structure.

Let us next suppose that another and outer layer of cells or vesicles develops upon some such lattice-work as has just been

* In all these latter cases we recognise a relation to, or extension of, the principle of Plateau's *bourrelet*, or van der Mensbrugghe's *masse annulaire*, or Gibbs's ring, of which we have had much to say.

described; and that instead of forming a second hexagonal lattice-work, the skeletal matter tends to be developed normally to the surface of the sphere, that is to say along the *radial* edges where the external vesicles (now compressed into hexagonal prisms) meet one another three by three. The result will be that, if the vesicles be removed, a series of radiating spicules will be left, directed outwards from the angles of the original polyhedron meshwork, all as is seen in Fig. 326. And it may further happen that these radiating skeletal rods branch at their outer ends into divergent rays, forming a triple fork, and corresponding (after the fashion

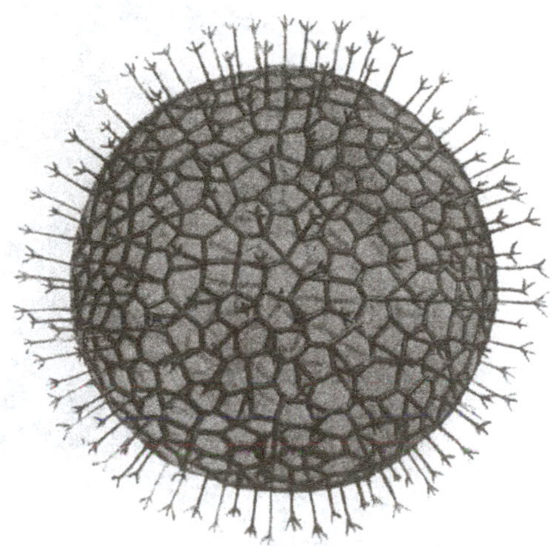

Fig. 326. *Aulastrum triceros* Hkl.

which we have already described as occurring in certain sponge-spicules) to the superficial furrows between the three adjacent cells; this is, as it were, a halfway stage between simple rods or radial spicules and the full completion of another sphere of latticed hexagons. Another possible case, among many, is when the large, uniform vesicles of the outer protoplasm are replaced by smaller vesicles, piled on one another in concentric layers. In this case the radial rods will no longer be straight, but will be bent zig-zag, with their angles in three vertical planes corresponding to the alternate contacts of the successive layers of cells (Fig. 327).

The solid skeleton is confined, in all these cases, to the boundary-lines, or edges, or grooves between adjacent cells or vesicles, but adsorptive energy may extend throughout the intervening walls. This happens in not a few Radiolaria, and in a certain group called the Nassellaria it produces geometrical forms of peculiar elegance and mathematical beauty.

When Plateau made the wire framework of a regular tetrahedron and dipped it in soap-solution, he obtained in an instant (as we well know) a beautifully symmetrical system of six films, meeting

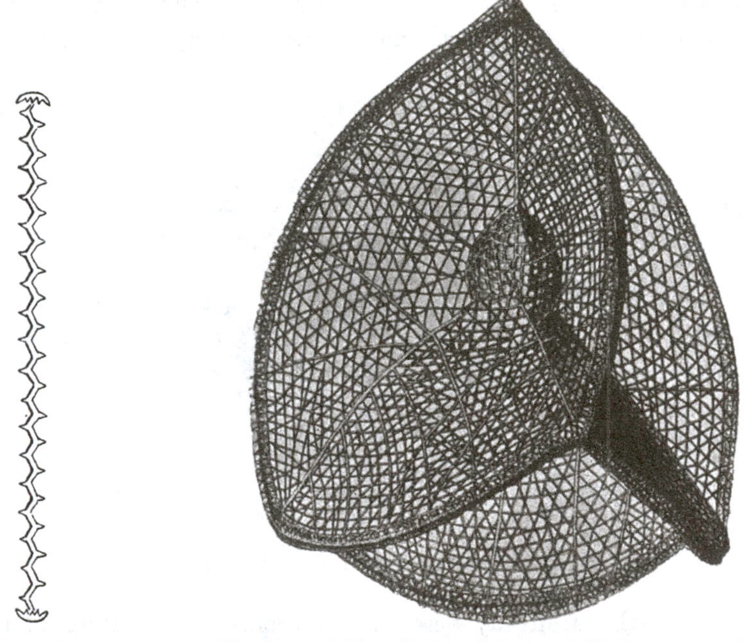

Fig. 327. Fig. 328. A Nassellarian skeleton, *Callimitra agnesae* Hkl (0·15 mm. diameter).

three by three in four edges, and these four edges running from the corners of the figure to its centre of symmetry. Here they meet, two by two, at the Maraldi angle; and the films meet three by three, to form the re-entrant solid angle which we have called a "Maraldi pyramid" in our account of the architecture of the honeycomb. The very same configuration is easily recognised in the minute siliceous skeleton of *Callimitra*. There are two discrepancies, neither of which need raise any difficulty. The figure is not a rectilinear but a *spherical tetrahedron*, such as might be formed by the boundary-edges of a tetrahedral cluster of four co-equal bubbles; and just as Plateau extended his experiment by blowing a small bubble in the centre of his tetrahedral system, so we have a central bubble also here.

This bubble may be of any size*; but its situation (if it be present at all) is always the same, and its shape is always such as to give the Maraldi angles at its own four corners. The tensions

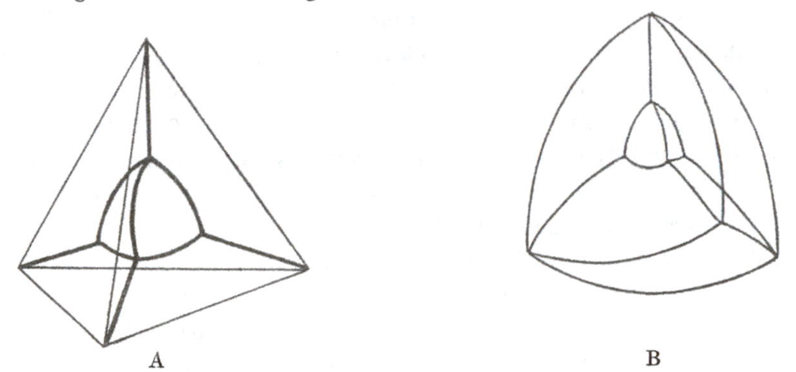

Fig. 329. Diagrammatic construction of *Callimitra*. A, a bubble suspended within a tetrahedral cage. B, another bubble within a skeleton of the former bubble.

of its own walls, and those of the films by which it is supported or slung, all balance one another. Hence the bubble appears in plane projection as a curvilinear equilateral triangle; and we have only got to convert this plane diagram into the corresponding solid to obtain the spherical tetrahedron we have been seeking to explain (Fig. 329).

We may make a simplified model (omitting the central bubble) of the tetrahedral skeleton of *Callimitra*, after the fashion of that of the bee's cell (p. 535). Take $OC = CD = DB$, and draw a circle with radius OB and diameter AB. Erect a perpendicular to AB at C, cutting the circle at E, F. AOE, AOF will be (as before) Maraldi angles of 109°; the arcs AE,

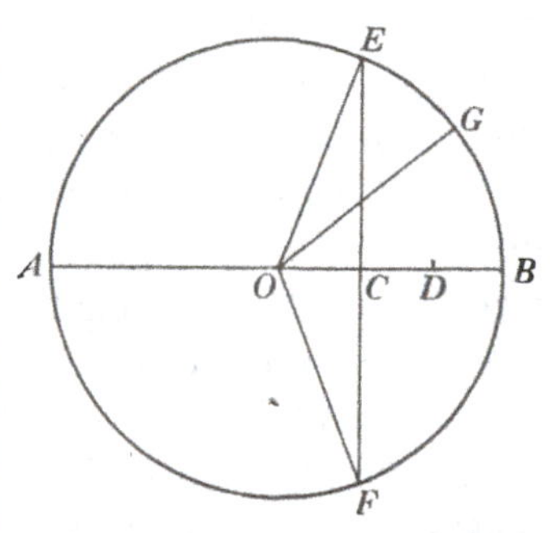

Fig. 330. Geometrical construction of *Callimitra*-skeleton.

* Plateau introduced the central bubble into his cube or tetrahedron by dipping the cage a second time, and so adding an extra face-film; under these circumstances the bubble has a definite magnitude.

AF will be edges of the spherical tetrahedron, and O will be the centre of symmetry. Make the angle $FOG = FOA = EOA$. Cut out the circle $EAFG$, and cut through the radius EO: fold at AO, FO, GO, and fasten together, using EOG for a flap. Make four such sheets, and fasten together back to back. The model will be much improved if little cusps be left at the corners in the cutting out.

The geometry of the little inner tetrahedron is not less simple and elegant. Its six edges and four faces are all equal. The films attaching it to the outer skeleton are all planes. Its faces are spherical, and each has its centre in the opposite corner. The edges are circular arcs, with cosine $\frac{1}{3}$; each is in a plane perpendicular to the chord of the arc opposite, and each has its centre in the middle of that chord. Along each edge the two intersecting spheres meet each other at an angle of 120°*.

This completes the elementary geometry of the figure; but one or two points remain to be considered.

We may notice that the outer edges of the little skeleton are thickened or intensified, and these thickened edges often remain whole or strong while the rest of the surfaces shew signs of imperfection or of breaking away; moreover, the four corners of the tetrahedron are not re-entrant (as in a group of bubbles) but a surplus of material forms a little point or cusp at each corner. In all this there is nothing anomalous, and nothing new. For we have already seen that it is at the margins or edges, and *a fortiori* at the corners, that the surface-energy reaches its maximum—with the double effect of accumulating protoplasmic material in the form of a Gibbs's ring or bourrelet, and of intensifying along the same lines the adsorptive secretion of skeletal matter. In some other tetrahedral systems analogous to *Callimitra*, the whole of the skeletal matter is concentrated along the boundary-edges, and none left to spread over the boundary-planes or interfaces: just as among our spherical Radiolaria it was at the boundary-edges of their many cells or vesicles, and often there alone, that skeletal formation occurred, and gave rise to the spherical skeleton and its meshwork

* For proof, see Lamarle, *op. cit.* pp. 6–8. Lamarle shewed that the sphere can be so divided in seven ways, but of these seven figures the tetrahedron alone is stable. The other six are the cube and the regular dodecahedron; prisms, triangular and pentagonal, with equilateral base and a certain ratio of base to height; and two polyhedra constructed of pentagons and quadrilaterals.

of hexagons. In the beautiful form which Haeckel calls *Archiscenium* the boundary edges disappear, the four edges converging on the median point are intensified, and only three of the six convergent facets are retained; but, much as the two differ in appearance, the geometry of this and of *Callimitra* remain essentially the same.

We learned also from Plateau that, just as a tetrahedral bubble can be inserted within the tetrahedral skeleton or cage, so may a cubical bubble be introduced within a cubical cage; and the edges of the inner cube will be just so curved as to give the Maraldi angles at the corners. We find among Haeckel's Radiolaria one (he calls it *Lithocubus geometricus*) which precisely corresponds to the skeleton

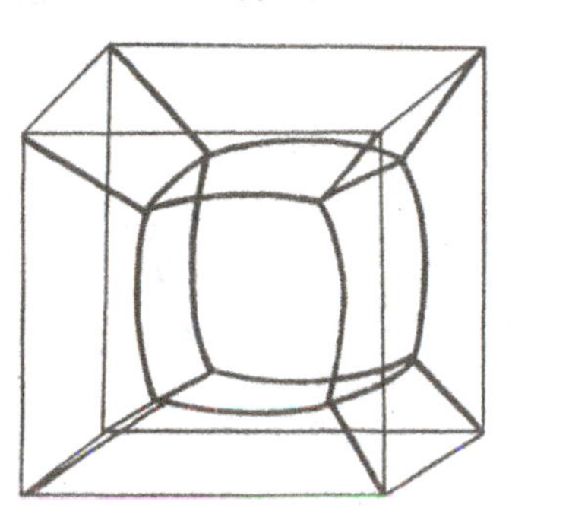

Fig. 331. A, bubble suspended within a cubical cage.
B, *Lithocubus geometricus* Hkl.

of this inner cubical bubble; and the little spokes or spikes which project from the corners are parts of the edges which once joined the corners of the enclosing figure to those of the bubble within (Fig. 331).

Again, if we construct a cage in the form of an equilateral triangular prism, and proceed as before, we shall probably see a vertical edge in the centre of the prism connecting two nodes near either end, in each of which the Maraldi figure is displayed. But if we gradually shorten our prism there comes a point where the two nodes disappear, a plane curvilinear triangle appears horizontally in the middle of the figure, and at each of its three corners four curved edges meet at the familiar angle. Here again we may insert a central bubble, which will now take the form of a curvilinear equilateral triangular prism; and Haeckel's *Prismatium tripodium* repeats this configuration (Fig. 332).

In a framework of two crossed rectangles, we may insert one bubble after another, producing a chain of superposed vesicles whose shapes vary as we alter the relative positions of the rectangular frames. Various species of *Triolampas, Theocyrtis*, etc. are more or less akin to these complicated figures of equilibrium. A very beautiful series of forms may be made by introducing successive bubbles within the film-system formed by a tetrahedron or a

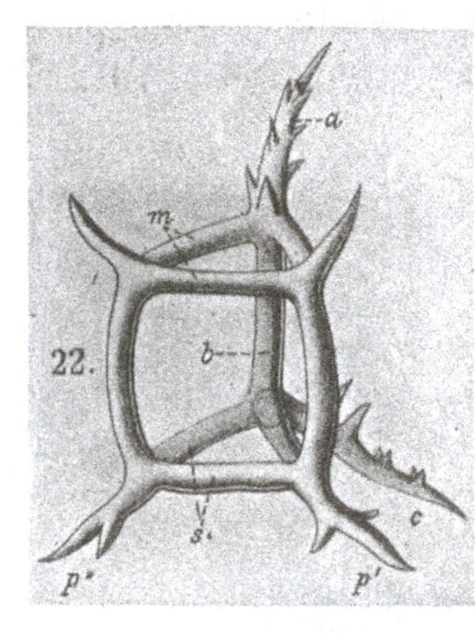

Fig. 332. *Prismatium tripodium* Hkl.

parallelepipedon. The shape and the curvature of the bubbles and of their suspensory films become extremely beautiful, and we have certain of them reproduced unmistakably in various Nassellarian genera, such as *Podocyrtis* and its allies.

In Fig. 333 we see a curious little skeletal structure or complex spicule, whose conformation is easily accounted for. Isolated spicules such as this form the skeleton in the genus *Dictyocha*, and occur scattered over the spherical surface of the organism (Fig. 334). The basket-shaped spicule has evidently been developed about a cluster of four cells or vesicles, lying in or on the surface of the organism, and therefore arranged, not in the three-dimensional, tetrahedral form of *Callimitra*, but in the manner in which four contiguous cells lying side by side in one plane normally set themselves, like the four cells of a segmenting egg: that is to say with an intervening "polar furrow," whose ends mark the meeting place, at equal angles, of the four cells in groups of three. The little projecting spokes, or spikes, which are set normally to the main basket-work, seem to be uncompleted portions of a larger basket, corresponding to a more numerous aggregation of cells. Similar but more complex forma-

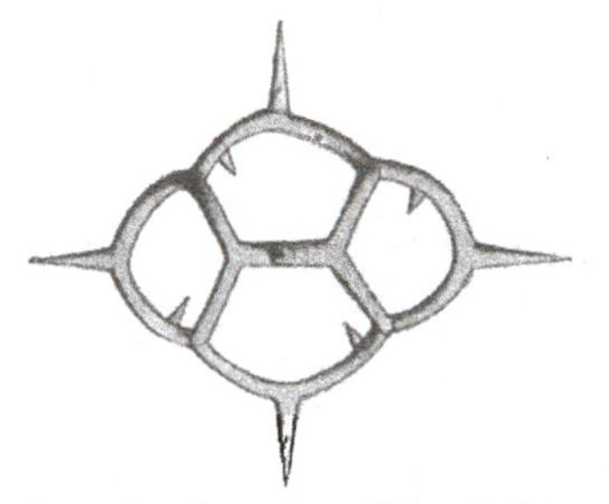

Fig. 333. An isolated portion of the skeleton of *Dityocha*.

tions, all explicable as basket-like frameworks developed around a cluster of cells, and adsorbed or secreted in the grooves common to adjacent cells or bubbles, are found in great variety.

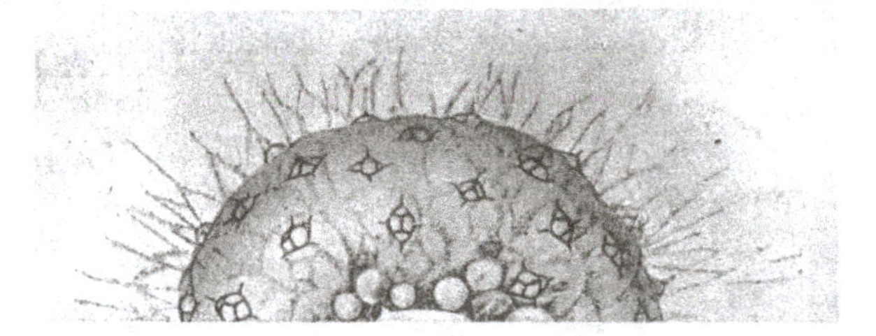

Fig. 334. *Dictyocha stapedia* Hkl.

The *Dictyocha*-spicule, laid down as a siliceous framework in the grooves between a few clustered cells, is too simple and natural to be confined to one group of animals. We have already seen it, as a calcareous spicule, in the holothurian genus *Thyone*, and we may find it again, in many various forms, in the protozoan group known as the Silicoflagellata*. Nothing can better illustrate the physico-mathematical character of these configurations than their

* See (*int. al.*) G. Deflandre, Les Silicoflagellés, etc., *Bull. Soc. Fr. de Microscopie*, I, p. 1, 1932: the figures in which article are mostly drawn from Ehrenberg's *Mikrogeologie*, 1854.

common occurrence in diverse groups of organisms. And the simple fact is, that we seem to know less and less of these things on the biological side, the more we come to understand their physical and

Fig. 335. Various species of *Distephanus* (Silicoflagellata). From Deflandre, after Ehrenberg.

mathematical characters. I have lost faith in Haeckel's four thousand "species" of Radiolaria.

In *Callimitra* itself, and elsewhere where the boundary-walls (and not merely their edges) are silicified, the skeletal matter is not

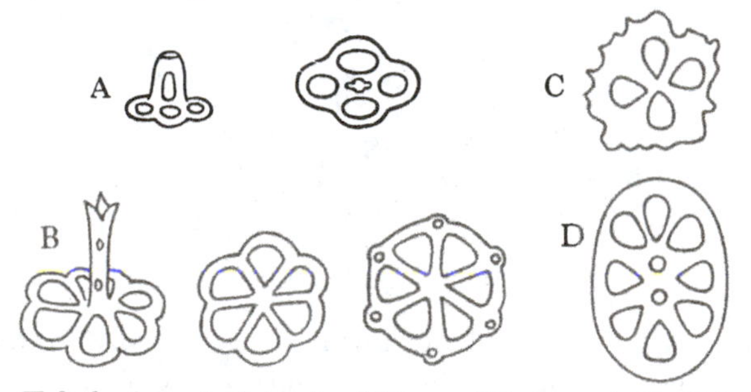

Fig. 336. Holothurian spicules. A, of *Thyone* (Mortensen); B, *Holothuria lactea* (Perrier); C, D, *Holothuria* and *Phyllophorus* (Deichman).

deposited in an even layer, like the waxen walls of a bee's cell, but in a close meshwork of fine curvilinear threads; and the curves seem to form three main series more or less closely related to the three edges of the partition. Sometimes (as may also be seen in our figure) the system is further complicated by a radial series running from the centre towards the free edge of each partition.

As to the former, their arrangement is such as would result if deposition or solidification had proceeded *in waves*, starting independently from each of the three boundary-edges of the little partition-wall, and something of this kind is doubtless what actually happened. We are reminded of the wave-like periodicity of the Liesegang phenomenon, and especially, perhaps, of the criss-cross rings which Liesegang observed in frozen gelatine (*supra*, p. 663). But there may be other explanations. For instance the film, liquid or other, which originally constituted the partition, might conceivably be thrown into *vibrations*, and then (like the dust upon a Chladni plate) minute particles in or on the film would tend to take up position in an orderly way, in relation to the nodal points or lines of the vibrating surface*. Some such hypothetical vibration may (to my thinking) account for the minute and varied and very beautiful patterns upon many diatoms, the resemblance of which patterns to the Chladni figures (in certain of their simpler cases) seems here and there striking and obvious. But I have not attempted to investigate the many special problems suggested by the diatom-skeleton.

The cusps at the four corners of the tetrahedral skeleton are a marked peculiarity of our Nassellarian shell, and we should by no means expect to see them in a skeleton formed at the boundary-edges of a simple tetrahedral pyramid of four bubbles or cells. But when we introduce another bubble into the centre of a system of four, then, as Plateau shewed, the tensions of its walls and of the surrounding partitions so balance one another that it becomes a regular curvilinear tetrahedron, or, as seen in plane projection (Fig. 337), a curvilinear, equilateral triangle, with prominent, not re-entrant angles. A drop of fluid tends to accumulate at each corner where four edges meet, and forms a bourrelet†; it is drawn out in the directions of the four films which impinge upon it, and

* Cf. Faraday's beautiful experiments, On the moving groups of particles found on vibrating elastic surfaces, etc., *Phil. Trans.* 1831, p. 299; *Researches*, 1859, pp. 314–358.

† The bourrelet is not only, as Plateau expresses it, a "surface of continuity," but we also recognise that it tends (so far as material is available for its production) to further lessen the free surface-area. On its relation to vapour-pressure and to the stability of foam, see FitzGerald's interesting note in *Nature*, Feb. 1, 1894 (*Works*, p. 309); and on its effect in thinning the soap-bubble to bursting-point, see Willard Gibbs, *Coll. Papers*, I, p. 307 *seq.*

so tends to assume in miniature the very same shape as the tetrahedron to whose corner it is attached. Out of these bourrelets, then, the cusps at the four corners of our little skeleton are formed.

A large and curiously beautiful family of radiolarian (or "polycystine") skeletons look, in a general way, like tiny helmets or *Pickelhauben*, with spike above, and three (or sometimes six) curved lobes, like helmet-straps, below. We recognise a family likeness, even a mathematical identity, between this figure and the last, for both alike are based on a tetrahedral symmetry: the body of the helmet corresponding to the inner vesicle of *Callimitra*, and the spike and the three straps to the four edges which ran out from the inner to the outer tetrahedron. In the one case an inner vesicle

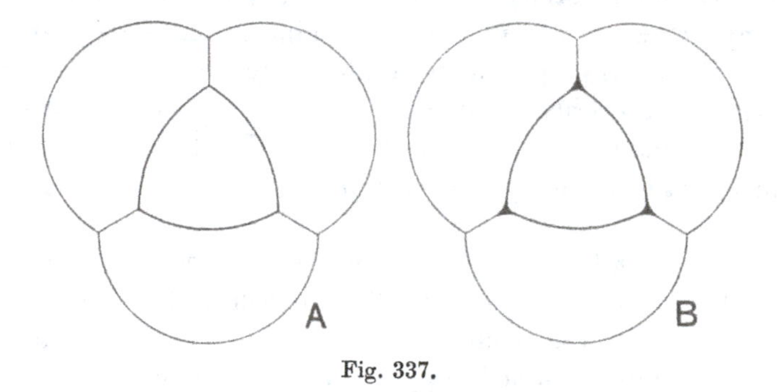

Fig. 337.

is surrounded by a tetrahedral figure whose outer walls, indeed, are absent, but its edges remain, and so do the walls connecting the outer and inner vesicles. In the other case the outer edges are gone, and so are the filmy partition-walls, save parts which correspond to the four internal edges between them*. There are apt to be two slight discrepancies. The helmet is often of somewhat complicated form, easily explained as due to the presence of two superposed bubbles instead of one. The other apparent anomaly is that the three helmet-straps are curved, while the corresponding edges are straight in Plateau's figure of the regular tetrahedron. But it is a paramount necessity (as we well know) for each set of four edges in a system of fluid films to meet in a point two and two *at the Maraldi angle*; just as it is necessary for the faces to meet

* Looking through Haeckel's very numerous figures, we see that now and then something more is left than the mere edges of the partition-walls.

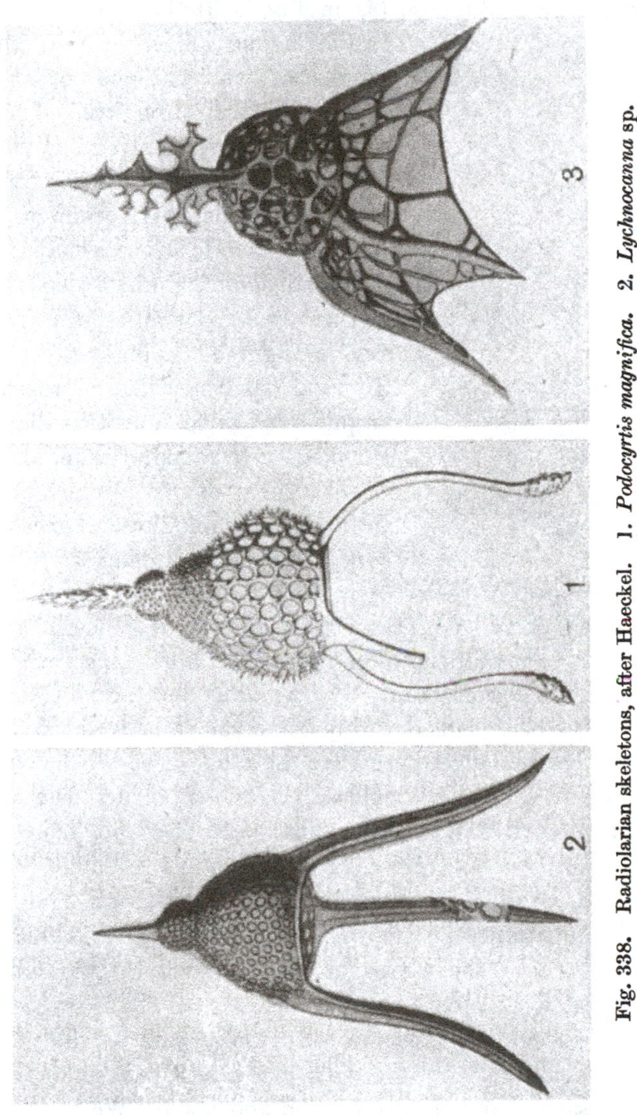

Fig. 338. Radiolarian skeletons, after Haeckel. 1. *Podocyrtis magnifica.* 2. *Lychnocanna* sp. 3. *Tripocyrtis* sp. (× c. 200 diam.).

at 120°; and faces and edges become curved, whenever necessary, in order that these conditions may be fulfilled. The need may arise in various ways. Suppose (as in Fig. 339) that our little central bubble be no longer in the centre of symmetry, but near one corner of the enclosing tetrahedron; the short edge running out from that corner will tend to remain straight (and so form the spike of the helmet); while the other three will each form an **S**-shaped curve, as a condition of making co-equal angles at their two extremities. An analogous case is figured in one of Sir David Brewster's papers where he repeats and amplifies some experiments of Plateau's*. He made a tetrahedral cage, and fitted it with three more wires, leading from the apex to the middle point of each basal edge. On dipping this into soap-solution, various complications were seen. At the apex, six films must not meet together, for no more than three surfaces may meet in an edge; intermediate or interstitial films make their appearance, with which we are not greatly concerned. But six films now ascend towards the apex from the base instead of three, and the three which come from the corners have a longer path than the other three which come from the mid-points of the basal edges; they *must* be curved in different degrees in order that all three may make at either end their co-equal angles. And if we now introduce a bubble (or two bubbles) into the interior of the system, we obtain the characteristic form of our helmet-shaped Radiolarian—the spike above, the single or double vesicle of the body, and the straps or lappets with their peculiar and characteristic curvatures.

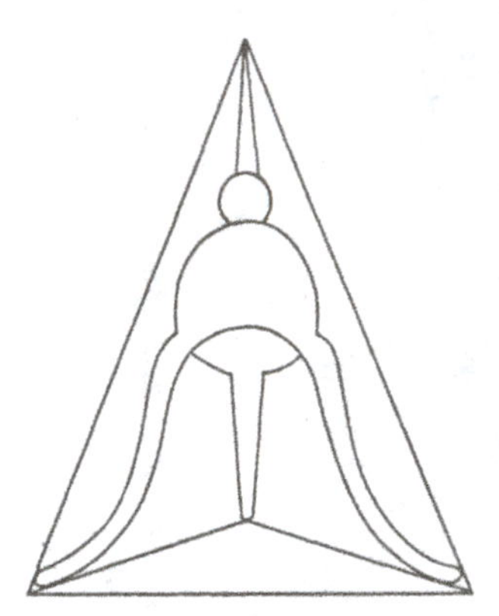

Fig. 339. Diagram of one of the helmet-shaped radiolaria, e.g. *Podocyrtis*: to shew its tetrahedral symmetry.

The little shell is perforated with many rounded holes, and it remains to account for these. They are required, so we are told, for the passage of pseudopodia; well and good. We have referred the *Dictyocha*-spicule and the hexagonal meshwork of *Aulonia* to froth-like associations of vesicles and to adsorption taking place

* On the figures of equilibrium in liquid films, *Trans. R.S.E.* XXIV, 1866.

between; so far so good, again. But the irregular lacework of the little helmets does not suit this explanation very well, and there may be yet other possibilities. We have already mentioned Tomlinson's "cohesion figures*." Experimenting with a great variety of substances, Tomlinson studied the innumerable ways in which drops, jets or floating films "cohere," disrupt or otherwise behave, under the resultant influences of surface-tension, cohesion, viscosity and friction; in one case a film runs out into a wavy or broken edge, in another it gives way here and there, and makes rents or holes in its surface. I take the little holes in our polycystine skeleton to be cohesion-figures, in Tomlinson's sense of the word—spots where the delicate film has given way and run into holes, and where surface-tension has rounded off the broken edges, and made the rents into rounded apertures.

In the foregoing examples of Radiolaria, the symmetry which the organism displays seems identical with that symmetry of forces which results from the play and interplay of surface-tensions in the whole system: this symmetry being displayed, in one class of cases, in a more or less spherical mass of froth, and in another class in a simpler aggregation of a few, otherwise isolated, vesicles. In either case skeletons are formed, in great variety, by one and the same kind of surface-action, namely by the adsorptive deposition of silica in walls and edges, corresponding to the manifold surfaces and interfaces of the system. But among the vast number of known Radiolaria, there are certain forms (especially among the Phaeodaria and Acantharia) which display a no less remarkable symmetry the origin of which is by no means clear, though surface-tension may play a part in its causation. Even this is doubtful; for the fact that three-way nodes are no longer to be seen at the junctions of the cells suggests that another law than that of minimal areas had been in action here. They are cases in which (as in some of those already described) the skeleton consists (1) of radiating spicular rods, definite in number and position, and (2) of interconnecting rods or plates, tangential to the more or less spherical body of the organism, whose form becomes, accordingly, that of a geometric, polyhedral solid. The great regularity, the numerical

* Cf. *supra*, p. 418

symmetries and the apparent simplicity of these latter forms makes of them a class apart, and suggests problems which have not been solved or even investigated.

The matter is partially illustrated by the accompanying figures (Fig. 340) from Haeckel's *Monograph of the Challenger Radiolaria**. In one of these we see a regular octahedron, in another a regular, or pentagonal, dodecahedron, in a third a regular icosahedron. In all cases the figure appears to be perfectly symmetrical, though neither the triangular facets of the octahedron and icosahedron, nor the pentagonal facets of the dodecahedron, are necessarily plane surfaces. In all of these cases, the radial spicules correspond to the corners of the figure; and they are, accordingly, six in number in the octahedron, twenty in the dodecahedron, and twelve in the icosahedron. If we add to these three figures the regular tetrahedron which we have just been studying, and the cube (which is represented, at least in outline, in the skeleton of the hexactinellid sponges), we have completed the series of the five regular polyhedra known to geometers, the *Platonic bodies*† of the older mathematicians. It is at first sight all the more remarkable that we should here meet with the whole five regular polyhedra, when we remember that, among the vast variety of crystalline forms known among minerals, the regular dodecahedron and icosahedron, simple as they are from the mathematical point of view, never occur. Not only do these latter never occur in crystallography, but (as is explained in textbooks of that science) it has been shewn that they cannot occur, owing to the fact that their indices (or numbers expressing the relation of the faces to the three primary axes) involve an irrational quantity: whereas it is a fundamental law of crystallography, involved in the whole theory of space-partitioning, that "the indices of any and every face of a crystal are small whole numbers‡." At the same time, an imperfect pentagonal dodeca-

* Of the many thousand figures in the hundred and forty plates of this beautifully illustrated book, there is scarcely one which does not depict some subtle and elegant *geometrical* configuration.

† They were known long before Plato: Πλάτων δὲ καὶ ἐν τούτοις πυθαγορίζει.

‡ If the equation of any plane face of a crystal be written in the form $hx+ky+lz=1$, then h, k, l are the indices of which we are speaking. They are the reciprocals of the parameters, or reciprocals of the distances from the origin at which the plane meets the several axes. In the case of the regular or pentagonal dodecahedron these indices are 2, $1+\sqrt{5}$, 0. Kepler described as follows, briefly

hedron, whose pentagonal sides are non-equilateral, is common among crystals. If we may safely judge from Haeckel's figures, the pentagonal dodecahedron of the Radiolarian (*Circorhegma*) is perfectly regular, and we may rest assured, accordingly, that it is not brought about by principles of space-partitioning similar to those which manifest themselves in the phenomenon of crystallisation. It will be observed that in all these radiolarian polyhedral shells, the surface of each external facet is formed of a minute hexagonal network, whose probable origin, in relation to a vesicular structure, is such as we have already discussed.

In certain allied Radiolaria of the family Acanthometridae (Fig. 341), which have twenty radial spines, the arrangement of these spines is commonly described in a somewhat singular way. The twenty spines are referred to five whorls of four spines each, arranged as parallel circles on the sphere, and corresponding to the equator, the tropics and the polar circles. This rule was laid down by the celebrated Johannes Müller, and has ever since been used and quoted as Müller's law*. But when we come to examine the figure, we find that Müller's law hardly does justice to the facts, and seems to overlook a simpler symmetry. We see in the first place that here, unlike our former cases, the twenty radial spines issue through the facets (and *all* the facets) of the polyhedron, instead of coming from its corners; and that our twenty spines correspond, therefore, not to the corners of a dodecahedron, but to the facets of some sort of an icosahedron. We see, in the next place, that this icosahedron is composed of faces of two kinds, hexagonal and pentagonal; and that the whole figure may be described as a hexagonal prism, whose twelve corners are truncated, and replaced by pentagonal facets. Both hexagons and pentagons appear to be equilateral, but if we

but adequately, the common characteristics of the dodecahedron and icosahedron: "Duo sunt corpora regularia, dodecaedron et icosaedron, quorum illud quinquangulis figuratur expresse, hoc triangulis quidem sed in quinquanguli formam coaptatis. Utriusque horum corporum ipsiusque adeo quinquanguli *structura perfici non potest sine proportione illa, quam hodierni geometrae divinam appellant*" (*De nive sexangula* (1611), Opera, ed. Fritsch, VII, p. 723). Here Kepler was dealing, somewhat after the manner of Sir Thomas Browne, with the mysteries of the quincunx, and also of the hexagon; and was seeking for an explanation of the mysterious or even mystical beauty of the 5-petalled or 3-petalled flower—*pulchritudinis aut proprietatis figurae, quae animam harum plantarum characterisavit.*

* See Johannes Müller, Ueber die Thalassicollen, Polycistinen und Acanthometren des Mittelmeeres, *Abh. d. Akad. Wiss. Berlin*, 1858, pp. 1–62, 11 pl.

try to construct a plane-sided polyhedron of this kind, we find it to be impossible; for into the angles between the six equatorial

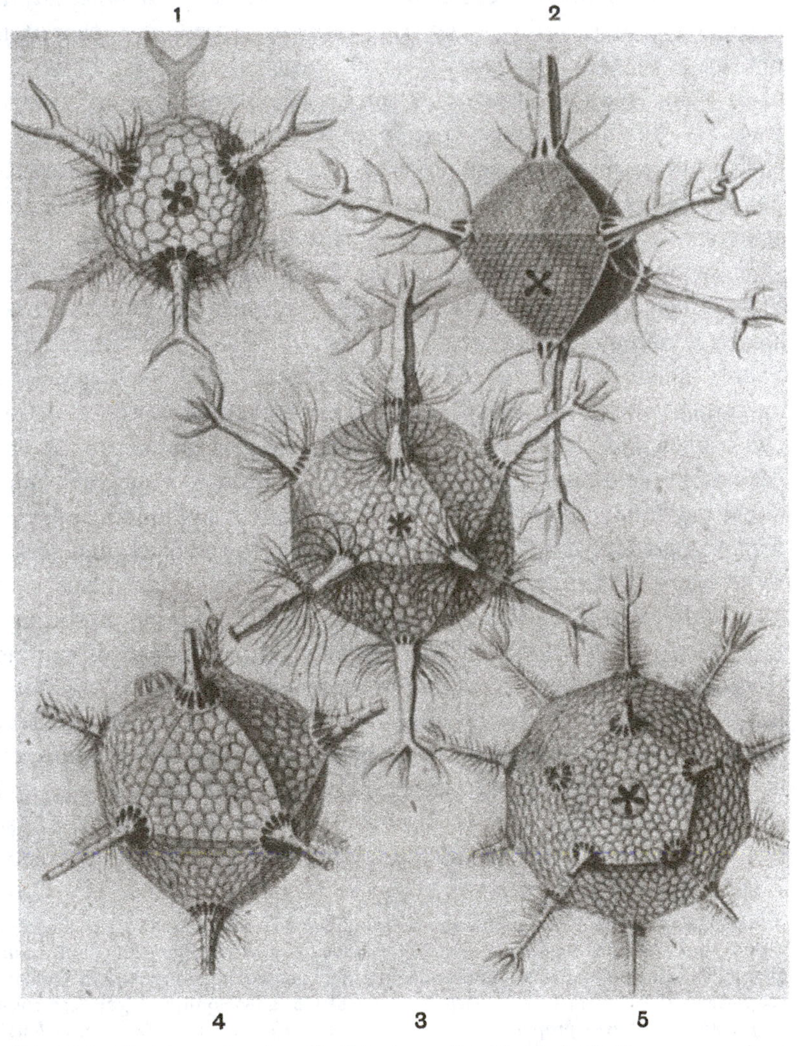

Fig. 340. Skeletons of various Radiolarians, after Haeckel. 1, *Circoporus sexfurcus*; 2, *C. octahedrus*; 3, *Circogonia icosahedra*; 4, *Circospathis novena*; 5, *Circorrhegma dodecahedra*.

regular hexagons six regular pentagons will not fit. The figure, however, can be easily constructed if we replace the straight edges (or some of them) by curves, the plane facets by slightly curved surfaces, or the regular by non-equilateral polygons*.

In some cases, such as Haeckel's *Phatnaspis cristata* (Fig. 342), we have an ellipsoidal body from which the spines emerge in the order described, but which is not obviously divided into facets. In Fig. 234 I have indicated the facets corresponding to the rays, and dividing the surface in the usual symmetrical way.

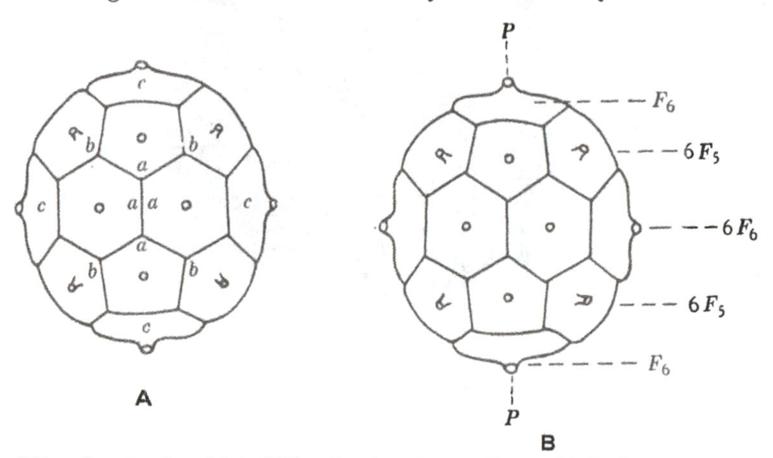

Fig. 341. *Dorataspis cristata* Hkl. A, viewed according to Müller's law: *a*, four polar plates; *b*, four intermediate or "tropical" plates; *c*, four equatorial plates. B, an alternative description: F_6, two polar and six equatorial hexagonal plates; F_5, two rows of six intermediate pentagonal plates.

About any polyhedron (within or without) we may describe another whose corners correspond to the sides, and whose sides to the corners, of the original figure; or the one configuration may be developed from the other by bevelling off, to a certain definite extent, the corners of the original polyhedron. The two figures, thus reciprocal to one another, form a "conjugate pair," and the principle is known as the "principle of duality" in polyhedra†. Of the regular solids, cube and octahedron, dodecahedron and

* Müller's interpretation was emended by Brandt, and what is known as Brandt's Law, viz. that the symmetry consists of two polar rays and three whorls of six each, coincides so far with the above description: save only that Brandt says plainly that the intermediate whorls stand equidistant between the equator and the poles, i.e. in latitude 45°, which, though not very far wrong, is geometrically inaccurate. But Brandt, if I understand him rightly, did not propose his "law" as a substitute for Müller's, but rather as a second law, applicable to a few special cases.

† First proved by Legendre, *Elém. de Géométrie*, VII, Prop. 25, 1794.

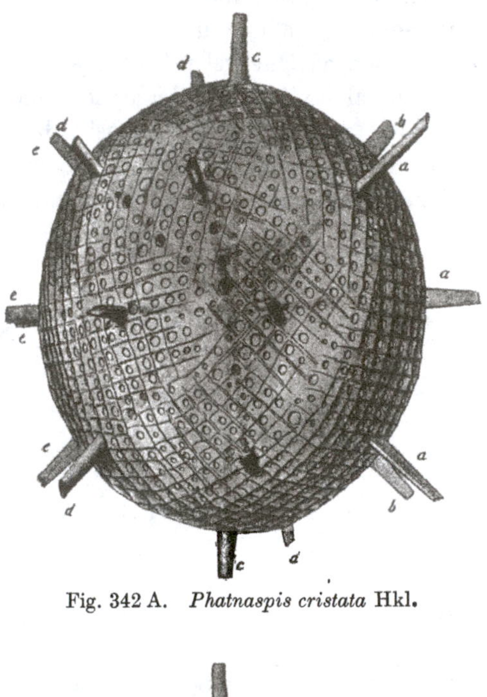

Fig. 342 A. *Phatnaspis cristata* Hkl.

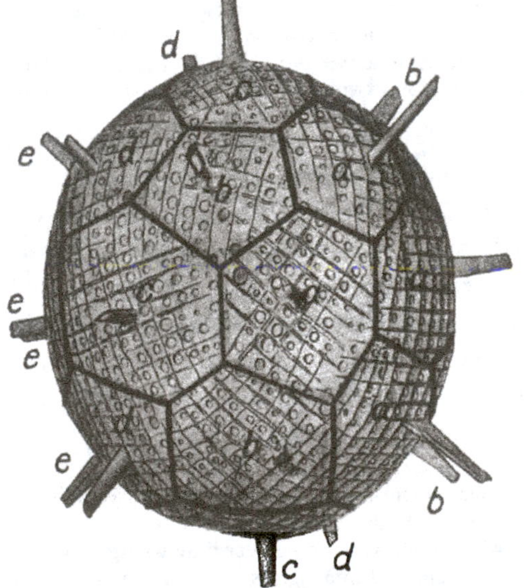

Fig. 342 B. The same, diagrammatic.

icosahedron, are conjugate pairs; but the first and simplest of all solid figures, the tetrahedron, has no conjugate but itself.

In our little shell of *Dorataspis*, the twenty spicules (as we have seen) spring from and correspond to the twenty facets of the polyhedron, twelve pentagonal and eight hexagonal, meeting at thirty-six corners, in all cases three by three; we may write the formula of the polyhedron, accordingly, as

$$12F_5 + 8F_6 + 36C_3.$$

If we now connect up the twenty spicules, three by three, we shall obtain thirty-six triangles, completely covering the figure; but we shall find that of the twenty corners twelve are surrounded by five, and eight by six triangles. The formula is now

$$36F_3 + 12C_5 + 8C_6,$$

and the two figures are fully reciprocal or conjugate.

I do not know of any radiolarian in which this configuration is to be found; nor does it seem a likely one, owing to the large and variable number of edges which meet in its corners. But we may have polyhedra related to, or derived from, one another in a less full and perfect degree. For instance, letting the twenty spicules of *Dorataspis* again serve as corners for the new figure, let *four facets* meet in each corner; or (which comes to the same thing) let each spicule give off four branches or offshoots, which shall meet their corresponding neighbours, and form the boundary-edges of a new network of facets. The result (Fig. 343) is a symmetrical figure, not geometrically perfect but elegant in its own way, which we recognise in a number of described forms*. It shews eight triangular and fourteen rhomboidal facets; and its formula is

$$8F_3 + 14F_4 + 20C_4.$$

Many subsidiary varieties may arise in turn: when, for instance, certain of the little branches fail to meet, or others grow large and widely confluent, always in symmetrical fashion.

We now see how in all such cases as these there is a *double symmetry* involved, that of two superimposed, and conjugate or semi-conjugate, figures. And the ambiguity which attends such descriptions as that which Johannes Müller embodied in his "law"

* Cf. W. Mielck, *Acanthometren aus Neu-Pommern*, Diss., Kiel, 1907.

seems due to a failure to recognise this twofold or alternative symmetry.

In all these latter cases it is the arrangement of the axial rods—the "polar symmetry" of the entire organism—which lies at the root of the matter; and which, if only we could account for it, would make it comparatively easy to explain the superficial configuration. But there are no obvious mechanical forces by which we can so explain this peculiar polarity. This at least is evident, that it arises in the central mass of protoplasm, which is the essential living portion of the organism as distinguished from that frothy

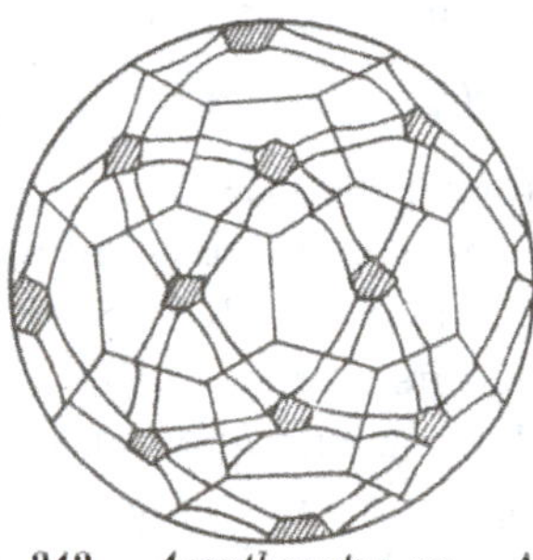

Fig. 343. *Acanthometra* sp. A derivative of the *Dorataspis* figure. After Mielck.

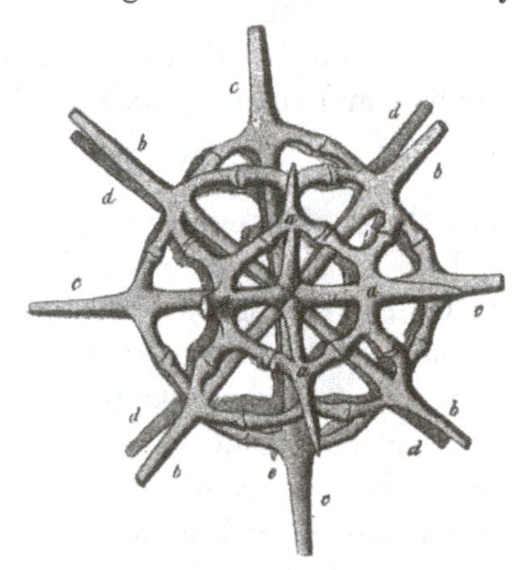

Fig. 344. *Phractaspis prototypus* Hkl.

peripheral mass whose structure has helped us to explain so many phenomena of the superficial or external skeleton. To say that the arrangement depends upon a specific polarisation of the cell is merely to refer the problem to other terms, and to set it aside for future solution. But it is possible that we may learn something about the lines in which *to seek for* such a solution by considering the case of Lehmann's "fluid crystals," and the light which they throw upon the phenomena of molecular aggregation.

The phenomenon of "fluid crystallisation" is found in a number of chemical bodies; it is exhibited at a specific temperature for each substance; and it would seem to be limited to bodies in which there is an elongated, or "long-chain" arrangement of the atoms in the molecule. Such bodies, at the appropriate temperature, tend to aggregate themselves into masses, which are sometimes spherical drops or globules (the so-called "spherulites"), and sometimes have the definite form of needle-like or prismatic crystals. In either case they remain liquid, and are also doubly refractive, polarising light in brilliant colours. Together with them are formed ordinary solid crystals, also with characteristic polarisation, and into such solid crystals all the fluid material ultimately turns. It seems that in these liquid crystals, though the molecules are freely mobile, just as are those of water, they are yet subject to, or endowed with, a "directive force," a force which confers upon them a definite configuration or "polarity," the "Gestaltungskraft" of Lehmann.

Such an hypothesis as this has been gradually extruded from the theories of mathematical crystallography*; and it has come to be understood that the symmetrical conformation of a homogeneous crystalline structure is sufficiently explained by the mere mechanical fitting together of appropriate structural units along the easiest and simplest lines of "close packing": just as a pile of oranges becomes definite, both in outward form and inward structural arrangement, without the play of any *specific* directive force. But while our conceptions of the tactical arrangement of crystalline molecules remain the same as before, and our hypotheses of "modes of packing" or of "space-lattices" remain as useful and as adequate as ever for the definition and explanation of the molecular arrangements, a new conception is introduced when we find something like such space-lattices maintained in what has hitherto been considered the molecular freedom of a liquid field; and Lehmann would persuade us, accordingly, to postulate a specific molecular force, or "Gestaltungskraft" (not unlike Kepler's "facultas formatrix"), to account for the phenomenon†.

Now just as some sort of specific "Gestaltungskraft" had been of old the *deus ex machina* accounting for all crystalline phenomena (*gnara totius geometriae, et in ea exercita*, as Kepler said), and as

* Cf. Tutton, *Crystallography*, 1911, p. 932.

† Kepler, if I understand him aright, saw his way to account for the shape of the bee's cell or the pomegranate-seed; and it was for want of any such mechanical explanation, and as little more than a confession of ignorance, that he fell back on a *facultas formatrix* to account for the six rays of the snow-crystal or the five petals of the flower. He was equally ready, unfortunately, to explain, by the same *facultas formatrix in aere*, the appearance of a plague of locusts or a swarm of flies.

such an hypothesis, after being dethroned and repudiated, has now fought its way back and claims a right to be heard, so it may be also in biology. We begin by an easy and general assumption of *specific properties*, by which each organism assumes its own specific form; we learn later (as it is the purpose of this book to shew) that throughout the whole range of organic morphology there are innumerable phenomena of form which are not peculiar to living things, but which are more or less simple manifestations of ordinary physical law. But every now and then we come to deep-seated signs of protoplasmic symmetry or polarisation, which *seem* to lie beyond the reach of the ordinary physical forces. It by no means follows that the forces in question are not essentially physical forces, more obscure and less familiar to us than the rest; and this would seem to be a great part of the lesson for us to draw from Lehmann's beautiful discovery. For Lehmann claims to have demonstrated, in non-living, chemical bodies, the existence of just such a determinant, just such a "Gestaltungskraft," as would be of infinite help to us if we might postulate it for the explanation (for instance) of our Radiolarian's axial symmetry. Further than this we cannot go; such analogy as we seem to see in the Lehmann phenomenon soon evades us, and refuses to be pressed home. The symmetry of crystallisation, which Haeckel tried hard to discover and to reveal in these and other organisms, resolves itself into remote analogies from which no conclusions can be drawn. Many a beautiful protozoan form has lent itself to easy physico-mathematical explanation; others, no less simple and no more beautiful prove harder to explain. That Nature keeps some of her secrets longer than others—that she tells the secret of the rainbow and hides that of the northern lights—is a lesson taught me when I was a boy.

A note on Polyhedra.

The theory of Polyhedra, Euler's *doctrina solidorum*, is a branch of geometry which deals with the more or less regular solids; and the rudiments of the theory may help us to study certain more or less symmetrical organic forms. Euler, a contemporary of Linnaeus, is the most celebrated of the many mathematicians who have carried this subject beyond where Pythagoras, Plato, Euclid and Archimedes had left it. He drew up a classification of polyhedral solids, using a binomial nomenclature based on the number of their corners or vertices, and sides or faces. Thus, for example, he called a figure with eight corners and seven faces *Octogonum heptaedrum*; and the analogy between this and Linnaeus's botanical classification and nomenclature—e.g. *Hexandria trigynia* and the rest—is very close and curious.

A simple theorem, of which Euler was vastly proud and which we still speak of as Euler's Law*, is fundamental to the theory of polyhedra. It tells us that in every polyhedron whatsoever, the faces and corners together outnumber the edges by two†:

$$C - E + F = 2 \qquad \ldots\ldots(1).$$

Another fundamental theorem follows. We know from Euclid that the three angles of a triangle are equal to two right angles; consequently, that in a polygon of C angles, the sum of the angles $= 2(C - 2)$ right angles. And there follows from this—but by no means expectedly—the analogous and extremely simple relation

* Euler, Elementa doctrinae solidorum, *Novi Comment. Acad. Sci. Imp. Petropol.* IV, p. 109 *seq.* (ad annos 1752 et 1753), 1758: "In omni solido hedris planis inclusum, aggregatum ex numero angulorum solidorum et ex numero hedrarum binario excedit numerum acierum." For a proof, see (*int. al.*) De Morgan, article Polyhedron in the *Penny Cyclopaedia*. There is reason to believe that Descartes was acquainted with this theorem between 1672 and 1676; cf. *Foucher de Careil, Œuvres inédites de Descartes,* Paris, II, p. 214. Cf. Baltzer, *Monatsber. Berlin. Akad.* 1861, p. 1043; and de Jonquières, *C.R.* 1890, p. 261. (The student will be struck by the *resemblance* between this formula and the phase rule of Willard Gibbs.)

† If we include, besides the corners, edges and faces (i.e. points, lines and surfaces) the solid figure itself, Euler's Law becomes

$$C - E + F - S = 1.$$

And in this form the theorem extends to n dimensions, as follows:

$$k_0 - k_1 + k_2 - k_3 + k_4 - \ldots = 1.$$

With equal beauty and simplicity, the *simplest* figure in each n-dimensional space is given as follows:

	k_0	k_1	k_2	k_3	k_4 etc.		
$n = 0$	1					$= 1$	(point)
1	2	-1				$= 1$	(line)
2	3	-3	$+1$			$= 1$	(triangle)
3	4	-6	$+4$	-1		$= 1$	(tetrahedron)
4	5	-10	$+10$	-5	$+1$	$= 1$	(pentahedroid)
etc.							

And, in a figure of n-dimensions, the sum of the plane angles $= 2^{(n-1)}(C - 2)$ right angles.

that in a polyhedron of C corners, the sum of the plane angles

$$= 4\,(C - 2) \text{ right angles} \qquad \ldots\ldots(2)^*.$$

Hence, if the polyhedron be isogonal, the sum of the plane angles at each corner

$$= \frac{4\,(C-2)}{C} \times 90^\circ, \text{ or } \left(4 - \frac{8}{C}\right) \text{ right angles} \qquad \ldots\ldots(3).$$

The five regular solids, or Platonic bodies—there can be no more—have been known from remote antiquity; they have their corners all alike and their faces all alike, they are isogonal and isohedral. Three of them, the tetrahedron, octahedron and icosahedron, have triangular faces; three of them, the tetrahedron, cube and dodecahedron, have trihedral or three-way corners. One or other of these, triangles or three-way corners, must (as we shall soon see) be present in every polyhedron whatsoever.

The semi-regular solids are regular in one respect or other, but not in both; they are *either* isogonal or isohedral—isohedral, when every face is an identical polygon and isogonal when at every corner the same set of faces is combined. The semi-regular *isogonal* solids, with all their corners alike but with two or more kinds of regular polygons for their faces, are thirteen in number—there can be no more; they were all described by Archimedes, and we call them by his name. One of them, with six square and eight hexagonal facets, derived by truncating the octahedron or the cube, we have found to be of peculiar interest, and it has become familiar to us as, of all homogeneous space-fillers, the one which encloses a given volume within a minimal area of surface. It is the *cubo-octahedron* of Kepler or of Fedorow, the *tetrakaidekahedron* of Kelvin, which latter name we commonly use.

Of semi-regular *isohedral* bodies, with all their sides alike (though no longer regular polygons) and their corners of two kinds or more, only one was known to antiquity; it is the rhombic dodecahedron, which is the crystalline form of the garnet, and appears in part

* On this remarkable parallel see Jacob Steiner, *Gesammelte Werke*, I, p. 97. It follows that the sum of the plane angles in a polyhedron, as in a plane polygon, is at once determined by the number of its corners: a result which delighted Euler, and led him to base his primary or generic classification of polyhedra on their corners rather than their sides.

again as a "space-filler" at the base of the bee's cell. A closely related rhombic icosahedron was known to Kepler; but it was left to Catalan* to discover, only some seventy-five years ago, that the isohedral bodies were thirteen in number, and were precisely comparable with and reciprocal to the Archimedean solids.

The semi-regular solids, both of Archimedes and of Catalan, are all, like the Platonic bodies, related to the sphere†, for a circumscribing sphere meets all the corners of an Archimedean solid, and an inscribed sphere touches all the faces of a solid of Catalan; and while the isogonal bodies can be constructed by various simple geometrical means, the general method of constructing the thirteen isohedral bodies is by dividing the sphere into so many similar and equal areas‡. It is a matter of spherical trigonometry rather than of simple geometry, and the problem, for that very reason, remained long unsolved.

The thirteen Archimedean bodies are derivable from the five Platonic bodies, in most cases easily, by so truncating their corners and their edges as to produce new and regularly polygonal faces in place of the old faces, corners and edges, and the possible number of faces in the new figure will be easily derived from the edges, corners and faces of the old. Part of the old faces will remain; each truncated corner will yield one new face; but each edge may be truncated, or bevelled, more than once, so as to yield one, two, or possibly three new faces. In short, if the faces, corners and edges of a regular solid be F, C, E, those of the Archimedean solids derivable from it (F_A) will be

$$F_A = F + mC + nE,$$

where $m = 0$ or 1, and $n = 0$, 1, or 2.

From the cube six Archimedean bodies may be derived, from the dodecahedron six, and from the tetrahedron one.

* *Journal de l'école impér. polytechnique*, XLI, pp. 1–71, 1865.

† It follows that the Chinese carved and perforated ivory balls, which are based on regular and symmetrical division of the sphere, can all be referred to one or another of the Platonic or Archimedean bodies.

‡ As a matter of fact, the Catalan bodies can be formed by adding to the Platonic bodies, just as (but not so easily as) the Archimedean bodies can be formed by truncating them.

The derivatives of the cube (with its six sides and eight corners) have the following numbers of sides:

$$\begin{aligned} F_A = F + C &= 6 + 8 = 14 \\ F + C + E &= 6 + 8 + 12 = 26 \\ F + C + 2E &= 6 + 8 + 24 = 38. \end{aligned}$$

The derivatives of the dodecahedron have, in like manner, 32, 62 or 92 sides; while the tetrahedron yields, by truncation of its four corners, a solid with eight sides.

The growth and form of crystals is a subject alien to our own, yet near enough to attract and tempt us. It is a curious thing (probably traceable to the Index Law of the crystallographer) that the Archimedean or isogonal bodies seldom occur and certainly play no conspicuous part in crystallography, while several of Catalan's isohedral figures are the characteristic forms of well-known minerals*.

Just as we pass from the Platonic to the Archimedean bodies by truncating the corners or edges of the former, so conversely, by producing their faces to a limit we obtain another family of figures —in all cases save the tetrahedron, which admits of no such extension; and the figures of this family are remarkable for the "twinned," or duplicate or multiple appearance which they present. If we extend the faces of an octahedron we get what looks like two tetrahedra, "twinned with" or interpenetrating one another; but there has been no interpenetration in the construction of this twin-like figure, only further accretion upon, and extension of, the facets of the octahedron. Among the higher polyhedra there are many figures which look, in a far more complicated way, like the twinning of simpler but still complicated forms; and these also have been constructed, not by interpenetration, but by the mere superposition of new parts on old.

An elementary, even a very elementary, knowledge of the theory of polyhedra becomes useful to the naturalist in various ways. Among organic structures we often find many-sided boxes (or what may be regarded as such), like the capsular seed-vessels of plants, the skeletons of certain Radiolaria, the shells of the Peridinia, the carapace of a tortoise, and a great many more. Or we may go

* E.g. the triakis, tetrakis and hexakis octahedra of fluor-spar. However the Archimedean tetrakaidekahedron ($6F_4\ 8F_6$) occurs in alum.

further and treat any cluster of cells, such as a segmenting ovum, as a species of polyhedron and study it from the point of view of Euler's Law and its associated theorems. We should have to include, as the geometer seldom does, the case of two-sided facets—facets with two corners and two curved sides or edges—like the "liths"* of a peeled orange; but the general formula would include these as a matter of course. On the other hand, we need very seldom consider any other than trihedral or three-way corners.

When we limit ourselves to polyhedra with trihedral corners the following formula applies:

$$4f_2 + 3f_3 + 2f_4 + f_5 \pm 0 . f_6 - f_7 - 2f_8 - \dots = 12 \quad \dots\dots(4).$$

That this formula applies to the tetrahedron with its four triangles, the cube with its six squares and the dodecahedron with its twelve pentagons, is at once obvious. The now familiar case of our four-celled egg with its polar furrows (Fig. 486, B, etc.) appears in two forms, according as the polar furrows run criss-cross or parallel. In the one case we have a curvilinear tetrahedron, in the other a figure with two two-sided and two four-sided facets; in either case the formula is obviously satisfied.

But the main lesson for us to learn is the broad, general principle that we cannot group as we please any number and sort of polygons into a polyhedron, but that the number and kind of facets in the latter is strictly limited to a narrow range of possibilities. For example, the case of *Aulonia* has already taught us that a polyhedron composed entirely of hexagons is a mathematical impossibility†; and the zero-coefficient which defines the number of hexagons in the above formula (4) is the mathematical statement of the fact‡. We can state it still more simply by the following corollary, likewise limited to the case of three-way corners:

$$(6 - n) F_n = 12.$$

* *Lith*, a useful Scottish word for a joint or segment. Cromwell, according to Carlyle, "gar'd kings ken they had a *lith* in their necks."

† That hexagons cannot enclose space, or form a "three-way graph," has been recognised as a significant fact in organic chemistry: where, for instance, it limits, somewhat unexpectedly, the ways in which a closed cyclol, or space-enclosing protein molecule, can be imagined to be built up. Cf. Dorothy Wrinch, in *Proc. R. S.* (A), No. 907, p. 510, 1937.

‡ Euler shewed at the same time the singular fact that no polyhedron can exist with seven edges.

This applies at once to the tetrahedron, the cube and the regular dodecahedron, and at once excludes the possibility of the closed hexagonal network.

We found in *Dorataspis* a closed shell consisting only of hexagons and pentagons; without counting these latter we know, by our formula, that they must be twelve in number, neither more nor less. Lord Kelvin's tetrakaidekahedron consists only of squares and hexagons; the squares are, and must be, six in number.

In a typical Peridinian, such as *Goniodoma*, there are twelve plates, all meeting by three-way nodes or corners; we know, and we have no difficulty in verifying the fact, that the twelve plates are all pentagonal.

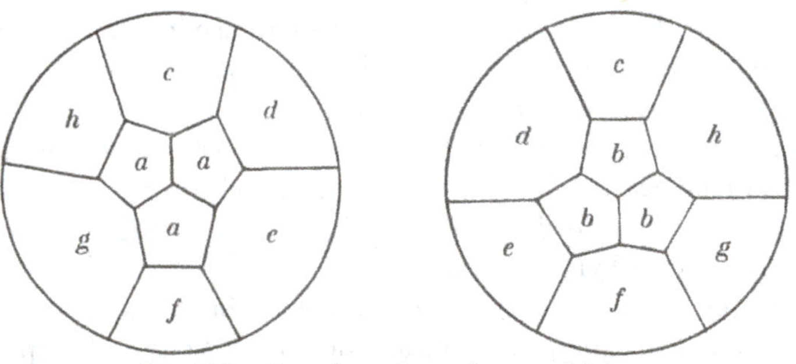

Fig. 345. *Goniodoma*, from above and below.

Without going beyond the elements of our subject we may want to extend our last formula, and remove the restriction to three-way corners under which it lay. We know that a tetrahedron has four triangles and four trihedral corners; that a cube has six squares and eight trihedral corners; an octahedron eight triangles and six four-way corners; an icosahedron twenty triangles and twelve five-way corners. By inspection of these numbers we are led to the following rule, and may establish it as a deduction from Euler's Law:

$$(f_3 + c_3) = 8 + 0\,(f_4 + c_4) + (f_5 + c_5) + 2\,(f_6 + c_6) + \text{etc.} \quad \ldots\ldots(5).$$

This important formula further illustrates the limitations to which all polyhedra are subject; for it shews us, among other things, that (if we neglect the exceptional case of dihedral facets or "liths") every polyhedron *must* possess either triangular faces or trihedral corners, and that these taken together are never less than eight in number.

We may add yet two more formulae, both related to the last, and all derivable, ultimately, from Euler's Law*:

$$3f_3 + 2f_4 + f_5 = 12 + 0 \,.\, c_3 + 2c_4 + 4c_5 + \ldots + 0 \,.\, f_6 + f_7 + 2f_8 \quad \ldots\ldots(6)$$

and

$$3c_3 + 2c_4 + c_5 = 12 + 0 \,.\, f_3 + 2f_4 + 4f_5 + \ldots + 0 \,.\, c_6 + c_7 + 2c_8 \quad \ldots(7).$$

These imply that in every polyhedron the triangular, quadrangular and pentagonal faces (*or* corners) must, taken together and multiplied as above, be at least twelve in number. Therefore no polyhedron can exist which has not a certain number of triangles, squares or pentagons in its composition; and the impossibility of a polyhedron consisting only of hexagons is demonstrated once again.

Formulae (5), (6) and (7) further shew us that not only is a three-way polyhedron of hexagons impossible, but also a four-way polyhedron of quadrangles, or one of six-way corners and triangular facets; all of which become the more obvious when we reflect that the plane angles meeting in each point or node must be, *on the average*, in the first case $3 \times 120°$, in the second $4 \times 90°$, and in the third $6 \times 60°$.

Lastly, having now considered the case of other than trihedral corners, we may learn a simple but very curious relation between the number of faces and corners, arising (like so much else) out of Euler's Law. In a polyhedron whose corners are all n-hedral, $nC = 2E$; therefore (by Euler) $nC/2 + 2 = F + C$; therefore $2F = (n-2)\,C + 4$.

Therefore, if

$$\left.\begin{aligned} n &= 3, \quad 2F = 4 + C \\ &= 4, \qquad\; = 4 + 2C \\ &= 5, \qquad\; = 4 + 3C \end{aligned}\right\} \quad \ldots\ldots(8).$$

Let us look again at the microscopic skeleton of *Dorataspis* (Fig. 341). We have seen that some of its facets are hexagonal, the rest pentagonal; there happen to be eight of the former, and therefore (as we now know) there *must* be twelve of the latter.

* Derivable from Euler together with the formulae for the "edge-counts," viz. $\Sigma nF_n = 2E$, and $\Sigma nC_n = 2E$; which merely mean that each edge separates two faces, and joins two corners.

We know also that, having no triangular facets, the polyhedral skeleton *must* possess trihedral corners; and these, moreover, must (by equation 5) be eight in number, plus the number of pentagons, plus twice the number of hexagons. The total,

$$8 + f_5 + 2f_6 = 8 + 12 + (2 \times 8) = 36,$$

is precisely the number of corners in the figure, all of them trihedral. We also know (from equation 2) that the sum of its plane angles

$$= 4\,(36 - 2) \times 90^\circ = 12{,}240^\circ,$$

which agrees with the sum of the angles of twelve pentagons and eight hexagons. The configuration, then, is a possible one.

So here and elsewhere an apparently infinite variety of form is defined by mathematical laws and theorems, and limited by the properties of space and number. And the whole matter is a running commentary on the cardinal fact that, under such *foedera Naturai* as Lucretius recognised of old, there are things which are possible, and things which are impossible, even to Nature herself.

CHAPTER X

A PARENTHETIC NOTE ON GEODESICS

We have made use in the last chapter of the mathematical principle of Geodesics (or Geodetics) in order to explain the conformation of a certain class of sponge-spicules; but the principle is of much wider application in morphology, and would seem to deserve attention which it has not yet received. The subject is not an easy one, and if we are to avoid mathematical difficulties we must keep within narrow bounds.

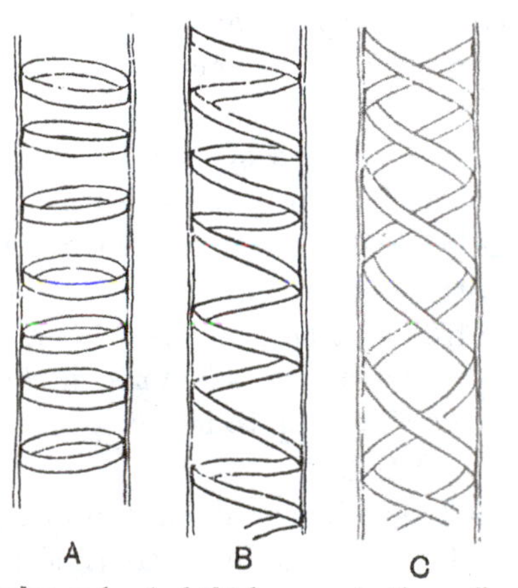

Fig. 346. Annular and spiral thickenings in the walls of plant-cells.

Defining, meanwhile, our geodesic line (as we have already done) as the shortest distance between two points on the surface of a solid of revolution, we find that the cylinder gives us some of the simplest of cases. Here it is plain that the geodesics are of three kinds: (1) a series of annuli around the cylinder, that is to say, a system of circles, in planes parallel to one another and at right angles to the axis of the cylinder (Fig. 346, A); (2) a series of straight lines parallel to the axis; and (3) a series of spiral curves winding round the wall of the cylinder (B, C). These three systems are all of frequent occurrence, and are illustrated in the local thickenings of the wall of the cylindrical cells or vessels of plants.

The spiral, or rather helical, geodesic is particularly common in cylindrical structures, and is beautifully shewn for instance in the spiral coil which stiffens the tracheal tubes of an insect, or the so-called tracheides of a woody stem. A like phenomenon is often witnessed in the splitting of a glass tube. If a crack appear in a test-tube it has a tendency to be prolonged in its own direction, and the more isotropic be the glass the more evenly will the split tend to follow the straight course in which it began. As a result, the crack often continues till our test-tube is split into a continuous spiral ribbon.

One may stretch a tape along a cylinder, but it no sooner swerves to one side than it begins to wind itself around; it is tracing its geodesic*.

In a circular cone, the spiral geodesic falls into closer and closer coils as the cone narrows, till it comes to the end, and then it winds back the same way; and a beautiful geodesic of this kind is exemplified in the sutural line of a spiral shell, such as *Turritella*, or in the striations which run parallel with the spiral suture. On a prolate spheroid, the coils of a spiral geodesic come closer together as they approach the ends of the long axis of the ellipse, and wind back and forward from one pole to the other†. We have a case of this kind in an *Equisetum*-spore, when the integument splits into the spiral "elaters," though the spire is not long enough to shew all its geodesic features in detail.

We begin to see that our first definition of a geodesic requires to be modified; for it is only subject to conditions that it is "the shortest distance between two points on the surface of the solid," and one of the commonest of these restricting conditions is that our geodesic may be constrained to go twice, or many times, round the surface on its way. In short, we may re-define a geodesic, as a curve drawn upon a surface such that, if we take any two *adjacent*

* It is not that the geodesic is rectified into a straight line; but that a straight line (the midline of the tape or ribbon) is converted into the geodesic on the given surface.

† In all these cases, $r \cos \alpha =$ a constant, where r is the radius of the circular section, and α the angle at which it is crossed by the geodesic. And the constant is measured by the smallest circle which the geodesic can reach.

points on the curve, the curve gives the shortest distance between them. It often happens, in the geodesic systems which we meet with in morphology, that two opposite spirals or rather helices run separate and distinct from one another, as in Fig. 346, C; and it is also common to find the two interfering with one another, and forming a criss-cross or reticulated arrangement. This indeed is a common source of reticulated patterns.

The microscopic and even ultramicroscopic structure of the cell-wall shews analogous configurations: as in the large cells of the alga *Valonia*, where the wall consists of many lamellae, each composed of parallel fibrillae running in spiral geodesics, and alternating in direction from one lamella to another. Here, and not less clearly in the young parenchyma of seedling oats, it is the long-chain cellulose molecules which follow a spiral course around the cell-wall, right-handed or left-handed as the case may be, and inclined more or less steeply according to the elongation of the cell. But these highly interesting questions of molecular, or micellar, structure lie beyond our scope*.

Among the ciliated infusoria, we have a variety of beautiful geodesic curves in the spiral patterns in which their cilia are arranged; though it is probable enough that in some complicated cases these are not simple geodesics, but developments of curves other than a straight line upon the surface of the organism. In other words, they seem to be instances of "geodesic curvature."

Lastly, an instructive case is furnished by the arrangement of the muscular fibres on the surface of a hollow organ, such as the heart or the stomach. Here we may consider the phenomenon from the point of view of mechanical efficiency, as well as from that of descriptive anatomy. In fact we have a right to expect that the muscular fibres covering such hollow organs will coincide with geodesic lines, in the sense in which we are using the term. For if we imagine a contractile fibre, or an elastic band, to be fixed by its two ends upon a curved surface, it is obvious that its first effort of contraction will tend to expend itself in accommodating

* Cf. (e.g.) C. Correns, Innere Struktur einiger Algenmembranen, *Beitr. zur Morphol. u. Physiol. d. Pflanzenzelle*, 1893, p. 260; W. T. Astbury and others, *Proc. R.S.* (B), CIX, p. 443, 1932, and other papers; G. van Iterson, jr., *Nature*, CXXXVIII, p. 364, 1936; R. D. Preston, *Proc. R.S.* (B), CXXV, p. 772, 1938; etc.

the band to the form of the surface, in "stretching it tight," or in other words in causing it to assume a direction which is the shortest possible line *upon the surface* between the two extremes: and it is only then that further contraction will have the effect of constricting the tube and so exercising pressure on its contents. Thus the muscular fibres, as they wind over the curved surface of an organ, arrange themselves automatically in geodesic curves: in precisely the same manner as we also automatically construct complex systems of geodesics whenever we wind a ball of wool or a spindle of tow, or when the skilful surgeon bandages a limb; indeed the surgeon must fold and crease his bandage if it is not to keep on geodesic lines. It is as a simple, necessary result of geodesic principles that we see those "figures-of-eight" produced, to which, in the case for instance of the heart-muscles, Pettigrew and other anatomists have ascribed peculiar importance. In the case of both heart and stomach we must look upon these organs as developed from a simple cylindrical tube, after the fashion of the glass-blower, as is further discussed on p. 1049 of this book, the modification of the simple cylinder consisting of various degrees of dilatation and of twisting. In the primitive undistorted cylinder, as in an artery or in the intestine, the muscles run in simple geodesic lines, and constitute the circular and longitudinal coats which form (or are said to form) the normal musculature of all tubular organs, or the cylindrical body of a worm. However, we can often recognise, in a small artery for instance, that the so-called circular fibres tend to take a slightly oblique or spiral course; and that the so-called *annular* muscle-fibres are really spirals is an old statement which may very likely be true*. If we consider each muscular fibre as an elastic strand embedded in the elastic membrane which constitutes the wall of the organ, it is evident that, whatever be the distortion suffered by the entire organ, the individual fibre will follow its own course, which will still, in a sense, be geodesic. But if the distortion be considerable, as for instance if the tube

* See A Discourse concerning the Spiral, instead of the supposed Annular, structure of the Fibres of the Intestins; discover'd and shewn by the Learn'd and Inquisitive Dr. William Cole to the Royal Society, *Phil. Trans.* XI, pp. 603–609, 1676. Cf. Eben J. Carey, Studies on the...small intestine, *Anat. Record*, XXI, pp. 189–215, 1921; F. T. Lewis, The spiral trend of intestinal muscle fibres, *Science*, LV, June 30, 1922.

become bent upon itself, or if at some point its walls bulge outwards in a diverticulum or pouch, then the old system of geodesics will only mark the shortest distance between two points more or less approximate to one another, and new systems of geodesics, peculiar to the new surface, will tend to appear, and link up points more remote from one another. This is evidently the case in the human stomach. We still have the systems, or their unobliterated remains, of circular and longitudinal muscles; but we also see two new systems of fibres, both obviously geodesic (or rather, when we look more closely, both parts of one and the same geodesic system), in the form of annuli encircling the pouch or diverticulum at the cardiac end of the stomach, and of oblique fibres taking a spiral course from the neighbourhood of the oesophagus over the sides of the organ.

In the heart we have a similar, but more complicated phenomenon. Its musculature consists, in great part, of the original simple system of circular and longitudinal muscles which enveloped the original arterial tubes, which tubes, after a process of local thickening, expansion, and especially *twisting*, came together to constitute the composite, or double, mammalian heart; and these systems of muscular fibres, geodesic to begin with, remain geodesic (in the sense in which we are using the word) after all the twisting which the primitive cylindrical tube or tubes have undergone. That is to say, these fibres still run their shortest possible course, from start to finish, over the complicated curved surface of the organ; and, as Borelli well understood, it is only because they do so that their contraction, or longitudinal shortening, is able to produce its direct effect in the contraction or systole of the heart*.

As a parenthetic corollary to the case of the spiral pattern upon the wall of a cylindrical cell, we may consider for a moment the spiral line which many small organisms tend to follow in their path

* The spiral fibres, or a large portion of them, constitute what Searle called "the rope of the heart" (Todd's *Cyclopaedia*, II, p. 621, 1836). The "twisted sinews of the heart" were known to early anatomists, and have been frequently and elaborately studied: for instance, by Gerdy (*Bull. Fac. Med. Paris*, 1820, pp. 40–148), and by Pettigrew (*Phil. Trans.* 1864), and again by J. B. Macallum (*Johns Hopkins Hospital Report*, IX, 1900) and by Franklin P. Mall (*Amer. Journ. Anat.* XI, 1911).

of locomotion*. A certain physiologist observed that an Amoeba, crawling within a narrow tube, wound its slow way in a spiral course instead of going straight along the tube. The creature was going nowhere in particular, but merely following the direction in which it had begun: in curious illustration of a familiar statement in the "dynamics of a particle," that a particle moving on a surface without constraint will describe geodesic lines.

But it is after a different fashion, and without any constraint to a surface, that the smaller ciliated organisms, such as the ciliate and flagellate infusoria, the rotifers, the swarm-spores of various Protista, and so forth, shew a tendency to pursue a spiral path in their ordinary locomotion. The means of locomotion which they possess in their cilia are at best somewhat primitive and inefficient; they have no apparent means of steering, or modifying their direction; and, if their course tended to swerve ever so little to one side, the result would be to bring them round and round again in an approximately circular path (such as a man astray on the prairie is said to follow), with little or no progress in a definite longitudinal direction. But as a matter of fact, by reason of a more or less unsymmetrical form of the body, all these creatures tend more or less to *rotate* about their long axis while they swim. And this axial rotation, just as in the case of a rifle-bullet, causes their natural swerve, which is always in the same direction as regards their own bodies, to be in a continually changing direction as regards space: in short, to make a spiral course around, and more or less near to, a straight axial line†.

In this short chapter we have touched on phenomena where form repeats itself, and mathematical analogies recur, in very different things and very different orders of magnitude. The spiral muscles of heart or stomach are the mechanical outcome of twists which these tubular organs have undergone in the course of their development, and come, accordingly, under the general category of organic

* Cf. Bütschli, "Protozoa," in Bronn's *Thierreich*, II, p. 848, III, p. 1785, etc., 1883–87; Jennings, *Amer. Nat.* XXXV, p. 369, 1901; Pütter, Thigmotaxie bei Protisten, *Arch. f. Anat. u. Phys.* (*Phys. Abth. Suppl.*), pp. 243–302, 1900.

† Cf. W. Ludwig, Ueber die Schraubenbahnen niederer Organismen, *Arch. f. vergl. Physiologie*, IX, 1919.

or embryological growth. But the spiral thickenings in the woody fibres of a plant are of another order of things, and lie in the region of molecular phenomena. The delicate spirals of the cell-wall of a cotton-hair are based on a complicated cellulose space-lattice, recalling Nägeli's micellar hypothesis in a new setting; and giving us a glimpse of organic growth after the very fashion of crystalline growth, that is to say from the starting-point of molecular structure and configuration*.

* W. Lawrence Balls, Determiners of cellulose structure as seen in the cell-wall of cotton-hairs, *Proc. R.S.* (B), xcv, pp. 72–89, 1923, and other papers. Cf. also Wilfred Robinson, Microscopical features of mechanical strains in timber, and the bearing of these on the structure of the cell-wall in plants, *Phil. Trans.* (B), ccx, pp. 49–82, 1920.

CHAPTER XI

THE EQUIANGULAR SPIRAL

THE very numerous examples of spiral conformation which we meet with in our studies of organic form are peculiarly adapted to mathematical methods of investigation. But ere we begin to study them we must take care to define our terms, and we had better also attempt some rough preliminary classification of the objects with which we shall have to deal.

In general terms, a Spiral is a curve which, starting from a point of origin, continually diminishes in curvature as it recedes from that point; or, in other words, whose *radius of curvature* continually increases. This definition is wide enough to include a number of different curves, but on the other hand it excludes at least one which in popular speech we are apt to confuse with a true spiral. This latter curve is the simple *screw*, or cylindrical *helix*, which curve neither starts from a definite origin nor changes its curvature as it proceeds. The "spiral" thickening of a woody plant-cell, the "spiral" thread within an insect's tracheal tube, or the "spiral" twist and twine of a climbing stem are not, mathematically speaking, *spirals* at all, but *screws* or *helices*. They belong to a distinct, though not very remote, family of curves.

Of true organic spirals we have no lack*. We think at once of horns of ruminants, and of still more exquisitely beautiful molluscan shells—in which (as Pliny says) *magna ludentis Naturae varietas.* Closely related spirals may be traced in the florets of a sunflower; a true spiral, though not, by the way, so easy of investigation, is seen in the outline of a cordiform leaf; and yet again, we can recognise typical though transitory spirals in a lock of hair, in a staple of wool†, in the coil of an elephant's trunk, in the "circling spires"

* A great number of spiral forms, both organic and artificial, are described and beautifully illustrated in Sir T. A. Cook's *Spirals in Nature and Art*, 1903, and *Curves of Life*, 1914.

† On this interesting case see, e.g. J. E. Duerden, in *Science*, May 25, 1934.

of a snake, in the coils of a cuttle-fish's arm, or of a monkey's or a chameleon's tail.

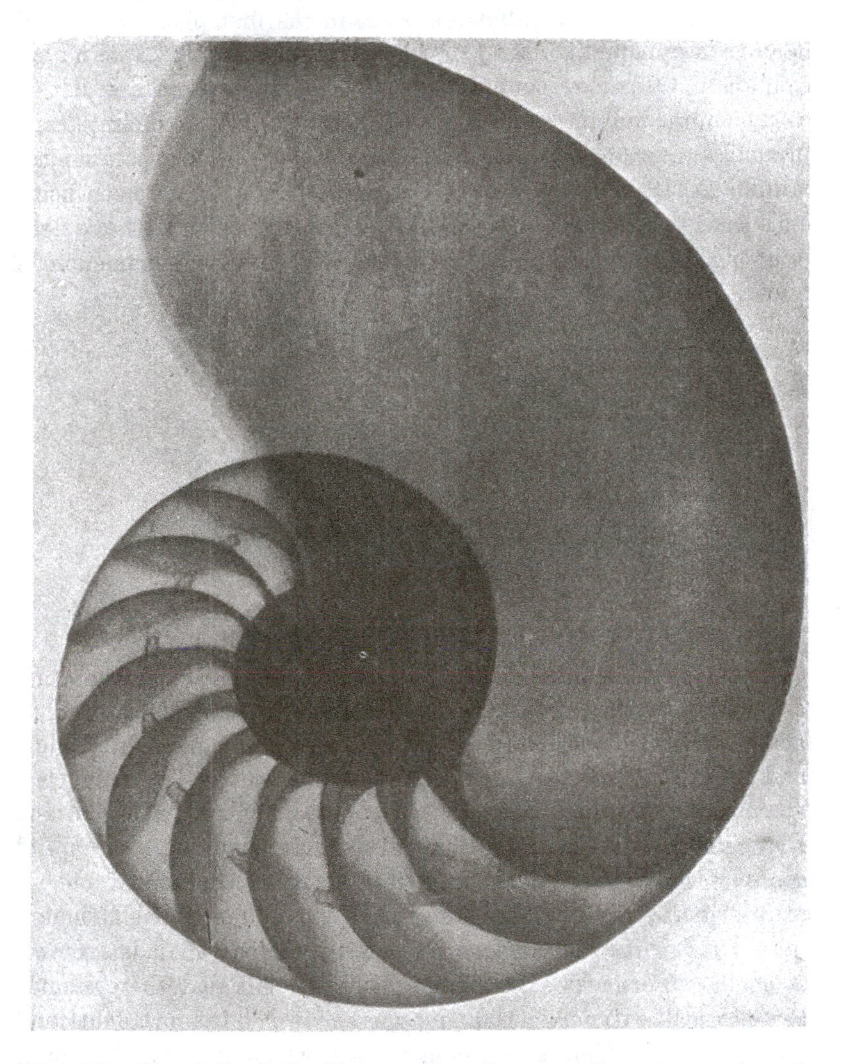

Fig. 347. The shell of *Nautilus pompilius*, from a radiograph: to shew the equiangular spiral of the shell, together with the arrangement of the internal septa. From Green and Gardiner, in *Proc. Malacol. Soc.* II, 1897.

Among such forms as these, and the many others which we might easily add to them, it is obvious that we have to do with things which, though mathematically similar, are biologically speaking fundamentally different; and not only are they biologically remote, but they are also physically different, in regard to the causes to which they are severally due. For in the first place, the spiral coil of the elephant's trunk or of the chameleon's tail is, as we have said, but a transitory configuration, and is plainly the result of certain muscular forces acting upon a structure of a definite, and normally an essentially different, form. It is rather a position, or an *attitude*, than a *form*, in the sense in which we have been using this latter term; and, unlike most of the forms which we have been studying, it has little or no direct relation to the phenomenon of growth.

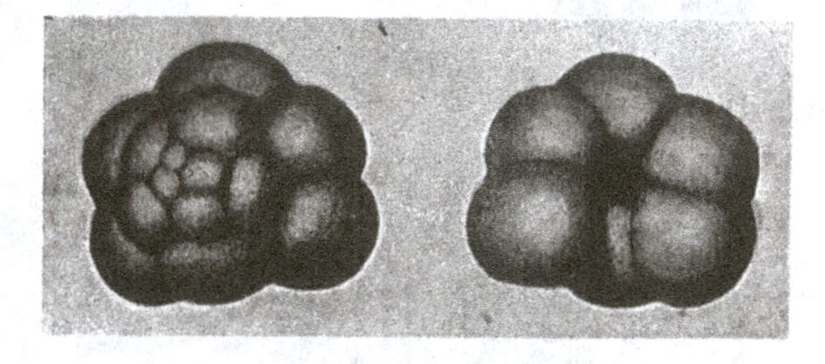

Fig. 348. A foraminiferal shell (*Pulvinulina*).

Again, there is a difference between such a spiral conformation as is built up by the separate and successive florets in the sunflower, and that which, in the snail or *Nautilus* shell, is apparently a single and indivisible unit. And a similar if not identical difference is apparent between the *Nautilus* shell and the minute shells of the Foraminifera which so closely simulate it: inasmuch as the spiral shells of these latter are composite structures, combined out of successive and separate chambers, while the molluscan shell, though it may (as in *Nautilus*) become secondarily subdivided, has grown as one continuous tube. It follows from all this that there cannot be a physical or dynamical, though there may well be a mathematical *law of growth*, which is common to, and which defines, the spiral form in *Nautilus*, in *Globigerina*, in the ram's horn, and in the inflorescence of the sunflower. Nature at least exhibits in them all "*un reflet des formes rigoureuses qu'étudie la géométrie**."

* Haton de la Goupillière, in the introduction to his important study of the *Surfaces Nautiloides, Annaes sci. da Acad. Polytechnica do Porto*, Coimbra, III, 1908.

Of the spiral forms which we have now mentioned, every one (with the single exception of the cordate outline of the leaf) is an example of the remarkable curve known as the equiangular or logarithmic spiral. But before we enter upon the mathematics of the equiangular spiral, let us carefully observe that the whole of the organic forms in which it is clearly and permanently exhibited, however different they may be from one another in outward appearance, in nature and in origin, nevertheless all belong, in a certain sense, to one particular class of conformations. In the great majority of cases, when we consider an organism in part or whole, when we look (for instance) at our own hand or foot, or contemplate an insect or a worm, we have no reason (or very little) to consider one part of the existing structure as *older* than another; through and through, the newer particles have been merged and commingled among the old; the outline, such as it is, is due to forces which for the most part are still at work to shape it, and which in shaping it have shaped it as a whole. But the horn, or the snail-shell, is curiously different; for in these the presently existing structure is, so to speak, partly old and partly new. It has been conformed by successive and continuous increments; and each successive stage of growth, starting from the origin, remains as an integral and unchanging portion of the growing structure.

We may go further, and see that horn and shell, though they belong to the living, are in no sense alive*. They are by-products of the animal; they consist of "formed material," as it is sometimes called; their growth is not of their own doing, but comes of living cells beneath them or around. The many structures which display the logarithmic spiral increase, or accumulate, rather than grow. The shell of nautilus or snail, the chambered shell of a foraminifer, the elephant's tusk, the beaver's tooth, the cat's claws or the canary-bird's—all these shew the same simple and very beautiful spiral curve. And all alike consist of stuff secreted or deposited by living cells; all grow, as an edifice grows, by accretion of accumulated

* For Oken and Goodsir the logarithmic spiral had a profound significance, for they saw in it a manifestation of life itself. For a like reason Sir Theodore Cook spoke of the *Curves of Life*; and Alfred Lartigues says (in his *Biodynamique générale*, 1930, p. 60): "Nous verrons la Conchyliologie apporter une magnifique contribution à la Stéréodynamique du tourbillon vital." The fact that the spiral is always formed of non-living matter helps to contradict these mystical conceptions.

material; and in all alike the parts once formed remain in being, and are thenceforward incapable of change.

In a slightly different, but closely cognate way, the same is true of the spirally arranged florets of the sunflower. For here again we are regarding serially arranged portions of a composite structure, which portions, similar to one another in form, *differ in age*; and differ also in magnitude in the strict ratio of their age. Somehow or other, in the equiangular spiral the *time-element* always enters in; and to this important fact, full of curious biological as well as mathematical significance, we shall afterwards return.

In the elementary mathematics of a spiral, we speak of the point of origin as the pole (O); a straight line having its extremity in the pole, and revolving about it, is called the radius vector; and a point (P), travelling along the radius vector under definite conditions of velocity, will then describe our spiral curve.

Of several mathematical curves whose form and development may be so conceived, the two most important (and the only two with which we need deal) are those which are known as (1) the equable spiral, or spiral of Archimedes, and (2) the equiangular or logarithmic spiral.

The former may be roughly illustrated by the way a sailor coils a rope upon the deck; as the rope is of uniform thickness, so in the whole spiral coil is each whorl of the same breadth as that which precedes and as that which follows it. Using its ancient definition, we may define it by saying, that "If a straight line revolve uniformly about its extremity, a point which likewise travels uniformly along it will describe the equable spiral*." Or, putting the same thing into our more modern words, "If, while the radius vector revolve uniformly about the pole, a point (P) travel with uniform velocity along it, the curve described will be that called the equable spiral, or spiral of Archimedes." It is plain that the spiral of Archimedes may be compared, but again roughly, to a *cylinder* coiled up. It is plain also that a radius ($r = OP$), made up of the successive and equal whorls, will increase in *arithmetical* progression: and will equal a certain constant quantity (a) multiplied

* Leslie's *Geometry of Curved Lines*, 1821, p. 417. This is practically identical with Archimedes' own definition (ed. Torelli, p. 219); cf. Cantor, *Geschichte der Mathematik*, I, p. 262, 1880.

by the whole number of whorls, (or more strictly speaking) multiplied by the whole angle (θ) through which it has revolved: so that $r = a\theta$. And it is also plain that the radius meets the curve (or its tangent) at an angle which changes slowly but continuously, and which tends towards a right angle as the whorls increase in number and become more and more nearly circular.

But, in contrast to this, in the equiangular spiral of the *Nautilus* or the snail-shell or *Globigerina*, the whorls continually increase in breadth, and do so in a steady and unchanging ratio. Our definition is as follows: "If, instead of travelling with a *uniform* velocity, our point move along the radius vector with a velocity *increasing as its distance from the pole*, then the path described is

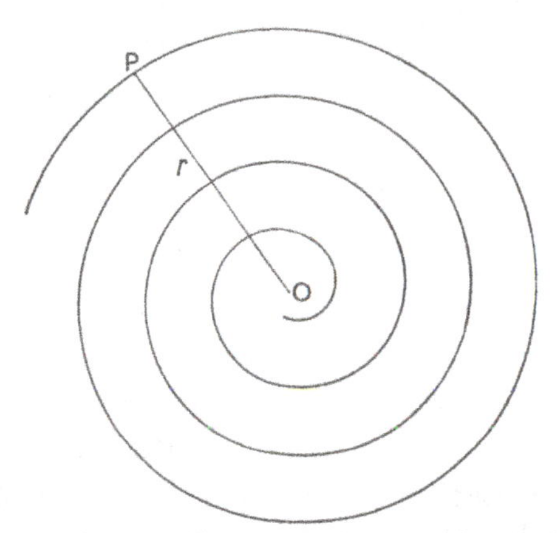

Fig. 349. The spiral of Archimedes.

called an equiangular spiral." Each whorl which the radius vector intersects will be broader than its predecessor in a definite ratio; the radius vector will increase in length in *geometrical* progression, as it sweeps through successive equal angles; and the equation to the spiral will be $r = a^{\theta}$. As the spiral of Archimedes, in our example of the coiled rope, might be looked upon as a coiled cylinder, so (but equally roughly) may the equiangular spiral, in the case of the shell, be pictured as a *cone* coiled upon itself; and it is the conical shape of the elephant's trunk or the chameleon's tail which makes them coil into a rough simulacrum of an equiangular spiral.

While the one spiral was known in ancient times, and was investigated if not discovered by Archimedes, the other was first recognised by Descartes, and discussed in the year 1638 in his letters to Mersenne*. Starting with the conception of a growing curve which should cut each radius vector at a constant angle—just as a circle does—Descartes shewed how it would necessarily follow that radii at equal angles to one another at the pole would be in continued proportion; that the same is therefore true of the parts cut off from a common radius vector by successive whorls or convolutions of the spire; and furthermore, that distances measured along the curve from its origin, and intercepted by any radii, as at B, C, are proportional to the lengths of these radii, OB, OC. It follows that

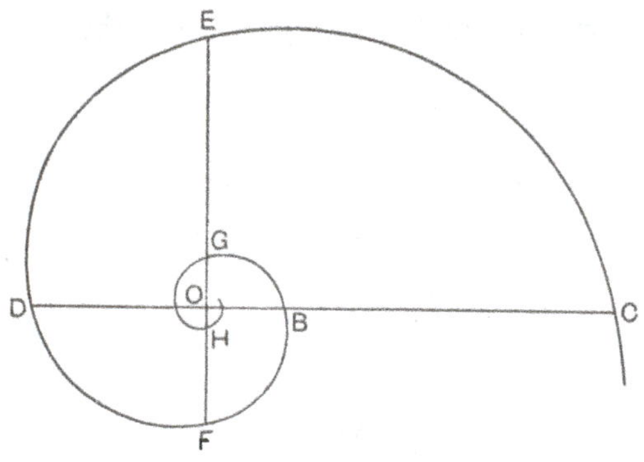

Fig. 350. The equiangular spiral.

the sectors cut off by successive radii, at equal vectorial angles, are similar to one another in every respect; and it further follows that the figure may be conceived as growing continuously without ever changing its shape the while.

If the whorls increase very slowly, the equiangular spiral will come to look like a spiral of Archimedes. The Nummulite is a case in point. Here we have a large number of whorls, very narrow, very close together, and apparently of equal breadth, which give rise to an appearance similar to that of our coiled rope. And, in a case of this kind, we might actually find that the whorls *were* of equal breadth, being produced (as is apparently the case in the Nummulite) not by any very slow and gradual growth in thickness of a continuous tube, but by a succession of similar cells or chambers laid on, round and round, determined as to their size by constant surface-tension conditions and therefore of unvarying dimensions. The Nummulite must always have a central core, or initial cell, around which the coil is not only wrapped, but out of which it springs; and this initial chamber corresponds to our a' in the expression $r = a' + a\theta \cot \alpha$.

* *Œuvres*, ed. Adam et Tannery, Paris, 1898, p. 360.

The many specific properties of the equiangular spiral are so interrelated to one another that we may choose pretty well any one of them as the basis of our definition, and deduce the others from it either by analytical methods or by elementary geometry. In algebra, when $m^x = n$, x is called the logarithm of n to the base m. Hence, in this instance, the equation $r = a^\theta$ may be written in the form $\log r = \theta \log a$, or $\theta = \log r / \log a$, or (since a is a constant) $\theta = k \log r$*. Which is as much as to say that (as Descartes discovered) the vector angles about the pole are proportional to the logarithms of the successive radii; from which circumstance the alternative name of the "logarithmic spiral" is derived†.

Moreover, for as many properties as the curve exhibits, so many names may it more or less appropriately receive. James Bernoulli called it the logarithmic spiral, as we still often do; P. Nicolas called it the geometrical spiral, because radii at equal polar angles are in geometrical progression; Halley, the proportional spiral, because the parts of a radius cut off by successive whorls are in continued proportion; and lastly, Roger Cotes, going back to Descartes' first description or first definition of all, called it the equiangular spiral‡. We may also recall Newton's remarkable demonstration that, had the force of gravity varied inversely as the *cube* instead of the *square* of the distance, the planets, instead of being bound to their

* Instead of $r = a^\theta$, we might write $r = r_0 a^\theta$; in which case r_0 is the value of r for zero value of θ.

† Of the two names for this spiral, equiangular and logarithmic, I used the latter in my first edition, but equiangular spiral seems to be the better name; for the constant angle is its most distinguishing characteristic, and that which leads to its remarkable property of continuous self-similarity. Equiangular spiral is its name in geometry; it is the analyst who derives from its geometrical properties its relation to the logarithm. The mechanical as well as the mathematical properties of this curve are very numerous. A Swedish admiral, in the eighteenth century, shewed an equiangular spiral (of a certain angle) to be the best form for an anchor-fluke (*Sv. Vet. Akad. Hdl.* XV, pp. 1–24, 1796), and in a parrot's beak it has the same efficiency. Macquorn Rankine shewed its advantages in the pitch of a cam or non-circular wheel (*Manual of Mechanics*, 1859, pp. 99–102; cf. R. C. Archibald, *Scripta Mathem.* III (4), p. 366, 1935).

‡ James Bernoulli, in *Acta Eruditorum*, 1691, p. 282; P. Nicolas, *De novis spiralibus*, Tolosae, 1693, p. 27; E. Halley, *Phil. Trans.* XIX, p. 58, 1696; Roger Cotes, *ibid.* 1714, and *Harmonia Mensurarum*, 1722, p. 19. For the further history of the curve see (e.g.) Gomes de Teixeira, *Traité des courbes remarquables*, Coimbre, 1909, pp. 76–86; Gino Loria, *Spezielle algebräische Kurven*, II, p. 60 *seq.*, 1911; R. C. Archibald (to whom I am much indebted) in *Amer. Mathem. Monthly*, XXV, pp. 189–193, 1918, and in Jay Hambidge's *Dynamic Symmetry*, 1920, pp. 146–157.

ellipses, would have been shot off in spiral orbits from the sun, the equiangular spiral being one case thereof.*

A singular instance of the same spiral is given by the route which certain insects follow towards a candle. Owing to the structure of their compound eyes, these insects do not look straight ahead but make for a light which they see abeam, at a certain angle. As they continually adjust their path to this constant angle, a spiral pathway brings them to their destination at last†.

In mechanical structures, *curvature* is essentially a mechanical phenomenon. It is found in flexible structures as the result of

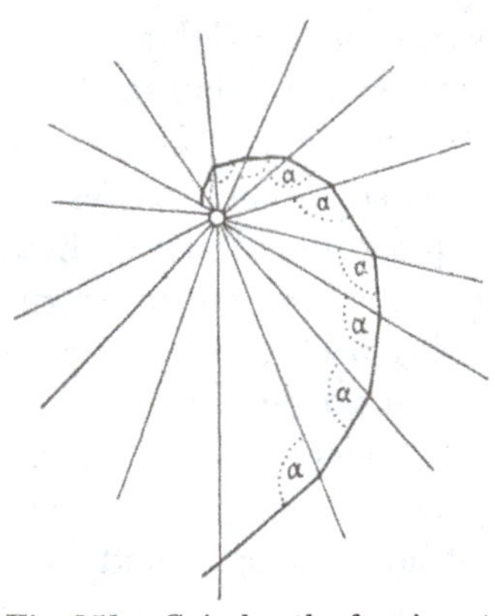

Fig. 351. Spiral path of an insect, as it draws towards a light. From Wigglesworth (after van Buddenbroek).

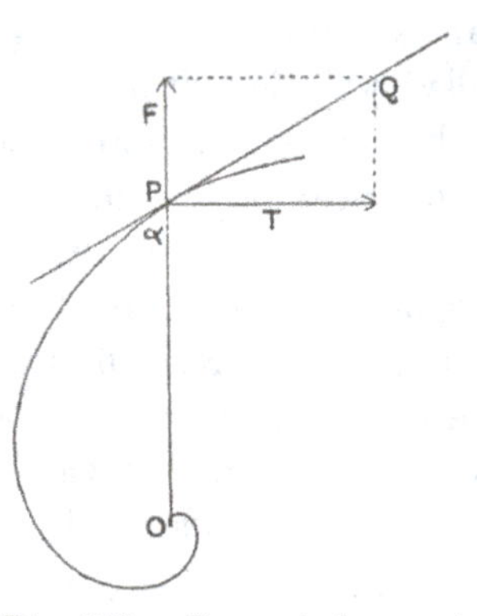

Fig. 352. Dynamical aspect of the equiangular spiral.

bending, or it may be introduced into the construction for the purpose of resisting such a bending-moment. But neither shell nor tooth nor claw are flexible structures; they have not been *bent* into their peculiar curvature, they have *grown* into it.

We may for a moment, however, regard the equiangular or logarithmic spiral of our shell from the dynamical point of view, by looking on *growth* itself as the force concerned. In the growing structure, let growth at any point P be resolved into a force F acting along the line joining P to a pole O, and a force T acting in a direction perpendicular to OP; and let the magnitude of these forces (or of these rates of growth) remain constant. It follows that

* *Principia*, I, 9; II, 15. On these "Cotes's spirals" see Tait and Steele, p. 147.

† Cf. W. Buddenbroek, *Sitzungsber. Heidelb. Akad.*, 1917; V. H. Wigglesworth, *Insect Physiology*, 1839, p. 167.

the resultant of the forces F and T (as PQ) makes a constant angle with the radius vector. But a constant angle between tangent and radius vector is a fundamental property of the "equiangular" spiral: the very property with which Descartes started his investigation, and that which gives its alternative name to the curve.

In such a spiral, radial growth and growth in the direction of the curve bear a constant ratio to one another. For, if we consider a consecutive radius vector, OP', whose increment as compared with OP is dr, while ds is the small arc PP', then $dr/ds = \cos\alpha = \text{constant}$.

In the growth of a shell, we can conceive no simpler law than this, namely, that it shall widen and lengthen in the same unvarying proportions: and this simplest of laws is that which Nature tends to follow. The shell, like the creature within it, grows in size *but does not change its shape*; and the existence of this constant relativity of growth, or constant similarity of form, is of the essence, and may be made the basis of a definition, of the equiangular spiral*.

Such a definition, though not commonly used by mathematicians, has been occasionally employed; and it is one from which the other properties of the curve can be deduced with great ease and simplicity. In mathematical language it would run as follows: "Any [plane] curve proceeding from a fixed point (which is called the pole), and such that the arc intercepted between any two radii at a given angle to one another is always similar to itself, is called an equiangular, or logarithmic, spiral."

In this definition, we have the most fundamental and "intrinsic" property of the curve, namely the property of continual similarity, and the very property by reason of which it is associated with organic growth in such structures as the horn or the shell. For it is peculiarly characteristic of the spiral shell, for instance, that it does not alter as it grows; each increment is similar to its predecessor, and the whole, after every spurt of growth, is just like what it was before. We feel no surprise when the animal which secretes the shell, or any other animal whatsoever, grows by such symmetrical expansion as to preserve its form unchanged; though even there, as we have already seen, the unchanging form denotes a nice balance between the rates of growth in various directions, which is

* See an interesting paper by W. A. Whitworth, The equiangular spiral, its chief properties proved geometrically, *Messenger of Mathematics* (1), I, p. 5, 1862. The celebrated Christian Wiener gave an explanation on these lines of the logarithmic spiral of the shell, in his highly original *Grundzüge der Weltordnung*, 1863.

but seldom accurately maintained for long. But the shell retains its unchanging form in spite of its *asymmetrical* growth; it grows at one end only, and so does the horn. And this remarkable property of increasing by *terminal* growth, but nevertheless retaining unchanged the form of the entire figure, is characteristic of the equiangular spiral, and of no other mathematical curve. It well deserves the name, by which James Bernoulli was wont to call it, of *spira mirabilis*.

We may at once illustrate this curious phenomenon by drawing the outline of a little *Nautilus* shell within a big one. We know, or we may see at once, that they are of precisely the same shape; so that, if we look at the little shell through a magnifying glass, it becomes identical with the big one. But we know, on the other hand, that the little *Nautilus* shell grows into the big one, not by growth or magnification in all parts and directions, as when the boy grows into the man, but by growing *at one end only*.

If we should want further proof or illustration of the fact that the spiral shell remains of the same shape while increasing in magnitude by its terminal growth, we may find it by help of our ratio $W : L^3$, which remains constant so long as the shape remains unchanged. Here are weights and measurements of a series of small land-shells (*Clausilia*):*

W (mgm.)	L (mm.)	$\sqrt[3]{W}/L$
50	14·4	2·56
53	15·1	2·49
56	15·2	2·52
56	15·2	2·52
56	15·4	2·44
58	15·5	2·50
61	16·4	2·40
63	16·0	2·49
67	16·0	2·54
69	16·1	2·56
	Mean	2·50

Though of all plane curves, this property of continued similarity is found only in the equiangular spiral, there are many rectilinear figures in which it may be shewn. For instance, it holds good of

* In 100 specimens of *Clausilia* the mean value of $\sqrt[3]{W}/L$ was found to be 2·517, the coefficient of variation 0·092, and the standard deviation 3·6. That is to say, over 90 per cent. grouped themselves about a mean value of 2·5 with a deviation of less than 4 per cent. Cf. C. Petersen, *Das Quotientengesetz*, 1921, p. 55.

any cone; for evidently, in Fig. 353, the little inner cone (represented in its triangular section) may become identical with the larger one either by magnification all round (as in *a*), or by an increment at one end (as in *b*); or for that matter on the rest of its surface, represented by the other two sides, as in *c*. All this is associated with the fact, which we have already noted, that the *Nautilus* shell is but a cone rolled up; that, in other words, the cone is but a particular variety, or "limiting case," of the spiral shell.

This singular property of continued similarity, which we see in the cone, and recognise as characteristic of the logarithmic spiral, would seem, under a more general aspect, to have engaged the particular attention of ancient mathematicians even from the days of Pythagoras, and so, with little doubt, from the still more ancient days of that Egyptian school whence he derived the foundations of

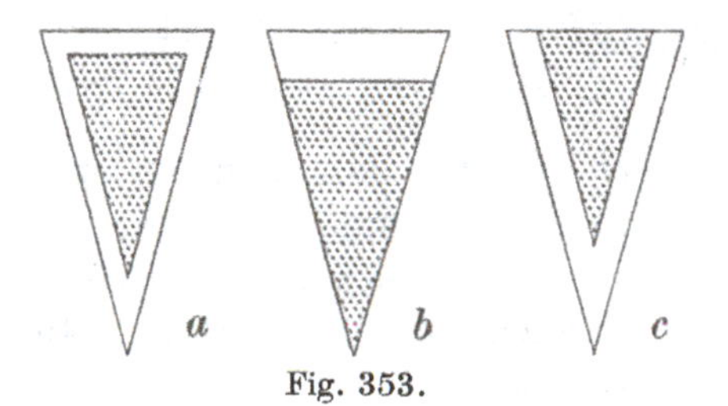

Fig. 353.

his learning*; and its bearing on our biological problem of the shell, however indirect, is close enough to deserve our very careful consideration.

There are certain things, says Aristotle, which suffer no alteration (save of magnitude) when they grow†. Thus if we add to a square an L-shaped portion, shaped like a carpenter's square, the resulting figure is still a square; and the portion which we have so added, with this singular result, is called in Greek a "gnomon."

Euclid extends the term to include the case of any parallelogram‡, whether rectangular or not (Fig. 354); and Hero of Alexandria

* I am well aware that the debt of Greek science to Egypt and the East is vigorously denied by many scholars, some of whom go so far as to believe that the Egyptians never had any science, save only some "rough rules of thumb for measuring fields and pyramids" (Burnet's *Greek Philosophy*, 1914, p. 5).

† *Categ.* 14, 15*a*, 30: ἔστι τινὰ αὐξανόμενα ἃ οὐκ ἀλλοιοῦται, οἷον τὸ τετράγωνον, γνώμονος περιτεθέντος, ηὔξηται μὲν ἀλλοιότερον δὲ οὐδὲν γεγένηται.

‡ Euclid (II, def. 2).

specifically defines a gnomon (as indeed Aristotle had implicitly defined it), as any figure which, being added to any figure whatsoever, leaves the resultant figure similar to the original. Included in this important definition is the case of numbers, considered geometrically; that is to say, the εἰδητικοὶ ἀριθμοί, which can be translated into *form*, by means of rows of dots or other signs (cf. Arist. *Metaph.* 1092 b 12), or in the pattern of a tiled floor: all according to "the mystical way of Pythagoras, and the secret

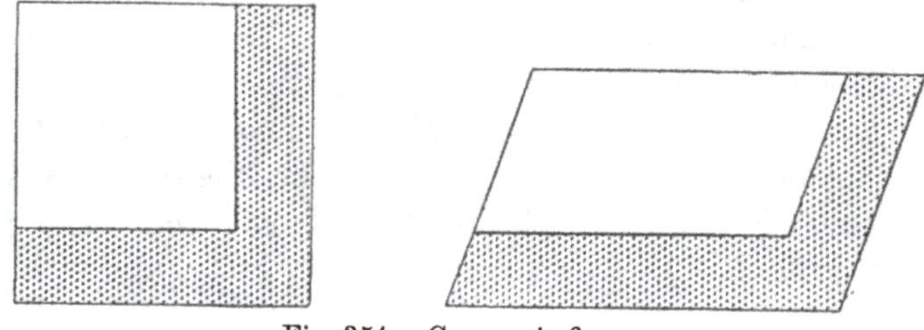

Fig. 354. Gnomonic figures.

magick of numbers." For instance, the triangular numbers, 1, 3, 6, 10 etc., have the natural numbers for their "differences"; and so the natural numbers may be called their gnomons, because they keep the triangular numbers still triangular. In like manner the square numbers have the successive odd numbers for their gnomons, as follows:

$$0 + 1 = 1^2$$
$$1^2 + 3 = 2^2$$
$$2^2 + 5 = 3^2$$
$$3^2 + 7 = 4^2 \quad \text{etc.}$$

And this gnomonic relation we may illustrate graphically (σχηματογραφεῖν) by the dots whose addition keeps the annexed figures perfect squares*:

There are other gnomonic figures more curious still. For example, if we make a rectangle (Fig. 355) such that the two sides are in the

* Cf. Treutlein, *Ztschr. f. Math. u. Phys.* (*Hist. litt. Abth.*), XXVIII, p. 209, 1883.

ratio of $1 : \sqrt{2}$, it is obvious that, on doubling it, we obtain a similar figure; for $1 : \sqrt{2} :: \sqrt{2} : 2$; and each half of the figure, accordingly, is now a gnomon to the other. Were we to make our paper of such a shape (say, roughly, 10 in. × 7 in.), we might fold and fold it, and the shape of folio, quarto and octavo pages would be all the same. For another elegant example, let us start with a rectangle (A) whose sides are in the proportion of the "divine" or "golden section*" that is to say as $1 : \frac{1}{2}(\sqrt{5} - 1)$, or, approximately, as 1 : 0·618.... The gnomon to this rectangle is the square (B) erected on its longer side, and so on successively (Fig. 356).

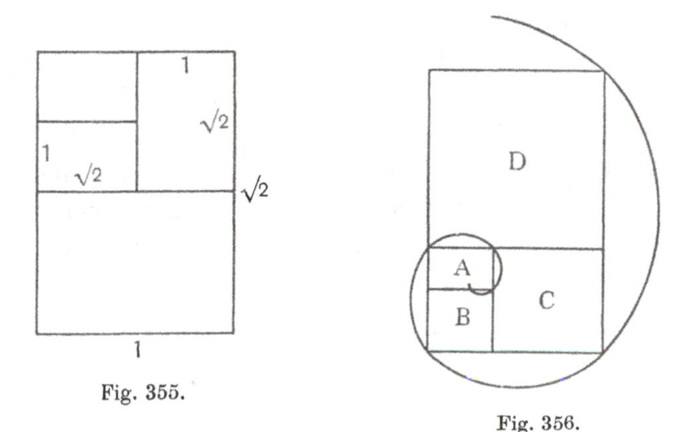

Fig. 355.

Fig. 356.

In any triangle, as Hero of Alexandria tells us, one part is always a gnomon to the other part. For instance, in the triangle ABC (Fig. 357), let us draw BD, so as to make the angle CBD equal to the angle A. Then the part BCD is a triangle similar to the whole triangle ABC, and ABD is a gnomon to BCD. A very elegant case is when the original triangle ABC is an isosceles triangle having one angle of 36°, and the other two angles, therefore, each equal to 72° (Fig. 358). Then, by bisecting one of the angles of the base, we subdivide the large isosceles triangle into two isosceles triangles, of which one is similar to the whole figure and the other is its gnomon†. There is good reason to believe that this triangle was especially studied by the Pythagoreans; for it lies at the root of

* Euclid, II, 11.

† This is the so-called *Dreifachgleichschenkelige Dreieck*; cf. Naber, *op. infra cit.* The ratio 1 : 0·618 is again not hard to find in this construction.

many interesting geometrical constructions, such as the regular pentagon, and its mystical "pentalpha," and a whole range of other curious figures beloved of the ancient mathematicians*: culminating

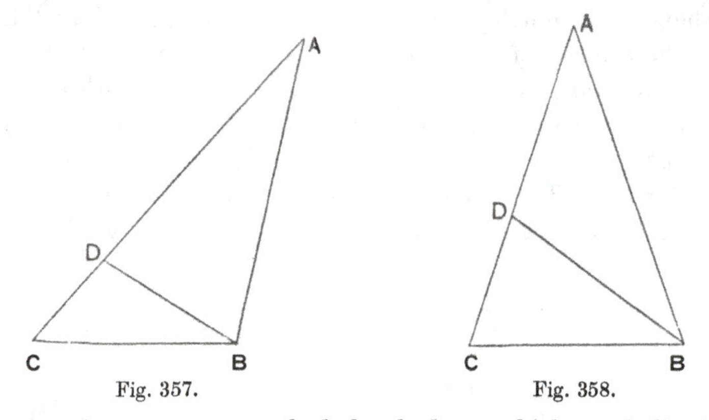

Fig. 357. Fig. 358.

in the regular, or pentagonal, dodecahedron, which symbolised the universe itself, and with which Euclidean geometry ends.

If we take any one of these figures, for instance the isosceles triangle which we have just described, and add to it (or subtract

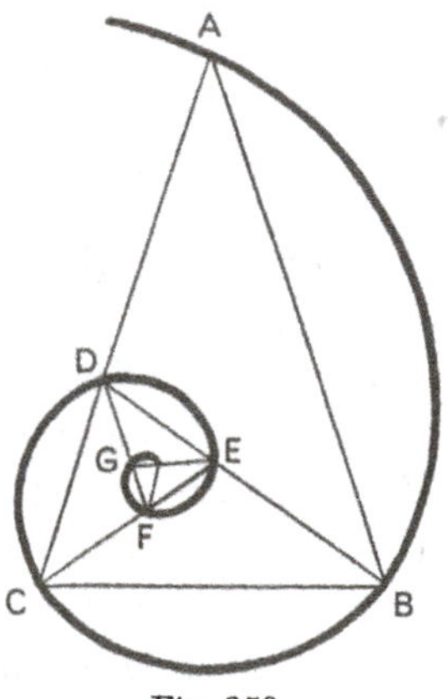

Fig. 359.

from it) in succession a series of gnomons, so converting it into larger and larger (or smaller and smaller) triangles all similar to the first, we find that the apices (or other corresponding points) of all these

* See, on the mathematical history of the gnomon, Heath's *Euclid*, I, *passim*, 1908; Zeuthen, *Théorème de Pythagore*, Genève, 1904; also a curious and interesting book, *Das Theorem des Pythagoras*, by Dr H. A. Naber, Haarlem, 1908.

triangles have their *locus* upon a equiangular spiral: a result which follows directly from that alternative definition of the equiangular spiral which I have quoted from Whitworth (p. 757).

If in this, or any other isosceles triangle, we take corresponding median lines of the successive triangles, by joining C to the mid-point (M) of AB, and D to the mid-point (N) of BC, then the pole of the spiral, or centre of similitude of ABC and BCD, is the point of intersection of CM and DN*.

Again, we may build up a series of right-angled triangles, each of which is a gnomon to the preceding figure; and here again, an equiangular spiral is the locus of corresponding points in these successive triangles. And lastly, whensoever we fill up space with a collection of equal and similar figures, as in Figs. 360, 361, there we can always discover a series of equiangular spirals in their successive multiples†.

Once more, then, we may modify our definition, and say that: "Any plane curve proceeding from a fixed point (or pole), and such that the vectorial area of any sector is always a gnomon to the whole preceding figure, is called an equiangular, or logarithmic, spiral." And we may now introduce this new concept and nomenclature into our description of the *Nautilus* shell and other related organic forms, by saying that: (1) if a growing structure be built up of successive parts, similar in form, magnified in geometrical progression, and similarly situated with respect to a centre of similitude, we can always trace through corresponding points a series of equiangular spirals; and (2) it is characteristic of the

* I owe this simple but novel construction, like so much else, to Dr G. T. Bennett.

† In each and all of these gnomonic figures we may now recognise a never-ending polygon, with equal angles at its corners, and with its successive sides in geometrical progression; and such a polygon we may look upon as the natural precursor of the equiangular spiral. If we call the exterior or "bending" angle of the polygon β, and the ratio of its sides λ, then the vertices lie on an equiangular spiral of angle α, given by $\log_e \lambda = \beta \cot \alpha$. In the spiral of Fig. 359 the constant angle is thus found to be about 75° 40′, in that of Fig. 355, 77° 40′, and in that of Fig. 356, 72° 50′.

The calculation is as follows. Taking, for example, the successive triangles of Fig. 359, the ratio (λ) of the sides, as $BC : AC$, is that of the golden section, 1 : 1·618. The external angle (β), as ADB, is 108°, or in radians 1·885. Then

$$\log 1{\cdot}618 = 0{\cdot}209, \text{ from which } \log_e 1{\cdot}618 = 0{\cdot}481$$

and

$$\cot \alpha = \frac{\log_e \lambda}{\beta} = \frac{0{\cdot}481}{1{\cdot}885} = 0{\cdot}255 = \cot 75^\circ\ 45'.$$

growth of the horn, of the shell, and of all other organic forms in which an equiangular spiral can be recognised, that *each successive increment of growth is similar, and similarly magnified, and similarly*

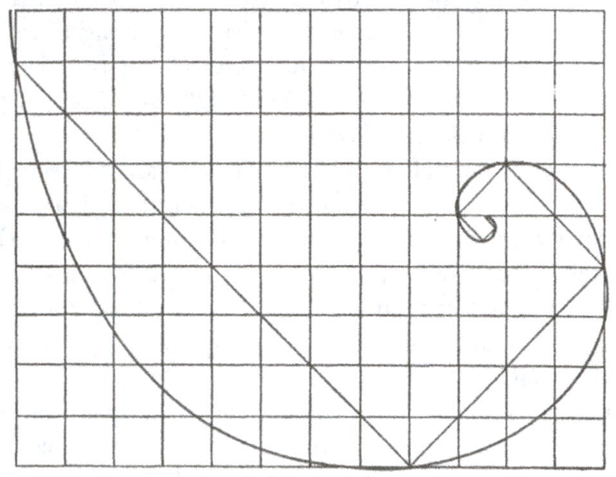

Fig. 360*. Logarithmic spiral derived from corresponding points in a system of squares.

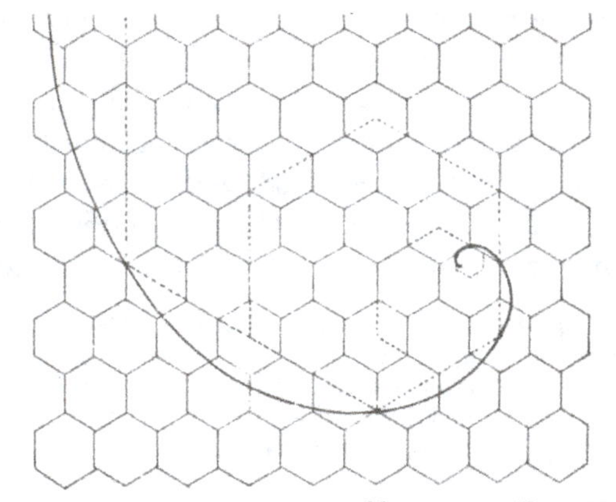

Fig. 361. The same in a system of hexagons. From Naber.

situated to its predecessor, and is in consequence a gnomon to the entire pre-existing structure. Conversely (3) it follows that in the spiral

* This diagram was at fault in my first edition (p. 512), as Dr G. T. Bennett shews me. The curve met its chords at equal angles at either end: whereas it ought to meet the further end at a lesser angle than the other, and ought in consequence to intersect the lines of the coordinate framework. The constant angle of this spiral is about 66° 11′ ($\tan \alpha = \pi/2 \log_e 2$).

outline of the shell or of the horn we can always inscribe an endless variety of other gnomonic figures, having no necessary relation, save as a mathematical accident, to the nature or mode of development of the actual structure*. But observe that the gnomons to a square may form increments of any size, and the same is true of the gnomons to a *Haliotis*-shell; but in the higher symmetry of a chambered *Nautilus*, or of the successive triangles in Fig. 359, growth goes on by a progressive series of gnomons, each one of which is the gnomon to another.

Fig. 362. A shell of *Haliotis*, with two of the many lines of growth, or generating curves, marked out in black: the areas bounded by these lines of growth being in all cases gnomons to the pre-existing shell.

Of these three propositions, the second is of great use and advantage for our easy understanding and simple description of the molluscan shell, and of a great variety of other structures whose mode of growth is analogous, and whose mathematical properties are therefore identical. We see that the successive chambers of a spiral *Nautilus* or of a straight *Orthoceras*, each whorl or part of a whorl of a periwinkle or other gastropod, each new increment of the operculum of a gastropod, each additional increment of an elephant's tusk, or each new chamber of a spiral foraminifer, has its leading characteristic at once described and its form so far explained by the

* For many beautiful geometrical constructions based on the molluscan shell, see S. Colman and C. A. Coan, *Nature's Harmonic Unity* (ch. IX, Conchology), New York, 1912.

simple statement that it constitutes a *gnomon* to the whole previously existing structure. And herein lies the explanation of that "time-element" in the development of organic spirals of which we have spoken already; for it follows as a simple corollary to this theory

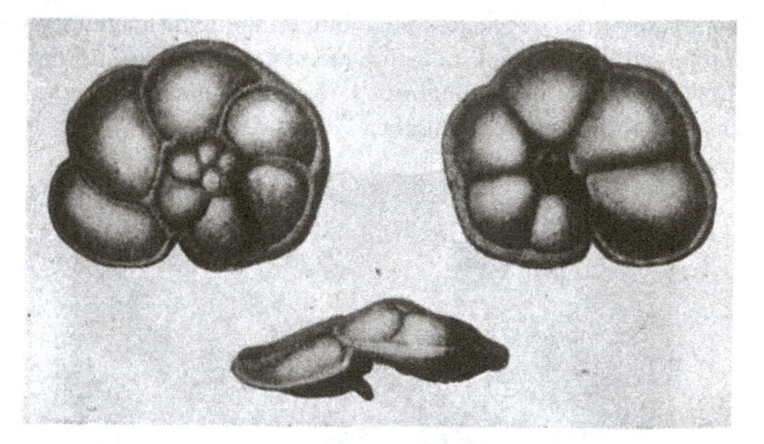

Fig. 363. A spiral foraminifer (*Pulvinulina*), to shew how each successive chamber continues the symmetry of, or constitutes a *gnomon* to, the rest of the structure.

of gnomons that we must never expect to find the logarithmic spiral manifested in a structure whose parts are simultaneously produced, as for instance in the margin of a leaf, or among the many curves that make the contour of a fish. But we most look for it wherever the organism retains, and still presents at a single view, the successive phases of preceding growth: the successive magnitudes attained, the successive outlines occupied, as growth pursued the even tenor of its way. And it follows from this that it is in the hard parts of organisms, and not the soft, fleshy, actively growing parts, that this spiral is commonly and characteristically found: not in the fresh mobile tisssue whose form is constrained merely by the active forces of the moment; but in things like shell and tusk, and horn and claw, visibly composed

Fig. 364. Another spiral foraminifer, *Cristellaria*.

of parts successively and permanently laid down. The shell-less molluscs are never spiral; the snail is spiral but not the slug*. In short, it is the shell which curves the snail, and not the snail which curves the shell. The logarithmic spiral is characteristic, not of the living tissues, but of the dead. And for the same reason it will always or nearly always be accompanied, and adorned, by a pattern formed of "lines of growth," the lasting record of successive stages of form and magnitude†.

The cymose inflorescences of the botanists are analogous in a curious and instructive way to the equiangular spiral.

In Fig. 365 B (which represents the *Cicinnus* of Schimper, or *cyme unipare scorpioide* of Bravais, as seen in the Borage), we begin with a primary shoot from which is given off, at a certain definite angle, a secondary shoot: and from that in turn, on the same side and at the same angle, another shoot, and so on. The deflection, or curvature, is continuous and progressive, for it is caused by no external force but only by causes intrinsic in the system. And the whole system is symmetrical: the angles at which the successive shoots are given off being all equal, and the lengths of the shoots diminishing *in constant ratio*. The result is that the successive shoots, or successive increments of growth, are tangents to a curve, and this curve is a true logarithmic spiral. Or in other words, we may regard each successive shoot as forming, or defining, a gnomon to the preceding structure. While in this simple case the successive shoots are depicted as lying in *a plane*, it may also happen that, in addition to their successive angular divergence from one another within that plane, they also tend to

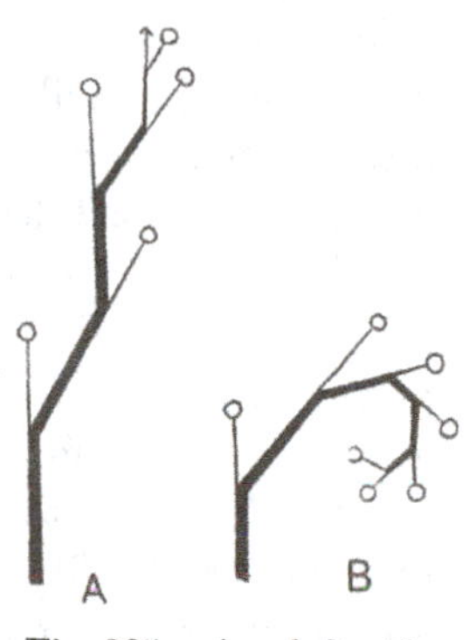

Fig. 365. A, a helicoid; B, a scorpioid cyme.

* Note also that *Chiton*, where the pieces of the shell are disconnected, shews no sign of spirality.

† That the invert to an equiangular spiral is identical with the original curve does not concern us in our study of organic form, but it is one of the most beautiful and most singular properties of the curve. It was this which led James Bernoulli, in imitation of Archimedes, to have the logarithmic spiral inscribed upon his tomb; and on John Goodsir's grave near Edinburgh the same symbol is reinscribed. Bernoulli's account of the matter is interesting and remarkable: "Cum autem ob proprietatem tam singularem tamque admirabilem mire mihi placeat spira haec mirabilis, sic ut ejus contemplatione satiari vix nequeam: cogitavi illam ad varias res symbolice repraesentandas non inconcinne adhiberi posse. Quoniam enim semper sibi et eandem spiram gignit, utcunque volvatur, evolvatur, radiet, hinc poterit esse vel sobolis parentibus per omnia similis Emblema: *Simillima Filia Matri*; vel (si rem aeternae veritatis Fidei mysteriis accommodare non est

diverge by successive equal angles *from* that plane of reference; and by this means, there will be superposed upon the equiangular spiral a twist or screw. And, in the particular case where this latter angle of divergence is just equal to 180°, or two right angles, the successive shoots will once more come to lie in a plane, but they will appear to come off from one another on *alternate* sides, as in Fig. 365 A. This is the *Schraubel* or *Bostryx* of Schimper, the *cyme unipare hélicoide* of Bravais. The equiangular spiral is still latent in it, as in the other; but is concealed from view by the deformation resulting from the helicoid. Many botanists did not recognise (as the brothers Bravais did) the mathematical significance of the latter case, but were led by the snail-like spiral of the scorpoid cyme to transfer the name "helicoid" to it*.

The spiral curve of the shell is, in a sense, a vector diagram of its own growth; for it shews at each instant of time the direction, radial and tangential, of growth, and the unchanging ratio of velocities in these directions. Regarding the *actual* velocity of growth in the shell, we know very little by way of experimental measurement; but if we make a certain simple assumption, then we may go a good deal further in our description of the equiangular spiral as it appears in this concrete case.

Let us make the assumption that *similar* increments are added to the shell in *equal* times; that is to say, that the amount of growth in unit time is measured by the areas subtended by equal angles. Thus, in the outer whorl of a spiral shell a definite area marked out by ridges, tubercles, etc., has very different linear dimensions to the corresponding areas of an inner whorl, but the symmetry of the figure implies that it subtends an equal angle with these; and it is reasonable to suppose that the successive regions, marked out in this way by successive natural boundaries or patterns, are produced in equal intervals of time.

prohibitum) ipsius aeternae generationis Filii, qui Patris veluti Imago, et ab illo ut Lumen a Lumine emanans, eidem ὁμοιούσιος existit, qualiscunque adumbratio. Aut, si mavis, quia Curva nostra mirabilis in ipsa mutatione semper sibi constantissime manet similis at numero eadem, poterit esse vel fortitudinis et constantiae in adversitatibus, vel etiam Carnis nostrae post varias alterationes et tandem ipsam quoque mortem, ejusdem numero resurrecturae symbolum: adeo quidem, ut si Archimedem imitandi hodiernum consuetudo obtineret, libenter Spiram hanc tumulo meo juberem incidi, cum Epigraphe, *Eadem numero mutata resurget*"; *Acta Eruditorum*, M. Maii, 1692, p. 213. Cf. L. Isely, Épigraphes tumulaires de mathématiciens, *Bull. Soc. Sci. nat. Neuchâtel*. XXVII, p. 171, 1899.

* The names of these structures have been often confused and misunderstood; cf. S. H. Vines, The history of the scorpioid cyme, *Journ. Bot.* (n.s.), X, pp. 3–9, 1881.

If this be so, the radii measured from the pole to the boundary of the shell will in each case be proportional to the velocity of growth at this point upon the circumference, and at the time when it corresponded with the outer lip, or region of active growth; and while the direction of the radius vector corresponds with the direction of growth in thickness of the animal, so does the tangent to the curve correspond with the direction, for the time being, of the animal's growth in length. The successive radii are a measure of the acceleration of growth, and the spiral curve of the shell itself, if the radius rotate uniformly, is no other than the *hodograph* of the growth of the contained organism*.

So far as we have now gone, we have studied the elementary properties of the equiangular spiral, including its fundamental property of *continued similarity*; and we have accordingly learned that the shell or the horn tends *necessarily* to assume the form of this mathematical figure, because in these structures growth proceeds by successive increments which are always similar in form, similarly situated, and of constant relative magnitude one to another. Our chief objects in enquiring further into the mathematical properties of the equiangular spiral will be: (1) to find means of confirming and verifying the fact that the shell (or other organic curve) is actually an equiangular spiral; (2) to learn how, by the properties of the curve, we may further extend our knowledge or simplify our descriptions of the shell; and (3) to understand the factors by which the characteristic form of any particular equiangular spiral is determined, and so to comprehend the nature of the specific or generic differences between one spiral shell and another.

Of the elementary properties of the equiangular spiral the following are those which we may most easily investigate in the concrete case of the molluscan shell: (1) that the polar radii whose vectorial angles are in arithmetical progression are themselves in geometrical progression; hence (2) that the vectorial angles are proportional to the *logarithms* of the corresponding radii; and (3) that the tangent at any point of an equiangular spiral makes a constant angle (called the *angle of the spiral*) with the polar radius vector.

* The hodograph of a logarithmic spiral (i.e. of a point which lies on a uniformly revolving radius and describes a logarithmic spiral) is likewise a logarithmic spiral: W. Walton, *Collection of Problems in Theoretical Mechanics* (3rd ed.), 1876, p. 296.

The first of these propositions may be written in a simpler form, as follows: radii which form equal angles about the pole of the equiangular spiral are themselves continued proportionals. That is to say, in Fig. 366, when the angle ROQ is equal to the angle QOP, then $OP : OQ :: OQ : OR$.

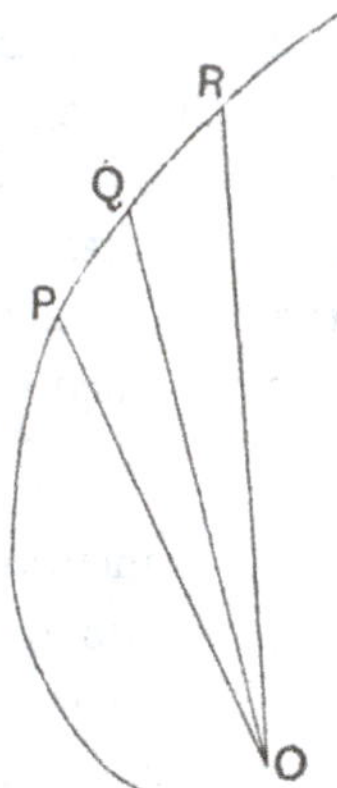

Fig. 366.

A particular case of this proposition is when the equal angles are each angles of 360°: that is to say when in each case the radius vector makes a complete revolution, and when, therefore, P, Q and R all lie upon the same radius.

It was by observing with the help of very careful measurement this continued proportionality, that Moseley was enabled to verify his first assumption, based on the general appearance of the shell, that the shell of *Nautilus* was actually an equiangular spiral, and this demonstration he was soon afterwards in a position to generalise by extending it to all spiral Ammonitoid and Gastropod mollusca*. For, taking a median transverse section of a *Nautilus pompilius*, and carefully measuring the successive breadths of the whorls (from the dark line which marks what was originally the outer surface, before it was covered up by fresh deposits on the part of the growing and advancing shell), Moseley found that "the distance of any two of its whorls measured upon a radius vector is one-third that of the two next whorls measured upon the same radius vector†. Thus (in

* The Rev. H. Moseley, On the geometrical forms of turbinated and discoid shells, *Phil. Trans.* 1838, Pt. I, pp. 351–370. Réaumur, in describing the snail-shell (*Mém. Acad. des Sci.* 1709, p. 378), had a glimpse of the same geometrical law: "Le diamètre de chaque tour de spirale, ou sa plus grande longueur, est à peu près double de celui qui la précède et la moitié de celui qui la suit." Leslie (in his *Geometry of Curved Lines*, 1822, p. 438) compared the "general form and the elegant *septa* of the *Nautilus*" to an equiangular spiral and a series of its involutes.

† It will be observed that here Moseley, speaking as a mathematician and considering the *linear* spiral, speaks of *whorls* when he means the linear boundaries, or lines traced by the revolving radius vector; while the conchologist usually applies the term *whorl* to the whole space between the two boundaries. As conchologists, therefore, we call the *breadth of a whorl* what Moseley looked upon as the *distance between two consecutive whorls*. But this latter nomenclature Moseley himself often uses. Observe also that Moseley gets a very good approximate result by his measurements "upon a radius vector," although he has to be content with a very rough determination of the pole.

Fig. 367), *ab* is one-third of *bc*, *de* of *ef*, *gh* of *hi*, and *kl* of *lm*. The curve is therefore an equiangular spiral."

The numerical ratio in the case of the *Nautilus* happens to be one of unusual simplicity. Let us take, with Moseley, a somewhat more complicated example.

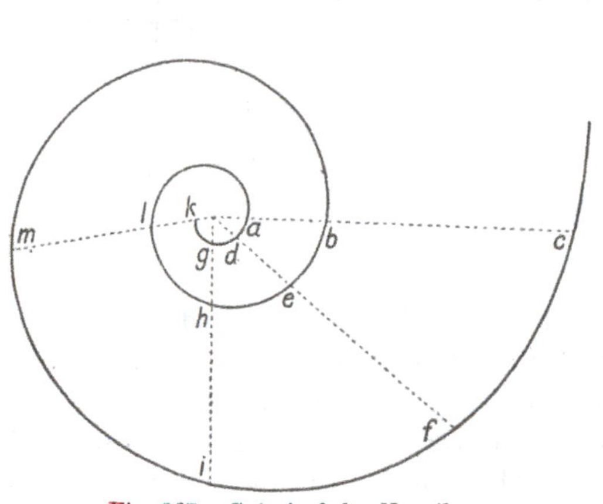

Fig. 367. Spiral of the *Nautilus*.

Fig. 368. *Turritella duplicata* (L.), Moseley's *Turbo duplicatus*. From Chenu. × ½.

From the apex of a large *Turritella* (*Turbo*) *duplicata** a line was drawn across its whorls, and their widths were measured upon it in succession, beginning with the last but one. The measure-

* In the case of "*Turbo*", and all other turbinate shells, we are dealing not with a plane logarithmic spiral, as in *Nautilus*, but with a "gauche" spiral, such that the radius vector no longer revolves in a plane perpendicular to the axis of the system, but is inclined to that axis at some constant angle (β). The figure still preserves its continued similarity, and may be called a logarithmic spiral in space; indeed it is commonly spoken of as a logarithmic spiral *wrapped upon a cone*, its pole coinciding with the apex of the cone. It follows that the distances of successive whorls of the spiral measured on the same straight line passing through the apex of the cone are in geometrical progression, and conversely; just as in the former case. But the ratio between any two consecutive interspaces (i.e. $R_3 - R_2/R_2 - R_1$) is now equal to $\epsilon^{2\pi \sin\beta \cot\alpha}$, β being the semi-angle of the enveloping cone. (Cf. Moseley, *Phil. Mag.* XXI, p. 300, 1842.)

ments were, as before, made with a fine pair of compasses and a diagonal scale. The sight was assisted by a magnifying glass. In a parallel column to the following admeasurements are the terms of a geometric progression, whose first term is the width of the widest whorl measured, and whose common ratio is 1·1804.

Turritella duplicata

Widths of successive whorls, measured in inches and parts of an inch	Terms of a geometrical progression, whose first term is the width of the widest whorl, and whose common ratio is 1·1804
1·31	1·310
1·12	1·110
0·94	0·940
0·80	0·797
0·67	0·675
0·57	0·572
0·48	0·484
0·41	0·410

The close coincidence between the observed and the calculated figures is very remarkable, and is amply sufficient to justify the conclusion that we are here dealing with a true logarithmic spiral*.

Nevertheless, in order to verify his conclusion still further, and to get partially rid of the inaccuracies due to successive small measurements, Moseley proceeded to investigate the same shell, measuring not single whorls but groups of whorls taken several at a time: making use of the following property of a geometrical progression, that "if μ represent the ratio of the sum of every even number (m) of its terms to the sum of half that number of terms, then the common ratio (r) of the series is represented by the formula

$$r = (\mu - 1)^{\frac{2}{m}}."$$

* Moseley, writing a hundred years ago, uses an obsolete nomenclature which is apt to be very misleading. His *Turbo duplicatus*, of Linnaeus, is now *Turritella duplicata*, the common large Indian *Turritella*, a slender, tapering shell with a very beautiful spiral, about six or seven inches long. But the operculum which he describes as that of *Turbo* does indeed belong to that genus, *sensu stricto*; it is the well-known calcareous operculum or "eyestone" of some such common species as *Turbo petholatus*. *Turritella* has a very different kind of operculum, a thin chitinous disc in the form of a close spiral coil, not nearly filling up the aperture of the shell. Moseley's *Turbo phasianus* is again no true *Turbo*, but is (to judge from his figure) *Phasianella bulimoides* Lam. =*P. australis* (Gmelin); and his *Buccinum subulatum* is *Terebra subulata* (L.).

Accordingly, Moseley made the following measurements, beginning from the second and third whorls respectively:

Width of		Ratio μ
Six whorls	Three whorls	
5·37	2·03	2·645
4·55	1·72	2·645
Four whorls	Two whorls	
4·15	1·74	2·385
3·52	1·47	2·394

"By the ratios of the two first admeasurements, the formula gives

$$r = (1\cdot645)^{\frac{1}{3}} = 1\cdot1804.$$

By the mean of the ratios deduced from the second two admeasurements, it gives

$$r = (1\cdot389)^{\frac{1}{2}} = 1\cdot1806.$$

"It is scarcely possible to imagine a more accurate verification than is deduced from these larger admeasurements, and we may with safety annex to the species *Turbo duplicatus* the characteristic number 1·18."

By similar and equally concordant observations, Moseley found for *Turbo phasianus* the characteristic ratio, 1·75; and for *Buccinum subulatum* that of 1·13.

From the measurements of *Turritella duplicata* (on p. 772), it is perhaps worth while to illustrate the logarithmic statement of the same thing: that is to say, the elementary fact, or corollary, that if the successive radii be in geometric progression, their logarithms will differ from one another by a constant amount.

Turritella duplicata

Widths of successive whorls	Logarithms of do.	Differences of logarithms	Ratios of successive widths
131	2·11727	—	—
112	2·04922	0·06805	1·170
94	1·97313	0·07609	1·191
80	1·90309	0·07004	1·175
67	1·82607	0·07702	1·194
57	1·75587	0·07020	1·175
48	1·68124	0·07463	1·188
41	1·61278	0·06846	1·171
	Mean	0·07207	1·1806

And 0·07207 is the logarithm of 1·1805.

Lastly, we may if we please, in this simple case, reduce the whole matter to arithmetic, and, dividing the width of each whorl by that of the next, see that these quotients are nearly identical, and that their mean value, or common ratio, is precisely that which we have already found.

We may shew, in the same simple fashion, by measurements of *Terebra* (Fig. 397), how the relative widths of successive whorls fall into a geometric progression, the criterion of a logarithmic spiral.

Measurements of a large specimen (15·5 *cm.*) *of* Terebra maculata, *along three several tangents* (*a*, *b*, *c*) *to the whorls.* (*After Chr. Peterson*, 1921.)

a		*b*		*c*	
Width (mm.)	Ratio	Width	Ratio	Width	Ratio
25		24·5		23	
	1·25		1·32		1·31
20		18·5		17·5	
	1·33		1·32		1·31
15		14		13·3	
	1·25		1·30		1·36
12		10·75		9·75	
	1·33		1·34		1·34
9		8		7·25	
Mean	1·29		1·32		1·33
	Mean ratio, 1·31				

The logarithmic spiral is not only very beautifully manifested in the molluscan shell*, but also, in certain cases, in the little lid or "operculum" by which the entrance to the tubular shell is closed after the animal has withdrawn itself within†. In the spiral shell of *Turbo*, for instance, the operculum is a thick calcareous structure, with a beautifully curved outline, which grows by successive increments applied to one portion of its edge, and shews, accordingly, a spiral line of growth upon its surface. The successive increments leave their traces on the surface of the operculum (Fig. 370), which traces have the form of curved lines in *Turbo*, and of straight lines

* It has even been proposed to use a logarithmic spiral in place of a table of logarithms. Cf. Ant. Favaro, *Statique graphique*, Paris, 1885; Hele-Shaw, in *Brit. Ass. Rep.* 1892, p. 403.

† Cf. Fred. Haussay, *Recherches sur l'opercule*, Diss., Paris, 1884.

in (e.g.) *Nerita* (Fig. 371); that is to say, apart from the side constituting the outer edge of the operculum (which side is always and of necessity curved) the successive increments constitute curvilinear

Fig. 369. Operculum of *Turbo.*

triangles in the one case, and rectilinear triangles in the other. The sides of these triangles are tangents to the spiral line of the

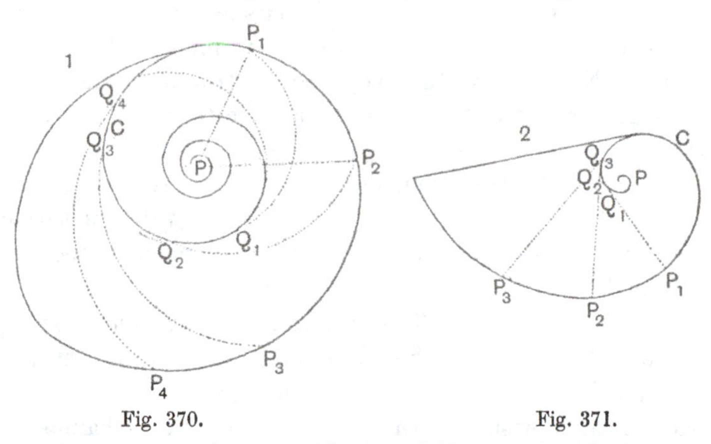

Fig. 370. Fig. 371.

Figs. 370, 371. Opercula of *Turbo* and *Nerita*. After Moseley.

operculum, and may be supposed to generate it by their consecutive intersections.

In a number of such opercula, Moseley measured the breadths

of the successive whorls along a radius vector*, just in the same way as he did with the entire shell in the foregoing cases; and here is one example of his results.

Operculum of Turbo *sp.; breadth (in inches) of successive whorls, measured from the pole*

Distance	Ratio	Distance	Ratio	Distance	Ratio	Distance	Ratio
0·24		0·16		0·2		0·18	
	2·28		2·31		2·30		2·30
0·55		0·37		0·6		0·42	
	2·32		2·30		2·30		2·24
1·28		0·85		1·38		0·94	

The ratio is approximately constant, and this spiral also is, therefore, a logarithmic spiral.

But here comes in a very beautiful illustration of that property of the logarithmic spiral which causes its whole shape to remain unchanged, in spite of its apparently unsymmetrical, or unilateral, mode of growth. For the mouth of the tubular shell, into which the operculum has to fit, is growing or widening on all sides: while the operculum is increasing, not by additions made at the same time all round its margin, but by additions made only on one side of it at each successive stage. One edge of the operculum thus remains unaltered as it advances into its new position, and comes to occupy a new-formed section of the tube, similar to but greater than the last. Nevertheless, the two apposed structures, the chamber and its plug, at all times fit one another to perfection. The mechanical problem (by no means an easy one) is thus solved: "How to shape a tube of a variable section, so that a piston driven along it shall, by one side of its margin, coincide continually with its surface as it advances, provided only that the piston be made at the same time continually to revolve in its own plane."

As Moseley puts it: "That the same edge which fitted a portion of the first less section should be capable of adjustment, so as to fit a portion of the next similar but greater section, supposes a geometrical provision in the curved form of the chamber of great

* As the successive increments evidently constitute similar figures, similarly related to the pole (P), it follows that their linear dimensions are to one another as the radii vectores drawn to similar points in them: for instance as PP_1, PP_2, which (in Fig. 370) are radii vectores drawn to the points where they meet the common boundary.

apparent complication and difficulty. But God hath bestowed upon this humble architect the practical skill of a learned geometrician, and he makes this provision with admirable precision in that curvature of the logarithmic spiral which he gives to the section of the shell. This curvature obtaining, he has only to turn his operculum slightly round in its own plane as he advances it into each newly formed portion of his chamber, to adapt one margin of it to a new and larger surface and a different curvature, leaving the space to be filled up by increasing the operculum wholly on the other margin." The fact is that self-similar or gnomonic growth is taking place both in the shell and its operculum; in both of them growth is in reference to a fixed centre, and to a fixed axis through that centre; and in both of them growth proceeds in geometric progression from the centre while rotation takes place in arithmetic progression about the axis. The same architecture which builds the house constructs the door. Moreover, not only are house and door governed by the same law of growth, but, growing together, door and doorway adapt themselves to one another.

The operculum of the gastropods varies from a more or less close-wound spiral, as in *Turritella*, *Trochus* or *Pleurotomaria*, to cases in which accretion takes place, by concentric (or more or less excentric) rings, all round. But these latter cases, so Mr Winckworth tells me, are not very common. *Paludina* and *Ampullaria* come near to having a concentric operculum, and so do some of the Murices, such as *M. tribulus*, and a few Turrids, and the genus *Helicina*; but even these opercula probably begin as spirals, adding on their gnomonic increments at one end or side, and only growing on all sides later on. There would seem to be a truly concentric operculum in the *Siphonium* group of *Vermetus*, where the spiral of the shell itself is lost, or nearly so; but it is usually overgrown with Melobesia, and hard to see.

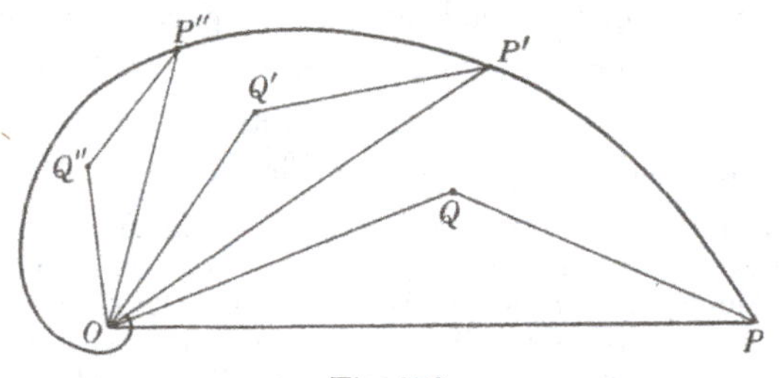

Fig. 372.

One more proposition, an all but self-evident one, we may make passing mention of here: If upon any polar radius vector OP, a triangle OPQ be drawn similar to a given triangle, the locus of the vertex Q will be a spiral similar to the original spiral. We may extend this proposition (as given by Whitworth) from the simple case of the triangle to any similar figures whatsoever; and see from it how every spot or ridge or tubercle repeated symmetrically from one radius vector (or one generating curve) to another becomes part of a spiral pattern on the shell.

Viewed in regard to its own fundamental properties and to those of its limiting cases, the equiangular spiral is one of the simplest of all known curves; and the rigid uniformity of the simple laws by which it is developed sufficiently account for its frequent manifestation in the structures built up by the slow and steady growth of organisms.

In order to translate into precise terms the whole form and growth of a spiral shell, we should have to employ a mathematical notation considerably more complicated than any that I have attempted to make use of in this book. But we may at least try to describe in elementary language the general method, and some of the variations, of the mathematical development of the shell. But here it is high time to observe that, while we have been speaking of the *shell* (which is a *surface*) as a logarithmic spiral (which is a *line*), we have been simplifying the case, in a provisional or preparatory way. The logarithmic spiral is but one factor in the case, albeit the chief or dominating one. The problem is one not of plane but of solid geometry, and the solid in question is described by the movement in space of a certain area, or closed curve*.

Let us imagine a closed curve in space, whether circular or elliptical or of some other and more complex specific form, not necessarily in a plane: such a curve as we see before us when we consider the mouth, or terminal orifice, of our tubular shell. Let

* For a more advanced study of the family of surfaces of which the *Nautilus* is a simple case, see M. Haton de la Goupillière (*op. cit.*). The turbinate shells represent a sub-family, which may be called that of the "surfaces cérithioides"; and "surfaces à front générateur" is a short title of the whole family. The form of the generating curve, its rate of expansion, the direction of its advance, and the angle which the generating front makes with the directrix, define, and give a wide extension to, the family. These parameters are all severally to be recognised in the growth of the living object; and they make of a collection of shells an unusually beautiful materialisation of the rigorous definitions of geometry.

us call this closed curve the "generating curve"; the surface which it bounds we may call (if need arise) the "generating front," and let us imagine some one characteristic point within this closed curve, such as its centre of gravity. Then, starting from a fixed origin, let this characteristic point describe an equiangular spiral

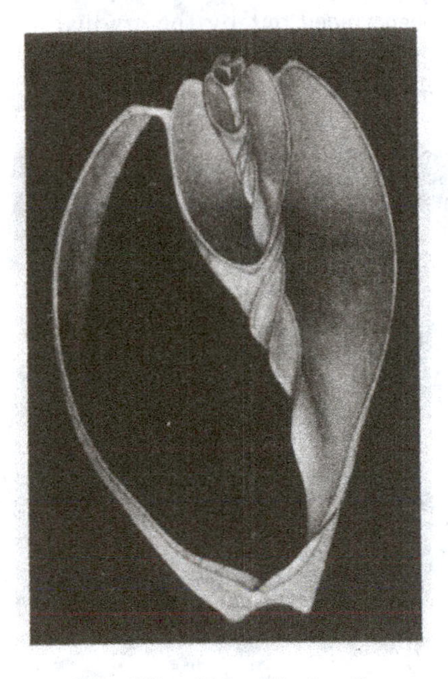

Fig. 373. *Melo ethiopicus* L.

in space about a fixed axis or "conductrix" (namely the axis of the shell), while at the same time the generating curve grows with each increment of rotation in such a way as to preserve the symmetry of the entire figure, with or without a simultaneous movement of translation along the axis.

The resulting shell may now be looked upon in either of two ways. It is, on the one hand, an *ensemble of similar closed curves*, spirally arranged in space, and gradually increasing in dimensions in proportion to the increase of their vector-angle from the pole*. In

* The plumber, the copper-smith and the glass-blower are at pains to conserve in every part of their tubular constructions, however these branch or bend, the constant form which their cross-sections ought to have. Throughout the spiral twisting of the shell, throughout the windings and branchings of the blood-vessels, the same uniformity is maintained.

other words, we can imagine our shell cut up into a system of rings, following one another in continuous spiral succession, from that terminal and largest one which constitutes the lip of the orifice of the shell. Or on the other hand, we may figure to ourselves the whole shell as made up of an *ensemble of spiral lines* in space, each spiral having been traced out by the gradual growth and revolution of a radius vector from the pole to a given point on the boundary of the generating curve.

Fig. 374. 1, *Harpa*; 2, *Dolium*. The ridges on the shell correspond in (1) to generating curves, in (2) to generating spirals.

Both systems of lines, the *generating spirals* (as these latter may be called), and the closed *generating curves* corresponding to successive margins or lips of the shell, may be easily traced in a great variety of cases. Thus, for example, in *Dolium*, *Eburna*, and a host of others, the generating spirals are beautifully marked out by ridges, tubercles or bands of colour. In *Trophon*, *Scalaria*, and (among countless others) in the Ammonites, it is the successive generating curves which more conspicuously leave their impress on the shell. And in not a few cases, as in *Harpa*, *Dolium perdix*, etc., both alike are conspicuous, ridges and colour-bands intersecting one another in a beautiful isogonal system.

In ordinary gastropods the shell is formed at or near the mantle-edge. Here, near the mantle-border, is a groove lined with a secretory epithelium which produces the horny cuticle or perio stracum of the shell*. A narrow zone of the mantle just behind this secretes lime abundantly, depositing it in a layer below the periostracum; and for some little way back more lime may be secreted, and pigment superadded from appropriate glands. Growth and secretion are periodic rather than continuous. Even in a snail-shell it is easy to see how the shell is built up of narrow annular increments; and many other shells record, in conspicuous colour-patterns, the alternate periods of rest and of activity which their pigment-glands have undergone.

The periodic accelerations and retardations in the growth of a shell are marked in various ways. Often we have nothing more than an increased activity from time to time at or near the mantle-edge—enough to give rise to slight successive ridges, each corresponding to a "generating curve" in the conformation of the shell. But in many other cases, as in *Murex*, *Ranella* and the like, the mantle-edge has its alternate phases of rest and of turgescence, its outline being plain and even in the one and folded and contorted in the other; and these recurring folds or pleatings of the edge leave their impress in the form of various ridges, ruffles or comb-like rows of spines upon the shell†.

In not a few cases the colour-pattern shews, or seems to shew, how some play of forces has fashioned and transformed the first elementary pattern of pigmentary drops or jets. As the bookbinder drops or dusts a little colour on a viscous fluid, and then produces the beautiful streamlines of his marbled papers by stirring

* That the shell grows by accretion at the mantle-edge was one of Réaumur's countless discoveries (*Mém. Acad. Roy. des Sc.* 1709, p. 364 *seq.*). It follows that the mathematical "generating curves," as Moseley chose them, correspond to the material increments of the shell.

† The periodic appearance of a ridge, or row of tubercles, or other ornament on the growing shell is illustrated or even exaggerated in the delicate "combs" of *Murex aculeatus*. Here normal growth is interrupted for the time being, the mantle-edge is temporarily folded and reflexed, and shell-substance is poured out into the folds.

and combing the colloid mass, so we may see, in the harp-shells or the volutes, how a few simple spots or lines have been drawn out into analogous wavy patterns by streaming movements during the formation of the shell.

In the complete mathematical formula for any given turbinate shell, we may include, with Moseley, factors for the following elements: (1) for the specific form of a section of the tube, or (as we have called it) the generating curve; (2) for the specific rate of growth of this generating curve; (3) for its inclination to the directrix, or to the axis; (4) for its specific rate of angular rotation about the pole, in a projection perpendicular to the axis; and (5) in turbinate (as opposed to nautiloid) shells, for its rate of screw-translation, parallel to the axis, as measured by the angle between a tangent to the whorls and the axis of the shell*. It seems a complicated affair; but it is only a pathway winding at a steady slope up a conical hill. This uniform gradient is traced by any given point on the generating curve while the vector angle increases in arithmetical progression, and the scale changes in geometrical progression; and a certain *ensemble*, or bunch, of these spiral curves in space constitutes the self-similar surface of the shell.

But after all this is not the only way, neither is it the easiest way, to approach our problem of the turbinate shell. The conchologist turned mathematician is apt to think of the generating curve by which the spiral surface is described as necessarily identical, or coincident, with the mouth or lip of the shell; for this is where growth actually goes on, and where the successive increments of shell-growth are visibly accumulated. But it does not follow that this particular generating curve is chosen for the best from the mathematical point of view; and the mathematician, unconcerned with the physiological side of the case and regardless of the succession of the parts in time, is free to choose any other generating curve which the geometry of the figure may suggest to him. We are following Moseley's example (as is usually done) when we think of no other generating curve but that which takes the form of a

* Note that this tangent touches the curve at a series of points, whorl by whorl, instead of at one only. Observe also that we may have various tangent-cones, all centred on the apex of the shell. In an open spiral, like a ram's horn, or a half-open spiral like the shell *Solarium*, we have two cones, one touching the outside, the other the inside of the shell.

frontal plane, outlined by the lip, and *sliding along* the axis while revolving round it; but the geometer takes a better and a simpler way. For, when of two similar figures in space one is derived from the other by a screw-displacement accompanied by change of scale—as in the case of a big whelk and a little whelk—there is a unique (apical) point which suffers no displacement; and if we choose for our generating curve a sectional figure centred on the apical point and passing through the axis of rotation, the whole development of the surface may be simply described as due to a rotation of this generating figure about the axis (z), together with a change of scale with the point 0 as centre of similitude. We need not, and now must not, think of a *slide* or *shear* as part of the operation; the translation along the axis is merely part and parcel of the *magnification* of the new generating curve. It follows that angular rotation in arithmetical progression, combined with change of scale (from 0) in geometrical progression, causes any arbitrary point on the generating curve to trace a path of uniform gradient round a circular cone, or in other words to describe a helico-spiral or gauche equiangular spiral in space. The spiral curve cuts all the straight-line generators of the cone at the same angle; and it further follows that the successive increments are, and the whole figure constantly remains, "self-similar"*.

Apart from the specific form of the generating curve, it is the ratios which happen to exist between the various factors, the ratio for instance between the growth-factor and the rate of angular revolution, which give the endless possibilities of permutation of form. For example, a certain rate of growth in the generating curve, together with a certain rate of vectorial rotation, will give us a spiral shell of which each successive whorl will just touch its predecessor and no more; with a slower growth-factor the whorls will stand asunder, as in a ram's horn; with a quicker growth-factor

* The equation to the surface of a turbinate shell is discussed by Moseley both in terms of polar and of rectangular coordinates, and the method of polar coordinates is used also by Haton de la Goupillière; but both accounts are subject to mathematical objection. Dr G. T. Bennett, choosing his generating curve (as described above) in the axial plane from which the vertical angles are measured (the plane $\theta = 0$), would state his equation in cylindrical coordinates, $f(za^\theta, ra^\theta) = 0$: that is to say in terms of z, conjointly with ordinary plane cylindrical coordinates.

each will cut or intersect its predecessor, as in an Ammonite or the majority of gastropods, and so on.

A similar relation of velocities suffices to determine the apical angle of the resulting cone, and give us the difference, for example, between the sharp, pointed cone of *Turritella*, the less acute one of *Fusus* or *Buccinum*, and the obtuse one of *Harpa* or of *Dolium*. In short it is obvious that *all* the differences of form which we observe between one shell and another are referable to matters of *degree*, depending, one and all, upon the relative magnitudes of the various factors in the complex equation to the curve. This is an immensely important thing. To learn that all the multitudinous shapes of shells, in their all but infinite variety, may be reduced to the variant properties of a single simple curve, is a great achievement. It exemplifies very beautifully what Bacon meant in saying that the forms or differences of things are simple and few, and the degrees and coordinations of these make all their variety*. And after such a fashion as this John Goodsir imagined that the naturalist of the future would determine and classify his shells, so that conchology should presently become, like mineralogy, a mathematical science†.

The paper in which, more than a hundred years ago, Canon Moseley‡ gave a simple mathematical account, on lines like these, of the spiral forms of univalve shells, is one of the classics of Natural History. But other students before, and sometimes long before, him had begun to recognise the same simplicity of form and structure. About the year 1818 Reinecke had declared *Nautilus* to be a well-defined geometrical figure, whose chambers followed

* For a discussion of this idea, and of the views of Bacon and of J. S. Mill, see J. M. Keynes, *op. cit.* p. 271.

† On the employment of mathematical modes of investigation in the determination of organic forms; in *Anatomical Memoirs*, II, p. 205, 1868 (posthumous publication).

‡ The Rev. Henry Moseley (1801–1872), of St John's College, Cambridge, Canon of Bristol, Professor of Natural Philosophy in King's College, London, was a man of great and versatile ability. He was father of H. N. Moseley, naturalist on board the *Challenger* and Professor of Zoology in Oxford; and he was grandfather of H. G. J. Moseley (1887–1915)—Moseley of the Moseley numbers—whose death at Gallipoli, long ere his prime, was one of the major tragedies of the Four Years War.

one another in a constant ratio or continued proportion*; and Leopold von Buch and others accepted and even developed the idea.

Long before, Swammerdam had grasped with a deeper insight the root of the whole matter; for, taking a few diverse examples, such as *Helix* and *Spirula*, he shewed that they and all other spiral shells whatsoever were referable to one common type, namely to that of a simple tube, variously curved according to definite mathematical laws; that all manner of ornamentation, in the way of spines, tuberosities, colour-bands and so forth, might be superposed upon them, but the type was one throughout and specific differences were of a geometrical kind. "Omnis enim quae inter eas animadvertitur differentia ex sola nascitur diversitate gyrationum: quibus si insuper externa quaedam adjunguntur ornamenta pinnarum, sinuum, anfractuum, planitierum, eminentiarum, profunditatum, extensionum, impressionum, circumvolutionum, colorumque: ...tunc deinceps facile est, quarumcumque Cochlearum figuras geometricas, curvosque, obliquos atque rectos angulos, ad unicam omnes speciem redigere: ad oblongum videlicet tubulum, qui vario modo curvatus, crispatus, extrorsum et introrsum flexus, ita concrevit†."

Nay more, we may go back yet another hundred years and find Sir Christopher Wren contemplating the architecture of a snail-shell, and finding in it the logarithmic spiral. For Wallis‡, after defining and describing this curve with great care and simplicity, tells us that Wren not only conceived the spiral shell to be a sort of cone or pyramid coiled round a vertical axis, but also saw that on the magnitude of *the angle of the spire* depended the specific form of the shell: "Hanc ipsam curvam...contemplatus est Wrennius noster. Nec tantum curvae longitudinem, partiumque ipsius, et

* J. C. M. Reinecke, *Maris protogaei Nautilos*, etc., Coburg, 1818, p. 17: "In eius forma, quae canalis spiram convoluti formam et proportiones simul subministrat, totius testae forma quoddammodo data est. Restaret solum scire, quota cujusque anfractus pars sequenti inclusa sit, ut testam geometrice construere possimus." Cf. Leopold von Buch, Ueber die Ammoniten in den älteren Gebir[illegible]-schichten, *Abh. Berlin. Akad., Phys. Kl.* 1830, pp. 135–158; *Ann. Sc. Nat.* XXVIII, pp. 5–43, 1833; cf. Elie de Beaumont, Sur l'enroulement des Ammonites, *Soc. Philom., Pr. verb.* 1841, pp. 45–48.

† *Biblia Naturae sive Historia Insectorum*, Leydae, 1737, p. 152.

‡ Joh. Wallis, *Tractatus duo, de Cycloide*, etc., Oxon., 1659, pp. 107, 108.

magnitudinem adjacentis plani; sed et, ipsius ope, Limacum et Conchiliorum domunculos metitur. Existimat utique, magna verisimilitudine, domunculos hosce non alios esse quam Pyramides convolutas: quarum Axis sit, istiusmodo Spiralis: non quidem in plano jacens, sed sensim in convolutione (circa erectum axim) assurgens: pro variis autem curvae, sive ad rectam circumductam sive ad subjacens planum, angulis, variae Conchiliorum formae enascantur. Atque hac hypothesi, mensurata Pyramide, metitur etiam ea conchiliorum spatia."

For some years after the appearance of Moseley's paper, a number of writers followed in his footsteps, and attempted in various ways to put his conclusions to practical use. For instance, d'Orbigny

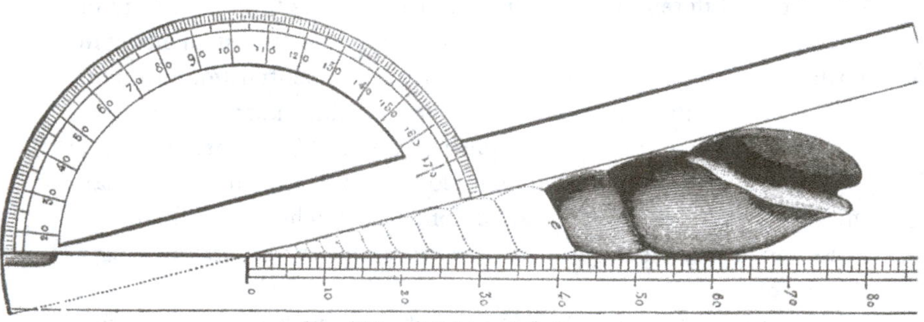

Fig. 375. d'Orbigny's helicometer.

devised a very simple protractor, which he called a Helicometer*, and which is represented in Fig. 375. By means of this little instrument the apical angle of the turbinate shell was immediately read off, and could then be used as a specific and diagnostic character. By keeping one limb of the protractor parallel to the side of the cone while the other was brought into line with the suture between two adjacent whorls, another specific angle, the "sutural angle," could in like manner be recorded. And, by the linear scale upon the instrument, the relative breadths of the consecutive whorls, and that of the terminal chamber to the rest of the shell, might

* Alcide d'Orbigny, *Bull. de la soc. géol. Fr.* XIII, p. 200, 1842; *Cours élém. de Paléontologie*, II, p. 5, 1851. A somewhat similar instrument was described by Boubée, in *Bull. soc. géol.* I, p. 232, 1831. Naumann's conchyliometer (*Poggend. Ann.* LIV, p. 544, 1845) was an application of the screw-micrometer; it was provided also with a rotating stage for angular measurement. It was adapted for the study of a discoid or ammonitoid shell, while d'Orbigny's instrument was meant for the study of a turbinate shell.

also, though somewhat roughly, be determined. For instance, in *Terebra dimidiata* the apical angle was found to be 13°, the sutural angle 109°, and so forth.

It was at once obvious that, in such a shell as is represented in Figs. 369 and 375 the entire outline (always excepting that of the immediate neighbourhood of the mouth) could be restored from a broken fragment. For if we draw our tangents to the cone, it follows from the symmetry of the figure that we can continue the projection of the sutural line, and so mark off the successive whorls, by simply drawing a series of consecutive parallels, and by then filling into the quadrilaterals so marked off a series of curves similar to one another, and to the whorls which are still intact in the broken shell. But the use of the helicometer soon shewed that it was by no means universally the case that one and the same cone was tangent to all the turbinate whorls; in other words, there was not always one specific apical angle which held good for the entire system. In the great majority of cases, it is true, the same tangent touches all the whorls, and is a straight line. But in others, as in the large *Cerithium nodosum*, such a line is slightly concave to the axis of the shell; and in the short spire of *Dolium*, for instance, the concavity is marked, and the apex of the spire is a distinct cusp. On the other hand, in *Pupa* and *Clausilia* the common tangent is convex to the axis of the shell.

So also is it, as we shall presently see, among the Ammonites: where there are some species in which the ratio of whorl to whorl remains, to all appearance, perfectly constant; others in which it gradually though only slightly increases; and others again in which it slightly and gradually falls away. It is obvious that, among the manifold possibilities of growth, such conditions as these are very easily conceivable. It is much more remarkable that, among these shells, the relative velocities of growth in various dimensions should be as constant as they are than that there should be an occasional departure from perfect regularity. In these latter cases the logarithmic law of growth is only approximately true. The shell is no longer to be represented simply as a cone which has been rolled up, but as a cone which (while rolling up) had grown trumpet-shaped, or conversely whose mouth had narrowed in, and which in longitudinal section is a curvilinear instead of a rectilinear triangle. But all that has happened is that a new factor, usually of small or all but imperceptible magnitude, has been introduced into the case; so that the ratio, $\log r = \theta \log \alpha$, is no longer constant but varies slightly, and in accordance with some simple law.

Some writers, such as Naumann* and Grabau, maintained that the molluscan spiral was no true logarithmic spiral, but differed from it specifically, and they gave it the name of *Conchospiral.* They said that the logarithmic spiral originates in a mathematical point, while the molluscan shell starts with a little embryonic shell, or central chamber (the "protoconch" of the conchologists), around which the spiral is subsequently wrapped. But this need not affect the logarithmic law of the shell as a whole; indeed we have already allowed for it by writing our equation in the form $r = ma^{\theta}$. And Grabau†, while he clung to Naumann's conchospiral against Moseley's logarithmic spiral, confessed that they were so much alike that ordinary measurements would seldom shew a difference between them.

There would seem, by the way, to be considerable confusion in the books with regard to the so-called "protoconch." In many cases it is a definite structure, of simple form, representing the more or less globular embryonic shell before it began to elongate into its conical or spiral form. But in many cases what is described as the "protoconch" is merely an empty space in the middle of the spiral coil, resulting from the fact that the actual spiral shell must have some magnitude to begin with, and that we cannot follow it down to its vanishing point in infinity. For instance, in the accompanying figure, the large space *a* is styled the protoconch, but it is the little bulbous or hemispherical chamber within it, at the end of the spire, which is the real beginning of the tubular shell. The form and magnitude of the space *a* are determined by the "angle of retardation," or ratio of rate of growth between the inner and outer curves of the spiral shell. They

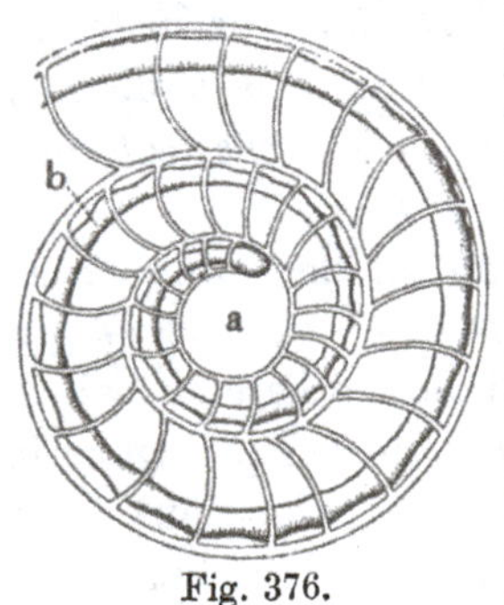

Fig. 376.

* C. F. Naumann, Beitrag zur Konchyliometrie, *Poggend. Ann.* L, p. 223, 1840; Ueber die Spiralen der Ammoniten, *ibid.* LI, p. 245, 1840; *ibid.* LIV, p. 541, 1845; etc. (See also p. 755.) Cf. also Lehmann, *Die von Seyfriedsche Konchyliensammlung und das Windungsgesetz von einigen Planorben,* Constanz, 1855.

† A. H. Grabau, Ueber die Naumannsche Conchospirale, und ihre Bedeutung für die Conchyliometrie, *Inauguraldiss.,* Leipzig, 1872; Ueber die Spiralen der Conchylien, etc., *Leipzig Progr.* No. 502, 1880; cf. *Sb. naturf. Gesellsch.* Leipzig, 1881, pp. 23–32.

are independent of the shape and size of the embryo, and depend only (as we shall see better presently) on the direction and relative rate of growth of the double contour of the shell*.

Now that we have dealt, in a general way, with some of the more obvious properties of the equiangular or logarithmic spiral, let us consider certain of them a little more particularly, keeping in view as our chief object of study the range of variation of the molluscan shell.

There is yet another equation to the logarithmic spiral, very commonly employed, and without the help of which we cannot get far. It is as follows: $r = e^{\theta \cot \alpha}$.

This follows directly from the fact that the angle α (the angle between the radius vector and the tangent to the curve) is constant.

For then,

$$\tan \alpha \ (= \tan \phi) = r \, d\theta / dr;$$

therefore $$dr/r = d\theta \cot \alpha,$$

and, integrating, $\log r = \theta \cot \alpha$,

or $$r = e^{\theta \cot \alpha}.$$

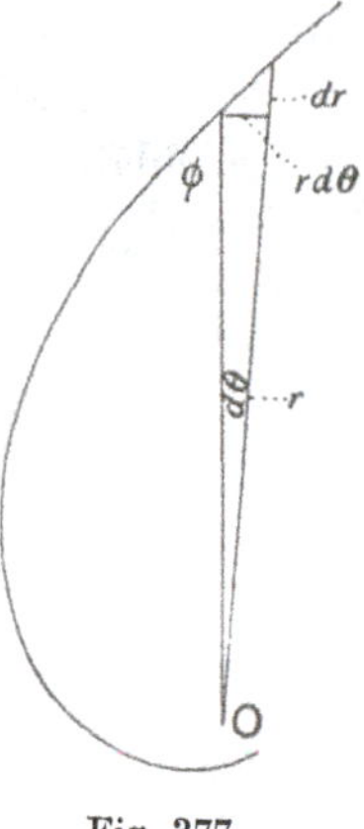

Fig. 377.

It is easy to see (we might indeed have noted it before) that the logarithmic spiral is but a plotting in polar coordinates of *increase by compound interest.* For if A be the "amount" of £1 in one year ($A = 1 + a$, where a is the rate of interest), and PA the amount of P in one year, then the whole amount, M, in t years is $M = PA^t$: this, provided that interest is payable once a year. But, as we are taught by algebra, and as we have seen in our study of growth, this formula becomes Pe^{at} when the intervals of time between the payments of interest decrease without limit, that is to say, when we may consider growth to be continuous. And this formula Pe^{at} is precisely that of our logarithmic spiral, when we represent the time

* J. F. Blake (cf. *infra*, p. 793) says of Naumann's formula: "By such a modification he hoped to bring the measurements of actual shells more into harmony with calculation. The errors of observation, however, are always greater than this change would correct—if founded on fact, which is doubtful; and all practical advantage is lost by the complication of the equations."

by a vector angle θ, and when for a, the particular rate of interest in the case, we write cot α, the constant measure of growth of the particular spiral.

As we have seen throughout our preliminary discussion, the two most important constants (or "specific characters," as the naturalist would say) in an equiangular or logarithmic spiral are (1) the magnitude of the angle of the spiral, or "constant angle" α, and (2) the rate of increase of the radius vector for any given angle of revolution, θ. But our two magnitudes, that of the constant angle and that of the ratio of the radii or breadths of whorl, are directly related to one another, so that we may determine either of them by measurement and calculate the other.

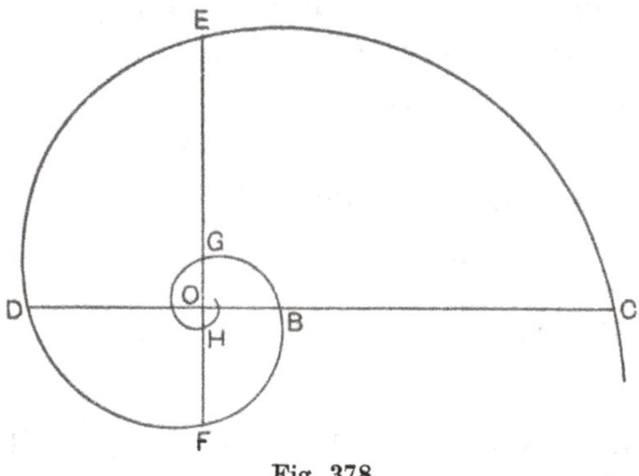

Fig. 378.

In any complete spiral, such as that of *Nautilus*, it is (as we have seen) easy to measure any two radii (r), or the breadths in a radial direction of any two whorls (W). We have then merely to apply the formula

$$\frac{r_{n+1}}{r_n} = e^{\theta \cot \alpha}, \quad \text{or} \quad \frac{W_{n+1}}{W_n} = e^{\theta \cot \alpha},$$

which we may simply write $r = e^{\theta \cot \alpha}$, etc., when one radius or whorl is regarded, for the purpose of comparison, as equal to unity.

Thus, in Fig. 378, OC/OE, or EF/BD, or DC/EF, being in each case radii, or diameters, at right angles to one another, are all equal to $e^{\frac{\pi}{2}\cot \alpha}$. While in like manner, EO/OF, EG/FH, or GO/HO, all equal $e^{\pi \cot \alpha}$; and BC/BA, or $CO/OB = e^{2\pi \cot \alpha}$.

As soon, then, as we have prepared tables for these values, the determination of the constant angle α in a particular shell becomes a very simple matter.

A complete table would be cumbrous, and it will be sufficient to deal with the simple case of the ratio between the breadths of adjacent, or immediately succeeding, whorls.

Here we have $r = e^{2\pi \cot \alpha}$, or $\log r = \log e \times 2\pi \times \cot \alpha$, from which we obtain the following figures*:

The shape of a nautiloid spiral

Ratio of breadth of each whorl to the next preceding $r/1$	Constant angle α
1·1	89° 8′
1·25	87 58
1·5	86 18
2·0	83 42
2·5	81 42
3·0	80 5
3·5	78 43
4·0	77 34
4·5	76 32
5·0	75 38
10·0	69 53
20·0	64 31
50·0	58 5
100·0	53 46
1000·0	42 17
10,000	34 19
100,000	28 37
1,000,000	24 28
10,000,000	21 18
100,000,000	18 50
1,000,000,000	16 52

We learn several interesting things from this short table. We see, in the first place, that where each whorl is about three times the breadth of its neighbour and predecessor, as is the case in *Nautilus*, the constant angle is in the neighbourhood of 80°; and hence also that, in all the ordinary ammonitoid shells, and in all the typically spiral shells of the gastropods†, the constant angle is also a large one, being very seldom less than 80°, and usually between 80° and 85°. In the next place, we see that with smaller

* It is obvious that the ratios of opposite whorls, or of radii 180° apart, are represented by the square roots of these values; and the ratios of whorls or radii 90° apart, by the square roots of these again.

† For the correction to be applied in the case of the helicoid, or "turbinate" shells, see p. 816.

angles the apparent form of the spiral is greatly altered, and the very fact of its being a spiral soon ceases to be apparent (Figs. 379, 380). Suppose one whorl to be an inch in breadth, then, if the angle of the spiral were 80°, the next whorl would (as we have just seen) be about three inches broad; if it were 70°, the next whorl would be nearly ten inches, and if it were 60°, the next whorl would be nearly four feet broad. If the angle were 28°, the next whorl would be a mile and a half in breadth; and if it were 17°, the next would be some 15,000 miles broad.

In other words, the spiral shells of gentle curvature, or of small constant angle, such as *Dentalium* or *Cristellaria*, are true equiangular spirals, just as are those of *Nautilus* or *Rotalia*: from

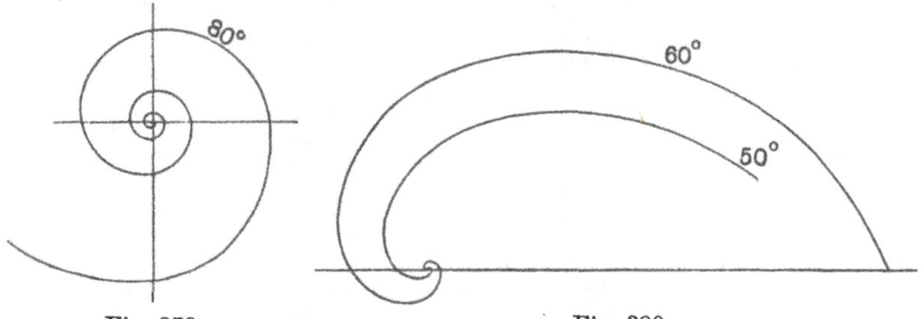

Fig. 379. Fig. 380.

which they differ only in degree, in the magnitude of an angular constant. But this diminished magnitude of the angle causes the spiral to dilate with such immense rapidity that, so to speak, it never comes round; and so, in such a shell as *Dentalium*, we never see but a small portion of a single whorl.

We might perhaps be inclined to suppose that, in such a shell as *Dentalium*, the lack of a visible spiral convolution was only due to our seeing but a small portion of the curve, at a distance from the pole, and when, therefore, its curvature had already greatly diminished. That is to say we might suppose that, however small the angle α, and however rapidly the whorls accordingly increased, there would nevertheless be a manifest spiral convolution in the immediate neighbourhood of the pole, as the starting point of the curve. But it is easy to see that it is not so. It is not that there cease to be convolutions of the spiral round the pole when α is a small angle; on the contrary, there are infinitely many, mathematically speaking. But as α diminishes, and $\cot \alpha$ increases towards infinity, the ratio between the breadth of one whorl and the next increases very rapidly. Our table shews us that even when α is no less than 40°, and our shell still looks strongly curved, one whorl is a thousandth part of the breadth of the next, and a thousandfold that of the one before; we cannot expect to see either of them under the materialised conditions of the actual shell. Our shells of small constant angle and gentle curvature, such as *Dentalium*, are accordingly as much as we can ever expect to see of their respective spirals.

The spiral whose constant angle is 45° is both a simple case and a mathematical curiosity; for, since the tangent of 45° is unity, we need merely write $r = e^{\theta}$; which is as much as to say that the natural logarithms of the radii give us, without more ado, the vector angles. In this spiral the ratio between the breadths of two consecutive whorls becomes $r = e^{2\pi} = e^{2\times3\cdot1416}$. Reducing this from Naperian to common logs, we have $\log r = 2\cdot729$; which tells us (by our tables) that the radius vector is multiplied about $535\frac{1}{2}$ times after a whole polar revolution; it is doubled after turning through a polar angle of less than 40°. Spirals of so low an angle as 45° are common enough in tooth and claw, but rare among molluscan shells; but one or two of the more strongly curved Dentaliums, like *D. elephantinum*, come near the mark. It is not easy to determine the pole, nor to measure the constant angle, in forms like these.

Let us return to the problem of how to ascertain, by direct measurement, the spiral angle of any particular shell. The method already employed is only applicable to complete spirals, that is to say to those in which the angle of the spiral is large, and furthermore it is inapplicable to portions, or broken fragments, of a shell. In the case of the broken fragment, it is plain that the determination of the angle is not merely of theoretic interest, but may be of great practical use to the conchologist as the one and only way by which he may restore the outline of the missing portions. We have a considerable choice of methods, which have been summarised by, and are partly due to, a very careful student of the Cephalopoda, the late Rev. J. F. Blake*.

(1) When an equiangular spiral rolls on a straight line, the pole traces another straight line at an angle to the first equal to the complement of the constant angle of the spiral; for the contact point is the instantaneous centre of the rotational movement, and the line joining it to the pole of the spiral is normal to the roulette path of that point. But the difficulty of determining the pole

* On the measurement of the curves formed by Cephalopods and other Mollusks, *Phil. Mag.* (5), VI, pp. 241–263, 1878.

(which is indeed asymptotic) makes this of little use as a method of determining the constant angle. It is, however, a beautiful property of the curve, and all the more interesting that Clerk Maxwell discovered it when he was a boy*.

(2) The following method is useful and easy when we have a portion of a single whorl, such as to shew both its inner and its outer edge. A broken whorl of an Ammonite, a curved shell such as *Dentalium*, or a horn of similar form to the latter, will fall under this head. We have merely to draw a tangent, *GEH*, to the outer whorl at any point *E*; then draw to the inner whorl a tangent parallel to *GEH*, touching the curve in some point *F*. The straight line joining the points of contact, *EF*, must evidently pass through the pole: and, accordingly, the angle *GEF* is the angle required. In shells which bear *longitudinal* striae or other ornaments, any pair of these will suffice for our purpose, instead of the actual boundaries of the whorl. But it is obvious that this method will be apt to fail us when the angle α is very small; and when, consequently, the points *E* and *F* are very remote.

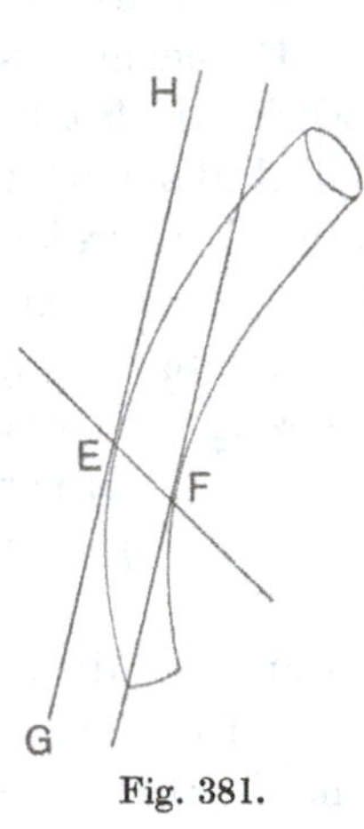

Fig. 381.

(3) In shells (or horns) shewing rings or other *transverse* ornamentation, we may take it that these ornaments are set at

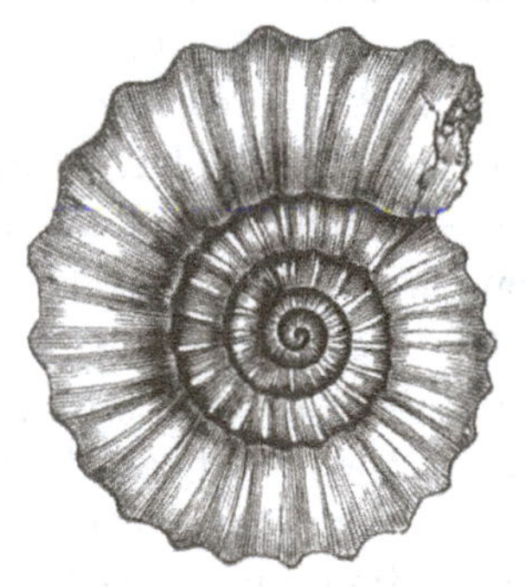

Fig. 382. An Ammonite, to shew corrugated surface-pattern.

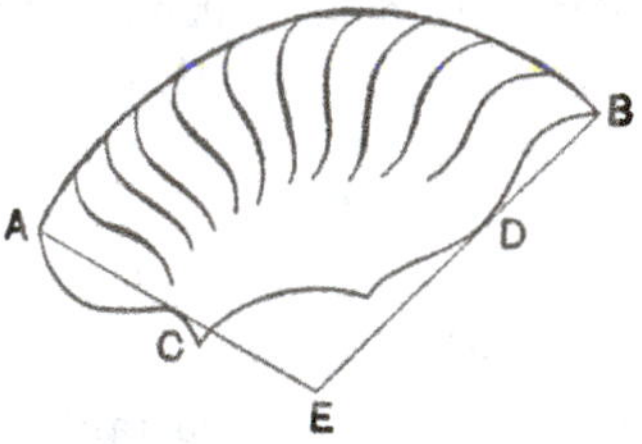

Fig. 383.

* Clerk Maxwell, On the theory of rolling curves, *Trans. R.S.E.* XVI, pp. 519–540, 1849; *Sci. Papers*, I, pp. 4–29.

a constant angle to the spire, and therefore to the radii. The angle (θ) between two of them, as AC, BD, is therefore equal to the angle θ between the polar radii from A and B, or from C and D; and therefore $BD/AC = e^{\theta \cot \alpha}$, which gives us the angle α in terms of known quantities.

(4) If only the outer edge be available, we have the ordinary geometrical problem—given an arc of an equiangular spiral, to find its pole and spiral angle. The methods we may employ depend (i) on determining directly the position of the pole, and (ii) on determining the radius of curvature.

The first method is theoretically simple, but difficult in practice; for it requires great accuracy in determining the points. Let AD, DB be two tangents drawn to the curve. Then a circle drawn through the points A, B, D will pass through the pole O, since the angles OAD, OBE (the supplement of OBD) are equal. The point O may be determined by the intersection of two such circles; and the angle DBO is then the angle, α, required.

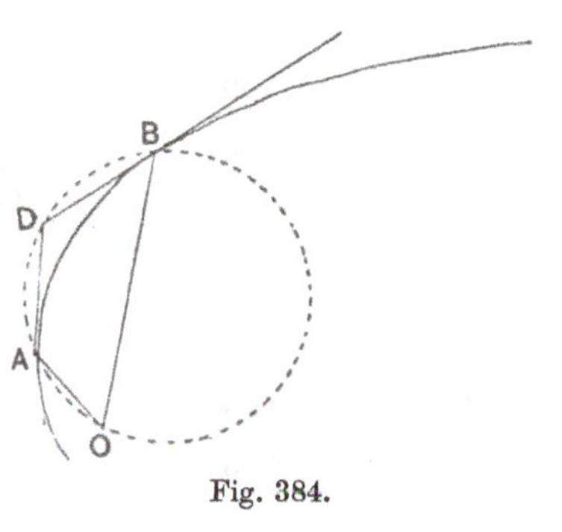

Fig. 384.

Or we may determine graphically, at two points, the radii of curvature $\rho_1 \rho_2$. Then, if s be the length of the arc between them (which may be determined with fair accuracy by rolling the margin of the shell along a ruler),

$$\cot \alpha = (\rho_1 - \rho_2)/s.$$

The following method*, given by Blake, will save actual determination of the radii of curvature.

Measure along a tangent to the curve the distance, AC, at which a certain small offset, CD, is made by the curve; and from another point B, measure the distance at which the curve makes an equal offset. Then, calling the offset μ; the arc AB, s; and AC, BE, respectively x_1, x_2, we have

$$\rho_1 = \frac{x_1^2 + \mu^2}{2\mu}, \text{ approximately,}$$

and

$$\cot \alpha = \frac{x_2^2 - x_1^2}{2\mu s}.$$

Of all these methods by which the mathematical constants, or specific characters, of a given spiral shell may be determined, the

* For an example of this method, see Blake, *loc. cit.* p. 251.

only one of which much use has been made is that which Moseley first employed, namely, the simple method of determining the relative breadths of the whorl at distances separated by some convenient vectorial angle such as 90°, 180°, or 360°.

Very elaborate measurements of a number of Ammonites have been made by Naumann*, by Grabau, by Sandberger†, and by Müller, among which we may choose a couple of cases for consideration‡. In the following table I have taken a portion of Grabau's

Ammonites intuslabiatus

Breadth of whorls (180° apart)	Ratio of breadth of successive whorls (360° apart)	The angle (α) as calculated
0·30 mm.	—	— —
0·30	1·333	87° 23′
0·40	1·500	86 19
0·45	1·500	86 19
0·60	1·444	86 39
0·65	1·417	86 49
0·85	1·692	85 13
1·10	1·588	85 47
1·35	1·545	86 2
1·70	1·630	85 33
2·20	1·441	86 40
2·45	1·432	86 43
3·15	1·735	85 0
4·25	1·683	85 16
5·30	1·482	86 25
6·30	1·519	86 12
8·05	1·635	85 32
10·30	1·416	86 50
11·40	1·252	87 57
12·90	—	— —
	Mean	86° 15′

* C. F. Naumann, Ueber die Spiralen von Conchylien, *Abh. k. sächs. Ges.* 1846, pp. 153–196; Ueber die cyclocentrische Conchospirale u. uber das Windungsgesetz von *Planorbis corneus*, *ibid.* I, pp. 171–195, 1849; Spirale von Nautilus u. *Ammonites galeatus*, *Ber. k. sächs. Ges.* II, p. 26, 1848; Spirale von *Amm. Ramsaueri*, *ibid.* XVI, p. 21, 1864. Oken, reviewing Naumann's work (in *Isis*, 1847, p. 867) foretold how some day the naturalist and the mathematician would each learn of the other: "Um die Sache zu Vollendung zu bringen wird der Mathematiker Zoolog und Physiolog, und diese Mathematiker werden müssen."

† G. Sandberger, *Clymenia subnautilina*, *Jahresber. d. Ver. f. Naturk. im Herzogth. Nassau*, 1855, p. 127; Spiralen des *Ammonites Amaltheus*, *A. Gaytani* und *Goniatites intumescens*, *Ztschr. d. d. Geolog. Gesellsch.* X, pp. 446–449, 1858. Also Müller, Beitrag zur Konchyliometrie, *Poggend. Ann.* LXXXVI, p. 533, 1850; *ibid.* XC, p. 323, 1853. These two authors upheld the logarithmic law against Naumann and Grabau.

‡ See also Chr. Petersen, *Das Quotientengesetz, eine biologisch-statistische Untersuchung*, 119 pp., Copenhagen, 1921; E. Sporn, Ueber die Gesetzmässigkeit im Baue der Muschelgehaüser, *Arch. f. Entw. Mech.* CVIII, pp. 228–242, 1926.

determinations of the breadth of the whorls in *Ammonites* (*Arcestes*) *intuslabiatus*; these measurements Grabau gives for every 45° of arc, but I have only set forth successive whorls measured along one diameter on both sides of the pole. The ratio between *alternate* measurements is therefore the same ratio as Moseley adopted, namely the ratio of breadth between *contiguous whorls* along a radius vector. I have then added to these observed values the corresponding calculated values of the angle α, as obtained from our usual formula.

There is considerable irregularity in the ratios derived from these measurements, but it will be seen that this irregularity only implies a variation of the angle of the spiral between about 85° and 87°; and the values fluctuate pretty regularly about the mean, which is 86° 15′. Considering the difficulty of measuring the whorls, especially towards the centre, and in particular the difficulty of determining with precise accuracy the position of the pole, it is clear that in such a case as this we are not justified in asserting that the law of the equiangular spiral is departed from.

Ammonites tornatus

Breadth of whorls (180° apart)	Ratio of breadth of successive whorls (360° apart)	The spiral angle (α) as calculated
0·25 mm.	—	— —
0·30	1·400	86° 56′
0·35	1·667	85 21
0·50	2·000	83 42
0·70	2·000	83 42
1·00	2·000	83 42
1·40	2·100	83 16
2·10	2·179	82 56
3·05	2·238	82 42
4·70	2·492	81 44
7·60	2·574	81 27
12·10	2·546	81 33
19·35	—	— —
Mean	2·11	83° 22′

In some cases, however, it is undoubtedly departed from. Here for instance is another table from Grabau, shewing the corresponding ratios in an Ammonite of the group of *Arcestes tornatus*. In this case we see a distinct tendency of the ratios to increase as we pass from the centre of the coil outwards, and consequently for the values of the angle α to diminish. The case is comparable to that of a cone with slightly curving sides: in which, that is to say, there is a slight acceleration of growth in a transverse as compared with the longitudinal direction.

In a tubular spiral, whether plane or helicoid, the consecutive whorls may either be (1) isolated and remote from one another; or (2) they may precisely meet, so that the outer border of one and the inner border of the next just coincide; or (3) they may overlap, the vector plane of each outer whorl cutting that of its immediate predecessor or predecessors.

Looking, as we have done, upon the spiral shell as being essentially a cone rolled up*, it is plain that, for a given spiral angle, intersection or non-intersection of the successive whorls will depend upon *the apical angle* of the original cone. For the wider the cone, the more will its inner border tend to encroach on the preceding whorl. But it is also plain that the greater the apical angle of the cone, and the broader, consequently, the cone itself, the greater difference will there be between the total *lengths* of its inner and outer borders. And, since the inner and outer borders are describing precisely the same spiral about the pole, we may consider the inner border as being *retarded* in growth as compared with the outer, and as being always identical with a smaller and earlier part of the latter.

If λ be the ratio of growth between the outer and the inner curve, then, the outer curve being represented by

$$r = ae^{\theta \cot \alpha},$$

the equation to the inner one will be

$$r' = a\lambda e^{\theta \cot \alpha},$$

or

$$r' = ae^{(\theta - \gamma) \cot \alpha},$$

* To speak of a cone "rolling up," and becoming a nautiloid spiral by doing so, is a rough and non-mathematical description; nor is it easy to see how a cone of wide angle could roll up, and yet remain a cone. But if (i) the centre of a sphere move along a straight line and its radius keep proportional to the distance the centre has moved, the sphere generates as its envelope a circular cone of which the straight line is the axis; and so, similarly, if (ii) the centre of a sphere move along an equiangular spiral and its radius keep proportional to the arc-distance along the spiral back to the pole, the sphere generates as its envelope a self-similar shell-surface, or nautiloid spiral.

and γ may then be called the angle of retardation, to which the inner curve is subject by virtue of its slower rate of growth.

Dispensing with mathematical formulae, the several conditions may be illustrated as follows:

In the diagrams (Fig. 385), $OP_1P_2P_3$, etc. represents a radius, on which P_1, P_2, P_3 are the points attained by the outer border of the tubular shell after as many entire consecutive revolutions. And P_1', P_2', P_3' are the points similarly intersected by the inner border; OP/OP' being always $= \lambda$, which is the ratio of growth, or "cutting-down factor." Then, obviously, (1) when OP_1 is less than

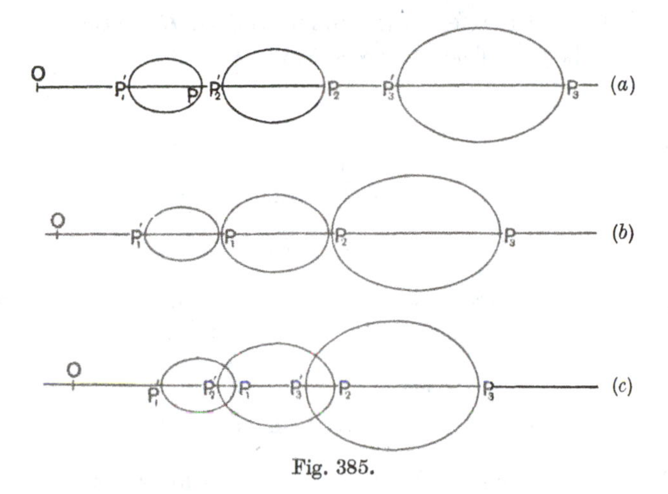

Fig. 385.

OP_2' the whorls will be separated by an interspace (a); (2) when $OP_1 = OP_2'$ they will be in contact (b), and (3) when OP_1 is greater than OP_2' there will be a greater or less extent of overlapping, that is to say of concealment of the surfaces of the earlier by the later whorls (c). And as a further case (4), it is plain that if λ be very large, that is to say if OP_1 be greater, not only than OP_2' but also than OP_3', OP_4', etc., we shall have complete, or all but complete, concealment by the last formed whorl of the whole of its predecessors. This latter condition is completely attained in *Nautilus pompilius*, and approached, though not quite attained, in *N. umbilicatus*; and the difference between these two forms, or "species," is constituted accordingly by a difference in the value of λ. (5) There is also a final case, not easily distinguishable externally from (4), where P' lies on the opposite side of the radius vector to P, and is therefore imaginary. This final condition is exhibited in *Argonauta*.

The limiting values of λ are easily ascertained.

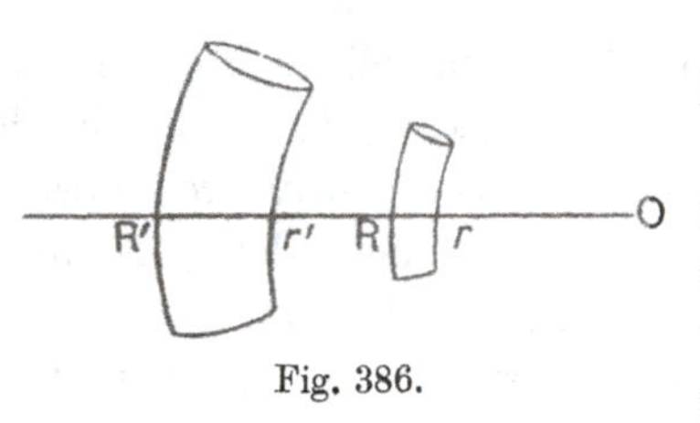

Fig. 386.

In Fig. 386 we have portions of two successive whorls, whose corresponding points on the same radius vector (as R and R') are, therefore, at a distance apart corresponding to 2π. Let r and r' refer to the inner, and R, R' to the outer sides of the two whorls. Then, if we consider

$$R = ae^{\theta \cot \alpha},$$

it follows that

$$R' = ae^{(\theta+2\pi) \cot \alpha},$$

$$r = \lambda ae^{\theta \cot \alpha} = ae^{(\theta-\gamma) \cot \alpha},$$

and

$$r' = \lambda ae^{(\theta+2\pi) \cot \alpha} = ae^{(\theta+2\pi-\gamma) \cot \alpha}.$$

Now in the three cases (a, b, c) represented in Fig. 385, it is plain that $r' \gtreqless R$, respectively. That is to say,

$$\lambda ae^{(\theta+2\pi) \cot \alpha} \gtreqless ae^{\theta \cot \alpha},$$

and

$$\lambda e^{2\pi \cot \alpha} \lesseqgtr 1.$$

The case in which $\lambda e^{2\pi \cot \alpha} = 1$, or $-\log \lambda = 2\pi \cot \alpha \log e$, is the case represented in Fig. 385, b: that is to say, the particular case, for each value of α, where the consecutive whorls just touch, without interspace or overlap. For such cases, then, we may tabulate the values of λ as follows:

Constant angle α of spiral	Ratio (λ) of rate of growth of inner border of tube, as compared with that of the outer border
89°	0·896
88	0·803
87	0·720
86	0·645
85	0·577
80	0·330
75	0·234
70	0·1016
65	0·0534

We see, accordingly, that in plane spirals whose constant angle lies, say, between 65° and 70°, we can only obtain contact between consecutive whorls if the rate of growth of the inner border of the tube be a small fraction—a tenth or a twentieth—of that of the outer border. In spirals whose constant angle is 80°, contact is attained when the respective rates of growth are, approximately, as 3 to 1; while in spirals of constant angle from about 85° to 89°, contact is attained when the rates of growth are in the ratio of from about $\frac{3}{5}$ to $\frac{9}{10}$.

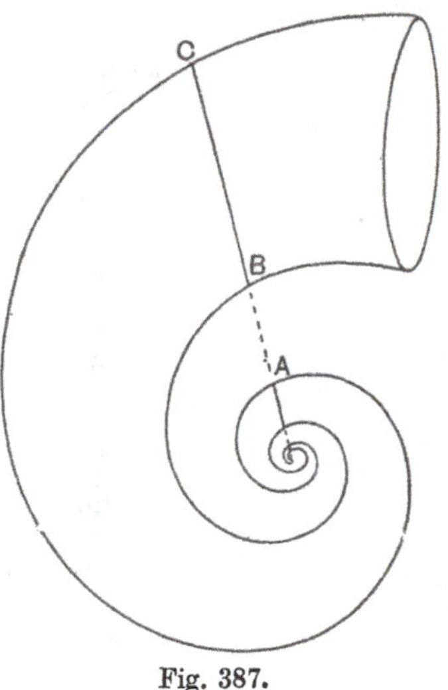

Fig. 387.

If on the other hand we have, for any given value of α, a value of λ greater or less than the value given in the above table, then we have, respectively, the conditions of separation or of overlap which are exemplified in Fig. 385, *a* and *c*. And, just as we have constructed this table for the particular case of simple contact, so we could construct similar tables for various degrees of separation or of overlap.

For instance, a case which admits of simple solution is that in which the interspace between the whorls is everywhere a mean proportional between the breadths of the whorls themselves (Fig. 387). In this case, let us call $OA = R$, $OC = R_1$, and $OB = r$. We then have

$$R_1 = OA = ae^{(\theta \cot \alpha)},$$

$$R_2 = OC = ae^{(\theta + 2\pi)\cot \alpha},$$

$$R_1 R_2 = ae^{2(\theta+\pi)\cot \alpha} = r^2 *.$$

And
$$r^2 = (1/\lambda)^2 \,.\, \epsilon^{2\theta \cot \alpha},$$

whence, equating,
$$1/\lambda = e^{\pi \cot \alpha}.$$

* It has been pointed out to me that it does not follow at once and obviously that, because the interspace AB is a mean proportional between the breadths of the adjacent whorls, therefore the whole distance OB is a mean proportional between OA and OC. This is a corollary which requires to be proved; but the proof is easy.

The corresponding values of λ are as follows:

Constant angle (α)	Ratio (λ) of rates of growth of outer and inner border, such as to produce a spiral with interspaces between the whorls, the breadth of which interspaces is a mean proportional between the breadths of the whorls themselves
90°	1·00 (imaginary)
89	0·95
88	0·89
87	0·85
86	0·81
85	0·76
80	0·57
75	0·43
70	0·32
65	0·23
60	0·18
55	0·13
50	0·090
45	0·063
40	0·042
35	0·026
30	0·016

As regards the angle of retardation, γ, in the formula

$$r' = \lambda e^{\theta \cot \alpha}, \quad \text{or} \quad r' = e^{(\theta - \gamma) \cot \alpha},$$

and in the case

$$r' = e^{(2\pi - \gamma) \cot \alpha}, \quad \text{or} \quad -\log \lambda = (2\pi - \gamma) \cot \alpha,$$

it is evident that when $\gamma = 2\pi$, that will mean that $\lambda = 1$. In other words, the outer and inner borders of the tube are identical, and the tube is constituted by one continuous line.

When λ is a very small fraction, that is to say when the rates of growth of the two borders of the tube are very diverse, then γ will tend towards infinity—tend that is to say towards a condition in which the inner border of the tube never grows at all. This condition is not infrequently approached in nature. I take it that *Cypraea* is such a case. But the nearly parallel-sided cone of *Dentalium*, or the widely separated whorls of *Lituites*, are cases where λ nearly approaches unity in the one case, and is still large in the other, γ being correspondingly small; while we can easily find cases where γ is very large, and λ is a small fraction, for instance in *Haliotis*, in *Calyptraea*, or in *Gryphaea*.

For the purposes of the morphologist, then, the main result of this last general investigation is to shew that all the various types of "open" and "closed" spirals, all the various degrees of separation or overlap of the successive whorls, are simply the outward expression of a varying ratio in the *rate of growth* of the outer as compared with the inner border of the tubular shell.

The foregoing problem of contact, or intersection, of successive whorls is a very simple one in the case of the discoid shell but a more complex one in the turbinate. For in the discoid shell contact will evidently take place when the retardation of the inner as compared with the outer whorl is just 360°, and the shape of the whorls need not be considered.

As the angle of retardation diminishes from 360°, the whorls stand further and further apart in an open coil; as it increases beyond 360°, they overlap more and more; and when the angle of retardation is infinite, that is to say when the true inner edge of the whorl does not grow at all, then the shell is said to be completely involute. Of this latter condition we have a striking example in *Argonauta*, and one a little more obscure in *Nautilus pompilius*.

In the turbinate shell the problem of contact is twofold, for we have to deal with the possibilities of contact on the *same* side of the axis (which is what we have dealt with in the discoid) and also with the new possibility of contact or intersection on the *opposite* side; it is this latter case which will determine the presence or absence of an open *umbilicus*. It is further obvious that, in the case of the turbinate, the question of contact or no contact will depend on the shape of the generating curve; and if we take the simple case where this generating curve may be considered as an ellipse, then contact will be found to depend on the angle which the major axis of this ellipse makes with the axis of the shell. The question becomes a complicated one, and the student will find it treated in Blake's paper already referred to.

When one whorl overlaps another, so that the generating curve cuts its predecessor (at a distance of 2π) on the same radius vector, the locus of intersection will follow a spiral line upon the shell, which is called the "suture" by conchologists. It is one of that *ensemble* of spiral lines in space of which, as we have seen, the whole shell may be conceived to be constituted; and we might call it a "contact-spiral," or "spiral of intersection." In discoid shells, such as an *Ammonite* or a *Planorbis*, or in *Nautilus umbilicatus*, there are obviously two such contact-spirals, one on each side of the shell, that is to say one on each side of a plane perpendicular to the axis. In turbinate shells such a condition is also possible, but is somewhat rare. We have it for instance in *Solarium perspectivum*, where the one contact-spiral is visible on the exterior of

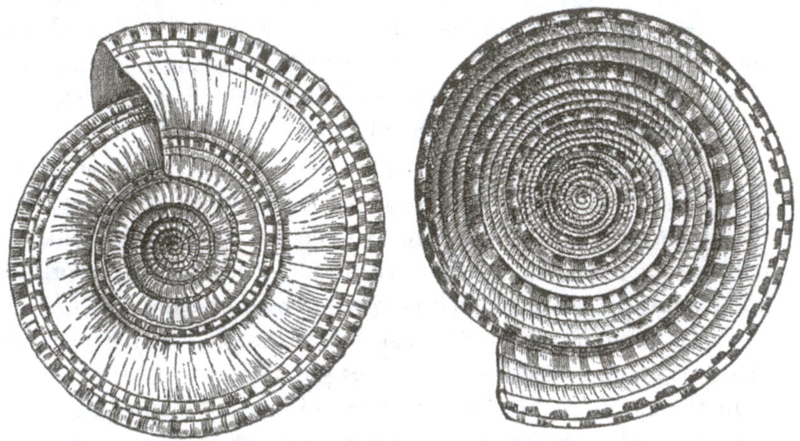

Fig. 388. *Solarium perspectivum.*

the shell, and the other lies internally, winding round the open cone of the umbilicus*; but this second contact-spiral is usually imaginary, or concealed within the whorls of the turbinated shell.

Fig. 389. *Haliotis tuberculata* L.; the ormer, or ear shell.

Fig. 390. *Scalaria pretiosa* L.; the wentletrap. From Cooke's *Spirals.*

Again, in *Haliotis*, one of the contact-spirals is non-existent, because of the extreme obliquity of the plane of the generating curve. In

* A beautiful construction: *stupendum Naturae artificium*, Linnaeus.

Scalaria pretiosa and in *Spirula** there is no contact-spiral, because the growth of the generating curve has been too slow in comparison with the vector rotation of its plane. In *Argonauta* and in *Cypraea* there is no contact-spiral, because the growth of the generating curve has been too quick. Nor, of course, is there any contact-spiral in *Patella* or in *Dentalium*, because the angle α is too small

Fig. 391. *Thatcheria mirabilis* Angas; from a radiograph by Dr A. Müller.

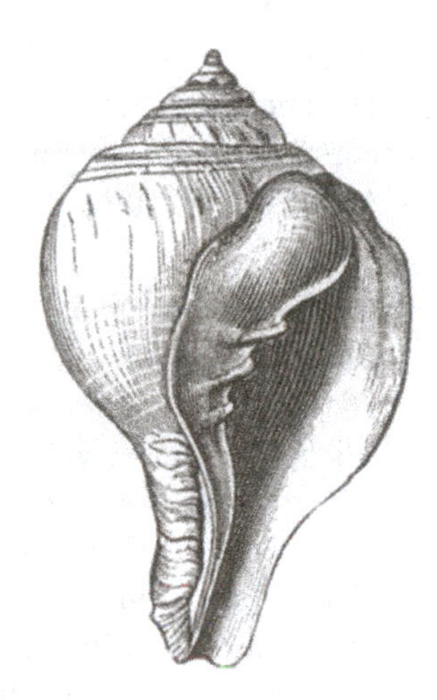

Fig. 392. *Turbinella napus* Lam.; an Indian chank-shell. From Chenu.

ever to give us a complete revolution of the spire. *Thatcheria mirabilis* is a peculiar and beautiful shell, in which the outline of the lip is sharply triangular, instead of being a smooth curve: with the result that the apex of the triangle forms a conspicuous "generating spiral", which winds round the shell and is more conspicuous than the suture itself.

In the great majority of helicoid or turbinate shells the innermost

* "It [*Spirula*] is curved so as its roundness is kept, and the Parts do not touch one another": R. Hooke, *Posthumous Works*, 1745, p. 284.

or axial portions of the whorls tend to form a solid axis or "columella"; and to this is attached the columellar muscle which on the one hand withdraws the animal within its shell, and on the other hand provides the controlling force or trammel, by which (in the gastropod) the growing shell is kept in its spiral course. This muscle is apt to leave a winding groove upon the columella (Fig. 373); now and then the muscle is split into strands or bundles, and then it leaves parallel grooves with ridges or pleats between, and the number of these folds or pleats may vary with the species, as in the Volutes, or even with race or locality. Thus, among the curiosities of conchology, the chank-shells on the Trincomali coast have four columellar folds or ridges; but all those from Tranquebar, just north of Adam's Bridge, have only three (Fig. 392)*.

The various forms of straight or spiral shells among the Cephalopods, which we have seen to be capable of complete definition by the help of elementary mathematics, have received a very complicated descriptive nomenclature from the palaeontologists. For instance, the straight cones are spoken of as *orthoceracones* or *bactriticones*, the loosely coiled forms as *gyroceracones* or *mimoceracones*, the more closely coiled shells, in which one whorl overlaps the other, as *nautilicones* or *ammoniticones*, and so forth. In such a series of forms the palaeontologist sees undoubted and unquestioned evidence of ancestral descent. For instance we read in Zittel's *Palaeontology*†: "The bactriticone obviously represents the primitive or primary radical of the Ammonoidea, and the mimoceracone the next or secondary radical of this order"; while precisely the opposite conclusion was drawn by Owen, who supposed that the straight chambered shells of such fossil Cephalopods as *Orthoceras* had been produced by the gradual unwinding of a coiled nautiloid shell‡. *The mathematical study of the forms of shells lends no support to these*

* Cf. R. Winckworth, *Proc. Malacol. Soc.* XXIII, p. 345, 1939.

† English edition, 1900, p. 537. The chapter is revised by Professor Alpheus Hyatt, to whom the nomenclature is largely due. For a more copious terminology, see Hyatt, *Phylogeny of an Acquired Characteristic*, 1894, p. 422 *seq.* Cf. also L. F. Spath, The evolution of the Cephalopoda, *Biol. Reviews*, VIII, pp. 418–462, 1933.

‡ This latter conclusion is adopted by Willey, *Zoological Results*, 1902, p. 747. Cf. also Graham Kerr, on *Spirula*: *Dana Reports*, No. 8, Copenhagen, 1931.

*or any suchlike phylogenetic hypotheses**. If we have two shells in which the constant angle of the spire be respectively 80° and 60°, that fact in itself does not at all justify an assertion that the one is more primitive, more ancient, or more "ancestral" than the other. Nor, if we find a third in which the angle happens to be 70°, does that fact entitle us to say that this shell is intermediate between the other two, in time, or in blood relationship, or in any other sense whatsoever save only the strictly formal and mathematical one. For it is evident that, though these particular arithmetical constants manifest themselves in visible and recognisable differences of form, yet they are not necessarily more deep-seated or significant than are those which manifest themselves only in difference of magnitude; and the student of phylogeny scarcely ventures to draw conclusions as to the relative antiquity of two allied organisms on the ground that one happens to be bigger or less, or longer or shorter, than the other.

At the same time, while it is obviously unsafe to rest conclusions upon such features as these, unless they be strongly supported and corroborated in other ways—for the simple reason that there is unlimited room for *coincidence*, or separate and independent attainment of this or that magnitude or numerical ratio—yet on the other hand it is certain that, in particular cases, the evolution of a race has actually involved gradual increase or decrease in some one or more numerical factors, magnitude itself included—that is to say increase or decrease in some one or more of the actual and relative velocities of growth. When we do meet with a clear and unmistakable series of such progressive magnitudes or ratios, manifesting themselves in a progressive series of "allied" forms, then we have the phenomenon of "*orthogenesis*." For orthogenesis is simply that phenomenon of continuous lines or series of form (and also of functional or physiological capacity),

* Phylogenetic speculation, fifty years ago the chief preoccupation of the biologist, has had its caustic critics. Cf. (*int. al.*) Rhumbler, in *Arch. f. Entw. Mech.* VII, p. 104, 1898: "Phylogenetische Speculationen...werden immer auf Anklang bei den Fachgenossen rechnen dürfen, sofern nicht ein anderer Fachgenosse auf demselben Gebiet mit gleicher Kenntniss der Dinge und mit gleicher Scharfsinn zufällig zu einer anderen Theorie gekommen ist....Die Richtigkeit 'guter' phylogenetischer Schlüsse lässt sich im schlimmsten Fälle anzweifeln, aber direkt widerlegen lasst sich in der Regel nicht."

which was the foundation of the Theory of Evolution, alike to Lamarck and to Darwin and Wallace; and which we see to exist whatever be our ideas of the "origin of species," or of the nature and origin of "functional adaptations." And to my mind, the mathematical (as distinguished from the purely physical) study of morphology bids fair to help us to recognise this phenomenon of orthogenesis in many cases where it is not at once patent to the eye; and, on the other hand, to warn us in many other cases that even strong and apparently complex resemblances in form may be capable of arising independently, and may sometimes signify no more than the equally accidental numerical coincidences which are manifested in identity of length or weight or any other simple magnitudes.

I have already referred to the fact that, while in general a very great and remarkable regularity of form is characteristic of the molluscan shell, yet that complete regularity is apt to be departed from. We have clear cases of such a departure in *Pupa*, *Clausilia* and various *Bulimi*, where the spire is not conical, but its sides are curved and narrow in.

The following measurements of three specimens of *Clausilia* shew a gradual change in the ratio to one another of successive whorls, or in other words a marked departure from the logarithmic law:

Clausilia lamellosa. (From Chr. Petersen*.)

	Width of successive whorls (mm.)				Ratios, or "quotients" of successive whorls			
	I	II	III		I	II	III	Mean
a	2·42	2·51	2·49	*a*/*b*	1·43	1·45	1·42	1·44
b	1·69	1·72	1·75	*b*/*c*	1·36	1·33	1·31	1·33
c	1·24	1·30	1·33	*c*/*d*	1·21	1·29	1·23	1·24
d	1·02	1·00	1·08	*d*/*e*	1·22	1·20	1·26	1·23
e	0·83	0·83	0·86					

In many ammonites, where the helicoid factor does not enter into the case, we have a clear illustration of how gradual and marked

* From Chr. Petersen, *Das Quotientengesetz*, p. 36. After making a careful statistical study of 1000 Clausilias, Peterson found the following mean ratios of the successive whorls, *a*/*b*, *b*/*c*, etc.: 1·37, 1·33, 1·27, 1·24, 1·22, 1·19.

changes in the spiral angle may be detected even in ammonites which present nothing abnormal to the eye. But let us suppose that the spiral angle increases somewhat rapidly; we shall then get a spiral with gradually narrowing whorls, which condition is characteristic of *Oekotraustes*, a subgenus of *Ammonites*. If on the other hand, the angle α gradually diminishes, and even falls away to zero, we shall have the spiral curve opening out, as it does in *Scaphites*, *Ancyloceras*

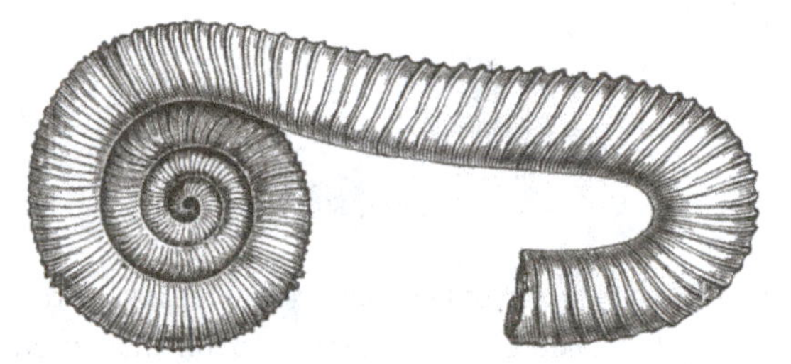

Fig. 393. An ammonitoid shell (*Macroscaphites*) to shew change of curvature.

and *Lituites*, until the spiral coil is replaced by a spiral curve so gentle as to seem all but straight. Lastly, there are a few cases, such as *Bellerophon expansus* and some *Goniatites*, where the outer spiral does not perceptibly change, but the whorls become more "embracing" or the whole shell more involute. Here it is the angle of retardation, the ratio of growth between the outer and inner parts of the whorl, which undergoes a gradual change.

In order to understand the relation of a close-coiled shell to its straighter congeners, to compare (for example) an *Ammonite* with an *Orthoceras*, it is necessary to estimate the length of the right cone which has, so to speak, been coiled up into the spiral shell. Our problem is, to find the length of a plane equiangular spiral, in terms of the radius and the constant angle α. Then, if *OP* be a radius vector, *OQ* a line of reference perpendicular to *OP*, and *PQ* a tangent to the curve, *PQ*, or sec α, is equal in length to the spiral arc *OP*. In other words, the arc measured from the pole is equal to the polar tangent*. And this is practically obvious: for

* Descartes made this discovery, and records it in a letter to Mersenne, 1638. The equiangular spiral was thus the first transcendental curve to be "rectified."

$PP'/PR' = ds/dr = \sec\alpha$, and therefore $\sec\alpha = s/r$, or the ratio of arc to radius vector.

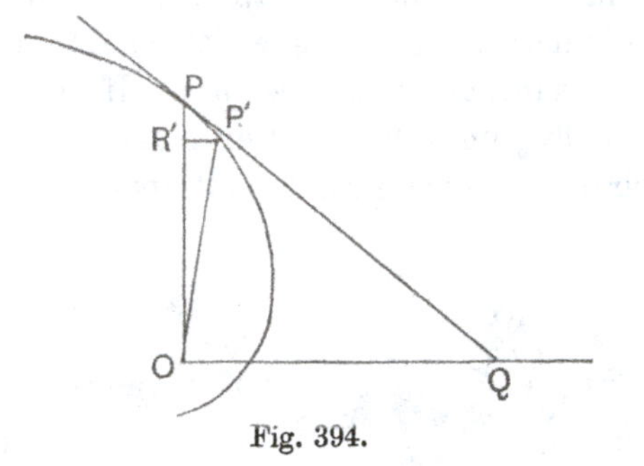

Fig. 394.

Accordingly, the ratio of l, the total length, to r, the radius vector up to which the total length is to be measured, is expressed by a simple table of secants; as follows:

α	l/r	α	l/r
5°	1·004	87°	19·1
10	1·015	88	28·7
20	1·064	89	57·3
30	1·165	89° 10′	68·8
40	1·305	20	85·9
50	1·56	30	114·6
60	2·0	40	171·9
70	2·9	50	343·8
75	3·9	55	687·5
80	5·8	59	3437·7
85	11·5	90	Infinite
86	14·3		

Putting the same table inversely, so as to shew the total length in terms of the radius, we have as follows:

Total length (in terms of the radius)	Constant angle
2	60°
3	70 31′
4	75 32
5	78 28
10	84 16
20	87 8
30	88 6
40	88 34
50	88 51
100	89 26
1000	89 56′ 36″
10,000	89 59 30

Accordingly, we see that (1), when the constant angle of the spiral is small, the shell (or for that matter the tooth, or horn or claw) is scarcely to be distinguished from a straight cone or cylinder; and this remains pretty much the case for a considerable increase of angle, say from 0° to 20° or more; (2) for a considerably greater increase of the constant angle, say to 50° or more, the shell would still only have the appearance of a gentle curve; (3) the characteristic close coils of the Nautilus or Ammonite would be typically represented only when the constant angle lies within a few degrees on either side of about 80°. The coiled up spiral of a Nautilus, with a constant angle of about 80°, is about six times the length of its radius vector, or rather more than three times its own diameter; while that of an Ammonite, with a constant angle of, say, from 85° to 88°, is from about six to fifteen times as long as its own diameter. And (4) as we approach an angle of 90° (at which point the spiral vanishes in a circle), the length of the coil increases with enormous rapidity. Our spiral would soon assume the appearance of the close coils of a Nummulite, and the successive increments of breadth in the successive whorls would become inappreciable to the eye.

The geometrical form of the shell involves many other beautiful properties, of great interest to the mathematician but which it is not possible to reduce to such simple expressions as we have been content to use. For instance, we may obtain an equation which shall express completely the surface of any shell, in terms of polar or of rectangular coordinates (as has been done by Moseley and by Blake), or in Hamiltonian vector notation*. It is likewise possible (though of little interest to the naturalist) to determine the area of a conchoidal surface or the volume of a conchoidal solid, and to find the centre of gravity of either surface or solid†. And Blake has further shewn, with considerable elaboration, how we may deal with the symmetrical distortion due to pressure which fossil shells are often found to have undergone, and how we may reconstitute by calculation their original undistorted form—a problem which, were the available methods only a little easier, would be

* Cf. H. W. L. Hime's *Outlines of Quaternions*, 1894, pp. 171–173.

† See Moseley, *op. cit.* p. 361 *seq.* Also, for more complete and elaborate treatment, Haton de la Goupillière, *op. cit.* 1908, pp. 5–46, 69–204.

very helpful to the palaeontologist; for, as Blake himself has shewn, it is easy to mistake a symmetrically distorted specimen of (for instance) an Ammonite for a new and distinct species of the same genus. But it is evident that to deal fully with the mathematical problems contained in, or suggested by, the spiral shell, would require a whole treatise, rather than a single chapter of this elementary book. Let us then, leaving mathematics aside, attempt to summarise, and perhaps to extend, what has been said about the general possibilities of form in this class of organisms.

The univalve shell: a summary

The surface of any shell, whether discoid or turbinate, may be imagined to be generated by the revolution about a fixed axis of a closed curve, which, remaining always geometrically similar to itself, increases its dimensions continually: and, since the scale of the figure increases in geometrical progression while the angle of rotation increases in arithmetical, and the centre of similitude remains fixed, the curve traced in space by corresponding points in the generating curve is, in all such cases, an equiangular spiral. In discoid shells, the generating figure revolves in a plane perpendicular to the axis, as in the Nautilus, the Argonaut and the Ammonite. In turbinate shells, it follows a skew path with respect to the axis of revolution, and the curve in space generated by any given point makes a constant angle to the axis of the enveloping cone, and partakes, therefore, of the character of a helix, as well as of a logarithmic spiral; it may be strictly entitled a helico-spiral. Such turbinate or helico-spiral shells include the snail, the periwinkle and all the common typical Gastropods.

When the envelope of the shell is a right cone—and it is seldom far from being so—then our helico-spiral is a loxodromic curve, and is obviously identical with a projection, parallel with the axis, of the logarithmic spiral of the base. As this spiral cuts all radii at a constant angle, so its orthogonal projection on the surface intersects all generatrices, and consequently all parallel circles, under a constant angle: this being the definition of a loxodromic curve on a surface of revolution. Guido Grandi describes this curve for the first time in a letter to Ceva, printed at the end of his *Demonstratio theorematum Hugenianorum circa...logarithmicam lineam*, 1701*.

* See R. C. Archibald, *op. cit.* 1918. Olivier discussed it again (*Rev. de géom. descriptive*, 1843) calling it a "conical equiangular" or "conical logarithmic"

The generating figure may be taken as any section of the shell, whether parallel, normal, or otherwise inclined to the axis. It is very commonly assumed to be identical with the mouth of the shell; in which case it is sometimes a plane curve of simple form; in other and more numerous cases, it becomes complicated in form and its boundaries do not lie in one plane: but in such cases as these we may replace it by its "trace," on a plane at some definite angle to the direction of growth, for instance by its form as it appears in a section through the axis of the helicoid shell. The generating curve is of very various shapes. It is circular in *Scalaria* or *Cyclostoma*, and in *Spirula*; it may be considered as a segment of a circle in *Natica* or in *Planorbis*. It is triangular in *Conus* or *Thatcheria*, and rhomboidal in *Solarium* or *Potamides*. It is very commonly more or less elliptical: the long axis of the ellipse being parallel to the axis of the shell in *Oliva* and *Cypraea*; all but perpendicular to it in many Trochi; and oblique to it in many well-marked cases, such as *Stomatella*, *Lamellaria*, *Sigaretus haliotoides* (Fig. 396) and *Haliotis*. In *Nautilus pompilius* it is approximately a semi-ellipse, and in *N. umbilicatus* rather more than a semi-ellipse, the long axis lying in both cases perpendicular to the axis of the shell*. Its form is seldom open to easy mathematical expression, save when it is an actual circle or

Fig. 395. Section of a spiral univalve, *Triton corrugatus* Lam. From Woodward.

spiral. Paul Serret (*Th. nouv...des lignes à double courbure*, 1860, p. 101) called it "*hélice cylindroconique*"; Haton de la Goupillière calls it a "*cônhélice*." It has also been studied by (*int. al.*) Tissot, *Nouv. ann. de mathém.* 1852; G. Pirondini, *Mathesis*, XIX, pp. 153–8, 1899; etc.

* In *Nautilus*, the "hood" has somewhat different dimensions in the two sexes, and these differences are impressed upon the shell, that is to say upon its "generating curve." The latter constitutes a somewhat broader ellipse in the male than in the female. But this difference is not to be detected in the young; in other words, the form of the generating curve perceptibly alters with advancing age. Somewhat similar differences in the shells of Ammonites were long ago suspected, by d'Orbigny, to be due to sexual differences. (Cf. Willey, *Natural Science*, VI, p. 411, 1895; *Zoological Results*, 1902, p. 742.)

ellipse; but an exception to this rule may be found in certain Ammonites, forming the group "Cordati," where (as Blake points out) the curve is very nearly represented by a cardioid, whose equation is $r = a\,(1 + \cos\theta)$.

When the generating curves of successive whorls cut one another, the line of intersection forms the conspicuous helico-spiral or loxodromic curve called the *suture* by conchologists.

The generating curve may grow slowly or quickly; its growth-factor is very slow in *Dentalium* or *Turritella*, very rapid in *Nerita*, or *Pileopsis*, or *Haliotis* or the Limpet. It may contain the axis in its plane, as in *Nautilus*; it may be parallel to the axis, as in the majority of Gastropods; or it may be inclined to the axis, as it is in a very marked degree in *Haliotis*. In fact, in *Haliotis* the generating

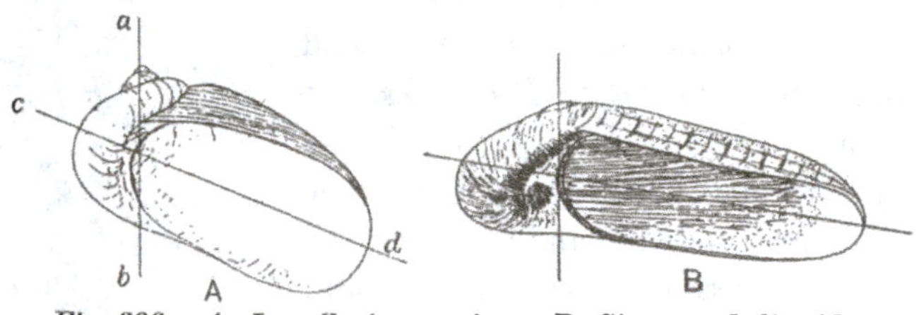

Fig. 396. A, *Lamellaria perspicua*; B, *Sigaretus haliotoides*. After Woodward.

curve is so oblique to the axis of the shell that the latter appears to grow by additions to one margin only (cf. Fig. 362), as in the case of the opercula of *Turbo* and *Nerita* referred to on p. 775; and this is what Moseley supposed it to do.

The general appearance of the entire shell is determined (apart from the form of its generating curve) by the magnitude of three angles; and these in turn are determined, as has been sufficiently explained, by the ratios of certain velocities of growth. These angles are (1) the constant angle of the equiangular spiral (α); (2) in turbinate shells, the enveloping angle of the cone, or (taking half that angle) the angle (β) which a tangent to the whorls makes with the axis of the shell; and (3) an angle called the "angle of retardation" (γ), which expresses the retardation in growth of the inner as compared with the outer part of each whorl, and therefore measures the extent to which one whorl overlaps, or the extent to which it is separated from, another.

The spiral angle (α) is very small in a limpet, where it is usually taken as $= 0°$; but it is evidently of a significant amount, though obscured by the shortness of the tubular shell. In *Dentalium* it is still small, but sufficient to give the appearance of a regular curve; it amounts here probably to about 30° to 40°. In *Haliotis* it is from about 70° to 75°; in *Nautilus* about 80°; and it lies between 80° and 85° or even more, in the majority of Gastropods*.

The case of *Fissurella* is curious. Here we have, apparently, a conical shell with no trace of spiral curvature, or (in other words) with a spiral angle which approximates to 0°; but in the minute embryonic shell (as in that of the limpet) a spiral convolution is distinctly to be seen. It would seem, then, that what we have to do with here is an unusually large growth-factor in the generating curve, causing the shell to dilate into a cone of very wide angle, the apical portion of which has become lost or absorbed, and the remaining part of which is too short to show clearly its intrinsic curvature. In the closely allied *Emarginula*, there is likewise a well-marked spiral in the embryo, which however is still manifested in the curvature of the adult, nearly conical, shell. In both cases we have to do with a very wide-angled cone, and with a high retardation-factor for its inner, or posterior, border. The series is continued, from the apparently simple cone to the complete spiral, through such forms as *Calyptraea*.

The angle α, as we have seen, is not always, nor rigorously, a constant angle. In some Ammonites it may increase with age, the whorls becoming closer and closer; in others it may decrease rapidly and even fall to zero, the coiled shell then straightening out, as in *Lituites* and similar forms. It diminishes somewhat, also, in many Orthocerata, which are slightly curved in youth but straight in age. It tends to increase notably in some common land-shells, the *Pupae* and *Bulimi*; and it decreases in *Succinea*.

Directly related to the angle α is the ratio which subsists between the breadths of successive whorls. The following table gives a few

* What is sometimes called, as by Leslie, the *angle of deflection* is the complement of what we have called the *spiral angle* (α), or obliquity of the spiral. When the angle of deflection is 6° 17′ 41″, or the spiral angle 83° 42′ 19″, the radiants, or breadths of successive whorls, are doubled at each entire circuit.

illustrations of this ratio in particular cases, in addition to those which we have already studied.

Ratio of breadth of consecutive whorls

Pointed Turbinates		Obtuse Turbinates and Discoids	
Telescopium fuscum ...	1·14	*Conus virgo*	1·25
Terebra subulata	1·16	‡*Clymenia laevigata*	1·33
**Turritella terebellata* ...	1·18	*Conus litteratus*	1·40
**Turritella imbricata*	1·20	*Conus betulinus*	1·43
Cerithium palustre	1·22	‡*Clymenia arietina*	1·50
Turritella duplicata	1·23	‡*Goniatites bifer*	1·50
Melanopsis terebralis ...	1·23	**Helix nemoralis*	1·50
Cerithium nodulosum ...	1·24	**Solarium perspectivum* ...	1·50
**Turritella carinata*	1·25	*Solarium trochleare* ...	1·62
Terebra crenulata	1·25	*Solarium magnificum* ...	1·75
Terebra maculata (Fig. 397)	1·25	**Natica aperta*	2·00
**Cerithium lignitarum* ...	1·26	*Euomphalus pentangulatus*	2·00
Terebra dimidiata	1·28	*Planorbis corneus*	2·00
Cerithium sulcatum	1·32	*Solaropsis pellis-serpent* ...	2·00
Fusus longissimus	1·34	*Dolium zonatum*	2·10
**Pleurotomaria conoidea* ...	1·34	‡*Goniatites carinatus* ...	2·50
Trochus niloticus (Fig. 398)	1·41	**Natica glaucina*	3·00
Mitra episcopalis	1·43	*Nautilus pompilius* ...	3·00
Fusus antiquus	1·50	*Haliotis excavatus*	4·20
Scalaria pretiosa	1·56	*Haliotis parvus*	6·00
Fusus colosseus	1·71	*Delphinula atrata*	6·00
Phasianella australis ...	1·80	*Haliotis rugoso-plicata* ...	9·30
Helicostyla polychroa ...	2·00	*Haliotis viridis*	10·00

Those marked * from Naumann; ‡ from Müller; the rest from Macalister†.

In the case of turbinate shells, we must take into account the angle β, in order to determine the spiral angle α from the ratio of the breadths of consecutive whorls; for the short table given on p. 791 is only applicable to discoid shells, in which the angle β is an angle of 90°. Our formula, as mentioned on p. 771, now becomes

$$R = \epsilon^{2\pi \sin \beta \cot \alpha}.$$

For this formula I have worked out the following table.

† Alex. Macalister, Observations on the mode of growth of discoid and turbinated shells, *Proc. R.S.* XVIII, pp. 529–532, 1870; *Ann. Mag. N.H.* (6), IV, p 160, 1870. Cf. also his Law of Symmetry as exemplified in animal form, *Journ. R. Dublin Soc.* 1869, p. 327.

Table shewing values of the spiral angle α corresponding to certain ratios of breadth of successive whorls of the shell, for various values of the apical semi-angle β

Ratio $R/1$	$\beta=5°$	10°	15°	20°	30°	40°	50°	60°	70°	80°	90°
1·1	80° 8′	85° 0′	86° 44′	87° 28′	88° 16′	88° 39′	88° 52′	89° 0′	89° 4′	89° 7′	89° 8′
1·25	67 51	78 27	82 11	84 5	85 56	86 50	87 21	87 39	87 50	87 56	87 58
1·5	53 30	69 37	76 0	79 21	82 39	84 16	85 13	85 44	86 4	86 15	86 18
2·0	38 20	57 35	66 55	73 11	77 34	80 16	81 52	82 45	83 18	83 37	83 42
2·5	30 53	50 0	60 35	67 0	73 45	77 13	79 19	80 26	81 11	81 35	81 42
3·0	26 32	44 50	56 0	63 0	70 45	74 45	77 17	78 35	79 28	79 56	80 5
3·5	23 37	41 5	52 25	59 50	68 15	72 45	75 35	77 2	78 1	78 33	78 43
4·0	21 35	38 10	49 35	57 15	66 10	71 3	74 9	75 42	76 47	77 22	77 34
4·5	20 0	36 0	47 15	55 5	64 25	69 35	72 54	74 33	75 43	76 20	76 35
5·0	18 45	34 10	45 20	53 15	62 55	68 15	71 48	73 31	74 45	75 25	75 38
10·0	13 25	25 20	35 15	43 5	53 45	60 20	64 57	67 4	68 42	69 35	69 53
20·0	10 25	20 0	28 30	35 45	46 25	53 25	58 52	61 10	63 6	64 10	64 31
50·0	8 0	15 35	22 35	28 50	38 45	45 55	52 1	54 18	56 28	57 42	58 6
100·0	6 50	13 20	19 30	25 5	34 20	41 15	47 35	49 45	52 3	53 20	53 46

From this table, by interpolation, we may easily fill in the approximate values of α, as soon as we have determined the apical angle β and measured the ratio R; as follows:

	R	β	α
Turritella sp.	1·12	7°	81°
Cerithium nodulosum ...	1·24	15	82
Conus virgo	1·25	70	88
Mitra episcopalis	1·43	16	78
Scalaria pretiosa	1·56	26	81
Phasianella australis ...	1·80	26	80
Solarium perspectivum ...	1·50	53	85
Natica aperta	2·00	70	83
Planorbis corneus ...	2·00	90	84
Euomphalus pentangulatus	2·00	90	84

We see from this that shells so different in appearance as *Cerithium*, *Solarium*, *Natica* and *Planorbis* differ very little indeed in the magnitude of the spiral angle α, that is to say in the relative velocities of radial and tangential growth. It is upon the angle β that the difference in their form mainly depends.

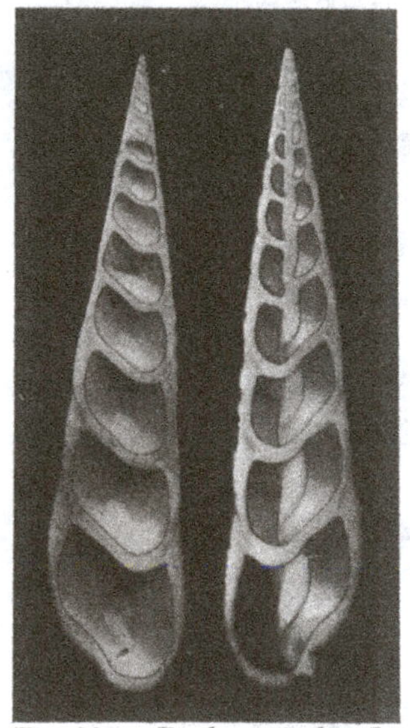

Fig. 397. *Terebra maculata* L.

The angle, or rather semi-angle (β), of the tangent cone may be taken as 90° in the discoid shells, such as *Nautilus* and *Planorbis*. It is still a large angle, of 70° or 75°, in *Conus* or in *Cymba*, somewhat less in *Cassis*, *Harpa*, *Dolium* or *Natica*; it is about 50° to 55° in the various species of *Solarium*, about 35° in the typical *Trochi*, such as *T. niloticus* or *T. zizyphinus*, and about 25° or 26° in *Scalaria pretiosa* and *Phasianella bulloides*; it becomes a very acute angle, of 15°, 10°, or even less, in *Eulima*, *Turritella* or *Cerithium*. The British species of '*Fusus*' form a series in which the apical angle ranges from about 28° in *F. antiquus*, through *F. Norvegicus*, *F. berniciensis*, *F. Turtoni*, *F. Islandicus*, to about 17° in *F. gracilis*. It varies much among the Cones; and the costly *Conus gloria-maris*, one of the great treasures of the conchologist, differs from its congeners in no important particular save in the somewhat "produced" spire, that is to say in the comparatively low value of the angle β.

A variation with advancing age of β is common, but (as Blake points out) it is often not to be distinguished or disentangled from an alteration of α. Whether alone, or combined with a change in α, we find it in all those many gastropods whose whorls cannot all be touched by the same enveloping cone, and whose spire is accordingly described as *concave* or *convex*. The former condition, as we have

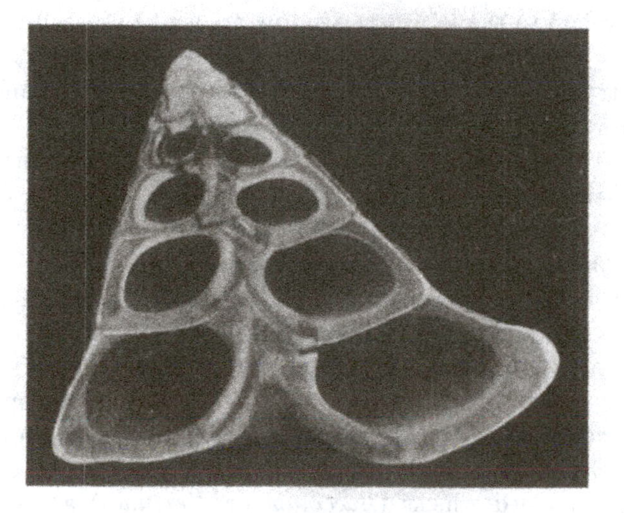

Fig. 398. *Trochus niloticus* L.

it in *Cerithium*, and in the cusp-like spire of *Cassis*, *Dolium* and some Cones, is much the commoner of the two*.

In the vast majority of spiral univalves the shell winds to the right, or turns clockwise, as we look along it in the direction in which the animal crawls and puts out its head. The thread of a carpenter's screw (except in China) runs the same way, and we call it a "right-handed screw." Save that it takes a right-handed movement to

* Many measurements of the linear dimensions of univalve shells have been made of late years, and studied by statistical methods in order to detect local races and other instances of variation and variability. But conchological statisticians seem to be content with some arbitrary linear ratio as a measure of "squatness" or the reverse; and the measurements chosen give little or no help towards the determination either of the apical or of the spiral angle. Cf. (e.g.) A. E. Boycott, Conchometry, *Proc. Malacol. Soc.* XVII, p. 8, 1928; C. Price-Jones, *ibid.* XIX, p. 146, 1930; etc. See also G. Duncker, Methode der Variations-Statistik, *Arch. f. Entw. Mech.* VIII, pp. 112–183, 1899.

drive in a "right-handed" screw, the terms right-handed and left-handed are purely conventional; and the mathematicians and the naturalists, unfortunately, use them in opposite ways. Thus the mathematicians call the snail-shell or the joiner's screw *leiotropic*; and Listing for one has much to say about lack of precision or even confusion on the part of the conchologists and the botanists, from Linnaeus downwards, in their attempts to deal with right-handed and left-handed spirals or screws*. The convolvulus twines to the right, the hop to the left; vine-tendrils are said to be mostly right-handed. At any rate, Clerk Maxwell spoke of *hop-spirals* and *vine-spirals*, trying to avoid the confusion or ambiguity of left and right. Some climbing plants are one and some the other; and the architect shews little preference, but builds his spiral staircases or twisted columns either way. But in all these, shells and all, the spiral runs *one way*; it is *isotropic*, while the fir-cone shews spirals running both ways at once, and we call them *heterotropic*, or *diadromic*.

When we find a "reversed shell," a whelk or a snail winding the wrong way, we describe it mathematically by the simple statement that the apical angle (β) has changed sign. Such left-handed shells occur as a well-known but rare abnormality; and the men who handle snails in the Paris market or whelks in Billingsgate keep a sharp look-out for them. In rare instances they become common. While left-handed whelks (*Buccinum* or *Neptunea*) are very rare nowadays, it was otherwise in the epoch of the Red Crag; for *Neptunea* was then extremely common, but right-handed specimens were as rare as left-handed are today. In the beautiful genus *Ampullaria*, or apple-snails, which inhabit tropical and sub-tropical rivers, there is unusual diversity; for the spire turns to the right in some species, and to the left in others, and again some are flat or "discoid," with no spire at all; and there are plenty of half-way stages, with right and left-handed spires of varying steepness or acuteness†; in short, within the limits of this singular genus the apical angle (β) may vary from about $\pm 35°$ to $\pm 125°$. But we need not imagine that the direction of growth actually changes over from right-handed to left-handed; it is enough to suppose

* See Listing's *Topologie*, p. 36; and cf. Clerk Maxwell's *Electricity and Magnetism*, I, p. 24.

† See figures in Arnold Lang's *Comparative Anatomy* (English translation), II, p. 161, 1902.

that the skew movement along the axis has changed its direction. For if I take a roll of tape and push the core out to one side or to the other, or if I keep the centre of the roll fixed and push the rim to the one side or to the other, I thereby convert the flat roll into a hollow cone, or (in other words) a plane into a gauche spiral. Whether we push one way or other, whether the spiral coil be plane or gauche, positively or negatively deformed, it remains right-handed or left-handed as the case may be; but it does change its direction as soon as we *turn it upside down*, or as soon as the animal does so in assuming its natural attitude. The linear spirals within and without the cone may change places but must remain congruent with one another; for they are merely the two edges of the ribbon, and as such are inseparable and identical twins. But of the shell itself we may reasonably say that a right-handed has given place to a left-handed spiral. Of these, the one is a mirror-image of the other; and the passing from one to the other through the plane of symmetry (which has no "handedness") is an operation which Listing called *perversion*. The flat or discoid apple-snails are like our roll of tape, which can be *converted* into a conical spire and *perverted* in one direction or the other; and in this genus, by a rare exception, it seems wellnigh as easy to depart one way as the other from the plane of symmetry. But why, in the general run of shells, all the world over, in the past and in the present, one direction of twist is so overwhelmingly commoner than the other, no man knows.

The phenomenon of reversal, or "sinistrality," has an interest of its own from the side of development and heredity. For careful study of certain pond-snails has shewn that dextral and sinistral varieties appear, not one by one, but by whole broods of the one sort or the other; a discovery which goes some way to account for the predominant left-handedness of *Fusus ambiguus* in the Red Crag. The right-handed, or ordinary form, is found to be "dominant" to the other; but the Mendelian heredity is of a curious and complicated kind. For the direction of the twist appears to be predetermined in the germ even prior to its fertilisation; and a left-handed pond-snail will produce a brood of left-handed young even when fertilised by a normal, or right-handed, individual*.

* See A. E. Boycott and others, Abnormal forms of *Limnaea peregra*...and their inheritance, *Phil. Trans.* (B), CCXXIX, p. 51, 1930; and other papers.

The angle of retardation (γ) is very small in *Dentalium* and *Patella*; it is very large in *Haliotis*; it becomes infinite in *Argonauta* and in *Cypraea*. Connected with the angle of retardation are the various possibilities of contact or separation, in various degrees, between adjacent whorls in the discoid shell, and between both adjacent and opposite whorls in the turbinate. But with these phenomena we have already dealt sufficiently.

The beautiful shell of the paper-nautilus (*Argonauta argo* L.) differs in sundry ways both from the Nautilus and from ordinary univalves. Only the female Argonaut possesses it; it is not attached to its owner, but is (so to speak) worn loose; it is rather a temporary cradle for the young than a true shell or bodily covering; and it is not secreted in the usual way, but is plastered on from the outside by two of the eight arms of the little Octopus to which it belongs. The shell shews a single whorl, or but little more; and the spiral is hard to measure, for this reason. It has been supposed by some to obey a law other than the logarithmic spiral. For my part I have made no special study of it, nor has any one else, to my knowledge, of recent years; but the simple fact that it conserves its shape as it grows, or that each increment is a gnomon to the rest, is enough to shew that this delicate and beautiful shell is mathematically, though not morphologically, homologous with all the others.

Of bivalve shells

Hitherto we have dealt only with univalve shells, and it is in these that all the mathematical problems connected with the spiral, or helico-spiral, configuration are best illustrated. But the case of the bivalve shell, whether of the lamellibranch or the brachiopod, presents no essential difference, save only that we have here to do with two conjugate spirals, whose two axes have a definite relation to one another, and some independent freedom of rotatory movement relatively to one another.

The bivalve or lamellibranch mollusca are very different creatures from the rest. The univalves or gastropods, like their cousins the cephalopods, go about their business and get their living in an ordinary way; but the bivalves are unintelligent, "acephalous" animals, and imbibe the invisible plankton-food which ciliary currents bring automatically to their mouths. There is something

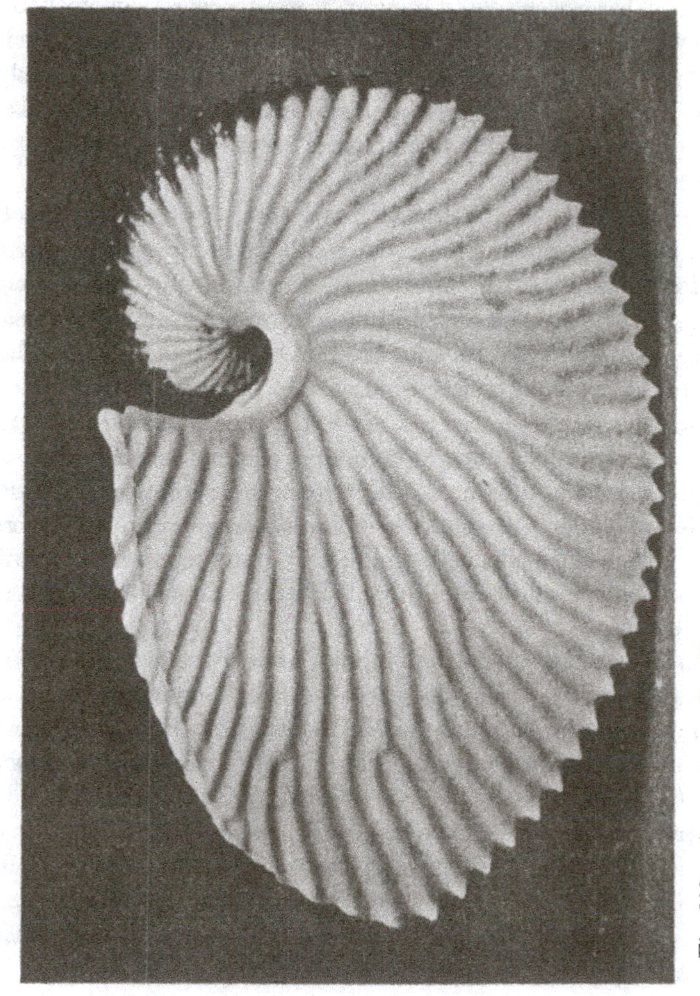

Fig. 399. *Argonauta argo* L. The paper-nautilus. From Cooke's *Spirals in Nature and Art.*

to be said for withdrawing them, as brachiopods and others have been withdrawn, from Cuvier's great class of the Mollusca. But whether bivalves and univalves be near relations or no is not the question. Both of them secrete a shell, and in both the shell grows by the successive addition of similar parts, gnomon after gnomon; so that in both the equiangular spiral makes, and is bound to make, its appearance. There is a mathematical analogy between the two; but it has no more bearing on zoological classification than has the still closer likeness between Nautilus and the nautiloid Foraminifera.

The generating curve is particularly well seen in the bivalve, where it simply constitutes what we call "the outline of the shell." It is for the most part a plane curve, but not always; for there are forms such as *Hippopus*, *Tridacna* and many Cockles, or *Rhynchonella* and *Spirifer* among the Brachiopods, in which the edges of the two valves interlock, and others, such as *Pholas*, *Mya*, etc., where they gape asunder. In such cases as these the generating curves, though not plane, are still conjugate, having a similar relation, but of opposite sign, to a median plane of reference or of projection. There are a few exceptional cases, e.g. *Arca* (*Parallelepipedon*) *tortuosa*, where there is no median plane of symmetry, but the generating curve, and therefore the outline of the shell itself, is a tortuous curve in three dimensions.

A great variety of form is exhibited among the bivalves by these generating curves. In many cases the curve or outline is all but circular, as in *Anomia*, *Sphaerium*, *Artemis*, *Isocardia*; it is nearly semicircular in *Argiope*; it is approximately elliptical in *Anodon*, *Lutraria*, *Orthis*; it may be called semi-elliptical in *Spirifer*; it is a nearly rectilinear triangle in *Lithocardium*, and a curvilinear triangle in *Mactra*. Many apparently diverse but more or less related forms may be shewn to be deformations of a common type, by a simple application of the mathematical theory of "transformations," which we shall have to study in a later chapter. In such a series as is furnished, for instance, by *Gervillea*, *Perna*, *Avicula*, *Modiola*, *Mytilus*, etc., a "simple shear" accounts for most, if not all, of the apparent differences.

Upon the surface of the bivalve shell we usually see with great clearness the "lines of growth" which represent the successive margins of the shell, or in other words the successive positions assumed during growth by the growing generating curve; and we have a good illustration, accordingly, of how it is characteristic of the generating curve that it should constantly increase, while never altering its geometric similarity.

Underlying these lines of growth, which are so characteristic of a molluscan shell (and of not a few other organic formations), there is, then, a law of growth which we may attempt to enquire into and which may be illustrated in various ways. The simplest cases are those in which we can study the lines of growth on a more or less flattened shell, such as the one valve of an oyster, a *Pecten* or a *Tellina*, or some such bivalve mollusc. Here around an origin, the so-called "umbo" of the shell, we have a series of curves, sometimes nearly circular, sometimes elliptical, often asymmetrical; and such curves are obviously not "concentric," though we are often apt to call them so, but have a common centre of similitude. This arrangement may be illustrated by various analogies. We might for instance compare it to a series of waves, radiating outwards from a point, through a medium which offered a resistance increasing, with the angle of divergence, according to some simple law. We may find another and perhaps a simpler illustration as follows:

In a simple and beautiful theorem, Galileo shewed that, if we imagine a number of inclined planes, or gutters, sloping downwards (in a vertical plane) at various angles from a common starting-point, and if we imagine a number of balls rolling each down its own gutter under the influence of gravity (and without hindrance from friction), then, at any given instant, the locus of all these moving bodies is a circle passing through the point of origin. For the acceleration along any one of the sloping paths, for instance AB (Fig. 400), is such that

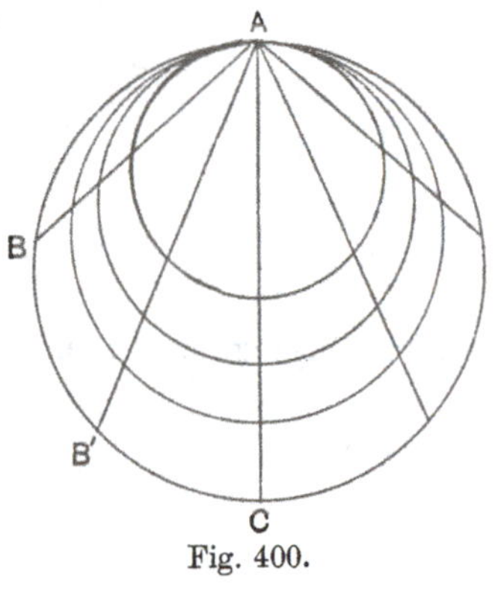

Fig. 400.

$$AB = \tfrac{1}{2}g\cos\theta\,.\,t^2$$
$$= \tfrac{1}{2}g\,.\,AB/AC\,.\,t^2.$$

Therefore $$t^2 = 2/g\,.\,AC.$$

That is to say, all the balls reach the circumference of the circle at the same moment as the ball which drops vertically from A to C.

Where, then, as often happens, the generating curve of the shell is approximately a circle passing through the point of origin, we may consider the acceleration of growth along various radiants to be governed by a simple mathematical law, closely akin to that simple law of acceleration which governs the movements of a falling body. And, *mutatis mutandis*, a similar definite law underlies the cases where the generating curve is continually elliptical, or where it assumes some more complex, but still regular and constant form.

It is easy to extend the proposition to the particular case where the lines of growth may be considered elliptical. In such a case we have $x^2/a^2 + y^2/b^2 = 1$, where a and b are the major and minor axes of the ellipse.

Or, changing the origin to the vertex of the figure,

$$\frac{x^2}{a^2} - \frac{2x}{a} + \frac{y^2}{b^2} = 0, \quad \text{giving} \quad \frac{(x-a)^2}{a^2} + \frac{y^2}{b^2} = 1.$$

Then, transferring to polar coordinates, where $r . \cos\theta = x$, $r . \sin\theta = y$, we have

$$\frac{r . \cos^2\theta}{a^2} - \frac{2\cos\theta}{a} + \frac{r . \sin^2\theta}{b^2} = 0,$$

which is equivalent to

$$r = \frac{2ab^2\cos\theta}{b^2\cos^2\theta + a^2\sin^2\theta},$$

or, simplifying, by eliminating the sine-function,

$$r = \frac{2ab^2\cos\theta}{(b^2 - a^2)\cos^2\theta + a^2}.$$

Obviously, in the case when $a = b$, this gives us the circular system which we have already considered. For other values, or ratios, of a and b, and for all values of θ, we can easily construct a table, of which the following is a sample:

Chords of an ellipse, whose major and minor axes (a, b) *are in certain given ratios*

θ	$a/b = 1/3$	1/2	2/3	1/1	3/2	2/1	3/1
0°	1·0	1·0	1·0	1·0	1·0	1·0	1·0
10	1·01	1·01	1·002	0·985	0·948	0·902	0·793
20	1·05	1·03	1·005	0·940	0·820	0·695	0·485
30	1·115	1·065	1·005	0·866	0·666	0·495	0·289
40	1·21	1·11	0·995	0·766	0·505	0·342	0·178
50	1·34	1·145	0·952	0·643	0·372	0·232	0·113
60	1·50	1·142	0·857	0·500	0·258	0·152	0·071
70	1·59	1·015	0·670	0·342	0·163	0·092	0·042
80	1·235	0·635	0·375	0·174	0·078	0·045	0·020
90	0·0	0·0	0·0	0·0	0·0	0·0	0·0

The ellipses which we then draw, from the values given in the table, are such as are shewn in Fig. 401 for the ratio $a/b = \frac{3}{1}$, and in Fig. 402 for the ratio $a/b = \frac{1}{2}$; these are fair approximations to the actual outlines, and to the actual arrangement of the lines of growth, in such forms as *Solecurtus* or *Cultellus*, and in *Tellina* or *Psammobia*. It is not difficult to introduce a constant into our equation to meet the case of a shell which is somewhat unsymmetrical on either side of the median axis. It is a somewhat more troublesome matter, however, to bring these configurations into relation with a "law of growth," as was so easily done in the case of the circular figure: in other words, to formulate a law of acceleration according to which points starting from the origin O, and moving along radial lines, would all lie, at any future epoch, on an ellipse passing through O; and this calculation we need not enter into.

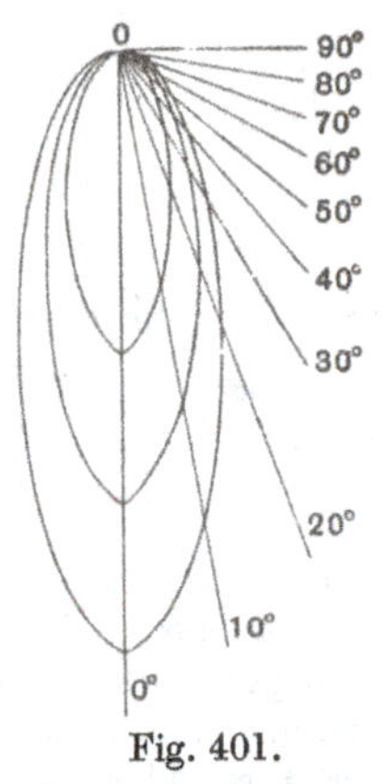

Fig. 401.

All that we are immediately concerned with is the simple fact that where a velocity, such as our rate of growth, varies with its direction—varies that is to say as a function of the angular divergence from a certain axis—then, in a certain simple case, we get lines of growth laid down as a system of coaxial circles, and, in somewhat less simple cases, we obtain a system of ellipses or of other more complicated coaxial figures, which may or may not be symmetrical on either side of the axis. Among our bivalve mollusca we shall find the lines of growth to be approximately circular in, for instance, *Anomia*; in *Lima* (e.g. *L. subauriculata*) we have a system of nearly symmetrical ellipses with the vertical axis about twice the transverse; in *Solen pellucidus*, we have again a system of lines of growth which are not far from being symmetrical ellipses,

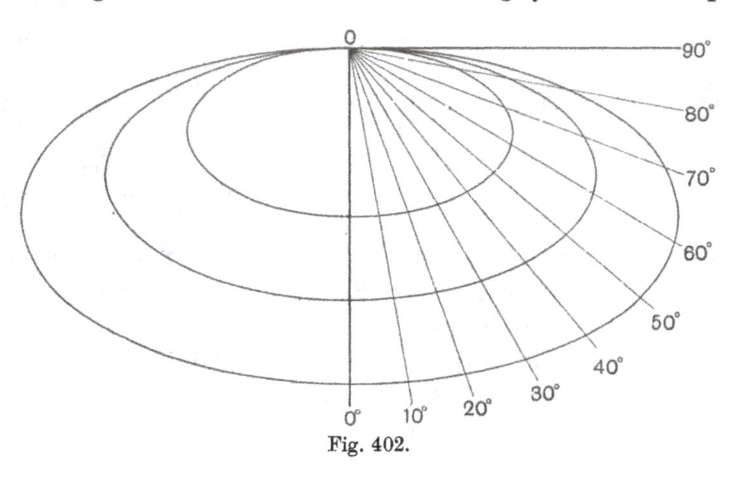

Fig. 402.

in which however the transverse is between three and four times as great as the vertical axis. In the great majority of cases, we have a similar phenomenon with the further complication of slight, but occasionally very considerable, lateral asymmetry.

In the above account of the mathematical form of the bivalve shell, we have supposed, for simplicity's sake, that the pole or origin of the system is at a point where all the successive curves touch one another. But such an arrangement is neither theoretically probable, nor is it actually the case; for it would mean that in a certain direction growth fell, not merely to a minimum, but to zero. As a matter of fact, the centre of the system (the "umbo" of the conchologists) lies not at the edge of the system, but very near to it; in other words, there is a certain amount of growth all round. But to take account of this condition would involve more troublesome mathematics, and it is obvious that the foregoing illustrations are a sufficiently near approximation to the actual case.

In certain little Crustacea (of the genus *Estheria*) the carapace takes the form of a bivalve shell, closely simulating that of a lamellibranchiate mollusc, and bearing lines of growth in all respects analogous to or even identical with those of the latter. The explanation is very curious and interesting. In ordinary Crustacea the carapace, like the rest of the chitinised and calcified integument, is shed off in successive moults, and is restored again as a whole. But in *Estheria* (and one or two other small crustacea) the moult is incomplete: the old carapace is retained, and the new, growing up underneath it, adheres to it like a lining, and projects beyond its edge: so that in course of time the margins of successive old carapaces appear as "lines of growth" upon the surface of the shell. In this mode of formation, then (but not in the usual one), we obtain a structure which "is partly old and partly new," and whose successive increments are all similar, similarly situated, and enlarged

Fig. 403. *Hemicardium inversum* Lam. From Chenu.

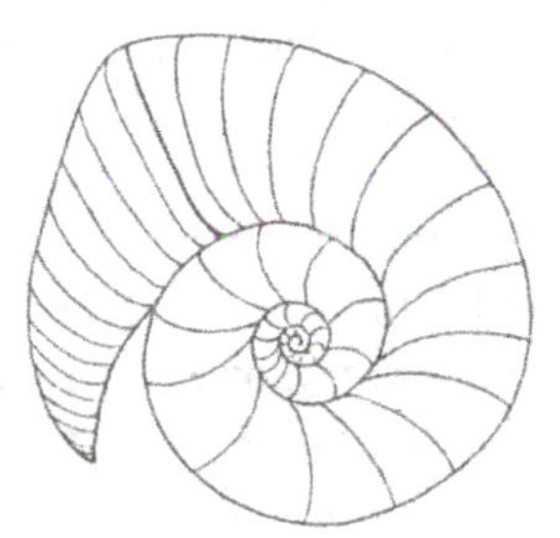

Fig. 404. *Caprinella adversa.* After Woodward.

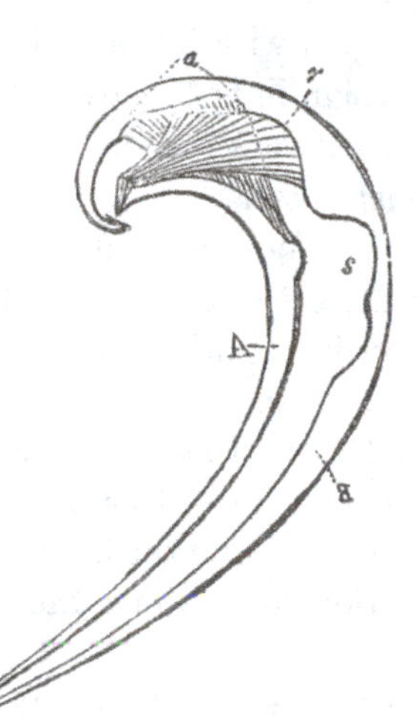

Fig. 405. Section of *Productus* (*Strophonema*) sp. From Woods.

in a continued progression. We have, in short, all the conditions appropriate and necessary for the development of a logarithmic spiral; and this logarithmic spiral (though it is one of small angle) gives its own character to the structure, and causes the little carapace to partake of the characteristic conformation of the molluscan shell.

Among the bivalves the spiral angle (α) is very small in the flattened shells, such as *Orthis*, *Lingula* or *Anomia*. It is larger, as a rule, in the Lamellibranchs than in the Brachiopods, but in the latter it is of considerable magnitude among the Pentameri.

Among the Lamellibranchs it is largest in such forms as *Isocardia* and *Diceras*, and in the very curious genus *Caprinella*; in all of these last-named genera its magnitude leads to the production of a spiral shell of several whorls, precisely as in the univalves. The angle is usually equal, but of opposite sign, in the two valves of the Lamellibranch, and usually of opposite sign but unequal in the two valves of the Brachiopod. It is very unequal in many Ostreidae, and especially in such forms as *Gryphaea*, or in *Caprinella*, which is a kind of exaggerated *Gryphaea*; in the cretaceous genus *Requienia*, the two valves of the shell closely resemble a turbinate gastropod with its flat calcified operculum. Occasionally it is of the same sign in both valves (that is to say, both valves curve the same way) as we see sometimes in *Anomia*, and better in *Productus* or *Strophonema*.

It will be observed, and it may not be difficult to explain, that the more the bivalve shell curves in the one direction the more it curves in the other; each valve tends to be spheroidal, or ellipsoidal, rather than cylindroidal. The cylindroidal form occurs, exceptionally, in *Solen*. But *Pecten*, *Gryphaea*, *Terebratula* are all cases of bivalve shells where one valve is flat and the other curved from *side to side*; and the flat valve tends to remain flat in the longitudinal direction also, while the curved valve grows into its logarithmic spiral.

In the genus *Gryphaea*, an oyster-like bivalve from the Jurassic, the creature lay on its side with its left valve downward, as oysters and scallops also do; and this valve adhered to the ground while the animal was young. The upper valve stays flat, and looks like a mere operculum; but the lower or deep valve grows into a more or less pronounced spiral. So is it also in the neighbouring genus *Pecten*, where *P. Jacobaeus* has its under-valve much deeper and more curved than, say, *P. opercularis*; but *Gryphaea incurva* is more spirally curved than any of these, and *G. arcuata* has a spiral angle very near to that of *Nautilus* itself. In both the spiral is a typical equiangular one, built up of a succession of gnomonic increments, which in turn depend on a constant ratio between the expansion of a generating figure and its rotation about a centre of similitude. *Rate of growth* is at the root of the whole matter. Now *Gryphaea*, like some Ammonites of which we spoke before, is one of those cases in which not only does the form of the shell vary, but geologists recognise, now and then, a *trend*, or progressive sequence of variation, from one stratum or one "horizon" to another. In short, *as time goes on*, we seem to see the shell growing thicker or wider, or more and more spirally curved, before our eyes. What meaning shall we give, what importance should we assign, to these changes, and what sort or grade of evolution do they imply? Some hold that these palaeontological features are "strictly comparable with those on which the geneticist bases his factorial studies"; and that as such they may shew "linkage of characters," as when "in the evolution of *Gryphaea* the area of attachment retrogresses as the arching progresses"*. These are debatable matters. But in so far as the changes depend on mere gradations of magnitude, they lead indeed to variety but fall short of the full concept of evolution. For to quote Aristotle once again (though we need not go to Aristotle to learn it): "some things shew increase but suffer no alteration; because increase is one thing and alteration is another."

The so-called "spiral arms" of *Spirifer* and many other Brachiopods are not difficult to explain. They begin as a single structure, in the form of a loop of shelly substance, attached to the dorsal valve of the shell, in the neighbourhood of the hinge, and forming a skeletal support for two ciliate and tentaculate arms. These grow to a considerable length, coiling up within the shell that they may do so. In *Terebratula* the loop remains short and simple, and is merely flattened and distorted somewhat by the restraining pressure of the ventral valve; but in *Spirifer*, *Atrypa*, *Athyris* and many more it forms a watchspring coil on either side, corresponding to the close-coiled arms of which it was the support and skeleton. In these curious and characteristic structures we see no sign of progressive

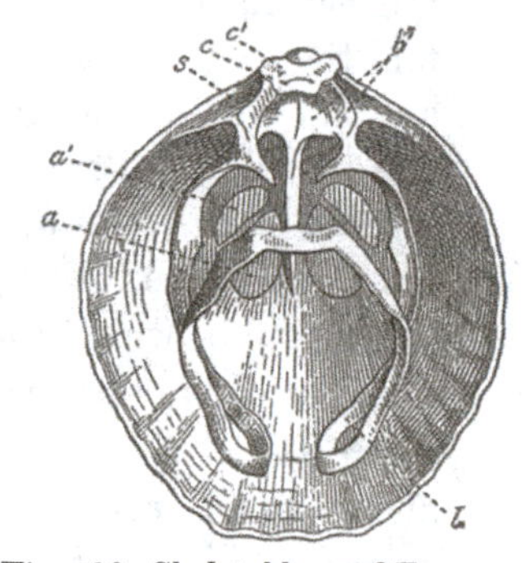

Fig. 406. Skeletal loop of *Terebratula*. From Woods.

* H. H. Swinnerton, Unit characters in fossils, *Biol. Reviews*, VII, pp. 321–335, 1932; cf. A. E. Truman, *Geol. Mag.* LIX, p. 258, LXI, p. 358, 1922–24.

growth, no successional increments, no "gnomons," no self-similarity in the figure. In short it has nothing to do with a logarithmic or equiangular spiral, but is a mere twist, or tapering helix, and it points now one way, now another. The cases in which the helicoid spires point towards, or point away from, the middle line are ascribed, in zoological classification, to particular "families" of Brachiopods, the former condition defining (or helping to define) the Atrypidae and the latter the Spiriferidae

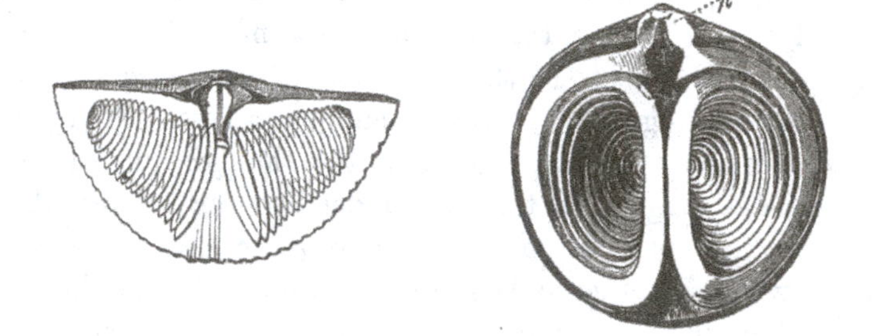

and Athyridae. It is obvious that the incipient curvature of the arms, and consequently the form and direction of the spirals, will be influenced by the surrounding pressures, and these in turn by the general shape of the shell. We shall expect, accordingly, to find the long outwardly directed spirals associated with shells which are transversely elongated, as *Spirifer* is; while the more rounded *Atrypa* will tend to the opposite condition. In a few cases, as in *Cyrtina* or *Reticularia*, where the shell is comparatively narrow but long, and where the uncoiled basal support of the arms is long also, the coils into which the latter grow are turned backwards, in the direction where there is most room for them. And in the few cases where the shell is very considerably flattened, the spirals (if they find room to grow at all) will be constrained to do so in a discoid or nearly discoid fashion, and this is actually the case in such flattened forms as *Koninckina* or *Thecidium*.

The shells of Pteropods

While mathematically speaking we are entitled to look upon the bivalve shell of the Lamellibranch as consisting of two distinct elements, each comparable to the entire shell of the univalve, we have no biological grounds for such a statement; for the shell arises from a single embryonic origin, and afterwards becomes split into portions which constitute the two separate valves. We can perhaps throw some indirect light upon this phenomenon, and upon several other phenomena connected with shell-growth, by a consideration of the simple conical or tubular shells of the Pteropods. The shells of the latter are in few cases suitable for simple mathematical investigation, but nevertheless they are of very considerable interest in connection with our general problem. The morphology of the Pteropods is by no means well understood, and in speaking of them

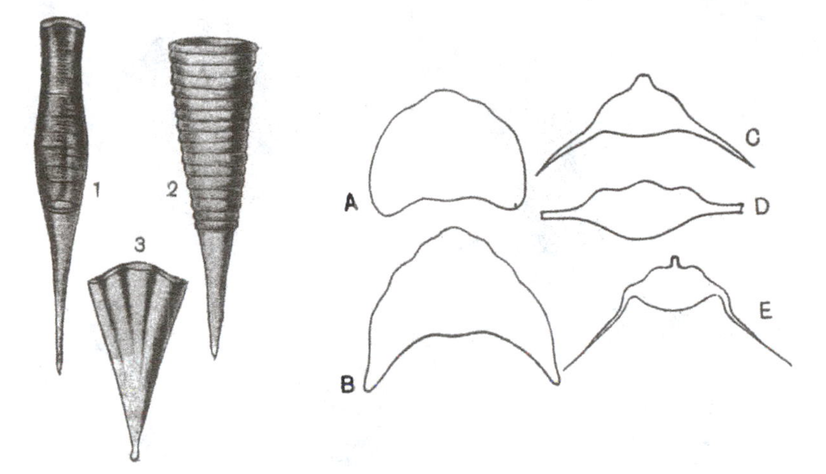

I will assume that there are still grounds for believing (in spite of Boas' and Pelseneer's arguments) that they are directly related to, or may at least be directly compared with, the Cephalopoda*.

The simplest shells among the Pteropods have the form of a tube, more or less cylindrical (*Cuvierina*), more often conical (*Creseis*, *Clio*); and this tubular shell (as we have already had occasion to remark, on p. 416), frequently tends, when it is very small and delicate, to assume the character of an unduloid. (In such a case it is more than likely that the tiny shell, or that portion of it which

* We need not assume a *close* relationship, nor indeed any more than such a one as permits us to compare the shell of a *Nautilus* with that of a Gastropod.

constitutes the unduloid, has not grown by successive increments or "rings of growth," but has developed as a whole.) A thickened "rib" is often, perhaps generally, present on the dorsal side of the little conical shell. In a few cases (*Limacina*, *Peraclis*) the tube becomes spirally coiled, in a normal equiangular spiral or helico-spiral.

In certain cases (e.g. *Cleodora*, *Hyalaea*) the tube or cone is curiously modified. In the first place, its cross-section, originally circular or nearly so, becomes flattened or compressed dorsoventrally; and

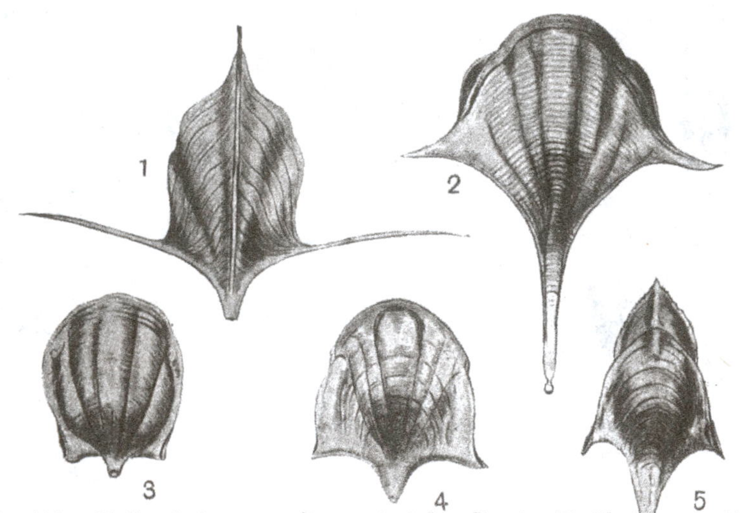

Fig. 411. Shells of thecosome Pteropods (after Boas). (1) *Cleodora cuspidata*; (2) *Hyalaea trispinosa*; (3) *H. globulosa*; (4) *H. uncinata*; (5) *H. inflexa*.

the angle, or rather edge, where dorsal and ventral walls meet, becomes more and more drawn out into a ridge or keel. Along the free margin, both of the dorsal and the ventral portion of the shell, growth proceeds with a regularly varying velocity, so that these margins, or lips, of the shell become regularly curved or markedly sinuous. At the same time, growth in a transverse direction proceeds with an acceleration which manifests itself in a curvature of the sides, replacing the straight borders of the original cone. In other words, the cross-section of the cone, or what we have been calling the generating curve, increases its dimensions more rapidly than its distance from the pole.

In the above figures, for instance in that of *Cleodora cuspidata*, the markings of the shell which represent the successive edges of the lip at former stages of growth furnish us at once with a "graph"

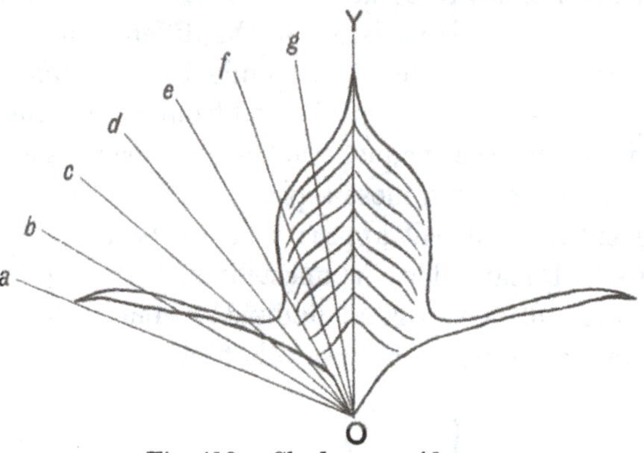

Fig. 412. *Cleodora cuspidata.*

of the varying velocities of growth as measured, radially, from the apex. We can reveal more clearly the nature of these variations in the following way, which is simply tantamount to converting our radial into rectangular coordinates. Neglecting curvature (if any)

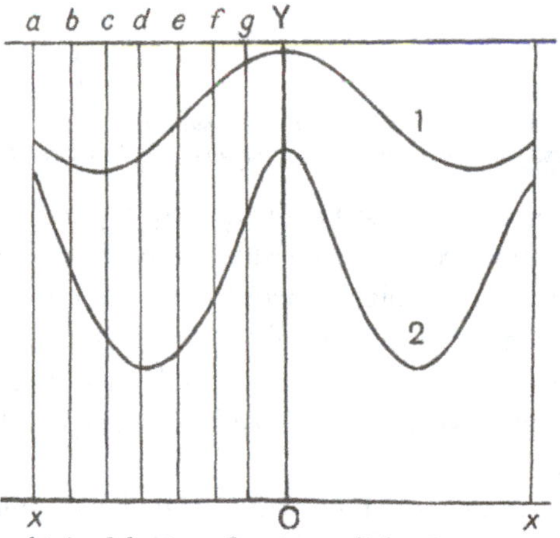

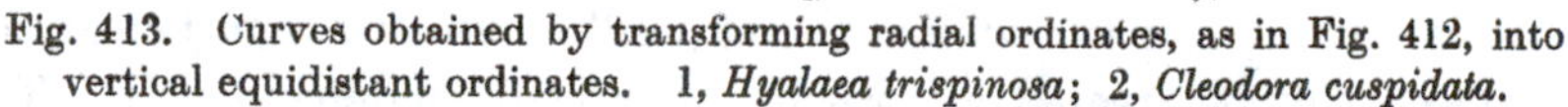

Fig. 413. Curves obtained by transforming radial ordinates, as in Fig. 412, into vertical equidistant ordinates. 1, *Hyalaea trispinosa*; 2, *Cleodora cuspidata.*

of the sides and treating the shell (for simplicity's sake) as a right cone, we lay off equal angles from the apex *O*, along the radii *Oa*, *Ob*, etc. If we then plot, as vertical equidistant ordinates, the magnitudes Oa, Ob ... OY, and again on to Oa', we obtain a diagram such as follows in Fig. 413: by help of which we not only see more clearly the way in which the growth-rate varies from point to point, but we also recognise better than before the nature of the law which governs this variation in the different species.

Furthermore, the young shell having become differentiated into a dorsal and a ventral part, marked off from one another by a lateral edge or keel, and the inequality of growth being such as to cause each portion to increase most rapidly in the median line, it follows that the entire shell will appear to have been split into a dorsal and a ventral plate, both connected with, and projecting from, what remains of the original undivided cone. Putting the same thing in other words, we may say that the generating figure, which

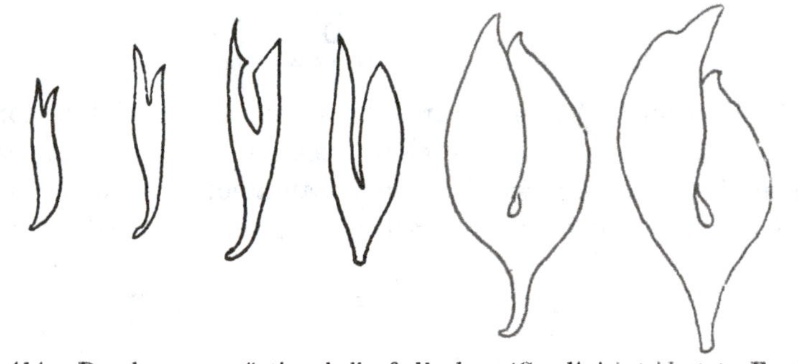

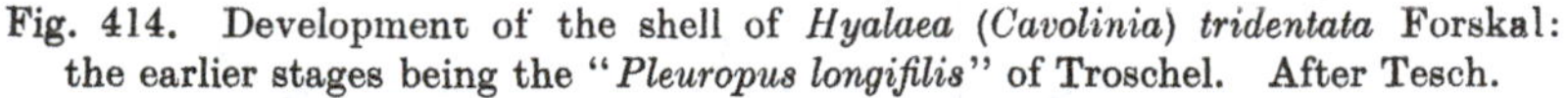

Fig. 414. Development of the shell of *Hyalaea* (*Cavolinia*) *tridentata* Forskal: the earlier stages being the "*Pleuropus longifilis*" of Troschel. After Tesch.

lay at first in a plane perpendicular to the axis of the cone, has now, by unequal growth, been sharply bent or folded, so as to lie approximately in two planes, parallel to the anterior and posterior faces of the cone. We have only to imagine the apical connecting portion to be further reduced, and finally to disappear or rupture, and we should have a *bivalve shell* developed out of the original simple cone.

In its outer and growing portion, the shell of our Pteropod now consists of two parts which, though still connected together at the apex, may be treated as growing practically independently. The shell is no longer a simple tube, or simple cone, in which regular inequalities of growth will lead to the development of a spiral; and this for the simple reason that we have now two opposite maxima of growth, instead of a maximum on the one side and a minimum on the other side of our tubular shell. As a matter of fact, the dorsal and the ventral plate tend to curve in opposite directions, towards the middle line, the dorsal curving ventrally and the ventral curving towards the dorsal side.

In the case of the Lamellibranch or the Brachiopod, it is quite possible for both valves to grow into more or less pronounced spirals, for the simple reason that they are *hinged* upon one another; and each growing edge, instead of being brought to a standstill by the growth of its opposite neighbour, is free to move out of the way, by the rotation about the hinge of the plane in which it lies.

But where there is no such hinge, as in the Pteropod, the dorsal and ventral halves of the shell (or dorsal and ventral valves, if we

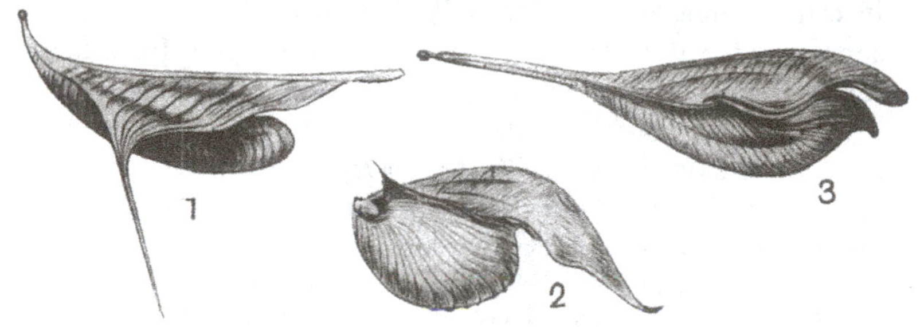

Fig. 415. Pteropod shells, from the side: (1) *Cleodora cuspidata*; (2) *Hyalaea longirostris*; (3) *H. trispinosa*. After Boas.

may call them so) would soon interfere with one another's progress if they curved towards one another (as they do in a cockle), and the development of a pair of conjugate spirals would become impossible. Nevertheless, there is obviously, in both dorsal and ventral valve, a *tendency* to the development of a spiral curve, that of the ventral valve being more marked than that of the larger and overlapping dorsal one, exactly as in the two unequal valves of *Terebratula*. In many cases (e.g. *Cleodora cuspidata*), the dorsal valve or plate, strengthened and stiffened by its midrib, is nearly straight, while the curvature of the other is well displayed. But the case will be materially altered and simplified if growth be arrested or retarded in either half of the shell. Suppose for instance that the dorsal valve grew so slowly that after a while, in comparison with the other, we might speak of it as being absent altogether:

or suppose that it merely became so reduced in relative size as to form no impediment to the continued growth of the ventral one; the latter would continue to grow in the direction of its natural curvature, and would end by forming a complete and coiled logarithmic spiral. It would be precisely analogous to the spiral shell of *Nautilus*, and, in regard to its ventral position, concave towards the dorsal side, it would even deserve to be called directly homologous with it. Suppose, on the other hand, that the ventral valve were to be greatly reduced, and even to disappear, the dorsal valve would then pursue its unopposed growth; and, were it to be markedly curved, it would come to form a logarithmic spiral, concave towards the ventral side, as is the case in the shell of *Spirula**. Were the dorsal valve to be destitute of any marked curvature (or in other words, to have but a low spiral angle), it would form a simple plate, as in the shells of *Sepia* or *Loligo*. Indeed, in the shells of these latter, and especially in that of *Sepia*, we seem to recognise a manifest resemblance to the dorsal plate of the Pteropod shell, as we have it (e.g.) in *Cleodora* or *Hyalaea*; the little "rostrum" of *Sepia* is but the apex of the primitive cone, and the rounded anterior extremity has grown according to a law precisely such as that which has produced the curved margin of the dorsal valve in the Pteropod. The ventral portion of the original cone is nearly, but not wholly, wanting; it is represented by the so-called posterior wall of the "siphuncular space." In many decapod cuttle-fishes also (e.g. *Todarodes*, *Illex*, etc.) we still see at the posterior end of the "pen" a vestige of the primitive cone, whose dorsal margin only has continued to grow; and the same phenomenon, on an exaggerated scale, is represented in the *Belemnites*.

It is not at all impossible that we may explain on the same lines the development of the curious "operculum" of the Ammonites. This consists of a single horny plate (*Anaptychus*), or of a thicker, more calcified plate divided into two symmetrical halves (*Aptychi*), often found inside the terminal chamber of the Ammonite, and occasionally to be seen lying *in situ*, as an operculum which partially closes the mouth of the shell; this structure is known to exist even

* Cf. Owen, "These shells [*Nautilus* and Ammonites] are revolutely spiral or coiled over the back of the animal, not involute like *Spirula*": *Palaeontology*, 1861, p. 97; cf. *Memoir on the Pearly Nautilus*, 1832; also *P.Z.S.* 1878, p. 955.

in connection with the early embryonic shell. In form the Anaptychus, or the pair of conjoined Aptychi, shew an upper and a lower border, the latter strongly convex, the former sometimes slightly concave, sometimes slightly convex, and usually shewing a median projection or slightly developed rostrum. From this rostral border the curves of growth start, and course round parallel to, finally constituting, the convex border. It is this convex border which fits into the free margin of the mouth of the Ammonite's shell, while the other is applied to and overlaps the preceding whorl of the spire. Now this relationship is precisely what we should expect, were we to imagine as our starting-point a shell similar to that of *Hyalaea*: in which however the dorsal part of the split cone had become separate from the ventral half, had remained flat, and had grown comparatively slowly, while at the same time it kept slipping forward over the growing and coiling spire into which the ventral half of the original shell develops*. In short, I think there is reason to believe, or at least to suspect, that we have in the shell and Aptychus of the Ammonites, two portions of a once united structure; of which other Cephalopods retain not both parts but only one or other, one as the ventrally situated shell of *Nautilus*, the other as the dorsally placed shell for example of *Sepia* or of *Spirula*.

In the case of the bivalve shells of the Lamellibranchs or of the Brachiopods, we have to deal with a phenomenon precisely analogous to the split and flattened cone of our Pteropods, save only that the primitive cone has been split into two portions, not incompletely, as in the Pteropod (*Hyalaea*), but completely, so as to form two separate valves. Though somewhat greater freedom is given to growth now that the two valves are separate and hinged, yet still the two valves oppose and hamper one another, so that in the longitudinal direction each is capable of only a moderate curvature. This curvature, as we have seen, is recognisable as an equiangular spiral, but only now and then does the growth of the spiral continue so far as to develop successive coils: as it does in a few symmetrical forms such as *Isocardia cor*; and as it does still more conspicuously in a few others, such as *Gryphaea* and *Caprinella*, where one of the

* The case of *Terebratula* or of *Gryphaea* would be closely analogous, if the smaller valve were less closely connected and co-articulated with the larger.

two valves is stunted, and the growth of the other is (relatively speaking) unopposed.

Of septa

Before we leave the subject of the molluscan shell, we have still another problem to deal with, in regard to the form and arrangement of the septa which divide up the tubular shell into chambers, in the Nautilus, the Ammonite and their allies.

The existence of septa in a nautiloid shell may probably be accounted for as follows. We have seen that it is a property of a cone that, while growing by increments at one end only, it conserves its original shape: therefore the animal within, which (though growing by a different law) also conserves its shape, will continue to fill the shell if it actually fills it to begin with: as does a snail or other Gastropod. But suppose that our mollusc fills a part only of a conical shell (as it does in the case of *Nautilus*); then, unless it alter its shape, it must move upward as it grows in the growing cone, until it comes to occupy a space similar in form to that which it occupied before: just, indeed, as a little ball drops far down into the cone, but a big one must stay farther up. Then, when the animal after a period of growth has moved farther up in the shell, the mantle-surface continues or resumes its secretory activity, and that portion which had been in contact with the former septum secretes a septum anew. In short, at any given epoch, the creature is not secreting a tube and a septum by separate operations, but is secreting a shelly case about its rounded body, of which case one part appears as the continuation of the tube, and the other part, merging with it by indistinguishable boundaries, appears as the septum*.

The various forms assumed by the septa in spiral shells† present us with a number of problems of great beauty, simple in their essence, but whose full investigation would soon lead us into difficult mathematics.

* "It has been suggested, and I think in some quarters adopted as a dogma, that the formation of successive septa [in *Nautilus*] is correlated with the recurrence of reproductive periods. This is not the case, since, according to my observations, propagation only takes place after the last septum is formed"; Willey, *Zoological Results*, 1902, p. 746.

† Cf. Henry Woodward, On the structure of camerated shells, *Pop. Sci. Rev.* XI, pp. 113–120, 1872.

We do not know how these septa are laid down in an Ammonite, but in the Nautilus the essential facts are clear*. The septum begins as a very thin cuticular membrane (composed of a substance called conchyolin), which is secreted by the skin, or mantle-surface, of the animal; and upon this membrane nacreous matter is gradually laid down on the mantle-side (that is to say between the animal's body and the cuticular membrane which has been thrown off from it), so that the membrane remains as a thin pellicle over the *hinder* surface of the septum, and so that, to begin with, the membranous septum is moulded on the flexible and elastic surface of the animal, within which the fluids of the body must exercise a uniform, or nearly uniform pressure.

Let us think, then, of the septa as they would appear in their uncalcified condition, formed of, or at least superposed upon, an elastic membrane. They must follow the general law, applicable to all elastic membranes under uniform pressure, that the tension varies inversely as the radius of curvature; and we come back once more to our old equation of Laplace and Plateau, that

$$P = T\left(\frac{1}{r} + \frac{1}{r'}\right).$$

Moreover, since the cavity below the septum is practically closed, and is filled either with air or with water, P will be constant over the whole area of the septum. And further, we must assume, at least to begin with, that the membrane constituting the incipient septum is homogeneous or isotropic.

Let us take first the case of a straight cone, of circular section, more or less like an *Orthoceras*; and let us suppose that the septum is attached to the shell in a plane perpendicular to its axis. The septum itself must then obviously be spherical. Moreover the extent of the spherical surface is constant, and easily determined. For obviously, in Fig. 417, the angle LCL' equals the supplement of the angle (LOL') of the cone; that is to say, the circle of contact subtends an angle at the centre of the spherical surface, which is constant, and which is equal to $\pi - 2\beta$. The case is not excluded where, owing to an asymmetry of tensions, the septum meets the

* See Willey, *op. cit.*, p. 749. Cf. also Bather, Shell-growth in Cephalopoda, *Ann. Mag. N.H.* (6), I, pp. 298–310, 1888; *ibid.* pp. 421–427, and other papers by Blake, Riefstahl, etc. quoted therein.

side walls of the cone at other than a right angle, as in Fig. 416; and here, while the septa still remain portions of spheres, the geometrical construction for the position of their centres is equally easy.

If, on the other hand, the attachment of the septum to the inner walls of the cone be in a plane oblique to the axis, then the outline of the septum will be an ellipse, but its surface will still be spheroidal. If

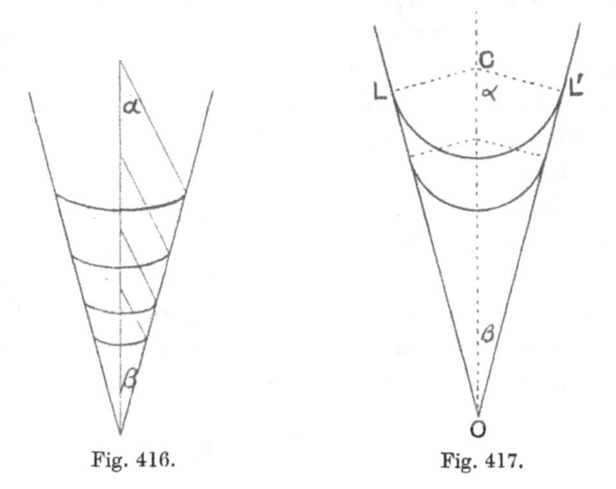

Fig. 416. Fig. 417.

the attachment of the septum be not in one plane, but forms a sinuous line of contact with the cone, then the septum will be a saddle-shaped surface, of great complexity and beauty. In all cases, provided only that the membrane be isotropic, the form assumed will be precisely that of a soap-bubble under similar conditions of attachment: that is to say, it will be (with the usual limitations or conditions) a surface of minimal area, and of constant mean curvature.

If our cone be no longer straight, but curved, then the septa will by symmetrically deformed in consequence. A beautiful and interesting case is afforded us by *Nautilus* itself. Here the outline of the septum, referred to a plane, is approximately bounded by two elliptic curves, similar and similarly situated, whose areas are to one another in a definite ratio, namely as

$$\frac{A_1}{A_2} = \frac{r_1 r'_1}{r_2 r'_2} = \epsilon^{-4\pi \cot \alpha},$$

and a similar ratio exists in Ammonites and all other close-whorled spirals, in which however we cannot always make the simple assumption of elliptical form. In a median section of *Nautilus*, we see each septum forming a tangent to the inner and to the outer wall, just as it did in a section of the straight *Orthoceras*; but the

Fig. 418. Section of *Nautilus*, shewing the contour of the septa in the median plane.

curvatures in the neighbourhood of these two points of contact are not identical, for they now vary inversely as the radii, drawn from the pole of the spiral shell. The contour of the septum in this median plane is a spiral curve—the conformal spiral transformation of the spherical septum of the rectilinear Orthoceratite.

But while the outline of the septum in median section is simple and easy to determine, the curved surface of the septum in its entirety is a very complicated matter, even in *Nautilus* which is one of the simplest of actual cases. For, in the first place, since the form of the septum, as seen in median section, is that of a logarithmic spiral, and as therefore its curvature is constantly

Fig. 419. Cast of the interior of *Nautilus*: to shew the contours of the septa at their junction with the shell-wall.

altering, it follows that, in successive *transverse* sections, the curvature is also constantly altering. But in the case of *Nautilus*, there are other aspects of the phenomenon, which we can illustrate, but only in part, in the following simple manner. Let us imagine a pack of cards, in which we have cut out of each card a similar concave arc of a logarithmic spiral, such as we actually see in the median section of the septum of a *Nautilus*. Then, while we hold the cards together, foursquare, in the ordinary position of the pack, we have a simple "ruled" surface, which in any longitudinal section has the form of a logarithmic spiral but in any transverse section is a straight horizontal line. If we shear or slide the cards upon one another, thrusting the middle cards of the pack forward in advance of the others, till the one end of the pack is a convex, and the other a concave, ellipse, the cut edges which combine to represent our septum will now form a curved surface of much greater complexity; and this is part, but not by any means all, of the deformation produced as a direct consequence of the form in *Nautilus* of the section of the tube within which the septum has to lie. The complex curvature of the surface will be manifested in a sinuous outline of the edge, or line of attachment of the septum to the tube, and will vary according to the configuration of the latter. In the case of *Nautilus*, it is easy to shew empirically (though not perhaps easy to demonstrate mathematically), that the sinuous or saddle-shaped contour of the "suture" (or line of attachment of the septum to the tube) is such as can be precisely accounted for in this manner; and we may find other forms, such as *Ceratites*, where the septal outline is only a little more sinuous, and still precisely analogous to that of *Nautilus*. It is also easy to see that, when the section of the tube (or "generating curve") is more complicated in form, when it is flattened, grooved, or otherwise ornamented, the curvature of the septum and the outline of its sutural attachment will become very complicated indeed*; but it will be comparatively simple in the case of the first few sutures of the young shell, laid down before any overlapping of whorls has taken place, and this comparative simplicity of the first-formed sutures is a marked feature among Ammonites†.

* The "lobes" and "saddles" which arise in this manner, and on whose arrangement the modern classification of the nautiloid and ammonitoid shells largely depends, were first recognised and named by Leopold von Buch, *Ann. Sci. Nat.* XXVII, XXVIII, 1829.

† Blake has remarked upon the fact (*op. cit.* p. 248) that in some Cyrtocerata we may have a curved shell in which the ornaments approximately run at a constant angular distance from the pole, while the septa approximate to a radial direction; and that "thus one law of growth is illustrated by the inside, and another by the outside." In this there is nothing at which we need wonder. It is merely a case where the generating curve is set very obliquely to the axis of the shell; but where the septa, which have no necessary relation to the *mouth* of the shell, take their places, as usual, at a certain definite angle to the *walls* of the tube. This relation of the septa to the walls of the tube arises after the tube itself is fully formed, and the obliquity of growth of the open end of the tube has no relation to the matter.

We have other sources of complication, besides those which are at once introduced by the sectional form of the tube. For instance, the siphuncle, or little inner tube which perforates the septa, exercises a certain amount of tension, sometimes evidently considerable, upon the latter: which tension is made manifest in *Spirula* (and slightly so even in *Nautilus*) by a dip in the septal floor where it meets the siphuncle. We can no longer, then, consider each septum as an isotropic surface under uniform pressure; and there may be other structural modifications, or inequalities, in that portion of the

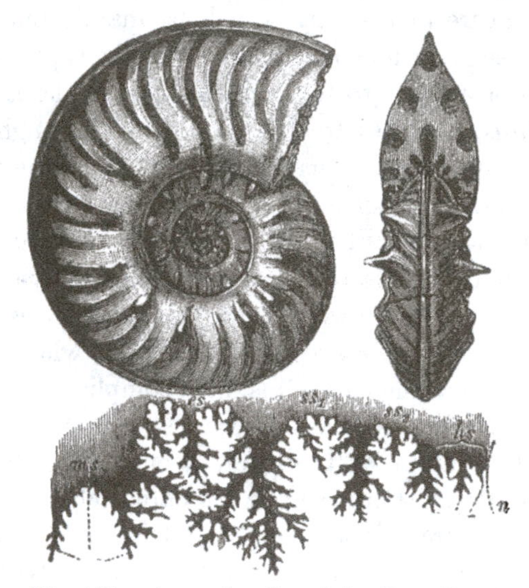

Fig. 420. *Ammonites Sowerbyi*. From Zittel.

animal's body with which the septum is in contact, and by which it is conformed. It is hardly likely, for all these reasons, that we shall ever attain to a full and particular explanation of the septal surfaces and their sutural outlines throughout the whole range of Cephalopod shells; but in general terms, the problem is probably not beyond the reach of mathematical analysis. The problem might be approached experimentally, after the manner of Plateau's experiments, by bending a wire into the complicated form of the suture-line, and studying the form of the liquid film which constitutes the corresponding surface *minimae areae*.

In certain Ammonites the septal outline is further complicated in another way. Superposed upon the usual sinuous outline, with its "lobes" and "saddles," we have here a minutely ramified, or arborescent outline, in which all the branches terminate in wavy, more or less circular arcs—looking just like the "landscape marble" from the Bristol Rhaetic. We have no difficulty in recognising in this a surface-tension phenomenon. The figures are precisely such as we can imitate (for instance) by pouring a few drops of milk upon a greasy plate, or of oil upon an alkaline solution*; they are what Charles Tomlinson called "cohesion figures."

Fig. 421. Suture-line of a Triassic Ammonite (*Pinacoceras*). From Zittel.

We must not forget that while the nautilus and the ammonite resemble one another, and are mathematically identical in their spiral curves, they are really very different things. The one is an external, the other an internal shell. The nautilus occupies the large terminal chamber of the many-chambered shell, and "Still as the spiral grew, He left the past year's dwelling for the new." But even the largest ammonites never contained the body of the animal, but lay hidden, as *Spirula* does, deep within the substance of the mantle. How the complicated septa and septal outlines of the ammonites are produced I do not know†.

We have very far from exhausted, we have perhaps little more than begun, the study of the logarithmic spiral and the associated curves which find exemplification in the multitudinous diversities of molluscan shells. But, with a closing word or two, we must now bring this chapter to an end.

* "The Fimbriae, or Edges, appeared on the Surface like the Outlines of some curious Foliage. This, upon Examination of them, I found to proceed from the Fulness of the Edges of the Diaphragms, whereby the Edges were waved or plaited somewhat in the manner of a Ruff" (R. Hooke, *op. cit.*).

† In certain rare cases the complicated sutural pattern of an ammonite is found *upside down*, but unchanged otherwise. Cf. Otto Haas, A case of inversion of suture lines in *Hysteroceras*, *Amer. Jl. of Sci.* CCXXXIX, p. 661, 1941.

In the spiral shell we have a problem, or a phenomenon, of growth, immensely simplified by the fact that each successive increment is no sooner formed than it is fixed irrevocably, instead of remaining in a state of flux and sharing in the further changes which the organism undergoes. In such a structure, then, we have certain primary phenomena of growth manifested in their original simplicity, undisturbed by secondary and conflicting phenomena. What actually *grows* is merely the lip of an orifice, where there is produced a ring of solid material, whose form we have discussed under the name of the generating curve; and this generating curve grows in magnitude without alteration of its form. Besides its increase in areal magnitude, the growing curve has certain strictly limited degrees of freedom, which define its motions in space. And, though we may know nothing whatsoever about the actual velocities of any of these motions, we do know that they are so correlated together that their *relative* velocities remain constant, and accordingly the form and symmetry of the whole system remain in general unchanged.

But there is a vast range of possibilities in regard to every one of these factors: the generating curve may be of various forms, and even when of simple form, such as an ellipse, its axes may be set at various angles to the system; the plane also in which it lies may vary, almost indefinitely, in its angle relatively to that of any plane of reference in the system; and in the several velocities of growth, of rotation and of translation, and therefore in the ratios between all these, we have again a vast range of possibilities. We have then a certain definite type, or group of forms, mathematically isomorphous, but presenting infinite diversities of outward appearance: which diversities, as Swammerdam said, *ex sola nascuntur diversitate gyrationum*; and which accordingly are seen to have their origin in differences of rate, or of magnitude, and so to be, essentially, neither more nor less than *differences of degree.*

In nature, we find these forms presenting themselves with but little relation to the character of the creature by which they are produced. Spiral forms of certain particular kinds are common to Gastropods and to Cephalopods, and to diverse families of each; while outside the class of molluscs altogether, among the Foraminifera and among the worms (as in *Spirorbis*, *Spirographis*,* and in the *Dentalium*-like shell of *Ditrupa*), we again meet with similar and corresponding spirals.

Again, we find the same forms, or forms which (save for external ornament) are mathematically identical, repeating themselves in all periods of the world's geological history; and we see them mixed up, one with another, irrespective of climate or local conditions, in the depths and on the shores of every sea. It is hard indeed (to my mind) to see in such a case as this where Natural Selection necessarily enters in, or to admit that it has had any share whatsoever in the production of these varied conformations. Unless indeed we use the term Natural Selection in a sense so wide as to deprive it of any purely biological significance; and so recognise as a sort of natural selection whatsoever nexus of causes suffices to differentiate between the likely and the unlikely, the scarce and the frequent, the easy and the hard: and leads accordingly, under the peculiar conditions, limitations and restraints which we call "ordinary circumstances," one type of crystal, one form of cloud, one chemical compound, to be of frequent occurrence and another to be rare*.

* Cf. Bacon, *Advancement of Learning*, Bk. II (p. 254): "Doth any give the reason, why some things in nature are so common and in so great mass, and others so rare and in so small quantity?"

CHAPTER XII

THE SPIRAL SHELLS OF THE FORAMINIFERA

WE have already dealt in a few simple cases with the shells of the Foraminifera*; and we have seen that wherever the shell is but a single unit or single chamber, its form may be explained in general by the laws of surface-tension: the argument (or assumption) being that the little mass of protoplasm which makes the simple shell behaves as a *fluid drop*, the form of which is perpetuated when the protoplasm acquires its solid covering. Thus the spherical Orbulinae and the flask-shaped Lagenae represent drops in equilibrium, under various conditions of freedom or constraint; while the irregular, amoeboid body of Astrorhiza is a manifestation not of equilibrium, but of a varying and fluctuating distribution of surface energy. When the foraminiferal shell becomes multilocular, the same general principles continue to hold; the growing protoplasm increases drop by drop, and each successive drop has its particular phenomena of surface energy, manifested at its fluid surface, and tending to confer upon it a certain place in the system and a certain shape of its own.

It is characteristic and even diagnostic of this particular group of Protozoa (1) that development proceeds by a well-marked alternation of rest and of activity—of activity during which the protoplasm increases, and of rest during which the shell is formed; (2) that the shell is formed at the outer surface of the protoplasmic organism, and tends to constitute a continuous or all but continuous covering; and it follows (3) from these two factors taken together that each successive increment is added on outside of and distinct from its predecessors, that the successive parts or chambers of the shell are of different and successive ages, so that one part of the shell is always relatively new, and the rest old in various grades of seniority.

The forms which we set together in the sister-group of Radiolaria are very differently characterised. Here the cells or vesicles of which each little composite organism is made up are but little

* Cf. pp. 420, 702, etc.

separated, and in no way walled off, from one another; the hard skeletal matter tends to be deposited in the form of isolated spicules or of little connected rods or plates, at the angles, the edges or the interfaces of the vesicles; the cells or vesicles form a coordinated and cotemporaneous rather than a successive series. In a word, the whole quasi-fluid protoplasmic body may be likened to a little mass of froth or foam: that is to say, to an aggregation of simultaneously formed drops or bubbles, whose physical properties and geometrical relations are very different from those of a system of

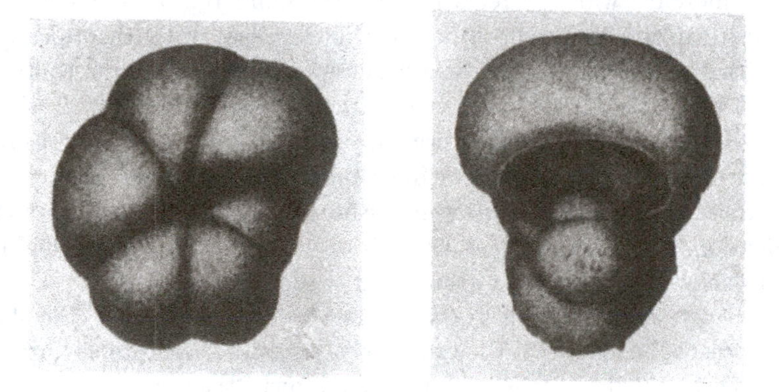

Fig. 422. *Hastigerina* sp.; to shew the "mouth."

drops or bubbles which are formed one after another, each solidifying before the next is formed.

With the actual origin or mode of development of the foraminiferal shell we are now but little concerned. The main factor is the adsorption, and subsequent precipitation at the surface of the organism, of calcium carbonate—the shell so formed being interrupted by pores or by some larger interspace or "mouth" (Fig. 422), which interruptions we may doubtless interpret as being due to unequal distributions of surface energy. In many cases the fluid protoplasm "picks up" sand-grains and other foreign particles, after a fashion which we have already described (p. 702); and it cements these together with more or less of calcareous material. The calcareous shell is a crystalline structure, and the micro-crystals of calcium carbonate are so set that their little prisms radiate outwards in each chamber through the thickness of the wall—which symmetry is subject to corresponding modification when the spherical chambers are more or less symmetrically deformed*.

In various ways the rounded, drop-like shells of the Foraminifera, both simple and compound, have been artificially imitated. Thus, if small globules of mercury be immersed in water in which a little chromic acid is allowed to dissolve, as the little beads of quicksilver become slowly covered with a crystalline coat of mercuric chromate they assume various forms reminiscent of the monothalamic Foraminifera. The mercuric chromate has a higher atomic volume than the mercury which it replaces, and therefore the fluid contents of the drop are under pressure, which increases with the thickness of the pellicle; hence at some weak spot in the latter the contents will presently burst forth, so forming a mouth to the little shell. Sometimes a long thread is formed, just as in *Rhabdammina linearis*; and sometimes unduloid swellings make their appearance on such a thread, just as in *R. discreta*. And again, by appropriate modifications of the experimental conditions, it is possible (as Rhumbler has shewn) to build up a chambered shell†.

In a few forms, such as *Globigerina* and its close allies, the shell is beset during life with excessively long and delicate calcareous spines or needles. It is only in oceanic forms that these are present, because only when poised in water can such delicate structures endure; in dead shells, such as we are much more familiar with, every trace of them is broken and rubbed away. The growth of these long needles may be partly explained (as we have already said on p. 675) by the phenomenon which Lehmann calls *orientirte Adsorption*—the tendency for a crystalline structure to grow by accretion, not necessarily in the outward form of a "crystal," but continuing in any direction or orientation which has once been impressed upon it: in this case the spicular growth is in direct continuation of the radial symmetry of the micro-crystalline

* In a few cases, according to Awerinzew and Rhumbler, where the chambers are added on in concentric series, as in Orbitolites, we have the crystalline structure arranged radially in the radial walls but tangentially in the concentric ones: whereby we tend to obtain, on a minute scale, a system of orthogonal trajectories, comparable to that which we shall presently study in connection with the structure of bone. Cf. S. Awerinzew, Kalkschale der Rhizopoden, *Z. f. w. Z.* LXXIV, pp. 478–490, 1903.

† L. Rhumbler, Die Doppelschalen von Orbitolites und anderer Foraminiferen, etc., *Arch. f. Protistenkunde*, I, pp. 193–296, 1902; and other papers. Also *Die Foraminiferen der Planktonexpedition*, I, pp. 50–56, 1911.

elements of the shell-wall. But the calcareous needles are secreted in, or *by*, no less long and delicate pseudopodia or "filopodia," and much has been learned since this book was written of the molecular, or micellar, orientation of the protoplasm in such filamentous structures; it is known that the long pseudopodia of the Foraminifera are doubly refractive, and it follows that their molecules are anisotropically arranged*. Whether the slender form and asymmetrical structure of calcareous rod and protoplasmic thread be independent phenomena, or merely two aspects of one and the same phenomenon, is a hard question, and not one for us to discuss. Nor can we profitably discuss (much as we should like to know) how far these patterns of molecular structure in threads, films and surface-pellicles affect the "fluidity" of the substance, and conflict with the capillary forces which influence its outward form. But we may safely say that the effects of surface-tension on cell-form have been so plainly seen all through this book that any counter-effects due to protoplasmic asymmetry must be phenomena of a second order, and inconspicuous on the whole. Over the whole surface of the shell of *Globigerina* the radiating spicules tend to occur in a hexagonal pattern, symmetrically grouped around the pores which perforate the shell. Rhumbler has suggested that this arrangement is due to diffusion-currents, forming little eddies about the base of the pseudopodia issuing from the pores: the idea being borrowed from Bénard, to whom is due the discovery of this type or order of vortices†. In one of Bénard's experiments a thin layer of paraffin is strewn with particles of graphite, then warmed to melting, whereupon each little solid granule becomes the centre of a vortex; by the interaction of these vortices the particles tend to be repelled to equal distances from one another, and in the end they are found to be arranged in a hexagonal pattern‡.

* Cf. W. J. Schmidt, *Die Bausteine des Tierkörpers in polarisiertem Lichte*, Bonn, 1924; Ueber den Feinbau der Filopodien; insb. ihre Doppelbrechung bei *Miliola*, *Protoplasma*, xxvii, p. 587, 1937; also D. L. Mackinnon, Optical properties of contractile organs in Heliozoa, *Jl. Physiol.* xxxviii, p. 254, 1909; R. O. Herzog, Lineare u. laminäre Feinstrukturen, *Kolloidzschr.* lxi, p. 280, 1932. See, for discussion and bibliography, L. E. Picken, *op. cit.*

† H. Bénard, Les tourbillons cellulaires, *Ann. de Chimie* (8), xxiv, 1901. Cf. also the pattern of cilia on an Infusorian, as figured by Bütschli in Bronn's *Protozoa*, iii, p. 1281, 1887.

‡ A similar hexagonal pattern is obtained by the mutual repulsion of floating magnets in Mr R. W. Wood's experiments, *Phil. Mag.* xlvi, pp. 162–164, 1898.

The analogy is plain between this experiment and those diffusion experiments by which Leduc produces his beautiful hexagonal systems of artificial cells, with which we have dealt in a previous chapter.

But let us come back to the shell itself, and consider particularly its spiral form. That the shell in the Foraminifera should tend towards a spiral form need not surprise us; for we have learned that one of the fundamental conditions of the production of a concrete spiral is just precisely what we have here, namely the development of a structure by means of successive graded increments superadded to its exterior, which then form part, successively, of a permanent and rigid structure. This condition is obviously forthcoming in the foraminiferal, but not at all in the radiolarian, shell. Our second fundamental condition of the production of a logarithmic spiral is that each successive increment shall be so posited and so conformed that its addition to the system leaves the form of the whole system unchanged. We have now to enquire into this latter condition; and to determine whether the successive increments, or successive chambers, of the foraminiferal shell actually constitute *gnomons* to the entire structure.

It is obvious enough that the spiral shells of the Foraminifera closely resemble true logarithmic spirals. Indeed so precisely do the minute shells of many Foraminifera repeat or simulate the spiral shells of *Nautilus* and its allies that to the naturalists of the early nineteenth century they were known as the *Céphalopodes microscopiques**, until Dujardin shewed that their little bodies comprised no complex anatomy of organs, but consisted merely of that slime-like organic matter which he taught us to call "sarcode," and which we learned afterwards from Schwann to speak of as "protoplasm."

One striking difference, however, is apparent between the shell of *Nautilus* and the little nautiloid or rotaline shells of the Foraminifera: namely that the septa in these latter, and in all other chambered Foraminifera, are convex outwards (Fig. 423), whereas they are concave outwards in *Nautilus* (Fig. 347) and in the rest of the chambered molluscan shells. The reason is perfectly simple.

* Cf. Alc. d'Orbigny, Tableau méthodique de la classe des Céphalopodes, *Ann. des Sci. Nat.* (1), VII, pp. 245–315, 1826; Félix Dujardin, Observations nouvelles sur les prétendus Céphalopodes microscopiques, *ibid.* (2), III, pp. 108, 109, 312–315, 1835; Recherches sur les organismes inférieurs, *ibid.* IV, pp. 343–377, 1835; etc.

In both cases the curvature of the septum was determined before it became rigid, and at a time when it had the properties of a fluid film or an elastic membrane. In both cases the actual curvature is determined by the tensions of the membrane and the pressures to which it was exposed. Now it is obvious that the extrinsic pressure which the tension of the membrane has to withstand is on opposite sides in the two cases. In *Nautilus*, the pressure to be resisted is that produced by the growing body of the animal, lying to the *outer side* of the septum, in the outer, wider portion of the tubular shell. In the Foraminifer the septum at the time of its formation was no septum at all; it was but a portion of the convex

Fig. 423. *Nummulina antiquior* R. and V. After V. von Moller.

surface of a drop—that portion namely which afterwards became overlapped and enclosed by the succeeding drop; and the curvature of the septum is concave towards the pressure to be resisted, which latter is *inside* the septum, being simply the hydrostatic pressure of the fluid contents of the drop. The one septum is, speaking generally, the reverse of the other; the organism, so to speak, is outside the one and inside the other; and in both cases alike, the septum tends to assume the form of a surface of minimal area, as permitted, or as defined, by all the circumstances of the case.

The logarithmic spiral is easily recognisable in typical cases* (and

* It is obvious that the actual *outline* of a foraminiferal, just as of a molluscan shell, may depart widely from a logarithmic spiral. When we say here, for short, that the shell *is* a logarithmic spiral, we merely mean that it is essentially related to one: that it can be inscribed in such a spiral, or that corresponding points (such, for instance, as the centres of gravity of successive chambers, or the extremities of successive septa) will be found to lie upon such a spiral.

especially where the spire makes more than one visible revolution about the pole), by its fundamental property of continued similarity: that is to say, by reason of the fact that the big many-chambered shell is of just the same shape as the smaller and younger shell—which phenomenon is as apparent and even obvious in the nautiloid Foraminifera, as in *Nautilus* itself: but nevertheless the nature of the curve must be verified by careful measurement, just as Moseley determined or verified it in his original study of *Nautilus* (cf. p. 770). This has accordingly been done, by various writers: and in the first instance by Valerian von Möller, in an elaborate study of *Fusulina*—a palaeozoic genus whose little shells have built up vast tracts of carboniferous limestone in European Russia*.

In this genus a growing surface of protoplasm may be conceived as wrapping round and round a small initial chamber, in such a way as to produce a fusiform or ellipsoidal shell—a transverse section of which reveals the close-wound spiral coil. The following are measurements of the successive whorls in a couple of species of this genus:—

	F. cylindrica Fischer		*F. Böcki* v. Möller	
	Breadth (in millimetres)			
Whorl	Observed	Calculated	Observed	Calculated
I	0·132	—	0·079	—
II	0·195	0·198	0·120	0·119
III	0·300	0·297	0·180	0·179
IV	0·449	0·445	0·264	0·267
V	—	—	0·396	0·401

In both cases the successive whorls are very nearly in the ratio of 1 : 1·5; and on this ratio the calculated values are based.

Here is another of von Möller's series of measurements of *F. cylindrica*, the measurements being those of opposite whorls—that is to say of whorls 180° apart:

Breadth (mm.)	0·096	0·117	0·144	0·176	0·216	0·264	0·323	0·395
Log. of do.	0·982	0·068	0·158	0·246	0·334	0·422	0·509	0·597
Diff. of logs.	—	0·086	0·090	0·088	0·088	0·088	0·087	0·088

The mean logarithmic difference is here 0·088, = log 1·225; or the mean difference of alternate logs (corresponding to a vector angle of 2π, i.e. to consecutive measurements along the *same* radius) is 0·176, = log 1·5, the same value as before. And this ratio of 1·5 between the breadths of successive whorls corresponds (as we see

* V. von Möller, Die spiral-gewundenen Foraminifera des rüssischen Kohlenkalks, *Mém. de l'Acad. Imp. Sci., St Pétersbourg* (7), xxv, 1878.

by our table on p. 791) to a constant angle of about 86°, or just such a spiral as we commonly meet with in the Ammonites (cf. p. 796).

In *Fusulina*, and in some few other Foraminifera (cf. Fig. 424, A), the spire seems to wind evenly on, with little or no external sign of the successive periods of growth, or successive chambers of the shell. The septa which mark off the chambers, and correspond to retardations or cessations in the periodicity of growth, are still to be found in sections of the shell of *Fusulina*, but they are somewhat irregular and comparatively inconspicuous; the measurements we have just spoken of are taken without reference to the segments or chambers, but only with reference to the whorls, or in other words with direct reference to the vectorial angle.

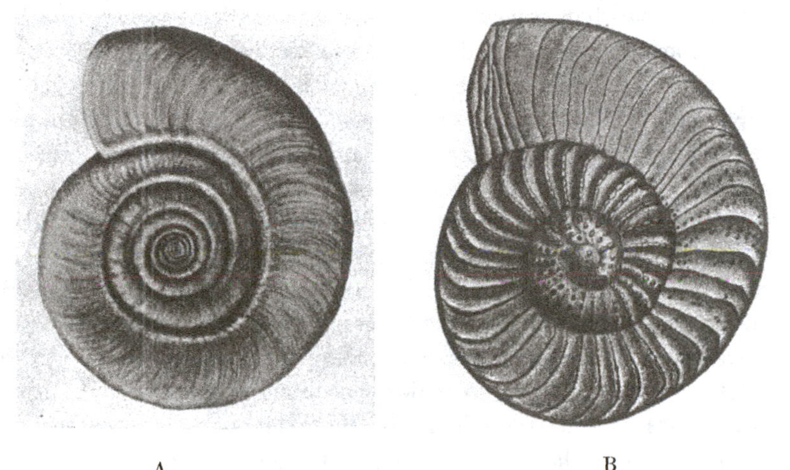

A B

Fig. 424. A, *Cornuspira foliacea* Phil.; B, *Operculina complanata* Defr.

The linear dimensions of successive chambers have been measured in a number of cases. Van Iterson* has done so in various Miliolinidae, with such results as the following:

Triloculina rotunda d'Orb.

No. of chamber	1	2	3	4	5	6	7	8	9	10
Breadth of chamber in μ	—	34	45	61	84	114	142	182	246	319
Breadth of chamber in μ, calculated	—	34	45	60	79	105	140	187	243	319

* G. van Iterson, *Mathem. u. mikrosk.-anat. Studien über Blattstellungen, nebst Betrachtungen über den Schalenbau der Miliolinen*, 331 pp., Jena, 1907.

Here the mean ratio of breadth of consecutive chambers may be taken as 1·323 (that is to say, the eighth root of 319/34); and the calculated values, as given above, are based on this determination. Again, Rhumbler has measured the linear dimensions of a number of rotaline forms, for instance *Pulvinulina menardi* (Fig. 363): in which common species he finds the mean linear ratio of consecutive chambers to be about 1·187. In both cases, and especially in the latter, the ratio is not strictly constant from chamber to chamber, but is subject to a small secondary fluctuation*.

When the linear dimensions of successive chambers are in continued proportion, then, in order that the whole shell may constitute a logarithmic spiral, it is necessary that the several chambers should subtend equal angles of revolution at the pole. In the case of the Miliolidae this is obviously the case (Fig. 425); for in this family the chambers lie in two rows (*Biloculina*), or three rows (*Triloculina*), or in some other small number of series: so that the angles subtended by them are large, simple fractions of the circular arc, such as 180° or 120°. In many of the nautiloid forms, such as *Cyclammina* (Fig. 426), the angles subtended, though of less magnitude, are still remarkably constant, as we may see by Fig. 427; where the angle subtended by each chamber is made equal to 20°, and this diagrammatic figure is not perceptibly different from the other. In some cases the subtended angle is less constant; and in these it would be necessary to equate the several linear dimensions with the corresponding vector angles, according to our equation $r = e^{\theta \cot \alpha}$. It is probable that, by so taking account of variations of θ, such variations of r as (according to Rhumbler's measurements) *Pulvinulina* and other genera appear to shew, would be found to diminish or even to disappear.

The law of increase by which each chamber bears a constant ratio of magnitude to the next may be looked upon as a simple

* Hans Przibram asserts that the linear ratio of successive chambers tends in many Foraminifera to approximate to 1·26, which $= \sqrt[3]{2}$; in other words, that the volumes of successive chambers tend to double. This Przibram would bring into relation with another law, viz. that insects and other arthropods tend to moult, or to metamorphose, just when they double their weights, or increase their linear dimensions in the ratio of $1 : \sqrt[3]{2}$. (Die Kammerprogression der Foraminiferen als Parallele zur Häutungsprogression der Mantiden, *Arch. f. Entw. Mech.* XXXIV, p. 680, 1813.) Neither rule seems to me to be well grounded (see above, p. 165).

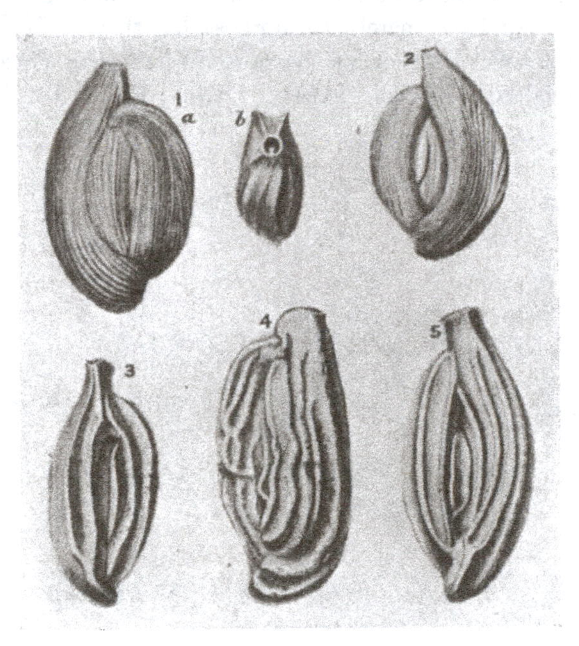

Fig. 425. 1, 2, *Miliolina pulchella* d'Orb.; 3–5, *M. linnaeana* d'Orb. After Brady.

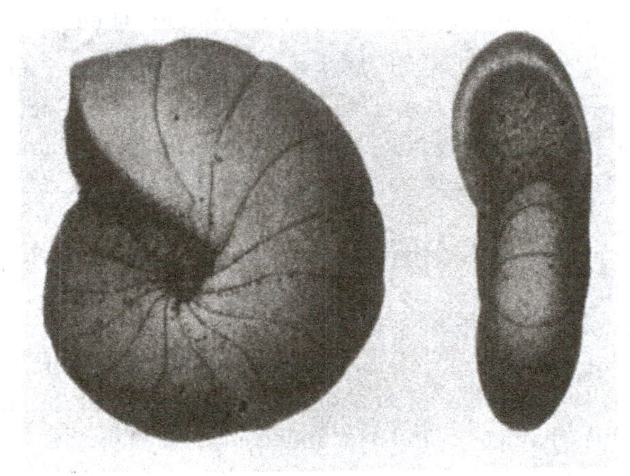

Fig. 426. *Cyclammina cancellata* Brady.

consequence of the structural uniformity or homogeneity of the organism; we have merely to suppose (as this uniformity would naturally lead us to do) that the rate of increase is at each instant proportional to the whole existing mass. For if V_0, V_1, etc. be

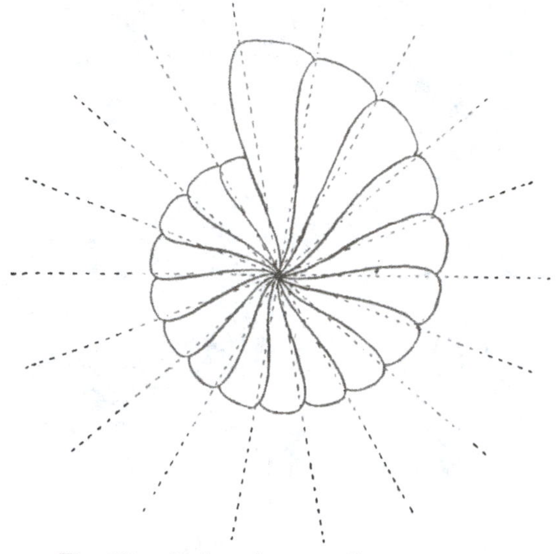

Fig. 427. *Cyclammina* sp. (Diagrammatic.)

the volumes of the successive chambers, let V_1 bear a constant proportion to V_0, so that $V_1 = qV_0$, and let V_2 bear the same proportion to the whole pre-existing volume: then

$$V_2 = q(V_0 + V_1) = q(V_0 + qV_0) = qV_0(1+q) \text{ and } V_2/V_1 = 1 + q.$$

This ratio of $1/(1+q)$ is easily shewn to be the constant ratio running through the whole series, from chamber to chamber; and if this ratio of volumes be constant, so also are the ratios of corresponding surfaces, and of corresponding linear dimensions, provided always that the successive increments, or successive chambers, are similar in form.

We have still to discuss the similarity of form and the symmetry of position which characterise the successive chambers, and which, together with the law of continued proportionality of size, are the distinctive characters and the indispensable conditions of a series of "gnomons."

The minute size of the foraminiferal shell or at least of each successive increment thereof, taken in connection with the fluid or semi-fluid nature of the protoplasmic substance, is enough to suggest that the molecular forces, and especially the force of surface-tension, must exercise a controlling influence over the form of the whole structure; and this suggestion, or belief, is already implied in our statement that each successive increment of growing protoplasm constitutes a separate *drop*. These "drops," partially concealed by their successors, but still shewing in part their rounded outlines, are easily recognisable in the various foraminiferal shells which are illustrated in this chapter.

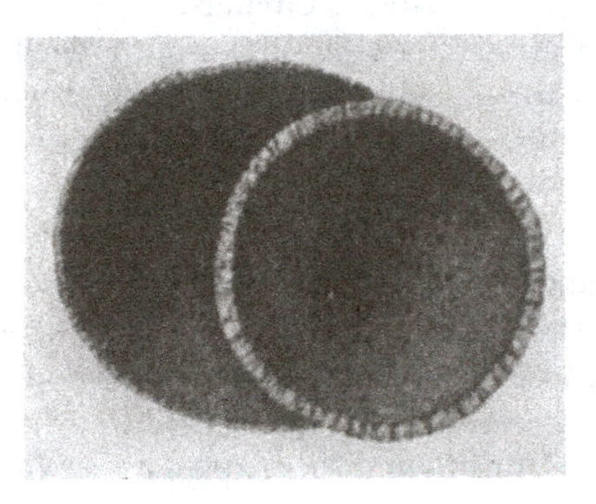

Fig. 428. *Orbulina universa* d'Orb.

The accompanying figure represents, to begin with, the spherical shell characteristic of the common, floating, oceanic *Orbulina*. In the specimen illustrated, a second chamber, superadded to the first, has arisen as a drop of protoplasm which exuded through the pores of the first chamber, accumulated on its surface, and spread over the latter till it came to rest in a position of equilibrium. We may take it that this position of equilibrium is determined, at least in the first instance, by the "law of the constant angle," which holds, or tends to hold, in all cases where the free surface of a given liquid is in contact with a given solid, in presence of another liquid or a gas. The corresponding equations are precisely the same as those which we have used in discussing the form of a drop (on p. 466); though some slight modification must be made in our definitions, inasmuch as the consideration of surface-*tension* is no longer appropriate at the solid surfaces, and the concept of surface-*energy* must take its place. Be that as it may, it is enough for us to observe that, in such a case as ours, when a given fluid (namely protoplasm) is in surface contact with a solid (viz. a calcareous shell), in presence of another fluid (sea-water), then the angle of contact, or angle by which the common surface (or interface) of the two liquids abuts against the solid wall, tends to be constant: and that being so, the drop will have a certain definite form, depending (*inter alia*) on the form of the surface with which it is in contact. After a period of rest, during which the surface of our second drop becomes rigid by calcification, a new period of growth will recur and a new drop of protoplasm be accumulated. Circumstances remaining the same, this new drop will meet the solid surface of the shell at the same angle as did the former one; and, the other forces at work on the system remaining the same, the form of the whole drop, or chamber, will be the same as before.

According to Rhumbler, this "law of the constant angle" is the fundamental principle in the mechanical conformation of the foraminiferal shell, and provides for the symmetry of form as well as of position in each succeeding drop of protoplasm: which form and position, once acquired, become rigid and fixed with the onset of calcification. But Rhumbler's explanation brings with it its own difficulties. It is by no means easy of verification, for on the very complicated curved surfaces of the shell it seems to me extraordinarily difficult to measure, or even to recognise, the actual angle of contact: of which angle of contact, by the way, but little is known, save only in the particular case where one of the three bodies is air, as when a surface of water is exposed to air and in contact with glass. It is easy moreover to see that in many of our Foraminifera the angle of contact, though it may be constant in homologous positions from chamber to chamber, is by no means constant at all points along the boundary of each chamber. In *Cristellaria*, for instance (Fig. 429), it would seem to be (and Rhumbler asserts that it actually is) about 90° on the outer side and only about 50° on the inner side of each septal partition; in *Pulvinulina* (Fig. 363), according to Rhumbler, the angles adjacent to the mouth are of 90°, and the opposite angles are of 60°, in each chamber. For these and other similar discrepancies Rhumbler would account by simply invoking the heterogeneity of the protoplasmic drop: that is to say, by assuming that the protoplasm has a different composition and different properties (including a very different distribution of surface-energy), at points near to and remote from the mouth of the shell. Whether the differences in angle of contact be as great as Rhumbler takes them to be, whether marked heterogeneities of the protoplasm occur, and whether these be enough to account for the differences of angle, I cannot tell. But it seems to me that we had better rest content with a general statement, and that Rhumbler has taken too precise and narrow a view.

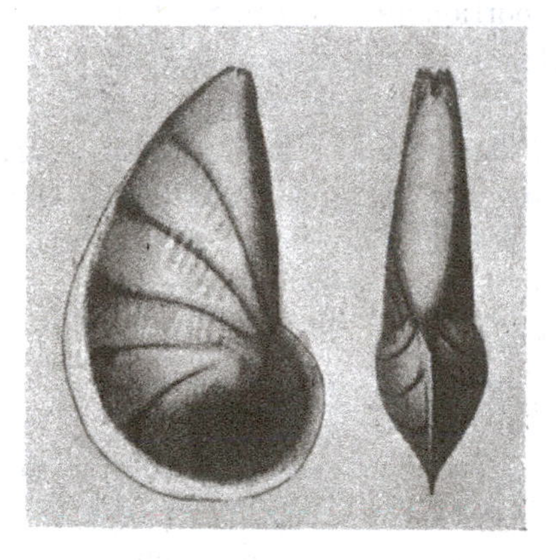

Fig. 429. *Cristellaria reniformis* d'Orb

In the molecular growth of a crystal, although we must of necessity assume that each molecule settles down in a position of minimum potential energy, we find it very hard indeed to explain precisely, even in simple cases and after all the labours of modern crystallographers, why this or that position is actually a place of minimum potential. In the case of our little Foraminifer (just as in the case of the crystal), let us then be content to assert that each drop or bead of protoplasm takes up a position of minimum potential energy, in relation to all the circumstances of the case; and let us not attempt, in the present state of our knowledge, to define that position of minimum potential by reference to angle of contact or any other particular condition of equilibrium. In most cases the whole exposed surface, on some portion of which the drop must come to rest, is an extremely complicated one, and the forces involved constitute a system which, in its entirety, is more complicated still; but from the symmetry of the case and the continuity of the whole phenomenon, we are entitled to believe that the conditions are just the same, or very nearly the same, time after time, from one chamber to another: as the one chamber is conformed so will the next tend to be, and as the one is situated relatively to the system so will its successor tend to be situated in turn. The physical law of minimum potential (including also the law of minimal area) is all that we need in order to explain, *in general terms*, the continued similarity of one chamber to another; and the physiological law of growth, by which a continued proportionality of size tends to run through the series of successive chambers, impresses the form of a logarithmic spiral upon this series of similar increments.

In each particular case the nature of the logarithmic spiral, as defined by its constant angle, will be chiefly determined by the rate of growth; that is to say by the particular ratio in which each new chamber exceeds its predecessor in magnitude. But shells having the same constant angle (α) may still differ from one another in many ways—in the general form and relative position of the chambers, in their extent of overlap, and hence in the actual contour and appearance of the shell; and these variations must correspond to particular distributions of energy within the system, which is governed as a whole by the law of minimum potential.

Our problem, then, becomes reduced to that of investigating the possible configurations which may be derived from the successive symmetrical apposition of similar bodies whose magnitudes are in continued proportion; and it is obvious, mathematically speaking, that the various possible arrangements all come under the head of the logarithmic spiral, together with the limiting cases which it includes. Since the difference between one such form and another depends upon the numerical value of certain coefficients of magnitude, it is plain that any one must tend to pass into any other by small and continuous gradations; in other words, that a *classification* of these forms must (like any classification whatsoever of logarithmic spirals or of any other mathematical curves) be theoretic or "artificial." But we may easily make such an artificial classification, and shall probably find it to agree, more or less, with the usual methods of classification recognised by biological students of the Foraminifera.

Firstly we have the typically spiral shells, which occur in great variety, and which (for our present purpose) we need hardly describe further. We may merely notice how in certain cases, for instance *Globigerina*, the individual chambers are little removed from spheres; in other words, the area of contact between the adjacent chambers is small. In such forms as *Cyclammina* and *Pulvinulina*, on the other hand, each chamber is greatly overlapped by its successor, and the spherical form of each is lost in a marked asymmetry. Furthermore, in *Globigerina* and some others we have a tendency to the development of a gauche spiral in space, as in so many of our univalve molluscan shells. The mathematical problem of how a shell should grow, under the assumptions which we have made, would probably find its most general statement in such a case as that of *Globigerina*, where the whole organism lives and grows freely poised in a medium whose density is little different from its own.

Fig. 430. *Discorbina bertheloti* d'Orb.

The majority of spiral forms, on the other hand, are plane or discoid spirals, and we may take it that in these cases some force has exercised a controlling influence, so as to keep all the chambers in a plane. This is especially the case in forms like *Rotalia* or *Discorbina* (Fig. 430), where the organism lives attached to a rock or a frond of sea-weed; for here (just as in the case of the coiled tubes which little worms such as *Serpula* and *Spirorbis* make, under similar conditions) the spiral disc is itself asymmetrical, its whorls being markedly flattened on their attached surfaces.

We may also conceive, among other conditions, the very curious case in which the protoplasm may entirely overspread the surface of the shell without reaching a position of equilibrium; in which case a new shell will be formed *enclosing* the old one, whether the old one be in the form of a single, solitary chamber, or have already attained to the form of a chambered or spiral shell. This is precisely what often happens in the case of *Orbulina*, when within the spherical shell we find a small, but perfectly formed, spiral "*Globigerina**."

The various Miliolidae (Fig. 425) only differ from the typical spiral, or rotaline forms, in the large angle subtended by each chamber, and the consequent abruptness of their inclination to each other. In these cases the *outward* appearance of a spiral tends to be lost; and it behoves us to recollect, all the more, that our spiral curve is not necessarily identical with the *outline* of the shell, but is always a line drawn through corresponding *points* in the successive chambers of the latter.

We reach a limiting case of the logarithmic spiral when the chambers are arranged in a straight line; and the eye will tend to associate with this limiting case the much more numerous forms in which the spiral angle is small, and the shell only exhibits a gentle curve with no succession of enveloping whorls. This constitutes the Nodosarian type (Fig. 134, p. 421); and here again, we must postulate some force which has tended to keep the chambers in a rectilinear series: such for instance as gravity, acting on a system of "hanging drops."

In *Textularia* and its allies (Fig. 431) we have a precise parallel to the helicoid cyme of the botanists (cf. p. 767): that is to say we have a screw translation, perpendicular to the plane of the underlying logarithmic spiral. In other words, in tracing a genetic spiral through the whole succession of chambers, we do so by a continuous vector rotation through successive angles of 180° (or 120° in some cases), while the pole moves along an axis perpendicular to the original plane of the spiral.

Another type is furnished by the "cyclic" shells of the Orbitolitidae, where small and numerous chambers tend to be added

* Cf. G. Schacko, Ueber *Globigerina*-Einschluss bei *Orbulina*, *Wiegmann's Archiv*, XLIX, p. 428, 1883; Brady, *Chall. Rep.* 1884, p. 607.

on round and round the system, so building up a circular flattened disc. This again we perceive to be, mathematically, a limiting case of the logarithmic spiral; the spiral has become wellnigh a circle and the constant angle is wellnigh 90°.

Lastly there are a certain number of Foraminifera in which, without more ado, we may simply say that the arrangement of the chambers is irregular, neither the law of constant ratio of magnitude nor that of constant form being obeyed. The chambers are heaped pell-mell upon one another, and such forms are known to naturalists as the Acervularidae.

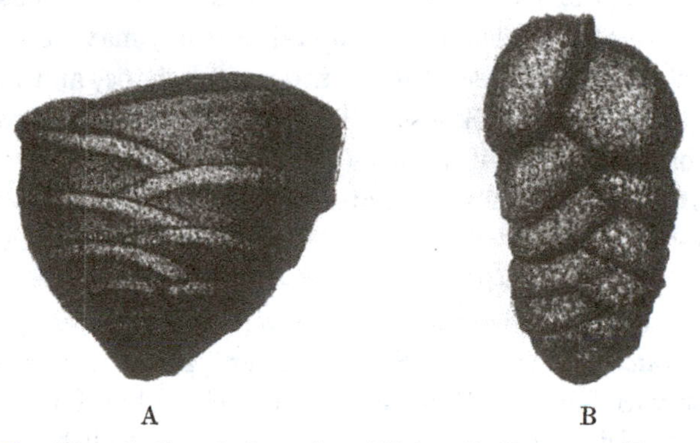

Fig. 431. A, *Textularia trochus* d'Orb. B, *T. concava* Karrer.

While in these last we have an extreme lack of regularity, we must not exaggerate the regularity or constancy which the more ordinary forms display. We may think it hard to believe that the simple causes, or simple laws, which we have described should operate, and operate again and again, in millions of individuals to produce the same delicate and complex conformations. But we are taking a good deal for granted if we assert that they do so, and in particular we are assuming, with very little proof, the "constancy of species" in this group of animals. Just as Verworn has shewn that the typical *Amoeba proteus*, when a trace of alkali is added to the water in which it lives, tends, by alteration of surface tensions, to protrude the more delicate pseudopodia characteristic of *A. radiosa*—and again when the water is rendered a little more alkaline, to turn apparently into the so-called *A. limax*—so it is evident that a very slight modification in the surface-energies concerned might tend to turn one so-called species into another among the Foraminifera. To what extent this process actually occurs, we do not know.

But that this, or something of the kind, does actually occur we can scarcely doubt. For example in the genus *Peneroplis*, the first portion of the shell consists of a series of chambers arranged in a spiral or nautiloid series; but as age advances the spiral is apt to be modified in various ways*. Sometimes the successive chambers grow rapidly broader, the whole shell becoming fan-shaped. Sometimes the chambers become narrower, till they no longer enfold the earlier chambers but only come in contact each with its immediate predecessor: the result being that the shell straightens out, and (taking into account the earlier spiral portion) may be described as crozier-shaped. Between these extremes of shape, and in regard to other variations of thickness or thinness, roughness or smoothness, and so on, there are innumerable gradations passing one into another and intermixed without regard to geographical distribution:—"wherever Peneroplides abound this wide variation exists, and nothing can be more easy than to pick out a number of striking specimens and give to each a distinctive name, but *in no other way can they be divided into 'species.'*†" Some writers have wondered at the peculiar variability of this particular shell‡; but for all we know of the life-history of the Foraminifera, it may well be that a great number of the other forms which we distinguish as separate species and even genera are no more than temporary manifestations of the same variability§.

* Cf. H. B. Brady, *Challenger Rep., Foraminifera*, 1884, p. 203, pl. XIII.

† Brady, *op. cit.* p. 206; Batsch, one of the earliest writers on Foraminifera, had already noticed that this whole series of ear-shaped and crozier-shaped shells was filled in by gradational forms; *Conchylien des Seesandes*, 1791, p. 4, pl. VI, fig. 15*a–f*. See also, in particular, Dreyer, *Peneroplis; eine Studie zur biologischen Morphologie und zur Speciesfrage*, Leipzig, 1898; also Eimer und Fickert, Artbildung und Verwandschaft bei den Foraminiferen, *Tübinger zqol. Arbeiten*, III, p. 35, 1899.

‡ Doflein, *Protozoenkunde*, 1911, p. 263: "Was diese Art veranlässt in dieser Weise gelegentlich zu variiren, ist vorläufig noch ganz räthselhaft."

§ In the case of *Globigerina*, some fourteen species (out of a very much larger number of described forms) were allowed by Brady (in 1884) to be distinct; and this list has been, I believe, rather added to than diminished. But these so-called species depend for the most part on slight differences of degree, differences in the angle of the spiral, in the ratio of magnitude of the segments, or in their area of contact one with another. Moreover with the exception of one or two "dwarf"

Conclusion

If we can comprehend and interpret on some such lines as these the form and mode of growth of the foraminiferal shell we may also begin to understand two striking features of the group, on the one hand the large number of diverse types or families which exist and the large number of species and varieties within each, and on the other the persistence of forms which in many cases seem to have undergone little change or none at all from the Cretaceous or even from earlier periods to the present day. In few other groups, perhaps only among the Radiolaria, do we seem to possess so nearly complete a picture of all possible transitions between form and form, and of the whole branching system of the evolutionary tree: as though little or nothing of it had ever perished, and the whole web of life, past and present, were as complete as ever. It leads one to imagine that these shells have grown according to laws so simple, so much in harmony with their material, with their environment, and with all the forces internal and external to which they are exposed, that none is better than another and none fitter or less fit to survive. It invites one also to contemplate the possibility of the lines of possible variation being here so narrow and determinate that identical forms may have come independently into being again and again.

While we can trace in the most complete and beautiful manner the passage of one form into another among these little shells, and ascribe them all at last (if we please) to a series which starts with the simple sphere of *Orbulina* or with the amoeboid body of *Astrorhiza*, the question stares us in the face whether this be an "evolution" which we have any right to correlate with historic *time*. The mathematician can trace one conic section into another, and "evolve" for example, through innumerable graded ellipses, the circle from the straight line: which tracing of continuous steps is a true "evolution," though time has no part therein. It was after this fashion that Hegel, and for that matter Aristotle himself, was an

forms, said to be limited to Arctic and Antarctic waters, there is no principle of geographical distribution to be discerned amongst them. A species found fossil in New Britain turns up in the North Atlantic; a species described from the West Indies is rediscovered at the ice-barrier of the Antarctic.

evolutionist—to whom evolution was a mental concept, involving order and continuity in thought but not an actual sequence of events in time. Such a conception of evolution is not easy for the modern biologist to grasp, and is harder still to appreciate. And so it is that even those who, like Dreyer* and like Rhumbler, study the foraminiferal shell as a physical system, who recognise that its whole plan and mode of growth is closely akin to the phenomena exhibited by fluid drops under particular conditions, and who explain the conformation of the shell by help of the same physical principles and mathematical laws—yet all the while abate no jot or tittle of the ordinary postulates of modern biology, nor doubt the validity and universal applicability of the concepts of Darwinian evolution. For these writers the *biogenetisches Grundgesetz* remains impregnable. The Foraminifera remain for them a great family tree, whose actual pedigree is traceable to the remotest ages; in which historical evolution has coincided with progressive change; and in which structural fitness for a particular function (or functions) has exercised its selective action and ensured "the survival of the fittest." By successive stages of historic evolution we are supposed to pass from the irregular *Astrorhiza* to a *Rhabdammina* with its more concentrated disc; to the forms of the same genus which consist of but a single tube with central chamber; to those where this chamber is more and more distinctly segmented; so to the typical many-chambered Nodosariae; and from these, by another definite advance and later evolution to the spiral Trochamminae. After this fashion, throughout the whole varied series of the Foraminifera, Dreyer and Rhumbler (following Neumayr) recognise so many successions of related forms, one passing into another and standing towards it in a definite relationship of ancestry or descent. Each evolution of form, from simpler to more complex, is deemed to have been attended by an advantage to the organism, an enhancement of its chances of survival or perpetuation; hence the historically older forms are on the whole structurally the simpler; or conversely, the simpler forms, such as the simple sphere, were the first to come into being in primeval seas; and finally, the gradual development and increasing complication of the individual

* F. Dreyer, Prinzipien der Gerüstbildung bei Rhizopoden, etc., *Jen. Zeitschr.* XXVI, pp. 204–468, 1892.

within its own lifetime is held to be at least a partial recapitulation of the unknown history of its race and dynasty*.

We encounter many difficulties when we try to extend such concepts as these to the Foraminifera. We are led for instance to assert, as Rhumbler does, that the increasing complexity of the shell, and of the manner in which one chamber is fitted on another, makes for advantage; and the particular advantage on which Rhumbler rests his argument is *strength.* Increase of strength, *die Festigkeitssteigerung*, is according to him the guiding principle in foraminiferal evolution, and marks the historic stages of their development in geologic time. But in days gone by I used to see the beach of a little Connemara bay bestrewn with millions upon millions of foraminiferal shells, simple Lagenae, less simple Nodosariae, more complex Rotaliae: all drifted by wave and gentle current from their sea-cradle to their sandy grave: all lying bleached and dead: one more delicate than another, but all (or vast multitudes of them) perfect and unbroken. And so I am not inclined to believe that niceties of form affect the case very much: nor in general that foraminiferal life involves a struggle for existence wherein breakage is a danger to be averted, and strength an advantage to be ensured†.

In the course of the same argument Rhumbler remarks that Foraminifera are absent from the coarse sands and gravels‡, as Williamson indeed had observed many years ago: so averting, or at least escaping, the dangers of concussion. But this is after all

* A difficulty arises in the case of forms (like *Peneroplis*) where the young shell appears to be more complex than the old, the first-formed portion being closely coiled while the later additions become straight and simple: "die biformen Arten verhalten sich, kurz gesagt, gerade umgekehrt als man nach dem biogenetischen Grundgesetz erwarten sollte," Rhumbler, *op. cit.* p. 33, etc.

† "Das Festigkeitsprinzip als *Movens* der Weiterentwicklung ist zu interessant und für die Aufstellung meines Systems zu wichtig um die Frage unerörtert zu lassen, warum diese Bevorzügung der Festigkeit stattgefunden hat. Meiner Ansicht nach lautet die Antwort auf diese Frage einfach, weil die Foraminiferen meistens unter Verhältnissen leben, die ihre Schalen in hohem Grade der Gefahr des Zerbrechens aussetzen; es muss also eine fortwahrende Auslese des Festeren stattfinden," Rhumbler, *op. cit.* p. 22.

‡ "Die Foraminiferen kiesige oder grobsandige Gebiete des Meeresbodens *nicht lieben*, u.s.w.": where the last two words have no particular meaning, save only that (as M. Aurelius says) "of things that use to be, we say commonly that they love to be."

a very simple matter of mechanical analysis. The coarseness or fineness of the sediment on the sea-bottom is a measure of the current: where the current is strong the larger stones are washed clean, where there is perfect stillness the finest mud settles down; and the light, fragile shells of the Foraminifera find their appropriate place, like every other graded sediment, in this spontaneous order of levigation.

The theorem of Organic Evolution is one thing; the problem of deciphering the lines of evolution, the order of phylogeny, the degrees of relationship and consanguinity, is quite another. Among the higher organisms we arrive at conclusions regarding these things by weighing much circumstantial evidence, by dealing with the resultant of many variations, and by considering the probability or improbability of many coincidences of cause and effect; but even then our conclusions are at best uncertain, our judgments are continually open to revision and subject to appeal, and all the proof and confirmation we can ever have is that which comes from the direct, but fragmentary evidence of palaeontology*.

But in so far as forms can be shewn to depend on the play of physical forces, and the variations of form to be directly due to simple quantitative variations in these, just so far are we thrown back on our guard before the biological conception of consanguinity, and compelled to revise the vague canons which connect classification with phylogeny.

The physicist explains in terms of the properties of matter, and classifies according to a mathematical analysis, all the drops and forms of drops and associations of drops, all the kinds of froth and foam, which he may discover among inanimate things; and his task ends there. But when such forms, such conformations and configurations, occur among *living* things, then at once the biologist introduces his concepts of heredity, of historical evolution, of succession in time, of recapitulation of remote ancestry in individual growth, of common origin (unless contradicted by direct evidence) of similar forms remotely separated by geographic space or geologic time, of fitness for a function, of adaptation to an environment, of higher and lower, of "better" and "worse." This is the fundamental

* In regard to the Foraminifera, "die Palaeontologie lásst uns leider an Anfang der Stammesgeschichte fast gänzlich im Stiche," Rhumbler, *op. cit.* p. 14.

difference between the "explanations" of the physicist and those of the biologist.

In the order of physical and mathematical complexity there is no question of the sequence of historic time. The forces that bring about the sphere, the cylinder or the ellipsoid are the same yesterday and to-morrow. A snow-crystal is the same to-day as when the first snows fell. The physical forces which mould the forms of *Orbulina*, of *Astrorhiza*, of *Lagena* or of *Nodosaria* to-day were still the same, and for aught we have reason to believe the physical conditions under which they worked were not appreciably different, in that yesterday which we call the Cretaceous epoch; or, for aught we know, throughout all that duration of time which is marked, but not measured, by the geological record.

In a word, the minuteness of our organism brings its conformation as a whole within the range of the molecular forces; the laws of its growth and form appear to lie on simple lines; what Bergson calls* the "ideal kinship" is plain and certain, but the "material affiliation" is problematic and obscure; and, in the end and upshot, it seems to me by no means certain that the biologist's usual mode of reasoning is appropriate to the case, or that the concept of continuous historical evolution must necessarily, or may safely and legitimately, be employed.

That things not only alter but improve is an article of faith, and the boldest of evolutionary conceptions. How far it be true were very hard to say; but I for one imagine that a pterodactyl flew no less well than does an albatross, and that Old Red Sandstone fishes swam as well and easily as the fishes of our own seas.

* The evolutionist theory, as Bergson puts it, "consists above all in establishing relations of ideal kinship, and in maintaining that wherever there is this relation of, so to speak, *logical* affiliation between forms, *there is also a relation of chronological succession between the species in which these forms are materialised*" (*Creative Evolution*, 1911, p. 26). Cf. *supra*, p. 412.

CHAPTER XIII

THE SHAPES OF HORNS, AND OF TEETH OR TUSKS: WITH A NOTE ON TORSION

We have had so much to say on the subject of shell-spirals that we must deal briefly with the analogous problems which are presented by the horns of sheep, goats, antelopes and other horned quadrupeds; and all the more, because these horn-spirals are on the whole less symmetrical, less easy of measurement than those of the shell, and in other ways also are less easy of investigation. Let us dispense altogether in this case with mathematics; and be content with a very simple account of the configuration of a horn.

There are three types of horn which deserve separate consideration: firstly, the horn of the rhinoceros; secondly, the horns of the sheep, the goat, the ox or the antelope, that is to say, of the so-called hollow-horned ruminants; and thirdly, the solid bony horns, or "antlers," which are characteristic of the deer.

The horn of the rhinoceros presents no difficulty. It is physiologically equivalent to a mass of consolidated hairs, and, like ordinary hair, it consists of non-living or "formed" material, continually added to by the living tissues at its base. In section the horn is elliptical, with the long axis fore-and-aft, or in some species nearly circular. Its longitudinal growth proceeds with a maximum velocity anteriorly, and a minimum posteriorly; and the ratio of these velocities being constant, the horn curves into the form of a logarithmic spiral in the manner that we have already studied. The spiral is of small angle, but in the longer-horned species, such as the great white rhinoceros (*Ceratorhinus*), the spiral curvature is distinctly recognised. As the horn occupies a median position on the head—a position, that is to say, of symmetry in respect to the field of force on either side—there is no tendency towards a lateral twist, and the horn accordingly develops as a *plane* logarithmic spiral. When two median horns coexist, the hinder one is much the smaller of the two: which is as much as to say that the force, or rate, of growth diminishes as we pass backwards, just as it does within the limits of the single horn. And accordingly, while both horns have *essentially* the same shape, the spiral curvature is less manifest in the second one, by the mere reason of its shortness.

The paired horns of the ordinary hollow-horned ruminants, such as the sheep or the goat, grow under conditions which are in some respects similar, but which differ in other and important respects from the conditions under which the horn grows in the rhinoceros. As regards its structure, the entire horn now consists of a bony core with a covering of skin; the inner, or dermal, layer of the latter is richly supplied with nutrient blood-vessels, while the outer layer, or epidermis, develops the fibrous or chitinous material, chemically and morphologically akin to a mass of cemented or consolidated hairs, which constitutes the "sheath" of the horn. A zone of active growth at the base of the horn keeps adding to this sheath, ring by ring, and the specific form of this annular zone may be taken as the "generating curve" of the horn*. Each horn no longer lies, as it does in the rhinoceros, in the plane of symmetry of the animal of which it forms a part; and the limited field of force concerned in the genesis and growth of the horn is bound, accordingly, to be more or less laterally asymmetrical. But the two horns are in symmetry one with another; they form "conjugate" spirals, one being the "mirror-image" of the other. Just as in the hairy coat of the animal each hair, on either side of the median "parting," tends to have a certain definite direction of its own, inclined away from the median axial plane of the whole system, so is it both with the bony core of the horn and with the consolidated mass of hairs or hair-like substance which constitutes its sheath; the primary axis of the horn is more or less inclined to, and may even be nearly perpendicular to, the axial plane of the animal.

The growth of the horny sheath is not continuous, but more or less definitely periodic: sometimes, as in the sheep, this periodicity is particularly well-marked, and causes the horny sheath to be com-

* In this chapter we keep to Moseley's way of regarding the equiangular spiral in space, of shell or horn, as generated by a certain figure which (*a*) grows, (*b*) revolves about an axis, and (*c*) is translated along or parallel to the said axis, all at certain appropriate and specific velocities. This method is simple, and even adequate, from the naturalist's point of view; but not so, or much less so, from the mathematician's, as we have found in the last chapter (p. 782).

posed of a series of all but separate rings, which are supposed to be formed year by year, and so to record the age of the animal*.

Just as Moseley sought for the true generating curve in the orifice, or "lip," of the molluscan shell, so we begin by assuming that in the spiral horn the generating curve corresponds to the lip or margin of one of the horny rings or annuli. This annular margin, or boundary of the ring, is usually a sinuous curve, not lying in a plane, but such as would form the boundary of an anticlastic surface of great complexity: to the meaning and origin of which phenomenon we shall return presently. But, as we have already seen in the case of the molluscan shell, the complexities of the lip itself, or of the corresponding lines of growth upon the shell, need

Fig. 432. The Argali sheep; *Ovis Ammon*. From Cook's *Spirals in Nature and Art*.

not concern us in our study of the development of the spiral: inasmuch as we may substitute for these actual boundary lines, their "trace," or projection on a plane perpendicular to the axis—in other words the simple outline of a transverse section of the whorl. In the horn, this transverse section is often circular or nearly so, as in the oxen and many antelopes: it now and then becomes of

* Cf. R. S. Hindekoper, *On the Age of the Domestic Animals*, Philadelphia and London, 1891, p. 173. In the case of the ram's horn, the assumption that the rings are annual is probably justified. In cattle they are much less conspicuous, but are sometimes well-marked in the cow; and in Sweden they are then called "calf-rings," from a belief that they record the number of offspring. That is to say, the growth of the horn is supposed to be retarded during gestation, and to be accelerated after parturition, when superfluous nourishment seeks a new outlet. (Cf. Lönnberg, *P.Z.S.* 1900, p. 689.)

somewhat complicated polygonal outline, as in a highland ram; but in many antelopes, and in most of the sheep, the outline is that of an isosceles or sometimes nearly equilateral triangle, a form which is typically displayed, for instance, in *Ovis Ammon*. The horn in this latter case is a trihedral prism, whose three faces are (1) an upper, or frontal face, in continuation of the plane of the frontal bone; (2) an outer, or orbital, starting from the upper margin of the orbit; and (3) an inner, or nuchal, abutting on the parietal bone*. Along these three faces, and their corresponding angles or edges, we can trace in the fibrous substance of the horn a series of homologous spirals, such as we have called in a preceding chapter the "*ensemble* of generating spirals" which define or constitute the surface.

The case of the horn differs in ways of its own from that of the molluscan shell. For one thing, the horn is always tubular—its generating curve is actually, as well as theoretically, a closed curve; there is no such thing as "involution," or the wrapping of one whorl within another, or successive intersection of the generating curve. Again, while the calcareous substance of the shell is laid down once for all, fixed and immovable, there is reason to believe that the young horn has, to begin with, a certain measure of flexibility, a certain freedom, even though it be slight, to bend or fold or wrinkle. And this being so, while it is no harder in the horn than in the shell to recognise the general field of force or general direction of growth, the actual conditions are somewhat more complex.

In some few cases, of which the male musk ox is one of the most notable, the horn is not developed in a continuous spiral curve. It changes its shape as growth proceeds; and this, as we have seen, is enough to show that it does not constitute a logarithmic spiral. The reason is that the bony exostoses, or horn-cores, about which the horny sheath is shaped and moulded, neither grow continuously nor even remain of constant size after attaining their full growth. But as the horns grow heavy the bony core is bent downwards by their weight, and so guides the growth of the horn in a new direction. Moreover as age advances, the core is further weakened and to a great extent absorbed: and the horny sheath or horn proper,

* Cf. Sir V. Brooke, On the large sheep of the Thian Shan, *P.Z.S.* 1875, p. 511.

deprived of its support, continues to grow, but in a flattened curve very different from its original spiral*. The chamois is a somewhat analogous case. Here the terminal, or oldest, part of the horn is curved; it tends to assume a spiral form, though from its comparative shortness it seems merely to be bent into a hook. But later on the bony core within, as it grows and strengthens, stiffens the horn and guides it into a straighter course or form. The same phenomenon of change of curvature, manifesting itself at the time when, or the place where, the horn is freed from the support of the internal core, is seen in a good many other antelopes (such as the

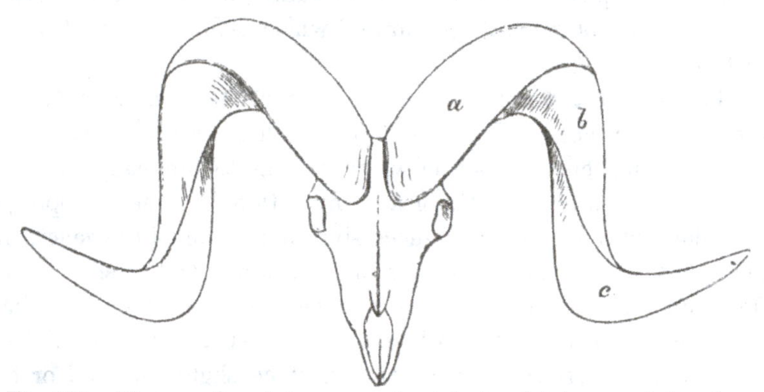

Fig. 433. Diagram of ram's horns. *a*, frontal; *b*, orbital; *c*, nuchal surface. After Sir Vincent Brooke, from *P.Z.S.*

hartebeest) and in many buffaloes; and the cases where it is most manifest appear to be those where the bony core is relatively short, or relatively weak. All these illustrate the cardinal difference between the growth of the horn and that of the bone below: the one dead, the other alive; the one adding and retaining its successive increments, the other mobile, plastic, and in continual flux throughout.

But in the great majority of horns we have no difficulty in recognising a continuous logarithmic spiral, nor in correlating it with an unequal rate of growth (parallel to the axis) on two opposite sides of the horn, the inequality maintaining a constant ratio as long as growth proceeds. In certain antelopes, such as the gemsbok, the spiral angle is very small, or in other words the horn

* Cf. E. Lönnberg, On the structure of the musk ox, *P.Z.S.* 1900, pp. 686–718.

is very nearly straight; in other species of the same genus *Oryx*, such as the Beisa antelope and the Leucoryx, a gentle curve (not unlike though generally less than that of a *Dentalium* shell) is evident; and the spiral angle, according to the few measurements I have made, is found to measure from about 20° to nearly 40°. In some of the large wild goats, such as the Scinde wild goat, we have

Fig. 434. Head of Arabian wild goat, *Capra sinaitica*. After Sclater, from *P.Z.S.*

a beautiful logarithmic spiral, with a constant angle of rather less than 70°; and we may easily arrange a series of forms, such for example as the Siberian ibex, the moufflon, *Ovis Ammon*, etc., and ending with the long-horned Highland ram: in which, as we pass from one to another, we recognise precisely homologous spirals with an increasing angular constant, the spiral angle being, for instance, about 75° or rather less in *Ovis Ammon*, and in the Highland ram a very little more. We have already seen that in the neighbourhood of 70° or 80° a small change of angle makes a marked difference in the appearance of the spire; and we know also that the actual length of the horn makes a very striking difference, for the spiral becomes especially conspicuous to the eye when horn or shell is long enough to shew several whorls, or at least a considerable part of one entire convolution.

Even in the simplest cases, such as the wild goats, the spiral is never a plane but always a *gauche* spiral: in greater or less degree there is always superposed upon the plane logarithmic spiral a helical spiral in space. Sometimes the latter is scarcely apparent, for the horn (though long, as in the said wild goats) is not nearly long enough to shew a complete convolution: at other times, as in the ram, and still better in many antelopes such as the koodoo, the corkscrew curve of the horn becomes its most characteristic feature. So we may study, as in the molluscan shell, the helicoid component of the spire—in other words the variation in what we have called (on p. 816) the angle β. This factor it is which, more than the constant angle of the logarithmic spiral, imparts a characteristic appearance to the various species of sheep, for instance to the various closely allied species of Asiatic wild sheep, or Argali. In all of these the constant angle of the logarithmic spiral is very much the same, but the enveloping angle of the cone differs greatly. Thus the long drawn out horns of *Ovis Poli*, four feet or more from tip to tip, differ conspicuously from those of *Ovis Ammon* or *O. hodgsoni*, in which a very similar logarithmic spiral is wound (as it were) round a much blunter cone.

Let us continue to dispense with mathematics, for the mathematical treatment of a gauche spiral is never very simple, and let us deal with the matter by experiment. We have seen that the generating curve, or transverse section, of a typical ram's horn is triangular in form. Measuring (along the curve of the horn) the length of the three edges of the trihedral structure in a specimen of *Ovis Ammon*, and calling them respectively the outer, inner, and hinder edges (from their position at the base of the horn, relatively to the skull), I find the outer edge to measure 80 cm., the inner 74 cm., and the posterior 45 cm.; let us say that, roughly, they are in the ratio of 9 : 8 : 5. Then, if we make a number of little cardboard triangles, equip each with three little legs (I make them of cork), whose relative lengths are as 9 : 8 : 5, and pile them up and stick them all together, we straightway build up a curve of double curvature precisely analogous to the ram's horn: except only that, in this first approximation, we have not allowed for the gradual increment (or decrement) of the triangular surfaces, that is to say, for the *tapering* of the horn due to the magnification of the generating curve.

In this case then, and in most other trihedral or three-sided horns, one of the three components, or three unequal velocities of growth, is of relatively small magnitude, but the other two are nearly equal one to the other; it would involve but little change for these latter to become precisely equal; and again but little to turn the balance of inequality the other way. But the immediate consequence of this altered ratio of growth would be that the horn would appear to wind the other way, as it does in the antelopes, and also in certain goats, e.g. the markhor, *Capra falconeri*.

For these two opposite directions of twist Dr Wherry has suggested a convenient nomenclature. When the horn winds so that we follow it from base to apex in the direction of the hands of a watch, it is customary to call it a "left-handed" spiral. Such a spiral we have in the horn on the left-hand side of a ram's head. Accordingly, Dr Wherry calls the condition *homonymous*, where, as in the sheep, a right-handed spiral is on the right side of the head, and a left-handed spiral on the left side; while he calls the opposite condition *heteronymous*, as we have it in the antelopes, where the right-handed twist is on the left side of the head, and the left-handed twist on the right-hand side. Among the goats, we may have either condition. Thus the domestic and most of the wild goats agree with the sheep; but in the markhor the twisted horns are heteronymous, as in the antelopes. The difference, as we have seen, is easily explained; and (very much as in the case of our opposite spirals in the apple-snail, referred to on p. 820) it has no very deep importance

Summarised then in a very few words, the argument by which we account for the spiral conformation of the horn is as follows: The horn elongates by dint of continual growth within a narrow zone, or annulus, at its base. If the rate of growth be identical on all sides of this zone, the horn will grow straight; if it be greater on one side than on the other, the horn will become curved; and it probably *will* be greater on one side than on the other, because each single horn occupies an unsymmetrical field with reference to the plane of symmetry of the animal. If the maximal and minimal velocities of growth be precisely at opposite sides of the zone of growth, the resultant spiral will be a plane spiral; but if they be not precisely or diametrically opposite, then the spiral will be a gauche spiral in space; and it is by no means likely that the maximum and minimum *will* occur at precisely opposite ends of a diameter, for no such plane of symmetry is manifested in the field of force to which the growing annulus corresponds or appertains.

Now we must carefully remember that the rates of growth of which we are here speaking are the net rates of longitudinal increment, in which increment the activity of the living cells in the zone of growth at the base of the horn is only one (though it is the fundamental) factor. In other words, if the horny sheath were continually being added to with equal rapidity all round its zone of active growth,

Fig. 435. Marco Polo's sheep: *Ovis Poli*. From Cook.

but at the same time had its elongation more retarded on one side than the other (prior to its complete solidification) by varying degrees of adhesion or membranous attachment to the bony core within, then the net result would be a spiral curve precisely such as would have arisen from initial inequalities in the rate of growth itself. It seems probable that this is an important factor, and sometimes even the chief factor in the case. The same phenomenon of attachment to the bony core, and the consequent friction or retardation with which the sheath slides over its surface, will lead to various subsidiary phenomena: among others to the presence of transverse folds or corrugations upon the horn, and to their unequal distribution upon its several faces or edges. And while it is perfectly true that nearly all the characters of the horn can be accounted for by unequal velocities of longitudinal growth upon its different sides, it is also plain that the actual field of force is a very complicated one indeed. For example, we can easily see (at least in the great majority of cases) that the direction of growth of the horny fibres of the sheath is by no means parallel to the axis of the core within; accordingly these fibres will tend to wind in a system of helicoid curves around the core, and not only this helicoid twist but any other tendency to spiral curvature on the part of the sheath will tend to be opposed or modified by the resistance of the core within. On the other hand living bone is a very plastic structure, and yields easily though slowly to any forces tending to its deformation; and so, to a considerable extent, the bony core itself will tend to be modelled by the curvature which the

Fig. 436. Head of *Ovis Ammon,* shewing St Venant's curves.

growing sheath assumes, and the final result will be determined by an equilibrium between these two systems.

While it is not very safe, perhaps, to lay down any general rule as to what horns are more and what are less spirally curved, I think it may be said that, on the whole, the thicker the horn the greater is its spiral curvature. It is the slender horns, of such forms as the Beisa antelope, which are gently curved, and it is the robust horns of goats or of sheep in which the curvature is more pronounced. Other things being the same, this is what we should expect to find; for it is where the transverse section of the horn is large that we may expect to find the more marked differences in the intensity of the field of force, whether of active growth or of retardation, on opposite sides or in different sectors thereof.

But there is yet another and a very remarkable phenomenon which we may discern in the growth of a horn when it takes the form of a curve of double curvature, namely, an effect of torsional strain; and this it is which gives rise to the sinuous "lines of growth," or sinuous boundaries of the separate horny rings, of which we have already spoken. It is not at first sight obvious that a mechanical strain of torsion is necessarily involved in the growth of the horn. In our experimental illustration (p. 880), we built up a twisted coil of separate elements, and no torsional strain attended the development of the system. So would it be if the horny sheath grew by successive annular increments, free save for their relation to one another and having no attachment to the solid core within. But as a matter of fact there is such an attachment, by subcutaneous connective tissue, to the bony core; and accordingly a torsional strain will be set up in the growing horny sheath, again provided that the forces of growth therein be directed more or less obliquely to the axis of the core; for a "couple" is thus introduced, giving rise to a strain which the sheath would not experience were it free (so to speak) to slip along, impelled only by the pressure of its own growth from below. And furthermore, the successive small increments of the growing horn (that is to say, of the horny sheath) are not instantaneously converted from living to solid and rigid substance; but there is an intermediate stage, probably long-continued, during which the new-formed horny substance in the neighbourhood of the zone of active growth is still plastic and capable of deformation.

Now we know, from the celebrated experiments of St Venant*, that in the torsion of an elastic body, other than a cylinder of circular section, a very remarkable state of strain is introduced. If the body be thus cylindrical (whether solid or hollow), then a twist leaves each circular section unchanged, in dimensions and in figure. But in all other cases, such as an elliptic rod or a prism of any particular sectional form, forces are introduced which act parallel to the axis of the structure, and which warp each section into a complex "anticlastic" surface. Thus in the case of a triangular and

* St Venant, De la torsion des prismes, avec des considérations sur leur flexion, etc., *Mém. des Savants Étrangers*, Paris, XIV, pp. 233–560, 1856. Karl Pearson dedicated part of his *History of the Theory of Elasticity* to the memory of this ingenious and original man. For a modern account of the subject see Love's *Elasticity* (2nd ed.), chap. XIV.

equilateral prism, such as is shewn in section in Fig. 437 A, if the part of the rod represented in the section be twisted by a force acting in the direction of the arrow, then the originally plane section will be warped as indicated in the diagram—where the full contour-lines represent elevation above, and the dotted lines represent depression below, the original level. On the external surface of the prism, then, contour-lines which were originally parallel and horizontal will be found warped into sinuous curves, such that, on each of the three faces, the curve will be convex upwards on one half, and concave upwards on the other half of the face. The ram's horn, and still better that of *Ovis Ammon*, is comparable to such a prism,

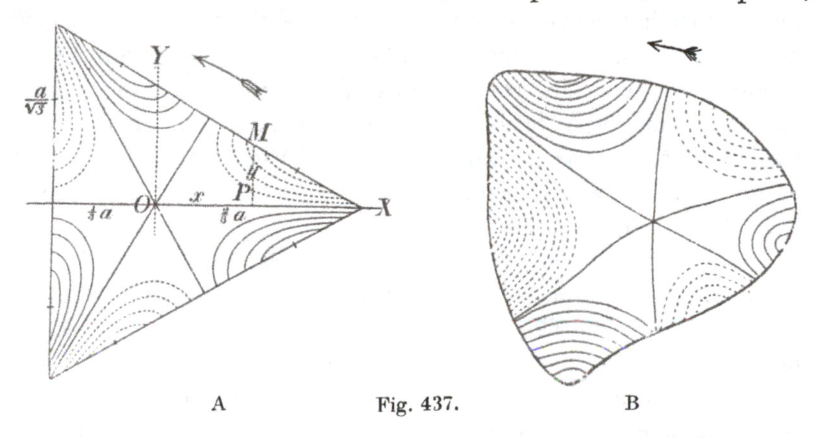

A Fig. 437. B

save that in section it is not quite equilateral, and that its three faces are not plane. The warping is therefore not precisely identical on the three faces of the horn; but, in the general distribution of the curves, it is in complete accordance with theory*. Similar anticlastic curves are well seen in many antelopes; but they are conspicuous by their absence in the *cylindrical* horns of oxen.

The better to illustrate this phenomenon, the nature of which is indeed obvious enough from a superficial examination of the horn, I made a plaster cast of one of the horny rings in a horn of *Ovis Ammon*, so as to get an accurate pattern of its sinuous edge: and then, filling the mould up with wet clay, I modelled an anticlastic

* The case of a thin conical shell under torsion is more complicated than either that of the cylinder or of a prismatic rod; and the more tapering horns doubtless deserve further study from this point of view. Cf. R. V. Southwell, On the torsion of conical shells, *Proc. R.S.* (A), CLXIII, pp. 337–355, 1937.

surface, such as to correspond as nearly as possible with the sinuous outline*. Finally, after making a plaster cast of this sectional surface, I drew its contour-lines (as shewn in Fig. 437 B) with the help of a simple form of spherometer. It will be seen that in great part this diagram is precisely similar to St Venant's diagram of the cross-section of a twisted triangular prism; and this is especially the case in the neighbourhood of the sharp angle of our prismatic section. That in parts the diagram is somewhat asymmetrical is not to be wondered at: and (apart from inaccuracies due to the somewhat rough means by which it was made) this asymmetry can be sufficiently accounted for by anisotropy of the material, by inequalities in thickness of different parts of the horny sheath, and especially (I think) by unequal distributions of rigidity due to the presence of the smaller corrugations of the horn. It is on account of these minor corrugations that in such horns as the Highland ram's, where they are strongly marked, the main St Venant effect is not nearly so well shewn as in smoother horns, such as those of *Ovis Ammon* and its congeners†.

The distribution of forces which manifest themselves in the growth and configuration of a horn is no simple nor merely superficial matter. One thing is coordinated with another; the direction of the axis of the horn, the form of its sectional boundary, the specific rates of growth in the mean spiral and at various parts of its periphery—all these play their parts, controlled in turn by the supply of nutriment which the character of the adjacent tissues and the distribution of the blood-vessels combine to determine. To suppose that this or that size or shape of horn has been produced or altered, acquired or lost, *by Natural Selection*, whensoever one type rather than another proved serviceable for defence or attack or any other purpose, is an hypothesis harder to define and to substantiate than some imagine it to be.

There are still one or two small matters to speak of before we leave these spiral horns. It is the way of sportsmen to keep record of big game by measuring the length along the curve of the horn and the span from tip to tip. Now if we study such measurements (as

* This is not difficult to do, with considerable accuracy, if the clay be kept well wetted or semi-fluid, and the smoothing be done with a large wet brush.

† The curves are well shewn in most of Sir V. Brooke's figures of the various species of Argali, in the paper quoted above, on p. 877.

they may be found in Mr Rowland Ward's book*), we shall soon see that the two measurements do not tally one with the other: but that a pair of horns, the longer when measured along the curve, may be the shorter from tip to tip, and *vice versa*. We might set this down to mere variability of form, but the true reason is simpler still. If the axes of the two horns stood straight out, at right angles to the median plane, then growth in length and in width of span would go on together. But if the two horns diverge at any lesser angle, then as the horns grow their spiral curvature will tend to bring their tips nearer and farther apart alternately.

There is one last, but not least curious property to be seen in a ram's horns. However large and heavy the horns may be—and in *Ovis Poli* 50 or 60 lb. is no unusual weight for the pair to grow to—the ram carries them with grace and ease, and they neither endanger his poise nor encumber his movements. The reason is that head and horns are very perfectly *balanced*, in such a way that no bending moment tends to turn the head up or down about its fulcrum in the atlas vertebra; if one puts two fingers into the foramen magnum one may lift up the heavy skull, and find it hang in perfect equilibrium. Moreover, the horns go on growing, but this equipoise is never lost nor changed; for the centre of gravity of the logarithmic spiral remains constant. There are other cases where heavy horns, well balanced as they doubtless are, yet visibly affect the set and balance of the head. The stag carries his head higher than a horse, and an Indian buffalo tilts his muzzle higher than a cow.

A further note upon torsion

The phenomenon of torsion, to which we have been thus introduced, opens up many wide questions in connection with form. Some of the associated phenomena are admirably illustrated in the case of climbing plants; but we can only deal with these still more briefly and parenthetically. The subject has been elaborately dealt with not only in Darwin's books†, but also by a great number of earlier and later writers. In "twining" plants, which constitute the greater number of "climbers," the essential phenomenon is a tendency of the growing shoot to revolve about a vertical axis—

* *Records of Big Game*, 9th edition, 1928.

† *Climbing Plants*, 1865 (2nd ed. 1875); *Power of Movement in Plants*, 1880.

a tendency long ago discussed by de Candolle and investigated by Palm, H. von Mohl and Dutrochet*. This tendency to revolution—circumvolution, as Darwin calls it, revolving nutation, as Sachs puts it—is very closely comparable to the process by which an antelope's horn (such as the koodoo's) acquires its spiral twist, and is due, in like manner, to inequalities in the rate of growth of the growing stem: with this difference between the two, that in the antelope's horn the zone of active growth is confined to the base of the horn, while in the climbing stem the same phenomenon is at work throughout the whole length of the growing structure. This growth is in the main due to "turgescence," that is to the extension, or elongation, of ready-formed cells through the imbibition of water; it is a phenomenon due to osmotic pressure. The particular stimulus to which these movements (that is to say, these inequalities of growth) have been ascribed can hardly be discussed here; but it was hotly debated fifty years ago and for many years thereafter, the point at issue being no other than whether direct physical causation, or the Darwinian concept of fitness or adaptation, should be invoked as an "explanation" of biological phenomena. The old *Naturphilosophie* had been inclined to look for spirals everywhere, and to attribute them to very simple causes: "Man wird nicht gross irren" (said Oken †) "wenn man sagt, alle Pflanzen entstehen als Spirale, und zwar weil sie feststehen und ein End gegen die Sonne kehren, die täglich einen Spiralgang um sie macht, u.s.w." When de Candolle saw a shoot curve under the influence of light (by heliotropism, as we are told to call it), he was content to regard the curvature as the result of different rates of growth on one side or other of the shoot, and these in turn as the direct result of differences of illumination. But by the Darwins, father and son, and by Sachs and by the Würzburg school, the curvature was ascribed to "irritability," a "stimulus" on one side of the shoot being followed by a "motor-reaction" on the other. The curvature was thus taken to be a "response" to external stimuli (such as light and gravity); and stimulus and response were supposed to have

* Palm, *Ueber das Winden der Pflanzen*, 1827; H. von Mohl, *Bau und Winden der Ranken*, etc., 1827; R. H. J. Dutrochet, Sur la volubilité des tiges de certains végétaux, et sur la cause de ce phénomène, *Ann. Sc. Nat.* (*Bot.*), II, pp. 156–167, 1844, and other papers.

† *Isis*, I, p. 222, 1817.

evolved together in the course of ages, to bring about something more and more fitted for survival in the struggle for existence. They were, in short, of the nature of *acquired habits*, rather than *physical phenomena*. But there was no gainsaying the fact that the immediate cause of curvature was inequality of growth on opposite sides*.

A simple stem growing upright in the dark, or in uniformly diffused light, would be in a position of equilibrium to a field of force radially symmetrical about its vertical axis. But this complete radial symmetry will not often occur; and the radial anomalies may be such as arise intrinsically from structural peculiarities in the stem itself, or externally to it by reason of unequal illumination or through various other localised forces. The essential fact, so far as we are concerned, is that in twining plants we have a very marked tendency to inequalities in longitudinal growth on different aspects of the stem—a tendency which is but an exaggerated manifestation of one which is more or less present, under certain conditions, in all plants whatsoever. Just as in the case of the ruminants' horns so we find here that this inequality may be, so to speak, positive or negative, the maximum lying to the one side or the other of the twining stem; and so it comes to pass that some climbers twine to the one side and some to the other: the hop and the honeysuckle following the sun, and the field-convolvulus twining in the reverse direction; there are also some, like the woody nightshade (*Solanum Dulcamara*), which twine indifferently either way.

Together with this circumnutatory movement, there is very generally to be seen an actual *torsion* of the twining stem—a twist, that is to say, about its own axis; and Mohl made the curious observation, confirmed by Darwin, that when a stem twines around a smooth cylindrical stick the torsion does not take place, save "only in that degree which follows as a mechanical necessity from the spiral winding": but that stems which had climbed around a rough stick were all more or less, and generally much, twisted. Here Darwin did not refrain from introducing that teleological argument which pervades his whole train of reasoning: "The stem," he says, "probably gains rigidity by being twisted (on the same

* On the whole controversy, see F. F. Blackman's obituary notice of Francis Darwin in *Proc. R.S.* (B), cx, 1932.

principle that a much twisted rope is stiffer than a slackly twisted one), and is thus indirectly benefited so as to be able to pass over inequalities in its spiral ascent, and to carry its own weight when allowed to revolve freely." The mechanical explanation would appear to be very simple, and such as to render the teleological hypothesis unnecessary. In the case of the roughened support, there is a temporary adhesion or "clinging" between it and the growing stem which twines around it; and a system of forces is thus set up, producing a "couple," just as it was in the case of the ram's or antelope's horn through direct adhesion of the bony core to the surrounding sheath. The twist is the direct result of this couple, and it disappears when the support is so smooth that no such force comes to be exerted.

Another important class of climbers includes the so-called "leaf-climbers." In these, some portion of the leaf, generally the petiole, sometimes (as in the fumitory) the elongated midrib, curls round a support; and a phenomenon of like nature occurs in many, though not all, of the so-called "tendril-bearers." Except that a different part of the plant, leaf or tendril instead of stem, is concerned in the twining process, the phenomenon here is strictly analogous to our former case; but in the resulting helix there is, as a rule, this obvious difference, that, while the twining stem, for instance of the hop, makes a slow revolution about its support, the typical leaf-climber makes a close, firm coil: the axis of the latter is nearly perpendicular and parallel to the axis of its support, while in the twining stem the angle between the two axes is comparatively small. Mathematically speaking, the difference merely amounts to this, that the component in the direction of the vertical axis is large in the one case, and the corresponding component is small, if not absent, in the other; in other words, we have in the climbing stem a considerable vertical component, due to its own tendency to grow in height, while this longitudinal or vertical extension of the whole system is not apparent, or little apparent, in the other cases. But from the fact that the twining stem tends to run obliquely to its support, and the coiling petiole of the leaf-climber tends to run transversely to the axis of its support, there immediately follows this marked difference, that the phenomenon of *torsion*, so manifest in the former case, will be absent in the latter.

There is one other phenomenon which meets us in the twining and twisted stem, and which is doubtless illustrated also, though not so well, in the antelope's horn; it is a phenomenon which forms the subject of a second chapter of St Venant's researches on the effects of torsional strain in elastic bodies. We have already seen how one effect of torsion, in for instance a prism, is to produce strains parallel to the axis, elevating parts and depressing other parts of each transverse section. But in addition to this, the same torsion has the effect of materially altering the form of the section itself, as we may easily see by twisting a square or oblong piece of india-rubber. If we start with a cylinder, such as a round piece of catapult india-rubber, and twist it on its own long axis, we have already seen that it suffers no other distortion; it still remains a cylinder, that is to say, it is still in section everywhere circular. But if it be of any other shape than cylindrical the case is different, for now the sectional shape tends to alter under the strain of torsion. Thus, if our rod be elliptical in section to begin with, it will, under torsion, become a more elongated ellipse; if it be square, its angles will become more prominent and its sides will curve inwards, till at length the square assumes the appearance of a four-pointed star with rounded angles. Furthermore, looking at the results of this process of modification, we find experimentally that the resultant figures are more easily twisted, less resistant to torsion, than were those from which we evolved them; and this is a very curious physical or mathematical fact. So a cylinder, which is especially resistant to torsion, is very easily bent or flexed; while projecting ribs or angles, such as an engineer makes in a bar or pillar of iron for the purpose of increasing its resistance to *bending*, actually make it much weaker than before (for the same amount of metal per unit length) in the way of resistance to *torsion*.

In the hop itself, and in a very considerable number of other twining and twisting stems, the ribbed or channelled form of the stem is a conspicuous feature. We may safely take it, (1) that such stems are especially susceptible of torsion; and (2) that the effect of torsion will be to intensify any such peculiarities of sectional outline which they may possess, though not to initiate them in an originally cylindrical structure. In the leaf-climbers the case does not present itself, for there, as we have seen, torsion itself is not, or is very slightly, manifested. There are very distinct traces of the phenomenon in the horns of certain antelopes, but the reason why it is not a more conspicuous feature of the antelope's horn or of the ram's is apparently a very simple one: namely, that the presence of the bony core within tends to check that deformation which is perpendicular, while it permits that which is parallel, to the axis of the horn.

Of deer's antlers

But let us return to our subject of the shapes of horns, and consider briefly our last class of these structures, namely the bony antlers of the elk and deer*. The problems which these present to us are very different from those which we have had to do with in the antelope or the sheep.

With regard to its structure, it is plain that the bony antler corresponds, upon the whole, to the bony core of the antelope's horn; while in place of the hard horny sheath of the latter, we have the soft "velvet," which every season covers the new growing antler, and protects the large nutrient blood-vessels by help of which the antler grows†. The main difference lies in the fact that in the one case the bony core, imprisoned within its sheath, is rendered incapable of branching and incapable also of lateral expansion, and the whole horn is only permitted to grow in length while retaining a sectional contour that is identical with (or but little altered from) that which it possesses at its growing base: but in the antler on the other hand no such restraint is imposed, and the living, growing fabric of bone is free to expand into a broad flat plate over which the blood-vessels run. In the immediate neighbourhood of the main blood-vessels growth will be most active, in the interspaces between it may wholly fail: with the result that we may have great notches cut out of the flattened plate, or may at length find it reduced to the

* For an elaborate study of antlers, see A. Rörig, *Arch. f. Entw. Mech.* x, pp. 525–644, 1900; xi, pp. 65–148, 225–309, 1901; C. Hoffmann, *Zur Morphologie der rezenten Hirsche,* 75 pp., 23 pls., 1901; also Sir Victor Brooke, On the classification of the Cervidae, *P.Z.S.* 1878, pp. 883–928. For a discussion of the development of horns and antlers, see H. Gadow, *P.Z.S.* 1902, pp. 206–222, and works quoted therein.

† Cf. L. Rhumbler, Ueber die Abhängigkeit des Geweihwachstums der Hirsche, speziell des Edelhirsches, vom Verlauf der Blutgefässe im Kalbengeweih, *Zeitschr. f. Forst. und Jagdwesen,* 1911, pp. 295–314.

form of a simple branching structure. The main point is that the "horn" is essentially an *axial rod*, while the "antler" is essentially an outspread *surface**. In other words, the whole configuration of an antler is more easily understood by conceiving it as a plate or a surface, more and more notched and scolloped till but a slender skeleton remains, than to look upon it the other way, namely as an axial stem (or beam) giving off branches (or tines), the interspaces between which latter may sometimes fill up to form a continuous surface.

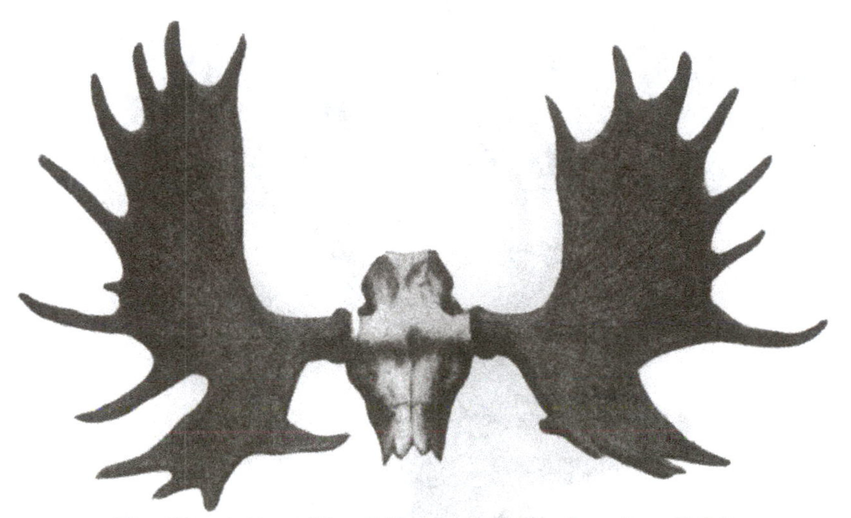

Fig. 438. Antlers of Swedish elk. After Lönnberg, from *P.Z.S.*

In a sense it matters very little whether we regard the broad plate-like antlers of the elk or the slender branching antlers of the stag as the more primitive type; for we are not concerned here with questions of hypothetical phylogeny, and even from the mathematical point of view it makes little or no difference whether we describe the plate as constituted by the interconnection of branches, or the branches as derived by the notching or incision

* The fact that in one very small deer, the little South American Coassus, the antler is reduced to a simple short spike, does not preclude the general distinction which I have drawn. In Coassus we have the beginnings of an antler, which has not yet manifested its tendency to expand; and in the many allied species of the American genus Cariacus, we find the expansion manifested in various simple modes of ramification or bifurcation.

of a plate. The important point for us is to recognise that (save for occasional slight irregularities) the branching system in the one *conforms* essentially to the curved plate or surface which we see plainly in the other. In short the arrangement of the branches is more or less comparable to that of the veins in a leaf, or to that of the blood-vessels as they course over the curved surface of an organ. It is a process of ramification, not, like that of a tree, in various planes, but strictly limited to a single surface. And just as the veins within a leaf are not necessarily confined (as they happen to

Fig. 439. Head and antlers of the Indian swamp-deer (*Cervus Duvauceli*). After Lydekker, from *P.Z.S.*

be in most ordinary leaves) to a *plane* surface, but, as in the petal of a tulip or the capsule of a poppy, may have to run their course within a curved surface, so does the analogy of the leaf lead us directly to the mode of branching which is characteristic of the antler. The surface to which the branches of the antler tend to be confined is a more or less spheroidal, or occasionally an ellipsoidal one; and furthermore, when we inspect any well-developed pair of antlers, such as those of a red deer, a sambur or a wapiti, we have no difficulty in seeing that the two antlers make up between them *a single surface*, and constitute a symmetrical figure, each half being the mirror-image of the other. It is what the ghillies call the "cup of the antler".

To put the case in another way, a pair of antlers (apart from occasional slight irregularities) tends to constitute a figure such that we could conceive an elastic sheet stretched over or round the entire system, and to form one continuous and even surface; and not only would the surface curvature be on the whole smooth and even, but the boundary of the surface would also tend to be an even curve: that is to say the tips of all the tines would approximately have their locus in a continuous curve.

It follows from this that if we want to make a simple model of a set of antlers, we shall be very greatly helped by taking some appropriate spheroidal surface as our groundwork or scaffolding. The best form of surface is a matter for trial and investigation in each particular case; but even in a sphere, by selecting appropriate areas thereof, we can obtain sufficient varieties of surface to meet all ordinary cases. With merely a bit of sculptor's clay or plasticine, we should be put hard to it to model the horns of a wapiti or a reindeer· but if we start with an orange (or a round florence flask) and lay our little tapered rolls of plasticine upon it, in simple natural curves, it is surprising to see how quickly and successfully we can imitate one type of antler after another. In either case, we shall be struck by the fact that our model may vary in its mode of branching within very considerable limits, and yet look perfectly natural; for the same wide range of variation is characteristic of the natural antlers themselves. As Sir V. Brooke says (*op. cit.* p. 892), "No two antlers are ever exactly alike; and the variation to which the antlers are subject is so great that in the absence of a large series they would be held to be indicative of several distinct species*." But all these many variations lie within a limited range, for they are all subject to our general rule that the entire structure is essentially confined to a single curved surface. A sheet of stiff paper makes an even simpler model. Fold it in two; cut a deer's head out of the double sheet, and leave a large oval where the antlers are to be; cut a few notches in this oval leaf, for the spaces between the tines (Fig. 440). The likeness to a pair of antlers seems remote to begin

* Cf. also the immense range of variation in elks' horns, as described by Lönnberg, *P.Z.S.* II, pp. 352–360, 1902.

with; but it is wonderfully improved as we separate the two antlers and give a twist to each, turning antler, tines and all, into the appropriate curved or twisted surface.

It is probable that in the curvatures both of the beam and of its tines, in the angles by which these latter meet the beam, and in the contours of the entire system, there are involved many elegant mathematical problems with which we cannot attempt to deal. Nor must we attempt meanwhile to enquire into the physical meaning or origin of these phenomena, for as yet the clue seems to be lacking and we should only heap one hypothesis upon another. That there is a complete contrast of mathematical properties between the horn and the antler is the main lesson with which, in the meantime, we must rest content.

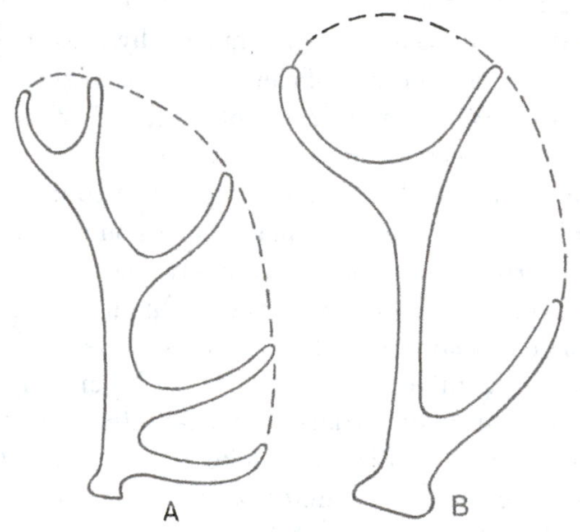

Fig. 440. Diagrams of antlers, before twisting into shape. A, Red-deer; B, Swamp-deer.

Of teeth, and of beak and claw

In a fashion similar to that manifested in the shell or the horn, we find the equiangular spiral to be implicit in a great many other organic structures where the phenomena of growth proceed in a similar way: that is to say, where about an axis there is some asymmetry leading to unequal rates of longitudinal growth, and where the structure is of such a kind that each new increment is added on as a permanent and unchanging part of the entire conformation. Nail and claw, beak and tooth, all come under this category. The logarithmic spiral *always* tends to manifest itself in such structures as these, though it usually only attracts our attention in elongated structures, where (that is to say) the radius vector has described a considerable angle. When the canary-bird's claws grow long from lack of use, or when the incisor tooth of a rabbit or a rat grows long by reason of disease or of injury of the opponent tooth against which it was wont to bite*, we know that the tooth or claw tends to grow into a spiral curve, and we speak of it as a malformation†. But there has been no fundamental change of form, only an abnormal increase in length; the elongated tooth or claw has the selfsame curvature which it had when it was short, but the spiral becomes more and more manifest the longer it grows. It is only natural, but nevertheless it is curious to see, how closely a rabbit's abnormally overgrown teeth come to resemble the tusks of swine or elephants, of which the normal state is one of hypertrophy. A curiously analogous case is that of the New Zealand Huia bird, in which the beak of the male is comparatively short and straight, while that of the female is long and curved; it is easy to see that there is a slight but identical curve also in the beak of the male, and that the beak of the female shews nothing but an extension or prolongation of the same. In the case of the more curved beaks, such as those of an eagle or a parrot, we may, if we please, determine the constant angle of the logarithmic spiral, just as we have done in the case of the *Nautilus* shell; and here again, as the bird grows older or the beak longer, the spiral nature of the curve becomes more and more apparent, as in the hooked beak of an old eagle, or in the great beak of a hyacinthine macaw.

Let us glance at one or two instances to illustrate the spiral curvature of teeth.

* Cf. John Hunter, *Natural History of the Human Teeth* (3rd ed.), 1808, p. 110: "Where a tooth has lost its opposite, it will in time become really so much longer than the rest as the others grow shorter by abrasion". Cf. James Murie, Notes on some diseased dental conditions in animals, *Tr. Odontol. Soc.* 1867–8, pp. 37–69, 257–298. We now know that a *Coenurus*-cyst in a rabbit's masseter muscle may twist the jaw sideways, so that the incisors fail to meet, and grow accordingly: H. A. Baylis, *Trans. R. Soc. Trop. Medicine*, XXXIII, p. 4, 1939.

† See Professor W. C. McIntosh's paper on "Abnormal teeth in certain mammals, especially in the rabbit," *Trans. R.S.E.* LVI, pp. 333–407, for a large collection of instances admirably illustrated.

A dentist knows that every tooth has a curvature of its own, and that in pulling the tooth he must follow the direction of the curve; but in an ordinary tooth this curvature is scarcely visible, and is least so when the diameter of the tooth is large compared with its length. In simple, more or less conical teeth, such as those of the dolphin, and in the more or less similarly shaped canines and incisors of mammals in general, the curvature of the tooth is particularly well seen. We see it in the little teeth of a hedgehog, and in the canines of a dog or a cat it is very obvious indeed. When the great canine of the carnivore becomes still further enlarged or elongated, as in *Machairodus*, it grows into the strongly curved sabre-tooth of that extinct tiger; and the boar's canine grows into the spiral tusk of wart-hog or babirussa. In rodents, it is the incisors which undergo elongation; their rate of growth differs, though but slightly, on the two sides of the axis, and by summation of these slight differences in the rapid growth of the tooth an unmistakable logarithmic spiral is gradually built up; we see it admirably in the beaver, or in the great ground-rat *Geomys*. The elephant is a similar case, save that the tooth or tusk remains, owing to comparative lack of wear, in a more perfect condition. In the rodent (save only in those abnormal cases mentioned on the last page) the tip, or first-formed part of the tooth wears away as fast as it is added to from behind; and in the grown animal, all those portions of the tooth near to the pole of the logarithmic spiral have long disappeared. In the elephant, on the other hand, we see, practically speaking, the whole unworn tooth, from point to root; and its actual tip nearly coincides with the pole of the spiral. If we assume (as with no great inaccuracy we may do) that the tip actually coincides with the pole, then we may very easily construct the continuous spiral of which the existing tusk constitutes a part; and by so doing, we see the short, gently curved tusk of our ordinary elephant growing gradually into the spiral tusk of the mammoth. No doubt, just as in the case of our molluscan shells, we have a tendency to variation, both individual and specific, in the constant angle of the spiral; some elephants, and some species of elephant, undoubtedly have a higher spiral angle than others. But in most cases, the angle would seem to be such that a spiral configuration would become very manifest indeed if only the tusk pursued its steady growth, unchanged otherwise in form, till it attained the dimensions which we meet with in the mammoth. In a species such as *Mastodon angustidens*, or *M. arvernensis*, the specific angle is low and the tusk comparatively straight; but the American mastodons and the existing species of elephant have tusks which do not differ appreciably, except in size, from the great spiral tusks of the mammoth, though from their comparative shortness the spiral is little developed and only appears to the eye as a gentle curve. Wherever the tooth is very long indeed, as in the mammoth or the beaver, the effect of some slight and all but inevitable lateral asymmetry in the rate of growth begins to shew itself: in other words, the spiral is seen to lie not absolutely in a plane, but to be a gauche curve, like a twisted horn. We see this condition very well in the huge canine tusks of the babirussa; it is a conspicuous feature in the mammoth, and it is more or less perceptible in any large tusk of the ordinary elephants.

The simplest of mammalian teeth are, like those of reptiles, conical buds which spring by single roots from a common origin: much as the pinnules of a compound leaf spring from a common petiole. A dolphin's teeth are typical of what Cope* called a *haplodont* dentition; a sloth's (whether degenerate or no) are no further advanced; canines remain unaltered throughout the mammalia, and incisors vary little save for some flattening due to crowding in a foreshortened jaw. Like the leaf and its pinnules, the tooth-germ buds and branches in endless ways; and we have no criterion of comparison (nor any right to expect it) between the individual cusps of a dog's, an elephant's and a horse's teeth, any more than between the several pinnules, cusps or leaflets of a rose, a maple and a horse-chestnut. The tooth-buds remain apart or coalesce in various numbers and degrees; and crowding, abrasion and mechanical pressure play a large part in the final arrangement and conformation†.

The dolphin's teeth, used only for prehension, do not impinge on one another, and stay sharp accordingly; those of the carnivores

* E. D. Cope, On the homologies and origin of the types of molar teeth in mammalian dentition, *Pr. Ac. N. S. Philad.* xxv, p. 371, 1873; *Journ. Ac. N. S. Philad.* (c), VIII, pp. 71–89, 1874.

† Cf. J. A. Ryder, Mechanical genesis of tooth forms, *Pr. Ac. N. S. Philad.* 1878, pp. 45–80.

interlock, rather than meet and oppose*; in herbivorous animals the molars grind one against another, and wear their crowns away. The teeth of ungulates have been studied with especial care by the palaeontologists on the basis of Cope's well-known tritubercular theory, and one is greatly daring who ventures to deal with them in a different way†. The case is neither plain nor easy. We are accustomed to speak of a "tooth" as a single unit, however complicated it may be; but we may err in doing so, and we encounter other difficulties in studying teeth whose crowns are worn away, and in interpreting the "patterns" which successive stages of wear and tear expose.

The elephant's molar is manifestly composite. We see on its worn surface a long succession of "enamel ridges," each marking a narrow ring or island, lying transversely, filled with dentine, surrounded by interstitial cement, and with a root or roots of its own. The molars develop one after another during the animal's lifetime; and each consists, to begin with, of so many separate island-elements, not yet cemented together nor worn down, each with its own roots, its own covering of enamel and its own transversely cuspidate crown. These are true dental units, the primitive individual "teeth", corresponding to the still simpler teeth of the dolphin; and they illustrate, and go far to confirm the view that the molar tooth is formed, both here and elsewhere, by "concrescence"‡. These *rudimenta dentium*, as old Patrick Blair called them, or *denticules* as Owen did, soon fuse together, and begin to wear down as soon as the great composite tooth rolls forward and emerges from the gum. As each denticule begins to wear away, it first appears as a transverse row of separate rings, the so-called *columns*, which represent the cusps of the original crown and vary in size, number and proportion with the species.

* This is precisely what Aristotle means when he describes the dog's teeth as *carcharodont*, or sharklike, i.e. interlocking—*καρχαρόδοντα γάρ ἐστιν ὅσα ἐπαλλάττει τὰς ὀδόντας τὰς ὀξεῖς*, H.A., ii, 501 a 18.

† See E. D. Cope, *loc. cit.* and H. F. Osborn, *passim.* Cf. also W. K. Gregory, A half century of trituberculy, *Proc. Am. Phil. Soc.* LXXIII, pp. 169–317, 1934, who says that "even the most complex molar patterns of the Ungulates are referable to the trituberculate type, in strict accord with the steps postulated by Cope and Osborn."

‡ A view held by Gaudry, Giebel, Kükental and others, but stoutly opposed by Cope and Osborn, who see in the molar tooth a single unit, complicated by "differentiation".

These columns soon fuse and vanish as the cusps wear down; and each denticule now appears as a continuous ring of enamel, within

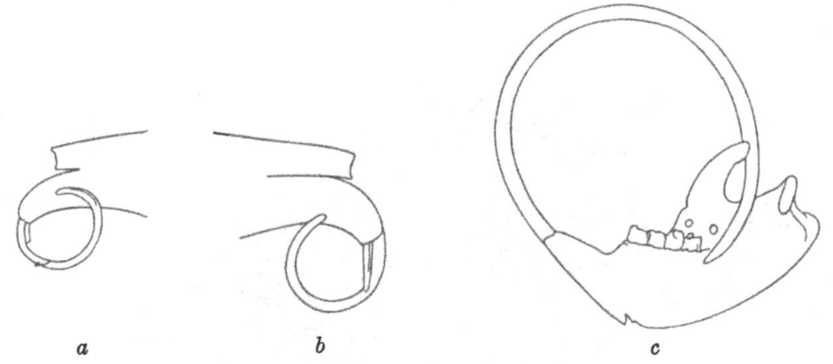

Fig. 441. Abnormal incisor teeth: *a*, *b*, of rabbit; *c*, of beaver. After McIntosh.

which the dentine is exposed and around which the cement accumulates*. A single great molar is made up of nearly a dozen of these

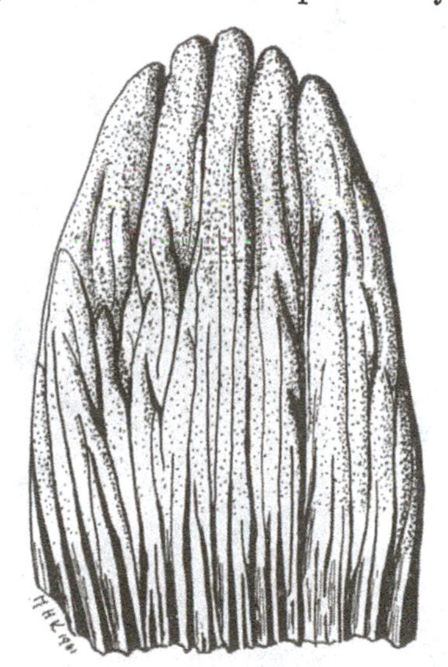

Fig. 442. A dental unit, or element of the composite molar tooth, of an Indian elephant. It consists of five "columns", terminating in yet unworn "cusps".

* See Blair's *Osteographia elephantina*, 1713, Tab. III, 19; also the figures in F. van Gaver's Étude de la tête d'un jeune Éléphant d'Asie, *Ann. Mus. Marseille*, XX, 1925. Cf. also L. Bolk, Zur Ontogenie des Elefantengebisses, *Odontologische Studien*, III, Leipzig, 1919.

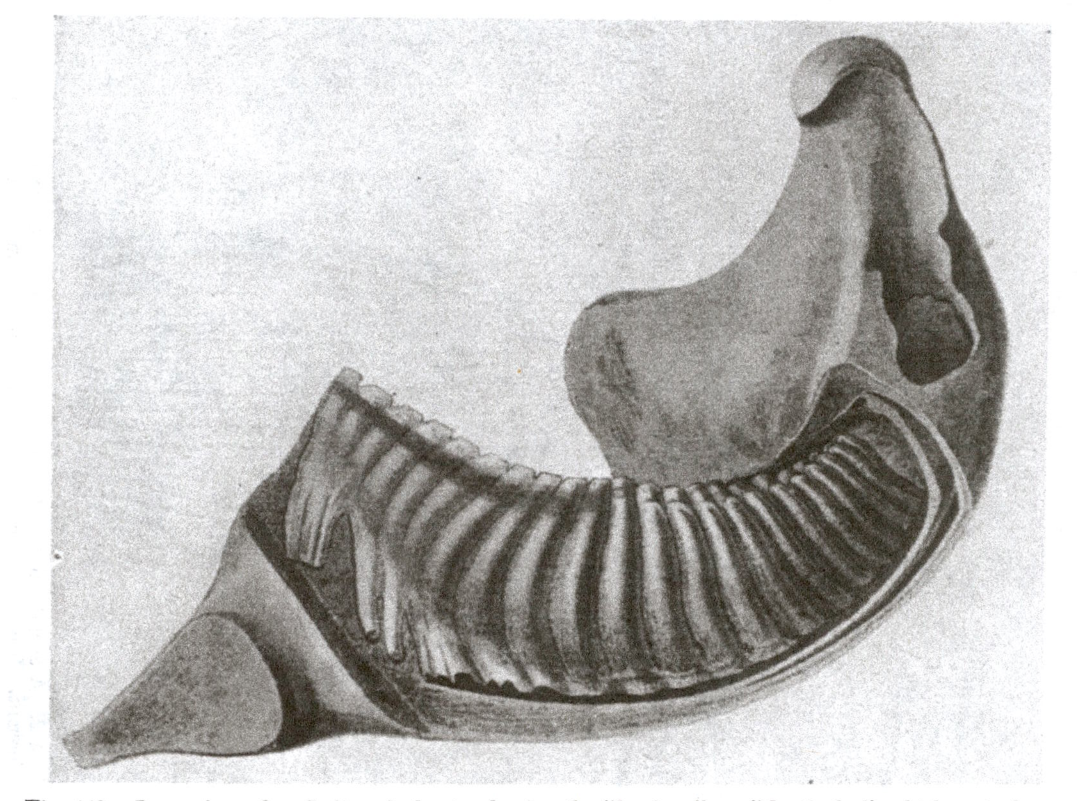

Fig. 443. Lower jaw of an Indian elephant: shewing the "laminae", or "denticules", which go to form a molar tooth. From John Corse, *Phil. Trans.* 1799.

elements in the African, and of twice as many (twice as much flattened or compressed) in the Indian elephant; in the mastodon they are much fewer and much larger, and their great tuberculated crowns never wholly wear away. In an old but admirable paper on the Indian elephant*, Mr John Corse says: "The number of teeth of which a grinder is composed varies from four to twenty-three, according as the elephant advances in years; so that a grinder, or case of teeth, in full-grown elephants, is more than sufficient to fill one side of the mouth....The same number of laminae generally fills the jaw of a young or of an old elephant; and from three till fifty years there are from ten to twelve teeth or laminae in use, in each side of either jaw, for the mastication of the food."

The molar teeth of a mouse, a hare or a capybara are likewise composite structures; they shew, precisely after the fashion of the elephant, successive narrow annular islands of enamel, with dentine within and cement between, all in varying degrees of independence or coalescence†.

The molars of a hippopotamus are composite but to a less degree; his upper molars have each two pair of roots, the last molar one root more. A block of dentine lying transversely to the jaw, with a pair of roots below and a pair of enamel-covered cusps above, is the unit of dentition, and is analogous to the young toothlet of the elephant.

In the horse and its kind the teeth are long and deeply sunk in the jaw, very much as in a rabbit or hare. Their length is made up not of root but of elongated crown, in which the deep valleys between the once high cusps are filled or flooded with cement; and these long crowns are soon worn down to an all but level surface,

* J. Corse, Observations on the different species of Asiatic elephant, and their mode of dentition, *Phil. Trans.* 1799, pp. 205–236; and cf. Owen's *Comp. Anat.* III, p. 361.

† The elephant (in my opinion) shews its likeness or affinity to the rodents throughout its whole anatomy, the metacromial process of its scapula being one conspicuous indication. *Hyrax* and *Elephas* are two isolated forms lying near the common origin of ungulates and rodents; the one lying rather to the ungulate side, the other to the rodent side, of the vague and indefinable border-line. On the relation of the rodent's dentition to the elephant's (a view strongly opposed by Dr W. K. Gregory), see M. Friant, Contribution à l'étude...des dents jugales chez les Mammifères, *Bull. Mus. Hist. Nat.* I, pp. 1–132, 1933.

in which the enamel-layer which covered the hills and lined the valleys is seen in sectional contour. A horse's incisor is the simplest case. On its worn surface we see an inner ring of enamel concentric with the enamel of the outer edge or surface of the tooth; cement fills up the inner ring, and dentine the space between. The tip of the tooth has sunk down, or been tucked in, till it forms a cement-filled lake on the top of the hill; the lake narrows in, and at last vanishes as the horse grows old and the tooth wears down; in the "aged" horse we see the "mark" no more. To recognise this lake or pit in the simple contours of the young incisor is an easy matter; but in the abraded molar the enamel-layer which once covered all its ups and downs forms a contour-line, or "curve of level," of great complexity. This contour-line alters as the levels change, and varies from one tooth to the next and from one year to another, so long as wear and tear continue. The geographer reads the lie of the land, with all its ups and downs, from a many-contoured map*, but the worn tooth shews us only one level and one contour at a time; we must eke out its scanty evidence by older and younger teeth in other phases or degrees of wear. The "pattern" of a horse's molar tooth is indeed so closely akin to a map-maker's contours that some of the terms he uses may be useful to us. He speaks, for instance, of *ridge-lines* and *course-lines*, *lignes de faîte* and *lignes de thalweg;* of a *gap*, or lowland way between two hills, in contrast to a *col* or *saddle* at the summit of a mountain-pass; or of a *gorge*, which is a narrow steep-sided valley; or a *scarp*, which is a long steep-faced hillside. We must take care all the while to see which side of our contour-line is positive or negative—on which side the ground slopes up and on which down. In our tooth we find that every enamel-contour has dentine on the one side and more or less cement on the other; the dentine belongs to the closed interior of the tooth itself, and on the other side of the enamel-line are spaces open to the world.

In a horse's molar we see the sinuous contours of two small lakes, remains of the two valleys which lay between the three transverse ridges of the compound tooth; and outside the enamel-edges of

* Contour-lines or *horizontals*, as some geographers prefer to call them, were invented by Buache, in 1752. These are discussed by Cayley, On contours and slope-lines, *Phil. Mag.* XVIII, pp. 294–8, 1859; and by Clerk Maxwell, On hills and dales, *ibid.* XL, pp. 421–7, 1870.

these contoured lakes is the dentinal substance of the tooth, surrounded again by the outer covering of enamel. The space between the outer and the inner contours is narrowed in each case at a certain point, suggesting a "col"; while it broadens out at other places, suggesting the former sites of cusps or hills. In neighbouring sections (B, C) we rise above the level of the cols, find a way open to the valleys, and see the separate transverse mountain-ranges (or lophs) of which the tooth is composed. The general plan

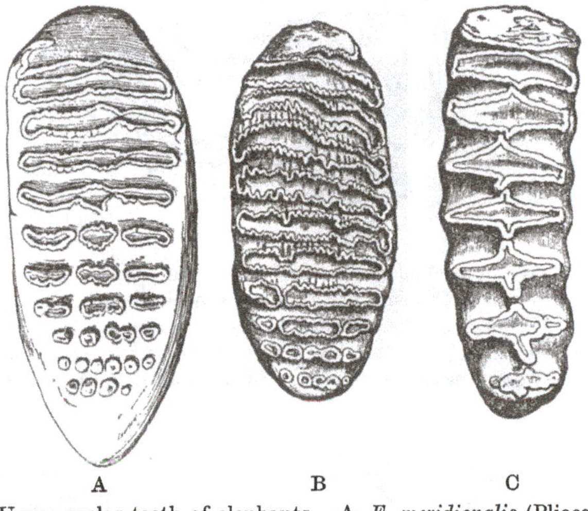

Fig. 444. Upper molar teeth of elephants. A, *E. meridionalis* (Pliocene), largest of elephants. B, Indian, and C, African elephants.

of this tooth is characteristic of the Perissodactyles; but the varying steepness of the hills, the depth of the valleys, and the amount of abrasion or erosion, lead to an infinite variety of patterns, varying with the species, with the age of the animal, and with the order of succession of the particular tooth; and so rendering it (as Osborn says) "one of the most difficult objects to define and describe in the whole field of vertebrate palaeontology*."

In *Elasmotherium* the hillsides are ridged and channelled, and their contours folded or sinuous accordingly. In *Rhinoceros* broad gaps replace the narrow cols, and certain jutting crags figure on

* H. F. Osborn, Equidae of the Oligocene, etc., *Mem. Amer. Mus. of N.H.* (n.s.) II, p. 3, 1918.

the contour-lines as the so-called *crochet* and *anticrochet*. In *Anchitherium*, erosion goes no farther than the summits of the several cusps or hill-tops.

These, to my thinking, are the few and simple lines on which we may study the architecture of the Perissodactyle tooth. But to say how far we may rely on the innumerable minor differences of pattern

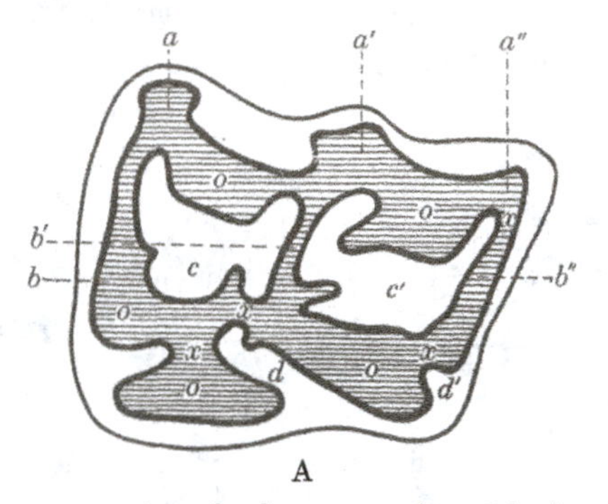

Fig. 445 A. Third upper molar of a horse. *a*, the *ectoloph*, with its three *styles*, separated by two *indents* (Owen); *b*, three transverse ridges, the *protoloph*, *mesoloph* and *metaloph* (Osborn); *c*, *c′*, two lakes, valleys or *fossettes*; *d*, *d′*, what Owen calls the *entries of the valleys*; *x*, *x′*, *cols*, where a less worn tooth would shew open roads or *passes*; *o*, *o*, *cusps* or *conules*, the sites of worn-down hills or hillocks.

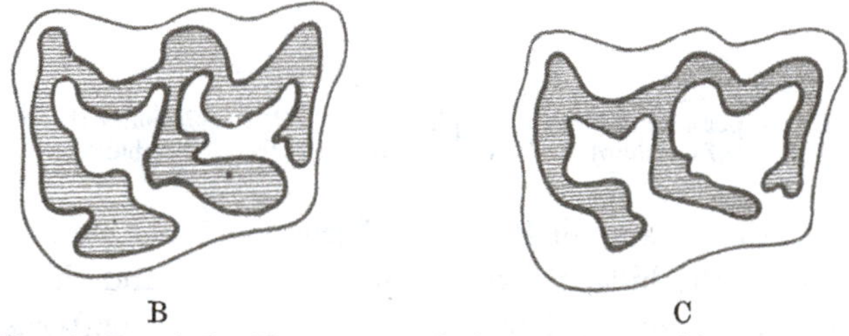

Fig. 445 B and C. The same tooth, but younger and less worn down than A. Diagrammatic.

as evidence of blood-relationship and evolutionary descent is quite another story, and deserves much more anxious consideration*.

* In the vast literature of mammalian dentition the following are conspicuous: R. Owen, *Odontography*, 1845; L. Rütimeyer, Zur Kenntniss der fossilen Pferde, und zu einer vergl. Odontographie der Hufthiere, *Verh. Naturf. Ges. Basel*, III, 1963; W. Leche, Zur Entwicklungsgeschichte des Zahnsystems der Säugetiere, *Bibl. Zool.* 1894–5, 160 pp.; E. D. Cope, On the trituberculate type of molar tooth in the Mammalia, *Proc. Amer. Philos. Soc.* XXI, pp. 324–326, 1885; W. K. Gregory, A half-century of trituberculy, *ibid.* LXXIII, pp. 161–317, 1934.

The "horn" or tusk of the narwhal is a very remarkable and a very anomalous thing. It is the only tooth in the creature's head to come to maturity; it grows to an immense and apparently

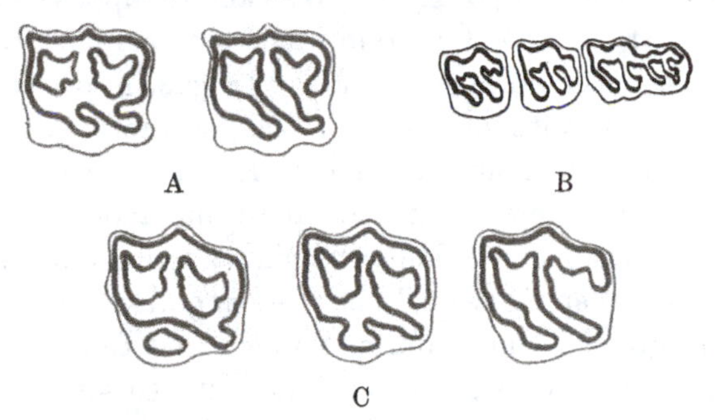

Fig. 446. Enamel patterns (diagrammatic) of certain fossil Equidae. A, *Protohippus*; B, *Hyohippus*; C, *Neohippus*.

unwieldy size, say to eight or even nine feet long; it never curves nor bends, but grows as straight as straight can be—a very singular and exceptional thing; it looks as though it were twisted, but really

Fig. 447. Enamel pattern (diagrammatic) of the upper molar teeth of *Rhinoceros*. The back-tooth (to the right-hand side) is the least worn, and its contour-line lies at the highest level.

carries on its straight axis a *screw* of several contiguous low-pitched threads; and (last and most anomalous thing of all) when, as happens now and then, two tusks are developed instead of one, one on either side, these two do not form a *conjugate* or symmetrical pair, they are not mirror-images of one another, but are *identical screws, with both threads running the same way**.

* The male narwhal carries the horn, the female being tuskless; but the whalers say that the rare two-horned specimens are all females. A famous two-horned skull in the Hamburg Museum is known to have belonged to a pregnant female. It was brought home in 1684, and is one of the oldest museum specimens in the world; the tusks measure 242 and 236 cm. During my thirty years' close acquaintance with the Dundee whalers, only four two-horned narwhals passed through my hands. Bateson (*Problems of Genetics*, 1913, p. 44) makes the curious remark that "the Narwhal's tusks, in being both twisted in the same direction, are highly anomalous, and are *comparable with pairs of twins*."

All ordinary teeth, as we have seen, have their own natural curvature, less or more, which becomes more manifest and conspicuous the longer they grow. We cannot suppose that the field of force (internal and external) in which the narwhal's tusk develops is so simple and uniform as to allow it to grow in perfect symmetry, year after year, without the least bias or intention toward either side; we must rather suppose that the resistances which the growing tusk encounters average out and cancel one another, and leave no one-sided resultant. The long, straight, tapering tooth is commonly said to have a "spiral twist," but there is no twist at all; the ivory is straight-grained and uniform, through and through. The tusk, in short, is a straight, right-handed, low-pitched screw or helix, with several threads; which threads, in the form of alternate grooves and ridges, wind evenly and continuously from one end of the tusk to the other, *even extending to its root*, deep-set in the socket or alveolus of the upper jaw.

How this composite spiral thread is formed is quite unknown. We have just seen that it is not due to any twisting of the dentinal axis of the tooth. That it is uniform and unbroken from end to end shews that the tooth somehow fashions it as a whole; and that it extends deep down within the alveolus is enough to shew that it is not impressed or graven on the tooth by any external agency. We note, as a minor feature, that the several grooves or ridges which constitute the composite thread have their individual or accidental differences; a broader or a narrower groove continues unchanged and recognisable from one end of the tooth to the other; in other words, whatever makes each ridge or groove goes on acting in the selfsame way, as long as growth goes on. A *screw* is made, in general, by compounding a translatory with a rotatory motion, and by bringing the latter into relation with the mould or matrix by which the thread is fashioned or imposed; and I cannot see how to avoid believing that the narwhal's tooth must revolve in like manner, very slowly on its longitudinal axis, all the while it grows—however strange, anomalous and hard to imagine such a mode of growth may be. We know that the tooth grows throughout life in its longitudinal direction, the open root and "permanent pulp" accounting for this; and only by a simultaneous and equally continuous rotation (so far as I can see) can we account for the perfect straightness of the tusk, for the grooving or "rifling" of the surface accompanied by no internal twist, for the extension of that rifling to the alveolar portion of the tusk within the jaw, and for the fact that the several associate grooves and ridges preserve their individual character as they pass along and wind their way around. A very slow rotation is all we need demand—say four or five complete revolutions of the tusk in the whole course of a lifetime.

The progress of a whale or dolphin through the water may be explained as the reaction to a wave which is caused to run from head to tail, the creature moving through the water somewhat slower than the wave travels. The same is true, so far, of a fish; but the wave tends to be in one plane in the fish, the dorsal and ventral fins helping to keep it so; while in the dolphin it may be said to be "circularly polarised," or resoluble into two oscillations in planes normal to one another, and caused by tail and tail-end swishing around in circular orbits which alter in phase from one transverse section to another. Just as in the case of a screw-propeller, or as in a torpedo (where it is specially corrected or compensated), this mode of action entails a certain waste of energy; it comes of the development of a "harmful moment," which tends to rotate the body about its axis, and to *screw* the animal along its course. A slight left-handed curvature of the dolphin's tail goes some little way towards correcting this tendency. M. Shuleikin's study of the kinematics of the dolphin*—a fine piece of work both on its experimental and its theoretical side—shows the dolphin to be a better swimmer than the fish, inasmuch as its speed of progression comes nearer to the velocity of the wave which is propagated along its body; the so-called "step," or fraction of the body-length travelled in a single period, is found to be about 0·7 in the dolphin, against 0·57 in a fast-swimming fish (tunny or mackerel).

Shuleikin makes the curious remark that the asymmetry of the skull (discernible in all Cetacea), which in the dolphin shews a screw-twist with a pitch about equal to the length of the body, acts as a compensatory check to the screw-component in the creature's movement of progression, and that "the till now obscure purpose

* Wassilev Shuleikin, Kinematics of a dolphin (Russian), *Bull. Acad. Sci. U.R.S.S.* (*Cl. sci. math. et phys.*), 1935, pp. 651–671; also, Dynamics, external and internal, of a fish, *ibid.* 1934, pp. 1151–1186. On the latter subject see James Gray, *Croonian Lecture*, 1940, and other papers.

of the skull's asymmetry" is accordingly explained. I should put this differently, and suggest that this counter-spirality of the skull is the direct *result* of the spiral component in locomotion. It implies, I take it, a lagging and incomplete response in the fore-part of the body to the rotatory impulse of the parts behind: or, in the plain words of the engineer, a *torque of inertia*.

This tendency, dimly seen in the dolphin's skull, is clearly demonstrated in the narwhal's "horn," and gives a complete explanation of its many singularities. The narwhal and its horn are joined together, and move together as one piece—nearly, but not quite! Stiff, straight and heavy, the great tusk has its centre of inertia well ahead of the animal, and far from the driving impulse of its tail. At each powerful stroke of the tail the creature not only darts forward, but twists or slews all of a sudden to one side; and the heavy horn, held only by its root, responds (so to speak) with difficulty. For at its slender base the "couple", by which it has to follow the twisting of the body, works at no small disadvantage. A "torque of inertia" is bound to manifest itself. The horn does not twist round in perfect synchronism with the animal; but the animal (so to speak) goes slowly, slowly, little by little, round its own horn! The play of motion, the lag, between head and horn is slight indeed; but it is repeated with every stroke of the tail. It is felt just at the growing root, the permanent pulp, of the tooth; and it puts a strain, or exercises a torque, at the very seat, and during the very process, of calcification.

Suppose that at every sweep of the tail there be a lag of no more than a fifth part of a second of arc* between the rotation of the tusk and of the body, that small amount would amply suffice to account, on a rough estimate of the age and of the activity of the animal, for as many turns of the screw as a fair-sized tusk is found to exhibit.

According to this explanation, or hypothesis, the slow rotation of the tusk corrects all tendency to flexure or curvature in one direction or another; the grooves and ridges which constitute the "thread" of the screw are the result of irregularities or inequalities within the alveolus, which "rifle" the tusk as it grows; and the

* Or say a hundred-thousandth part of the angle subtended by a minute on the clock.

identity of direction in the two horns of a pair is at once accounted for.

Beautiful as the spiral pattern of the tusk is, it obviously falls short, in regularity and elegance, of what we find, for instance, in a long tapering *Terebra* or *Turritella*, or any other spiral gasteropod shell. In the narwhal we have, as we suppose, only a *general* and never a precise agreement between rate of torsion and rate of growth; for these two velocities—of translation and rotation—are separate and independent, and their resultant keeps fairly steady but no more. In the snail-shell, on the other hand, actual tissue-growth is the common cause of both longitudinal and torsional displacements, and the resultant spiral is very perfect and regular.

Before we leave the teeth, let us note that their extreme tightness in their sockets is a remarkable thing. A thin "periodontal membrane," less than 0·25 mm. thick, fills up the space between tooth and socket; and this membrane, elastic, homogeneous and incompressible, is analogous to the thin layer of viscous liquid dealt with in modern theories of lubrication. The equilibrium of the system, the tightness of the fit, the displacement of the tooth under given forces, and the conditions of stress and strain in the membrane, are all open to mathematical treatment; distributions of pressure can be assigned to the tooth, a centre of rotation can be found, a critical load can be approximately determined, and the pressures calculated at various points. If the membrane thickens the tooth loosens; its freedom of movement or range of displacement varies with the cube of the thickness of the membrane, and is at most exceedingly small*.

* J. L. Synge, The tightness of the teeth, etc., *Phil. Trans.* (A), CCXXXI, pp. 435–477, 1933.

CHAPTER XIV

ON LEAF-ARRANGEMENT, OR PHYLLOTAXIS

THE beautiful configurations produced by the orderly arrangement of leaves or florets on a stem have long been an object of admiration and curiosity; and not the least curious feature of the case is the limited, even the small number of possible arrangements which we observe and recognise. Leonardo da Vinci would seem, as Sir Theodore Cook tells us, to have been the first to record his thoughts upon this subject; but the old Greek and Egyptian geometers are not likely to have left unstudied or unobserved the spiral traces of the leaves upon a palm-stem, or the spiral order of the petals of a lotus or the florets in a sunflower. For so, as old Nehemiah Grew says, "from the contemplation of Plants, men might first be invited to Mathematical Enquirys*."

The spiral leaf-order has been regarded by many learned botanists as involving a fundamental law of growth, of the deepest and most far-reaching importance; while others, such as Sachs, have looked upon the whole doctrine of "phyllotaxis" as "a sort of geometrical or arithmetical playing with ideas," and "the spiral theory as a mode of view gratuitously introduced into the plant." Sachs even went so far as to declare this doctrine to be "in direct opposition to scientific investigation, and based upon the idealism of the Naturphilosophie"—the mystical biology of Oken and his school.

The essential facts of the case are not difficult to understand; but the theories built upon them are so varied, so conflicting, and sometimes so obscure, that we must not attempt to submit them to detailed analysis and criticism. There are said to be two chief ways by which we may approach the question, according to whether we regard as the more fundamental and typical, one or other of two chief modes in which the phenomenon presents itself. That is to say, we may hold that the phenomenon is displayed in its essential

* N. Grew, *The Anatomy of Plants*, 1682, p. 152.

simplicity by the corkscrew spirals, or helices, which mark the position of the leaves on a cylindrical stem or tapering fir-cone; or, on

Fig. 448. A giant sunflower, *Helianthus maximus*. From H. A. Naber, after M. Brocard.

the other hand, we may be more attracted by, and may regard as of greater importance, the spirals traced by the curving rows of florets in the discoidal inflorescence of a sunflower. Whether one way or the other be the better, or even whether one be not positively correct and the other radically wrong, has been vehemently debated; but as a matter of fact they are, both mathematically and biologically, inseparable and even identical phenomena. For the face of the sunflower is but a shortened stem, and the curves upon its

Fig. 449. A cauliflower, its composite inflorescence shewing spiral patterns of the first and second order.

surfaces are but the projection on a plane of a more elongated inflorescence.

We speak, as botanists are wont to do, of these spirals of sunflower, cauliflower and the rest as logarithmic spirals, but not without hesitation. They doubtless resemble the logarithmic or equiangular spiral, but different spirals may look much alike; and these are ill-suited to the careful admeasurement and rigorous verification which Moseley gave to the spirals of his molluscan shells*. But in the sunflower, to judge by the eye, the spirals remain self-similar as they grow; each fresh increment forms, or seems to form, a *gnomon* to what went before; each new floret falls into line as part of a continuous and self-similar curve: and this goes a long way to justify our use of the familiar term logarithmic, or equiangular spiral. But the leaf-arrangement or the inflorescence are far less simple than the shell. The shell grew as one continuous and indivisible whole; its tip is the oldest part, it remains the smallest part, and the spiral tube expands continuously as it goes on. But each floret of the sunflower has its own separate and individual growth; the oldest is also the largest, and the youngest is the least; and as younger and younger florets are added on, the spiral advances in the direction of its own focus, or its own little end. And the conditions may be less simple still in other cases, as in the fir-cone itself.

The spiral tesselation of the fir-cone was carefully studied in the middle of the eighteenth century by the celebrated Bonnet, with the help of Calandrini the mathematician. Memoirs published about 1835, by Schimper and Braun, greatly amplified Bonnet's investigations, and introduced a nomenclature which still holds its own in botanical textbooks. Naumann and the brothers Bravais are among those who continued the investigation in the years immediately following, and Hofmeister, in 1868, gave an admirable account and summary of the work of these and many other writers†.

* Thus Dr A. H. Church, in his *Interpretation of Phyllotaxis Phenomena*, 1920, p. 3, begins by saying that "angular measurements on actual plant-specimens... can never hope to come within a range of accuracy admitting of an error of less than half a degree, while precise mathematical theory soon begins to tabulate minutes and seconds."

† Besides papers referred to below, and many others quoted in Sachs's *Botany* and elsewhere, the following are important: Alex. Braun, Vergl. Untersuchung über die Ordnung der Schuppen an den Tannenzapfen, etc., *Nova Acta Acad. Car. Leop.* XV, pp. 199–401, 1831; C. F. Schimper's Vorträge über die Moglichkeit eines wissenschaftlichen Verständnisses der Blattstellung, etc., *Flora,* XVIII, pp. 145–191, 737–756, 1835; C. F. Schimper, Geometrische Anordnung der um eine Achse peripherischen Blattgebilde, *Verhandl. Schweiz. Ges.* 1836, pp. 113–117; L. and A. Bravais, Essai sur la disposition des feuilles curvisériées, *Ann. Sci. Nat.* (2), VII, pp. 42–110, 1837; Sur la disposition symétrique des inflorescences, *ibid.* pp. 193–221, 291–348, VIII, pp. 11–42, 1838; Sur la disposition générale des feuilles rectisériées, *ibid.* XII, pp. 5–41, 65–77, 1839; *Mémoire sur la disposition géométrique des feuilles et des inflorescences,* Paris, 1838; Zeising, *Normalverhältniss*

The surface of a pine-cone shews a crowded assemblage of woody scales, close-packed and pressed together in such a way that each has a quadrangular, rhomboidal form*. Each scale forms part of, and marks the intersection of, two linear series; these run upwards in a spiral course, one in one direction and one in the other, and are called accordingly *diadromous* spirals. In the little cones of the Scotch Fir (*Pinus silvestris*), the whose assemblage of scales may be looked on as forming five linear series, or spiral bands, running side by side the one way, or as eight such series running the other. But these two sets are far from being all the spirals which we can trace upon the cone. Sometimes the packing is closer still, especially if the cone be long and slender. Then each scale tends to come in contact with six others, and so to become roughly hexagonal; we recognise a third spiral series besides the other two, and this new series is found to consist of thirteen rows. But let us disregard for the moment this perplexing phenomenon of a cone composed of so many series of scales, five, eight or thirteen in number as we happen to look at them; and try to find a single series in which every scale takes part. We are in no way limited to the fir-cone, which is a somewhat special case; but may consider, in a very general way, the case of any leafy stem.

Starting from some given level and proceeding upwards, let us mark the position of some one leaf (A) upon the cylindrical stem.

der chemischen und morphologischen Proportionen, Leipzig, 1856; C. F. Naumann, Ueber den Quincunx als Gesetz der Blattstellung bei Sigillaria, etc., *Neues Jahrb. f. Miner.* 1842, pp. 410–417; T. Lestiboudois, *Phyllotaxie anatomique*, Paris, 1848; G. Henslow, Phyllotaxis, or the arrangement of leaves according to mathematical laws, *Jl. Victoria Inst.* VI, pp. 129–140, 1873; On the origin of the prevailing systems of Phyllotaxis, *Tr. Linn. Soc.* (*Bot.*), I, pp. 37–45, 1880. J. Wiesner, Bemerkungen über rationale und irrationale Divergenzen, *Flora*, LVIII, pp. 113–115, 139–143, 1875; H. Airy, On leaf arrangement, *Proc. R.S.* XXI, p. 176, 1873; S. Schwendener, *Mechanische Theorie der Blattstellungen*, Leipzig, 1878; F. Delpino, *Causa meccanica della filotasse quincunciale*, Genova, 1880; *Teoria generale di Filotasse*, *ibid.* 1883; S. Günther, Das mathematische Grundgesetz im Bau des Pflanzenkorpers, *Kosmos*, IV, pp. 270–284, 1879; F. Ludwig, Wichtige Abschnitte aus der mathematischen Botanik, *Zeitschr. f. mathem. u. naturw. Unterricht*, XIV, p. 161, 1883; Weiteres über Fibonacci-Kurven und die numerische Variation der gesammten Bluthenstände der Kompositen, *Botan. Cblt.* LXVIII, p. 1, 1896; Alex. Dickson, Phyllotaxis of *Lepidodendron* and *Knossia*, *Jl. Bot.* IX, p. 166, 1871. For a historical account of the earlier literature, see Casimir de Candolle's *Considérations générales sur l'étude de la phyllotaxie*, Genève, 1881.

* Cf. *supra*, p. 515.

Another, and a younger leaf (B) will be found standing at a certain distance *around* the stem, and a certain distance *along* the stem, from the first. The former distance may be expressed as a fractional "divergence" (such as two-fifths of the circumference of the stem) as the botanists describe it, or by an "angle of azimuth" (such as $\phi = 144°$) as the mathematician would be more likely to state it. The position of B relatively to A may be determined, not only by this angle ϕ, in the horizontal plane, but also by an angle of slope (θ), or merely by linear distance from its basal plane; for the height of B above the level of A, in comparison with the diameter of the cylinder, will obviously make a great difference in the appearance of the whole system. But this matter botanical students have not concerned themselves with; in other words, their studies have been limited (or mainly limited) to the relation of the leaves to one another in *azimuth*—in other words, to the angle ϕ and its multiples.

Whatever relation we have found between A and B, let precisely the same relation subsist between B and C: and so on. Let the growth of the system, that is to say, be continuous and uniform; it is then evident that we have the elementary conditions for the development of a simple cylindrical helix; and this "primary helix" or "genetic spiral" we can now trace, winding round and round the stem, through A, B, C, etc. But if we can trace such a helix through A, B, C, it follows from the symmetry of the system, that we have only to join A to some other leaf to trace another spiral helix, such, for instance, as A, C, E, etc.; parallel to which will run another and similar one, namely in this case B, D, F, etc. And these spirals will run in the opposite direction to the spiral ABC*

In short, the existence of one helical arrangement of points implies and involves the existence of another and then another helical pattern, just as, in the pattern of a wall-paper, our eye travels from one linear series to another.

A modification of the helical system will be introduced when, instead of the leaves appearing, or standing, in singular succession, we get two or more appearing simultaneously upon the same level. If there be two such, then we shall have two generating spirals

* For the spiral ACE to be different from ABC, the angle of divergence, or angle of azimuth for one step, must exceed 90°, so that the nearer way from A to C is backwards; otherwise the spiral ACE is $ABCDE$, or ABC over again.

precisely equivalent to one another; and we may call them A, B, C, etc., and A', B', C', and so on. These are the cases which we call "whorled" leaves; or in the simplest case, where the whorl consists of two opposite leaves only, we call them "decussate."

Among the phenomena of phyllotaxis, two points in particular have been found difficult of explanation, and have aroused discussion. These are (1), the presence of the logarithmic spirals such as we have already spoken of in the sunflower; and (2) the fact that, as regards the number of the helical or spiral rows, certain numerical coincidences are apt to recur again and again, to the exclusion of others, and so to become characteristic features of the phenomenon.

As to the first of these, we have seen that the curves resemble, and sometimes closely resemble, the logarithmic spiral; but that they are, strictly speaking, logarithmic is neither proved nor capable of proof. That they appear as spiral curves (whether equable or logarithmic) is then a mere matter of mathematical "deformation." The stem which we have begun to speak of as a cylinder is not strictly so, inasmuch as it tapers off towards its summit. The curve which winds evenly around this stem is, accordingly, not a true helix, for that term is confined to the curve which winds evenly around the *cylinder*: it is a curve in space which (like the spiral curve we have studied in our turbinate shells) partakes of the characters of a helix and of a spiral, and which is in fact a spiral with its pole drawn out of its originál plane by a force acting in the direction of the axis. If we imagine a tapering cylinder, or cone, projected by vertical projection on a plane, it becomes a circular disc; and a helix described about the cone becomes in the disc a spiral described about a pole which corresponds to the apex of our cone. In like manner we may project an identical spiral in space upon such surfaces as (for instance) a portion of a sphere or of an ellipsoid; and in all these cases we preserve the spiral configuration, which is the more clearly brought into view the more we reduce the vertical component by which it was accompanied. The converse is equally true, and equally obvious, namely that any spiral traced upon a circular disc or spheroidal surface will be transformed into a corresponding spiral helix when the plane or spheroidal disc is extended into an elongated cone approximating to a cylinder. This mathematical conception is translated, in botany, into actual fact. The fir-cone may be looked upon as a cylindrical axis contracted at both ends, until it becomes approximately an ellipsoidal solid of revolution, generated about the long axis of the ellipse; and the semi-ellipsoidal capitulum of the teasel, the more or less hemispherical one of the thistle, and the flattened but still convex one of the sunflower, are all beautiful and successive deformations of what is typically a long, conical, and all but cylindrical stem. On the other hand, every stem as it grows out into its long cylindrical shape is but a deformation of the little spheroidal or ellipsoidal or conical surface which was its forerunner in the bud.

This identity of the helical spirals around the stem with spirals projected on a plane was clearly recognised by Hofmeister, who was accustomed to represent his diagrams of leaf-arrangement either in one way or the other, though not in a strictly geometrical projection*.

According to Mr A. H. Church†, who has dealt carefully and elaborately with the whole question of phyllotaxis, the spirals such as we see in the disc of the sunflower have a far greater importance and a far deeper meaning than this brief treatment of mine would accord to them: and Sir Theodore Cook, in his book on the *Curves of Life*, adopted and helped to expound and popularise Mr Church's investigations.

Mr Church, regarding the problem as one of "uniform growth," easily arrives at the conclusion that, *if* this growth can be conceived as taking place symmetrically about a central point or "pole," the uniform growth would then manifest itself in logarithmic spirals, including of course the limiting cases of the circle and straight line. With this statement I have little fault to find; it is in essence identical with much that I have said in a previous chapter. But other statements of Mr Church's, and many theories woven about them by Sir T. Cook and himself, I am less able to follow. Mr Church tells us that the essential phenomenon in the sunflower disc is a series of orthogonally intersecting logarithmic spirals. Unless I wholly misapprehend Mr Church's meaning, I should say that this

* *Allgemeine Morphologie der Gewächse*, 1868, p. 442, etc.

† *Relation of Phyllotaxis to Mechanical Laws*, Oxford, 1901–1903; cf. *Ann. Bot.* xv, p. 481, 1901.

is very far from essential. The spirals intersect isogonally, but orthogonal intersection would be only one particular case, and in all probability a very infrequent one, in the intersection of logarithmic spirals developed about a common pole. Again on the analogy of the hydrodynamic lines of force in certain vortex movements, and of similar lines of force in certain magnetic phenomena, Mr Church proceeds to argue that the energies of life follow lines comparable to those of electric energy, and that the logarithmic spirals of the sunflower are, so to speak, lines of equipotential*. And Sir T. Cook remarks that this "theory, if correct, would be fundamental for all forms of growth, though it would be more easily observed in plant construction than in animals." But the physical analogies are remote, and the deductions I am not able to follow.

Mr Church sees in phyllotaxis an organic mystery, a something for which we are unable to suggest any precise cause: a phenomenon which is to be referred, somehow, to waves of growth emanating from a centre, but on the other hand not to be explained by the division of an apical cell, or any other histological factor. As Sir T. Cook puts it, "at the growing point of a plant where the new members are being formed, there is simply *nothing to see.*"

But it is impossible to deal satisfactorily, in brief space, either with Mr Church's theories, or my own objections to them†. Let it suffice to say that I, for my part, see no subtle mystery in the matter, other than what lies in the steady production of similar growing parts, similarly situated, at similar successive intervals of time. If such be the case, then we are bound to have in consequence

* "The proposition is that the genetic spiral is a logarithmic spiral, homologous with the line of current-flow in a spiral vortex; and that in such a system the action of orthogonal forces will be mapped out by other orthogonally intersecting logarithmic spirals—the 'parastichies'"; Church, *op. cit.* I, p. 42.

† Mr Church's whole theory, if it be not based upon, is interwoven with, Sachs's theory of the orthogonal intersection of cell-walls, and the elaborate theories of the symmetry of a growing point or apical cell which are connected therewith. According to Mr Church, "the law of the orthogonal intersection of cell-walls at a growing apex may be taken as generally accepted" (p. 32); but I have taken a very different view of Sachs's law, in the eighth chapter of the present book. With regard to his own and Sachs's hypotheses, Mr Church makes the following curious remark (p. 42): "Nor are the hypotheses here put forward more imaginative than that of the paraboloid apex of Sachs which remains incapable of proof, or his construction for the apical cell of *Pteris* which does not satisfy the evidence of his own drawings."

a series of symmetrical patterns, whose nature will depend upon the form of the entire surface. If the surface be that of a cylinder, we shall have a system, or systems, of spiral helices: if it be a plane with an infinitely distant focus, such as we obtain by "unwrapping" our cylindrical surface, we shall have straight lines; if it be a plane containing the focus within itself, or if it be any other symmetrical surface containing the focus, then we shall have a system of logarithmic spirals. The appearance of these spirals is sometimes spoken of as a "subjective" phenomenon, but the description is inaccurate: it is a purely mathematical phenomenon, an inseparable secondary result of other arrangements which we, for the time being, regard as primary. When the bricklayer builds a factory chimney, he lays his bricks in a certain steady, orderly way, with no thought of the spiral patterns to which this orderly sequence inevitably leads, and which spiral patterns are by no means "subjective." The designer of a wall-paper not only has no intention of producing a pattern of criss-cross lines, but on the contrary he does his best to avoid them; nevertheless, so long as his design is a symmetrical one, the criss-cross intersections inevitably come. And as the train carries us past an orchard we see not one single symmetrical configuration, but a multiplicity of collineations among the trees.

Let us, however, leave this discussion, and return to the facts of the case.

Our second question, which relates to the numerical coincidences so familiar to all students of phyllotaxis, is not to be set and answered in a word.

Let us, for simplicity's sake, avoid consideration of simultaneous or whorled leaf origins, and consider only the more frequent cases where a single "genetic spiral" can be traced throughout the entire system.

It is seldom that this primary, genetic spiral catches the eye, for the leaves which immediately succeed one another in this genetic order are usually far apart on the circumference of the stem, and it is only in close-packed arrangements that the eye readily apprehends the continuous series. Accordingly in such a case as a fir-cone, for instance, it is certain of the secondary spirals or "parastichies" which catch the eye; and among fir-cones, we can easily count these, and we find them to be on the whole very constant in number, according to the species.

Thus in many cones, such as those of the Norway spruce, we can trace five rows of scales winding steeply up the cone in one direction, and three rows winding less steeply the other way; in certain other species, such as the common larch, the normal number is eight rows in the one direction and five in the other; while in the American larch we have again three in the one direction and five in the other. It not seldom happens that two arrangements grade into one another on different parts of one and the same cone. Among other cases in which such spiral series are readily visible we have, for instance, the crowded leaves of the stone-crops and mesembryanthemums, and (as we have said) the crowded florets of the composites. Among these we may find plenty of examples in which the numbers of the serial rows are similar to those of the fir-cones; but in some cases, as in the daisy and others of the smaller composites, we shall be able to trace thirteen rows in one direction and twenty-one in the other, or perhaps twenty-one and thirty-four; while in a great big sunflower we may find (in one and the same species) thirty-four and fifty-five, fifty-five and eighty-nine, or even as many as eighty-nine and one hundred and forty-four. On the other hand, in an ordinary "pentamerous" flower, such as a ranunculus, we may be able to trace, in the arrangement of its sepals, petals and stamens, shorter spiral series, three in one direction and two in the other; and the scales on the little cone of a Cypress shew the same numerical simplicity. It will be at once observed that these arrangements manifest themselves in connection with very different things, in the orderly interspacing of single leaves and of entire florets, and among all kinds of leaf-like structures, foliage-leaves, bracts, cone-scales, and the various parts or members of the flower. Again we must be careful to note that, while the above numerical characters are by much the most common, so much so as to be deemed "normal," many other combinations are known to occur.

The arrangement, as we have seen, is apt to vary when the entire structure varies greatly in size, as in the disc of the sunflower. It is also subject to less regular variation within one and the same species, as can always be discovered when we examine a sufficiently large sample of fir-cones. For instance, out of 505 cones of the

Norway spruce, Beal* found 92 per cent. in which the spirals were in five and eight rows; in 6 per cent. the rows were four and seven, and in 4 per cent. they were four and six. In each case they were nearly equally divided as regards direction; for instance, of the 467 cones shewing the five-eight arrangement, the five-series ran in right-handed spirals in 224 cases, and in left-handed spirals in 243.

Omitting the "abnormal" cases, such as we have seen to occur in a small percentage of our cones of the spruce, the arrangements which we have just mentioned may be set forth as follows (the fractional number used being simply an abbreviated symbol for the number of associated helices or parastichies which we can count running in the opposite directions): 2/3, 3/5, 5/8, 8/13, 13/21, 21/34, 34/55, 55/89, 89/144. Now these numbers form a very interesting series, which happens to have a number of curious mathematical properties†. We see, for instance, that the denominator of each

* *Amer. Naturalist,* VII, p. 449, 1873.

† This celebrated series corresponds to the continued fraction $\frac{1}{1+}\frac{1}{1+}$ etc., and converges to 1·618..., the numerical equivalent of the *sectio divina,* or "*Golden Mean.*" The series of numbers, 1, 1, 2, 3, 5, 8, ..., of which each is the sum of the preceding two, was used by Leonardo of Pisa (*c.* 1170–1250), nicknamed *Fi Bonacci,* or *filius bonassi,* in his *Liber Abbaci,* a work dedicated to *magister meus, summus philosophus,* Michael Scot (*Scritti,* I, pp. 283–284, 1857). This learned man was educated in Morocco, where his father was clerk or dragoman to Pisan merchants; and he is said to have been the first to bring the Arabic numerals, or "*novem figurae Indorum,*" into Europe. The Fibonacci numbers were first so-called by Eduard Lucas, *Bollettino di Bibliogr. e Storia dei Sci. Matem. e Fis.* X, p. 129, 1877. The general expression for the series

$$u_n = \frac{1}{\sqrt{5}}\left\{\left(\frac{1+\sqrt{5}}{2}\right)^n - \left(\frac{1-\sqrt{5}}{2}\right)^n\right\},$$

was known to Euler and to Daniel Bernoulli (*Comm. Acad. Sci. Imp. Petropol.* 1732, p. 90), and was rediscovered by Binet, *C.R.* XVIII, p. 563, 1843; XIX, p. 939, 1844) and by Lamé, *ibid.* XIX, p. 867, after whom it is sometimes called Lamé's series. But the Greeks were familiar with the series 2, 3: 5, 7 : 12, 17, etc.; which converges to $\sqrt{2}$, as the other does to the Golden Mean; and so closely related are the two series, that it seems impossible that the Greeks could have known the one and remained ignorant of the other. (See a paper of mine, on "Excess and Defect, etc.," in *Mind,* XXXVIII, No. 149, 1928.)

The Fibonacci (or Lamé) series was well known to Kepler, who, in his paper *De nive sexangula* (1611, cf. *supra,* p. 695), discussed it in connection with the form of the dodecahedron and icosahedron, and with the ternary or quinary symmetry of the flower. (Cf. F. Ludwig, Kepler über das Vorkommen der Fibonaccireihe im Pflanzenreich, *Bot. Centralbl.* LXVIII, p. 7, 1896.) Professor William Allman, Professor of Botany in Dublin (father of the historian of Greek geometry), fraction is the numerator of the next; and further, that each successive numerator, or denominator, is the sum of the preceding two. Our immediate problem, then, is to determine, if possible, how these numerical coincidences come about, and why these particular numbers should be so commonly met with as to be considered "normal" and characteristic features of the general phenomenon of phyllotaxis. The following account is based on a short paper by Professor P. G. Tait*.

Of the two following diagrams, Fig. 450 represents the general case, and Fig. 451 a particular one, for the sake of possibly greater simplicity. Both diagrams represent a portion of a branch, or fir-cone, regarded as cylindrical, and unwrapped to form a plane surface. A, a, at the two ends of the base-line, represent the same initial leaf or scale; O is a leaf which can be reached from A by m steps

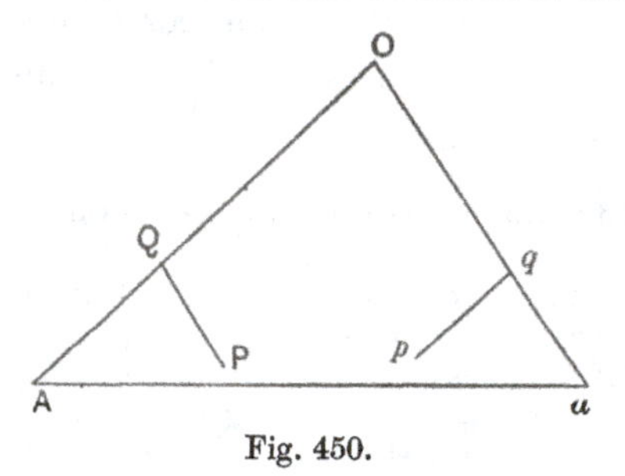

Fig. 450.

in a right-hand spiral (developed into the straight line AO), and by n steps from a in a left-handed spiral aO. Now it is obvious in our fir-cone, that we can include *all* the scales upon the cone by taking so many spirals in the one direction, and again include them all by so

speculating on the same facts, put forward the curious suggestion that the cellular tissue of the dicotyledons, or exogens, would be found to consist of dodecahedra, and that of the monocotyledons or endogens of icosahedra (*On the mathematical connection between the parts of vegetables*: abstract of a Memoir read before the Royal Society in the year 1811 (privately printed, n.d.). Cf. De Candolle, *Organogenie végétale*, I, p. 534. See also C. E. Wasteels, Over de Fibonaccigetalen, *3de Natuur. Congres, Antwerpen*, 1899, pp. 25–37; R. C. Archibald, in Jay Hambidge's *Dynamic Symmetry*, 1920, pp. 146–157; and, on the many mathematical properties of the series, L. E. Dickson, *Theory of Numbers*, I, pp. 393–411, 1919.

Of these famous and fascinating numbers a mathematical friend writes to me: "All the romance of continued fractions, linear recurrence relations, surd approximations to integers and the rest, lies in them, and they are a source of endless curiosity. How interesting it is to see them striving to attain the unattainable, the golden ratio, for instance; and this is only one of hundreds of such relations."

* *Proc. R.S.E.* VII, p. 391, 1872.

many in the other. Accordingly, in our diagrammatic construction, the spirals AO and aO *must*, and always *can*, be so taken that m spirals parallel to aO, and n spirals parallel to AO, shall separately include all the leaves upon the stem or cone.

If m and n have a common factor, l, it can easily be shewn that the arrangement is composite, and that there are l fundamental, or genetic spirals, and l leaves (including A) which are situated

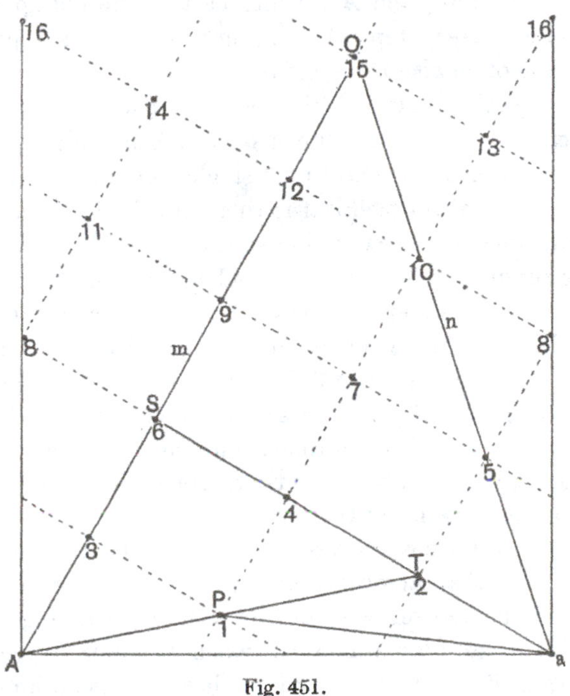

Fig. 451.

exactly on the line Aa. That is to say, we have here a *whorled* arrangement, which we have agreed to leave unconsidered in favour of the simpler case. We restrict ourselves, accordingly, to the cases where there is but one genetic spiral, and when *therefore* m and n are prime to one another.

Our fundamental, or genetic, spiral, as we have seen, is that which passes from A (or a) to the leaf which is situated nearest to the base-line Aa. The fundamental spiral will thus be right-handed (A, P, etc.) if P, which is nearer to A than to a, be this leaf—left-handed if it be p. That is to say, we make it a convention that we shall always, for our fundamental spiral, run round the system, from one leaf to the next, *by the shortest way*.

Now it is obvious, from the symmetry of the figure (as further shewn in Fig. 451), that, besides the spirals running along AO and aO, we have a series running *from the steps on* aO to the steps on AO. In other words we can find a leaf (S) upon AO, which, like the leaf O, is reached directly by a spiral series from A and from a, such that aS includes n steps, and AS (being part of the old spiral line AO) now includes $m-n$ steps. And, since m and n are prime to one another (for otherwise the system would have been a composite or whorled one), it is evident that we can continue this process of convergence until we come down to a 1, 1 arrangement, that is to say to a leaf which is reached by a single step, in opposite directions from A and from a, which leaf is therefore the first leaf, next to A, of the fundamental or generating spiral.

If our original lines along AO and aO contain, for instance, 13 and 8 steps respectively (i.e. $m = 13$, $n = 8$), then our next series, observable in the same cone, will be 8 and (13 − 8) or 5; the next 5 and (8 − 5) or 3; the next 3, 2; and the next 2, 1; leading to the ultimate condition of 1, 1. These are the very series which we have found to be common, or normal; and so far as our investigation has yet gone, it has proved to us that, if one of these exists, it entails, *ipso facto*, the presence of the rest.

In following down our series, according to the above construction, we have seen that at every step we have changed direction, the longer and the shorter sides of our triangle changing places every time. Let us stop for a moment, when we come to the 1, 2 series, or AT, aT of Fig. 451. It is obvious that there is nothing to prevent us making a new 1, 3 series if we please, by continuing the generating spiral through three leaves, and connecting the leaf so reached directly with our initial one. But in the case represented in Fig. 451, it is obvious that these two series (A, 1, 2, 3, etc., and a, 3, 6, etc.) will be running in the same direction; i.e. they will both be right-handed, or both left-handed spirals. The simple meaning of this is that the third leaf of the generating spiral was distant from our initial leaf by *more than the circumference* of the cylindrical stem; in other words, that there were more than two, but *less than three* leaves in a single turn of the fundamental spiral.

Less than two there can obviously never be. When there are exactly two, we have the simplest of all possible arrangements, namely that in which the leaves are placed alternately on opposite sides of the stem. When there are more than two, but less than three, we have the elementary condition for the production of the series which we have been considering, namely 1, 2; 2, 3; 3, 5, etc. To put the latter part of this argument in more precise language, let us say that: If, in our descending series, we come to steps 1 and t, where t is determined by the condition that 1 and $t+1$ would give spirals both right-handed, or both left-handed; it follows that there are less than $t+1$ leaves in a single turn of the fundamental spiral. And, determined in this manner, it is found in the great majority of cases, in fir-cones, and a host of other examples of phyllotaxis, that $t=2$. In other words, in the great majority of cases, we have what corresponds to an arrangement next in order of simplicity to the simplest case of all: next, that is to say, to the arrangement which consists of opposite and alternate leaves.

"These simple considerations," as Tait says, "explain completely the so-called mysterious appearance of terms of the recurring series 1, 2, 3, 5, 8, 13, etc.* The other natural series, usually but misleadingly represented by convergents to an infinitely extended continuous fraction, are easily explained, as above, by taking $t=3$, 4, 5, etc., etc." Many examples of these latter series have been recorded, as more or less rare abnormalities, by Dickson† and other writers.

We have now learned, among other elementary facts, that wherever any one system of spiral steps is present, certain others invariably and of necessity accompany it, and are definitely related to it. In any diagram, such as Fig. 451, in which we represent our leaf-arrangement by means of uniform and regularly interspaced dots, we can draw one series of spirals after another, and one as easily

* The necessary existence of these recurring spirals is also proved, in a somewhat different way, by Leslie Ellis, On the theory of vegetable spirals, in *Mathematical and other Writings*, 1863, pp. 358–372. Leslie Ellis, Whewell's brother-in-law, was a man of great originality. He is best remembered, perhaps, for his views on the Theory of Probabilities (cf. J. M. Keynes, *Treatise on Probabilities*, 1921, p. 92), and for his association with Stebbing as editor of Bacon.

† *Proc. R.S.E.* VII, p. 397, 1872; *Trans. R.S.E.* XXVI, pp. 505–520, 1872.

as another. In a fire-cone one particular series, or rather two conjugate series, are always conspicuous, but the related series may be sought and found with little difficulty. The spruce-fir is commonly said to have a phyllotaxis of 8/13; but we may count still

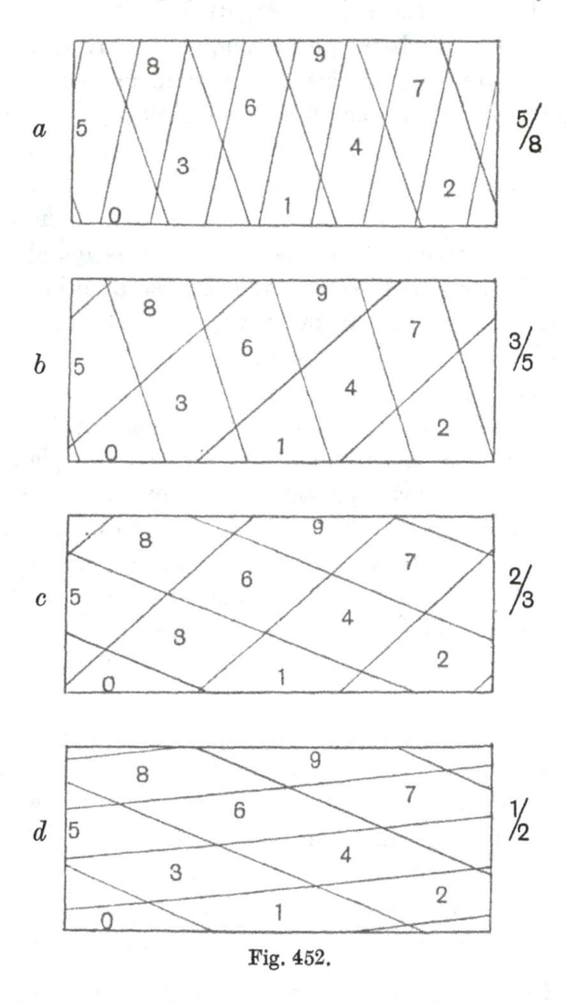

Fig. 452.

steeper and nearly vertical rows of scales to the number of 13/21; and if we take pains to number all the scales consecutively, we may find the lower series, 5/8, 3/5, and even 1/2, with ease and certainty.

The phenomenon is illustrated by Fig. 452, *a–d*. The ground-plan of all these diagrams is identically the same. The generating spiral in each case represents a divergence of 3/8, or 135° of azimuth; and the points succeed one another at the same successional distances parallel to the axis. The rectangular outlines, which correspond to the exposed surface of the leaves or cone-scales, are of equal area, and of equal number. Nevertheless the appearances presented by these diagrams are very different; for in one the eye catches a 5/8 arrangement, in another a 3/5; and so on, down to an arrangement of 1/1. The mathematical side of this very curious phenomenon I have not attempted to investigate. But it is quite obvious that, in a system within which various spirals are implicitly contained, the conspicuousness of one set or another does not depend upon angular divergence. It depends on the relative proportions in length and breadth of the leaves themselves; or, more strictly speaking, on the ratio of the diagonals of the rhomboidal figure by which each leaf-area is circumscribed. When, as in the fir-cone, the scales by mutual compression conform to these rhomboidal outlines, their inclined edges at once guide the eye in the direction of some one particular spiral; and we shall not fail to notice that in such cases the usual result is to give us arrangements corresponding to the middle diagrams in Fig. 452, which are the configurations in which the quadrilateral outlines approach most nearly to a rectangular form, and give us accordingly the least possible ratio (under the given conditions) of sectional boundary-wall to surface area.

The manner in which one system of spirals may be caused to slide, so to speak, into another, has been ingeniously demonstrated by Schwendener on a mechanical model, consisting essentially of a framework which can be opened or closed to correspond with one another of the above series of diagrams*.

The same curious fact, that one Fibonacci series leads to, or involves the rest, is further shewn, in a very simple way, in the following diagrammatic Table (p. 930). It shews, in the first instance, the numerical order of the scales on a fir-cone, in so-called 5/8 phyllotaxis; that is to say, it represents the cone *unwrapped*, with the two principal spirals lying along the axes of a rectangular system. Starting from 0, the abscissae increase by 5, the ordinates by 8; or, in other words, any given number $m = 5x + 8y$; it is

* A common form of pail-shaped waste-paper basket, with wide rhomboidal meshes of cane, is well-nigh as good a model as is required.

easy, then, to number the entire system. The generating spiral, 0, 1, 2, 3, . . . , and the various secondary Fibonacci spirals, are then easily recognised.

A Fibonacci series, unwrapped from a cone or cylinder. $m = 5x + 8y$.

1	6	11	**16**	21	26	31	36
$\overline{7}$	$\overline{2}$	3	**8**	13	18	23	28
$\overline{\mathbf{15}}$	$\overline{\mathbf{10}}$	$\overline{\mathbf{5}}$	**0**	**5**	**10**	**15**	**20**
$\overline{23}$	$\overline{18}$	$\overline{13}$	$\overline{\mathbf{8}}$	$\overline{3}$	2	7	12
$\overline{31}$	$\overline{26}$	$\overline{21}$	$\overline{\mathbf{16}}$	$\overline{11}$	$\overline{6}$	$\overline{1}$	4

The place of the first scale in each series is then found to be as follows:

Series	1	2	3	5	8	13	21	34	55
$\frac{x}{y} =$	$\frac{-3}{2}$	$\frac{2}{-1}$	$\frac{-1}{1}$	$\frac{1}{0}$	$\frac{0}{1}$	$\frac{1}{1}$	$\frac{1}{2}$	$\frac{2}{3}$	$\frac{3}{5}$

And this is the Fibonacci series over again. We also see how the several spirals, of which these are the beginnings, alternate to the right and left of an asymptotic line, where $x/y = 0{\cdot}618 \ldots$.

The Fibonacci numbers, so conspicuous in the fir-cone, make their appearance also in the flower. The commonest of floral numbers are 3 and 5; among the Composites we find 8 ray-florets in the single dahlia, 13 in the ragwort, 21 in the ox-eye daisy or the marigold. In the last two, heads with 34 ray-florets are apt to be produced at certain times or in certain places*; and in *C. segetum* these florets are said to vary in a bimodal curve of frequency, with a high maximum at 13 and a lower at 21†. The simplest explanation (though perhaps it does not go far) is to suppose that a ligulate floret terminates, or tends to terminate, each of the principal spiral series. But among the higher numbers these numerical relations are only approximate, and the whole matter rests, so far, on somewhat scanty evidence.

The determination of the precise angle of divergence of two consecutive leaves of the generating spiral does not enter into the above general investigation (though Tait gives, in the same paper, a method

* Cf. G. Henslow, On the origin of dimerous and trimerous whorls among the flowers of Dicotyledons, *Trans. Linn. Soc.* (*Bot.*) (2), VII, p. 161, 1908.

† Cf. A. Gravis, *Éléments de Physiologie végétale*, 1921, p. 122.

by which it may be easily determined); and the very fact that it does not so enter shews it to be essentially unimportant. The determination of so-called "orthostichies," or precisely vertical successions of leaves, is also unimportant. We have no means, other than observation, of determining that one leaf is vertically above another, and spiral series such as we have been dealing with will appear, whether such orthostichies exist, whether they be near or remote, or whether the angle of divergence be such that no precise vertical superposition ever occurs. And lastly, the fact that the successional numbers, expressed as fractions, 1/2, 2/3, 3/5, represent a convergent series, whose final term is equal to 0·61803..., the *sectio aurea* or "golden mean" of unity, is seen to be a mathematical coincidence, devoid of biological significance; it is but a particular case of Lagrange's theorem that the roots of every numerical equation of the second degree can be expressed by a periodic continued fraction. The same number has a multitude of curious arithmetical properties. It is the final term of all similar series to that with which we have been dealing, such for instance as 1/3, 3/4, 4/7, etc., or 1/4, 4/5, 5/9, etc. It is a number beloved of the circle-squarer, and of all those who seek to find, and then to penetrate, the secrets of the Great Pyramid. It is deep-set in the regular pentagon and dodecahedron, the triumphs of Pythagorean or Euclidean geometry. It enters (as the chord of an angle of 36°) into the thrice-isosceles triangle of which we have spoken on p. 762; it is a number which becomes (by the addition of unity) its own reciprocal—its properties never end. To Kepler (as Naber tells us) it was a symbol of Creation, or Generation. Its recent application to biology and art-criticism by Sir Theodore Cook and others is not new. Naber's book, already quoted, is full of it. Zeising*, in 1854, found in it the key to all

* A. Zeising, *Neue Lehre von der Proportion des menschlichen Körpers aus einem bisher unerkannt gebliebenen die ganze Natur und Kunst durchdringenden morphologischen Grundgesetze entwickelt*, Leipzig, 1854, 457 pp.; *ibid. Deutsche Vierteljahrsschrift*, 1868, p. 261; also, posthumously, *Der Goldene Schnitt*, Leipzig, 1884, 24 pp. Cf. S. Gunther, Adolph Zeising als Mathematiker, *Ztschr. f. Math. u. Physik. (Hist. Lit. Abth.)*, XXI, pp. 157–165, 1876; also F. X. Pfeiffer, Die Proportionen des goldenen Schnittes an den Blättern u. Stengelen der Pflanzen, *Ztschr. f. math. u. naturw. Unterricht*, XV, pp. 325–338, 1885. For other references, see R. C. Archibald, *op. cit.* Among modern books on similar lines, the following are curious, interesting and beautiful (whether we agree with them or not): Jay Hambidge, *Dynamic Symmetry*, Yale, 1920; C. Arthur Coan, *Nature's Harmonic Unity*, New York, 1912.

morphology, and the same writer, later on, declared it to dominate both architecture and music. But indeed, to use Sir Thomas Browne's words (though it was of another number that he spoke): "To enlarge this contemplation into all the mysteries and secrets accommodable unto this number, were inexcusable Pythagorisme."

That this number has any serious claim at all to enter into the biological question of phyllotaxis seems to depend on the assertion, first made by Chauncey Wright*, that, if the successive leaves of the fundamental spiral be placed at the particular azimuth which divides the circle in this "sectio aurea," then no two leaves will ever be superposed†; and thus we are said to have "the most thorough and rapid distribution of the leaves round the stem, each new or higher leaf falling over the angular space between the two older ones which are nearest in direction, so as to divide it in the same ratio (K), in which the first two or any two successive ones divide the circumference. Now 5/8 and all successive fractions differ inappreciably from K." To this view there are many simple objections. In the first place, even 5/8, or 0·625, is but a moderately close approximation to the "golden mean"; and furthermore, the arrangements by which a better approximation is got, such as 8/13, 13/21, and the very close approximations such as 34/55, 55/89, 89/144, etc., are comparatively rare, while the much less close approximations of 3/5 or 2/3, or even 1/2, are extremely common. Again, the general type of argument such as that which asserts that the plant is "aiming at" something which we may call an "ideal angle" is one which cannot commend itself to a plain student of physical science: nor is the hypothesis rendered more acceptable when Sir T. Cook qualifies it by telling us that "all that a plant can do is to vary, to make blind shots at constructions, or to 'mutate' as it is now termed: and the most suitable of these constructions will in the long run be isolated by the action of Natural Selection." Thirdly, we must not suppose the Fibonacci numbers

* On the uses and origin of the arrangement of leaves in plants, *Mem. Amer. Acad.* IX, p. 380, 1871, Cambridge, Mass. Cf. J. Wiesner, *Ueber die Beziehungen der Stellungsverhältnisse der Laubblätter zur Beleuchtung*, Wien, 1902.

† This is what Ruskin spoke of as "the vacant space"; *Mod. Painters*, V, chap. VI, p. 44, 1860. Leonardo had in like manner explained the leaf-arrangement as serving to let air pass between the leaves, keep one from overshadowing another, and let rain-drops fall from the one leaf to the one below.

to have any *exclusive* relation to the Golden Mean; for arithmetic teaches us that, beginning with any two numbers whatsoever, we are led by successive summations toward one out of innumerable series of numbers whose ratios one to another converge to the Golden Mean*. Fourthly, the supposed isolation of the leaves, or their most complete "distribution to the action of the surrounding atmosphere" is manifestly very little affected by any conditions which are confined to the angle of azimuth. For if it be (so to speak) Nature's object to set them farther apart than they actually are, to give them freer exposure to the air or to the sunlight than they actually have, then it is surely manifest that the simple way to do so is to elongate the axis, and to set the leaves farther apart, lengthways on the stem. This has at once a far more potent effect than any nice manipulation of the "angle of divergence."

Lastly, and this seems the simplest, the most cogent and most unanswerable objection of them all, if it be indeed desirable that no leaf should be superimposed above another, the one condition necessary is that the common angle of azimuth should *not* be a rational multiple of a right angle—should not be equivalent to $\frac{m}{n}\left(\frac{\pi}{2}\right)$. One irrational angle is as good as another: there is no special merit in any one of them, not even in the *ratio divina*. We come then without more ado to the conclusion that while the Fibonacci series stares us in the face in the fir-cone, it does so for mathematical reasons; and its supposed usefulness, and the hypothesis of its introduction into plant-structure through natural selection, are matters which deserve no place in the plain study of botanical phenomena. As Sachs shrewdly recognised years ago, all such speculations as these hark back to a school of mystical idealism.

* Thus, instead of beginning with 1, 1, let us begin 1, 7. The summation-series is then 1, 7, 8, 15, 23, 38, 61, 99, 160, 259, ..., etc.; and 99/160 = 0·618... and 259/160 = 1·619...; and so on. But after all, the old Fibonacci numbers are not far away. For we may write the new series in the form:

$$\begin{aligned} &7\ (0,1,1,2,3,5,8,\ldots)\\ +\ &1\ (1,0,1,1,2,3,5,\ldots). \end{aligned}$$

CHAPTER XV

ON THE SHAPES OF EGGS, AND OF CERTAIN OTHER HOLLOW STRUCTURES

THE eggs of birds and all other hard-shelled eggs, such as those of the tortoise and the crocodile, are simple solids of revolution; but they differ greatly in form, according to the configuration of the plane curve by the revolution of which the egg is, in a mathematical sense, generated. Some few eggs, such as those of the owl, the penguin, or the tortoise, are spherical or very nearly so; a few more, such as the grebe's, the cormorant's or the pelican's, are approximately ellipsoidal, with symmetrical or nearly symmetrical ends, and somewhat similar are the so-called "cylindrical" eggs of the megapodes and the sand-grouse; the great majority, like the hen's egg, are "ovoid," a little blunter at one end than the other; and some, by an exaggeration of this lack of antero-posterior symmetry, are blunt at one end but characteristically pointed at the other, as is the case with the eggs of the guillemot and puffin, the sandpiper, plover and curlew. It is an obvious but by no means negligible fact that the egg, while often pointed, is never flattened or discoidal; it is a prolate, but never an oblate, spheroid. Its oval outline has one maximal and two minimal radii of curvature, one minimum being less than the other. The evolute to a curve often emphasises, even exaggerates, its features; and the evolutes to a series of eggs (i.e. to their generating curves) are more conspicuously different than the eggs themselves (Fig. 453)*.

The careful study and collection of birds' eggs would seem to have begun with the Count de Marsigli†, the same celebrated naturalist

* Cf. A. Mallock, On the shapes of birds' eggs, *Nature*, CXVI, p. 311, 1925. The evolute may be easily if somewhat roughly drawn by erecting perpendiculars on a sufficient number of tangents to the curve. The evolute then appears as an *envelope*, the perpendiculars all being tangents to it.

† *De avibus circa aquas Danubii vagantibus et de ipsarum nidis* (Vol. v of the *Danubius Panonico-Mysicus*), Hagae Com. 1726. Count Giuseppi Ginanni, or Zinanni, came soon afterwards with his book *Delle uove e dei nidi degli uccelli*, Venezia, 1737.

who first studied the "flowers" of the coral, and who wrote the *Histoire physique de la mer*; and the specific form as well as the colour and other attributes of the egg have been again and again discussed, and not least by the many dilettanti naturalists of the eighteenth century who soon followed in Marsigli's footsteps*.

We need do no more than mention Aristotle's belief, doubtless old in his time, that the more pointed egg produces the male chicken, and the blunter egg the hen; though this theory survived into modern times† and still lingers on (cf. p. 943). Several naturalists, such as Günther (1772) and Bühle (1818), have taken the trouble to disprove it by experiment. A more modern and more generally accepted

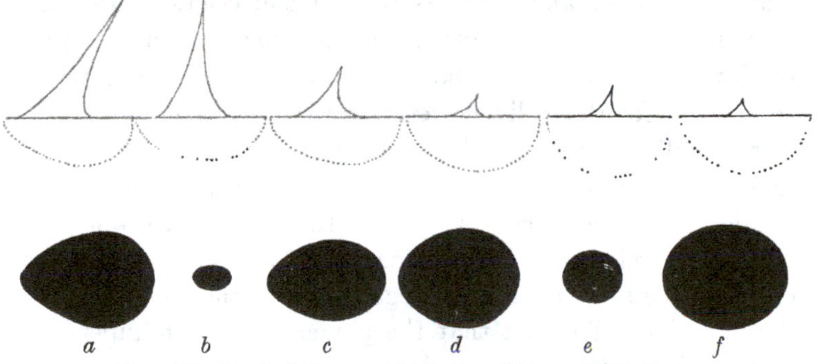

Fig. 453. Typical forms of birds' eggs: from A. Mallock.

The figures below are pinhole photographs of the eggs. The upper figures (drawn to a uniform scale) shew the generating curves and their evolutes.

a Green plover. *c* Crow. *e* Kingfisher.
b Humming-bird. *d* Pheasant. *f* Owl.

explanation has been that the form of the egg is in direct relation to that of the bird which has to be hatched within—a view that would seem to have been first set forth by Naumann and Bühle, in their great treatise on eggs‡, and adopted by Des Murs§ and many other well-known writers.

In a treatise by de Lafresnaye||, an elaborate comparison is made

* But Sir Thomas Browne had a collection of eggs at Norwich in 1671, according to Evelyn.

† Cf. Lapierre, in Buffon's *Histoire Naturelle*, ed. Sonnini, 1800.

‡ *Eier der Vogel Deutschlands*, 1818–28 (*cit.* Des Murs, p. 36).

§ *Traité d'Oologie*, 1860.

|| F. de Lafresnaye, Comparaison des œufs des oiseaux avec leurs squelettes, comme seul moyen de reconnaître la cause de leurs différentes formes, *Rev. Zool.* 1845, pp. 180–187, 239–244.

between the skeleton and the egg of various birds, to shew, for instance, how those birds with a deep-keeled sternum laid rounded eggs, which alone could accommodate the form of the young. According to this view, that "Nature had foreseen*" the form adapted to and necessary for the growing embryo, it was easy to correlate the owl with its spherical egg, the diver with its elliptical one, and in like manner the round egg of the tortoise and the elongated one of the crocodile, with the shape of the creatures which had afterwards to be hatched therein. A few writers, such as Thienemann†, looked at the same facts the other way, and asserted that the form of the egg was determined by that of the bird by which it was laid and in whose body it had been conformed.

In more recent times, other theories, based upon the principles of Natural Selection, have been current and very generally accepted to account for these diversities of form. The pointed, conical egg of the guillemot is generally supposed to be an adaptation, advantageous to the species in the circumstances under which the egg is laid; the pointed egg is less apt than a spherical one to roll off the narrow ledge of rock on which this bird is said to lay its solitary egg, and the more pointed the egg, so much the fitter and likelier is it to survive. The fact that the plover or the sandpiper, breeding in very different situations, lay eggs that are also conical, elicits another explanation, to the effect that here the conical form permits the many large eggs to be packed closely under the mother bird‡. Whatever truth there be in these apparent adaptations to existing circumstances, it is only by a very hasty logic that we can accept them as a *vera causa*, or adequate explanation of the facts; and it is obvious that in the bird's egg we have an admirable case for direct investigation of the mechanical or physical significance of its form§.

* Cf. Des Murs, p. 67: "Elle devait encore penser au moment où ce germe aurait besoin de l'espace nécessaire à son accroissement, à ce moment où...il devra remplir exactement l'intervalle circonscrit par sa fragile prison, etc."

† F. A. L. Thienemann, *Syst. Darstellung der Fortpflanzung der Vögel Europas*, Leipzig, 1825–38.

‡ Cf. Newton's *Dictionary of Birds*, 1893, p. 191; Szielasko, Gestalt der Vogeleier, *Journ. f. Ornith.* LIII, pp. 273–297, 1905.

§ Jacob Steiner suggested a Cartesian oval, $r + mr' = c$, as a general formula for all eggs (cf. Fechner, *Ber. sächs. Ges.* 1849, p. 57); but this formula (which fails in such a case as the guillemot) is purely empirical, and has no mechanical foundation.

Of all the many naturalists of the eighteenth and nineteenth centuries who wrote on the subject of eggs, only two (so far as I am aware) ascribed the form of the egg to direct mechanical causes. Günther*, in 1772, declared that the more or less rounded or pointed form of the egg is a mechanical consequence of the pressure of the oviduct at a time when the shell is yet unformed or unsolidified; and that accordingly, to explain the round egg of the owl or the kingfisher, we have only to admit that the oviduct of these birds is somewhat larger than that of most others, or less subject to violent contractions. This statement contains, in essence, the whole story of the mechanical conformation of the egg. A hundred and twenty years after, Dr J. Ryder of Philadelphia gave, as near as may be, the same explanation†.

Let us consider, very briefly, the conditions to which the egg is subject in its passage down the oviduct.

(1) The "egg," as it enters the oviduct, consists of the yolk only, enclosed in its vitelline membrane. As it passes down the first portion of the oviduct the white is gradually superadded, and becomes in turn surrounded by the "shell-membrane." About this latter the shell is secreted, rapidly and at a late period: the egg having meanwhile passed on into a wider portion of the oviducal tube, called (by loose analogy, as Owen says) the "uterus." Here the egg assumes its permanent form, here it ultimately becomes rigid, and it is to this portion of the oviduct that our argument principally refers.

(2) Both the yolk and the entire egg tend to fill completely their respective membranes, and, whether this be due to growth or imbibition on the part of the contents or to contraction on the part of the surrounding membranes, the resulting tendency is for both yolk and egg to be, in the first instance, spherical, unless or until distorted by external pressure.

(3) The egg is subject to pressure within the oviduct, which is an elastic, muscular tube, along the walls of which pass peristaltic

* F. C. Günther, *Sammlung von Nestern und Eyern verschiedener Vögel*, Nürnb. 1772. Cf. also Raymond Pearl, Morphogenetic activity of the oviduct, *J. Exp. Zool.* VI, pp. 339–359, 1909.

† J. Ryder, The mechanical genesis of the form of the fowl's egg, *Proc. Amer. Philosoph. Soc.* Philadelphia, XXXI, pp. 203–209, 1893; cf. A. S. Packard, Inheritance of acquired characters, *Proc. Amer. Acad.* 1894, p. 360.

waves of contraction. These muscular contractions may be described as the contraction of successive annuli of muscle, giving annular (or radial) pressure to successive portions of the egg; they drive the egg forward against the frictional resistance of the tube, while tending at the same time to distort its form. While nothing is known, so far as I am aware, of the muscular physiology of the oviduct, it is well known in the case of the intestine that the presence of an obstruction leads to the development of violent contractions in its rear, which waves of contraction die away, and are scarcely if at all propagated in advance of the obstruction; indeed in normal intestinal peristalsis a wave of relaxation travels close ahead of the wave of constriction.

(4) The egg is, to all intents and purposes, a solid of revolution; in other words, its transverse sections are all but perfect circles, so nearly perfect that, chucked in the lathe, an egg "runs true." This may be taken to shew that the direct pressure of the oviduct, whether elastic or muscular, is large compared with the weight of the egg. Even in ostrich eggs, where if anywhere gravitational deformation should be found, the greatest and least equatorial diameters do not differ by 1 per cent., and sometimes by less than one part in a thousand*.

(5) It is known by observation that a hen's egg is always laid blunt end foremost†.

(6) It can be shewn, at least as a very common rule, that those eggs which are most unsymmetrical, or most tapered off posteriorly, are also eggs of a large size relatively to the parent bird. The guillemot is a notable case in point, and so also are the curlews, sandpipers, phaleropes and terns. We may accordingly presume that the more pointed eggs are those that are large relatively to the tube or oviduct through which they have to pass, or, in other words, are those which are subject to the greatest pressure while

* Cf. Mallock, *op. cit.*

† This was known to Albertus Magnus, though his explanation was wrong, "Ova autem habentia duos colores non sunt omnino penitus rotunda, sed ex una parte sunt acuta habentia angulum sphericum acutum, sicut sunt composita ex duobus semispheris, in una parte extensis ad angulum acutum et in alia parte sphericis non extensis in loco ubi est polus ovi....Et in exitu ovi acutus angulus exit ultimo, eo quod ipse porrectus est ad interiora matricis versus parietem ubi ovum cum matrice continuatur in sui generatione" (*De animalibus*, lib. XVII, tract. 1, c. 3).

being forced along. So general is this relation that we may go still further, and presume with great plausibility in the few exceptional cases (of which the apteryx is the most conspicuous) where the egg is relatively large though not markedly unsymmetrical, that in these cases the oviduct itself is in all probability large (as Günther had suggested) in proportion to the size of the bird. In the case of the common fowl we can trace a direct relation between the size and shape of the egg, for the first eggs laid by a young pullet are usually smaller, and at the same time are much more nearly spherical than the later ones; and, moreover, some breeds of fowls lay proportionately smaller eggs than others, and on the whole the former eggs tend to be rounder than the latter*.

We may now proceed to enquire more particularly how the form of the egg is controlled by the pressures to which it is subjected.

The egg, just prior to the formation of the shell, is, as we have seen, a fluid body, tending to a spherical shape and *enclosed within a membrane*.

Our problem, then, is: Given an incompressible fluid, contained in a deformable capsule, which is either (*a*) entirely inextensible, or (*b*) slightly extensible, and which is placed in a long elastic tube the walls of which are radially contractile, to determine the shape under some given distribution of pressure. We may assume, at least to begin with, that the shell-membrane is homogeneous and isotropic—uniform in all parts and in all directions.

If the capsule be spherical, inextensible, and completely filled with the fluid, absolutely no deformation can take place. The few eggs that are actually or approximately spherical, such as those of the tortoise or the owl, may thus be alternatively explained as cases where little or no deforming pressure has been applied prior to the solidification of the shell, or else as cases where the capsule was so

* In so far as our explanation involves a shaping or moulding of the egg by the uterus or oviduct (an agency supplemented by the proper tensions of the egg), it is curious to note that this is very much the same as that old view of Telesius regarding the formation of the embryo (*De rerum natura*, VI, cc. 4 and 10), which he had inherited from Galen, and of which Bacon speaks (*Nov. Org.* cap. 50; cf. Ellis's note). Bacon expressly remarks that "Telesius should have been able to shew the like formation in the shells of eggs." This old theory of embryonic modelling survives in our usage of the term "matrix" for a "mould."

little capable of extension and so completely filled as to preclude the possibility of deformation.

If the capsule be not spherical, but be inextensible, then only such deformation can take place as tends to make the shape more nearly spherical; and as the surface area is thereby decreased, the envelope must either shrink or pucker. In other words, an incompressible fluid contained in an inextensible envelope cannot be deformed without puckering of the envelope.

But let us next assume, as the condition by which this result may be avoided, that the envelope is to some extent extensible and that deformation is so far permitted. It is obvious that, on the presumption that the envelope is only moderately extensible, the whole structure can only be distorted to a moderate degree away from the spherical or spheroidal form.

At all points the shape is determined by the law of the distribution of *radial pressure within the given region of the tube*, surface friction helping to maintain the egg in position. If the egg be under pressure from the oviduct, but without any marked component either in a forward or backward direction, the egg will be compressed in the middle, and will tend more or less to the form of a cylinder with spherical ends. The eggs of the grebe, cormorant, or crocodile may be supposed to receive their shape in such circumstances.

When the egg is subject to the peristaltic contraction of the oviduct during its formation, then from the nature and direction of motion of the peristaltic wave the pressure will be greatest somewhere behind the middle of the egg; in other words, the tube is converted for the time being into a more conical form, and the simple result follows that the anterior end of the egg becomes the broader and the posterior end the narrower.

The peristalsis of the oviduct thus plays a double part, in propelling the egg down the oviduct and in impressing on it its ovoid form; but the whole process is a very slow one, for the hen's oviduct is only a few inches long, and the egg is some ten or twelve hours upon its way. We shall consider presently certain shells which may be regarded as so many drops or vesicles deformed by gravity; that is a statical problem. Compared with it the problem of the egg is a dynamical one; and yet it becomes a quasi-statical one, because the action is so very slow. It is an action without lag and without momentum; and the question, common in dynamical problems, of the relation between the period of the application of the force and the free period of response or adjustment to it need not concern us at all.

Again, the case of the egg is somewhat akin to a hydrodynamical problem; for as it lies in the oviduct we may look on it as a stationary body round which waves are flowing, with the same result as when a body moves through a fluid at rest. Thus we may treat it as a hydrodynamical problem, but a very simple one—simplified by the absence of all eddies and every form of turbulence; and we come to look on the egg as a *streamlined* structure, though its streamlines are of a very simple kind.

The mathematical statement of the case begins as follows: In our egg, consisting of an extensible membrane filled with an incompressible fluid and under external pressure, the equation of the envelope is $p_n + T(1/r + 1/r') = P$, where p_n is the normal component of external pressure at a point where r and r' are the radii of curvature, T is the tension of the envelope, and P the internal fluid pressure. This is simply the equation of an elastic surface where T represents the coefficient of elasticity; in other words, a flexible elastic shell has the same mathematical properties as our fluid, membrane-covered egg. And this is the identical equation which we have already had so frequent occasion to employ in our discussion of the forms of cells; save only that in these latter we had chiefly to study the tension T (i.e. the surface-tension of the semi-fluid cell) and had little or nothing to do with the factor of external pressure (p_n), which in the case of the egg becomes of chief importance.

To enquire how an elastic sphere or spheroid will be deformed in passing down a peristaltic tube is an ill-defined and indeterminate problem; but we can study the effect produced in the shape of any particular egg, and so far infer the forces which have been in action. We need only study a single meridian of the egg, inasmuch as we have found it to be a solid of revolution. At successive points along this meridian, let us determine the amount of curvature, that is to say the principal radii of curvature, in latitude and longitude, in the Gaussian formula $P = p_n + T(1/r + 1/r')$: or, as we may write it if we have any reason to doubt the uniformity or isotropy of the membrane, $T/r + T'/r'$. The sum of these curvatures varies from point to point; the internal or hydrodynamical pressure, P, is constant; and therefore the external pressure, p_n, varies from point to point with the curvature, and is a direct function of the shape of the egg.

Some few eggs, such as the owl's and the kingfisher's, are so nearly spherical that we are apt to speak of them as spheres; but they are all prolate more or less, and no egg is so nearly circular in meridional section as all eggs are in their circles of latitude. When the egg is all but spherical that shape may be due (as we have seen) to various causes: to a relatively small size of the egg, allowing it to descend the tube under a minimum of peristaltic pressure; perhaps to an unusually strong shell-membrane, resistant of deformation; in general terms, to a possible diminution of p_n, or a possible increase of T. But all eggs have approximately spherical ends, and the big anterior end of the large conical eggs of plover or curlew or guillemot is conspicuously so. Here the egg projects into the wide cavity of the uncontracted oviduct, external or peristaltic pressure does not exist, the shell-membrane has to resist internal pressure without further external support, and the resultant spherical curvature is an indication of the uniformity, or isotropy, of the membrane. The lesser of the two spherical ends, that is to say the posterior end, has by much the greater curvature, and the tension there is correspondingly great. It would seem that the membrane ought to be thicker or stronger at this pointed end than elsewhere, but it is not known to be so. In any case, it is just here, in this presumably weakest part, that we are most apt to find the irregularities and deformities of misshapen eggs.

Within the egg lies the yolk, and the yolk is invariably spherical or very nearly so, whatever be the form of the entire egg. The reason is simple, and lies in the fact that the fluid yolk is itself enclosed within another membrane, between which and the shell-membrane lies the fluid albumin, which transmits a uniform hydrostatic pressure to the yolk*. The lack of friction between the yolk-membrane and the white of the egg is indicated by the well-known fact that the "germinal spot" on the surface of the yolk is always

* In like manner, the cell-nucleus is "usually globular, except in certain specialised tissues, or when it degenerates" (Darlington). Whether it possesses a *membrane* is matter in dispute, but it at all events possesses a *surface*, with a phase-difference between it and the surrounding cytoplasm. Cf. above, p. 295.

found uppermost, however we may place and wherever we may open the egg; that is to say, the yolk easily rotates within the egg, bringing its lighter pole uppermost.

In its passage down the oviduct the egg is not merely thrust but also *screwed* along; and its spiral course leaves traces on wellnigh all its structure save the shell. When we have broken the shell of a hard-boiled egg the shell-membrane below peels off in spiral strips, and even the white tends to flake off in layers, spirally. In the fresh unboiled egg two knotted cords—the treadles or *chalazae*—are connected with the yolk, and lie fore-and-aft of it, loose in the albumen. These represent the free ends of a yolk-membrane, which got caught in the constricted oviduct while the yolk between them was being screwed along: very much as we may wrap an apple in a handkerchief, hold the two ends fast, and twirl the apple round.

These, then, are the general principles involved in, and illustrated by, the configuration of an egg; and they take us as far as we can safely go without actual quantitative determination, in each particular case, of the forces concerned*.

In certain cases among the invertebrates, we again find instances of hard-shelled eggs which have obviously been moulded by the oviduct, or so-called "ootype," in which they have lain: and not merely in such a way as to shew the effects of peristaltic pressure upon a uniform elastic envelope, but so as to impress upon the egg the more or less irregular form of the cavity within which it had been for a time contained and compressed. After this fashion is explained the curious form of the egg in *Bilharzia* (*Schistosoma*) *haematobium*, a formidable parasitic worm to which is due a disease wide-spread in Africa and Arabia, and an especial scourge of the Mecca pilgrims. The egg in this worm is provided at one end with a little spine, which is explained as having been moulded within a little funnel-shaped expansion of the uterus, just where it communicates with the common duct leading from the ovary and yolk-gland. Owing to some anatomical difference in the uterus, the little

* It is a common but unfounded belief among poultry-men that shape and size are related to the sex of the egg, the longer eggs producing mostly male chicks. That there is no such correlation between sex on the one hand and weight, length or shape on the other, has been clearly demonstrated. Cf. M. A. Jull and J. P. Quinn, *Journ. Agr. Research*, XXIX, pp. 195–201, 1924.

spine may be at the end or towards the side of the egg: and this visible difference has led to the recognition of a new species, *S. mansoni**. In a third species, *S. japonicum*, the egg is described as bulging into a so-called "calotte," or bubble-like convexity at the end opposite to the spine. This, I think, may, with very little doubt, be ascribed to hardening of the egg-shell having taken place just at the time when partial relief from pressure was being experienced by the egg in the neighbourhood of the dilated orifice of the oviduct.

This case of Bilharzia is not, from our present point of view, a very important one, but nevertheless it is interesting. It ascribes to a mechanical cause a curious peculiarity of form; and it shews, by reference to this mechanical principle, how two simple mechanical modifications of the same thing may not only seem very different to the systematic naturalist's eye, but may actually lead to the recognition of a new species, with its own geographical distribution, and its own pathogenic characteristics.

On the form of sea-urchins

As a corollary to the problem of the bird's egg, we may consider for a moment the forms assumed by the shells of the sea-urchins. These latter are commonly divided into two classes—the Regular and the Irregular Echinids. The regular sea-urchins, save in slight details which do not affect our problem, have a complete axial symmetry. The axis of the animal's body is vertical, with mouth below and the intestinal outlet above; and around this axis the shell is built as a symmetrical system. It follows that in horizontal section the shell is everywhere circular, and we need only consider its form as seen in vertical section or projection. The irregular urchins (very inaccurately so-called) have the anal extremity of the body removed from its central, dorsal situation; and it follows that they have now a single plane of symmetry, about which the organism, shell and all, is bilaterally symmetrical. We need not concern ourselves in detail with the shapes of their shells, which may be very simply interpreted, by the help of radial coordinates, as deformations of the circular or "regular" type.

* L. W. Sambon, *Proc. Zool. Soc.* 1907 (I), p. 283; also in *Journ. Trop. Med. and Hygiene*, Sept. 15, 1926.

The sea-urchin shell consists of a membrane, stiffened into rigidity by calcareous deposits, which constitute a beautiful skeleton of separate, neatly fitting "ossicles." The rigidity of the shell is more apparent than real, for the entire structure is, in a sluggish way, plastic; inasmuch as each little ossicle is capable of growth, and the entire shell grows by increments to each and all of these multitudinous elements, whose individual growth involves a certain amount of freedom to move relatively to one another; in a few cases the ossicles are so little developed that the whole shell appears soft and flexible. The viscera of the animal occupy but a small part of the space within the shell, the cavity being mainly filled by a large quantity of watery fluid, whose density must be very near to that of the external sea-water.

Apart from the fact that the sea-urchin continues to grow, it is plain that we have here the same general conditions as in the egg-shell, and that the form of the sea-urchin is subject to a similar equilibrium of forces. But there is this important difference, that an external muscular pressure (such as the oviduct administers during the consolidation of the egg-shell) is now lacking. In its place we have the steady continuous influence of gravity, and there is yet another force which in all probability we require to take into consideration.

While the sea-urchin is alive, an immense number of delicate "tube-feet," with suckers at their tips, pass through minute pores in the shell, and, like so many long cables, moor the animal to the ground. They constitute a symmetrical system of forces, with one resultant downwards, in the direction of gravity, and another outwards in a radial direction; and if we look upon the shell as originally spherical, both will tend to depress the sphere into a flattened cake. We need not consider the radial component, but may treat the case as that of a spherical shell symmetrically depressed under the influence of gravity. This is precisely the condition which we have to deal with in a drop of liquid lying on a plate; the form of which is determined by its own uniform surface-tension, plus gravity, acting against the internal hydrostatic pressure. Simple as this system is, the full mathematical investigation of the form of a drop is not easy, and we can scarcely hope that the systematic study of the Echinodermata will ever be conducted by methods based on Laplace's differential equation*; but we have little difficulty in seeing that the various forms represented in a series of sea-urchin shells are no other than those which we may easily and perfectly imitate in drops.

In the case of the drop of water (or of any other particular liquid) the specific surface-tension is always constant, and the pressure varies inversely as the radius of curvature; therefore the smaller the drop the more nearly is it able to conserve the spherical form, and the larger the drop the more does it become flattened under gravity†. We can imitate this phenomenon by using india-rubber balls filled with water, of different sizes; the little ones will remain very nearly spherical, but the larger will fall down "of their own weight," into the form of more and more flattened cakes; and we see the same thing when we let drops of heavy oil (such as the orthotoluidene spoken of on p. 370) fall through a tall column of water, the little ones remaining round, and the big ones getting more and more flattened as they sink. In the case of the sea-urchin, the same series of forms may be assumed to occur, irrespective of size, through variations in T, the specific tension, or "strength" of the enveloping shell. Accordingly we may study, entirely from this point of view, such a series as the following (Fig. 454). In a very few cases, such as the fossil *Palaeechinus*, we have an approximately spherical shell, that is to say a shell so strong that the influence of gravity becomes negligible as a cause of deformation, just as (to compare small things with great) the surface tension of mercury is so high that small drops of it seem perfectly spherical‡. The ordinary species of *Echinus* begin to display a pronounced depression, and this reaches its maximum in such soft-shelled flexible forms as *Phormosoma*. On the general question I took the oppor-

* Cf. Bashforth and Adams, *Theoretical Forms of Drops, etc.*, Cambridge, 1883.

† The drops must be spherical, or very nearly so, to produce a rainbow. But the bow is said to be always better defined near the top than down below; which seems to shew that the lower and larger raindrops are the less perfect spheres. (Cf. T. W. Backhouse, *Symons's M. Met. Mag.* 1879, p. 25.) For the small round droplets in the cloud tend to cannon off one another, and remain small and spherical. But when there comes a difference of potential between cloud and cloud, or between earth and sky, then the spherules become distorted, one droplet coalesces with another, and the big drops begin to fall.

‡ Cf. A. Ferguson, On the theoretical shape of large bubbles and drops, *Phil. Mag.* (6), xxv, pp. 507–520, 1913.

tunity of consulting Mr C. R. Darling, who is an acknowledged expert in drops, and he at once agreed with me that such forms as are represented in Fig. 454 are no other than diagrammatic illustrations of various kinds of drops, "most of which can easily be reproduced in outline by the aid of liquids of approximately equal density to water, although some of them are fugitive." He found a difficulty in the case of the outline which represents *Asthenosoma*, but the reason for the anomaly is obvious; the flexible shell has flattened

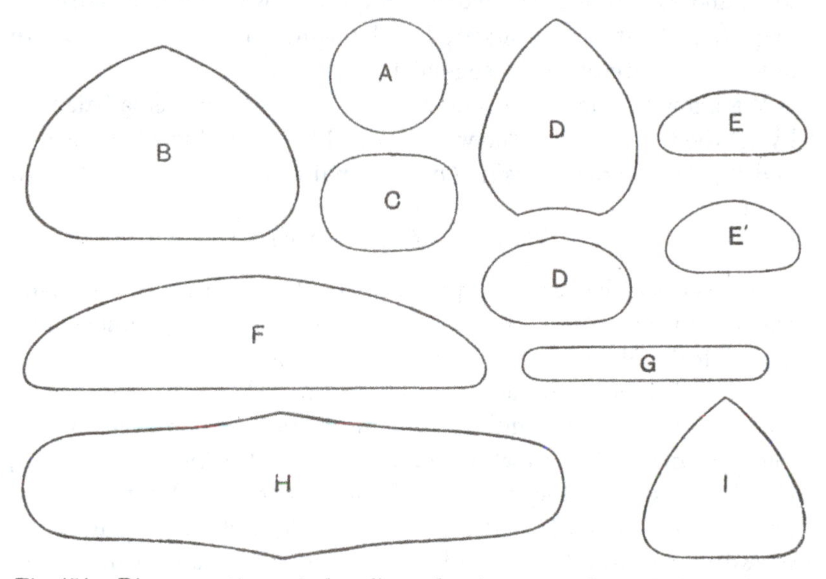

Fig. 454. Diagrammatic vertical outlines of various sea-urchins: A, *Palaeechinus*; B, *Echinus acutus*; C, *Cidaris*; D, D′, *Coelopleurus*; E, E′, *Genicopatagus*; F, *Phormosoma luculenter*; G, *P. tenuis*; H, *Asthenosoma*; I, *Urechinus*.

down until it has come in contact with the hard skeleton of the jaws, or "Aristotle's lantern," within, and the curvature of the outline is accordingly disturbed. The elevated, conical shells such as those of *Urechinus* and *Coelopleurus* evidently call for some further explanation; for there is here some cause at work to elevate, rather than to depress the shell. Mr Darling tells me that these forms "are nearly identical in shape with globules I have frequently obtained, in which, on standing, bubbles of gas rose to the summit and pressed the skin upwards, without being able to escape." The same condition may be at work in the sea-urchin; but a similar tendency would also be manifested by the presence in the upper part of the shell of any accumulation of substance lighter than water, such as is actually present in the masses of fatty, oily eggs.

On the form and branching of blood-vessels

Passing to what may seem a very different subject, we may investigate a number of interesting points in connection with the form and structure of the blood-vessels, and we shall find ourselves helped, at least in the outset, by the same equations as those we have used in studying the egg-shell.

We know that the fluid pressure (P) within the vessel is balanced by (1) the tension (T) of the wall, divided by the radius of curvature, and (2) the external pressure (p_n), normal to the wall: according to our formula

$$P = p_n + T\,(1/r + 1/r').$$

If we neglect the external pressure, that is to say any support which may be given to the vessel by the surrounding tissues, and if we deal only with a cylindrical vein or artery, this formula becomes simplified to the form $P = T/R$. That is to say, under constant pressure, the tension varies as the radius. But the tension, per unit area of the vessel, depends upon the thickness of the wall, that is to say on the amount of membranous and especially of muscular tissue of which it is composed. Therefore, so long as the pressure is constant, the thickness of the wall should vary as the radius, or as the diameter, of the blood-vessel.

But it is not the case that the pressure is constant, for it gradually falls off, by loss through friction, as we pass from the large arteries to the small; and accordingly we find that while, for a time, the cross-sections of the larger and smaller vessels are symmetrical figures, with the wall-thickness proportional to the size of the tube, this proportion is gradually lost, and the walls of the small arteries, and still more of the capillaries, become exceedingly thin, and more so than in strict proportion to the narrowing of the tube.

In the case of the heart we have, within each of its cavities, a pressure which, at any given moment, is constant over the whole wall-area, but the thickness of the wall varies very considerably. For instance, in the left ventricle the apex is by much the thinnest portion, as it is also that with the greatest curvature. We may assume, therefore (or at least suspect), that the formula, $t\,(1/r + 1/r') = C$, holds good; that is to say, that the thickness (t) of the wall varies inversely as the mean curvature. This may be tested experimentally, by dilating a heart with alcohol under a known pressure, and then measuring the thickness of the walls in various parts after the whole organ has become hardened. By this means it is found that, for each of the cavities, the law holds good with great accuracy*. Moreover, if we begin by dilating the right ventricle and then dilate the left in like manner, until the whole heart is equally and symmetrically dilated, we find (1) that we have had to use a pressure in the left ventricle from six to seven times as great as in the right ventricle, and (2) that the thickness of the walls is just in the same proportion†.

Many problems of a hydrodynamical kind arise in connection with the flow of blood through the blood-vessels; and while these are of primary importance to the physiologist they interest the morphologist in so far as they bear on questions of structure and form. As an example of such mechanical problems we may take the conditions which go to determine the manner of branching of an artery, or the angle at which its branches are given off; for, as John Hunter said‡, "To keep up a circulation sufficient for the part, and no more, Nature has varied the angle of the origin of the arteries accordingly." This is a vastly important theme, and leads us a deal farther than does the problem, petty in comparison, of the shape of an egg. For the theorem which John Hunter has set forth in these simple words is no other than that "principle of minimal work" which is fundamental in physiology, and which some have deemed the very criterion

* R. H. Woods, On a physical theorem applied to tense membranes, *Journ. of Anat. and Phys.* XXVI, pp. 362–371, 1892. A similar investigation of the tensions in the uterine wall, and of the varying thickness of its muscles, was attempted by Haughton in his *Animal Mechanics*, 1873, pp. 151–158.

† This corresponds with a determination of the normal pressures (in systole) by Knohl, as being in the ratio of 1 : 6·8.

‡ *Essays*, edited by Owen, I, p. 134, 1861. The subject greatly interested Keats. See his *Notebook*, edited by M. B. Forman, 1932, p. 7; and cf. *Keats as a Medical Student*, by Sir Wm Hale-White, in *Guy's Hospital Reports*, LXXIII, pp. 249–262, 1925.

of "organisation*." For the principle of Lagrange, the "principle of virtual work," is the key to physiological equilibrium, and physiology itself has been called a problem in maxima and minima†.

This principle, overflowing into morphology, helps to bring the morphological and the physiological concepts together. We have dealt with problems of maxima and minima in many simple configurations, where form alone seemed to be in question; and we meet with the same principle again wherever work has to be done and mechanism is at hand to do it. That this mechanism is the best possible under all the circumstances of the case, that its work is done with a maximum of efficiency and at a minimum of cost, may not always lie within our range of quantitative demonstration, but to believe it to be so is part of our common faith in the perfection of Nature's handiwork. All the experience and the very instinct of the physiologist tells him it is true; he comes to use it as a postulate, or *methodus inveniendi*, and it does not lead him astray. The discovery of the circulation of the blood was implicit in, or followed quickly after, the recognition of the fact that the valves of heart and veins are adapted to a one-way circulation; and we may begin likewise by assuming a perfect fitness or adaptation in all the minor details of the circulation.

As part of our concept of organisation we assume that the cost of operating a physiological system is a minimum, what we mean by *cost* being measurable in calories and ergs, units whose dimensions are equivalent to those of *work*. The circulation teems with illustrations of this great and cardinal principle. "To keep up a circulation sufficient for the part and no more" Nature has not only varied the angle of branching of the blood-vessels to suit her purpose, she has regulated the dimensions of every branch and stem and twig and capillary; the normal operation of the heart is perfection itself, even the amount of oxygen which enters and leaves the capillaries is such that the work involved in its exchange and transport is a minimum. In short, oxygen transport is the main object of the circulation, and it seems that through all the trials and errors of

* Cf. Cecil D. Murray, The physiological principle of minimal work, in the vascular system, and the cost of blood-volume, *Proc. Acad. Nat. Sci.* XII, pp. 207–214, 1926; The angle of branching of the arteries, *Journ. Gen. Physiol.* IX, pp. 835–841, 1926; On the branching-angles of trees, *ibid.* X, p. 725, 1927.

† By Dr F. H. Pike, quoted by C. D. Murray.

growth and evolution an efficient mode of transport has been attained. To prove that it is the very best of all possible modes of transport may be beyond our powers and beyond our needs; but to assume that it is *perfectly economical* is a sound working hypothesis*. And by this working hypothesis we seek to understand the form and dimensions of this structure or that, in terms of the work which it has to do.

The general principle, then, is that the form and arrangement of the blood-vessels is such that the circulation proceeds with a minimum of effort, and with a minimum of wall-surface, the latter condition leading to a minimum of friction and being therefore included in the first. What, then, should be the angle of branching, such that there shall be the least possible loss of energy in the course of the circulation? In order to solve this problem in any particular case we should obviously require to know (1) how the loss of energy depends upon the distance travelled, and (2) how the loss of energy varies with the diameter of the vessel. The loss of energy is evidently greater in a narrow tube than in a wide one, and greater, obviously, in a long journey than a short. If the large artery, AB, gives off a comparatively narrow branch leading to P (such as CP, or DP), the route ACP is evidently shorter than ADP, but on the other hand, by the latter path, the blood has tarried longer in the wide vessel AB, and has had a shorter course in the narrow branch.

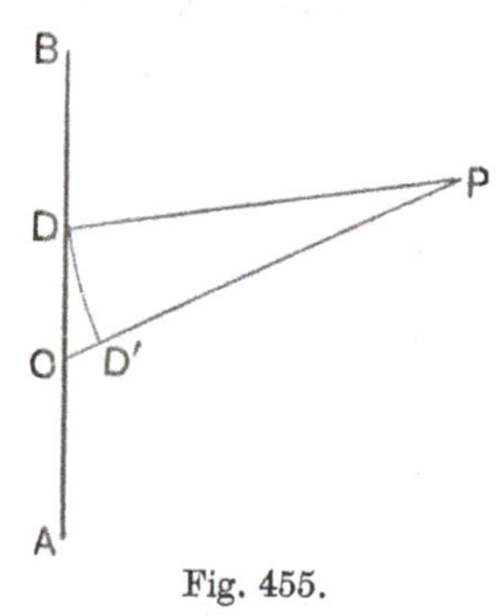

Fig. 455.

The relative advantage of the two paths will depend on the loss of energy in the portion CD, as compared with that in the alternative portion CD', the one being short and narrow, the other long and wide. If we ask, then, which factor is the more important, length

* Cf. A. W. Volkmann, *Die Haemodynamik nach Versuchen*, Leipzig, 1850 (a work of great originality); G. Schwalbe, Ueber...die Gestaltung des Arteriensystems, *Jen. Zeitschr.* XII, p. 267, 1878; W. Hess, Eine mechanischbedingte Gesetzmassigkeit im Bau des Blutgefässsystems, *A. f. Entw. Mech.* XVI, p. 632, 1903; Ueber die peripherische Regulierung der Blutzirkulation, *Pfluger's Archiv*, CLXVIII, pp. 439–490, 1917; R. Thoma, Die mittlere Durchflussmengen der Arterien des Menschen als Funktion des Gefässradius, *ibid.* CLXXXIX, pp. 282–310, CXCIII, pp. 385–406, 1921–22; E. Blum, Querschnittsbeziehungen zwischen Stamm u. Ästen im Arteriensystem, *ibid.* CLXXV, pp. 1–19, 1919.

or width, we may safely take it that the question is one of degree; and that the factor of width will become the more important of the two wherever artery and branch are markedly unequal in size. In other words, it would seem that for small branches a large angle of bifurcation, and for large branches a small one, is always the better. Roux has laid down certain rules in regard to the branching of arteries, which correspond with the general conclusions which we have just arrived at. The most important of these are as follows: (1) If an artery bifurcates into two equal branches, these branches come off at equal angles to the main stem. (2) If one of the two branches be smaller than the other, then the main branch, or continuation of the original artery, makes with the latter a smaller angle than does the smaller or "lateral" branch. And (3) all branches which are so small that they scarcely seem to weaken or diminish the main stem come off from it at a large angle, from about 70° to 90°.

We may follow Hess in a further investigation of the phenomenon. Let *AB* be an artery, from which a branch has to be given off so as to reach *P*, and let *ACP*, *ADP*, etc., be alternative courses which the branch may follow: *CD*, *DE*, etc., in the diagram, being equal distances ($= l$) along *AB*. Let us call the angles *PCD*, *PDE*, x_1, x_2, etc.: and the distances *CD'*, *DE'*, by which each branch exceeds the next in length, we shall call l_1, l_2, etc. Now it is evident that, of the courses shewn, *ACP* is the shortest which the blood can take, but it is also that by which its transit through the narrow branch is the longest. We may reduce its transit through the narrow branch more and more, till we come to *CGP*, or rather to a point where the branch comes off at right angles to the main stem; but in so doing we very considerably increase the whole distance travelled. We may take it that there will be some intermediate point which will strike the balance of advantage.

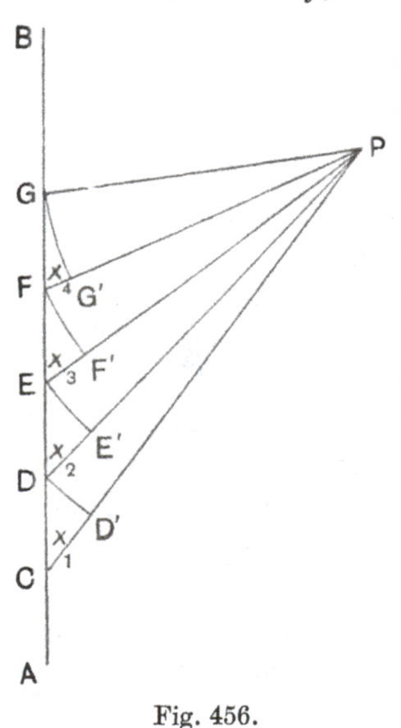

Fig. 456.

Now it is easy to shew that if, in Fig. 456, the route ADP and AEP (two contiguous routes) be equally favourable, then any other route on either side of these, such as ACP or AFP, must be less favourable than either. Let ADP and AEP, then, be equally favourable; that is to say, let the loss of energy which the blood suffers in its passage along these two routes be equal. Then, if we make the distance DE very small, the angles x_2 and x_3 are nearly equal, and may be so treated. And again, if DE be very small, then $DE'E$ becomes a right angle, and l_2 (or DE') $= l \cos x_2$. But if L be the loss of energy per unit distance in the wide tube AB, and L' be the corresponding loss of energy in the narrow tube DP, etc., then $lL = l_2 L'$, because, as we have assumed, the loss of energy on the route DP is equal to that on the whole route DEP. Therefore $lL = lL' \cos x_2$, and $\cos x_2 = L/L'$. That is to say, the most favourable angle of branching will be such that the cosine of the angle is equal to the ratio of the loss of energy which the blood undergoes, per unit of length, in the main vessel, as compared with that which it undergoes in the branch. The path of a ray of light from one refractive medium to another is an analogous but much more famous problem; and the analogy becomes a close one when we look upon the branching artery as the special case of "grazing incidence."

After thus dealing with the most suitable angle of branching, we have still to consider the appropriate cross-section of the branches compared with the main trunk, for instance in the special case where a main artery bifurcates into two. That the sectional area of the two branches may together equal the area of the parent trunk, it is (of course) only necessary that the diameters of trunk and branch should be as $\sqrt{2} : 1$, or (say) as $14 : 10$, or (still more roughly) as $10 : 7$; and in the great vessels, this simple ratio comes very nearly true. We have, for instance, the following measurements of the common iliac arteries, into which the abdominal aorta subdivides:

*Internal diameter of abdominal arteries**

Aorta abdom. (mm.)	15·2	12·0	14·1	13·9
Iliaca comm. d.	10·8	8·8	10·4	8·6
Iliaca comm. s.	10·7	8·6	9·5	10·0
Mean of do.	10·8	8·7	10·0	9·3
Ratio	71	72	69	67 p.c.
			Av. 70 p.c.	

* From R. Thoma, *op. cit.* p. 388.

But the increasing surface of the branches soon means increased friction, and a slower pace of the blood travelling through; and therefore the branches must be more capacious than at first appears. It becomes a question not of capacity but of resistance; and in general terms the answer is that the ratio of resistance to cross-section shall be equal in every part of the system, before and after bifurcation, as a condition of least possible resistance in the whole system; the total cross-section of the branches, therefore, must be greater than that of the trunk in proportion to the increased resistance.

An approximate result, familiar to students of hydrodynamics, is that the resistance is a minimum, and the condition an optimum, when the cross-section of the main stem is to the sum of the cross-sections of the branches as $1 : \sqrt[3]{2}$, or $1 : 1{\cdot}26$. Accordingly, in the case of a blood-vessel bifurcating into two equal branches, the diameter of each should be to that of the main stem (approximately) as

$$\sqrt{\frac{1{\cdot}26}{2}} : 1, \quad \text{or (say) } 8 : 10.$$

While these statements are so far true, and while they undoubtedly cover a great number of observed facts, yet it is plain that, as in all such cases, we must regard them not as a complete explanation, but as *factors* in a complicated phenomenon: not forgetting that (as one of the most learned of all students of the heart and arteries, Dr Thomas Young, said in his Croonian lecture*) all such questions as these, and all matters connected with the muscular and elastic powers of the blood-vessels, "belong to the most refined departments of hydraulics"; and Euler himself had commented on the "in-

* On the functions of the heart and arteries, *Phil. Trans.* 1809, pp. 1–31, cf. 1808, pp. 164–186; *Collected Works*, I, pp. 511–534, 1855. The same lesson is conveyed by all such work as that of Volkmann, E. H. Weber and Poiseuille. Cf. Stephen Hales's *Statical Essays*, II, *Introduction*: "Especially considering that they [i.e. animal Bodies] are in a manner framed of one continued Maze of innumerable Canals, in which Fluids are incessantly circulating, some with great Force and Rapidity, others with very different Degrees of rebated Velocity: Hence, etc." Even Leonardo had brought his knowledge of hydrodynamics to bear on the valves of the heart and the vortex-like eddies of the blood. Cf. J. Playfair McMurrich, *L. da Vinci, the Anatomist*, 1930, p. 165; etc. How complicated the physiological aspect of the case becomes may be judged by Thoma's papers quoted above.

superable difficulties" of this sort of problem*. Some other explanation must be sought in order to account for a phenomenon which particularly impressed John Hunter's mind, namely the gradually altering angle at which the successive intercostal arteries are given off from the thoracic aorta: the special interest of this case arising from the regularity and symmetry of the series, for "there is not another set of arteries in the body whose origins are so much the same, whose offices are so much the same, whose distances from their origin to the place of use, and whose uses [? sizes]† are so much the same."

The mechanical and hydrodynamical aspect of the circulation was as plain to John Hunter's mind as it had been to William Harvey or to Stephen Hales, or as it was afterwards to Thomas Young; but it was not always plain to other men. When a turtle's heart has been removed from its body, the blood may still be seen moving in the capillaries for some short while thereafter; and Haller, seeing this, "attributed it to some unknown power which he conceived to be exerted by the solid tissues on the blood and also by the globules of the blood on each other; to which power, until further investigation should elucidate its nature, he gave the name of *attraction.*" So said William Sharpey, the father of modern English physiology; and Sharpey went on to say that "many physiologists accordingly maintain the existence of a peculiar propulsive power in the coats of the capillary vessels different from contractility, or that the globules of the blood are possessed of the power of spontaneous motion." Alison, great physician and famous vitalist, "extended this view, in so far as he regards the motion of the blood in the capillaries as one of the effects produced by what he calls vital attraction and repulsion, powers which he conceives to be general attributes of living matter." But Sharpey's own clear insight so far overcame his faith in Alison that he found it "not impossible that a certain degree of agitation might be occasioned in the blood by the elastic resilience of the vessels reacting on it, after the distending force of the heart has been withdrawn"; and, in short, that the evidence in the case did not "warrant the

* In a tract entitled *Principia pro motu sanguinis per arterias determinando Op. posth.* XI, pp. 814–823, 1862.

† "Sizes" is Owen's editorial emendation, which seems amply justified.

assumption of a peculiar power acting on the blood, of whose existence in the animal economy we have as yet no other evidence*."

Sir Charles Bell, whose anatomical skill was great but his mathematical insight small, drew the conclusion, of no small historic interest, that "the laws of hydraulics, though illustrative, are not strictly applicable to the explanation of the circulation of the blood, nor to the actions of the living frame." He goes on to say: "Although we perceive admirable mechanism in the heart, and in the adjustment of the tubes on hydraulic principles: and although the arteries and veins have form, calibre and curves suited to the conveyance of fluid, according to our knowledge of hydraulic engines: yet the laws of life, or of physiology, are essential to the explanation of the circulation of the blood. And this conclusion we draw, not only from the extent and minuteness of the vessels, but also from the peculiar nature of the blood itself. Life is in both, and a mutual influence prevails†." This peculiar form of vitalism savours more of Bichat and the French school than of the teaching of John Hunter or Thomas Young. It is precisely that idea of "organic control" or "organic coordination," which the physiologists are always reluctant to accept, always unwilling to abandon: which is said to be inherent in every process or operation of the body, and to differentiate biology from all the physical sciences: and of which in our own day Haldane has been the chief and great protagonist. But it is a subject with which this book is not concerned.

To conclude, we may now approach the question of economical size of the blood-vessels in a broader way. They must not be too small, or the work of driving blood through them will be too great;

* See Sharpey's article on *Cilia*, in Todd's *Cyclopaedia*, I, p. 637, 1836; also Allen Thomson's admirable article on the *Circulation*, *ibid.* p. 672. Alison's views were based not only on Haller, but largely on Dr James Black's *Essay on the Capillary Circulation*, London, 1825.

† *Practical Essays*, 1842, p. 88. When Sir Charles Bell declared that hydraulic principles were not enough, but that "the laws of life" were needed to explain the circulation of the blood, he was right from his point of view. He was slow to see, and unwilling to admit, that hydrodynamical principles suffice to explain a large, essential part of the problem; but as a physiologist he had every reason to know that that part was not the whole. He may have had many things in mind: the arrest of the circulation in inflammation, as we see it in a frog's web; that a cut artery bleeds to death while a torn one does not bleed at all; that blood does not coagulate when stagnant within its own vessels—a fact which, as John Hunter said, "has ever appeared to me the most interesting fact in physiology."

they must not be too large, or they will hold more blood than is needed—and blood is a costly thing. We rely once more on Poiseuille's Law*, which tells us the amount of work done in causing so much fluid to flow through a tube against resistance, the said resistance being measured by the viscosity of the fluid, the coefficient of friction and the dimensions of the tube; but we have also to account for the blood itself, whose maintenance requires a share of the bodily fuel, and whose cost per c.c. may (in theory at least) be expressed in calories, or in ergs per day. The total cost, then, of operating a given section of artery will be measured by (1) the work done in overcoming its resistance, and (2) the work done in providing blood to fill it; we have come again to a differential equation, leading to an equation of maximal efficiency. The general result† is as follows: it can be made a quantitative one by introducing known experimental values. Were blood a cheaper thing than it is we might expect all arteries to be uniformly larger than they are, for thereby the burden on the heart (the flow remaining equal) would be greatly reduced—thus if the blood-vessels were doubled in diameter, and their volume thereby quadrupled, the work of the heart would be reduced to one-sixteenth. On the other hand, were blood a scarcer and still costlier fluid, narrower blood-vessels would hold the available supply; but a larger and stronger heart would be needed to overcome the increased resistance.

* Owing to faulty determination of the fall of pressure in the capillaries, Poiseuille's equation used to be deemed inapplicable to them; but Krogh's recent work removes, or tends to remove, the inconsistency (*Anatomy and Physiology of the Capillaries*, Yale University Press, 1922).

† Cf. C. D. Murray, *op. cit.* p. 211.

CHAPTER XVI

ON FORM AND MECHANICAL EFFICIENCY

There is a certain large class of morphological problems of which we have not yet spoken, and of which we shall be able to say but little. Nevertheless they are so important, so full of deep theoretical significance, and so bound up with the general question of form and its determination as a result of growth, that an essay on growth and form is bound to take account of them, however imperfectly and briefly. The phenomena which I have in mind are just those many cases where *adaptation* in the strictest sense is obviously present, in the clearly demonstrable form of mechanical fitness for the exercise of some particular function or action which has become inseparable from the life and well-being of the organism.

When we discuss certain so-called "adaptations" to outward circumstance, in the way of form, colour and so forth, we are often apt to use illustrations convincing enough to certain minds but unsatisfying to others—in other words, incapable of demonstration. With regard to coloration, for instance, it is by colours "cryptic," "warning," "signalling," "mimetic," and so on*, that we prosaically expound, and slavishly profess to justify, the vast Aristotelian synthesis that Nature makes all things with a purpose and "does nothing in vain." Only for a moment let us glance at some few instances by which the modern teleologist accounts for this or that manifestation of colour, and is led on and on to beliefs and doctrines to which it becomes more and more difficult to subscribe.

Some dangerous and malignant animals are said (in sober earnest) to wear a perpetual war-paint, in order to "remind their enemies that they had better leave them alone†." The wasp and the hornet,

* For a more elaborate classification, into colours cryptic, procryptic, anticryptic, apatetic, epigamic, sematic, episematic, aposematic, etc., see Poulton's *Colours of Animals* (Int. Scientific Series, LXVIII, 1890; cf. also R. Meldola, Variable protective colouring in insects, *P.Z.S.* 1873, pp. 153–162; etc. The subject is well and fully set forth by H. B. Cott, *Adaption coloration in Animals*, 1940.

† Dendy, *Evolutionary Biology*, 1912, p. 336.

in gallant black and gold, are terrible as an army with banners; and the Gila Monster (the poison-lizard of the Arizona desert) is splashed with scarlet—its dread and black complexion stained with heraldry more dismal. But the wasp-like livery of the noisy, idle hover-flies and drone-flies is but stage armour, and in their tinsel suits the little counterfeit cowardly knaves mimic the fighting crew.

The jewelled splendour of the peacock and the humming-bird, and the less effulgent glory of the lyre-bird and the Argus pheasant, are ascribed to the unquestioned prevalence of vanity in the one sex and wantonness in the other*.

The zebra is striped that it may graze unnoticed on the plain, the tiger that it may lurk undiscovered in the jungle; the banded Chaetodont and Pomacentrid fishes are further bedizened to the hues of the coral-reefs in which they dwell†. The tawny lion is yellow as the desert sand; but the leopard wears its dappled hide to blend, as it crouches on the branch, with the sun-flecks peeping through the leaves.

The ptarmigan and the snowy owl, the arctic fox and the polar bear, are white among the snows; but go he north or go he south, the raven (like the jackdaw) is boldly and impudently black.

The rabbit has his white scut, and sundry antelopes their piebald flanks, that one timorous fugitive may hie after another, spying the warning signal. The primeval terrier or collie-dog had brown spots over his eyes that he might seem awake when he was sleeping‡: so that an enemy might let the sleeping dog lie, for the singular reason that he imagined him to be awake. And a flock of flamingos,

* Delight in beauty is one of the pleasures of the imagination; there is no limit to its indulgence, and no end to the results which we may ascribe to its exercise. But as for the particular "standard of beauty" which the bird (for instance) admires and selects (as Darwin says in the *Origin*, p. 70, edit. 1884), we are very much in the dark, and we run the risk of arguing in a circle; for wellnigh all we can safely say is what Addison says (in the 412th *Spectator*)—that each different species "is most affected with the beauties of its own kind.... Hinc merula in nigro se oblectat nigra marito;... hinc noctua tetram Canitiem alarum et glaucos miratur ocellos."

† Cf. T. W. Bridge, *Cambridge Natural History* (Fishes), VII, p. 173, 1904; also K. v. Frisch, Ueber farbige Anpassung bei Fische, *Zool. Jahrb.* (*Abt. Allg. Zool.*), XXXII, pp. 171–230, 1914. But Reighard, in what Raymond Pearl calls "one of the most beautiful experimental studies of natural selection which has ever been made," found no relation between the colours of coral-reef fishes and their elimination by natural enemies (*Carnegie Inst. Publication* 103, pp. 257–325, 1908).

‡ *Nature*, L, p. 572; LI, pp. 33, 57, 533, 1894–95.

wearing on rosy breast and crimson wings a garment of invisibility, fades away into the sky at dawn or sunset like a cloud incarnadine*.

To buttress the theory of natural selection the same instances of "adaptation" (and many more) are used, as in an earlier but not distant age testified to the wisdom of the Creator and revealed to simple piety the immediate finger of God. In the words of a certain learned theologian†, "The free use of final causes to explain what seems obscure was temptingly easy....Hence the finalist was often the man who made a liberal use of the *ignava ratio*, or lazy argument: when you failed to explain a thing by the ordinary process of causality, you could 'explain' it by reference to some purpose of nature or of its Creator. This method lent itself with dangerous facility to the well-meant endeavours of the older theologians to expound and emphasise the beneficence of the divine purpose." *Mutatis mutandis*, the passage carries its plain message to the naturalist.

The fate of such arguments or illustrations is always the same. They attract and captivate for awhile; they go to the building of a creed, which contemporary orthodoxy‡ defends under its severest penalties: but the time comes when they lose their fascination, they somehow cease to satisfy and to convince, their foundations

* They are "wonderfully fitted for 'vanishment' against the flushed, rich-coloured skies of early morning and evening...their chief feeding-times"; and "look like a real sunset or dawn, repeated on the opposite side of the heavens—either east or west as the case may be" (Thayer, *Concealing-coloration in the Animal Kingdom*, New York, 1909, pp. 154–155). This hypothesis, like the rest, is not free from difficulty. Twilight is apt to be short in the homes of the flamingo; moreover, Mr Abel Chapman watched them on the Guadalquivir *feeding by day*, as I also have seen them at Walfisch Bay.

† Principal Galloway, *Philosophy of Religion*, 1914, p. 344.

‡ Professor D. M. S. Watson, addressing the British Association in 1929 on *Adaptation*, parted company with what I had called *contemporary orthodoxy* in 1917. Speaking of such morphological differences as "have commonly been assumed to be of an adaptive nature," he said: "That these structural differences are adaptive is for the most part pure assumption....There is no branch of zoology in which assumption has played a greater part, or evidence a less part, than in the study of such presumed adaptations." Hume, in his *Dialogue concerning Natural Religion*, shewed similar caution: "Steps of a stair are plainly constructed that human legs may use them in mounting; and this inference is certain and infallible. Human legs are also contrived for walking and mounting; and this inference, I allow, is not altogether so certain."

are discovered to be insecure, and in the end no man troubles to controvert them*.

But of a very different order from all such "adaptations" as these are those very perfect adaptations of form which, for instance, fit a fish for swimming or a bird for flight. Here we are far above the region of mere hypothesis, for we have to deal with questions of mechanical efficiency where statical and dynamical considerations can be applied and established in detail. The naval architect learns a great part of his lesson from the stream-lining of a fish†; the yachtsman learns that his sails are nothing more than a great bird's wing, causing the slender hull to *fly* along‡; and the mathematical study of the stream-lines of a bird, and of the principles underlying the areas and curvatures of its wings and tail, has helped to lay the very foundations of the modern science of aeronautics.

We know, for example, how in strict accord with theory (it was George Cayley who explained it first) the wing, whether of bird or insect, stands stiff along its "leading edge," like the mast before the sail; and how, conversely, it thins out exquisitely fine along its rear or "trailing edge," where sharp discontinuity favours the formation of uplifting eddies. And we see how, alike in the flying wing, in the penguin's swimming wing and in the whale's flipper, the same design of stiff fore-edge and thin fine trailing edge, both curving away evenly to meet at the tip, is continually exemplified.

We learn how lifting power not only depends on area but has a linear factor besides, such that a long narrow wing is more stable and effective both for speedy and for soaring flight than a short and broad one of equal area; and how in this respect the hawkmoth differs from the butterfly, the swallow from the thrush. We are taught how every wing, and every kite or sail, must have a certain

* The influence of environment on coloration is one thing, and the hypothesis of protective colouring is quite another. That arctic animals are often white, and desert animals sandy-hued or isabelline, are simple and undisputed facts; but such field-naturalists as Theodore Roosevelt, Selous (in his *African Nature Notes*), Buxton (*Animal Life in Deserts*), and Abel Chapman (*Savage Sudan*, and *Retrospect*) reject with one accord the theory of colour-protection.

† No creature shews more perfect stream-lining than a fur-seal swimming. Every curve is a *continuous* curve, the very ears and eye-slits and whiskers falling into the scheme, and the flippers folding close against the body.

‡ Cf. Manfred Curry, *Yacht Racing, and the Aerodynamics of Sails*, London, 1928.

amount of arch* or "belly," slight in the rapid fliers, deeper in the slow, flattened in the strong wind, bulging in the gentler breeze; and how advantageous is all possible stiffening of sail or wing, and why accordingly the yachtsman inserts "battens" and the Chinaman bamboos in his sail. We are shewn by Lilienthal himself how a powerful eddy, the so-called "ram," forms under the fore-edge, and is sometimes caught in a pocket of the bird's under wing-coverts and made use of as a forward drive.

We have lately learned how the gaps or slots between the primary wing-feathers of a crow, and a slight power of the wing-feathers to twist, like the slats of a Venetian blind, play their necessary

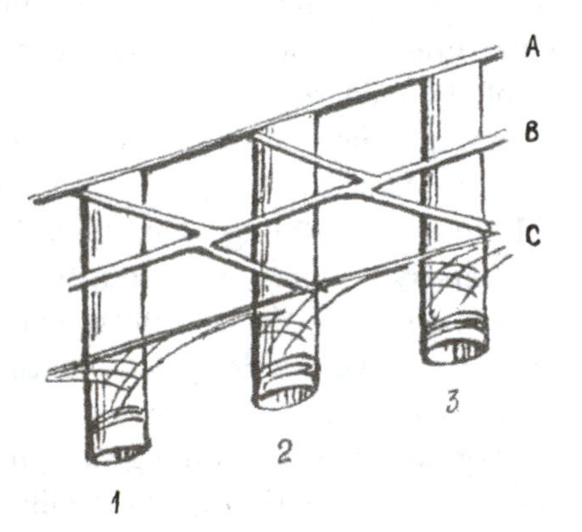

Fig. 457. Ligaments in a swan's wing. 1, 2, 3, remiges; A, B, longitudinal ligaments, with their oblique branches; C, small subcutaneous ligaments. From Marey, after Pettigrew.

part under certain conditions in the perfect working of the machine. Nothing can be simpler than the mechanism by which all this is done. Delicate ligaments run along the base of the wing from feather to feather, and send a branch to every quill (Fig. 457); by these, as the wing extends, the quills are raised into their places, and kept at their due and even distances apart. Not only that, but every separate ligamentous strand curls a little way round its feather where it is inserted into it; and thereby the feather is not only elevated into its place, but is given the little twist which brings it to its proper and precise obliquity.

* On the *curvatura veli*, cf. J. Bernoulli, *Acta Erudit. Lips.* 1692, p. 202. Studied also by Eiffel, *Résistance de l'air et l'aviation*, Paris, 1910.

All this is part of the automatic mechanism of the wing, than which there is nothing prettier in all anatomy. The triceps muscle, massed on the shoulder, extends the elbow-joint in the usual way. But another muscle (a long flexor of the wrist) has its origin above the elbow and its insertion below the wrist. It passes over two joints; it is what German anatomists call a *zweigelenkiger Muskel*. It transmits to the one joint the movements of the other; and in birds of powerful flight it becomes less and less muscular, more and more tendinous, and so more and more completely automatic. The wing itself is kept light; its chief muscle is far back on the shoulder; a contraction of that remote muscle throws the whole wing into gear. The little ligaments we have been speaking of are so linked up with the rest of the mechanism that we have only to hold a bird's wing by the arm-bone and extend its elbow-joint, to see the whole wing spring into action, with every joint extended, and every feather tense and in its place*.

Again on the fore-edge of the wing there lies a tiny mobile "thumb," whose little tuft of stiff, strong feathers forms the so-called "bastard wing." We used to look on it as a "vestigial organ," a functionless rudiment, a something which from ancient times had "lagged superfluous on the stage"; until a man of genius saw that it was just the very thing required to break the leading vortices, keep the flow stream-lined at a larger angle of incidence than before, and thereby help the plane to land. So he invented, to his great profit and advantage, the "slotted wing."

We learn many and many another interesting thing. How a stiff "comb" along the leading edge, a broad soft fringe along the trailing edge (the fringe acting as a damper and preventing "fluttering"), and a soft, downy upper surface of the wing, all help as silencers, and give the owl her noiseless flight. How the wing-loading of the owls is lower than in any other birds, lower even than in the eagles; and how owl and eagle have power to spare to carry easily their prey of mouse or mountain-hare. How the deep terminal wing-slots aid the heavy rook or heron in their slow

* The mole's forelimb has a somewhat similar action, by which, as the arm and hand are pulled violently backward, the claws are powerfully and automatically flexed for digging. The "suspensory ligament" of the horse, which is, or was, a short flexor of the digit, is an analogous mechanism. See a paper of mine On the nature and action of certain ligaments, *Journ. of Anat. and Physiol.* xviii, pp. 406–410, 1884.

flapping flight; how the sea-gull does not need them, for his load is lighter and his wings move slowly. There is never a discovery made in the theory of aerodynamics but we find it adopted already by Nature, and exemplified in the construction of the wing*.

We may illustrate some few of the principles involved in the construction of the bird's wing with a half-sheet of paper, whose laws, as it planes or glides downwards, Clerk Maxwell explained many years ago†.

To improve this first and roughest of models, we see that its leading edge had better be as long as possible, and that sharp

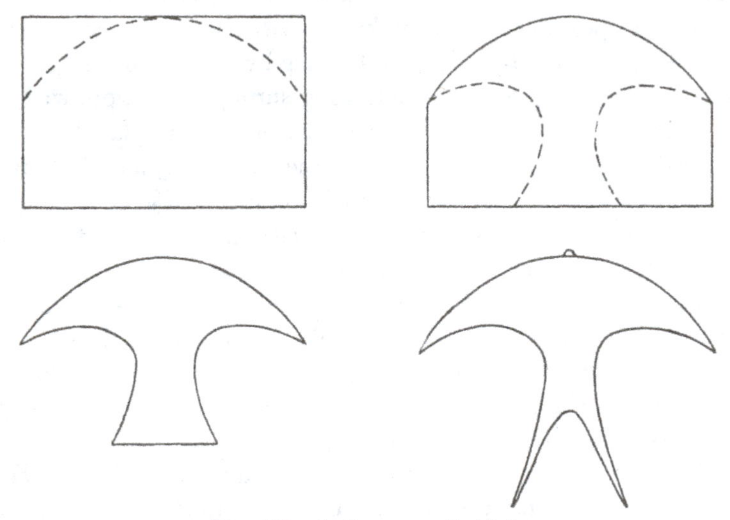

Fig. 458. A diagrammatic bird.

corners are bound to cause disturbance; let us get rid of the corners and turn the leading edge into a continuous curve (Fig. 458). The leading edge is now doing most of the work, and the area within and behind is doing little good. Vorticoid air-currents are beating down on either side on this inner area; moreover, air is "sliding out" below, and tending to curl round the tip and edges of the

* Note that the aeroplane copies the beetle rather than the bird, as Lilienthal himself points out, in *Vom Gleitflug zu Segelflug*, Berlin, 1923.

† On a particular case of the descent of a heavy body in a resisting medium, *Camb. and Dublin Math. Journ.* IX, pp. 145–148, 1854; *Sci. Papers*, I, pp. 115–118. This elegant and celebrated little paper was written by Clerk Maxwell while an undergraduate at Trinity College, Cambridge.

wing, all with so much waste of energy*. In short the broad wing is less efficient than the narrow; and on either side of our sheet of paper we may cut out a portion which is useless and in the way: for the same reasons we may cut out the middle part of the tail, which also is doing more harm than good†.

Hard as the problem is, and harder as it becomes, we may venture on. In aeronautics, as in hydrodynamics, we try to determine the resistances encountered by bodies of various shapes, moving through various fluids at various speeds; and in so doing we learn the enormous, the paramount importance of "stream-lining." There would be no need for stream-lining in a "perfect fluid," but in air or water it makes all the difference in the world. Stream-lining implies a shape round which the medium streams so smoothly that resistance is at last practically *nil*; there only remains the slight "skin-friction," which can be reduced or minimised in various ways. But the least imperfection of the stream-lining leads to whirls and "pockets" of dead water or dead air, which mean large resistance and waste of energy. The converse and more general problem soon emerges, of how in natural objects stream-lining comes to be; and whether or no the more or less stream-lined shape tends to be impressed on a deformable or plastic body by its own steady motion through a fluid‡. The principle of least action, the "loi de repos," is enough to suggest that the stream will tend to impress its stream-lines on the plastic body, causing it to yield or "give," until it ends by offering a minimum of resistance; and experiment goes some way to support the hypothesis. A bubble of mercury, poised in a tube through which air is blown, assumes a stream-lined shape, in so far as the forces due to the moving current avail against the

* This is why "slotting" so improves the broad wing of the crow. See on this and other matters, R. R. Graham's papers on Safety devices in the wings of birds, in *British Birds*, XXIV, 1930.

† Pettigrew shewed long ago (*Tr. R.S.E.* XXVI, p. 361, 1872) that the wing-area (in insects) "is usually greatly in excess of what is absolutely required for flight," and that the posterior or trailing edge could be largely trimmed away without the power of flight being at all diminished. We see how in the swallow this trailing edge is "trimmed away" till a bare minimum is left, and how (at least for a certain kind of flight) the wing is thereby greatly improved.

‡ Cf. Enoch Farrer, The shape assumed by a deformable body immersed in a moving fluid, *Journ. Franklin Inst.* 1921, pp. 737–756; also Vaughan Cornish, on *Waves of Sand and Snow*.

other forces of restraint. We have seen how the egg is automatically stream-lined, after a simple fashion, by the muscular pressure which drives it on its way. The contours of a snowdrift, of a wind-swept sand-dune, even of the flame of a lamp, shew endless illustrations of stream-lines or eddy-curves which the stream itself imposes, and which are oftentimes of great elegance and complexity. Always the stream tends to mould the bodies it streams over, facilitating its own flow; and the same principle must somehow come into play, at least as a contributory factor, in the making of a fish or of a bird. But it is obvious in both of these that even though the stream-lining be perfected in the individual it is also an inheritance of the race; and the twofold problem of accumulated inheritance, and of perfect structural adaptation, confronts us once again and passes all our understanding*.

When, after attempting to comprehend the exquisite adaptation of the swallow or the albatross to the navigation of the air, we try to pass beyond the empirical study and contemplation of such perfection of mechanical fitness, and to ask how such fitness came to be, then indeed we may be excused if we stand wrapt in wonderment, and if our minds be occupied and even satisfied with the conception of a final cause. And yet all the while, with no loss of wonderment nor lack of reverence, do we find ourselves constrained to believe that somehow or other, in dynamical principles and natural law, there lie hidden the steps and stages of physical causation by which the material structure was so shapen to its ends†.

The problems associated with these phenomena are difficult at every stage, even long before we approach to the unsolved secrets of causation; and for my part I confess I lack the requisite knowledge for even an elementary discussion of the form of a fish, or of an insect, or of a bird. But in the form of a bone we have a problem

* Mechanical perfection has often little to do with immunity from accident or with capacity to survive. Legs and wings of locust or mayfly are indescribably perfect for their brief spell of life and narrow sphere of toil; but they may be torn asunder in a moment, and whole populations perish in an hour. Careful of the type, but careless of the single life, Nature seems ruthless and indiscriminate in the sacrifice of these little lives.

† Cf. Professor Flint, in his Preface to Affleck's translation of Janet's *Causes finales*: "We are, no doubt, still a long way from a mechanical theory of organic growth, but it may be said to be the *quaesitum* of modern science, and no one can say that it is a chimaera."

of the same kind and order, so far simplified and particularised that we may to some extent deal with it, and may possibly even find, in our partial comprehension of it, a partial clue to the principles of causation underlying this whole class of phenomena.

Before we speak of the form of a bone, let us say a word about the mechanical properties of the material of which it is built*, in relation to the strength it has to manifest or the forces it has to resist: understanding always that we mean thereby the properties of fresh or living bone, with all its organic as well as inorganic constituents, for dead, dry bone is a very different thing. In all the structures raised by the engineer, in beams, pillars and girders of every kind, provision has to be made, somehow or other, for strength of two kinds, strength to resist compression or crushing, and strength to resist tension or pulling asunder. The evenly loaded column is designed with a view to supporting a downward pressure, the wire-rope, like the tendon of a muscle, is adapted only to resist a tensile stress; but in many or most cases the two functions are very closely inter-related and combined. The case of a loaded beam is a familiar one; though, by the way, we are now told that it is by no means so simple as it looks, and indeed that "the stresses and strains in this log of timber are so complex that the problem has not yet been solved in a manner that reasonably accords with the known strength of the beam as found by actual experiment†." However, be that as it may, we know, roughly, that when the beam is loaded in the middle and supported at both ends, it tends to be bent into an arc, in which condition its lower fibres are being stretched, or are undergoing a tensile

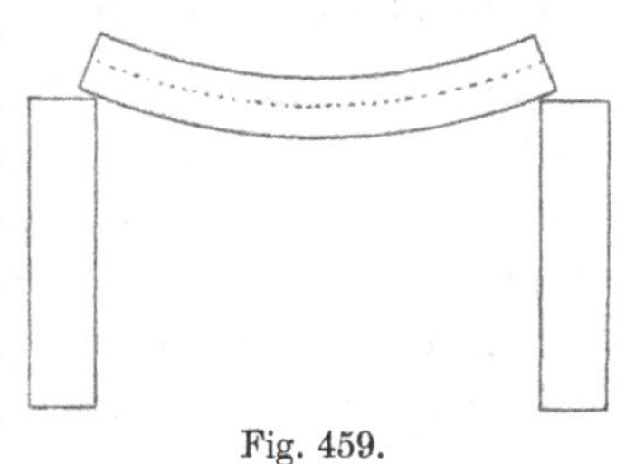

Fig. 459.

* Cf. Sir Donald MacAlister, How a bone is built, *Engl. Ill. Mag.* 1884.

† Professor Claxton Fidler, *On Bridge Construction*, p. 22 (4th ed.), 1909; cf. (*int. al.*) Love's *Elasticity*, p. 20 (*Historical Introduction*), 2nd ed., 1906, where the bending of the beam, and the distortion or warping of its cross-section, are studied after the manner of St Venant, in his Memoir on Torsion (1855). How complex the question has become may be judged from such papers as Price, On the structure of wood in relation to its elastic properties, *Phil. Trans.* (A), CCVIII, 1928; or D. B. Smith and R. V. Southwell, On the stresses induced by flexure in a deep rectangular beam, *Proc. R.S.* (A), CXLIII, pp. 271–285, 1934.

stress, while its upper fibres are undergoing compression. It follows that in some intermediate layer there is a "neutral zone," where the fibres of the wood are subject to no stress of either kind.

The phenomenon of a compression-member side by side with a tension-member may be illustrated in many simple ways. Ruskin (in *Deucalion*) describes it in a glacier. He then bids us warm a stick of sealing-wax and bend it in a horseshoe: "you will then see, through a lens of moderate power, the most exquisite facsimile of glacier fissures produced by extension on its convex surface, and as faithful an image of glacier surge produced by compression on its concave side." A still more beautiful way of exhibiting the distribution of strain is to use gelatin, into which bubbles of gas have been introduced with the help of sodium bicarbonate. A bar of such gelatin, when bent into a hoop, shews on the one side the bubbles elongated by tension and on the other those shortened by compression*.

In like manner a vertical pillar, if unevenly loaded (as for instance the shaft of our thigh-bone normally is), will tend to bend, and so to endure compression on its concave, and tensile stress upon its convex side. In many cases it is the business of the engineer to separate out, as far as possible, the pressure-lines from the tension-lines, in order to use separate modes of construction, or even different materials for each. In a suspension-bridge, for instance, a great part of the fabric is subject to tensile strain only, and is built throughout of ropes or wires; but the massive piers at either end of the bridge carry the weight of the whole structure and of its load, and endure all the "compression-strains" which are inherent in the system. Very much the same is the case in that wonderful arrangement of struts and ties which constitute, or complete, the skeleton of an animal. The "skeleton," as we see it in a Museum, is a poor and even a misleading picture of mechanical efficiency†. From the engineer's point of view, it is a diagram shewing all the compression-lines, but by no means all the tension-lines of the construction; it shews all the struts, but few of the ties, and perhaps we might even say *none* of the principal

* Cf. Emil Hatschek, Gestalt und Orientirung von Gasblasen in Gelen, *Kolloid-Ztschr.* XV, pp. 226–234, 1914.

† In preparing or "macerating" a skeleton, the naturalist nowadays carries on the process till nothing is left but the whitened bones. But the old anatomists, whose object was not the study of "comparative morphology" but the wider theme of comparative physiology, were wont to macerate by easy stages; and in many of their most instructive preparations the ligaments were intentionally left in connection with the bones, and as part of the "skeleton."

ones; it falls all to pieces unless we clamp it together, as best we can, in a more or less clumsy and immobilised way. But in life, that fabric of struts is surrounded and interwoven with a complicated system of ties—"its living mantles jointed strong, With glistering band and silvery thong*": ligament and membrane, muscle and tendon, run between bone and bone; and the beauty and strength of the mechanical construction lie not in one part or in another, but in the harmonious concatenation which all the parts, soft and hard, rigid and flexible, tension-bearing and pressure-bearing, make up together†.

However much we may find a tendency, whether in Nature or art, to separate these two constituent factors of tension and compression, we cannot do so completely; and accordingly the engineer seeks for a material which shall, as nearly as possible, offer equal resistance to both kinds of strain‡.

From the engineer's point of view, bone may seem weak indeed; but it has the great advantage that it is very nearly as good for a tie as for a strut, nearly as strong to withstand rupture, or tearing asunder, as to resist crushing. The strength of timber varies with the kind, but it always stands up better to tension than to compression, and wrought iron, with its greater strength, does much the same; but in cast-iron there is a still greater discrepancy the other way, for it makes a good strut but a very bad tie indeed. Mild steel, which has displaced the old-fashioned wrought iron in all engineering constructions, is not only a much stronger material, but it also possesses, like bone, the two kinds of strength in no very great relative disproportion§.

* See Oliver Wendell Holmes' *Anatomist's Hymn*.

† In a few anatomical diagrams, for instance in some of the drawings in Schmaltz's *Atlas der Anatomie des Pferdes*, we may see the system of "ties" diagrammatically inserted in the figure of the skeleton. Cf. W. K. Gregory, On the principles of quadrupedal locomotion, *Ann. N. Y. Acad. of Sciences*, XXII, p. 289, 1912.

‡ The strength of materials is not easy to discuss, and is still harder to tabulate. The wide range of qualities in each material, in timber the wide differences according to the direction in which the block is cut, and in all cases the wide difference between yield-point and fracture-point, are some of the difficulties in the way of a succinct statement.

§ In the modern device of "reinforced concrete," blocks of cement and rods of steel are so combined together as to resist both compression and tension in due or equal measure.

When the engineer constructs an iron or steel girder, to take the place of the primitive wooden beam, we know that he takes advantage of the elementary principle we have spoken of, and saves weight and economises material by leaving out as far as possible all the middle portion, all the parts in the neighbourhood of the "neutral zone"; and in so doing he reduces his girder to an upper and lower "flange," connected together by a "web," the whole resembling, in cross-section, an **I** or an **ꟷ**.

But it is obvious that, if the strains in the two flanges are to be equal as well as opposite, and if the material be such as cast-iron or wrought-iron, one or other flange must be made much thicker than the other in order that they may be equally strong*; and if at times the two flanges have, as it were, to change places, or play each other's parts, then there must be introduced a margin of safety by making both flanges thick enough to meet that kind of stress in regard to which the material happens to be weakest. There is great economy, then, in any material which is, as nearly as possible, equally strong in both ways; and so we see that, from the engineer's or contractor's point of view, bone is a good and suitable material for purposes of construction.

The I or the H-girder or rail is designed to resist bending in one particular direction, but if, as in a tall pillar, it be necessary to resist bending in all directions alike, it is obvious that the tubular or cylindrical construction best meets the case; for it is plain that this hollow tubular pillar is but the I-girder turned round every way, in a "solid of revolution," so that on any two opposite sides compression and tension are equally met and resisted, and there is now no need for any substance at all in the way of web or "filling" within the hollow core of the tube. And it is not only in the supporting pillar that such a construction is useful; it is appropriate in every case where *stiffness* is required, where bending has to be resisted. A sheet of paper becomes a stiff rod when you roll it up, and hollow tubes of thin bent wood withstand powerful thrusts in aeroplane construction. The long bone of a bird's wing has little or no weight to carry, but it has to withstand powerful bending

* This principle was recognized as soon as iron came into common use as a structural material. The great suspension bridges only became possible, in Telford's hands, when wrought iron became available.

moments; and in the arm-bone of a long-winged bird, such as an albatross, we see the tubular construction manifested in its perfection, the bony substance being reduced to a thin, perfectly cylindrical, and almost empty shell*. The quill of the bird's feather, the hollow shaft of a reed, the thin tube of the wheat-straw bearing its heavy burden in the ear, are all illustrations which Galileo used in his account of this mechanical principle†; and the working of his practical mind is exemplified by this catalogue of varied instances which one demonstration suffices to explain.

The same principle is beautifully shewn in the hollow body and tubular limbs of an insect or a crustacean; and these complicated and elaborately jointed structures have doubtless many constructional lessons to teach us. We know, for instance, that a thin cylindrical tube, under bending stress, tends to flatten before it buckles, and also to become "lobed" on the compression side of the bend; and we often recognise both of these phenomena in the joints of a crab's leg‡.

Two points, both of considerable importance, present themselves here, and we may deal with them before we go further on. In the first place, it is not difficult to see that in our bending beam the stress is greatest at its middle; if we press our walking-stick hard against the ground, it will tend to snap midway. Hence, if our cylindrical column be exposed to strong bending stresses, it will be prudent and economical to make its walls thickest in the middle and thinning off gradually towards the ends; and if we look at a longitudinal section of a thigh-bone, we shall see that this is just what Nature has done. The presence of a "danger-point" has been avoided, and the thickness of the walls becomes nothing less than

* Marsigli (*op. cit.*) was acquainted with the hollow wing-bones of the pelican; and Buffon deals with the whole subject in his *Discours sur la nature des oiseaux*.

† Galileo, *Dialogues concerning Two New Sciences* (1638), Crew and Salvio's translation, New York, 1914, p. 150; *Opere*, ed. Favaro, VIII, p. 186. (According to R. A. Millikan, "we owe our present day civilisation to Galileo.") Cf. Borelli, *De Motu Animalium*, I, prop. CLXXX, 1685. Cf. also P. Camper, La structure des os dans les oiseaux, *Opp.* III, p. 459, ed. 1803; A. Rauber, Galileo über Knochenformen, *Morphol. Jahrb.* VII, pp. 327, 328, 1881; Paolo Enriques, Della economia di sostanza nelle osse cave, *Arch. f. Entw. Mech.* XX, pp. 427–465, 1906. Galileo's views on the mechanism of the human body are also discussed by O. Fischer, in his article on *Physiologische Mechanik*, in the *Encycl. d. mathem. Wissenschaften*, 1904.

‡ Cf. L. G. Brazier, On the flexure of thin cylindrical shells, etc., *Proc. R.S.* (A), CXVI, p. 104, 1927.

a diagram, or "graph," of the bending-moments from one point to another along the length of the bone.

The second point requires a little more explanation. If we imagine our loaded beam to be supported at one end only (for instance, by being built into a wall), so as to form what is called a "bracket" or "cantilever," then we can see, without much difficulty, that the lines of stress in the beam run somewhat as in the accompanying diagram. Immediately under the load, the "compression-lines" tend to run vertically downward, but where the bracket is fastened to the wall there is pressure directed horizontally against the wall in the lower part of the surface of attachment; and the vertical beginning and the horizontal end of these pressure-lines must be continued into one another in the form of some even mathematical curve—which, as it happens, is part of a parabola. The tension-lines are identical in form with the compression-lines, of which they constitute the "mirror-image"; and where the two systems intercross they do so at right angles, or "orthogonally" to one another. Such systems of stress-lines as these we shall deal with again; but let us take note here of the important though well-nigh obvious fact, that while in the beam they both unite to carry the load, yet it is often possible to weaken one set of lines at the expense of the other, and in some cases to do altogether away with one set or the other. For example, when we replace our end-supported beam by a curved bracket, bent upwards or downwards as the case may be, we have evidently cut away in the one case the greater part of the tension-lines, and in the other the greater part of the compression-lines. And if instead of bridging a stream with our beam of wood we bridge it with a rope, it is evident that this new construction contains all the tension-lines, but none of the compression-lines of the old. The biological interest connected with this principle lies chiefly in the mechanical construction of the rush or the straw, or any other typically cylindrical stem. The material of which the stalk is constructed is very weak to withstand compression, but parts of it have a very great tensile strength. Schwendener, who was both botanist and engineer, has elaborately investigated the factor of strength in the cylindrical

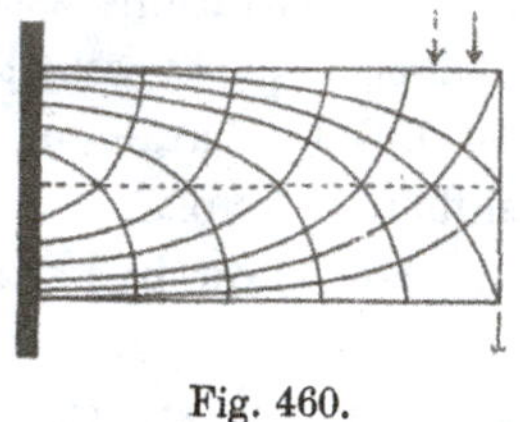

Fig. 460.

stem, which Galileo was the first to call attention to. Schwendener* shewed that its strength was concentrated in the little bundles of "bast-tissue," but that these bast-fibres had a tensile strength per square mm. of section not less, up to the limit of elasticity, than that of steel-wire of such quality as was in use in his day.

For instance, we see in the following table the load which various fibres, and various wires, were found capable of sustaining, not up to the breaking-point but up to the "elastic limit," or point beyond which complete recovery to the original length took place no longer after release of the load.

	Stress, or load in gms. per sq. mm., at Limit of Elasticity	Do., in tons per sq. inch	Strain, or amount of stretching, per mille
Secale cereale	15–20	9·4–12·5	4·4
Lilium auratum	19	11·8	7·6
Phormium tenax	20	12·5	13·0
Papyrus antiquorum	20	12·5	15·2
Molinia coerulea	22	13·8	11·0
Pincenectia recurvata	25	15·6	14·5
Copper wire	12·1	7·6	1·0
Brass ,,	13·3	8·5	1·35
Iron ,,	21·9	13·7	1·0
Steel ,,	24·6*	15·4	1·2

* This figure should be considerably higher for the best modern steel.

In other respects, it is true, the plant-fibres were inferior to the wires; for the former broke asunder very soon after the limit of elasticity was passed, while the iron-wire could stand, before snapping, about twice the load which was measured by its limit of elasticity: in the language of a modern engineer, the bast-fibres had a low "yield-point," little above the elastic limit. Nature seems content, as Schwendener puts it, if the strength of the fibre be ensured up to the elastic limit; for the equilibrium of the structure is lost as soon as that limit is passed, and it then matters little how far off the actual breaking-point may be†. But nevertheless, within certain limits, plant-fibre and wire were just as good and strong one

* S. Schwendener, *Das mechanische Princip im anatomischen Bau der Monocotyleen*, Leipzig, 1874; Zur Lehre von der Festigkeit der Gewächse, *Sb. Berlin. Akad.* 1884, pp. 1045–1070.

† The great extensibility of the plant-fibre is due to the spiral arrangement of the ultramicroscopic micellae of which the bast-fibre is built up: the spiral untwisting as the fibre stretches, in a right or left-hand spiral according to the species. Cf. C. Steinbruck, Die Micellartheorie auf botanischem Gebiete, *Biol. Centralbl.* 1925, p. 1.

as the other. And then Schwendener proceeds to shew, in many beautiful diagrams, the various ways in which these strands of strong tensile tissue are arranged in various stems: sometimes, in the simpler cases, forming numerous small bundles arranged in a peripheral ring, not quite at the periphery, for a certain amount of space has to be left for living and active tissue; sometimes in a sparser ring of larger and stronger bundles; sometimes with these bundles further strengthened by radial balks or ridges; sometimes with all the fibres set close together in a continuous hollow cylinder. In the case figured in Fig. 461, Schwendener calculated that the resistance to bending was at least twenty-five times as great as it would have been had the six main bundles been brought close together in a solid core. In many cases the centre of the stem is altogether empty; in all other cases it is filled with soft tissue, suitable for various functions, but never such as to confer mechanical rigidity. In a tall conical stem, such as that of a palm-tree, we can see not only these principles in the construction of the cylindrical trunk, but we can observe, towards the apex, the bundles of fibre curving over and intercrossing orthogonally with one another, exactly after the fashion of our stress-lines in Fig. 460; but of course, in this case, we are still dealing with tensile members, the opposite bundles taking on in turn, as the tree sways, the alternate function of resisting tensile strain*.

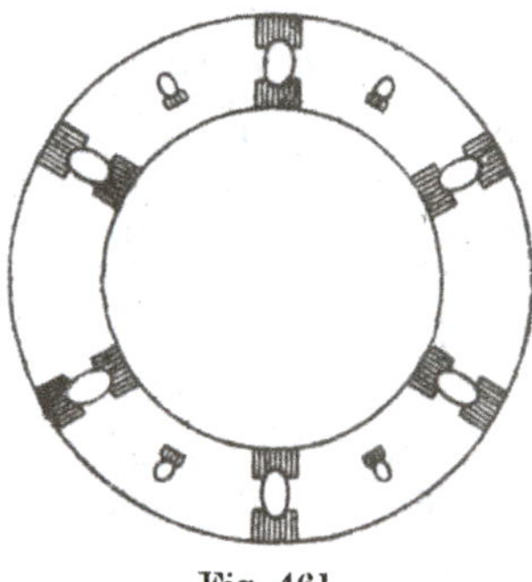

Fig. 461.

* For further botanical illustrations, see (*int. al.*) R. Hegler, Einfluss der Zugkräften auf die Festigkeit und die Ausbildung mechanischer Gewebe in Pflanzen, *SB. sächs. Ges. d. Wiss.* 1891, p. 638; Einfluss des mechanischen Zuges auf das Wachstum der Pflanze, *Cohn's Beiträge*, VI, pp. 383–432, 1893; O. M. Ball, Einfluss von Zug auf die Ausbildung der Festigkeitsgewebe, *Jahrb. d. wiss. Bot.* XXXIX, pp. 305–341, 1903; L. Kny, Einfluss von Zug und Druck auf die Richtung der Scheidewände in sich teilenden Pflanzenzellen, *Ber. d. bot. Gesellsch.* XIV, pp. 378–391, 1896; Sachs, Mechanomorphose und Phylogenie, *Flora*, LXXVIII, 1894; cf. also Pflüger, Einwirkung der Schwerkraft, etc., über die Richtung der Zelltheilung, *Archiv*, XXXIV, 1884; G. Haberlandt's *Physiological Plant Anatomy*, tr. by Montagu Drummond, 1914, pp. 150–213. On the engineering side of the case, see Angus R. Fulton, Experiments to show how failure under stress occurs in timber, etc., *Trans. R.S.E.* XLVIII, pp. 417–440, 1912; Fulton shews (*int. al.*) that "the initial cause of fracture in timbers lies in the medullary rays."

The Forth Bridge, from which the anatomist may learn many a lesson, is built of tubes, which correspond even in detail to the structure of a cylindrical branch or stem. The main diagonal struts are tubes twelve feet in diameter, and within the wall of each of these lie six T-shaped "stiffeners," corresponding precisely to the fibro-vascular bundles of Fig. 461; in the same great tubular struts the tendency to "buckle" is resisted, just as in the jointed stem of a bamboo, by "stiffening rings," or perforated diaphragms set twenty feet apart within the tube. We may draw one more curious, albeit parenthetic, comparison. An engineering construction, no less than the skeleton of plant or animal, has *to grow*; but the living thing is in a sense complete during every phase of its existence, while the engineer is often hard put to it to ensure sufficient strength in his unfinished and imperfect structure. The young twig stands more upright than the old, and between winter and summer the weight of leafage affects all the curving outlines of the tree. A slight upward curvature, a matter of a few inches, was deliberately given to the great diagonal tubes of the bridge during their piecemeal construction; and it was a triumph of engineering foresight to see how, like the twig, as length and weight increased, they at last came straight and true.

Let us now come, at last, to the mechanical structure of bone, of which we find a well-known and classical illustration in the various bones of the human leg. In the case of the tibia, the bone is somewhat widened out above, and its hollow shaft is capped by an almost flattened roof, on which the weight of the body directly rests. It is obvious that, under these circumstances, the engineer would find it necessary to devise means for supporting this flat roof, and for distributing the vertical pressures which impinge upon it to the cylindrical walls of the shaft.

In the long wing-bones of a bird the hollow of the bone is empty, save for a thin layer of living tissue lining the cylinder of bone; but in our own bones, and all weight-carrying bones in general, the hollow space is filled with marrow, blood-vessels and other tissues; and amidst these living tissues lies a fine lattice-work of little interlaced "trabeculae" of bone, forming the so-called "cancellous tissue." The older anatomists were content to describe this cancellous tissue as a sort of spongy network or irregular honeycomb*; but at length its orderly construction began to be perceived, and attempts were made to find a meaning or "purpose" in the arrangement. Sir Charles Bell had a glimpse of the truth when he asserted† that "this minute lattice-work, or the cancelli which constitute the interior structure of bone, have still reference to the forces acting on the bone"; but he did not succeed in shewing what these forces are, nor how the arrangement of the cancelli is related to them.

Jeffries Wyman, of Boston, came much nearer to the truth in a paper long neglected and forgotten‡. He gives the gist of the whole matter in two short paragraphs: "1. The cancelli of such bones as assist in supporting the weight of the body are arranged either in the direction of that weight, or in such a manner as to support and brace those cancelli which are in that direction. In a mechanical point of view they may be regarded in nearly all these bones as a series of 'studs' and 'braces.' 2. The direction of these fibres in some of the bones of the human skeleton is characteristic and, it is believed, has a definite relation to the erect position which is naturally assumed by man alone." A few years afterwards the story was told again, and this time with convincing accuracy. It was shewn by Hermann Meyer (and afterwards in greater detail by Julius Wolff and others) that the trabeculae, as seen in a longitudinal section of the femur, spread in beautiful curving lines from the head to the hollow shaft of the bone; and that these linear bundles are crossed by others, with so nice a regularity of arrangement that each intercrossing is as nearly as possible an orthogonal one: that is to say, the one set of fibres or cancelli cross the other everywhere at right angles. A great engineer, Professor Culmann of Zürich, to whom by the way we owe the whole modern method of "graphic statics," happened (in the year 1866) to come into his colleague Meyer's dissecting-room, where the anatomist was contemplating

* Sir John Herschel described a bone as a "framework of the most curious carpentry: in which occurs not a single straight line nor any known geometrical curve, yet all evidently systematic, and constructed by rules which defy our research" (*On the Study of Natural Philosophy*, 1830, p. 203).

† In *Animal Mechanics, or Proofs of Design in the Animal Frame*, 1827.

‡ Animal mechanics: on the cancellated structure of some of the bones of the human body, *Boston Soc. of Nat. Hist.* 1849. Reprinted, together with Sir C. Bell's work, by Morrill Wyman, Cambridge, Mass., 1902.

the section of a bone*. The engineer, who had been busy designing a new and powerful crane, saw in a moment that the arrangement of the bony trabeculae was nothing more nor less than a diagram of the lines of stress, or directions of tension and compression, in the loaded structure: in short, that Nature was strengthening the bone in precisely the manner and direction in which strength was required; and he is said to have cried out, "That's my crane!"

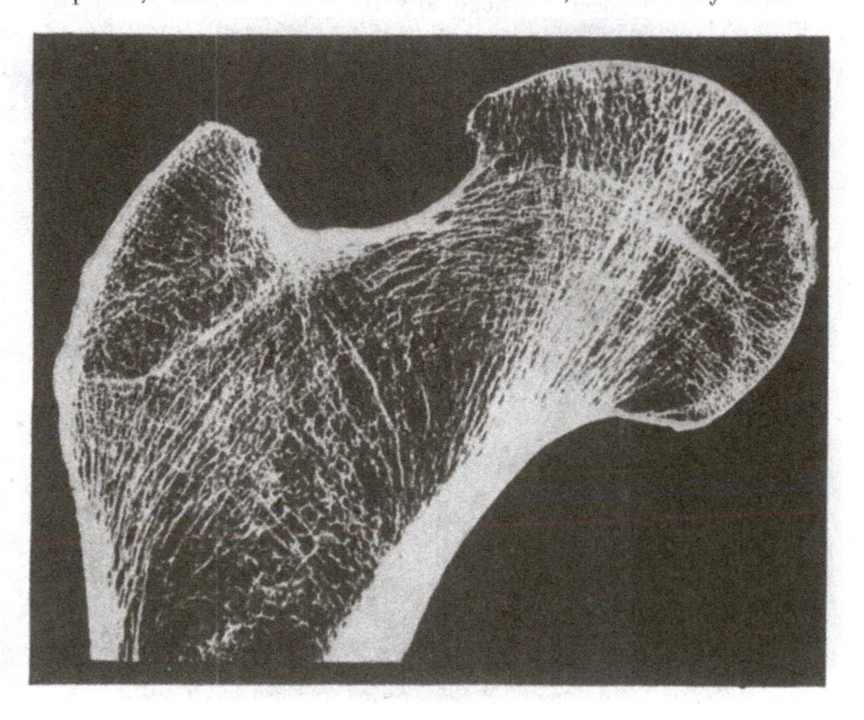

Fig. 462. Head of the human femur in section. After Schäfer, from a photo by Professor A. Robinson.

In the accompanying diagram of Culmann's crane-head, we recognise a simple modification, due entirely to the curved shape of the structure, of the still simpler lines of tension and compression which we have already seen in our end-supported beam, as represented in Fig. 460. In the shaft of the crane the concave or inner side,

* The first metatarsal, rather than the femur, is said to have been the bone which Meyer was demonstrating when Culmann first recognised the orthogonal intercrossing of the cancelli in tension and compression; cf. A. Kirchner, Architektur der Metatarsalien des Menschen, *Arch. f. Entw. Mech.* XXIV, pp. 539–616, 1907.

overhung by the loaded head, is the "compression-member"; the outer side is the "tension-member"; the pressure-lines, starting from the loaded surface, gather themselves together, always in the direction of the resultant pressure, till they form a close bundle running down the compressed side of the shaft: while the tension-lines, running upwards along the opposite side of the shaft, spread out through the head, orthogonally to, and linking together, the system of compression-lines. The head of the femur (Fig. 462) is

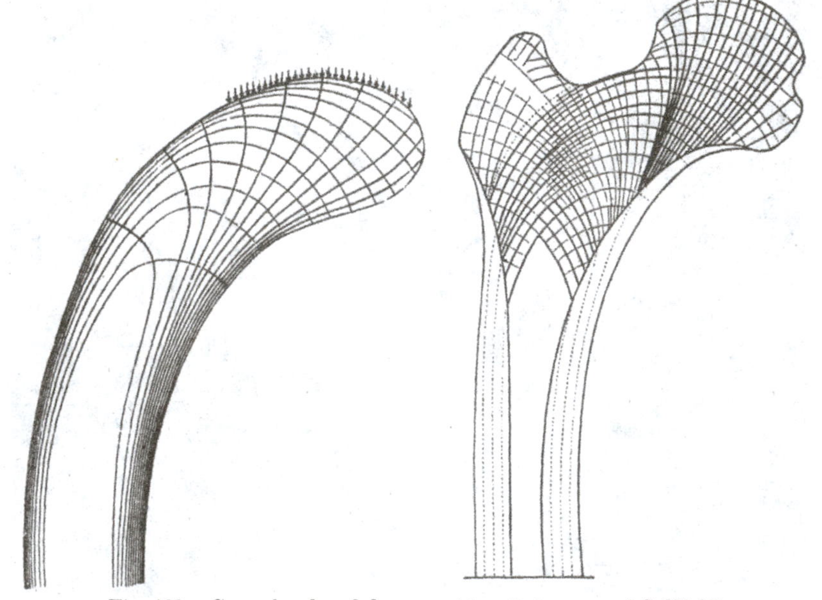

Fig. 463. Crane-head and femur. After Culmann and J. Wolff.

a little more complicated in form and a little less symmetrical than Culmann's diagrammatic crane, from which it chiefly differs in the fact that its load is divided into two parts, that namely which is borne by the head of the bone, and that smaller portion which rests upon the great trochanter; but this merely amounts to saying that a *notch* has been cut out of the curved upper surface of the structure, and we have no difficulty in seeing that the anatomical arrangement of the trabeculae follows precisely the mechanical distribution of compressive and tensile stress or, in other words, accords perfectly with the theoretical stress-diagram of the crane.

The lines of stress are bundled close together along the sides of the shaft, and lost or concealed there in the substance of the solid wall of bone; but in and near the head of the bone, a peripheral shell of bone does not suffice to contain them, and they spread out through the central mass in the actual concrete form of bony trabeculae*.

Mutatis mutandis, the same phenomenon may be traced in any other bone which carries weight and is liable to flexure; and in the *os calcis* and the tibia, and more or less in all the bones of the lower limb, the arrangement is found to be very simple and clear.

Thus, in the *os calcis*, the weight resting on the head of the bone has to be transmitted partly through the backward-projecting heel to the ground, and partly forwards through its articulation with

* Among other works on the mechanical construction of bone see: Bourgery, *Traité de l'anatomie* (*I. Ostéologie*), 1832 (with admirable illustrations of trabecular structure); L. Fick, *Die Ursachen der Knochenformen*, Gottingen, 1857; H. Meyer, Die Architektur der Spongiosa, *Arch. f. Anat. und Physiol.* XLVII, pp. 615–628, 1867; *Statik u. Mechanik des menschlichen Knochengerüstes*, Leipzig, 1873; H. Wolfermann, Beitrag zur K. der Architektur der Knochen, *Arch. f. Anat. und Physiol.* 1872, p. 312; J. Wolff, Die innere Architektur der Knochen, *Arch. f. Anat. und Phys.* L, 1870; *Das Gesetz der Transformation bei Knochen*, 1892; Y. Dwight, The significance of bone-architecture, *Mem. Boston Soc. N.H.* IV, p. 1, 1886; V. von Ebner, Der feinere Bau der Knochensubstanz, *Wiener Bericht*, LXXII, 1875; Anton Rauber, *Elastizitat und Festigkeit der Knochen*, Leipzig, 1876; O. Meserer, *Elast. u. Festigk. d. menschlichen Knochen*, Stuttgart, 1880; Sir Donald MacAlister, How a bone is built, *English Illustr. Mag.* 1884, pp. 640–649; Rasumowsky, Architektonik des Fussskelets, *Int. Monatsschr. f. Anat.* 1889, p. 197; Zschokke, *Weitere Unters. uber das Verhältnis der Knochenbildung zur Statik und Mechanik des Vertebratenskelets*, Zurich, 1892; W. Roux, *Ges. Abhandlungen über Entwicklungsmechanik der Organismen, Bd. I, Funktionelle Anpassung*, Leipzig, 1895; J. Wolff, Die Lehre von der funktionellen Knochengestalt, *Virchow's Archiv*, CLV, 1899; R. Schmidt, Vergl. anat. Studien über den mechanischen Bau der Knochen und seine Vererbung, *Z. f. w. Z.* LXV, p. 65, 1899; B. Solger, Der gegenwärtige Stand der Lehre von der Knochenarchitektur, in Moleschott's *Unters. z. Naturlehre des Menschen*. XVI, p. 187, 1899; H. Triepel, Die Stossfestigkeit der Knochen, *Arch. f. Anat. und Phys.* 1900; Gebhardt, Funktionellwichtige Anordnungsweisen der feineren und gröberen Bauelemente des Wirbelthierknochens, etc., *Arch. f. Entw. Mech.* 1900–10; Revenstorf, Ueber die Transformation der Calcaneus-architektur, *Arch. f. Entw. Mech.* XXIII, p. 379, 1907; H. Bernhardt, *Vererbung der inneren Knochenarchitektur beim Menschen, und die Teleologie bei J. Wolff*, Inaug. Diss., München, 1907; Herm. Triepel, Die trajectoriellen Structuren (in *Einf. in die Physikalische Anatomie*, 1908); A. F. Dixon, Architecture of the cancellous tissue forming the upper end of the femur, *Journ. Anat. and Phys.* (3), XLIV, pp. 223–230, 1910; A. Benninghoff, Ueber Leitsystem der Knochencompacta; Studien zur Architektur der Knochen, *Beitr. z. Anat. funktioneller Systeme*, I, 1930.

the cuboid bone, to the arch of the foot. We thus have, very much as in a triangular roof-tree, two compression-members sloping apart from one another; and these have to be bound together by a "tie" or tension-member, corresponding to the third, horizontal member of the truss.

It is a simple corollary, confirmed by observation, that the trabeculae have a very different distribution in animals whose actions and attitudes are materially different, as in the aquatic mammals, such as the beaver and the seal*. And in much less extreme cases there are lessons to be learned from a study of the

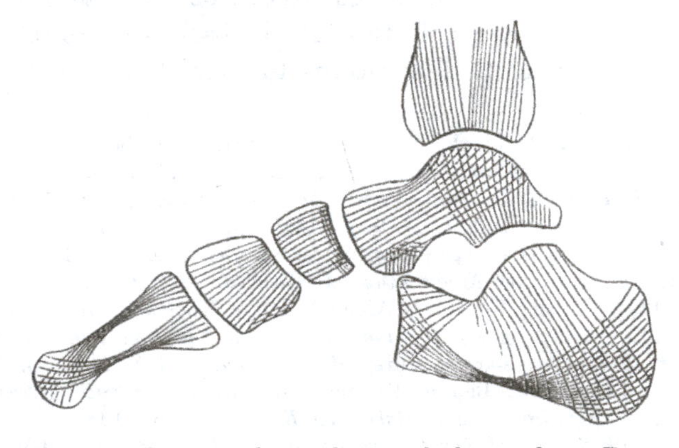

Fig. 464. Diagram of stress-lines in the human foot. From Sir D. MacAlister, after H. Meyer.

same bone in different animals, as the loads alter in direction and magnitude. The gorilla's heelbone resembles man's, but the load on the heel is much less, for the erect posture is imperfectly achieved: in a common monkey the heel is carried high, and consequently the direction of the trabeculae is still more changed. The bear walks on the sole of his foot, though less perfectly than does man, and the lie of the trabeculae is plainly analogous in the two; but in the bear more powerful strands than in the *os calcis* of man transmit the load forward to the toes, and less of it through the heel to the ground. In the leopard we see the full effect of tip-toe, or digitigrade, progression. The long hind part (or tuberosity) of

* Cf. G. de M. Rudolf, Habit and the architecture of the mammalian femur, *Journ. Anatomy*, LVI, pp. 139–146, 1922.

the heel is now more a mere lever than a pillar of support; it is little more than a stiffened rod, with compression-members and tension-members in opposite bundles, inosculating orthogonally at the two ends*.

In the bird the small bones of the hand, dwarfed as they are in size, have still a deal to do in carrying the long primary flight-feathers, and in forming a rigid axis for the terminal part of the wing. The simple tubular construction, which answers well for the long, slender arm-bones, does not suffice where a still more efficient stiffening is required. In all the mechanical side of anatomy

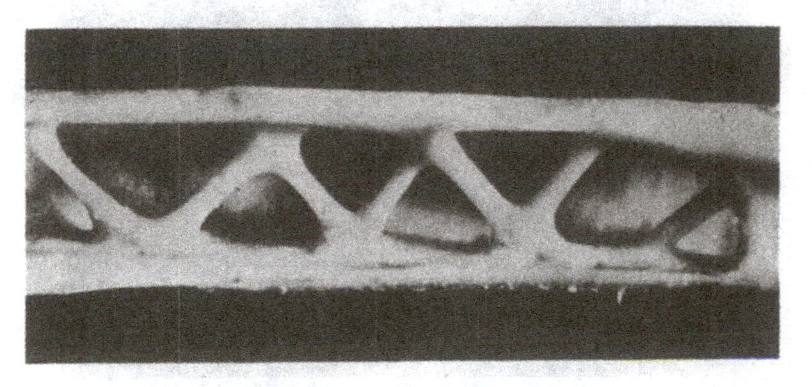

Fig. 465. Metacarpal bone from a vulture's wing; stiffened after the manner of a Warren's truss. From O. Prochnow, *Formenkunst der Natur*.

nothing can be more beautiful than the construction of a vulture's metacarpal bone, as figured here (Fig. 465). The engineer sees in it a perfect Warren's truss, just such a one as is often used for a main rib in an aeroplane. Not only so, but the bone is better than the truss; for the engineer has to be content to set his V-shaped struts all in one plane, while in the bone they are put, with obvious but inimitable advantage, in a three-dimensional configuration.

So far, dealing wholly with the stresses and strains due to tension and compression, we have omitted to speak of a third very important factor in the engineer's calculations, namely what is known as "shearing stress." A shearing force is one which produces

* Cf. Fr. Weidenreich, Ueber formbestimmende Ursachen am Skelett, und die Erblichkeit der Knochenform, *Arch. f. Entw. Mech.* LI, pp. 438–481, 1922.

"angular distortion" in a figure, or (what comes to the same thing) which tends to cause its particles to slide over one another. A shearing stress is a somewhat complicated thing, and we must try to illustrate it (however imperfectly) in the simplest possible way. If we build up a pillar, for instance, of flat horizontal slates, or of a pack of cards, a vertical load placed upon it will produce compression, but will have no tendency to cause one card to slide, or shear, upon another; and in like manner, if we make up a cable of parallel wires and, letting it hang vertically, load it evenly with

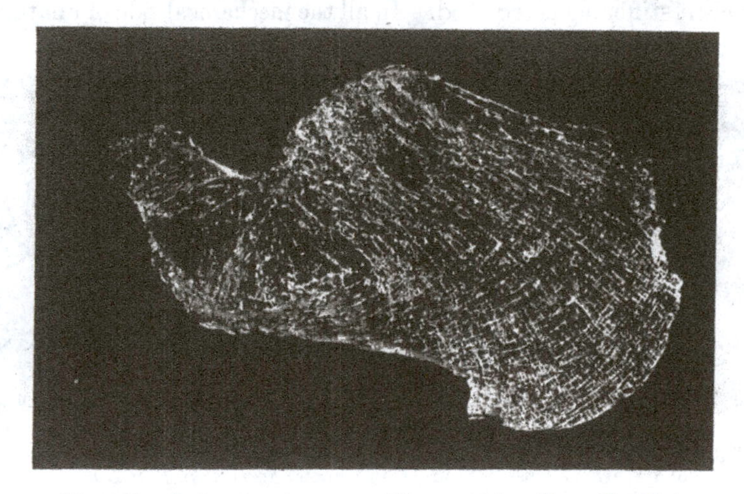

Fig. 466. Trabecular structure of the os calcis. From MacAlister.

a weight, again the tensile stress produced has no tendency to cause one wire to slip or shear upon another. But the case would have been very different if we had built up our pillar of cards or slates lying obliquely to the lines of pressure, for then at once there would have been a tendency for the elements of the pile to slip and slide asunder, and to produce what the geologists call "a fault" in the structure.

Somewhat more generally, if AB be a bar, or pillar, of cross-section a under a direct load P, giving a direct and uniformly distributed stress per unit area $=p$, then the whole pressure $P=pa$. Let CD be an oblique section, inclined at an angle θ to the cross-section; the pressure on CD will evidently be $=pa \cos \theta$. But at any point O in CD, the pressure P may be resolved into the shearing force Q acting along CD, and the direct force N perpendicular to it: where $N=P \cos \theta = pa \cos \theta$, and $Q=P \sin \theta = pa \sin \theta$. The shearing force

Q upon $CD = q \,.\,$area of CD, which is $= q \,.\, a/\cos\theta$. Therefore $qa/\cos\theta = pa\sin\theta$, therefore $q = p\sin\theta\cos\theta = \frac{1}{2}p\sin 2\theta$. Therefore when $\sin 2\theta = 1$, that is, when $\theta = 45°$, q is a maximum, and $= p/2$; and when $\sin 2\theta = 0$, that is when $\theta = 0°$ or $90°$, then q vanishes altogether.

This is as much as to say, that under this form of loading there is no shearing stress along or perpendicular to the lines of principal stress, or along the lines of maximum compression or tension; but shear has a definite value on all other planes, and a maximum value when it is inclined at 45° to the cross-section. This may be further illustrated in various simple ways. When we submit a cubical block of iron to compression in the testing machine, it does not tend to give way by crumbling all to pieces, but always disrupts by shearing, and along some plane approximately at 45° to the axis of compression; this is known as Coulomb's Theory of Fracture, and, while subject to many qualifications, it is still an important first approximation to the truth. Again, in the beam which we have already considered under a bending moment, we know that if we substitute for it a pack of cards, they will be strongly sheared on one another; and the shearing stress is greatest in the "neutral zone," where neither tension nor compression is manifested: that is to say in the line which cuts at equal angles of 45° the orthogonally intersecting lines of pressure and tension.

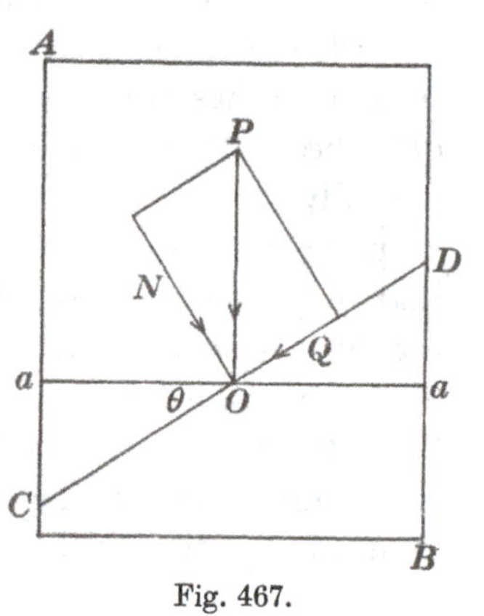

Fig. 467.

In short we see that, while shearing *stresses* can by no means be got rid of, the danger of rupture or breaking-down under shearing stress is lessened the more we arrange the materials of our construction along the pressure-lines and tension-lines of the system; for *along these lines* there is no shear*.

To apply these principles to the growth and development of our bone, we have only to imagine a little trabecula (or group of trabeculae) being secreted and laid down fortuitously in any direction within the substance of the bone. If it lie in the direction of one of the pressure-lines, for instance, it will be in a position of comparative

* It is also obvious that a free surface is always a region of zero-shear.

equilibrium, or minimal disturbance; but if it be inclined obliquely to the pressure-lines, the shearing force will at once tend to act upon it and move it away. This is neither more nor less than what happens when we comb our hair, or card a lock of wool: filaments lying in the direction of the comb's path remain where they were; but the others, under the influence of an oblique component of pressure, are sheared out of their places till they too come into coincidence with the lines of force. So straws show how the wind blows—or rather how it has been blowing. For every straw that lies askew to the wind's path tends to be sheared into it; but as soon as it has come to lie the way of the wind it tends to be disturbed no more, save (of course) by a violence such as to hurl it bodily away.

In the biological aspect of the case, we must always remember that our bone is not only a living, but a highly plastic structure; the little trabeculae are constantly being formed and deformed, demolished and formed anew. Here, for once, it is safe to say that "heredity" need not and cannot be invoked to account for the configuration and arrangement of the trabeculae: for we can see them at any time of life in the making, under the direct action and control of the forces to which the system is exposed. If a bone be broken and so repaired that its parts lie somewhat out of their former place, so that the pressure- and tension-lines have now a new distribution, before many weeks are over the trabecular system will be found to have been entirely remodelled, so as to fall into line with the new system of forces. And as Wolff pointed out, this process of reconstruction extends a long way off from the seat of injury, and so cannot be looked upon as a mere accident of the physiological process of healing and repair; for instance, it may happen that, after a fracture of the *shaft* of a long bone, the trabecular meshwork is wholly altered and reconstructed within the distant *extremities* of the bone. Moreover, in cases of transplantation of bone, for example when a diseased metacarpal is repaired by means of a portion taken from the lower end of the ulna, with astonishing quickness the plastic capabilities of the bony tissue are so manifested that neither in outward form nor inward structure can the old portion be distinguished from the new.

Herein then lies, so far as we can discern it, a great part at least of the physical causation of what at first sight strikes us as a purely functional adaptation: as a phenomenon, in other words, whose physical cause is as obscure as its final cause or end is apparently manifest.

Partly associated with the same phenomenon, and partly to be looked upon (meanwhile at least) as a fact apart, is the very important physiological truth that a condition of *strain*, the result of a *stress*, is a direct stimulus to growth itself. This indeed is no less than one of the cardinal facts of theoretical biology. The soles of our boots wear thin, but the soles of our feet grow thick, the more we walk upon them: for it would seem that the living cells are "stimulated" by pressure, or by what we call "exercise," to increase and multiply. The surgeon knows, when he bandages a broken limb, that his bandage is doing something more than merely keeping the parts together: and that the even, constant pressure which he skilfully applies is a direct encouragement of growth and an active agent in the process of repair. In the classical experiments of Sédillot*, the greater part of the shaft of the tibia was excised in some young puppies, leaving the whole weight of the body to rest upon the fibula. The latter bone is normally about one-fifth or sixth of the diameter of the tibia; but under the new conditions, and under the "stimulus" of the increased load, it grew till it was as thick or even thicker than the normal bulk of the larger bone. Among plant tissues this phenomenon is very apparent, and in a somewhat remarkable way; for a strain caused by a constant or increasing weight (such as that in the stalk of a pear while the pear is growing and ripening) produces a very marked increase of *strength* without any necessary increase of bulk, but rather by some histological, or molecular, alteration of the tissues. Hegler, Pfeffer, and others have investigated this subject, by loading the young shoot of a plant nearly to its breaking point, and then redetermining the breaking-strength after a few days. Some young shoots of the sunflower were found to break with a strain of 160 gm.; but when loaded with 150 gm., and retested after two days, they were able to support 250 gm.; and being again loaded with something short

* Sédillot, De l'influence des fonctions sur la structure et la forme des organes, *C.R.* LIX, p. 539, 1864; cf. LX, p. 97, 1865; LXVIII, p. 1444, 1869.

of this, by next day they sustained 300 gm., and a few days later even 400 gm.*

The kneading of dough is an analogous phenomenon. The viscosity and perhaps other properties of the stuff are affected by the strains to which we have submitted it, and may thus be said to depend not only on the nature of the substance but on its history†. It is a long way from this simple instance, but we stretch across it easily in imagination, to the experimental growth of a nerve-fibre within a mass of clotted lymph: where, when we draw out the clot in one direction or another we lay down traction-lines, or tension-lines, and make of them a path for growth to follow‡.

Such experiments have been amply confirmed, but so far as I am aware we do not know much more about the matter: we do not know, for instance, how far the change is accompanied by increase in number of the bast-fibres, through transformation of other tissues; or how far it is due to increase in size of these fibres; or whether it be not simply due to strengthening of the original fibres by some molecular change. But I should be much inclined to suspect that this last had a good deal to do with the phenomenon. We know nowadays that a railway axle, or any other piece of steel, is weakened by a constant succession of frequently interrupted strains; it is said to be "fatigued," and its strength is restored by a period of rest. The converse effect of continued strain in a uniform direction may be illustrated by a homely example. The confectioner takes a mass of boiled sugar or treacle (in a particular molecular condition determined by the temperature to which it has been raised), and draws the soft sticky mass out into a rope; and then, folding it up lengthways, he repeats the process again and again. At first the rope is pulled out of the ductile mass without difficulty; but as the work goes on it gets harder to do, until all the man's force is used to stretch the rope. Here we have the phenomenon

* *Op. cit.* Hegler's results are criticised by O. M. Ball, Einfluss von Zug auf die Ausbildung der Festigungsgewebe, *Jb. d. wiss. Botanik*, XXXIX, pp. 305–341, 1903, and by H. Keller, Einfluss von Belastung und Lage auf die Ausbildung des Gewebes in Fruchtstielen, *Inaug. Diss.* Kiel, 1904.

† Cf. R. K. Schofield and G. W. S. Blair, On dough, *Proc. R.S.* (A), CXXXVIII, p. 707; CXXXIX, p. 557, 1932–33; also Nadai and Wahl's *Plasticity*, 1931. For analogous properties of hairs and fibres, see Shorter, *Journ. Textile Inst.* XV, 1824; etc.

‡ Cf. Ross Harrison's *Croonian Lecture*, 1933.

of increasing strength, following mechanically on a rearrangement of molecules, as the original isotropic condition is transmuted more and more into molecular asymmetry or anisotropy; and the rope apparently "adapts itself" to the increased strain which it is called on to bear, all after a fashion which at least suggests a parallel to the increasing strength of the stretched and weighted fibre in the plant. For increase of strength by rearrangement of the particles we have already a rough illustration in our lock of wool or hank of tow. The tow will carry but little weight while its fibres are tangled and awry: but as soon as we have carded or "hatchelled" it out, and brought all its long fibres parallel and side by side, we make of it a strong and useful cord*.

But the lessons which we learn from dough and treacle are nowadays plain enough in steel and iron, and become immensely more important in these. For here again plasticity is associated with a certain capacity for structural rearrangement, and increased strength again results therefrom. Elaborate processess of rolling, drawing, bending, hammering, and so on, are regularly employed to toughen and strengthen the material. The "mechanical structure" of solids has become an important subject. And when the engineer talks of repeated loading, of elastic fatigue, of hysteresis, and other phenomena associated with plasticity and strain, the physiological analogues of these physical phenomena are perhaps not far away.

In some such ways as these, then, it would seem that we may coordinate, or hope to coordinate, the phenomenon of growth with certain of the beautiful structural phenomena which present themselves to our eyes as "provisions," or mechanical adaptations†, for the display of strength where strength is most required. That is to say the origin, or causation, of the phenomenon would seem to lie partly in the tendency of growth to be accelerated under strain: and partly in the automatic effect of shearing strain, by which it tends to displace parts which grow obliquely to the direct lines of tension and of pressure, while leaving those in place which happen to lie parallel or perpendicular to those lines: an automatic effect

* Cf. Sir Charles Bell's *Animal Mechanics*, chap. v, "Of the tendons compared with cordage."

† So P. Enriques (*op. cit. supra*, p. 5), writing on the economy of material in the construction of a bone, admits that "una certa impronta di teleologismo qua e là è rimasta, mio malgrado, in questo scritto."

which we can probably trace as working on all scales of magnitude, and as accounting therefore for the rearrangement of minute particles in the metal or the fibre, as well as for the bringing into line of the fibres within the plant, or of the trabeculae within the bone.

But we may now attempt to pass from the study of the individual bone to the much wider and not less beautiful problems of mechanical construction which are presented to us by the skeleton as a whole. Certain problems of this class are by no means neglected by writers on anatomy, and many have been handed down from Borelli, and even from older writers. For instance, it is an old tradition of anatomical teaching to point out in the human body examples of the three orders of levers*; again, the principle that the limb-bones tend to be shortened in order to support the weight of a very heavy animal is well understood by comparative anatomists, in accordance with Euler's law, that the weight which a column liable to flexure is capable of supporting varies inversely as the square of its length; and again, the statical equilibrium of the body, in relation for instance to the erect posture of man, has long been a favourite theme of the philosophical anatomist. But the general method, based upon that of graphic statics, to which we have been introduced in our study of a bone, has not, so far as I know, been applied to the general fabric of the skeleton. Yet it is plain that each bone plays a part in relation to the whole body, analogous to that which a little trabecula, or a little group of trabeculae, plays within the bone itself: that is to say, in the normal distribution of forces in the body the bones tend to follow the lines of stress, and especially the pressure-lines. To demonstrate this in a comprehensive way would doubtless be difficult; for we should be dealing with a framework of very great complexity, and should have to take account of

* E.g. (1) the head, nodding backwards and forwards on a fulcrum, represented by the atlas vertebra, lying between the weight and the power; (2) the foot, raising on tip-toe the weight of the body against the fulcrum of the ground, where the weight is between the fulcrum and the power, the latter being represented by the *tendo Achillis*; (3) the arm, lifting a weight in the hand, with the power (i.e. the biceps muscle) between the fulcrum and the weight. (The second case, by the way, has been much disputed; cf. Haycraft in Schäfer's *Textbook of Physiology*, 1900, p. 251.) Cf. (*int. al.*) G. H. Meyer, *Statik u. Mechanik der menschlichen Knochengerüste*, 1873, pp. 13–25.

a great variety of conditions*. This framework is complicated as we see it in the skeleton, where (as we have said) it is only, or chiefly, the *struts* of the whole fabric which are represented; but to understand the mechanical structure in detail, we should have to follow out the still more complex arrangement of the *ties*, as represented by the muscles and ligaments, and we should also require much detailed information as to the weights of the various parts and as to the other forces concerned. Without these latter data we can only treat the question in a preliminary and imperfect way. But, to take once again a small and simplified part of a big problem, let us think of a quadruped (for instance, a horse) in a standing posture, and see whether the methods and terminology of the engineer may not help us, as they did in regard to the minute structure of the single bone. And let us note in passing that the "standing posture," whether on two legs or on four, is no very common thing; but is (so to speak), with all its correlated anatomy, a privilege of the few.

Standing four-square upon its fore-legs and hind-legs, with the weight of the body suspended between, the quadruped at once suggests to us the analogy of a bridge, carried by its two piers. And if it occurs to us, as naturalists, that we never look at a standing quadruped without contemplating a bridge, so, conversely, a similar idea has occurred to the engineer; for Professor Fidler, in this *Treatise on Bridge-Construction*, deals with the chief descriptive part of his subject under the heading of "The Comparative Anatomy of Bridges†." The designation is most just, for in studying the various types of bridges we are studying a series of well-planned *skeletons*‡; and (at the cost of a little pedantry)

* Our problem is analogous to Thomas Young's problem of the best disposition of the timbers in a wooden ship (*Phil. Trans.* 1814, p. 303). He was not long of finding that the forces which act upon the fabric are very numerous and very variable, and that the best mode of resisting them, or best structural arrangement for ultimate strength, becomes an immensely complicated problem.

† By a bolder metaphor Fontenelle said of Newton that he had "fait l'anatomie de la lumière."

‡ In like manner, Clerk Maxwell could not help employing the term "skeleton" in defining the mathematical conception of a "frame," constituted by points and their interconnecting lines: in studying the equilibrium of which, we consider its different points as mutually acting on each other with forces whose directions are those of the lines joining each pair of points. Hence (says Maxwell), "in order to exhibit the mechanical action of the frame in the most elementary manner, we may

we might go even further, and study (after the fashion of the anatomist) the "osteology" and "desmology" of the structure, that is to say the bones which are represented by "struts," and the ligaments, etc., which are represented by "ties." Furthermore after the methods of the comparative anatomist, we may classify the families, genera and species of bridges according to their distinctive mechanical features, which correspond to certain definite conditions and functions.

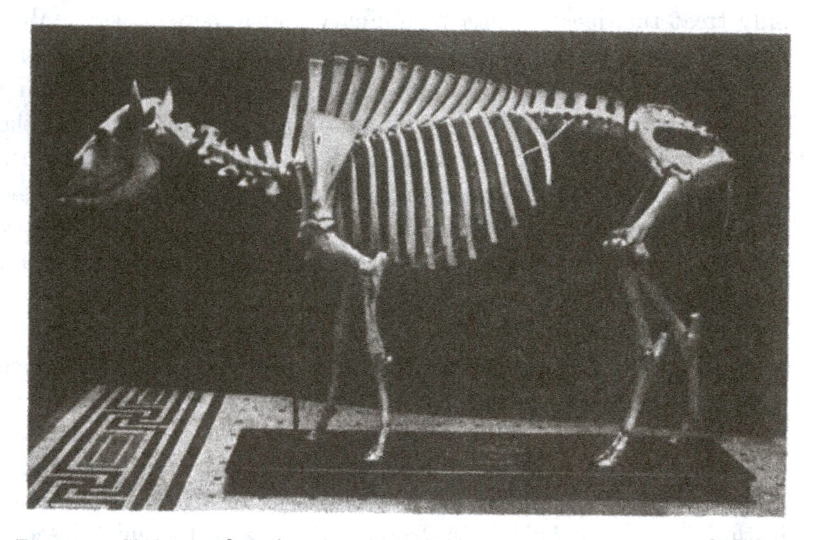

Fig. 468. Skeleton of an American bison. (An unusually well-mounted skeleton, of American workmanship, now in the Anatomical Museum of Edinburgh University.)

In more ways than one, the quadrupedal bridge is a remarkable one; and perhaps its most remarkable peculiarity is that it is a jointed and flexible bridge, remaining in equilibrium under considerable and sometimes great modifications of its curvature, such as we see, for instance, when a cat humps or flattens her back. The fact that *flexibility* is an essential feature in the quadrupedal

draw it as a *skeleton*, in which the different points are joined by straight lines, and we may indicate by numbers attached to these lines the tensions or compressions in the corresponding pieces of the frame" (*Trans. R.S.E.* XXVI, p. 1, 1870). It follows that the diagram so constructed represents a "diagram of forces," in this limited sense that it is geometrical as regards the position and direction of the forces, but arithmetical as regards their magnitude. It is to just such a diagram that the animal's skeleton tends to approximate.

bridge, while it is the last thing which an engineer desires and the first which he seeks to provide against, will impose certain important limiting conditions upon the design of the skeletal fabric. But let us begin by considering the quadruped at rest, when he stands upright and motionless upon his feet, and when his legs exercise no function save only to carry the weight of the whole body. So far as that function is concerned, we might now perhaps compare the horse's legs with the tall and slender piers of some railway bridge; but it is obvious that these jointed legs are ill-adapted to receive the *horizontal thrust* of any *arch* that may be placed atop of them. Hence it follows that the curved backbone of the horse, which appears to cross like an arch the span between his shoulders and his flanks, cannot be regarded as an *arch*, in the

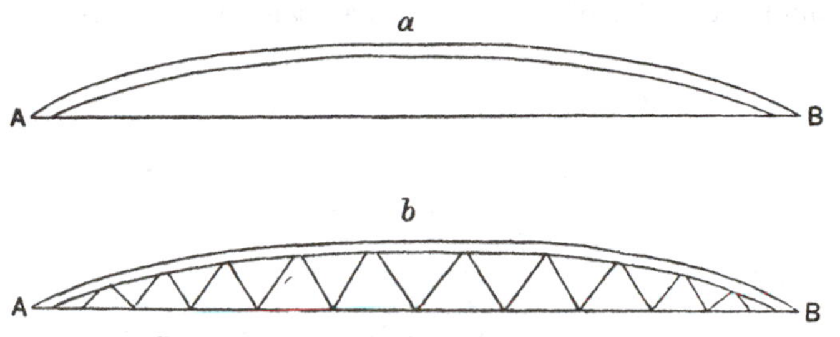

Fig. 469. *a*, tied arch; *b*, bowstring girder.

engineer's sense of the word. It resembles an arch in *form*, but not in *function*, for it cannot act as an arch unless it be held back at each end (as every arch is held back) by *abutments* capable of resisting the horizontal thrust; and these necessary abutments are not present in the structure. But in various ways the engineer can modify his superstructure so as to supply the place of these *external* reactions, which in the simple arch are obviously indispensable. Thus, for example, we may begin by inserting a straight steel tie, *AB* (Fig. 469), uniting the ends of the curved rib *AaB*; and this tie will supply the place of the external reactions, converting the structure into a "tied arch," such as we may see in the roofs of many railway stations. Or we may go on to fill in the space between arch and tie by a "web-system," converting it into what the engineer describes as a "parabolic bowstring girder" (Fig. 469 *b*). In either case, the structure becomes an independent "detached girder," supported at each end but not otherwise fixed, and consisting essentially of an upper compression-member, AaB, and a lower tension-member, AB. But again, in the skeleton of the quadruped, *the necessary tie, AB, of the simple bow-girder is not to be found*; and it follows that these comparatively simple types of bridge do not correspond to, nor do they help us to understand, the type of bridge which Nature has designed in the skeleton of the quadruped. Nevertheless if we try to look, as an engineer would look, at the actual design of the animal skeleton and the actual distribution of its load, we find that the one is most admirably adapted to the other, according to the strict principles of engineering construction. The structure is not an arch, nor a tied arch, nor a bowstring girder: but it is strictly and beautifully comparable to the main girder of a double-armed cantilever bridge.

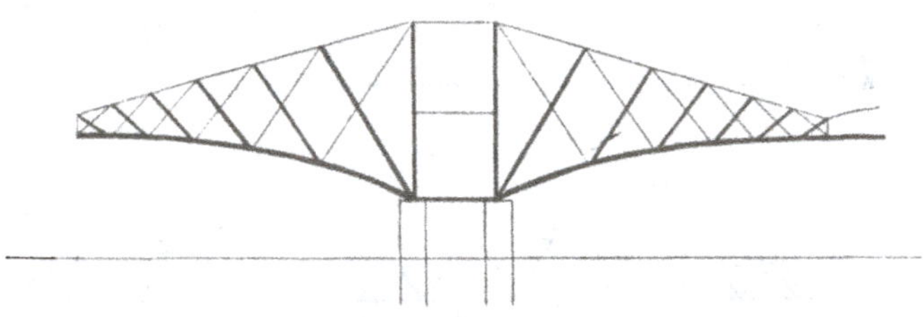

Fig. 470. A two-armed cantilever of the Forth Bridge. Thick lines, compression-members (bones); thin lines, tension-members (ligaments).

Obviously, in our quadrupedal bridge, the superstructure does not terminate (as it did in our former diagram) at the two points of support, but it extends beyond them, carrying the head at one end and sometimes a heavy tail at the other, upon projecting arms or "cantilevers."

In a typical cantilever bridge, such as the Forth Bridge (Fig. 470), a certain simplification is introduced. For each pier carries, in this case, its own double-armed cantilever, linked by a short connecting girder to the next, but so jointed to it that no weight is transmitted from one cantilever to another. The bridge in short is *cut* into separate sections, practically independent of one another; at the joints a certain amount of bending is not precluded, but shearing strain is evaded; and each pier carries only its own load. By this arrangement the engineer finds that design and construction are alike simplified and facilitated. In the horse or the ox, it is obvious that the two piers of the bridge, that is to say the fore-legs and the hind-legs, do not bear (as they do in the Forth Bridge) separate and independent loads, but the whole system forms a continuous structure. In this case, the calculation of the loads will be a little more difficult and the corresponding design of the structure a little more complicated. We shall accordingly simplify our problem very considerably if, to begin with, we look upon the quadrupedal skeleton as constituted of two separate systems, that is to say of two balanced cantilevers, one supported on the fore-legs and the other on the hind; and we may deal afterwards with the fact that these two cantilevers are not independent, but are bound up in one common field of force and plan of construction.

In both horse and ox it is plain that the two cantilever systems into which we may thus analyse the quadrupedal bridge are unequal in magnitude and importance. The fore-part of the animal is much bulkier than its hind-quarters, and the fact that the fore-legs carry, as they so evidently do, a greater weight than the hind-legs has long been known and is easily proved; we have only to walk a horse on to a weigh-bridge, weigh first his fore-legs and then his hind-legs, to discover that what we may call his front half weighs a good deal more than what is carried on his hind feet, say about three-fifths of the whole weight of the animal.

The great (or anterior) cantilever then, in the horse, is constituted by the heavy head and still heavier neck on one side of that pier which is represented by the fore-legs, and by the dorsal vertebrae carrying a large part of the weight of the trunk upon the other side; and this weight is so balanced over the fore-legs that the cantilever, while "anchored" to the other parts of the structure, transmits but little of its weight to the hind-legs, and the amount so transmitted will vary with the attitude of the head and with the position of any artificial load*. Under certain conditions, as when the head is thrust well forward, it is evident that the hind-legs will be actually relieved of a portion of the comparatively small load which is their normal share.

* When the jockey crouches over the neck of his race-horse, and when Tod Sloan introduced the "American seat," the avowed object in both cases is to relieve the hind-legs of weight, and so leave them free for the work of propulsion. On the share taken by the hind-limbs in this latter duty, and other matters, cf. Stillman, *The Horse in Motion*, 1882, p. 69.

But here we pass from the statical problem to the dynamical, from the horse at rest to the horse in motion, from the observed fact that weight lies mainly over the fore-legs to the question of what advantage is gained by such a distribution of the load. Taking the hind-legs as the main propulsive agency, as we may now safely do, the moment of propulsion is about the hind-hooves; then (as we see in Fig. 471) we may take the weight, $W = A \sin \alpha$, and the propulsive force, $f = A \cos \alpha$, and $\frac{W}{f} = \frac{H}{L}$, $WL = fH$ being the balanced condition. From the statical point of view the load must balance over the fore-legs; from the dynamical point of view it might well lie even farther forward. And when the jockey crouches

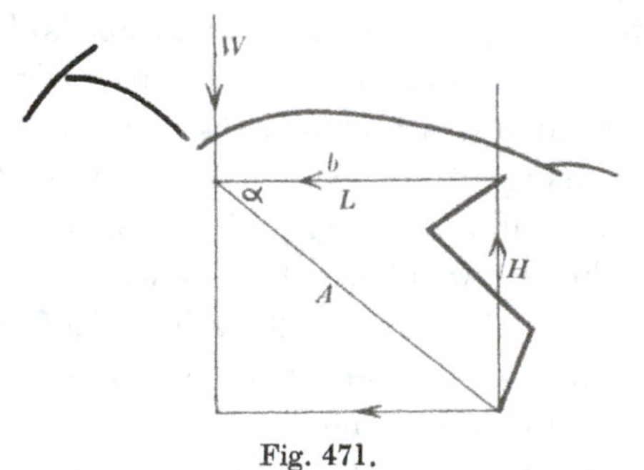

Fig. 471.

over the horse's neck, and when Tod Sloan introduced the "American seat," both shew a remarkable, though perhaps unconscious, insight into the dynamical proposition.

Our next problem is to discover, in a rough and approximate way, some of the structural details which the balanced load upon the double cantilever will impress upon the fabric.

Working by the methods of graphic statics, the engineer's task is, in theory, one of great simplicity. He begins by drawing in outline the structure which he desires to erect; he calculates the stresses and bending-moments necessitated by the dimensions and load on the structure; he draws a new diagram *representing these forces*, and he designs and builds his fabric on the lines of this statical diagram. He does, in short, precisely what we have seen *nature* doing in the case of the bone. For if we had begun, as it were, by blocking out the femur roughly, and considering its position and dimensions, its means of support and the load which it has to bear, we could have proceeded at once to draw the system of stress-lines which must occupy that field of force: and to precisely

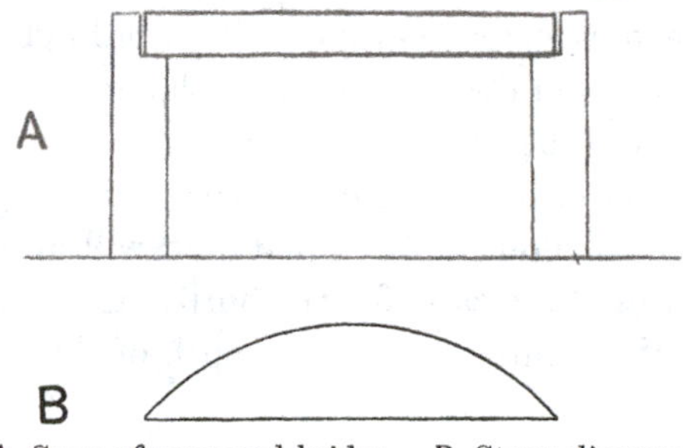

Fig. 472. A, Span of proposed bridge. B, Stress diagram, or diagram of bending-moments*.

those stress-lines has Nature kept in the building of the bone, down to the minute arrangement of its trabeculae.

The essential function of a bridge is to stretch across a certain span, and carry a certain definite load; and this being so, the chief

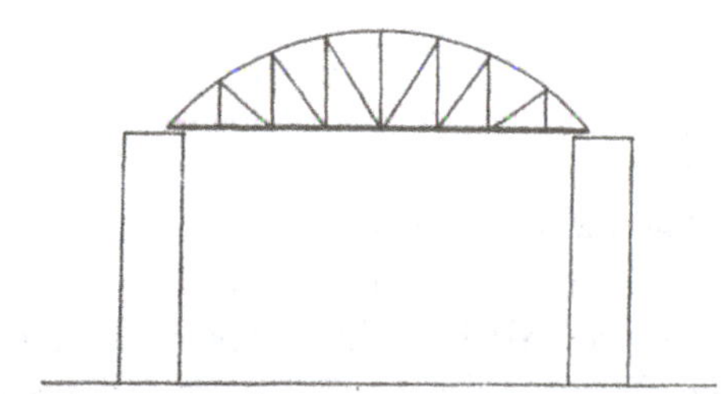

Fig. 473. The bridge constructed, as a parabolic girder.

problem in the designing of a bridge is to provide due resistance to the "bending-moments" which result from the load. These bending-moments will vary from point to point along the girder,

* This and the following diagrams are borrowed and adapted from Professor Fidler's *Bridge Construction.* We may reflect with advantage on Clerk Maxwell's saying that "the use of diagrams is a particular instance of that method of symbols which is so powerful an aid in the advancement of science"; and on his explanation that "a diagram differs from a picture in this respect that in a diagram no attempt is made to represent those factors of the actual material system which are not the special objects of our study."

and taking the simplest case of a uniform load, whether supported at one or both ends, they will be represented by points on a parabola. If the girder be of uniform depth and section, that is to say if its two flanges, respectively under tension and compression, be equal and parallel to one another, then the stress upon these flanges will vary as the bending-moments, and will accordingly be very severe in the middle and will dwindle towards the ends. But if we make the *depth* of the girder everywhere proportional to the bending-moments, that is to say if we copy in the girder the outlines of the bending-moment diagram, then our design will automatically meet the circumstances of the case, for the horizontal stress in each flange will now be uniform throughout the length of the girder. In short,

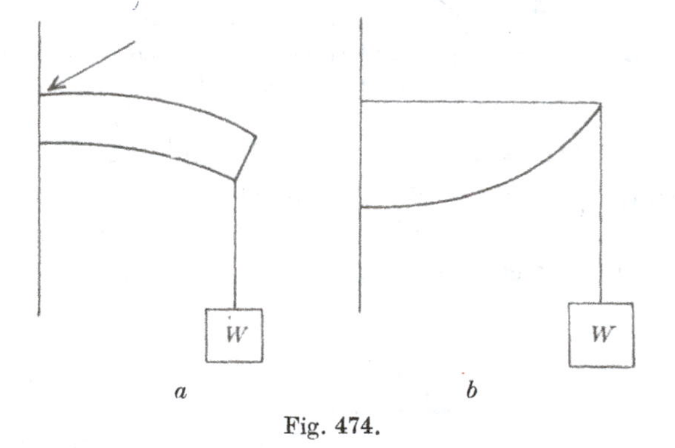

Fig. 474.

in Professor Fidler's words, "Every diagram of moments represents the outline of a framed structure which will carry the given load with a uniform horizontal stress in the principal members."

In the above diagrams (Fig. 474, *a*, *b*) (which are taken from the original ones of Culmann), we see at once that the loaded beam or bracket (*a*) has a "danger-point" close to its fixed base, that is to say at the point remotest from its load. But in the parabolic bracket (*b*) there is no danger-point at all, for the dimensions of the structure are made to increase *pari passu* with the bending-moments: stress and resistance vary together. Again in Fig. 475, we have a simple span (A), with its stress diagram (B); and in (C) we have the corresponding parabolic girder, whose stresses are now uniform throughout. In fact we see that, by a process of conversion, the stress diagram in each case becomes the structural diagram in the other*. Now all this is but the modern rendering of one of Galileo's most famous propositions. In the Dialogue which we have already quoted more than once†, Sagredo says "It would be a fine thing if one could discover the proper shape to give a solid in order to make it equally resistant at every point, in which case a load placed at the middle would not produce fracture more easily

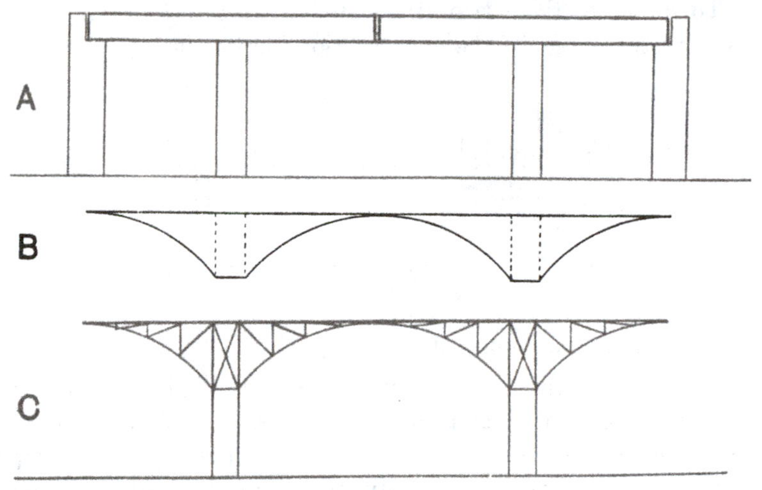

Fig. 475.

than if placed at any other point‡." And Galileo (in the person of Salviati) first puts the problem into its more general form; and then shews us how, by giving a parabolic outline to our beam, we have its simple and comprehensive solution. It was such teaching as this that led R. A. Millikan to say that "we owe our present-day civilisation to Galileo."

* The method of constructing *reciprocal diagrams*, of which one should represent the outlines of a frame and the other the system of forces necessary to keep it in equilibrium, was first indicated in Culmann's *Graphische Statik*; it was greatly developed soon afterwards by Macquorn Rankine (*Phil. Mag.* Feb. 1864, and *Applied Mechanics, passim*), to whom the application of the principle to engineering practice is mainly due. See also Fleeming Jenkin, On the practical application of reciprocal figures to the calculation of strains in framework, *Trans. R.S.E.* xxv, pp. 441–448, 1869; and Clerk Maxwell, *ibid.* xxvi, p. 9, 1870, and *Phil. Mag.* April 1864.

† *Dialogues concerning Two New Sciences* (1638); Crew and Salvio's translation p. 140 *seq.*

‡ As in the great case of the Eiffel Tower, *supra*, p. 29.

In the case of our cantilever bridge, we shew the primitive girder in Fig. 475, A, with its bending-moment diagram (B); and it is evident that, if we turn this diagram upside down, it will still be illustrative, just as before, of the bending-moments from point to point: for as yet it is merely a diagram, or graph, of relative magnitudes.

To either of these two stress diagrams, direct or inverted, we may fit the design of the construction, as in Figs. 475, C and 476.

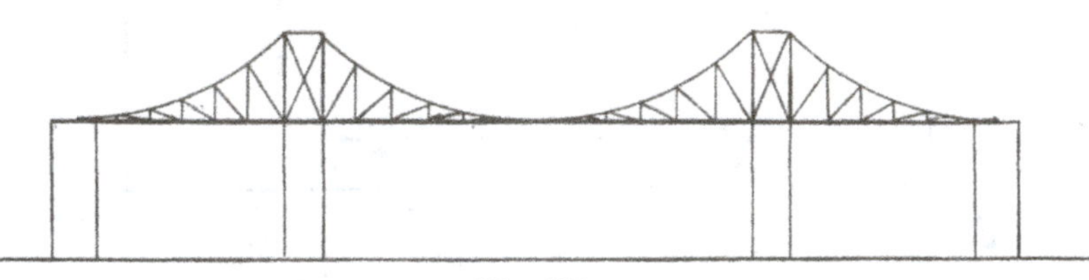

Fig. 476.

Now in different animals the amount and distribution of the load differ so greatly that we can expect no single diagram, drawn from the comparative anatomy of bridges, to apply equally well to all the cases met with in the comparative anatomy of quadrupeds; but nevertheless we have already gained an insight into the general principles of "structural design" in the quadrupedal bridge.

In our last diagram the upper member of the cantilever is under tension; it is represented in the quadruped by the *ligamentum nuchae* on the one side of the cantilever, and by the supraspinous ligaments of the dorsal vertebrae on the other. The compression-member is similarly represented, on both sides of the cantilever, by the vertebral column, or rather by the *bodies* of the vertebrae; while the web, or "filling," of the girders, that is to say the upright or sloping members which extend from one flange to the other, is represented on the one hand by the spines of the vertebrae, and on the other hand by the oblique interspinous ligaments and muscles—that is to say, by compression-members and tension-members inclined in opposite directions to one another. The high spines over the quadruped's withers are no other than the high struts which rise over the supporting piers in the parabolic girder, and correspond to the position of the maximal bending-moments. The fact that these tall vertebrae of the withers usually slope backwards, sometimes steeply, in a quadruped, is easily and obviously explained*. For each vertebra tends to act as a "hinged lever," and its spine, acted on by the tensions transmitted by the ligaments on either side, takes up its position as the diagonal of the parallelogram of forces to which it is exposed.

It happens that in these comparatively simple types of cantilever bridge the whole of the parabolic curvature is transferred to one or other of the principal members, either the tension-member or the compression-member as the case may be. But it is of course equally permissible to have both members curved, in opposite directions. This, though not exactly the case in the Forth Bridge, is approximately so; for here the main compression-member is curved or arched, and the main tension-member slopes downwards on either side from its maximal height above the piers. In short, the Forth Bridge (Fig. 470) is a nearer approach than either of the other bridges which we have illustrated to the plan of the quadrupedal skeleton; for the main compression-member almost exactly recalls the form of the backbone, while the main tension-member, though not so closely similar to the supraspinous and nuchal ligaments, corresponds to the plan of these in a somewhat simplified form.

We may now pass without difficulty from the two-armed cantilever supported on a single pier, as it is in each separate section of the Forth Bridge, or as we have imagined it to be in the fore-quarters of a horse, to the condition which actually exists in a quadruped, when a two-armed cantilever has its load distributed over two separate piers. This is not precisely what an engineer calls a "continuous" girder, for that term is applied to a girder which, as a continuous structure, has three supports and crosses two or more spans, while here there is only one. But nevertheless, this girder

* The form and direction of the vertebral spines have been frequently and elaborately described; cf. (e.g.) H. Gottlieb, Die Anticlinie der Wirbelsäule der Saugethiere, *Morphol. Jahrb.* LXIX, pp. 179–220, 1915, and many works quoted therein. According to Morita, Ueber die Ursachen der Richtung und Gestalt der thoracalen Dornfortsätze der Säugethierwirbelsäule (*ibi cit.* p. 201), various changes take place in the direction or inclination of these processes in rabbits, after section of the interspinous ligaments and muscles. These changes seem to be very much what we should expect, on simple mechanical grounds. See also O. Fischer, *Theoretische Grundlagen für eine Mechanik der lebenden Körper*, Leipzig, 1906, pp. x, 372.

is *effectively* continuous from the head to the tip of the tail; and at each point of support (A and B) it is subjected to the negative bending-moment due to the overhanging load on each of the projecting cantilever arms AH and BT. The diagram of bending-moments will (according to the ordinary conventions) lie below the base line (because the moments are negative), and must take some such form as that shewn in the diagram: for the girder must suffer its greatest bending stress not at the centre, but at the two points of support A and B, where the moments are measured by the vertical ordinates. It is plain that this figure only differs from a representation of *two* independent two-armed cantilevers in the fact that there is no point midway in the span where the bending-moment vanishes, but only a region between the two piers in which it tends to diminish.

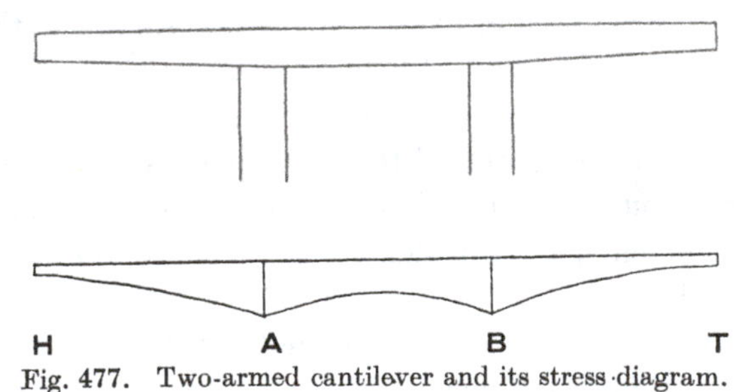

Fig. 477. Two-armed cantilever and its stress-diagram.

The diagram effects a graphic summation of the positive and negative moments, but its form may assume various modifications according to the method of graphic summation which we choose to adopt; and it is obvious also that the form of the diagram may assume many modifications of detail according to the actual distribution of the load. In all cases the essential points to be observed are these: firstly that the girder which is to resist the bending-moments induced by the load must possess its two principal members—an upper tension-member or tie, represented by ligament (whose tension doubtless varies along its length), and a lower compression-member represented by bone: these members being united by a web represented by the vertebral spines with their interspinous ligaments, and being placed one above the other in the order named because the moments are negative; secondly we observe that the depth of the web, or distance apart of the principal members—that is to say the height of the vertebral spines—must be proportional to the bending-moment at each point along the length of the girder.

In the case of an animal carrying most of his weight upon his fore-legs, as the horse or the ox do, the bending-moment diagram will be unsymmetrical, after the fashion of Fig. 478, the precise form depending on the distribution of weights and distances.

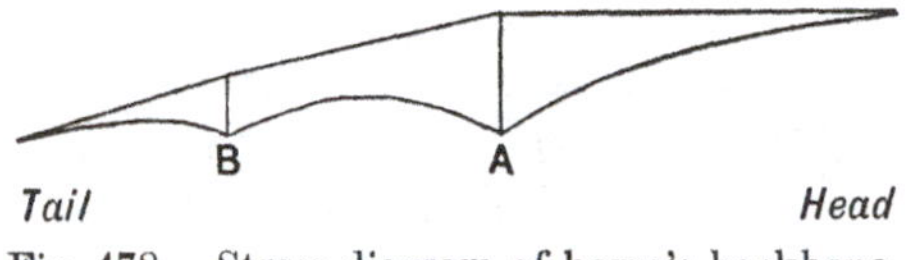

Fig. 478. Stress-diagram of horse's backbone.

On the other hand the Dinosaur, with his light head and enormous tail would give us a moment-diagram with the opposite kind of asymmetry, the greatest bending stress being now found over the haunches, at *B* (Fig. 479). A glance at the skeleton of Diplodocus will shew us the high vertebral spines over the loins, in precise correspondence with the requirements of this diagram: just as in the horse, under the opposite conditions of load, the highest vertebral spines are those of the withers, that is to say those of the posterior cervical and anterior dorsal vertebrae.

We have now not only dealt with the general resemblance, both in structure and in function, of the quadrupedal backbone with its associated ligaments to a double-armed cantilever girder, but we have begun to see how the characters of the vertebral system must differ in different quadrupeds, according to the conditions imposed by the varying distribution of the load: and in particular how the height of the vertebral spines which constitute the web will be in a definite relation, as regards magnitude and position, to the bending-moments induced thereby. We should require much detailed information as to the actual weights of the several parts of the body before we could follow out quantitatively the mechanical efficiency of each type of skeleton; but in an approximate way what we have already learnt will enable us to trace many interesting correspondences between structure and function in this particular part of comparative anatomy. We must, however, be careful to note that the great cantilever system is not of necessity constituted by the vertebral column and its ligaments alone, but that the pelvis, firmly united as it is to the sacral vertebrae, and stretching backwards far beyond the acetabulum, becomes an intrinsic part of the system; and helping (as it does) to carry the load of the abdominal viscera, it constitutes a great portion of the posterior cantilever arm, or even its chief portion in cases where the size and weight of the tail are insignificant, as is the case in the majority of terrestrial mammals.

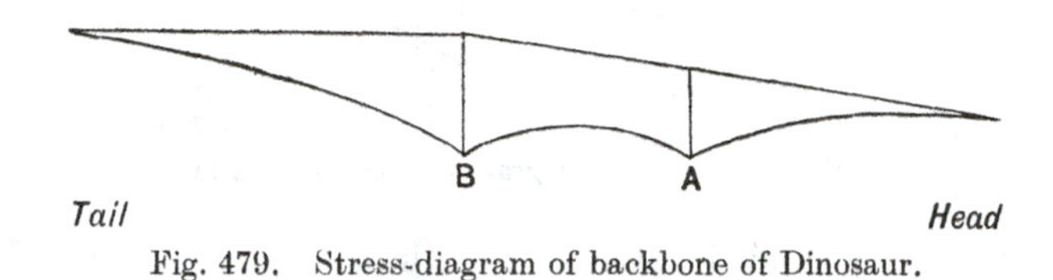

Fig. 479. Stress-diagram of backbone of Dinosaur.

We may also note here, that just as a bridge is often a "combined" or composite structure, exhibiting a combination of principles in its construction, so in the quadruped we have, as it were, another girder supported by the same piers to carry the viscera; and consisting of an inverted parabolic girder, whose compression-member is again constituted by the backbone, its tension-member by the line of the sternum and the abdominal muscles, while the ribs and intercostal muscles play the part of the web or filling.

A very few instances must suffice to illustrate the chief variations in the load, and therefore in the bending-moment diagram, and therefore also in the plan of construction, of various quadrupeds. But let us begin by setting forth, in a few cases, the actual weights which are borne by the fore-limbs and the hind-limbs, in our quadrupedal bridge*.

	Gross weight ton	Gross weight cwt.	On fore-feet cwt.	On hind-feet cwt.	% on fore-feet	% on hind-feet
Camel (Bactrian)	–	14·25	9·25	4·5	67·3	32·7
Llama	–	2·75	1·75	0·875	66·7	33·3
Elephant (Indian)	1	15·75	20·5	14·75	58·2	41·8
Horse	–	8·25	4·75	3·5	57·6	42·4
Horse (large Clydesdale)	–	15·5	8·5	7·0	54·8	45·2

* I owe the first four of these determinations to the kindness of Sir P. Chalmers Mitchell, who had them made for me at the Zoological Society's Gardens; while the great Clydesdale carthorse was weighed for me by a friend in Dundee.

It will be observed that in all these animals the load upon the fore-feet preponderates considerably over that upon the hind, the preponderance being rather greater in the elephant than in the horse, and markedly greater in the camel and the llama than in the other two. But while these weights are helpful and suggestive, it is obvious that they do not go nearly far enough to give us a full insight into the constructional diagram to which the animals are conformed. For such a purpose we should require to weigh the total load, not in two portions but in many; and we should also have to take close account of the general form of the animal, of the relation between that form and the distribution of the load, and of the actual directions of each bone and ligament by which the forces of compression and tension were transmitted. All this lies beyond us for the present; but nevertheless we may consider,

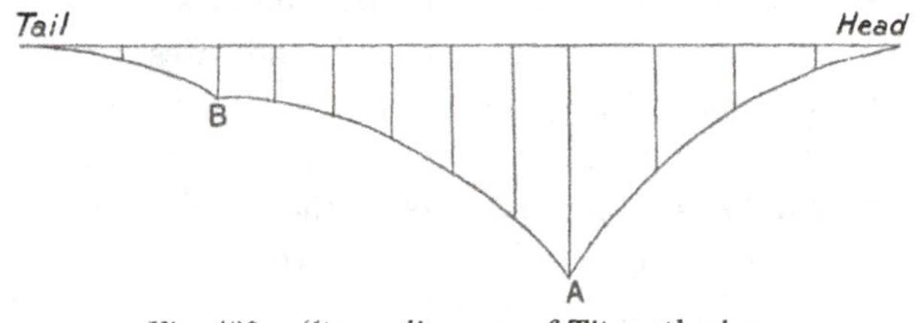

Fig. 480. Stress-diagram of Titanotherium.

very briefly, the principal cases involved in our enquiry, of which the above animals form a partial and preliminary illustration.

(1) Wherever we have a heavily loaded anterior cantilever arm, that is to say whenever the head and neck represent a considerable fraction of the whole weight of the body, we tend to have large bending-moments over the fore-legs, and correspondingly high spines over the vertebrae of the withers. This is the case in the great majority of four-footed terrestrial animals, the chief exceptions being found in animals with comparatively small heads but large and heavy tails, such as the anteaters or the Dinosaurian reptiles, and also (very naturally) in animals such as the crocodile, where the "bridge" can scarcely be said to be developed, for the long heavy body sags down to rest upon the ground. The case is sufficiently exemplified by the horse, and still more notably by the stag, the ox, or the pig. It is illustrated in the skeleton of a bison (Fig. 468), or

in the accompanying diagram of the conditions in the great extinct Titanotherium.

(2) In the elephant and the camel we have similar conditions, but slightly modified. In both cases, and especially in the latter, the weight on the fore-quarters is relatively large; and in both cases the bending-moments are all the larger, by reason of the length and forward extension of the camel's neck and the forward position of the heavy tusks of the elephant. In both cases the dorsal spines are large, but they do not strike us as exceptionally so; but in both cases, and especially in the elephant, they slope backwards in a marked degree. Each spine, as already explained, must in all cases assume the position of the diagonal in the parallelogram of forces defined by the tensions acting on it at its extremity; for it constitutes a "hinged lever," by which the bending-moments on either side are automatically balanced; and it is plain that the more the spine slopes backwards the more it indicates a relatively large strain thrown upon the great ligament of the neck, and a relief of strain upon the more directly acting, but weaker, ligaments of the back and loins. In both cases, the bending-moments would seem to be more evenly distributed over the region of the back than, for instance, in the stag, with its light hind-quarters and heavy load of antlers: and in both cases the high "girder" is considerably prolonged, by an extension of the tall spines backwards in the direction of the loins. When we come to such a case as the mammoth, with its immensely heavy and immensely elongated tusks, we perceive at once that the bending-moments over the fore-legs are now very severe; and we see also that the dorsal spines in this region are much more conspicuously elevated than in the ordinary elephant.

(3) In the case of the giraffe we have, without doubt, a very heavy load upon the fore-legs, though no weighings are at hand to define the ratio; but as far as possible this disproportionate load would seem to be relieved by help of a downward as well as backward thrust, through the sloping back to the unusually low hind-quarters. The dorsal spines of the vertebrae are very high and strong, and the whole girder-system very perfectly formed. The elevated rather than protruding position of the head lessens the anterior bending-moment as far as possible, but it leads to a strong

compressional stress transmitted almost directly downwards through the neck: in correlation with which we observe that the bodies of the cervical vertebrae are exceptionally large and strong, and steadily increase in size and strength from the head downwards.

(4) In the kangaroo, the fore-limbs are entirely relieved of their load, and accordingly the tall spines over the withers, which were so conspicuous in all heavy-headed *quadrupeds*, have now completely vanished. The creature has become bipedal, and body and tail form the extremities of *a single* balanced cantilever, whose maximal bending-moments are marked by strong, high lumbar and sacral vertebrae, and by iliac bones of peculiar form, of exceptional strength and nearly upright position.

Precisely the same condition is illustrated in the Iguanodon, and better still by reason of the great bulk of the creature and of the heavy load which falls to be supported by the great cantilever and by the hind-legs which form its piers. The long heavy body and neck require a balance-weight (as in the kangaroo) in the form of a long heavy tail; and the double-armed cantilever, so constituted, shews a beautiful parabolic curvature in the graded heights of the whole series of vertebral spines, which rise to a maximum over the haunches and die away slowly towards the neck and towards the tip of the tail.

(5) In the case of some of the great American fossil reptiles such as Diplodocus, it has always been a more or less disputed question whether or not they assumed, like Iguanodon, an erect, bipedal attitude. In all of them we see an elongated pelvis, and, in still more marked degree, we see elevated spinous processes of the vertebrae over the hind-limbs; in all of them we have a long heavy tail, and in most of them we have a marked reduction in size and weight both of the fore-limb and of the head itself. The great size of these animals is not of itself a proof against the erect attitude; because it might well have been accompanied by an aquatic or partially submerged habitat, and the crushing stress of the creature's huge bulk proportionately relieved. But we must consider each such case in the whole light of its own evidence; and it is easy to see that, just as the quadrupedal mammal may carry the greater part but not all of its weight upon its fore-limbs, so a heavy-tailed reptile may carry the greater part upon its hind-limbs, without

this process going so far as to relieve its fore-limbs of all weight whatsoever. This would seem to be the case in such a form as Diplodocus, and also in Stegosaurus, whose restoration by Marsh is doubtless substantially correct*. The fore-limbs, though comparatively small, are obviously fashioned for support, but the weight which they have to carry is far less than that which the hind-limbs bear. The head is small and the neck short, while on the other hand the hind-quarters and the tail are big and massive. The backbone bends into a great double-armed cantilever, culminating over the pelvis and the hind-limbs, and here furnished with its highest and strongest spines to separate the tension-member from the com-

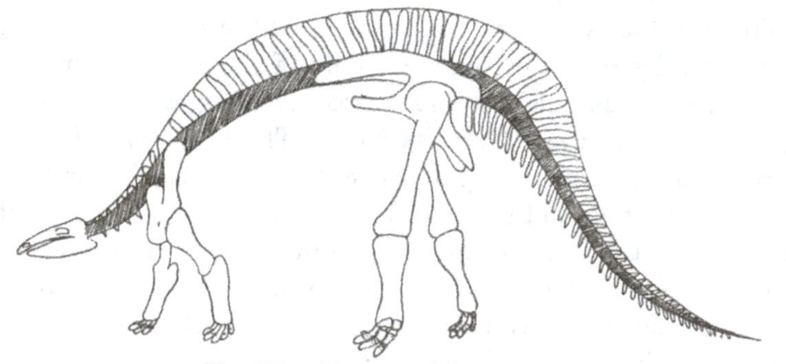

Fig. 481. Diagram of Stegosaurus.

pression-member of the girder. The fore-legs form a secondary supporting pier to this great continuous cantilever, the greater part of whose weight is poised upon the hind-limbs alone.

(6) In the slender body of a weasel, neither head nor tail is such as to form an efficient cantilever; and though the lithe body is arched in active exercise, our parallel of the bridge no longer works well. What else to compare it with is far from clear; but the mechanism has some resemblance (perhaps) to an elastic spring. Animals of this habit of body are all small; their bodily weight is a light burden, and gravity becomes an ineffectual force.

* This pose of Diplodocus, and of other Sauropodous reptiles, has been much discussed. Cf. (*int. al.*) O. Abel, *Abh. k. k. zoöl. bot. Ges. Wien*, v, 1909–10 (60 pp.); Tornier, *SB. Ges. Naturf. Fr. Berlin*, 1909, pp. 193–209; O. P. Hay, *Amer. Nat.* Oct. 1908; *Tr. Wash. Acad. Sci.* XLII, pp. 1–25, 1910; Holland, *Amer. Nat.* May 1910, pp. 259–283; Matthew, *ibid.* pp. 547–560; C. W. Gilmore (*Restoration of Stegosaurus*), *Pr. U.S. Nat. Museum*, 1915.

(7) An abnormal and very curious case is that of the sloth, which hangs by hooked hands and feet, head downwards, from high branches in the Brazilian forest. The vertebrae are unusually numerous, they are all much alike one to another, and (as we might well suppose) the whole pensile chain of vertebrae hangs in what closely approximates to a catenary curve*.

(8) We find a highly important corollary in the case of aquatic animals. For here the effect of gravity is neutralised; we have neither piers nor cantilevers; and we find accordingly in all aquatic mammals of whatsoever group—whales, seals or sea-cows—that the high arched vertebral spines over the withers, or corresponding structures over the hind-limbs, have both entirely disappeared.

But in the whale or dolphin (and not less so in the aquatic bird), *stiffness* must be ensured in order to enable the muscles to act against the resistance of the water in the act of swimming; and accordingly Nature must provide against bending-moments irrespective of gravity. In the dolphin, at any rate as regards its tail-end, the conditions will be not very different from those of a column or beam with fixed ends, in which, under deflection, there will be two points of contrary flexure, as at *C*, *D*, in Fig. 482.

Here, between *C* and *D* we have a varying bending-moment, represented by a continuous curve with its maximal elevation midway between the points of inflection. And correspondingly, in our dolphin, we have a continuous series of high dorsal spines, rising to a maximum about the middle of the animal's body, and falling to nil at some distance from the end of the tail. It is their business (as usual) to keep the tension-member, represented by the strong supraspinous ligaments, wide apart from the compression-member, which is as usual represented by the backbone itself. But in our diagram we see that on the farther side of *C* and *D* we have a *negative* curve of bending-moments, or bending-moments in a contrary direction. Without enquiring how these stresses are precisely met

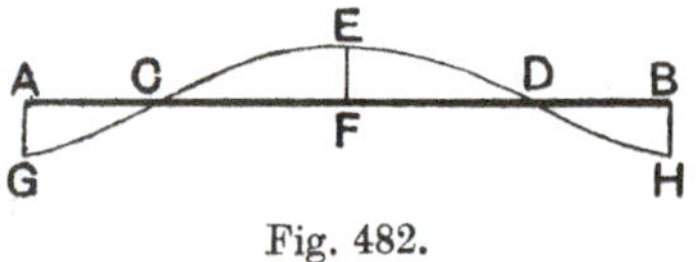

Fig. 482.

* A *heavy* cord, or a cord carrying equal weights for equal distances along its line, hangs in a catenary: imagine it frozen and inverted, and we have an arch, carrying the same sort of load, and under compression only. On the other hand, a flexible cable (itself of negligible weight), carrying a uniform load along the line of its horizontal projection, hangs in the form of a parabola.

towards the dolphin's head (where the coalesced cervical vertebrae suggest themselves as a partial explanation), we see at once that towards the tail they are met by the strong series of chevron-bones, which in the caudal region, where tall *dorsal* spines are no longer needed, take their place *below* the vertebrae, in precise correspondence with the bending-moment diagram. In many cases other than these aquatic ones, when we have to deal with animals with long and heavy tails (like the Iguanodon and the kangaroo of which we have already spoken), we are apt to meet with similar, though usually shorter chevron-bones; and in all these cases we may see without difficulty that a negative bending-moment in the vertical direction has to be resisted or controlled.

In the dolphin we may find an illustration of the fact that not only is it necessary to provide for rigidity in the vertical direction but often also in the horizontal, where a tendency to bending must be resisted on either side. This function is effected in part by the ribs with their associated muscles, but they extend but a little way and their efficacy for this purpose can be but small. We have, however, behind the region of the ribs and on either side of the backbone a strong series of elongated and flattened transverse processes, forming a web for the support of a tension-member in the usual form of ligament, and so playing a part precisely analogous to that performed by the dorsal spines in the same animal. In an ordinary fish, such as a cod or a haddock, we see precisely the same thing:

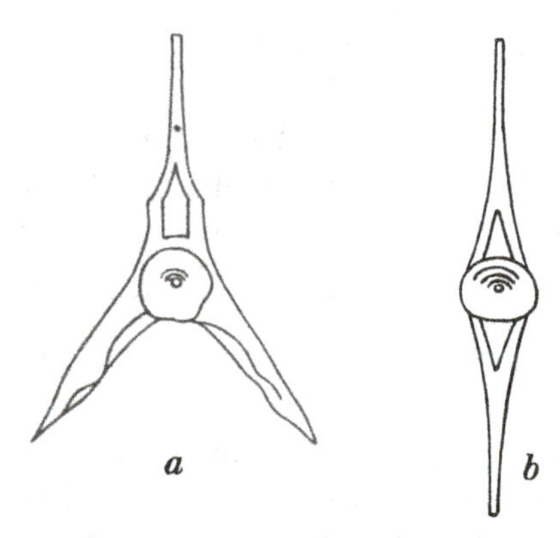

Fig. 483. *a*, dorsal and *b*, caudal vertebrae of haddock.

the backbone is stiffened by the indispensable help of its *three series* of ligament-connected processes, the dorsal and the two transverse series; but there are no such stiffeners in the eel. When we come to the region of the tail, where rigidity gives place to lateral flexibility, the three stiffeners give place to two—the dorsal and haemal spines of the caudal vertebrae. And here we see that the three series of processes, or struts, tend (when all three are present) to be arranged well-nigh at equal angles, of 120°, with one another, giving the greatest and most uniform strength of which such a system is capable. On the other hand, in a flat fish, such as a plaice, where from the natural mode of progression it is necessary that the backbone should be flexible in one direction while stiffened in another, we find the whole outline of the fish comparable to that of a double bowstring girder, the compression-member being (as usual) the backbone itself, the tension-member on either side being constituted by the interspinous ligaments and muscles, while the web or filling is very beautifully represented by the long and evenly graded neural and haemal spines, which spring symmetrically up and down from each individual vertebra.

In the skeleton of the flat fishes, the web of the otherwise perfect parabolic girder has to be cut away and encroached on to make room for the viscera. When the body is long and the vertebrae many, as in the sole, the space required is small compared with the length of the girder, and the strength of the latter is not much impaired. In the shorter, rounder kinds with fewer vertebrae, like the turbot, the visceral cavity is large compared with the length of the fish, and its presence would seem to weaken the girder very seriously. But Nature repairs the breach by framing in the hinder part of the space with a strong curved bracket or angle-iron, which takes the place very efficiently of the bony struts which have been cut away.

The main result at which we have now arrived, in regard to the construction of the vertebral column and its associated parts, is that we may look upon it as a certain type of *girder*, whose depth is everywhere very nearly proportional to the height of the corresponding ordinate in the diagram of moments: just as it is in a girder designed by a modern engineer. In short, after the nineteenth or twentieth century engineer has done his best in framing the design of a big cantilever, he may find that some of his best ideas had, so to speak, been anticipated ages ago in the fabric of the great saurians and the larger mammals.

But it is possible that the modern engineer might be disposed to criticise the skeleton girder at two or three points; and in particular he might think the girder, as we see it for instance in Diplodocus or Stegosaurus, not deep enough for carrying the animal's enormous weight of some twenty tons. If we adopt a much greater depth (or ratio of depth to length) as in the modern cantilever, we shall greatly increase the *strength* of the structure; but at the same time we should greatly increase its *rigidity*, and this is precisely what, in the circumstances of the case, it would seem that Nature is bound to avoid. We need not suppose that the great saurian was by any means active and limber; but a certain amount of activity and flexibility he was bound to have, and in a thousand ways he would find the need of a backbone that should be *flexible* as well as *strong*. Now this opens up a new aspect of the matter and is the beginning of a long, long story, for in every direction this double requirement of strength and flexibility imposes new conditions upon the design. To represent all the correlated quantities we should have to construct not only a diagram of moments but also a diagram of elastic deflection and its so-called "curvature"; and the engineer would want to know something more about the *material* of the ligamentous tension-member—its flexibility, its modulus of elasticity in direct tension, its elastic limit, and its safe working stress.

In various ways our structural problem is beset by "limiting conditions." Not only must rigidity be associated with flexibility, but also stability must be ensured in various positions and attitudes; and the primary function of support or weight-carrying must be combined with the provision of *points d'appui* for the muscles concerned in locomotion. We cannot hope to arrive at a numerical or quantitative solution of this complicate problem, but we have found it possible to trace it out in part towards a qualitative solution. And speaking broadly we may certainly say that in each case the problem has been solved by Nature herself, very much as she solves the difficult problems of minimal areas in a system of soap-bubbles; so that each animal is fitted with a backbone adapted to his own individual needs, or (in other words) corresponding to the mean resultant of the many stresses to which as a mechanical system it is exposed.

The mechanical construction of a bird is a more elaborate affair than a quadruped's, inasmuch as it has a double part to play, the bird's whole weight being borne now by its legs and now by its wings. As it stands on the ground our bird is a balanced cantilever, carried on two legs as on a pier, the cantilever being constituted by the pelvic bones, drawn out fore and aft and firmly welded to a long stretch of vertebral column. The centre of gravity is kept in a line passing through the acetabulum, and the long toes help to preserve an unstable but well-adjusted equilibrium. One arm of the cantilever carries head, neck and wings, the other, the shorter arm, carries the abdomen; but the whole weight of the viscera hangs in the abdomen as in a bag, and on the other hand head and neck are kept small and light, and their purchase on the fulcrum is under constant modification and control. A stork or a heron is continually balancing itself;

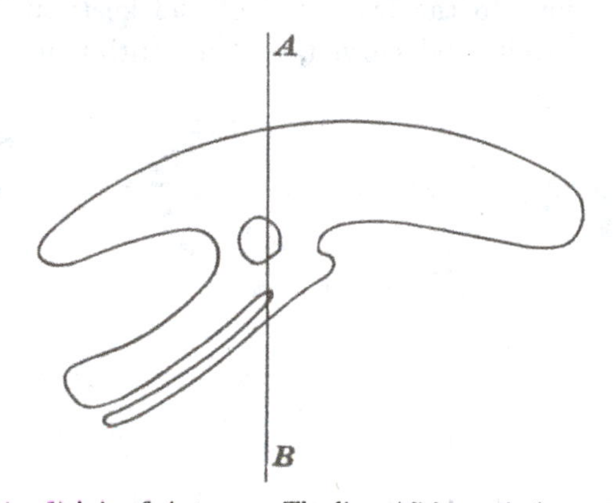

Fig. 484. Pelvis of *Apteryx*. The line *AB* is vertical, or nearly so, in the standing posture of the bird.

as the beak is thrust forward a leg stretches back, as the bird walks along its whole body sways in keeping. No less elegant is the perfect balance of the same birds at rest—the heron standing on one leg, even on a tree top, the flamingo also on one long leg, with its neck close coiled and its head tucked amongst the feathers.

The approximately parabolic form of the great pelvic cantilever is best seen in the ostrich and other running birds, but more commonly the strength of the cantilever is got in other ways. Usually, as in the fowl, it consists of a thin shell of bone curved over like the bonnet of a motor car and stiffened, or "cambered," by ridges converging on the acetabulum. A doubled sheet of paper, cut roughly to the shape of the pelvis and then pinched up into folds on either side, as in Fig. 485, will serve as a model of the skeletal cantilever and shew how its limp surface is stiffened by the folds.

Save in the ostrich, and a few other flightless birds, the breast-bone or sternum is a broad, flat bone, produced into a deep, descending ridge or "keel." Very firmly fixed to the sternum on either side is a short strong bone, the coracoid; attached to it again, and bending backwards over the ribs, is the scapula; and at the junction of scapula and coracoid is the socket, or glenoid cavity, for the wing. The clavicles, fused into a "merry-thought," run from near the glenoid cavity to the front end of the keel; in strong-flying birds they are stout and curved, and a continuous curve sweeps

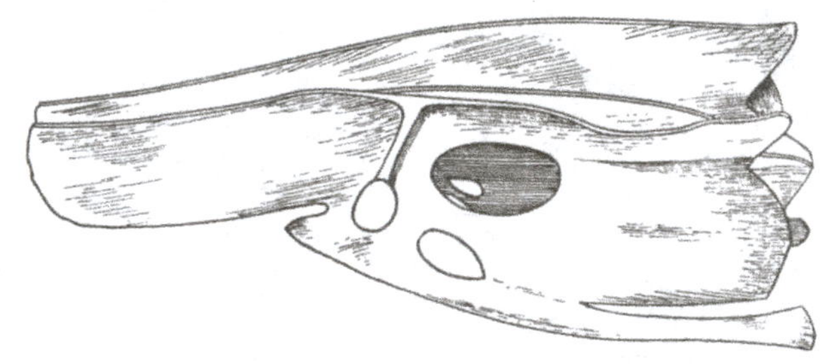

Fig. 485. Rough paper model of a fowl's pelvis.

round from scapula to sternal keel. The keel is commonly explained as necessary to give space enough for the attachment of the muscles of flight, but this explanation is inadequate, even untrue; for one thing, the great pectoral muscle springs from the edge, not from the broad surface of the keel. The keel is essentially *a flange*, and, as in a piece of T iron, adds immensely to the strength and stiffness of the construction*; that it tends to give the fibres of the muscle more stretch and play, and a straighter pull on the arm-bone to which they run, is a secondary advantage. Strong as they are, these bones are exquisitely light and thin. A great frigate-bird, with a 7-foot span of wings, weighs a little over a couple of pounds, and all its bones weigh about four ounces. The bones weigh less than the feathers†.

* T irons, if I am not mistaken, were among the many inventions of Robert Stephenson, in his construction of the Menai tubular bridge a hundred years ago.

† Cf. R. C. Murphy, *Natural History*, Oct. 1939.

While the bird stands on the ground its backbone is, as in ourselves, its skeletal axis, and it, including the great cantilever associated with

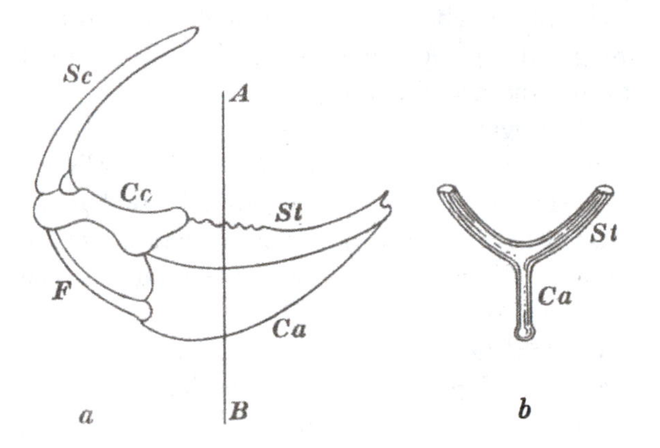

Fig. 485. *a*, Sternum and shoulder-girdle of a skua gull. *b*, section of do., through the line *AB*. *St*, sternum; *Ca*, its carina or keel; *Co*, coracoid bone; *Sc*, scapula; *F*, merry-thought or furcula.

it, carries, and transmits to the legs, the whole weight of the body. But as soon as a bird spreads its wings and rests upon the air, legs,

Fig. 487. A flying bird. Sternum, shoulder-girdle and wings combine to support the body; and all the rest lies as a dead weight thereon.

backbone, cantilever and all become merely so much weight to be carried; and the whole rests, as on a floor, on the strong, stiff platform made of sternum and shoulder-girdle, which the wings (so to speak) take hold of and support, and which is now, mechanically speaking, the axis of the body. The bird has two points of suspension, which it uses alternately: the one through the two acetabula, the other through the glenoid cavities and the outstretched wings. Glenoid cavity and acetabulum are but a little way apart, and the bird swings its weight over from one to the other easily and smoothly. At first sight it seems a curious feature of the bird's skeleton that breast-bone, shoulder-girdle, wings and all are but very slightly attached to the rest of the body, and to what we look on, usually, as its main axis of support; the only skeletal attachment is by the framework of the ribs, and these are slight and slender. The fact is that the two skeletal axes, the backbone and the breast-bone, have their separate and independent roles, and each is but loosely connected with the other.

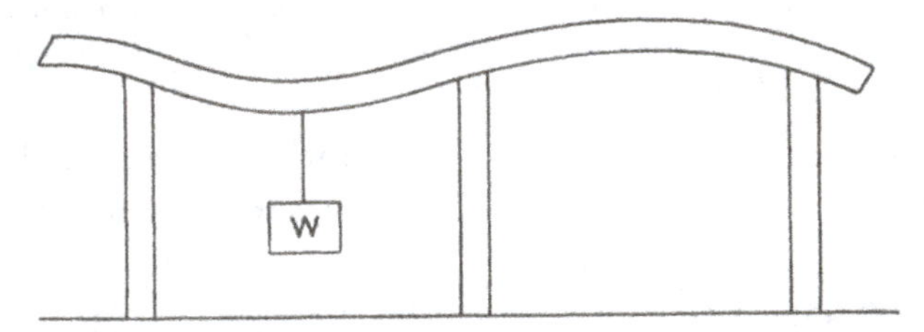

Fig. 488. Diagram of a continuous girder.

The curvature of the bird's neck is very beautiful: one curve leads on to another; and indeed the bird's whole axial skeleton, from head to tail, is one even and continuous curve. Where a bridge crosses the gap between two piers, it sags as the load passes over; where successive girders cross successive gaps, each sags in its turn under the travelling load. But suppose one *continuous* girder to cross two gaps; it bends in a more complicated way, and one half tends to bend up while the other is sagging down. We cannot analyse the whole field of force to which the bird is subject, but we realise that it is a *continuous* field, in which what the engineer calls a "continuous girder" has its great part to play. The continuous girder is apt to sag and bend and sway in an erratic fashion unless its ends be firm and secure, and the bird's head must, of necessity, be under some analogous control; the semicircular canals are the potent factors in equilibrium, and the bird "keeps a level head" by their help and guidance. Then, between the level of the skull and the level of the great pelvic cantilever a continuous field of force governs and defines the S-shaped curvature. Man's vertebral column shews, *mutatis mutandis*, the same phenomenon of continuous but alternating curvature. The dorsal region is, of necessity, concave towards the cavity of the chest, and as a simple consequence the cervical and lumbar regions curve the other way.

The typically aquatic birds, such as swim under water as penguins and divers do, have characteristic features and adaptations of their own. Just as the cantilever girder becomes obsolete in the aquatic mammal so does it tend to weaken and disappear in the aquatic bird. There is a marked contrast between the high-arched strongly built pelvis in the ostrich or the hen, and the long, thin, comparatively straight and apparently weakly bone which represents it in a diver, a grebe or a penguin. Wings large enough for air would be an obstruction under water, and small wings are enough; for they have to produce thrust only, not lift, and the former is but a small fraction of the latter load. The feet also are now mainly concerned with the same forward thrust, and we begin to see how the long narrow pelvis gives just the *point d'appui* which that thrust requires.

The woodcock, as ornithologists are aware, shews us an osteological paradox, which is commonly described by saying that this bird's ear is in front of its eye! If we hold a woodcock's skull level, *beak and all*, this indeed seems to be the case, but no woodcock does so. Standing or flying, the woodcock holds its beak pointing downwards, and its skull is then level, like that of other birds; in other words, its beak is not in a line with the basi-cranial axis, as a guillemot's is, but bends sharply downwards. When the axis of the skull is horizontal, the beak points downwards at an angle of nearly 60°, and the auditory aperture is then as much behind the eye as in other birds.

There is a certain other principle much to the fore in the construction of the skeleton, well known to the designer of a hydroplane or "flying boat," and not wholly neglected by the bridge-building engineer; it is the principle of non-rigid, flexible or *elastic stability**. A homely comparison between a basket and a tin-can tells us in a moment what it means, and shews us some at least of its peculiar advantages. This method of construction helps to *distribute* the load,

* India-rubber has great elastic stability. It is *not* compressible, but is almost as incompressible as water itself, as J. D. Forbes discovered a century ago.

bridges over points or areas where pressure might be unduly concentrated or confined, adapts itself to a sudden impact or concentrated stress, helps to lessen or to guard against *shock*, and imparts to the whole structure a quality which we may call, for short, *resiliency*.

The engineer finds it easiest of attainment when his principal members are in tension; hence elastic movement and resilience are apt to be conspicuous in a suspension-bridge. One way and another, resilience shews to perfection in a bird. The **S**-shaped curve of the neck carrying the light weight of the head, the zig-zag flexures of the legs bearing the balanced burden of the body, the supple basket

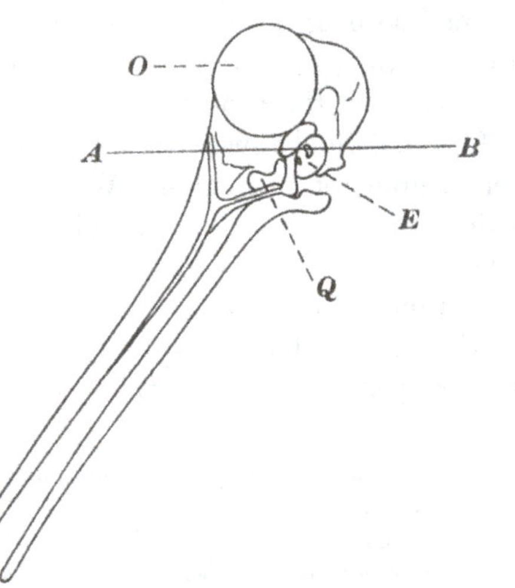

Fig. 489. A woodcock's skull, in (or nearly in) the natural attitude. *A*, *B*, the *basis cranii*; *E*, auditory meatus; *O*, orbit; *Q*, quadrate bone.

of the ribs, each rib in two halves one flexed on the other, all these are such as to make the whole framework act like an elastic spring, absorbing every shock as the bird lights on or rises from the ground. Bird, beast and man exhibit this resilience, each in its degree; a springy step is part of the joy of youth, and its loss is one of the first infirmities of age.

Nature's engineering is marvellous in our eyes, and our finest work is narrow in scope and clumsy in execution compared to her construction and design. But following her example, wittingly or unwittingly, our own problems evolve and our ambitions enlarge towards the conception of an "organised structure." In such triumphs of modern mechanism as a torpedo, a racing aeroplane, a high-speed railway-train, the whole construction is knit together in a new way. It finds its streamlined outline in what seems to be a simple and natural way; it is solid and robust, it is graceful as well as strong; it is no longer a bundle of parts, it has become an organic whole: its likeness, even its outward likeness, to a living organism has become patent and clear.

Throughout this short discussion of the principles of construction, we see the same general principles at work in the skeleton as a whole as we recognised in the plan and construction of an individual bone. That is to say, we see a tendency for material to be laid down just in the lines of *stress*, and so to evade thereby the distortions and disruptions due to *shear*. In these phenomena there lies a definite law of growth, whatever its ultimate expression or explanation may come to be. Let us not press either argument or hypothesis too far: but be content to see that skeletal form, as brought about by growth, is to a very large extent determined by mechanical considerations, and tends to manifest itself as a diagram, or reflected image, of mechanical stress. If we fail, owing to the immense complexity of the case, to unravel all the mathematical principles involved in the construction of the skeleton, we yet gain something, and not a little, by applying this method to the familiar objects of anatomical study: *obvia conspicimus, nubem pellente mathesi**.

Before we leave this subject of mechanical adaptation, let us dwell once more for a moment upon the considerations which arise from our conception of a field of force, or field of stress, in which tension and compression (for instance) are inevitably combined, and are met by the materials naturally fitted to resist them. It has been remarked over and over again how harmoniously the whole organism hangs together, and how throughout its fabric one part is related and fitted to another in strictly functional correlation. But this conception, though never denied, is sometimes apt to be forgotten in the course of that process of more and more minute

* The motto was Macquorn Rankine's, in 1857; cf. *Trans. R.S.E.* xxvi, p. 715, 1872.

analysis by which, for simplicity's sake, we seek to unravel the intricacies of a complex organism.

As we analyse a thing into its parts or into its properties, we tend to magnify these, to exaggerate their apparent independence, and to hide from ourselves (at least for a time) the essential integrity and individuality of the composite whole. We divide the body into its organs, the skeleton into its bones, as in very much the same fashion we make a subjective analysis of the mind, according to the teachings of psychology, into component factors: but we know very well that judgment and knowledge, courage or gentleness, love or fear, have no separate existence, but are somehow mere manifestations, or imaginary coefficients, of a most complex integral. And likewise, as biologists, we may go so far as to say that even the bones themselves are only in a limited and even a deceptive sense, separate and individual things. The skeleton begins as a *continuum*, and a *continuum* it remains all life long. The things that link bone with bone, cartilage, ligaments, membranes, are fashioned out of the same primordial tissue, and come into being *pari passu* with the bones themselves. The entire fabric has its soft parts and its hard, its rigid and its flexible parts; but until we disrupt and dismember its bony, gristly and fibrous parts one from another, it exists simply as a "skeleton," as one integral and individual whole.

A bridge was once upon a time a loose heap of pillars and rods and rivets of steel. But the identity of these is lost, just as if they were fused into a solid mass, when once the bridge is built; their separate functions are only to be recognised and analysed in so far as we can analyse the stresses, the tensions and the pressures, which affect this part of the structure or that; and these forces are not themselves separate entities, but are the resultants of an analysis of the whole field of force. Moreover when the bridge is broken it is no longer a bridge, and all its strength is gone. So is it precisely with the skeleton. In it is reflected a field of force: and keeping pace, as it were, in action and interaction with this field of force, the whole skeleton and every part thereof, down to the minute intrinsic structure of the bones themselves, is related in form and in position to the lines of force, to the resistances it has to encounter; for by one of the mysteries of biology, resistance begets resistance, and where pressure falls there growth springs up in strength to meet it. And, pursuing the same train of thought, we see that all this is true not of the skeleton alone but of the whole fabric of the body. Muscle and bone, for instance, are inseparably associated and connected; they are moulded one with another; they come into being together, and act and react together*. We may study them apart, but it is as a concession to our weakness and to the narrow outlook of our minds. We see, dimly perhaps but yet with all the assurance of conviction, that between muscle and bone there can be no change in the one but it is correlated with changes in the other; that through and through they are linked in indissoluble association; that they are only separate entities in this limited and subordinate sense, that they are *parts* of a whole which, when it loses its composite integrity, ceases to exist.

The biologist, as well as the philosopher, learns to recognise that the whole is not merely the sum of its parts. It is this, and much more than this. For it is not a bundle of parts but an organisation of parts, of parts in their mutual arrangement, fitting one with another, in what Aristotle calls "a single and indivisible principle of unity"; and this is no merely metaphysical conception, but is in biology the fundamental truth which lies at the basis of Geoffroy's (or Goethe's) law of "compensation," or "balancement of growth."

Nevertheless Darwin found no difficulty in believing that "natural selection will tend in the long run to reduce *any part* of the organisation, as soon as, through changed habits, it becomes superfluous: without by any means causing some other part to be largely developed in a corresponding degree. And conversely, that natural selection may perfectly well succeed in largely developing an organ without requiring as a necessary compensation the reduction of some adjoining part†." This view has been developed into a doctrine of the "independence of single characters" (not to be confused with the germinal "unit characters" of Mendelism), especially by the palaeontologists. Thus Osborn asserts a "principle of hereditary correlation," combined with a "principle of *hereditary separability*,

* John Hunter was seldom wrong; but I cannot believe that he was right when he said (*Scientific Works*, ed. Owen, I, p. 371), "The bones, in a mechanical view, appear to be the first that are to be considered. We can study their shape, connections, number, uses, etc., *without considering any other part of the body*."

† *Origin of Species*, 6th ed. p. 118.

whereby the body is a colony, a mosaic, of single individual and separable characters*." I cannot think that there is more than a very small element of truth in this doctrine. As Kant said, "die Ursache der Art der Existenz bei jedem Theile eines lebenden Körpers *ist im Ganzen enthalten.*" And, according to the trend or aspect of our thought, we may look upon the coordinated parts, now as related and fitted *to the end or function of* the whole, and now as related to or resulting *from the physical causes* inherent in the entire system of forces to which the whole has been exposed, and under whose influence it has come into being†.

In John Hunter's day the anatomist studied every bone of the skeleton in its own place, in order to discover its useful purpose and understand its mechanical perfection. The morphologist of a hundred years later preferred to study an isolated bone from many animals, collar-bones or shoulder-blades by themselves, apart from the field of force in which their work was done, in the search for signs of blood-relationship and common ancestry. Truth lies both ways; immediate use and old inheritance are blended in Nature's handiwork as in our own. In the marble columns and architraves of a Greek temple we still trace the timbers of its wooden prototype, and see beyond these the tree-trunks of a primeval sacred grove; roof and eaves of a pagoda recall the sagging mats which roofed an

* *Amer. Naturalist*, April, 1915, p. 198, etc. Cf. *infra*, p. 1036.

† Driesch saw in "Entelechy" that something which differentiates the whole from the sum of its parts in the case of the organism: "The organism, we know, is a system the single constituents of which are inorganic in themselves; only the whole constituted by them in their typical order or arrangement owes its specificity to 'Entelechy'" (*Gifford Lectures*, 1908, p. 229): and I think it could be shewn that many other philosophers have said precisely the same thing. So far as the argument goes, I fail to see how *this* Entelechy is shewn to be peculiarly or specifically related to the *living* organism. The conception (at the bottom of General Smuts's '*Holism*') that the whole is *always* something very different from its parts is a very ancient doctrine. The reader will perhaps remember how, in another vein, the theme is treated by Martinus Scriblerus (Huxley quoted it once, for his own ends): "In every Jack there is a *meat-roasting* Quality, which neither resides in the fly, nor in the weight, nor in any particular wheel of the Jack, but is the result of the whole composition; etc., etc." Indeed it was at that very time, in the early eighteenth century, that the terms *organism* and *organisation* were coming into use, to connote that harmonious combination of parts "qui conspirent toutes ensembles à produire cet effet général que nous nommons la vie" (Buffon). Cf. Ch. Robin, Recherches sur l'origine et le sens des termes organisme et organisation, *Jl. de l'Anat.* LX, pp. 1–55, 1880.

earlier edifice; Anglo-Saxon land-tenure influences the planning of our streets, and the cliff-dwelling and the cave-dwelling linger on in the construction of our homes! So we see enduring traces of the past in the living organism—landmarks which have lasted on through altered functions and altered needs; and yet at every stage new needs are met and new functions effectively performed.

When we consider (for instance) the several bones in a fish's shoulder-girdle—clavicle, supra-clavicle, post-clavicle, post-temporal and so on—and recognise these in this fish or that under countless minor transformations, we have something which is not only wide-

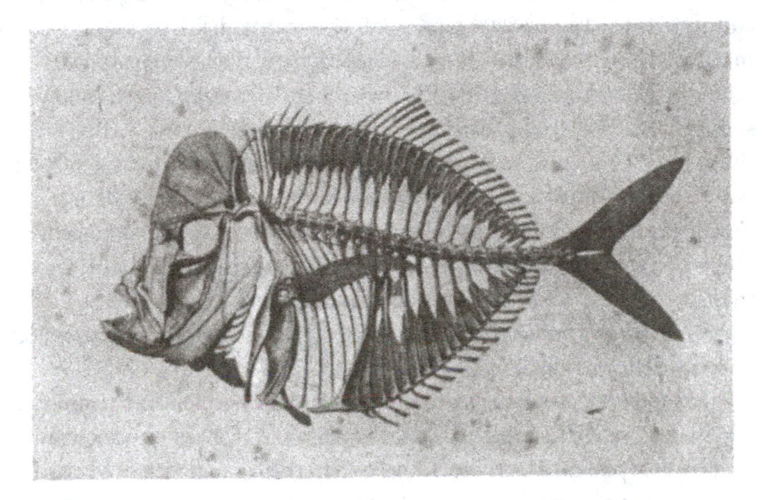

Fig. 490. Skeleton of moonfish, *Vomer* sp. From L. Agassiz.

spread but is rooted in antiquity, and whose full significance seems beyond our reach. But take the skeleton of some particular fish, a moonfish or a John Dory will do very well, and look at its shoulder-girdle from the mechanical point of view. It is a deal more than is needed for the support of the small, weak pectoral fin; but another function, and its perfect adaptation for that function, are not hard to see. The flattened body of the fish is built (as we have seen also in the plaice) on the plan of a parabolic girder; but out of this girder a great gap has had to be cut, to hold the viscera. The great shoulder-girdle serves to strengthen and complete the girder, to bind its upper and lower members together, and to compensate for the part which has been taken away. It fulfils this function by various means; by the way in which the two sides of the girdle are conjoined into a single arch; by its strong attachment to the head, and again to the pelvis, and through the latter to the chain of ossicles which bound or constitute the abdominal border of the fish; and a large part of the stress upon the shoulder-girdle proper is taken up, or relieved, by the strong post-clavicular bones, which form a supplementary arch running downwards from the clavicle (just where it begins to incline forward), straight to the ventral border, to be firmly attached there to the ventral ossicles. Similarly we notice at the hinder border of the abdominal cavity, a strong curved bone running from the anterior part of the ventral fin to a solid attachment with the vertebral column, stiffening the ventral part, and helping the shoulder-girdle to restore full strength to the girder after it had been reduced, so to speak, to the brink of inevitable collapse. The skull itself is not only streamlined with the rest of the body, but is an intrinsic part of the whole engineering construction. The lines of stress run simply and clearly through the skeleton, and a bone can no longer teach us its full and proper lesson after we have taken it apart. To look on the hereditary or evolutionary factor as *the guiding principle* in morphology is to give to that science a one-sided and fallacious simplicity*.

It would seem to me that the mechanical principles and phenomena which we have dealt with in this chapter are of no small importance to the morphologist, all the more when he is inclined to direct his study of the skeleton exclusively to the problem of phylogeny; and especially when, according to the methods of modern comparative morphology, he is apt to take the skeleton to pieces, and to draw from the comparison of a series of scapulae, humeri, or individual vertebrae, conclusions as to the descent and relationship of the animals to which they belong.

It would, I dare say, be an exaggeration to see in every bone nothing more than a resultant of immediate and direct physical or mechanical conditions; for to do so would be to deny the existence,

* The extreme evolutionary, or phylogenetic, aspect of morphology was being questioned even forty years ago. "Where we once thought we detected relationships we now know we were often being misled, and the old-time supposition that mere community of structure is necessarily an index of community of origin has gone to the wall" (G. B. Howes, in *Nature*, Jan. 10, 1901).

in this connection, of a principle of heredity. And though I have tried throughout this book to lay emphasis on the direct action of causes other than heredity, in short to circumscribe the employment of the latter as a working hypothesis in morphology, there can still be no question whatsoever but that heredity is a vastly important as well as a mysterious thing; it is *one* of the great factors in biology, however we may attempt to figure to ourselves, or howsoever we may fail even to imagine, its underlying physical explanation. But I maintain that it is no less an exaggeration if we tend to neglect these direct physical and mechanical modes of causation altogether, and to see in the characters of a bone merely the results of variation and of heredity, and to trust, in consequence, to those characters as a sure and certain and unquestioned guide to affinity and phylogeny. Comparative anatomy has its physiological side, which filled men's minds in John Hunter's day, and in Owen's day; it has its classificatory and phylogenetic aspect, which all but filled men's minds in the early days of Darwinism; and we can lose sight of neither aspect without risk of error and misconception.

It is certain that the question of phylogeny, always difficult, becomes especially so in cases where a great change of physical or mechanical conditions has come about, and where accordingly the former physical and physiological constraints are altered or removed. The great depths of the sea differ from other habitations of the living, not least in their eternal quietude. The fishes which dwell therein are quaint and strange; their huge heads, prodigious jaws, and long tails and tentacles are, as it were, gross exaggerations of the common and conventional forms. We look in vain for any purposeful cause or physiological explanation of these enormities; and are left under a vague impression that life has been going on in the security of all but perfect equilibrium, and that the resulting forms, liberated from many ordinary constraints, have grown with unusual freedom*.

To discuss these questions at length would be to enter on a discussion of Lamarck's philosophy of biology, and of many other things besides. But let us take one single illustration. The affinities of the whales constitute, as will be readily admitted, a very hard problem in phylogenetic classification. We know now that the

* Cf. *supra*, p. 423.

extinct Zeuglodons are related to the old Creodont carnivores, and thereby (though distantly) to the seals*; and it is supposed, but it is by no means so certain, that in turn they are to be considered as representing, or as allied to, the ancestors of the modern toothed whales†. The proof of any such a contention becomes, to my mind, extraordinarily difficult and complicated; and the arguments commonly used in such cases may be said (in Bacon's phrase) to allure, rather than to extort assent. Though the Zeuglodons were aquatic animals, we do not know, and we have no right to suppose or to assume, that they swam after the fashion of a whale (any more than the seal does), that they dived like a whale, or leaped like a whale. But the fact that the whale does these things, and the way in which he does them, is reflected in many parts of his skeleton—perhaps more or less in all: so much so that the lines of stress which these actions impose are the very plan and working-diagram of great part of his structure. That the Zeuglodon has a scapula like that of a whale is to my mind no necessary argument that he is akin by blood-relationship to a whale: that his dorsal vertebrae are very different from a whale's is no conclusive argument that such blood-relationship is lacking. The former fact goes a long way to prove that he used his flippers very much as a whale does; the latter goes still farther to prove that his general movements and equilibrium in the water were totally different. The whale may be descended from the Carnivora, or might for that matter, as an older school of naturalists believed, be descended from the Ungulates; but whether or no, we need not expect to find in him the scapula, the pelvis or the vertebral column of the lion or of the cow, for it would be physically impossible that he could live the life he does with any one of them. In short, when we hope to find the missing links between a whale and his terrestrial ancestors, it must be not by means of conclusions drawn from a scapula, an axis, or

* See (*int. al.*) my paper On the affinities of *Zeuglodon* in *Studies from the Museum of University College, Dundee*, 1889.

† "There can be no doubt that Fraas is correct in regarding this type (*Procetus*) as an annectant form between the Zeuglodonts and the Creodonta, but, although the origin of the Zeuglodonts is thus made clear, it still seems to be by no means so certain as that author believes, that they may not themselves be the ancestral forms of the Odontoceti" (Andrews, *Tertiary Vertebrata of the Fayum*, 1906, p. 235).

even from a tooth, but by the discovery of forms so intermediate in their general structure as to indicate an organisation and, *ipso facto*, a mode of life, intermediate between the terrestrial and the Cetacean form. There is no valid syllogism to the effect that *A* has a flat curved scapula like a seal's, and *B* has a flat curved scapula like a seal's: and therefore *A* and *B* are related to the seals and to each other; it is merely a flagrant case of an "undistributed middle." But there is validity in an argument that *B* shews in its general structure, extending over this bone and that bone, resemblances both to *A* and to the seals: and that therefore he may be presumed to be related to both, in his hereditary habits of life and in actual kinship by blood. It is cognate to this argument that (as every palaeontologist knows) we find clues to affinity more easily, that is to say with less confusion and perplexity, in certain structures than in others. The deep-seated rhythms of growth which, as I venture to think, are the chief basis of morphological heredity, bring about similarities of form which endure in the absence of conflicting forces; but a new system of forces, introduced by altered environment and habits, impinging on those particular parts of the fabric which lie within this particular field of force, will assuredly not be long of manifesting itself in notable and inevitable modifications of form. And if this be really so, it will further imply that modifications of form will tend to manifest themselves, not so much in small and *isolated* phenomena, in this part of the fabric or in that, in a scapula for instance or a humerus: but rather in some slow, *general*, and more or less uniform or graded modification, spread over a number of correlated parts, and at times extending over the whole, or over great portions, of the body. Whether any such general tendency to widespread and correlated transformation exists, we shall attempt to discuss in the following chapter.

CHAPTER XVII

ON THE THEORY OF TRANSFORMATIONS, OR THE COMPARISON OF RELATED FORMS

In the foregoing chapters of this book we have attempted to study the inter-relations of growth and form, and the part which the physical forces play in this complex interaction; and, as part of the same enquiry, we have tried in comparatively simple cases to use mathematical methods and mathematical terminology to describe and define the forms of organisms. We have learned in so doing that our own study of organic form, which we call by Goethe's name of Morphology, is but a portion of that wider Science of Form which deals with the forms assumed by matter under all aspects and conditions, and, in a still wider sense, with forms which are theoretically imaginable.

The study of form may be descriptive merely, or it may become analytical. We begin by describing the shape of an object in the simple words of common speech: we end by defining it in the precise language of mathematics; and the one method tends to follow the other in strict scientific order and historical continuity. Thus, for instance, the form of the earth, of a raindrop or a rainbow, the shape of the hanging chain, or the path of a stone thrown up into the air, may all be described, however inadequately, in common words; but when we have learned to comprehend and to define the sphere, the catenary, or the parabola, we have made a wonderful and perhaps a manifold advance. The mathematical definition of a "form" has a quality of precision which was quite lacking in our earlier stage of mere description; it is expressed in few words or in still briefer symbols, and these words or symbols are so pregnant with meaning that thought itself is economised; we are brought by means of it in touch with Galileo's aphorism (as old as Plato, as old as Pythagoras, as old perhaps as the wisdom of the Egyptians), that "the Book of Nature is written in characters of Geometry*."

* Cf. Plutarch, *Symp.* viii, 2, on the meaning of Plato's aphorism ("if it actually was Plato's"): πῶς Πλάτων ἔλεγε τὸν θεὸν ἀεὶ γεωμετρεῖν.

We are apt to think of mathematical definitions as too strict and rigid for common use, but their rigour is combined with all but endless freedom. The precise definition of an ellipse introduces us to all the ellipses in the world; the definition of a "conic section" enlarges our concept, and a "curve of higher order" all the more extends our range of freedom*. By means of these large limitations, by this controlled and regulated freedom, we reach through mathematical analysis to mathematical synthesis. We discover homologies or identities which were not obvious before, and which our descriptions obscured rather than revealed: as for instance, when we learn that, however we hold our chain, or however we fire our bullet, the contour of the one or the path of the other is always mathematically homologous.

Once more, and this is the greatest gain of all, we pass quickly and easily from the mathematical concept of form in its statical aspect to form in its dynamical relations: we rise from the conception of form to an understanding of the forces which gave rise to it; and in the representation of form and in the comparison of kindred forms, we see in the one case a diagram of forces in equilibrium, and in the other case we discern the magnitude and the direction of the forces which have sufficed to convert the one form into the other. Here, since a change of material form is only effected by the movement of matter†, we have once again the support of the schoolman's and the philosopher's axiom, *Ignorato motu, ignoratur Natura.*"

* So said Gustav Theodor Fechner, the author of Fechner's Law, a hundred years ago. (Ueber die mathematische Behandlung organischer Gestalten und Processe, *Berichte d. k. sachs. Gesellsch., Math.-phys. Cl.*, Leipzig, 1849, pp. 50–64.) Fechner's treatment is more purely mathematical and less physical in its scope and bearing than ours, and his paper is but a short one, but the conclusions to which he is led differ little from our own. Let me quote a single sentence which, together with its context, runs precisely on the lines which we have followed in this book: "So ist also die mathematische Bestimmbarkeit im Gebiete des Organischen ganz eben so gut vorhanden als in dem des Unorganischen, und in letzterem eben solchen oder äquivalenten Beschränkungen unterworfen als in ersterem; und nur sofern die unorganischen Formen und das unorganische Geschehen sich einer einfacheren Gesetzlichkeit mehr nähern als die organischen, kann die Approximation im unorganischen Gebiet leichter und weiter getrieben werden als im organischen. Dies wäre der ganze, sonach rein relative, Unterschied." Here, in a nutshell, is the gist of the whole matter.

† "We can *move* matter, that is all we can do to it" (Oliver Lodge).

There is yet another way—we learn it of Henri Poincaré—to regard the function of mathematics, and to realise why its laws and its methods *are bound* to underlie all parts of physical science. Every natural phenomenon, however simple, is really composite, and every visible action and effect is a summation of countless subordinate actions. Here mathematics shews her peculiar power, to combine and to generalise. The concept of an average, the equation to a curve, the description of a froth or cellular tissue, all come within the scope of mathematics for no other reason than that they are summations of more elementary principles or phenomena. Growth and Form are throughout of this composite nature; therefore the laws of mathematics are bound to underlie them, and her methods to be peculiarly fitted to interpret them.

In the morphology of living things the use of mathematical methods and symbols has made slow progress; and there are various reasons for this failure to employ a method whose advantages are so obvious in the investigation of other physical forms. To begin with, there would seem to be a psychological reason, lying in the fact that the student of living things is by nature and training an observer of concrete objects and phenomena and the habit of mind which he possesses and cultivates is alien to that of the theoretical mathematician. But this is by no means the only reason; for in the kindred subject of mineralogy, for instance, crystals were still treated in the days of Linnaeus as wholly within the province of the naturalist, and were described by him after the simple methods in use for animals and plants: but as soon as Haüy shewed the application of mathematics to the description and classification of crystals, his methods were immediately adopted and a new science came into being.

A large part of the neglect and suspicion of mathematical methods in organic morphology is due (as we have partly seen in our opening chapter) to an ingrained and deep-seated belief that even when we seem to discern a regular mathematical figure in an organism, the sphere, the hexagon, or the spiral which we so recognise merely resembles, but is never entirely explained by, its mathematical analogue; in short, that the details in which the figure differs from its mathematical prototype are more important and more interesting than the features in which it agrees; and even that the peculiar aesthetic pleasure with which we regard a living thing is somehow bound up with the departure from mathematical regularity which it manifests as a peculiar attribute of life. This view seems to me to involve a misapprehension. There is no such essential difference between these phenomena of organic form and those which are manifested in portions of inanimate matter*. The mathematician knows better than we do the value of an approximate result†. The child's skipping-rope is but an approximation to Huygens's catenary curve—but in the catenary curve lies the whole gist of the matter. We may be dismayed too easily by contingencies which are nothing short of irrelevant compared to the main issue; there is a *principle of negligibility*. Someone has said that if Tycho Brahé's instruments had been ten times as exact there would have been no Kepler, no Newton, and no astronomy.

If no chain hangs in a perfect catenary and no raindrop is a perfect sphere, this is for the reason that forces and resistances other than the main one are inevitably at work. The same is true of organic form, but it is for the mathematician to unravel the conflicting forces which are at work together. And this process of investigation may lead us on step by step to new phenomena, as it has done in physics, where sometimes a knowledge of form leads us to the interpretation of forces, and at other times a knowledge of the forces at work guides us towards a better insight into form. After the fundamental advance had been made which taught us that the world

* M. Bergson repudiates, with peculiar confidence, the application of mathematics to biology; cf. *Creative Evolution*, p. 21, "Calculation touches, at most, certain phenomena of organic destruction. Organic creation, on the contrary, the evolutionary phenomena which properly constitute life, we cannot in any way subject to a mathematical treatment." Bergson thus follows Bichat: "C'est peu connaître les fonctions animales que de vouloir les soumettre au moindre calcul, parceque leur instabilité est extrême. Les phénomènes restent toujours les mêmes, et c'est ce qui nous importe; mais leurs variations, en plus ou en moins, sont sans nombre" (*La Vie et la Mort*, p. 257).

† When we make a 'first approximation' to the solution of a physical problem, we usually mean that we are solving one part while neglecting others. Geometry deals with *pure forms* (such as a straight line), defined by a single law; but these are few compared with the *mixed forms*, like the surface of a polyhedron, or a segment of a sphere, or any ordinary mechanical construction or any ordinary physical phenomenon. It is only in a purely mathematical treatment of physics that the "single law" can be dealt with alone, and the approximate solution dispensed with accordingly.

was round, Newton shewed that the forces at work upon it must lead to its being imperfectly spherical, and in the course of time its oblate spheroidal shape was actually verified. But now, in turn, it has been shewn that its form is still more complicated, and the next step is to seek for the forces that have deformed the oblate spheroid. As Newton somewhere says, "Nature delights in transformations."

The organic forms which we can define more or less precisely in mathematical terms, and afterwards proceed to explain and to account for in terms of force, are of many kinds, as we have seen; but nevertheless they are few in number compared with Nature's all but infinite variety. The reason for this is not far to seek. The living organism represents, or occupies, a field of force which is never simple, and which as a rule is of immense complexity. And just as in the very simplest of actual cases we meet with a departure from such symmetry as could only exist under conditions of *ideal* simplicity, so do we pass quickly to cases where the interference of numerous, though still perhaps very simple, causes leads to a resultant complexity far beyond our powers of analysis. Nor must we forget that the biologist is much more exacting in his requirements, as regards form, than the physicist; for the latter is usually content with either an ideal or a general description of form, while the student of living things must needs be specific. Material things, be they living or dead, shew us but a shadow of mathematical perfection*. The physicist or mathematician can give us perfectly satisfying expressions for the form of a wave, or even of a heap of sand; but we never ask him to define the form of any particular wave of the sea, nor the actual form of any mountain-peak or hill.

In this there lies a certain justification for a saying of Minot's, of the greater part of which, nevertheless, I am heartily inclined to disapprove. "We biologists," he says, "cannot deplore too frequently or too emphatically the great mathematical delusion by which men often of great if limited ability have been misled into becoming advocates of an erroneous conception of accuracy. The delusion is that no science is accurate until its results can be expressed mathematically. The error comes from the assumption that mathematics can express complex relations. Unfortunately mathematics have a very limited scope, and are based upon a few extremely rudimentary

* Cf. Haton de la Goupillière, *op. cit.*: "On a souvent l'occasion de saisir dans la nature *un reflet* des formes rigoureuses qu'étudie la géometrie."

experiences, which we make as very little children and of which no adult has any recollection. The fact that from this basis men of genius have evolved wonderful methods of dealing with numerical relations should not blind us to another fact, namely, that the observational basis of mathematics is, psychologically speaking, very minute compared with the observational basis of even a single minor branch of biology....While therefore here and there the mathematical methods may aid us, *we need a kind and degree of accuracy of which mathematics is absolutely incapable*....With human minds constituted as they actually are, we cannot anticipate that there will ever be a mathematical expression for any organ or even a single cell, although formulae will continue to be useful for dealing now and then with isolated details..." (*op. cit.* p. 19, 1911). It were easy to discuss and criticise these sweeping assertions, which perhaps had their origin and parentage in an *obiter dictum* of Huxley's, to the effect that "Mathematics is that study which knows nothing of observation, nothing of experiment, nothing of induction, nothing of causation" (*cit.* Cajori, *Hist. of Elem. Mathematics*, p. 283). But Gauss, "rex mathematicorum," called mathematics "a science of the eye"; and Sylvester assures us that "most, if not all, of the great ideas of modern mathematics have had their origin in observation" (*Brit. Ass. Address*, 1869, and *Laws of Verse*, p. 120, 1870.

Réaumur said the same thing two hundred years ago (*Mém.* I, p. 49, 1734). Maupertuis, he said, was both naturalist and mathematician; and all his mathematics "n'ont en rien affaibli son gout pour les insectes, personne peut-être n'a plus d'amour pour eux." He goes on to say: "L'esprit d'observation qu'on regarde comme le caractere d'esprit essentiel aux naturalistes, est également necessaire pour faire des progrès en quelque science que ce soit. C'est l'esprit d'observation qui fait appercevoir ce qui a échappé aux autres, qui fait saisir des rapports qui sont entre des choses qui semblent differentes, ou qui fait trouver les differences qui sont entre celles qui paroissent semblables. On ne résoud les problemes les plus épineux de Geometrie qu'après avoir sçû observer des rapports qui ne se découvrent qu'à un esprit penetrant, et extrêmement attentif. Ce sont des observations qui mettent en état de résoudre les problemes de physique comme ceux d'histoire naturelle, car l'histoire naturelle a ses problemes à résoudre, et elle n'en a même que trop qui ne sont pas résolus." It is in a deeper sense than this, however, that the modern physicist looks on mathematics as an "empirical" science, and no longer a matter of pure intuition, or "reine Anschauung." Cf. Max Born, on Some philosophical aspects of modern physics, *Proc. R.S.E.* LVII, pp. 1–18, 1936.

For one reason or another there are very many organic forms which we cannot describe, still less define, in mathematical terms: just as there are problems even in physical science beyond the mathematics of our age. We never even seek for a formula to define this fish or that, or this or that vertebrate skull. But we may already use mathematical language to describe, even to define in general terms, the shape of a snail-shell, the twist of a horn, the outline of a leaf, the texture of a bone, the fabric of a skeleton, the stream-lines of fish or bird, the fairy lace-work of an insect's wing. Even to do this we must learn from the mathematician to eliminate and to discard; to keep the type in mind and leave the single case, with all its accidents, alone; and to find in this sacrifice of what matters little and conservation of what matters much one of the peculiar excellences of the method of mathematics*.

In a very large part of morphology, our essential task lies in the comparison of related forms rather than in the precise definition of each; and the *deformation* of a complicated figure may be a phenomenon easy of comprehension, though the figure itself have to be left unanalysed and undefined. This process of comparison, of recognising in one form a definite permutation or *deformation* of another, apart altogether from a precise and adequate understanding of the original "type" or standard of comparison, lies within the immediate province of mathematics, and finds its solution in the elementary use of a certain method of the mathematician. This method is the Method of Coordinates, on which is based the Theory of Transformations†.

I imagine that when Descartes conceived the method of coordinates, as a generalisation from the proportional diagrams of the artist and the architect, and long before the immense possibilities of this analysis could be foreseen, he had in mind a very simple purpose; it was perhaps no more than to find a way of translating the *form* of a curve (as well as the position of a point) into *numbers* and into *words*. This is precisely what we do, by the method of coordinates, every time we study a statistical curve; and conversely, we translate numbers into form whenever we "plot a curve," to illustrate a table of mortality, a rate of growth, or the daily variation of temperature or barometric pressure. In precisely the same way

* Cf. W. H. Young, The mathematical method and its limitations, *Congresso dei Matematici*, Bologna, 1928.

† The mathematical Theory of Transformations is part of the Theory of Groups, of great importance in modern mathematics. A distinction is drawn between Substitution-groups and Transformation-groups, the former being discontinuous, the latter continuous—in such a way that within one and the same group each transformation is infinitely little different from another. The distinction among biologists between a mutation and a variation is curiously analogous.

it is possible to inscribe in a net of rectangular coordinates the outline, for instance, of a fish, and so to translate it into a table of numbers, from which again we may at pleasure reconstruct the curve.

But it is the next step in the employment of coordinates which is of special interest and use to the morphologist; and this step consists in the alteration, or deformation, of our system of coordinates, and in the study of the corresponding transformation of the curve or figure inscribed in the coordinate network.

Let us inscribe in a system of Cartesian coordinates the outline of an organism, however complicated, or a part thereof: such as a fish, a crab, or a mammalian skull. We may now treat this complicated figure, in general terms, as a function of x, y. If we submit our rectangular system to deformation on simple and recognised lines, altering, for instance, the direction of the axes, the ratio of x/y, or substituting for x and y some more complicated expressions, then we obtain a new system of coordinates, whose deformation from the original type the inscribed figure will precisely follow. In other words, we obtain a new figure which represents the old figure under a more or less homogeneous *strain*, and is a function of the new coordinates in precisely the same way as the old figure was of the original coordinates x and y.

The problem is closely akin to that of the cartographer who transfers identical data to one projection or another*; and whose object is to secure (if it be possible) a complete correspondence, *in each small unit of area*, between the one representation and the other. The morphologist will not seek to draw his organic forms in a new and artificial projection; but, in the converse aspect of the problem, he will enquire whether two different but more or less obviously related forms can be so analysed and interpreted that each may be shewn to be a transformed representation of the other. This once demonstrated, it will be a comparatively easy task (in all probability) to postulate the direction and magnitude of the force capable of effecting the required transformation. Again, if such a simple alteration of the system of forces can be proved adequate to meet the case, we may find ourselves able to dispense with many

* Cf. (e.g.) Tissot, *Mémoire sur la représentation des surfaces, et les projections des cartes géographiques*, Paris, 1881.

widely current and more complicated hypotheses of biological causation. For it is a maxim in physics that an effect ought not to be ascribed to the joint operation of many causes if few are adequate to the production of it: *Frustra fit per plura, quod fieri potest per pauciora.*

We might suppose that by the combined action of appropriate forces any material form could be transformed into any other: just as out of a "shapeless" mass of clay the potter or the sculptor models his artistic product; or just as we attribute to Nature herself the power to effect the gradual and successive transformation of the simple germ into the complex organism. But we need not let these considerations deter us from our method of comparison of *related* forms. We shall strictly limit ourselves to cases where the transformation necessary to effect a comparison shall be of a simple kind, and where the transformed, as well as the original, coordinates shall constitute an harmonious and more or less symmetrical system. We should fall into deserved and inevitable confusion if, whether by the mathematical or any other method, we attempted to compare organisms separated far apart in Nature and in zoological classification. We are limited, both by our method and by the whole nature of the case, to the comparison of organisms such as are manifestly related to one another and belong to the same zoological class. For it is a grave sophism, in natural history as in logic, to make a transition into another kind*.

Our enquiry lies, in short, just within the limits which Aristotle himself laid down when, in defining a "genus," he shewed that (apart from those superficial characters, such as colour, which he called "accidents") the essential differences between one "species" and another are merely differences of proportion, of relative magnitude, or (as he phrased it) of "excess and defect." "Save only for a difference in the way of excess or defect, the parts are identical in the case of such animals as are of one and the same genus; and by 'genus' I mean, for instance, Bird or Fish." And again: "Within the limits of the same genus, as a general rule, most of the parts exhibit differences...in the way of multitude or fewness, magnitude

* The saying *heterogenea comparari non possunt* is discussed by Coleridge in his *Aids to Reflexion.*

or parvitude, in short, in the way of excess or defect. For 'the more' and 'the less' may be represented as 'excess' and 'defect'*." It is precisely this difference of relative magnitudes, this Aristotelian "excess and defect" in the case of form, which our coordinate method is especially adapted to analyse, and to reveal and demonstrate as the main cause of what (again in the Aristotelian sense) we term "specific" differences.

The applicability of our method to particular cases will depend upon, or be further limited by, certain practical considerations or qualifications. Of these the chief, and indeed the essential, condition is, that the form of the entire structure under investigation should be found to vary in a more or less uniform manner, after the fashion of an approximately homogeneous and isotropic body. But an imperfect isotropy, provided always that some "principle of continuity" run through its variations, will not seriously interfere with our method; it will only cause our transformed coordinates to be somewhat less regular and harmonious than are those, for instance, by which the physicist depicts the motions of a perfect fluid, or a theoretic field of force in a uniform medium.

Again, it is essential that our structure vary in its entirety, or at least that "independent variants" should be relatively few. That independent variations occur, that localised centres of diminished or exaggerated growth will now and then be found, is not only probable but manifest; and they may even be so pronounced as to appear to constitute new formations altogether. Such independent variants as these Aristotle himself clearly recognised: "It happens further that some have parts which others have not; for instance, some [birds] have spurs and others not, some have crests, or combs, and others not; but, as a general rule, most parts and those that go to make up the bulk of the body are either identical with one another, or differ from one another in the way of contrast and of excess and defect. For 'the more' and 'the less' may be represented as 'excess' or 'defect'†."

If, in the evolution of a fish, for instance, it be the case that its

* *Historia Animalium* I, 1.

† Aristotle's argument is even more subtle and far-reaching; for the differences of which he speaks are not merely those between one bird and another, but between them all and the very type itself, or Platonic "idea" of a bird.

several and constituent parts—head, body and tail, or this fin and that fin—represent so many independent variants, then our coordinate system will at once become too complex to be intelligible; we shall be making not one comparison but several separate comparisons, and our general method will be found inapplicable. Now precisely this independent variability of parts and organs—here, there, and everywhere within the organism—would appear to be implicit in our ordinary accepted notions regarding variation; and, unless I am greatly mistaken, it is precisely on such a conception of the easy, frequent, and normally independent variability of parts that our conception of the process of natural selection is fundamentally based. For the morphologist, when comparing one organism with another, describes the differences between them point by point, and "character" by "character*." If he is from time to time constrained to admit the existence of "correlation" between characters (as a hundred years ago Cuvier first shewed the way), yet all the while he recognises this fact of correlation somewhat vaguely, as a phenomenon due to causes which, except in rare instances, he can hardly hope to trace; and he falls readily into the habit of thinking and talking of evolution as though it had proceeded on the lines of his own descriptions, point by point, and character by character†.

With the "characters" of Mendelian genetics there is no fault to be found; tall and short, rough and smooth, plain or coloured are opposite tendencies or contrasting qualities, in plain logical contradistinction. But when the morphologist compares one animal with another, point by point or character by character, these are too often the mere outcome of artificial dissection and analysis. Rather is the living body one integral and indivisible whole, in

* Cf. *supra*, p. 1020.

† Cf. H. F. Osborn, On the origin of single characters, as observed in fossil and living animals and plants, *Amer. Nat.* XLIX, pp. 193–239, 1915 (and other papers); *ibid.* p. 194, "Each individual is composed of a vast number of somewhat similar new or old characters, each character has its independent and separate history, each character is in a certain stage of evolution, each character is correlated with the other characters of the individual....The real problem has always been that of the origin and development of characters. Since the *Origin of Species* appeared, the terms variation and variability have always referred to single characters; if a species is said to be variable, we mean that a considerable number of the single characters or groups of characters of which it is composed are variable," etc.

which we cannot find, when we come to look for it, any strict dividing line even between the head and the body, the muscle and the tendon, the sinew and the bone. Characters which we have differentiated insist on integrating themselves again; and aspects of the organism are seen to be conjoined which only our mental analysis had put asunder. The coordinate diagram throws into relief the integral solidarity of the organism, and enables us to see how simple a certain kind of *correlation* is which had been apt to seem a subtle and a complex thing.

But if, on the other hand, diverse and dissimilar fishes can be referred as a whole to identical functions of very different coordinate systems, this fact will of itself constitute a proof that variation has proceeded on definite and orderly lines, that a comprehensive "law of growth" has pervaded the whole structure in its integrity, and that some more or less simple and recognisable system of forces has been in control. It will not only shew how real and deep-seated is the phenomenon of "correlation," in regard to form, but it will also demonstrate the fact that a correlation which had seemed too complex for analysis or comprehension is, in many cases, capable of very simple graphic expression. This, after many trials, I believe to be in general the case, bearing always in mind that the occurrence of independent or localised variations must sometimes be considered.

We are dealing in this chapter with the forms of related organisms, in order to shew that the differences between them are as a general rule simple and symmetrical, and just such as might have been brought about by a slight and simple change in the system of forces to which the living and growing organism was exposed. Mathematically speaking, the phenomenon is identical with one met with by the geologist, when he finds a bed of fossils squeezed flat or otherwise symmetrically deformed by the pressures to which they, and the strata which contain them, have been subjected. In the first step towards fossilisation, when the body of a fish or shellfish is silted over and buried, we may take it that the wet sand or mud exercises, approximately, a hydrostatic pressure—that is to say a pressure which is uniform in all directions, and by which the form of the buried object will not be appreciably changed. As the strata consolidate and accumulate, the fossil organisms which they contain will tend to be flattened by the vast superincumbent load, just as the stratum which contains them will also be compressed and will have its molecular arrangement more or less modified*. But the deformation due to direct vertical pressure in a horizontal stratum is not nearly so striking as are the deformations produced by the oblique or shearing stresses to which inclined

* Cf. Sorby, *Quart. Journ. Geol. Soc.* (*Proc.*), 1879, p. 88.

and folded strata have been exposed, and by which their various "dislocations" have been brought about. And especially in mountain regions, where these dislocations are especially numerous and complicated, the contained fossils are apt to be so curiously and yet so symmetrically deformed (usually by a simple shear) that they may easily be interpreted as so many distinct and separate "species*." A great number of described species, and here and there a new genus (as the genus *Ellipsolithes* for an obliquely deformed *Goniatite* or *Nautilus*), are said to rest on no other foundation†.

If we begin by drawing a net of rectangular equidistant coordinates (about the axes x and y), we may alter or *deform* this network in various ways, several of which are very simple indeed. Thus (1) we may alter the dimensions of our system, extending it along one or

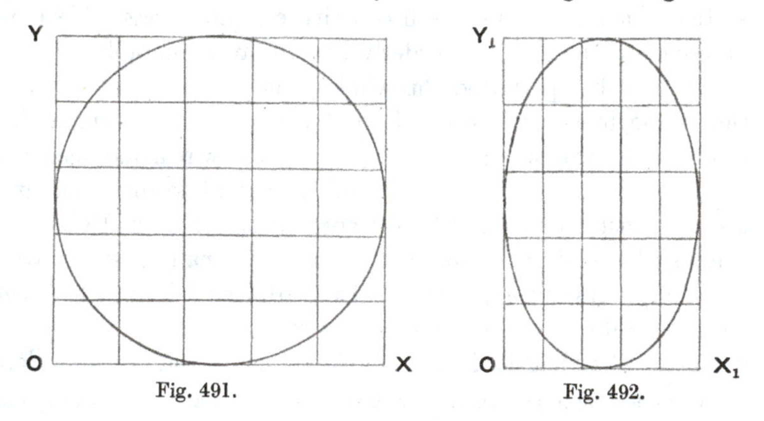

Fig. 491. Fig. 492.

other axis, and so converting each little square into a corresponding and proportionate oblong (Figs. 491, 492). It follows that any figure which we may have inscribed in the original net, and which we transfer to the new, will thereby be *deformed* in strict proportion to the deformation of the entire configuration, being still defined by corresponding points in the network and being throughout in conformity with the original figure. For instance, a circle inscribed

* Cf. Alc. D'Orbigny, *Cours élém. de Paléontologie*, etc., I, pp. 144–148, 1849; see also Daniel Sharpe, On slaty cleavage, *Q.J.G.S.* III, p. 74, 1847.

† Thus *Ammonites erugatus*, when compressed, has been described as *A. planorbis*: cf. J. F. Blake, *Phil. Mag.* (5), VI, p. 260, 1878. Wettstein has shewn that several species of the fish-genus *Lepidopus* have been based on specimens artificially deformed in various ways: Ueber die Fischfauna des Tertiären Glarnerschiefers, *Abh. Schw. Palaeont. Gesellsch.* XIII, 1886 (see especially pp. 23–38, pl. I). The whole subject, interesting as it is, has been little studied; both Blake and Wettstein deal with it mathematically.

in the original "Cartesian" net will now, after extension in the y-direction, be found elongated into an ellipse. In elementary mathematical language, for the original x and y we have substituted x_1 and cy_1, and the equation to our original circle, $x^2 + y^2 = a^2$, becomes that of the ellipse, $x_1{}^2 + c^2 y_1{}^2 = a^2$.

If I draw the cannon-bone of an ox (Fig. 493, A), for instance, within a system of rectangular coordinates, and then transfer the same drawing, point for point, to a system in which for the x of the original diagram we substitute $x' = 2x/3$, we obtain a drawing (B) which is a very close approximation to the cannon-bone of the sheep. In other words, the main (and perhaps the only) difference

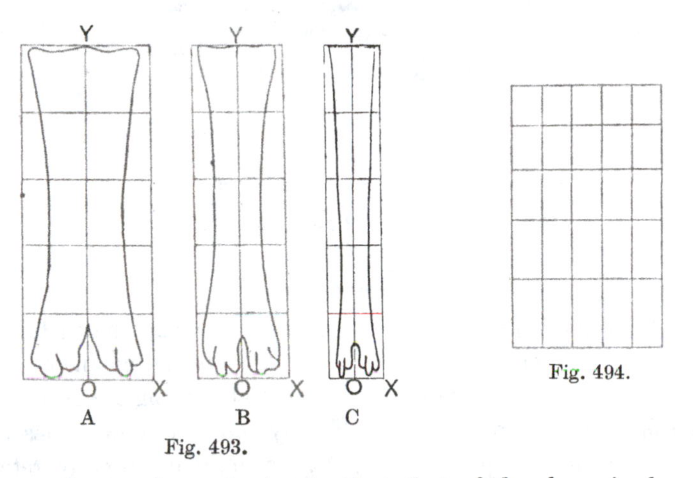

Fig. 493.

Fig. 494.

between the two bones is simply that that of the sheep is elongated along the vertical axis as compared with that of the ox, in the proportion of 3/2. And similarly, the long slender cannon-bone of the giraffe (C) is referable to the same identical type, subject to a reduction of breadth, or increase of length, corresponding to $x'' = x/3$.

(2) The second type is that where extension is not equal or uniform at all distances from the origin: but grows greater or less, as, for instance, when we stretch a *tapering* elastic band. In such cases, as I have represented it in Fig. 494, the ordinate increases logarithmically, and for y we substitute ϵ^y. It is obvious that this logarithmic extension may involve both abscissae and ordinates, x becoming ϵ^x while y becomes ϵ^y. The circle in our original figure is now deformed into some such shape as that of Fig. 495. This

method of deformation is a common one, and will often be of use to us in our comparison of organic forms.

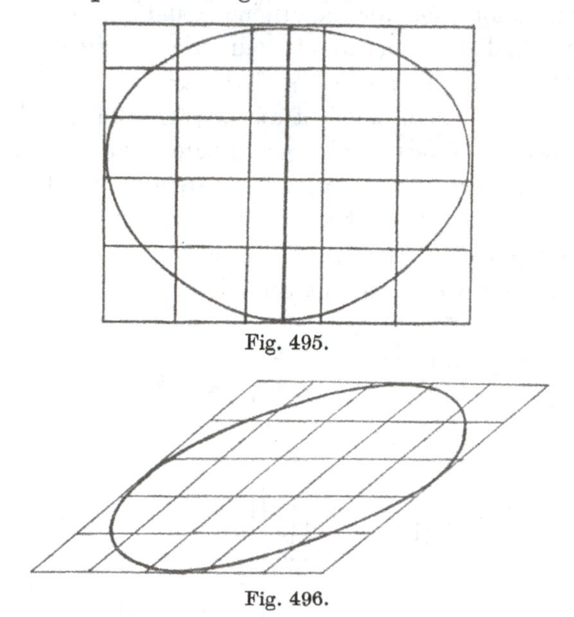

Fig. 495.

Fig. 496.

(3) Our third type is the "simple shear," where the rectangular coordinates become "oblique," their axes being inclined to one another at a certain angle ω. Our original rectangle now becomes such a figure as that of Fig. 496. The system may now be described in terms of the oblique axes X, Y; or may be directly referred to new rectangular coordinates ξ, η by the simple transposition $x = \xi - \eta \cot \omega$, $y = \eta \operatorname{cosec} \omega$.

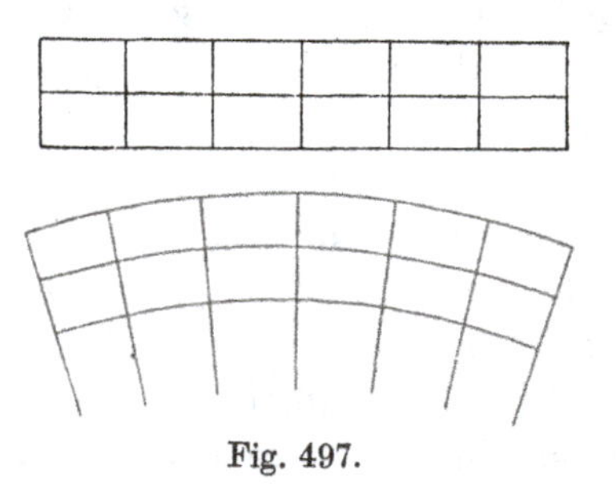

Fig. 497.

(4) Yet another important class of deformations may be represented by the use of radial coordinates, in which one set of lines are

represented as radiating from a point or "focus," while the other set are transformed into circular arcs cutting the radii orthogonally. These radial coordinates are especially applicable to cases where there exists (either within or without the figure) some part which is supposed to suffer no deformation; a simple illustration is afforded by the diagrams which illustrate the flexure of a beam (Fig. 497). In biology these coordinates will be especially applicable in cases where the growing structure includes a "node," or point where growth is absent or at a minimum; and about which node the rate of growth may be assumed to increase symmetrically. Precisely

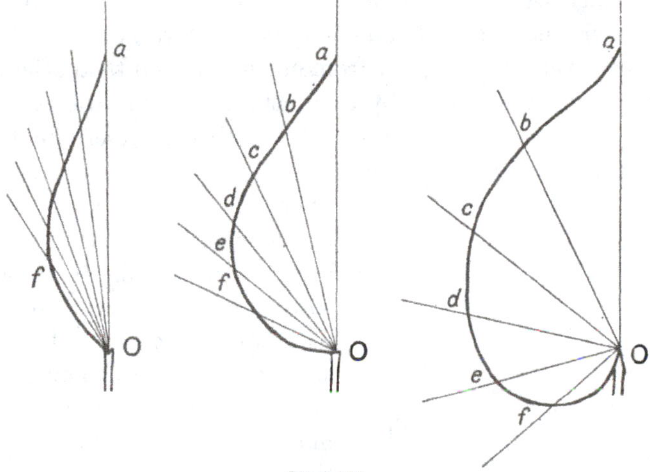

Fig. 498.

such a case is furnished us in a leaf of an ordinary dicotyledon. The leaf of a typical monocotyledon—such as a grass or a hyacinth, for instance—grows continuously from its base, and exhibits no node or "point of arrest." Its sides taper off gradually from its broad base to its slender tip, according to some law of decrement specific to the plant; and any alteration in the relative velocities of longitudinal and transverse growth will merely make the leaf a little broader or narrower, and will effect no other conspicuous alteration in its contour. But if there once come into existence a node, or "locus of no growth," about which we may assume growth—which in the hyacinth leaf was longitudinal and transverse—to take place radially and transversely to the radii, then we shall soon see the sloping sides of the hyacinth leaf give place to a more typical and "leaf-like" shape. If we alter the ratio between the radial and tangential velocities of growth—in other words, if we increase the angles between corresponding radii—we pass successively through the various configurations which the botanist describes as the lanceolate, the ovate, and the cordiform leaf. These successive changes may to some extent, and in appropriate cases, be traced as the individual leaf grows to maturity; but as a much more general rule, the balance of forces, the ratio between radial and tangential velocities of growth, remains so nicely and constantly balanced that the leaf increases in size without conspicuous modification of form. It is rather what we may call a long-period variation, a tendency for the relative velocities to alter from one generation to another, whose result is brought into view by this method of illustration.

There are various corollaries to this method of describing the form of a leaf which may be here alluded to. For instance, the so-called unsymmetrical leaf* of a begonia, in which one side of the leaf may be merely ovate while the other has a cordate outline, is seen to be really a case of *unequal*, and not truly asymmetrical, growth on either side of the midrib. There is nothing more mysterious in its conformation than, for instance, in that of a forked twig in which one limb of the fork has grown longer than the other. The case of the begonia leaf is of sufficient interest to deserve illustration, and in Fig. 499 I have outlined a leaf of the large *Begonia daedalea*. On the smaller left-hand side of the leaf I have taken at random three points a, b, c, and have measured the angles, AOa, etc.,

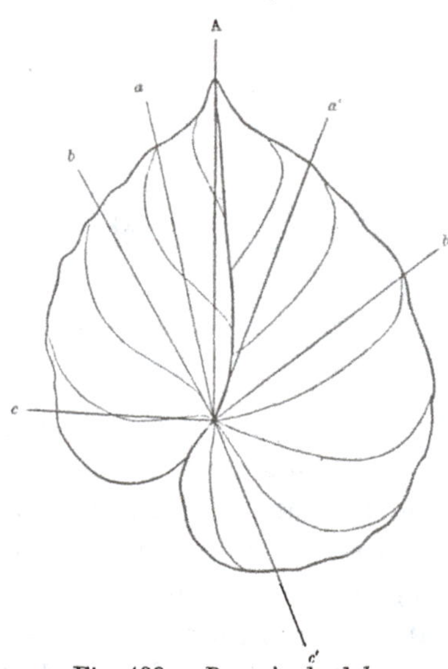

Fig. 499. *Begonia daedalea.*

* Cf. Sir Thomas Browne, in *The Garden of Cyrus*: "But why ofttimes one side of the leaf is unequall unto the other, as in Hazell and Oaks, why on either side the master vein the lesser and derivative channels stand not directly opposite, not at equall angles, respectively unto the adverse side, but those of one side do often exceed the other, as the Wallnut and many more, deserves another enquiry."

which the radii from the hilus of the leaf to these points make with the median axis. On the other side of the leaf I have marked the points a', b', c', such that the radii drawn to this margin of the leaf are equal to the former, Oa' to Oa, etc. Now if the two sides of the leaf are mathematically similar to one another, it is obvious that the respective angles should be in continued proportion, i.e. as AOa is to AOa', so should AOb be to AOb'. This proves to be very nearly the case. For I have measured the three angles on one side, and one on the other, and have then compared, as follows, the calculated with the observed values of the other two:

	AOa	AOb	AOc	AOa'	AOb'	AOc'
Observed values	12°	28·5°	88°	—	—	157°
Calculated ,,	—	—	—	21·5°	51·1°	—
Observed ,,	—	—	—	20	52	—

The agreement is very close, and what discrepancy there is may be amply accounted for, firstly, by the slight irregularity of the sinuous margin of the leaf; and secondly, by the fact that the true axis or midrib of the leaf is not straight but slightly curved, and therefore that it is curvilinear and not rectilinear triangles which we ought to have measured. When we understand these few points regarding the peripheral curvature of the leaf, it is easy to see that its principal veins approximate closely to a beautiful system of isogonal coordinates. It is also obvious that we can easily pass, by a process of shearing, from those cases where the principal veins start from the base of the leaf to those where they arise successively from the midrib, as they do in most dicotyledons.

It may sometimes happen that the node*, or "point of arrest," is at the upper instead of the lower end of the leaf-blade; and occasionally there is a node at both ends. In the former case, as we have it in the daisy, the form of the leaf will be, as it were, inverted, the broad, more or less heart-shaped, outline appearing at the upper end, while below the leaf tapers gradually downwards to an ill-defined base. In the latter case, as in *Dionaea*, we obtain a leaf equally expanded, and similarly ovate or cordate, at both ends. We may notice, lastly, that the shape of a solid fruit, such as an apple or a cherry, is a solid of revolution, developed from similar curves and to be explained on the same principle. In the

* "Node," in the botanical, not the mathematical, sense.

cherry we have a "point of arrest" at the base of the berry, where it joins its peduncle, and about this point the fruit (in imaginary section) swells out into a cordate outline; while in the apple we have two such well-marked points of arrest, above and below, and about both of them the same conformation tends to arise. The bean and the human kidney owe their "reniform" shape to precisely the same phenomenon, namely, to the existence of a node or "hilus," about which the forces of growth are radially and symmetrically arranged. When the seed is small and the pod roomy, the seed may grow round, or nearly so, like a pea; but it is flattened and bean-shaped, or elliptical like a kidney-bean, when compressed within a narrow and elongated pod. If the original seed have any simple pattern, of the nature for instance of meridians or parallels of latitude, it is easy to see how these will suffer a conformal transformation, corresponding to the deformation of the sphere*.

We might go farther, and farther than we have room for here, to illustrate the shapes of leaves by means of radial coordinates, and even to attempt to define them by polar equations. In a former chapter we learned to look upon the curve of sines as an easy, gradual and natural transition—perhaps the simplest and most natural of all—from minimum to corresponding maximum, and so on alternately and continuously; and we found the same curve going round like the hands of a clock, when plotted on radial coordinates and (so to speak) prevented from leaving its place. Either way it represents a "simple harmonic motion." Now we have just seen an ordinary dicotyledonous leaf to have a "point of arrest," or zero-growth in a certain direction, while in the opposite direction towards the tip it has grown with a maximum velocity. This progress from zero to maximum suggests one-half of the sine-curve; in other words, if we look on the outline of the leaf as a vector-diagram of its own growth, at rates varying from zero to zero in a complete circuit of 360°, this suggests, as a possible and very simple case, the plotting of $r = \sin \theta/2$. Doing so, we obtain a curve (Fig. 500) closely resembling what the botanists call a *reniform* (or kidney-shaped) leaf, that is to say, with a cordate outline at the base formed of two "auricles," one on either side, and then rounded

* *Vide supra*, p. 524.

off with no projecting apex*. The ground-ivy and the dog-violet (Fig. 501) illustrate such a leaf; and sometimes, as in the violet, the veins of the leaf show similar curves congruent with the outer edge. Moreover the violet is a good example of how the reniform leaf may be drawn out more and more into an acute and ovate form.

From $\sin \theta/2$ we may proceed to any other given fraction of θ, and plot, for instance, $r = \sin 5\theta/3$, as in Fig. 502; which now no longer represents a single leaf but has become a diagram of the

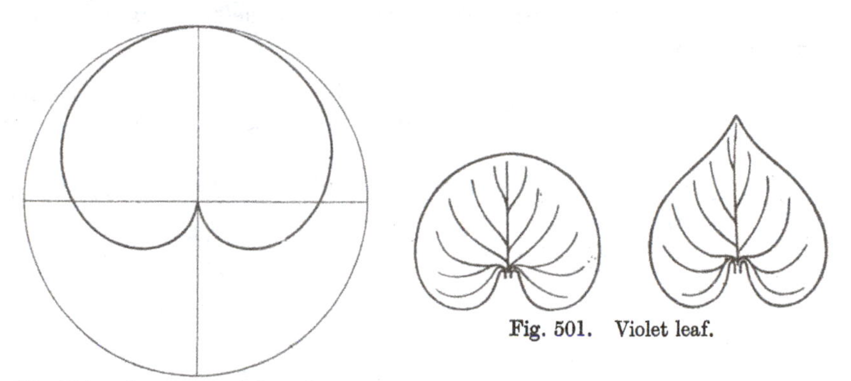

Fig. 501. Violet leaf.

Fig. 500. Curve resembling the outline of a reniform leaf: $r = \sin \theta/2$.

five petals of a pentamerous flower. Abbot Guido Grandi, a Pisan mathematician of the early eighteenth century, drew such a curve and pointed out its botanical analogies; and we still call the curves of this family "Grandi's curves†."

The gamopetalous corolla is easily transferred to polar coordinates, in which the radius vector now consists of two parts, the one a constant, the other expressing the amplitude (or half-amplitude) of the sine-curve; we may write the formula $r = a + b \cos n\theta$. In Fig. 503 $n = 5$; in this figure, if the radius of the outermost circle be taken as unity, the outer of the two sinuous curves has $a:b$ as

* Fig. 500 illustrates the whole leaf, but only shows one-half of the sine-curve. The rest is got by reflecting the moiety already drawn in the horizontal axis ($\theta = \pi/2$).

† Dom. Guido Grandus, *Flores geometrici ex rhodonearum et cloeliarum curvarum descriptione resultantes...*, Florentiae, 1728. Cf. Alfred Lartigue, *Biodynamique générale*, Paris, 1930—a curious but eccentric book.

9:1, and the inner curve as 3:1; while the five petals become separate when $a = b$, and the formula reduces to $r = \cos^2 \dfrac{5\theta}{2}$.

In Fig. 504 we have what looks like a first approximation to a horse-chestnut leaf. It consists of so many separate leaflets, akin to the five petals in Fig. 503; but these are now inscribed in (or have a *locus* in) the cordate or reniform outline of Fig. 500. The new curve is, in short, a composite one; and its general formula is $r = \sin \theta/2 . \sin n\theta$. The small size of the two leaflets adjacent to the petiole is characteristic of the curve, and helps to explain the development of "stipules."

Fig. 502. Grandi's curves based on $r = \sin \frac{5}{3}\theta$, and illustrating the five petals of a simple flower.

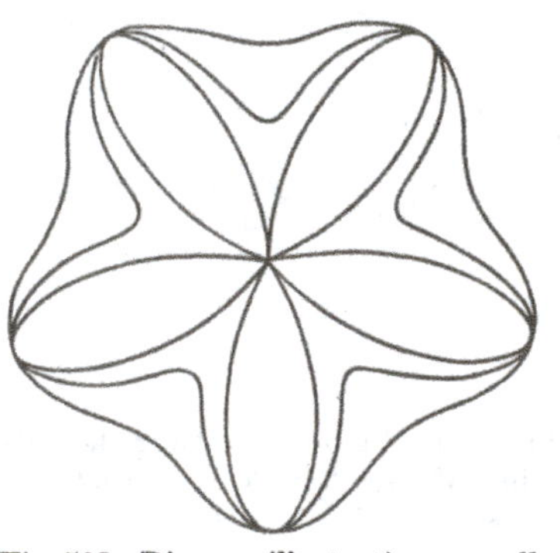

Fig. 503. Diagram illustrating a corolla of five petals, or of five lobes, are based on the equation $r = a + b \cos \theta$.

In this last case we have combined one curve with another, and the doing so opens out a new range of possibilities. On the outline of the simple leaf, whether ovate, lanceolate or cordate, we may superpose secondary sine-curves of lesser period and varying amplitude, after the fashion of a Fourier series; and the results will vary from a mere crenate outline to the digitate lobes of an ivy-leaf, or to separate leaflets such as we have just studied in the horse-chestnut. Or again, we may inscribe the separate petals of Fig. 505 within a spiral curve, equable or equiangular as the case may be; and then, continuing the series on and on, we shall obtain a figure resembling the clustered leaves of a stonecrop, or the petals of a water-lily or other polypetalous flower.

Most of the transformations which we have hitherto considered (other than that of the simple shear) are particular cases of a general transformation, obtainable by the method of conjugate functions and equivalent to the projection of the original figure on a new plane. Appropriate transformations, on these general lines, provide for the cases of a coaxial system where the Cartesian coordinates are replaced by coaxial circles, or a confocal system in which they are replaced by confocal ellipses and hyperbolas.

Fig. 504. Outline of a compound leaf, like a horse-chestnut, based on a composite sine-curve, of the form $r = \sin \theta/2 \,.\, \sin n\theta$.

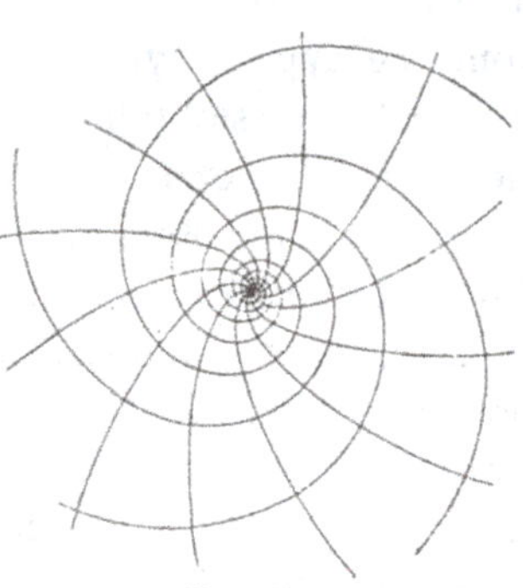

Fig. 505.

Yet another curious and important transformation, belonging to the same class, is that by which a system of straight lines becomes transformed into a conformal system of logarithmic spirals: the straight line $Y - AX = c$ corresponding to the logarithmic spiral $\theta - A \log r = c$ (Fig. 505). This beautiful and simple transformation lets us at once convert, for instance, the straight conical shell of the Pteropod or the *Orthoceras* into the logarithmic spiral of the Nautiloid; it involves a mathematical symbolism which is but a slight extension of that which we have employed in our elementary treatment of the logarithmic spiral.

These various systems of coordinates, which we have now briefly considered, are sometimes called "isothermal coordinates," from the fact that, when employed in this particular branch of physics, they perfectly represent the phenomena of the conduction of heat, the contour lines of equal temperature appearing, under appropriate conditions, as the orthogonal lines of the coordinate system. And it follows that the "law of growth" which our biological analysis by means of orthogonal coordinate systems presupposes, or at least foreshadows, is one according to which the organism grows or develops along *stream-lines*, which may be defined by a suitable mathematical transformation.

When the system becomes no longer orthogonal, as in many of the following illustrations—for instance, that of *Orthagoriscus* (Fig. 526)—then the transformation is no longer within the reach of comparatively simple mathematical analysis. Such departure from the typical symmetry of a "stream-line" system is, in the first instance, sufficiently accounted for by the simple fact that the developing organism is very far from being homogeneous and isotropic, or, in other words, does not behave like a perfect fluid. But though under such circumstances our coordinate systems may be no longer capable of strict mathematical analysis, they will still indicate *graphically* the relation of the new coordinate system to the old, and conversely will furnish us with some guidance as to the "law of growth," or play of forces, by which the transformation has been effected.

Before we pass from this brief discussion of transformations in general, let us glance at one or two cases in which the forces applied are more or less intelligible, but the resulting transformations are, from the mathematical point of view, exceedingly complicated.

The "marbled papers" of the bookbinder are a beautiful illustration of visible "stream-lines." On a dishful of a sort of semi-liquid gum the workman dusts a few simple lines or patches of colouring matter; and then, by passing a comb through the liquid, he draws the colour-bands into the streaks, waves, and spirals which constitute the marbled pattern, and which he then transfers to sheets of paper laid down upon the gum. By some such system of shears, by the effect of unequal traction or unequal growth in various directions and superposed on an originally simple pattern, we may account for the not dissimilar marbled patterns which we recognise, for instance, on a large serpent's skin. But it must be remarked, in the case of the marbled paper, that though the method of application of the forces is simple, yet in the aggregate the system of forces set up by the many teeth of the comb is exceedingly complex, and its complexity is revealed in the complicated "diagram of forces" which constitutes the pattern.

To take another and still more instructive illustration. To turn one circle (or sphere) into two circles (or spheres) would be, from the point of view of the mathematician, an extraordinarily difficult transformation; but, physically speaking, its achievement may be extremely simple. The little round gourd grows naturally, by its symmetrical forces of expansive growth, into a big, round, or somewhat oval pumpkin or melon*. But the Moorish husbandman ties a rag round its middle, and the same forces of growth, unaltered save for the presence of this trammel, now expand the globular structure into two superposed and connected globes. And again, by varying the position of the encircling band, or by applying several such ligatures instead of one, a great variety of artificial forms of "gourd" may be, and actually are, produced. It is clear, I think, that we may account for many ordinary biological processes of development or transformation of form by the existence of trammels or lines of constraint, which limit and determine the action of the expansive forces of growth that would otherwise be uniform and symmetrical. This case has a close parallel in the operations of the glass-blower, to which we have already, more than once, referred in passing†. The glass-blower starts his operations with a *tube*, which he first closes at one end so as to form a hollow vesicle, within which his blast of air exercises a uniform pressure on all sides; but the spherical conformation which this uniform expansive force would naturally tend to produce is modified into all kinds of forms by the trammels or resistances set up as the workman lets one part or another of his bubble be unequally heated or cooled. It was Oliver Wendell

* Analogous structural differences, especially in the fibrovascular bundles, help to explain the differences between (e.g.) a smooth melon and a cantelupe, or between various elongate, flattened and globular varieties. These breed true to type, and obey, when crossed. the laws of Mendelian inheritance. Cf. E. W. Sinnett, Inheritance of fruit-shape in Cucurbita, *Botan. Gazette*, LXXIV, pp. 95–103, 1922, and other papers.

† Where gourds are common, the glass-blower is still apt to take them for a prototype, as the prehistoric potter also did. For instance, a tall, annulated Florence oil-flask is an exact but no longer a conscious imitation of a gourd which has been converted into a bottle in the manner described.

Holmes who first shewed this curious parallel between the operations of the glass-blower and those of Nature, when she starts, as she so often does, with a simple tube*. The alimentary canal, the arterial system including the heart, the central nervous system of the vertebrate, including the brain itself, all begin as simple tubular structures. And with them Nature does just what the glass-blower does, and, we might even say, no more than he. For she can expand the tube here and narrow it there; thicken its walls or thin them; blow off a lateral offshoot or caecal diverticulum; bend the tube, or twist and coil it; and infold or crimp its walls as, so to speak, she pleases. Such a form as that of the human stomach is easily explained when it is regarded from this point of view; it is simply an ill-blown bubble, a bubble that has been rendered lopsided by a trammel or restraint along one side, such as to prevent its symmetrical expansion—such a trammel as is produced if the glass-blower lets one side of his bubble get cold, and such as is actually present in the stomach itself in the form of a muscular band.

The Florence flask, or any other handiwork of the glass-blower, is always beautiful, because its graded contours are, as in its living analogues, a picture of the graded forces by which it was conformed. It is an example of mathematical beauty, of which the machine-made, moulded bottle has no trace at all. An alabaster bottle is different again. It is no longer an unduloid figure of equilibrium. Turned on a lathe, it is a solid of revolution, and not without beauty; but it is not near so beautiful as the blown flask or bubble.

The gravitational field is part of the complex field of force by which the form of the organism is influenced and determined. Its share is seldom easy to define, but there is a resultant due to gravity in hanging breasts and tired eyelids and all the sagging wrinkles of the old. Now and then we see gravity at work in the normal construction of the body, and can describe its effect on form in a general, or qualitative, way. Each pair of ribs in man forms a hoop which droops of its own weight in front, so flattening the chest, and at the same time twisting the rib on either hand near its point of suspension†. But in the dog each costal hoop is dragged

* Cf. *Elsie Venner*, chap. II.

† See T. P. Anderson Stuart, How the form of the thorax is partly determined by gravitation, *Proc. R.S.* XLIX, p.143, 1891.

straight downwards, into a vertical instead of a transverse ellipse, and is even narrowed to a point at the sternal border.

We may now proceed to consider and illustrate a few permutations or transformations of organic form, out of the vast multitude which are equally open to this method of enquiry.

We have already compared in a preliminary fashion the metacarpal or cannon-bone of the ox, the sheep, and the giraffe (Fig. 493); and we have seen that the essential difference in form between these three bones is a matter of relative length and breadth, such that, if we reduce the figures to an identical standard of length (or identical values of y), the breadth (or value of x) will be approximately two-thirds that of the ox in the case of the sheep and one-third that of the ox in the case of the giraffe. We may easily, for the sake of closer comparison, determine these ratios more accurately, for instance, if it be our purpose to compare the different racial varieties within the limits of a single species. And in such cases, by the way, as when we compare with one another various breeds or races of cattle or of horses, the ratios of length and breadth in this particular bone are extremely significant*.

If, instead of limiting ourselves to the cannon-bone, we inscribe the entire foot of our several Ungulates in a coordinate system, the same ratios of x that served us for the cannon-bones still give us a first approximation to the required comparison; but even in the case of such closely allied forms as the ox and the sheep there is evidently something wanting in the comparison. The reason is that the relative elongation of the several parts, or individual bones, has not proceeded equally or proportionately in all cases; in other words, that the equations for x will not suffice without some simultaneous modification of the values of y (Fig. 506). In such a case it may be found possible to satisfy the varying values of y by some logarithmic

* This significance is particularly remarkable in connection with the development of speed, for the metacarpal region is the seat of very important leverage in the propulsion of the body. In a certain Scottish Museum there stand side by side the skeleton of an immense carthorse (celebrated for having drawn all the stones of the Bell Rock Lighthouse to the shore), and a beautiful skeleton of a racehorse, long supposed to be the actual skeleton of Eclipse. When I was a boy my grandfather used to point out to me that the cannon-bone of the little racer is not only relatively, but actually, longer than that of the great Clydesdale.

or other formula; but, even if that be possible, it will probably be somewhat difficult of discovery or verification in such a case as the present, owing to the fact that we have too few well-marked points of correspondence between the one object and the other, and that especially along the shaft of such long bones as the cannon-bone of the ox, the deer, the llama, or the giraffe there is a complete lack of easily recognisable corresponding points. In such a case a brief tabular statement of apparently corresponding values of y, or of those obviously corresponding values which coincide with the boundaries of the several bones of the foot, will, as in the following example, enable us to dispense with a fresh equation.

		a	*b*	*c*	*d*
y (Ox) ...	0	18	27	42	100
y' (Sheep) ...	0	10	19	36	100
y'' (Giraffe) ...	0	5	10	24	100

This summary of values of y', coupled with the equations for the value of x, will enable us, from any drawing of the ox's foot, to construct a figure of that of the sheep or of the giraffe with remarkable accuracy.

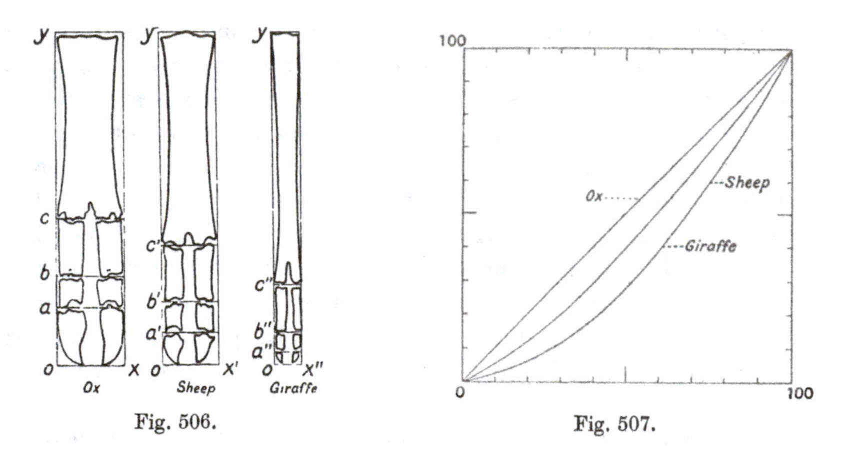

Fig. 506.

Fig. 507.

That underlying the varying amounts of extension to which the parts or segments of the limb have been subject there is a law, or principle of continuity, may be discerned from such a diagram as the above (Fig. 507), where the values of y in the case of the ox are plotted as a straight line, and the corresponding values for the sheep (extracted from the above table) are seen to form a more or less regular and even curve. This simple graphic result implies the existence of a comparatively simple equation between y and y'.

An elementary application of the principle of coordinates to the study of proportion, as we have here used it to illustrate the varying proportions of a bone, was in common use in the sixteenth and seventeenth centuries by artists in their study of the human form. The method is probably much more ancient, and may even be classical*; it is fully described and put in practice by Albert Dürer in his *Geometry*, and especially in his *Treatise on Proportion*†. In this latter work, the manner in which the human figure, features, and facial expression are all transformed and modified by slight variations in the relative magnitude of the parts is admirably and copiously illustrated (Fig. 508).

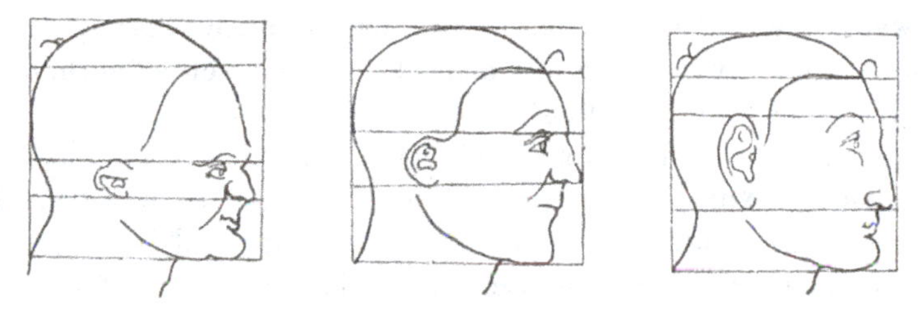

Fig. 508. (After Albert Dürer.)

In a tapir's foot there is a striking difference, and yet at the same time there is an obvious underlying resemblance, between the middle toe and either of its unsymmetrical lateral neighbours. Let us take the median terminal phalanx and inscribe its outline in a net of rectangular equidistant coordinates (Fig. 509, *a*). Let us then make a similar network about axes which are no longer at right angles, but inclined to one another at an angle of about 50° (*b*).

* Cf. Vitruvius, III, 1.

† *Les quatres livres d'Albert Dürer de la proportion des parties et pourtraicts des corps humains*, Arnheim, 1613, folio (and earlier editions). Cf. also Lavater, *Essays on Physiognomy*, III, p. 271, 1799; also H. Meige, La géométrie des visages d'après Albert Dürer, *La Nature*, Dec. 1927. On Dürer as mathematician, cf. Cantor, II, p. 459; S. Günther, *Die geometrische Näherungsconstructione Albrecht Dürers*, Ansbach, 1866; H. Staigmuller, *Dürer als Mathematiker*, Stüttgart, 1891.

If into this new network we fill in, point for point, an outline precisely corresponding to our original drawing of the middle toe, we shall find that we have already represented the main features of the adjacent lateral one. We shall, however, perceive that our new diagram looks a little too bulky on one side, the inner side, of the lateral toe. If now we substitute for our equidistant ordinates,

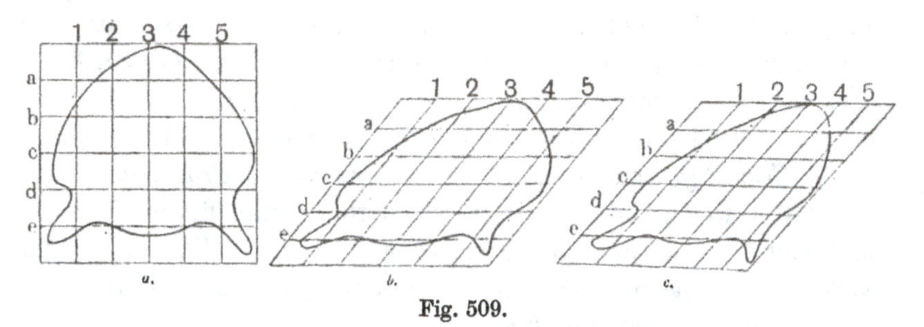

Fig. 509.

ordinates which get gradually closer and closer together as we pass towards the median side of the toe, then we shall obtain a diagram which differs in no essential respect from an actual outline copy of the lateral toe (*c*). In short, the difference between the outline of the middle toe of the tapir and the next lateral toe may be almost completely expressed by saying that if the one be represented by rectangular equidistant coordinates, the other will be represented by oblique coordinates, whose axes make an angle of 50°, and in

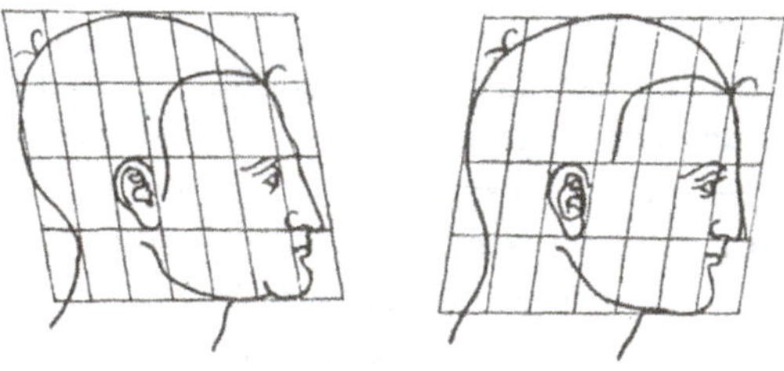

Fig. 510. (After Albert Dürer.)

which the abscissal interspaces decrease in a certain logarithmic ratio. We treated our original complex curve or projection of the tapir's toe as a function of the form $F(x, y) = 0$. The figure of the tapir's lateral toe is a precisely identical function of the form $F(e^x, y_1) = 0$, where x_1, y_1 are oblique coordinate axes inclined to one another at an angle of 50°.

Dürer was acquainted with these oblique coordinates also, and I have copied two illustrative figures from his book*.

In Fig. 511 I have sketched the common Copepod *Oithona nana*, and have inscribed it in a rectangular net, with abscissae three-fifths the length of the ordinates. Side by side (Fig. 512) is drawn a very different Copepod, of the genus *Sapphirina*; and about it is drawn a network such that each coordinate passes (as nearly as possible) through points corresponding to those of the former figure. It will be seen that two differences are apparent. (1) The values of y in Fig. 512 are large in the upper part of the figure, and diminish

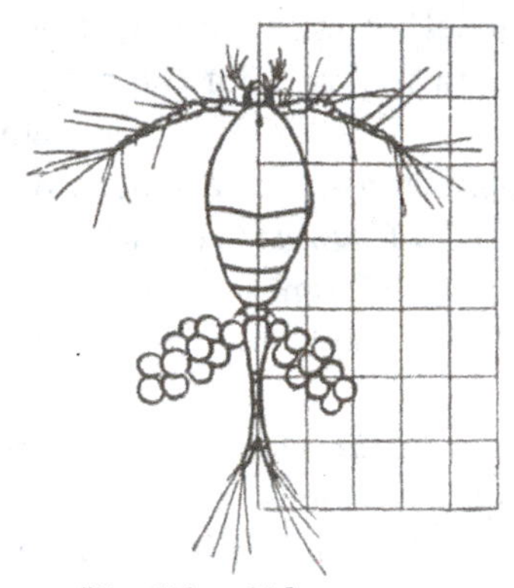

Fig. 511. *Oithona nana.*

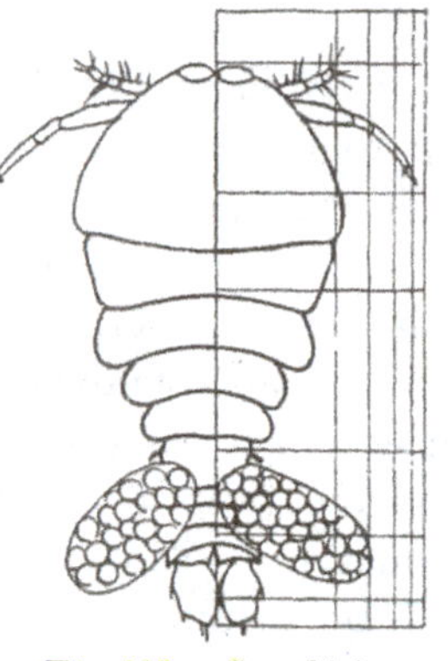

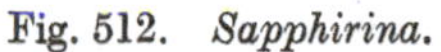

Fig. 512. *Sapphirina.*

rapidly towards its base. (2) The values of x are very large in the neighbourhood of the origin, but diminish rapidly as we pass towards either side, away from the median vertical axis; and it is probable that they do so according to a definite, but somewhat complicated,

* It was these very drawings of Dürer's that gave to Peter Camper his notion of the "facial angle." Camper's method of comparison was the very same as ours, save that he only drew the axes, without filling in the network, of his coordinate system; he saw clearly the essential fact, that the skull *varies as a whole*, and that the "facial angle" is the index to a general deformation. "The great object was to shew that natural differences might be reduced to rules, of which the direction of the facial line forms the *norma* or canon; and that these directions and inclinations are always accompanied by correspondent form, size and position of the other parts of the cranium," etc.; from Dr T. Cogan's preface to Camper's work *On the Connexion between the Science of Anatomy and the Arts of Drawing, Painting and Sculpture* (1768?), quoted in Dr R. Hamilton's Memoir of Camper, in *Lives of Eminent Naturalists (Nat. Libr.)*, Edinburgh, 1840. See also P. Camper, *Dissertation sur les différences réelles que presentent les Traits du Visage chez les hommes de différents pays et de différents âges*, Paris, 1791 (*op. posth.*); cf. P. Topinard, Études sur Pierre Camper, et sur l'angle facial dit de Camper, *Rev. d'Anthropol.* ii. 1874.

ratio. If, instead of seeking for an actual equation, we simply tabulate our values of x and y in the second figure as compared with the first (just as we did in comparing the feet of the Ungulates), we get the dimensions of a net in which, by simply projecting the figure of *Oithona*, we obtain that of *Sapphirina* without further trouble, e.g.:

x (*Oithona*)	0	3	6	9	12	15	—
x' (*Sapphirina*)	0	8	10	12	13	14	—
y (*Oithona*)	0	5	10	15	20	25	30
y' (*Sapphirina*)	0	2	7	3	23	32	40

In this manner, with a single model or type to copy from, we may record in very brief space the data requisite for the production of approximate outlines of a great number of forms. For instance, the difference, at first sight immense, between the attenuated body of a *Caprella* and the thick-set body of a *Cyamus* is obviously little, and is probably nothing more than a difference of relative magnitudes, capable of tabulation by numbers and of complete expression by means of rectilinear coordinates.

The Crustacea afford innumerable instances of more complex deformations. Thus we may compare various higher Crustacea with one another, even in the case of such dissimilar forms as a lobster and a crab. It is obvious that the whole body of the former is elongated as compared with the latter, and that the crab is relatively broad in the region of the carapace, while it tapers off rapidly towards its attenuated and abbreviated tail. In a general way, the elongated rectangular system of coordinates in which we may inscribe the outline of the lobster becomes a shortened triangle in the case of the crab. In a little more detail we may compare the outline of the carapace in various crabs one with another: and the comparison will be found easy and significant, even, in many cases, down to minute details, such as the number and situation of the marginal spines, though these are in other cases subject to independent variability.

If we choose, to begin with, such a crab as *Geryon* (Fig. 513, 1) and inscribe it in our equidistant rectangular coordinates, we shall see that we pass easily to forms more elongated in a transverse direction, such as *Matuta* or *Lupa* (5), and conversely, by transverse compression, to such a form as *Corystes* (2). In certain other cases the carapace conforms to a triangular diagram, more or less curvilinear, as in Fig. 513, 4, which represents the genus *Paralomis*. Here we can easily see that the posterior border is transversely elongated as compared with that of *Geryon*, while at the same time the anterior

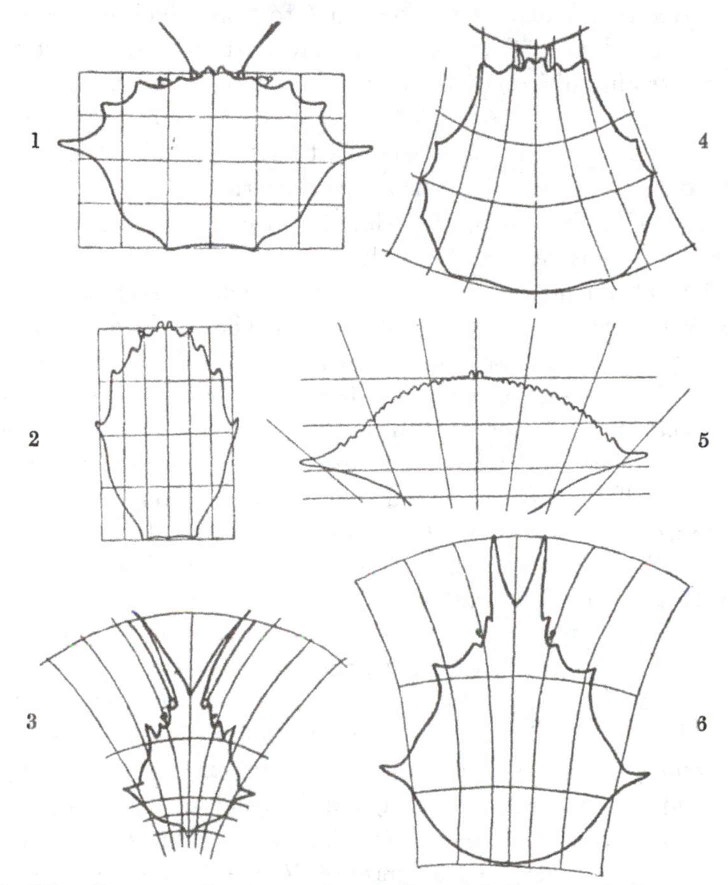

Fig. 513. Carapaces of various crabs. 1, *Geryon*; 2, *Corystes*; 3, *Scyramathia*; 4, *Paralomis*; 5, *Lupa*; 6, *Chorinus*.

part is longitudinally extended as compared with the posterior. A system of slightly curved and converging ordinates, with orthogonal and logarithmically interspaced abscissal lines, as shewn in the figure, appears to satisfy the conditions.

In an interesting series of cases, such as the genus *Chorinus*, or *Scyramathia*, and in the spider-crabs generally, we appear to have just the converse of this. While the carapace of these crabs presents a somewhat triangular form, which seems at first sight more or less similar to those just described, we soon see that the actual posterior border is now narrow instead of broad, the broadest part of the carapace corresponding precisely, not to that which is broadest in *Paralomis*, but to that which was broadest in *Geryon*; while the most striking difference from the latter lies in an antero-posterior lengthening of the forepart of the carapace, culminating in a great elongation of the frontal region, with its two spines or "horns." The curved ordinates here converge posteriorly and diverge widely in front (Fig. 513, 3 and 6), while the decremental interspacing of the abscissae is very marked indeed.

We put our method to a severer test when we attempt to sketch an entire and complicated animal than when we simply compare corresponding parts such as the carapaces of various Malacostraca, or related bones as in the case of the tapir's toes. Nevertheless, up to a certain point, the method stands the test very well. In other words, one particular mode and direction of variation is often (or even usually) so prominent and so paramount throughout the entire organism, that one comprehensive system of coordinates suffices to give a fair picture of the actual phenomenon. To take another illustration from the Crustacea, I have drawn roughly in Fig. 514, 1 a little amphipod of the family Phoxocephalidae (*Harpinia* sp.). Deforming the coordinates of the figure into the curved orthogonal* system in Fig. 514, 2, we at once obtain a very fair representation of an allied genus, belonging to a different family of amphipods, namely *Stegocephalus*. As we proceed further from our type our coordinates will require greater deformation, and the resultant figure will usually be somewhat less accurate. In Fig. 514, 3 I shew a network, to which, if we transfer our diagram of *Harpinia* or of *Stegocephalus*, we shall obtain a tolerable representation of the aberrant genus *Hyperia*†, with its narrow abdomen, its reduced pleural lappets, its great eyes, and its inflated head.

* Similar coordinates are treated of by Lamé, *Leçons sur les coordonnées curvilignes*, Paris, 1859.

† For an analogous, but more detailed comparison, see H. Mogk, Versuch einer Formanalyse bei Hyperiden, *Int. Rev. d. ges. Hydrobiol.*, etc., XIV, pp. 276–311, 1923; XVII, pp. 1–98, 1926.

The hydroid zoophytes constitute a "polymorphic" group, within which a vast number of species have already been distinguished; and the labours of the systematic naturalist are constantly adding to the number. The specific distinctions are for the móst part based, not upon characters directly presented by the living animal, but upon the form, size and arrangement of the little cups, or "calycles," secreted and inhabited by the little individual polyps

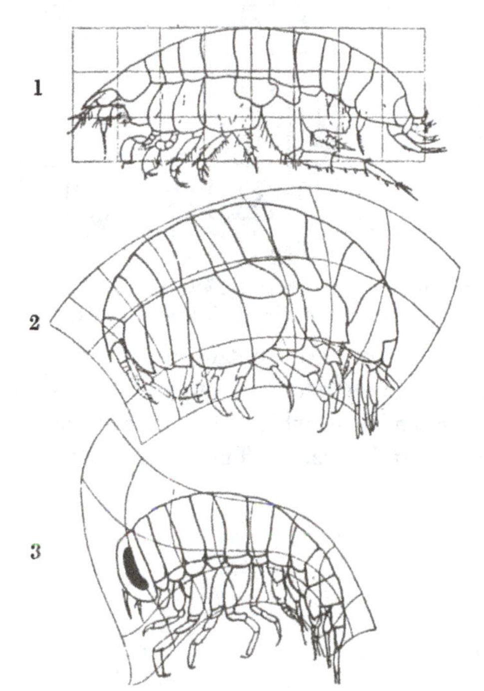

Fig. 514. 1, *Harpinia plumosa* Kr.; 2, *Stegocephalus inflatus* Kr.; 3, *Hyperia galba.*

which compose the compound organism. The variations, which are apparently infinite, of these conformations are easily seen to be a question of relative magnitudes, and are capable of complete expression, sometimes by very simple, sometimes by somewhat more complex, coordinate networks.

For instance, the varying shapes of the simple wineglass-shaped cups of the Campanularidae are at once sufficiently represented and compared by means of simple Cartesian coordinates (Fig. 515). In the two allied families of Plumulariidae and Aglaopheniidae the calycles are set unilaterally upon a jointed stem, and small cup-like structures (holding rudimentary polyps) are associated with the large calycles in definite number and position. These small calyculi

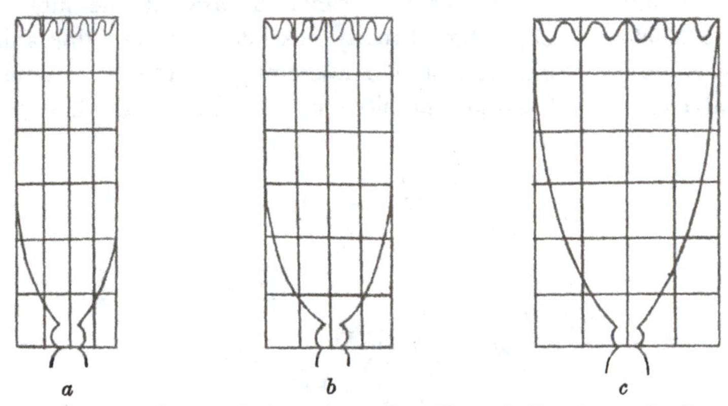

Fig. 515. *a*, *Campanularia macroscyphus* Allm.; *b*, *Gonothyraea hyalina* Hincks; *c*, *Clytia Johnstoni* Alder.

are variable in number, but in the great majority of cases they accompany the large calycle in groups of three—two standing by its upper border, and one, which is especially variable in form and magnitude, lying at its base. The stem is liable to flexure and,

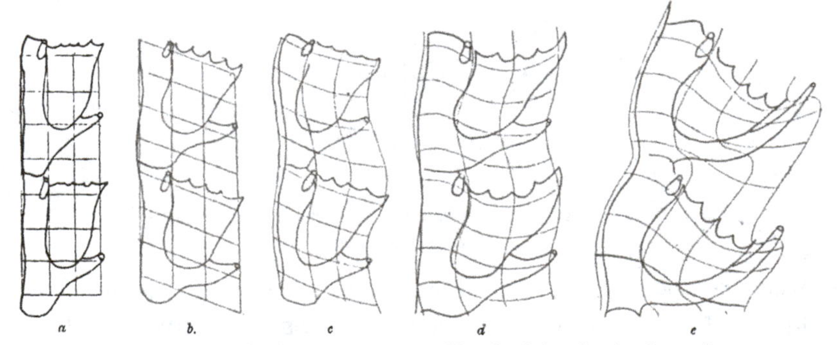

Fig. 516. *a*, *Cladocarpus crenatus* F.; *b*, *Aglaophenia pluma* L.; *c*, *A. rhynchocarpa* A.; *d*, *A. cornuta* K.; *e*, *A. ramulosa* K.

in a high degree, to extension or compression; and these variations extend, often on an exaggerated scale, to the related calycles. As a result we find that we can draw various systems of curved or sinuous coordinates, which express, all but completely, the configuration of the various hydroids which we inscribe therein (Fig. 516).

The comparative smoothness of denticulation of the margin of the calycle, and the number of its denticles, constitutes an independent variation, and requires separate description; we have already seen (p. 391) that this denticulation is in all probability due to a particular physical cause.

Among countless other invertebrate animals which we might illustrate, did space and time permit, we should find the bivalve molluscs shewing certain things extremely well. If we start with a more or less oblong shell, such as *Anodon* or *Mya* or *Psammobia*, we can see how easily it may be transformed into a more circular or orbicular, but still closely related form; while on the other hand a simple shear is well-nigh all that is needed to transform the oblong *Anodon* into the triangular, pointed *Mytilus*, *Avicula* or *Pinna*. Now suppose we draw the shell of *Anodon* in the usual rectangular coordinates, and deform this network into the corresponding oblique coordinates of *Mytilus*, we may then proceed to draw within the same two nets the anatomy of the same two molluscs. Then of the two adductor muscles, coequal in *Anodon*, one becomes small, the other large, when transferred to the oblique network of *Mytilus*; at the same time the foot becomes stunted and the siphonal aperture enlarged. In short, having "transformed" one shell into the other we may perform an identical transformation on their contained anatomy: and so (provided the two are not too distantly related) deduce the bodily structure of the one from our knowledge of the other, to a first but by no means negligible approximation.

Among the fishes we discover a great variety of deformations, some of them of a very simple kind, while others are more striking and more unexpected. A comparatively simple case, involving a simple shear, is illustrated by Figs. 517 and 518. The one represents, within Cartesian coordinates, a certain little oceanic fish known as *Argyropelecus Olfersi*. The other represents precisely the same outline, transferred to a system of oblique coordinates whose axes are inclined at an angle of 70°; but this is now (as far as can be seen on the scale of the drawing) a very good figure of an allied fish, assigned to a different genus, under the name of *Sternoptyx diaphana*. The deformation illustrated by this case of *Argyropelecus* is precisely analogous to the simplest and commonest kind of deformation to which fossils are subject (as we have seen on p. 811) as the result of shearing-stresses in the solid rock.

Fig. 519 is an outline diagram of a typical Scaroid fish. Let us deform its rectilinear coordinates into a system of (approximately) coaxial circles, as in Fig. 520, and then filling into the new system,

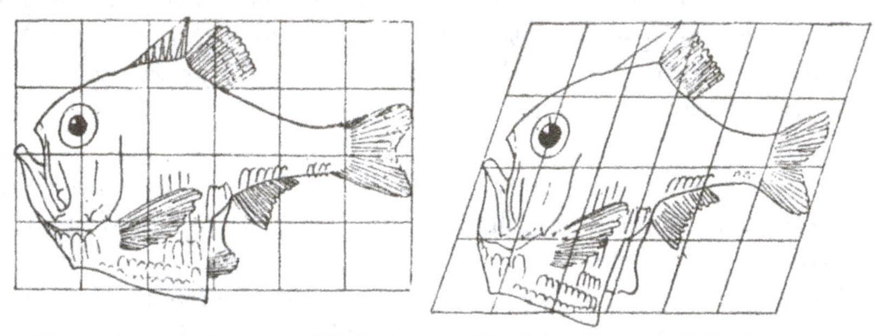

Fig. 517. *Argyropelecus Olfersi*. Fig. 518. *Sternoptyx diaphana*.

space by space and point by point, our former diagram of *Scarus*, we obtain a very good outline of an allied fish, belonging to a neighbouring family, of the genus *Pomacanthus*. This case is all the more interesting, because upon the body of our *Pomacanthus* there are striking colour bands, which correspond in direction very closely

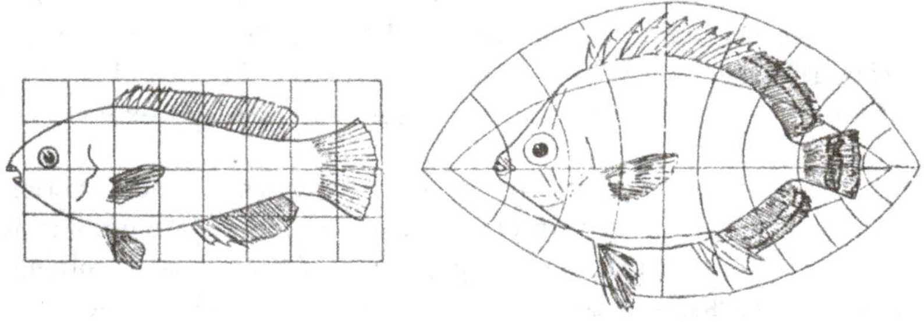

Fig. 519. *Scarus* sp. Fig. 520. *Pomacanthus*.

to the lines of our new curved ordinates. In like manner, the still more bizarre outlines of other fishes of the same family of Chaetodonts will be found to correspond to very slight modifications of similar coordinates; in other words, to small variations in the values of the constants of the coaxial curves.

In Figs. 521–524 I have represented another series of Acanthopterygian fishes, not very distantly related to the foregoing. If we start this series with the figure of *Polyprion*, in Fig. 521, we see that the outlines of *Pseudopriacanthus* (Fig. 522) and of *Sebastes* or *Scorpaena* (Fig. 523) are easily derived by substituting a system

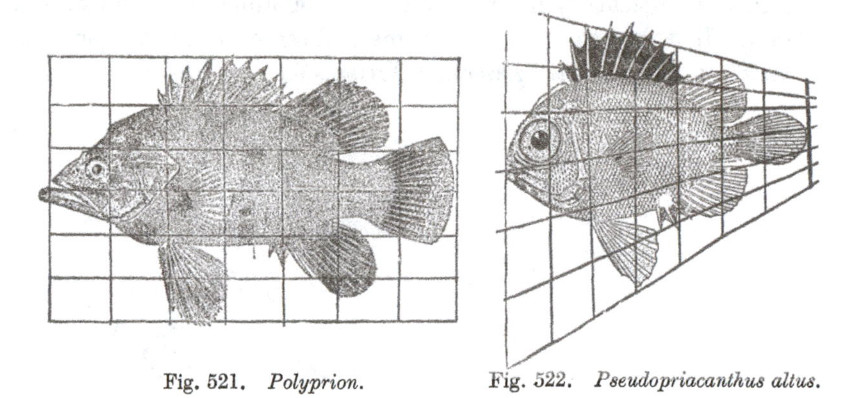

Fig. 521. *Polyprion*. Fig. 522. *Pseudopriacanthus altus*.

of triangular, or radial, coordinates for the rectangular ones in which we had inscribed *Polyprion*. The very curious fish *Antigonia capros*, an oceanic relative of our own boar-fish, conforms closely to the peculiar deformation represented in Fig. 524.

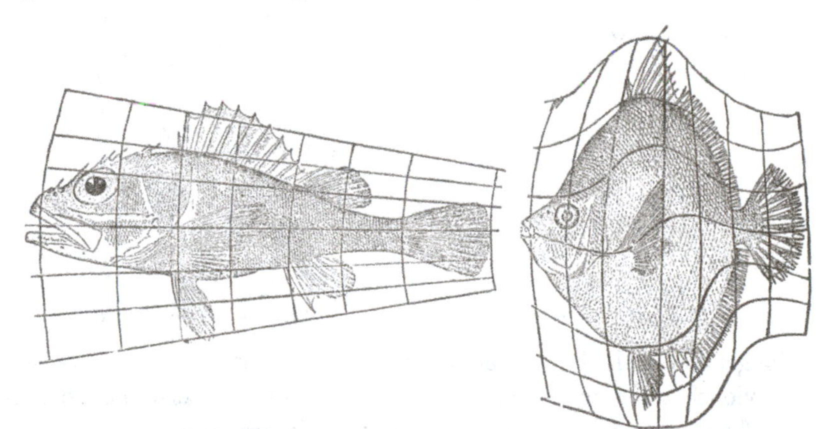

Fig. 523. *Scorpaena* sp. Fig. 524. *Antigonia capros*.

Fig. 525 is a common, typical *Diodon* or porcupine-fish, and in Fig. 526 I have deformed its vertical coordinates into a system of concentric circles, and its horizontal coordinates into a system of curves which, approximately and provisionally, are made to resemble a system of hyperbolas*. The old outline, transferred in its integrity to the new network, appears as a manifest representation of the closely allied, but very different looking, sunfish, *Orthagoriscus mola*. This is a particularly instructive case of deformation or transformation. It is true that, in a mathematical sense, it is not a perfectly satisfactory or perfectly regular deformation, for the system is no

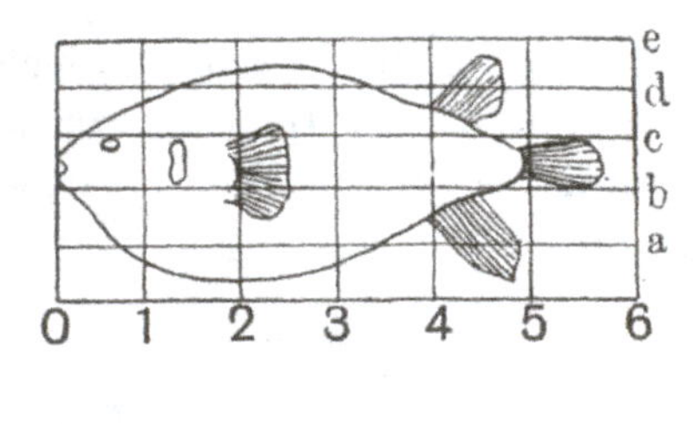

Fig. 525. *Diodon*.

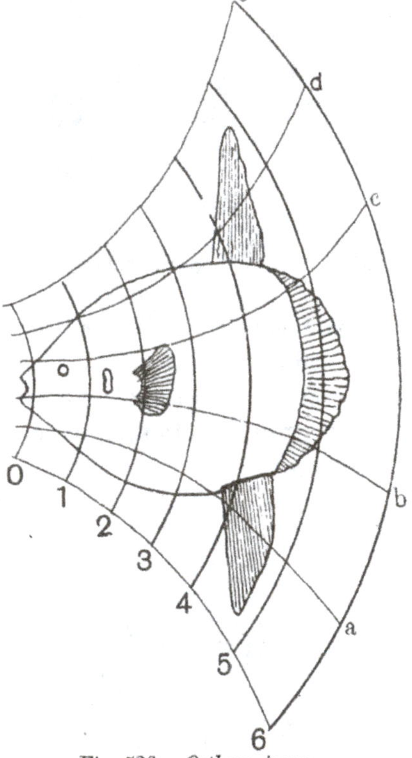

Fig. 526. *Orthagoriscus*.

longer isogonal; but nevertheless, it is symmetrical to the eye, and obviously approaches to an isogonal system under certain conditions of friction or constraint. And as such it accounts, by one single integral transformation, for all the apparently separate and distinct external differences between the two fishes. It leaves the parts

* The coordinate system of Fig. 526 is somewhat different from that which I first drew and published. It is not unlikely that further investigation will further simplify the comparison, and shew it to involve a still more symmetrical system.

near to the origin of the system, the whole region of the head, the opercular orifice and the pectoral fin, practically unchanged in form, size and position; and it shews a greater and greater apparent modification of size and form as we pass from the origin towards the periphery of the system.

In a word, it is sufficient to account for the new and striking contour in all its essential details, of rounded body, exaggerated dorsal and ventral fins, and truncated tail. In like manner, and using precisely the same coordinate networks, it appears to me possible to shew the relations, almost bone for bone, of the skeletons of the two fishes; in other words, to reconstruct the skeleton of the one from our knowledge of the skeleton of the other, under the guidance of the same correspondence as is indicated in their external configuration.

The family of the crocodiles has had a special interest for the evolutionist ever since Huxley pointed out that, in a degree only second to the horse and its ancestors, it furnishes us with a close and almost unbroken series of transitional forms, running down in continuous succession from one geological formation to another. I should be inclined to transpose this general statement into other terms, and to say that the Crocodilia constitute a case in which, with unusually little complication from the presence of independent variants, the trend of one particular mode of transformation is visibly manifested. If we exclude meanwhile from our comparison a few of the oldest of the crocodiles, such as *Belodon*, which differ more fundamentally from the rest, we shall find a long series of genera in which we can refer not only the changing contours of the skull, but even the shape and size of the many constituent bones and their intervening spaces or "vacuities," to one and the same simple system of transformed coordinates. The manner in which the skulls of various Crocodilians differ from one another may be sufficiently illustrated by three or four examples.

Let us take one of the typical modern crocodiles as our standard of form, e.g. *C. porosus*, and inscribe it, as in Fig. 527, *a*, in the usual Cartesian coordinates. By deforming the rectangular network into a triangular system, with the apex of the triangle a little way in front of the snout, as in *b*, we pass to such a form as *C. americanus*.

By an exaggeration of the same process we at once get an approximation to the form of one of the sharp-snouted, or longirostrine, crocodiles, such as the genus *Tomistoma*; and, in the species figured, the oblique position of the orbits, the arched contour of the occipital border, and certain other characters suggest a certain amount of curvature, such as I have represented in the diagram (Fig. 527, *b*), on the part of the horizontal coordinates. In the still more elongated skull of such a form as the Indian Gavial, the whole skull has undergone a great longitudinal extension, or, in other words, the ratio of x/y is greatly diminished; and this extension is not uniform, but is at a maximum in the region of the nasal and maxillary bones.

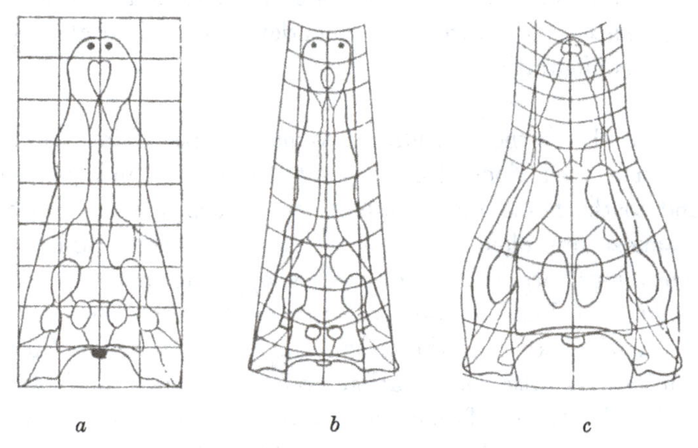

Fig. 527. *a*, *Crocodilus porosus*; *b*, *C. americanus*; *c*, *Notosuchus terrestris.*

This especially elongated region is at the same time narrowed in an exceptional degree, and its excessive narrowing is represented by a curvature, convex towards the median axis, on the part of the vertical ordinates. Let us take as a last illustration one of the Mesozoic crocodiles, the little *Notosuchus*, from the Cretaceous formation. This little crocodile is very different from our type in the proportions of its skull. The region of the snout, in front of and including the frontal bones, is greatly shortened; from constituting fully two-thirds of the whole length of the skull in *Crocodilus*, it now constitutes less than half, or, say, three-sevenths of the whole; and the whole skull, and especially its posterior part, is curiously compact, broad, and squat. The orbit is unusually large. If in the diagram of this skull we select a number of points obviously corresponding to points where our rectangular coordinates intersect particular bones or other recognisable features in our typical crocodile, we shall easily discover that the lines joining these points in *Notosuchus* fall into such a coordinate network as that which is represented in Fig. 527, *c*. To all intents and purposes, then, this not very complex system, representing one harmonious "deformation," accounts for *all* the differences between the two figures, and is sufficient to enable one at any time to reconstruct a detailed drawing, bone for bone, of the skull of *Notosuchus* from the model furnished by the common crocodile.

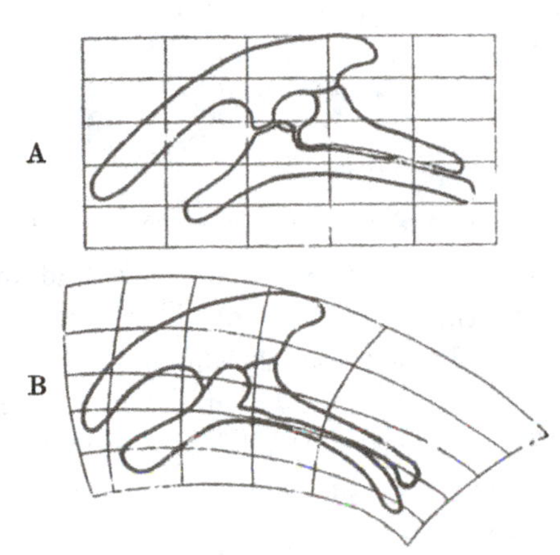

Fig. 528. Pelvis of (A) *Stegosaurus*; (B) *Camptosaurus*.

The many diverse forms of Dinosaurian reptiles, all of which manifest a strong family likeness underlying much superficial diversity, furnish us with plentiful material for comparison by the method of transformations. As an instance, I have figured the pelvic bones of *Stegosaurus* and of *Camptosaurus* (Fig. 528, *a*, *b*) to shew that, when the former is taken as our Cartesian type, a slight curvature and an approximately logarithmic extension of the x-axis brings us easily to the configuration of the other. In the original specimen of *Camptosaurus* described by Marsh*, the anterior portion of the iliac bone is missing; and in Marsh's restoration this part of the bone is drawn as though it came somewhat abruptly to a sharp point. In my figure I have completed this missing part

* *Dinosaurs of North America*, pl. LXXXI, etc., 1896.

of the bone in harmony with the general coordinate network which is suggested by our comparison of the two entire pelves; and I venture to think that the result is more natural in appearance, and more likely to be correct than was Marsh's conjectural restoration. It would seem, in fact, that there is an obvious field for the employment of the method of coordinates in this task of reproducing missing

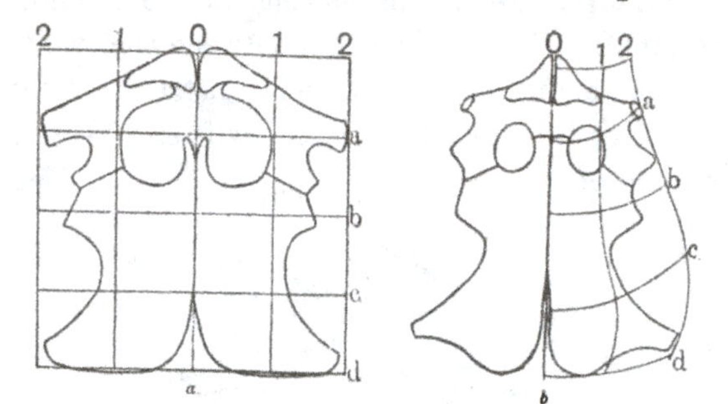

Fig. 529. Shoulder-girdle of *Cryptocleidus*. *a*, young; *b*, adult.

portions of a structure to the proper scale and in harmony with related types. To this subject we shall presently return.

In Fig. 529, *a*, *b*, I have drawn the shoulder-girdle of *Cryptocleidus*, a Plesiosaurian reptile, half-grown in the one case and full-grown in the other. The change of form during growth in this region of the body is very considerable, and its nature is well brought out

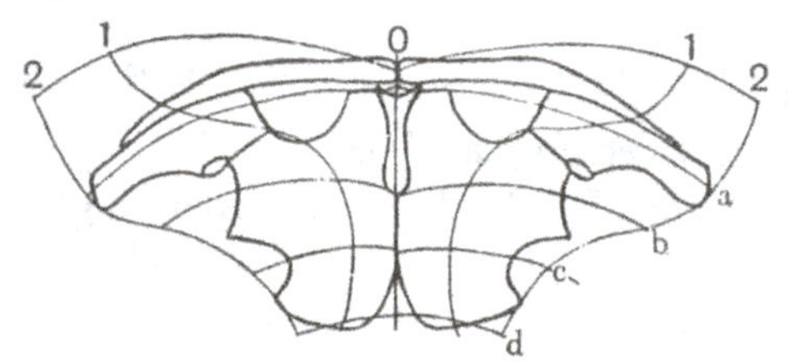

Fig. 530. Shoulder-girdle of *Ichthyosaurus*.

by the two coordinate systems. In Fig. 530 I have drawn the shoulder-girdle of an Ichthyosaur, referring it to *Cryptocleidus* as a standard of comparison. The interclavicle, which is present in *Ichthyosaurus*, is minute and hidden in *Cryptocleidus*; but the numerous other differences between the two forms, chief among which is the great elongation in *Ichthyosaurus* of the two clavicles, are all seen by our diagrams to be part and parcel of one general and systematic deformation.

Before we leave the group of reptiles we may glance at the very strangely modified skull of *Pteranodon*, one of the extinct flying reptiles, or Pterosauria. In this very curious skull the region of the jaws, or beak, is greatly elongated and pointed; the occipital bone is drawn out into an enormous backwardly directed crest; the posterior part of the lower jaw is similarly produced backwards; the orbit is small; and the quadrate bone is strongly inclined downwards and forwards. The whole skull has a configuration which stands, apparently, in the strongest possible contrast to that of a more normal Ornithosaurian such as *Dimorphodon*. But if we inscribe the latter in Cartesian coordinates (Fig. 531, *a*), and refer

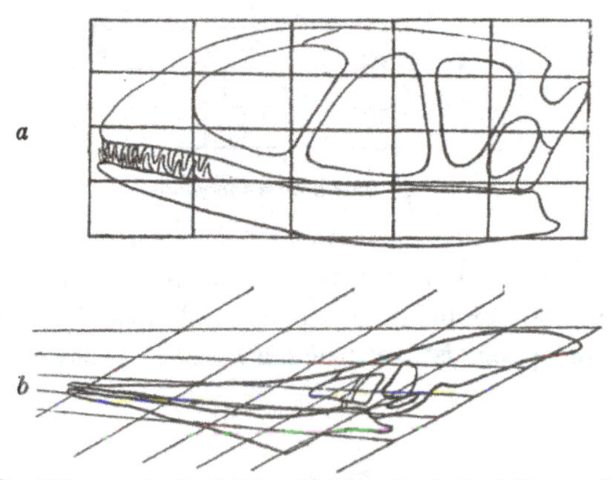

Fig. 531. *a*, skull of *Dimorphodon*; *b*, skull of *Pteranodon*.

our *Pteranodon* to a system of oblique coordinates (*b*), in which the two coordinate systems of parallel lines become each a pencil of diverging rays, we make manifest a correspondence which extends uniformly throughout all parts of these very different-looking skulls.

We have dealt so far, and for the most part we shall continue to deal, with our coordinate method as a means of comparing one known structure with another. But it is obvious, as I have said, that it may also be employed for drawing hypothetical structures, on the assumption that they have varied from a known form in some definite way. And this process may be especially useful, and will be most obviously legitimate, when we apply it to the particular case of representing intermediate stages between two forms which are actually known to exist, in other words, of reconstructing the transitional stages through which the course of evolution must have successively travelled if it has brought about the change from some ancestral type to its presumed descendant. Some years ago I sent my friend, Mr Gerhard Heilmann of Copenhagen, a few of

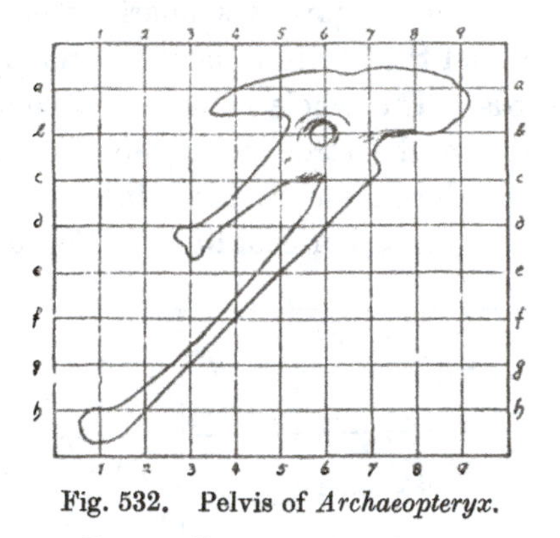

Fig. 532. Pelvis of *Archaeopteryx*.

my own rough coordinate diagrams, including some in which the pelves of certain ancient and primitive birds were compared one with another. Mr Heilmann, who is both a skilled draughtsman and an able morphologist, returned me a set of diagrams which are

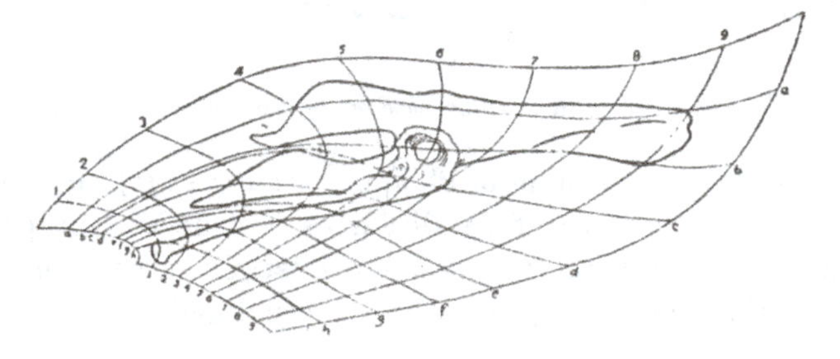

Fig. 533. Pelvis of *Apatornis*.

a vast improvement on my own, and which are reproduced in Figs. 532–537. Here we have, as extreme cases, the pelvis of *Archaeopteryx*, the most ancient of known birds, and that of *Apatornis*, one of the fossil "toothed" birds from the North American Cretaceous formations—a bird shewing some resemblance to the modern terns. The pelvis of *Archaeopteryx* is taken as our type, and referred accordingly to Cartesian coordinates (Fig. 532); while the corresponding coordinates of the very different pelvis of *Apatornis* are represented in Fig. 533. In Fig. 534 the outlines of these two

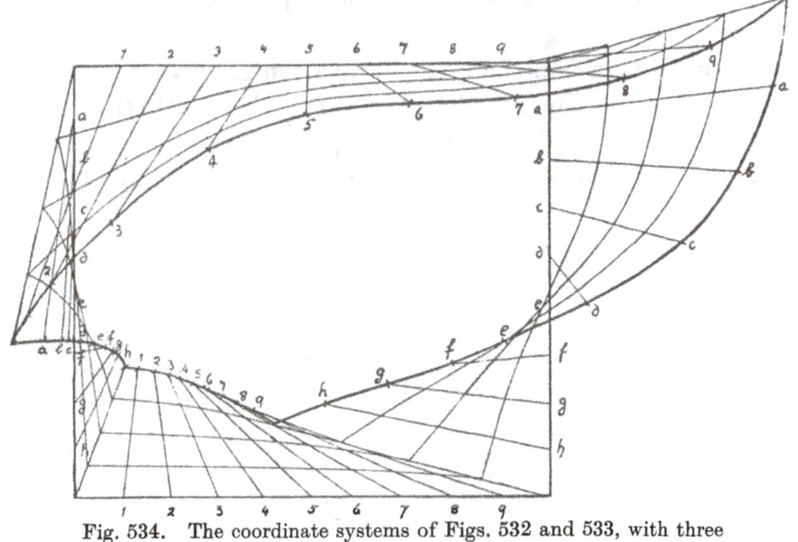

Fig. 534. The coordinate systems of Figs. 532 and 533, with three intermediate systems interpolated.

coordinate systems are superposed upon one another, and those of three intermediate and equidistant coordinate systems are interpolated between them. From each of these latter systems,

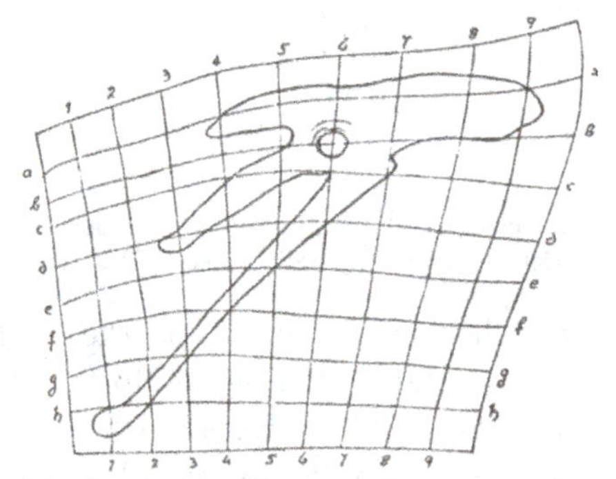

Fig. 535. The first intermediate coordinate network, with its corresponding inscribed pelvis.

so determined by direct interpolation, a complete coordinate diagram is drawn, and the corresponding outline of a pelvis is found from each of these systems of coordinates, as in Figs. 535, 536. Finally, in Fig. 537 the complete series is represented, beginning with the known pelvis of *Archaeopteryx*, and leading up by our three intermediate hypothetical types to the known pelvis of *Apatornis*.

Among mammalian skulls I will take two illustrations only, one drawn from a comparison of the human skull with that of the higher apes, and another from the group of Perissodactyle Ungulates, the group which includes the rhinoceros, the tapir, and the horse.

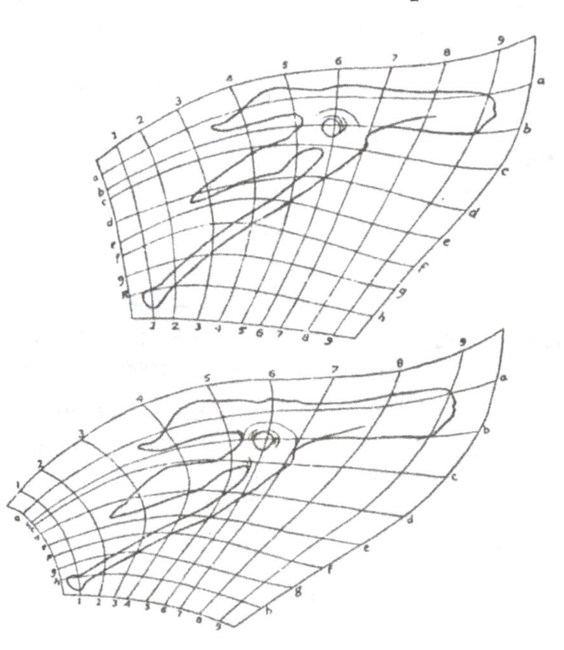

Fig. 536. The second and third intermediate coordinate networks, with their corresponding inscribed pelves.

Let us begin by choosing as our type the skull of *Hyrachyus agrarius* Cope, from the Middle Eocene of North America, as figured by Osborn in his Monograph of the Extinct Rhinoceroses* (Fig. 538). The many other forms of primitive rhinoceros described in the monograph differ from *Hyrachyus* in various details—in the characters of the teeth, sometimes in the number of the toes, and so forth; and they also differ very considerably in the general appearance of the skull. But these differences in the conformation

* *Mem. Amer. Mus. of Nat. Hist.* I, III, 1898. .

of the skull, conspicuous as they are at first sight, will be found easy to bring under the conception of a simple and homogeneous transformation, such as would result from the application of some not very complicated stress. For instance, the corresponding coordinates of *Aceratherium tridactylum*, as shewn in Fig. 539, indicate that the

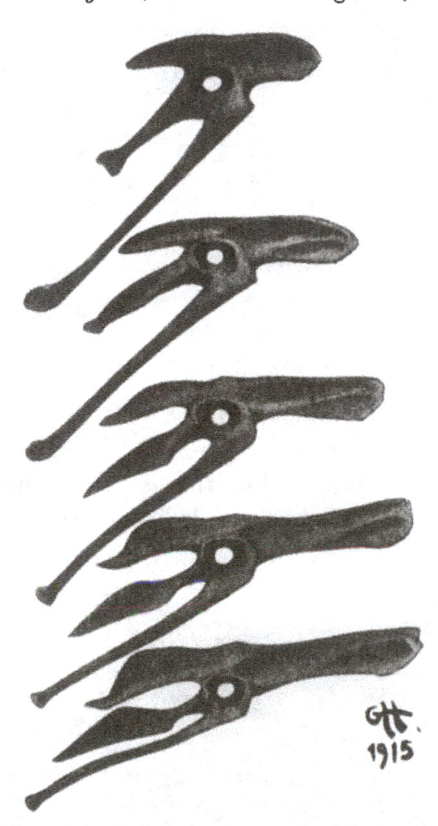

Fig. 537. The pelvis of *Archaeopteryx* and of *Apatornis*, with three transitional types interpolated between them.

essential difference between this skull and the former one may be summed up by saying that the long axis of the skull of *Aceratherium* has undergone a slight double curvature, while the upper parts of the skull have at the same time been subject to a vertical expansion, or to growth in somewhat greater proportion than the lower parts. Precisely the same changes, on a somewhat greater scale, give us the skull of an existing rhinoceros.

Among the species of *Aceratherium*, the posterior, or occipital, view of the skull presents specific differences which are perhaps more conspicuous than those furnished by the side view; and these

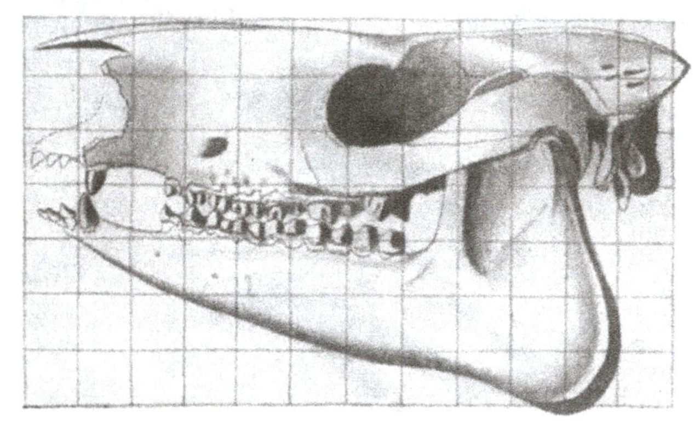

Fig. 538. Skull of *Hyrachyus agrarius*. After Osborn.

differences are very strikingly brought out by the series of conformal transformations which I have represented in Fig. 540. In this case it will perhaps be noticed that the correspondence is not always quite accurate in small details. It could easily have been made

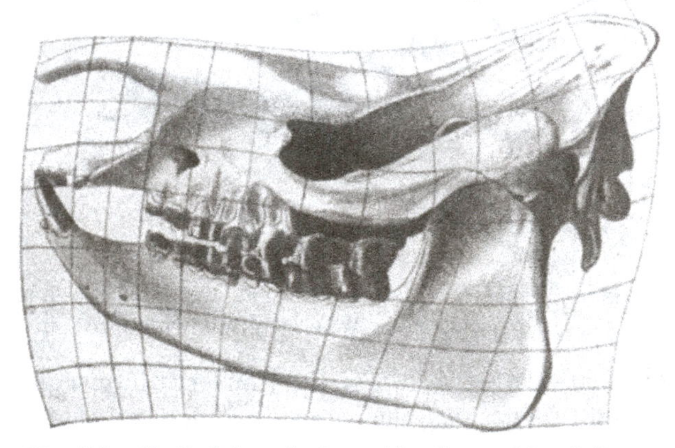

Fig. 539. Skull of *Aceratherium tridactylum*. After Osborn.

much more accurate by giving a slightly sinuous curvature to certain of the coordinates. But as they stand, the correspondence indicated is very close, and the simplicity of the figures illustrates all the better the general character of the transformation.

By similar and not more violent changes we pass easily to such allied forms as the Titanotheres (Fig. 541); and the well-known series of species of *Titanotherium*, by which Professor Osborn has

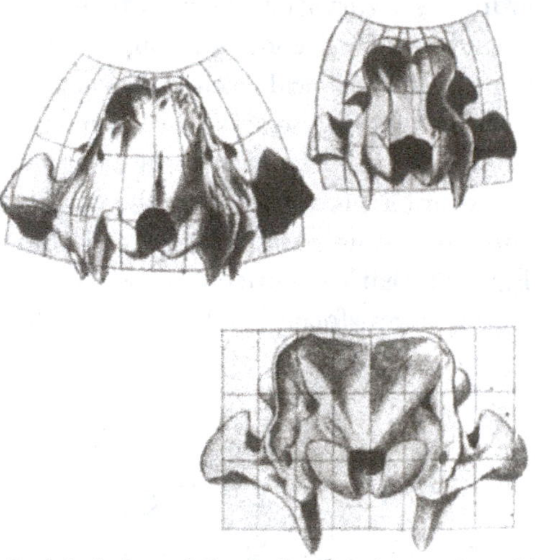

Fig. 540. Occipital view of the skulls of various extinct rhinoceroses (*Aceratherium* spp.). After Osborn.

illustrated the evolution of this genus, constitutes a simple and suitable case for the application of our method.

But our method enables us to pass over greater gaps than these, and to discern the general, and to a very large extent even the

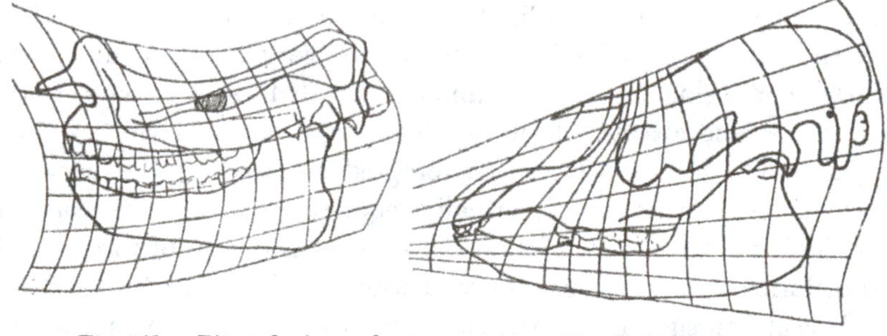

Fig. 541. *Titanotherium robustum*. Fig. 542. Tapir's skull.

detailed, resemblances between the skull of the rhinoceros and those of the tapir or the horse. From the Cartesian coordinates in which we have begun by inscribing the skull of a primitive rhinoceros, we pass to the tapir's skull (Fig. 542), firstly, by converting the rectangular into a triangular network, by which we represent the depression of the anterior and the progressively increasing elevation of the posterior part of the skull; and secondly, by giving to the vertical ordinates a curvature such as to bring about a certain longitudinal compression, or condensation, in the forepart of the skull, especially in the nasal and orbital regions.

The conformation of the horse's skull departs from that of our primitive Perissodactyle (that is to say our early type of rhinoceros, *Hyrachyus*) in a direction that is nearly the opposite of that taken by *Titanotherium* and by the recent species of rhinoceros. For we perceive, by Fig. 543, that the horizontal coordinates, which in these latter cases become transformed into curves with the concavity

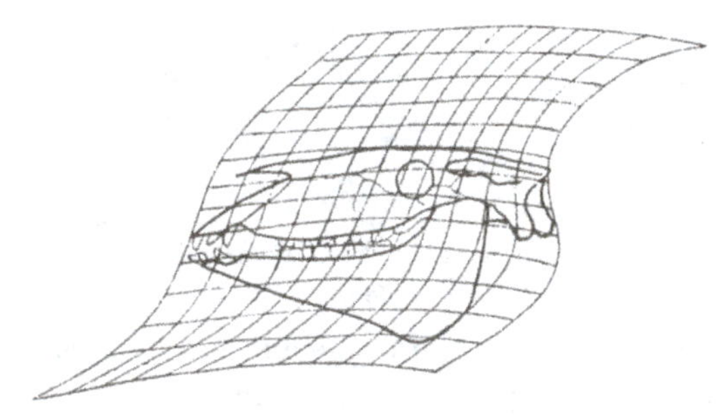

Fig. 543. Horse's skull.

upwards, are curved, in the case of the horse, in the opposite direction. And the vertical ordinates, which are also curved, somewhat in the same fashion as in the tapir, are very nearly equidistant, instead of being, as in that animal, crowded together anteriorly. Ordinates and abscissae form an oblique system, as is shewn in the figure. In this case I have attempted to produce the network beyond the region which is actually required to include the diagram of the horse's skull, in order to shew better the form of the general transformation, with a part only of which we have actually to deal.

It is at first sight not a little surprising to find that we can pass, by a cognate and even simpler transformation, from our Perissodactyle skulls to that of the rabbit; but the fact that we can easily do so is a simple illustration of the undoubted affinity which exists between the Rodentia, especially the family of the Leporidae, and the more primitive Ungulates. For my part, I would go further; for I think there is strong reason to believe that the Perissodactyles are more closely related to the Leporidae than the former are to the other Ungulates, or than the Leporidae are to the rest of the Rodentia. Be that as it may, it is obvious from Fig. 544 that the rabbit's skull

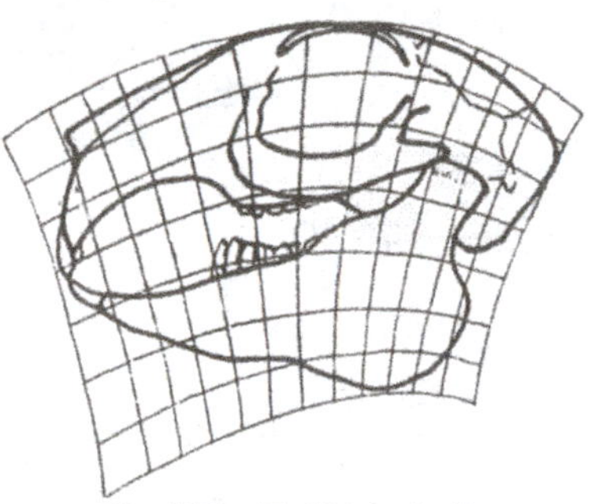

Fig. 544. Rabbit's skull.

conforms to a system of coordinates corresponding to the Cartesian coordinates in which we have inscribed the skull of *Hyrachyus*, with the difference, firstly, that the horizontal ordinates of the latter are

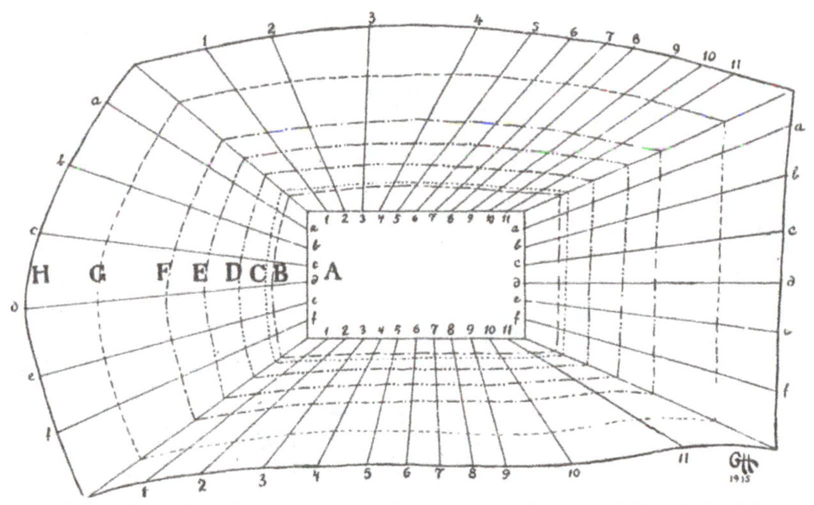

Fig. 545. A, outline diagram of the Cartesian coordinates of the skull of *Hyracotherium* or *Eohippus*, as shewn in Fig. 546, A. H, outline of the corresponding projection of the horse's skull. B–G, intermediate, or interpolated, outlines.

transformed into equidistant curved lines, approximately arcs of circles, with their concavity directed downwards; and secondly, that the vertical ordinates are transformed into a pencil of rays approxi-

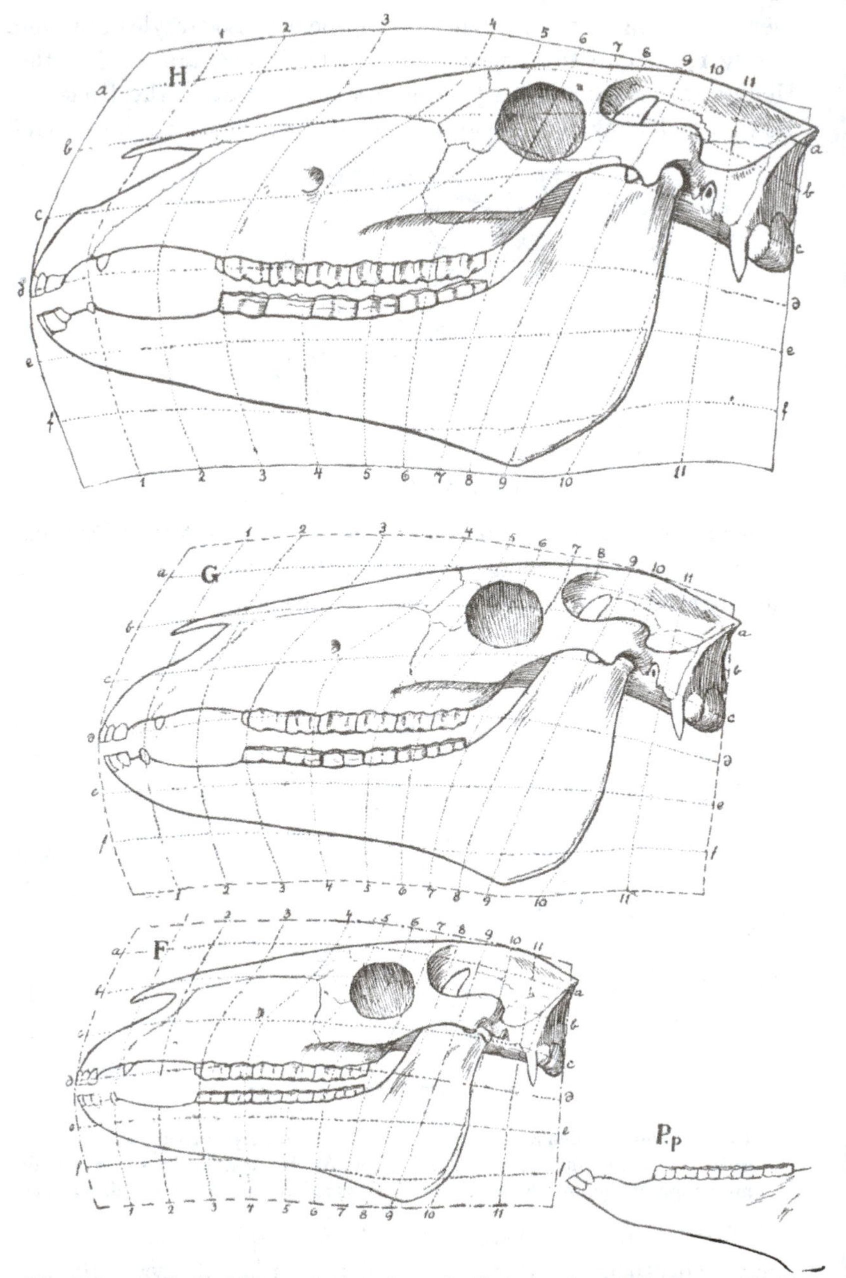

Fig. 546. A, skull of *Hyracotherium*, from the Eocene, after W. B. Scott; H, skull of horse, represented as a coordinate transformation of that of *Hyracotherium*,

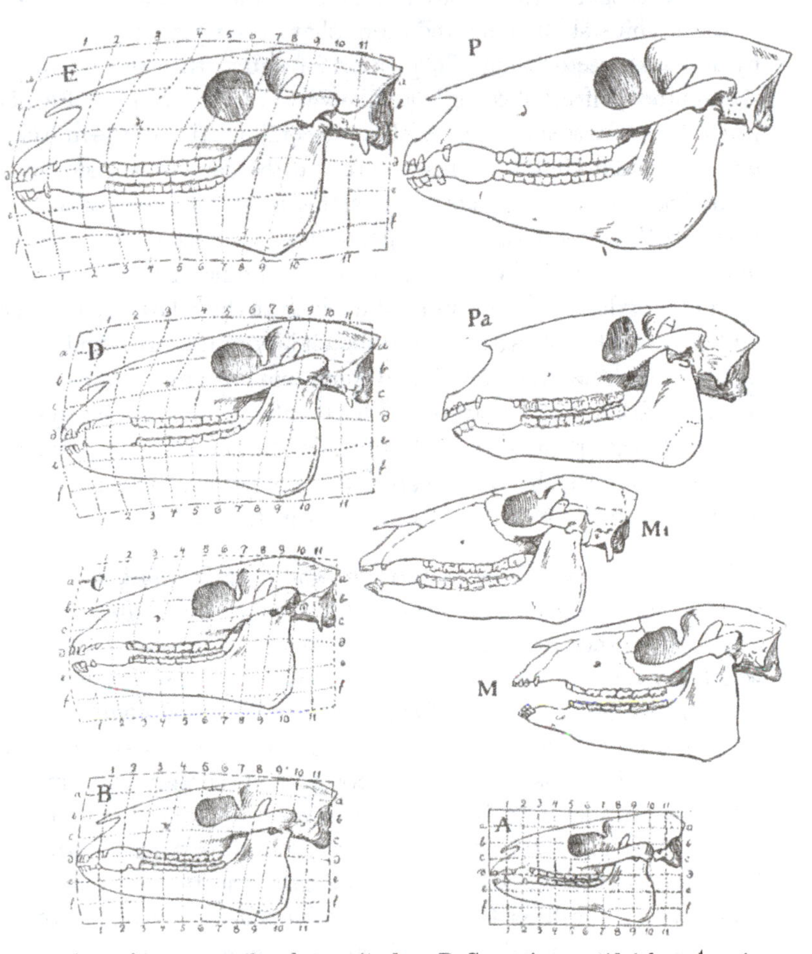

and to the same scale of magnitude; B–G, various artificial or imaginary types, reconstructed as intermediate stages between A and H; M, skull of *Mesohippus*, from the Oligocene, after Scott, for comparison with C; P, skull of *Protohippus*, from the Miocene, after Cope, for comparison with E; Pp, lower jaw of *Protohippus placidus* (after Matthew and Gidley), for comparison with F; Mi, *Miohippus* (after Osborn), Pa, *Parahippus* (after Peterson), shewing resemblance, but less perfect agreement, with C and D.

mately orthogonal to the circular arcs. In short, the configuration of the rabbit's skull is derived from that of our primitive rhinoceros by the unexpectedly simple process of submitting the latter to a strong and uniform flexure in the downward direction (cf. Fig. 538, p. 1074). In the case of the rabbit the configuration of the individual bones does not conform quite so well to the general transformation as it does when we are comparing the several Perissodactyles one with another; and the chief departures from conformity will be found in the size of the orbit and in the outline of the immediately surrounding bones. The simple fact is that the relatively enormous eye of the rabbit constitutes an independent variation, which cannot be brought into the general and fundamental transformation, but must be dealt with separately. The enlargement of the eye, like the modification in form and number of the teeth, is a separate phenomenon, which supplements but in no way contradicts our general comparison of the skulls taken in their entirety.

Before we leave the Perissodactyla and their allies, let us look a little more closely into the case of the horse and its immediate relations or ancestors, doing so with the help of a set of diagrams which I again owe to Mr Gerard Heilmann*. Here we start afresh, with the skull (Fig. 546, A) of *Hyracotherium* (or *Eohippus*), inscribed in a simple Cartesian network. At the other end of the series (H) is a skull of *Equus*, in its own corresponding network; and the intermediate stages (B–G) are all drawn by direct and simple interpolation, as in Mr Heilmann's former series of drawings of *Archaeopteryx* and *Apatornis*. In this present case, the relative magnitudes are shewn, as well as the forms, of the several skulls. Alongside of these reconstructed diagrams are set figures of certain extinct "horses" (Equidae or Palaeotheriidae), and in two cases, viz. *Mesohippus* and *Protohippus* (M, P), it will be seen that the actual fossil skull coincides in the most perfect fashion with one of the hypothetical forms or stages which our method shews to be implicitly involved in the transition from *Hyracotherium* to *Equus*†. In a third case, that of *Parahippus* (Pa), the correspondence (as Mr Heilmann

* These and also other coordinate diagrams will be found in Mr G. Heilmann's beautiful and original book *Fuglenes Afstamning*, 398 pp., Copenhagen, 1916; see especially pp. 368–380.

† Cf. Zittel, *Grundzüge d. Palaeontologie*, 1911, p. 463.

points out) is by no means exact. The outline of this skull comes nearest to that of the hypothetical transition stage D, but the "fit" is now a bad one; for the skull of *Parahippus* is evidently a longer, straighter and narrower skull, and differs in other minor characters besides. In short, though some writers have placed *Parahippus* in the direct line of descent between *Equus* and *Eohippus*, we see at once that there is no place for it there, and that it must, accordingly, represent a somewhat divergent branch or offshoot of the Equidae*. It may be noticed, especially in the case of *Protohippus* (P), that the configuration of the angle of the jaw does not tally quite so accurately with that of our hypothetical diagrams as do other parts of the skull. As a matter of fact, this region is somewhat variable, in different species of a genus, and even in different individuals of the same species; in the small figure (Pp) of *Protohippus placidus* the correspondence is more exact.

In considering this series of figures we cannot but be struck, not only with the regularity of the succession of "transformations," but also with the slight and inconsiderable differences which separate each recorded stage from the next, and even the two extremes of the whole series from one another. These differences are no greater (save in regard to actual magnitude) than those between one human skull and another, at least if we take into account the older or remoter races; and they are again no greater, but if anything less, than the range of variation, racial and individual, in certain other human bones, for instance the scapula†.

The variability of this latter bone is great, but it is neither surprising nor peculiar; for it is linked with all the considerations of

* Cf. W. B. Scott (*Amer. Journ. of Science*, XLVIII, pp. 335–374, 1894), "We find that any mammalian series at all complete, such as that of the horses, is remarkably continuous, and that the progress of discovery is steadily filling up what few gaps remain. So closely do successive stages follow upon one another that it is sometimes extremely difficult to arrange them all in order, and to distinguish clearly those members which belong in the main line of descent, and those which represent incipient branches. Some phylogenies actually suffer from an embarrassment of riches."

† Cf. T. Dwight, The range of variation of the human scapula, *Amer. Nat.* XXI, pp. 627–638, 1887. Cf. also Turner, *Challenger Rep.* XLVII, on Human Skeletons, p. 86, 1886: "I gather both from my own measurements, and those of other observers, that the range of variation in the relative length and breadth of the scapula is very considerable in the same race, so that it needs a large number of bones to enable one to obtain an accurate idea of the mean of the race."

mechanical efficiency and functional modification which we dealt with in our last chapter. The scapula occupies, as it were, a focus in a very important field of force; and the lines of force converging on it will be very greatly modified by the varying development of

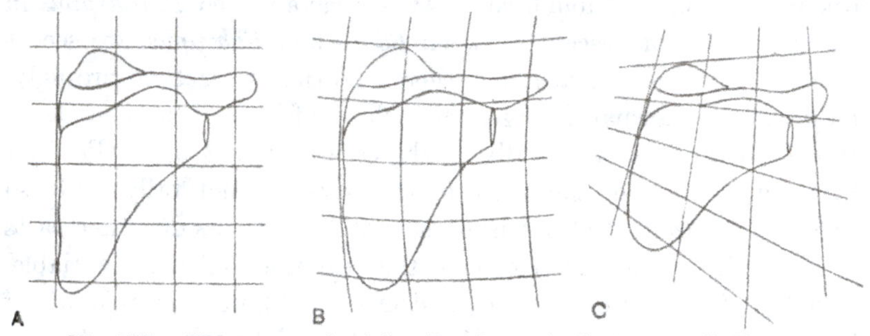

Fig. 547. Human scapulae (after Dwight). A, Caucasian; B, Negro; C, North American Indian (from Kentucky Mountains).

the muscles over a large area of the body and of the uses to which they are habitually put.

Let us now inscribe in our Cartesian coordinates the outline of a human skull (Fig. 548), for the purpose of comparing it with the skulls of some of the higher apes. We know beforehand that the main differences between the human and the simian types depend

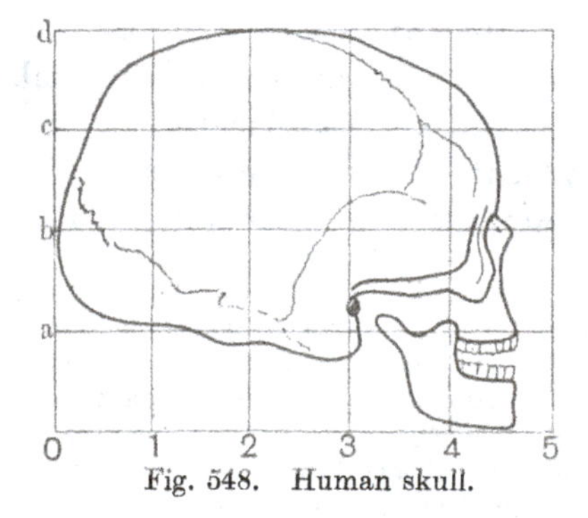

Fig. 548. Human skull.

upon the enlargement or expansion of the brain and braincase in man, and the relative diminution or enfeeblement of his jaws. Together with these changes, the "facial angle" increases from an oblique angle to nearly a right angle in man, and the configuration of every constituent bone of the face and skull undergoes an alteration. We do not know to begin with, and we are not shewn by the ordinary methods of comparison, how far these various changes form part of one harmonious and congruent transformation, or whether we are to look, for instance, upon the changes undergone by the frontal, the occipital, the maxillary, and the mandibular

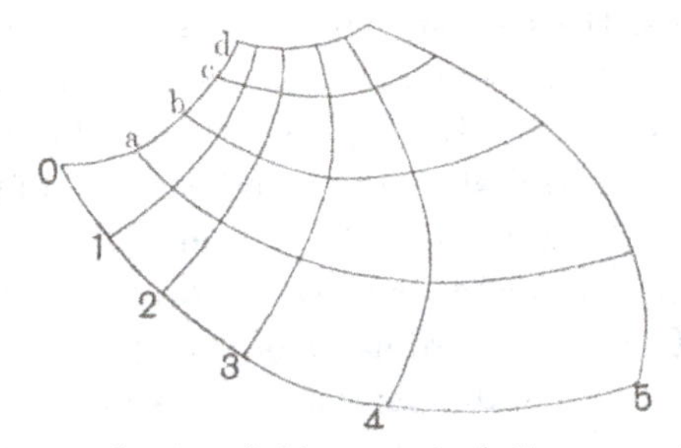

Fig. 549. Coordinates of chimpanzee's skull, as a projection of the Cartesian coordinates of Fig. 548.

regions as a congeries of separate modifications or independent variants. But as soon as we have marked out a number of points in the gorilla's or chimpanzee's skull, corresponding with those which our coordinate network intersected in the human skull, we find that these corresponding points may be at once linked up by smoothly curved lines of intersection, which form a new system of coordinates

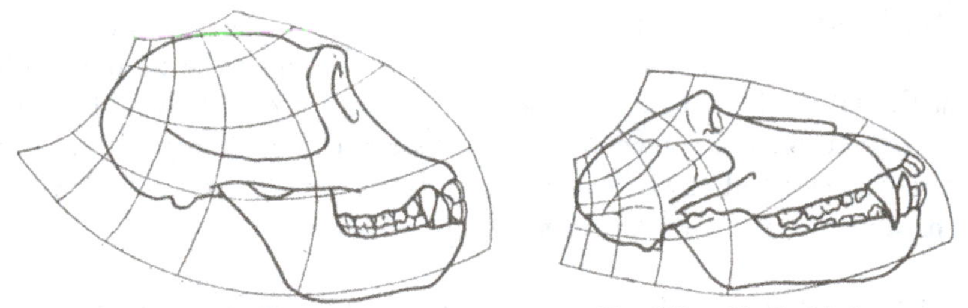

Fig. 550. Skull of chimpanzee.

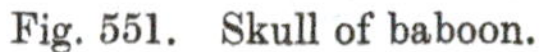

Fig. 551. Skull of baboon.

and constitute a simple "projection" of our human skull. The network represented in Fig. 549 constitutes such a projection of the human skull on what we may call, figuratively speaking, the "plane" of the chimpanzee; and the full diagram in Fig. 550 demonstrates the correspondence. In Fig. 551 I have shewn the similar deformation in the case of a baboon, and it is obvious that the transformation is of precisely the same order, and differs only in an increased intensity or degree of deformation*. These anthropoid skulls, then, which we can transform one into another by a "continuous transformation," are admirable examples of what Listing called "topological similitude."

In both dimensions, as we pass from above downwards and from behind forwards, the corresponding areas of the network are seen to increase in a gradual and approximately logarithmic order in the lower as compared with the higher type of skull; and, in short, it becomes at once manifest that the modifications of jaws, brain-case, and the regions between are all portions of one continuous and integral process. It is of course easy to draw the inverse diagrams, by which the Cartesian coordinates of the ape are transformed into curvilinear and non-equidistant coordinates in man†.

From this comparison of the gorilla's or chimpanzee's with the human skull we realise that an inherent weakness underlies the anthropologist's method of comparing skulls by reference to a small number of axes. The most important of these are the "facial" and "basicranial" axes, which include between them the "facial angle." But it is, in the first place, evident that these axes are merely the principal axes of a system of coordinates, and that their restricted and isolated use neglects all that can be learned from the filling in of the rest of the coordinate network. And, in the second place, the "facial axis," for instance, as ordinarily used in the anthropological comparison of one human skull with another, or of the human skull with the gorilla's, is in all cases treated as a straight line; but our investigation has shewn that rectilinear axes only meet the case in the simplest and most closely related transformations; and that, for instance, in the anthropoid skull no rectilinear axis is homologous

* The empirical coordinates which I have sketched in for the chimpanzee as a conformal transformation of the Cartesian coordinates of the human skull look as if they might find their place in an equipotential elliptic field. They are indeed closely analogous to some already figured by MM. Y. Ikada and M. Kuwaori, Some conformal representations by means of the elliptic integrals, *Sci. Papers Inst. Phys. Research, Tokyo*, XXVI, pp. 208–215, 1936: e.g. pl. XXXI*b*.

† Speaking of "diagrams in pairs," and doubtless thinking of his own "reciprocal diagrams," Clerk Maxwell says (in his article *Diagrams* in the *Encyclopaedia Britannica*): "The method in which we simultaneously contemplate two figures, and recognise a correspondence between certain points in the one figure and certain points in the other, is one of the most powerful and fertile methods hitherto known in science....It is sometimes spoken of as the method or principle of duality."

with a rectilinear axis in a man's skull, but what is a straight line in the one has become a certain definite curve in the other.

Mr Heilmann tells me that he has tried, but without success, to obtain a transitional series between the human skull and some prehuman, anthropoid type, which series (as in the case of the Equidae) should be found to contain other known types in direct linear sequence. It appears impossible, however, to obtain such a series, or to pass by successive and continuous gradations through such forms as Mesopithecus, Pithecanthropus, *Homo neanderthalensis*, and the lower or higher races of modern man. The failure is not the fault of our method. It merely indicates that no one straight line of descent, or of consecutive transformation, exists; but on the contrary, that among human and anthropoid types, recent and extinct, we have to do with a complex problem of divergent, rather than of continuous, variation. And in like manner, easy as it is to correlate the baboon's and chimpanzee's skulls severally with that of man, and easy as it is to see that the chimpanzee's skull is much nearer to the human type than is the baboon's, it is also not difficult to perceive that the series is not, strictly speaking, continuous, and that neither of our two apes lies *precisely* on the same direct line or sequence of deformation by which we may hypothetically connect the other with man.

After easily transforming our coordinate diagram of the human skull into a corresponding diagram of ape or of baboon, we may effect a further transformation of man or monkey into dog no less easily; and we are thereby encouraged to believe that any two mammalian skulls may be compared with, or transformed into, one another by this method. There is something, an essential and indispensable something, which is common to them all, something which is the subject of all our transformations, and remains *invariant* (as the mathematicians say) under them all. In these transformations of ours every point may change its place, every line its curvature, every area its magnitude; but on the other hand every point and every line continues to exist, and keeps its relative order and position throughout all distortions and transformations. A series of points, a, b, c, along a certain line persist as corresponding points a', b', c', however the line connecting them may lengthen or bend; and as with points, so with lines, and so also with areas. Ear, eye and nostril, and all the other great landmarks of cranial anatomy, not only continue to exist but retain their relative order and position throughout all our transformations.

We can discover a certain *invariance*, somewhat more restricted than before, between the mammalian skull and that of fowl, frog or even herring. We have still something common to them all; and using another mathematical term (somewhat loosely perhaps) we may speak of the *discriminant characters* which persist unchanged, and continue to form the subject of our transformation. But the method, far as it goes, has its limitations. We cannot fit both beetle and cuttlefish into the same framework, however we distort it; nor by any coordinate transformation can we turn either of them into one another or into the vertebrate type. They are

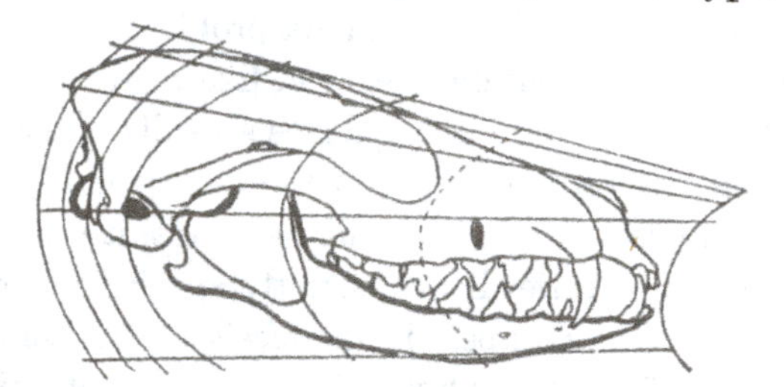

Fig. 552. Skull of dog, compared with the human skull of Fig. 548.

essentially different; there is nothing about them which can be legitimately compared. Eyes they all have, and mouth and jaws; but what we call by these names are no longer in the same order or relative position; they are no longer the same thing, there is no *invariant* basis for transformation. The cuttlefish eye seems as perfect, optically, as our own; but the lack of an invariant relation of position between them, or lack of true homology between them (as we naturalists say), is enough to shew that they are unrelated things, and have come into existence independently of one another.

As a final illustration I have drawn the outline of a dog's skull (Fig. 552), and inscribed it in a network comparable with the Cartesian network of the human skull in Fig. 548. Here we attempt to bridge over a wider gulf than we have crossed in any of our former comparisons. But, nevertheless, it is obvious that our method still holds good, in spite of the fact that there are various specific differences, such as the open or closed orbit, etc., which have to be separately described and accounted for. We see that the chief essential differences in plan between the dog's skull and the man's lie in the fact that, relatively speaking, the former tapers away in front, a triangular taking the place of a rectangular conformation; secondly, that, coincident with the tapering off, there is a progressive elongation, or pulling out, of the whole forepart of the skull; and lastly, as a minor difference, that the straight vertical ordinates of the human skull become curved, with their convexity directed forwards, in the dog. While the net result is that in the dog, just as in the chimpanzee, the brain-pan is smaller and the jaws are larger than in man, it is now conspicuously evident that the coordinate network of the ape is by no means intermediate between those which fit the other two. The mode of deformation is on different lines; and, while it may be correct to say that the chimpanzee and the baboon are more brute-like, it would be by no means accurate to assert that they are more dog-like, than man.

In this brief account of coordinate transformations and of their morphological utility I have dealt with plane coordinates only, and have made no mention of the less elementary subject of coordinates in three-dimensional space. In theory there is no difficulty whatsoever in such an extension of our method; it is just as easy to refer the form of our fish or of our skull to the rectangular coordinates x, y, z, or to the polar coordinates ξ, η, ζ, as it is to refer their plane projections to the two axes to which our investigation has been confined. And that it would be advantageous to do so goes without saying, for it is the shape of the solid object, not that of the mere drawing of the object, that we want to understand; and already we have found some of our easy problems in solid geometry leading us (as in the case of the form of the bivalve and even of the univalve shell) quickly in the direction of coordinate analysis and the theory of conformal transformations. But this extended theme I have not attempted to pursue, and it must be left to other times, and to other hands. Nevertheless, let us glance for a moment at the sort of simple cases, the simplest possible cases, with which such an investigation might begin; and we have found our plane coordinate systems so easily and effectively applicable to certain fishes that we may seek among them for our first and tentative introduction to the three-dimensional field.

It is obvious enough that the same method of description and analysis which we have applied to one plane, we may apply to another: drawing by observation, and by a process of trial and error, our various cross-sections and the coordinate systems which seem best to correspond. But the new and important problem which now emerges is to *correlate* the deformation or transformation which we discover in one plane with that which we have observed in another: and at length, perhaps, after grasping the general principles of such correlation, to forecast approximately what is likely to take place in the third dimension when we are acquainted with two, that is to say, to determine the values along one axis in terms of the other two.

Let us imagine a common "round" fish, and a common "flat" fish, such as a haddock and a plaice. These two fishes are not as nicely adapted for comparison by means of plane coordinates as some which we have studied, owing to the presence of essentially unimportant, but yet conspicuous differences in the position of the eyes, or in the number of the fins—that is to say in the manner in which the continuous dorsal fin of the plaice appears in the haddock to be cut or scolloped into a number of separate fins. But speaking broadly, and apart from such minor differences as these, it is manifest that the chief factor in the case (so far as we at present see) is simply the broadening out of the plaice's body, as compared with the haddock's, in the dorso-ventral direction, that is to say, along the y axis; in other words, the ratio x/y is much less (and indeed little more than half as great) in the haddock than in the plaice. But we also recognise at once that while the plaice (as compared with the haddock) is expanded in one direction, it is also flattened, or thinned out, in the other: y increases, but z diminishes, relatively to x. And furthermore, we soon see that this is a common or even a general phenomenon. The high, expanded body in our Antigonia or in our sun-fish or in a John Dory is at the same time flattened or *compressed* from side to side, in comparison with the related fishes which we have chosen as standards of reference or comparison; and conversely, such a fish as the skate, while it is expanded from side to side in comparison with a shark or dogfish, is at the same time flattened or *depressed* in its vertical section. We hasten to enquire whether there be any simple relation of *magnitude* discernible between these twin factors of expansion and compression; and the very fact that the two dimensions of breadth and depth tend to vary inversely assures us that, in the general process of deformation, the volume and the area of cross-section are less affected than are those two linear dimensions. Some years ago, when I was studying the weight-length coefficient in fishes (of which we have already spoken in chapter III), that is to say the coefficient k in the formula $W = kL^3$, I was not a little surprised to find that k (let us call it in this case k_l) was all but identical in two such different looking fishes as the haddock and the plaice: thus indicating that these two fishes have approximately the same *volume* when they are equal in *length*; or, in other words, that the extent to which the plaice has broadened is *just about compensated for* by the extent to which it has also got flattened or thinned. In short, if we might conceive of a haddock being transformed directly into a plaice, a very large part of the change would be accounted for by supposing the round fish to be "rolled out" into the flat one, as a baker rolls a piece of dough. This is, as it were, an extreme case of the *balancement des organes*, or "compensation of parts."

We must not forget, while we consider the "deformation" of a fish, that the fish, like the bird, is subject to certain strict limitations of form. What we happen to have found in a particular case was observed fifty years ago, and brought under a general rule, by a naval engineer studying fishes from the shipbuilder's point of view. Mr Parsons* compared the contours and the sectional areas of a number of fishes and of several whales; and he found the sectional areas to be always very much the same at the same proportional distances from the front end of the body†. Increase in depth was balanced (as we also have found) by diminution of breadth; and the magnitude of the "entering angle" presented to the water by the advancing fish was fairly constant. Moreover, according to Parsons, the position of the greatest cross-section is fixed for all species, being situated at 36 per cent. of the length behind the

* H. de B. Parsons, Displacements and area-curves of fish, *Trans. Amer. Soc. of Mechan. Engineers*, IX, pp. 679–695, 1888.

† That is to say, if the areas of cross-section be plotted against their distances from the front end of the body, the results are very much alike for all the species examined. See also Selig Hecht, Form and growth in fishes, *Journ. Morph.* XXVII, pp. 379–400, 1916.

snout. We need not stop to consider such extreme cases as the eel or the globefish (*Diodon*), whose ways of propulsion and locomotion are materially modified. But it is certainly curious that no sooner do we try to correlate deformation in one direction with deformation in another, than we are led towards a broad generalisation, touching on hydrodynamical conditions and the limitations of form and structure which are imposed thereby.

Our simple, or simplified, illustrations carry us but a little way, and only half prepare us for much harder things. But interesting as the whole subject is we must meanwhile leave it alone; recognising, however, that if the difficulties of description and representation could be overcome, it is by means of such coordinates in space that we should at last obtain an adequate and satisfying picture of the processes of deformation and the directions of growth.

A Note on Pattern

We have had so much to do with the study of Form that *pattern* has been wellnigh left out of the account, although it is part of the same story. Like any other aspect of form, pattern is correlated with growth, and even determined by it. A feather, for example, which is equally and equidistantly striped to begin with, may have this simple striping transformed into a more complex pattern by the unequal but *graded* elongation of the feather. We need not go farther than the zebra for a characteristic pattern of stripes, nor need we seek a better illustration of how a common pattern may vary in related species.

A zebra's stripes may be broad or narrow, uniform or alternately dark and pale—these are minor or secondary diversities; but the pattern of the stripes shews more conspicuous differences than these, though the differences remain of a simple kind. A zebra's stripes fall into several series. One set covers the neck, including the mane, and extends backwards over the body and forwards on to the face; and these "body-stripes" are all that the extinct Quagga possessed. On the head they are interrupted by the ears and eyes, and end at a definite vertex on the forehead: from which, however, they run down the face in pairs, of which the first pair of all may or may not coalesce into a single median stripe (Fig. 553). A second series runs up the foreleg, and where it meets the body we have the problem of how best to fit the horizontal leg-stripes and the vertical body-stripes together. There is only one way. A pair of body-stripes diverge apart and the upper leg-stripes fit in between, becoming at the same time chevron-shaped so as to adapt themselves to the space they have come to occupy. The stripes of the forelegs, and their manner of fitting on to the body-stripes, vary very little in the several species or varieties.

A third series of stripes ascends the hindlegs, in a fashion identical to begin with for all, but open to modification where these leg-stripes spread over the haunches; for here there may be great

Fig. 553. Zebra's head, to shew how the body-stripes extend to the face. From A. Rzasnicki.

differences in the extent to which the leg-stripes compete with and interfere with, or (so to speak) encroach upon, the stripes of the body. The typical *Equus zebra* is easily recognised by the so-called "gridiron" on its rump; this is a dorsal continuation of the body-stripes, extending to the tail, but sharply cut off on either side by the stripes ascending from the leg (Fig. 554, C). In Burchell's zebra the hindleg-stripes encroach still farther on the body, and even reach up to the rump, so that the gridiron is entirely cut away*.

* Ward's zebra and Grant's zebra are varieties of *Equus zebra*, the former with a very strong "gridiron," the latter with a mere vestige of the same: which is as much as to say that the leg-stripes encroach little in the one, and much in the other, on the hindmost body-stripes. Chapman's zebra is a form of *E. Burchelli*, with well-striped legs and faint intermediate striping. Cf. W. Ridgeway, on The differentiation of the three species of Zebra, *P.Z.S.* 1909, pp. 547–563; also (*int. al.*) Adolf Rzasnicki, *Zebry*, Warsaw, 1931.

In the Abyssinian *Equus Grevyi*, all the stripes are very numerous, narrow and close-set. The body-stripes refuse, as it were, to be encroached on or obliterated by those of the hindlegs; which latter are merely intercalated between them, chevron fashion, wedging

Fig. 554. Zebra patterns. A, B, *Equus Burchelli*; C, *E. zebra*; D, *E. Grevyi*.

in between the body-stripes as the foreleg-stripes are wont to do. It follows that in the middle of the haunch, over the region of the hip-joint, there is in this species a characteristic "focus," where the leg-stripes fit in between the lumbar and the caudal sections of the body-stripes. We may now add, as a fourth and last series, common to all kinds, the few stripes which surround the lips on either side, and wedge in between the stripes upon the face.

Conclusion

There is one last lesson which coordinate geometry helps us to learn; it is simple and easy, but very important indeed. In the study of evolution, and in all attempts to trace the descent of the animal kingdom, fourscore years' study of the *Origin of Species* has had an unlooked-for and disappointing result. It was hoped to begin with, and within my own recollection it was confidently believed, that the broad lines of descent, the relation of the main branches to one another and to the trunk of the tree, would soon be settled, and the lesser ramifications would be unravelled bit by bit and later on. But things have turned out otherwise. We have long known, in more or less satisfactory detail, the pedigree of horses, elephants, turtles, crocodiles and some few more; and our conclusions tally as to these, again more or less to our satisfaction, with the direct evidence of palaeontological succession. But the larger and at first sight simpler questions remain unanswered; for eighty years' study of Darwinian evolution has not taught us how birds descend from reptiles, mammals from earlier quadrupeds, quadrupeds from fishes, nor vertebrates from the invertebrate stock. The invertebrates themselves involve the selfsame difficulties, so that we do not know the origin of the echinoderms, of the molluscs, of the coelenterates, nor of one group of protozoa from another. The difficulty is not always quite the same. We may fail to find the actual links between the vertebrate groups, but yet their resemblance and their relationship, real though indefinable, are plain to see; there are gaps between the groups, but we can see, so to speak, across the gap. On the other hand, the breach between vertebrate and invertebrate, worm and coelenterate, coelenterate and protozoon, is in each case of another order, and is so wide that we cannot see across the intervening gap at all.

This failure to solve the cardinal problem of evolutionary biology is a very curious thing; and we may well wonder why the long pedigree is subject to such breaches of continuity. We used to be told, and were content to believe, that the old record was of necessity imperfect—we could not expect it to be otherwise; the story was hard to read because every here and there a page had been lost or torn away, like some *hiatus valde deflendus* in an ancient manuscript. But there is a deeper reason. When we begin to draw comparisons between our algebraic curves and attempt to transform one into another, we find ourselves limited by the very nature of the case to curves having some tangible degree of relation to one another; and these "degrees of relationship" imply a *classification* of mathematical forms, analogous to the classification of plants or animals in another part of the *Systema Naturae*.

An algebraic curve has its fundamental formula, which defines the family to which it belongs; and its parameters, whose quantitative variation admits of infinite variety within the limits which the formula prescribes. With some extension of the meaning of parameters, we may say the same of the families, or genera, or other classificatory groups of plants and animals. We cross a boundary every time we pass from family to family, or group to group. The passage is easy at first, and we are led, *along definite lines*, to more and more subtle and elegant comparisons. But we come in time to forms which, though both may still be simple, yet stand so far apart that direct comparison is no longer legitimate. We never think of "transforming" a helicoid into an ellipsoid, or a circle into a frequency-curve. So it is with the forms of animals. We *cannot* transform an invertebrate into a vertebrate, nor a coelenterate into a worm, by any simple and legitimate deformation, nor by anything short of reduction to elementary principles.

A "principle of discontinuity," then, is inherent in all our classifications, whether mathematical, physical or biological; and the infinitude of possible forms, always limited, may be further reduced and discontinuity further revealed by imposing conditions—as, for example, that our parameters must be whole numbers, or proceed by *quanta*, as the physicists say. The lines of the spectrum, the six families of crystals, Dalton's atomic law, the chemical elements themselves, all illustrate this principle of discontinuity. In short, nature proceeds *from one type to another* among organic as well as inorganic forms; and these types vary according to their own parameters, and are defined by physico-mathematical conditions of possibility. In natural history Cuvier's "types" may not be perfectly chosen nor numerous enough, but *types* they are; and to seek for stepping-stones across the gaps between is to seek vain, for ever.

This is no argument against the theory of evolutionary descent. It merely states that formal resemblance, which we depend on as our trusty guide to the affinities of animals within certain bounds or grades of kinship and propinquity, ceases in certain other cases to serve us, because under certain circumstances it ceases to exist. Our geometrical analogies weigh heavily against Darwin's conception of endless small continuous variations; they help to show that discontinuous variations are a natural thing, that "mutations"—or sudden changes, greater or less—are bound to have taken place, and new "types" to have arisen, now and then. Our argument indicates, if it does not prove, that such mutations, occurring on a comparatively few definite lines, or plain alternatives, of physico-mathematical possibility, are likely to repeat themselves: that the "higher" protozoa, for instance, may have sprung not from or through one another, but severally from the simpler forms; or that the worm-type, to take another example, may have come into being again and again.

EPILOGUE

In the beginning of this book I said that its scope and treatment were of so prefatory a kind that of other preface it had no need; and now, for the same reason, with no formal and elaborate conclusion do I bring it to a close. The fact that I set little store by certain postulates (often deemed to be fundamental) of our present-day biology the reader will have discovered and I have not endeavoured to conceal. But it is not for the sake of polemical argument that I have written, and the doctrines which I do not subscribe to I have only spoken of by the way. My task is finished if I have been able to shew that a certain mathematical aspect of morphology, to which as yet the morphologist gives little heed, is interwoven with his problems, complementary to his descriptive task, and helpful, nay essential, to his proper study and comprehension of Growth and Form. *Hic artem remumque repono.*

And while I have sought to shew the naturalist how a few mathematical concepts and dynamical principles may help and guide him, I have tried to shew the mathematician a field for his labour—a field which few have entered and no man has explored. Here may be found homely problems, such as often tax the highest skill of the mathematician, and reward his ingenuity all the more for their trivial associations and outward semblance of simplicity. *Haec utinam excolant, utinam exhauriant, utinam aperiant nobis Viri mathematice docti**.

That I am no skilled mathematician I have had little need to confess. I am "advanced in these enquiries no farther than the threshold"; but something of the use and beauty of mathematics I think I am able to understand. I know that in the study of material things, number, order and position are the threefold clue to exact knowledge; that these three, in the mathematician's hands, furnish the "first outlines for a sketch of the Universe"; that by square and circle we are helped, like Emile Verhaeren's carpenter, to conceive "Les lois indubitables et fécondes Qui sont la règle et la clarté du monde."

For the harmony of the world is made manifest in Form and

* So Boerhaave, in his *Oratio de Usu Ratiocinii Mechanici in Medicina* (1703).

Number, and the heart and soul and all the poetry of Natural Philosophy are embodied in the concept of mathematical beauty. A greater than Verhaeren had this in mind when he told of "the golden compasses prepared In God's eternal store." A greater than Milton had magnified the theme and glorified Him "that sitteth upon the circle of the earth," saying: He hath measured the waters in the hollow of his hand, and meted out heaven with the span, and comprehended the dust of the earth in a measure.

Moreover, the perfection of mathematical beauty is such (as Colin Maclaurin learned of the bee), that whatsoever is most beautiful and regular is also found to be most useful and excellent.

Not only the movements of the heavenly host must be determined by observation and elucidated by mathematics, but whatsoever else can be expressed by number and defined by natural law. This is the teaching of Plato and Pythagoras, and the message of Greek wisdom to mankind. So the living and the dead, things animate and inanimate, we dwellers in the world and this world wherein we dwell—*πάντα γα μὰν τὰ γιγνωσκόμενα*—are bound alike by physical and mathematical law. "Conterminous with space and coeval with time is the kingdom of Mathematics; within this range her dominion is supreme; otherwise than according to her order nothing can exist, and nothing takes place in contradiction to her laws." So said, some sixty years ago, a certain mathematician*; and Philolaus the Pythagorean had said much the same.

But with no less love and insight has the science of Form and Number been appraised in our own day and generation by a very great Naturalist indeed†—by that old man eloquent, that wise student and pupil of the ant and the bee, who died while this book was being written; who in his all but saecular life had tasted of the firstfruits of immortality; who curiously conjoined the wisdom of antiquity with the learning of today; whose Provençal verse seems set to Dorian music; in whose plainest words is a sound as of bees' industrious murmur; and who, being of the same blood and marrow with Plato and Pythagoras, saw in Number *le comment et le pourquoi des choses*, and found in it *la clef de voûte de l'Univers*.

* William Spottiswoode, in his presidential address to the British Association at Dublin in 1878.

† Henri Fabre.

INDEX

A CATALOG OF SELECTED
DOVER BOOKS
IN SCIENCE AND MATHEMATICS

Mathematics

FUNCTIONAL ANALYSIS (Second Corrected Edition), George Bachman and Lawrence Narici. Excellent treatment of subject geared toward students with background in linear algebra, advanced calculus, physics and engineering. Text covers introduction to inner-product spaces, normed, metric spaces, and topological spaces; complete orthonormal sets, the Hahn-Banach Theorem and its consequences, and many other related subjects. 1966 ed. 544pp. 6⅛ x 9¼. 0-486-40251-7

ASYMPTOTIC EXPANSIONS OF INTEGRALS, Norman Bleistein & Richard A. Handelsman. Best introduction to important field with applications in a variety of scientific disciplines. New preface. Problems. Diagrams. Tables. Bibliography. Index. 448pp. 5⅜ x 8½. 0-486-65082-0

VECTOR AND TENSOR ANALYSIS WITH APPLICATIONS, A. I. Borisenko and I. E. Tarapov. Concise introduction. Worked-out problems, solutions, exercises. 257pp. 5⅝ x 8¼. 0-486-63833-2

AN INTRODUCTION TO ORDINARY DIFFERENTIAL EQUATIONS, Earl A. Coddington. A thorough and systematic first course in elementary differential equations for undergraduates in mathematics and science, with many exercises and problems (with answers). Index. 304pp. 5⅜ x 8½. 0-486-65942-9

FOURIER SERIES AND ORTHOGONAL FUNCTIONS, Harry F. Davis. An incisive text combining theory and practical example to introduce Fourier series, orthogonal functions and applications of the Fourier method to boundary-value problems. 570 exercises. Answers and notes. 416pp. 5⅜ x 8½. 0-486-65973-9

COMPUTABILITY AND UNSOLVABILITY, Martin Davis. Classic graduate-level introduction to theory of computability, usually referred to as theory of recurrent functions. New preface and appendix. 288pp. 5⅜ x 8½. 0-486-61471-9

ASYMPTOTIC METHODS IN ANALYSIS, N. G. de Bruijn. An inexpensive, comprehensive guide to asymptotic methods–the pioneering work that teaches by explaining worked examples in detail. Index. 224pp. 5⅜ x 8½ 0-486-64221-6

APPLIED COMPLEX VARIABLES, John W. Dettman. Step-by-step coverage of fundamentals of analytic function theory–plus lucid exposition of five important applications: Potential Theory; Ordinary Differential Equations; Fourier Transforms; Laplace Transforms; Asymptotic Expansions. 66 figures. Exercises at chapter ends. 512pp. 5⅜ x 8½. 0-486-64670-X

INTRODUCTION TO LINEAR ALGEBRA AND DIFFERENTIAL EQUATIONS, John W. Dettman. Excellent text covers complex numbers, determinants, orthonormal bases, Laplace transforms, much more. Exercises with solutions. Undergraduate level. 416pp. 5⅜ x 8½. 0-486-65191-6

RIEMANN'S ZETA FUNCTION, H. M. Edwards. Superb, high-level study of landmark 1859 publication entitled "On the Number of Primes Less Than a Given Magnitude" traces developments in mathematical theory that it inspired. xiv+315pp. 5⅜ x 8½. 0-486-41740-9

Physics

OPTICAL RESONANCE AND TWO-LEVEL ATOMS, L. Allen and J. H. Eberly. Clear, comprehensive introduction to basic principles behind all quantum optical resonance phenomena. 53 illustrations. Preface. Index. 256pp. 5⅜ x 8½. 0-486-65533-4

QUANTUM THEORY, David Bohm. This advanced undergraduate-level text presents the quantum theory in terms of qualitative and imaginative concepts, followed by specific applications worked out in mathematical detail. Preface. Index. 655pp. 5⅜ x 8½. 0-486-65969-0

ATOMIC PHYSICS (8th EDITION), Max Born. Nobel laureate's lucid treatment of kinetic theory of gases, elementary particles, nuclear atom, wave-corpuscles, atomic structure and spectral lines, much more. Over 40 appendices, bibliography. 495pp. 5⅜ x 8½. 0-486-65984-4

A SOPHISTICATE'S PRIMER OF RELATIVITY, P. W. Bridgman. Geared toward readers already acquainted with special relativity, this book transcends the view of theory as a working tool to answer natural questions: What is a frame of reference? What is a "law of nature"? What is the role of the "observer"? Extensive treatment, written in terms accessible to those without a scientific background. 1983 ed. xlviii+172pp. 5⅜ x 8½. 0-486-42549-5

AN INTRODUCTION TO HAMILTONIAN OPTICS, H. A. Buchdahl. Detailed account of the Hamiltonian treatment of aberration theory in geometrical optics. Many classes of optical systems defined in terms of the symmetries they possess. Problems with detailed solutions. 1970 edition. xv + 360pp. 5⅜ x 8½. 0-486-67597-1

PRIMER OF QUANTUM MECHANICS, Marvin Chester. Introductory text examines the classical quantum bead on a track: its state and representations; operator eigenvalues; harmonic oscillator and bound bead in a symmetric force field; and bead in a spherical shell. Other topics include spin, matrices, and the structure of quantum mechanics; the simplest atom; indistinguishable particles; and stationary-state perturbation theory. 1992 ed. xiv+314pp. 6⅛ x 9¼. 0-486-42878-8

LECTURES ON QUANTUM MECHANICS, Paul A. M. Dirac. Four concise, brilliant lectures on mathematical methods in quantum mechanics from Nobel Prize-winning quantum pioneer build on idea of visualizing quantum theory through the use of classical mechanics. 96pp. 5⅜ x 8½. 0-486-41713-1

THIRTY YEARS THAT SHOOK PHYSICS: THE STORY OF QUANTUM THEORY, George Gamow. Lucid, accessible introduction to influential theory of energy and matter. Careful explanations of Dirac's anti-particles, Bohr's model of the atom, much more. 12 plates. Numerous drawings. 240pp. 5⅜ x 8½. 0-486-24895-X

ELECTRONIC STRUCTURE AND THE PROPERTIES OF SOLIDS: THE PHYSICS OF THE CHEMICAL BOND, Walter A. Harrison. Innovative text offers basic understanding of the electronic structure of covalent and ionic solids, simple metals, transition metals and their compounds. Problems. 1980 edition. 582pp. 6⅛ x 9¼. 0-486-66021-4

CATALOG OF DOVER BOOKS

A TREATISE ON ELECTRICITY AND MAGNETISM, James Clerk Maxwell. Important foundation work of modern physics. Brings to final form Maxwell's theory of electromagnetism and rigorously derives his general equations of field theory. 1,084pp. 5⅜ x 8½. Two-vol. set. Vol. I: 0-486-60636-8 Vol. II: 0-486-60637-6

QUANTUM MECHANICS: PRINCIPLES AND FORMALISM, Roy McWeeny. Graduate student-oriented volume develops subject as fundamental discipline, opening with review of origins of Schrödinger's equations and vector spaces. Focusing on main principles of quantum mechanics and their immediate consequences, it concludes with final generalizations covering alternative "languages" or representations. 1972 ed. 15 figures. xi+155pp. 5⅜ x 8½. 0-486-42829-X

INTRODUCTION TO QUANTUM MECHANICS With Applications to Chemistry, Linus Pauling & E. Bright Wilson, Jr. Classic undergraduate text by Nobel Prize winner applies quantum mechanics to chemical and physical problems. Numerous tables and figures enhance the text. Chapter bibliographies. Appendices. Index. 468pp. 5⅜ x 8½. 0-486-64871-0

METHODS OF THERMODYNAMICS, Howard Reiss. Outstanding text focuses on physical technique of thermodynamics, typical problem areas of understanding, and significance and use of thermodynamic potential. 1965 edition. 238pp. 5⅜ x 8½. 0-486-69445-3

THE ELECTROMAGNETIC FIELD, Albert Shadowitz. Comprehensive undergraduate text covers basics of electric and magnetic fields, builds up to electromagnetic theory. Also related topics, including relativity. Over 900 problems. 768pp. 5⅝ x 8¼. 0-486-65660-8

GREAT EXPERIMENTS IN PHYSICS: FIRSTHAND ACCOUNTS FROM GALILEO TO EINSTEIN, Morris H. Shamos (ed.). 25 crucial discoveries: Newton's laws of motion, Chadwick's study of the neutron, Hertz on electromagnetic waves, more. Original accounts clearly annotated. 370pp. 5⅜ x 8½. 0-486-25346-5

EINSTEIN'S LEGACY, Julian Schwinger. A Nobel Laureate relates fascinating story of Einstein and development of relativity theory in well-illustrated, nontechnical volume. Subjects include meaning of time, paradoxes of space travel, gravity and its effect on light, non-Euclidean geometry and curving of space-time, impact of radio astronomy and space-age discoveries, and more. 189 b/w illustrations. xiv+250pp. 8⅜ x 9¼. 0-486-41974-6

STATISTICAL PHYSICS, Gregory H. Wannier. Classic text combines thermodynamics, statistical mechanics and kinetic theory in one unified presentation of thermal physics. Problems with solutions. Bibliography. 532pp. 5⅜ x 8½. 0-486-65401-X

Paperbound unless otherwise indicated. Available at your book dealer, online at **www.doverpublications.com**, or by writing to Dept. GI, Dover Publications, Inc., 31 East 2nd Street, Mineola, NY 11501. For current price information or for free catalogues (please indicate field of interest), write to Dover Publications or log on to **www.doverpublications.com** and see every Dover book in print. Dover publishes more than 500 books each year on science, elementary and advanced mathematics, biology, music, art, literary history, social sciences, and other areas.

ANTIOCH UNIVERSITY LOS ANGELES
40000992